Common Functional Groups and Linkages in Biochemistry

Compound Name	Structure[a]	Functional Group
Amine[b]	RNH_2 or RNH_3^+ R_2NH or $R_2NH_2^+$ R_3N or R_3NH^+	$-N\Big<$ or $-\overset{+}{N}-$ (amino group)
Alcohol	ROH	$-OH$ (hydroxyl group)
Thiol	RSH	$-SH$ (sulfhydryl group)
Ether	ROR	$-O-$ (ether linkage)
Aldehyde	$R-\overset{\overset{O}{\|\|}}{C}-H$	$-\overset{\overset{O}{\|\|}}{C}-$ (carbonyl group), $R-\overset{\overset{O}{\|\|}}{C}-$ (acyl group)
Ketone	$R-\overset{\overset{O}{\|\|}}{C}-R$	$-\overset{\overset{O}{\|\|}}{C}-$ (carbonyl group), $R-\overset{\overset{O}{\|\|}}{C}-$ (acyl group)
Carboxylic acid[b] (Carboxylate)	$R-\overset{\overset{O}{\|\|}}{C}-OH$ or $R-\overset{\overset{O}{\|\|}}{C}-O^-$	$-\overset{\overset{O}{\|\|}}{C}-OH$ (carboxyl group) or $-\overset{\overset{O}{\|\|}}{C}-O^-$ (carboxylate group)
Ester	$R-\overset{\overset{O}{\|\|}}{C}-OR$	$-\overset{\overset{O}{\|\|}}{C}-O-$ (ester linkage)
Amide	$R-\overset{\overset{O}{\|\|}}{C}-NH_2$ $R-\overset{\overset{O}{\|\|}}{C}-NHR$ $R-\overset{\overset{O}{\|\|}}{C}-NR_2$	$-\overset{\overset{O}{\|\|}}{C}-N\Big<$ (amido group)
Imine[b]	$R{=}NH$ or $R{=}NH_2^+$ $R{=}NR$ or $R{=}NHR^+$	$\Big>C{=}N-$ or $\Big>C{=}\overset{+}{N}\Big<^{H}$ (imino group)
Phosphoric acid ester[b]	$R-O-\overset{\overset{O}{\|\|}}{\underset{\underset{OH}{\|}}{P}}-OH$ or $R-O-\overset{\overset{O}{\|\|}}{\underset{\underset{O^-}{\|}}{P}}-O^-$	$-O-\overset{\overset{O}{\|\|}}{\underset{\underset{OH}{\|}}{P}}-O-$ (phosphoester linkage) $-\overset{\overset{O}{\|\|}}{\underset{\underset{OH}{\|}}{P}}-OH$ or $-\overset{\overset{O}{\|\|}}{\underset{\underset{O^-}{\|}}{P}}-O^-$ (phosphoryl group, P_i)
Diphosphoric acid ester[b]	$R-O-\overset{\overset{O}{\|\|}}{\underset{\underset{OH}{\|}}{P}}-O-\overset{\overset{O}{\|\|}}{\underset{\underset{OH}{\|}}{P}}-OH$ or $R-O-\overset{\overset{O}{\|\|}}{\underset{\underset{O^-}{\|}}{P}}-O-\overset{\overset{O}{\|\|}}{\underset{\underset{O^-}{\|}}{P}}-O^-$	$-O-\overset{\overset{O}{\|\|}}{\underset{\underset{OH}{\|}}{P}}-O-\overset{\overset{O}{\|\|}}{\underset{\underset{OH}{\|}}{P}}-O-$ (phosphoanhydride linkage) $-\overset{\overset{O}{\|\|}}{\underset{\underset{OH}{\|}}{P}}-O-\overset{\overset{O}{\|\|}}{\underset{\underset{OH}{\|}}{P}}-OH$ or $-\overset{\overset{O}{\|\|}}{\underset{\underset{O^-}{\|}}{P}}-O-\overset{\overset{O}{\|\|}}{\underset{\underset{O^-}{\|}}{P}}-O^-$ (diphosphoryl group, pyrophosphoryl group, PP_i)

[a] R represents any carbon-containing group. In a molecule with more than one R group, the groups may be the same or different.

[b] Under physiological conditions, these groups are ionized and hence bear a positive or negative charge.

Third
Edition

Essential
Biochemistry

Third Edition

Essential Biochemistry

CHARLOTTE W. PRATT
Seattle Pacific University

KATHLEEN CORNELY
Providence College

WILEY

VICE PRESIDENT & EXECUTIVE PUBLISHER	Kaye Pace
VICE PRESIDENT, PUBLISHER	Petra Recter
SPONSORING EDITOR	Joan Kalkut
EDITORIAL ASSISTANT	Susan Tsui
SENIOR MARKETING MANAGER	Kristine Ruff
MARKETING ASSISTANT	Andrew Ginsberg
CONTENT MANAGER	Juanita Thompson
PRODUCTION EDITOR	Sandra Dumas
CONTENT EDITOR	Alyson Rentrop
PHOTO DEPARTMENT MANAGER	Hilary Newman
PHOTO RESEARCHER	Teri Stratford
DESIGN DIRECTOR	Harry Nolan
SENIOR DESIGNER	Wendy Lai
COVER DESIGNER	Thomas Nery
PRODUCTION MANAGEMENT SERVICES	Suzanne Ingrao

Cover Photo: ©Volker Steger/Science Source/Photo Researchers, Inc.

This book was typeset in 10.5/12 Adobe Garamond at Aptara and printed and bound by Courier/Kendallville. The cover was printed by Courier/Kendallville.

Founded in 1807, John Wiley & Sons, Inc. has been a valued source of knowledge and understanding for more than 200 years, helping people around the world meet their needs and fulfill their aspirations. Our company is built on a foundation of principles that include responsibility to the communities we serve and where we live and work. In 2008, we launched a Corporate Citizenship Initiative, a global effort to address the environmental, social, economic, and ethical challenges we face in our business. Among the issues we are addressing are carbon impact, paper specifications and procurement, ethical conduct within our business and among our vendors, and community and charitable support. For more information, please visit our website: www.wiley.com/go/citizenship.

The paper in this book was manufactured by a mill whose forest management programs include sustained yield-harvesting of its timberlands. Sustained yield harvesting principles ensure that the number of trees cut each year does not exceed the amount of new growth.

This book is printed on acid-free paper.

ISBN 13 978-1118-08350-5
ISBN 13 978-1118-44168-8

Printed in the United States of America.

10 9 8 7 6 5 4 3

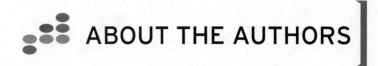

Charlotte Pratt received a B.S. in biology from the University of Notre Dame and a Ph.D. in biochemistry from Duke University. She is a protein chemist who has conducted research in blood coagulation and inflammation at the University of North Carolina at Chapel Hill. She is currently Associate Professor in the Biology Department at Seattle Pacific University. Her interests include molecular evolution, enzyme action, and the relationship between metabolic processes and disease. She has written numerous research and review articles, has worked as a textbook editor, and is a co-author, with Donald Voet and Judith G. Voet, of *Fundamentals of Biochemistry*, published by John Wiley & Sons, Inc.

Kathleen Cornely holds a B.S. in chemistry from Bowling Green (Ohio) State University, an M.S. in biochemistry from Indiana University, and a Ph.D. in nutritional biochemistry from Cornell University. Her experimental research has included a wide range of studies of protein purification and chemical modification. She currently serves as a Professor of Chemistry and Biochemistry at Providence College where she has taught courses in biochemistry, organic chemistry, and general chemistry. Her recent research efforts have focused on chemical education, particularly the use of case studies and guided inquiry in biochemistry education. She serves on the editorial board of *Biochemistry and Molecular Biology Education* and is a member of the American Society for Biochemistry and Molecular Biology.

BRIEF CONTENTS

CONTENTS

part three METABOLISM

part four GENETIC INFORMATION

APPLICATIONS AND ONLINE EXERCISES

Many fascinating aspects of biochemistry do not fit easily into the pages of a textbook. For this reason, some material has been placed in boxes that do not interrupt the flow of the text. *Biochemistry Note* boxes cover biochemical applications, human health, nutrition, and other topics that can enrich student appreciation of biochemistry. Biochemistry Notes include a question or two to help students see how the topic relates to what they have already learned. In addition, *Clinical Connection* boxes explore in detail the biochemistry behind certain human diseases and are accompanied by a set of questions for students to test their understanding. Finally, *Bioinformatics Projects* use a guided-tour approach and a set of questions and activities to introduce students to the databases and online tools used by researchers.

BIOCHEMISTRY NOTES

CLINICAL CONNECTIONS

BIOINFORMATICS PROJECTS

Several years ago, we set out to write a short biochemistry textbook that combined succinct, clear chapters with extensive problem sets. We believed that students would benefit from a modern approach involving broad but not overwhelming coverage of biochemical facts, focusing on the chemistry behind biology, and providing students with practical knowledge and problem-solving opportunities. Our experience in the classroom tells us that effective learning also requires students to become as fully engaged with the material as possible. To that end, we have embraced a strategy of posing questions in different ways throughout each chapter, so that students will not simply read and memorize but will explore and discover on their own—a truer reflection of how biochemists approach their work in the laboratory or clinic.

As always, we view our textbook as a guidebook for students, providing a solid foundation in biochemistry, presenting complete, up-to-date information, and showing the practical aspects of biochemistry as it applies to human health, nutrition, and disease. We hope that students will develop a sense of familiarity and comfort as they encounter new material, explore it, and test their understanding through problem solving.

New to This Edition

Learning is an active rather than passive activity, and one of the most powerful mechanisms for encouraging students to assume responsibility for their learning is to prompt them to examine the material in different ways. This edition includes a variety of questions that require students to review what they already know, to take a closer look at diagrams, to derive their own explanations, and to apply their skills to novel situations. Some features of this question-based approach are outlined here.

Chapter opening "Do you Remember?" sections. Students often have difficulty recalling previously learned material and relating it to the new topics they encounter. New chapter openers provide context with a set of 4 to 6 *Do You Remember* questions to remind students about relevant, foundational material they have seen in previous chapters.

Chapter opening **questions** are answered within that chapter

Do You Remember? reminds students about relevant material from previous chapters.

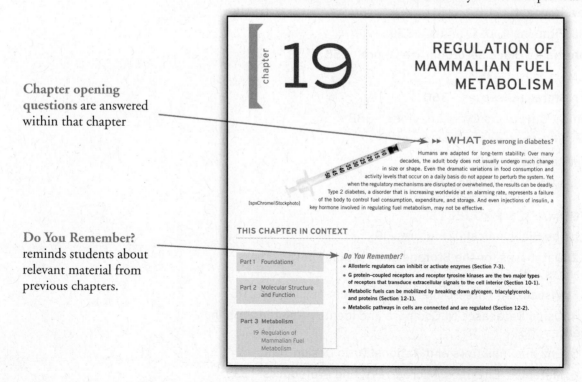

Art program. Revisions to the art program have been made with students in mind. Our goal has been to maximize the clarity and improve the pedagogical value of the illustrations that accompany the text.

- In this edition, we focused on adding visual impact in illustrations by providing more accurate three-dimensional depictions of proteins and membranes.

- **Figure Questions** are now included in many captions to prompt students to inspect the diagram more closely, compare it to similar figures, or draw connections to other topics.

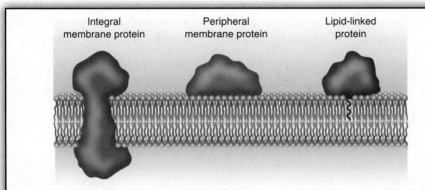

Integral membrane protein   Peripheral membrane protein   Lipid-linked protein

Figure 8-6 Types of membrane proteins. This schematic cross-section of a membrane shows an integral protein spanning the width of the membrane, a peripheral membrane protein associated with the membrane surface, and a lipid-linked protein whose attached hydrophobic tail is incorporated into the lipid bilayer.

? **Which type of membrane protein would be easiest to separate from the lipid bilayer?**

Figure Questions encourage students to more fully engage the material.

- Many of the molecular graphics were re-rendered at higher resolution and to show newly discovered structural features. A number of illustrations were provided by the researchers themselves, providing students a glimpse of how biochemists create models to communicate their new discoveries.

Sample Calculations with Practice Problems. Students need to develop their quantitative skills to understand particular aspects of biochemistry. This edition includes 6 new Sample Calculations, for a total of 24. In addition, each Sample Calculation now includes a set of Practice Problems to provide additional opportunities for students to master quantitative skills. Students can quickly check their answers in the Appendix before moving on.

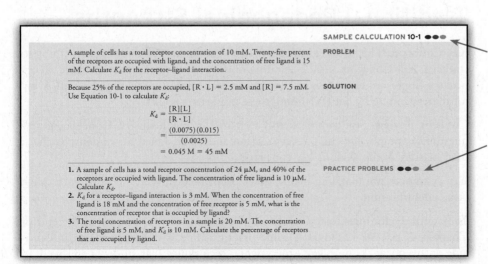

SAMPLE CALCULATION 10-1 ●●●

A sample of cells has a total receptor concentration of 10 mM. Twenty-five percent of the receptors are occupied with ligand, and the concentration of free ligand is 15 mM. Calculate K_d for the receptor–ligand interaction.

PROBLEM

Because 25% of the receptors are occupied, $[R \cdot L] = 2.5$ mM and $[R] = 7.5$ mM. Use Equation 10-1 to calculate K_d:

SOLUTION

$$K_d = \frac{[R][L]}{[R \cdot L]}$$

$$= \frac{(0.0075)(0.015)}{(0.0025)}$$

$$= 0.045 \text{ M} = 45 \text{ mM}$$

1. A sample of cells has a total receptor concentration of 24 μM, and 40% of the receptors are occupied with ligand. The concentration of free ligand is 10 μM. Calculate K_d.
2. K_d for a receptor–ligand interaction is 3 mM. When the concentration of free ligand is 18 mM and the concentration of free receptor is 5 mM, what is the concentration of receptor that is occupied by ligand?
3. The total concentration of receptors in a sample is 20 mM. The concentration of free ligand is 5 mM, and K_d is 10 mM. Calculate the percentage of receptors that are occupied by ligand.

PRACTICE PROBLEMS ●●●

Sample Calculation models the application of key equations to real data.

Practice Problems provide opportunities to check progress.

End-of-chapter problems. Students need ample opportunities to practice and apply their knowledge. This edition includes 36% more problems than the previous edition, bringing the total number of end-of-chapter problems to 1380. The problems, encompassing a variety of types, are grouped by section and are designed to prompt students to recall information, apply it to novel situations, draw connections between different topics, and relate new material to material learned previously or in other courses. Many problems are case studies based on data from research publications and clinical reports. To promote students' independence and to build confidence in their problem-solving skills, virtually all problems are arranged as successive pairs that address the same or related topics. The solution to the first (odd-numbered) problem is included in an appendix, while the solution to the second (even-numbered) problem is available only at the instructor's discretion. The problem sets can therefore also be used for class activities or assigned as homework.

Students are more engaged when they process information that is relevant to their own lives. Furthermore, we understand that many biochemistry students pursue careers in related fields and want to better understand how the subject matter they are learning in biochemistry relates to their intended study of nutrition, pharmacology, biotechnology, exercise science, and so on. Students also need an introduction to what is new and exciting in the field of biochemistry.

Biochemistry Notes. 46 Biochemistry Note boxes (15 of them new to this edition) demonstrate how biochemistry is integrated into all aspects of students' daily lives, addressing a wide range to topics including carbon monoxide poisoning, antifungal drugs, cellulosic biofuels, dietary guidelines, and antibiotics. Each box includes at least one question for students to consider.

Clinical Connections. A total of 15 Clinical Connection boxes, most of them new to this edition, explore the biochemical causes of diseases, their symptoms, and their treatment. Topics include diseases related to protein misfolding, vitamin deficiencies, antidepressants, alcohol toxicity, and metabolism of cancer cells. Questions at the end of each Clinical Connection box help students develop their analytical skills. Answers are provided in an appendix, so students can easily check their understanding as they proceed.

Up-to-date topics. New material for this edition focuses on developments in genomic analyses (Chapter 3), quantitative PCR (Chapter 3), mass spectrometry for protein analysis (Chapter 4), receptor protein structure (Chapter 10), the ATP synthase mechanism (Chapter 15), the role of probiotic organisms (Chapter 19), treatments for diabetes (Chapter 19), the histone code (Chapter 20), and eukaryotic ribosome structure (Chapter 22).

Traditional Pedagogical Strengths

Key Concepts. The information conveyed in biochemistry can be overwhelming to students. Key Concepts at the beginning of each section prompt students to recognize the important takeaways or concepts in each section, providing the scaffolding for understanding by better defining these important points.

Concept Review. Students need to "keep up" in biochemistry. Concept Review questions appear at the end of every section and are designed for students to check their mastery of the section's key concepts.

Key sentences summarizing main points are printed in italics to assist with quick visual identification.

Tools and Techniques Sections. These portions of text, appearing at the end of Chapters 2, 3, and 4, showcase some practical aspects of biochemistry and provide an overview of some key experimental techniques that students are likely to encounter in their reading or laboratory experience, such as preparation of buffers, techniques for manipulating nucleic acids (DNA sequencing, PCR, recombinant DNA, genetically modified organisms, and gene therapy), and protein analysis (chromatographic separations, protein sequencing, and techniques for determining protein three-dimensional structure).

Metabolism overview figures. Metabolism is a challenging subject for students. An overview figure outlining the major metabolic pathways is introduced in Chapter 12 and revisited in subsequent chapters on metabolism in order to help students place individual metabolic pathways into a broader context. Chapters 13 and 17, which focus heavily on multiple pathways, include an additional summary figure as a study aid.

Summary. Chapter summaries, organized by major section headings, highlight important concepts to guide students to focus on the most important points within each section.

Key terms are in boldface. Their definitions are also included in the **Glossary**.

Bioinformatics Projects. A great deal of biochemistry can be done outside the laboratory, using databases and programs available to researchers as well as students. The six exercises in this edition focus on various aspects of bioinformatics (the biochemical literature, nucleotide sequences, protein structures, protein evolution, drug design, and metabolic pathways), In these in-depth exercises, designed to introduce students to the online databases and software tools of bioinformatics, students are given detailed instructions for accessing and manipulating different types of information as well as specific questions to answer and suggestions for further exploration.

An annotated list of **Selected Readings** following each chapter includes recent short papers, mostly reviews, that students are likely to find useful as sources of additional information.

Organization

We have chosen to focus on aspects of biochemistry that tend to receive little coverage in other courses or present a challenge to many students. Thus, in this textbook, we devote proportionately more space to topics such as acid–base chemistry, enzyme mechanisms, enzyme kinetics, oxidation–reduction reactions, oxidative phosphorylation, photosynthesis, and the enzymology of DNA replication, transcription, and translation. At the same time, we appreciate that students can become overwhelmed with information. To counteract this tendency, we have intentionally left out some details, particularly in the chapters on metabolic pathways, in order to emphasize some general themes, such as the stepwise nature of pathways, their evolution, and their regulation.

In keeping with our guidebook approach to biochemistry, we have attempted to place some information in its broader biological context by telling a story. For example, in Chapter 9, the generation of a nerve impulse ties together information about membrane permeability, transport, and fusion. In Chapter 17, different aspects of lipid metabolism are linked to atherosclerosis.

The 22 chapters of *Essential Biochemistry* are relatively short so that students can spend less time reading and more time extending their learning through active problem-solving. Most of the problems require some analysis rather than simple recall of facts. Many problems based on research data provide students a glimpse of the "real world" of science and medicine.

Although each chapter of *Essential Biochemistry, Third Edition* is designed to be self-contained so that it can be covered at any point in the syllabus, the 22 chapters are organized into four parts that span the major themes of biochemistry, including

Metabolic Overview Figures provide "big picture" context.

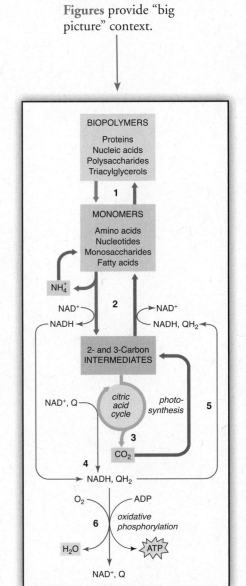

some chemistry background, structure–function relationships, the transformation of matter and energy, and how genetic information is stored and made accessible.

Part 1 of the textbook includes an introductory chapter and a chapter on water. Students with extensive exposure to chemistry can use this material for review. For students with little previous experience, these two chapters provide the chemistry background they will need to appreciate the molecular structures and metabolic reactions they will encounter later.

Part 2 begins with a chapter on the genetic basis of macromolecular structure and function (Chapter 3, From Genes to Proteins). This is followed by chapters on protein structure (Chapter 4) and protein function (Chapter 5), with coverage of myoglobin and hemoglobin, and cytoskeletal and motor proteins. An explanation of how enzymes work (Chapter 6) precedes a discussion of enzyme kinetics (Chapter 7), an arrangement that allows students to grasp the importance of enzymes and to focus on the chemistry of enzyme-catalyzed reactions before delving into the more quantitative aspects of enzyme kinetics. A chapter on lipid chemistry (Chapter 8, Lipids and Membranes) is followed by two chapters that discuss critical biological functions of membranes (Chapter 9, Membrane Transport, and Chapter 10, Signaling). The section ends with a chapter on carbohydrate chemistry (Chapter 11), completing the survey of molecular structure and function.

Part 3 begins with an introduction to metabolism that provides an overview of fuel acquisition, storage, and mobilization as well as the thermodynamics of metabolic reactions (Chapter 12). This is followed, in traditional fashion, by chapters on glucose and glycogen metabolism (Chapter 13); the citric acid cycle (Chapter 14); electron transport and oxidative phosphorylation (Chapter 15); the light and dark reactions of photosynthesis (Chapter 16); lipid catabolism and biosynthesis (Chapter 17); and pathways involving nitrogen-containing compounds, including the synthesis and degradation of amino acids, the synthesis and degradation of nucleotides, and the nitrogen cycle (Chapter 18). The final chapter of Part 2 explores the integration of mammalian metabolism, with extensive discussions of hormonal control of metabolic pathways and disorders of fuel metabolism (Chapter 19).

Part 4, the management of genetic information, includes three chapters, covering DNA replication and repair (Chapter 20), transcription (Chapter 21), and protein synthesis (Chapter 22). Because these topics are typically also covered in other courses, Chapters 20–22 emphasize the relevant biochemical details, such as topoisomerase action, nucleosome structure, mechanisms of polymerases and other enzymes, structures of accessory proteins, proofreading strategies, and chaperone-assisted protein folding.

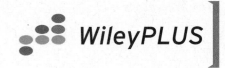

Teaching and Learning Resources

WileyPLUS for Essential Biochemistry, Third Edition—A Powerful Teaching and Learning Solution

WileyPLUS is an innovative, research-based online environment for effective teaching and learning. *WileyPLUS* builds student confidence because it takes the guesswork out of studying by providing students with a clear roadmap: what to do, how to do it, and whether they did it right. Students will take more initiative so instructors will have greater impact on their achievement in the classroom and beyond.

What do students receive with *WileyPLUS*?

- The complete digital textbook, saving students up to 60% off the cost of a printed text.

- Question assistance, including links to relevant sections in the online digital textbook.

- Immediate feedback and proof of progress, 24/7.

- Integrated multimedia resources that address your students' unique learning styles, levels of proficiency, and levels of preparation. These resources, including visualizations that are crucial for understanding biochemistry, provide multiple study paths and encourage active learning with opportunities to practice and review.

WileyPLUS Student Resources

Guided Explorations. A set of 25 self-contained presentations, many with narration, employ extensive animated computer graphics to enhance student understanding of key topics.

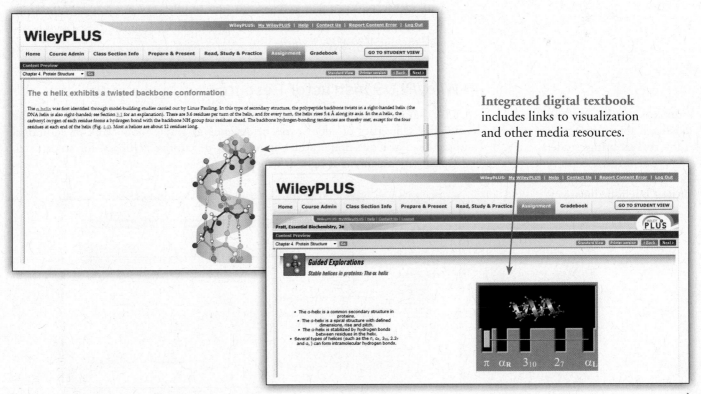

Integrated digital textbook includes links to visualization and other media resources.

Interactive Exercises. A set of 21 molecular structures from the text have been rendered in Jmol, a browser-independent viewer for manipulating structures in three dimensions, and paired with questions designed to facilitate comprehension of concepts. A tutorial for using Jmol is also provided.

Animated Figures. A set of 34 figures from the text, illustrating various concepts, techniques, and processes, are presented as brief animations to facilitate learning.

These media resources are intended to enrich the learning process for students, especially those who rely heavily on visual information. Whereas some resources, particularly the Interactive Exercises and Animated Figures, are brief and could easily be incorporated into an instructor's classroom lecture, all the resources are ideal for student self-study, allowing students to proceed at their own pace or back up and review as needed.

Chapter 0 Chemistry Refresher. To ensure that students have mastered the necessary prerequisite content from both General Chemistry and Organic Chemistry, and to eliminate the burden on instructors to review this material in lecture, *Wiley-PLUS* now includes a complete chapter of core General and Organic Chemistry topics with corresponding assignments. Chapter 0 is available to students and can be assigned in *WileyPLUS* to gauge understanding of the core topics required to succeed in Biochemistry.

What do instructors receive with *WileyPLUS*?

- Reliable resources that reinforce course goals inside and outside of the classroom.

- The ability to easily identify those students who are falling behind by tracking their progress and offering assistance easily, even before they come to office hours. *WileyPLUS* simplifies and automates such tasks as student performance assessment, creating assignments, scoring student work, keeping grades, and more.

- Media-rich course materials and assessment content that allow you to customize your classroom presentation with a wealth of resources and functionality from PowerPoint slides to a database of rich visuals. You can even add your own materials to your *WileyPLUS* course.

WileyPLUS Instructor Resources

Exercises Over 800 conceptually based questions by Rachel Milner and Adrienne Wright, University of Alberta can be assigned as graded homework or additional practice. Each question features immediate, descriptive feedback for students to explain why an answer is right or wrong.

Students receive immediate feedback about answer, while Link to Text brings student to pertinent text section to better understand why an answer might be incorrect.

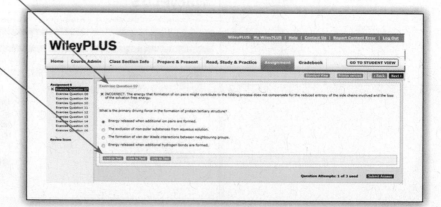

Image PowerPoint Slides contain all images and tables in the text, optimized for viewing onscreen.

Digital Image Library with images from the text, also available in JPEG format, can be used to customize presentations and provide additional visual support for quizzes and exams.

Lecture Power Points by Mary Peek, Georgia Tech University, contain lecture outlines, Key Concepts, and relevant figures from the text.

Test Bank by Scott Lefler, Arizona State University, contains over 1100 questions, with each multiple choice question keyed to the relevant section in the text and rated by difficulty level. A computerized version of the Test Bank is also available.

Prelecture Questions and Practice Quizzes by Steven Vik, Southern Methodist University, accompany each chapter and consist of multiple-choice, true/false, and fill-in-the blank questions, with instant feedback to help students master concepts.

Classroom Response Questions ("Clicker Questions") by Gail Grabner, University of Texas at Austin, consist of interactive questions for classroom response systems to facilitate classroom participation and discussion. These questions can also be used by instructors as prelecture questions that help gauge students' knowledge of overall concepts, while addressing common misconceptions.

Bioinformatics Projects A set of newly updated exercises by Paul Craig, Rochester Institute of Technology, cover the contents and uses of databases related to nucleic acids, protein sequences, protein structures, enzyme inhibition, and other topics. The exercises use real data sets, pose specific questions, and prompt students to obtain information from online databases and to access the software tools for analyzing such data.

Additional Student Resources include

Animated Exercises guide students through key topics and present related questions that challenge their understanding of these topics

Interactive Periodic Table provides students with key information on the elements in the Periodic Table in an interactive environment

Wiley Encyclopedia of Chemical Biology links bring students to selected articles in this reference work for the widely expanding field of chemical biology. Relevant articles will facilitate deeper research and encourage additional reading.

Annotated List of Selected Readings includes recent short papers, mostly reviews, that students are likely to find useful as sources of additional information.

ACKNOWLEDGMENTS

We would like to thank everyone who helped develop *Essential Biochemistry, Third Edition* including Biochemistry Editor Joan Kalkut, Senior Media Editor Geraldine Osnato, Content Editor Aly Rentrop, the production management services of Ingrao Associates, Production Editor Sandra Dumas, Designer (interior) Wendy Lai, Designer (cover) Tom Nery, Photo Researcher Teri Stratford, and Editorial Assistant Susan Tsui.

We also thank all the reviewers who provided essential feedback on manuscript and media, corrected errors, and made valuable suggestions for improvements that have been so important in the writing and development of *Essential Biochemistry, 3e*.

Arkansas
Anne Grippo, *Arkansas State University*

Arizona
Matthew Gage, *Northern Arizona University*
Scott Lefler, *Arizona State University, Tempe*
Allan Scruggs *Arizona State University, Tempe*

California
Elaine Carter, *Los Angeles City College*
Daniel Edwards, *California State University, Chico*
Pavan Kadandale, *University of California, Irvine*
Rakesh Mogul, *California State Polytechnic University, Pomona*

Colorado
Andrew Bonham, *Metropolitan State University of Denver*
Johannes Rudolph, *University of Colorado*

Connecticut
Matthew Fisher, *Saint Vincent's College*

Florida
David Brown, *Florida Gulf Coast University*

Georgia
Chavonda Mills, *Georgia College*
Mary E. Peek, *Georgia Tech University*
Rich Singiser, *Clayton State University*

Hawaii
Jon-Paul Bingham, *University of Hawaii-Manoa, College of Tropical Agriculture and Human Resources*

Illinois
Jon Friesen, *Illinois State University*
Constance Jeffrey, *University of Illinois, Chicago*
Gary Roby, *College of DuPage*
Lisa Wen, *Western Illinois University*

Indiana
Mohammad Qasim, *Indiana University-Purdue University*

Kansas
Peter Gegenheimer, *The University of Kansas*
Ramaswamy Krishnamoorth, *Kansas State University*

Louisiana
James Moroney, *Louisiana State University*
Jeffrey Temple, *Southeastern Louisiana University*

Maine
Robert Gundersen, *University of Maine, Orono*

Massachusetts
Jeffry Nicols, *Worcester State University*

Michigan
Kathleen Foley, *Michigan State University*
Deborah Heyl-Clegg, *Eastern Michigan University*
Jon Stoltzfus, *Michigan State University*
Mark Thomson, *Ferris State University*

Minnesota
Sandra Olmsted, *Augsburg College*

Missouri
Karen Bame, *University of Missouri, Kansas City*
Nuran Ercal, *Missouri University of Science & Technology*

Nebraska
Jodi Kreiling, *University of Nebraska, Omaha*

New Jersey
Yufeng Wei, *Seton Hall University*
Bryan Spiegelberg, *Rider University*

New York
Wendy Pogozelski, *SUNY Geneseo*
Susan Rotenberg *Queens College of CUNY*

Ohio
Edward Merino, *University of Cincinnati*
Heeyoung Tai, *Miami University*
Lai-Chu Wu, *The Ohio State University*

Oregon
Jeannine Chan, *Pacific University*

Pennsylvania

Mahrukh Azam, *West Chester University of Pennsylvania*

David Edwards, *University of Pittsburgh School of Pharmacy*

Robin Ertl, *Marywood University*

Amy Hark, *Muhlenberg College*

Justin Huffman, *Pennsylvania State University, Altoona*

Sandra Turch-Dooley, *Millersville University*

Rhode Island

Lenore Martin, *University of Rhode Island*

Erica Oduaranm *Roger Williams University*

South Carolina

Weiguo Cao, *Clemson University*

Ramin Radfar, *Wofford College*

Paul Richardson, *Coastal Carolina University*

Kerry Smith, *Clemson University*

Tennessee

Meagan Mann, *Austin Peay State University*

Texas

Johannes Bauer, *Southern Methodist University*

Gail Grabner, *University of Texas, Austin*

Marcos Oliveira, *Feik School of Pharmacy, University of the Incarnate Word*

Utah

Craig Thulin, *Utah Valley University*

We are also grateful to the many people who provided reviews that guided preparation of the earlier editions of our book.

Arizona

Allan Bieber, *Arizona State University*

Matthew Gage, *Northern Arizona University*

Richard Posner, *Northern Arizona State University*

California

Gregg Jongeward, *University of the Pacific*

Paul Larsen, *University of California, Riverside*

Colorado

Paul Azari, *Colorado State University*

Illinois

Lisa Wen, *Western Illinois University*

Indiana

Brenda Blacklock, *Indiana University-Purdue University Indianapolis*

Todd Hrubey, *Butler University*

Christine Hrycyna, *Purdue University*

Scott Pattison, *Ball State University*

Iowa

Don Heck, *Iowa State University*

Michigan

Marilee Benore, *University of Michigan*

Kim Colvert, *Ferris State University*

Melvin Schindler, *Michigan State University*

Minnesota

Tammy Stobb, *St. Cloud State University*

Mississippi

Jeffrey Evans, *University of Southern Mississippi*

James R. Heitz, *Mississippi State University*

Nebraska

S. Madhavan, *University of Nebraska*

Russell Rasmussen, *Wayne State College*

New Mexico

Beulah Woodfin, *University of New Mexico*

Oklahoma

Charles Crittell, *East Central University*

Oregon

Steven Sylvester, *Oregon State University*

Pennsylvania

Jeffrey Brodsky, *University of Pittsburgh*

Michael Sypes, *Pennsylvania State University*

South Carolina

Carolyn S. Brown, *Clemson University*

Texas

James Caras, *Austin, Texas*

Paige Caras, *Austin, Texas*

David W. Eldridge, *Baylor University*

Edward Funkhouser, *Texas A&M University*

Barrie Kitto, *University of Texas at Austin*

Richard Sheardy, *Texas Woman's University*

Linette Watkins, *Southwest Texas State University*

Wisconsin

Sandy Grunwald, *University of Wisconsin La Crosse*

Canada

Isabelle Barrette-Ng, *University of Calgary*

Many of the molecular graphics that illustrate this book were created using publicly available coordinates from the Protein Data Bank (www.rcsb.org). The figures were rendered with the Swiss-Pdb Viewer [Guex, N., Peitsch, M.C., SWISS-MODEL and the Swiss-Pdb Viewer: an environment for comparative protein modeling, *Electrophoresis* **18**, 2714–2723 (1997); program available at spdbv.vital-it.ch/] and Pov-Ray (available at www.povray.org) or with PyMol [developed by Warren DeLano; available at www.pymol.org/].

GUIDE TO MEDIA RESOURCES

FOR THE STUDENT: HOW TO USE THIS BOOK

Welcome to Biochemistry! Your success in this course will depend to a great extent on your willingness to take an active role in your education. Learning biochemistry requires more than simply reading the textbook, although we recommend that as a first step! *Essential Biochemistry, Third Edition* has been designed and written with you in mind, and we urge you to take advantage of all it has to offer.

Biochemical knowledge is cumulative; it is not something that can be learned all at once. We advise you to keep up with your reading and other assignments so that you have plenty of time to reflect, ask questions, and, if necessary, seek help from your instructor. As you read each chapter of the textbook, make sure you understand how it fits into the course syllabus. Use the study aids provided in the textbook: First check out the **Do You Remember** questions to be sure you are prepared. Note the list of **Key Concepts** at the start of each text section. Use the **Concept Review** and **Chapter Summary** to check your understanding. Take time to answer the figure questions and the questions at the ends of Biochemistry Note and Clinical Connection boxes. Be sure to view the **online resources** that expand on material covered in the textbook. These exercises include animations of dynamic biochemical processes and interactive molecular graphics. You can enrich your understanding of biochemistry by exploring the exercises and answering the questions they pose.

As you study, note the **key sentences** that are highlighted in italics. Be able to define the **key terms** in boldface, and test your knowledge by answering the questions in *WileyPLUS*. Most importantly, solve the **problems** at the end of each chapter. You should make every effort to complete all the problems without looking at the solutions. Developing problem-solving skills will facilitate your understanding of biochemistry and will help pave the way to success in any future academic or career endeavor.

Finally, take advantage of the additional resources available—such as the list of Selected Readings, Bioinformatics Exercises, and articles from the Wiley Encyclopedia of Chemical Biology—if you need help, are curious about biochemistry, or need up-to-date information as a starting point for a class project.

In writing *Essential Biochemistry*, we endeavored to select topics that would provide a solid introduction to modern biochemistry, which is a vast and ever-changing field. We realize that most students who use this book will not become biochemists. Nevertheless, it is our hope that you will come to understand the major themes of biochemistry and see how they relate to current and future developments in science and medicine.

Charlotte W. Pratt
Kathleen Cornely

Third
Edition

Essential
Biochemistry

chapter 1

THE CHEMICAL BASIS OF LIFE

[© Biophoto Associates/Photo Researchers, Inc.]

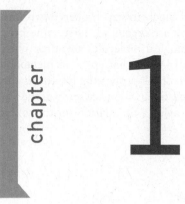

▸▸ DO living systems obey the laws of thermodynamics?

The rules that describe matter and energy suggest that chemical systems tend to reach a state of equilibrium, in the process becoming more disordered. Yet we know that living systems, such as the *Anabaena* cells shown here, are not at equilibrium and are highly ordered. In this chapter, we'll introduce the molecular components of cells and explore the laws that apply to the chemical reactions that occur in cells, so you'll see that living systems obey the same rules that govern nonliving systems.

THIS CHAPTER IN CONTEXT

Part 1 Foundations
1 The Chemical Basis of Life

Part 2 Molecular Structure and Function

Part 3 Metabolism

Part 4 Genetic Information

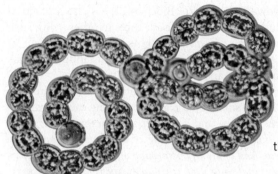

This first chapter offers a preview of the study of biochemistry, broken down into three sections that reflect how topics in this book are organized. First come brief descriptions of the four major types of small biological molecules and their polymeric forms. Next is a summary of the thermodynamics that apply to metabolic reactions. Finally, there is a discussion of the origin of self-replicating life-forms and their evolution into modern cells. These short discussions introduce some of the key players and major themes of biochemistry and provide a foundation for the topics that will be encountered in subsequent chapters.

1-1 What Is Biochemistry?

Biochemistry is the scientific discipline that seeks to explain life at the molecular level. It uses the tools and terminology of chemistry to describe the various attributes of living organisms. Biochemistry offers answers to such fundamental questions as "What are we made of?" and "How do we work?" Biochemistry is also a practical science: It generates powerful techniques that underlie advances in other fields, such as genetics, cell biology, and immunology; it offers insights into the treatment of diseases such as cancer and diabetes; and it improves the efficiency of industries such as wastewater treatment, food production, and drug manufacturing.

Biochemistry has traditionally been a science of reductionism; that is, it attempts to explain the whole by breaking it into smaller parts and examining each part separately. For biochemists, this means isolating and characterizing an organism's component molecules (**Fig. 1-1**). *A thorough understanding of each molecule's physical structure and chemical reactivity helps lead to an understanding of how molecules cooperate and combine to form larger functional units and, ultimately, the intact organism.*

Nevertheless, a holistic approach is indispensable for unraveling the secrets of nature. Just as a clock completely disassembled no longer resembles a clock, information about a multitude of biological molecules does not necessarily reveal how an organism lives. Some molecular interactions are too complex to be teased apart in the laboratory, so it may be necessary to examine a cultured organism to see how it fares when a particular molecule is modified or absent. In addition, so much is known about so many molecules that the sheer volume of data—much of it collected in online databases—requires a computer to access and analyze it using the tools of

Figure 1-1 Levels of organization in a living organism. Biochemistry focuses on the structures and functions of molecules. Interactions between molecules give rise to higher-order structures (for example, organelles), which may themselves be components of larger entities, leading ultimately to the entire organism.

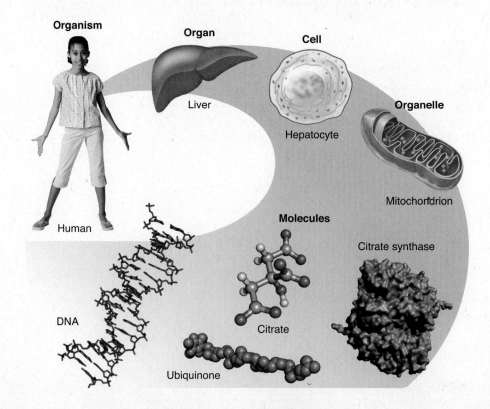

bioinformatics (see Bioinformatics Project 1, The Biochemical Literature). Thus, a biochemist's laboratory is as likely to hold racks of test tubes as flasks of bacteria or computers. The material included in this book reflects both traditional and modern approaches to biochemical understanding.

Chapters 3 through 22 of this book are divided into three groups that roughly correspond to three major themes of biochemistry:

1. *What are living organisms made of?* Some molecules are responsible for the physical shapes of cells. Others carry out various activities in the cell. (For convenience, we use *cell* interchangeably with *organism* since the simplest living entity is a single cell.) In all cases, the structure of a molecule is intimately linked to its function. Understanding a molecule's structural characteristics is therefore an important key to understanding its functional significance.

2. *How do organisms acquire and use energy?* The ability of a cell to carry out metabolic reactions—to synthesize its constituents and to move, grow, and reproduce—requires the input of energy. A cell must extract this energy from the environment and spend it or store it in a manageable form.

3. *How does an organism maintain its identity across generations?* Modern human beings look much like they did 100,000 years ago. Certain bacteria have persisted for millions, if not billions, of years. In all organisms, information specifying a cell's structural composition and functional capacity must be faithfully maintained and transmitted each time the cell divides.

Even within its own lifetime, a cell may dramatically alter its shape or metabolic activities, but it does so within certain limits. Throughout this book we will examine how regulatory processes define an organism's ability to respond to changing internal and external conditions. In addition, we will examine the diseases that can result from a molecular defect in any of the cell's attributes—its structural components, its metabolism, or its ability to follow genetic instructions.

1-2 Biological Molecules

Even the simplest organisms contain a staggering number of different molecules, yet this number represents only an infinitesimal portion of all the molecules that are chemically possible. For one thing, *only a small subset of the known elements are found in living systems* (Fig. 1-2). The most abundant of these are C, N, O, and H, followed by Ca, P, K, S, Cl, Na, and Mg. Certain **trace elements** are also present in very small quantities.

Virtually all the molecules in a living organism contain carbon, so biochemistry can be considered to be a branch of organic chemistry. In addition, biological molecules are constructed from H, N, O, P, and S. Most of these molecules belong to one of a few structural classes, which are described below. Similarly, *the chemical reactivity of biomolecules is limited relative to the reactivity of all chemical compounds.* A few of the functional groups and intramolecular linkages that are common in biochemistry are listed in Table 1-1. It may be helpful to review them, because we will refer to them throughout this book.

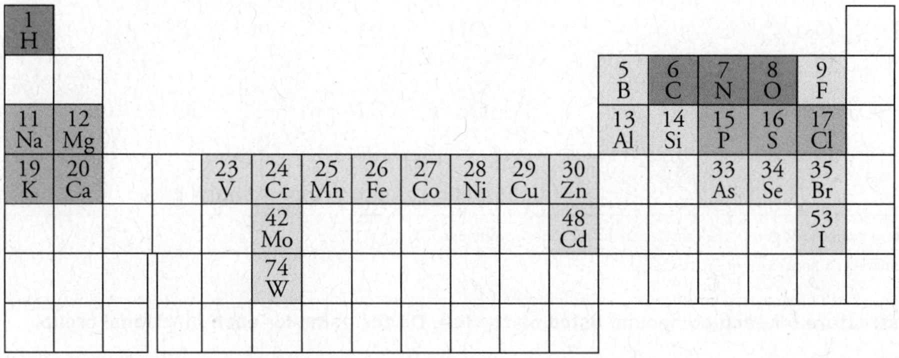

KEY CONCEPTS
- Biological molecules are composed of a subset of all possible elements and functional groups.
- Cells contain four major types of biological molecules and three major types of polymers.

1 H																	
												5 B	6 C	7 N	8 O	9 F	
11 Na	12 Mg											13 Al	14 Si	15 P	16 S	17 Cl	
19 K	20 Ca		23 V	24 Cr	25 Mn	26 Fe	27 Co	28 Ni	29 Cu	30 Zn				33 As	34 Se	35 Br	
			42 Mo							48 Cd						53 I	
			74 W														

Figure 1-2 Elements found in biological systems. The most abundant elements are most darkly shaded; trace elements are most lightly shaded. Not every organism contains every trace element. Biological molecules primarily contain H, C, N, O, P, and S.

[TABLE]

Compo...

Ami...

	Structure[a]	Functional Group
	RNH_2 or RNH_3^+ R_2NH or $R_2NH_2^+$ R_3N or R_3NH^+	$-N\!\!<$ or $^+N\!\!-$ (amino group)
	ROH	$-OH$ (hydroxyl group)
	RSH	$-SH$ (sulfhydryl group)
	ROR	$-O-$ (ether linkage)
Aldehyde	$R-\overset{\overset{\displaystyle O}{\|}}{C}-H$	$-\overset{\overset{\displaystyle O}{\|}}{C}-$ (carbonyl group), $R-\overset{\overset{\displaystyle O}{\|}}{C}-$ (acyl group)
Ketone	$R-\overset{\overset{\displaystyle O}{\|}}{C}-R$	$-\overset{\overset{\displaystyle O}{\|}}{C}-$ (carbonyl group), $R-\overset{\overset{\displaystyle O}{\|}}{C}-$ (acyl group)
Carboxylic acid[b] (Carboxylate)	$R-\overset{\overset{\displaystyle O}{\|}}{C}-OH$ or $R-\overset{\overset{\displaystyle O}{\|}}{C}-O^-$	$-\overset{\overset{\displaystyle O}{\|}}{C}-OH$ (carboxyl group) or $-\overset{\overset{\displaystyle O}{\|}}{C}-O^-$ (carboxylate group)
Ester	$R-\overset{\overset{\displaystyle O}{\|}}{C}-OR$	$-\overset{\overset{\displaystyle O}{\|}}{C}-O-$ (ester linkage)
Amide	$R-\overset{\overset{\displaystyle O}{\|}}{C}-NH_2$ $R-\overset{\overset{\displaystyle O}{\|}}{C}-NHR$ $R-\overset{\overset{\displaystyle O}{\|}}{C}-NR_2$	$-\overset{\overset{\displaystyle O}{\|}}{C}-N\!\!<$ (amido group)
Imine[b]	$R{=}NH$ or $R{=}NH_2^+$ $R{=}NR$ or $R{=}NHR^+$	$>\!C{=}N-$ or $>\!C{=}\overset{+}{N}\!\!<^H$ (imino group)
Phosphoric acid ester[b]	$R-O-\overset{\overset{\displaystyle O}{\|}}{\underset{\underset{\displaystyle OH}{\|}}{P}}-OH$ or $R-O-\overset{\overset{\displaystyle O}{\|}}{\underset{\underset{\displaystyle O^-}{\|}}{P}}-O^-$	$-O-\overset{\overset{\displaystyle O}{\|}}{\underset{\underset{\displaystyle OH}{\|}}{P}}-O-$ (phosphoester linkage) $-\overset{\overset{\displaystyle O}{\|}}{\underset{\underset{\displaystyle OH}{\|}}{P}}-OH$ or $-\overset{\overset{\displaystyle O}{\|}}{\underset{\underset{\displaystyle O^-}{\|}}{P}}-O^-$ (phosphoryl group, P_i)
Diphosphoric acid ester[b]	$R-O-\overset{\overset{\displaystyle O}{\|}}{\underset{\underset{\displaystyle OH}{\|}}{P}}-O-\overset{\overset{\displaystyle O}{\|}}{\underset{\underset{\displaystyle OH}{\|}}{P}}-OH$ or $R-O-\overset{\overset{\displaystyle O}{\|}}{\underset{\underset{\displaystyle O^-}{\|}}{P}}-O-\overset{\overset{\displaystyle O}{\|}}{\underset{\underset{\displaystyle O^-}{\|}}{P}}-O^-$	$-O-\overset{\overset{\displaystyle O}{\|}}{\underset{\underset{\displaystyle OH}{\|}}{P}}-O-\overset{\overset{\displaystyle O}{\|}}{\underset{\underset{\displaystyle OH}{\|}}{P}}-O-$ (phosphoanhydride linkage) $-\overset{\overset{\displaystyle O}{\|}}{\underset{\underset{\displaystyle OH}{\|}}{P}}-O-\overset{\overset{\displaystyle O}{\|}}{\underset{\underset{\displaystyle OH}{\|}}{P}}-OH$ or $-\overset{\overset{\displaystyle O}{\|}}{\underset{\underset{\displaystyle O^-}{\|}}{P}}-O-\overset{\overset{\displaystyle O}{\|}}{\underset{\underset{\displaystyle O^-}{\|}}{P}}-O^-$ (diphosphoryl group, pyrophosphoryl group, PP_i)

[a]R represents any carbon-containing group. In a molecule with more than one R group, the groups may be the same or different.

[b]Under physiological conditions, these groups are ionized and hence bear a positive or negative charge.

? **Cover the Structure column and draw the structure for each compound listed on the left. Do the same for each functional group.**

Cells contain four major types of biomolecules

Most of the cell's small molecules can be divided into four classes. Although each class contains many members, *they are united under a single structural or functional definition.* Identifying a particular molecule's class may help predict its chemical properties and possibly its role in the cell.

1. Amino Acids

The simplest compounds are the **amino acids,** so named because they contain an amino group ($-NH_2$) and a carboxylic acid group ($-COOH$). Under physiological conditions, these groups are actually ionized to $-NH_3^+$ and $-COO^-$. The common amino acid alanine—like other small molecules—can be depicted in different ways, for example, by a structural formula, a ball-and-stick model, or a space-filling model (**Fig. 1-3**). Other amino acids resemble alanine in basic structure, but instead of a methyl group ($-CH_3$), they have another group—called a side chain or R group—that may also contain N, O, or S; for example,

Asparagine

Cysteine

2. Carbohydrates

Simple **carbohydrates** (also called **monosaccharides** or just sugars) have the formula $(CH_2O)_n$, where n is ≥ 3. Glucose, a monosaccharide with six carbon atoms, has the formula $C_6H_{12}O_6$. It is sometimes convenient to draw it as a ladder-like chain (*left*); however, glucose forms a cyclic structure in solution (*right*):

Glucose

In the representation of the cyclic structure, the darker bonds project in front of the page and the lighter bonds project behind it. In many monosaccharides, one or more hydroxyl groups are replaced by other groups, but the ring structure and multiple $-OH$ groups of these molecules allow them to be easily recognized as carbohydrates.

3. Nucleotides

A five-carbon sugar, a nitrogen-containing ring, and one or more phosphate groups are the components of **nucleotides.** For example, adenosine triphosphate (ATP)

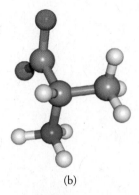

(a)

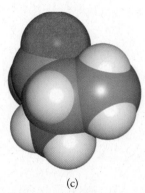

(b)

(c)

Figure 1-3 Representations of alanine. The structural formula (a) indicates all the atoms and the major bonds. Some bonds, such as the C—O and N—H bonds, are implied. Because the central carbon atom has tetrahedral geometry, its four bonds do not lie flat in the plane of the paper: The horizontal bonds actually extend slightly above the plane of the page, and the vertical bonds extend slightly behind it. This tetrahedral arrangement is more accurately depicted in the ball-and-stick model (b). Here, the atoms are color-coded by convention: C is gray, N is blue, O is red, and H is white. This ball-and-stick representation reveals the identities of the atoms and their positions in space but does not indicate their relative size or electrical charge. In a space-filling model (c), each atom is presented as a sphere whose radius corresponds to the distance of closest approach by another atom. This model most accurately depicts the actual size of the molecule, but it obscures some of its atoms and linkages.

contains the nitrogenous group adenine linked to the monosaccharide ribose, to which a triphosphate group is also attached:

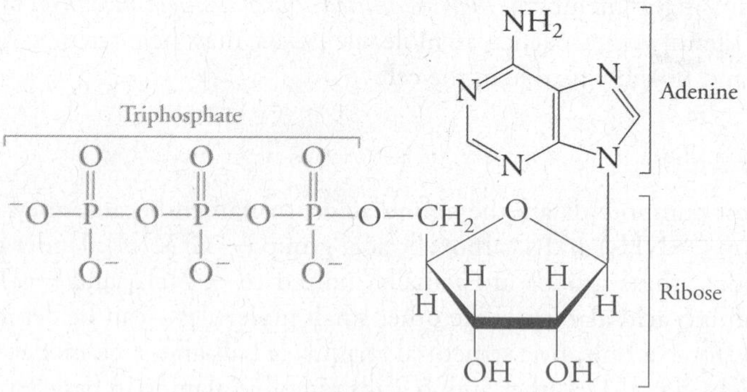

Adenosine triphosphate (ATP)

The most common nucleotides are mono-, di-, and triphosphates containing the nitrogenous ring compounds (or "bases") adenine, cytosine, guanine, thymine, or uracil (abbreviated A, C, G, T, and U).

4. Lipids

The fourth major group of biomolecules consists of the **lipids.** These compounds cannot be described by a single structural formula since they are a diverse collection of molecules. However, they all have in common a tendency to be poorly soluble in water because the bulk of their structure is hydrocarbon-like. For example, palmitic acid consists of a highly insoluble chain of 15 carbons attached to a carboxylic acid group, which is ionized under physiological conditions. The anionic lipid is therefore called palmitate.

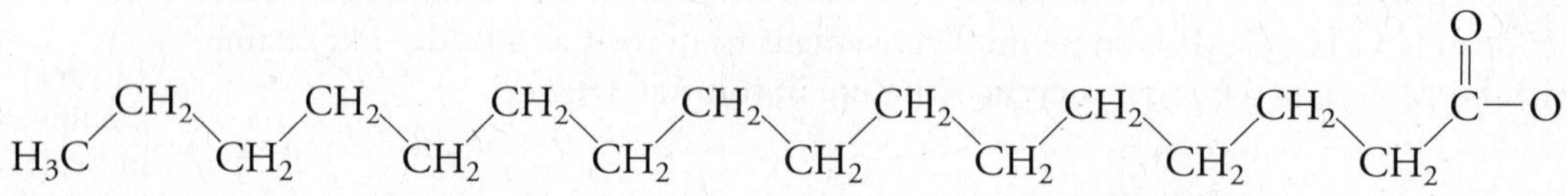

Palmitate

Cholesterol, although it differs significantly in structure from palmitate, is also poorly soluble in water because of its hydrocarbon-like composition.

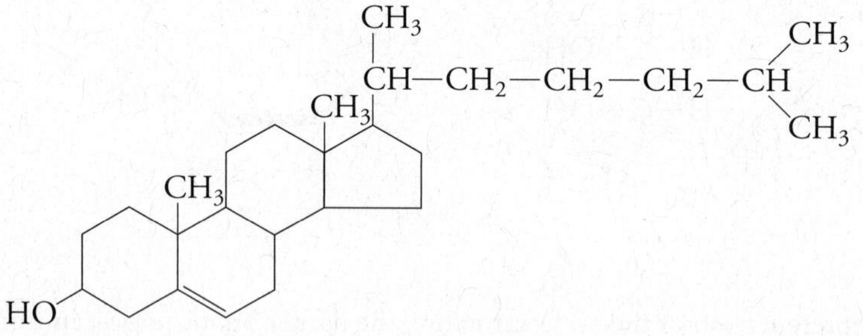

Cholesterol

Cells also contain a few other small molecules that cannot be easily classified into the groups above or that are constructed from molecules belonging to more than one group.

There are three major kinds of biological polymers

In addition to small molecules consisting of relatively few atoms, organisms contain macromolecules that may consist of thousands of atoms. Such huge molecules are not synthesized in one piece but are built from smaller units. This is a universal feature of

BOX 1-A ⬢ **BIOCHEMISTRY NOTE**

7
Biological Molecules

Units Used in Biochemistry

Convention rules the biochemist's choice of terms to quantify objects on a molecular scale. For example, the mass of a molecule can be expressed in atomic mass units; however, the masses of biological molecules—especially very large ones—are typically given without units. Here it is understood that the mass is expressed relative to one-twelfth the mass of an atom of the common carbon isotope ^{12}C (12.011 atomic mass units). Occasionally, units of daltons (D) are used (1 dalton = 1 atomic mass unit), often with the prefix kilo, k (kD). This is useful for macromolecules such as proteins, many of which have masses in the range from 20,000 (20 kD) to over 1,000,000 (1000 kD).

The standard metric prefixes are also necessary for expressing the minute concentrations of biomolecules in living cells. Concentrations are usually given as moles per liter ($mol \cdot L^{-1}$ or M), with the appropriate prefix such as m, μ, or n:

mega (M)	10^6	nano (n)	10^{-9}
kilo (k)	10^3	pico (p)	10^{-12}
milli (m)	10^{-3}	femto (f)	10^{-15}
micro (μ)	10^{-6}		

For example, the concentration of the sugar glucose in human blood is about 5 mM, but many intracellular molecules are present at concentrations of μM or less.

Distances are customarily expressed in angstroms, Å (1 Å = 10^{-10} m) or in nanometers, nm (1 nm = 10^{-9} m). For example, the distance between the centers of carbon atoms in a C—C bond is about 1.5 Å, and the diameter of a DNA molecule is about 20 Å.

⬢ **Question:** The diameter of a typical spherical bacterial cell is about 1 μM. What is the cell's volume?

---

nature: *A few kinds of building blocks can be combined in different ways to produce a wide variety of larger structures.* This is advantageous for a cell, which can get by with a limited array of raw materials. In addition, the very act of chemically linking individual units (**monomers**) into longer strings (**polymers**) is a way of encoding information (the sequence of the monomeric units) in a stable form. Biochemists use certain units of measure to describe both large and small molecules (Box 1-A).

Amino acids, monosaccharides, and nucleotides each form polymeric structures with widely varying properties. In most cases, the individual monomers become covalently linked in head-to-tail fashion:

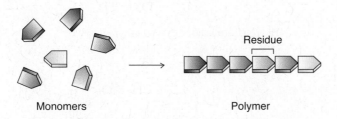

Monomers   Polymer

The linkage between monomeric units is characteristic of each type of polymer. The monomers are called **residues** after they have been incorporated into the polymer. Strictly speaking, lipids do not form polymers, although they do tend to aggregate to form larger structures.

1. Proteins

Polymers of amino acids are called **polypeptides** or **proteins.** Twenty different amino acids serve as building blocks for proteins, which may contain many hundreds of amino acid residues. The amino acid residues are linked to each other by amide bonds called **peptide bonds.** A peptide bond (*arrow*) links the two residues in a dipeptide (the side chains of the amino acids are represented by R_1 and R_2).

$$\overset{+}{H_3N}-\underset{\underset{H}{|}}{\overset{\overset{R_1}{|}}{C}}-\underset{}{\overset{\overset{O}{\parallel}}{C}}-\underset{\underset{H}{|}}{\overset{\overset{H}{|}}{N}}-\underset{\underset{H}{|}}{\overset{\overset{R_2}{|}}{C}}-\overset{\overset{O}{\parallel}}{C}-O^-$$

Because the side chains of the 20 amino acids have different sizes, shapes, and chemical properties, the exact **conformation** (three-dimensional shape) of the polypeptide chain depends on its amino acid composition and sequence. For example, the small polypeptide endothelin, with 21 residues, assumes a compact shape in which the polymer bends and folds to accommodate the functional groups of its amino acid residues (**Fig. 1-4**).

The 20 different amino acids can be combined in almost any order and in almost any proportion to produce myriad polypeptides, all of which have unique three-dimensional shapes. This property makes proteins as a class the most structurally variable and therefore the most functionally versatile of all the biopolymers. Accordingly, *proteins perform a wide variety of tasks in the cell, such as mediating chemical reactions and providing structural support.*

2. Nucleic Acids

Polymers of nucleotides are termed **polynucleotides** or **nucleic acids,** better known as DNA and RNA. Unlike polypeptides, with 20 different amino acids available for polymerization, each nucleic acid is made from just four different nucleotides. For example, the residues in RNA contain the bases adenine, cytosine, guanine, and uracil, whereas the residues in DNA contain adenine, cytosine, guanine, and thymine. Polymerization involves the phosphate and sugar groups of the nucleotides, which become linked by **phosphodiester bonds.**

(a)

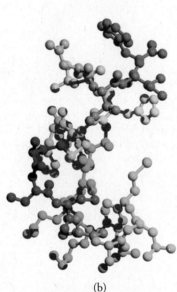

(b)

Figure 1-4 Structure of human endothelin. The 21 amino acid residues of this polypeptide, shaded from blue to red, form a compact structure. In (a), each amino acid residue is represented by a sphere. The ball-and-stick model (b) shows all the atoms except hydrogen. [Structure (pdb 1EDN) determined by B. A. Wallace and R. W. Jones.]

In part because nucleotides are much less variable in structure and chemistry than amino acids, nucleic acids tend to have more regular structures than proteins. *This is in keeping with their primary role as carriers of genetic information, which is contained in their sequence of nucleotide residues rather than in their three-dimensional shape* (**Fig. 1-5**). Nevertheless, some nucleic acids do bend and fold into compact globular shapes, as proteins do.

3. Polysaccharides

Polysaccharides usually contain only one or a few different types of monosaccharide residues, so even though a cell may synthesize dozens of different kinds of monosaccharides, most of its polysaccharides are homogeneous polymers. This tends to limit their potential for carrying genetic information in the sequence of their residues (as nucleic acids do) or for adopting a large variety of shapes and mediating chemical reactions (as proteins do). On the other hand, *polysaccharides perform essential cell functions by serving as fuel-storage molecules and by providing structural support.* For example, plants link the monosaccharide glucose, which is a fuel for virtually all cells, into the polysaccharide starch for long-term storage. The glucose residues are linked by **glycosidic bonds** (the bond is shown in red in this disaccharide):

Glucose monomers are also the building blocks for cellulose, the extended polymer that helps make plant cell walls rigid **(Fig. 1-6)**. The starch and cellulose polymers differ in the arrangement of the glycosidic bonds between glucose residues.

The brief descriptions of biological polymers given above are generalizations, meant to convey some appreciation for the possible structures and functions of these macromolecules. *Exceptions to the generalizations abound.* For example, some small polysaccharides encode information that allows cells bearing the molecules on their

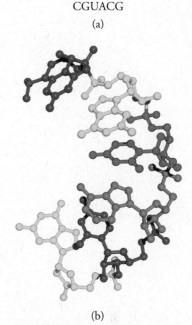

CGUACG

(a)

(b)

Figure 1-5 Structure of a nucleic acid. (a) Sequence of nucleotide residues, using one-letter abbreviations. (b) Ball-and-stick model of the polynucleotide, showing all atoms except hydrogen (this structure is a six-residue segment of RNA). [Structure (pdb ARF0108) determined by R. Biswas, S. N. Mitra, and M. Sundaralingam.]

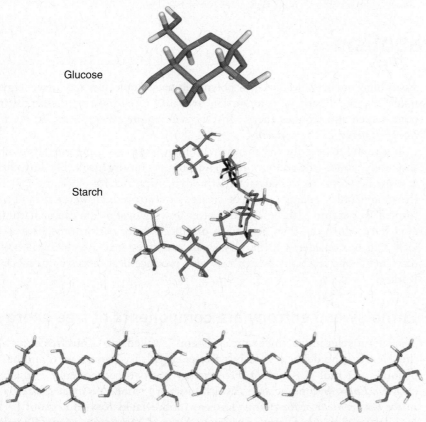

Glucose

Starch

Figure 1-6 Glucose and its polymers. Both starch and cellulose are polysaccharides containing glucose residues. They differ in the type of chemical linkage between the monosaccharide units. Starch molecules have a loose helical conformation, whereas cellulose molecules are extended and relatively stiff.

[**TABLE 1-2**] Functions of Biopolymers

Biopolymer	Encode Information	Carry Out Metabolic Reactions	Store Energy	Support Cellular Structures
Proteins	—	✔	✓	✔
Nucleic acids	✔	✓	—	✓
Polysaccharides	✓	—	✔	✔

✔ major function
✓ minor function

surfaces to recognize each other. Likewise, some nucleic acids perform structural roles, for example, by serving as scaffolding in ribosomes, the small particles where protein synthesis takes place. Under certain conditions, proteins are called on as fuel-storage molecules. A summary of the major and minor functions of proteins, polysaccharides, and nucleic acids is presented in Table 1-2.

CONCEPT REVIEW
- Which six elements are most abundant in biological molecules?
- Name the common functional groups and linkages shown in Table 1-1.
- Give the structural or functional definitions for amino acids, monosaccharides, nucleotides, and lipids.
- Why is it efficient for macromolecules to be polymers?
- What is the relationship between a monomer and a residue?
- Give the structural definitions and major functions of proteins, polysaccharides, and nucleic acids.

1-3 Energy and Metabolism

KEY CONCEPTS
- The free energy of a system is determined by its enthalpy and entropy.
- Living organisms obey the laws of thermodynamics.

Assembling small molecules into polymeric macromolecules requires energy. And unless the monomeric units are readily available, a cell must synthesize the monomers, which also requires energy. In fact, *cells require energy for all the functions of living, growing, and reproducing.*

It is useful to describe the energy in biological systems using the terminology of thermodynamics (the study of heat and power). An organism, like any chemical system, is subject to the laws of thermodynamics. According to the first law of thermodynamics, energy cannot be created or destroyed. However, it can be transformed. For example, the energy of a river flowing over a dam can be harnessed as electricity, which can then be used to produce heat or perform mechanical work. Cells can be considered to be very small machines that use chemical energy to drive metabolic reactions, which may also produce heat or carry out mechanical work.

Enthalpy and entropy are components of free energy

The energy relevant to biochemical systems is called the Gibbs free energy (after the scientist who defined it) or just **free energy.** It is abbreviated G and has units of joules per mol ($J \cdot mol^{-1}$). Free energy has two components: enthalpy and entropy. **Enthalpy** (abbreviated H, with units of $J \cdot mol^{-1}$) *is taken to be equivalent to the heat content of the system.* **Entropy** (abbreviated S, with units of $J \cdot K^{-1} \cdot mol^{-1}$) *is a measure of how the energy is dispersed within that system.* Entropy can therefore be considered to be a measure of the system's disorder or randomness, because the more ways a system's components can be arranged, the more dispersed its energy. For example, consider a pool table at the start of a game when all 15 balls are arranged in one neat triangle (a state of high order or low entropy). After play has begun, the balls are scattered across the table, which is now in a state of disorder and high entropy (Fig. 1-7).

Free energy, enthalpy, and entropy are related by the equation

$$G = H - TS \qquad \text{[1-1]}$$

where T represents temperature in Kelvin (equivalent to degrees Celsius plus 273). Temperature is a coefficient of the entropy term because entropy varies with temperature; the entropy of a substance increases when it is warmed because more thermal energy has been dispersed within it. The enthalpy of a chemical system can be measured, although with some difficulty, but it is next to impossible to measure a system's entropy because this would require counting all the possible arrangements of its components or all the ways its energy could be spread out among them. Therefore, it is more practical to deal with *changes* in these quantities (change is indicated by the Greek letter delta, Δ) so that

$$\Delta G = \Delta H - T\Delta S \qquad \text{[1-2]}$$

Biochemists can measure how the free energy, enthalpy, and entropy of a system differ before and after a chemical reaction. For example, some chemical reactions are accompanied by the release of heat to the surroundings ($H_{after} - H_{before} = \Delta H < 0$), whereas others absorb heat from the surroundings ($\Delta H > 0$). Similarly, the entropy change, $S_{after} - S_{before} = \Delta S$, can be positive or negative. When ΔH and ΔS for a process are known, Equation 1-2 can be used to calculate the value of ΔG at a given temperature (see Sample Calculation 1-1).

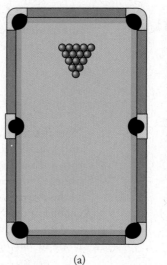
(a)

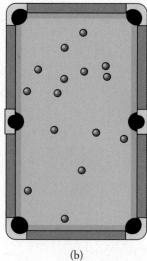
(b)

Figure 1-7 Illustration of entropy. Entropy is a measure of the dispersal of energy in a system, so it reflects the system's randomness or disorder. (a) Entropy is low when all the balls are arranged in a single area of the pool table. (b) Entropy is high after the balls have been scattered, because there are now a large number of different possible arrangements of the balls on the table.

? Compare the entropy of a ball of yarn before and after a cat has played with it.

SAMPLE CALCULATION 1-1 ●●●

PROBLEM

Use the information below to calculate the change in enthalpy and the change in entropy for the reaction A → B.

	Enthalpy (kJ · mol⁻¹)	Entropy (J · K⁻¹ · mol⁻¹)
A	60	22
B	75	97

SOLUTION

$$\begin{aligned} \Delta H &= H_B - H_A \\ &= 75\ \text{kJ} \cdot \text{mol}^{-1} - 60\ \text{kJ} \cdot \text{mol}^{-1} \\ &= 15\ \text{kJ} \cdot \text{mol}^{-1} \\ &= 15{,}000\ \text{J} \cdot \text{mol}^{-1} \end{aligned}$$

$$\begin{aligned} \Delta S &= S_B - S_A \\ &= 97\ \text{J} \cdot \text{K}^{-1} \cdot \text{mol}^{-1} \\ &\quad -22\ \text{J} \cdot \text{K}^{-1} \cdot \text{mol}^{-1} \\ &= 75\ \text{J} \cdot \text{K}^{-1} \cdot \text{mol}^{-1} \end{aligned}$$

PRACTICE PROBLEMS ●●●

1. The change in enthalpy for a reaction is 8 kJ · mol⁻¹. Is heat absorbed or given off during the reaction?
2. A reacting system undergoes a change from ordered to disordered. Is the entropy change positive or negative?

ΔG is less than zero for a spontaneous process

A china cup dropped from a great height will break, but the pieces will never reassemble themselves to restore the cup. The thermodynamic explanation is that the broken pieces have less free energy than the intact cup. *In order for a process to occur, the overall change in free energy (ΔG) must be negative.* For a chemical reaction, this means that the free energy of the products must be less than the free energy of the reactants:

$$\Delta G = G_{products} - G_{reactants} < 0 \qquad \text{[1-3]}$$

When ΔG is less than zero, the reaction is said to be **spontaneous** or **exergonic**. A **nonspontaneous** or **endergonic** reaction has a free energy change greater than zero; in this case, the reverse reaction is spontaneous.

$$A \rightarrow B \qquad B \rightarrow A$$
$$\Delta G > 0 \qquad \Delta G < 0$$
$$\text{Nonspontaneous} \quad \text{Spontaneous}$$

Note that thermodynamic spontaneity does not indicate how *fast* a reaction occurs, only whether it will occur as written. (The rate of a reaction depends on other factors, such as the concentrations of the reacting molecules, the temperature, and the presence of a catalyst.) When a reaction, such as $A \rightarrow B$, is at equilibrium, the rate of the forward reaction is equal to the rate of the reverse reaction, so there is no net change in the system. In this situation, $\Delta G = 0$.

A quick examination of Equation 1-2 reveals that *a reaction that occurs with a decrease in enthalpy and an increase in entropy is spontaneous at all temperatures because ΔG is always less than zero.* These results are consistent with everyday experience. For example, heat moves spontaneously from a hot object to a cool object, and items that are neatly arranged tend to become disordered, never the other way around. (This is a manifestation of the second law of thermodynamics, which states that energy tends to spread out.) Accordingly, reactions in which the enthalpy increases and entropy decreases do not occur. If enthalpy and entropy both increase or both decrease during a reaction, the value of ΔG then depends on the temperature, which governs whether the $T\Delta S$ term of Equation 1-2 is greater than or less than the ΔH term. This means that a large increase in entropy can offset an unfavorable (positive) change in enthalpy. Conversely, the release of a large amount of heat ($\Delta H < 0$) during a reaction can offset an unfavorable decrease in entropy (see Sample Calculation 1-2).

●●● **SAMPLE CALCULATION 1-2**

PROBLEM

Use the information given in Sample Calculation 1-1 to determine whether the reaction $A \rightarrow B$ is spontaneous at 25°C.

SOLUTION

Substitute the values for ΔH and ΔS, calculated in Sample Calculation 1-1, into Equation 1-2. To express the temperature in Kelvin, add 273 to the temperature in degrees Celsius: $273 + 25 = 298$ K.

$$\begin{aligned}\Delta G &= \Delta H - T\Delta S \\ &= 15{,}000 \text{ J}\cdot\text{mol}^{-1} - 298 \text{ K} (75 \text{ J}\cdot\text{K}^{-1}\cdot\text{mol}^{-1}) \\ &= 15{,}000 - 22{,}400 \text{ J}\cdot\text{mol}^{-1} \\ &= -7400 \text{ J}\cdot\text{mol}^{-1} \\ &= -7.4 \text{ kJ}\cdot\text{mol}^{-1}\end{aligned}$$

Because ΔG is less than zero, the reaction is spontaneous. Even though the change in enthalpy is unfavorable, the large increase in entropy makes ΔG favorable.

●●● **PRACTICE PROBLEMS**

3. If $\Delta H = -15{,}000$ J $\cdot$ mol^{-1} and $\Delta S = -75$ J $\cdot$ K^{-1} $\cdot$ mol^{-1}, would the reaction be spontaneous at 25°C?

4. Use the information given in Sample Calculation 1-1 to determine the temperature at which $\Delta G = 0$.

Life is thermodynamically possible

In order to exist, life must be thermodynamically spontaneous. Does this hold at the molecular level? When analyzed in a test tube (***in vitro,*** literally "in glass"), many of a cell's metabolic reactions have free energy changes that are less than zero, but some reactions do not. Nevertheless, the nonspontaneous reactions are able to proceed ***in vivo*** (in a living organism) because they occur in concert with other reactions that

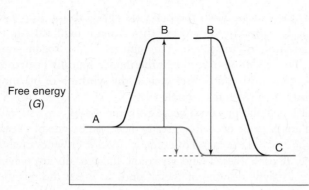

Figure 1-8 Free energy changes in coupled reactions. A nonspontaneous reaction, such as A → B, which has a positive value of ΔG, can be coupled to another reaction, B → C, which has a negative value of ΔG and is therefore spontaneous. The reactions are coupled because the product of the first reaction, B, is a reactant for the second reaction.

? **Which reaction occurs spontaneously in reverse: C → B, B → A, or C → A?**

are thermodynamically favorable. Consider two reactions *in vitro*, one nonspontaneous ($\Delta G > 0$) and one spontaneous ($\Delta G < 0$):

$$A \dashrightarrow B \qquad \Delta G = +15 \text{ kJ} \cdot \text{mol}^{-1} \quad \text{(nonspontaneous)}$$
$$B \rightarrow C \qquad \Delta G = -20 \text{ kJ} \cdot \text{mol}^{-1} \quad \text{(spontaneous)}$$

When the reactions are combined, their ΔG values are added, so the overall process has a negative change in free energy:

$$A + B \rightarrow B + C \qquad \Delta G = (15 \text{ kJ} \cdot \text{mol}^{-1}) + (-20 \text{ kJ} \cdot \text{mol}^{-1})$$
$$A \rightarrow C \qquad \Delta G = -5 \text{ kJ} \cdot \text{mol}^{-1}$$

This phenomenon is shown graphically in **Figure 1-8.** In effect, the unfavorable "uphill" reaction A → B is pulled along by the more favorable "downhill" reaction B → C.

Cells couple unfavorable metabolic processes with favorable ones so that the net change in free energy is negative. Note that it is permissible to add ΔG values because the free energy, G, depends only on the initial and final states of the system, without regard to the specific chemical or mechanical work that occurred in going from one state to the other.

Most macroscopic life on earth today is sustained by the energy of the sun (this was not always the case, nor is it true of all organisms). In photosynthetic organisms, such as green plants, light energy excites certain molecules so that their subsequent chemical reactions occur with a net negative change in free energy. These thermodynamically favorable (spontaneous) reactions are coupled to the unfavorable synthesis of monosaccharides from atmospheric CO_2 **(Fig. 1-9).** In this process, the carbon is **reduced.** Reduction, the gain of electrons, is accomplished by the addition of hydrogen or the removal of oxygen (the oxidation states of carbon are reviewed in Table 1-3). The plant—or an animal that eats

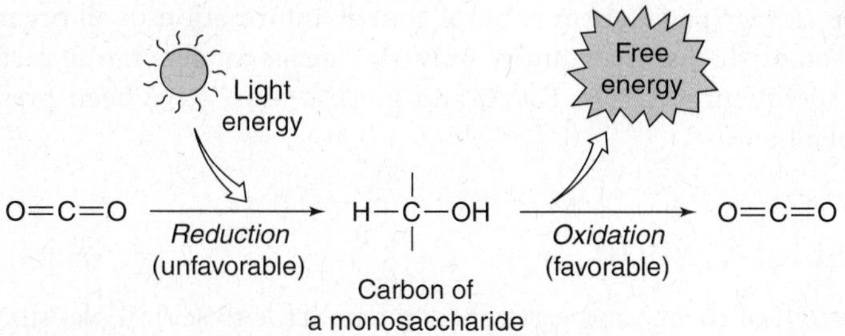

Figure 1-9 Reduction and reoxidation of carbon compounds. The sun provides the free energy to convert CO_2 to reduced compounds such as monosaccharides. The reoxidation of these compounds to CO_2 is thermodynamically spontaneous, so free energy can be made available for other metabolic processes. Note that free energy is not actually a substance that is physically released from a molecule.

[**TABLE 1-3**]	
Oxidation States of Carbon	
Compound[a]	**Formula**
Carbon dioxide *most oxidized* (*least reduced*)	$O{=}C{=}O$
Acetic acid	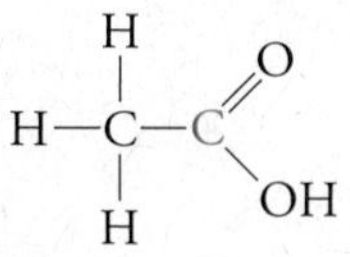 H—C—C with H, and C double bond O, OH
Carbon monoxide	$C{\equiv}O$
Formic acid	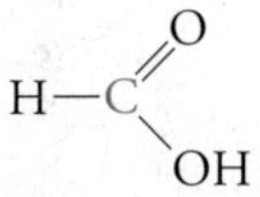 H—C double bond O, OH
Acetone	H—C—C—C—H with H's, middle C double bond O
Acetaldehyde	H—C—C with H's, C double bond O, H
Formaldehyde	H—C double bond O, H
Acetylene	$H{-}C{\equiv}C{-}H$
Ethanol	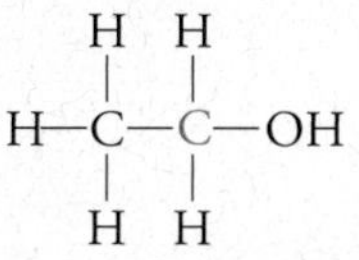 H—C—C—OH with H's
Ethene	C=C with H's
Ethane	H—C—C—H with H's
Methane *least oxidized* (*most reduced*)	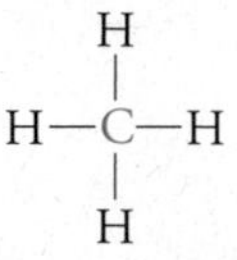 H—C—H with H's

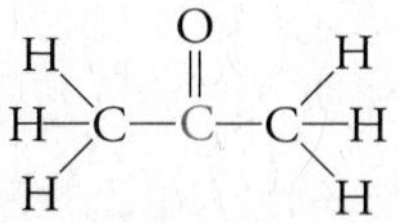

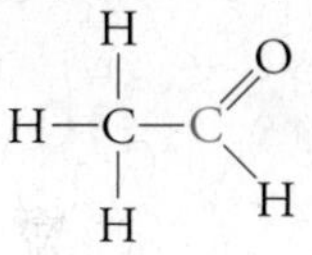

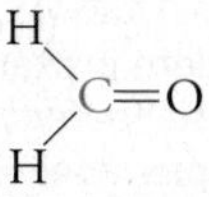

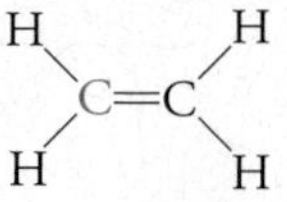

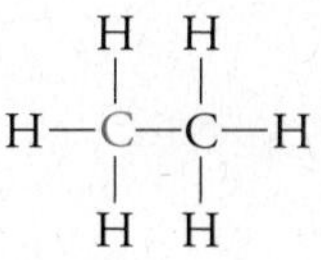

[a]Compounds are listed in order of decreasing oxidation state of the red carbon atom.

the plant—can then break down the monosaccharide to use it as a fuel to power other metabolic activities. In the process, the carbon is **oxidized**—it loses electrons through the addition of oxygen or the removal of hydrogen—and ultimately becomes CO_2. The oxidation of carbon is thermodynamically favorable, so it can be coupled to energy-requiring processes such as the synthesis of building blocks and their polymerization to form macromolecules.

Virtually all metabolic processes occur with the aid of catalysts called **enzymes,** most of which are proteins (a catalyst greatly increases the rate of a reaction without itself undergoing any net change). For example, specific enzymes catalyze the formation of peptide, phosphodiester, and glycosidic linkages during polymer synthesis. Other enzymes catalyze cleavage of these bonds to break the polymers into their monomeric units.

A living organism—with its high level of organization of atoms, molecules, and larger structures—represents a state of low entropy relative to its surroundings. Yet the organism can maintain this thermodynamically unfavorable state as long as it continually obtains free energy from its food. This is the answer to the question posed at the start of the chapter: Living organisms do indeed obey the laws of thermodynamics. When the organism ceases to obtain a source of free energy from its surroundings or exhausts its stored food, the chemical reactions in its cells reach equilibrium ($\Delta G = 0$), which results in death.

▶▶ **DO** living systems obey the laws of thermodynamics?

> **CONCEPT REVIEW**
> - What do enthalpy and entropy mean and how are they related to free energy?
> - What does the value of ΔG reveal about a biochemical process?
> - How can thermodynamically unfavorable reactions proceed *in vivo*?
> - Why must an organism have a steady supply of food?
> - Describe the cycle of carbon reduction and oxidation in photosynthesis and in the breakdown of a compound such as a monosaccharide.

1-4 The Origin and Evolution of Life

> **KEY CONCEPTS**
> - Modern prokaryotic and eukaryotic cells apparently evolved from simpler nonliving systems.
> - The three domains of life are bacteria, archaea, and eukarya.

Every living cell originates from the division of a parental cell. Thus, the ability to **replicate** (make a replica or copy of itself) is one of the universal characteristics of living organisms. *In order to leave descendants that closely resemble itself, a cell must contain a set of instructions—and the means for carrying them out—that can be transmitted from generation to generation.* Over time, the instructions change gradually, so that species also change, or evolve. By carefully examining an organism's genetic information and the cellular machinery that supports it, biochemists can draw some conclusions about the organism's relationship to more ancient life-forms. The history of evolution is therefore contained not just within the fossil record but also in the molecular makeup of all living cells. For example, nucleic acids participate in the storage and transmission of genetic information in all organisms, and the oxidation of glucose is an almost universal means for generating metabolic free energy. Consequently, DNA, RNA, and glucose must have been present in the ancestor of all cells.

The prebiotic world

A combination of theory and experimental data leads to several plausible scenarios for the emergence of life from nonbiological (prebiotic) materials on the early earth. In one scenario, inorganic compounds such as H_2, H_2O, NH_3, and CH_4—which may have been present in the early atmosphere—could have given rise to simple biomolecules, such as amino acids, when sparked by lightning. Laboratory experiments with the same raw materials and electrical discharges to simulate lightning do

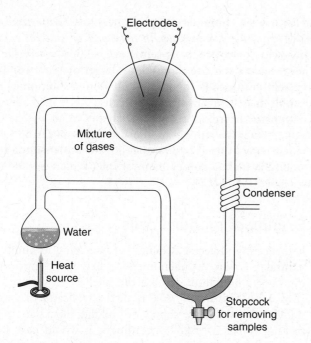

Figure 1-10 Laboratory synthesis of biological molecules. A mixture of gases—H_2, H_2O, NH_3, and CH_4—is subject to an electrical discharge. Newly formed compounds, such as amino acids, accumulate in the aqueous phase as water vapor condenses. Samples of the reaction products can be removed via the stopcock.

in fact yield these molecules (**Fig. 1-10**). Other experiments suggest that hydrogen cyanide (HCN), formaldehyde (HCOH), and phosphate could have been converted to nucleotides with a similarly modest input of energy.

Over time, simple molecular building blocks could have accumulated and formed larger structures, particularly in shallow waters where evaporation would have had a concentrating effect. Eventually, conditions would have been ripe for the assembly of functional, living cells. Charles Darwin proposed that life might have arisen in some "warm little pond"; however, the early earth was probably a much more violent place, with frequent meteorite impacts and volcanic activity.

In an alternative scenario, supported by studies of the metabolism of some modern bacteria, the first cells could have developed at deep-sea hydrothermal vents, some of which are characterized by temperatures as high as 350°C and clouds of gaseous H_2S and metal sulfides (giving them the name "black smokers"; **Fig. 1-11**). In the laboratory, incubating a few small molecules in the presence of iron sulfide and nickel sulfide at 100°C yields acetic acid, an organic compound with a newly formed C—C bond:

$$CH_3SH \ + \ CO \ + \ H_2O \xrightarrow{\text{FeS, NiS}} CH_3COOH + H_2S$$

Methyl   Carbon   Acetic acid
thiol   monoxide

Under similar conditions, amino acids spontaneously form short polypeptides. Although the high temperatures that are necessary for their synthesis also tend to break them down, these compounds would have been stable in the cooler water next to the hydrothermal vent.

Regardless of how they formed, *the first biological building blocks would have had to polymerize.* This process might have been stimulated when the organic molecules—often bearing anionic (negatively charged) groups—aligned themselves on a cationic (positively charged) mineral surface.

Figure 1-11 A hydrothermal vent. Life may have originated at these "black smokers," where high temperatures, H_2S, and metal sulfides might have stimulated the formation of biological molecules. [B. Murton/Southhampton Oceanography Centre/Science Photo Library/ Photo Researchers.]

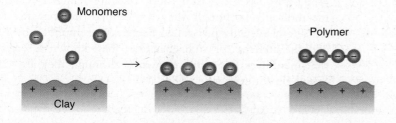

In fact, in the laboratory, common clay promotes the polymerization of nucleotides into RNA. *Primitive polymers would have had to gain the capacity for self-replication.* Otherwise, no matter how stable or chemically versatile, such molecules would never have given rise to anything larger or more complicated: The probability of assembling a fully functional cell from a solution of thousands of separate small molecules is practically nil. Because RNA in modern cells represents a form of genetic information and participates in all aspects of expressing that information, it may be similar to the first self-replicating biopolymer. It might have made a copy of itself by first making a **complement,** a sort of mirror image, that could then make a complement of *itself,* which would be identical to the original molecule (**Fig. 1-12**).

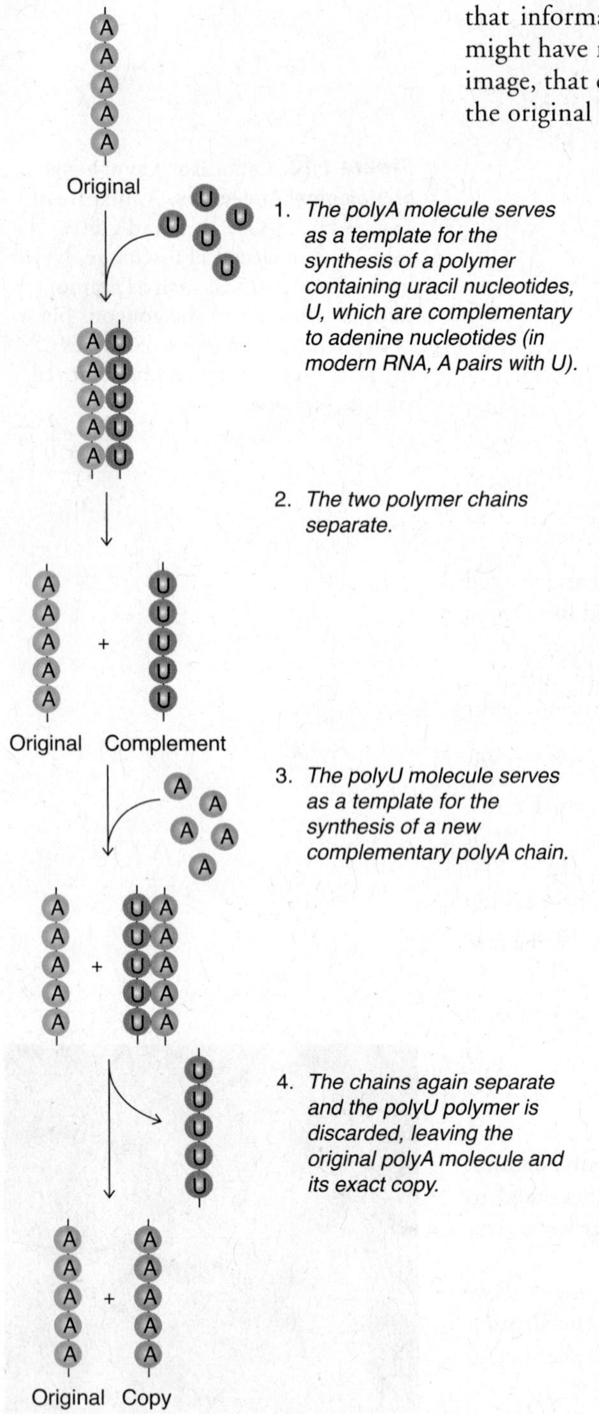

Origins of modern cells

A replicating molecule's chances of increasing in number depend on **natural selection,** the phenomenon whereby the entities best suited to the prevailing conditions are the likeliest to survive and multiply (Box 1-B). This would have favored a replicator that was chemically stable and had a ready supply of building blocks and free energy for making copies of itself. Accordingly, it would have been advantageous to become enclosed in some sort of membrane that could prevent valuable small molecules from diffusing away. Natural selection would also have favored replicating systems that developed the means for synthesizing their own building blocks and for more efficiently harnessing sources of free energy.

The first cells were probably able to "fix" CO_2—that is, convert it to reduced organic compounds—using the free energy released in the oxidation of readily available inorganic compounds such as H_2S or Fe^{2+}. Vestiges of these processes can be seen in modern metabolic reactions that involve sulfur and iron.

Later, photosynthetic organisms similar to present-day cyanobacteria (also called blue-green algae) used the sun's energy to fix CO_2:

$$CO_2 + H_2O \rightarrow (CH_2O) + O_2$$

The concomitant oxidation of H_2O to O_2 dramatically increased the concentration of atmospheric O_2, about 2.4 billion years ago, and made it possible for **aerobic** (oxygen-using) organisms to avail themselves of this powerful oxidizing agent. The **anaerobic** origins of life are still visible in the most basic metabolic reactions of modern organisms; these reactions proceed in the absence of oxygen. Now that the earth's atmosphere contains about 18% oxygen, anaerobic organisms have not disappeared, but they have been restricted to microenvironments where O_2 is scarce, such as the digestive systems of animals or underwater sediments.

The earth's present-day life-forms are of two types, which are distinguished by their cellular architecture:

1. *Prokaryotes are small unicellular organisms that lack a discrete nucleus and usually contain no internal membrane systems.* This group comprises two subgroups that are remarkably different metabolically, although they are similar in appearance: the eubacteria (usually just called **bacteria**), exemplified by *E. coli,* and the **archaea** (or archaebacteria), best known as organisms that inhabit extreme environments, although they are actually found almost everywhere (**Fig. 1-13**).

1. The polyA molecule serves as a template for the synthesis of a polymer containing uracil nucleotides, U, which are complementary to adenine nucleotides (in modern RNA, A pairs with U).

2. The two polymer chains separate.

3. The polyU molecule serves as a template for the synthesis of a new complementary polyA chain.

4. The chains again separate and the polyU polymer is discarded, leaving the original polyA molecule and its exact copy.

Figure 1-12 Possible mechanism for the self-replication of a primitive RNA molecule. For simplicity, the RNA molecule is shown as a polymer of adenine nucleotides, A.

? Draw a diagram showing how polyU would be replicated.

BOX 1-B ⬡ **BIOCHEMISTRY NOTE**

17
The Origin and Evolution of Life

How Does Evolution Work?

Documenting evolutionary change is relatively straightforward, but the mechanisms whereby evolution occurs are prone to misunderstanding. Populations change over time, and new species arise as a result of natural selection. Selection operates on individuals, but its effects can be seen in a population only over a period of time. Most populations are collections of individuals that share an overall genetic makeup but also exhibit small variations due to random alterations (mutations) in their genetic material as it is passed from parent to offspring. In general, the survival of an individual depends on how well suited it is to the particular conditions under which it lives.

Individuals whose genetic makeup grants them the greatest rate of survival have more opportunities to leave offspring with the same genetic makeup. Consequently, their characteristics become widespread in a population, and, over time, the population appears to adapt to its environment. A species that is well suited to its environment tends to persist; a poorly adapted species fails to reproduce and therefore dies out.

Because evolution is the result of random variations and changing probabilities for successful reproduction, it is inherently random and unpredictable. Furthermore, natural selection acts on the raw materials at hand. It cannot create something out of nothing but must operate in increments. For example, the insect wing did not suddenly appear in the offspring of a wingless parent but most likely developed bit by bit, over many generations, by modification of a gill or heat-exchange appendage. Each step of the wing's development would have been subject to natural selection, eventually making an individual that bore the appendage more likely to survive, perhaps by being able to first glide and then actually fly in pursuit of food or to evade predators.

Although we tend to think of evolution as an imperceptibly slow process, occurring on a geological time scale, it is ongoing and accessible to observation and quantification in the laboratory. For example, under optimal conditions, the bacterium *Escherichia coli* requires only about 20 minutes to produce a new generation. In the laboratory, a culture of *E. coli* cells can progress through about 2500 generations in a year (in contrast, 2500 human generations would require about 60,000 years). Hence, it is possible to subject a population of cultured bacterial cells to some "artificial" selection—for example, by making an essential nutrient scarce—and observe how the genetic composition of the population changes over time as it adapts to the new conditions.

● **Question:** Why can't acquired (rather than genetic) characteristics serve as the raw material for evolution?

2. ***Eukaryotic** cells are usually larger than prokaryotic cells and contain a nucleus and other membrane-bounded cellular compartments* (such as mitochondria, chloroplasts, and endoplasmic reticulum). Eukaryotes may be unicellular or multicellular. This group (also called the **eukarya**) includes microscopic organisms as well as familiar macroscopic plants and animals (**Fig. 1-14**).

By analyzing the sequences of nucleotides in certain genes that are present in all species, it is possible to construct a diagram that indicates how the bacteria, archaea, and eukarya are related. *The number of sequence differences between two groups of organisms indicates how long ago they diverged from a common ancestor:* Species with similar sequences have a longer shared evolutionary history than species with dissimilar sequences. This sort of analysis has produced the evolutionary tree shown in **Figure 1-15**.

The evolutionary history of eukaryotes is complicated by the fact that eukaryotic cells exhibit characteristics of both bacteria and archaea. Eukaryotic cells also contain organelles that are almost certainly the descendants of free-living cells. Specifically, the chloroplasts of plant cells, which carry out photosynthesis, closely resemble the

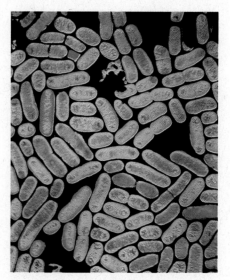

Figure 1-13 Prokaryotic cells. These single-celled *Escherichia coli* bacteria lack a nucleus and internal membrane systems.
[E. Gray/Science Photo Library/Photo Researchers.]

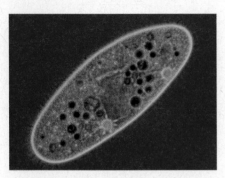

Figure 1-14 **A eukaryotic cell.** The paramecium, a one-celled organism, contains a nucleus and other membrane-bounded compartments. [Dr. David Patterson/Science Photo Library/Photo Researchers.]

? Compare this cell to the bacterial cells shown in Figure 1-13.

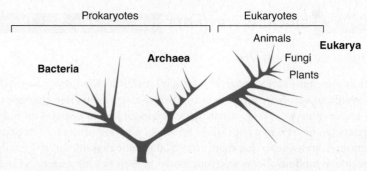

Figure 1-15 **Evolutionary tree based on nucleotide sequences.** This diagram reveals that the bacteria separated before the archaea and eukarya diverged. Note that the closely spaced fungi, plants, and animals are actually more similar to each other than are many groups of prokaryotes. [After Wheelis, M. L., Kandler, O., and Woese, C. R., *Proc. Natl. Acad. Sci. USA* 89, 2930–2934 (1992).]

photosynthetic cyanobacteria. The mitochondria of plant and animal cells, which are the site of much of the eukaryotic cell's aerobic metabolism, resemble certain bacteria. In fact, both chloroplasts and mitochondria contain their own genetic material and protein-synthesizing machinery.

It is likely that an early eukaryotic cell developed gradually from a mixed population of prokaryotic cells. Over many generations of living in close proximity and sharing each other's metabolic products, some of these cells became incorporated within a single larger cell. This arrangement would account for the mosaic-like character of modern eukaryotic cells (**Fig. 1-16**).

At some point, cells in dense populations might have traded their individual existence for a colonial lifestyle. This would have allowed for a division of labor as cells became specialized and would have eventually produced multicellular organisms.

The earth currently sustains about 9 million different species (although estimates vary widely). Perhaps some 500 million species have appeared and vanished over the course of evolutionary history. It is unlikely that the earth harbors more than a few mammals that have yet to be discovered, but new microbial species are routinely described. And although the number of known prokaryotes (about 10,000) is much less than the number of known eukaryotes (for example, there are about 900,000 known species of insects), prokaryotic metabolic strategies are amazingly varied. Nevertheless, by documenting characteristics that are common to all species, we can derive far-reaching conclusions about what life is made of, what sustains it, and how it has developed over the eons.

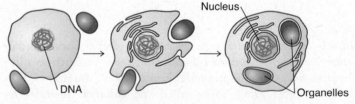

Figure 1-16 **Possible origin of eukaryotic cells.** The close association of different kinds of free-living cells gradually led to the modern eukaryotic cell, which appears to be a mosaic of bacterial and archaeal features and contains organelles that resemble whole bacterial cells.

CONCEPT REVIEW
- Explain how simple prebiotic compounds could give rise to biological monomers and polymers.
- Why is replication a requirement for life?
- Why were the first organisms anaerobic, and why are they now relatively scarce?
- Describe the differences between prokaryotes and eukaryotes.
- Why do eukaryotic cells appear to be mosaics?

[SUMMARY]

1-2 Biological Molecules

• The most abundant elements in biological molecules are H, C, N, O, P, and S, but a variety of other elements are also present in living systems.
• The major classes of small molecules in cells are amino acids, monosaccharides, nucleotides, and lipids. The major types of biological polymers are proteins, nucleic acids, and polysaccharides.

1-3 Energy and Metabolism

• Free energy has two components: enthalpy (heat content) and entropy (disorder). Free energy decreases in a spontaneous process.

• Life is thermodynamically possible because unfavorable endergonic processes are coupled to favorable exergonic processes.

1-4 The Origin and Evolution of Life

• The earliest cells may have evolved in concentrated solutions of molecules or near hydrothermal vents.
• Eukaryotic cells contain membrane-bounded organelles. Prokaryotic cells, which are smaller and simpler, include the bacteria and the archaea.

[GLOSSARY TERMS]

bioinformatics	polynucleotide	*in vivo*
trace element	nucleic acid	reduction
amino acid	phosphodiester bond	oxidation
carbohydrate	polysaccharide	enzyme
monosaccharide	glycosidic bond	replication
nucleotide	free energy (G)	complement
lipid	enthalpy (H)	natural selection
monomer	entropy (S)	aerobic
polymer	ΔG	anaerobic
residue	spontaneous process	prokaryote
polypeptide	exergonic reaction	bacteria
protein	nonspontaneous process	archaea
peptide bond	endergonic reaction	eukaryote
conformation	*in vitro*	eukarya

BIOINFORMATICS ▶ **PROJECT 1**   Learn to use the PubMed database to search for published papers on a given topic and to retrieve information from published reports.

THE BIOCHEMICAL LITERATURE

[PROBLEMS]

1-2 Biological Molecules

1. Identify the functional groups in the following molecules:

(a)
$$^{+}H_3N-CH-\overset{\overset{\displaystyle O}{\|}}{C}-O^-$$
$$CH_2$$
$$C=O$$
$$NH_2$$

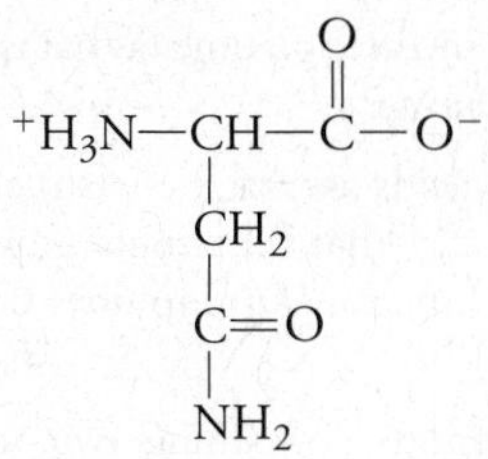

(b)
$$COOH$$
$$C=O$$
$$CH_2$$
$$H-C-OH$$
$$CH_3-\overset{\overset{\displaystyle O}{\|}}{C}-NH-C-H$$
$$HO-C-H$$
$$H-C-OH$$
$$H-C-OH$$
$$CH_2OH$$

(c)
$$CH_2-O-\overset{\overset{\displaystyle O}{\|}}{C}-(CH_2)_{14}-CH_3$$
$$HO-CH \quad\quad O$$
$$CH_2-O-\overset{}{\underset{\overset{\displaystyle |}{OH}}{P}}-OH$$

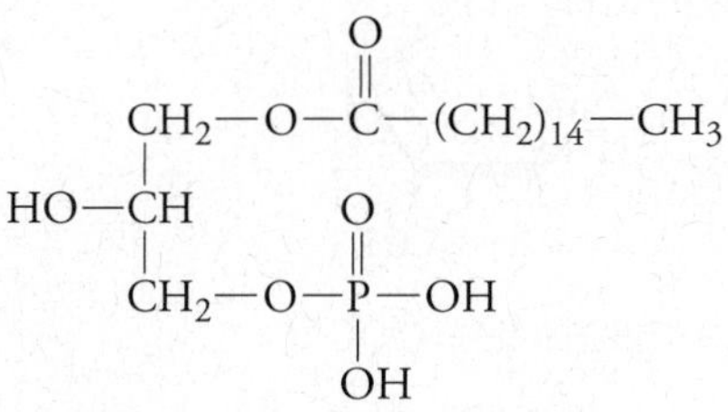

(d)
$$^{+}H_3N-CH-\overset{\overset{\displaystyle O}{\|}}{C}-O^-$$
$$CH_2$$
$$SH$$

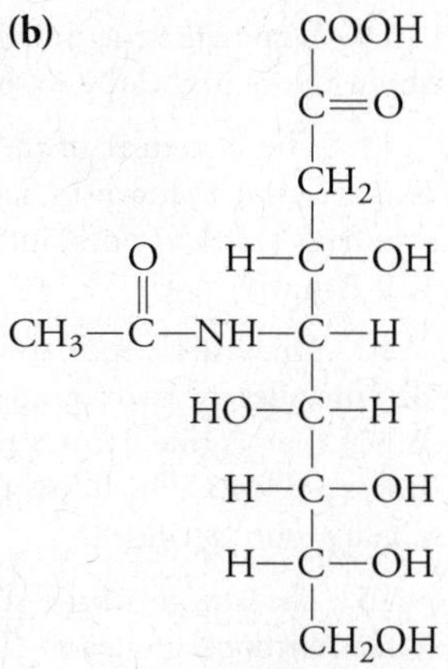

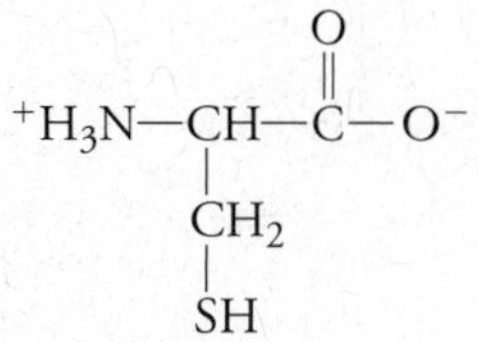

(e)

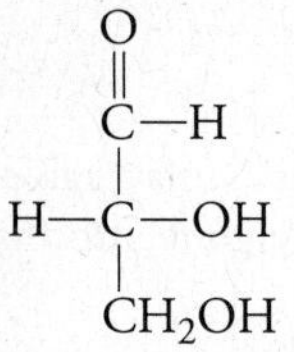

$$\begin{array}{c} O \\ \parallel \\ C-H \\ | \\ H-C-OH \\ | \\ CH_2OH \end{array}$$

(f)

$$\begin{array}{c} O \\ \parallel \\ C-O^- \\ | \\ HN \end{array}$$

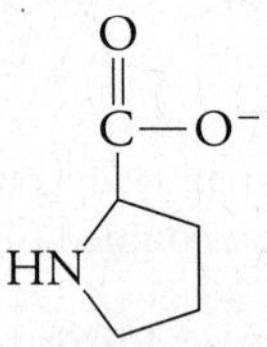

2. The structures of several vitamins are shown below. Identify the functional groups in each vitamin.

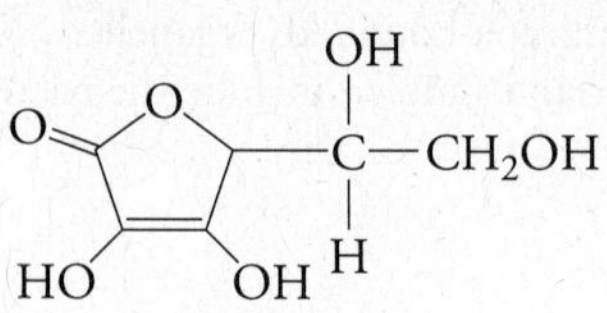

Vitamin C

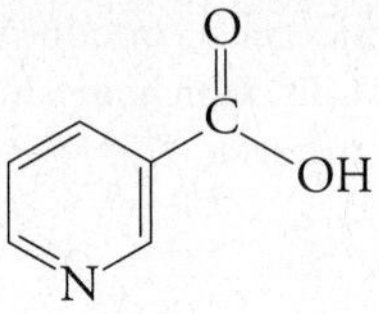

Nicotinic acid (niacin)

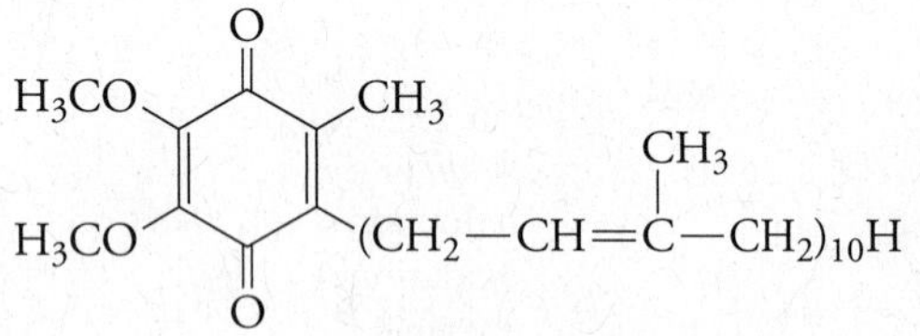

Coenzyme Q

3. Name the four types of small biological molecules. Which three are capable of forming polymeric structures? What are the names of the polymeric structures that are formed?

4. To which of the four classes of biomolecules do the following compounds belong?

(a)

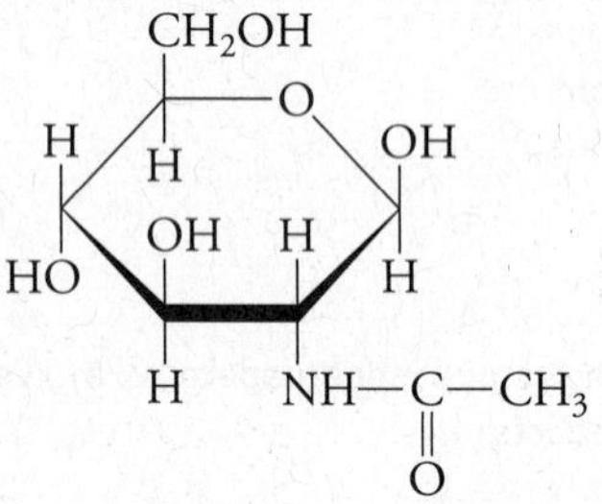

(b)

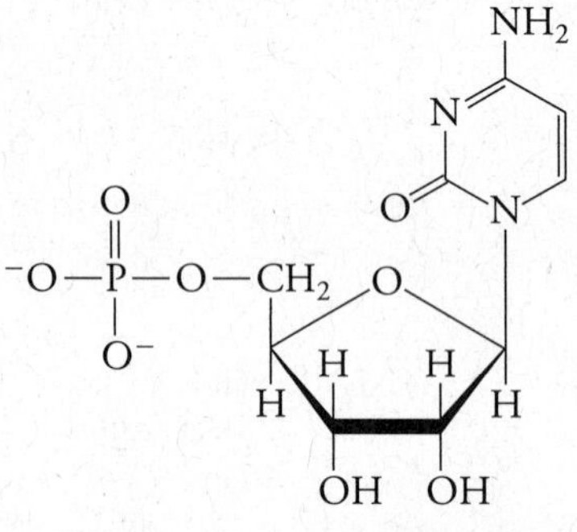

(c) $HS-CH_2-CH_2-CH-COO^-$ with $\overset{|}{NH_3^+}$

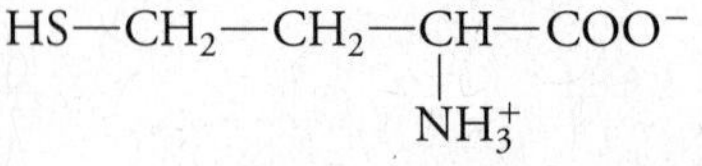

(d)

$R-\overset{O}{\underset{\parallel}{C}}-O-$

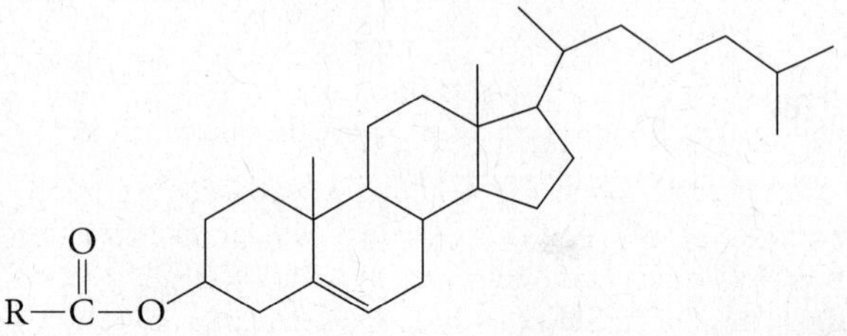

5. The nutritive quality of food can be analyzed by measuring the chemical elements it contains. Most foods are mixtures of the three major types of molecules: fats (lipids), carbohydrates, and proteins.
 (a) What elements are present in fats?
 (b) What elements are present in carbohydrates?
 (c) What elements are present in proteins?

6. A compound that is present in many foods has the formula $C_{44}H_{86}O_8NP$. To which class of molecules does this compound belong? Explain your answer.

7. A healthy diet must include some protein.
 (a) Assuming you had a way to measure the amount of each element in a sample of food, which element would you measure in order to tell whether the food contained protein?
 (b) The structures of three compounds are shown below. Based on your answer to part (a), which of the three compounds would you add to a food sample so that it would appear to contain more protein? Explain.
 (c) Which of the three compounds would already be present in a food sample that actually did contain protein? Explain.

(a) **(b)** 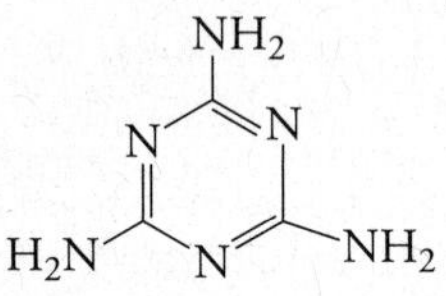 **(c)**

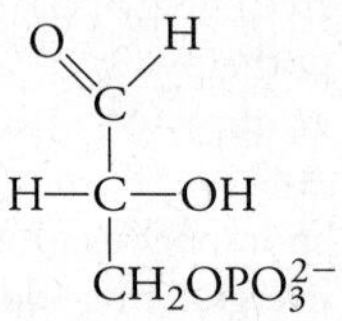

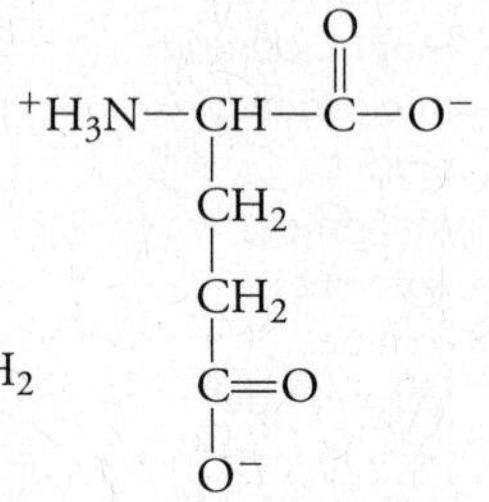

8. The structure of the compound urea is shown. Urea is a waste product of metabolism excreted by the kidneys into the urine. Why do doctors tell patients with kidney damage that they should consume a low-protein diet?

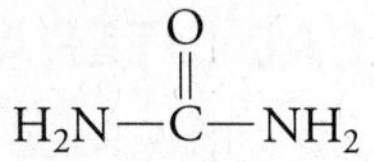

$$H_2N-\overset{O}{\underset{\parallel}{C}}-NH_2$$
Urea

9. There are 20 different amino acids that occur in proteins (see Fig. 4-2). Each has the same basic structure with the exception of the R group, which is unique to each amino acid. What functional groups are present in all amino acids?

10. Draw the structure of the amino acid alanine. What is special about the central carbon atom of alanine?

11. The structures of the amino acids asparagine (Asn) and cysteine (Cys) are shown in Section 1-2. What functional group does Asn have that Cys does not? What functional group does Cys have that Asn does not?

12. Draw a dipeptide (a polypeptide containing two residues) that includes the two amino acids Asn and Cys shown in Section 1-2. When the peptide bond between the two residues is formed, which atoms are lost? Which functional groups are lost? Which new functional group is formed?

13. The "straight-chain" structure of glucose is shown in Section 1-2. What functional groups are present in the glucose molecule?

14. Consider the monosaccharide fructose.
(a) How does its molecular formula differ from that of glucose?
(b) How does its structure differ from the structure of glucose?

$$
\begin{array}{c}
CH_2OH \\
| \\
C=O \\
| \\
HO-C-H \\
| \\
H-C-OH \\
| \\
H-C-OH \\
| \\
CH_2OH
\end{array}
$$

Fructose

15. The structures of the nitrogenous bases uracil and cytosine are shown below. How do their functional groups differ?

Uracil   Cytosine

16. What are the structural components of the biological molecules called nucleotides?

17. Compare the solubilities in water of alanine, glucose, palmitate, and cholesterol, and explain your reasoning.

18. Cell membranes are largely hydrophobic structures. Which compound will pass through a membrane more easily, glucose or 2,4-dinitrophenol? Explain.

2,4-Dinitrophenol

19. What polymeric molecule forms a more regular structure, DNA or protein? Explain this observation in terms of the cellular roles of the two different molecules.

20. What are the two major biological roles of polysaccharides?

21. Pancreatic amylase digests the glycosidic bonds that link glucose residues together in starch. Would you expect this enzyme to digest the glycosidic bonds in cellulose as well? Explain why or why not.

22. The complete digestion of starch in mammals yields 4 kilocalories per gram (see Problem 21). What is the energy yield for cellulose?

1-3 Energy and Metabolism

23. What is the sign of the entropy change for each of the following processes?
(a) Water freezes.
(b) Water evaporates.
(c) Dry ice sublimes.
(d) Sodium chloride dissolves in water.

(e) Several different types of lipid molecules assemble to form a membrane.

24. Does entropy increase or decrease in the following reactions in aqueous solution?

(a)
$$
\begin{array}{c}
COO^- \\
| \\
C=O \ \ + \ CO_2(g) \longrightarrow \\
| \\
CH_3
\end{array}
\quad
\begin{array}{c}
COO^- \\
| \\
C=O \\
| \\
CH_2 \\
| \\
COO^-
\end{array}
$$

(b)
$$
\begin{array}{c}
COO^- \\
| \\
C=O \ \ + \ H^+ \longrightarrow \\
| \\
CH_3
\end{array}
\quad
\begin{array}{c}
H \\
| \\
C=O \ \ + \ CO_2(g) \\
| \\
CH_3
\end{array}
$$

25. Which has the greater entropy, a polymeric molecule or a mixture of its constituent monomers?

26. How does the entropy change when glucose undergoes combustion?

$$C_6H_{12}O_6 + 6\,O_2 \rightarrow 6\,CO_2 + 6\,H_2O$$

27. A soccer coach keeps a couple of instant cold packs in her bag in case one of her players suffers a muscle injury. Instant cold packs are composed of a plastic bag containing a smaller water bag and solid ammonium nitrate. In order to activate the cold pack, the bag is kneaded until the smaller water bag breaks, which allows the released water to dissolve the ammonium nitrate. The equation for the dissolution of ammonium nitrate in water is shown below. How does the cold pack work?

$$NH_4NO_3(s) \xrightarrow{H_2O} NH_4^+(aq) + NO_3^-(aq)$$
$$\Delta H = 26.4 \ kJ \cdot mol^{-1}$$

28. Campers carry hot packs with them, especially when camping during the winter months or at high altitudes. Their design is similar to that described in Problem 27, except that calcium chloride is used in place of the ammonium nitrate. The equation for the dissolution of calcium chloride in water is shown below. How does the hot pack work?

$$CaCl_2(s) \xrightarrow{H_2O} Ca^{2+}(aq) + 2Cl^-(aq) \quad \Delta H = -81 \ kJ \cdot mol^{-1}$$

29. For the reaction in which reactant A is converted to product B, tell whether this process is favorable at (a) 4°C and (b) 37°C.

	$H\,(kJ \cdot mol^{-1})$	$S\,(J \cdot K^{-1} \cdot mol^{-1})$
A	54	22
B	60	43

30. For a given reaction, the value of ΔH is 15 kJ · mol^{-1} and the value of ΔS is 51 J·K^{-1} · mol^{-1}. Above what temperature will this reaction be spontaneous?

31. The hydrolysis of pyrophosphate at 25°C is spontaneous. The enthalpy change for this reaction is −14.3 kJ · mol^{-1}. What is the sign and the magnitude of ΔS for this reaction?

32. Phosphoenolpyruvate donates a phosphate group to ADP to produce pyruvate and ATP. The ΔG value for this reaction at 25°C is −63 kJ · mol^{-1} and the value of ΔS is 190 J · K^{-1} · mol^{-1}. What is the value of ΔH? Is the reaction exothermic or endothermic?

33. Which of the following processes are spontaneous?
(a) a reaction that occurs with any size decrease in enthalpy and any size increase in entropy
(b) a reaction that occurs with a small increase in enthalpy and a large increase in entropy
(c) a reaction that occurs with a large decrease in enthalpy and a small decrease in entropy
(d) a reaction that occurs with any size increase in enthalpy and any size decrease in entropy

34. When a stretched rubber band is allowed to relax, it feels cooler.
(a) Is the enthalpy change for this process positive or negative?
(b) Since a stretched rubber band spontaneously relaxes, what can you conclude about the entropy change during relaxation?

35. Urea (NH_2CONH_2) dissolves readily in water; that is, this is a spontaneous process. The beaker containing the dissolved compound is cold to the touch. What conclusions can you make about the sign of the **(a)** enthalpy change and **(b)** entropy change for this process?

36. Phosphofructokinase catalyzes the transfer of a phosphate group (from ATP) to fructose-6-phosphate to produced fructose-1,6-bisphosphate. The ΔH value for this reaction is -9.5 kJ $\cdot$ mol^{-1} and the ΔG is -17.2 kJ $\cdot$ mol^{-1}.
(a) Is the reaction exothermic or endothermic?
(b) What is the value of ΔS for the reaction? Does this reaction proceed with an increase or decrease in entropy?
(c) Which component makes a greater contribution to the free energy value—the ΔH or ΔS value? Comment on the significance of this observation.

37. Glucose can be converted to glucose-6-phosphate:

$$\text{glucose} + \text{phosphate} \rightleftharpoons \text{glucose-6-phosphate} + H_2O$$
$$\Delta G = 13.8 \text{ kJ} \cdot \text{mol}^{-1}$$

(a) Is this reaction favorable? Explain.
(b) Suppose the synthesis of glucose-6-phosphate is coupled with the hydrolysis of ATP. Write the overall equation for the coupled process and calculate the ΔG of the coupled reaction. Is the conversion of glucose to glucose-6-phosphate favorable under these conditions? Explain.

$$\text{ATP} + H_2O \rightleftharpoons \text{ADP} + \text{phosphate} \quad \Delta G = -30.5 \text{ kJ} \cdot \text{mol}^{-1}$$

38. Glyceraldehyde-3-phosphate (GAP) is converted to 1,3-bisphosphoglycerate (1,3BPG) as shown below.

$$\text{GAP} + P_i + \text{NAD}^+ \rightleftharpoons 1,3\text{BPG} + \text{NADH}$$
$$\Delta G = +6.7 \text{ kJ} \cdot \text{mol}^{-1}$$

(a) Is this reaction spontaneous? The reaction shown above is coupled to the following reaction in which 1,3BPG is converted to 3-phosphoglycerate (3PG):

$$1,3\text{BPG} + \text{ADP} \rightleftharpoons 3\text{PG} + \text{ATP} \quad \Delta G = -18.8 \text{ kJ} \cdot \text{mol}^{-1}$$

(b) Write the equation for the overall conversion of GAP to 3PG. Is the coupled reaction favorable?

39. Place these molecules in order from the most oxidized to the most reduced.

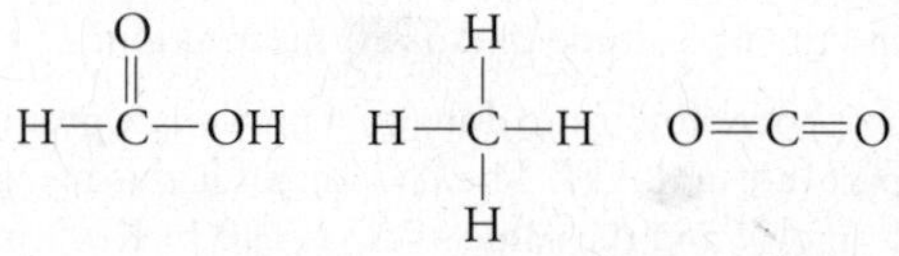

A $\quad$ B $\quad$ C

40. Identify the process described in the statements below as an oxidation or reduction process.
(a) Monosaccharides are synthesized from carbon dioxide by plants during photosynthesis.
(b) An animal eats the plant and breaks down the monosaccharide in order to obtain energy for cellular processes.

41. Given the following reactions, tell whether the reactant is being oxidized or reduced. Reactions may not be balanced.

(a) $CH_3-(CH_2)_{14}-C(=O)-O^- \longrightarrow 8\ CH_3-C(=O)-S-CoA$

(b) (reaction diagram)

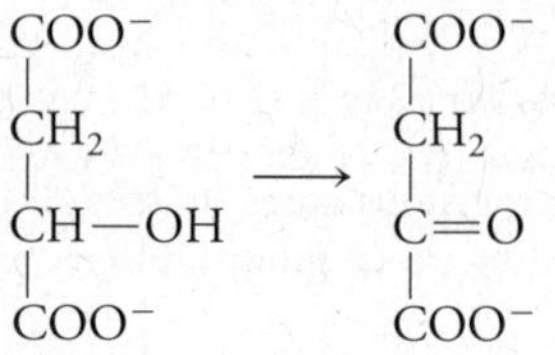

(c) (reaction diagram)

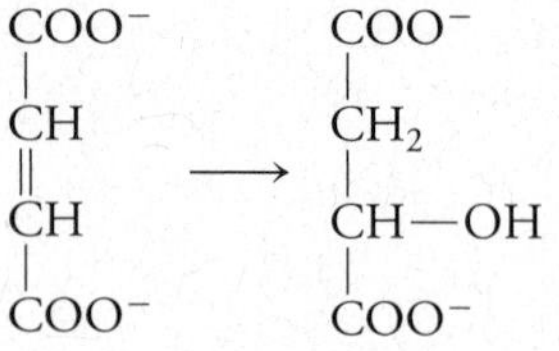

(d) (reaction diagram) $+ H_2 \longrightarrow 2\ ^+H_3N-CH-C(=O)-O^-$

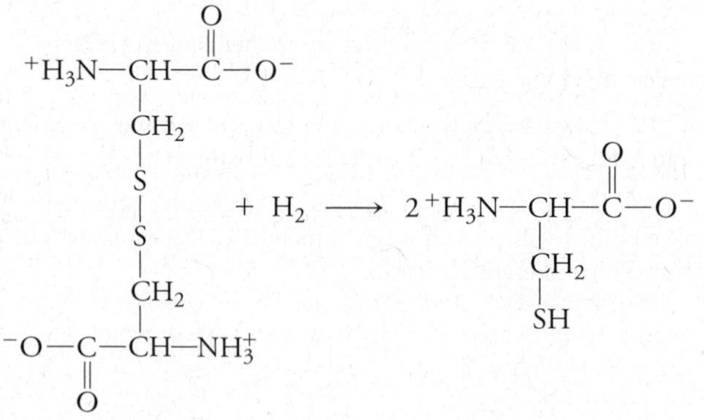

42. For each of the reactions in Problem 41, tell whether an oxidizing agent or a reducing agent is needed to accomplish the reaction.

43. In some cells, lipids such as palmitate (shown in Section 1-2), rather than monosaccharides, serve as the primary metabolic fuel.
(a) Consider the oxidation state of palmitate's carbon atoms and explain how it fits into a scheme such as the one shown in Figure 1-9.
(b) On a per-carbon basis, which would make more free energy available for metabolic reactions: palmitate or glucose?

44. Which yields more free energy when completely oxidized, stearate or α-linolenate?

$$H_3C-(CH_2)_{16}-COO^-$$
Stearate

$$H_3C-CH_2-(CH=CHCH_2)_3-(CH_2)_6-COO^-$$
α-Linolenate

1-4 The Origin and Evolution of Life

45. In the 1920s, Oparin and Haldane independently suggested that the energy of lightning storms in the prebiotic world could have transformed gases in the early atmosphere to small organic molecules. In 1953, Miller and Urey conducted an experiment demonstrating that this scenario is plausible. They subjected a mixture of water, methane, ammonia, and hydrogen gas to electric discharge and refluxed the mixture so that any compounds formed would dissolve and accumulate in the water. At the end of a week, they analyzed the solution and found glycine, alanine, lactic acid, urea, and other amino acids and small organic acids. What was the significance of this experiment?

46. In order to give rise to more highly complex structures, what capabilities did the first biological molecules have to have?

47. Why is molecular information so important for classifying and tracing the evolutionary relatedness of bacterial species but less important for vertebrate species?

48. The first theories to explain the similarities between bacteria and mitochondria or chloroplasts suggested that an early eukaryotic cell actually engulfed but failed to fully digest a free-living prokaryotic cell. Why is such an event unlikely to account for the origin of mitochondria or chloroplasts?

49. Draw a simple evolutionary tree that shows the relationships between species A, B, and C based on the DNA sequences given here.

Species A	T C G T C G A G T C
Species B	T G G A C T A G C C
Species B	T G G A C C A G C C

50. A portion of the evolutionary tree for a flu virus is shown here. Different strains are identified by an H followed by a number.

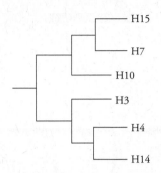

(a) Identify two pairs of closely related flu strains.
(b) Which strain(s) is(are) most closely related to strain H3?

[SELECTED READINGS]

Koonin, E. V., The origin and early evolution of eukaryotes in the light of phylogenomics, *Genome Biol.* **11**, 209 (2010). [Discusses several models for the origin of eukaryotic cells.]

Koshland, D. E., Jr., The seven pillars of life, *Science* **295**, 2215–2216 (2002). [Describes some of the essential attributes of all organisms, including a DNA program, ability to mutate, compartmentalization, need for energy, ability to regenerate, adaptability, and seclusion.]

Mora, C., Tittensor, D. P., Adl, S., Simpson, A. G. B., and Worm, B., How many species are there on earth and in the ocean? *PLoS Biol* **9(8):** e1001127. doi:10.1371/journal.pbio.1001127 (2011). [Shows how statistical analysis of databases can be used to estimate the total number of known and not-yet-described species.]

Nee, S., More than meets the eye, *Nature* **429**, 804–805 (2004). [A brief commentary about appreciating the metabolic diversity of microbial life.]

Nisbet, E. G., and Sleep, N. H., The habitat and nature of early life, *Nature* **409**, 1083–1091 (2001). [Explains some of the hypotheses regarding the early earth and the origin of life, including the possibility that life originated at hydrothermal vents.]

Tinoco, I., Jr., Sauer, K., Wang, J. C., and Puglisi, J., *Physical Chemistry: Principles and Applications in Biological Sciences* (4th ed.), Chapters 2–5, Prentice Hall (2002). [This and other physical chemistry textbooks present the basic equations of thermodynamics.]

chapter

2 AQUEOUS CHEMISTRY

[stuartbur/iStockphoto]

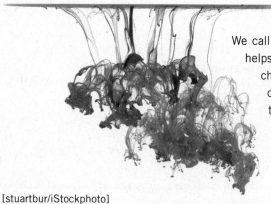

▸▸ WHY do so many substances dissolve in water?

We call water the medium of life because water surrounds biological molecules and helps determine their shapes and chemical reactivity. But why is this so? In this chapter we will see how water interacts with other substances, such as the green dye shown here—either through electrostatic or hydrophobic effects—in ways that are not possible for other types of solvents.

THIS CHAPTER IN CONTEXT

Part 1 Foundations

2 Aqueous Chemistry

Part 2 Molecular Structure and Function

Part 3 Metabolism

Part 4 Genetic Information

Do You Remember?

- **Biological molecules are composed of a subset of all possible elements and functional groups (Section 1-2).**
- **The free energy of a system is determined by its enthalpy and entropy (Section 1-3).**

Living organisms can be found in virtually every portion of the earth that contains liquid water. In the polar ice caps, prokaryotes and small eukaryotes survive in the spaces between ice crystals (**Fig. 2-1**). The hot waters near hydrothermal vents host the prokaryote *Pyrolobus fumarii*, which grows best at a temperature of 105°C and can tolerate 113°C. Living organisms have even been discovered several kilometers below the earth's surface, but only where water is present.

Water is a fundamental requirement for life, so it is important to understand the structural and chemical properties of water. Not only are most biological molecules surrounded by water, but their molecular structure is in part governed by how their component groups interact with water. And water plays a role in how these molecules assemble to form larger structures or undergo chemical transformation. In fact, water itself—or its H^+ and OH^- constituents—participates directly in many biochemical processes. Therefore, an examination of water is a logical prelude to exploring the structures and functions of biomolecules in the following chapters.

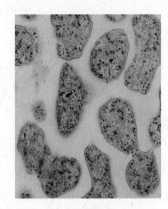

Figure 2-1 *Methanococcoides burtonii.* These bacterial cells from an Antarctic lake survive at temperatures as low as $-2.5°C$. [M. Rohde, GBF/Science Photo Library/Photo Researchers.]

2-1 Water Molecules Form Hydrogen Bonds

What is the nature of the substance that accounts for about 70% of the mass of most organisms? The human body, for example, is about 60% by weight water, most of it in the extracellular fluid (the fluid surrounding cells) and inside cells:

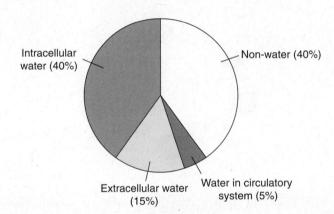

Intracellular water (40%)

Non-water (40%)

Extracellular water (15%)

Water in circulatory system (5%)

KEY CONCEPTS
- The polar water molecule forms hydrogen bonds with other molecules.
- Noncovalent forces, including hydrogen bonds, ionic interactions, and van der Waals forces, act on biological molecules.
- Water dissolves both ionic and polar substances.

In an individual H_2O molecule, the central oxygen atom forms covalent bonds with two hydrogen atoms, leaving two unshared pairs of electrons. The molecule therefore has approximately tetrahedral geometry, with the oxygen atom at the center of the tetrahedron, the hydrogen atoms at two of the four corners, and electrons at the other two corners (**Fig. 2-2**).

As a result of this electronic arrangement, *the water molecule is* **polar**; that is, it has an uneven distribution of charge. The oxygen atom bears a partial negative charge (indicated by the symbol δ^-), and each hydrogen atom bears a partial positive charge (indicated by the symbol δ^+):

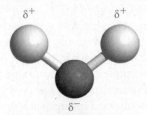

This polarity is the key to many of water's unique physical properties.

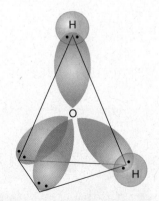

Figure 2-2 Electronic structure of the water molecule. Four electron orbitals, in an approximately tetrahedral arrangement, surround the central oxygen. Two orbitals participate in bonding to hydrogen (gray), and two contain unshared electron pairs.

Neighboring water molecules tend to orient themselves so that each partially positive hydrogen is aligned with a partially negative oxygen:

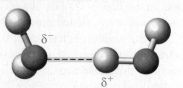

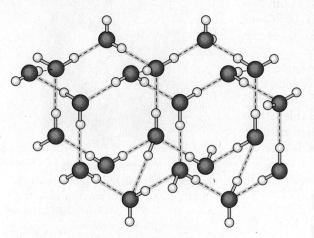

Figure 2-3 **Structure of ice.** Each water molecule acts as a donor for two hydrogen bonds and an acceptor for two hydrogen bonds, thereby interacting with four other water molecules in the crystal. (Only two layers of water molecules are shown here.)

? **Identify a hydrogen bond donor and acceptor in this structure.**

This interaction, shaded yellow here, is known as a **hydrogen bond.** Traditionally shown as a simple electrostatic attraction between oppositely charged particles, the hydrogen bond is now known to have some covalent character. This means that the bond has directionality, or a preferred orientation.

Each water molecule can potentially participate in four hydrogen bonds, since it has two hydrogen atoms to "donate" to a hydrogen bond and two pairs of unshared electrons that can "accept" a hydrogen bond. In ice, a crystalline form of water, each water molecule does indeed form hydrogen bonds with four other water molecules (**Fig. 2-3**). This regular, lattice-like structure breaks down when the ice melts.

In liquid water, each molecule can potentially form hydrogen bonds with up to four other water molecules, but each bond has a lifetime of only about 10^{-12} s. As a result, *the structure of water is continually flickering as water molecules rotate, bend, and reorient themselves.* Theoretical calculations and spectroscopic data suggest that water molecules participate in only two strong hydrogen bonds, one as a donor and one as an acceptor, generating transient hydrogen-bonded clusters such as the six-membered ring shown here:

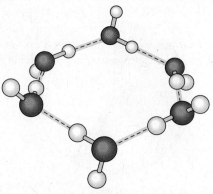

Because of its ability to form hydrogen bonds, water is highly cohesive. This accounts for its high surface tension, which allows certain insects to walk on water (**Fig. 2-4**). The cohesiveness of water molecules also explains why water remains a liquid, whereas molecules of similar size, such as CH_4 and H_2S, are gases at room temperature (25°C). At the same time, water is less dense than other liquids because hydrogen bonding demands that individual molecules not just approach each other but interact with a certain orientation. This geometrical constraint also explains why ice floats; for other materials, the solid is denser than the liquid.

Hydrogen bonds are one type of electrostatic force

Powerful covalent bonds define basic molecular constitutions, but much weaker noncovalent bonds, including hydrogen bonds, govern the final three-dimensional shapes of molecules and how they interact with each other. For example, about $460 \text{ kJ} \cdot \text{mol}^{-1}$ (110 kcal · mol^{-1}) of energy is required to break a covalent O—H bond. But a hydrogen bond in water has a strength of only about $20 \text{ kJ} \cdot \text{mol}^{-1}$ (4.8 kcal · mol^{-1}). Other noncovalent interactions are weaker still.

Among the noncovalent interactions that occur in biological molecules are electrostatic interactions between charged groups such as carboxylate (—COO$^-$) and amino (—NH$_3^+$) groups. These **ionic interactions** are intermediate in strength to covalent bonds and hydrogen bonds (**Fig. 2-5**).

Figure 2-4 **A water strider supported by the surface tension of water.**
[Hermann Eisenbeiss/Photo Research, Inc.]

Hydrogen bonds, despite their partial covalent nature, are classified as a type of electrostatic interaction. At about 1.8 Å, they are longer and hence weaker than a covalent O—H bond (which is about 1 Å long). However, a completely noninteracting O and H would approach no closer than about 2.7 Å, which is the sum of their **van der Waals radii** (the van der Waals radius of an isolated atom is the distance from its nucleus to its effective electronic surface).

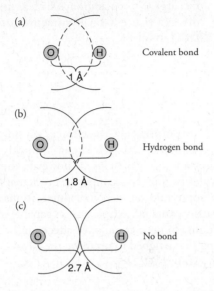

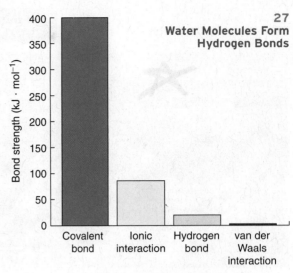

Figure 2-5 Relative strengths of bonds in biological molecules.

Hydrogen bonds usually involve N—H, O—H, and S—H groups as hydrogen donors and the electronegative N, O, or S atoms as hydrogen acceptors (**electronegativity** is a measure of an atom's affinity for electrons; Table 2-1). *Water, therefore, can form hydrogen bonds not just with other water molecules but with a wide variety of other compounds that bear N-, O-, or S-containing functional groups.*

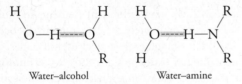

Likewise, these functional groups can form hydrogen bonds among themselves. For example, the complementarity of bases in DNA and RNA is determined by their ability to form hydrogen bonds with each other:

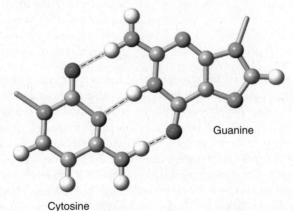

Other electrostatic interactions occur between particles that are polar but not actually charged, for example, two carbonyl groups:

$$\text{C}=\overset{\delta^-}{\text{O}}\text{-----}\overset{\delta^+}{\text{C}}=\text{O}$$

TABLE 2-1	
Electronegativities of Some Elements	
Element	**Electronegativity**
C	2.55
F	3.98
H	2.20
N	3.04
O	3.44

Figure 2-6 The cumulative effect of small forces. Just as the fictional giant Gulliver was restrained by many small tethers at the hands of the tiny Lilliputians, the structures of macromolecules are constrained by the effects of many weak noncovalent interactions. [Hulton Archive/Getty Images.]

These forces, called **van der Waals interactions,** are usually weaker than hydrogen bonds. The interaction shown on the previous page, between two strongly polar groups, is known as a **dipole–dipole interaction** and has a strength of about 9 kJ · mol^{-1}. Very weak van der Waals interactions, called **London dispersion forces,** occur between nonpolar molecules as a result of small fluctuations in their distribution of electrons that create a temporary separation of charge. Nonpolar groups such as methyl groups can therefore experience a small attractive force, in this case about 0.3 kJ · mol^{-1}:

Not surprisingly, these forces act only when the groups are very close, and their strength quickly falls off as the groups draw apart. If the groups approach too closely, however, their van der Waals radii collide and a strong repulsive force overcomes the attractive force.

Although hydrogen bonds and van der Waals interactions are individually weak, biological molecules usually contain multiple groups capable of participating in these intermolecular interactions, so their cumulative effect can be significant (**Fig. 2-6**). Drug molecules are typically designed to optimize the weak interactions that govern their therapeutic activity (Box 2-A).

BOX 2-A 🔷 **BIOCHEMISTRY NOTE**

Why Do Some Drugs Contain Fluorine?

As mentioned in Section 1-2, the most abundant elements in biological molecules are H, C, N, O, P, and S. Fluorine only rarely appears in naturally occurring organic compounds. Why, then, do about one-quarter of all drug molecules, including the widely prescribed Prozac (fluoxetine, an antidepressant; Box 9-C), fluorouracil (an anticancer agent; Section 7-3), and Ciprofloxacin (an antibacterial agent; Section 20-1), contain F?

Prozac (Fluoxetine)

In designing an effective drug, pharmaceutical scientists often intentionally introduce F in order to alter the drug's chemical or biological properties without significantly altering its shape. Fluorine can take the place of hydrogen in a chemical structure, but with its high electronegativity (see Table 2-1), F behaves much more like O than H. Consequently, transforming a relatively inert C—H group into an electron-withdrawing C—F group can decrease the basicity of nearby amino groups (see Section 2-3). Fewer positive charges in a drug allow it to more easily pass through membranes to enter cells and exert its biological effect.

In addition, the polar C—F bond can participate in hydrogen bonding (C—F · · · H—C) or other dipole–dipole interactions (such as C—F · · · C=O), potentially augmenting the intermolecular attraction between a drug and its target molecule in the body. Better binding usually means that the drug will be effective at lower concentrations and will have fewer side effects.

🔷 **Question:** Identify the hydrogen-bonding groups in Prozac.

Water dissolves many compounds

The answer to the question we asked at the start of the chapter is that, unlike most other solvent molecules, water molecules are able to form hydrogen bonds and participate in other electrostatic interactions with a wide variety of compounds. Water has a relatively high **dielectric constant,** which is a measure of a solvent's ability to diminish the electrostatic attractions between dissolved ions (Table 2-2). The higher the dielectric constant of the solvent, the less able the ions are to associate with each other. The polar water molecules surround ions (for example, the Na^+ and Cl^- ions from the salt NaCl) by aligning their partial charges with the oppositely charged ions. Because the interactions between the polar water molecules and the ions are stronger than the attractive forces between the Na^+ and Cl^- ions, the salt dissolves (the dissolved particle is called a **solute**). Each solute ion surrounded by water molecules (shown at right) is said to be **solvated** (or **hydrated,** to indicate that the solvent is water).

Biological molecules that bear polar or ionic functional groups are also readily solubilized, in this case because the groups can form hydrogen bonds with the solvent water molecules. Glucose, for example, with its six hydrogen-bonding oxygens, is highly soluble in water:

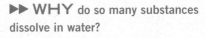

▶▶ **WHY** do so many substances dissolve in water?

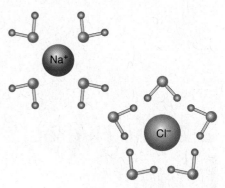

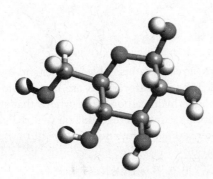

The concentration of glucose in human blood is about 5 mM. In a solution of 5 mM glucose in water, there are about 10,000 water molecules for every glucose molecule (the water molecules are present at a concentration of about 55.5 M). However, biological molecules are never found alone in such dilute conditions *in vivo*, because a large number of small molecules, large polymers, and macromolecular aggregates collectively form a solution that is more like a hearty stew than a thin, watery soup (Fig. 2-7).

Inside a cell, the spaces between molecules may be only a few Å wide, enough room for only two water molecules to fit. This allows solute molecules, each with a coating of properly oriented water molecules, to slide past each other. This thin coating, or shell, of water may be enough to keep molecules from coming into

Figure 2-7 Portion of a *Dictyostelium* cell visualized by cryoelectron tomography. In this technique, the cells are rapidly frozen so that they retain their fine structure, and two-dimensional electron micrographs taken from different angles are merged to re-create a three-dimensional image. The red structures are filaments of the protein actin, ribosomes and other macromolecular complexes are colored green, and membranes are blue. Small molecules (not visible) fill the spaces between these larger cell components. [Courtesy Wolfgang Baumeister, Max Planck Institute for Biochemistry.]

[TABLE 2-2] Dielectric Constants for Some Solvents at Room Temperature

Solvent	Dielectric Constant
Formamide ($HCONH_2$)	109
Water	80
Methanol (CH_3OH)	33
Ethanol (CH_3CH_2OH)	25
1-Propanol ($CH_3CH_2CH_2OH$)	20
1-Butanol ($CH_3CH_2CH_2CH_2OH$)	18
Benzene (C_6H_6)	2

? **Compare the hydrogen-bonding ability of these solvents.**

van der Waals contact (van der Waals interactions are weak but attractive), thereby helping maintain the cell's contents in a crowded but fluid state.

> **CONCEPT REVIEW**
> • Why is a water molecule polar?
> • What is a hydrogen bond, and why does it form between water molecules?
> • Describe the structure of liquid water.
> • Describe the nature and relative strength of ionic interactions, hydrogen bonds, and van der Waals interactions.
> • What happens when an ionic substance dissolves?
> • Explain why water is a more effective solvent than ammonia or methanol.

2-2 The Hydrophobic Effect

> **KEY CONCEPTS**
> • The hydrophobic effect, which is driven by entropy, excludes nonpolar substances from water.
> • Amphiphilic molecules form micelles or bilayers.

Glucose and other readily hydrated substances are said to be **hydrophilic** (water-loving). In contrast, a compound such as dodecane (a C_{12} alkane),

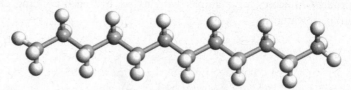

which lacks polar groups, is relatively insoluble in water and is said to be **hydrophobic** (water-fearing). Although pure hydrocarbons are rare in biological systems, many biological molecules contain hydrocarbon-like portions that are insoluble in water.

When a nonpolar substance such as vegetable oil (which consists of hydrocarbon-like molecules) is added to water, it does not dissolve but forms a separate phase. In order for the water and oil to mix, free energy must be added to the system (for example, by stirring vigorously or applying heat). Why is it thermodynamically unfavorable to dissolve a hydrophobic substance in water? One possibility is that enthalpy is required to break the hydrogen bonds among solvent water molecules in order to create a "hole" into which a nonpolar molecule can fit. Experimental measurements, however, show that the free energy barrier (ΔG) to the solvation process depends much more on the entropy term (ΔS) than on the enthalpy term (ΔH; recall from Chapter 1 that $\Delta G = \Delta H - T\Delta S$; Equation 1-2). This is because when a hydrophobic molecule is hydrated, it becomes surrounded by a layer of water molecules that cannot participate in normal hydrogen bonding but instead must align themselves so that their polar ends are not oriented toward the nonpolar solute. *This constraint on the structure of water represents a loss of entropy in the system*, because now the highly mobile water molecules have lost some of their freedom to rapidly form, break, and re-form hydrogen bonds with other water molecules (**Fig. 2-8**). The loss

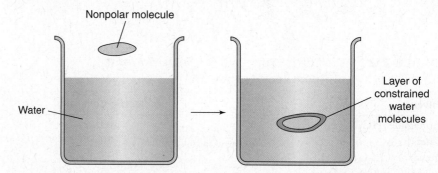

Figure 2-8 Hydration of a nonpolar molecule. When a nonpolar molecule (green) is added to water, the system loses entropy because the water molecules surrounding the nonpolar solute (orange) lose their freedom to form hydrogen bonds. The loss of entropy is a property of the entire system, not just the water molecules nearest the solute, because these molecules are continually changing places with water molecules from the rest of the solution. The loss of entropy presents a thermodynamic barrier to the hydration of a nonpolar solute.

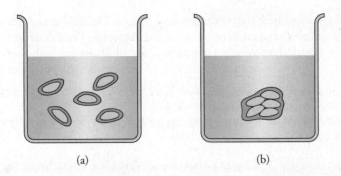

(a)   (b)

Figure 2-9 Aggregation of nonpolar molecules in water. (a) The individual hydration of dispersed nonpolar molecules (green) decreases the entropy of the system because the hydrating water molecules (orange) are not as free to form hydrogen bonds. (b) Aggregation of the nonpolar molecules increases the entropy of the system, since the number of water molecules required to hydrate the aggregated solutes is less than the number of water molecules required to hydrate the dispersed solute molecules. This increase in entropy accounts for the spontaneous aggregation of nonpolar substances in water.

? Explain why it is incorrect to describe the behavior shown in part (b) in terms of "hydrophobic bonds."

of entropy is not due to the formation of a frozen "cage" of water molecules around the nonpolar solute, as commonly pictured, because in liquid water, the solvent molecules are in constant motion.

When a large number of nonpolar molecules are introduced into a sample of water, they do not disperse and become individually hydrated, each surrounded by a layer of water molecules. Instead, the nonpolar molecules tend to clump together, removing themselves from contact with water molecules. (This explains why small oil droplets coalesce into one large oily phase.) Although the entropy of the nonpolar molecules is thereby reduced, this thermodynamically unfavorable event is more than offset by the increase in the entropy of the water molecules, which regain their ability to interact freely with other water molecules (Fig. 2-9).

[The exclusion of nonpolar substances from an aqueous solution is known as the **hydrophobic effect.**] It is a powerful force in biochemical systems, even though it is not a bond or an attractive interaction in the conventional sense. The nonpolar molecules do not experience any additional attractive force among themselves; they aggregate only because they are driven out of the aqueous phase by the unfavorable entropy cost of individually hydrating them. The hydrophobic effect governs the structures and functions of many biological molecules. For example, the polypeptide chain of a protein folds into a globular mass so that its hydrophobic groups are in the interior, away from the solvent, and its polar groups are on the exterior, where they can interact with water. Similarly, the structure of the lipid membrane that surrounds all cells is maintained by the hydrophobic effect acting on the lipids.

Amphiphilic molecules experience both hydrophilic interactions and the hydrophobic effect

Consider a molecule such as the fatty acid palmitate:

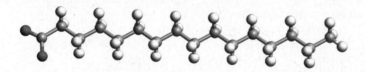

The hydrocarbon "tail" of the molecule (on the right) is nonpolar, while its carboxylate "head" (on the left) is strongly polar. Molecules such as this one, which have both hydrophobic and hydrophilic portions, are said to be **amphiphilic** or **amphipathic.** What happens when amphiphilic molecules are added to water? In general, *the polar groups of amphiphiles orient themselves toward the solvent molecules and are therefore hydrated, while the nonpolar groups tend to aggregate due to the hydrophobic effect.* As a result, the amphiphiles may form a spherical **micelle,** a particle with a solvated surface and a hydrophobic core (Fig. 2-10).

Depending in part on the relative sizes of the hydrophilic and hydrophobic portions of the amphiphiles, the molecules may form a sheet rather than a spherical micelle. The amphiphilic lipids that provide the structural basis of biological membranes form two-layered sheets called **bilayers,** in which a hydrophobic layer is

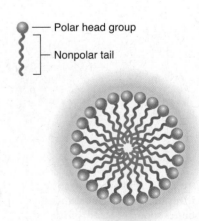

— Polar head group

— Nonpolar tail

Figure 2-10 A micelle formed by amphiphilic molecules. The hydrophobic tails of the molecules aggregate, out of contact with water, due to the hydrophobic effect. The polar head groups are exposed to and can interact with the solvent water molecules.

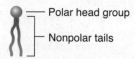

Polar head group

Nonpolar tails

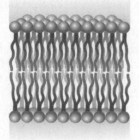

Figure 2-11 **A lipid bilayer.** The amphiphilic lipid molecules form two layers so that their polar head groups are exposed to the solvent while their hydrophobic tails are sequestered in the interior of the bilayer, away from water. The likelihood of amphiphilic molecules forming a bilayer rather than a micelle depends in part on the sizes and nature of the hydrophobic and hydrophilic groups. One-tailed lipids tend to form micelles (see Fig. 2-10), and two-tailed lipids tend to form bilayers.

? **Indicate where a sodium ion and a benzene molecule would be located.**

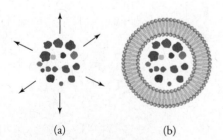

(a)   **(b)**

Figure 2-12 **A bilayer prevents the diffusion of polar substances.** (a) Solutes spontaneously diffuse from a region of high concentration to a region of low concentration. (b) A lipid barrier, which presents a thermodynamic barrier to the passage of polar substances, prevents the diffusion of polar substances out of the inner compartment (it also prevents the inward diffusion of polar substances from the external solution).

sandwiched between hydrated polar surfaces (**Fig. 2-11**). The structures of biological membranes are discussed in more detail in Chapter 8. The formation of micelles or bilayers is thermodynamically favored because the hydrogen-bonding capacity of the polar head groups is satisfied through interactions with solvent water molecules, and the nonpolar tails are sequestered from the solvent.

The hydrophobic core of a lipid bilayer is a barrier to diffusion

To eliminate its solvent-exposed edges, a lipid bilayer tends to close up to form a **vesicle,** shown cut in half:

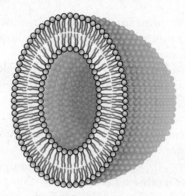

Many of the subcellular compartments (organelles) in eukaryotic cells have a similar structure.

When the vesicle forms, it traps a volume of the aqueous solution. Polar solutes in the enclosed compartment tend to remain there because they cannot easily pass through the hydrophobic interior of the bilayer. The energetic cost of transferring a hydrated polar group through the nonpolar lipid tails is too great. (In contrast, small nonpolar molecules such as O_2 can pass through the bilayer relatively easily.)

Normally, substances that are present at high concentrations tend to diffuse to regions of lower concentration (this movement "down" a concentration gradient is a spontaneous process driven by the increase in entropy of the solute molecules). *A barrier such as a bilayer can prevent this diffusion* (**Fig. 2-12**). This helps explain why cells, which are universally enclosed by a membrane, can maintain their specific concentrations of ions, small molecules, and biopolymers even when the external concentrations of these substances are quite different (**Fig. 2-13**). The solute composition of intracellular compartments and other biological fluids is carefully regulated. Not surprisingly, organisms spend a considerable amount of metabolic energy to maintain the proper concentrations of water and salts, and losses of one or the other must be compensated (Box 2-B).

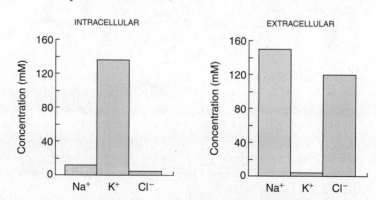

Figure 2-13 **Ionic composition of intracellular and extracellular fluid.** Human cells contain much higher concentrations of potassium than of sodium or chloride; the opposite is true of the fluid outside the cell. The cell membrane helps maintain the concentration differences.

BOX 2-B ⬡ BIOCHEMISTRY NOTE

Sweat, Exercise, and Sports Drinks

Animals, including humans, generate heat, even at rest, due to their metabolic activity. Some of this heat is lost to the environment by radiation, convection, conduction, and—in terrestrial animals—the vaporization of water. Evaporation has a significant cooling effect because about 2.5 kJ of heat is given up for every gram (mL) of water lost. In humans and certain other animals, an increase in skin temperature triggers the activity of sweat glands, which secrete a solution containing (in humans) about 50 mM Na^+, 5 mM K^+, and 45 mM Cl^-. The body is cooled as the sweat evaporates from its surface.

The evaporation of water accounts for a small portion of a resting body's heat loss, but sweating is the main mechanism for dissipating heat generated when the body is highly active. During vigorous exercise or exertion at high ambient temperatures, the body may experience a fluid loss of up to 2 L per hour. Athletic training not only improves the performance of the muscles and cardiopulmonary system, it also increases the capacity for sweating so that the athlete begins to sweat at a lower skin temperature and loses less salt in the secretions of the sweat glands. But regardless of training, a fluid loss representing more than 2% of the body's weight may impair cardiovascular function. In fact, "heat exhaustion" in humans is usually due to dehydration rather than an actual increase in body temperature.

Numerous studies have concluded that athletes seldom drink enough before or during exercise. Ideally, fluid intake should match the losses due to sweat, and the rate of intake should keep pace with the rate of sweating. So what should the conscientious athlete drink? For activities lasting less than about 90 minutes, especially when periods of high intensity alternate with brief periods of rest, water alone is sufficient. Commercial sports drinks containing carbohydrates can replace the water lost as sweat and also provide a source of energy. However, this carbohydrate boost may be an advantage only during prolonged sustained activity, such as during a marathon, when the body's own carbohydrate stores are depleted. A marathon runner or a manual laborer in the hot sun might benefit from the salt found in sports drinks, but most athletes don't need the supplemental salt (although it does make the carbohydrate solution more palatable). A normal diet usually contains enough Na^+ and Cl^- to offset the losses in sweat.

⬤ **Question:** Compare the ion concentrations of sweat and extracellular fluid.

CONCEPT REVIEW
- Why do polar molecules dissolve more easily than nonpolar substances in water?
- What does entropy have to do with the solubility of nonpolar substances in water?
- Can a molecule be both hydrophilic and hydrophobic?
- Why is a lipid bilayer a barrier to the diffusion of polar molecules?

2-3 Acid-Base Chemistry

Water is not merely an inert medium for biochemical processes; it is an active participant. Its chemical reactivity in biological systems is in part a result of its ability to ionize. This can be expressed in terms of a chemical equilibrium:

$$H_2O \rightleftharpoons H^+ + OH^-$$

The products of water's dissociation are a hydrogen ion or proton (H^+) and a hydroxide ion (OH^-).

KEY CONCEPTS
- Water ionizes to form H^+ and OH^-.
- An acid's pK value describes its tendency to ionize.
- The pH of a solution of acid depends on the pK and the concentrations of the acid and its conjugate base.

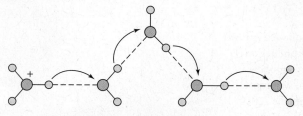

Figure 2-14 Proton jumping. A proton associated with one water molecule (as a hydronium ion, at left) appears to jump rapidly through a network of hydrogen-bonded water molecules.

Aqueous solutions do not actually contain lone protons. Instead, the H^+ can be visualized as combining with a water molecule to produce a **hydronium ion** (H_3O^+):

However, the H^+ is somewhat delocalized, so it probably exists as part of a larger, fleeting structure such as

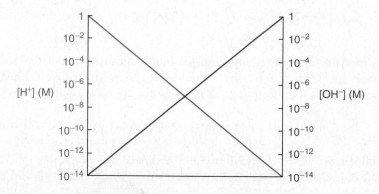

Because a proton does not remain associated with a single water molecule, it appears to be relayed through a hydrogen-bonded network of water molecules (**Fig. 2-14**). This rapid **proton jumping** means that the effective mobility of H^+ in water is much greater than the mobility of other ions that must physically diffuse among water molecules. Consequently, acid–base reactions are among the fastest biochemical reactions.

$[H^+]$ and $[OH^-]$ are inversely related

Pure water exhibits only a slight tendency to ionize, so the resulting concentrations of H^+ and OH^- are actually quite small. According to the law of mass action, the ionization of water can be described by a dissociation constant, K, which is equivalent to the concentrations of the reaction products divided by the concentration of un-ionized water:

$$K = \frac{[H^+][OH^-]}{[H_2O]} \qquad \text{[2-1]}$$

The square brackets represent the molar concentrations of the indicated species.

Because the concentration of H_2O (55.5 M) is so much greater than $[H^+]$ or $[OH^-]$, it is considered to be constant, and K is redefined as K_w, the **ionization constant of water:**

$$K_w = K[H_2O] = [H^+][OH^-] \qquad \text{[2-2]}$$

K_w is 10^{-14} at 25°C. In a sample of pure water, $[H^+] = [OH^-]$, so $[H^+]$ and $[OH^-]$ must both be equal to 10^{-7} M:

$$K_w = 10^{-14} = [H^+][OH^-] = (10^{-7} M)(10^{-7} M) \qquad \text{[2-3]}$$

Since *the product of $[H^+]$ and $[OH^-]$ in any solution must be equal to 10^{-14}*, a hydrogen ion concentration greater than 10^{-7} M is balanced by a hydroxide ion concentration less than 10^{-7} M (**Fig. 2-15**).

Figure 2-15 Relationship between $[H^+]$ and $[OH^-]$. The product of $[H^+]$ and $[OH^-]$ is K_w, which is equal to 10^{-14}. Consequently, when $[H^+]$ is greater than 10^{-7} M, $[OH^-]$ is less than 10^{-7} M, and vice versa.

A solution in which $[H^+] = [OH^-] = 10^{-7}$ M is said to be **neutral;** a solution with $[H^+] > 10^{-7}$ M ($[OH^-] < 10^{-7}$ M) is **acidic;** and a solution with $[H^+] < 10^{-7}$ M ($[OH^-] > 10^{-7}$ M) is **basic.** To more easily describe such solutions, the hydrogen ion concentration is expressed as a **pH:**

$$pH = -\log[H^+] \qquad \textbf{[2-4]}$$

Accordingly, a neutral solution has a pH of 7, an acidic solution has a pH < 7, and a basic solution has a pH > 7 (**Fig. 2-16**). Note that because the pH scale is logarithmic, *a difference of one pH unit is equivalent to a 10-fold difference in [H⁺].* The so-called physiological pH, the normal pH of human blood, is a near-neutral 7.4. The pH values of some other body fluids are listed in Table 2-3. Acid–base chemistry is also a concern outside the laboratory (Box 2-C).

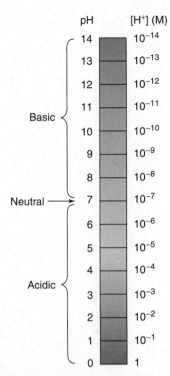

Figure 2-16 Relationship between pH and [H⁺]. Because pH is equal to $-\log [H^+]$, the greater the $[H^+]$, the lower the pH. A solution with a pH of 7 is neutral, a solution with a pH < 7 is acidic, and a solution with a pH > 7 is basic.

? **What is the difference in H⁺ concentration between a solution at pH 4 and a solution at pH 8?**

![BOX 2-C] **BIOCHEMISTRY NOTE**

Atmospheric CO₂ and Ocean Acidification

The human-generated increase in atmospheric carbon dioxide that is contributing to global warming is also impacting the chemistry of the world's oceans. Atmospheric CO_2 dissolves in water and reacts with it to generate carbonic acid. The acid immediately dissociates to form protons (H⁺) and bicarbonate (HCO_3^-):

$$CO_2 + H_2O \rightleftharpoons H_2CO_3 \rightleftharpoons H^+ + HCO_3^-$$

The addition of hydrogen ions from CO_2-derived carbonic acid therefore leads to a decrease in the pH. Currently, the earth's oceans are slightly basic, with a pH of approximately 8.0. It has been estimated that over the next 100 years, the ocean pH will drop to about 7.8. Although the oceans act as a CO_2 "sink" that helps mitigate the increase in atmospheric CO_2, the increase in acidity in the marine environment represents an enormous challenge to organisms that must adapt to the new conditions.

Many marine organisms, including mollusks, many corals, and some plankton, use dissolved carbonate ions (CO_3^{2-}) to construct protective shells of calcium carbonate ($CaCO_3$). However, carbonate ions can combine with H⁺ to form bicarbonate:

$$CO_3^{2-} + H^+ \rightleftharpoons HCO_3^-$$

Consequently, the increase in ocean acidity could decrease the availability of carbonate and thereby slow the growth of shell-building organisms. This not only would affect the availability of shellfish for human consumption but also would impact huge numbers of unicellular organisms at the base of the marine food chain. It is possible that acidification of the oceans could also dissolve existing calcium carbonate–based materials, such as coral reefs:

$$CaCO_3 + H^+ \rightleftharpoons HCO_3^- + Ca^{2+}$$

This could have disastrous consequences for these species-rich ecosystems.

● **Question:** Paradoxically, some marine organisms appear to benefit from increased atmospheric CO_2. Write an equation that describes how increased bicarbonate concentrations in seawater could promote shell growth.

[**TABLE 2-3**]

pH Values of Some Biological Fluids

Fluid	pH
Pancreatic juice	7.8–8.0
Blood	7.4
Saliva	6.4–7.0
Urine	5.0–8.0
Gastric juice	1.5–3.0

The pH of a solution can be altered

The pH of a sample of water can be changed by adding a substance that affects the existing balance between $[H^+]$ and $[OH^-]$. Adding an acid increases the concentration of $[H^+]$ and decreases the pH; adding a base has the opposite effect. *Biochemists define an **acid** as a substance that can donate a proton and a **base** as a substance that can accept a proton.* For example, adding hydrochloric acid (HCl) to a sample of water increases the hydrogen ion concentration ($[H^+]$ or $[H_3O^+]$) because the HCl donates a proton to water:

$$HCl + H_2O \rightarrow H_3O^+ + Cl^-$$

Note that in this reaction, H_2O acts as a base that accepts a proton from the added acid.

Similarly, adding the base sodium hydroxide (NaOH) increases the pH (decreases $[H^+]$) by introducing hydroxide ions that can recombine with existing hydrogen ions:

$$NaOH + H_3O^+ \rightarrow Na^+ + 2\,H_2O$$

In this reaction, H_3O^+ is the acid that donates a proton to the added base. The final pH of the solution depends on how much H^+ (for example, from HCl) has been introduced or how much H^+ has been removed from the solution by its reaction with a base (for example, the OH^- ion of NaOH). Substances such as HCl and NaOH are known as "strong" acids and bases because they ionize completely in water. The Na^+ and Cl^- ions are called spectator ions and do not affect the pH. Calculating the pH of a solution of strong acid or base is straightforward (see Sample Calculation 2-1).

●●● **SAMPLE CALCULATION 2-1**

PROBLEM

Calculate the pH of 1 L of water to which is added **(a)** 10 mL of 5.0 M HCl or **(b)** 10 mL of 5.0 M NaOH.

SOLUTION

(a) The final concentration of HCl is $\dfrac{(0.01\text{ L})(5.0\text{ M})}{1.01\text{ L}} = 0.050$ M

Since HCl dissociates completely, the added $[H^+]$ is equal to [HCl], or 0.050 M (the existing hydrogen ion concentration, 10^{-7} M, can be ignored because it is much smaller).

$$pH = -\log[H^+]$$
$$= -\log 0.050$$
$$= 1.3$$

(b) The final concentration of NaOH is 0.050 M. Since NaOH dissociates completely, the added $[OH^-]$ is 0.050 M. Use Equation 2-2 to calculate $[H^+]$.

$$K_w = 10^{-14} = [H^+][OH^-]$$
$$[H^+] = 10^{-14}/[OH^-]$$
$$= 10^{-14}/(0.050\text{ M})$$
$$= 2.0 \times 10^{-13}\text{ M}$$
$$pH = -\log[H^+]$$
$$= -\log(2.0 \times 10^{-13})$$
$$= 12.7$$

●●● **PRACTICE PROBLEMS**

1. Calculate the pH of 500 mL of water after the addition of 50 mL of 25 mM HCl.
2. Calculate the pH of 250 mL of water after mixing with 250 mL of 5 mM NaOH.

Most biologically relevant acids and bases, unlike HCl and NaOH, do not dissociate completely when added to water. In other words, proton transfer to or from water is not complete. Therefore, the final concentrations of the acidic and basic species (including water itself) must be expressed in terms of an equilibrium. For example, acetic acid partially ionizes, or donates only some of its protons to water:

$$CH_3COOH + H_2O \rightleftharpoons CH_3COO^- + H_3O^+$$

The equilibrium constant for this reaction takes the form

$$K = \frac{[CH_3COO^-][H_3O^+]}{[CH_3COOH][H_2O]} \qquad \textbf{[2-5]}$$

Because the concentration of H_2O is much higher than the other concentrations, it is considered constant and is incorporated into the value of K, which is then formally known as K_a, the **acid dissociation constant:**

$$K_a = K[H_2O] = \frac{[CH_3COO^-][H^+]}{[CH_3COOH]} \qquad \textbf{[2-6]}$$

The acid dissociation constant for acetic acid is 1.74×10^{-5}. The larger the value of K_a, the more likely the acid is to ionize; that is, the greater its tendency to donate a proton to water. The smaller the value of K_a, the less likely the compound is to donate a proton.

Acid dissociation constants, like hydrogen ion concentrations, are often very small numbers. Therefore, it is convenient to transform the K_a to a **pK** value as follows:

$$pK = -\log K_a \qquad \textbf{[2-7]}$$

For acetic acid,

$$pK = -\log(1.74 \times 10^{-5}) = 4.76 \qquad \textbf{[2-8]}$$

The larger an acid's K_a, the smaller its pK and the greater its "strength" as an acid.

Consider an acid such as the ammonium ion, NH_4^+:

$$NH_4^+ \rightleftharpoons NH_3 + H^+$$

Its K_a is 5.62×10^{-10}, which corresponds to a pK of 9.25. This indicates that the ammonium ion is a relatively weak acid, a compound that tends not to donate a proton. On the other hand, ammonia (NH_3), which is the **conjugate base** of the acid NH_4^+, readily accepts a proton. The pK values of some compounds are listed in Table 2-4. A **polyprotic acid,** a compound with more than one acidic hydrogen, has a pK value for each dissociation (called pK_1, pK_2, etc.). The first proton dissociates with the lowest pK value. Subsequent protons are less likely to dissociate and so have higher pK values.

The pH of a solution of acid is related to the pK

When an acid (represented as the proton donor HA) is added to water, the final hydrogen ion concentration of the solution depends on the acid's tendency to ionize:

$$HA \rightleftharpoons A^- + H^+$$

[**TABLE 2.4**] pK Values of Some Acids

Name	Formula[a]	pK
Trifluoroacetic acid	CF_3COOH	0.18
Phosphoric acid	H_3PO_4	2.15[b]
Formic acid	$HCOOH$	3.75
Succinic acid	$HOOCCH_2CH_2COOH$	4.21[b]
Acetic acid	CH_3COOH	4.76
Succinate	$HOOCCH_2CH_2COO^-$	5.64[c]
Thiophenol	C_6H_5SH	6.60
Phosphate	$H_2PO_4^-$	6.82[c]
N-(2-acetamido)-2-amino-ethanesulfonic acid (ACES)	$H_2NCOCH_2\overset{+}{N}H_2CH_2CH_2SO_3^-$	6.90
Imidazolium ion	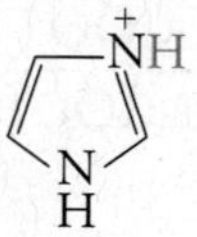	7.00
p-Nitrophenol	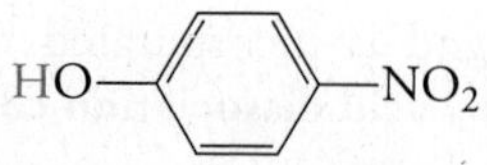	7.24
N-2-hydroxyethylpiperazine-N'-2-ethanesulfonic acid (HEPES)	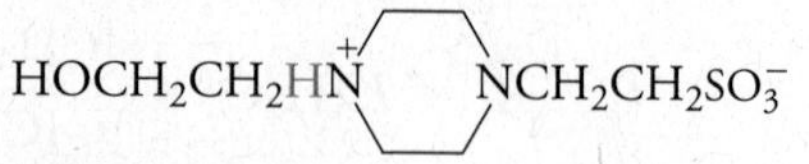	7.55
Glycinamide	$^+H_3NCH_2CONH_2$	8.20
Tris(hydroxymethyl)-aminomethane (Tris)	$(HOCH_2)_3C\overset{+}{N}H_3$	8.30
Boric acid	H_3BO_3	9.24
Ammonium ion	NH_4^+	9.25
Phenol	C_6H_5OH	9.90
Methylammonium ion	$CH_3NH_3^+$	10.60
Phosphate	HPO_4^{2-}	12.38[d]

[a] The acidic hydrogen is highlighted in red; [b] pK_1; [c] pK_2; [d] pK_3.

In other words, the final pH depends on the equilibrium between HA and A^-,

$$K_a = \frac{[A^-][H^+]}{HA} \qquad \textbf{[2-9]}$$

so that

$$[H^+] = K_a \frac{[HA]}{[A^-]} \qquad \textbf{[2-10]}$$

We can express $[H^+]$ as a pH, and K_a as a pK, which yields

$$-\log[H^+] = -\log K_a - \log\frac{[HA]}{[A^-]} \qquad \textbf{[2-11]}$$

or

$$pH = pK + \log\frac{[A^-]}{[HA]} \qquad \textbf{[2-12]}$$

Equation 2-12 is known as the **Henderson–Hasselbalch equation.** It relates the pH of a solution to the pK of an acid and the concentration of the acid (HA) and its conjugate base (A^-). This equation makes it possible to perform practical calculations to predict the pH of a solution (see Sample Calculation 2-2) or the concentrations of an acid and its conjugate base at a given pH (see Sample Calculation 2-3).

SAMPLE CALCULATION 2-2

PROBLEM

Calculate the pH of a 1-L solution to which has been added 6.0 mL of 1.5 M acetic acid and 5.0 mL of 0.4 M sodium acetate.

SOLUTION

First, calculate the final concentrations of acetic acid (HA) and acetate (A^-). The final volume of the solution is 1 L + 6 mL + 5 mL = 1.011 L.

$$[HA] = \frac{(0.006\ L)(1.5\ M)}{1.011\ L} = 0.0089\ M$$

$$[A^-] = \frac{(0.005\ L)(0.4\ M)}{1.011\ L} = 0.0020\ M$$

Next, substitute these values into the Henderson–Hasselbalch equation using the pK for acetic acid given in Table 2-4:

$$pH = pK + \log\frac{[A^-]}{[HA]}$$

$$pH = 4.76 + \log\frac{0.0020}{0.0089}$$

$$= 4.76 - 0.65$$

$$= 4.11$$

PRACTICE PROBLEMS

3. Calculate the pH of a 1-L solution to which has been added 25 mL of 10 mM acetic acid and 25 mL of 30 mM sodium acetate.
4. Calculate the pH of a 500-mL solution to which has been added 10 mL of 50 mM boric acid and 20 mL of 20 mM sodium borate.

SAMPLE CALCULATION 2-3

PROBLEM

Calculate the concentration of formate in a 10-mM solution of formic acid at pH 4.15.

SOLUTION

The solution of formic acid contains both the acid species (formic acid) and its conjugate base (formate). Use the Henderson–Hasselbalch equation to determine the ratio of formate (A^-) to formic acid (HA) at pH 4.15, using the pK value given in Table 2-4.

$$pH = pK + \log\frac{[A^-]}{[HA]}$$

$$\log\frac{[A^-]}{[HA]} = pH - pK = 4.15 - 3.75 = 0.40$$

$$\frac{[A^-]}{[HA]} = 2.51 \text{ or } [A^-] = 2.51[HA]$$

Since the total concentration of formate and formic acid is 0.01 M, $[A^-] + [HA] = 0.01$ M, and $[HA] = 0.01\ M - [A^-]$. Therefore,

$$[A^-] = 2.51[HA]$$

$$[A^-] = 2.51(0.01\ M - [A^-])$$

$$[A^-] = 0.0251\ M - 2.51[A^-]$$

$$3.51[A^-] = 0.0251\ M$$

$$[A^-] = 0.0072\ M \text{ or } 7.2\ mM$$

PRACTICE PROBLEMS

5. Calculate the concentration of acetate in a 50-mM solution of acetic acid at pH 5.0.
6. Calculate the concentration of phosphoric acid in a 50-mM solution of phosphate at pH 3.0.

The Henderson–Hasselbalch equation indicates that *when the pH of a solution of acid is equal to the pK of that acid, then the acid is half dissociated; that is, exactly half of the molecules are in the protonated HA form and half are in the unprotonated A⁻ form.* You can prove to yourself that when $[A^-] = [HA]$, the log term of the Henderson–Hasselbalch equation becomes zero (log 1 = 0), and pH = pK. When the pH is far below the pK, the acid exists mostly in the HA form; when the pH is far above the pK, the acid exists mostly in the A^- form. Note that A^- signifies a deprotonated acid; if the acid (HA) bears a positive charge to begin with, the dissociation of a proton yields a neutral species, still designated A^-.

Knowing the ionization state of an acidic substance at a given pH can be critical. For example, a drug that has no net charge at pH 7.4 may readily enter cells, whereas a drug that bears a net positive or negative charge at that pH may remain in the bloodstream and be therapeutically useless (see Sample Calculation 2-4).

●●● SAMPLE CALCULATION 2-4

PROBLEM

Determine which molecular species of phosphoric acid predominates at pH values of **(a)** 1.5, **(b)** 4, **(c)** 9, and **(d)** 13.

SOLUTION

From the pK values in Table 2-4, we know that:

Below pH 2.15, the fully protonated H_3PO_4 species predominates.

At pH 2.15, $[H_3PO_4] = [H_2PO_4^-]$.

Between pH 2.15 and 6.82, the $H_2PO_4^-$ species predominates.

At pH 6.82, $[H_2PO_4^-] = [HPO_4^{2-}]$.

Between pH 6.82 and 12.38, the HPO_4^{2-} species predominates.

At pH 12.38, $[HPO_4^{2-}] = [PO_4^{3-}]$.

Above pH 12.38, the fully deprotonated PO_4^{3-} species predominates.

Therefore, the predominant species at the indicated pH values are **(a)** H_3PO_4, **(b)** $H_2PO_4^-$, **(c)** HPO_4^{2-}, and **(d)** PO_4^{3-}.

●●● PRACTICE PROBLEMS

7. Which species of phosphoric acid predominates at pH 6?
8. Which species of phosphoric acid predominates at pH 8?
9. At what pH will $[NH_3] = [NH_4^+]$?
10. Over what pH range will the $HOOCCH_2CH_2COO^-$ form of succinic acid predominate?

The pH-dependent ionization of biological molecules is also key to understanding their structures and functions. *Many of the functional groups on biological molecules act as acids and bases.* Their ionization states depend on their respective pK values and on the pH ($[H^+]$) of their environment. For example, at physiological pH, a polypeptide bears multiple ionic charges because its carboxylic acid (—COOH) groups are ionized to carboxylate (—COO⁻) groups and its amino (—NH₂) groups are protonated (—NH₃⁺). This is because the pK values for the carboxylic acid groups are about 4, and the pK values for the amino groups are above 10. Consequently, below pH 4, both the carboxylic acid and amino groups are mostly protonated; above pH 10, both groups are mostly deprotonated.

pH < 4	4 < pH < 10	pH > 10
—COOH	—COO⁻	—COO⁻
—NH₃⁺	—NH₃⁺	—NH₂

Note that a compound containing a —COO⁻ group is sometimes still called an "acid," even though it has already given up its proton. Similarly, a "basic" compound may already have accepted a proton.

2-4 Tools and Techniques: Buffers

When a strong acid such as HCl is added to pure water, all the added acid contributes directly to a decrease in pH. But when HCl is added to a solution containing a weak acid in equilibrium with its conjugate base (A$^-$), the pH does not change so dramatically, because some of the added protons combine with the conjugate base to re-form the acid and therefore do not contribute to an increase in [H$^+$].

$$HCl \rightarrow H^+ + Cl^- \quad \text{large increase in [H}^+\text{]}$$
$$HCl + A^- \rightarrow HA + Cl^- \quad \text{small increase in [H}^+\text{]}$$

Conversely, when a strong base (such as NaOH) is added to the solution of weak acid/conjugate base, some of the added hydroxide ions accept protons from the acid to form H$_2$O and therefore do not contribute to a decrease in [H$^+$].

$$NaOH \rightarrow Na^+ + OH^- \quad \text{large decrease in [H}^+\text{]}$$
$$NaOH + HA \rightarrow Na^+ + A^- + H_2O \quad \text{small decrease in [H}^+\text{]}$$

The weak acid/conjugate base system (HA/A$^-$) acts as a **buffer** against the added acid or base by preventing the dramatic changes in pH that would otherwise occur.

The buffering activity of a weak acid, such as acetic acid, can be traced by titrating the acid with a strong base (**Fig. 2-17**). At the start of the titration, all the acid is present in its protonated (HA) form. As base (for example, NaOH) is added, protons begin to dissociate from the acid, producing A$^-$. The continued addition of base eventually causes all the protons to dissociate, leaving all the acid in its conjugate base (A$^-$) form. At the midpoint of the titration, exactly half the protons have dissociated, so [HA] = [A$^-$] and pH = pK (Equation 2-12). The broad, flat shape of the titration curve shown in Figure 2-17 indicates that the pH does not change drastically with added acid or base when the pH is near the pK. *The effective buffering capacity of an acid is generally taken to be within one pH unit of its pK.* For acetic acid (pK = 4.76), this would be pH 3.76–5.76.

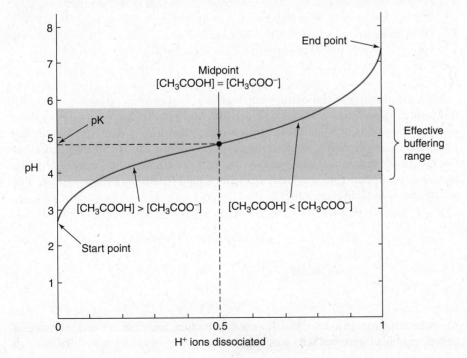

Figure 2-17 Titration of acetic acid. At the start point (before base is added), the acid is present mainly in its CH$_3$COOH form. As small amounts of base are added, protons dissociate until, at the midpoint of the titration (where pH = pK), [CH$_3$COOH] = [CH$_3$COO$^-$]. The addition of more base causes more protons to dissociate until nearly all the acid is in the CH$_3$COO$^-$ form (the end point). The shaded area indicates the effective buffering range of acetic acid. Within one pH unit of the pK, additions of acid or base do not greatly perturb the pH of the solution. ⊕ **See Animated Figure. Titration curves for acetic acid, phosphate, and ammonia.**

? **Sketch the titration curve for ammonia.**

Biochemists nearly always perform experiments in buffered solutions in order to maintain a constant pH when acidic or basic substances are added or when chemical reactions produce or consume protons. Without buffering, fluctuations in pH would alter the ionization state of the molecules under study, which might then behave differently. Before biochemists appreciated the importance of pH, experimental results were often poorly reproducible, even within the same laboratory.

A buffer solution is typically prepared from a weak acid and the salt of its conjugate base (see Sample Calculation 2-5). The two are mixed together in the appropriate ratio, according to the Henderson–Hasselbalch equation, and the final pH is adjusted if necessary by adding a small amount of concentrated HCl or NaOH. In addition to choosing a buffering compound with a pK value near the desired pH, a biochemist must consider other factors, including the compound's solubility, stability, toxicity to cells, reactivity with other molecules, and cost.

●●● SAMPLE CALCULATION 2-5

PROBLEM

How many mL of a 2.0-M solution of boric acid must be added to 600 mL of a solution of 10 mM sodium borate in order for the pH to be 9.45?

SOLUTION

Rearrange the Henderson–Hasselbalch equation to isolate the [A⁻]/[HA] term:

$$pH = pK + \log \frac{[A^-]}{[HA]}$$

$$\log \frac{[A^-]}{[HA]} = pH - pK$$

$$\frac{[A^-]}{[HA]} = 10^{(pH - pK)}$$

Substitute the known pK (from Table 2-4) and the desired pH:

$$\frac{[A^-]}{[HA]} = 10^{(9.45 - 9.24)} = 10^{0.21} = 1.62$$

The starting solution contains $(0.6 \text{ L})(0.01 \text{ mol} \cdot \text{L}^{-1}) = 0.006$ moles of borate (A⁻). The amount of boric acid (HA) needed is $0.006 \text{ mol}/1.62 = 0.0037$ mol.

Since the stock boric acid is 2.0 M, the volume of boric acid to be added is $(0.0037 \text{ mol})/(2.0 \text{ mol} \cdot \text{L}^{-1}) = 0.0019$ L or 1.9 mL.

●●● PRACTICE PROBLEMS

11. How many mL of a 5.0-M solution of sodium borate must be added to a 200-mL solution of 50 mM boric acid in order for the pH to be 9.6?
12. How many mL of a 1-M solution of imidazolium chloride must be added to a 500-mL solution of 10 mM imidazole in order for the pH to be 6.5?

One commonly used laboratory buffer system that mimics physiological conditions contains a mixture of dissolved NaH_2PO_4 and Na_2HPO_4 for a total phosphate concentration of 10 mM. The Na⁺ ions are spectator ions and are usually not significant because the buffer solution usually also contains about 150 mM NaCl (see Fig. 2-13). In this "phosphate-buffered saline," the equilibrium between the two species of phosphate ions can "soak up" added acid (producing more $H_2PO_4^-$) or added base (producing more HPO_4^{2-}).

$$pK = 6.82$$
$$H_2PO_4^- \rightleftharpoons H^+ + HPO_4^{2-}$$

Acid added $H_2PO_4^- \Longleftarrow H^+ + HPO_4^{2-}$

Base added $H_2PO_4^- \Longrightarrow H^+ + HPO_4^{2-}$

This phenomenon illustrates **Le Châtelier's principle,** which states that a change in concentration of one reactant will shift the concentrations of other reactants in order to restore equilibrium. In the human body, the major buffering systems involve bicarbonate, phosphate, and other ions (Box 2-D).

CONCEPT REVIEW
- How does a solution of a weak acid and its conjugate base minimize changes in pH when a strong acid or base is added to the solution?
- What is the useful pH range of a buffer?

BOX 2-D CLINICAL CONNECTION

Acid-Base Balance in Humans

The cells of the human body typically maintain an internal pH of 6.9–7.4. The body does not normally have to defend itself against strong inorganic acids, but many metabolic processes generate acids, which must be buffered so that they do not cause the pH of blood to drop below its normal value of 7.4. The functional groups of proteins and phosphate groups can serve as biological buffers; however, the most important buffering system involves CO_2 (itself a product of metabolism) in the blood plasma (plasma is the fluid component of blood).

CO_2 reacts with water to form carbonic acid, H_2CO_3

$$CO_2 + H_2O \rightleftharpoons H_2CO_3$$

This freely reversible reaction is accelerated *in vivo* by the enzyme carbonic anhydrase, which is present in most tissues and is particularly abundant in red blood cells. Carbonic acid ionizes to bicarbonate, HCO_3^- (see Box 2-C):

$$H_2CO_3 \rightleftharpoons H^+ + HCO_3^-$$

so that the overall reaction is

$$CO_2 + H_2O \rightleftharpoons H^+ + HCO_3^-$$

The pK for this process is 6.1 (the ionization of HCO_3^- to CO_3^{2-} occurs with a pK of 10.3 and is therefore not significant at physiological pH).

Although a pK of 6.1 appears to be just outside the range of a useful physiological buffer (which would be within one pH unit of 7.4), the effectiveness of the bicarbonate buffer system is augmented by the fact that excess hydrogen ions can not only be buffered but can also be eliminated from the body. This is possible because after the H^+ combines with HCO_3^- to re-form H_2CO_3, which rapidly equilibrates with $CO_2 + H_2O$, some of the CO_2 can be given off as a gas in the lungs. If it becomes necessary to retain more H^+ to maintain a constant pH, breathing can be adjusted so that less gaseous CO_2 is lost during exhalation.

Changes in pulmonary function can adequately adjust blood pH on the order of minutes to hours; however, longer-term adjustments of hours to days are made by the kidneys, which use a variety of mechanisms to excrete or retain H^+, bicarbonate, and other ions. In fact, the kidneys play a major role in the buffering of metabolic acids. Normal metabolic activity generates acids as the result of the degradation of amino acids, the incomplete oxidation of glucose and fatty acids, and the ingestion of acidic groups in the form of phosphoproteins and phospholipids. The HCO_3^- required to buffer these acids is initially filtered out of the bloodstream in the kidneys, but the kidneys actively reclaim this bicarbonate before it is lost in the urine. Most bicarbonate reabsorption is accomplished by a system in which H^+ leaves the kidney cells in exchange for Na^+ (step 1 in the figure below). The expelled H^+ combines with HCO_3^- in the filtrate, forming CO_2 (step 2). Because it is nonpolar, the CO_2 can diffuse into the kidney cell, where it is converted back to $H^+ + HCO_3^-$ (step 3).

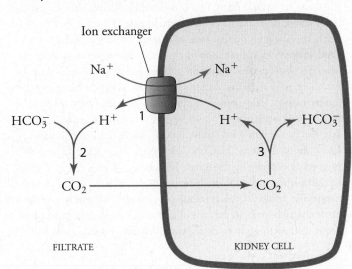

FILTRATE   KIDNEY CELL

In addition to reabsorbing filtered HCO_3^-, the kidneys also generate additional HCO_3^- to offset losses due to the buffering of metabolic acids and the exhalation of CO_2. Metabolic activity in the kidney cells produces CO_2, which is converted to $H^+ + HCO_3^-$. The cells actively secrete the H^+, which is lost via the urine (step 1 in the figure below), accounting for the mildly acidic pH of normal urine. The bicarbonate remaining in the cell is returned to the bloodstream in exchange for Cl^- (step 2).

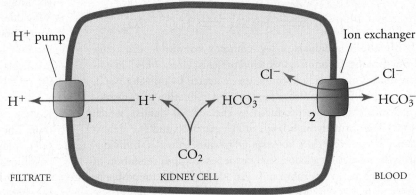

FILTRATE   KIDNEY CELL   BLOOD

Additional HCO_3^- accumulates as a result of the metabolism of the amino acid glutamine in the kidneys.

$$\begin{array}{c} COO^- \qquad\qquad\qquad O \\ | \qquad\qquad\qquad\qquad \parallel \\ H-C-CH_2-CH_2-C \\ | \qquad\qquad\qquad\qquad \backslash \\ {}^+NH_3 \qquad\qquad\qquad NH_2 \end{array}$$

Glutamine

The two amino groups are removed as ammonia (NH_3), which is ultimately excreted in the urine. Because the ammonium ion (NH_4^+) has a pK value of 9.25, nearly all the ammonia molecules become protonated at physiological pH. The consumption of protons from carbonic acid (ultimately, CO_2) leaves an excess of HCO_3^-.

Certain medical conditions can disrupt normal acid–base balance, leading to **acidosis** (blood pH less than 7.35) or **alkalosis** (blood pH greater than 7.45). The activities of the lungs and kidneys, as well as other organs, may contribute to the imbalance or respond to help correct the imbalance.

The most common disorder of acid–base chemistry is **metabolic acidosis,** which is caused by the accumulation of the acidic products of metabolism and can develop during shock, starvation, severe diarrhea (in which bicarbonate-rich digestive fluid is lost), certain genetic diseases, and renal failure (in which damaged kidneys eliminate too little acid). Metabolic acidosis can interfere with cardiac function and oxygen delivery and contribute to central nervous system depression. Despite its many different causes, one common symptom of metabolic acidosis is rapid, deep breathing. The increased ventilation helps compensate for the acidosis by "blowing off" more acid in the form of CO_2 derived from H_2CO_3. However, this mechanism also impairs O_2 uptake by the lungs. Metabolic acidosis can be treated by administering sodium bicarbonate ($NaHCO_3$). In chronic metabolic acidosis, the mineral component of bone serves as a buffer, which leads to the loss of calcium, magnesium, and phosphate and ultimately to osteoporosis and fractures.

Normal conditions

$$H^+ + HCO_3^- \rightleftharpoons H_2CO_3 \rightleftharpoons H_2O + CO_2$$

Excess acid

$$H^+ + HCO_3^- \rightarrow H_2CO_3 \rightleftharpoons H_2O + CO_2$$
$$H^+ + HCO_3^- \rightleftharpoons H_2CO_3 \rightarrow H_2O + CO_2$$
$$H^+ + HCO_3^- \rightleftharpoons H_2CO_3 \rightleftharpoons H_2O + CO_2$$

Insufficient acid

$$H^+ + HCO_3^- \rightleftharpoons H_2CO_3 \leftarrow H_2O + CO_2$$
$$H^+ + HCO_3^- \leftarrow H_2CO_3 \rightleftharpoons H_2O + CO_2$$
$$H^+ + HCO_3^- \rightleftharpoons H_2CO_3 \rightleftharpoons H_2O + CO_2$$

Metabolic alkalosis, a less common condition, can result from prolonged vomiting, which represents a loss of HCl in gastric fluid. This specific disorder can be treated by infusing NaCl. Metabolic alkalosis is also caused by overproduction of mineralocorticoids (hormones produced by the adrenal glands), which leads to abnormally high levels of H^+ excretion and Na^+ retention. This disorder does not respond to saline infusion. Individuals with metabolic alkalosis may experience apnea (cessation of breathing) for 15 s or more and cyanosis (blue coloration) due to inadequate oxygen uptake. Both these symptoms reflect the body's effort to compensate for the high blood pH by minimizing the loss of CO_2 from the lungs.

What happens when abnormal lung function is the *cause* of an acid–base imbalance? **Respiratory acidosis** can result from impaired pulmonary function caused by airway blockage, asthma (constriction of the airways), and emphysema (loss of alveolar tissue). In all cases, the kidneys respond by adjusting their activity, mainly by increasing the synthesis of the enzymes that break down glutamine in order to produce NH_3. Excretion of NH_4^+ helps correct the acidosis. However, renal compensation of respiratory acidosis takes hours to days (the time course for adjusting enzyme levels), so this condition is best treated by restoring pulmonary function through bronchodilation, supplemental oxygen, or mechanical ventilation (assisted breathing).

Some of the diseases that impair lung function (for example, asthma) can also contribute to **respiratory alkalosis,** but this relatively rare condition is more often caused by hyperventilation brought on by fear or anxiety. Unlike the other acid–base disorders, this form of respiratory alkalosis is seldom life-threatening.

● **Questions:**

1. Impaired pulmonary function can contribute to respiratory acidosis. Using the appropriate equations, explain how the failure to eliminate sufficient CO_2 through the lungs leads to acidosis.

2. Metabolic acidosis often occurs in patients with impaired circulation from cardiac arrest. Mechanical hyperventilation (a standard treatment for acidosis) cannot be used with these patients because they often have acute lung injury (ALI). A group of physicians at San Francisco General Hospital advocated using tris(hydroxymethyl)aminomethane (Tris) to treat the metabolic acidosis of these patients.

   (a) How would mechanical hyperventilation help alleviate acidosis in patients who do not have ALI?

   (b) Why would treatment with sodium bicarbonate be effective in treating metabolic acidosis? Why would this treatment be unacceptable for patients with ALI?

   (c) Explain how Tris works to treat metabolic acidosis. Why is this treatment acceptable for patients with ALI?

$$\begin{array}{c} CH_2OH \\ | \\ HOH_2C-C-NH_2 \\ | \\ CH_2OH \end{array}$$

Tris(hydroxymethyl)aminomethane

[From Kallet, R. H., et al., *Am. J. Respir. Crit. Care Med.* **161,** 1149–1153 (2000).]

3. An individual who develops alkalosis by hyperventilating is encouraged to breathe into a paper bag for several minutes. Why does this treatment correct the alkalosis?

4. Ammonia produced in kidney cells can diffuse out of the cell, but only in its unprotonated (NH_3) form.

   (a) Explain.

   (b) The NH_4^+ ion, however, can exit the cell in exchange for another ion. What ion is most likely involved?

5. Kidney cells excrete H^+ and reabsorb HCO_3^- from the filtrate. The movement of each of these ions is thermodynamically unfavorable. However, the movement of each becomes possible when it is coupled to another, thermodynamically favorable ion transport process. Explain how the movement of other ions drives the transport of H^+ and HCO_3^-.

6. In uncontrolled diabetes, the body converts fats to the so-called ketone bodies acetoacetate and 3-hydroxybutyrate, which accumulate in the bloodstream.

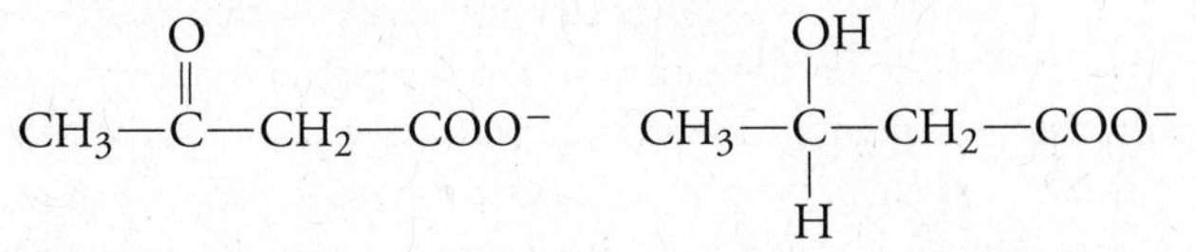

$$CH_3\!-\!\overset{\displaystyle O}{\overset{\|}{C}}\!-\!CH_2\!-\!COO^- \qquad CH_3\!-\!\overset{\displaystyle OH}{\underset{\displaystyle H}{\overset{|}{\underset{|}{C}}}}\!-\!CH_2\!-\!COO^-$$

Acetoacetate　　　　　　　3-Hydroxybutyrate

Do the ketone bodies contribute to acidosis or alkalosis? How might the body compensate for this acid–base imbalance?

7. Kidney cells have a carbonic anhydrase on their external surface as well as an intracellular carbonic anhydrase. What are the functions of these two enzymes?

8. Explain why the lungs can rapidly compensate for metabolic acidosis, whereas the kidneys are slow to compensate for respiratory acidosis.

[SUMMARY]

2-1 Water Molecules Form Hydrogen Bonds

- Water molecules are polar; they form hydrogen bonds with each other and with other polar molecules bearing hydrogen bond donor or acceptor groups.
- The electrostatic forces acting on biological molecules also include ionic interactions and van der Waals interactions.
- Water dissolves polar and ionic substances.

2-2 The Hydrophobic Effect

- Nonpolar (hydrophobic) substances tend to aggregate rather than disperse in water in order to minimize the decrease in entropy that would be required for water molecules to surround each nonpolar molecule. This is the hydrophobic effect.
- Amphiphilic molecules, which contain both polar and nonpolar groups, may aggregate to form micelles or bilayers.

2-3 Acid–Base Chemistry

- The dissociation of water produces hydroxide ions (OH^-) and protons (H^+) whose concentration can be expressed as a pH value. The pH of a solution can be altered by adding an acid (which donates protons) or a base (which accepts protons).
- The tendency for a proton to dissociate from an acid is expressed as a pK value.
- The Henderson–Hasselbalch equation relates the pH of a solution of a weak acid and its conjugate base to the pK and the concentrations of the acid and base.

2-4 Tools and Techniques: Buffers

- A buffered solution, which contains an acid and its conjugate base, resists changes in pH when more acid or base is added.

[GLOSSARY TERMS]

polarity	hydrophobic effect	base
hydrogen bond	amphiphilic	acid dissociation constant (K_a)
ionic interaction	amphipathic	pK
van der Waals radius	micelle	conjugate base
electronegativity	bilayer	polyprotic acid
van der Waals interaction	vesicle	Henderson–Hasselbalch equation
dipole–dipole interaction	hydronium ion	buffer
London dispersion forces	proton jumping	Le Châtelier's principle
dielectric constant	ionization constant of water (K_w)	acidosis
solute	neutral solution	alkalosis
solvation	acidic solution	metabolic acidosis
hydration	basic solution	metabolic alkalosis
hydrophilic	pH	respiratory acidosis
hydrophobic	acid	respiratory alkalosis

[PROBLEMS]

2-1 Water Molecules Form Hydrogen Bonds

1. The H—C—H bond angle in the perfectly tetrahedral CH_4 molecule is 109°. Explain why the H—O—H bond angle in water is only about 104.5°.

2. Each C=O bond in CO_2 is polar, yet the whole molecule is nonpolar. Explain.

3. Is ammonia a polar molecule?

4. Consider the following molecules and their melting points listed below. How can you account for the differences in melting points among these molecules of similar size?

	Molecular weight (g · mol^{-1})	Melting point (°C)
Water, H_2O	18.0	0
Ammonia, NH_3	17.0	−77
Methane, CH_4	16.0	−182

5. Identify the hydrogen bond acceptor and donor groups in the following molecules. Use an arrow to point toward each acceptor and away from each donor.

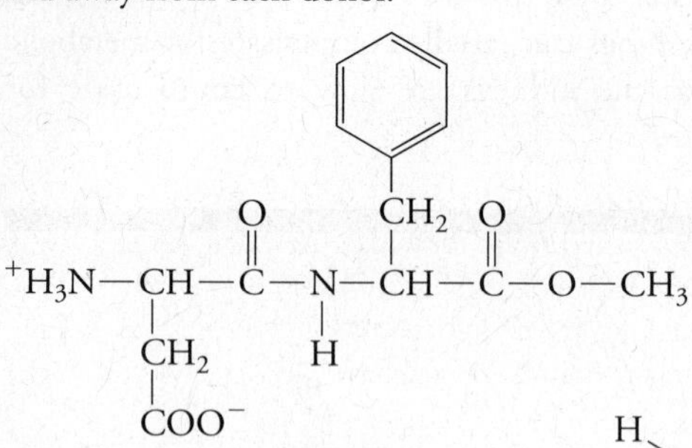

$$^+H_3N-CH-C-N-CH-C-O-CH_3$$

Aspartame

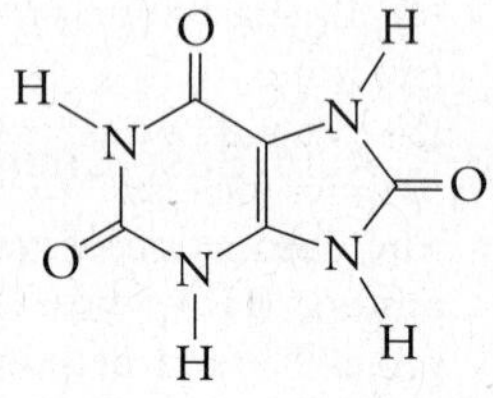

Sulfanilamide

Uric acid

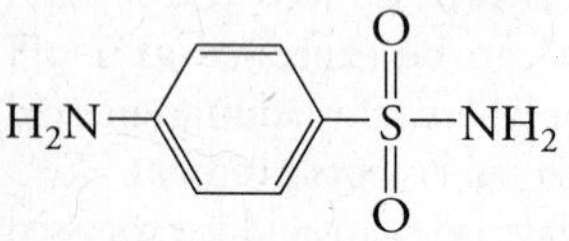

6. Do intermolecular hydrogen bonds form in the compounds below? Draw the hydrogen bonds where appropriate.

(a) H_3C-C-H

(b)

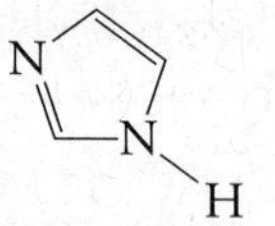

(c) H_3C-CH_2-OH

(d) H_3C-CH_2-Cl and H_2O

(e) $H_3C-CH_2-O-CH_2-CH_3$ and

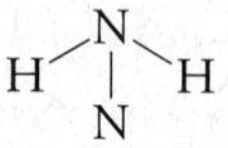

7. What are the most important intermolecular interactions in the following molecules?

(a) $H_3C-C-CH_3$

(b) $H_3C-C-NH_2$

(c) $H_3C-CH_2-CH_3$

(d) CsCl

8. Examine Table 2-1.
(a) Rank the five atoms listed in order of increasing electronegativity.
(b) What is the relationship between an atom's electronegativity and its ability to participate in hydrogen bonding?

9. Rank the melting points of the following compounds:

(a) $H_3C-CH_2-O-CH_3$

(b) $H_3C-C-NH_2$

(c) $H_2N-C-NH_2$

(d) $H_3C-(CH_2)_3-CH_3$

(e) H_3C-CH_2-CH

10. Classify the compounds listed in Problem 9 as soluble, slightly soluble, or insoluble in water and explain your answer.

11. Water is unusual in that its solid form is less dense than its liquid form. This means that when a pond freezes in the winter, ice is found as a layer on top of the pond, not on the bottom. What are the biological advantages of this?

12. Ice skaters skate not on the solid ice surface itself but on a thin film of water that forms between the skater's blade and the ice. What unique property of water makes ice skating possible?

13. Ammonium sulfate, $(NH_4)_2SO_4$, is a water-soluble salt. Draw the structures of the hydrated ions that form when ammonium sulfate dissolves in water.

14. The amino acid glycine is sometimes drawn as structure A below. However, the structure of glycine is more accurately represented by structure B. Glycine has the following properties: white crystalline solid, high melting point, and high water solubility. Why does structure B more accurately represent the structure of glycine than structure A?

$$H_2N-CH_2-COOH \qquad ^+H_3N-CH_2-COO^-$$
Structure A \qquad\qquad Structure B

15.
(a) Water has a surface tension that is nearly three times greater than that of ethanol. Explain.
(b) The surface tension of water has been shown to decrease with increasing temperature. Explain.

16. Explain why water forms nearly spherical droplets on the surface of a freshly waxed car. Why doesn't water bead on a clean windshield?

17. Which of the four primary alcohols listed in Table 2-2 would be the best solvent for ammonium ions? What can you conclude about the polarity of these solvents, which all contain an —OH group that can form hydrogen bonds?

18. Editors of a popular sporting magazine recommend sponging skin down with isopropyl alcohol after a noontime bike ride. Why is this a quick alternative to a shower?

19. A typical spherical bacterial cell with a diameter of 1 μm contains 1000 molecules of a certain protein.
(a) What is the protein's concentration in units of mM? (*Hint:* The volume of a sphere is $4\pi r^3/3$.)
(b) How many molecules of glucose does the cell contain if the internal glucose concentration is 5 mM?

20. Practitioners of homeopathic medicine believe that good health can be restored by administering a small amount of a harmful substance that stimulates the body's natural healing processes. The remedy is typically prepared by serially diluting an animal or plant extract. In a so-called 30× dilution, the active substance is diluted 10-fold, 30 times in succession.
(a) If the substance is initially present at a concentration of 1 M, what is its concentration after a 30× dilution?

(b) A typical homeopathic dose is a few drops. How many molecules of the active substance would be present in 1 mL?

(c) When confronted with the results of an exercise such as the one in part (b), proponents of homeopathy claim that a molecule can leave an imprint, or memory of itself, on the surrounding water molecules. What information presented in this chapter might be used to support this claim? To refute it?

2-2 The Hydrophobic Effect

21. Consider the structures of the molecules below. Are these molecules polar, nonpolar, or amphiphilic? Which are capable of forming micelles? Which are capable of forming bilayers?

(a) $H_3C-(CH_2)_{11}-N^+-CH_2COO^-$
with CH_3 groups above and below the N^+

(b) $H_3C-(CH_2)_{11}-CH_3$

(c) $H_3C-N^+-CH_3$ with CH_3 above and below, Cl^-

(d) $CH_2-O-\overset{O}{\overset{\|}{C}}-(CH_2)_{11}-CH_3$
$HC-O-C-(CH_2)_{11}-CH_3$
$HO-CH_2 \quad O$

(e) $H_3C-CH-COO^-$
with OH below

22. The compound bis-(2-ethylhexyl)sulfosuccinate (abbreviated AOT) is capable of forming "reverse" micelles in the hydrocarbon solvent isooctane (2,2,4-trimethylpentane). Scientists have investigated the use of reverse micelles for extracting water-soluble proteins. A two-phase system is formed: the hydrocarbon phase containing the reverse micelles and the water phase containing the protein. After a certain period of time, the protein is transferred to the reverse micelle.

(a) Draw the structure of the reverse micelle that AOT would form in isooctane.

(b) Where would the protein be located in the reverse micelle?

$$H_3C-(CH_2)_3-\overset{CH_2CH_3}{\overset{|}{CH}}-CH_2-O-\overset{O}{\overset{\|}{C}}-CH_2$$
$$H_3C-(CH_2)_3-\overset{|}{CH}-CH_2-O-\overset{|}{C}-CH-\overset{|}{S}-O^-$$
$$\overset{|}{CH_2CH_3} \qquad \overset{\|}{O} \quad \overset{\|}{O}$$

Bis-(2-ethylhexyl)sulfosuccinate (AOT)

23. Many household soaps are amphiphilic substances, often the salts of long-chain fatty acids, that form water-soluble micelles. An example is sodium dodecyl sulfate (SDS above, right).

(a) Identify the polar and nonpolar portions of the SDS molecule.

(b) Draw the structure of the micelle formed by SDS.

(c) Explain how the SDS micelles "wash away" water-insoluble substances such as cooking grease.

$$H_3C-(CH_2)_{11}-O-\overset{O}{\overset{\|}{\underset{\|}{S}}}-O^-Na^+$$
with O above and below the S

Sodium dodecyl sulfate (SDS)

24. A device sometimes called a laundry ball is advertised as a replacement for environmentally harmful detergent. The laundry ball is a rubber-coated, baseball-sized plastic sphere containing fluid that is said to serve as an organizing template for water molecules.

(a) Evaluate this claim based on your knowledge of water structure.

(b) Magnets inside the laundry ball are said to rip apart clusters of water molecules to make it easier for individual water molecules to access and remove dirt molecules. If true, would this action help wash away dirt?

(c) The instructions recommend that a laundry ball, rather than the usual detergent, be added to a load of laundry. Why do the instructions recommend washing with hot rather than cold water?

25. Just as a dissolved substance tends to move spontaneously down its concentration gradient, water also tends to move from an area of high concentration (low solute concentration) to an area of low concentration (high solute concentration), a process known as osmosis.

(a) Explain why a lipid bilayer would be a barrier to osmosis.

(b) Why are isolated human cells placed in a solution that typically contains about 150 mM NaCl? What would happen if the cells were placed in pure water?

26. Fresh water is obtained from seawater in a desalination process using reverse osmosis. In reverse osmosis, seawater is forced through a membrane; the salt remains on one side of the membrane and fresh water is produced on the other side. In what ways does reverse osmosis differ from osmosis as described in Problem 25?

27. Which of the following substances might be able to cross a bilayer? Which substances could not? Explain your answer.

(a) CO_2

(b)

$$\begin{array}{c} CHO \\ H-OH \\ HO-H \\ H-OH \\ H-OH \\ CH_2OH \end{array}$$

Glucose

(c)

(structure of 2,4-dinitrophenol with OH, two NO_2 groups)

2,4-Dinitrophenol (DNP)

(d) Ca^{2+}

28. Drug delivery is a challenging problem because the drug molecules must be sufficiently water soluble to dissolve in the blood, but also sufficiently nonpolar to pass through the cell membrane for delivery. A medicinal chemist proposes to encapsulate a water-soluble drug into a vesicle (see Section 2-2). How does this strategy facilitate the delivery of the drug to its target cell?

29. A specialized protein pump in the red blood cell membrane exports Na^+ ions and imports K^+ ions in order to maintain the sodium and potassium ion concentrations shown in Figure 2-13. Does the movement of these ions occur spontaneously, or is this an energy-requiring process? Explain.

48

30. Estimate the amount of Na^+ lost in sweat during 15 minutes of vigorous exercise. What is the mass of potato chips (200 mg Na^+ per ounce) you would have to consume in order to replace the lost sodium?

31. The bacterium *E. coli* has the ability to change both its water content and its cytoplasmic K^+ concentration to adapt to changes in the solute concentration of its growth medium. (The solute concentration is referred to as *osmolarity*—a solution with a high concentration of solute has a high osmolarity and a solution with a low concentration of solute has a low osmolarity.) The bacterial cell consists of a cytoplasmic compartment bounded by a membrane, which is in turn surrounded by a cell wall. The cell wall is porous and can accommodate the passage of water and ions. (The membrane is not as porous but does allow passage of water and ions.) In addition, the cell wall is also fairly elastic and can stretch to accommodate increases in the cytoplasmic compartment volume.

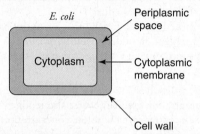

Under nongrowing conditions, *E. coli* responds to changes in osmolarity by regulating its cytoplasmic water content. What would happen to the cytoplasmic volume if *E. coli* were grown in a high-osmolarity medium? In a low-osmolarity medium?

32. Under growth conditions, *E. coli* also regulates cytoplasmic K^+ content (in addition to water) in response to growth medium osmolarity (see Problem 31). This prevents the large changes in cell volume that would occur in response to changes in growth medium osmolarity if only water content were regulated. How might *E. coli* regulate both the cytoplasmic concentrations of K^+ and water when grown in a low-osmolarity medium? What happens when *E. coli* is grown in a high-osmolarity medium?

2-3 Acid-Base Chemistry

33. Compare the concentrations of H_2O and H^+ in a sample of pure water at pH 7.0 at 25°C.

34. Like all equilibrium constants, the value of K_w is temperature dependent. K_w is 1.47×10^{-14} at 30°C. What is "neutral" pH at this temperature?

35. What is the pH of a solution of 1.0×10^{-9} M HCl?

36. What is the pH of a solution of 1.0×10^{-9} M NaOH?

37. Explain why H_3O^+ is the strongest acid that can exist in a biological system.

38. Draw a diagram similar to Fig. 2-14 showing how a hydroxide ion appears to jump through an aqueous solution.

39. Calculate $[H^+]$ for saliva (pH ~6.6) and urine (pH ~5.5).

40. Fill in the blanks of the following table:

	Acid, base, or neutral?	pH	$[H^+]$ (M)	$[OH^-]$ (M)
A	_____	5.60	_____	
B	_____	_____	_____	4.5×10^{-7}
C	Neutral	_____	_____	
D	_____	_____	2.1×10^{-3}	_____

41. Calculate the pH of 500 mL of water to which **(a)** 20 mL of 1.0 M HNO_3 or **(b)** 15 mL of 1.0 M KOH has been added.

42. What is the pH of 1.0 L of water to which **(a)** 1.5 mL of 3.0 M HCl or **(b)** 1.5 mL of 3.0 M NaOH has been added?

43. Several hours after a meal, partially digested food leaves the stomach and enters the small intestine, where pancreatic juice is added. How does the pH of the partially digested mixture change as it passes from the stomach into the intestine (see Table 2-3)?

44. The pH of urine has been found to be correlated with diet. Acidic urine results when meat and dairy products are consumed, because the oxidation of sulfur-containing amino acids in the proteins produces protons. Consumption of fruits and vegetables leads to alkaline urine, because these foods contain plentiful amounts of potassium and magnesium carbonate salts. Why would the presence of these salts result in alkaline urine?

45. Give the conjugate base of the following acids:
(a) $HC_2O_4^-$
(b) HSO_3^-
(c) $H_2PO_4^-$
(d) HCO_3^-
(e) $HAsO_4^{2-}$
(f) HPO_4^{2-}
(g) HO_2^-

46. Give the conjugate acid of the species listed in Problem 45.

47. Identify the acidic hydrogens in the following compounds.

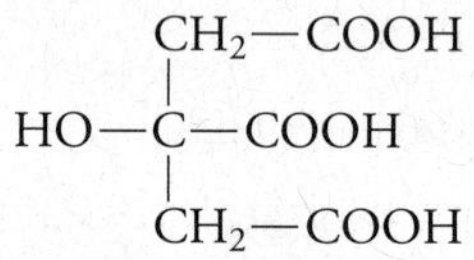

Citric acid 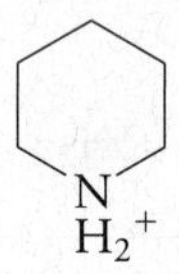 Piperidine

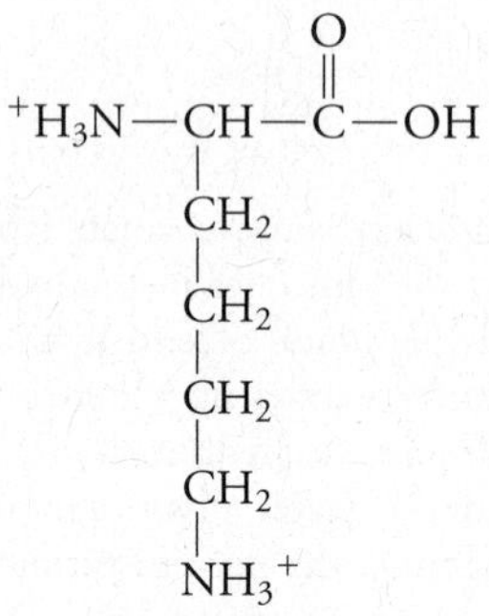

Lysine 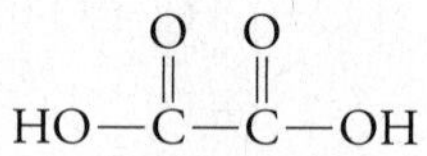 Oxalic acid 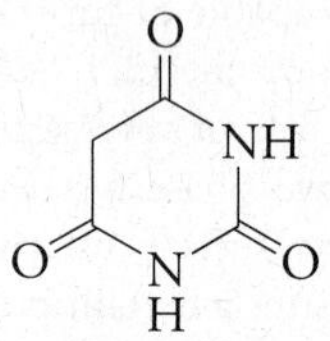 Barbituric acid

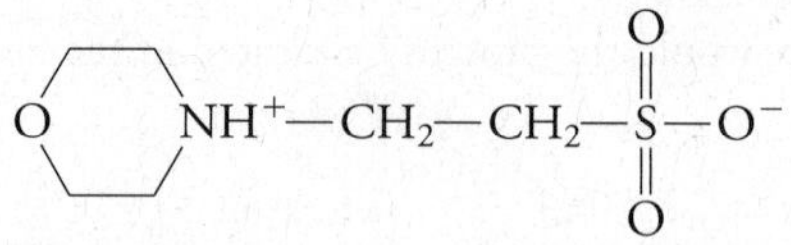

4-Morphine ethanesulfonic acid (MES)

48. The structure of pyruvic acid is shown below.
(a) Draw the structure of pyruvate.
(b) Using what you have learned about acidic functional groups, which form of this compound is likely to predominate in the cell at pH 7.4? Explain.

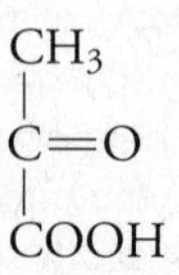

Pyruvic acid

49. The structure of a neutral form of lysine is shown below. Use the information presented in Section 2-3 to draw the structure of this compound at pH 7.

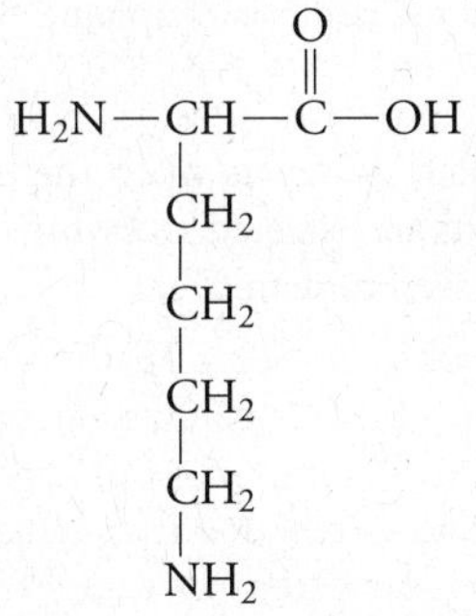

$$H_2N-\overset{\displaystyle O}{\underset{\displaystyle CH_2}{\overset{\displaystyle |}{C}}}H-\overset{\displaystyle ||}{C}-OH$$

$$\begin{array}{c} CH_2 \\ | \\ CH_2 \\ | \\ CH_2 \\ | \\ CH_2 \\ | \\ NH_2 \end{array}$$

50. Rank the following according to their strength as acids:

	Acid	K_a	pK
A	Citrate		4.76
B	Succinic acid	6.17×10^{-5}	___
C	Succinate	2.29×10^{-6}	___
D	Formic acid	1.78×10^{-4}	___
E	Citric acid		3.13

51. The pK of $CH_3CH_2NH_3^+$ is 10.7. Would the pK of $FCH_2CH_2NH_3^+$ be higher or lower?

52. The amino acid glycine (H_2N-CH_2-COOH) has pK values of 2.35 and 9.78. Indicate the structure and net charge of the molecular species that predominates at pH 2, 7, and 10.

2-4 Tools and Techniques: Buffers

53. Which would be a more effective buffer at pH 8.0 (see Table 2-4)?
(a) 10 mM HEPES buffer or 10 mM glycinamide buffer
(b) 10 mM Tris buffer or 20 mM Tris buffer
(c) 10 mM boric acid or 10 mM sodium borate

54. Which would be a more effective buffer at pH 5.0 (see Table 2-4)?
(a) 10 mM acetic acid buffer or 10 mM HEPES buffer
(b) 10 mM acetic acid buffer or 20 mM acetic acid buffer
(c) 10 mM acetic acid or 10 mM sodium acetate

55. Explain why metabolically generated CO_2 does not accumulate in tissues but is quickly converted to carbonic acid by the action of carbonic anhydrase in red blood cells.

56. The pH of blood is maintained within a narrow range (7.35–7.45). Carbonic acid, H_2CO_3, participates in blood buffering.
(a) Write the equations for the dissociation of the two ionizable protons.
(b) The pK for the first ionizable proton is 6.35; the pK for the second ionizable proton is 10.33. Use this information to identify the weak acid and the conjugate base present in the blood.

57. Phosphoric acid, H_3PO_4, has three ionizable protons.
(a) Sketch the titration curve. Indicate the pK values and the species that predominate in each area of the curve.
(b) Which two phosphate species are present in the blood at pH 7.4?
(c) Which two phosphate species would be used to prepare a buffer solution of pH 11?

58. The structure of acetylsalicylic acid (aspirin) is shown. Is aspirin more likely to be absorbed (pass through a lipid membrane) in the stomach (pH ~2) or in the small intestine (pH ~8)? Explain.

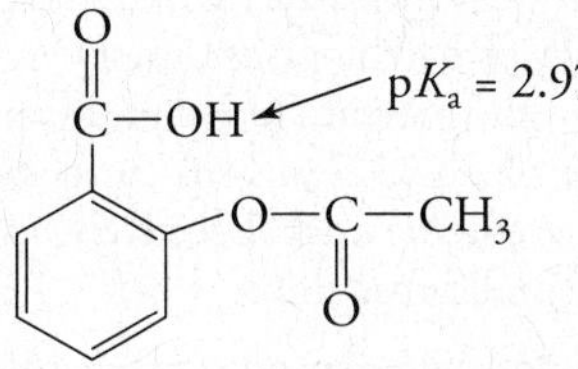

Acetylsalicylic acid (aspirin)

59. A solution is made by mixing 50 mL of a stock solution of 2.0 M K_2HPO_4 and 25 mL of a stock solution of 2.0 M KH_2PO_4. The solution is diluted to a final volume of 200 mL. What is the final pH of the solution?

60. Calculate the ratio of imidazole to the imidazolium ion in a solution at pH 7.4.

61. What is the volume (in mL) of glacial acetic acid (17.4 M) that would have to be added to 500 mL of a solution of 0.20 M sodium acetate in order to achieve a pH of 5.0?

62. What is the mass of NaOH that would have to be added to 500 mL of a solution of 0.20 M acetic acid in order to achieve a pH of 5.0?

63. An experiment requires the buffer HEPES, pH = 8.0 (see Table 2-4).
(a) Write an equation for the dissociation of HEPES in water. Identify the weak acid and the conjugate base.
(b) What is the effective buffering range for HEPES?
(c) The buffer will be prepared by making 1.0 L of a 0.10 M solution of HEPES. Hydrochloric acid will be added until the desired pH is achieved. Describe how you will make 1.0 L of 0.10 M HEPES. (HEPES is supplied by a chemical company as a sodium salt with a molecular weight of 260.3 g · mol^{-1}.)
(d) What is the volume (in mL) of a stock solution of 6.0 M HCl that must be added to the 0.1 M HEPES to achieve the desired pH of 8.0? Describe how you will make the buffer.

64. One liter of a 0.10 M Tris buffer (see Table 2-4) is prepared and adjusted to a pH of 8.2.
(a) Write the equation for the dissociation of Tris in water. Identify the weak acid and the conjugate base.
(b) What is the effective buffering range for Tris?
(c) What are the concentrations of the conjugate acid and weak base at pH 8.2?
(d) What is the ratio of conjugate base to weak acid if 1.5 mL of 3.0 M HCl is added to 1.0 L of the buffer? What is the new pH? Has the buffer functioned effectively? Compare the pH change to that of Problem 42a in which the same amount of acid was added to the same volume of pure water.
(e) What is the ratio of conjugate base to weak acid if 1.5 mL of 3.0 M NaOH is added to 1.0 L of the buffer? What is the new pH? Has the buffer functioned effectively? Compare the pH change to that of Problem 42b in which the same amount of base was added to the same volume of pure water.

65. One liter of a 0.1 M Tris buffer (see Table 2-4) is prepared and adjusted to a pH of 2.0.
(a) What are the concentrations of the conjugate base and weak acid at this pH?
(b) What is the pH when 1.5 mL of 3.0 M HCl is added to 1.0 L of the buffer? Has the buffer functioned effectively? Explain.

50

(c) What is the pH when 1.5 mL of 3.0 M NaOH is added to 1.0 L of the buffer? Has the buffer functioned effectively? Explain.

66. A patient who has taken an overdose of aspirin is brought into the emergency room for treatment. She suffers from respiratory alkalosis, and the pH of her blood is 7.5. Determine the ratio of HCO_3^- to H_2CO_3 in the patient's blood at this pH. How does this compare to the ratio of HCO_3^- to H_2CO_3 in normal blood? Can the H_2CO_3/HCO_3^- system work effectively as a buffer in this patient under these conditions?

67. Metabolic acidosis is a general term that describes a number of metabolic disorders in the body that result in a lowering of the blood pH from 7.4 to 7.35 or below. The kidney plays a vital role in regulating blood pH. The kidney can either excrete or reabsorb various ions, including phosphate, $H_2PO_4^-$; ammonium, NH_4^+; or bicarbonate, HCO_3^-. Which ions are excreted and which ions are reabsorbed in metabolic acidosis? Explain, using relevant chemical equations.

68. Metabolic alkalosis occurs when the blood pH rises to 7.45 or greater. Which ions are excreted and which ions are reabsorbed in metabolic alkalosis (see Problem 67)?

[SELECTED READINGS]

Ellis, R. J., Macromolecular crowding: obvious but underappreciated, *Trends Biochem. Sci.* **26,** 597–603 (2001). [Discusses how the large number of macromolecules in cells could affect reaction equilibria and reaction rates.]

Gerstein, M., and Levitt, M., Simulating water and the molecules of life, *Sci. Am.* **279(11),** 101–105 (1998). [Describes the structure of water and how water interacts with other molecules.]

Halperin, M. L., and Goldstein, M. B., *Fluid, Electrolyte, and Acid–Base Physiology: A Problem-Based Approach* (3rd ed.), W. B. Saunders (1999). [Includes extensive problem sets with explanations of basic science as well as clinical effects of acid–base disorders.]

Kropman, M. F., and Bakker, H. J., Dynamics of water molecules, *Science* **291,** 2118–2120 (2001). [Describes how water molecules in solvation shells move more slowly than bulk water molecules.]

Yucha, C., Renal regulation of acid–base balance, *Nephrol. Nursing J.* **31,** 201–206 (2004). [A short review of physiological buffer systems and the role of the kidney.]

3

FROM GENES
TO PROTEINS

▶▶ **HOW** do researchers decipher the information in DNA?

Virtually all biological materials, such as this blood sample, contain DNA. *Even minute amounts of biological fluids can yield traces of DNA.* Because the sequence of nucleotides in a DNA molecule is a form of information, the ability to read and interpret the sequence makes it possible to identify an individual or diagnose a disease. For example, analysis of fetal DNA in the maternal circulation can reveal chromosomal abnormalities as well as the presence of sequence variants that can cause disease. Yet the techniques for reading DNA sequences, described in this chapter, also generate challenges in compiling vast quantities of data and understanding how cells themselves store and use that information.

[Lewis Wright/iStockphoto.]

THIS CHAPTER IN CONTEXT

Part 1 Foundations

Part 2 Molecular Structure and Function
3 From Genes to Proteins

Part 3 Metabolism

Part 4 Genetic Information

Do You Remember?

- Cells contain four major types of biological molecules and three major types of polymers (Section 1-2).

- Modern prokaryotic and eukaryotic cells apparently evolved from simpler nonliving systems (Section 1-4).

- Noncovalent forces, including hydrogen bonds, ionic interactions, and van der Waals forces, act on biological molecules (Section 2-1).

All the structural components of cells and the machinery that carries out the cell's activities are ultimately specified by the cell's genetic material—DNA. Therefore, before examining other types of biological molecules and their metabolic transformations, we must consider the nature of DNA, including its chemical structure and how its biological information is organized and expressed. The Tools and Techniques section of this chapter includes some of the methods used to study and manipulate DNA in the laboratory.

3-1 DNA Is the Genetic Material

KEY CONCEPTS
- DNA and RNA are polymers of nucleotides, each of which consists of a purine or pyrimidine base, deoxyribose or ribose, and phosphate.
- A DNA molecule contains two antiparallel strands that wind around each other to form a double helix in which A and T bases in opposite strands, and C and G bases in opposite strands, pair through hydrogen bonding.
- Double-stranded nucleic acids are denatured at high temperatures; at lower temperatures, complementary polynucleotides anneal.

Gregor Mendel was certainly not the first to notice that an organism's characteristics (for example, flower color or seed shape in pea plants) were passed to its progeny, but in 1865 he was the first to describe their predictable patterns of inheritance. By 1903, Mendel's inherited factors (now called genes) were recognized as belonging to **chromosomes** (a word that means "colored bodies"), which are visible by light microscopy (**Fig. 3-1**).

Nucleic acids had been discovered in 1869 by Friedrich Miescher, who isolated this material from the white blood cells in pus on surgical bandages. However, when it became clear that chromosomes were composed of both proteins and nucleic acids, nucleic acids were dismissed as possible carriers of genetic information due to their lack of complexity: They contained only four different types of structural units, called **nucleotides.** In contrast, proteins contained 20 different types of amino acids and exhibited great diversity in composition, size, and shape—attributes that seemed more appropriate for carriers of genetic information.

Years later, microbiologists showed that a substance from a dead pathogenic (disease-causing) strain of the bacterium *Streptococcus pneumoniae* could transform cells from a normal strain to the pathogenic type. In 1944, Oswald Avery, Colin MacLeod, and Maclyn McCarty showed that this transforming substance was **deoxyribonucleic acid (DNA),** but their results did not garner much attention. Another seven years went by until Alfred Hershey and Martha Chase demonstrated that in **bacteriophages** (viruses that infect bacterial cells and that consist only of protein and DNA), the DNA, not the protein, was the infectious agent (**Fig. 3-2**).

By this time, DNA was known to contain chains of polymerized nucleotides—abbreviated A, C, G, and T—but these were thought to occur as simple repeating tetranucleotides, for example,

$$—ACGT\text{-}ACGT\text{-}ACGT\text{-}ACGT—$$

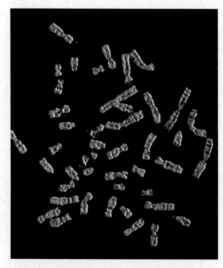

Figure 3-1 Human chromosomes from amniocentesis. In this image, the chromosomes have been stained with fluorescent dyes. [Dr. P. Boyer/Photo Researchers, Inc.]

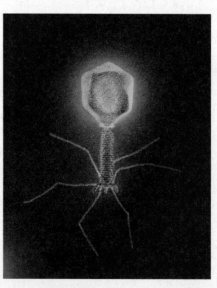

Figure 3-2 A T-type bacteriophage. The phage consists mostly of a protein coat surrounding a molecule of DNA. Alfred Hershey and Martha Chase identified the DNA as the infectious agent. [Dept. of Microbiology, Biozentrum/Science Photo Library/Photo Researchers.]

When Erwin Chargaff showed in 1950 that the nucleotides in DNA were not all present in equal numbers and that the nucleotide composition varied among species, it became apparent that DNA might be complex enough to be the genetic material after all, and the race was on to decipher its molecular structure.

The DNA structure ultimately elucidated by James Watson and Francis Crick in 1953 incorporated Chargaff's observations. Specifically, Chargaff noted that the amount of A is equal to the amount of T, the amount of C is equal to the amount of G, and the total amount of A + G is equal to the total amount of C + T. *Chargaff's "rules" could be satisfied by a molecule with two **polynucleotide** strands (polymers of nucleotides) in which A and C in one strand pair with T and G in the other.*

Nucleic acids are polymers of nucleotides

Each nucleotide of DNA includes a nitrogen-containing **base.** The bases adenine (A) and guanine (G) are known as **purines** because they resemble the organic compound purine:

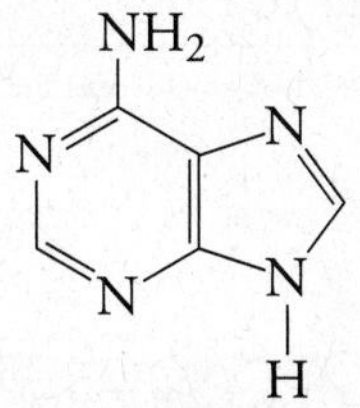

Adenine 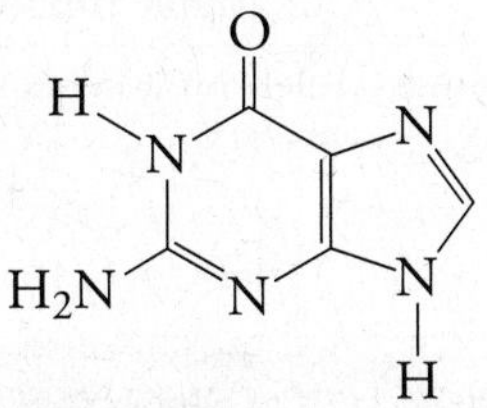   Guanine 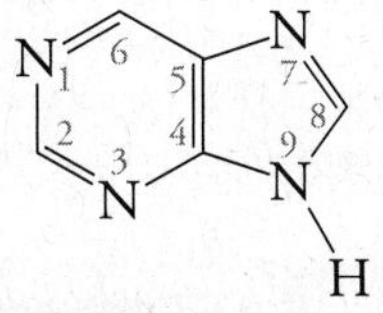   Purine

The bases cytosine (C) and thymine (T) are known as **pyrimidines** because they resemble the organic compound pyrimidine:

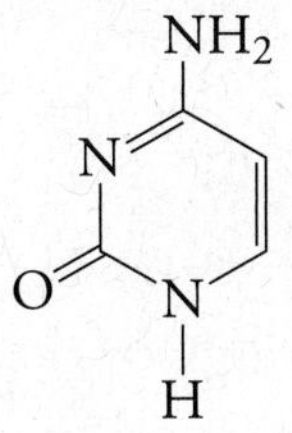

Cytosine 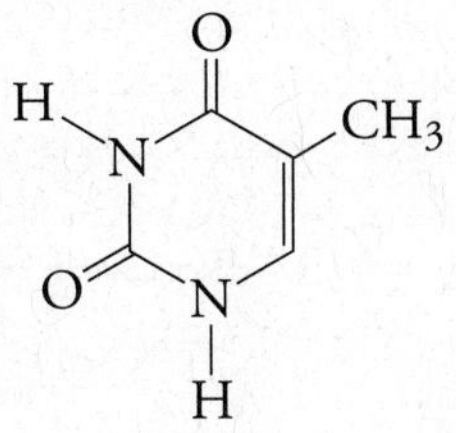   Thymine 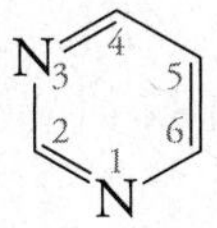   Pyrimidine

Ribonucleic acid (RNA) contains the pyrimidine uracil (U) rather than thymine:

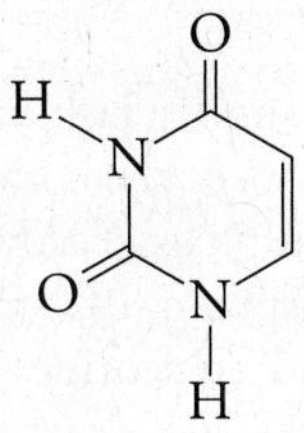

Uracil

so that DNA contains the bases A, C, G, and T, whereas RNA contains A, C, G, and U.

Linking atom N9 in a purine or atom N1 in a pyrimidine to a five-carbon sugar forms a **nucleoside.** *In DNA, the sugar is 2'-deoxyribose; in RNA, the sugar is ribose* (the sugar atoms are numbered with primes to distinguish them from the atoms of the attached bases).

Ribose   2'-Deoxyribose

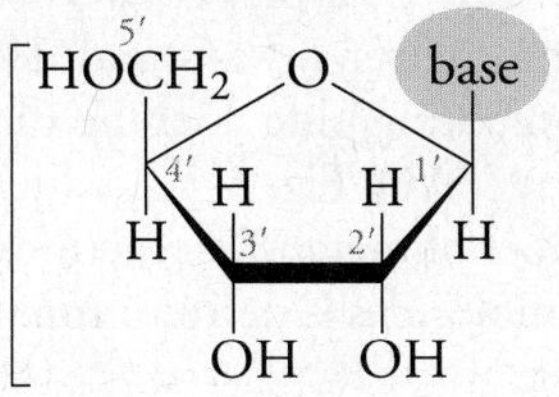

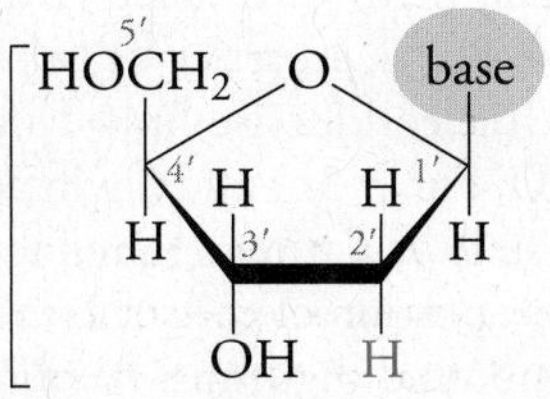

[TABLE 3-1] Nucleic Acid Bases, Nucleosides, and Nucleotides

Base	Nucleoside[a]	Nucleotides[a]
Adenine (A)	Adenosine	Adenylate; adenosine monophosphate (AMP) adenosine diphosphate (ADP) adenosine triphosphate (ATP)
Cytosine (C)	Cytidine	Cytidylate; cytidine monophosphate (CMP) cytidine diphosphate (CDP) cytidine triphosphate (CTP)
Guanine (G)	Guanosine	Guanylate; guanosine monophosphate (GMP) guanosine diphosphate (GDP) guanosine triphosphate (GTP)
Thymine (T)[b]	Thymidine	Thymidylate; thymidine monophosphate (TMP) thymidine diphosphate (TDP) thymidine triphosphate (TTP)
Uracil (U)[c]	Uridine	Uridylate; uridine monophosphate (UMP) uridine diphosphate (UDP) uridine triphosphate (UTP)

[a]Nucleosides and nucleotides containing 2′-deoxyribose rather than ribose may be called deoxynucleosides and deoxynucleotides. The nucleotide abbreviation is then preceded by "d."

[b]Thymine is found in DNA but not in RNA.

[c]Uracil is found in RNA but not in DNA.

A nucleotide is a nucleoside to which one or more phosphate groups are linked, usually at C5′ of the sugar. Depending on whether there are one, two, or three phosphate groups, the nucleotide is known as a nucleoside monophosphate, nucleoside diphosphate, or nucleoside triphosphate and is represented by a three-letter abbreviation, for example,

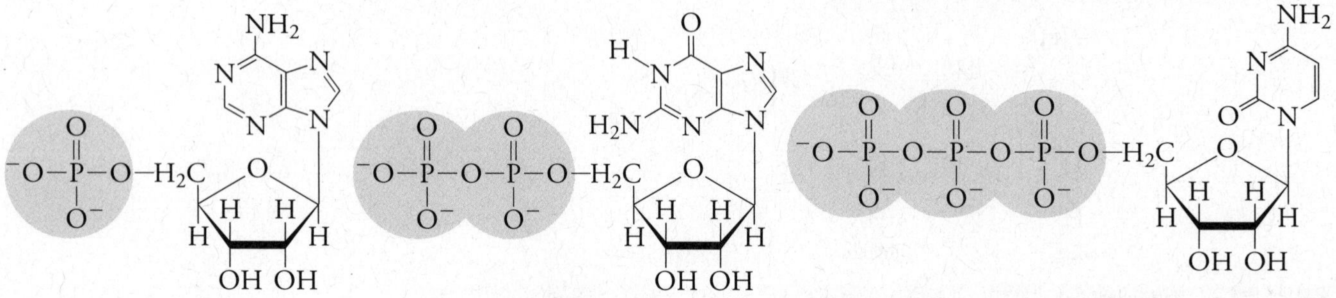

Adenosine monophosphate (AMP)

Guanosine diphosphate (GDP)

Cytidine triphosphate (CTP)

Deoxynucleotides are named in a similar fashion, and their abbreviations are preceded by "d." The deoxy counterparts of the compounds shown above would therefore be deoxyadenosine monophosphate (dAMP), deoxyguanosine diphosphate (dGDP), and deoxycytidine triphosphate (dCTP). The names and abbreviations of the common bases, nucleosides, and nucleotides are summarized in Table 3-1.

Some nucleotides have other functions

In addition to serving as the building blocks for DNA and RNA, nucleotides perform a variety of functions in the cell. They are involved in energy transduction, intracellular signaling, and regulation of enzyme activity. Some nucleotide derivatives are essential players in the metabolic pathways that synthesize biomolecules or degrade them in order to "capture" free energy. For example, coenzyme A (CoA; **Fig. 3-3a**) is a carrier of other molecules during their synthesis and degradation. Two nucleotides are linked in the compounds nicotinamide adenine dinucleotide (NAD; Fig. 3-3b) and flavin adenine dinucleotide (FAD; Fig. 3-3c), which undergo reversible oxidation and reduction during a number of metabolic reactions. Interestingly, a portion of the structures of each of these molecules is derived from a **vitamin**, a compound that must be obtained from the diet.

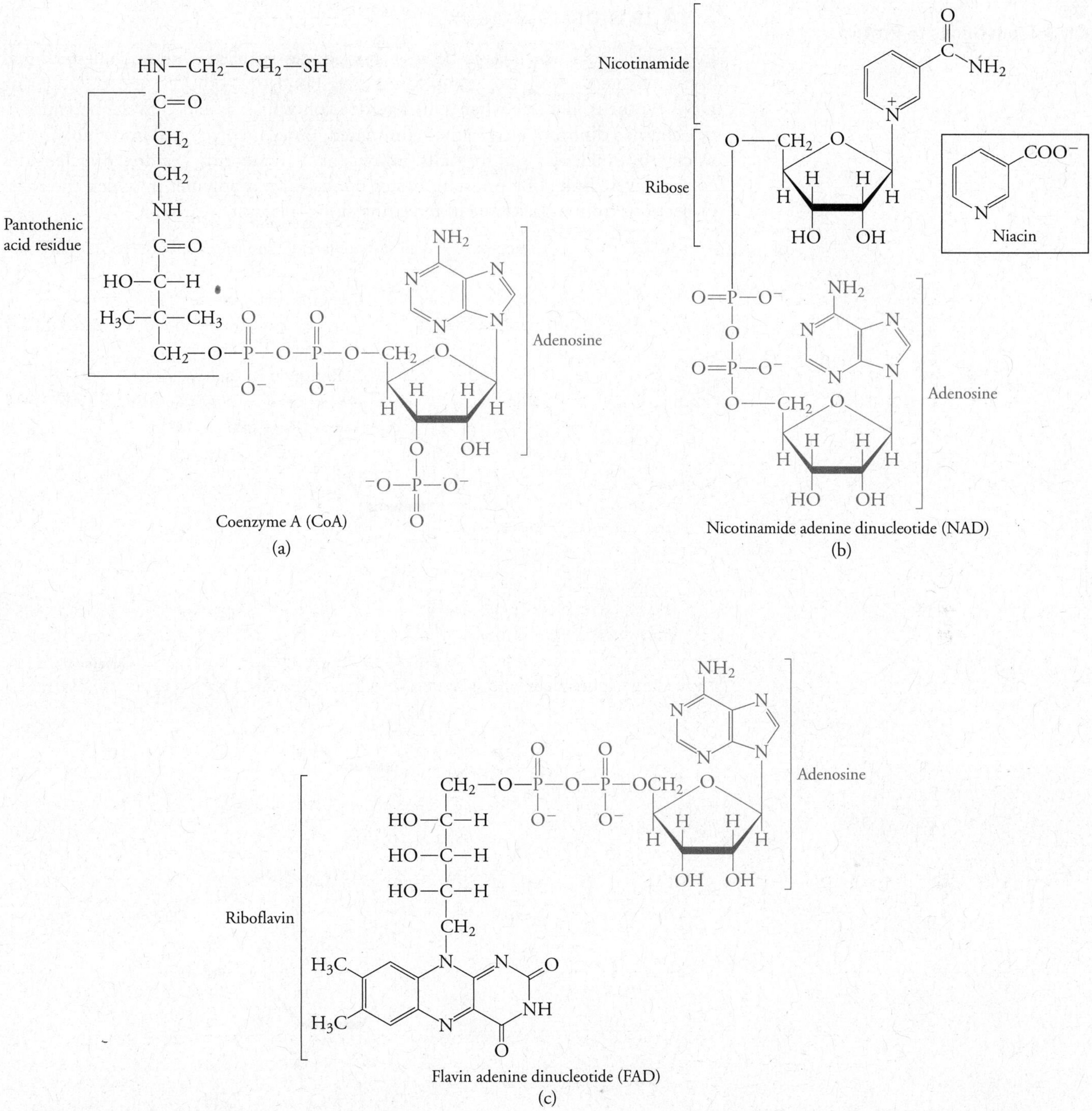

Figure 3-3 Some nucleotide derivatives. The adenosine group of each of these compounds is shown in red. Note that each also contains a vitamin derivative. (a) Coenzyme A (CoA) contains a residue of pantothenic acid (pantothenate), also known as vitamin B_5. The sulfhydryl group is the site of attachment of other groups. (b) The nicotinamide group of nicotinamide adenine dinucleotide (NAD) is a derivative of the vitamin niacin (also called nicotinic acid or vitamin B_3; see inset) and undergoes oxidation and reduction. The related compound nicotinamide adenine dinucleotide phosphate (NADP) contains a phosphoryl group at the adenosine C2′ position. (c) Oxidation and reduction of flavin adenine dinucleotide (FAD) occurs at the riboflavin group (also known as vitamin B_2).

? **Identify the nitrogenous base(s) and sugar(s) in each structure.**

DNA is a double helix

In a nucleic acid, the linkage between nucleotides is called a **phosphodiester bond** because a single phosphate group forms ester bonds to both C5′ and C3′. During DNA synthesis in a cell, when a nucleoside triphosphate is added to the polynucleotide chain, a diphosphate group is eliminated. Once incorporated into a polynucleotide, the nucleotide is formally known as a nucleotide **residue.** Nucleotides consecutively linked by phosphodiester bonds form a polymer in which the bases project out from a backbone of repeating sugar–phosphate groups.

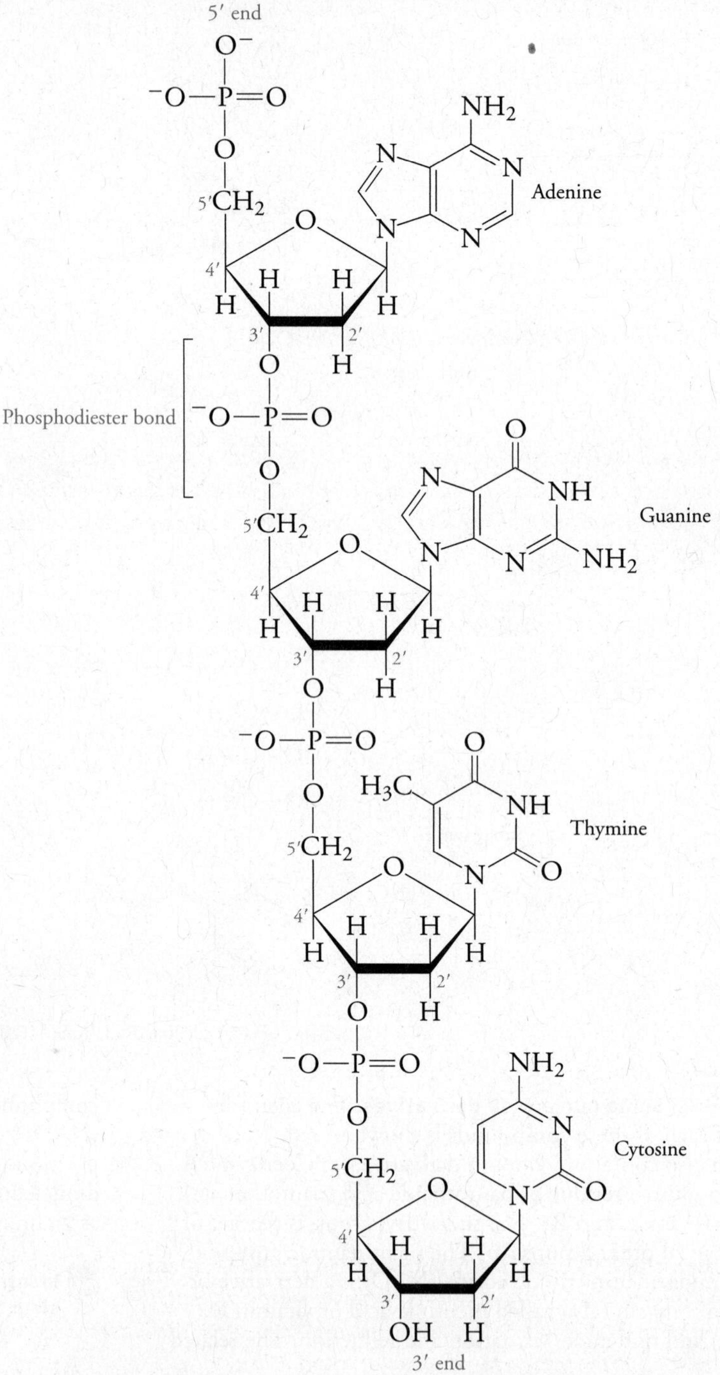

The end of the polymer that bears a phosphate group attached to C5′ is known as the **5′ end,** and the end that bears a free OH group at C3′ is the **3′ end.** By convention, the base sequence in a polynucleotide is read from the 5′ end (on the left) to the 3′ end (on the right).

DNA contains two polynucleotide strands whose bases pair through hydrogen bonding (hydrogen bonds are discussed in Section 2-1). Two hydrogen bonds link adenine and thymine, and three hydrogen bonds link guanine and cytosine:

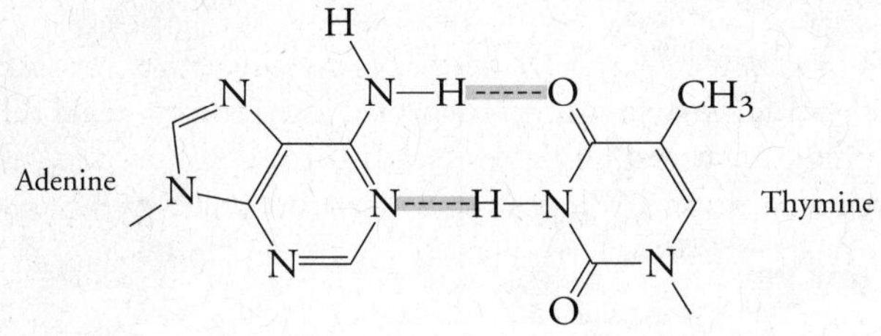

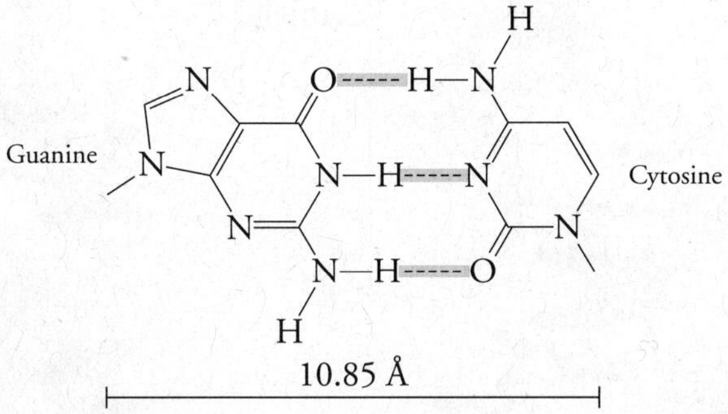

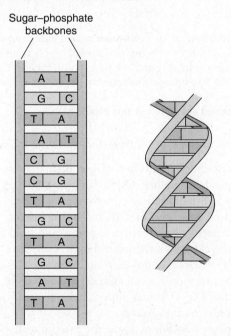

10.85 Å

*All the **base pairs**, which consist of a purine and a pyrimidine, have the same molecular dimensions* (about 11 Å wide). Consequently, the **sugar–phosphate backbones** of the two strands of DNA are separated by a constant distance, regardless of whether the base pair is A:T, G:C, T:A, or C:G.

Although the DNA is shown here as a ladder-like structure (*left*), with the two sugar–phosphate backbones as the vertical supports and the base pairs as the rungs, the two strands of DNA twist around each other to generate the familiar double helix (*right*). This conformation allows successive base pairs, which are essentially planar, to stack on top of each other with a center-to-center distance of only 3.4 Å. In fact, Watson and Crick derived this model for DNA not just from Chargaff's rules but also from Rosalind Franklin's studies of the diffraction (scattering) of an X-ray beam by a DNA fiber, which suggested a helix with a repeating spacing of 3.4 Å.

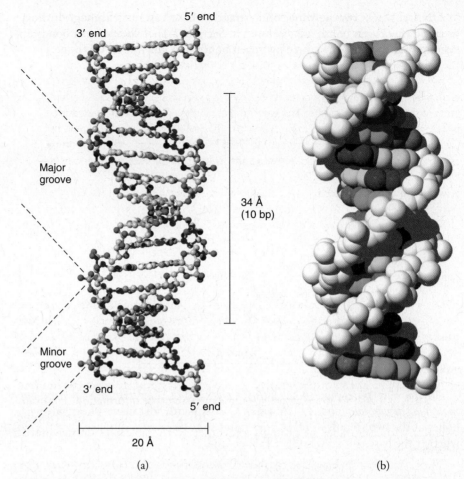

5′ end
3′ end
Major groove
34 Å (10 bp)
Minor groove
3′ end
5′ end
20 Å
(a)
(b)

Figure 3-4 Model of DNA. (a) Ball-and-stick model with atoms colored: C gray, O red, N blue, and P gold (H atoms are not shown). (b) Space-filling model with the sugar–phosphate backbone in gray and the bases color-coded: A green, C blue, G yellow, and T red. ⊕ See Interactive Exercise. **Three-dimensional structure of DNA.**

? **How many nucleotides are shown in this double helix?**

The major features of the DNA molecule include the following (**Fig. 3-4**):

1. The two polynucleotide strands are **antiparallel;** that is, their phosphodiester bonds run in opposite directions. One strand has a 5′ → 3′ orientation, and the other has a 3′ → 5′ orientation.
2. The DNA "ladder" is twisted in a right-handed fashion. (If you climbed the DNA helix as if it were a spiral staircase, you would hold the outer railing—the sugar–phosphate backbone—with your right hand.)
3. The diameter of the helix is about 20 Å, and it completes a turn about every 10 base pairs, which corresponds to an axial distance of about 34 Å.
4. The twisting of the DNA "ladder" into a helix creates two grooves of unequal width, the **major** and **minor grooves.**
5. The sugar–phosphate backbones define the exterior of the helix and are exposed to the solvent. The negatively charged phosphate groups bind Mg^{2+} cations *in vivo*, which helps minimize electrostatic repulsion between these groups.
6. The base pairs are located in the center of the helix, approximately perpendicular to the helix axis.
7. The base pairs stack on top of each other, so the core of the helix is solid (see Fig. 3-4b). Although the planar faces of the base pairs are not accessible to the solvent, their edges are exposed in the major and minor grooves (this allows certain DNA-binding proteins to recognize specific bases).

In nature, DNA seldom assumes a perfectly regular conformation because of small sequence-dependent irregularities. For example, base pairs can roll or twist like propeller blades, and the helix may wind more tightly or loosely at certain nucleotide sequences. DNA-binding proteins may take advantage of these small variations to locate their specific binding sites, and they in turn may further distort the DNA helix by causing it to bend or partially unwind.

The size of a DNA segment is expressed in units of base pairs (**bp**) or kilobase pairs (1000 bp, abbreviated **kb**). Most naturally occurring DNA molecules comprise thousands to millions of base pairs. A short single-stranded polymer of nucleotides is usually called an **oligonucleotide** (*oligo* is Greek for "few"). In a cell, nucleotides are polymerized by the action of enzymes known as **polymerases.** The phosphodiester bonds linking nucleotide residues can be broken by the action of **nucleases.** An **exonuclease** removes a residue from the end of a polynucleotide chain, whereas an **endonuclease** cleaves at some other point along the chain. Polymerases and nucleases are usually specific for either DNA or RNA. In the absence of these enzymes, the structures of nucleic acids are remarkably stable.

RNA is single-stranded

RNA, which is a single-stranded polynucleotide, has greater conformational freedom than DNA, whose structure is constrained by the requirements of regular base-pairing between its two strands. *An RNA strand can fold back on itself so that base pairs form between complementary segments of the same strand.* Consequently, RNA molecules tend to assume intricate three-dimensional shapes (**Fig. 3-5**). Unlike DNA, whose regular structure is suited for the long-term storage of genetic information, RNA can assume more active roles in expressing that information. For example, the molecule shown in Figure 3-4, which carries the amino acid phenylalanine, interacts with a number of proteins and other RNA molecules during protein synthesis.

The residues of RNA are also capable of base-pairing with a complementary single strand of DNA to produce an RNA–DNA hybrid double helix. A double helix involving RNA is wider and flatter than the standard DNA helix (its diameter is about 26 Å, and it makes one helical turn every 11 residues). In addition, its base pairs are inclined to the helix axis by about 20° (**Fig. 3-6**). These structural differences relative to the standard DNA helix primarily reflect the presence of the 2′ OH groups in RNA.

A double-stranded DNA helix can adopt this same helical conformation; it is known as **A-DNA.** The standard DNA helix shown in Figure 3-4 is known as **B-DNA.** Other conformations of DNA have been described, and there is evidence that they exist *in vivo,* at least for certain nucleotide sequences, but their functional significance is not completely understood.

DNA can be denatured and renatured

The pairing of polynucleotide strands in a double-stranded nucleic acid is possible because bases in each strand form hydrogen bonds with complementary bases in the other strand: A is the complement of T (or U), and G is the complement of C. However, the structural stability of the DNA helix does not depend significantly on hydrogen bonding between complementary bases. (If the strands were separated, the bases could still satisfy their hydrogen-bonding requirements by forming hydrogen bonds with solvent water molecules.) Instead, *stability depends mostly on **stacking interactions,** which are a form of van der Waals interaction, between adjacent base pairs.* A view down the helix axis shows that stacked base pairs do not overlap exactly, due to the winding of the helix (**Fig. 3-7**). Although individual stacking interactions are weak, they are additive along the length of a DNA molecule.

The stacking interactions between neighboring G:C base pairs are stronger than those of A:T base pairs (this is not related to the fact that G:C base pairs have one more hydrogen bond than A:T base pairs). Consequently, *a DNA helix that is rich in G and C is harder to disrupt than DNA with a high proportion of A and T.* These differences can be quantified in the **melting temperature (T_m)** of the DNA.

Figure 3-5 A transfer RNA molecule. This 76-nucleotide single-stranded RNA molecule folds back on itself so that base pairs form between complementary segments. [Structure (pdb 4TRA) determined by E. Westhoff, P. Dumas, and D. Moras.]

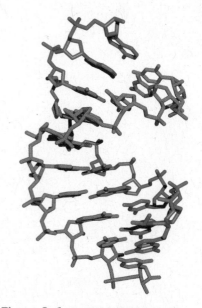

Figure 3-6 An RNA-DNA hybrid helix. In a double helix formed by one strand of RNA (red) and one strand of DNA (blue), the planar base pairs are tilted and the helix does not wind as steeply as in a standard DNA double helix (compare with Fig. 3-4). [Structure (pdb 1FIX) determined by N. C. Horton and B. C. Finzel.]

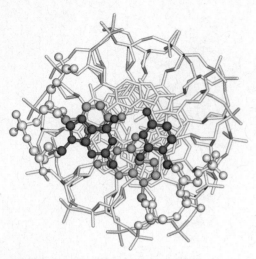

Figure 3-7 **Axial view of DNA base pairs.** A view down the central axis of the DNA helix shows the overlap of neighboring base pairs (only the first two nucleotide pairs are highlighted).

? **Locate the base and sugar in the blue nucleotides.**

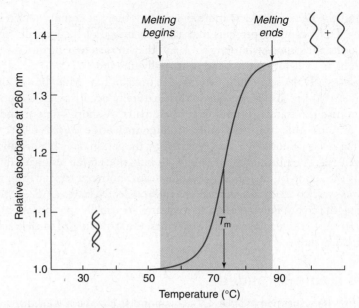

Figure 3-8 **A DNA melting curve.** Thermal denaturation (melting, or strand separation) of DNA results in an increase in ultraviolet absorbance relative to the absorbance at 25°C. The melting point, T_m, of the DNA sample is defined as the midpoint of the melting curve.

To determine the melting point of a sample of DNA, the temperature is slowly increased. At a sufficiently high temperature, the base pairs begin to unstack, hydrogen bonds break, and the two strands begin to separate. This process continues as the temperature rises, until the two strands come completely apart. The melting, or **denaturation,** of the DNA can be recorded as a melting curve **(Fig. 3-8)** by monitoring an increase in the absorbance of ultraviolet (260-nm) light (the aromatic bases absorb more light when unstacked). The midpoint of the melting curve (that is, the temperature at which half the DNA has separated into single strands) is the T_m. Table 3-2 lists the GC content and the melting point of the DNA from different species. Since manipulating DNA in the laboratory frequently requires the thermal separation of paired DNA strands, it is sometimes helpful to know the DNA's GC content.

When the temperature is lowered slowly, denatured DNA can **renature;** that is, *the separated strands can re-form a double helix by reestablishing hydrogen bonds between the complementary strands and by restacking the base pairs.* The maximum rate of renaturation occurs at about 20–25°C below the melting temperature. If the DNA is cooled too rapidly, it may not fully renature because base pairs may form randomly between short complementary segments. At low temperatures, the improperly paired segments are frozen in place since they do not have enough thermal energy to melt apart and find their correct complements **(Fig. 3-9).** The rate of renaturation of denatured DNA depends on the length of the double-stranded molecule: Short segments come together **(anneal)** faster than longer segments because the bases in each strand must locate their partners along the length of the complementary strand.

The ability of short single-stranded nucleic acids (either DNA or RNA) to hybridize with longer polynucleotide chains is the basis for a number of useful laboratory techniques (described in detail in Section 3-4). For example, an oligonucleotide **probe**

[TABLE 3-2] GC Content and Melting Points of DNA

Source of DNA	GC Content (%)	T_m (°C)
Dictyostelium discoideum (fungus)	23.0	79.5
Clostridium butyricum (bacterium)	37.4	82.1
Homo sapiens	40.3	86.5
Streptomyces albus (bacterium)	72.3	100.5

[Data from Brown, T. A. (ed.), *Molecular Biology LabFax*, vol. I., Academic Press (1998), pp. 233–237.]

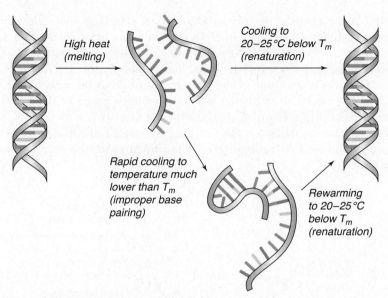

Figure 3-9 **Renaturation of DNA.** DNA strands that have been melted apart can renature at a temperature of 20–25°C below the T_m. At much lower temperatures, base pairs may form between short complementary segments within and between the single strands. Correct renaturation is possible only if the sample is rewarmed so that the improperly paired strands can separate and reanneal.

that has been tagged with a radioactive isotope or a fluorescent group can be used to detect the presence of a complementary nucleic acid sequence in a complex mixture.

> **CONCEPT REVIEW**
> - What are the relationships among purines, pyrimidines, nucleosides, nucleotides, and nucleic acids?
> - Describe the arrangement of the base pairs and sugar-phosphate backbones in DNA.
> - What did Chargaff's rules reveal about the structure of DNA?
> - How do DNA and RNA differ?
> - Describe the molecular events in DNA denaturation and renaturation.

3-2 Genes Encode Proteins

The complementarity of the two strands of DNA is essential for its function as the storehouse of genetic information, since this information must be **replicated** (copied) for each new generation. As first suggested by Watson and Crick, the separated strands of DNA direct the synthesis of complementary strands, thereby generating two identical double-stranded molecules (**Fig. 3-10**). The parental strands are said to act as templates for the assembly of the new strands because their sequence of nucleotides determines the sequence of nucleotides in the new strands. Thus, genetic information—in the form of a sequence of nucleotide residues—is transmitted each time a cell divides.

A similar phenomenon is responsible for the **expression** of that genetic information, a process in which the information is used to direct the synthesis of proteins that carry out the cell's activities. First, *a portion of the DNA, a **gene**, is **transcribed** to produce a complementary strand of RNA; then the RNA is translated into protein.* This paradigm, known as the central dogma of molecular biology, was formulated by Francis Crick. It can be shown schematically as

KEY CONCEPT
- The biological information encoded by a sequence of DNA is transcribed to RNA and then translated into the amino acid sequence of a protein.

⊕ See Guided Exploration. Overview of transcription and translation.

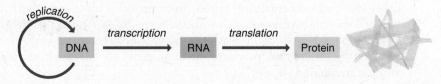

Even in the simplest organisms, DNA is an enormous molecule, and many organisms contain several different DNA molecules (for example, the chromosomes of eukaryotes). An organism's complete set of genetic information is called its **genome.** A genome may comprise several hundred to perhaps 35,000 genes.

To transcribe a gene, one of the two strands of DNA serves as a template for an RNA polymerase to synthesize a complementary strand of RNA. The RNA therefore has the same sequence (except for the substitution of U for T) and the same 5′ → 3′ orientation as the nontemplate strand of DNA. This strand of DNA is often called the **coding strand** (the template strand is called the **noncoding strand**).

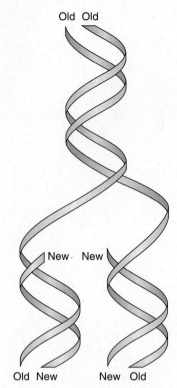

Figure 3-10 DNA replication. The double helix unwinds so that each parental strand can serve as a template for the synthesis of a new complementary strand. The result is two identical double-helical DNA molecules.

....

? **Label the 5′ and 3′ end of each strand.**

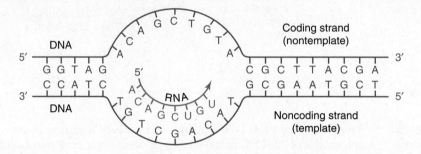

The transcribed RNA is known as **messenger RNA (mRNA)** because it carries the same genetic message as the gene.

The mRNA is translated in the **ribosome,** a cellular particle consisting of protein and **ribosomal RNA (rRNA).** At the ribosome, small molecules called **transfer RNA (tRNA),** which carry amino acids, recognize sequential sets of three bases (known as **codons**) in the mRNA through complementary base-pairing (a tRNA molecule is shown in Fig. 3-5). The ribosome covalently links the amino acids carried by successive tRNAs to form a protein. *The protein's amino acid sequence therefore ultimately depends on the nucleotide sequence of the DNA.*

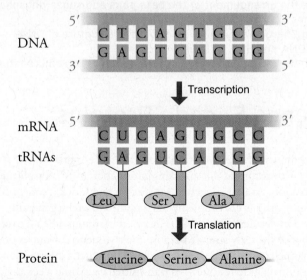

The correspondence between amino acids and mRNA codons is known as the **genetic code.** There are a total of 64 codons: 3 of these are "stop" signals that terminate translation, and the remaining 61 represent, with some redundancy, the 20 standard amino acids found in proteins. Table 3-3 shows which codons specify which amino acids. In theory, knowing a gene's nucleotide sequence should be equivalent to knowing the amino acid sequence of the protein encoded by the gene. However, as we will see, genetic information is often "processed" at several points before the protein reaches its mature form. Keep in mind that the rRNA and tRNA required for protein synthesis, as well as other types of RNA, are also encoded by genes. The "products" of these genes are the result of transcription without translation.

[**TABLE 3-3**] The Standard Genetic Code[a]

First Position (5′ end)	Second Position				Third Position (3′end)
	U	**C**	**A**	**G**	
U	UUU Phe	UCU Ser	UAU Tyr	UGU Cys	U
	UUC Phe	UCC Ser	UAC Tyr	UGC Cys	C
	UUA Leu	UCA Ser	UAA Stop	UGA Stop	A
	UUG Leu	UCG Ser	UAG Stop	UGG Trp	G
C	CUU Leu	CCU Pro	CAU His	CGU Arg	U
	CUC Leu	CCC Pro	CAC His	CGC Arg	C
	CUA Leu	CCA Pro	CAA Gln	CGA Arg	A
	CUG Leu	CCG Pro	CAG Gln	CGG Arg	G
A	AUU Ile	ACU Thr	AAU Asn	AGU Ser	U
	AUC Ile	ACC Thr	AAC Asn	AGC Ser	C
	AUA Ile	ACA Thr	AAA Lys	AGA Arg	A
	AUG Met	ACG Thr	AAG Lys	AGG Arg	G
G	GUU Val	GCU Ala	GAU Asp	GGU Gly	U
	GUC Val	GCC Ala	GAC Asp	GGC Gly	C
	GUA Val	GCA Ala	GAA Glu	GGA Gly	A
	GUG Val	GCG Ala	GAG Glu	GGG Gly	G

[a]The 20 amino acids are abbreviated; Ala, alanine; Arg, arginine; Asn, asparagine; Asp, aspartate; Cys, cysteine; Gly, glycine; Gln, glutamine; Glu, glutamate; His, histidine; Ile, isoleucine; Leu, leucine; Lys, lysine; Met, methionine; Phe, phenylalanine; Pro, proline; Ser, serine; Thr, threonine; Trp, tryptophan; Tyr, tyrosine; and Val, valine.

? How many amino acids would be uniquely specified by a genetic code that consisted of just the first two nucleotides in each codon?

A mutated gene can cause disease

Because an organism's genetic material influences the organism's entire repertoire of activities, it is vitally important to unravel the sequence of nucleotides in that organism's DNA, even by examining one gene at a time. Thousands of genes have been identified through studies of the genes' protein products, and millions more have been catalogued through genome-sequencing projects (discussed below in Section 3-3). Although the functions of many genes are not yet understood, some genes have come to light through the study of inherited diseases. In a traditional approach, researchers have used the defective protein associated with a particular disease to track down the relevant genetic defect. For example, the variant hemoglobin protein that causes sickle cell anemia results from the substitution of the amino acid glutamate by valine. In the gene for that protein chain, the normal GAG codon has been **mutated** (altered) to GTG.

Normal gene · · · ACT CCT GAG GAG AAG · · ·
Protein · · · Thr – Pro – Glu – Glu – Lys · · ·

Mutated gene · · · ACT CCT GTG GAG AAG · · ·
Protein · · · Thr – Pro – Val – Glu – Lys · · ·

More modern approaches begin with analysis of the DNA to discover the genetic changes that lead to disease. The first successful application of this method identified the cystic fibrosis gene, that is, the gene whose mutation cause the disease (Box 3-A).

Over 3000 genes have been linked to specific monogenetic diseases, such as sickle cell anemia and cystic fibrosis. In many cases, a variety of different mutations have been catalogued for each disease gene, which explains in part why symptoms of the disease vary between individuals. The database known as OMIM (Online Mendelian Inheritance in Man; http://www.ncbi.nlm.nih.gov/omim) contains information on thousands of genetic variants, including the clinical features of the resulting disorder and its biochemical basis. The Genetic Testing Registry (http://www.ncbi.nlm.nih.gov/gtr/) is a database of the diseases that can be detected through analysis of DNA, carried out by either clinical or research laboratories.

BOX 3-A   CLINICAL CONNECTION

Discovery of the Cystic Fibrosis Gene

About 1 in 3000 babies in the United States is born with **cystic fibrosis (CF),** the most common inherited disease in individuals of northern European extraction. The most serious symptom of CF is the obstruction of the airways by thick, sticky mucus, which tends to create an ideal environment for bacterial growth. Individuals with CF may also suffer from impaired secretion of digestive enzymes from the pancreas, which contributes to malnutrition and poor growth. Historically, individuals with CF died in childhood, but a variety of treatments, including the use of antibiotics to prevent lung infections, have now extended survival well into adulthood.

Before the era of DNA testing, one of the diagnostic signs of CF was high chloride concentrations in sweat (according to medieval folklore, a baby who tasted salty when kissed was predicted to die soon). But neither this characteristic nor other symptoms, such as the thick mucus in the airways, pointed unequivocally to a defect in any known protein. Consequently, the search for the genetic basis for CF required a strategy that did not use a protein as its starting point.

To find the cystic fibrosis gene, researchers analyzed DNA from affected individuals, who had two copies of the defective CF gene, and from family members who were asymptomatic carriers and had one normal and one defective copy of the gene. Individuals with one or two copies of the defective CF gene shared two other genetic features that can be detected in a laboratory test. These two **DNA markers** were used to define a chromosomal region that was likely to contain the cystic fibrosis gene. In particular, one DNA segment on chromosome 7 appeared to be present in a number of mammalian species, which suggested that the segment contained an essential gene (about 98% of mammalian DNA does not encode any protein). The researchers then deduced the sequence of nucleotides in this region of DNA, ultimately identifying a stretch of about 250,000 bp as the CF gene.

As is the case for nearly all mammalian genes, only certain portions of the CF gene directly correspond to a protein product, because segments of the mRNA molecule transcribed from the gene are excised, an event called **splicing,** before the message is translated into protein (splicing is discussed further in Section 21-3). In addition, sequences at each end of the mRNA are not translated. After splicing, the mRNA is only 6129 nucleotides long. Of this molecule, 4440 nucleotides (or 4440 ÷ 3 = 1480 codons) specify the 1480 amino acid residues of the protein product.

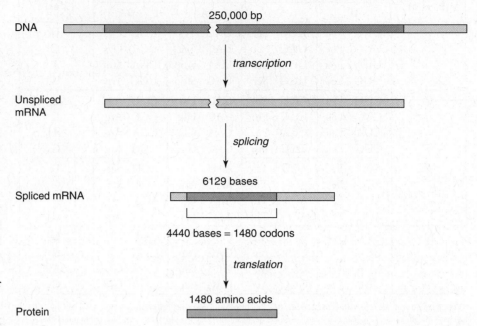

Matching every three bases in the derived mRNA sequence with the appropriate amino acid (see Table 3-3) yielded the amino acid sequence of the protein.

Additional sequencing studies showed that in about 70% of CF patients, the gene is missing three nucleotides. This results in the deletion of a single phenylalanine (Phe) residue at position 508 (the 508th amino acid residue in the encoded protein):

	504	505	506	507	508	509	510	511	512
Normal gene	···GAA	AAT	ATC	ATC	TTT	GGT	GTT	TCC	TAT···
Protein	···Glu	Asn	Ile	Ile	Phe	Gly	Val	Ser	Tyr···

	504	505	506	507	508	509	510	511	512
Mutated gene	···GAA	AAT	ATC	AT–	––T	GGT	GTT	TCC	TAT···
Protein	···Glu	Asn	Ile	Ile		Gly	Val	Ser	Tyr···

Note that although the nucleotide deletion affects codons 507 and 508, the redundancy of the genetic code means that the isoleucine (Ile) at position 507 is not affected (because codons ATC and ATT both specify Ile). The protein lacking Phe 508 is abnormally processed by the cell, so very little is present in functional form.

So what does the CF gene do? The putative function of the cystic fibrosis gene product was identified by its sequence similarity to a large family of proteins involved in the transport of substances across cell membranes (recall from Section 2-2 that only nonpolar substances can spontaneously traverse a lipid bilayer; all other substances require a protein transporter). Each member of this protein family has one or two segments that position the protein in the membrane. The CF protein also contains an additional domain thought to play a regulatory role. Accordingly, the protein was named the cystic fibrosis transmembrane conductance regulator (CFTR). When the CFTR gene was introduced into different cell types, its function could be studied. The CFTR protein is, in fact, a membrane protein that acts as a channel to allow Cl⁻ to exit the cell

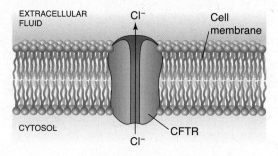

CFTR also appears to regulate Na⁺ uptake by the cell. Consequently, a defective or absent CFTR protein disrupts the normal distribution of Na⁺ and Cl⁻. In the CF lung, the concentrations of the ions are low in the extracellular space. As a result, the water that would normally be drawn by high concentrations of these ions is absent. In a normal lung, the extracellular fluid is thin and watery, but in the CF lung, the fluid is thick and viscous. In the sweat gland, a defective CFTR alters the transport of Na⁺ and Cl⁻, causing the salty sweat that is diagnostic of CF.

● **Questions:**

1. The Phe-deletion mutation described on the facing page causes a severe form of CF. Other types of mutations in the CF gene produce milder forms of the disease that may not be detected until adulthood. Explain.
2. One portion of the normal CF gene has the sequence

$$\cdots \text{AAT ATA GAT ACA G} \cdots$$

   In some individuals with cystic fibrosis, this portion of the gene has the sequence

$$\cdots \text{AAT AGA TAC AG} \cdots$$

   How has the DNA sequence changed and how does this affect the encoded protein?
3. Most genetic diseases that limit survival to reproductive age are relatively rare; CF is an exception. The prevalence of the defective CF gene suggests that it may have had benefits during human evolution. Some pathogens (disease-causing organisms) use the normal CFTR as an entry point for infecting cells. Explain why individuals with one normal and one defective CF gene are more likely to survive (and pass on their genes) than individuals with two normal or two defective copies of the CF gene.
4. Would cystic fibrosis be a good candidate for treatment by gene therapy (Section 3-4)?

CONCEPT REVIEW
- How does DNA encode genetic information and how is this information expressed?
- What is the relationship between the nucleotide sequence in a gene and the amino acid sequence of a protein?
- List some reasons why knowing a gene's sequence might be useful.

3-3 Genomics

The ability to sequence large tracts of DNA has made it possible to study entire genomes, from the small DNA molecules of parasitic bacteria to the enormous multichromosome genomes of plants and mammals. Sequence data are customarily deposited in a public database such as GenBank. The data can be accessed electronically in order to compare a given sequence to sequences from other genes (see Bioinformatics Project 2).

Some of the thousands of organisms whose genomes have been partially or fully sequenced are listed in Table 3-4. This list includes species that are widely used as model organisms for different types of biochemical studies (Fig. 3-11).

KEY CONCEPTS
- The genomes of different species vary in size and number of genes.
- Genes can be identified by their nucleotide sequences.
- Analysis of genetic data can provide information about gene function and risk of disease.

Gene number is roughly correlated with organismal complexity

Not surprisingly, *organisms with the simplest lifestyles tend to have the least amount of DNA and the fewest genes.* For example, *M. genitalium* and *H. influenzae* (see Table 3-4) are human parasites that depend on their host to provide nutrients; these organisms do not contain as many genes as free-living bacteria such as *Synechocystis* (a photosynthetic bacterium). Multicellular organisms generally have even more DNA and more genes, presumably to support the activities of their many specialized

[TABLE 3-4] Genome Size and Gene Number of Some Organisms

Organism	Genome Size (kb)	Number of Genes
Bacteria		
Mycoplasma genitalium	580	525
Haemophilus influenzae	1,830	1,789
Synechocystis PCC6803	3,947	3,618
Escherichia coli	4,643	4,630
Archaea		
Methanocaldococcus jannaschii	1,740	1,830
Archaeoglobus fulgidus	2,178	2,486
Fungi		
Saccharomyces cerevisiae (yeast)	12,071	6,281
Plants		
Arabidopsis thaliana	119,146	33,323
Oryza sativa (rice)	382,151	30,294
Zea mays (corn)	~2,046,000	~32,000
Animals		
Caenorhabditis elegans (nematode)	100,268	21,175
Drosophila melanogaster (fruit fly)	139,466	15,016
Homo sapiens	3,102,000	~21,000

[Data from NCBI Genome Project.]

? **What is the relationship between genome size and gene number in prokaryotes? How does this differ in eukaryotes?**

Figure 3-11 Some model organisms.
(a) *Escherichia coli,* a normal inhabitant of the mammalian digestive tract, is a metabolically versatile bacterium that tolerates both aerobic and anaerobic conditions. (b) Baker's yeast, *Saccharomyces cerevisiae,* is one of the simplest eukaryotic organisms, with just over 6000 genes. (c) *Caenorhabditis elegans* is a small (1-mm) and transparent roundworm. As a multicellular organism, it bears genes not found in unicellular organisms. (d) The plant kingdom is represented by *Arabidopsis thaliana,* which has a short generation time and readily takes up foreign DNA. [Dr. Kari Lounatmaa/Science Photo Library/Photo Researchers; Andrew Syred/Science Photo Library/Photo Researchers; Sinclair Stammers/Science Photo Library/Photo Researchers; Dr. Jeremy Burgess/Science Photo Library/Photo Researchers.]

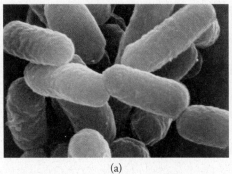
(a)

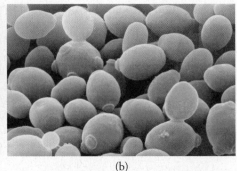
(b)

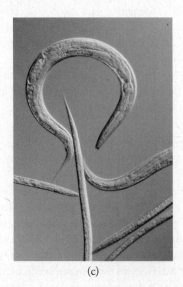
(c)

(d)

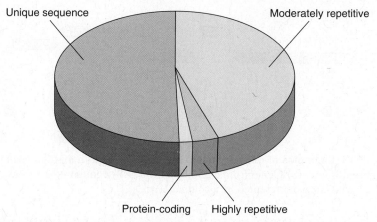

Unique sequence
Moderately repetitive
Protein-coding
Highly repetitive

Figure 3-12 Coding and noncoding portions of the human genome. Approximately 1.5% of the genome codes for proteins. Moderately repetitive sequences make up 45% of the genome and highly repetitive sequences about 3%, so that roughly half of the human genome consists of unique DNA sequences of unknown function. Up to 80% of the genome may be transcribed, however.

cell types. Interestingly, humans contain about as many genes as nematodes, suggesting that organismal complexity results not just from the raw number of genes but from how the genes are transcribed and translated into protein. Note that humans and many other organisms are **diploid** (having two sets of genetic information, one from each parent), so that each human cell contains roughly 6.2 billion base pairs of DNA. For simplicity, genetic information usually refers to the **haploid** state, equivalent to one set of genetic instructions.

In prokaryotic genomes, all but a few percent of the DNA represents genes for proteins and RNA. The proportion of noncoding DNA generally increases with the complexity of the organism. For example, about 30% of the yeast genome, about half of the *Arabidopsis* genome, and over 98% of the human genome is noncoding DNA. Although up to 80% of the human genome may actually be transcribed to RNA, *the protein-coding segments account for only about 1.5% of the total* (**Fig. 3-12**).

Much of the noncoding DNA consists of repeating sequences with no known function. The presence of repetitive DNA helps explain why certain very large genomes actually include only a modest number of genes. For example, the maize (corn) and rice genomes contain about the same number of genes, but the maize genome is as much as 10 times larger than the rice genome. Over half of the maize genome appears to be composed of **transposable elements,** short segments of DNA that are copied many times and inserted randomly into the chromosomes.

The human genome contains several types of repetitive DNA sequences, including the inactive remnants of transposable elements. About 45% of human DNA consists of **moderately repetitive sequences,** which are blocks of hundreds or thousands of nucleotides scattered throughout the genome. The most numerous of these are present in hundreds of thousands of copies. **Highly repetitive sequences** account for another 3% of the human genome. These segments of 2 to 10 bases are present in millions of copies. They are repeated tandemly (side by side), sometimes thousands of times. The number of repeats of a given sequence often varies between individuals, even in the same family, so this information can be analyzed to produce a DNA "fingerprint" (see Section 3-4).

Some so-called noncoding DNA in fact consists of genes for RNA molecules, which appear to play a variety of roles in regulating the expression of protein-coding genes (Chapter 21). Comparisons of mammalian genomes indicate that as much as 6% of the human genome appears to have changed relatively little during evolution, suggesting that it has some essential function.

Genes are identified by comparing sequences

For many genomes, the exact number of genes has not yet been determined, and different methods for identifying genes yield different estimates. For example, a

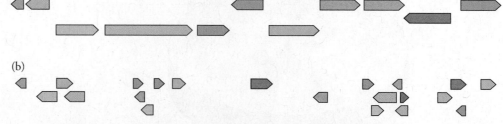

Figure 3-13 Examples of genome maps. (a) Genes located in a 10-kb span of the *E. coli* chromosome. (b) Genes from a 2500-kb gene-rich segment of the mouse genome. Each gene is represented by a colored block.

computer can scan a DNA sequence for an **open reading frame (ORF),** that is, a stretch of nucleotides that can potentially be transcribed or translated. For a protein-coding gene, the ORF begins with a "start" codon: ATG in the coding strand of DNA, which corresponds to AUG in RNA (see Table 3-3). This codon specifies methionine, the initial residue of all newly synthesized proteins. The ORF ends with one of the three "stop" codons: DNA coding sequences of TAA, TAG, or TGA, which correspond to the three mRNA stop codons (see Table 3-3). Other so-called *ab initio* ("from the beginning") gene-identifying methods scan the DNA for other features that characterize the beginnings and endings of genes. These methods tend to overestimate the number of genes.

Another method for identifying genes in a genome relies on sequence comparisons with known genes (and thereby probably underestimates the true number of genes). Such genome-to-genome comparisons are possible because of the universal nature of the genetic code and the relatedness of all organisms through evolution (Section 1-4). *Genes with similar functions in different species tend to have similar sequences;* such genes are said to be **homologous.** Even an inexact match can still indicate a protein's functional category, such as enzyme or hormone receptor, although its exact role in the cell may not be obvious. Genes that appear to lack counterparts in other species are known as **orphan genes.** At present, the number of known genes exceeds the number of known gene products (proteins and RNA molecules). This is hardly surprising, given that some genes are expressed at low levels or generate products that have not yet been detected through conventional biochemical isolation approaches. About 20% of the genes in the well-studied organism *E. coli* have not yet been assigned functions.

Genome maps, such as the ones shown in Figure 3-13, indicate the placement and orientation of genes on a chromosome. Arrows pointing in opposite directions represent genes encoded by different strands of the double-stranded chromosome. Note that mammalian genes are typically much longer than bacterial genes (27 kb on average), since they contain sequences that are spliced out of the transcript before translation. In addition, the spaces between genes are much larger in the mammalian genome.

Gene-mapping projects have uncovered some interesting aspects of evolution, including **horizontal gene transfer.** This occurs when a gene is transferred between species rather than from parent to offspring of the same species (vertical gene transfer). Horizontal gene transfer may be mediated by viruses, which can pick up extra DNA as they insert and excise themselves from the host's chromosomes. This activity can generate, for example, what appears to be a mammalian gene inside a bacterial genome. The ease with which many bacterial organisms trade their genes has given rise to the idea that groups of bacteria should be viewed as a continuum of genomic variations instead of separate species with discrete genomes.

Genomic data can be linked to disease

Genomics, the study of genomes, has a number of practical applications. For one thing, *the number of genes and their putative functions provide a rough snapshot of the*

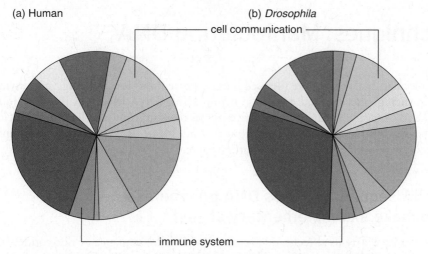

(a) Human (b) *Drosophila*
— cell communication —

— immune system —

Figure 3-14 Functional classification of genes. This diagram is based on 17,181 human genes (a) and 9837 *Drosophila* genes (b), grouped according to the biochemical function of the gene product. Humans devote a larger proportion of genes to cell communication (25.4% versus 18.6% in *Drosophila*) and to the immune system (15.3% versus 10.2% in *Drosophila*). [Data from the Protein Analysis through Evolutionary Relationships classification system, www.pantherdb.org/.]

metabolic capabilities of a given organism. For example, humans and fruit flies differ in the number of genes that code for cell-signaling pathways and immune system functions (**Fig. 3-14**). An unusual number of genes belonging to one category might indicate some unusual biological property in an organism. This sort of knowledge could be useful for developing drugs to inhibit the growth of a pathogenic organism according to its unique metabolism.

Genomic analysis also reveals variations among individuals, some of which can be linked to an individual's chance of developing a particular disease. In addition to genetic changes that are clearly associated with a single-gene disorder, millions more sequence variations have been catalogued. On average, *the DNA of any two humans differs at 3 million sites, or about once every thousand base pairs.* These **single-nucleotide polymorphisms** (**SNPs,** instances where the DNA sequence differs among individuals) are compiled in databases. Some of the factors that can alter DNA are discussed in Section 20-5).

Researchers have attempted to correlate SNPs with disorders, such as cardiovascular disease or cancer, that likely depend on the contributions of many genes. **Genome-wide association studies (GWAS)** have identified, for example, 39 sites that are associated with type 2 diabetes and 71 that are associated with Crohn's disease, an autoimmune disorder. The risk tied to any particular genetic variant is low, but the entire set of variations can explain up to 50% of the heritability of the disease. Although the SNPs are only proxies for disease genes, these data should provide a starting point for researchers to explore the DNA near the SNPs to discover the genes that are directly involved in the disease. Several commercial enterprises offer individual genome-sequencing services, but until genetic information can be reliably translated into effective disease-prevention or treatment regimens, the practical value of "personal genomics" is somewhat limited.

CONCEPT REVIEW
- Describe the rough correlation between gene number and organismal lifestyle.
- How are genes identified?
- Why is it useful to identify homologs of human genes?
- What is the value of genome-wide association studies? What are their limitations?

KEY CONCEPTS
- A DNA molecule can be sequenced or amplified by using DNA polymerase to make a copy of a template strand.
- Linking together DNA fragments produces recombinant DNA molecules that can be used to study gene function, to express genes in host organisms, and to engineer genes for commercial and therapeutic purposes.

⊕ **See Guided Exploration. DNA sequence determination by the chain-terminator method.**

▶▶ **HOW** do researchers decipher the information in DNA?

Molecular biologists have devised clever procedures for isolating, amplifying, sequencing, altering, and otherwise manipulating DNA. Many of the techniques take advantage of naturally occurring enzymes that cut, copy, and link nucleic acids. These techniques also exploit the ability of nucleic acids to interact with complementary molecules purified from natural sources or synthesized in the laboratory.

DNA sequencing uses DNA polymerase to make a complementary strand

The most widely used technique for determining the sequence of nucleotides in a segment of DNA was developed by Frederick Sanger in 1975. The Sanger sequencing technique is also known as **dideoxy DNA sequencing** because it uses dideoxy nucleotides, that is, nucleotides lacking both 2′ and 3′ hydroxyl groups:

$$\text{P}-\text{P}-\text{P}-\text{OCH}_2 \quad \text{O} \quad \text{base}$$

2′,3′-Dideoxynucleoside triphosphate
(ddNTP)

A dideoxynucleoside triphosphate can be abbreviated **ddNTP,** where N represents any of the four bases.

The sequencing protocol is outlined in **Figure 3-15.** First, the DNA is denatured to separate its two strands. The DNA is then incubated with a mixture of all four deoxynucleoside triphosphates (dATP, dCTP, dGTP, and dTTP) and a bacterial DNA polymerase, an enzyme that catalyzes the polymerization of the nucleotides in the order determined by their base-pairing with a single-stranded DNA template. Because the DNA polymerase cannot begin a new nucleotide strand but can only extend a preexisting chain, a short single-stranded **primer** that base pairs with the template strand is added to the mixture. Keep in mind that the reaction mixture actually contains millions of molecules of the template, the primer, and the polymerase.

The reaction mixture also includes small amounts of four dideoxynucleotides (ddATP, ddCTP, ddGTP, and ddTTP), each of which is tagged with a different fluorescent dye. *As the DNA polymerase proceeds to synthesize a new DNA chain, it occasionally adds one of the ddNTPs in place of the corresponding dNTP.* This halts further extension of the DNA chain because the newly incorporated ddNTP, which lacks a 3′ OH group, cannot form the 3′ portion of a 3′–5′ phosphodiester bond to the next nucleotide (see Section 3-1).

The concentrations of the ddNTPs in the reaction mixture are lower than the concentrations of their corresponding **dNTPs,** so the DNA polymerase can assemble new chains of varying length before the random incorporation of a ddNTP halts further polymerization. The result is a population of truncated DNA strands, each capped by a fluorescent ddNTP residue. *The fragments differ in length by a single nucleotide since they all started from identical primers.*

The reaction products are subjected to **electrophoresis,** a procedure in which the molecules move through a gel-like matrix such as agarose or polyacrylamide under the influence of an electric field. Because all the DNA segments have a uniform charge density, they are separated on the basis of their size (the smallest molecules move fastest). The separated molecules are excited by a laser so that the fluorescent dye

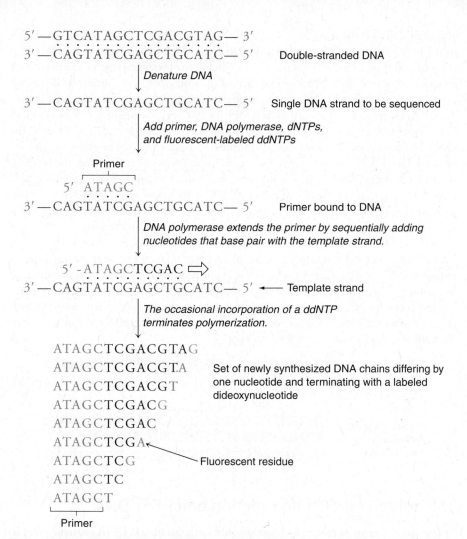

5′ —GTCATAGCTCGACGTAG— 3′
3′ —CAGTATCGAGCTGCATC— 5′   Double-stranded DNA

↓ *Denature DNA*

3′ —CAGTATCGAGCTGCATC— 5′   Single DNA strand to be sequenced

↓ *Add primer, DNA polymerase, dNTPs,*
and fluorescent-labeled ddNTPs

Primer
5′ ATAGC
3′ —CAGTATCGAGCTGCATC— 5′   Primer bound to DNA

↓ *DNA polymerase extends the primer by sequentially adding*
nucleotides that base pair with the template strand.

5′ -ATAGCTCGAC ⟹
3′ —CAGTATCGAGCTGCATC— 5′ ← Template strand

↓ *The occasional incorporation of a ddNTP*
terminates polymerization.

ATAGCTCGACGTAG
ATAGCTCGACGTA
ATAGCTCGACGT   Set of newly synthesized DNA chains differing by
ATAGCTCGACG   one nucleotide and terminating with a labeled
ATAGCTCGAC   dideoxynucleotide
ATAGCTCGA ←
ATAGCTCG   ⟵ Fluorescent residue
ATAGCTC
ATAGCT
Primer

Figure 3-15 **Dideoxy DNA sequencing procedure.**

attached to each dideoxynucleotide residue emits its characteristic color (**Fig. 3-16**). The order of appearance of the colors corresponds to the order of nucleoides in the newly synthesized DNA. Keep in mind that this sequence is complementary to the DNA strand that was used as a template in the sequencing reaction. With automation, a single reaction can yield a sequence of 400–1000 nucleotides.

Newer sequencing procedures also use DNA polymerase to generate a complementary DNA strand. In the **pyrosequencing** technique, molecules of the template DNA are immobilized on a plastic surface, a primer and DNA polymerase are added,

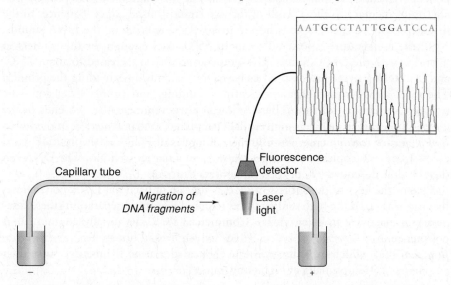

AATGCCTATTGGATCCA

Fluorescence
detector

Capillary tube

Migration of →   Laser
DNA fragments   light

−   +

Figure 3-16 **Results of dideoxy sequencing.** In electrophoresis, the DNA fragments generated by dideoxy sequencing move through a thin tube under an electric field and are detected by their fluorescence. The peaks represent fragments that differ in size by one nucleotide, and the color corresponds to the dideoxynucleotide at the end of the fragment. Thus, the successive peaks reveal the nucleotide sequence.

and then a dNTP substrate is introduced. If DNA polymerase adds that nucleotide to the new DNA strand, pyrophosphate (the "diphosphate" portion of the dNTP) is released and triggers a chemical reaction involving the firefly enzyme luciferase, which generates a flash of light. Solutions of each of the four dNTPs are successively washed across the immobilized DNA template, and a detector records whether light is produced in the presence of a particular dNTP. In this way, the sequence of nucleotides complementary to the template strand can be deduced. In a similar approach, the DNA to be sequenced is introduced into microscopic wells positioned atop an ion-sensitive layer. When DNA polymerase adds the appropriate nucleotide, the reaction releases a hydrogen ion, which is detected as a voltage change. Pyrosequencing and related techniques can "read" stretches of 300–500 nucleotides, somewhat shorter than the sequences revealed by dideoxy sequencing, but the small scale of the fluid-delivery system allows many templates to be sequenced simultaneously.

Because DNA samples are frequently too large to be sequenced in a single pass, the DNA must be broken into a number of overlapping segments that are individually sequenced. The nucleotide sequence of the entire DNA can then be reconstructed by computer analysis of the overlapping sequences. Although this piece-by-piece approach can be carried out systematically, it is usually more efficient to generate a large number of DNA fragments in a random, or "shotgun," fashion and then examine the sequenced pieces for overlaps.

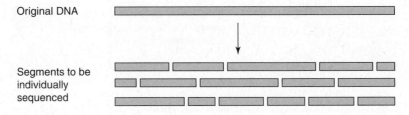

Original DNA

Segments to be individually sequenced

The polymerase chain reaction amplifies DNA

⊕ See Guided Exploration. PCR and site-directed mutagenesis.

At one time, scientists requiring large amounts of a particular DNA sequence had to laboriously isolate the target DNA and clone (copy) it in living cells, as described later in this section. That changed in 1985 with the invention of the **polymerase chain reaction (PCR)** by Kary Mullis. Although PCR cannot entirely replace cloning as a research tool, it provides a relatively easy and rapid way to amplify a segment of DNA. *One of the advantages of PCR over traditional cloning techniques is that the starting material need not be pure* (this makes the technique ideal for analyzing complex mixtures such as tissues or biological fluids).

Like DNA sequencing, PCR uses DNA polymerase to make copies of a particular DNA sequence (**Fig. 3-17**). The reaction mixture contains the DNA sample, a DNA polymerase, all four deoxynucleotides, and two oligonucleotide primers that are complementary to the 3′ ends of the two strands of the target sequence. In the first step of PCR, the sample is heated to 90–95°C to separate the DNA strands. Next, the temperature is lowered to about 55°C, cool enough for the primers to hybridize with the DNA strands. The temperature is then increased to about 75°C, and the DNA polymerase synthesizes new DNA strands by extending the primers. The three steps—strand separation, primer binding, and primer extension—are repeated as many as 40 times. Because the primers represent the two ends of the target DNA, this sequence is preferentially amplified so that *it doubles in concentration with each reaction cycle.* For example, 20 cycles of PCR can theoretically yield $2^{20} = 1,048,576$ copies of the target sequence in a matter of hours. The DNA can then be cloned, sequenced, or used for another purpose.

One of the keys to the success of PCR is the use of bacterial DNA polymerases that can withstand the high temperatures required for strand separation (these temperatures inactivate most enzymes). Commercial PCR kits usually contain DNA polymerase from *Thermus aquaticus* ("Taq," which lives in hot springs) or *Pyrococcus furiosus* ("Pfu," which inhabits geothermally heated marine sediments), since their enzymes perform optimally at high temperatures.

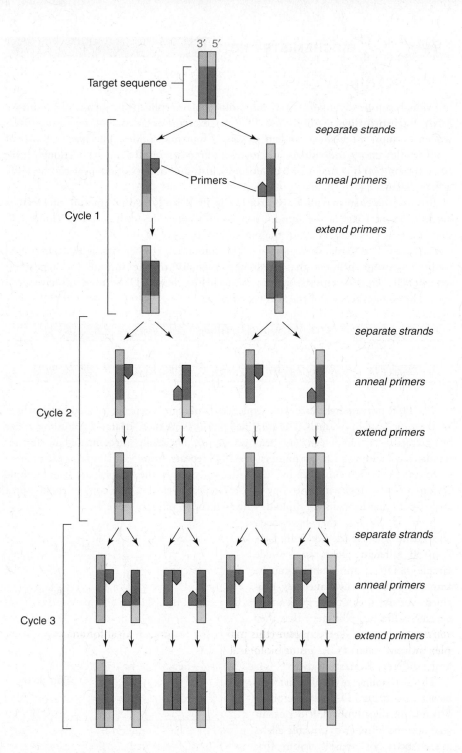

Figure 3-17 **The polymerase chain reaction.** Each cycle consists of separation of DNA strands, binding of primers to the 3′ ends of the target sequence, and extension of the primers by DNA polymerase. The target DNA doubles in concentration with each cycle.

? Indicate the temperature at which each step takes place.

One of the limitations of PCR is that choosing appropriate primers requires some knowledge of the DNA sequence to be amplified; otherwise, the primers will not anneal with complementary sequences in the DNA and the polymerase will not be able to make any new DNA strands. However, since no new DNA is synthesized *unless* the primers can bind to the DNA sequence, *PCR can be used to verify the presence of that sequence.* In fact, PCR is the most efficient way to detect certain organisms that are difficult to grow in the laboratory. For obvious reasons, PCR is also preferred over culturing as a way to detect the presence of highly contagious bacteria and viruses. Blood banks use PCR to test donations for the human immunodeficiency virus (HIV), hepatitis viruses, and West Nile virus. Scientists have used PCR to amplify DNA from ancient bones and, in the forensics laboratory, to analyze DNA of more recent origin (Box 3-B).

If merely detecting a specific DNA sequence does not provide enough information, **quantitative PCR (qPCR,** also known as real-time PCR) can be employed.

BOX 3-B   BIOCHEMISTRY NOTE

DNA Fingerprinting

Individuals can be distinguished by examining polymorphisms in their DNA. Current **DNA fingerprinting** methods use PCR to examine segments of repetitive DNA sequences, most often short tandem repeats of four nucleotides. The exact number of repeats varies among individuals, and each set of repeats, or **allele,** is small enough (usually less than 500 bp) that alleles differing by just one four-residue repeat can be easily differentiated.

Because the first step of fingerprinting is PCR, only a tiny amount of DNA is needed—about 1 μg, or the amount present on a coffee cup or a licked envelope. And since the target segment is short, the purity and integrity of the DNA sample is usually not an issue. The **locus,** or region of DNA containing the repeats, is PCR-amplified using fluorescent primers that are complementary to the unique (nonrepeating) sequences flanking the repeats. In the example below, the two DNA segments have seven and eight tandem repeats of the AATG sequence.

| | | | AATG | AATG | AATG | AATG | AATG | AATG | AATG | | |

| | | | AATG | AATG | AATG | AATG | AATG | AATG | AATG | AATG | | |

The PCR primers hybridize with sequences flanking the repeats (shaded blue), which are the same in all individuals. The amplified products are then separated according to size by electrophoresis and detected by their fluorescence. The results are compared to reference standards containing a known number of AATG repeats, from 5 to 10 in this example.

Sample A comes from an individual with two copies of the 7-repeat allele and sample B from an individual with one copy of the 7-repeat allele and one copy of the 8-repeat allele (recall that humans are diploid, with two copies of each "gene").

Each of the loci that have been selected for forensic use generally have 7 to 30 different alleles. In a single sample of DNA, multiple loci can be amplified by PCR simultaneously, provided that the sizes of the PCR products are sufficiently different that they will not overlap during electrophoresis. Alternatively, each PCR primer can bear a different fluorescent dye.

The probability of two individuals having matching DNA fingerprints depends on the number of loci examined and the number of possible alleles at each locus. For example, assume that one locus has 20 alleles and that each allele has a frequency in the population of 5% (1 in 20, or 1/20). Another locus has 10 alleles, and each has a frequency of 10% (1 in 10, or 1/10). The probability that two individuals would match at both sites is $1/20 \times 1/10 = 1/200$ (the probabilities of independent events are multiplied). If multiple loci are examined, the probabilities can reach the range of 1 in a million or more. For this reason, most courts now consider DNA sequences to be unambiguous identifiers of individuals.

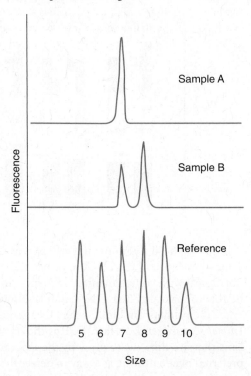

● **Question:** Would a child's DNA fingerprint match the parents' DNA fingerprints?

In this technique, the polymerase chain reaction continually generates new DNA sequences that bind to fluorescent probes, so the amount of the DNA sequence can be monitored over time (rather than at the end of many reaction cycles, as in standard PCR). This approach is useful for quantifying infectious agents such as bacteria and viruses. Analysis of fetal DNA from the mother's blood relies on the accuracy of qPCR to detect the small increase in concentration of a particular stretch of DNA that the fetus may have inherited from its mother, since circulating maternal DNA is about 10 times more abundant than the circulating fetal DNA.

qPCR methods are also used to assess the level of gene expression in cells: Cellular mRNA is first reverse-transcribed to DNA, then a specific DNA sequence is amplified by PCR. A gene's level of expression is sometimes reported relative to that of a gene that encodes a protein such as actin, which is typically produced at constant levels in cells and can therefore serve as a sort of benchmark.

Restriction enzymes cut DNA at specific sequences

Naturally occurring DNA molecules tend to break from mechanical stress during laboratory manipulations. However, such randomly fragmented DNA is not always useful, so researchers take advantage of enzymes that cut DNA in defined ways. Bacteria produce DNA-cleaving enzymes known as **restriction endonucleases** (or **restriction enzymes**) that catalyze the breakage of phosphodiester bonds at or near specific nucleotide sequences. These enzymes can thereby destroy foreign DNA that enters the cell (for example, bacteriophage DNA). In this way, the bacterial cell "restricts" the growth of the phage. The bacterial cell protects its own DNA from endonucleolytic digestion by methylating it (adding a —CH₃ group) at the same sites recognized by its restriction endonucleases. In the laboratory, the most useful restriction enzymes are those that cleave *at* (rather than just *near*) the recognition site. Hundreds of these enzymes have been examined; some are listed in Table 3-5 along with their recognition sequences and cleavage sites.

Restriction enzymes typically recognize a 4- to 8-base sequence that is identical, when read in the same $5' \rightarrow 3'$ direction, on both strands. DNA with this form of symmetry is said to be **palindromic** (words such as *madam* and *noon* are palindromes). One restriction enzyme isolated from *E. coli* is known as EcoRI (the first three letters are derived from the genus and species names). Its recognition sequence is

$$5' — G\overset{\downarrow}{A}ATTC— 3'$$
$$3' — CTTA\underset{\uparrow}{A}G— 5'$$

The arrows indicate the phosphodiester bonds that are cleaved. Note that the sequence reads the same on both strands.

Because the EcoRI cleavage sites are symmetrical but staggered, the enzyme generates DNA fragments with single-stranded extensions known as **sticky ends:**

—G   AATTC—
—CTTAA   G—

In contrast, the *E. coli* restriction enzyme known as EcoRV cleaves both strands of DNA at the center of its 6-bp recognition sequence, so that the resulting DNA fragments have **blunt ends:**

$$5' — GAT\overset{\downarrow}{A}TC— 3' \qquad\qquad — GAT \quad ATC—$$
$$3' — CTA\underset{\uparrow}{T}AG— 5' \qquad\longrightarrow\qquad — CTA \quad TAG—$$

Restriction enzymes have many uses in the laboratory. For example, they are indispensable for reproducibly breaking large pieces of DNA into smaller pieces of manageable size. **Restriction digests** of well-characterized DNA molecules, such as the 48,502 bp *E. coli* bacteriophage λ, yield **restriction fragments** of predictable size (**Fig. 3-18**).

[TABLE **3-5**]

Recognition and Cleavage Sites of Some Restriction Endonucleases

Enzyme	Recognition/ Cleavage Site[a]
AluI	AG | CT
MspI	C | CGG
AsuI	G | GNCC[b]
EcoRI	G | AATTC
EcoRV	GAT | ATC
PstI	CTGCA | G
SauI	CC | TNAGG
NotI	GC | GGCCGC

[a]The sequence of one of the two DNA strands is shown. The vertical bar indicates the cleavage site.

[b]N represents any nucleotide.

[An exhaustive source of information on restriction enzymes is available through the Restriction Enzyme Database: rebase.neb.com/rebase/rebase.html.]

23130
9416
6557
4361
2322
2027

Figure 3-18 Digestion of bacteriophage λ DNA by the restriction enzyme HindIII. The restriction enzyme cleaves the DNA to produce eight fragments of defined size, six of which are large enough to be separated by electrophoresis in an agarose gel. The DNA was applied to the top of the gel, and the fragments were visualized by binding a fluorescent dye. The numbers indicate the number of base pairs in each fragment. [Reprinted from www.neb.com, http://www.neb.com; © 2012 with permission from New England Biolabs.]

? **Explain why the bands at the top appear brighter than the bands at the bottom.**

DNA fragments are joined to produce recombinant DNA

Recombinant DNA technology (also known as **genetic engineering** or **molecular cloning**) is a set of methods for cleaving, joining, and copying DNA fragments *in vivo* (PCR copies DNA *in vitro*). Briefly, a fragment of DNA is combined with another DNA molecule to produce a recombinant DNA molecule:

1. A fragment of DNA of the appropriate size is generated by the action of a restriction enzyme, by PCR, or by chemical synthesis.
2. The fragment is incorporated into another DNA molecule.
3. The recombinant DNA is introduced into cells, where it replicates.
4. Cells containing the desired DNA are identified.

Restriction endonucleases are valuable tools for genetic engineers. When different samples of DNA are digested with the same sticky end–generating restriction endonuclease, all the fragments have identical sticky ends. If the fragments are mixed together, the sticky ends can find their complements and re-form base pairs. The discontinuities in the sugar–phosphate backbone can then be mended by a **DNA ligase** (an enzyme that forms new phosphodiester bonds between adjacent nucleotide residues). These cutting-and-pasting reactions allow a segment of DNA to be excised from a chromosome and inserted into a carrier DNA molecule, such as a circular **plasmid,** that has been cut by the same restriction enzyme. DNA ligase seals the breaks in the nucleotide strands, leaving an unbroken double-stranded recombinant DNA molecule consisting of the plasmid with a foreign DNA insert (**Fig. 3-19**).

Plasmids are small, circular DNA molecules present in many bacterial cells. A single cell may contain multiple copies of a plasmid, which replicates independently of the bacterial chromosome and usually does not contain genes essential for the host's normal activities. However, plasmids often do carry genes for specialized functions, such as resistance to certain antibiotics (these genes often encode proteins that inactivate the antibiotics). An antibiotic resistance gene allows the **selection** of cells that harbor the plasmid: Only cells that contain the plasmid can survive in the presence of the antibiotic. Growing large quantities of plasmid-laden cells is one way to produce large amounts of the foreign DNA insert. (It can be removed later by treating the plasmid with the same restriction enzyme used to insert the foreign DNA.) A piece of DNA that is amplified in this way is said to be cloned. The plasmid that contains the foreign DNA is called a **cloning vector.** Note that a **clone** is simply an identical copy of an original. The term is used to refer either to a gene that has been amplified, as described above, or to a cell or organism that is genetically identical to its parent.

An example of a cloning vector is shown in **Figure 3-20.** This plasmid contains a gene (called *amp*^R) for resistance to the antibiotic ampicillin and a gene (called *lacZ*) encoding the enzyme β-galactosidase, which catalyzes the hydrolysis of certain galactose derivatives. The *lacZ* gene has been engineered to contain several restriction sites, any one of which can be used as an insertion point for a piece of foreign DNA with compatible sticky ends. Interrupting the *lacZ* gene with foreign DNA prevents the synthesis of the β-galactosidase protein.

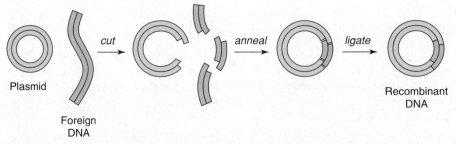

Figure 3-19 Production of a recombinant DNA molecule. A small circular plasmid and a sample of DNA are cut with the same restriction enzyme, generating complementary sticky ends, so that the fragment of foreign DNA can be ligated into the plasmid. ⊕ **See Animated Figure. Construction of a recombinant DNA molecule.**

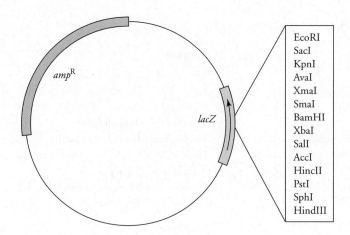

Figure 3-20 Map of a cloning vector. This 2743-bp circular DNA molecule, called pGEM-3Z, has a gene for resistance to ampicillin so that bacterial cells containing the plasmid can be selected by their ability to grow in the presence of the antibiotic. The plasmid also has a site comprising recognition sequences for 14 different restriction enzymes. Inserting a foreign DNA segment at this site interrupts the *lacZ* gene, which encodes the enzyme β-galactosidase.

Colonies of bacterial cells harboring the intact plasmid can be detected when their β-galactosidase cleaves a galactose derivative that generates a blue dye:

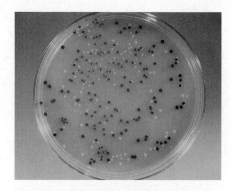

Colonies of bacterial cells in which a foreign DNA insert has interrupted the *lacZ* gene are unable to cleave the galactose derivative and therefore do not turn blue (**Fig. 3-21**). This method for identifying (screening) bacterial cells containing plasmids with the insert is known as **blue-white screening.** A single white colony can then be removed from the culture plate and grown in order to harvest its recombinant DNA for sequencing or some other purpose. Bacterial cells that lack the plasmid entirely would also be white. However, these cells are eliminated by including ampicillin in the culture medium, which kills cells that don't contain the *amp*^{R} gene.

A large number of cloning vectors have been developed to accommodate different sizes of DNA inserts (Table 3-6). Vectors also differ by the cell type in which the vector can grow (for example, bacterial, fungal, insect, or mammalian cells) and in the strategy used to select cells containing the foreign DNA.

Cloned genes yield valuable products

If a gene that has been isolated and cloned in a host cell is also expressed (transcribed and translated into protein), it may affect the metabolism of that cell. The functions of some gene products have been assessed in this way. Sometimes a specific combination of vector and host cell are chosen so that large quantities of the gene product can be isolated from the cultured cells or from the medium in which they grow. This is an economical method for producing certain proteins that are difficult to obtain directly from human tissues (Table 3-7).

After a gene has been isolated and cloned, it can be specifically altered in order to alter the amino acid sequence of the encoded protein. **Site-directed mutagenesis** (also called *in vitro* **mutagenesis**) mimics the natural process of evolution and allows predictions about the structural and functional roles of particular amino acids in a protein to be rigorously tested in the laboratory. One technique for site-directed mutagenesis is a variation of PCR in which the primers are oligonucleotides whose sequences are identical to a portion of the gene of interest except for one or a few bases corresponding to the codon(s) to be altered. The primers can hybridize to the

Figure 3-21 Culture dish used in blue-white screening. Blue colonies arise from cells whose plasmids have an intact β-galactosidase gene. White colonies arise from cells whose plasmids contain an insert that interrupts the β-galactosidase gene. [Courtesy S. Kopczak and D. P. Snustad, University of Minnesota.]

[**TABLE 3-6**]

Types of Cloning Vectors

Vector	Size of DNA Insert (kb)
Plasmid	<20
Cosmid	40–45
Bacterial artificial chromosome	40–400
Yeast artificial chromosome	400–1000

[**TABLE 3-7**] Some Recombinant Protein Products

Protein	Purpose
Insulin	Treat insulin-dependent diabetes
Growth hormone	Treat certain growth disorders in children
Erythropoietin	Stimulate production of red blood cells; useful in kidney dialysis
Coagulation factors IX and X	Treat hemophilia b and other bleeding disorders
Tissue plasminogen activator	Promote clot lysis following myocardial infarction or stroke
Colony stimulating factor	Promote white blood cell production after bone marrow transplant

wild-type (naturally occurring) gene sequence if there are only a few mismatched bases. Multiple rounds of DNA polymerase–catalyzed primer extension yield many copies of the gene with the desired sequence (**Fig. 3-22**). The mutagenized gene can then be inserted into a vector for expression in a host cell.

Figure 3-22 Site-directed mutagenesis. ⊕ See Animated Figure. Site-directed mutagenesis.

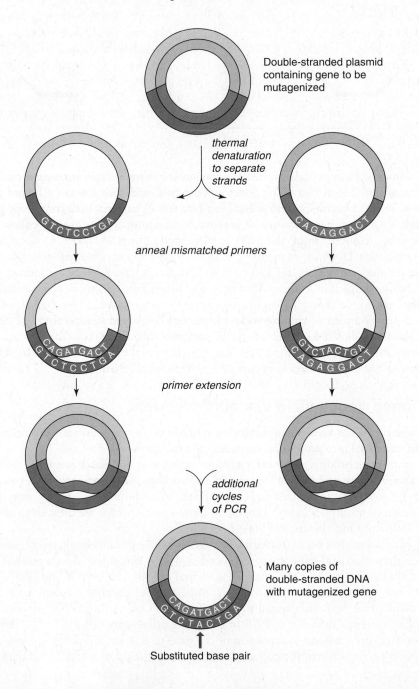

Double-stranded plasmid containing gene to be mutagenized

thermal denaturation to separate strands

anneal mismatched primers

primer extension

additional cycles of PCR

Many copies of double-stranded DNA with mutagenized gene

Substituted base pair

Genetically modified organisms have practical applications

Introducing a foreign gene into a single host cell via a cloning vector alters the genetic makeup of that cell and all its descendants. Producing a **transgenic organism** whose cells all contain the foreign gene is more difficult. In mammals, the DNA must be injected into fertilized eggs, which are then implanted in a foster mother. Some of the resulting embryos' cells (possibly including their reproductive cells) will contain the foreign gene. When the animals mature, they must be bred in order to yield offspring whose cells are all transgenic.

Transgenic plants are produced by introducing recombinant DNA into a few cells, which can often develop into an entire plant whose cells all contain the foreign DNA. Desirable traits such as resistance to insect pests have been introduced into several important crop species. For example, about two-thirds of the U.S. corn (maize) crop is genetically modified to produce a protein that is toxic to plant-eating insects. So-called Bt corn plants have been engineered to express an insecticidal toxin from the bacterium *Bacillus thuringiensis*. However, the extensive acreage devoted to the transgenic corn (also known as a genetically modified, or GM, food) raises some concerns: It increases the selective pressure on insects to evolve resistance to the toxin, and it increases the likelihood that the toxin gene will be transferred to other plant species, with disastrous effects on the insects that feed on them.

Transgenic plants have also been engineered for better nutrition. For example, researchers have developed strains of rice with foreign genes, from other plant species, that encode enzymes necessary to synthesize β-carotene (an orange pigment that is the precursor of vitamin A) and a gene for the iron-storage protein ferritin (**Fig. 3-23**). The genetically modified rice strains are intended to help alleviate vitamin A deficiencies (which afflict some 400 million people) and iron deficiencies (an estimated 30% of the world's population suffers from iron deficiency).

Using techniques similar to those for producing transgenic animals, researchers can produce animals in which a particular gene has been inactivated. These "gene knockouts" can serve as animal models for human diseases. For example, mice lacking the cystic fibrosis gene have been used to study the development of the disease and to test potential treatments.

Figure 3-23 Genetically modified rice. The plant has been genetically engineered to produce more β-carotene, the precursor of vitamin A. As a result, the rice grains are yellow rather than white. [© phloen/Alamy.]

Gene therapy can cure some human diseases

*The goal of **gene therapy** is to introduce a functional gene into an individual in order to compensate for an existing malfunctioning gene.* The first successful gene therapy trials treated children with severe combined immunodeficiency (SCID), a normally fatal condition caused by the absence of the enzyme adenosine deaminase or the absence of a particular receptor protein. Bone marrow cells were removed from each patient, genetically modified, and infused into the patient, where they differentiated into functional immune system cells. The "correct" gene sequence was generated by site-directed mutagenesis and then introduced into the bone marrow cells via a viral vector. Vectors derived from viruses are effective because viruses have evolved to be efficient gene-delivery systems for mammalian cells.

Some of the diseases treated by gene therapy are listed in Table 3-8. Despite some 30 years of research effort, significant challenges remain. For example, the viral

[TABLE 3-8] Some Hereditary Diseases Treated by Gene Therapy

Disease	Symptoms
Adrenoleukodystrophy	Neurodegeneration
Hemophilia	Bleeding
Leber's congenital amaurosis	Blindness
Severe combined immunodeficiency (SCID)	Immunodeficiency
β-Thalassemia	Anemia
Wiskott-Aldrich syndrome	Immunodeficiency

vectors used in gene therapy may elicit an immune response that destroys the vector before it can deliver its cargo or, as occurred in one case, kills the patient. Moreover, viruses tend to insert themselves into the host cell's chromosomes somewhat at random, which may interrupt the function of other genes. This has led to several cases of leukemia (uncontrolled growth of an immune system cell) in gene therapy recipients. An additional limitation of the current gene therapy approaches is that most viruses can accommodate a foreign gene no larger than about 8 kb, whereas some human genes have a total length of hundreds of kb (the CF gene is 250 kb). Some genes also require additional regulatory sequences, at some distance from the coding segments, in order for the gene to be expressed in a tissue-specific and developmentally appropriate manner. Whereas a person's immune system can be "re-seeded" using a few altered cells that continue to multiply, other types of tissues do not readily take up foreign DNA or cannot proliferate to an extent that would permit the introduced gene to compensate for the defective gene.

CONCEPT REVIEW
- Summarize the steps of dideoxy DNA sequencing. How does pyrosequencing differ from the dideoxy approach?
- How do researchers reconstruct long DNA sequences?
- Describe the polymerase chain reaction (PCR). Why is PCR a step in DNA fingerprinting?
- Why are restriction endonucleases useful for constructing recombinant DNA?
- How can cells containing recombinant DNA molecules be identified by selection or screening?
- Describe how a cloned gene can be used to produce an altered protein, to generate a transgenic organism, or to cure a disease.

[SUMMARY]

3-1 DNA Is the Genetic Material

- The genetic material in virtually all organisms consists of DNA, a polymer of nucleotides. A nucleotide contains a purine or pyrimidine base linked to a ribose group (in RNA) or a deoxyribose group (in DNA) that also bears one or more phosphate groups.
- DNA contains two antiparallel helical strands of nucleotides linked by phosphodiester bonds. Each base pairs with a complementary base in the opposite strand: A with T and G with C. The structure of RNA, which is single-stranded and contains U rather than T, is more variable.
- Nucleic acid structures are stabilized primarily by stacking interactions between bases. The separated strands of DNA can reanneal.

3-2 Genes Encode Proteins

- The central dogma summarizes how the sequence of nucleotides in DNA is transcribed into RNA, which is then translated into protein according to the genetic code.
- The sequence of nucleotides in a segment of DNA can reveal mutations that cause disease.

3-3 Genomics

- Genomes contain variable amounts of repetitive and other forms of noncoding DNA in addition to genes, which are identified by their sequence characteristics or similarity to other genes.

- Genetic variations can be linked to human diseases even when specific disease genes have not been identified.

3-4 Tools and Techniques: Manipulating DNA

- The sequence of nucleotides in a segment of DNA is commonly determined by the dideoxy method, in which labeled complementary copies of a DNA strand are synthesized. The presence of dideoxynucleotides, which cannot support further synthesis, generates a set of fragments of different lengths that are separated and analyzed to deduce the sequence of the original template strand.
- Large amounts of a particular DNA segment can be obtained by the polymerase chain reaction, in which a DNA polymerase makes complementary copies of a selected segment of DNA.
- DNA molecules can be reproducibly fragmented by the action of restriction enzymes, which cleave DNA at specific sequences.
- DNA fragments can be joined to each other to generate recombinant DNA molecules that are then cloned (copied) in host cells.
- A cloned gene can be manipulated so that large amounts of its protein product can be produced. The sequence of the gene may first be altered by site-directed mutagenesis.
- Genes can be introduced into foreign hosts to produce transgenic organisms. In gene therapy, a normal gene is introduced in order to cure a genetic disease.

[GLOSSARY TERMS]

chromosome	melting temperature (T_m)	horizontal gene transfer
nucleic acid	denaturation	genomics
nucleotide	renaturation	single-nucleotide polymorphism (SNP)
DNA (deoxyribonucleic acid)	anneal	genome-wide association study (GWAS)
bacteriophage	probe	dideoxy DNA sequencing
polynucleotide	replication	ddNTP
base	gene	primer
purine	gene expression	dNTP
pyrimidine	transcription	electrophoresis
RNA (ribonucleic acid)	translation	pyrosequencing
nucleoside	genome	polymerase chain reaction (PCR)
deoxynucleotide	coding strand	quantitative PCR (qPCR)
vitamin	noncoding strand	restriction endonuclease
phosphodiester bond	messenger RNA (mRNA)	palindrome
residue	ribosome	DNA fingerprinting
5′ end	ribosomal RNA (rRNA)	allele
3′ end	transfer RNA (tRNA)	locus
base pair	codon	sticky ends
sugar–phosphate backbone	genetic code	blunt ends
antiparallel	mutation	restriction digest
major groove	cystic fibrosis (CF)	restriction fragment
minor groove	DNA marker	recombinant DNA technology
bp	splicing	DNA ligase
kb	diploid	plasmid
oligonucleotide	haploid	selection
polymerase	transposable element	cloning vector
nuclease	moderately repetitive DNA	clone
exonuclease	highly repetitive DNA	blue-white screening
endonuclease	open reading frame (ORF)	site-directed mutagenesis
A-DNA	homologous genes	transgenic organism
B-DNA	orphan gene	gene therapy
stacking interactions	genome map	

BIOINFORMATICS ▶ **PROJECT 2**   Learn to locate nucleotide sequences, examine their encoded information, and identify similarities to other sequences.

DATABASES FOR THE STORAGE AND "MINING" OF GENOME SEQUENCES

[PROBLEMS]

3-1 DNA Is the Genetic Material

1. The identification of DNA as the genetic material began with Griffith's "transformation" experiment conducted in 1928. Griffith worked with *Pneumococcus*, an encapsulated bacterium that causes pneumonia. Wild-type *Pneumococcus* forms smooth colonies when plated on agar and causes death when injected into mice. A mutant *Pneumococcus* lacking the enzymes needed to synthesize the polysaccharide capsule (required for virulence) forms rough colonies when plated on agar and does not cause death when injected into mice. Griffith found that heat-treated wild-type *Pneumococcus* did not cause death when injected into the mice because the heat treatment destroyed the polysaccharide capsule. However, if Griffith mixed heat-treated wild-type *Pneumococcus* and the mutant unencapsulated *Pneumococcus* together and injected this mixture, the mice died. Even more surprisingly, upon autopsy,

Griffith found live, encapsulated *Pneumococcus* bacteria in the mouse tissue. Griffith concluded that the mutant *Pneumococcus* had been "transformed" into disease-causing *Pneumococcus*, but he could not explain how this occurred. Using your current knowledge of how DNA works, explain how the mutant *Pneumococcus* became transformed.

2. In 1944, Avery, MacLeod, and McCarty set out to identify the chemical agent capable of transforming mutant unencapsulated *Pneumococcus* to the deadly encapsulated form. They isolated a viscous substance with the chemical and physical properties of DNA and showed that it was capable of transformation. Transformation could still occur if proteases (enzymes that degrade proteins) or ribonucleases (enzymes that degrade RNA) were added prior to the experiment. What did these treatments tell the investigators about the molecular identity of the transforming factor?

3. In 1952, Alfred Hershey and Martha Chase carried out experiments using bacteriophages, which consist of nucleic acid enclosed by a protein capsid (coat). Hershey and Chase first labeled the bacteriophages with the radioactive isotopes ^{35}S and ^{32}P. Because proteins contain sulfur but not phosphorus, and DNA contains phosphorus but not sulfur, each type of molecule was separately labeled. The radiolabeled bacteriophages were allowed to infect the bacteria, and then the preparation was treated to separate the empty capsids (ghosts) from the bacterial cells. The ghosts were found to contain most of the ^{35}S label, whereas 30% of the ^{32}P was found in the new bacteriophages produced by the infected cells. What does this experiment reveal about the roles of bacteriophage DNA and protein?

4. In February 1953 (two months before Watson and Crick published their paper describing DNA as a double helix), Linus Pauling and Robert Corey published a paper in which they proposed that DNA adopts a triple-helical structure. In their model, the three chains are tightly packed together. The phosphate groups reside on the inside of the triple helix, while the nitrogenous bases are located on the outside. They proposed that the triple helix was stabilized by hydrogen bonds between the interior phosphate groups. What are the flaws in this model?

5. In some organisms, DNA is modified by addition of methyl groups. Draw the structure of 5-methylcytosine.

6. In certain pathogenic bacteria, the methylation of certain adenines in DNA is required in order for the bacteria to cause disease.
(a) Draw the structure of N^6-methyladenine.
(b) Why might scientists be interested in the fact that other bacteria produce an N^6-DNA methyltransferase (an enzyme that catalyzes the transfer of methyl groups to adenine)?

7. Certain strains of *E. coli* incorporate the nitrogenous base shown here into nucleotides. For which base is this one a substitute?

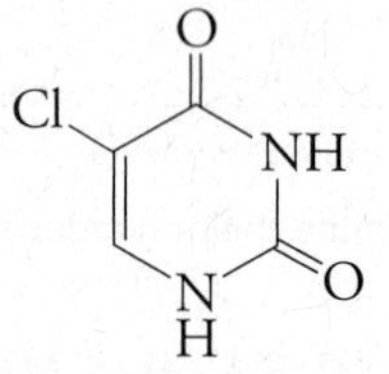

8. An *E. coli* culture is grown in the presence of the base shown in Problem 7. A control culture is grown in the absence of this modified base. Compare the masses of the DNA isolated from *E. coli* in these two cultures.

9. Describe the chemical difference between uracil and thymine.

10. (a) What kind of linkage joins the two nucleotides in the dinucleotides NAD and FAD (see Figure 3-3)?
(b) How do the adenosine groups in FAD and CoA differ?

11. Draw a CA (ribo)dinucleotide and label the phosphodiester bond. How would the structure differ if it were DNA?

12. Many cellular signaling pathways involve the conversion of ATP to cyclic AMP, in which a single phosphate group is esterified to both C3′ and C5′. Draw the structure of cyclic AMP.

13. Do Chargaff's rules hold true for RNA? Explain why or why not.

14. A diploid organism with a 30,000-kb haploid genome contains 19% T residues. Calculate the number of A, C, G, and T residues in the DNA of each cell in this organism.

15. A well-studied bacteriophage has 97,004 bases in its complete genome.
(a) There are 24,182 G residues in the genome. Calculate the number of C, A, and T residues.
(b) Why does GenBank report a total of 48,502 bases for this bacteriophage genome?

16. The complete genome of a virus contains 1578 T residues, 1180 G residues, 1609 A residues, and 1132 C residues. What can you conclude about the structure of the viral genome, given this information?

17. Identify the base pair highlighted in blue in Figure 3-7.

18. The adenine derivative hypoxanthine can base pair with cytosine, adenine, and uracil. Show the structures of these base pairs.

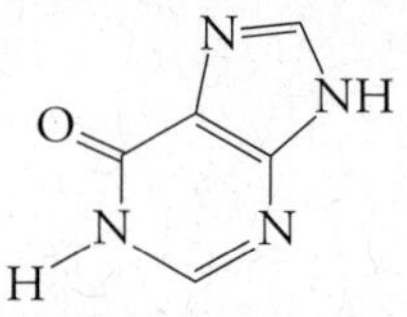

Hypoxanthine

19. Explain whether the following statement is true or false: Because a G:C base pair is stabilized by three hydrogen bonds, whereas an A:T base pair is stabilized by only two hydrogen bonds, GC-rich DNA is harder to melt than AT-rich DNA.

20. Hydrogen bonding does not make a significantly large contribution to the overall stability of the DNA molecule. Explain.

21. How can the hydrophobic effect (Section 2-2) explain why DNA adopts a helical structure?

22. (a) Would you expect proteins to bind to the major groove or the minor groove of DNA? Explain.
(b) Eukaryotic DNA is packaged with histones, small proteins with a high lysine and arginine content. Why do histones have a high affinity for DNA?

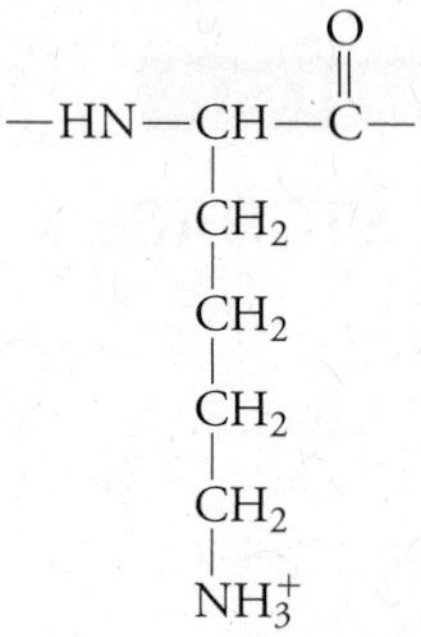

Lysine

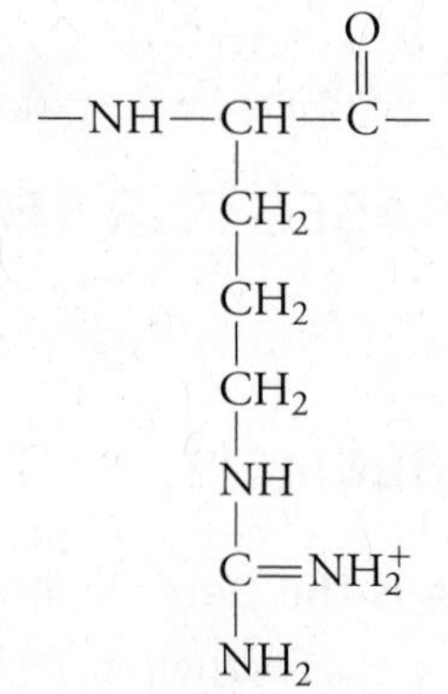

Arginine

23. Draw melting curves that would be obtained from the DNA of *Dictyostelium discoideum* and *Streptomyces albus* (see Table 3-2).

24. Using Table 3-2 as a guide, estimate the melting temperature of the DNA from an organism whose genome contains equal amounts of all four nucleotides.

25. What might you find in comparing the GC content of DNA from *Thermus aquaticus* or *Pyrococcus furiosus* and DNA from bacteria in a typical backyard pond?

26. Explain why the melting temperature of a sample of double-helical DNA increases when the Na^+ concentration increases.

27. You have a short piece of synthetic RNA that you want to use as a probe to identify a gene in a sample of DNA. The RNA probe has a tendency to hybridize with sequences that are only weakly complementary. Should you increase or decrease the temperature to improve your chances of tagging the correct sequence?

28. In the laboratory technique known as fluorescence *in situ* hybridization (FISH), a fluorescent oligonucleotide probe is allowed to hybridize with a cell's chromosomes, which are typically spread on a microscope slide. Explain why the chromosome preparation must be heated before the probe is added to it.

3-2 Genes Encode Proteins

29. Discuss the shortcomings of the following definitions for *gene:*
(a) A gene is the information that determines an inherited characteristic such as flower color.
(b) A gene is a segment of DNA that encodes a protein.
(c) A gene is a segment of DNA that is transcribed in all cells.

30. The semiconservative nature of DNA replication (as shown in Fig. 3-10) was proposed by Watson and Crick in 1953, but it wasn't experimentally verified until 1958. Meselson and Stahl grew bacteria on a "heavy nitrogen" source (ammonium chloride containing the isotope ^{15}N) for many generations so that virtually every nitrogen atom in the bacterial DNA was the ^{15}N isotope. This resulted in DNA that was denser than normal. The food source was abruptly switched to one containing only ^{14}N. Bacteria were harvested and the DNA isolated by density gradient centrifugation.
(a) What is the density of the DNA of the first-generation daughter DNA molecules? Explain.
(b) What is the density of the DNA isolated after two generations? Explain.

31. A segment of the coding strand of a gene is shown below.

ACACCATGGTGCATCTGACT

(a) Write the sequence of the complementary strand that DNA polymerase would make.
(b) Write the sequence of the mRNA that RNA polymerase would make from the gene segment.

32. A portion of a gene is shown below.

5'-ATGATTCGCCTCGGGGCTCCCCAGTCGCTGGTGCT-
3'-TACTAAGCGGAGCCCCGAGGGGTCAGCGACCACGA-
GCTGACGCTGCTCGTCG-3'
CGACTGCGACGAGCAGC-5'

The sequence of the mRNA transcribed from this gene has the following sequence:

5'-AUGAUUCGCCUCGGGGCUCCCCAGUCGCU-
GGUGCUGCUGACGCUGCUCGUCG-3'

(a) Identify the coding and noncoding strands of the DNA.
(b) Explain why only the coding strands of DNA are commonly published in databanks.

33. In the early 1960s, Marshall Nirenberg deciphered the genetic code by designing an experiment in which he synthesized a polynucleotide strand consisting solely of U residues, then added this strand to a test tube containing all of the components needed for protein synthesis.
(a) What polypeptide was produced by this "cell-free" system?
(b) What polypeptides were produced when poly(A), poly(C), and poly(G) were added to the cell-free system?

34. Har Gobind Korana, using a new polynucleotide synthesis method that he developed, extended Nirenberg's work by synthesizing polynucleotides with precisely defined sequences.
(a) What polypeptide(s) would be produced if a poly(GU) were added to the cell-free system described in Problem 33?
(b) Do these results help to decipher the identities of the codons involved?

35. An open reading frame (ORF) is a portion of the genome that potentially codes for a protein. A given nucleotide sequence of mRNA potentially has three different reading frames, only one of which is correct (the selection of the correct ORF will be discussed more fully in Section 22-3). A portion of the gene for a type II human collagen is shown. What are the sequences of amino acids that can potentially be translated from each of the three possible reading frames?

AGGTCTTCAGGGAATGCCTGGCGAGAGGGGAGCAGCT-
GGTATCGCTGGGCCCAAAGGC

36. Refer to Problem 35. Collagen's amino acid sequence consists of repeating triplets in which every third amino acid is glycine. Does this information assist you in your identification of the correct reading frame?

37. One form of the disease adrenoleukodystrophy (ALD) is caused by the substitution of serine for asparagine in the ALD protein. List the possible single-nucleotide alterations in the DNA of the ALD gene that could cause this genetic disease.

38. In another form of adrenoleukodystrophy, a CGA codon in the ALD gene is converted to a UGA codon. Explain how this mutation affects the ALD protein.

39. A mutation occurs when there is a base change in the DNA sequence. Some base changes do not lead to changes in the amino acid sequence of the resulting protein. Explain why.

40. Is it possible for the same segment of DNA to encode two different proteins? Explain.

41. A portion of the nucleotide sequence from the DNA coding strand of the chick ovalbumin gene is shown. What is the partial amino acid sequence of the encoded protein?

CTCAGAGTTCACCATGGGCTCCATCGGTGCAGCAA-
GCATGGAA—(1104 bp)—TTCTTTGGCAGATGTGTTT-
CCCCTTAAAAAGAA

42. A type of gene therapy called RNA interference (RNAi) is being investigated to treat Huntington's disease. This disease is the result of a mutation in the DNA that results in the synthesis of a nervous system protein with an altered amino acid sequence. The mutated protein forms clumps, which cause nervous system defects. To treat this disease, scientists synthesize short sequences of RNA (siRNA, or small interfering RNA) that form base pairs with the mRNA that codes for the mutated protein.
(a) Design an siRNA that will interfere with the synthesis of the protein shown in Problem 41.
(b) Explain how the addition of the siRNA will prevent the synthesis of the protein.
(c) What are the difficulties that must be overcome in order for RNA interference to be an effective technique for treating the disease?

3-3 Genomics

43. The genome of the bacterium *Carsonella ruddii* contains 159 kb of DNA with 182 ORFs. What can you conclude about the habitat or lifestyle of this bacterium?

44. In theory, both strands of DNA can code for proteins; that is, genes can be overlapping. Propose an explanation for why overlapping genes are more commonly observed in prokaryotes than in eukaryotes.

45. For many years, biologists and others have claimed that humans and chimpanzees are 98% identical at the level of DNA. Both the human and chimp genomes, which are roughly the same size, have now been sequenced, and the data reveal approximately 35 million nucleotide differences between the two species. How does this number compare to the original claim?

46. When genomes of various organisms were sequenced, biologists expected that the DNA content (the C-value) would be positively correlated with organismal complexity. But no such correlation has been demonstrated. In fact, some amphibians have DNA content that is similar to that found in humans. The *C-value paradox* is the term that refers to this puzzling lack of correlation between DNA content and organismal complexity. What questions do biologists need to ask as they attempt to solve the paradox?

47. A partial sequence of a newly discovered bacteriophage is shown below.
(a) Identify the longest open reading frame (ORF).
(b) Assuming that the ORF has been correctly identified, where is the most likely start site?

TATGGGATGGCTGAGTACAGCACGTTGAATGAGGCGAT-
GGCCGCTGGTGATG

48. The bacteriophage described in Problem 47 contains 59 kb and 105 ORFs. None of the ORFs code for tRNAs. How does the bacteriophage replicate its DNA and synthesize the structural proteins necessary to replicate itself?

49. If each person's genome contains a SNP every 300 nucleotides or so, how many SNPs are in that person's genome?

50. Assuming that genes and SNPs are distributed evenly throughout the human genome, estimate how many protein-coding genes are likely to differ between two individuals.

51. A genome-wide association study was carried out to identify the SNPs located on chromosome 1 that were correlated with an intestinal disease. The location of three genes on chromosome 1 (between positions 6.3×10^7 and 6.6×10^7) is shown in the figure below. A $-\log_{10}$ P-value of 7 or greater is assumed to be associated with the disease.
(a) What are the locations on the chromosome that show the strongest correlation with the disease?
(b) Which genes contain SNPs associated with the disease and which do not?

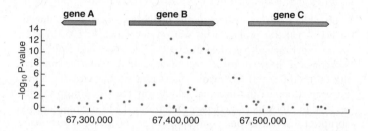

52. Use the information in the figure below to determine the chromosomal locations for the SNPs that are most closely associated with a colon disease.

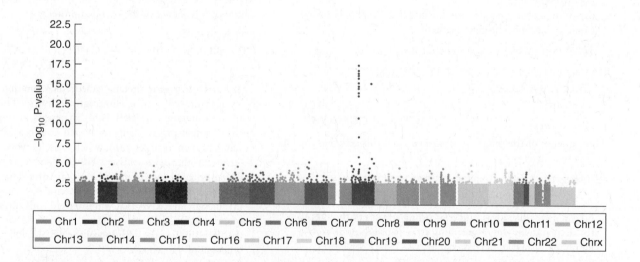

3-4 Tools and Techniques: Manipulating DNA

53. Design a 10-bp primer that could be used to determine the following sequence of DNA:

5′-AGTCGATCCCTGATCGTACGCTACGGTAACGT-3′

54. The following fragments were produced by using the "shotgun" method. Show how the fragments should be aligned to determine the sequence of the DNA.

ACCGTGTTTCCGACCG

ATTGTTCCCACAGACCG

CGGCGAAGCATTGTTCC

TTGTTCCCACAGACCGTG

55. The primer used in sequencing a cloned DNA segment often includes the recognition sequence for a restriction endonuclease. Explain.

56. Heat-stable DNA polymerases are sometimes used for dideoxy sequencing at high temperatures, especially when the template DNA has a high content of G + C. Explain.

57. Could you perform PCR with an ordinary DNA polymerase, that is, one that is destroyed by high temperatures? What modifications would you make in the PCR protocol?

58. Examine the sequence of the protein in Solution 41. Assume that the corresponding DNA sequence is *not* known. Using the amino acid sequence as a guide, design a pair of nine-base deoxynucleotide primers that could be used for PCR amplification of the protein-coding portion of the gene. (*Hint:* DNA polymerase can extend a primer only from its 3′ end.) How many different pairs of primers could you choose from?

59. Which restriction enzymes in Table 3-5 generate sticky ends? Blunt ends?

60. A sample of the DNA segment shown below is treated with the restriction enzyme MspI. Next, the sample is incubated with an exonuclease that acts only on single-stranded polynucleotides. What mononucleotides will be present in the reaction mixture?

TGCTTAGCCGGAACGA

ACGAATCGGCCTTGCT

61. Which is more likely to be called a "rare cutter": a restriction enzyme with a four-base recognition sequence or a restriction enzyme with an eight-base recognition sequence?

62. Restriction enzymes are used to construct restriction maps of DNA. These are diagrams of specific DNA molecules that show the sites where the restriction enzymes cleave the DNA. To construct a restriction map, purified samples of the DNA are treated with restriction enzymes, either alone or in combination, and then the reaction products are separated by agarose gel electrophoresis, a technique that separates the DNA fragments based on size. (The smaller fragments travel more quickly and are found near the bottom of the gel, whereas the larger fragments are found near the top of the gel.)

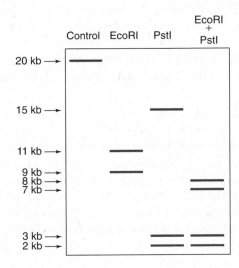

Use the results of the agarose gel electrophoresis separation (above) to construct a restriction map for the sample of DNA.

63. The search for a gene sometimes starts with a DNA library, which is a set of cloned DNA fragments representing all the sequences in an organism's genome. If you were to construct a human DNA library to search for novel genes, would you choose to clone the DNA fragments in plasmids or in yeast artificial chromosomes? Explain.

64. A researcher trying to identify the gene for a known protein might begin by looking closely at the protein's sequence in order to design a single-stranded oligonucleotide probe that will hybridize with the DNA of the gene. Why would the researcher focus on a segment of the protein containing a methionine (Met) or tryptophan (Trp) residue (see Table 3-3)?

65. A group of investigators is interested in studying the gp41 protein from the human immunodeficiency virus (HIV). In order to do this, they use site-directed mutagenesis to synthesis a series of truncated proteins. A partial sequence of the gp41 protein is shown. Design an 18 bp "mismatched primer" (see Fig. 3-22) that could be used to synthesize a truncated protein that would terminate after the Leu residue at position 700.

700
↓

··· WYIKLFIMIVGGLVGLRIVFAVLSIVNRVRGGYSP ···

66. Refer to the DNA sequence shown in Problem 41.
(a) Using the coding sequence for this gene, design two 18-bp PCR primers that could be used to amplify this gene.
(b) Suppose you wanted to add EcoRI restriction sites to each end of the gene. How would you modify the sequences of the PCR primers you designed in (a) to amplify the gene?

[SELECTED READINGS]

Collins, F. S., Cystic fibrosis: molecular biology and therapeutic implications, *Science* **256,** 774–779 (1992). [Describes the discovery of the CFTR gene and its protein product.]

Dickerson, R. E., DNA structure from A to Z, *Methods Enzymol.* **211,** 67–111 (1992). [Describes the various crystallographic forms of DNA.]

Fischer, A., and Cavazzana-Calvo, M., Gene therapy of inherited diseases, *Lancet* **371,** 2044–2047 (2008). [Discusses strategies, progress, and ongoing challenges in human gene therapy.]

Galperin, M. Y., and Fernández-Suárez, X. M., The 2012 Nucleic Acids Research Database Issue and the online Molecular Biology Database Collection, *Nucleic Acids Res.* **40,** D1–D8 (2012). [Briefly describes some new additions to a set of 1380 databases related to molecular biology. Also available at http://nar.oxfordjournals.org/.]

International Human Genome Sequencing Consortium, Initial sequencing and analysis of the human genome, *Nature* **409,** 860–921 (2001) and Venter, J. C., et al., The sequence of the human genome, *Science* **291,** 1304–1351 (2001). [These and other papers in the same issues of *Nature* and *Science* describe the data that constitute the draft sequence of the human genome and discuss how this information can be used in understanding biological function, evolution, and human health.]

Lander, E. S., Initial impact of the sequencing of the human genome, *Nature* **470,** 187–197 (2011). [Reviews genome structure and provides insights into its variation and links to human diseases.]

Thieffry, D., Forty years under the central dogma, *Trends Biochem. Sci.* **23,** 312–316 (1998). [Traces the origins, acceptance, and shortcomings of the idea that nucleic acids contain biological information.]

4 PROTEIN STRUCTURE

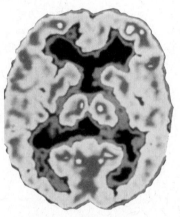

[Dr. Robert Friedland/Photo Researchers, Inc.]

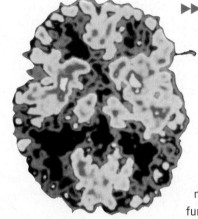

▶▶ **HOW** do faulty protein structures lead to disorders such as Alzheimer's disease?

The effects of this neurodegenerative condition are clearly visible when comparing scans of the brain of an Alzheimer's disease patient (*right*) and a normal brain (*left*). In this chapter, we'll examine protein structure, beginning with the amino acids that form polypeptide chains. We'll also see how proteins are stabilized by weak noncovalent forces, including the hydrophobic effect. When some protein structures are disrupted, the molecules tend to aggregate, eventually impairing cellular functions and producing the symptoms of disorders such as Alzheimer's disease.

THIS CHAPTER IN CONTEXT

Part 1 Foundations

Part 2 Molecular Structure and Function

    4 Protein Structure

Part 3 Metabolism

Part 4 Genetic Information

Do You Remember?

- Cells contain four major types of biological molecules and three major types of polymers (Section 1-2).
- Noncovalent forces, including hydrogen bonds, ionic interactions, and van der Waals forces, act on biological molecules (Section 2-1).
- The hydrophobic effect, which is driven by entropy, excludes nonpolar substances from water (Section 2-2).
- An acid's pK value describes its tendency to ionize (Section 2-3).
- The biological information encoded by a sequence of DNA is transcribed to RNA and then translated into the amino acid sequence of a protein (Section 3-2).

Proteins are the workhorses of the cell. They provide structural stability and motors for movement; they form the molecular machinery for harvesting free energy and using it to carry out other metabolic activities; they participate in the expression of genetic information; and they mediate communication between the cell and its environment. In subsequent chapters, we will describe in more detail these protein-driven phenomena, but for now we will focus on protein structure.

Proteins come in many shapes and sizes (**Fig. 4-1**). The essence of their biological function is their interaction with other molecules, including other proteins. We will look first at the amino acid components of proteins. Next comes a discussion of how the protein backbone folds and how the backbone plus all the side chains assume a unique three-dimensional shape. The Tools and Techniques section of this chapter examines some of the procedures for purifying and sequencing proteins and determining their three-dimensional shapes.

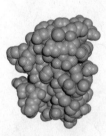

Plastocyanin (poplar)
Shuttles electrons as part of the apparatus for converting light energy to chemical energy
(more in Section 16-2)

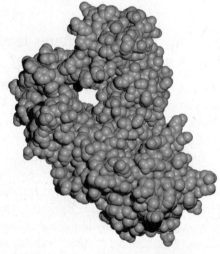

DNA polymerase (*E. coli* Klenow fragment)
Synthesizes a new DNA chain using an existing DNA strand as a template (more in Section 20-2)

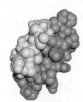

Insulin (pig)
Released from the pancreas to signal the availability of the metabolic fuel glucose
(more in Section 19-2)

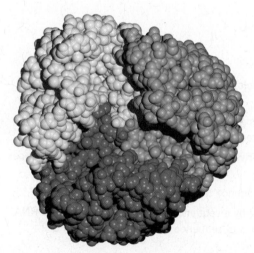

Maltoporin (*E. coli*)
Permits sugars to cross the bacterial cell membrane (more in Section 9-2)

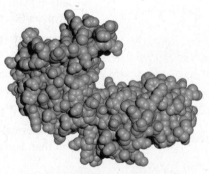

Phosphoglycerate kinase (yeast)
Catalyzes one of the central reactions in metabolism
(more in Section 13-1)

Figure 4-1 A gallery of protein structures. These space-filling models are all shown at approximately the same scale. In proteins that consist of more than one chain of amino acids, the chains are shaded differently. [Structure of insulin (pdb 1ZNI) determined by M. G. W. Turkenburg, J. L. Whittingham, G. G. Dodson, E. J. Dodson, B. Xiao, and G. A. Bentley; structure of maltoporin (pdb 1MPM) determined by R. Dutzler and T. Schirmer; structure of phosphoglycerate kinase (pdb 3PGK) determined by P. J. Shaw, N. P. Walker, and H. C. Watson; structure of DNA polymerase (pdb 1KFS) determined by C. A. Brautigan and T. A. Steitz; and structure of plastocyanin (pdb 1PND) determined by B. A. Fields, J. M. Guss, and H. C. Freeman.]

4-1 Proteins Are Chains of Amino Acids

A **protein** is a biological molecule that consists of one or more **polypeptides,** which are chains of polymerized amino acids. A cell may contain dozens of different amino acids, but only 20 of these—called the "standard" amino acids—are commonly found in proteins. As introduced in Section 1-2, an **amino acid** is a small molecule containing an amino group ($-NH_3^+$) and a carboxylate group ($-COO^-$) as well as a side chain of variable structure, called an **R group:**

$$\begin{array}{c} COO^- \\ | \\ H-C-R \\ | \\ NH_3^+ \end{array}$$

Note that at physiological pH, the carboxyl group is unprotonated and the amino group is protonated, so an isolated amino acid bears both a negative and a positive charge.

The 20 amino acids have different chemical properties

The identities of the R groups distinguish the 20 standard amino acids. The R groups can be classified by their overall chemical characteristics as hydrophobic, polar, or charged, as shown in **Figure 4-2**, which also includes the one- and three-letter codes for each amino acid. These compounds are formally called **α-amino acids** because the amino and carboxylate (acid) groups are both attached to a central carbon atom known as the α carbon (abbreviated **Cα**).

Nineteen of the twenty standard amino acids are asymmetric, or chiral, molecules. Their **chirality,** or handedness (from the Greek *cheir,* "hand"), results from the asymmetry of the alpha carbon. The four different substituents of Cα can be arranged in two ways. For alanine, a small amino acid with a methyl R group, the possibilities are

$$\begin{array}{cc} COO^- & COO^- \\ | & | \\ H_3N^+-C_\alpha-H & H-C_\alpha-NH_3^+ \\ | & | \\ CH_3 & CH_3 \end{array}$$

You can use a simple model-building kit to satisfy yourself that the two structures are not identical. They are nonsuperimposable mirror images, like right and left hands.

The amino acids found in proteins all have the form on the left. For historical reasons, these are designated L amino acids (from the Greek *levo,* "left"). Their mirror images, which rarely occur in proteins, are the D amino acids (from *dextro,* "right").

Molecules related by mirror symmetry are physically indistinguishable and are usually present in equal amounts in synthetic preparations. However, the two forms behave differently in biological systems (Box 4-A).

It is advisable to become familiar with the structures of the standard amino acids, since their side chains ultimately help determine the three-dimensional shape of the protein as well as its chemical reactivity.

The Hydrophobic Amino Acids

As their name implies, *the hydrophobic amino acids have essentially nonpolar side chains that interact very weakly or not at all with water.* The aliphatic (hydrocarbon-like) side chains of alanine (Ala), valine (Val), leucine (Leu), isoleucine (Ile), and phenylalanine (Phe) obviously fit into this group. Although methionine (Met, with an S atom) and tryptophan (Trp, with an NH group) include atoms that can form hydrogen bonds, the bulk of their side chains is nonpolar. Proline (Pro) is unique among the amino acids because its aliphatic side chain is also covalently linked to its amino group.

KEY CONCEPTS
- The 20 amino acids differ in the chemical characteristics of their R groups.
- Amino acids are linked by peptide bonds to form a polypeptide.
- A protein's structure may be described at four levels, from primary to quaternary.

BOX 4-A   BIOCHEMISTRY NOTE

Does Chirality Matter?

The importance of chirality in biological systems was brought home in the 1960s when pregnant women with morning sickness were given the sedative thalidomide, which was a mixture of right- and left-handed forms. The active form of the drug has the structure shown below.

Thalidomide

Tragically, its mirror image, which was also present, caused severe birth defects, including abnormally short or absent limbs.

Although the mechanisms of action of the two forms of thalidomide are not well understood, different responses to the two forms can be rationalized. An organism's ability to distinguish chiral molecules results from the handedness of its molecular constituents. For example, proteins contain all L amino acids, and polynucleotides coil in a right-handed helix (see Fig. 3-4). The lessons learned from thalidomide have made drugs more costly to develop and test but have also made them safer.

Question: Which of the 20 amino acids is not chiral?

In proteins, the hydrophobic amino acids are almost always located in the interior of the molecule, among other hydrophobic groups, where they do not interact with water. And because they lack reactive functional groups, the hydrophobic side chains do not directly participate in mediating chemical reactions.

The Polar Amino Acids

The side chains of the polar amino acids can interact with water because they contain hydrogen-bonding groups. Serine (Ser), threonine (Thr), and tyrosine (Tyr) have hydroxyl groups; cysteine (Cys) has a thiol group; and asparagine (Asn) and glutamine (Gln) have amide groups. All these amino acids, along with histidine (His, which bears a polar imidazole ring), can be found on the solvent-exposed surface of a protein, although they also occur in the protein interior, provided that their hydrogen-bonding

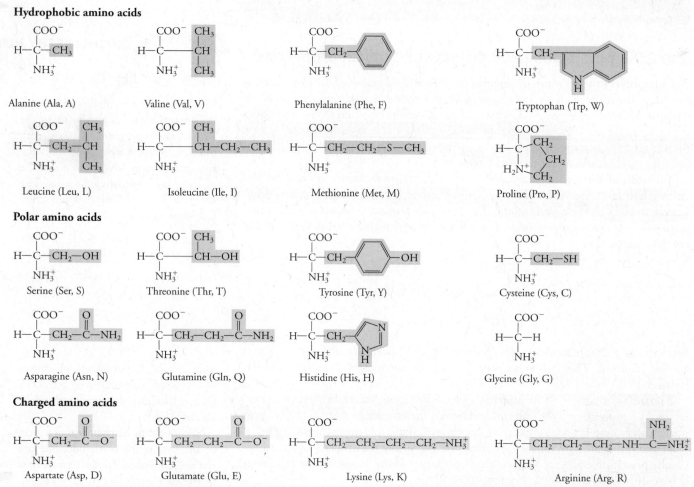

Figure 4-2 Structures and abbreviations of the 20 standard amino acids. The amino acids can be classified according to the chemical properties of their R groups as hydrophobic, polar, or charged. The side chain (R group) of each amino acid is shaded. The three-letter code is usually the first three letters of the amino acid's name. The one-letter code is derived as follows: If only one amino acid begins with a particular letter, that letter is used: C = cysteine, H = histidine, I = isoleucine, M = methionine, S = serine, and V = valine. If more than one amino acid begins with a particular letter, the letter is assigned to the most abundant amino acid: A = alanine, G = glycine, L = leucine, P = proline, and T = threonine. Most of the others are phonetically suggestive: D = aspartate ("asparDate"), F = phenylalanine ("Fenylalanine"), N = asparagine ("asparagiNe"), R = arginine ("aRginine"), W = tryptophan ("tWyptophan"), and Y = tyrosine ("tYrosine"). The rest are assigned as follows: E = glutamate (near D, aspartate), K = lysine, and Q = glutamine (near N, asparagine). The carbon atoms of amino acids are sometimes assigned Greek letters, beginning with Cα, the carbon to which the R group is attached. Thus, glutamate has a γ-carboxylate group, and lysine has an ε-amino group.

? **Identify the functional groups in each amino acid. Refer to Table 1-1.**

requirements are satisfied by their proximity to other hydrogen bond donor or acceptor groups. Glycine (Gly), whose side chain consists of only an H atom, cannot form hydrogen bonds but is included with the polar amino acids because it is neither hydrophobic nor charged.

Depending on the presence of nearby groups that increase their polarity, *some of the polar side chains can ionize at physiological pH values*. For example, the neutral (basic) form of His can accept a proton to form an imidazolium ion (an acid):

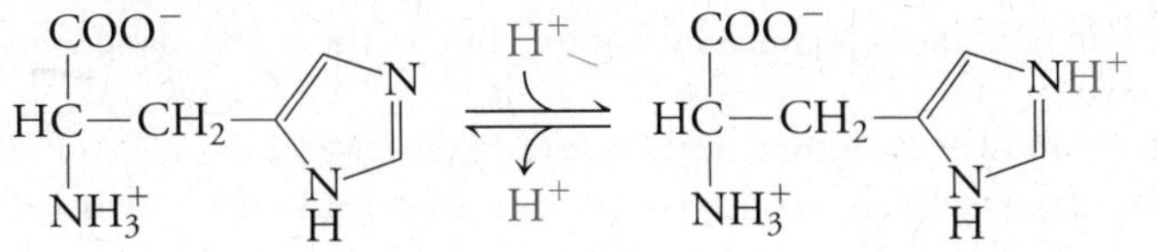

As we will see, the ability of His to act as an acid or a base gives it great versatility in catalyzing chemical reactions.

Similarly, the thiol group of Cys can be deprotonated, yielding a thiolate anion:

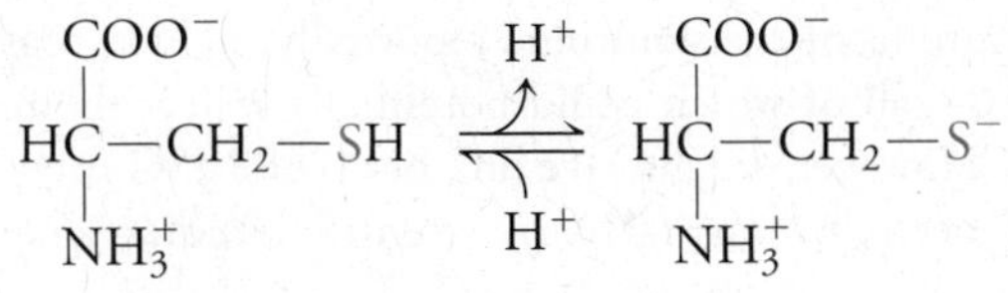

Occasionally, cysteine's thiol group undergoes oxidation with another thiol group, such as another Cys side chain, to form a **disulfide bond:**

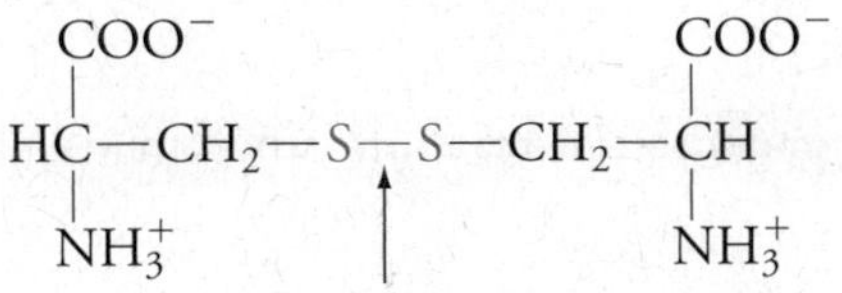

Disulfide bond

In rare cases, the very weakly acidic hydroxyl groups of Ser, Thr, and Tyr ionize to yield hydroxide groups that can act as strong bases in chemical reactions.

The Charged Amino Acids

Four amino acids have side chains that are virtually always charged under physiological conditions. Aspartate (Asp) and glutamate (Glu), which bear carboxylate groups, are negatively charged. Lysine (Lys) and arginine (Arg) are positively charged. These side chains are usually located on the protein's surface, where their charged groups can be surrounded by water molecules or interact with other polar or ionic substances.

Although it is convenient to view amino acids merely as the building blocks of proteins, many amino acids play key roles in regulating physiological processes (Box 4-B).

Peptide bonds link amino acids in proteins

The polymerization of amino acids to form a polypeptide chain involves the condensation of the carboxylate group of one amino acid with the amino

BOX 4-B **BIOCHEMISTRY NOTE**

Monosodium Glutamate

A number of amino acids and compounds derived from them function as signaling molecules in the nervous system (we will look at some of these in more detail in Section 18-2). Among the amino acids with signaling activity is glutamate, which most often operates as an excitatory signal and is necessary for learning and memory. Because glutamate is abundant in dietary proteins and because the human body can manufacture it, glutamate deficiency is rare. But is there any danger in eating too much glutamate?

Glutamate binds to receptors on the tongue that register the taste of umami—one of the five human tastes, along with sweet, salty, sour, and bitter. By itself, the umami taste is not particularly pleasing, but when combined with other tastes, it imparts a sense of savoriness and induces salivation. For this reason, glutamate in the form of monosodium glutamate (MSG) is sometimes added to processed foods as a flavor enhancer. For example, a low-salt food item can be made more appealing by adding MSG to it.

According to some popular accounts, "Chinese restaurant syndrome" can be attributed to the consumption of excess MSG added to prepared foods or present in soy sauce (MSG is also naturally present in many other foods, including cheese and tomatoes). The symptoms of the syndrome reportedly include muscle tingling, head-ache, and drowsiness—all of which could potentially reflect the role of glutamate in the nervous system. However, a definitive link between MSG intake and neurological symptoms has not been demonstrated in scientific studies and therefore remains mostly anecdotal.

● **Question:** Draw the structure of MSG. Which group is ionized in order to pair with a single sodium ion? What is the structure of magnesium glutamate? Would it have the same physiological effects as MSG?

group of another (a **condensation reaction** is one in which a water molecule is eliminated):

Peptide bond

The resulting amide bond linking the two amino acids is called a **peptide bond.** The remaining portions of the amino acids are called amino acid **residues.** In a cell, peptide bond formation is carried out in several steps involving the ribosome and ad-ditional RNA and protein factors (Section 22-3). Peptide bonds can be broken, or **hydrolyzed,** by the action of **exo-** or **endopeptidases** (enzymes that act from the end or the middle of the chain, respectively).

By convention, a chain of amino acid residues linked by peptide bonds is written or drawn so that the residue with a free amino group is on the left (this end of the

[TABLE 4-1] pK Values of Ionizable Groups in Polypeptides

Group[a]		pK
C-terminus	—COOH	3.5
Asp	—CH$_2$—C(=O)—OH	3.9
Glu	—CH$_2$—CH$_2$—C(=O)—OH	4.1
His	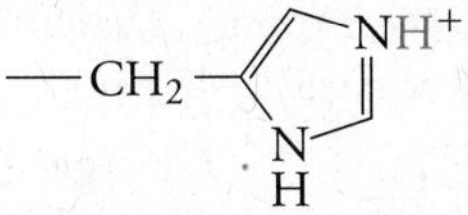 —CH$_2$— (imidazole with NH$^+$)	6.0
Cys	—CH$_2$—SH	8.4
N-terminus	—NH$_3^+$	9.0
Tyr	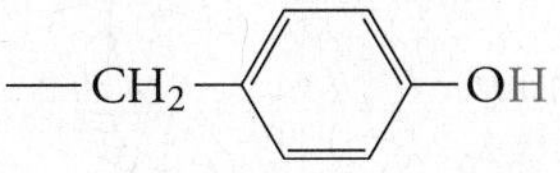 —CH$_2$— (phenyl) —OH	10.5
Lys	—CH$_2$—CH$_2$—CH$_2$—CH$_2$—NH$_3^+$	10.5
Arg	—CH$_2$—CH$_2$—CH$_2$—NH—C(NH$_2$)=NH$_2^+$	12.5

[a]The ionizable proton is indicated in red.

polypeptide is called the **N-terminus**) and the residue with a free carboxylate group is on the right (this end is called the **C-terminus**):

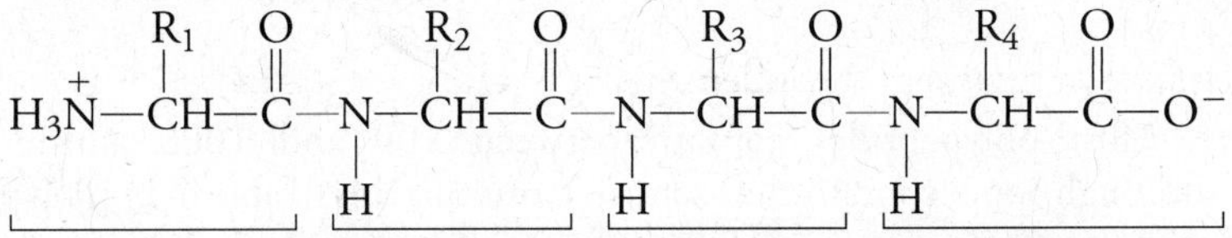

Note that, except for the two terminal groups, the charged amino and carboxylate groups of each amino acid are eliminated in forming peptide bonds. *The electrostatic properties of the polypeptide therefore depend primarily on the identities of the side chains (R groups) that project out from the polypeptide **backbone.***

The pK values of all the charged and ionizable groups in amino acids are given in Table 4-1 (recall from Section 2-3 that a pK value is a measure of a group's tendency to ionize). Thus, it is possible to calculate the net charge of a protein at a given pH (see Sample Calculation 4-1). At best, this value is only an estimate, since the side chains of polymerized amino acids do not behave as they do in free amino acids. This is because of the electronic effects of the peptide bond and other functional groups that may be brought into proximity when the polypeptide chain folds into a three-dimensional shape. The chemical properties of a side chain's immediate neighbors, its **microenvironment,** may alter its polarity, thereby altering its tendency to lose or accept a proton.

Nevertheless, *the chemical and physical properties of proteins depend on their constituent amino acids,* so proteins exhibit different behaviors under given laboratory conditions. These differences can be exploited to purify a protein, that is, to isolate it from a mixture containing other molecules (see Section 4-5).

PROBLEM

Estimate the net charge of the polypeptide chain below at physiological pH (7.4) and at pH 5.0.

Ala–Arg–Val–His–Asp–Gln

SOLUTION

The polypeptide contains the following ionizable groups, whose pK values are listed in Table 4-1: the N-terminus (pK = 9.0), Arg (pK = 12.5), His (pK = 6.0), Asp (pK = 3.9), and the C-terminus (pK = 3.5).

At pH 7.4, the groups whose pK values are less than 7.4 are mostly deprotonated, and the groups with pK values greater than 7.4 are mostly protonated. The polypeptide therefore has a net charge of 0:

Group	Charge
N-terminus	+1
Arg	+1
His	0
Asp	−1
C-terminus	−1
net charge	0

At pH 5.0, His is likely to be protonated, giving the polypeptide a net charge of +1:

Group	Charge
N-terminus	+1
Arg	+1
His	+1
Asp	−1
C-terminus	−1
net charge	+1

1. Estimate the net charge of a Glu–Tyr dipeptide at pH 6.0.
2. Estimate the net charge of an Asp–Asp–Asp tripeptide at pH 7.0.
3. Estimate the net charge of a His–Lys–Glu tripeptide at pH 8.0.

Most polypeptides contain between 100 and 1000 amino acid residues, although some contain thousands of amino acids (Table 4-2). Polypeptides smaller than about 40 residues are often called **oligopeptides** (*oligo* is Greek for "few") or just **peptides.** Since there are 20 different amino acids that can be polymerized to form polypeptides, even peptides of similar size can differ dramatically from each other, depending on their complement of amino acids.

[**TABLE 4-2**]   Composition of Some Proteins

Protein	Number of Amino Acid Residues	Number of Polypeptide Chains	Molecular Mass (D)
Insulin (bovine)	51	2	5733
Rubredoxin (*Pyrococcus*)	53	1	5878
Myoglobin (human)	153	1	17,053
Phosphorylase kinase (yeast)	416	1	44,552
Hemoglobin (human)	574	4	61,972
Reverse transcriptase (HIV)	986	2	114,097
Nitrite reductase (*Alcaligenes*)	1029	3	111,027
C-reactive protein (human)	1030	5	115,160
Pyruvate decarboxylase (yeast)	1112	2	121,600
Immunoglobulin (mouse)	1316	4	145,228
Ribulose bisphosphate carboxylase (spinach)	5048	16	567,960
Glutamine synthetase (*Salmonella*)	5628	12	621,600
Carbamoyl phosphate synthetase (*E. coli*)	5820	8	637,020

The potential for sequence variation is enormous. For a modest-sized polypeptide of 100 residues, there are 20^{100} or 1.27×10^{130} possible amino acid sequences. This number is clearly unattainable in nature, since there are only about 10^{79} atoms in the universe, but it illustrates the tremendous structural variability of proteins.

Unraveling the amino acid sequence of a protein may be relatively straightforward if its gene has been sequenced (see Section 3-3). In this case, it is just a matter of reading successive sets of three nucleotides in the DNA as a sequence of amino acids in the protein. However, this exercise may not be accurate if the gene's mRNA is spliced before being translated or if the protein is hydrolyzed or otherwise covalently modified immediately after it is synthesized. And, of course, nucleic acid sequencing is of no use if the protein's gene has not been identified. The alternative is to use a technique such as mass spectrometry to directly determine the protein's amino acid sequence (Section 4-5).

The amino acid sequence is the first level of protein structure

The sequence of amino acids in a polypeptide is called the protein's **primary structure.** There are as many as four levels of structure in a protein (Fig. 4-3). Under physiological conditions, a polypeptide very seldom assumes a linear extended conformation but instead folds up to form a more compact shape, usually consisting of

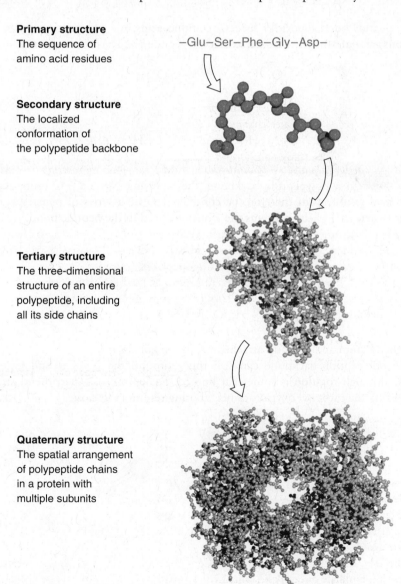

Primary structure
The sequence of
amino acid residues

−Glu−Ser−Phe−Gly−Asp−

Secondary structure
The localized
conformation of
the polypeptide backbone

Tertiary structure
The three-dimensional
structure of an entire
polypeptide, including
all its side chains

Quaternary structure
The spatial arrangement
of polypeptide chains
in a protein with
multiple subunits

Figure 4-3 Levels of protein structure in hemoglobin. [Structure of human hemoglobin (pdb 2HHB) determined by G. Fermi and M. F. Perutz.]

several layers. The local folding arrangement of the polypeptide backbone (exclusive of the side chains) is known as **secondary structure.** The complete three-dimensional conformation of the polypeptide, including its backbone atoms and all its side chains, is the polypeptide's **tertiary structure.** In a protein that consists of more than one polypeptide chain, the **quaternary structure** refers to the spatial arrangement of all the chains. In the following sections we will consider the second, third, and fourth levels of protein structure.

> **CONCEPT REVIEW**
> - Draw the structures and give the one- and three-letter abbreviations for the 20 standard amino acids.
> - Divide the 20 amino acids into groups that are hydrophobic, polar, and charged.
> - Which polar amino acids are sometimes charged?
> - Describe the four levels of protein structure.

4-2 Secondary Structure: The Conformation of the Peptide Group

KEY CONCEPTS
- The polypeptide backbone has limited conformational flexibility.
- The α helix and β sheet are common secondary structures characterized by hydrogen bonding between backbone groups.

In the peptide bond that links successive amino acids in a polypeptide chain, the electrons are somewhat delocalized so that the peptide bond has two resonance forms:

Due to this partial (about 40%) double-bond character, there is no rotation around the C—N bond. In a polypeptide backbone, the repeating N—Cα—C units of the amino acid residues can therefore be considered to be a series of planar peptide groups (where each plane contains the atoms involved in the peptide bond):

Here the H atom and R group attached to Cα are not shown.

The polypeptide backbone can still rotate around the N—Cα and Cα—C bonds, although rotation is somewhat limited. For example, a sharp bend at Cα could bring the carbonyl oxygens of neighboring residues too close:

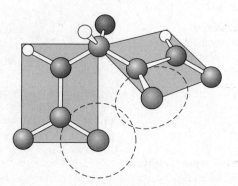

Here the atoms are color-coded: C gray, O red, N blue, and H white.

As the resonance structures on the opposite page indicate, the groups involved in the peptide bond are strongly polar, with a tendency to form hydrogen bonds. The backbone amide groups are hydrogen bond donors, and the carbonyl oxygens are hydrogen bond acceptors. *Under physiological conditions, the polypeptide chain folds so that it can satisfy as many of these hydrogen-bonding requirements as possible.* At the same time, *the polypeptide backbone must adopt a conformation (a secondary structure) that minimizes steric strain.* In addition, side chains must be positioned in a way that minimizes their steric interference. Two kinds of secondary structure commonly found in proteins meet these criteria: the α helix and the β sheet.

The α helix exhibits a twisted backbone conformation

The **α helix** was first identified through model-building studies carried out by Linus Pauling. In this type of secondary structure, the polypeptide backbone twists in a right-handed helix (the DNA helix is also right-handed; see Section 3-1 for an explanation). There are 3.6 residues per turn of the helix, and for every turn, the helix rises 5.4 Å along its axis. In the α helix, the carbonyl oxygen of each residue forms a hydrogen bond with the backbone NH group four residues ahead. The backbone hydrogen-bonding tendencies are thereby met, except for the four residues at each end of the helix (**Fig. 4-4**). Most α helices are about 12 residues long.

Like the DNA helix, whose side chains fill the helix interior (see Figure 3-4b), the α helix is solid—the atoms of the polypeptide backbone are in van der Waals contact. However, in the α helix, the side chains extend outward from the helix (**Fig. 4-5**).

The β sheet contains multiple polypeptide strands

Pauling, along with Robert Corey, also built models of the **β sheet.** This type of secondary structure consists of aligned strands of polypeptide whose hydrogen-bonding requirements are met by bonding between neighboring strands. The strands of a β sheet can be arranged in two ways (**Fig. 4-6**): In a **parallel β sheet,** neighboring chains run in the same direction; in an **antiparallel β sheet,** neighboring chains run in opposite directions. Each residue forms two hydrogen bonds with a neighboring strand, so all hydrogen-bonding requirements are met, except in the first and last strands of the sheet.

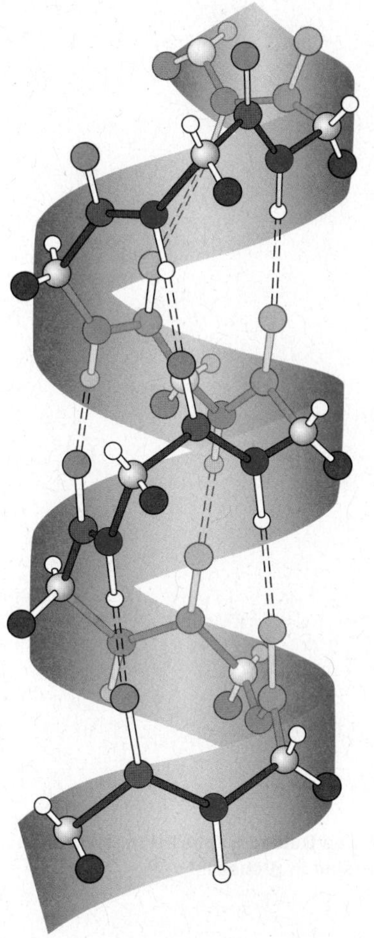

Figure 4-4 **The α helix.** In this conformation, the polypeptide backbone twists in a right-handed fashion so that hydrogen bonds (dashed lines) form between C=O and N—H groups four residues farther along. Atoms are color-coded: Cα light gray, carbonyl C dark gray, O red, N blue, side chain purple, H white. [Based on a drawing by Irving Geis.] ⊕ **See Animated Figure. The α helix.**

? How many amino acid residues are shown here? How many hydrogen bonds?

⊕ **See Guided Exploration. Secondary structures in proteins.**

⊕ **See Guided Exploration. Stable helices in proteins: the α helix.**

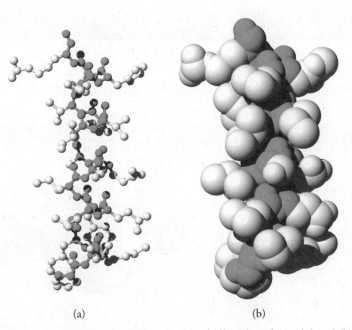

(a)   (b)

Figure 4-5 An α helix from myoglobin. In (a) a ball-and-stick model and (b) a space-filling model of residues 100–118 of myoglobin, the backbone atoms are green and the side chains are gray. [Structure of sperm whale myoglobin (pdb 1MBD) determined by S. E. V. Phillips.]

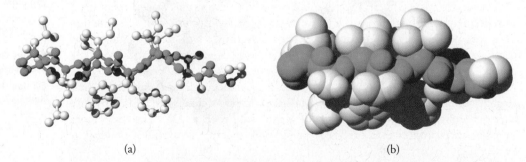

Parallel   Antiparallel

Figure 4-6 β sheets. In a parallel β sheet and an antiparallel β sheet, the polypeptide backbone is extended. In both types of β sheet, hydrogen bonds form between the amide and carbonyl groups of adjacent strands. The H and R attached to Cα are not shown. Note that the strands are not necessarily separate polypeptides but may be segments of a single chain that loops back on itself. ⊕ **See Animated Figure. β sheets.**

⊕ **See Guided Exploration. Hydrogen bonding in β sheets.**

? How many amino acid residues are shown in each chain?

A single β sheet may contain from 2 to more than 12 polypeptide strands, with an average of 6 strands, and each strand has an average length of 6 residues. In a β sheet, the amino acid side chains extend from both faces (Fig. 4-7).

Proteins also contain irregular secondary structure

α Helices and β sheets are classified as **regular secondary structures,** because *their component residues exhibit backbone conformations that are the same from one residue to the next.* In fact, these elements of secondary structure are easily recognized in the

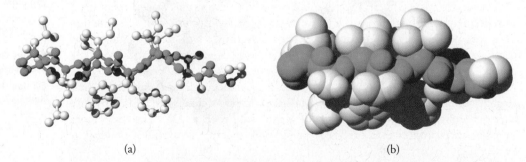

(a)   (b)

Figure 4-7 Side view of two parallel strands of a β sheet. In (a) a ball-and-stick model and (b) a space-filling model of a β sheet from carboxypeptidase A, the backbone atoms are green. The amino acid side chains (gray) point alternately to each side of the β sheet. [Structure of carboxypeptidase (pdb 3CPA) determined by W. N. Lipscomb.]

three-dimensional structures of a huge variety of proteins, regardless of their amino acid composition. Of course, depending on the identities of the side chains and other groups that might be present, α helices and β sheets may be slightly distorted from their ideal conformations. For example, the final turn of some α helices becomes "stretched out" (longer and thinner than the rest of the helix).

In every protein, elements of secondary structure (individual α helices or strands in a β sheet) are linked together by polypeptide loops of various sizes. A loop may be a relatively simple hairpin turn, as in the connection of two antiparallel β strands (which are shown below as flat arrows; *left*). Or it may be quite long, especially if it joins successive strands in a parallel β sheet (*right*):

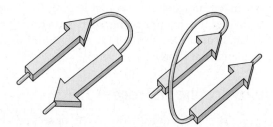

Usually, the loops that link β strands or α helices consist of residues with **irregular secondary structure;** that is, the polypeptide does not adopt a defined secondary structure in which successive residues have the same backbone conformation. Note that "irregular" does not mean "disordered": The peptide backbone almost always adopts a single, unique conformation. Most proteins contain a combination of regular and irregular secondary structure. On average, 31% of residues are in α helices, 28% are in β sheets, and most of the remainder are in irregular loops of different sizes.

CONCEPT REVIEW
- Which amino acid atoms constitute the polypeptide backbone?
- What limits the rotation of the backbone?
- Summarize the structural features of α helices and β sheets.
- Distinguish regular and irregular secondary structure.
- Why does virtually every protein contain loops?

4-3 Tertiary Structure and Protein Stability

The three-dimensional shape of a protein, known as its tertiary structure, includes its regular and irregular secondary structure (that is, the overall folding of its peptide backbone) as well as the spatial arrangement of all its side chains. In a fully folded protein under physiological conditions, virtually every atom has a designated place.

One of the most powerful techniques for probing the atomic structures of macromolecules, including proteins, is X-ray crystallography (Section 4-5). The structure of myoglobin, the first protein structure to be determined by X-ray crystallography, came to light in 1958 through the efforts of John Kendrew, who painstakingly determined the conformation of every backbone and side-chain group. Kendrew's results—coming just a few years after Watson and Crick had published their elegant model of DNA—were a bit of a disappointment. The myoglobin structure lacked the simplicity and symmetry of a molecule such as DNA and was more irregular and complex than expected (myoglobin structure is examined in detail in Section 5-1).

The structure of another well-studied protein, the enzyme triose phosphate isomerase, is shown in **Figure 4-8a**. This enzyme is a **globular protein** (**fibrous proteins**, in contrast, are usually highly elongated; examples of these are presented in Section 5-2). The proteins shown in Figure 4-1 are all globular proteins. The tertiary structure of

KEY CONCEPTS
- A folded polypeptide assumes a shape with a hydrophilic surface and a hydrophobic core.
- Protein folding and protein stabilization depend on noncovalent forces.
- Some proteins can adopt more than one stable conformation.

triose phosphate isomerase can be simplified by showing just the peptide backbone (Fig. 4-8b). Alternatively, the structure can be represented by a ribbon that passes through Cα of each residue (Fig. 4-8c). This rendering makes it easier to identify elements of secondary structure.

Commonly used systems for analyzing protein structures are based on the presence of secondary structural elements. For example, four classes are recognized by the CATH system (the name refers to a hierarchy of organizational levels: Class, Architecture, Topology, and Homology). Proteins may contain mostly α structure, mostly β structure, a combination of α and β, or very few secondary structural elements. Examples of each class are shown in **Figure 4-9**. Structural data—including the three-dimensional coordinates for each atom—for tens of thousands of proteins and other macromolecules are available in online databases. The use of software for visualizing and manipulating such structures provides valuable insight into molecular structure and function (see Bioinformatics Project 3, Visualizing Three-Dimensional Protein Structures).

Proteins have hydrophobic cores

Globular proteins typically contain at least two layers of secondary structure. This means that the protein has definite surface and core regions. On the protein's surface, some backbone and side-chain groups are exposed to the solvent; in the core,

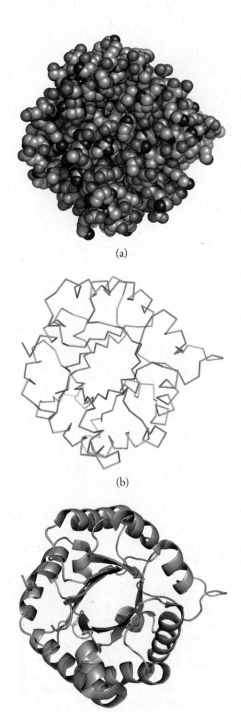

(a)

(b)

(c)

Figure 4-8 Triose phosphate isomerase. (a) Space-filling model. All atoms (except H) are shown (C gray, O red, N blue). (b) Polypeptide backbone. The trace connects the α carbons of successive amino acid residues. (c) Ribbon diagram. The ribbon represents the overall conformation of the backbone.

[Structure (pdb 1YPI) determined by T. Alber, E. Lolis, and G. A. Petsko.]

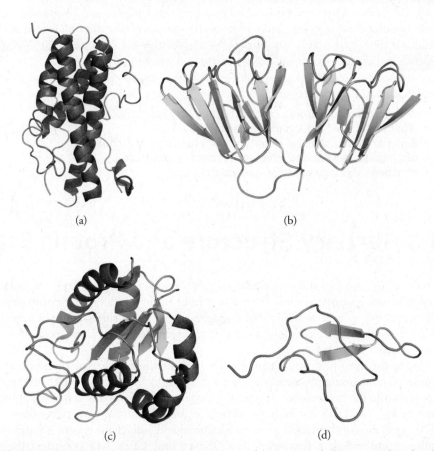

(a)   (b)

(c)   (d)

Figure 4-9 Classes of protein structure. In each protein, α helices are colored red, and β strands are colored yellow. (a) Growth hormone, an all-α protein. [Structure (pdb 1HGU) determined by L. Chantalat, N. Jones, F. Korber, J. Navaza, and A. G. Pavlovsky.] (b) γB-Crystallin, an all-β protein. [Structure (pdb 1GCS) determined by S. Najmudin, P. Lindley, C. Slingsby, O. Bateman, D. Myles, S. Kumaraswamy, and I. Glover.] (c) Flavodoxin, an α/β protein. [Structure (pdb 1CZN) determined by W. W. Smith, K. A. Pattridge, C. L. Luschinsky, and M. L. Ludwig.] (d) Tachystatin, a protein with little secondary structure. [Structure (pdb 1CIX) determined by N. Fujitani, S. Kawabata, T. Osaki, Y. Kumaki, M. Demura, K. Nitta, and K. Kawano.]

these groups are sequestered from the solvent. In other words, *the protein comprises a hydrophilic surface and a hydrophobic core.*

A polypeptide segment that has folded into a single structural unit with a hydrophobic core is often called a **domain.** Some small proteins consist of a single domain. Larger proteins may contain several domains, which may be structurally similar or dissimilar (Fig. 4-10).

The core of a small hydrophobic protein (or a domain) is typically rich in regular secondary structure. This is because the formation of α helices and β sheets, which are internally hydrogen-bonded, minimizes the hydrophilicity of the polar backbone groups. Irregular secondary structures (loops) are more often found on the surface of the protein (or domain), where the polar backbone groups can form hydrogen bonds to water molecules.

The requirement for a hydrophobic core and a hydrophilic surface also places constraints on amino acid sequence. The location of a particular side chain in a protein's tertiary structure is related to its hydrophobicity: *The greater a residue's hydrophobicity, the more likely it is to be located in the protein interior.* In the protein interior, side chains pack together, leaving very little empty space or space that could be occupied by a water molecule.

Table 4-3 lists two scales for assessing the hydrophobicity of amino acid side chains. Such information is useful for predicting the locations of amino acid residues within proteins. For example, highly hydrophobic residues such as Phe and Met are almost always buried. Polar side chains, like hydrogen-bonding backbone groups, can participate in hydrogen bonding in the protein interior, which helps "neutralize" their polarity and allows them to be buried in a nonpolar environment. When a charged residue occurs in the protein interior, it is almost always located near another residue with the opposite charge, so the two groups can interact electrostatically to form an **ion pair.** By color-coding the amino acid residues of myoglobin according to their hydrophobicity, it is easy to see that hydrophobic side chains cluster in the interior while hydrophilic side chains predominate on the surface (Fig. 4-11).

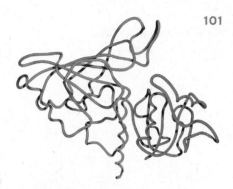

Figure 4-10 A two-domain protein. In this model of glyceraldehyde-3-phosphate dehydrogenase, the small domain is red, and the large domain is green. [Structure (pdb 1GPD) determined by D. Moras, K. W. Olsen, M. N. Sabesan, M. Buehner, G. C. Ford, and M. G. Rossmann.]

? **Which of the proteins in Figure 4-9 consists of two domains?**

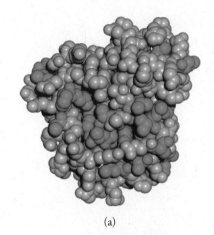

(a)

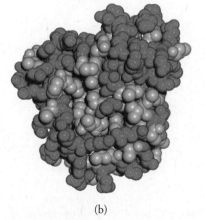

(b)

Figure 4-11 Hydrophobic and hydrophilic residues in myoglobin. (a) All hydrophobic side chains (Ala, Ile, Leu, Met, Phe, Pro, Trp, and Val) are colored green. These residues are located mostly in the protein interior. (b) All polar and charged side chains are colored red. These are located primarily on the protein surface.

[**TABLE 4-3**] Hydrophobicity Scales

Residue	Scale A[a]	Scale B[b]
Phe	2.8	3.7
Met	1.9	3.4
Ile	4.5	3.1
Leu	3.8	2.8
Val	4.2	2.6
Cys	2.5	2.0
Trp	−0.9	1.9
Ala	1.8	1.6
Thr	−0.7	1.2
Gly	−0.4	1.0
Ser	−0.8	0.6
Pro	−1.6	−0.2
Tyr	−1.3	−0.7
His	−3.2	−3.0
Gln	−3.5	−4.1
Asn	−3.5	−4.8
Glu	−3.5	−8.2
Lys	−3.9	−8.8
Asp	−3.5	−9.2
Arg	−4.5	−12.3

[a]Scale A is from Kyte, J., and Doolittle, R.F., *J. Mol. Biol.* **157,** 105–132 (1982).

[b]Scale B is from Engelman, D.M., Steitz, T.A., and Goldman, A., *Annu. Rev. Biophys. Chem.* **15,** 321–353 (1986).

(a) Folded

(b) Unfolded

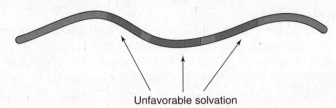

Unfavorable solvation

Figure 4-12 The hydrophobic effect in protein folding. (a) In a folded protein, hydrophobic regions (represented by blue segments of the polypeptide chain) are sequestered in the protein interior. (b) Unfolding the protein exposes these segments to water. This arrangement is energetically unfavorable, because the presence of the hydrophobic groups interrupts the hydrogen-bonded network of water molecules.

Protein structures are stabilized mainly by the hydrophobic effect

Surprisingly, the fully folded conformation of a protein is only marginally more stable than its unfolded form. The difference in thermodynamic stability amounts to about $0.4 \text{ kJ} \cdot \text{mol}^{-1}$ per amino acid, or about $40 \text{ kJ} \cdot \text{mol}^{-1}$ for a 100-residue polypeptide. This is equivalent to the amount of free energy required to break two hydrogen bonds (about $20 \text{ kJ} \cdot \text{mol}^{-1}$ each). This quantity seems incredibly small, considering the number of potential interactions among all of a protein's backbone and side-chain atoms. Nevertheless, most proteins do fold into a unique and stable three-dimensional arrangement of atoms.

The largest force governing protein structure is the hydrophobic effect (introduced in Section 2-2), which causes nonpolar groups to aggregate in order to minimize their contact with water. The hydrophobic effect is driven by the increase in entropy of the solvent water molecules, which would otherwise have to order themselves around each hydrophobic group. As we have seen, hydrophobic side chains are located predominantly in the interior of a protein. This arrangement stabilizes the folded polypeptide backbone, since unfolding it or extending it would expose the hydrophobic side chains to the solvent (**Fig. 4-12**).

Hydrogen bonding by itself is not a major determinant of protein stability, because in an unfolded protein, polar groups could just as easily form energetically equivalent hydrogen bonds with water molecules. Instead, hydrogen bonding may help the protein fine-tune a folded conformation that is already largely stabilized by the hydrophobic effect.

Cross-links help stabilize proteins

Many folded polypeptides appear to be held in place by cross-links of various kinds, the most common being ion pairs, disulfide bonds, and inorganic ions such as zinc. Do these cross-links help stabilize the protein?

An ion pair can form between oppositely charged side chains or the N- and C-terminal groups (**Fig. 4-13**). Although the resulting electrostatic interaction is strong, it does not contribute much to protein stability. This is because the favorable free energy of the electrostatic interaction is offset by the loss of entropy when the side chains become fixed in the ion pair. For a buried ion pair, there is the additional energetic cost of desolvating the charged groups in order for them to enter the hydrophobic core.

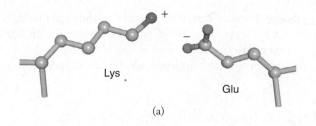

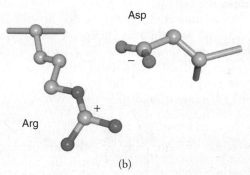

Figure 4-13 Examples of ion pairs in myoglobin. (a) The ε-amino group of Lys 77 interacts with the carboxylate group of Glu 18. (b) The carboxylate group of Asp 60 interacts with Arg 45. The atoms are color-coded: C gray, N blue, and O red. Note that these intramolecular interactions occur between side chains that are near each other in the protein's tertiary structure but are far apart in the primary structure.

? **Identify residues that could form two other types of ion pairs.**

Disulfide bonds (shown in Section 4.1) can form within and between polypeptide chains. Experiments show that even when the Cys residues of certain proteins are chemically blocked, the proteins may still fold and function normally. This suggests that disulfide bonds are not essential for stabilizing these proteins. In fact, disulfides are rare in intracellular proteins, since the cytoplasm is a reducing environment. They are more plentiful in proteins that are secreted to an extracellular (oxidizing) environment (**Fig. 4-14**). Here, the bonds may help prevent protein unfolding under relatively harsh extracellular conditions.

Domains containing cross-links called **zinc fingers** are common in DNA-binding proteins. These structures consist of 20–60 residues with one or two Zn^{2+} ions. The Zn^{2+} ions are tetrahedrally coordinated by the side chains of Cys and/or His and sometimes Asp or Glu (**Fig. 4-15**). Protein domains this size are

Figure 4-14 Disulfide bonds in lysozyme, an extracellular protein. This 129-residue enzyme from hen egg white contains eight Cys residues (yellow), which form four disulfide bonds that link different sites on the polypeptide backbone. [Structure (pdb 1E8L) determined by H. Schwalbe, S. B. Grimshaw, A. Spencer, M. Buck, J. Boyd, C. M. Dobson, C. Redfield, and L. J. Smith.]

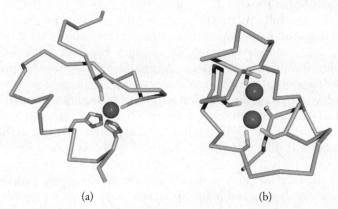

Figure 4-15 Zinc fingers. (a) A zinc finger with one Zn^{2+} (purple sphere) coordinated by two Cys and two His residues, from *Xenopus* transcription factor IIIA. (b) A zinc finger with two Zn^{2+} coordinated by six Cys residues, from the yeast transcription factor GAL4. [Structure of transcription factor IIIA (pdb 1TF6) determined by R. T. Nolte, R. M. Conlin, S. C. Harrison, and R. S. Brown; structure of GAL4 (pdb 1D66) determined by R. Marmorstein and S. Harrison.]

too small to assume a stable tertiary structure without a metal ion cross-link. Zinc is an ideal ion for stabilizing proteins: It can interact with ligands (S, N, or O) provided by several amino acids, and it has only one oxidation state (unlike Cu or Fe ions, which readily undergo oxidation–reduction reactions under cellular conditions).

Protein folding begins with the formation of secondary structures

The crowded nature of the cell interior (see Fig. 2-7) demands that proteins and other macromolecules assume compact shapes. In the cell, a newly synthesized polypeptide begins to fold as soon as it emerges from the ribosome, so part of the polypeptide may adopt its mature tertiary structure before the entire chain has been synthesized. It is difficult to monitor this process in the cell, so studies of protein folding *in vitro* usually use full-length polypeptides that have been chemically unfolded (**denatured**) and then allowed to refold (**renature**). In the laboratory, proteins can be denatured by adding highly soluble substances such as salts or urea (NH_2—CO—NH_2). Large amounts of these solutes interfere with the structure of the solvent water, thereby attenuating the hydrophobic effect and causing the protein to unfold. When the solutes are removed, the proteins renature.

Protein renaturation experiments demonstrate that *protein folding is not a random process.* That is, the protein does not just happen upon its most stable tertiary structure (its **native structure**) by trial and error but approaches this conformation through one or a few alternate pathways. During this process, small elements of secondary structure form first, then these coalesce under the influence of the hydrophobic effect to produce a mass with a hydrophobic core. Finally, small rearrangements yield the native tertiary structure (**Fig. 4-16**).

All the information required for a protein to fold is contained in its amino acid sequence. Unfortunately, there are no completely reliable methods for predicting how a polypeptide chain will fold. In fact, it is difficult to determine whether a given amino acid sequence will form an α helix, β sheet, or irregular secondary structure. This presents a formidable obstacle to assigning three-dimensional structures—and possible functions—to the burgeoning number of proteins identified through genome sequencing (see Section 3-3).

In the laboratory, certain small proteins can be repeatedly denatured and renatured, but in the cell, protein folding is more complicated and may require the assistance of other proteins. Some of these are known as **molecular chaperones** and are described in more detail in Section 22-4. Proteins that do not fold properly—often due to genetic mutations that substitute one amino acid for another—can lead to a variety of diseases (Box 4-C).

The pathway to full functionality may require additional steps beyond polypeptide folding. Some proteins contain several polypeptide chains, which must fold individually before assembling. In addition, many proteins undergo **processing** before reaching their mature forms. Depending on the protein, this might mean removal of some amino acid residues or the covalent attachment of another group such as a lipid, carbohydrate, or phosphate group (**Fig. 4-17**). The attached groups usually have a discrete biological function and may also help stabilize the folded conformation of the protein. Finally, some proteins become functional only after associating with metal ions or a specific organic molecule.

Some proteins can adopt more than one conformation. A protein's native structure is not necessarily rigid and inflexible. In fact, some minor movement, primarily the result of bending and stretching of individual bonds, is required for most proteins to carry out their biological functions. In addition, some proteins—as many as 50% in eukaryotes—may include highly flexible extended segments that are rich in hydrophilic amino acids. These **intrinsically unstructured proteins** or protein segments might be particularly well suited for regulatory roles, in which one protein must interact with several others.

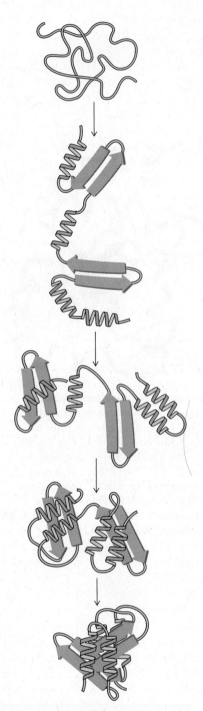

Figure 4-16 Model of protein folding. In this hypothetical example, the polypeptide first forms elements of secondary structure (α helices and β sheets). These coalesce into a single globular mass, and small adjustments generate the final stable tertiary structure. [After Goldberg, M. E., *Trends Biochem. Sci.* **10**, 389 (1985).]

BOX 4-C ⬢ CLINICAL CONNECTION

Protein Misfolding and Disease

▶▶ **HOW** do faulty protein structures lead to disorders such as Alzheimer's disease? At the start of the chapter, we posed a question about how faulty protein structures cause disease. The answer lies in the cell's failure to deal with misfolded proteins. Normally, the chaperones that help new proteins to fold can also help misfolded proteins to refold. If the protein cannot be salvaged in this way, it is usually degraded to its component amino acids.

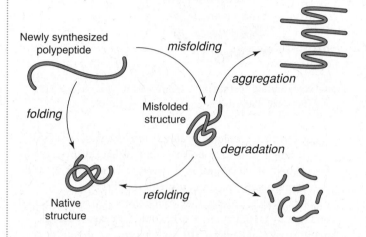

The operation of this quality control system explains what happens in the most common form of cystic fibrosis (see Box 3-A): A mutated form of the CFTR protein folds incorrectly and therefore never reaches its intended cellular destination.

Other diseases result when misfolded proteins are not immediately refolded or degraded but instead aggregate, often as long insoluble fibers. Although the fibers can occur throughout the body, they appear to be deadliest when they occur in the brain. Among the human diseases characterized by such fibrous deposits are Alzheimer's disease, Parkinson's disease, and the transmissible spongiform encephalopathies, which lead to neurological abnormalities and loss of neurons (nerve cells). The aggregated proteins, a different type in each disease, are commonly called **amyloid deposits** (a name originally referring to their starchlike appearance).

Alzheimer's disease, the most common neurodegenerative disease, is accompanied by both intracellular "tangles" and extracellular "plaques."

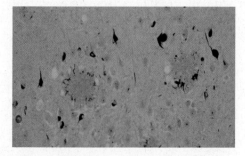

Amyloid deposits (large red areas) surrounded by intracellular tangles (smaller dark shapes). [Courtesy Dennis Selkoe and Marcia Podlisny, Harvard University Medical School.]

The fibrous material inside cells is made of a protein called tau, which is involved in the assembly of microtubules, a component of the cytoskeleton (Section 5-2). Tau deposits also appear in some other neurodegenerative diseases, and tau's role in Alzheimer's disease is not clear. The extracellular amyloid material in Alzheimer's disease consists primarily of a protein called amyloid-β, a 40- or 42-residue fragment of the much larger amyloid precursor protein, which is a membrane protein. Proteases called β-secretase and γ-secretase cleave the precursor protein to generate amyloid-β. Normal brain tissue contains some extracellular amyloid-β, but neither its function nor the function of its precursor protein is completely understood. However, excess amyloid-β is clearly linked to Alzheimer's disease.

The neurodegeneration of Alzheimer's disease may begin many years before symptoms such as memory loss appear, and studies of the disease in animal models indicate that both of these signs can be detected before amyloid fibers accumulate in the brain. These and other experimental results support the hypothesis that the early stages of amyloid-β misfolding and aggregation are toxic to neurons and are the ultimate cause of Alzheimer's disease. The accumulation of extracellular fibers may actually be a protective mechanism to deal with excessive amyloid-β production.

In Parkinson's disease, neurons in a portion of the brain accumulate fragments of a protein known as α-synuclein. Like amyloid-β, α-synuclein's function is not fully known, but it appears to play a role in neurotransmission. α-Synuclein is a small soluble protein (140 amino acid residues) with an extended conformation, part of which appears to form α helices upon binding to other molecules. The intrinsic disorder of the protein may contribute to its propensity to form amyloid deposits, which are characterized by a high content of β secondary structure. Accumulation of this material is associated with the death of neurons, leading to the typical symptoms of Parkinson's disease, which include tremor, muscular rigidity, and slow movements. Mutations in the gene for α-synuclein that lead to increased expression of the protein or promote its self-aggregation are responsible for some hereditary forms of Parkinson's disease.

Mad cow disease is the best known of the **transmissible spongiform encephalopathies (TSEs),** a group of disorders that also includes scrapie in sheep and Creutzfeldt–Jakob disease in humans. These fatal diseases, which give the brain a spongy appearance, were once thought to be caused by a virus. However, extensive investigation has revealed that the infectious agent is a protein called a **prion.** Interestingly, normal human brain tissue contains the same protein, named PrP^C (C for cellular), which occurs on neural cell membranes and appears to play a role in normal brain function. The scrapie form of the prion protein, PrP^{Sc}, has the same 253–amino acid sequence as PrP^C but includes more β secondary structure. Introduction of PrP^{Sc} into cells apparently triggers PrP^C, which contains more α-helical structure, to assume the PrP^{Sc} conformation and thereby aggregate with it.

How is a prion disease transmitted? One route seems to be by ingestion, as illustrated by the incidence of bovine TSE in cows consuming feed prepared from scrapie-infected sheep. (This feeding practice has been discontinued, but not before a number of people—who presumably consumed the infected beef—developed a variant form of Creutzfeldt–Jakob disease). PrP^{Sc} must be absorbed, without being digested, and transported to the central nervous system. Here, it causes neurodegeneration, possibly through the loss of PrP^{C} as it converts to PrP^{Sc} or through the toxic effects of the PrP^{Sc} as it accumulates.

Despite the lack of sequence or structural similarities among amyloid-β, α-synuclein, and PrP^{Sc}, their misfolded forms are all rich in β structure, and this seems to be the key to the formation of amyloid deposits. Studies of a fungal polypeptide similar to PrP^{Sc} show how amyloid formation might occur. In its original state, the protein is mostly α-helical. A segment of polypeptide shifts to an all-β conformation, which allows single molecules to stack on top of each other in parallel to form a triangular fiber stabilized by interchain hydrogen bonding. In this model, five aggregated polypeptides are shown. The vertical arrow indicates the long axis of the amyloid fiber.

Presumably, other amyloid-forming proteins undergo similar conformational changes. Significant aggregation may not occur until the protein concentration reaches some critical threshold (this may explain why the amyloid diseases, even the TSEs, take years to develop). After a few β structures have assembled, they act to trigger further protein misfolding, and the amyloid fibers rapidly propagate. Once formed, the fibers are resistant to degradation by proteases.

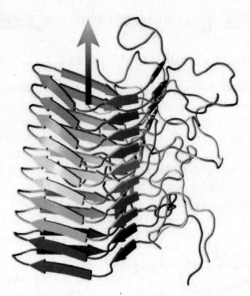

Model for amyloid formation. [From Meier, B., *et al.*, *Science*, 319, 1523–1526 (2008). Reprinted with permission from AAAS.]

● Questions:

1. The gene for the amyloid precursor protein is located on chromosome 21. Individuals with Down syndrome (trisomy 21) have three rather than two copies of chromosome 21. Explain why individuals with Down syndrome tend to exhibit Alzheimer's-like symptoms by middle age.

2. Mice that have been genetically engineered to lack the PrP gene (but are normal in all other ways) do not develop spongiform encephalopathy after being inoculated with PrP^{Sc}, which causes disease in normal mice. What do these results reveal about the relationship between PrP and TSE? What do they reveal about the cellular function of PrP?

3. Myoglobin is a protein that contains mostly α helices and no β strands (see Fig. 5-1). **(a)** Would you expect myoglobin to form amyloid fibers? **(b)** Under certain laboratory conditions, myoglobin can be induced to form amyloid fibers. What does this suggest about a polypeptide's ability to adopt β secondary structure?

4. Discuss the role of amino acid side chains in the formation of an amyloid fiber.

In some cases, a protein's conformational flexibility includes two stable alternatives in dynamic equilibrium. A change in cellular conditions, such as pH or oxidation state, or the presence of a binding partner can tip the balance toward one conformation or the other (Fig. 4-18). Because structural biologists have traditionally examined proteins with fixed, stable shapes, newer approaches to studying protein structure will most likely reveal more about intrinsically unstructured proteins and proteins with more than one conformational variant.

(a)

$H_3C-(CH_2)_{14}-\overset{\displaystyle O}{\overset{\displaystyle \|}{C}}-S-CH_2-\overset{\displaystyle NH}{\underset{\displaystyle C=O}{CH}}$

(b)

(c)

$^-O-\overset{\displaystyle O}{\underset{\displaystyle O}{\overset{\displaystyle \|}{P}}}-O-CH_2-\overset{\displaystyle NH}{\underset{\displaystyle C=O}{CH}}$

Figure 4-17 Some covalent modifications of proteins. (a) A 16-carbon fatty acid (palmitate, in red) is linked by a thioester bond to a Cys residue. **(b)** A chain of several carbohydrate units (here only one sugar residue is shown, in red) is linked to the amide N of an Asn side chain. **(c)** A phosphoryl group (red) is esterified to a Ser side chain.

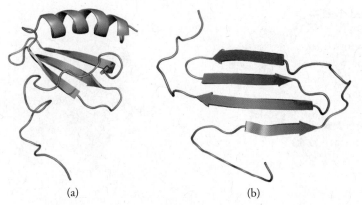

(a)   (b)

Figure 4-18 A protein with two stable conformations. One form of the protein lymphotactin consists of a three-stranded β sheet and an β helix (a). This form interconverts rapidly with an alternate form (b) with an all-β structure. [Structures (a, pdb 1J9O and b, pdb 2JP1) determined by B. F. Volkman.]

CONCEPT REVIEW

- How do the surface and core of a protein differ? Which residues are likely to be found in each region?
- What forces stabilize protein structure?
- How does an unfolded polypeptide reach its native conformation?

4-4 Quaternary Structure

Most proteins, especially those with molecular masses greater than 100 kD, consist of more than one polypeptide chain. The individual chains, called **subunits,** may all be identical, in which case the protein is known as a **homodimer, homotetramer,** and so on. If the chains are not all identical, the prefix *hetero-* is used. The spatial arrangement of these polypeptides is known as the protein's quaternary structure.

The forces that hold subunits together are similar to those that determine the tertiary structures of the individual polypeptides. The interface (the area of contact) between two subunits is often hydrophobic, with closely packed side chains. Hydrogen bonds, ion pairs, and disulfide bonds dictate the exact geometry of the interacting subunits.

Among the most common quaternary structures in proteins are symmetrical arrangements of two or more identical subunits (**Fig. 4-19**). Even in proteins with nonidentical subunits, the symmetry is based on groups of subunits. For example, hemoglobin, a heterotetramer with two αβ units, can be considered to be a dimer of dimers (see Fig. 4-19c). A few multimeric proteins are known to adopt more than one possible quaternary structure and apparently shift among the alternatives by dissociating, adopting different tertiary structures, and reassembling.

The advantages of multisubunit protein structure are many. For starters, the cell can construct extremely large proteins through the incremental addition of small building blocks (we will see examples of this in the next chapter). This is clearly an asset for certain structural proteins that—due to their enormous size—cannot be synthesized all in one piece or must be assembled outside the cell. Moreover, the impact of the inevitable errors in transcription and translation can be minimized if the affected polypeptide is small and readily replaced.

Finally, the interaction between subunits in a multisubunit protein affords an opportunity for the subunits to influence each other's behavior or work cooperatively. The result is a way of regulating function that is not possible in single-subunit proteins or in multisubunit proteins whose subunits each operate independently. In Chapter 5, we will examine the cooperative behavior of hemoglobin, which has four interacting oxygen-binding sites.

KEY CONCEPT

- Proteins containing more than one polypeptide chain have quaternary structure.

CONCEPT REVIEW

- What are the advantages of quaternary structure?

Figure 4-19 Some proteins with quaternary structure. The alpha carbon backbone of each polypeptide is shown. (a) Nitrite reductase, an enzyme with three identical subunits, from *Alcaligenes.* [Structure (pdb 1AS8) determined by M. E. P Murphy, E. T. Adams, and S. Turley.] (b) *E. coli* fumarase, a homotetrameric enzyme. [Structure (pdb 1FUQ) determined by T. Weaver and L. Banaszak.] (c) Human hemoglobin, a heterotetramer with two α subunits (blue) and 2 β subunits (red). [Structure (pdb 2HHB) determined by G. Fermi and M. F. Perutz.] (d) Bacterial methane hydroxylase, whose two halves (*right* and *left* in this image) each contain three kinds of subunits. [Structure (pdb 1MMO) determined by A. C. Rosenzweig, C. A. Frederick, S. J. Lippard, and P. Nordlund.]

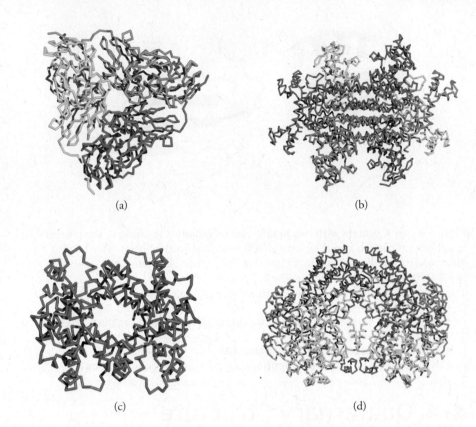

(a)

(b)

(c)

(d)

? **Which of the proteins shown in Fig. 4-1 has quaternary structure?**

4-5 Tools and Techniques: Analyzing Protein Structure

KEY CONCEPTS
- Chromatography is a technique for separating molecules on the basis of size, charge, or specific binding behavior.
- The sequence of amino acids in a polypeptide can be determined by mass spectrometry.
- The three-dimensional arrangement of atoms in a protein can be deduced by measuring the diffraction of X-rays or electrons or by analyzing nuclear magnetic resonance.

Like nucleic acids (Section 3-4), proteins can be purified and analyzed in the laboratory. In this section, we will examine some commonly used approaches to isolating proteins and determining their sequence of amino acids and their three-dimensional structures.

Chromatography takes advantage of a polypeptide's unique properties

As described in Section 4-1, a protein's amino acid sequence determines its overall chemical characteristics, including its size, shape, charge, and ability to interact with other substances. A variety of laboratory techniques have been devised to exploit these features in order to separate proteins from other cellular components. One of the most powerful techniques is **chromatography.** Originally performed with solvents moving across paper, chromatography now typically uses a column packed with a porous matrix (the stationary phase) and a buffered solution (the mobile phase) that percolates through the column. *Proteins or other solutes pass through the column at different rates, depending on how they interact with the stationary phase.*

In **size-exclusion chromatography** (also called gel filtration chromatography), the stationary phase consists of tiny beads with pores of a characteristic size. If a solution containing proteins of different sizes is applied to the top of the column, the proteins will move through the column as fluid drips out the bottom. Larger proteins will be excluded from the spaces inside the beads and will pass through the column faster than smaller proteins, which will spend time inside the beads. The proteins gradually become separated and can be recovered by collecting the solution that exits the column (Fig. 4-20).

(a)   (b)

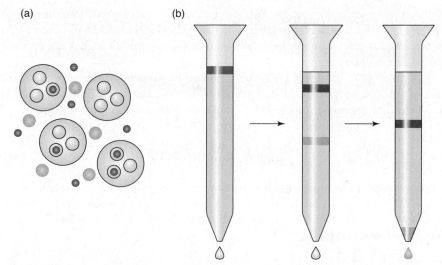

Figure 4-20 Size-exclusion chromatography. (a) Small molecules (blue) can enter the spaces inside the porous beads of the stationary phase, while larger molecules (gold) are excluded. (b) When a mixture of proteins (green) is applied to the top of a size-exclusion column, the large proteins (gold) migrate more quickly than small proteins (blue) through the column and are recovered by collecting the material that flows out the bottom of the column. In this way, a mixture of proteins can be separated according to size. ⊕ **See Animated Figure. Gel filtration chromatography.**

A protein's net charge at a particular pH can be exploited for its purification by **ion exchange chromatography.** In this technique, the solid phase typically consists of beads derivatized with positively charged diethylaminoethyl (DEAE) groups or negatively charged carboxymethyl (CM) groups:

$$\textbf{DEAE} \quad \bullet \!-\! CH_2\!-\!CH_2\!-\!NH^+ \genfrac{}{}{0pt}{}{CH_2-CH_3}{CH_2-CH_3}$$

$$\textbf{CM} \quad \bullet \!-\! CH_2\!-\!COO^-$$

Negatively charged proteins will bind tightly to the DEAE groups, while uncharged and positively charged proteins pass through the column. The bound proteins can then be dislodged by passing a high-salt solution through the column so that the dissolved ions can compete with the protein molecules for binding to DEAE groups **(Fig. 4-21).** Alternatively, the pH of the solvent can be decreased

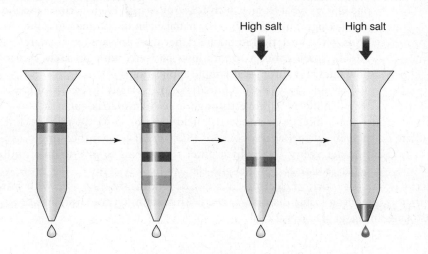

High salt   High salt

Figure 4-21 Ion exchange chromatography. When a mixture of proteins (brown) is applied to the top of a positively charged anion exchange column (e.g., a DEAE matrix), negatively charged proteins (red) bind to the matrix, while uncharged and cationic proteins (orange and purple) flow through the column. The desired protein can be dislodged by applying a high-salt solution (whose anions compete with the protein for binding to the DEAE groups). In cation exchange chromatography, negatively charged proteins bind to an anionic matrix (e.g., one containing carboxymethyl groups). ⊕ **See Animated Figure. Ion exchange chromatography.**

so that the bound protein's anionic groups become protonated, loosening their hold on the DEAE matrix. Similarly, positively charged proteins will bind to CM groups and can subsequently be dislodged by solutions with a higher salt concentration or a higher pH.

The success of ion exchange can be enhanced by knowing something about the protein's net charge (see Sample Calculation 4-1) or its **isoelectric point, pI,** the pH at which it carries no net charge. For a molecule with two ionizable groups, the pI lies between the pK values of those two groups:

$$pI = {}^1\!/_2\,(pK_1 + pK_2) \qquad \qquad \textbf{[4-1]}$$

Calculating the pI of an amino acid is relatively straightforward (Sample Calculation 4-2). However, a protein may contain many ionizable groups, so although its pI can be estimated from its amino acid composition, its pI is more accurately determined experimentally.

●●● SAMPLE CALCULATION 4-2

PROBLEM	Estimate the isoelectric point of arginine.
SOLUTION	In order for arginine to have no net charge, its α-carboxyl group must be unprotonated (negatively charged), its α-amino group must be unprotonated (neutral), and its side chain must be protonated (positively charged). Because protonation of the α-amino group or deprotonation of the side chain would change the amino acid's net charge, the pK values of these groups (9.0 and 12.5) should be used with Equation 4-1:

$$pI = {}^1\!/_2\,(9.0 + 12.5) = 10.75$$

●●● PRACTICE PROBLEMS

4. Estimate the pI of alanine.
5. Estimate the pI of glutamate.

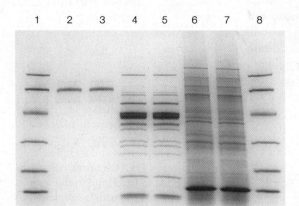

Figure 4-22 SDS-PAGE. Following electrophoresis, the gel was stained with Coomassie blue dye so that each protein is visible as a blue band. Lanes 1 and 8 contain proteins of known molecular mass that serve as standards for estimating the masses of other proteins. [Courtesy of Bio-Rad Laboratories, ©2012.]

Other binding behaviors can be adapted for chromatographic separations. For example, a small molecule can be immobilized on the chromatographic matrix, and proteins that can specifically bind to that molecule will stick to the column while other substances exit the column without binding. This technique, called **affinity chromatography,** is a particularly powerful separation method because its takes advantage of a protein's unique ability to interact with another molecule, rather than one of its general features such as size or charge. **High-performance liquid chromatography (HPLC)** is the name given to chromatographic separations, often analytical in nature rather than preparative, that are carried out in closed columns under high pressure, with precisely controlled flow rates and automated sample application.

Proteins are sometimes analyzed or isolated by electrophoresis, in which molecules move through a gel-like matrix such as polyacrylamide under the influence of an electric field (Section 3-4). In sodium dodecyl sulfate polyacrylamide gel electrophoresis **(SDS-PAGE),** both the sample and the gel contain the detergent SDS, which binds to proteins to give them a uniform density of negative charges. When the electric field is applied, the proteins all move toward the positive electrode at a rate depending on their size, with smaller proteins migrating faster than larger ones. After staining, the proteins are visible as bands in the gel **(Fig. 4-22).**

Mass spectrometry reveals amino acid sequences

A standard approach to sequencing a protein has several steps:

1. *The sample of protein to be sequenced is purified* (for example, by chromatographic or other methods) so that it is free of other proteins.
2. If the protein contains more than one kind of polypeptide chain, *the chains are separated* so that each can be individually sequenced. In some cases, this requires breaking (reducing) disulfide bonds.
3. *Large polypeptides must be broken into smaller pieces (<100 residues) that can be individually sequenced.* Cleavage can be accomplished chemically, for example, by treating the polypeptide with cyanogen bromide (CNBr), which cleaves the peptide bond on the C-terminal side of Met residues. Cleavage can also be accomplished with **proteases** (another term for peptidases) that hydrolyze specific peptide bonds. For example, the protease trypsin cleaves the peptide bond on the C-terminal side of the positively charged residues Arg and Lys:

⊕ **See Guided Exploration. Protein sequence determination.**

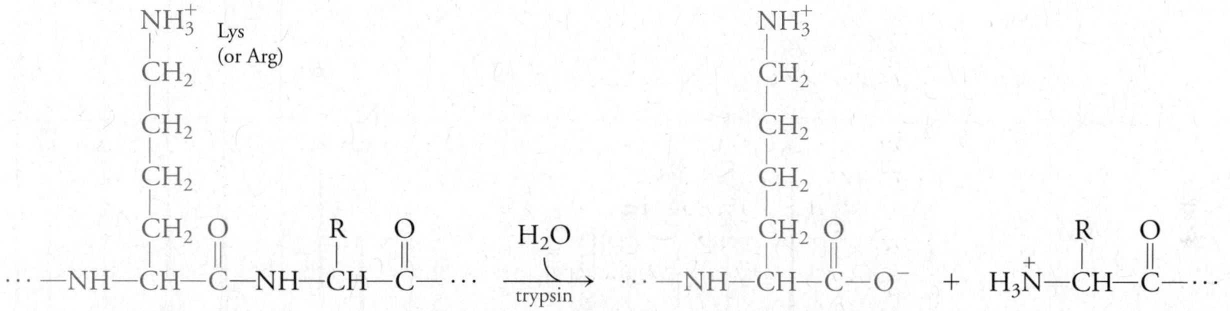

Some commonly used proteases and their preferred cleavage sites are listed in Table 4-4.
4. Each peptide is sequenced. In a classic procedure known as ***Edman degradation,*** *the N-terminal residue of a peptide is chemically derivatized, cleaved off, and identified; the process is then repeated over and over so that the peptide's sequence can be deduced, one residue at a time.* Alternatively, each peptide can be sequenced by mass spectrometry.
5. To reconstruct the sequence of the intact polypeptide, *a different set of fragments that overlaps the first is generated* so that the two sets of sequenced fragments can be aligned.

[**TABLE 4-4**]

Specificities of Some Proteases

Protease	Residue Preceding Cleaved Peptide Bond[a]
Chymotrypsin	Phe, Trp, Tyr
Elastase	Ala, Gly, Ser, Val
Thermolysin	Ile, Met, Phe, Trp, Tyr, Val
Trypsin	Arg, Lys

[a]Cleavage does not occur if the following residue is Pro.

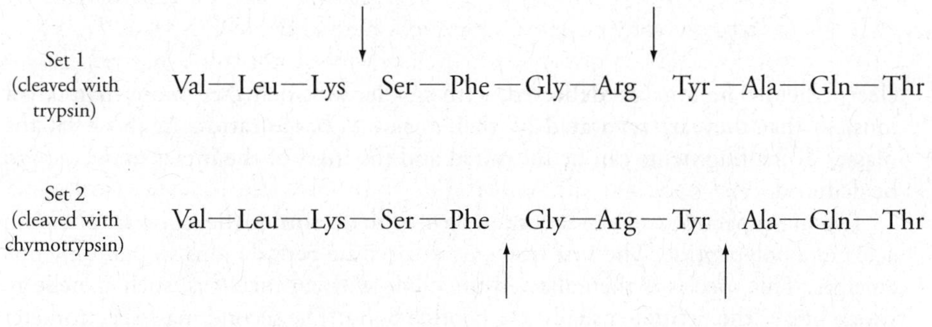

A more efficient approach for analyzing protein structure measures the sizes of peptide fragments by **mass spectrometry.** In standard mass spectrometry, a solution of the protein (or another macromolecule) is sprayed from a tiny nozzle at high voltage. This yields droplets of positively charged molecular fragments, from which the solvent quickly evaporates. Each gas-phase ion then passes through an

⊕ **See Animated Figure. Generating overlapping peptides to determine amino acid sequence.**

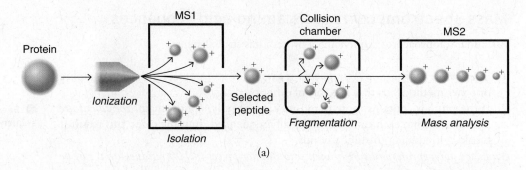

(a)

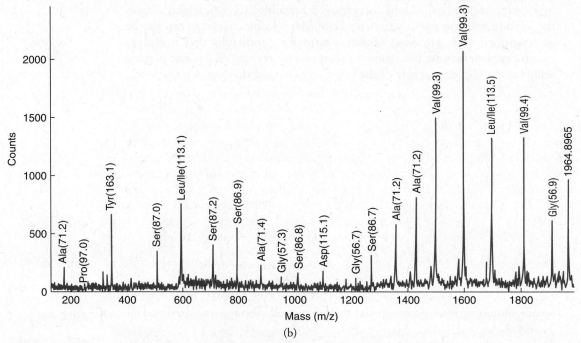

(b)

Figure 4-23 Peptide sequencing by mass spectrometry. (a) A solution of charged peptides is sprayed into the first mass spectrometer (MS1). One peptide ion is selected to enter the collision chamber to be fragmented. The second mass spectrometer (MS2) then measures the mass-to-charge ratio of the ionic peptide fragments. The peptide sequence is determined by comparing the masses of increasingly larger fragments. (b) An example of peptide sequencing by mass spectrometry. The difference in mass of each successive peak identifies each residue, allowing the amino acid sequence to be read from right to left. [From Keough, T., *et al.*, *Proc. Nat. Acad. Sci.* 96, 7131–7136 (1999). Reprinted with permission of PNAS.]

electric field. The ions are deflected, with smaller ions deflected more than larger ions, so that they are separated by their mass-to-charge ratios. In this way, the masses of the fragments can be measured and the mass of the intact molecule can be deduced.

Two mass spectrometers in series can be used to determine the sequence of amino acids in a polypeptide. The first instrument sorts the peptide ions so that only one emerges. This species is then allowed to collide with an inert gas, such as helium, which breaks the peptide, usually at a peptide bond. The second mass spectrometer then measures the mass-to-charge ratios of the peptide fragments (**Fig. 4-23**). *Because successively larger fragments all bear the same charge but differ by one amino acid, and because the mass of each of the 20 amino acids is known, the sequence of amino acids in a set of fragments can be deduced.* Although mass spectrometry is not practical for sequencing large polypeptides, even a partial sequence may be valuable (Box 4-D). Mass spectrometry can also reveal additional details of protein structure, such as covalently modified amino acid residues.

Mass Spectrometry Applications

Mass spectrometry has been used for several decades in clinical and forensics laboratories to identify normal metabolic compounds as well as toxins and drugs (both therapeutic and illicit). Instruments that detect traces of explosives at airport security checkpoints also use mass spectrometry, which is rapid, sensitive, and reliable. The analysis of small compounds by mass spectrometry is relatively straightforward. It is much more challenging, however, to analyze large molecules in complex mixtures for the purpose of diagnosing a disease such as cancer or tracking its effects on human tissues.

Fluids such as blood contain so many different proteins, with concentrations ranging over many order of magnitude, that it is difficult to detect rare proteins against a backdrop of very abundant ones such as serum albumin and immunoglobulins, which together account for about 75% of the plasma proteins. Urinalysis by mass spectrometry holds more promise, as urine normally contains relatively few proteins and none with a mass greater than about 15,000 D. Even so, over 2000 different proteins are detectable in urine.

One common approach for identifying proteins in biological samples is to fractionate the mixture by electrophoresis, then extract the separated proteins from the gel, partially digest them with a protease, and subject them to mass spectrometry. The resulting pattern of peaks, a "fingerprint," can be compared to a database to identify the protein. Even when tandem mass spectrometers are used, it may not be necessary to completely sequence each polypeptide. A partial sequence of just a few amino acids is often sufficient to identify the parent protein. Of course, the availability of complete genomic sequences makes this approach possible, and a number of software programs have been developed to translate mass spectral data into query sequences for searching sequence databases.

⬤ **Question:** Explain why it is possible that a 5-residue sequence can uniquely identify a protein whose total length is 200 residues. (*Hint*: compare the total number of possible pentapeptide sequences to the actual number of pentapeptide fragments in the polypeptide.)

Protein structures are determined by X-ray crystallography, electron crystallography, and NMR spectroscopy

Most proteins are too small to be directly visualized, even by electron microscopy, but their atomic structures are accessible to high-energy probes in the form of X-rays. **X-Ray crystallography** is performed on samples of protein that have been induced to form crystals. A protein preparation must be exceptionally pure in order to crystallize without imperfections. A protein crystal, often no more than 0.5 mm in diameter, usually contains 40–70% water by volume and is therefore more gel-like than solid (**Fig. 4-24**).

When bombarded with a narrow beam of X-rays, the electrons of the atoms in the crystal scatter the X-rays, which reinforce and cancel each other to produce a **diffraction pattern** of light and dark spots that can be captured electronically or on a piece of X-ray film (**Fig. 4-25**).

Mathematical analysis of the intensities and positions of the diffracted X-rays yields a three-dimensional map of electron density in the crystallized molecule. The level of detail of the image depends in part on the quality of the crystal. Slight conformational variations among the crystallized protein molecules often limit resolution to about 2 Å. However, this is usually sufficient to trace the polypeptide backbone and discern the general shapes of the side chains (**Fig. 4-26**). When Kendrew determined the X-ray structure of myoglobin, its amino acid sequence was not known. Today, protein crystallographers usually take advantage of amino acid sequence information to simplify the task of elucidating protein structures.

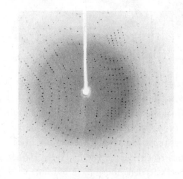

Figure 4-24 Crystals of the protein streptavidin. [Courtesy I. Le Trong and R. E. Stenkamp, University of Washington.]

Figure 4-25 An X-ray diffraction pattern. [Courtesy Isolde Le Trong, David Teller, and Ronald Stenkamp, University of Washington.]

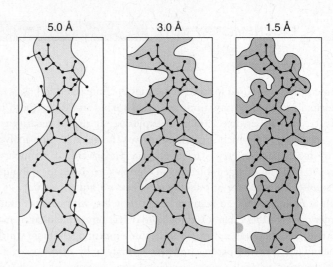

Figure 4-26 Protein structure at different resolutions. In this example, the green areas represent electron density, and the protein structure is superimposed in black. As the resolution improves, it becomes easier to trace the pattern of the protein backbone (a portion of an α helix). [Based on a drawing by Wayne Hendrickson.]

Are X-ray structures accurate? One might expect a crystallized molecule to be utterly immobile, quite different from a protein in solution, which undergoes bending and stretching movements among its many bonds. But, in fact, crystallized proteins retain some of their ability to move. They can sometimes bind small molecules that diffuse into the crystal (which is about half water) and can sometimes mediate chemical reactions. These activities would not be possible if the structure of the crystallized protein did not closely approximate the structure of the protein in solution. Finally, nuclear magnetic resonance (NMR) methods (see below) for determining the structures of small proteins in solution appear to yield results consistent with X-ray crystallographic data.

Proteins that are difficult to crystallize, such as membrane proteins with extensive hydrophobic regions, can sometimes be analyzed by **electron crystallography,** a technique in which electron beams rather than X-rays probe the protein's structure. Protein samples, which do not have to be in a crystalline array, are placed in an electron microscope, and diffraction information is collected from many different angles. In this way, the three-dimensional structure of the protein can be reconstructed. Because electrons interact more strongly with atoms than X-rays do, the sample is susceptible to radiation damage. This can be minimized by rapidly freezing the sample and collecting data at the temperature of liquid nitrogen (–196°C), a method called **cryoelectron microscopy.** The structures of some macromolecular complexes, including the ribosome (Section 22-2), have been visualized using cryoelectron microscopy.

Proteins in solution can be analyzed by **nuclear magnetic resonance (NMR) spectroscopy,** which takes advantage of the ability of atomic nuclei (most commonly, hydrogen) to resonate in an applied magnetic field according to their interactions with nearby atoms. An NMR spectrum consists of numerous peaks that can be analyzed to reveal the distances between two H atoms that are close together in space or are covalently connected through one or two other atoms. These measurements, along with information about the protein's amino acid sequence, are used to construct a three-dimensional model of the protein. Due to the inherent imprecision of the measurements, NMR spectroscopy typically yields a set of closely related structures, which may convey a sense of the protein's natural conformational flexibility (**Fig. 4-27**).

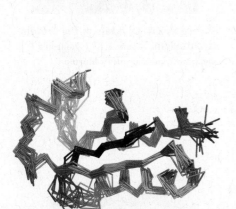

Figure 4-27 The NMR structure of glutaredoxin. Twenty structures for this 85-residue protein, all compatible with the NMR data, are shown as α-carbon traces, colored from the N-terminus (blue) to the C-terminus (red). [Structure (pdb 1EGR) determined by P. Sodano, T. H. Xia, J. H. Bushweller, O. Bjornberg, A. Holmgren, M. Billeter, and K. Wuthrich.]

CONCEPT REVIEW
- What types of chromatography would separate proteins of different size? Of different charge?
- Summarize the strategy used to determine the sequence of a large protein.
- Compare the physical state of the protein to be analyzed by X-ray crystallography, electron crystallography, and NMR spectroscopy.

[SUMMARY]

4-1 Proteins Are Chains of Amino Acids

- The 20 amino acid constituents of proteins are differentiated by the chemical properties of their side chains, which can be roughly classified as hydrophobic, polar, or charged.
- Amino acids are linked by peptide bonds to form a polypeptide.

4-2 Secondary Structure: The Conformation of the Peptide Group

- Protein secondary structure includes the α helix and β sheet, in which hydrogen bonds form between backbone carbonyl and amide groups. Irregular secondary structure has no regularly repeating conformation.

4-3 Tertiary Structure and Protein Stability

- The three-dimensional shape (tertiary structure) of a protein includes its backbone and all side chains. A protein may contain all α, all β, or a mix of α and β structures.
- A globular protein has a hydrophobic core and is stabilized primarily by the hydrophobic effect. Ion pairing, disulfide bonds, and other cross-links may also help stabilize proteins.

- A denatured protein may refold to achieve its native structure. In a cell, chaperones assist protein folding.

4-4 Quaternary Structure

- Proteins with quaternary structure have multiple subunits.

4-5 Tools and Techniques: Analyzing Protein Structure

- In the laboratory, proteins can be purified by chromatographic techniques that take advantage of the proteins' size, charge, and ability to bind other molecules.
- The sequence of amino acids in a polypeptide can be determined by chemically derivatizing and removing them in Edman degradation or by measuring the mass-to-charge ratio of peptide fragments in mass spectrometry.
- X-Ray crystallography, electron crystallography, and NMR spectroscopy provide information about the three-dimensional of proteins at the atomic level.

[GLOSSARY TERMS]

protein	tertiary structure	prion
polypeptide	quaternary structure	intrinsically unstructured protein
amino acid	α helix	subunit
R group	β sheet	homo-
α-amino acid	parallel β sheet	hetero-
Cα	antiparallel β sheet	chromatography
chirality	regular secondary structure	size-exclusion chromatography
disulfide bond	irregular secondary structure	ion exchange chromatography
condensation reaction	globular protein	pI
peptide bond	fibrous protein	affinity chromatography
residue	domain	HPLC
hydrolysis	ion pair	SDS-PAGE
exopeptidase	zinc finger	protease
endopeptidase	denaturation	Edman degradation
N-terminus	renaturation	mass spectrometry
C-terminus	native structure	X-ray crystallography
backbone	molecular chaperone	diffraction pattern
microenvironment	processing	electron crystallography
oligopeptide	amyloid deposits	cryoelectron microscopy
peptide	transmissible spongiform	NMR spectroscopy
primary structure	encephalopathy	
secondary structure	(TSE)	

....

BIOINFORMATICS ▶ **PROJECT 3**   Learn to use Jmol or PyMOL to view and manipulate structures obtained from the Protein Data Bank.

VISUALIZING THREE-DIMENSIONAL PROTEIN STRUCTURES

....

[PROBLEMS]

4-1 Proteins Are Chains of Amino Acids

1. Amino acids (except for glycine) are chiral; that is, they contain a chiral carbon and thus can form mirror-image isomers, or enantiomers. Biochemists use D and L to distinguish enantiomers instead of the *RS* system used by chemists.

(a) The structure of L-alanine is shown. Label the chiral carbon and draw the structure of D-alanine.

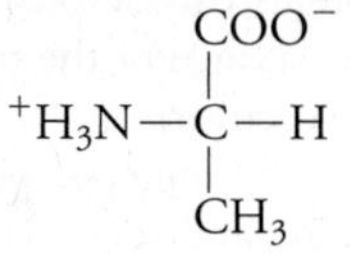

$$\begin{array}{c} COO^- \\ | \\ {}^+H_3N-C-H \\ | \\ CH_3 \end{array}$$

L-Alanine

(b) The vast majority of proteins in living systems consist of L amino acids. However, D amino acids are found in some short bacterial peptides that make up cell walls. Why might this be advantageous to the bacteria?

2. How many chiral carbons does threonine have? How many stereoisomers are possible? Draw the stereoisomers of threonine.

3. Which of the 20 standard amino acids are
(a) cyclic
(b) aromatic
(c) sometimes charged at physiological pH
(d) technically not hydrophobic, polar, or charged
(e) basic
(f) acidic
(g) sulfur-containing

4. The side chains of asparagine and glutamine can undergo hydrolysis in aqueous acid. Draw the reactants and products for the deamination reaction. Name the products.

5. Rank the solubility of the following amino acids in water at pH 7: Trp, Arg, Ser, Val, Thr.

6. What is more soluble, a solution of free histidine or a solution of the histidine tetrapeptide described in Problem 11?

7. At what pH would an amino acid bear both a COOH and an NH_2 group?

8. Most amino acids have melting points around 300°C. They are soluble in water but insoluble in nonpolar organic solvents. How do these observations support your answer to Problem 7?

9. The pK values of the amino and carboxyl groups in free amino acids differ from the pK values of the N- and C-termini of polypeptides. Explain.

10. Histones are basic proteins that bind to DNA. What amino acids are found in abundance in histones and why? What important intermolecular interactions form between histones and DNA?

11. Estimate the net charge of a His–His–His–His tetrapeptide at pH 6.0.

12. Certain bacteria synthesize cyclic tetrapeptides.
(a) Estimate the net charge at pH 7.0 of such a compound that consists of two Pro and two Tyr residues.
(b) If the peptide were linear rather than cyclic, what would be its net charge?

13. Biochemists sometimes link a recombinant protein to a protein known as green fluorescent protein (GFP), which was first purified from bioluminescent jellyfish. The fluorophore in GFP (shown below) is a derivative of three consecutive amino acids that undergo cyclization of the polypeptide chain and an oxidation. Identify
(a) the three residues and the bonds that result from the
(b) cyclization and **(c)** oxidation reactions.

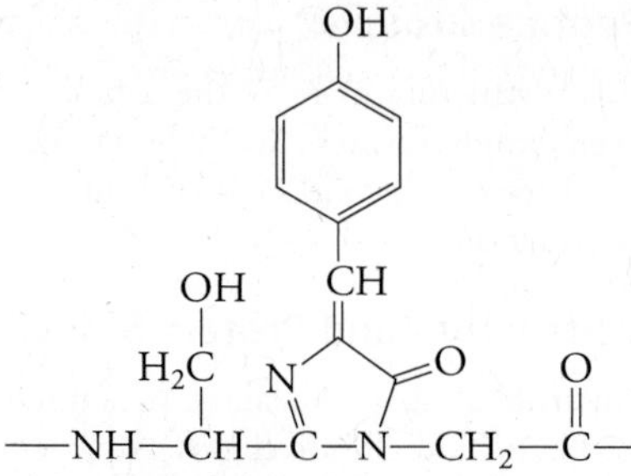

14. Draw the structure of the dipeptide Lys–Glu at pH 7.0. Label the following:
(a) peptide bond, **(b)** N-terminus, **(c)** C-terminus, **(d)** an α-amino group and an ε-amino group, **(e)** an α-carboxylate group and a γ-carboxylate group.

15. The artificial non-nutritive sweetener aspartame is a dipeptide with the sequence Asp–Phe. The carboxyl terminus is methylated. Draw the structure of aspartame at pH 7.0.

16. The zebrafish is a good animal model to use to test the actions of various drugs. The zebrafish produces a novel heptapeptide that binds to opiate receptors in the brain. The sequence of the peptide is Tyr–Gly–Gly–Phe–Met–Gly–Tyr. Draw the structure of the heptapeptide at pH 7.

17. Glutathione is a Cys-containing tripeptide found in red blood cells that functions to remove hydrogen peroxide and organic peroxides that can irreversibly damage hemoglobin and cell membranes.
(a) The sequence of glutathione is γ-Glu–Cys–Gly. The γ-Glu indicates that the first peptide bond forms between the γ-carboxyl group of the Glu side chain and the α-amino group of Cys. Draw the structure of glutathione.
(b) Glutathione is sometimes abbreviated as GSH to show the importance of the Cys side chain in reactions. For example, two molecules of GSH react with an organic peroxide as shown below. The glutathione is oxidized to GSSG, and the organic peroxide is reduced to a less harmful alcohol. Draw the structure of GSSG.

$$2\ GSH + R-O-O-H \xrightarrow[\text{peroxidase}]{\text{glutathione}} GSSG + ROH + H_2O$$

18. Why are the terms *protein* and *polypeptide* not interchangeable?

19. A fusion protein can be synthesized in which the protein carries a specific tag. These tags are useful in identifying the proteins in various types of experiments. Fusion proteins with the FLAG epitope bind to specific anti-FLAG antibodies and carry this extra sequence: Asp–Tyr–Lys–Asp–Asp–Asp–Asp–Lys. Draw the structure of the FLAG epitope.

20. The biotech company that sells anti-FLAG epitopes provides the information that fusion proteins carrying the FLAG epitope display the sequence on the surface of the protein (see Problem 19). Examine the sequence and explain why the company can make this claim.

21. You have isolated a tripeptide containing Arg, His, and Pro. How many different sequences are possible for this tripeptide?

22. You have isolated a tetrapeptide with an unknown sequence, and after hydrolyzing its peptide bonds, you recover Ala, Glu, Lys, and Thr. How many different sequences are possible for this tetrapeptide?

23. Read the following sentences and identify each statement as describing the primary, secondary, tertiary, or quaternary structure of the protein:
(a) The shape of myoglobin is roughly spherical.
(b) Hemoglobin consists of four polypeptide chains.
(c) About one-third of the amino acid residues in collagen are glycines.
(d) The lysozyme molecule contains regions of helical structure.

24. Which level of protein architecture is discussed in Problem 13?

4-2 Secondary Structure: The Conformation of the Peptide Group

25. Draw the structure of the peptide group and draw hydrogen bonds that form between functional groups of the peptide group and water.

26. The structure of the peptide bond is drawn in Section 4.1 in the *trans* configuration in which the carbonyl group and the amide hydrogen are on opposite sides of the peptide bond. Draw the structure of the peptide bond in the *cis* configuration. Which configuration is more likely to be found in proteins and why?

27. Compare and contrast the structures of the DNA helix and the α helix.

28. The α helix is stabilized by hydrogen bonds that are directed along the length of the helix. The peptide carbonyl group of the *n*th residue forms a hydrogen bond with the peptide —NH group of the (*n* + 4)th residue. Draw the structure of this hydrogen bond.

29. Proline is known as a helix disrupter; it sometimes appears at the beginning or the end of an α helix but never in the middle. Explain.

30. Interestingly, glycine, like proline (see Problem 29), is not usually found in an α helix, but for the opposite reason. Explain why glycine residues are not usually found in α helices.

31. In 1967, Schiffer and Edmundson developed a tool called a helical wheel that is still widely used today. A helical wheel is used to visualize an α helix in which the angle of rotation between two consecutive amino acid residues is 100°. Because the α helix consists of 3.6 residues per turn, the pattern repeats after five turns and 18 residues. In the final representation the view is down the helical axis.

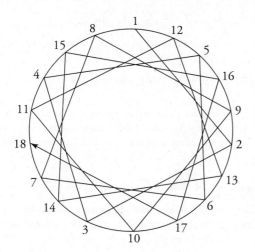

The sequence of a domain of the gp160 protein (found in the envelope of HIV) is shown below, using one-letter codes for the amino acids. Plot this sequence on the helical wheel. What do you notice about the types of amino acid residues on either side of the wheel?

MRVKEKYQHLWRWGWRWG

32. A 24-residue peptide called Pandinin 2 was isolated from scorpion venom and was found to have both antimicrobial and hemolytic properties. The sequence of the first 18 residues of this peptide is shown below. **(a)** Use the helical wheel shown in Problem 31 to plot the sequence of this peptide. **(b)** Propose a hypothesis that explains why the peptide has the ability to lyse bacteria and red blood cells.

FWGALAKGALKLIPSLFS

4-3 Tertiary Structure and Protein Stability

33. Examine Figure 4-8 (triose phosphate isomerase). To which of the CATH classes does this protein belong?

34. Choose the amino acid in the following pairs that would be more likely to appear on the solvent-exposed surface of a protein:
(a) Lys or Leu
(b) Ser or Ala
(c) Phe or Tyr
(d) Trp or Gln
(e) Asn or Ile

35. In a receptor protein, a Glu residue is altered to an Ala residue in a site-directed mutagenesis experiment. Binding of the ligand to this receptor decreases 100-fold. What does this experiment tell you about the interaction between the ligand and the receptor?

36. Scientists studying the enzyme PEP carboxylase from corn conducted site-directed mutagenesis studies in which they synthesized three different mutant forms of the enzyme. For each mutant, a lysine was converted to a different amino acid—an Asn, a Glu, or an Arg. Which substitution would you expect would have the least effect on enzymatic activity? Which substitution would have the greatest effect? Explain.

37. Draw two amino acid side chains that can interact with each other via the following intermolecular interactions:
(a) ion pair
(b) hydrogen bond
(c) van der Waals interaction (London dispersion forces)

38. A type of muscular dystrophy, called severe childhood autosomal recessive muscular dystrophy (SCARMD), results from a mutation in the gene for a 50-kD muscle protein. The defective protein leads to muscle necrosis. Detailed studies of this protein have revealed that an arginine residue at position 98 has been mutated to a histidine. Why might replacing an arginine with histidine result in a defective protein?

39. Laboratory techniques for randomly linking together amino acids typically generate an insoluble polypeptide, yet a naturally occurring polypeptide of the same length is usually soluble. Explain.

40. Proteins can be unfolded, or denatured, by agents that alter the balance of weak noncovalent forces that maintain the native conformation. How would the following agents cause a protein to denature? Be specific about the type of intermolecular forces that would be affected.
(a) heat
(b) pH

(c) amphiphilic detergents

(d) reducing agents such as 2-mercaptoethanol (HSCH$_2$CH$_2$OH)

41. In 1957, Christian Anfinsen carried out a denaturation experiment with ribonuclease. Ribonuclease, a pancreatic enzyme that catalyzes the digestion of RNA, consists of a single chain of 124 amino acids cross-linked by four disulfide bonds. Urea and 2-mercaptoethanol were added to a solution of ribonuclease, which caused it to unfold, or denature. The loss of tertiary structure resulted in a loss of biological activity. When the denaturing agent (urea) and the reducing agent (mercaptoethanol) were simultaneously removed, the ribonuclease spontaneously folded back up to its native conformation and regained full enzymatic activity in a process called renaturation. What is the significance of this experiment?

42. Insulin consists of two chains, a shorter A chain and a longer B chain. The two chains are held together with disulfide bonds. *In vivo*, insulin is processed from proinsulin, a single polypeptide chain. The C chain is removed from proinsulin to form insulin.

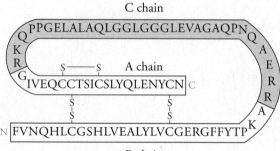

(a) A denaturation/renaturation experiment similar to the one carried out by Anfinsen with ribonuclease (see Problem 41) was carried out using insulin. However, in contrast to Anfinsen's results, approximately 2–4% of the activity of the native protein was recovered when the urea and 2-mercaptoethanol were removed by dialysis (this is the level of activity to be expected if the disulfide bridges formed randomly). When the experiment was repeated with proinsulin, 60% of the activity was restored upon renaturation. Explain these observations.

(b) The investigators noted that the refolding of proinsulin depended on pH. For example, when the proinsulin was incubated in pH 7.5 buffer, only 10% of the proinsulin was renatured. But at pH 10.5, 60% of the proinsulin was renatured. Explain these observations. (*Hint:* The pK value of the cysteine side chain in proinsulin is similar to the pK value of the sulfhydryl group in free cysteine.)

43. In the mid-1980s, scientists noted that if cells were incubated at 42°C instead of the normal 37°C, the synthesis of a group of proteins dramatically increased. For lack of a better name, the scientists called these heat-shock proteins. It was later determined that the heat-shock proteins were chaperones. Why do you think that the cell would increase its synthesis of chaperones when the temperature is increased?

44. A protein engineering laboratory studying monoclonal antibody proteins characterized the thermal stability of these proteins by measuring their melting temperature (T_m), which is defined as the temperature at which the proteins are half unfolded. The investigators found that there was a positive correlation between T_m and the proteins' —SH molar ratio. In other words, proteins with a higher content of —SH were more thermally stable. Explain the reason for this observation.

45. Spectrin is a protein found in the cytoskeleton of the red blood cell. The cytoskeleton consists of proteins anchored to the cytosolic surface of the membrane and gives the cell the strength and flexibility to squeeze through capillaries. The spectrin protein chains consist of repeating segments that form α-helical bundles. Recently, a mutant spectrin was isolated in which a glutamine was replaced with proline. What would be the effect of this mutation on the spectrin protein? What are the consequences for the red blood cell?

46. During evolution, why do insertions, deletions, and substitutions of amino acids occur more often in loops than in elements of regular secondary structure?

4-4 Quaternary Structure

47. The restriction endonucleases EcoRI and EcoRV are dimeric (two-subunit) enzymes (see Section 3-4). Based on how these proteins interact with DNA, do you expect them to be homo- or heterodimeric?

48. A protein with two identical subunits can often be rotated 180° (halfway) around its axis so as to generate an identical structure; such a protein is said to have rotational symmetry. Why is it not possible for a protein to have mirror symmetry (that is, its halves would be related as if reflected in a mirror)?

49. Glutathione transferase consists of a dimer of two identical subunits. The dimer is in equilibrium with its constituent monomers. Site-directed mutagenesis studies were carried out on this protein in which two arginine residues were mutated to glutamines and two aspartates were mutated to asparagines. These substitutions caused the equilibrium to shift in favor of the monomeric form of the enzyme. Where were the arginines and aspartates likely to be found on the protein and what is their role in stabilizing the dimeric form of the enzyme?

50. A tetrameric protein dissociates into dimers when the detergent sodium dodecyl sulfate (SDS) is added to a solution of the protein. But the dimers are termed SDS-resistant because they do not further dissociate into monomers in the presence of the detergent. What intermolecular forces might be acting at the dimer–dimer interface? Are the intermolecular forces acting at the monomer–monomer interface different? Explain.

4-5 Tools and Techniques: Analyzing Protein Structure

51. Estimate the pI of a Ser–Ile dipeptide.

52. Estimate the pI of a Gly–Tyr–Val tripeptide.

53. A certain protein has a pI of 4.3. What types of amino acids are likely to be relatively abundant in this protein?

54. What can you conclude about the net charge of any protein at a pH less than 3.5?

55. In your laboratory, you plan to use ion exchange chromatography to separate the peptide shown below (using one-letter codes) from a mixture of different peptides at pH 7.0. Should you choose a matrix containing DEAE or CM groups?

Peptide: GLEKSLVRLGDVQPSLGKESRAKKFQRQ

56. The protein allergen from peanuts, a protein called Ara h8, was recently purified and characterized. The investigators initially had difficulty separating Ara h8 from a similar protein in peanuts called Ara h6 because the two proteins were of similar size and had nearly identical pI values. Separation of the two proteins was finally achieved when it was noted that the Ara h6 protein contained 10 cysteine residues involved in disulfide bridges, whereas Ara h8 contained no cysteines. The protein mixture was treated with a reducing agent, dithiothreitol (DTT), and then treated with iodoacetic acid (ICH_2COOH, a reagent that adds to, or alkylates, an —SH group and releases free iodine). The mixture was then loaded onto an anion exchange column and the two proteins were successfully separated.

(a) Show the structural changes that occur when Cys residues are exposed to DTT followed by iodoacetic acid.

(b) Draw a plausible elution profile (protein concentration versus solvent volume) for the separation of the proteins by anion exchange chromatography.

(c) Explain why this treatment resulted in successful separation of the two proteins.

57. The sequence of kassinin, a tachykinin dodecapeptide from the African frog *Kassina senegalensis*, was determined. A single round of Edman degradation identifies Asp as the N-terminus. A second sample of the peptide is treated with chymotrypsin. Two fragments are released with the following amino acid compositions: fragment I (Gly, Leu, Met, Val) and fragment II (Asp2, Gln, Lys, Phe, Pro, Ser, Val). Next, a third sample of peptide is treated with trypsin, which results in two fragments with the following amino acid compositions: fragment III (Asp, Pro, Lys, Val) and fragment IV (Asp, Gln, Gly, Leu, Met, Phe, Ser, Val). Treatment of another sample with elastase yields a single Gly residue and three fragments—fragment V (Leu, Met), fragment VI (Asp, Lys, Pro, Ser, Val), and fragment VII, which was sequenced: Asp–Gln–Phe–Val. A fourth sample was treated with CNBr, but the dodecapeptide was not cleaved. What is the sequence of the dodecapeptide?

58. The ubiquitin protein from the malaria parasite was treated with chymotrypsin, and the resulting fragments were sequenced. A second sample of the polypeptide was treated with trypsin, and the fragments were sequenced. What is the sequence of the polypeptide?

Chymotrypsin fragments	Trypsin fragments
AGKQLEDGRTLSDY	LR
IPPDQQRLIF	AK
VKTLTGKTITLDVEPSDTIEN- VKAKIQDKEGI	EGI
NIQKESTLHLVLRLRGGMQIF	IQDK
	LIFAGK
	QLEDGR
	TLTGK
	IPPDQQR
	GGMQIFVK
	TLSDYNIQK
	ESTLHLVLR
	TITLDVEPSDTIENVK

59. Before sequencing, a protein whose two identical polypeptide chains are linked by a disulfide bond must be reduced and alkylated (to chemically block them). Why should reduction and alkylation also be performed for a single polypeptide chain that includes an intramolecular disulfide bond?

60. You must cleave the following peptide into smaller fragments. Which of the proteases listed in Table 4-4 would theoretically yield the most fragments? The fewest?

NMTQGRCKPVNTFVHEPLVDVQNVCFKE

61. Seedlings use seed storage proteins as an important nitrogen source during germination. The amino acid sequence of a seed storage protein named BN was determined in the following manner. The BN protein was first treated with 2-mercaptoethanol to reduce any disulfide bridges. This treatment revealed that BN consists of two chains, a light chain and a heavy chain. Next, the two chains were separated and three separate samples of each chain were treated with different proteases. The fragments obtained were individually sequenced by Edman degradation. The sequence is shown below (the first five residues of the light chain are missing due to a blocked amino terminus).

Light chain
RIPKCRKEFQQAQHLRACQQWLHKQANQSGGGPS

Heavy chain
PQGPQQRPPLLQQCCNEKHQEEPLCVCPTLKGASKAVRQ-
QIRQQGQQQGQQGQQLQREISRIYQTATHLPRVCNIPRV-
SICPFQKTMPGP

(a) Why was it necessary to carry out a minimum of two different proteolytic cleavages of the protein using different proteases?

(b) One of the enzymes used by the investigators was trypsin. Indicate the sequences of the fragments that would result from trypsin digestion.

(c) Choose a second protease to cleave both the light and heavy chains into smaller fragments. What protease did you choose, and why? Indicate the sequences of the fragments that would result from digestion by the protease you chose.

62. A peptide sequence linked to the amino terminal of an antibody was sequenced by purifying its mRNA and then using the genetic code to determine the protein sequence. The amino acid sequence of this peptide is shown below:

METDTLLLWVLLLWVPGSTG

Why would sequencing using the traditional methods (e.g., enzymatic cleavage) be difficult to accomplish with this particular peptide?

63. In prokaryotes, the error rate in protein synthesis may be as high as 5×10^{-4} per codon. How many polypeptides containing 500 residues would you expect to contain an amino acid substitution?

64. How many polypeptides containing 2000 residues would you expect to contain an amino acid substitution (see Problem 63)?

65. The mass of each amino acid residue is shown below. Explain why mass spectrometry, which is highly accurate, cannot distinguish Leu and Ile.

Residue	Mass (D)
Ala	71.0
Arg	156.1
Asn	114.0
Asp	115.0
Cys	103.0
Gln	128.1
Glu	129.0
Gly	57.0
His	137.1
Ile	113.1
Leu	113.1
Lys	128.1
Met	131.0

Residue	Mass (D)
Phe	147.1
Pro	97.1
Ser	87.0
Thr	101.0
Trp	186.1
Tyr	163.1
Val	99.1

66. In sequencing by mass spectrometry, not every peptide bond may break. **(a)** If cleavage between two Gly residues does not occur, which amino acid would be identified in place of the two glycines? **(b)** What amino acid would be identified if a bond between Ser and Val did not break (see Problem 65 for amino acid masses)?

67. The peptide shown here is subjected to mass spectrometry to determined its sequence. If all its peptide bonds (but no others) are broken, what is the mass of the smallest fragment? Assume that the only charged group is the N-terminus.

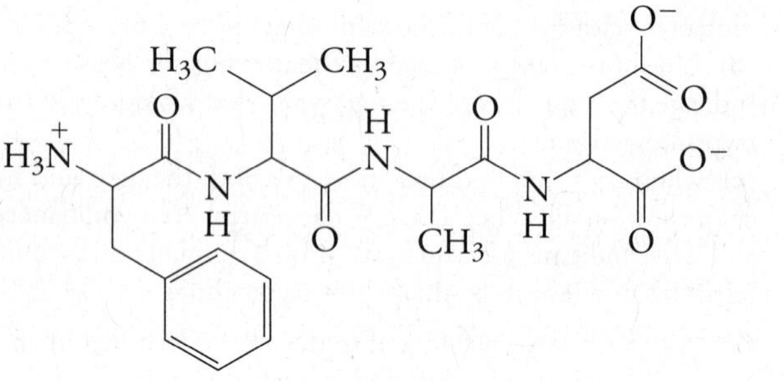

68. For the peptide described in Problem 67, determine the difference in mass between the smallest and the next-smallest fragment.

69. X-ray crystallographic analysis of a protein crystal sometimes fails to reveal the positions of the first few residues of a polypeptide chain. Explain.

70. X-ray crystallography yields models showing the positions of atoms such as C, N, and O. Hydrogen atoms are not detected in most electron density maps and must be added back in their inferred positions. Add back the H atoms in the polypeptide structure below.

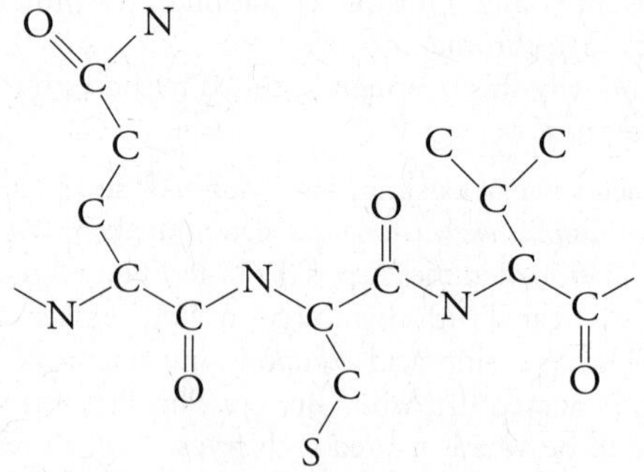

[SELECTED READINGS]

Branden, C., and Tooze, J., *Introduction to Protein Structure* (2nd ed.), Garland Publishing (1999). [A well-illustrated book with chapters introducing amino acids and protein structure, plus chapters on specific proteins categorized by their structure and function.]

Cuff, A. L., Sillitoe, I., Lewis, T., Redfern, O. C., Garratt, R., Thornton, J., and Orengo, C. A., The CATH classification revisited—architectures reviewed and new ways to characterize structural divergence in superfamilies, *Nuc. Acids Res.* **37,** D310–D314 (2009). [Summarizes a widely used approach to classifying protein structures.]

Goodsell, D. S., Visual methods from atoms to cells, *Structure* **13,** 347–354 (2005). [Discusses the challenges of presenting different features of molecular structures.]

Hartl, F. U., Bracher, A., and Hayer-Hartl, M., Molecular chaperones in protein folding and proteostasis, *Nature* **475,** 324–332 (2011). [Includes summaries of protein folding pathways and the fate of misfolded proteins.]

Proteopedia, www.proteopedia.org [This interactive resource showcases tutorials on the three-dimensional structures of proteins, nucleic acids, and other molecules.]

Richardson, J. S., Richardson, D. C., Tweedy, N. B., Gernert, K. M., Quinn, T. P., Hecht, M. H., Erickson, B. W., Yan, Y., McClain, R. D., Donlan, M. E., and Surles, M. C., Looking at proteins: representations, folding, packing, and design, *Biophys. J.* **63,** 1186–1209 (1992). [A highly readable review showing a variety of ways of visualizing protein folding patterns.]

Soto, C., and Estrada, L. D., Protein misfolding and neurodegeneration, *Arch. Neurol.* **65,** 184–189 (2008). [Summarizes the evidence for misfolded proteins as the cause of Alzheimer's and Parkinson's diseases, TSEs, and other disorders.]

Street, J. M., and Dear, J. W., The application of mass-spectrometry-based protein, biomarker discovery to theragnostics, *Br. J. Clin. Pharmacol.* **69,** 367–378 (2010). [Discusses how mass spectrometry can be used to discover and identify proteins associated with diseases.]

5

PROTEIN FUNCTION

▶▶ **CAN** proteins move?

The traditional approach to studying proteins focuses on their architecture, such as α helices and β sheets, suggesting that proteins are like sculpture: beautiful perhaps, but somewhat rocklike. Indeed, a common technique for determining protein structure requires that the protein molecules first form a crystal. The static images of proteins that appear on the pages of this textbook further contribute to the impression that proteins are not particularly dynamic. This chapter highlights several types of proteins, describing how they move—not just by bonds stretching and flexing but also by dramatic movements that are integral to the proteins' physiological functions. The swimming movements of the parasite *Giardia,* for example, depend on the activity of motor proteins in its flagella.

[Alliance Images/Alamy Limited]

THIS CHAPTER IN CONTEXT

Part 1   Foundations

**Part 2 Molecular Structure
and Function**

   5 Protein Function

Part 3   Metabolism

Part 4   Genetic Information

Do You Remember?

● **Noncovalent forces, including hydrogen bonds, ionic interactions, and
van der Waals forces, act on biological molecules (Section 2-1).**

● **A protein's structure may be described at four levels, from primary to quaternary
(Section 4-1).**

● **Some proteins can adopt more than one stable conformation (Section 4-3).**

● **Proteins containing more than one polypeptide chain have quaternary structure
(Section 4-4).**

Every protein, with its unique three-dimensional structure, can perform some unique function for the organism that produces it. For example, the hormone insulin (a 51-residue protein) docks with its receptor (a much larger membrane protein) to trigger certain intracellular responses. And nearly all enzymes are proteins that interact with specific molecules and mediate their chemical transformation. This chapter begins by examining myoglobin, an intracellular oxygen-binding protein that gives vertebrate muscle a reddish color, and hemoglobin, a major protein of red blood cells, which transports O_2 from the lungs to other tissues. These two proteins have been studied for many decades and provide a wealth of information about protein function.

Myoglobin and hemoglobin are globular proteins, but many of the most abundant proteins are fibrous proteins that are elongated and may assemble to form extended structures that determine the shape and other physical attributes of cells and entire organisms. These structural proteins include collagen, an extracellular matrix protein, and a variety of proteins that constitute the intracellular scaffolding known as the **cytoskeleton.** Other than forming fibrous networks, these proteins have little in common, exhibiting a variety of secondary, tertiary, and quaternary structural characteristics related to their distinct physiological functions.

Whereas the supportive role of fibrous proteins in cellular architecture may seem obvious, it turns out that many of the dynamic functions of cells are also intricately tied to the cytoskeleton. The movements of cells and the movements of organelles within cells reflect the action of **motor proteins** that operate along tracks provided by cytoskeletal fibers. Motor proteins provide some additional lessons in protein structure and function due to the mechanism whereby they use chemical energy (the energy of ATP) to do mechanical work, that is, molecular movement.

5-1 Myoglobin and Hemoglobin: Oxygen-Binding Proteins

KEY CONCEPTS
- O_2 binds to the heme group of myoglobin such that binding is half-maximal when the oxygen concentration is equal to the dissociation constant.
- The similarities in structure and sequence between myoglobin and hemoglobin indicate a common evolutionary origin.
- O_2 can bind cooperatively to hemoglobin as the protein shifts from the deoxy to the oxy conformation.
- The Bohr effect and BPG modulate hemoglobin function *in vivo*.

Myoglobin is a relatively small protein with a compact shape about $44 \times 44 \times 25$ Å (Fig. 5-1a). Myoglobin lacks β structure entirely, and all but 32 of its 153 amino acids are part of eight α helices, which range in length from 7 to 26 residues and are labeled A through H (Fig. 5-1b). Hemoglobin is a tetrameric protein whose four subunits each resemble myoglobin.

The fully functional myoglobin molecule contains a polypeptide chain plus the iron-containing porphyrin derivative known as **heme** (shown below). The heme is a type of **prosthetic group,** an organic compound that allows a protein to carry out some function that the polypeptide alone cannot perform—in this case, binding oxygen.

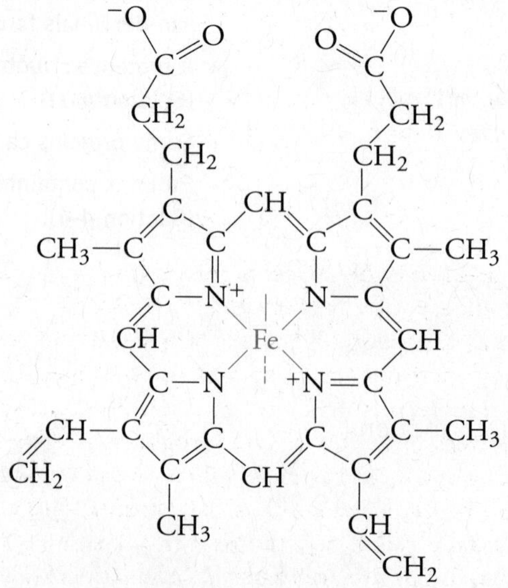

The planar heme is tightly wedged into a hydrophobic pocket between helices E and F of myoglobin. It is oriented so that its two nonpolar vinyl ($-CH=CH_2$) groups are buried and its two polar propionate ($-CH_2-CH_2-COO^-$) groups are exposed to the solvent. The central iron atom, with six possible coordination bonds, is liganded by four N atoms of the porphyrin ring system. A fifth ligand is provided by a His residue of myoglobin known as His F8 (the eighth residue of helix F). Molecular oxygen (O_2) can bind reversibly to form the sixth coordination bond. (This is what allows certain heme-containing proteins, such as myoglobin and hemoglobin, to function physiologically as oxygen carriers.) Residue His E7 (the seventh residue of helix E) forms a hydrogen bond to the O_2 molecule (Fig. 5-2). By itself, heme is not an effective oxygen carrier because the central Fe(II) (or Fe^{2+}) atom is easily oxidized to the ferric Fe(III) (or Fe^{3+}) state, which cannot bind O_2. Oxidation does not readily take place when the heme is part of a protein such as myoglobin or hemoglobin.

Oxygen binding to myoglobin depends on the oxygen concentration

The muscles of diving mammals are especially rich in myoglobin. At one time, myoglobin was believed to be an oxygen-storage protein—which would be advantageous during a long dive—but it most likely just facilitates oxygen diffusion through muscle cells or binds other small molecules such as nitric oxide (NO).

Myoglobin's O_2-binding behavior can be quantified. To begin, the reversible binding of O_2 to myoglobin (Mb) is described by a simple equilibrium

$$Mb + O_2 \rightleftharpoons MbO_2$$

with a dissociation constant, K:

$$K = \frac{[Mb][O_2]}{[MbO_2]} \qquad \text{[5-1]}$$

where the square brackets indicate molar concentrations. (Note that biochemists tend to describe binding phenomena in terms of dissociation constants, sometimes given as K_d, which are the reciprocals of the association constants, K_a, used by chemists.) The proportion of the total myoglobin molecules that have bound O_2 is called the **fractional saturation** and is abbreviated Y:

$$Y = \frac{[MbO_2]}{[Mb] + [MbO_2]} \qquad \text{[5-2]}$$

Since $[MbO_2]$ is equal to $[Mb][O_2]/K$ (Equation 5-1, rearranged),

$$Y = \frac{[O_2]}{K + [O_2]} \qquad \text{[5-3]}$$

O_2 is a gas, so its concentration is expressed as pO_2, the **partial pressure of oxygen**, in units of torr (where 760 torr = 1 atm). Thus,

$$Y = \frac{pO_2}{K + pO_2} \qquad \text{[5-4]}$$

In other words, *the amount of O_2 bound to myoglobin (Y) is a function of the oxygen concentration (pO_2) and the affinity of myoglobin for O_2 (K).*

A plot of fractional saturation (Y) versus pO_2 yields a hyperbola (Fig. 5-3). As the O_2 concentration increases, more and more O_2 molecules bind to the heme groups of myoglobin molecules until, at very high O_2 concentrations, virtually all the myoglobin molecules have bound O_2. Myoglobin is then said to be **saturated** with oxygen. The oxygen concentration at which myoglobin is half-saturated—that is, the concentration of O_2 at which Y is half-maximal—is equivalent to K. For convenience, K is usually called p_{50}, the oxygen pressure at 50% saturation. For human myoglobin, p_{50} is 2.8 torr (see Sample Calculation 5-1).

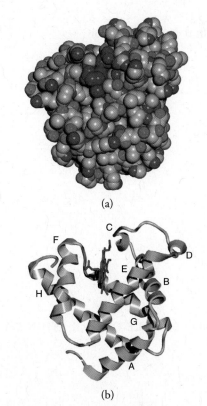

(a)

(b)

Figure 5-1 Models of myoglobin structure. (a) Space-filling model. All atoms (except H) are shown (C gray, O red, N blue). The heme group, where oxygen binds, is purple. (b) Ribbon diagram with the eight α helices labeled A–H. [Structure of myoglobin (pdb 1MBO) determined by S. E. V. Phillips.]

? **To which class of the CATH system (Section 4-3) does myoglobin belong?**

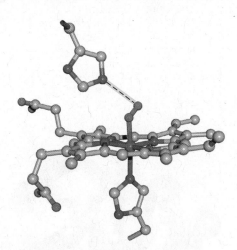

Figure 5-2 Oxygen binding to the heme group of myoglobin. The central Fe(II) atom of the heme group (purple) is liganded to four porphyrin N atoms and to the N of His F8 below the porphyrin plane. O_2 (red) binds reversibly to the sixth coordination site, above the porphyrin plane. Residue His E7 forms a hydrogen bond to O_2.

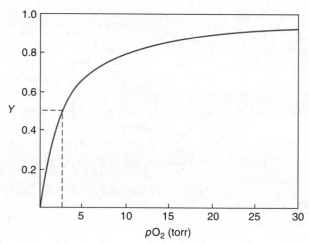

Figure 5-3 Myoglobin oxygen-binding curve. The relationship between the fractional saturation of myoglobin (Y) and the oxygen concentration (pO_2) is hyperbolic. When $pO_2 = K = 2.8$ torr, myoglobin is half-saturated ($Y = 0.5$).

●●● SAMPLE CALCULATION 5-1

PROBLEM

Calculate the fractional saturation of myoglobin when $pO_2 = 1$ torr, 10 torr, and 100 torr.

SOLUTION

Use Equation 5-4 and let $K = 2.8$ torr.

When $pO_2 = 1$ torr, $Y = \dfrac{1}{2.8 + 1} = 0.26$

When $pO_2 = 10$ torr, $Y = \dfrac{10}{2.8 + 10} = 0.78$

When $pO_2 = 100$ torr, $Y = \dfrac{100}{2.8 + 100} = 0.97$

●●● **PRACTICE PROBLEMS**

1. Calculate the fractional saturation of myoglobin when $pO_2 = 5.6$ torr.
2. How would the value of Y in Practice Problem 1 change if $K = 1.4$ torr?
3. At what pO_2 value would myoglobin be 75% saturated with O_2?

Myoglobin and hemoglobin are related by evolution

Hemoglobin is a heterotetramer containing two α chains and two β chains. Each of these subunits, called a **globin,** looks a lot like myoglobin. The hemoglobin α chains, the hemoglobin β chains, and myoglobin have remarkably similar tertiary structures (**Fig. 5-4**). All have a heme group in a hydrophobic pocket, a His F8 that ligands the Fe(II) ion, and a His E7 that forms a hydrogen bond to O_2.

Somewhat surprisingly, the amino acid sequences of the three globin polypeptides are only 18% identical. **Figure 5-5** shows the aligned sequences, with the necessary gaps (for example, the hemoglobin α chain has no D helix). The lack of striking sequence similarities among these proteins highlights an important principle of protein three-dimensional structure: *Certain tertiary structures—for example, the backbone folding pattern of a globin polypeptide—can accommodate a variety of amino acid sequences. In fact, many proteins with completely unrelated sequences adopt similar structures.*

Figure 5-4 Tertiary structures of myoglobin and the α and β chains of hemoglobin. Backbone traces of α globin (blue) and β globin (red) are aligned with myoglobin (green) to show their structural similarity. The heme group of myoglobin is shown in gray. [Structure of hemoglobin (pdb 2HHB) determined by G. Fermi and M. F. Perutz.]

? **Which portion of globin structure shows the most variability? The least? Explain.**

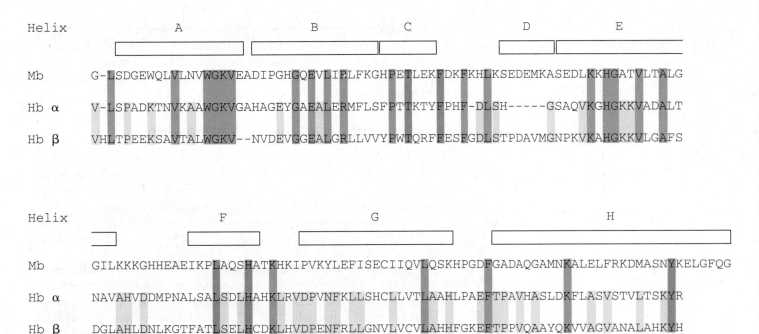

Helix   A   B   C   D   E

Mb   G-LSDGEWQLVLNVWGKVEADIPGHGQEVLIRLFKGHPETLEKFDKFKHLKSEDEMKASEDLKKHGATVLTALG

Hb α   V-LSPADKTNVKAAWGKVGAHAGEYGAEALERMFLSFPTTKTYFPHF-DLSH-----GSAQVKGHGKKVADALT

Hb β   VHLTPEEKSAVTALWGKV--NVDEVGGEALGRLLVVYPWTQRFFESFGDLSTPDAVMGNPKVKAHGKKVLGAFS

Helix   F   G   H

Mb   GILKKKGHHEAEIKPLAQSHATKHKIPVKYLEFISECIIQVLQSKHPGDFGADAQGAMNKALELFRKDMASNYKELGFQG

Hb α   NAVAHVDDMPNALSALSDLHAHKLRVDPVNFKLLSHCLLVTLAAHLPAEFTPAVHASLDKFLASVSTVLTSKYR

Hb β   DGLAHLDNLKGTFATLSELHCDKLHVDPENFRLLGNVLVCVLAHHFGKEFTPPVQAAYQKVVAGVANALAHKYH

Figure 5-5 The amino acid sequences of myoglobin and the hemoglobin α and β chains. The sequence of human myoglobin (Mb) and the human hemoglobin (Hb) chains are written so that their helical segments (bars labeled A through H) are aligned. Residues that are identical in the α and β globins are shaded yellow; residues identical in myoglobin and the α and β globins are shaded blue, and residues that are invariant in all vertebrate myoglobin and hemoglobin chains are shaded purple. The one-letter abbreviations for amino acids are given in Figure 4-2. [After Dickerson, R. E., and Geis, I., *Hemoglobin*, pp. 68–69, Benjamin/Cummings (1983).]

? **Can you identify positions occupied by structurally similar amino acids in all three globins?**

Clearly, the globins are **homologous proteins** that have evolved from a common ancestor through genetic mutation (see Section 3-3). The α and β chains of human hemoglobin share a number of residues; some of these are also identical in human myoglobin. A few residues are found in all vertebrate hemoglobin and myoglobin chains. The **invariant residues,** those that are identical in all the globins, are essential for the structure and/or function of the proteins and cannot be replaced by other residues. Some positions are under less selective pressure to maintain a particular amino acid match and can be **conservatively substituted** by a similar amino acid (for example, Ile for Leu or Ser for Thr). Still other positions are **variable,** meaning that they can accommodate a variety of residues, none of which is critical for the protein's structure or function. *By looking at the similarities and differences in sequences among evolutionarily related proteins such as the globins, it is possible to deduce considerable information about elements of protein structure that are central to protein function* (see Bioinformatics Project 4).

Sequence analysis also provides a window on the course of globin evolution, since *the number of sequence differences roughly corresponds to the time since the genes diverged.* An estimated 1.1 billion years ago, a single globin gene was duplicated, possibly by aberrant genetic recombination, leaving two globin genes that then could evolve independently (Fig. 5-6). Over time, the gene sequences diverged by mutation. One gene became the myoglobin gene. The other coded for a monomeric hemoglobin, which is still found in some primitive vertebrates such as the lamprey (an organism that originated about 425 million years ago). Subsequent duplication of the hemoglobin gene and additional sequence changes yielded the α and β globins, which made possible the evolution of a tetrameric hemoglobin (whose structure is abbreviated $\alpha_2\beta_2$). Additional gene duplications and mutations produced the ζ chain (from the α chain) and the γ and ε chains (from the β chain). In fetal mammals, hemoglobin

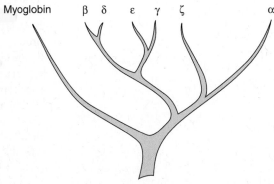

Figure 5-6 Evolution of the globins. Duplication of a primordial globin gene allowed the separate evolution of myoglobin and a monomeric hemoglobin. Additional duplications among the hemoglobin genes gave rise to six different globin chains that combine to form tetrameric hemoglobin variants at various times during development.

has the composition $\alpha_2\gamma_2$, and early human embryos synthesize a $\zeta_2\varepsilon_2$ hemoglobin. In primates, a recent duplication of the β chain gene has yielded the δ chain. An $\alpha_2\delta_2$ hemoglobin occurs as a minor component (about 2%) of adult human hemoglobin. At present, the δ chain appears to have no unique biological function, but it may eventually evolve one.

Oxygen binds cooperatively to hemoglobin

A milliliter of human blood contains about 5 billion red blood cells, each of which is packed with about 300 million hemoglobin molecules. Consequently, blood can carry far more oxygen than a comparable volume of pure water. The oxygen-carrying capacity of the blood can be quickly assessed by measuring the hematocrit (the percentage of the blood volume occupied by red blood cells, which ranges from about 40% (in women) to 45% (in men). Individuals with anemia, too few red blood cells, can sometimes be treated to increase red blood cell production (Box 5-A).

The hemoglobin in red blood cells, like myoglobin, binds O_2 reversibly, but it does not exhibit the simple behavior of myoglobin. A plot of fractional saturation (Y) versus pO_2 for hemoglobin is sigmoidal (S-shaped) rather than hyperbolic (Fig. 5-7). Furthermore, hemoglobin's overall oxygen affinity is lower than that of myoglobin: Hemoglobin is half-saturated at an oxygen pressure of 26 torr (p_{50} = 26 torr), whereas myoglobin is half-saturated at 2.8 torr.

Why is hemoglobin's binding curve sigmoidal? At low O_2 concentrations, hemoglobin appears to be reluctant to bind the first O_2, but as the pO_2 increases, O_2 binding increases sharply, until hemoglobin is almost fully saturated. A look at the binding curve in reverse shows that at high O_2 concentrations, oxygenated hemoglobin is reluctant to give up its first O_2, but as the pO_2 decreases, all the O_2 molecules are easily given up. This behavior suggests that the binding of the first O_2 increases the affinity of the remaining O_2-binding sites. Apparently, *hemoglobin's four heme groups are not independent but communicate with each other in order to work in a unified fashion.* This is known as **cooperative** binding behavior. In fact, the fourth O_2 taken up by hemoglobin binds with about 100 times greater affinity than the first.

Hemoglobin's relatively low oxygen affinity and its cooperative binding behavior are the keys to its physiological function (see Fig. 5-7). In the lungs, where the pO_2

BOX 5-A ⬡ BIOCHEMISTRY NOTE

Erythropoietin Boosts Red Blood Cell Production

Erythropoietin (EPO) is a 165-residue protein hormone that signals the bone marrow to produce more red blood cells (also known as erythrocytes). EPO is synthesized primarily by the kidneys, so individuals with kidney disease are often deficient in EPO and develop anemia. Treating these individuals with recombinant EPO can increase red blood cell production to near-normal levels.

The results of EPO treatment have not been lost on athletes, particularly in endurance sports such as cycling; some competitors have found that taking EPO increases their red cell counts, thereby increasing their ability to deliver oxygen to their muscles. The illegal use of EPO to enhance athletic performance is difficult to detect. For one thing, EPO is already present in the body, and an increase in red blood cells by itself is not proof of EPO doping. A clever abuser may even take iron supplements to mask the telltale drop in stored iron that accompanies the increased production of erythrocytes. Successfully detecting the presence of recombinant EPO takes advantage of subtle differences between the natural and recombinant hormones. Although they have the same amino acid sequence, recombinant EPO produced by cultured cells is not derivatized with carbohydrates in the same manner as EPO produced naturally by the kidneys.

● **Questions:** Would measurement of mean red blood cell volume or red blood cell number provide the same information as the hematocrit? Why might it be hazardous to take EPO to increase red blood cell levels *above* normal?

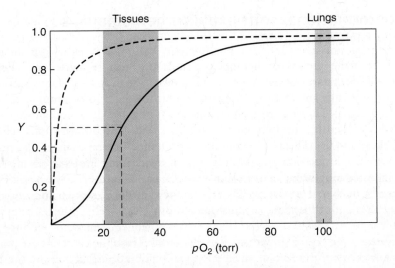

Figure 5-7 Oxygen binding to hemoglobin. The relationship between fractional saturation (Y) and oxygen concentration (pO_2) is sigmoidal. The pO_2 at which hemoglobin is half-saturated ($p50$) is 26 torr. For comparison, myoglobin's O_2-binding curve is indicated by the dashed line. The difference in oxygen affinity between hemoglobin and myoglobin ensures that O_2 bound to hemoglobin in the lungs is released to myoglobin in the muscles. This oxygen-delivery system is efficient because the tissue pO_2 corresponds to the part of the hemoglobin binding curve where the O_2 affinity falls off most sharply. ⊕ **See Animated Figure. Oxygen-binding curve of hemoglobin.**

? Could this O_2-delivery system still operate if the $p50$ for myoglobin was twice as high?

is about 100 torr, hemoglobin is about 95% saturated with O_2. In the tissues, where the pO_2 is only about 20 to 40 torr, hemoglobin's oxygen affinity drops off rapidly (it is only about 55% saturated when the pO_2 is 30 torr). Under these conditions, *the O_2 released from hemoglobin is readily taken up by myoglobin in muscle cells, since myoglobin's affinity for O_2 is much higher,* even at the lower oxygen pressure. Myoglobin can therefore relay O_2 from red blood cells in the capillaries to the muscle cell's mitochondria, where it is consumed in the oxidative reactions that sustain muscle activity. Agents such as carbon monoxide, which interferes with O_2 binding to hemoglobin, prevent the efficient delivery of O_2 to cells (Box 5-B).

BOX 5-B ⬡ BIOCHEMISTRY NOTE

Carbon Monoxide Poisoning

The affinity of hemoglobin for carbon monoxide is about 250 times higher than its affinity for oxygen. However, the concentration of CO in the atmosphere is only about 0.1 ppm (parts per million by volume), compared to an O_2 concentration of about 200,000 ppm. Normally, only about 1% of the hemoglobin molecules in an individual are in the carboxyhemoglobin (Hb · CO) form, probably as a result of endogenous production of CO in the body (CO acts as a signaling molecule, although its physiological role is not well understood).

Danger arises when the fraction of carboxyhemoglobin rises, which can occur when individuals are exposed to high levels of environmental CO. For example, the incomplete combustion of fuels, as occurs in gas-burning appliances and vehicle engines, releases CO. The concentration of CO can rise to about 10 ppm in these situations and to as high as 100 ppm in highly polluted urban areas. The concentration of carboxyhemoglobin may reach 15% in some heavy smokers, although the symptoms of CO poisoning are usually not apparent.

CO toxicity, which occurs when the concentration of carboxyhemoglobin rises above about 25%, causes neurological impairment, usually dizziness and confusion. High doses of CO, which cause carboxyhemoglobin levels to rise above 50%, can trigger coma and death. When CO is bound to some of the heme groups of hemoglobin, O_2 is not able to bind to those sites because its low affinity means that it cannot displace the bound CO. In addition, the carboxyhemoglobin molecule remains in a high-affinity conformation, so that even if O_2 does bind to some of the hemoglobin heme groups in the lungs, O_2 release to the tissues is impaired. The effects of mild CO poisoning are largely reversible through the administration of O_2. But because the CO remains bound to hemoglobin with a half-life of several hours, recovery is slow.

⬢ **Question:** Sketch the oxygen-binding curves of hemoglobin and carboxyhemoglobin [Hb · (CO)$_2$].

▶▶ CAN proteins move?

A conformational shift explains hemoglobin's cooperative behavior

The four heme groups of hemoglobin must be able to sense one another's oxygen-binding status so that they can bind or release their O_2 in concert. But the four heme groups are 25 to 37 Å apart, too far for them to communicate via an electronic signal. Therefore, the signal must be mechanical. In a mechanism worked out by Max Perutz, the four globin subunits undergo conformational changes when they bind O_2. This is one example of protein structural changes that provide an answer to the question posed at the start of the chapter: Proteins *can* move.

In **deoxyhemoglobin** (hemoglobin without any bound O_2), the heme Fe ion has five ligands, so the porphyrin ring is somewhat dome-shaped and the Fe lies about 0.6 Å out of the plane of the porphyrin ring. As a result, the heme group is bowed slightly toward His F8 (**Fig. 5-8**). When O_2 binds to produce **oxyhemoglobin**, the Fe—now with six ligands—moves into the center of the porphyrin plane. This movement of the Fe ion pulls His F8 farther toward the heme group, and this in turn drags the entire F helix so that it moves up to 1 Å. The F helix cannot move in this manner unless the entire protein alters its conformation, culminating in the rotation of one αβ unit relative to the other. Consequently, *hemoglobin has two quaternary structures, corresponding to the oxy and deoxy states.*

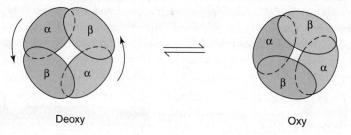

Deoxy   Oxy

The shift in conformation between the oxy and deoxy states primarily involves rotation of one αβ unit relative to the other. Oxygen binding decreases the size of the central cavity between the four subunits and alters some of the contacts between subunits. The two conformational states of hemoglobin are

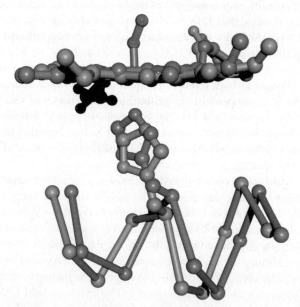

Figure 5-8 Conformational changes in hemoglobin upon O_2 binding. In deoxyhemoglobin (blue), the porphyrin ring is slightly bowed down toward His F8 (shown in ball-and-stick form). The remainder of the F helix is represented by its alpha carbon atoms. In oxyhemoglobin (purple), the heme group becomes planar, pulling His F8 and its attached F helix upward. The bound O_2 is shown in red. ⊕ **See Animated Figure. Movements of heme and F helix in hemoglobin.**

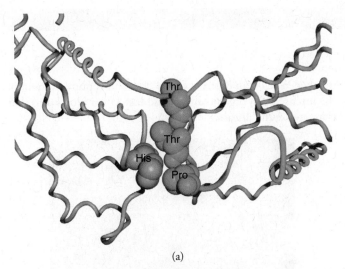

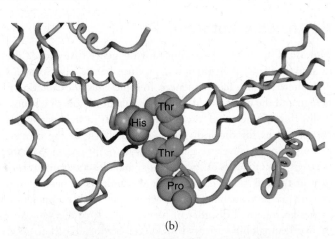

(a)

(b)

Figure 5-9 Some of the subunit interactions in hemoglobin. The interactions between the αβ units of hemoglobin include contacts between side chains. The relevant residues are shown in space-filling form. (a) In deoxyhemoglobin, a His residue on the β chain (blue, left) fits between a Pro and a Thr residue on the α chain (green, right). (b) Upon oxygenation, the His residue moves between two Thr residues. An intermediate conformation (between the deoxy and oxy conformations) is disallowed in part because the highlighted side chains would experience strain. [Structure of human deoxyhemoglobin (pdb 2HHB) determined by G. Fermi and M. F. Perutz; structure of human oxyhemoglobin (pdb 1HHO) determined by B. Shaanan.]

formally known as **T** (for "tense") and **R** (for "relaxed"). The T state corresponds to deoxyhemoglobin, and the R state corresponds to oxyhemoglobin.

Deoxyhemoglobin is reluctant to bind the first O_2 molecule because the protein is in the deoxy (T) conformation, which is unfavorable for O_2 binding (the Fe atom lies out of the heme plane). However, once O_2 has bound, probably to the α chain in each αβ pair, the entire tetramer switches to the oxy (R) conformation as the Fe atom and the F helix move. An intermediate conformation is not possible, because the contacts between the αβ units do not allow it (**Fig. 5-9**). Molecular dynamics studies suggest that hemoglobin does not instantaneously snap from one conformation to the other but instead undergoes fluctuations in tertiary structure that precede the shift in quaternary structure.

Subsequent O_2 molecules bind with higher affinity because the protein is already in the oxy (R) conformation, which is favorable for O_2 binding. Similarly, oxyhemoglobin tends to retain its bound O_2 molecules until the oxygen pressure drops significantly. Then some O_2 is released, triggering the change to the deoxy (T) conformation. This decreases the affinity of the remaining bound O_2 molecules, making it easier for hemoglobin to unload its bound oxygen. Because measurements of O_2 binding reflect the average behavior of many individual hemoglobin molecules, the result is a smooth curve (as shown in Fig. 5-7).

Hemoglobin and many other proteins with multiple binding sites are known as **allosteric proteins** (from the Greek *allos,* meaning "other," and *stereos,* meaning "space"). In these proteins, *the binding of a small molecule (called a **ligand**) to one site alters the ligand-binding affinity of the other sites.* In principle, the ligands need not be identical, and their binding may either increase or decrease the binding activity of the other sites. In hemoglobin, the ligands are all oxygen molecules, and O_2 binding to one subunit of the protein increases the O_2 affinity of the other subunits.

H⁺ ions and bisphosphoglycerate regulate oxygen binding to hemoglobin *in vivo*

Decades of study have revealed the detailed chemistry behind hemoglobin's activity (and have also revealed how molecular defects can lead to disease; Box 5-C). The conformational change that transforms deoxyhemoglobin to oxyhemoglobin alters the microenvironments of several ionizable groups in the protein, including the two N-terminal amino groups of the α subunits and the two His residues near the

BOX 5-C ⬡ CLINICAL CONNECTION

Hemoglobin Mutations

Mutations in the DNA sequence for the genes that encode the α and β chains of hemoglobin produce hemoglobin proteins with altered amino acid sequences. In some cases, the mutation is benign and the hemoglobin molecules function more or less normally. But in other cases, the mutation results in serious physical complications for the individual, as the ability of the mutant hemoglobin to deliver oxygen to cells is compromised. Mutant hemoglobins are often unstable, which may also result in destruction of the red blood cell (**anemia**). Hundreds of hemoglobin variants are known, and about 5% of the world's population carries an inherited disorder of hemoglobin. One of the better-known hemoglobin variants is sickle cell hemoglobin (known as hemoglobin S or Hb S). Individuals with two copies of the defective gene develop sickle cell anemia, a debilitating disease that predominantly affects populations of African descent.

The discovery of the molecular defect that causes sickle cell anemia was a groundbreaking event in biochemistry. The disease was first described in 1910, but for many years there was no direct evidence that sickle cell anemia—or any genetic disease—was the result of an alteration in the molecular structure of a protein. Then in 1949, Linus Pauling (who was already on his way to discovering the α helix) showed that hemoglobin from patients with sickle cell anemia had a different electrical charge than hemoglobin from healthy individuals. Eight years later, in 1957, Vernon Ingram identified a single amino acid difference: Glu at position A3 in the β chain is replaced by Val in sickle cell hemoglobin. This was the first evidence that an alteration in a gene caused an alteration in the amino acid sequence of the corresponding polypeptide. The mutation is described in Section 3-2.

In normal hemoglobin, the switch from the oxy to the deoxy conformation exposes a hydrophobic patch on the protein surface between the E and F helices. The hydrophobic Val residues on hemoglobin S are optimally positioned to bind to this patch. This intermolecular association leads to the rapid aggregation of hemoglobin S molecules to form long rigid fibers.

These fibers physically distort the red blood cell, producing the familiar sickle shape. Because hemoglobin S aggregation occurs only among deoxyhemoglobin S molecules, sickling tends to occur when the red blood cells pass through oxygen-poor capillaries. The misshapen cells can obstruct blood flow and rupture, leading to the intense pain, organ damage, and loss of red blood cells that characterize the disease.

Normal red blood cells (*top*) and a sickled cell (*bottom*). [From Andrew Skred/Science Photo Library/Photo Researchers and Jacki Lewin, Royal Free Hospital/Science Photo Library/Photo Researchers.]

The high frequency of the gene for sickle cell anemia (that is, the mutated β globin gene) was at first puzzling: Genes that lead to disabling diseases tend to be rare because individuals with two copies of the gene usually die before they can pass the gene to their offspring. However, carriers of the sickle cell variant appear to have a selective advantage. They are protected against malaria, a parasitic disease that afflicts about 225 million people and kills about 1 million each year, mostly children. In fact, the sickle cell hemoglobin variant is common in regions of the world where malaria is endemic. In heterozygotes (individuals with one normal and one defective β globin gene), only about 2% of red blood cells undergo sickling. Sickling does not directly lead to elimination of *Plasmodium falciparum*, the intracellular protozoan that causes malaria. However, the presence of hemoglobin S triggers production of heme oxygenase, an enzyme whose reaction products protect against the damage caused by the parasite.

The **thalassemias** result from genetic defects that reduce the rate of synthesis of the α or β globin chains. These disorders are prevalent in the Mediterranean area (the name comes from the Greek word *thalassa*, meaning "sea") and in South Asia. Depending on the nature of the mutation, individuals with thalassemia may experience mild to severe anemia and their red blood cells may be smaller than normal. But like heterozygotes for hemoglobin S, individuals with thalassemia are apparently more resistant to malaria.

The accompanying table lists some hemoglobin residues that are critical for normal function; their mutation produces clinical symptoms. Note that defective hemoglobins are usually named for the place where they were first observed or characterized.

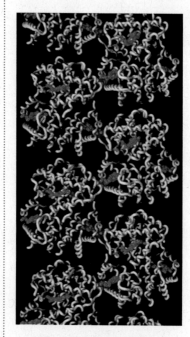

In this model of polymerized hemoglobin S molecules, the heme groups are red and the mutant Val residues are blue. [From W. Royer and D. Harrington, *J. Mol. Biol.* 272, 398–407 (1992).]

Chain	Position	Amino Acid	Role	Significance
α	44	Pro	Participates in the formation of the $\alpha_1\beta_2$ interface in the deoxy form but not the oxy form.	Stabilizes the deoxy form.
α	141 (C-terminus)	Arg	Its COO^- forms an ion pair with Lys 127 and its side chain forms an ion pair with Asp 126 in the deoxy form.	Stabilizes the deoxy form.
β	82	Lys	Forms an ion pair to BPG in the central cavity.	Stabilizes the deoxy form.
β	146	His	The side-chain imidazole ring forms an ion pair with Asp 94. It also forms an ion pair with BPG in the central cavity.	Stabilizes the deoxy form.

● **Questions:**

1. In the mutant hemoglobin Hb Ohio (β142Ala → Asp), the substitution of Asp for Ala results in the displacement of the G helix relative to the H helix in the β chain. This decreases the stability of the β146His–β94Asp ion pair. Draw an oxygen-binding curve that compares the relative p_{50} values of normal hemoglobin (Hb A) and Hb Ohio. What is the effect of the decreased stability of the His–Asp ion pair on Hb Ohio?

2. Investigators of mutant human hemoglobins often subject the proteins to cellulose acetate electrophoresis at pH 8.5, a pH at which most hemoglobins are negatively charged. The proteins are applied at the negative pole and migrate toward the positive pole when the current is turned on. The more negatively charged the hemoglobin, the faster it migrates to the positive pole. Results for normal (Hb A) and sickle cell hemoglobin (Hb S) are shown. Draw a diagram that shows the results of electrophoresis of Hb A and Hb Ohio (β 142Ala → Asp).

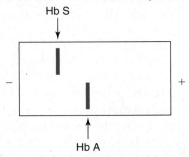

3. Hb Milledgeville (α44Pro → Leu) results in a mutated hemoglobin with altered oxygen affinity. Explain how the oxygen affinity is altered.

4. Hb Providence (β44Lys → Asn) results from a single point mutation of the DNA.
   (a) What is the change in the DNA that occurred to produce the mutant hemoglobin?
   (b) Compare the oxygen affinities of Hb Providence and Hb A.
   (c) There are actually two forms of Hb Providence in affected individuals. Hb Providence in which the β44Lys has been replaced with Asn is referred to as Hb Providence Asn. This mutant hemoglobin can undergo deamidation to produce Hb Providence Asp. Draw the reaction that converts Hb Providence Asn to Hb Providence Asp.
   (d) Draw a diagram that shows the cellulose acetate electrophoresis results for Hb A, Hb Providence Asn, and Hb Providence Asp.
   (e) Compare the oxygen affinities of Hb Providence Asp and Hb Providence Asn.

5. Hb Syracuse (β146His → Pro) is a mutant hemoglobin with altered oxygen affinity.
   (a) Draw a diagram showing the cellulose acetate electrophoresis results for Hb Syracuse and Hb A.
   (b) Evaluate the ability of Hb Syracuse to respond to normal allosteric effectors of hemoglobin. How is the oxygen affinity affected as a result?

C-terminus of the β subunits. As a result, these groups become more acidic and release H^+ when O_2 binds to the protein:

$$Hb \cdot H^+ + O_2 \rightleftharpoons Hb \cdot O_2 + H^+$$

Therefore, increasing the pH of a solution of hemoglobin (decreasing $[H^+]$) favors O_2 binding by "pushing" the reaction to the right, as written above. Decreasing the pH (increasing $[H^+]$) favors O_2 dissociation by "pushing" the reaction to the left. *The reduction of hemoglobin's oxygen-binding affinity when the pH decreases is known as the **Bohr effect**.*

The Bohr effect plays an important role in O_2 transport *in vivo*. Tissues release CO_2 as they consume O_2 in respiration. The dissolved CO_2 enters red blood cells, where it is rapidly converted to bicarbonate (HCO_3^-) by the action of the enzyme carbonic anhydrase (see Box 2-D):

$$CO_2 + H_2O \rightleftharpoons HCO_3^- + H^+$$

LUNGS   BLOOD   TISSUES

$$O_2 \longrightarrow Hb \cdot O_2 \longrightarrow Hb \cdot O_2 \longrightarrow O_2$$

$$Hb \qquad\qquad Hb$$

$$Hb \cdot H^+ \longleftarrow Hb \cdot H^+$$

$$H^+ \qquad\qquad H^+$$

$$CO_2 \longleftarrow HCO_3^- \longleftarrow HCO_3^- \longleftarrow CO_2$$

Figure 5-10 Oxygen transport and the Bohr effect. Hemoglobin picks up O_2 in the lungs. In the tissues, H^+ derived from the metabolic production of CO_2 decreases hemoglobin's affinity for O_2, thereby promoting O_2 release to the tissues. Back in the lungs, hemoglobin binds more O_2, releasing the protons, which recombine with bicarbonate to re-form CO_2. ⊕ **See Animated Figure. The Bohr effect.**

? **Write the net equation for the process shown in the diagram.**

The H^+ released in this reaction induces hemoglobin to unload its O_2 (**Fig. 5-10**). In the lungs, the high concentration of oxygen promotes O_2 binding to hemoglobin. This causes the release of protons that can then combine with bicarbonate to re-form CO_2, which is breathed out.

Red blood cells use one additional mechanism to fine-tune hemoglobin function. These cells contain a three-carbon compound, 2,3-bisphosphoglycerate (BPG):

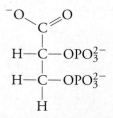

2,3-Bisphosphoglycerate (BPG)

BPG binds in the central cavity of hemoglobin, but only in the T (deoxy) state. The five negative charges in BPG interact with positively charged groups in deoxyhemoglobin; in oxyhemoglobin, these cationic groups have moved away and the central cavity is too narrow to accommodate BPG. Thus, *the presence of BPG stabilizes the deoxy conformation of hemoglobin.* Without BPG, hemoglobin would bind O_2 too tightly to release it to cells. In fact, hemoglobin stripped of its BPG *in vitro* exhibits very strong O_2 affinity, even at low pO_2 (**Fig. 5-11**).

The fetus takes advantage of this chemistry to obtain O_2 from its mother's hemoglobin. Fetal hemoglobin has the subunit composition $\alpha_2\gamma_2$. In the γ chains, position H21 is not His (as it is in the mother's β chain) but Ser. His H21 bears one of the positive charges important for binding BPG in adult hemoglobin. The absence of this interaction in fetal hemoglobin reduces BPG binding. Consequently, *hemoglobin in fetal red blood cells has a higher O_2 affinity than adult hemoglobin, which helps transfer O_2 from the maternal circulation across the placenta to the fetus.*

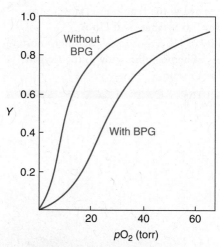

Figure 5-11 Effect of BPG on hemoglobin. BPG binds to deoxyhemoglobin but not to oxyhemoglobin. It therefore reduces hemoglobin's O_2 affinity by stabilizing the deoxy conformation. ⊕ **See Animated Figure. Effect of BPG and CO_2 on hemoglobin.**

- Why does myoglobin require a prosthetic group?
- What is the relationship between myoglobin's fractional saturation and the oxygen concentration?
- How can the sequences of homologous proteins provide information about residues that are essential or nonessential for a protein's function?
- How do the different O_2 affinities of myoglobin and hemoglobin relate to their physiological roles?
- Explain the structural basis for hemoglobin's cooperative O_2-binding behavior.
- How do the Bohr effect and BPG regulate O_2 transport *in vivo*?

5-2 Structural Proteins

A typical eukaryotic cell contains three types of cytoskeletal proteins that form fibers extending throughout the cell (Fig. 5-12): These are **microfilaments** (actin filaments, with a diameter of about 70 Å), **intermediate filaments** (with a diameter of about 100 Å), and **microtubules** (with a diameter of about 240 Å). In large multicellular organisms, fibers of the protein collagen provide structural support extracellularly. Bacterial cells also contain proteins that form structures similar to microfilaments and microtubules. In the following discussion, note how the structure of each protein influences the overall structure and flexibility of the fiber as well as the fiber's ability to disassemble and reassemble.

Microfilaments are made of actin

A major portion of the eukaryotic cytoskeleton consists of microfilaments, or polymers of actin. In many cells, a network of actin filaments supports the plasma membrane and therefore determines cell shape (see Fig. 2-7 and Fig. 5-12). Certain proteins cross-link individual actin polymers to help form bundles of microfilaments, thereby increasing their strength.

Monomeric actin is a globular protein with about 375 amino acids (Fig. 5-13). On its surface is a cleft in which adenosine triphosphate (ATP) binds. The adenosine group slips into a pocket on the protein, and the ribose hydroxyl groups and the phosphate groups form hydrogen bonds with the protein.

KEY CONCEPTS
- Globular actin subunits associate in a double chain to form a microfilament.
- The growth and regression of actin filaments can change a cell's shape.
- Microtubules are hollow tubes built from tubulin dimers.
- Intermediate filaments are long-lasting fibrous proteins consisting of coiled α helices.
- Three left-handed Gly-rich helical polypeptides form the collagen triple helix.

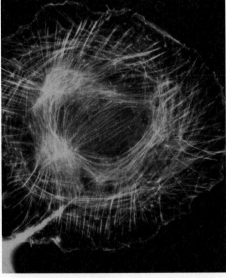
Actin filaments

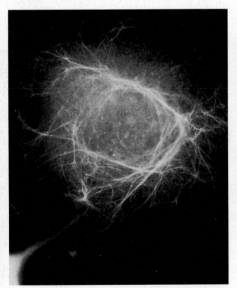
Intermediate filaments

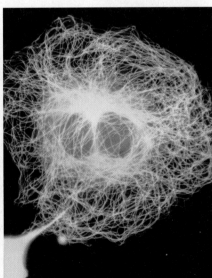
Microtubules

Figure 5-12 Distribution of cytoskeletal fibers in a single cell. To make these micrographs, each type of fiber was labeled with a fluorescent probe that binds specifically to one type of cytoskeletal protein. Note how the distribution of actin filaments differs somewhat from that of intermediate filaments and microtubules. [Courtesy J. Victor Small, Austrian Academy of Sciences, Vienna, Austria.]

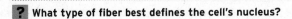

? **What type of fiber best defines the cell's nucleus?**

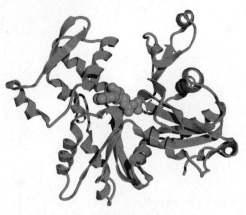

Figure 5-13 Actin monomer. This protein assumes a globular shape with a cleft where ATP (green) binds. [Structure of rabbit actin (pdb 1J6Z) determined by L. R. Otterbein, P. Graceffa, and R. Dominguez.]

(−) end

(+) end

Figure 5-14 Model of an actin filament. The structure of F-actin was determined from X-ray diffraction data and computer model-building. Fourteen actin subunits are shown (all are different colors except the central actin subunit, whose two halves are blue and gray). [Courtesy Ken Holmes, Max Planck Institute for Medical Research.]

Polymerized actin is sometimes referred to as **F-actin** (for filamentous actin, to distinguish it from **G-actin,** the globular monomeric form). *The actin polymer is actually a double chain of subunits in which each subunit contacts four neighboring subunits* (Fig. 5-14). Each actin subunit has the same orientation (for example, all the nucleotide-binding sites point up in Fig. 5-14), so the assembled fiber has a distinct polarity. The end with the ATP site is known as the **(−) end,** and the opposite end is the **(+) end.**

Initially, polymerization of actin monomers is slow because actin dimers and trimers are unstable. But once a longer polymer has formed, subunits add to both ends. Addition is usually much more rapid at the (+) end (hence its name) than at the (−) end (Fig. 5-15).

Actin polymerization is driven by the hydrolysis of ATP (splitting ATP by the addition of water) to produce ADP + inorganic phosphate (P_i):

Adenosine triphosphate (ATP)

Adenosine diphosphate (ADP)

Inorganic phosphate (P_i)

This reaction is catalyzed by F-actin but not by G-actin. Consequently, *most of the actin subunits in a microfilament contain bound ADP.* Only the most recently added subunits still contain ATP. Because ATP-actin and ADP-actin assume slightly different conformations, proteins that interact with microfilaments may be able to distinguish rapidly polymerizing (ATP-rich) microfilaments from longer-established (ADP-rich) microfilaments.

Actin filaments continuously extend and retract

Microfilaments are dynamic structures. *Polymerization of actin monomers is a reversible process, so the polymer undergoes constant shrinking and growing as subunits add to and dissociate from one or both ends of the microfilament* (see Fig. 5-15). When the net rate of addition of subunits to one end of a microfilament matches the net rate of removal of subunits at the other end, the polymer is said to be **treadmilling** (Fig. 5-16).

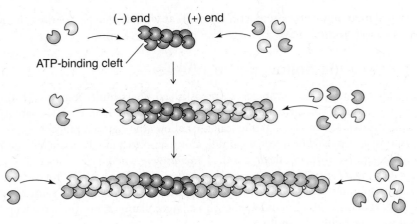

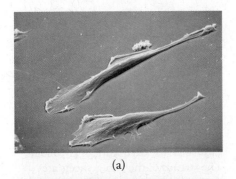

Figure 5-15 Actin filament assembly. A microfilament grows as subunits add to its ends. Subunits usually add more rapidly to the (+) end, which therefore grows faster than the (−) end. Actual microfilaments are much longer than depicted here.

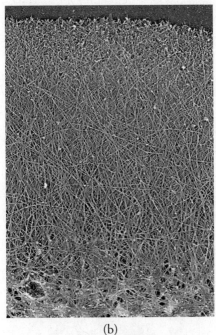

Calculations suggest that under cellular conditions, the equilibrium between monomeric actin and polymeric actin favors the polymer. However, the growth of microfilaments *in vivo* is limited by capping proteins that bind to and block further polymerization at the (+) or (−) ends. A process that removes a microfilament cap will target growth to the uncapped end. New microfilament growth can also occur as branches form along existing microfilaments.

A supply of actin monomers to support microfilament growth in one area must come at the expense of microfilament disassembly elsewhere. In a cell, certain proteins sever microfilaments by binding to a polymerized actin subunit and inducing a small structural change that weakens actin–actin interactions and thereby increases the likelihood that the microfilament will break at that point. Actin subunits can then dissociate from the newly exposed ends unless they are subsequently capped.

Capping, branching, and severing proteins, along with other proteins whose activity is sensitive to extracellular signals, regulate the assembly and disassembly of actin filaments. A cell containing a network of actin filaments can therefore change its shape as the filaments lengthen in one area and regress in another. Certain cells use this system to move. When a cell crawls along a surface, actin polymerization extends its "leading" edge, while depolymerization helps retract its "trailing" edge (**Fig. 5-17a**). The high density of growing filament ends at the leading edge of the cell (Fig. 5-17b) illustrates how the rapid formation and outward extension of actin filaments can modulate cell shape and drive cell locomotion. Not only do actin filaments provide structural support and generate cell movement by assembly and disassembly, they also participate in generating tensile force. This system is well

Figure 5-17 Actin filament dynamics in cell crawling. (a) Scanning electron micrograph of crawling cells. The leading edges of the cells (*lower left*) are ruffled where they have become detached from the surface and are in the process of extending. The trailing edges or tails of the cells, still attached to the surface (*upper right*), are gradually pulled toward the leading edge. The rate of actin polymerization is greatest at the leading edge. [Courtesy Guenter Albrecht-Buehler.] (b) Organization of actin filaments in a fish epithelial cell. At the leading edge of the cell (*top*), the filaments form a dense and highly branched network. Deeper within the cell (*bottom*), the filaments are sparser. [Courtesy Tatyana Svitkina, Northwestern University Medical School.]

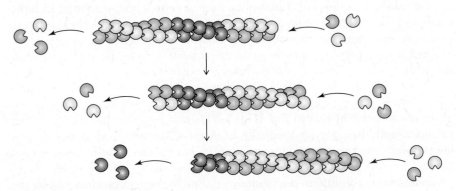

Figure 5-16 Actin filament treadmilling. Net assembly at one end balances net dissociation at the other end.

developed in muscle cells, where actin filaments are an essential part of the contractile apparatus (see Section 5-3).

Tubulin forms hollow microtubules

Like actin filaments, microtubules are cytoskeletal fibers built from small globular protein subunits. Consequently, they share with actin filaments the ability to assemble and disassemble on a time scale that allows the cell to rapidly change shape in response to external or internal stimuli. Compared to a microtubule, however, an actin filament is a thin and flexible rod. *A microtubule is about three times thicker and much more rigid because it is constructed as a hollow tube.* Consider the following analogy: A metal rod with the dimensions of a pencil is easily bent. The same quantity of metal, fashioned into a hollow tube with a larger diameter but the same length, is much more resistant to bending.

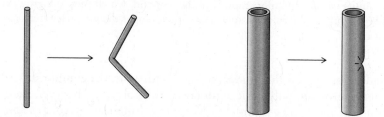

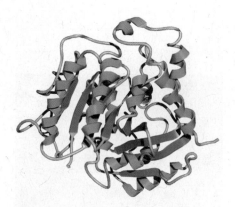

Figure 5-18 Structure of β-tubulin. The strands of the two β sheets are shown in blue, and the 12 α helices that surround them are green. [Structure of pig tubulin (pdb 1TUB) determined by E. Nogales and K. H. Downing.]

Bicycle frames, plant stems, and bones are built on this same principle. Cells use hollow microtubules to reinforce other elements of the cytoskeleton (see Fig. 5-12), to construct cilia and flagella, and to align and separate pairs of chromosomes during mitosis.

The basic structural unit of a microtubule is the protein tubulin. Two monomers, known as α-tubulin and β-tubulin, form a dimer, and a microtubule grows by the addition of tubulin dimers. Each tubulin monomer contains about 450 amino acids, 40% of them identical in α- and β-tubulin. The core of tubulin consists of a four-stranded and a six-stranded β sheet surrounded by 12 α helices (**Fig. 5-18**).

Each tubulin subunit includes a nucleotide-binding site. Unlike actin, tubulin binds a guanine nucleotide, either guanosine triphosphate (GTP) or its hydrolysis product, guanosine diphosphate (GDP). When the dimer forms, the α-tubulin GTP-binding site becomes buried in the interface between the monomers. The nucleotide-binding site in β-tubulin remains exposed to the solvent (**Fig. 5-19**). After the tubulin dimer is incorporated into a microtubule and another dimer binds on top of it, the β-tubulin nucleotide-binding site is also sequestered from solvent. The GTP is then hydrolyzed, but the resulting GDP remains bound to β-tubulin because it cannot diffuse away (the GTP in the α-tubulin subunit remains where it is and is not hydrolyzed).

Assembly of a microtubule begins with the end-to-end association of tubulin dimers to form a short linear **protofilament.** Protofilaments then align side-to-side in a curved sheet, which wraps around on itself to form a hollow tube of 13 protofilaments (**Fig. 5-20**). The microtubule extends as tubulin dimers add to both ends. Like a microfilament, the microtubule is polar and one end grows more rapidly. *The (+) end, terminating in β-tubulin, grows about twice as fast as the (−) or α-tubulin end because tubulin dimers bind preferentially to the (+) end.*

Disassembly of a microtubule also takes place at both ends but occurs more rapidly at the (+) end. Under conditions that favor depolymerization, the ends of the microtubule appear to fray (**Fig. 5-21**). This suggests that tubulin dimers do not simply dissociate individually from the microtubule ends but that the interactions between protofilaments weaken before the tubulin dimers come loose.

Under certain conditions, microtubule treadmilling can occur when tubulin subunits add to the (+) end as fast as they leave the (−) end. *In vivo,* the (−) ends are often anchored to some sort of organizing center in the cell. This means that most

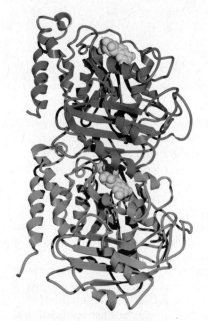

Figure 5-19 The tubulin dimer. The guanine nucleotide (gold) in the α-tubulin subunit (*bottom*) is inaccessible in the dimer, whereas the nucleotide in the β-tubulin subunit (*top*) is more exposed to the solvent.

(a)

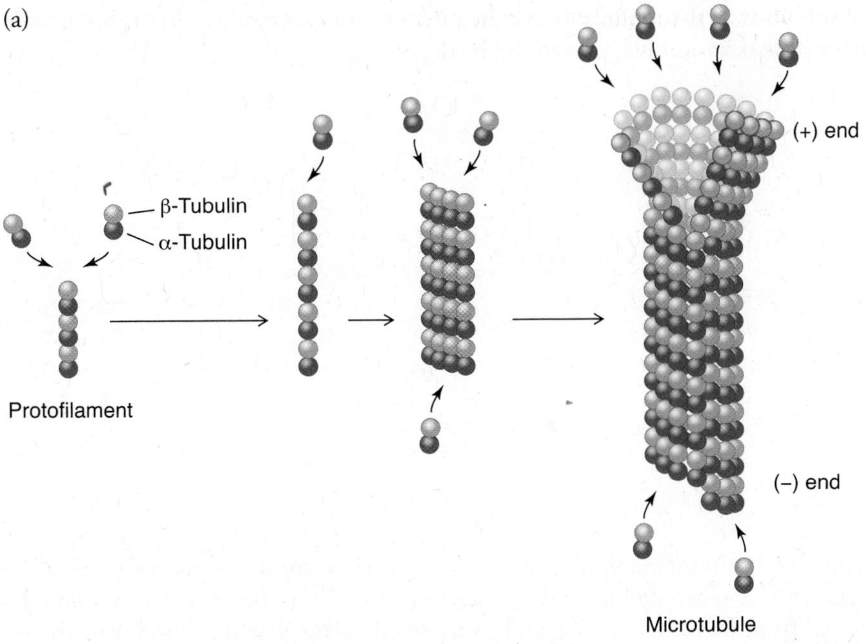

β-Tubulin
α-Tubulin

(+) end

(−) end

Protofilament

Microtubule

(b)

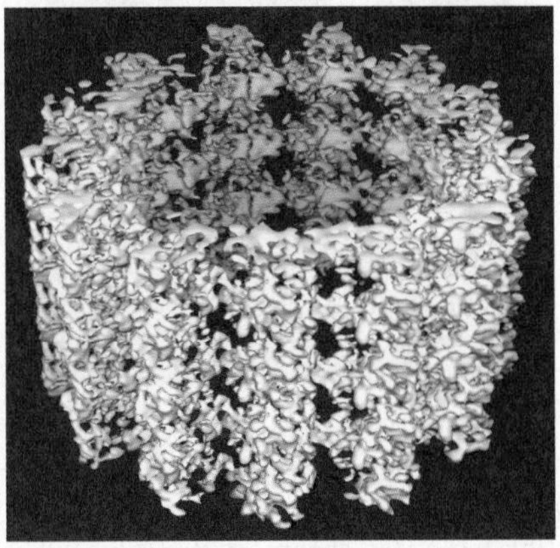

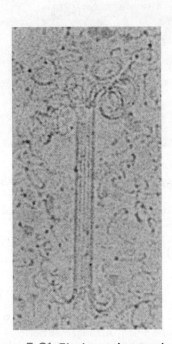

Figure 5-20 **Assembly of a microtubule.** (a) αβ Dimers of tubulin initially form a linear protofilament. Protofilaments associate side by side, ultimately forming a tube. Tubulin dimers can add to either end of the microtubule, but growth is about twice as fast at the (+) end. (b) Cryoelectron microscopy view of a microtubule. [Courtesy Kenneth Downing, Lawrence Berkeley National Laboratory.]

? **Compare a microtubule and an actin filament (Fig. 5-14) in terms of strength and speed of assembly.**

microtubule growth and regression occur at the (+) end. Microtubule dynamics are also regulated by proteins that cross-link microtubules and promote or prevent depolymerization.

Some drugs affect microtubules

Compounds that interfere with microtubule dynamics can have drastic physiological effects. One reason is that during mitosis, chromosomes separate along a spindle made of microtubules (**Fig. 5-22**). The drug colchicine, a product of the meadow saffron plant, causes microtubules to depolymerize, thereby blocking cell division.

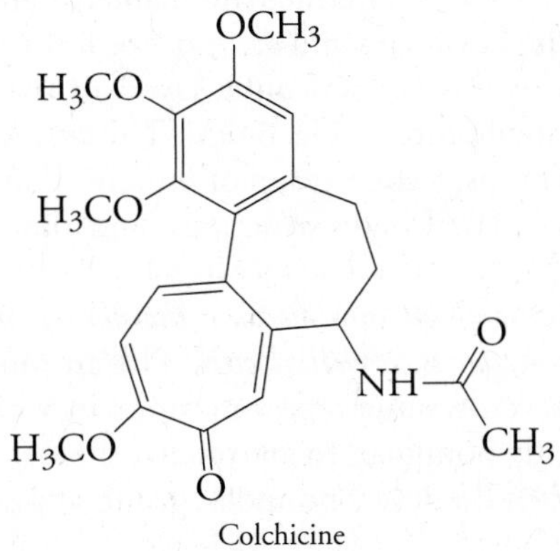

OCH₃

H₃CO

H₃CO

H₃CO

O

NH

O

CH₃

Colchicine

Colchicine binds at the interface between α- and β-tubulin in a dimer, facing the inside of the microtubule cylinder. The bound drug may induce a slight conformational change that weakens the lateral contacts between protofilaments. If enough colchicine is present, microtubules shorten and eventually disappear. Colchicine was first used over 2000 years ago to treat **gout** (inflammation stemming from the precipitation of uric acid in the joints) because it inhibits the action of the white blood cells that mediate inflammation.

Figure 5-21 **Electron micrograph of a depolymerizing microtubule.** The ends of protofilaments apparently curve away from the microtubule and separate before tubulin dimers dissociate. [Courtesy Ronald Milligan, The Scripps Research Institute.]

Taxol binds to β-tubulin subunits in a microtubule, but not to free tubulin, so it stabilizes the microtubule, preventing its depolymerization.

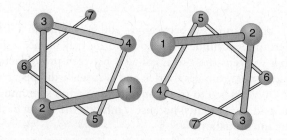

Taxol

Figure 5-22 Microtubules in a dividing cell. During mitosis, microtubules (green fluorescence) link replicated chromosomes (blue fluorescence) to two points at opposite sides of the cell. Disassembly of the microtubules, along with the action of motor proteins, draws the chromosomes apart before the cell splits in half. [Courtesy Alexey Khodjakov and Conly L. Rieder, Wadsworth Center, Albany.]

The taxol–tubulin interaction appears to include close contacts between taxol's phenyl groups and hydrophobic residues such as Phe, Val, and Leu. Taxol was originally extracted from the slow-growing and endangered Pacific yew tree, but it can also be purified from more renewable sources or chemically synthesized. *Taxol is used as an anticancer agent because it blocks cell division and is therefore toxic to rapidly dividing cells such as tumor cells.*

Keratin is an intermediate filament

In addition to actin filaments and microtubules, eukaryotic cells—particularly those in multicellular organisms—contain intermediate filaments. With a diameter of about 100 Å, these fibers are intermediate in thickness to actin filaments and microtubules. *Intermediate filaments are exclusively structural proteins.* They play no part in cell motility, and unlike actin filaments and microtubules, they have no associated motor proteins. However, they do interact with actin filaments and microtubules via cross-linking proteins.

Intermediate filament proteins as a group are much more heterogeneous than the highly conserved actin and tubulin. For example, humans have about 65 intermediate filament genes. The lamins are the intermediate filaments that help form the nuclear lamina in animal cells, a 30–100-Å-thick network inside the nuclear membrane that helps define the nuclear shape and may play a role in DNA replication and transcription. In many cells, intermediate filaments are much more abundant than microfilaments or microtubules and are most prominent in the dead remnants of epidermal cells—that is, in the hard outer layers of the skin—where they may account for 85% of the total protein (**Fig. 5-23**). The best-known intermediate filament proteins are the keratins, a large group of proteins that include the "soft" keratins, which help define internal body structures, and the "hard" keratins of skin, hair, and claws.

*The basic structural unit of an intermediate filament is a dimer of α helices that wind around each other—that is, a **coiled coil**.* The amino acid sequence in such a structure consists of seven-residue repeating units in which the first and fourth residues are predominantly nonpolar. In an α helix, these nonpolar residues line up along one side (**Fig. 5-24**). Because a nonpolar group appears on average every 3.5

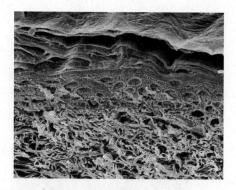

Figure 5-23 Scanning electron micrograph of sectioned human skin. The layers of dead epidermal cells at the top consist mostly of keratin. [Science Photo Library/Photo Researchers.]

Figure 5-24 Arrangement of residues in a coiled coil. This view down the axis of two seven-residue α helices shows that amino acids at positions 1 and 4 line up on one side of each helix. Nonpolar residues occupying these positions form a hydrophobic strip along the sides of the helices.

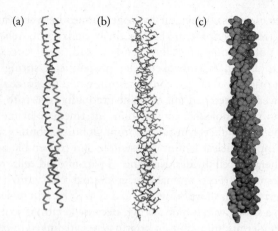

Figure 5-25 Three views of a coiled coil. These models show a segment of the coiled coil from the protein tropomyosin. (a) Backbone model. (b) Stick model. (c) Space-filling model. Each α-helical chain contains 100 residues. The nonpolar strips along each helix contact each other, so the two helices wind around each other in a gentle left-handed coil. [Structure of tropomyosin (pdb 1C1G) determined by F. G. Whitby and G. N. Phillips, Jr.]

? Which nonpolar residues would be most likely to appear at positions 1 and 4? Which would be least likely?

residues but there are 3.6 residues per α-helical turn, the strip of nonpolar residues actually winds slightly around the surface of the helix. Two helices whose nonpolar strips contact each other therefore adopt a coiled structure with a left-handed twist (**Fig. 5-25**).

Each intermediate filament subunit contains a stretch of α helix flanked by nonhelical regions at the N- and C-termini. Two of these polypeptides interact in register (parallel and with ends aligned) to form a coiled coil. The dimers then associate in a staggered antiparallel arrangement to form higher-order fibrous structures (**Fig. 5-26**). The fully assembled intermediate filament may consist of 16 to 32 polypeptides in cross-section. Note that no nucleotides are required for intermediate filament assembly. The N- and C-terminal domains may help align subunits during polymerization and interact with proteins that cross-link intermediate filaments to other cell components. Keratin fibers themselves are cross-linked through disulfide bonds between Cys residues on adjacent chains.

An animal hair—for example, sheep's wool or the hair on your head—consists almost entirely of keratin filaments (**Fig. 5-27**). Hair resists deformation but can

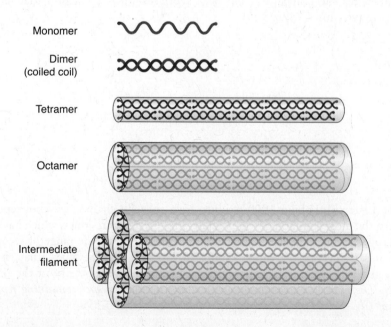

Monomer

Dimer
(coiled coil)

Tetramer

Octamer

Intermediate
filament

Figure 5-26 Model of an intermediate filament. Pairs of polypeptides form coiled coils. These dimers associate to form tetramers and so on, ultimately producing an intermediate filament composed of 16 to 32 polypeptides in cross-section. Although drawn as straight rods here, the intermediate filament and its component structures are probably all twisted around each other in some way, much like a man-made rope or cable.

be stretched. Tensile stress breaks the hydrogen bonds between carbonyl and amide groups four residues apart in the keratin α helix. The helices can then be pulled until the polypeptides are fully extended. Additional force causes the polypeptide chain to break. If unbroken, the protein can spring back—at least partially—to its original α-helical conformation when the force is removed. This is why a wool sweater stretched out of shape gradually reverts to its former style.

Perhaps due to their cable-like construction, intermediate filaments undergo less remodeling than cellular fibers constructed from globular subunits such as actin and tubulin. For example, nuclear lamins disassemble and reassemble only once during the cell cycle, when the cell divides. Keratins, as part of dead cells, remain intact for years. In the innermost layers of animal skin, epidermal cells synthesize large amounts of keratin. As layers of cells move outward and die, their keratin molecules are pushed together to form a strong waterproof coating. Because keratin is so important for the integrity of the epidermis, mutations in keratin genes are linked to certain skin disorders. In diseases such as epidermolysis bullosa simplex (EBS), cells rupture when subject to normal mechanical stress. The result is separation of epidermal layers, which leads to blistering. The most severe cases of the disease arise from mutations in the most highly conserved regions of the keratin molecules, near the ends of the α-helical regions.

Figure 5-27 Scanning electron micrograph of a human hair. [Tony Brain/ Science Photo Library/Photo Researchers.]

Collagen is a triple helix

Unicellular organisms can get by with just a cytoskeleton, but multicellular animals must have a way to hold their cells together according to some characteristic body plan. Large animals—especially nonaquatic ones—must also support the body's weight. This support is provided by collagen, which is the most abundant animal protein. It plays a major structural role in the extracellular matrix (the material that helps hold cells together), in connective tissue within and between organs, and in bone. Its name was derived from the French word for *glue* (at a time when glue was derived from animal connective tissue).

There are at least 28 types of collagen with different three-dimensional structures and physiological functions. The most familiar is the collagen from animal bones and tendons, which forms thick, ropelike fibers (**Fig. 5-28**). This type of collagen is a trimeric molecule about 3000 Å long but only 15 Å wide. As in all forms of collagen, the polypeptide chains have an unusual amino acid composition and an unusual conformation. Except in the extreme N- and C-terminal regions of the polypeptides (which are cleaved off once the protein exits the cell), every third amino acid is Gly, and about 30% of the remaining residues are proline and hydroxyproline (Hyp). Hyp residues result from the hydroxylation of Pro residues after the polypeptide has been synthesized, in a reaction that requires ascorbate (vitamin C; Box 5-D).

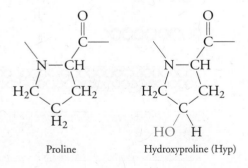

Proline   Hydroxyproline (Hyp)

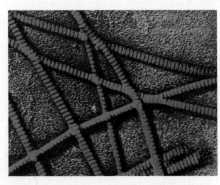

Figure 5-28 Electron micrograph of collagen fibers. Thousands of aligned collagen proteins yield structures with a diameter of 500 to 2000 Å. [J. Gross/ Science Photo Library/Photo Researchers.]

For a stretch of about 1000 residues, each collagen chain consists of repeating triplets, the most common of which is Gly–Pro–Hyp. Gly residues, which have only a hydrogen atom for a side chain, can normally adopt a wide range of secondary structures. However, the **imino groups** of Pro and Hyp residues (that is, their connected side chains and amino groups) constrain the geometry of the peptide group. *The most stable conformation for a polypeptide sequence containing repeating units of Gly –Pro –Hyp is a narrow left-handed helix* (**Fig. 5-29a**).

BOX 5-D ⬢ BIOCHEMISTRY NOTE

141
Structural Proteins

Vitamin C Deficiency Causes Scurvy

Ascorbate (vitamin C)

In the absence of ascorbate (vitamin C), collagen contains too few hydroxyproline residues and hydroxylated lysyl residues, so the resulting collagen fibers are relatively weak. Ascorbate also participates in enzymatic reactions involved in fatty acid breakdown and production of certain hormones. The symptoms of ascorbate deficiency include poor wound healing, loss of teeth, and easy bleeding—all of which can be attributed to abnormal collagen synthesis—as well as lethargy and depression.

Historically, ascorbate deficiency, known as scurvy, was common in sailors on long voyages where fresh fruit was unavailable. A remedy, in the form of a daily ration of limes, was discovered in the mid-eighteenth century. Unfortunately, citrus juice, which was also widely administered, proved much less effective in preventing scurvy because ascorbate is destroyed by heating and by prolonged exposure to air. For this reason, factors such as exercise and good hygiene were also believed to prevent scurvy.

Fruit is not the only source of ascorbate. Most animals—with the exception of bats, guinea pigs, and primates—produce ascorbate, so a diet containing fresh meat can supply sufficient ascorbate, which is vital in locations, such as the far North, where fruit is not available.

Given the presence of ascorbate in many foods, a true deficiency is rare in adults. Scurvy does still occur, however, as a side effect of general malnutrition or in individuals consuming odd diets. Fortunately, the symptoms of scurvy, which is otherwise fatal, can be easily reversed by administering ascorbate or consuming fresh food.

⬤ **Question:** Refer to the structure of ascorbate to explain why it does not accumulate in the body but instead is readily lost via the kidneys.

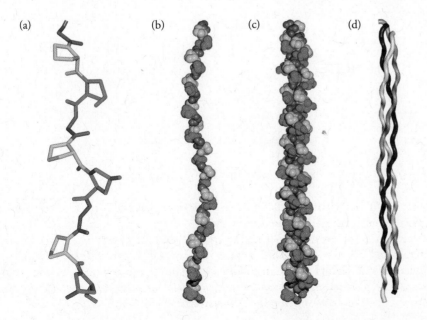

Figure 5-29 Collagen structure.
(a) A sequence of repeating Gly–Pro–Hyp residues adopts a secondary structure in which the polypeptide forms a narrow left-handed helix. The residues in this stick model are color-coded: Gly gray, Pro orange, Hyp red. H atoms are not shown. (b) Space-filling model of a single collagen polypeptide. (c) Space-filling model of the triple helix. (d) Backbone trace showing the three polypeptides in different shades of gray. Each polypeptide has a left-handed twist, but the triple helix has a right-handed twist. [Model of collagen (pdb 2CLG) constructed by J. M. Chen.]

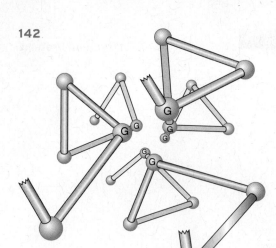

Figure 5-30 Cross-section of the collagen triple helix. In this view, looking down the axis of the three-chain molecule, each ball represents an amino acid, and the bars represent peptide bonds. Gly residues (which lack side chains and are marked by "G") are located in the center of the triple helix, whereas the side chains of other residues point outward from the triple helix.

? Compare this axial view of a collagen triple helix to the axial view of a coiled coil (Fig. 5-24).

In collagen, three polypeptides wind around each other to form a right-handed **triple helix** (Fig. 5-29b–d). The chains are parallel but staggered by one residue so that Gly appears at every position along the axis of the triple helix. The Gly residues are all located in the center of the helix, whereas all other residues are on the periphery. A look down the axis of the triple helix shows why Gly— but no other residue—occurs in the center of the helix (**Fig. 5-30**). The side chain of any other residue would be too large to fit. In fact, replacing Gly with Ala, the next-smallest amino acid, greatly perturbs the structure of the triple helix.

The collagen triple helix is stabilized through hydrogen bonding. One set of interactions links the backbone N—H group of each Gly residue to a backbone C=O group in another chain. The geometry of the triple helix prevents the other backbone N—H and C=O groups from forming hydrogen bonds with each other, but they are able to interact with a highly ordered network of water molecules surrounding the triple helix like a sheath.

Collagen molecules are covalently cross-linked

Trimeric collagen molecules assemble in the endoplasmic reticulum. After they are secreted from the cell, they are trimmed by proteases and align side-to-side and end-to-end to form the enormous fibers visible by electron microscopy (see Fig. 5-28). The fibers are strengthened by several kinds of cross-links. Because collagen polypeptides contain almost no cysteine, these links are not disulfide bonds. Instead, *the cross-links are covalent bonds between side chains that have been chemically modified following polypeptide synthesis.* For example, one kind of cross-link requires the enzyme-catalyzed oxidation of two Lys side chains, which then react to form a covalent bond. The number of these and other types of cross-links tends to increase with age, which explains why meat from older animals is tougher than meat from younger animals.

$$\vdots$$
$$C=O$$
$$CH-(CH_2)_3-CH_2-\overset{+}{N}H_3 \qquad H_3\overset{+}{N}-CH_2-(CH_2)_3-CH$$
$$NH \qquad\qquad Lys \qquad\qquad\qquad Lys \qquad\qquad NH$$
$$\vdots \qquad\qquad\qquad\qquad\qquad\qquad\qquad\qquad C=O$$

$$\downarrow \qquad\qquad\qquad\qquad\qquad \downarrow$$

$$\vdots \qquad\qquad\qquad\qquad\qquad\qquad\qquad\qquad \vdots$$
$$C=O \qquad\qquad O \qquad\qquad\qquad O \qquad\qquad C=O$$
$$CH-(CH_2)_3-CH \qquad\qquad HC-(CH_2)_3-CH$$
$$NH \qquad\qquad\qquad\qquad\qquad\qquad\qquad\qquad\qquad NH$$
$$\vdots \qquad\qquad\qquad\qquad\qquad\qquad\qquad\qquad \vdots$$

$$\underbrace{\qquad\qquad\qquad\qquad\qquad\qquad}$$
$$\downarrow$$

$$\vdots \qquad\qquad\qquad\qquad\qquad\qquad \vdots$$
$$C=O \qquad\qquad\qquad\qquad\qquad C=O$$
$$CH-(CH_2)_2-C=CH-(CH_2)_3-CH$$
$$NH \qquad\qquad\qquad CH \qquad\qquad NH$$
$$\vdots \qquad\qquad\qquad\quad O \qquad\qquad\quad \vdots$$

Collagen fibers have tremendous tensile strength. On a per-weight basis, collagen is stronger than steel. Not all types of collagen form thick linear fibers, however. Many nonfibrillar collagens form sheetlike networks of fibers that support layers of cells in tissues. Often, several types of collagen are found together. Not surprisingly, defects in collagen affect a variety of organ systems (Box 5-E).

CONCEPT REVIEW

- What is the basis of actin filament polarity?
- How does actin filament remodeling affect the cell's shape?
- How does a microtubule differ structurally from an actin filament?
- Why does normal physiological function require regulation of the assembly and disassembly of cytoskeletal elements?
- How do the structures of intermediate filaments such as keratin differ from the structures of actin filaments and microtubules?
- How do two α helices form a coiled coil?
- How does the structure of collagen differ from the structure of keratin or the structures of fibers with globular subunits?
- What is the structural importance of Gly residues in collagen?
- How are collagen fibers stabilized and cross-linked?

BOX 5-E ⬡ **CLINICAL CONNECTION**

Genetic Collagen Diseases

Connective tissue, such as cartilage and bone, consists of cells embedded in a matrix containing proteins (mainly collagen) and a space-filling "ground substance" (mostly polysaccharides; see Section 11-3). The polysaccharides, which are highly hydrated, are resilient and return to their original shape after being compressed. The collagen fibers are strong and relatively rigid, resisting tensile (stretching) forces. Together, the polysaccharides and collagen give ligaments (which attach bone to bone) and tendons (which join muscles to bones) the appropriate degree of resistance and flexibility. The connective tissues that surround muscles and organs contain collagen fibers arranged in sheetlike networks with similar physical properties.

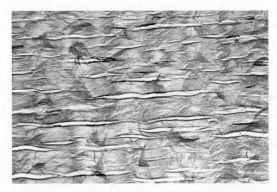

The regular arrays of collagen fibers (*horizontal bands*) in tendon allow the tissue to resist tension. Fibroblasts, collagen-producing cells, occupy the spaces between the fibers. Cartilage typically lacks blood vessels. [Mark Nielsen.]

In bone, the extracellular matrix is supplemented by minerals, mainly hydroxyapatite, $Ca(PO_4)_3(OH)$. This form of calcium phosphate forms extensive white crystals that account for up to 50% of the mass of bone. By itself, calcium phosphate is brittle. Yet bone is thousands of times stronger than the crystals because it is a composite structure in which the mineral is interspersed with collagen fibers. This layered arrangement dissipates stresses so that the bone remains strong but can "give" a little under pressure without shattering. Bone tissue also includes passageways for small blood vessels.

During development, most of the skeleton takes shape as cartilaginous tissue becomes mineralized, forming hard bone. The repair of a fractured bone follows a similar process in which fibroblasts moving to the injured site synthesize large amounts of collagen, chondroblasts produce cartilage, and osteoblasts gradually replace the cartilage with bony tissue. Mature bone continues to undergo remodeling, which begins with the release of enzymes and acid from cells known as osteoclasts. The enzymes digest collagen and other extracellular matrix components, while the low pH helps dissolve calcium phosphate. Osteoblasts then fill the void with new bone material. Some bones are remodeled faster than others, and the system responds to physical demands so that the bone becomes stronger and thicker when subjected to heavy loads. This is the basis for orthodontics: Teeth are realigned by placing stress on the bone in the tooth sockets for a period long enough for remodeling to take place.

The importance of collagen for the structure and function of connective tissues means that irregularities in the collagen protein itself or in the enzymes that process collagen molecules can lead to serious physical abnormalities. Hundreds of collagen-related mutations have been identified. Because most tissues contain more than one type of collagen, the physiological manifestations of collagen mutations are highly variable.

Defects in collagen type I (the major form in bones and tendons) cause the congenital disease **osteogenesis imperfecta.** The primary symptoms of the disease include bone fragility leading to easy fracture, long-bone deformation, and abnormalities of the skin and teeth.

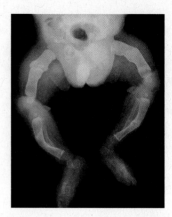

X-Ray of a child with a moderately severe case of osteogenesis imperfecta. [ISM/Phototake.]

Collagen type I, a trimeric molecule, contains two different types of polypeptide chains. Therefore, the severity of the disease depends in part on whether one or two chains in a collagen molecule are affected. Furthermore, *the location and nature of the mutation determine whether the abnormal collagen retains some normal function.* For example, in one severe form of osteogenesis imperfecta, a 599-base deletion in a collagen gene represents the loss of a large portion of triple helix. The resulting protein is unstable and is degraded intracellularly. Milder cases of osteogenesis imperfecta result from amino acid substitutions, for example, the replacement of Gly by a bulkier residue. Other amino acid changes may slow intracellular processing and excretion of collagen polypeptides, which affects the assembly of collagen fibers. Osteogenesis imperfecta affects about one in 10,000 people.

Mutations in collagen type II, a form found in cartilage, lead to osteoarthritis. This genetic disease, which becomes apparent in childhood, is distinct from the osteoarthritis that can develop later in life, often after years of wear and tear on the joints. Defects in the proteins that process collagen extracellularly and help assemble collagen fibers lead to disorders such as dermatosparaxis, which is characterized by extreme skin fragility.

Ehlers–Danlos syndrome results from abnormalities in collagen type III, a molecule that is abundant in most tissues but is scarce in skin and bone. Symptoms of this phenotypically variable disorder include easy bruising, thin or elastic skin, and joint hyperextensibility. In one form of the disease, which is accompanied by a high risk for arterial rupture, the molecular defect is a mutation in a collagen type III gene. In another form of the disease, in which individuals often suffer from scoliosis (curvature of the spine), the collagen genes are normal. In these cases, the disease results from a deficiency of lysyl oxidase, the enzyme that modifies Lys residues so that they can participate in collagen cross-links. Ehlers–Danlos syndrome is both rarer and less severe than osteogenesis imperfecta, with many affected individuals surviving to adulthood.

● Questions:

1. Explain why a broken bone, which is hard, can heal within a few weeks, whereas a torn ligament (which is softer) may take months to heal.
2. Osteoporosis is the weakening of bones that commonly occurs with aging. Why is exercise recommended as a strategy for preventing fractures in the elderly?
3. Osteogenesis imperfecta is a genetic disease, but most cases result from new mutations rather than ones that are passed from parent to child. Explain.
4. Because networks of collagen fibers lend resilience to the deep (living) layers of skin, some "anti-aging" cosmetics contain purified collagen. Explain why this exogenous (externally supplied) collagen is unlikely to attenuate the skin wrinkling that accompanies aging.

5-3 Motor Proteins

KEY CONCEPTS
- The motor protein myosin couples the steps of ATP hydrolysis to conformational changes, resulting in muscle contraction.
- Kinesin transports cargo by moving processively along a microtubule track.

▶▶ **CAN** proteins move?

An assortment of molecular machinery acts on structural proteins—actin filaments and microtubules—to generate the movements that allow cells to reorganize their contents, change their shape, and even crawl or swim. For example, muscle contraction is accomplished by the motor protein myosin pulling on actin filaments. Eukaryotic cells that move via the wavelike motion of cilia or flagella rely on the motor protein dynein, which acts by bending a bundle of microtubules. Intracellular transport is carried out by motor proteins that move along microfilament and microtubule tracks. In this section, we focus on two well-characterized motor proteins, myosin and kinesin, which provide additional examples of the ways that proteins can move.

Myosin has two heads and a long tail

There are at least 20 different types of myosin, which is present in nearly all eukaryotic cells. Myosin works with actin to produce movement by transducing chemical energy (in the form of ATP) to mechanical energy. Muscle myosin, with a total

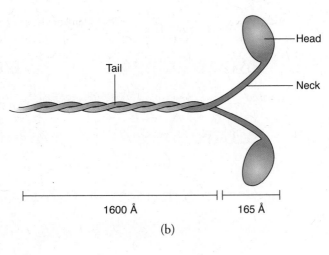

(b)

(a)

Figure 5-31 Structure of muscle myosin. (a) Electron micrograph. [Courtesy John Trinick, University of Leeds.] (b) Drawing of a myosin molecule. Myosin's two globular heads are connected via necks to myosin's tail, where the polypeptide chains form a coiled coil.

? In what part of the protein are large hydrophobic residues most likely to occur?

molecular mass of about 540 kD, consists mostly of two large polypeptides that form two globular heads attached to a long tail (**Fig. 5-31**). Each head includes a binding site for actin and a binding site for an adenine nucleotide. In the tail region, the two polypeptides twist around each other to form a single rodlike coiled coil (the same structural motif that occurs in intermediate filaments). The "neck" that joins each myosin head to the tail region consists of an α helix about 100 Å long, around which are wrapped two small polypeptides (**Fig. 5-32**). These so-called light chains help stiffen the neck helix so that it can act as a lever.

Each myosin head can interact noncovalently with a subunit in an actin filament, but the two heads act independently, and only one head binds to the actin filament at a given time. *In a series of steps that include protein conformational changes and the hydrolysis of ATP, the myosin head releases its bound actin subunit and rebinds another subunit closer to the (+) end of the actin filament.* Repetition of this reaction cycle allows myosin to progressively walk along the length of the actin filament.

In a muscle cell, hundreds of myosin tails associate to form a **thick filament** with the head domains sticking out (**Fig. 5-33**). These heads act as cross-bridges to **thin filaments,** which each consist of an actin filament and actin-binding proteins that regulate the accessibility of the actin subunits to myosin heads. When a muscle contracts, the multitude of myosin heads individually bind and release actin, like rowers working asynchronously, which causes the thin and thick filaments to slide past each other (**Fig. 5-34**). Because of the arrangement of filaments in the muscle cell, the action of myosin on actin results in an overall shortening of the muscle. This phenomenon is commonly called contraction, but the muscle does not undergo any compression

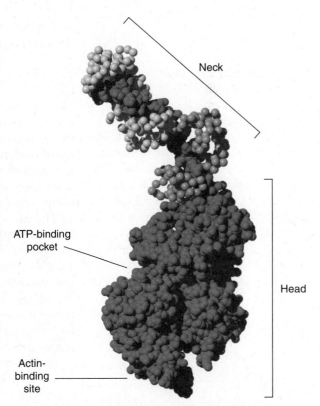

Figure 5-32 Myosin head and neck region. The myosin neck forms a molecular lever between the head domain and the tail. Two light chains (lighter shades of purple) help stabilize the α-helical neck. The actin-binding site is at the far end of the myosin head. ATP binds in a cleft near the middle of the head. Only the alpha carbons of the light chains are visible in this model. [Structure of chicken myosin (pdb 2MYS) determined by I. Rayment and H. M. Holden.]

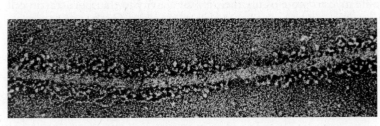

Figure 5-33 Electron micrograph of a thick filament. The heads of many myosin molecules project laterally from the thick rod formed by the aligned myosin tails. [From Trinick, J., and Elliott, A., *J. Mol. Biol.* 131, 15 (1977).]

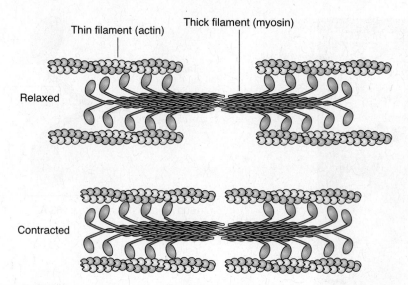

Figure 5-34 Movement of thin and thick filaments during muscle contraction.

and its volume remains constant—it actually becomes thicker around the middle. A shortening on the order of 20% of a muscle's length is typical; 40% is extreme.

Myosin operates through a lever mechanism

How does myosin work? The key to its mechanism is the hydrolysis of the ATP that is bound to the myosin head. Although the ATP-binding site is about 35 Å away from the actin-binding site, *the conversion of ATP to ADP + P_i triggers conformational changes in the myosin head that are communicated to the actin-binding site as well as to the lever (the neck region). The chemical reaction of ATP hydrolysis thereby drives the physical movement of myosin along an actin filament.* In other words, the free energy released by hydrolysis of ATP is transformed into mechanical work.

The four steps of the myosin–actin reaction sequence are shown in **Figure 5-35**. Note how each event at the nucleotide-binding site correlates with a conformational change related to either actin binding or bending of the lever. An α helix makes an ideal lever because it can be quite long. It is also relatively incompressible, so it can pull the coiled-coil myosin tail along with it. Altogether, the lever swings by about 70° relative to the myosin head. The return of the lever to its original conformation (step 4 of the reaction cycle) is the force-generating step. When adjusted for the difference in mass, the myosin–actin system has a power output comparable to a typical automobile.

Each cycle of ATP hydrolysis moves the myosin head by an estimated 50–100 Å. Since individual actin subunits are spaced about 55 Å apart along the thin filament, the myosin head advances by at least one actin subunit per reaction cycle. Because the reaction cycle involves several steps, some of which are essentially irreversible (such as ATP → ADP + P_i), the entire cycle is unidirectional.

Myosin operates in many cells, not just in muscles. For example, myosin works with actin during **cytokinesis** (the splitting of the cell into two halves following mitosis), and some myosin proteins use their motor activity to transport certain cell components along microfilament tracks. Myosin molecules may also act as tension rods to cross-link the microfilaments of the cytoskeleton. This is one reason why mutations in myosin in the sensory cells of the ear cause deafness and other abnormalities (Box 5-F).

Kinesin is a microtubule-associated motor protein

Many cells also contain motor proteins, such as kinesin, that move along microtubule tracks. There are several different types of kinesins; we will describe the prototypical one, also known as conventional kinesin.

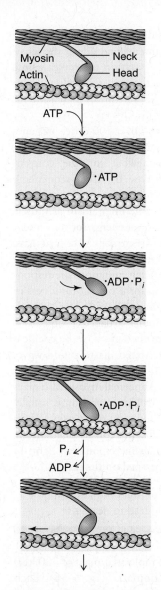

1. *The reaction sequence begins with a myosin head bound to an actin subunit of the thin filament. ATP binding alters the configuration of the myosin head so that it releases actin.*

2. *The rapid hydrolysis of ATP to ADP + P_i triggers a conformational change that rotates the myosin lever and increases the affinity of myosin for actin.*

3. *Myosin binds to an actin subunit farther along the thin filament.*

4. *Binding to actin causes P_i and then ADP to be released. As these reaction products exit, the myosin lever returns to its original position. This causes the thin filament to move relative to the thick filament (the power stroke).*

ATP replaces the lost ADP to repeat the reaction cycle.

Figure 5-35 The myosin–actin reaction cycle. For simplicity, only one myosin head is shown. ● See Animated Figure. Mechanism of force generation in muscle.

? **Write an equation that describes the reaction catalyzed by myosin.**

BOX 5-F ⬡ **BIOCHEMISTRY NOTE**

Myosin Mutations and Deafness

Inside the cochlea, the spiral-shaped organ of the inner ear, are thousands of hair cells, each of which is topped with a bundle of bristles known as **stereocilia.**

[P. Motta/Science Photo Library/Photo Researchers.]

Each stereocilium contains several hundred cross-linked actin filaments and is therefore extremely rigid, except at its base, where there are fewer actin filaments. Sound waves deflect the stereocilia at the base, initiating an electrical signal that is transmitted to the brain.

(continued on the next page)

Myosin molecules probably help control the tension inside each stereocilium, so the ratcheting activity of the myosin motors along the actin filaments may adjust the sensitivity of the hair cells to different degrees of stimulus. Other myosin molecules whose tails bind certain cell constituents may use their motor activity to redistribute these substances along the length of the actin filaments. Abnormalities in any of these proteins could interfere with normal hearing.

About half of all cases of deafness have a genetic basis, and over 100 different genes have been linked to deafness. One of these codes for myosin type VIIa, which is considered to be an "unconventional" myosin because it differs somewhat from the "conventional" myosin of skeletal muscle (also known as myosin type II). Gene-sequencing studies indicate that myosin VIIa has 2215 residues and forms a dimer with two heads and a long tail. Its head domains probably operate by the same mechanism as muscle myosin, converting the chemical energy of ATP into mechanical energy for movement relative to an actin filament. Over a hundred mutations in the myosin VIIa gene have been identified, including premature stop codons, amino acid substitutions, and deletions—all of which compromise the protein's function. Such mutations are responsible for many cases of **Usher syndrome,** the most common form of deaf-blindness in the United States. Usher syndrome is characterized by profound hearing loss, retinitis pigmentosa (which leads to blindness), and sometimes vestibular (balance) problems.

The congenital deafness of Usher syndrome results from the failure of the cochlear hair cells to develop properly. The unresponsiveness of the stereocilia to sound waves probably also accounts for their inability to respond normally to the movement of fluid in the inner ear, which is necessary for maintaining balance. Abnormal myosin also plays a role in the blindness that often develops in individuals with Usher syndrome, usually by the second or third decade. The intracellular transport function of myosin VIIa is responsible for distributing bundles of pigment in the retina. In retinitis pigmentosa, retinal neurons gradually lose their ability to transmit signals in response to light; in advanced stages of the disease, pigment actually becomes clumped on the retina.

● **Question:** Not all mutations are associated with a loss of function. Explain how a myosin mutation that *increases* the rate of ADP release could lead to deafness.

Kinesin, like myosin, is a relatively large protein (with a molecular mass of 380 kD) and has two large globular heads and a coiled-coil tail domain (**Fig. 5-36**). Each 100-Å-long head consists of an eight-stranded β sheet flanked by three α helices on each side and includes a tubulin-binding site and a nucleotide-binding site. The light chains, situated at the opposite end of the protein, bind to proteins in the

(a)

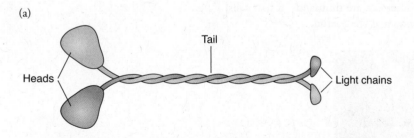

(b)

Figure 5-36 Structure of kinesin. (a) Diagram of the molecule. (b) Model of the head and neck region. Each globular head, which contains α helices and β sheets, connects to an α helix that winds around its counterpart to form a coiled coil. Two light chains at the end of the coiled-coil tail can interact with a membranous vesicle, the "cargo." [Structure of rat kinesin (pdb 3KIN) determined by F. Kozielski, S. Sack, A. Marx, M. Thormahlen, E. Schonbrunn, V Biou, A. Thompson, E.-M. Mandelkow, and E. Mandelkow.]

? Compare the structure of kinesin to the structure of myosin (Fig. 5-31).

membrane shell of a **vesicle.** The vesicle and its contents become kinesin's cargo. *Kinesin moves its cargo toward the (+) end of a microtubule by stepping along the length of a single protofilament.* Other microtubule-associated motor proteins appear to use a similar mechanism but move toward the (−) end of the microtubule.

The motor activity of kinesin requires the chemical energy of ATP hydrolysis. However, *kinesin cannot follow the myosin lever mechanism because its head domains are not rigidly fixed to its neck regions.* In kinesin, a relatively flexible polypeptide segment joins each head to an α helix that eventually becomes part of the coiled-coil tail (see Fig. 5-36b). (Recall that in myosin, the lever is a long α helix that extends from the head to the coiled-coil region and is stiffened by the two light chains; see Fig. 5-32.) Nevertheless, the relative flexibility of kinesin's neck is critical for its function.

Kinesin's two heads are not independent but work in a coordinated fashion so that the two heads alternately bind to successive β-tubulin subunits along a proto-filament, as if walking. Conformational changes elicited by ATP binding and hydrolysis are relayed to other regions of the molecule (**Fig. 5-37**). *This transforms the free energy of ATP into the mechanical movement of kinesin.* Each ATP-binding event

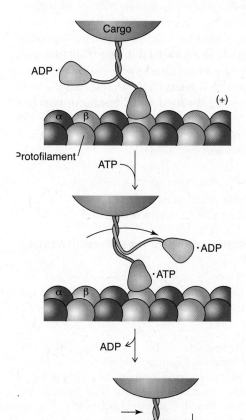

1. *ATP binding to the leading head (orange) induces a conformational change in which the neck docks against the head. This movement swings the trailing head (yellow) forward by 180° toward the (+) end of the microtubule. This is the force-generating step.*

2. *The new leading head (yellow) quickly binds to a tubulin subunit and releases its ADP. This step moves kinesin's cargo forward along the protofilament.*

3. *In the trailing head (orange), ATP is hydrolyzed to ADP + P_i. The P_i diffuses away, and the trailing head begins to detach from the microtubule.*

4. *ATP binds to the leading head to repeat the reaction cycle.*

Figure 5-37 The kinesin reaction cycle. The cycle begins with one kinesin head bound to a tubulin subunit in a protofilament of a microtubule. The trailing head has ADP in its nucleotide-binding site. For clarity, the relative size of the neck region is exaggerated.

? **What prevents kinesin from moving backward?**

yanks the trailing head forward by about 160 Å, so the net movement of the at-tached cargo is about 80 Å, or the length of a tubulin dimer.

Kinesin is a processive motor

As in the myosin–actin system (see Fig. 5-35), the kinesin–tubulin reaction cycle proceeds in only one direction. Although most molecular movement is associated with ATP binding, ATP hydrolysis is a necessary part of the reaction cycle. The slowest step of the reaction cycle shown in Figure 5-37 is the dissociation of the trailing kinesin head from the microtubule. ATP binding to the leading head may help promote trailing-head release as part of the forward-swing step. Because the free head quickly rebinds tubulin, the kinesin heads spend most of their time bound to the microtubule track.

One consequence of kinesin's almost constant hold on a microtubule is that many—perhaps 100 or more—cycles of ATP hydrolysis and kinesin advancement can occur before the motor dissociates from its microtubule track. *Kinesin is therefore said to have high* **processivity**. A motor protein such as myosin, which dissociates from an actin filament after a single stroke, is not processive.

In a muscle cell, low processivity is permitted because the many myosin–actin interactions occur more or less simultaneously to cause the thin and thick filaments to slide past each other (see Fig. 5-34). *High processivity is advantageous for a transport engine such as kinesin, because its cargo (which is relatively large and bulky) can be moved long distances without being lost.* Consider the need for efficient transport in a neuron. Neurotransmitters and membrane components are synthesized in the cell body, where ribosomes are located, but must be moved to the end of the axon, which may be several meters long in some cases (**Fig. 5-38**).

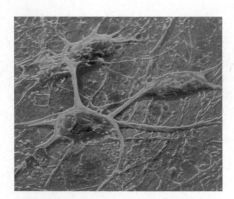

Figure 5-38 Electron micrograph of neurons. Microtubule-associated motor proteins move cargo between the cell body and the ends of the axon and other cell processes. [CNRI/Science Photo Library/Photo Researchers.]

> **CONCEPT REVIEW**
> • Describe the steps of the myosin-actin reaction cycle.
> • Which events at the nucleotide-binding site correlate with protein conformational changes?
> • Compare and contrast the myosin-actin and kinesin-tubulin reaction cycles.
> • Why is kinesin a processive motor?

[SUMMARY]

5-1 Myoglobin and Hemoglobin: Oxygen-Binding Proteins

• Myoglobin contains a heme prosthetic group that reversibly binds oxygen. The amount of O_2 bound depends on the O_2 concentration and on myoglobin's affinity for oxygen.
• Hemoglobin's α and β chains are homologous to myoglobin, indicating a common evolutionary origin.
• O_2 binds cooperatively to hemoglobin with low affinity, so hemoglobin can efficiently bind O_2 in the lungs and deliver it to myoglobin in the tissues.
• Hemoglobin is an allosteric protein whose four subunits alternate between the T (deoxy) and R (oxy) conformations in response to O_2 binding to the heme groups. The deoxy conformation is favored by low pH (the Bohr effect) and by the presence of BPG.

5-2 Structural Proteins

• The microfilament elements of a cell's cytoskeleton are built from ATP-binding globular actin subunits that polymerize as a double chain. Polymerization is reversible, so actin filaments undergo growth and regression. Their dynamics may be modified by proteins that mediate filament capping, branching, and severing.
• GTP-binding tubulin dimers polymerize to form a hollow microtubule. Polymerization is more rapid at one end, and the

microtubule can disassemble rapidly by fraying. Drugs that affect microtubule dynamics interfere with cell division.
• The intermediate filament keratin contains two long α-helical chains that coil around each other so that their hydrophobic residues are in contact. Keratin filaments associate and are cross-linked to form semipermanent structures.
• Collagen polypeptides contain a large amount of proline and hydroxyproline and include a Gly residue at every third position. Each chain forms a narrow left-handed helix, and three chains coil around each other to form a right-handed triple helix with Gly residues at its center. Covalent cross-links strengthen collagen fibers.

5-3 Motor Proteins

• Myosin, a protein with a coiled-coil tail and two globular heads, interacts with actin filaments to perform mechanical work. ATP-driven conformational changes allow the myosin head to bind, release, and rebind actin. This mechanism, in which myosin acts as a lever, is the basis of muscle contraction.
• The motor protein kinesin has two globular heads connected by flexible necks to a coiled-coil tail. Kinesin transports vesicular cargo along the length of a microtubule by a processive stepping mechanism that is driven by conformational changes triggered by ATP binding and hydrolysis.

[GLOSSARY TERMS]

cytoskeleton	oxyhemoglobin	treadmilling
motor protein	T state	protofilament
heme	R state	gout
prosthetic group	allosteric protein	coiled coil
Y	ligand	imino group
pO_2	anemia	triple helix
saturation	thalassemia	osteogenesis imperfecta
p_{50}	Bohr effect	Ehlers–Danlos syndrome
globin	microfilament	thick filament
homologous proteins	intermediate filament	thin filament
invariant residue	microtubule	stereocilia
conservative substitution	F-actin	Usher syndrome
variable residue	G-actin	cytokinesis
cooperative binding	(−) end	vesicle
deoxyhemoglobin	(+) end	processivity

BIOINFORMATICS ▶ **PROJECT 4**   Learn to use database tools to compare and align protein sequences and construct phylogenetic trees.

USING DATABASES TO COMPARE AND INDENTIFY RELATED PROTEIN SEQUENCES

[PROBLEMS]

5-1 Myoglobin and Hemoglobin: Oxygen-Binding Proteins

1. Explain why globin alone or heme alone is not effective as an oxygen carrier.

2. Myoglobin transfers oxygen obtained from hemoglobin to mitochondrial proteins involved in the catabolism of metabolic fuels to produce energy for the muscle cell. Using what you have learned in this chapter about myoglobin and hemoglobin, what can you conclude about the structures of these mitochondrial proteins?

3. Myoglobin most effectively facilitates oxygen diffusion through muscle cells when the intracellular oxygen partial pressure is comparable to the p_{50} value of myoglobin. Explain why.

4. Heart and muscle cells, where myoglobin resides, maintain an intracellular oxygen partial pressure of about 2.5 torr. Explain why a small change in oxygen partial pressure of 1 torr in either direction results in a dramatic change in myoglobin oxygen binding.

5. Oxygenated myoglobin is bright red, whereas deoxygenated myoglobin is a purplish color. Myoglobin in which the Fe^{2+} has been oxidized to Fe^{3+} is referred to as met-myoglobin and appears brown. The sixth ligand in met-myoglobin is water rather than oxygen.

(a) When a fresh, uncooked beef roast is sliced in two, the surface of the meat first appears purplish but then turns red. Explain why.

(b) Why does meat turn brown when it is cooked?

(c) Why might a retailer choose to package meat using oxygen-permeable plastic wrap instead of a vacuum-sealed package?

6. Myoglobins from three species of fish—tuna, bonito, and mackerel—were purified, and their oxygen binding affinities were measured at 20°C. The results are shown in the figure above right. (a) Use the graph to estimate the p_{50} values for each species. (b) Which species of fish has the highest oxygen-binding affinity at the experimental temperature? The least?

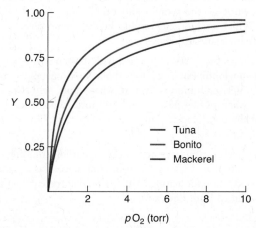

7. The oxygen-binding site in myoglobin, shown in Figure 5-2 and in the figure below, resides in a heme pocket in which a valine residue provides one of the pocket boundaries. How is oxygen binding affected when the Val residue is mutated to an Ile?

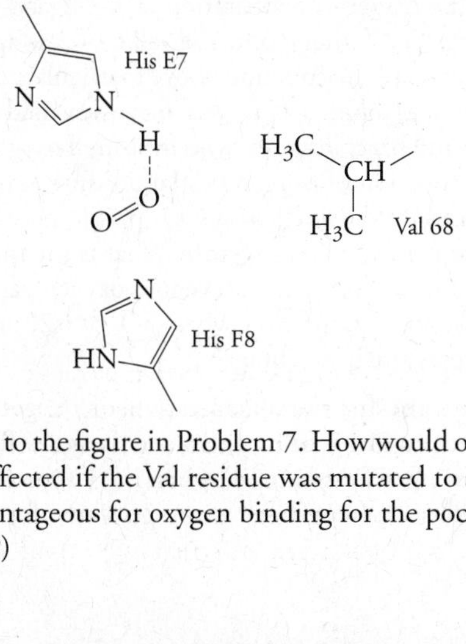

His E7

H

O═O

H_3C — CH — H_3C Val 68

His F8

8. Refer to the figure in Problem 7. How would oxygen-binding affinity be affected if the Val residue was mutated to a Ser? (*Hint:* Is it more advantageous for oxygen binding for the pocket to be polar or nonpolar?)

9. Hexacoordinate Fe(II) in heme is bright red. Pentacoordinate Fe(II) is blue. Explain how these electronic changes account for the different colors of arterial (scarlet) and venous (purple) blood.

10. Carbon monoxide, CO, binds to hemoglobin about 250 times as tightly as oxygen. The resulting complex is a red color that is much brighter than that observed when oxygen is bound. What symptoms would you expect to observe in a patient with CO poisoning?

11. Below is a list of the first 10 residues of the B helix in myoglobin from different organisms. Based on this information, which positions **(a)** cannot tolerate substitution, **(b)** can tolerate conservative substitution, and **(c)** are highly variable?

Position	Human	Chicken	Alligator	Turtle	Tuna	Carp
1	D	D	K	D	D	D
2	I	I	L	L	Y	F
3	P	A	P	S	T	E
4	G	G	E	A	T	G
5	H	H	H	H	M	T
6	G	G	G	G	G	G
7	Q	H	H	Q	G	G
8	E	E	E	E	L	E
9	V	V	V	V	V	V
10	L	L	I	I	L	L

12. Invariant residues are those that are essential for the structure and/or function of the protein and cannot be replaced by other residues. Which amino acid residue in the globin chain is most likely to be invariant? Explain.

13. Hemoglobin is 50% saturated with oxygen when $pO_2 = 26$ torr. If hemoglobin exhibited hyperbolic binding (as myoglobin does) with 50% saturation at 26 torr, what would be the fractional saturation when $pO_2 = 30$ torr and 100 torr? What does this tell you about the physiological importance of hemoglobin's sigmoidal oxygen-binding curve?

14. But hemoglobin does not exhibit hyperbolic binding, as we assumed in Problem 13; its oxygen-binding curve has a sigmoidal shape. The equation for hemoglobin's sigmoidal oxygen dissociation curve is modified from Equation 5-4:

$$Y = \frac{(pO_2)^n}{(p_{50})^n + (pO_2)^n}$$

The quantity n is known as the Hill coefficient, and its value increases with the degree of cooperation of the ligand (in this case, oxygen) binding. For hemoglobin, the value of n is approximately equal to 3. We can use the equation above to calculate the fractional saturation of hemoglobin. (This was first measured in 1904 by Christian Bohr and his colleagues, who measured oxygen binding in dog blood as a function of oxygen partial pressure at different concentrations of carbon dioxide.) At a CO_2 partial pressure of 5 torr, the p_{50} value for hemoglobin is 15 torr. What is the fractional saturation when $pO_2 = 25$ torr, a typical venous oxygen partial pressure? What is the fractional saturation when $pO_2 = 120$ torr, a typical oxygen partial pressure in the lungs?

15. Calculate the fractional saturation of oxygen binding to hemoglobin measured by Bohr and his colleagues for $pO_2 = 25$ torr and 120 torr (see Problem 14) when the CO_2 partial pressure is 80 torr. Under these conditions, the p_{50} value for hemoglobin is 40 torr.

16. Compare your answers to Problems 14 and 15. How much oxygen is delivered to tissues when the pCO_2 is 5 torr? How does this compare to the amount of oxygen delivered to tissues when the pCO_2 is 40 torr? What role does CO_2 play in assisting oxygen delivery to tissues?

17. The grelag goose lives year-round in the Indian plains, while its close relative, the bar-headed goose, spends the summer months in the Tibetan lake region. The p_{50} value for bar-headed goose hemoglobin is less than the p_{50} for grelag goose hemoglobin. What is the significance of this adaptation?

18. Drinking a few drops of a preparation called "vitamin O," which consists of oxygen and sodium chloride dissolved in water, purportedly increases the concentration of oxygen in the body.
(a) Use your knowledge of oxygen transport to evaluate this claim.
(b) Would vitamin O be more or less effective if it was infused directly into the bloodstream?

19. Highly active muscle generates lactic acid by respiration so fast that the blood passing through the muscle actually experiences a drop in pH from about 7.4 to about 7.2. Under these conditions, hemoglobin releases about 10% more O_2 than it does at pH 7.4. Explain.

20. *Plasmodium falciparum*, the protozoan that causes malaria, slightly decreases the pH of the red blood cells it infects. Invoke the Bohr effect to explain why *Plasmodium*-infected cells are more likely to undergo sickling in individuals with the Hb S variant.

21. About two dozen histidine residues in hemoglobin are involved in binding the protons produced by cellular metabolism. In this manner, hemoglobin contributes to buffering in the blood, and the imidazole groups able to bind and release protons contribute to the Bohr effect. One important contributor to the Bohr effect is His 146 on the β chain of hemoglobin, whose side chain is in close proximity to the side chain of Asp 94 in the deoxy form of hemoglobin but not the oxy form.
(a) What kind of interaction occurs between Asp 94 and His 146 in deoxyhemoglobin?
(b) The proximity of Asp 94 alters the pK value of the imidazole ring of His. In what way?

22. Refer to the data given in Problems 14 and 15. Compare the p_{50} values for hemoglobin when the pCO_2 is 5 torr and 80 torr. What is the significance of the different p_{50} values?

23. Lampreys are among the world's most primitive vertebrates and are therefore interesting organisms to study. It has been shown that lamprey hemoglobin forms a tetramer when deoxygenated, but when the hemoglobin becomes oxygenated, the tetramers dissociate into monomers. The binding of oxygen to lamprey hemoglobin is influenced by pH; as in human hemoglobin, the deoxygenated form is favored when the pH decreases. Glutamate residues on the surface of the monomers play an important role. How might glutamate residues influence the equilibrium between the monomer and tetramer form with changes in pH?

24. While most humans are able to hold their breath for only a minute or two, crocodiles can stay submerged under water for an hour or longer. This adaptation allows the crocodiles to kill their prey by drowning. In 1995, Nagai and colleagues suggested that bicarbonate (HCO_3^-) binds to crocodile hemoglobin to promote oxygen dissociation in a manner similar to BPG in humans and, in so doing, promotes delivery of a large fraction of bound oxygen to tissues. Information gathered in these experiments may allow scientists to design effective blood replacements.

(a) Consider the hypothesis that bicarbonate serves as an allosteric modulator of oxygen binding to hemoglobin in crocodiles. What is the source of HCO_3^- in crocodile tissues?
(b) Draw oxygen-binding curves for crocodile hemoglobin in the presence and absence of bicarbonate. Which conditions increase the p_{50} value for crocodile hemoglobin? What does this tell you about the oxygen-binding affinity of hemoglobin under those conditions?
(c) The investigators found that the bicarbonate binding site on crocodile hemoglobin was located at the $\alpha_1\beta_2$ subunit interface, where the two subunits slide with respect to each other during the oxy $\leftrightarrow$ deoxy transition. Based on their results, the researchers modeled a stereochemically plausible binding site that included the phenolate anions of Tyr 41β and Tyr 42α and the ε-amino group of Lys 38β. What interactions might form between these amino acid side chains and bicarbonate?
(d) Fish use ATP or GTP as allosteric effectors to encourage hemoglobin to release its bound oxygen. What structural characteristics do all of these different allosteric effectors have in common and how would the molecules bind to hemoglobin?

25. In the developing fetus, fetal hemoglobin (Hb F) is synthesized beginning at the third month of gestation and continuing until birth. After the baby is born, the concentration of Hb F declines and is replaced with adult hemoglobin (Hb A) as synthesis of the γ chain declines and synthesis of the β chain increases. By the time the baby is six months old, 98% of its hemoglobin is Hb A.

(a) In the graph below, which curve represents fetal hemoglobin?

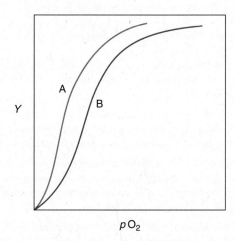

(b) Why does Hb F have a higher oxygen affinity than Hb A? Why does this provide an advantage to the developing fetus?

26. The drug hydroxyurea can be used to treat sickle cell anemia, although it is not used often because of undesirable side effects. Hydroxyurea is thought to function by stimulating the afflicted person's synthesis of fetal hemoglobin. In a clinical study, patients who took hydroxyurea showed a 50% reduction in frequency of hospital admissions for severe pain, and there was also a decrease in the frequency of fever and abnormal chest X-rays. Why would this drug alleviate the symptoms of sickle cell anemia?

27. The oxygen-binding curves for normal hemoglobin (Hb A) and a mutant hemoglobin (Hb Great Lakes) are shown in the figure.

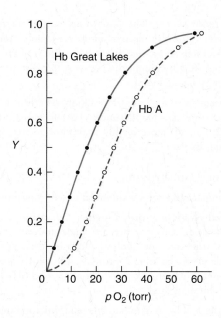

(a) Compare the shapes of the curves for the two hemoglobins and comment on their significance.
(b) Which hemoglobin has a higher affinity for oxygen when $pO_2 = 20$ torr?
(c) Which hemoglobin has a higher affinity for oxygen when $pO_2 = 75$ torr?
(d) Which hemoglobin is most efficient at delivering oxygen from arterial blood ($pO_2 = 75$ torr) to active muscle ($pO_2 = 20$ torr)?

28. Rank the stability of the following mutant hemoglobins:

Hb Hammersmith	β42Phe → Ser
Hb Bucuresti	β42Phe → Leu
Hb Sendagi	β42Phe → Val
Hb Bruxelles	β42Phe → 0 (Phe is deleted)

29. People who live at or near sea level undergo *high-altitude acclimatization* when they go up to moderately high altitudes of about 5000 m.
(a) The low concentration of oxygen at these altitudes initially induces hypoxia, which results in hyperventilation. Explain why this occurs. What happens to blood pH during hyperventilation?
(b) After a period of weeks, the alveolar pCO_2 decreases and the 2,3-bisphosphoglycerate concentration increases. Explain these observations.

30. Animals indigenous to high altitudes, such as yaks, llamas, and alpacas, have adapted to the low oxygen concentrations at high altitudes because they have high-affinity hemoglobins, achieved through a combination of globin-chain mutations and persistence of fetal hemoglobin (i.e., fetal hemoglobin is synthesized by mature animals). This is a true evolutionary adaptation and is different from the high-altitude acclimatization described in Problem 29.
(a) Compare the p_{50} values of the high-altitude animals with the p_{50} values of their low-altitude counterparts.
(b) How does the continued synthesis of fetal hemoglobin help these animals survive at high altitudes?

31. A patient with a high-affinity hemoglobin mutant was the subject of a paper published in 1970s. The subject's α chain was normal, but a mutation in the β chain occurred that resulted in the replacement of a lysine by an asparagine at position 144. (The normal C-terminal sequence of the β chain is –Lys 144–Tyr

145–His 146–COO⁻). The C-terminal portion of the β chain, which resides in the central cavity of the hemoglobin molecule, normally forms interactions that affect the position of the F helix that contains His F8. How might the Lys → Asn substitution result in a higher-affinity hemoglobin?

32. Patients with the types of mutations described in Problem 31 are able to undergo high-altitude acclimatization (described in Problem 29) much more readily than "lowlanders" with normal hemoglobin. Explain why.

5-2 Structural Proteins

33. Compare and contrast globular and fibrous proteins.

34. Of the various proteins highlighted in this chapter (actin, tubulin, keratin, collagen, myosin, and kinesin), **(a)** which are exclusively structural (not involved in cell shape changes); **(b)** which are considered to be motor proteins; **(c)** which are not motor proteins but can undergo structural changes; and **(d)** which contain nucleotide-binding sites?

35. Explain why microfilaments and microtubules are polar whereas intermediate filaments are not.

36. In order to obtain crystals suitable for X-ray crystallography, G-actin was first mixed with another protein.
(a) Why was this step necessary?
(b) What additional information was required to solve the structure of G-actin?

37. Phalloidin, a peptide isolated from a poisonous mushroom, binds to F-actin but not G-actin. How does the addition of phalloidin affect cell motility?

38. Phalloidin (see Problem 37) is used as an imaging tool by covalently attaching a fluorescent tag to it. What structures can be visualized in cells treated with fluorescently labeled phalloidin? What is not visible?

39. How could a microtubule-binding protein distinguish a rapidly growing microtubule from one that was growing more slowly?

40. Why would fraying be a faster mechanism for microtubule disassembly than dissociation of tubulin dimers?

41. Microtubule ends with GTP bound tend to be straight, whereas microtubule ends with GDP bound are curved and tend to fray. An experiment was carried out in which monomers of β-tubulin were allowed to polymerize in the presence of a nonhydrolyzable analog of GTP called guanylyl-(αβ)-methylenediphosphonate (GMP · CPP). Compare the stability of this polymer with one polymerized in the presence of GTP.

GMP · CPP

42. Explain why colchicine (which promotes microtubule depolymerization) and paclitaxel (taxol, which prevents depolymerization) both prevent cell division.

43. A recent review listed literally dozens of compounds that are being investigated as possible therapeutic agents for treating cancer. Tubulin was the target for all of the compounds. Why is tubulin a good target for anticancer drugs?

44. The three-dimensional structure of paclitaxel bound to the tubulin dimer has been solved. Paclitaxel binds to a pocket in β-tubulin on the inner surface of the microtubule and causes a conformational change that counteracts the effects of GTP hydrolysis. How does the binding of paclitaxel affect the stability of the microtubule?

45. Gout is a disease characterized by the deposition of sodium urate crystals in the synovial (joint) fluid. The crystals are ingested by neutrophils, which are white blood cells that circulate in the bloodstream before they crawl through tissues to reach sites of injury. Upon ingestion of the crystals, a series of biochemical reactions lead to a release of mediators from the neutrophils that cause inflammation, resulting in joint pain. Colchicine is used as a drug to treat gout because colchicine can inhibit neutrophil mobility. How does colchicine accomplish this?

46. The microtubules involved in cell division are less stable than microtubules found in axon extensions of nerve cells. Why is this the case?

47. Vinblastine, a compound in the periwinkle, has been shown to affect microtubule assembly by stabilizing the (+) ends and destabilizing the (−) ends of microtubules. How does vinblastine affect the formation of the mitotic spindle during mitosis?

48. Why would you expect vinblastine (Problem 47) to have a more dramatic effect on rapidly dividing cells such as cancer cells?

49. Hydrophobic residues usually appear at the first and fourth positions in the seven-residue repeats of polypeptides that form coiled coils.
(a) Why do polar or charged residues usually appear in the remaining five positions?
(b) Why is the sequence Ile–Gln–Glu–Val–Glu–Arg–Asp more likely than the sequence Trp–Gln–Glu–Tyr–Glu–Arg–Asp to appear in a coiled coil?

50. Globular proteins are typically constructed from several layers of secondary structure, with a hydrophobic core and a hydrophilic surface. Is this true for a fibrous protein such as keratin?

51. Straight hair can be curled and curly hair can be straightened by exposing wet hair to a reducing agent, repositioning the hair with rollers, then exposing the hair to an oxidizing agent. Explain how this procedure alters the shape of the hair.

52. "Hard" keratins such as those found in hair, horn, and nails have a high sulfur content, whereas "soft" keratins found in the skin have a lower sulfur content. Explain the structural basis for this observation.

53. Describe the primary, secondary, tertiary, and quaternary structures of **(a)** actin and **(b)** collagen. Why is it difficult to assign the four structural categories to these proteins?

54. A nutrition website advocates the consumption of collagen-rich foods such as chicken skin, shark fins, and pigs' feet as part of a "wrinkle-free" diet. Is the consumption of foods rich in collagen required for proper collagen biosynthesis? Do you think this diet is likely to decrease the appearance of wrinkles?

55. The highly pathogenic bacterium *Clostridium perfringens* causes gangrene, a disease that results in the destruction of animal tissue. The bacterium secretes two types of collagenases that cleave the peptide bonds linking Gly and Pro residues. A biochemical company sells a partially purified preparation of these collagenases to investigators who are interested in culturing cells derived from bone, cartilage, muscle, or endothelial tissue. How does this enzyme preparation assist the investigator in obtaining cells for culturing?

56. An enzyme found in tadpoles cleaves the bond indicated in the figure below. What kind of enzyme is this, and what is its role in the metamorphosis of the tadpole to the adult frog?

$$—Gly—Pro—Gln—Gly \downarrow Ile—Ala$$

57. Members of the collagen family are often characterized by their melting temperature (T_m). The collagen is subjected to increasing temperature and monitored for structural changes. At high temperatures, the intermolecular forces holding the three chains together in the triple helix are disrupted and the chains come apart, resulting in denatured collagen, or gelatin (a substance that is liquid at high temperatures but solidifies upon cooling). Because the intermolecular forces holding the collagen strands together are cooperative, the collagen triple helix tends to come apart all at once. The T_m is defined as that temperature at which the collagen is half-denatured and is used as a criterion for collagen stability. T_m measurements for collagen from two different species are presented in the table below. One is from rat; the other is from sea urchin. **(a)** Assign each collagen to the proper organism, and **(b)** explain the correlation between imino acid content and T_m.

Collagen	Number of Hyp per 1000 residues	Number of Pro per 1000 residues	T_m (°C)
A	48.5	81.3	27.0
B	68.5	111	38.5

58. A series of collagen-like peptides, each containing 30 amino acid residues, were synthesized in order to study the important interactions among the three chains in the triple helix. Peptide 1 is (Pro–Hyp–Gly)$_{10}$, peptide 2 is (Pro–Hyp–Gly)$_4$–Glu–Lys–Gly–(Pro–Hyp–Gly)$_5$, and peptide 3 is Gly–Lys–Hyp–Gly–Glu–Hyp–Gly–Pro–Lys–Gly–Asp–Ala–(Gly–Ala–Hyp)$_2$–(Gly–Pro–Hyp)$_4$. Their structural properties and melting temperatures (see Problem 57) are summarized in the table below. Note that "imino acid content" refers to the combined content of Pro and Hyp.

Peptide	Forms trimer	Imino acid content	T_m (°C)	
1	Yes	67%	pH = 1	61
			pH = 7	58
			pH = 11	60
2	Yes	60%	pH = 1	44
			pH = 7	46
			pH = 1	49
3	Yes	30%	pH = 7	18
			pH = 7	26.5
			pH = 13	19

(a) Rank the stability of the three collagen-like peptides. What is the reason for the observed stability?
(b) Compare the T_m values of peptide 3 at the various pH values. Why does peptide 3 have a maximum T_m value at pH = 7? What interactions are primarily responsible?

(c) Compare the T_m values of peptide 1 at the various pH values. Why is there less of a difference of T_m values at different pH values for peptide 1 than for peptide 3?
(d) Consider the answers to the above questions. Which factor is more important in stabilizing the structure of the collagen-like peptides: ion pairing or imino acid content?

59. The deep-sea hydrothermal vent worm *Riftia pachyptila* lives under extreme conditions of high temperature, low oxygen content, and drastic temperature changes. The worm has a thick collagen-containing cuticle that protects it from its harsh environment. Recently the structure of this collagen was investigated. Sequence analyses indicated that the collagen had the customary Gly–X–Y triplet but that hydroxyproline occurred only in the X position and that Y was often a glycosylated threonine (a galactose sugar residue was covalently attached via a condensation reaction between a hydroxyl group on the galactose and the hydroxyl group of the threonine side chain). Experiments were carried out using synthetic peptides in order to evaluate the stability of each. Results are shown in the table.

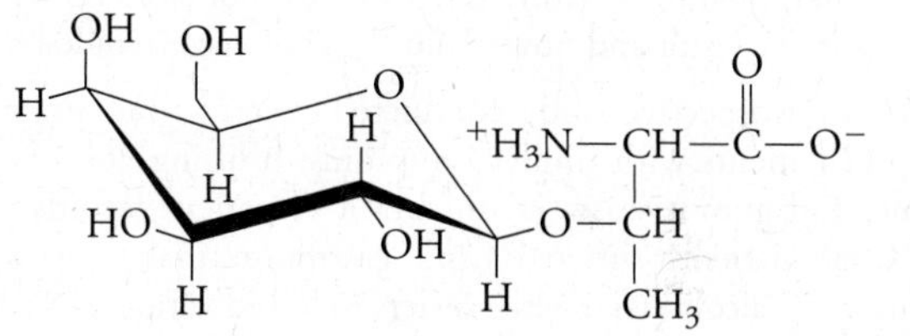

Gal-Thr

Synthetic peptide	Forms a triple helix?	T_m (°C)
(Pro–Pro–Gly)$_{10}$	Yes	41
(Pro–Hyp–Gly)$_{10}$	Yes	60
(Gly–Pro–Thr)$_{10}$	No	N/A
(Gly–Pro–Thr(Gal))$_{10}$	Yes	41

(a) Compare the melting temperatures of (Pro–Pro–Gly)$_{10}$ and (Pro–Hyp–Gly)$_{10}$. What is the structural basis for the difference?
(b) Compare the melting temperature of (Pro–Pro–Gly)$_{10}$ and (Gly–Pro–Thr(Gal))$_{10}$ and provide an explanation.
(c) Why was (Gly–Pro–Thr)$_{10}$ included by the investigators?

60. Several investigators have sought to explain why hydroxyproline-containing collagen molecules have increased stability. Some suggested that the hydroxyl group on the pyrrolidine ring of Hyp participates in a network of hydrogen bonds with "bridging" water molecules. However, this has been called into question by others, who have noted that triple helices made of (Pro–Pro–Gly)$_{10}$ and (Pro–Hyp–Gly)$_{10}$ are stable in methanol. To investigate the question of stability further, one group of investigators synthesized a synthetic collagen that contains 4-fluoroproline (Flp), which resembles hydroxyproline except that fluorine substitutes for the hydroxyl group. The melting points of three synthetic collagen molecules were measured and are shown in the table.

Synthetic peptide	Forms a triple helix?	T_m (°C)
(Pro–Pro–Gly)$_{10}$	Yes	41
(Pro–Hyp–Gly)$_{10}$	Yes	60
(Pro–Flp–Gly)$_{10}$	Yes	91

(a) Compare the structure of hydroxyproline and fluoroproline. Why do you suppose the investigators chose fluorine?
(b) Compare the melting points of the three synthetic collagens. What factor contributes the most to the stability of the molecule?

61. The triple helix of collagen is stabilized through hydrogen bonding. Draw an example of one of these hydrogen bonds, as described in the text.

62. Because collagen molecules are difficult to isolate from the connective tissue of animals, studies of collagen structure often use synthetic peptides. Would it be practical to try to purify collagen molecules from cultures of bacterial cells that have been engineered to express collagen genes?

63. Gelatin is a food product obtained from the thermal degradation of collagen. Powdered gelatin is soluble in hot water and forms a gel upon cooling to room temperature. Why is gelatin nutritionally inferior to other types of protein?

64. Papaya and pineapple contain proteolytic enzymes that can degrade collagen.
(a) Cooking meat with these fruits helps to tenderize the meat. How does this work?
(b) A cook decides to flavor a gelatin dessert with fresh papaya and pineapple. The fruits are added to a solution of gelatin in hot water, then the mixture is allowed to cool (see Problem 63). What is the result and how could the cook have avoided this?

65. A retrospective study conducted over 25 years examined a group of patients with similar symptoms: bruising, joint swelling, fatigue, and gum disease. Examination of patient records showed that some patients suffered from gastrointestinal diseases, poor dentition, or alcoholism, while others followed various food fads.
(a) What disease afflicts this diverse group of patients?
(b) Explain why the patients suffer from the symptoms listed.
(c) Explain why your diagnosis is consistent with the patients' medical histories.

66. Radioactively labeled [^{14}C]-proline is incorporated into collagen in cultured fibroblasts. The radioactivity is detected in the collagen protein. But collagen synthesized in the presence of [^{14}C]-hydroxyproline is not radiolabeled. Explain why.

67. Lysyl hydroxylase, like prolyl hydroxylase, requires ascorbate. Draw the product of the lysyl hydroxylase reaction, a 5-hydroxylysyl residue.

68. In a collagen polypeptide consisting mostly of repeating Gly–Pro–Hyp sequences, which residue(s) is (are) most likely to be substituted by 5-hydroxylysine?

69. The effect of the drug minoxidil was measured in cultured human skin fibroblasts, and the data are shown in the table below. Fibroblasts were incubated with [^{3}H]-proline or [^{3}H]-lysine to allow the cells to incorporate the radioactively labeled amino acids into collagen chains. Following treatment with minoxidil (controls were untreated), the cells were harvested and then subjected to enzyme treatment, which hydrolyzed all cellular protein. The amount of [^{3}H]-proline or [^{3}H]-lysine incorporated into collagen was then measured in terms of ^{3}H$_2$O release.

Culture conditions	Prolyl hydroxylase activity (cpm, ^{3}H$_2$O/mg protein)	Lysyl hydroxylase activity (cpm, ^{3}H$_2$O/mg protein)
Control	13,056	12,402
Minoxidil	13,242	1,936

(a) What is the effect of minoxidil on the cultured fibroblasts?
(b) Why would this drug be effective in treating fibrosis (a skin condition associated with an accumulation of collagen)?
(c) What are the potential dangers in the long-term use of minoxidil to treat hair loss?

70. A small portion of the middle of the gene for the α1(II) chain of type II collagen (the entire gene contains 3421 base pairs) is shown below as the sequence of the nontemplate, or coding, strand of the DNA. Also shown is the gene isolated from the skin fibroblasts of a patient with a mutant α1(II) gene. The patient was a neonate delivered at 38 weeks who died shortly after birth. The neonate had shortened limbs and a large head, with a flat nasal bridge and short bones and ribs. The symptoms are consistent with the disease hypochondrogenesis. Analysis of tissue taken from the neonate revealed a single point mutation in the α1(II) collagen gene.

Normal α1(II) collagen gene:
   … TAACGGCGAGAAGGGAGAAGTTGGACCTCCT…

Mutant α1(II) collagen gene:
   …TAACGGCGAGAAGGCAGAAGTTGGACCTCCT…

(a) What is the sequence of the protein encoded by both the normal and mutant genes? (Note that you must first determine the correct open reading frame.) What is the change in the protein sequence?
(b) How would the T_m values for normal collagen and mutant collagen compare?
(c) Why did the patient die shortly after birth?

5-3 Motor Proteins

71. Is myosin a fibrous protein or a globular protein? Explain.

72. Myosin type V is a two-headed myosin that operates as a transport motor to move its attached cargo along actin filaments. Its mechanism is similar to that of muscle myosin, but it acts processively, like kinesin. The reaction cycle diagrammed here begins with both myosin V heads bound to an actin filament. ADP is bound to the leading head, and the nucleotide-binding site in the trailing head is empty.

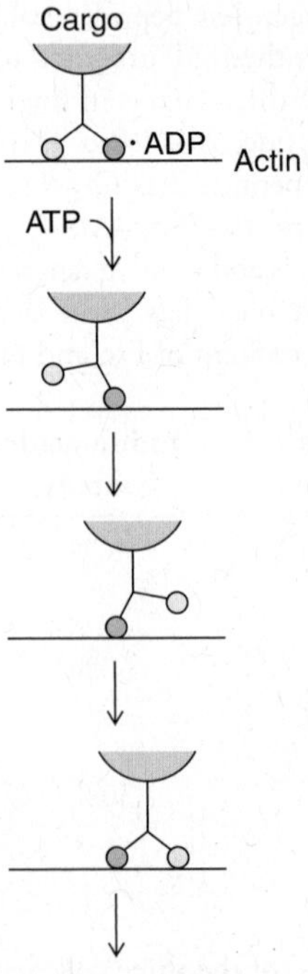

Based on your knowledge of the muscle myosin reaction cycle (Fig. 5-35), propose a reaction mechanism for myosin V, starting with the entry of ATP. How does each step of the ATP hydrolysis reaction correspond to a conformational change in myosin V?

73. Early cell biologists, examining living cells under the microscope, observed that the movement of certain cell constituents was rapid, linear, and targeted (that is, directed toward a particular point).

(a) Why are these qualities inconsistent with diffusion as a mechanism for redistributing cell components?

(b) List the minimum requirements for an intracellular transport system that is rapid, linear, and targeted.

74. Explain why the movement of myosin along an actin filament is "hopping," whereas the movement of kinesin along a microtubule is "walking."

75. Rigor mortis is the stiffening of a corpse that occurs in the hours following death. Using what you know about the mechanism of motor proteins, explain what causes this stiffness.

76. (a) Why is rigor mortis of concern to those in the meat industry?

(b) How could a forensic pathologist use the concept of rigor mortis to investigate the circumstances of an individual's death?

77. In individuals with muscular dystrophy, a mutation in a gene for a structural protein in muscle fibers leads to progressive muscle weakening beginning in early childhood. Explain why muscular dystrophy patients typically also exhibit abnormal bone development.

78. Explain why a one-headed kinesin motor would be ineffective.

[SELECTED READINGS]

Brodsky, B., and Persikov, A. V., Molecular structure of the collagen triple helix, *Adv. Protein Chem.* **70**, 310–339 (2005). [Discusses collagen structure and diseases associated with collagen mutations.]

Endow, S. A., Kinesin motors as molecular machines, *Bioessays* **25**, 1212–1219 (2003).

Gardner, M. K., Hunt, A. J., Goodson, H. V., and Odde, D. J., Microtubule assembly dynamics: new insights at the nanoscale, *Curr. Opin. Cell Biol.* **20**, 64–70 (2008). [Describes some of the techniques used to characterize microtubule formation.]

O'Connell, C. B., Tyska, M. J., and Mooseker, M. S., Myosin at work: motor adaptations for a variety of cellular functions, *Biochim. Biophys. Acta* **1773**, 615–630 (2007).

Oshima, R. G., Intermediate filaments: a historical perspective, *Exp. Cell Res.* **313**, 1981–1994 (2007). [Describes what is known about various types of intermediate filaments, including keratin.]

Reisler, E., and Egelman, E. H., Actin structure and function: what we still do not understand, *J. Biol. Chem.* **282**, 36113–36137 (2007). [Summarizes some areas of research on actin structure and its assembly into filaments.]

Schechter, A. N., Hemoglobin research and the origins of molecular medicine, *Blood* **112**, 3927–3938 (2008). [Reviews hemoglobin structure and function, including hemoglobin S.]

Wilson, M. T., and Reeder, B. J., Oxygen-binding haem proteins, *Exp. Physiol.* **93**, 128–132 (2007). [Discusses the lesser-known functions of hemoglobin, myoglobin, and some of their relatives.]

6

HOW ENZYMES WORK

▶▶ **DOES** the lock-and-key model explain enzyme action?

In 1894, long before the first enzyme structure had been determined and several decades before it had been shown that enzymes are proteins, Emil Fischer proposed that a substrate molecule fits into the enzyme active site like a key in a lock. This simple idea has been widely embraced, in part because it appears to explain the exquisite substrate specificity of enzymes. Yet the lock model does not explain why some substances can fit into the keyhole but not undergo any chemical reaction. Once you learn more about how enzymes work, you will appreciate why Fischer's lock-and-key model has been revised.

[iSci/iStockphoto]

THIS CHAPTER IN CONTEXT

Do You Remember?

- Living organisms obey the laws of thermodynamics (Section 1-3).
- Noncovalent forces, including hydrogen bonds, ionic interactions, and van der Waals forces, act on biological molecules (Section 2-1).
- An acid's pK value describes its tendency to ionize (Section 2-3).
- The 20 amino acids differ in the chemical characteristics of their R groups (Section 4-1).
- Some proteins can adopt more than one stable conformation (Section 4-3).

We have already seen how protein structures relate to their physiological functions. We are now ready to examine enzymes, which directly participate in the chemical reactions by which matter and energy are transformed by living cells. In this chapter, we examine the fundamental features of enzymes, including the thermodynamic underpinnings of their activity. We describe various mechanisms by which enzymes accelerate chemical reactions, focusing primarily on the digestive enzyme chymotrypsin to illustrate how different structural features influence catalytic activity. The following chapter continues the discussion of enzymes by describing how enzymatic activity is quantified and how it can be regulated.

6-1 What Is an Enzyme?

In vitro, under physiological conditions, the peptide bond that links amino acids in peptides and proteins has a half-life of about 20 years. That is, after 20 years, about half of the peptide bonds in a given sample of a peptide will have been broken down through **hydrolysis** (cleavage by water):

$$\cdots -N-CH-C-N-CH-C- \cdots + H_2O$$

$$\downarrow$$

$$\cdots -N-CH-C{\stackrel{O}{_{O^-}}} + H-N-CH-C- \cdots$$

Obviously, the long half-life of the peptide bond is advantageous for living organisms, since many of their structural and functional characteristics depend on the integrity of proteins. On the other hand, many proteins—some hormones, for example—must be broken down very rapidly so that their biological effects can be limited. Clearly, an organism must be able to accelerate the rate of peptide bond hydrolysis.

In general, there are three ways to increase the rate of a chemical reaction, including hydrolysis:

1. *Increasing the temperature (adding energy in the form of heat).* Unfortunately, this is not very practical, since the vast majority of organisms cannot regulate their internal temperature and thrive only within relatively narrow temperature ranges. Furthermore, an increase in temperature accelerates all chemical reactions, not just the desired reaction.
2. *Increasing the concentrations of the reacting substances.* Higher concentrations of **reactants** increase the likelihood that they will encounter each other in order to react. But a cell may contain tens of thousands of different types of molecules, space is limited, and many essential reactants are scarce inside as well as outside the cell.
3. *Adding a **catalyst**.* A catalyst is a substance that participates in the reaction yet emerges at the end in its original form. A huge variety of chemical catalysts are known. For example, the catalytic converter in an automobile engine contains a mixture of platinum and palladium that accelerates the conversion of carbon monoxide and unburned hydrocarbons to the relatively harmless carbon dioxide. *Living systems use catalysts called **enzymes** to increase the rates of chemical reactions.*

Most enzymes are proteins, but a few are made of RNA (these are called **ribozymes** and are described more fully in Section 21-3). One of the best-studied enzymes is chymotrypsin, a digestive protein that is synthesized in the pancreas and secreted

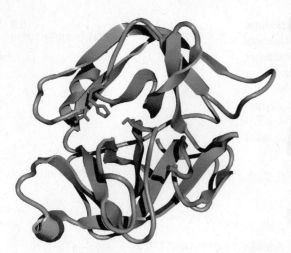

Figure 6-1 Ribbon model of chymotrypsin. The polypeptide chain (gray) folds into two domains. Three residues essential for the enzyme's activity are shown in red. [Structure (pdb 4CHA) determined by H. Tsukada and D. M. Blow.]

into the small intestine, where it helps break down dietary proteins. Perhaps because it can be purified in relatively large quantities from the pancreas of cows, chymotrypsin was one of the first enzymes to be crystallized (it is also widely used in the laboratory; for example, see Table 4-4). Chymotrypsin's 241 amino acid residues form a compact two-domain structure (Fig. 6-1). Hydrolysis of polypeptide substrates takes place in a cleft between the two domains, near the side chains of three residues (His 57, Asp 102, and Ser 195). This area of the enzyme is known as the **active site.** The active sites of nearly all known enzymes are located in similar crevices on the enzyme surface.

Chymotrypsin catalyzes the hydrolysis of peptide bonds at a rate of about 190 per second, which is about 1.7×10^{11} times faster than in the absence of a catalyst. This is also orders of magnitude faster than would be possible with a simple chemical catalyst. In addition, *chymotrypsin and other enzymes act under mild conditions* (atmospheric pressure and physiological temperature), whereas many chemical catalysts require extremely high temperatures and pressures for optimal performance.

Chymotrypsin's catalytic power is not unusual: *Rate enhancements of 10^8 to 10^{12} are typical of enzymes* (Table 6-1 gives the rates of some enzyme-catalyzed reactions). Of course, the slower the rate of the uncatalyzed reaction, the greater the opportunity for rate enhancement by an enzyme (see, for example, orotidine-5′-monophosphate decarboxylase in Table 6-1). Interestingly, even relatively fast reactions are subject to enzymatic catalysis in biological systems. For example, the conversion of CO_2 to carbonic acid in water

$$CO_2 + H_2O \rightleftharpoons H_2CO_3$$

has a half-time of 5 seconds (half the molecules will have reacted within 5 seconds). This reaction is accelerated over a millionfold by the enzyme carbonic anhydrase (see Table 6-1).

Another feature that sets enzymes apart from nonbiological catalysts is their **reaction specificity.** Most enzymes are highly specific for their reactants (called **substrates**) and products. The functional groups in the active site of an enzyme are so carefully arranged that the enzyme can distinguish its substrates from among many others that are similar in size and shape and can then mediate a single chemical reaction involving those substrates. This reaction specificity stands in marked contrast to the permissiveness of most organic catalysts, which can act on many different kinds of substrates and, for a given substrate, sometimes yield more than one product.

Chymotrypsin and some other digestive enzymes are somewhat unusual in acting on a relatively broad range of substrates and, at least in the laboratory, catalyzing several types of reactions. For instance, chymotrypsin catalyzes the hydrolysis of the peptide bond following almost any large nonpolar residue such as Phe, Trp, or Tyr. It can also catalyze the hydrolysis of other amide bonds and ester bonds. This behavior

[TABLE 6-1]   Rate Enhancements of Enzymes

Enzyme	Half-Time (uncatalyzed)[a]		Uncatalyzed Rate (s⁻¹)	Catalyzed Rate (s⁻¹)	Rate Enhancement (catalyzed rate/ uncatalyzed rate)
Orotidine-5′-monophosphate decarboxylase	78,000,000	years	2.8×10^{-16}	39	1.4×10^{17}
Staphylococcal nuclease	130,000	years	1.7×10^{-13}	95	5.6×10^{14}
Adenosine deaminase	120	years	1.8×10^{-10}	370	2.1×10^{12}
Chymotrypsin	20	years	1.0×10^{-9}	190	1.7×10^{11}
Triose phosphate isomerase	1.9	years	4.3×10^{-6}	4,300	1.0×10^9
Chorismate mutase	7.4	hours	2.6×10^{-5}	50	1.9×10^6
Carbonic anhydrase	5	seconds	1.3×10^{-1}	1,000,000	7.7×10^6

[a]The half-times of very slow reactions were estimated by extrapolating from measurements made at very high temperatures.

[Data mostly from Radzicka, R., and Wolfenden, R., *Science* **267,** 90–93 (1995).]

has proved to be convenient for quantifying the activity of purified chymotrypsin. An artificial substrate such as *p*-nitrophenylacetate (an ester) is readily hydrolyzed by the action of chymotrypsin (the name of the enzyme appears next to the reaction arrow to indicate that it participates as a catalyst):

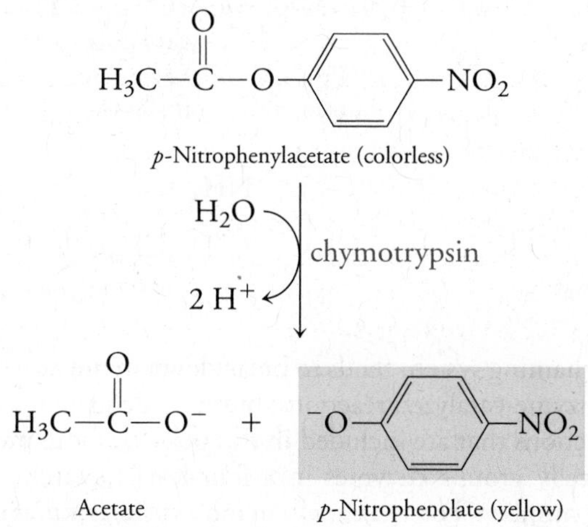

p-Nitrophenylacetate (colorless)

Acetate   *p*-Nitrophenolate (yellow)

The *p*-nitrophenolate reaction product is bright yellow, so the progress of the reaction can be easily monitored in a spectrophotometer at 410 nm.

Finally, enzymes differ from nonbiological catalysts in that *the activities of many enzymes are regulated so that the organism can respond to changing conditions or follow genetically determined developmental programs.* For this reason, biochemists seek to understand *how* enzymes work as well as *when* and *why.* These aspects of enzyme behavior are fairly well understood for chymotrypsin, which makes it an ideal subject to showcase the fundamentals of enzyme activity.

Enzymes are usually named after the reaction they catalyze

The enzymes that catalyze biochemical reactions have been formally classified into six subgroups according to the type of reaction carried out (Table 6-2). Basically, *all biochemical reactions involve either the addition of some substance to another, or its removal, or the rearrangement of that substance.* Keep in mind that although the substrates of many biochemical reactions appear to be quite large (for example, proteins or nucleic acids), the action really involves just a few chemical bonds and a few small groups (sometimes H_2O or even just an electron).

The name of an enzyme frequently provides a clue to its function. In some cases, an enzyme is named by incorporating the suffix -*ase* into the name of its substrate. For example, fumarase is an enzyme that acts on fumarate (Reaction 7 in the citric acid cycle; see Section 14-2). Chymotrypsin can similarly be called a proteinase, a protease, or a peptidase. Most enzyme names contain more descriptive words (also ending in -*ase*) to indicate the nature of the reaction catalyzed by that enzyme. For example, pyruvate decarboxylase catalyzes the removal of a CO_2 group from pyruvate:

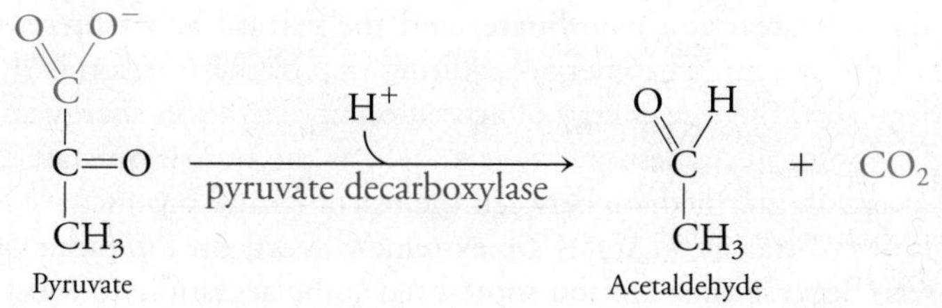

Pyruvate   Acetaldehyde

[TABLE 6-2]

Enzyme Classification

Class of Enzyme	Type of Reaction Catalyzed
1. Oxidoreductases	Oxidation–reduction reactions
2. Transferases	Transfer of functional groups
3. Hydrolases	Hydrolysis reactions
4. Lyases	Group elimination to form double bonds
5. Isomerases	Isomerization reactions
6. Ligases	Bond formation coupled with ATP hydrolysis

Alanine aminotransferase catalyzes the transfer of an amino group from alanine to an α-keto acid:

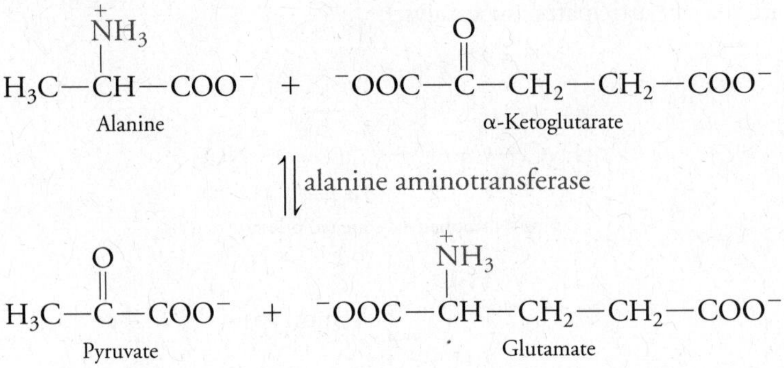

$$H_3C-\overset{\overset{+}{N}H_3}{\underset{|}{C}H}-COO^- \;+\; {}^-OOC-\overset{\overset{O}{\|}}{C}-CH_2-CH_2-COO^-$$

Alanine   α-Ketoglutarate

$$\big\Updownarrow\text{ alanine aminotransferase}$$

$$H_3C-\overset{\overset{O}{\|}}{C}-COO^- \;+\; {}^-OOC-\overset{\overset{+}{N}H_3}{\underset{|}{C}H}-CH_2-CH_2-COO^-$$

Pyruvate   Glutamate

Such a descriptive naming system tends to break down in the face of the many thousands of known enzyme-catalyzed reactions, but it is adequate for the small number of well-known reactions that are included in this book. A more precise classification scheme systematically groups enzymes in a four-level hierarchy and assigns each enzyme a unique number. For example, chymotrypsin is known as EC 3.4.21.1 (*EC* stands for Enzyme Commission, part of the nomenclature committee of the International Union of Biochemistry and Molecular Biology; the EC database can be accessed at http://enzyme.expasy.org).

Keep in mind that even within an organism, more than one protein may catalyze a given chemical reaction. Multiple enzymes catalyzing the same reaction are called **isozymes.** Although they usually share a common evolutionary origin, isozymes differ in their catalytic properties. Consequently, the various isozymes that are expressed in different tissues or at different developmental stages can perform slightly different metabolic functions.

> **CONCEPT REVIEW**
> • How do enzymes differ from nonbiological catalysts?
> • Explain why an enzyme's common name may not reveal its biological role.

6-2 The Chemistry of Catalysis

KEY CONCEPTS
- The height of the activation energy barrier determines the rate of a reaction.
- An enzyme provides a lower-energy pathway from reactants to products.
- Enzymes accelerate chemical reactions using acid–base catalysis, covalent catalysis, and metal ion catalysis.
- Chymotrypsin's catalytic triad participates in acid–base and covalent catalysis.

In a biochemical reaction, the reacting species must come together and undergo electronic rearrangements that result in the formation of products. In other words, some old bonds break and new bonds form. Let us consider an idealized transfer reaction in which compound A—B reacts with compound C to form two new compounds, A and B—C:

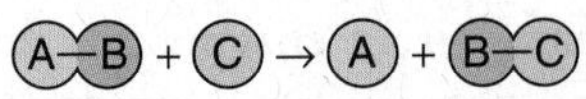

$$\boxed{A\!-\!B} + \boxed{C} \rightarrow \boxed{A} + \boxed{B\!-\!C}$$

In order for the first two compounds to react, they must approach closely enough for their constituent atoms to interact. Normally, atoms that approach too closely repel each other. But if the groups have sufficient free energy, they can pass this point and react with each other to form products. The progress of the reaction can be depicted on a diagram (**Fig. 6-2**) in which the horizontal axis represents the progress of the reaction (the **reaction coordinate**) and the vertical axis represents the free energy (G) of the system. The energy-requiring step of the reaction is shown as an energy barrier, called the **free energy of activation** or **activation energy** and symbolized $\Delta G^{\ddagger}$. The point of highest energy is known as the **transition state** and can be considered a sort of intermediate between the reactants and products.

The lifetimes of transition states are extremely short, on the order of 10^{-14} to 10^{-13} seconds. Because they are too short-lived to be accessible to most analytical

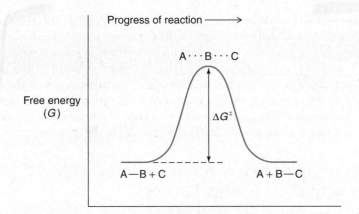

Progress of reaction ⟶

A···B···C

$\Delta G^{\ddagger}$

Free energy (G)

A—B + C   A + B—C

Reaction coordinate

Figure 6-2 Reaction coordinate diagram for the reaction A—B + C → A + B—C. The progress of the reaction is shown on the horizontal axis, and free energy is shown on the vertical axis. The transition state of the reaction, represented as A···B···C, is the point of highest free energy. The free energy difference between the reactants and the transition state is the free energy of activation ($\Delta G^{\ddagger}$).

techniques, the transition states of many reactions cannot be identified with absolute certainty. However, it is useful to visualize the transition state as a molecular species in the process of breaking old bonds and forming new bonds. For the reaction above, we can represent this as A····B····C. The reactants require free energy ($\Delta G^{\ddagger}$) to reach this point (for example, some energy is required to break existing bonds and bring other atoms together to begin forming new bonds). The analogy of going uphill in order to undergo a reaction is appropriate.

The height of the activation energy barrier determines the rate of a reaction (the amount of product formed per unit time). The higher the activation energy barrier, the less likely the reaction is to occur (the slower it is). Although the reactant molecules have varying free energies, very few of them have enough free energy to reach the transition state during a given time interval. But the lower the energy barrier, the more likely the reaction is to occur (the faster it is), because more reactant molecules happen to have enough free energy to achieve the transition state during the same time interval. Note that the transition state, at the peak, can potentially roll down either side of the free energy hill. Therefore, not all the reactants that get together to form a transition state actually proceed all the way to products; they may return to their original state. Similarly, the products (A and B—C in this case) can react, pass through the same transition state (A····B····C), and yield the original reactants (A—B and C).

In nature, the free energies of the reactants and products of a chemical reaction are seldom identical, so the reaction coordinate diagram looks more like **Figure 6-3a.** When the products have a lower free energy than the reactants, then the overall free

(a)   (b)

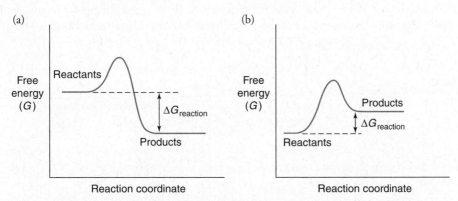

Free energy (G)   Reactants   Free energy (G)

$\Delta G_{\text{reaction}}$ Products

Products   Reactants $\Delta G_{\text{reaction}}$

Reaction coordinate   Reaction coordinate

Figure 6-3 Reaction coordinate diagram for a reaction in which reactants and products have different free energies. The free energy change for the reaction (ΔG) is equivalent to $G_{\text{products}} - G_{\text{reactants}}$. (a) When the free energy of the reactants is greater than that of the products, the free energy change for the reaction is negative, so the reaction proceeds spontaneously. (b) When the free energy of the products is greater than that of the reactants, the free energy change for the reaction is positive, so the reaction does not proceed spontaneously (however, the reverse reaction does proceed).

? **In a cell, some enzyme-catalyzed reactions proceed in both the forward and reverse directions. Sketch a reaction coordinate diagram for such a reaction.**

energy change of the reaction ($\Delta G_{reaction}$, or $G_{products} - G_{reactants}$) is less than zero. A negative free energy change indicates that a reaction proceeds spontaneously as written. Note that "spontaneously" does not mean "quickly." A reaction with a negative free energy change is thermodynamically favorable, but the height of the activation energy barrier ($\Delta G^{\ddagger}$) determines how fast the reaction actually occurs. If the products have greater free energy than the reactants (Fig. 6-3b), then the overall free energy change for the reaction ($\Delta G_{reaction}$) is greater than zero. This reaction does not proceed as written (because it goes "uphill"), but it does proceed in the reverse direction.

A catalyst provides a reaction pathway with a lower activation energy barrier

A catalyst, whether inorganic or enzymatic, decreases the activation energy barrier ($\Delta G^{\ddagger}$) of a reaction (Fig. 6-4). It does so by interacting with the reacting molecules such that they are more likely to assume the transition state. A catalyst speeds up the reaction because as more reacting molecules achieve the transition state per unit time, more molecules of product can form per unit time. (An increase in temperature increases the rate of a reaction for a similar reason: The input of thermal energy allows more molecules to achieve the transition state per unit time.) Thermodynamic calculations indicate that lowering $\Delta G^{\ddagger}$ by about 5.7 kJ · mol^{-1} accelerates the reaction 10-fold. A rate increase of 10^6 requires lowering $\Delta G^{\ddagger}$ by six times this amount, or about 34 kJ · mol^{-1}.

An enzyme catalyst does not alter the net free energy change for a reaction; it merely provides a pathway from reactants to products that passes through a transition state that has lower free energy than the transition state of the uncatalyzed reaction. *An enzyme, therefore, lowers the height of the activation energy barrier ($\Delta G^{\ddagger}$) by lowering the energy of the transition state.* The hydrolysis of a peptide bond is always thermodynamically favorable, but the reaction occurs quickly only when a catalyst (such as the protease chymotrypsin) is available to provide a lower-energy route to the transition state.

Enzymes use chemical catalytic mechanisms

The idea that living organisms contain agents that can promote the change of one substance into another has been around at least since the early nineteenth century, when scientists began to analyze the chemical transformations carried out by organisms such as yeast. However, it took some time to appreciate that these catalytic agents were not part of some "vital force" present only in intact, living organisms. In 1878, the word *enzyme* was coined to indicate that there was something *in* yeast

Figure 6-4 Effect of a catalyst on a chemical reaction. Here, the reactants proceed through a transition state symbolized X‡ during their conversion to products. A catalyst lowers the free energy of activation ($\Delta G^{\ddagger}$) for the reaction so that $\Delta G^{\ddagger}_{cat}$, $<\Delta G^{\ddagger}_{uncat}$. Lowering the free energy of the transition state (X‡) accelerates the reaction because more reactants are able to achieve the transition state per unit time.

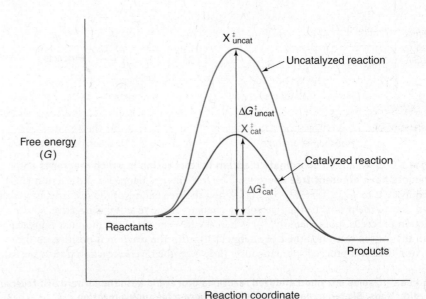

(Greek *en* = "in," *zyme* = "yeast"), rather than the yeast itself, that was responsible for breaking down (fermenting) sugar. In fact, the action of enzymes can be explained in purely chemical terms. What we currently know about enzyme mechanisms rests solidly on a foundation of knowledge about simple chemical catalysts.

In an enzyme, certain functional groups in the enzyme's active site perform the same catalytic function as small chemical catalysts. In some cases, the amino acid side chains of an enzyme cannot provide the required catalytic groups, so a tightly bound **cofactor** participates in catalysis. For example, many oxidation–reduction reactions require a metal ion cofactor, since a metal ion can exist in multiple oxidation states, unlike an amino acid side chain. Some enzyme cofactors are organic molecules known as **coenzymes,** which may be derived from vitamins. Enzymatic activity still requires the protein portion of the enzyme, which helps position the cofactor and reactants for the reaction (this situation is reminiscent of myoglobin and hemoglobin, where the globin and heme group together function to bind oxygen; see Section 5-1). Some coenzymes, termed cosubstrates, enter and exit the active site as substrates do; a tightly bound coenzyme that remains in the active site between reactions is called a prosthetic group (**Fig. 6-5**).

There are three basic kinds of chemical catalytic mechanisms used by enzymes: acid–base catalysis, covalent catalysis, and metal ion catalysis. We will examine each of these, using model reactions to illustrate some of their fundamental features.

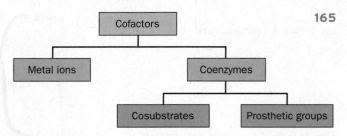

Figure 6-5 Types of enzyme cofactors.

1. Acid–Base Catalysis

Many enzyme mechanisms include **acid–base catalysis,** in which a proton is transferred between the enzyme and the substrate. This mechanism of catalysis can be further divided into **acid catalysis** and **base catalysis.** Some enzymes use one or the other; many use both. Consider the following model reaction, the tautomerization of a ketone to an enol (**tautomers** are interconvertible isomers that differ in the placement of a hydrogen and a double bond):

$$
\begin{array}{ccc}
\begin{matrix} R \\ | \\ C=O \\ | \\ CH_2 \\ | \\ H \end{matrix}
& \rightleftharpoons &
\left[\begin{matrix} R \\ | \\ C \text{---} O^- \\ \vdots \\ CH_2 \\ \vdots \\ H^+ \end{matrix} \right]
& \rightleftharpoons &
\begin{matrix} R \\ | \\ C-O-H \\ \| \\ CH_2 \end{matrix} \\
\text{Ketone} & & \text{Transition state} & & \text{Enol}
\end{array}
$$

Here the transition state is shown in square brackets to indicate that it is an unstable, transient species. The dotted lines represent bonds in the process of breaking or forming. The uncatalyzed reaction occurs slowly because formation of the carbanion-like transition state has a high activation energy barrier (a **carbanion** is a compound in which the carbon atom bears a negative charge).

If a catalyst (symbolized H—A) donates a proton to the ketone's oxygen atom, it reduces the unfavorable carbanion character of the transition state, thereby lowering its energy and hence lowering the activation energy barrier for the reaction:

$$
\begin{array}{ccc}
\begin{matrix} R \\ | \\ C=O \\ | \\ CH_2 \\ | \\ H \end{matrix} + H-A
& \rightleftharpoons &
\left[\begin{matrix} R \\ | \\ C \text{---} O^- \cdots H^+ \cdots A^- \\ \vdots \\ CH_2 \\ \vdots \\ H^+ \end{matrix} \right]
& \rightleftharpoons &
\begin{matrix} R \\ | \\ C-O-H \\ \| \\ CH_2 \end{matrix} + H-A
\end{array}
$$

This is an example of acid catalysis, since the catalyst acts as an acid by donating a proton. Note that *the catalyst is returned to its original form at the end of the reaction.*

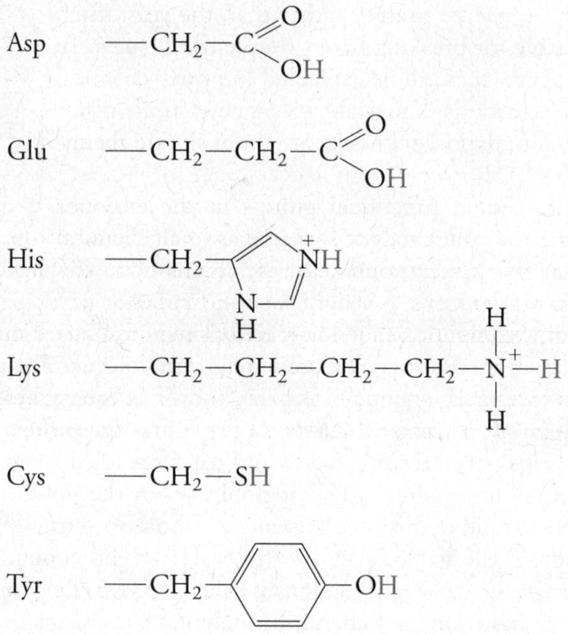

Figure 6-6 Amino acid side chains that can act as acid-base catalysts. These groups can act as acid or base catalysts, depending on their state of protonation in the enzyme's active site. The side chains are shown in their protonated forms, with the acidic proton highlighted.

? At neutral pH, which of these side chains most likely function as acid catalysts? Which most likely function as base catalysts? (*Hint:* See Table 4-1.)

The same keto-enol tautomerization reaction shown above can be accelerated by a catalyst that can accept a proton, that is, by a base catalyst. Here, the catalyst is shown as :B, where the dots represent unpaired electrons:

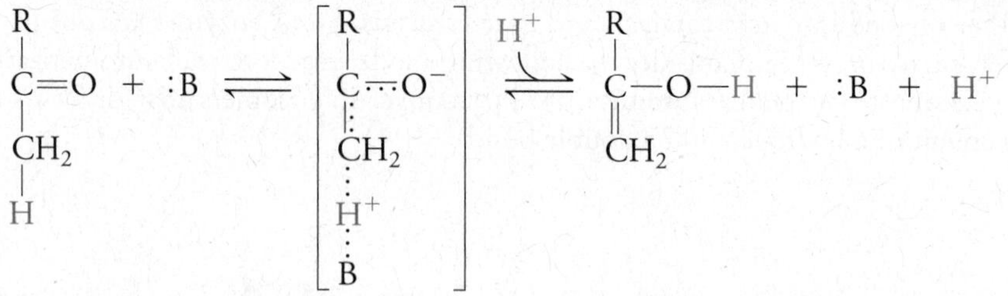

Base catalysis lowers the energy of the transition state and thereby accelerates the reaction.

In enzyme active sites, several amino acid side chains can potentially act as acid or base catalysts. These are the groups whose pK values are in or near the physiological pH range. The residues most commonly identified as acid–base catalysts are shown in **Figure 6-6**. Because the catalytic functions of these residues depend on their state of protonation or deprotonation, the catalytic activity of the enzyme may be sensitive to changes in pH.

2. Covalent Catalysis

In **covalent catalysis**, the second major chemical reaction mechanism used by enzymes, a covalent bond forms between the catalyst and the substrate during formation of the transition state. Consider as a model reaction the decarboxylation of acetoacetate. In this reaction, the movement of electron pairs among atoms is indicated by red curved arrows (Box 6-A).

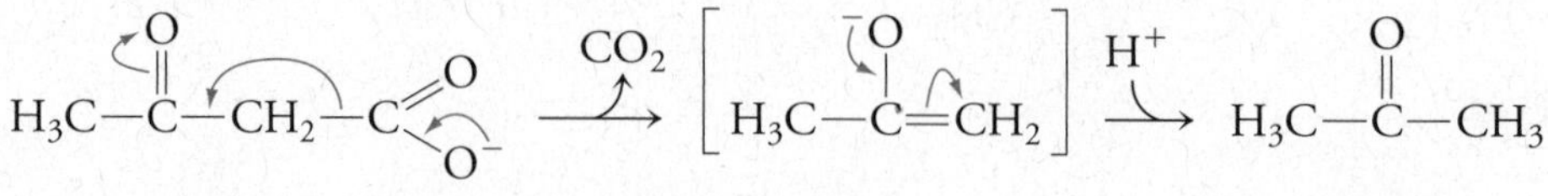

Acetoacetate   Enolate   Acetone

BOX 6-A ⬡ **BIOCHEMISTRY NOTE**

167
The Chemistry of Catalysis

Depicting Reaction Mechanisms

While it is often sufficient to draw the structures of a reaction's substrates and products, a full understanding of the reaction mechanism requires knowing what the electrons are doing. Recall that a covalent bond forms when two atoms share a pair of electrons, and a vast number of biochemical reactions involve breaking and forming covalent bonds. Although single-electron reactions also occur in biochemistry, we will focus on the more common two-electron reactions.

The curved arrow convention shows how electrons are rearranged during a reaction. The arrow emanates from the original location of an electron pair. This can be either an unshared electron pair, on an atom such as N or O, or the electrons of a covalent bond. The curved arrow points to the final location of the electron pair. For example, to show a bond breaking:

$$X-Y \longrightarrow X^+ + Y^-$$

and to show a bond forming:

$$X^+ + \;:Y^- \longrightarrow X-Y$$

A familiarity with Lewis dot structures and an understanding of electronegativity (see Section 2-1) is helpful for identifying electron-rich groups (often the source of electrons during a reaction) and electron-poor groups (where the electrons often end up).

A typical biochemical reaction requires several curved arrows, for example,

● **Question:** Draw arrows to show the electron movements in the reaction $2\,H_2O \rightarrow OH^- + H_3O^+$.

The transition state, an enolate, has a high free energy of activation. This reaction can be catalyzed by a primary amine (RNH_2), which reacts with the carbonyl group of acetoacetate to form an **imine,** a compound containing a C=N bond (this adduct is known as a **Schiff base**):

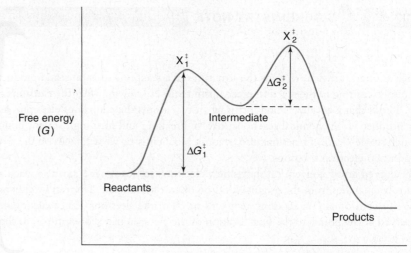

Figure 6-7 Diagram for a reaction accelerated by covalent catalysis. Two transition states flank the covalent intermediate. The relative heights of the activation energy barriers (to achieve the two transition states, $X_1^{\ddagger}$ and $X_2^{\ddagger}$) vary depending on the reaction.

? **Explain why the free energy of the reaction intermediate must be greater than the free energy of either the reactants or products.**

In this covalent intermediate, the protonated nitrogen atom acts as an electron sink to reduce the enolate character of the transition state in the decarboxylation reaction:

Finally, the Schiff base decomposes, which regenerates the amine catalyst and releases the product, acetone:

In enzymes that use covalent catalysis, an electron-rich group in the enzyme active site forms a covalent adduct with a substrate. This covalent complex can sometimes be isolated; it is much more stable than a transition state. *Enzymes that use covalent catalysis undergo a two-part reaction process, so the reaction coordinate diagram contains two energy barriers with the reaction intermediate between them* (**Fig. 6-7**).

Many of the same groups that make good acid–base catalysts (see Fig. 6-6) also make good covalent catalysts because they contain unshared electron pairs (**Fig. 6-8**). Covalent catalysis is often called nucleophilic catalysis because the catalyst is a **nucleophile,** that is, an electron-rich group in search of an electron-poor center (a compound with an electron deficiency is known as an **electrophile**).

3. Metal Ion Catalysis

Metal ion catalysis occurs when metal ions participate in enzymatic reactions by mediating oxidation–reduction reactions, as mentioned earlier, or by promoting the reactivity of other groups in the enzyme's active site through electrostatic effects. A protein-bound metal ion can also interact directly with the reacting substrate. For example, during the conversion of acetaldehyde to ethanol as catalyzed by the liver

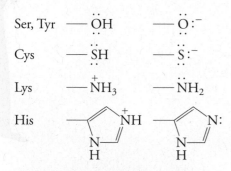

Figure 6-8 Protein groups that can act as covalent catalysts. In their deprotonated forms (*right*), these groups act as nucleophiles. They attack electron-deficient centers to form covalent intermediates.

enzyme alcohol dehydrogenase, a zinc ion stabilizes the developing negative charge on the oxygen atom during formation of the transition state:

$$H_3C-\overset{O}{\underset{H}{C}} \rightleftharpoons \left[H_3C-\overset{O^{--}\cdots Zn^{2+}}{\underset{H}{C}} \right] \overset{2\,H\cdot}{\rightleftharpoons} H_3C-\overset{OH}{\underset{H}{\overset{|}{C}}}-H$$

Acetaldehyde   Ethanol

The catalytic triad of chymotrypsin promotes peptide bond hydrolysis

Chymotrypsin uses both acid–base catalysis and covalent catalysis to accelerate peptide bond hydrolysis. These activities depend on three active-site residues whose identities and catalytic importance have been the object of intense study since the 1960s. Two of chymotrypsin's catalytic residues were identified using a technique called **chemical labeling.** When chymotrypsin is incubated with the compound diisopropylphosphofluoridate (DIPF), one of its 27 Ser residues (Ser 195) becomes covalently tagged with the diisopropylphospho (DIP) group, and the enzyme loses activity.

$$-CH_2-OH \quad + \quad \underset{\underset{CH(CH_3)_2}{\overset{\overset{CH(CH_3)_2}{|}}{\overset{O}{|}}}{F-\overset{\overset{||}{P}=O}{|}} \quad \overset{HF}{\longrightarrow} \quad -CH_2-O-\underset{\underset{CH(CH_3)_2}{\overset{\overset{CH(CH_3)_2}{|}}{\overset{O}{|}}}{\overset{\overset{||}{P}=O}{|}}$$

Ser 195   Diisopropylphosphofluoridate   DIP-Ser 195
(active)   (DIPF)   (inactive)

This observation provided strong evidence that Ser 195 is essential for catalysis. Chymotrypsin is therefore known as a **serine protease.** It is one of a large family of enzymes that use the same Ser-dependent catalytic mechanism. A similar labeling technique was used to identify the catalytic importance of His 57. The third residue involved in catalysis by chymotrypsin—Asp 102—was identified only after the fine structure of chymotrypsin was visualized through X-ray crystallography.

The hydrogen-bonded arrangement of the Asp, His, and Ser residues of chymotrypsin and other serine proteases is called the **catalytic triad** (**Fig. 6-9**). The substrate's **scissile bond** (the bond to be cleaved by hydrolysis) is positioned near Ser 195 when the substrate binds to the enzyme. The side chain of Ser is not normally a strong enough nucleophile to attack an amide bond. However, His 57, acting as a base catalyst, abstracts a proton from Ser 195 so that the oxygen can act as a covalent catalyst. Asp 102 promotes catalysis by stabilizing the resulting positively charged imidazole group of His 57.

Chymotrypsin-catalyzed peptide bond hydrolysis actually occurs in two phases that correspond to the formation and breakdown of a covalent reaction intermediate. The steps of catalysis are detailed in **Figure 6-10**. Nucleophilic attack of Ser 195 on the substrate's carbonyl carbon leads to an unstable tetrahedral intermediate in which the carbonyl carbon assumes tetrahedral geometry. This structure then collapses to an intermediate in which the N-terminal portion of the substrate remains covalently attached to the enzyme. The second part of the reaction, during which the oxygen of a water molecule attacks the carbonyl carbon, also includes an unstable tetrahedral intermediate. Although the enzyme-catalyzed reaction requires multiple steps, the net reaction is the same as the uncatalyzed reaction shown in Section 6-1.

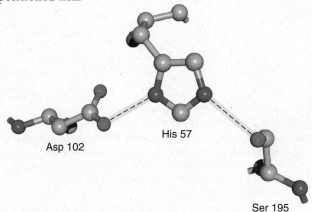

His 57

Asp 102

Ser 195

Figure 6-9 The catalytic triad of chymotrypsin. Asp 102, His 57, and Ser 195 are arrayed in a hydrogen-bonded network. Atoms are color-coded (C gray, N blue, O red), and the hydrogen bonds are shaded yellow.

? **Add hydrogen atoms to the three side chains.**

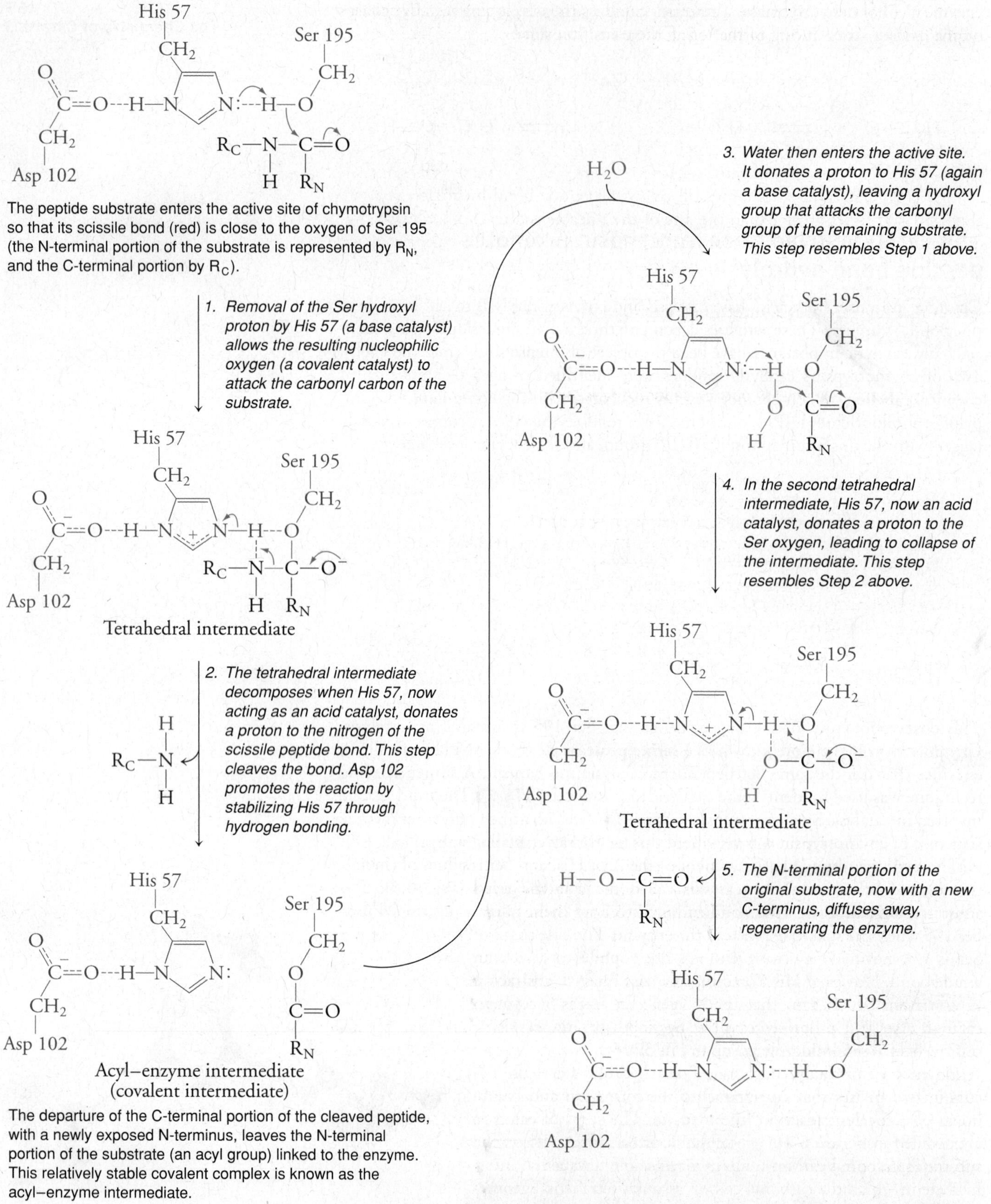

The peptide substrate enters the active site of chymotrypsin so that its scissile bond (red) is close to the oxygen of Ser 195 (the N-terminal portion of the substrate is represented by R_N, and the C-terminal portion by R_C).

1. Removal of the Ser hydroxyl proton by His 57 (a base catalyst) allows the resulting nucleophilic oxygen (a covalent catalyst) to attack the carbonyl carbon of the substrate.

Tetrahedral intermediate

2. The tetrahedral intermediate decomposes when His 57, now acting as an acid catalyst, donates a proton to the nitrogen of the scissile peptide bond. This step cleaves the bond. Asp 102 promotes the reaction by stabilizing His 57 through hydrogen bonding.

Acyl–enzyme intermediate (covalent intermediate)

The departure of the C-terminal portion of the cleaved peptide, with a newly exposed N-terminus, leaves the N-terminal portion of the substrate (an acyl group) linked to the enzyme. This relatively stable covalent complex is known as the acyl–enzyme intermediate.

3. Water then enters the active site. It donates a proton to His 57 (again a base catalyst), leaving a hydroxyl group that attacks the carbonyl group of the remaining substrate. This step resembles Step 1 above.

4. In the second tetrahedral intermediate, His 57, now an acid catalyst, donates a proton to the Ser oxygen, leading to collapse of the intermediate. This step resembles Step 2 above.

Tetrahedral intermediate

5. The N-terminal portion of the original substrate, now with a new C-terminus, diffuses away, regenerating the enzyme.

Figure 6-10 The catalytic mechanism of chymotrypsin and other serine proteases. The two tetrahedral intermediates are believed to resemble the transition states leading to and from the acyl–enzyme intermediate.

? Choose one step of the reaction and identify all the bonds that break and form in that step.

The roles of Asp, His, and Ser in peptide bond hydrolysis, as catalyzed by chymotrypsin and other members of the serine protease family, have been tested through site-directed mutagenesis (see Section 3-4). Replacing the catalytic Asp with another residue decreases the rate of substrate hydrolysis about 5000-fold. Adding a methyl group to His by chemical labeling (so that it can't accept or donate a proton) has a similar effect. Replacing the catalytic Ser with another residue decreases enzyme activity about a millionfold. Surprisingly, replacing all three catalytic residues—Asp, His, and Ser—through site-directed mutagenesis does not completely abolish protease activity: The modified enzyme still catalyzes peptide bond hydrolysis at a rate about 50,000 times greater than the rate of the uncatalyzed reaction. Clearly, chymotrypsin and its relatives rely on the acid–base catalysis and covalent catalysis carried out by the Asp–His–Ser catalytic triad, but these enzymes must have additional catalytic mechanisms that allow them to achieve reaction rates 10^{11} times greater than the rate of the uncatalyzed reaction.

CONCEPT REVIEW
- Why does a free energy barrier separate reactants and products in a chemical reaction?
- What is the relationship between the height of the activation energy barrier and the rate of the reaction?
- How does an enzyme lower the activation energy of a reaction?
- How can an acid or base accelerate a reaction?
- How does a covalent catalyst accelerate a reaction?
- Why must a reaction that proceeds via covalent catalysis have two transition states?
- How do metal ions accelerate reactions?
- How can chemical labeling identify an enzyme's catalytic residues?
- Which residues constitute chymotrypsin's catalytic triad?
- How do these residues participate in peptide bond hydrolysis as catalyzed by chymotrypsin?

6-3 The Unique Properties of Enzyme Catalysts

If only a few residues in an enzyme directly participate in catalysis (for example, Asp, His, and Ser in chymotrypsin), why are enzymes so large? One obvious answer is that the catalytic residues must be precisely aligned in the active site, so a certain amount of surrounding structure is required to hold them in place. This view supports Fischer's **lock-and-key model** for enzyme action. But how can an active site that perfectly accommodates substrates also accommodate products before they are released from the enzyme? Moreover, how can the lock-and-key model explain why an enzyme inhibitor can bind tightly in the active site but not react? The answer is that enzymes, like other proteins (see Section 4-3), are not rigid molecules but instead can flex while binding substrates. In other words, the enzyme–substrate interaction must be more dynamic than a key in a lock, more like a hand in a glove. In some cases, the enzyme can physically distort the substrate as it binds, in effect pushing it toward a higher-energy conformation closer to the reaction's transition state. Current theories attribute much of the catalytic power of enzymes to these and other specific interactions between enzymes and their substrates.

KEY CONCEPT
- The catalytic activity of enzymes also depends on transition state stabilization, proximity and orientation effects, induced fit, and electrostatic catalysis.

⊕ **See Guided Exploration. Catalytic mechanism of serine proteases.**

▶▶ **DOES** the lock-and-key model explain enzyme action?

Enzymes stabilize the transition state

The lock-and-key model does contain a grain of truth, a principle first formulated by Linus Pauling in 1946. He proposed that *an enzyme increases the reaction rate not by binding tightly to the substrates but by binding tightly to the reaction's transition state* (that is, substrates that have been strained toward the structures of the products). In other words, the tightly bound key of the lock-and-key model is the transition state, not the substrate. In an enzyme, tight binding (stabilization) of the transition state occurs in addition to acid–base, covalent, or metal ion catalysis. In general, transition state stabilization is accomplished through the close complementarity in shape and charge between the active site and the transition state. A nonreactive substance that mimics the transition state can therefore bind tightly to the enzyme and block its catalytic activity (enzyme inhibition is discussed further in Section 7-3).

Transition state stabilization appears to be an important part of the chymotrypsin reaction. In this case, the two tetrahedral intermediates, which resemble the transition state (see Fig. 6-10), are stabilized through interactions that do not occur at any other point in the reaction. Rate acceleration is believed to result from an increase in both the number and strength of the bonds that form between active site groups and the substrate in the transition state.

1. During formation of the tetrahedral intermediate, the planar peptide group of the substrate changes its geometry, and the carbonyl oxygen, now an anion, moves into a previously unoccupied cavity near the Ser 195 side chain. In this cavity, called the **oxyanion hole,** the substrate oxygen ion can form two new hydrogen bonds with the backbone NH groups of Ser 195 and Gly 193 (Fig. 6-11). The backbone NH group of the substrate residue preceding the scissile bond forms another hydrogen bond to Gly 193 (not shown in Fig. 6-11). Thus, the transition state is stabilized (its energy lowered) by three hydrogen bonds that cannot form when the enzyme first binds its substrate. The stabilizing effect of these three new hydrogen bonds could account for a significant portion of chymotrypsin's catalytic power, since the energy of a standard hydrogen bond is about 20 kJ · mol^{-1} and the reaction rate increases 10-fold for every decrease in $\Delta G^{\ddagger}$ of 5.7 kJ · mol^{-1}.

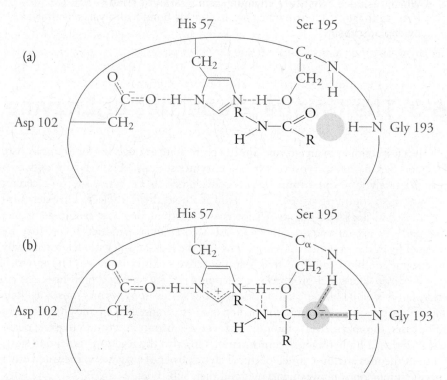

Figure 6-11 Transition state stabilization in the oxyanion hole. (a) The chymotrypsin active site is shown with the oxyanion hole shaded in pink. The carbonyl carbon of the peptide substrate has trigonal geometry, so the carbonyl oxygen cannot occupy the oxyanion hole. (b) Nucleophilic attack by the oxygen of Ser 195 on the substrate carbonyl group leads to a transition state, in which the carbonyl carbon assumes tetrahedral geometry. At this point, the substrate's anionic oxygen (the oxyanion) can move into the oxyanion hole, where it forms hydrogen bonds (shaded yellow) with two enzyme backbone groups. ⊕ See Animated Figure. Effect of preferential transition state binding.

2. NMR studies, which can identify individual hydrogen-bonding interactions, suggest that the hydrogen bond between Asp 102 and His 57 becomes shorter during formation of the two tetrahedral intermediates (**Fig. 6-12**). Such a bond is called a **low-barrier hydrogen bond** because the hydrogen is shared equally between the original donor and acceptor atoms (in a standard hydrogen bond, the proton still "belongs" to the donor atom and there is an energy barrier for its full transfer to the acceptor atom). A decrease in bond length from ~2.8 Å to ~2.5 Å in forming the low-barrier hydrogen bond is accompanied by a three- to fourfold increase in bond strength. The low-barrier hydrogen bond that forms during catalysis in chymotrypsin helps stabilize the transition state and thereby accelerate the reaction.

Efficient catalysis depends on proximity and orientation effects

Enzymes increase reaction rates by bringing reacting groups into close proximity so as to increase the frequency of collisions that can lead to a reaction. Furthermore, when substrates bind to an enzyme, their translational and rotational motions are frozen out so that they can be oriented properly for reaction (**Fig. 6-13**). These **proximity and orientation effects** likely explain some of the residual activity of chymotrypsin whose catalytic residues have been altered. Nevertheless, *an enzyme must be more than a template for assembling and aligning reacting groups.*

The active-site microenvironment promotes catalysis

In nearly all cases, an enzyme's active site is somewhat removed from the solvent, with its catalytic residues in some sort of cleft or pocket on the enzyme surface. Upon binding substrates, some enzymes undergo a pronounced conformational change so that they almost fully enclose the substrates. Daniel Koshland has called this phenomenon **induced fit.** Many studies of enzyme activity, supplemented by X-ray crystallographic data, support this feature of enzyme action, further highlighting the shortcomings of the lock-and-key model. In fact, techniques that assess protein structure on millisecond time scales suggest that conformational fluctuations may be critical for all stages of an enzyme-catalyzed reaction, not just substrate binding.

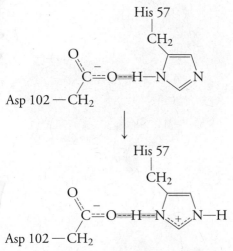

Figure 6-12 Formation of a low-barrier hydrogen bond during catalysis in chymotrypsin. The Asp 102—His 57 hydrogen bond becomes shorter and stronger, so the imidazole proton comes to be shared equally between the O of Asp and the N of histidine in a low-barrier hydrogen bond.

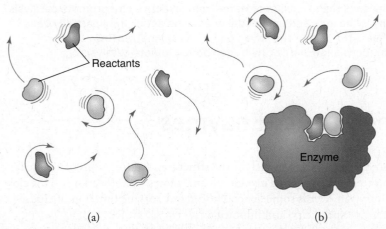

(a)   (b)

Figure 6-13 Proximity and orientation effects in catalysis. In order to react, two groups must come together and collide with the correct orientation. (a) Reactants in solution are separated in space and have translational and rotational motions that must be overcome. (b) When the reactants bind to an enzyme, their motion is limited, and they are held in close proximity and with the correct alignment for a productive reaction.

A classic case of induced fit occurs in hexokinase, which catalyzes the phosphorylation of glucose by ATP (Reaction 1 of glycolysis; Section 13-1):

Glucose + ATP $\xrightarrow{\text{hexokinase}}$ Glucose-6-phosphate + ADP

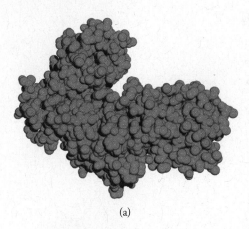

(a)

(b)

Figure 6-14 Conformational changes in hexokinase. (a) The enzyme consists of two lobes connected by a hinge region. The active site is located in a cleft between the lobes. (b) When glucose (not shown) binds to the active site, the enzyme lobes swing together, enclosing glucose and preventing the entry of water. [Structure of "open" hexokinase (pdb 2YHX) determined by T. A. Steitz, C. M. Anderson, and R. E. Stenkamp; structure of "closed" hexokinase (pdb 1HKG) determined by W. S. Bennett Jr. and T. A. Steitz.]

The enzyme consists of two hinged lobes with the active site located between them **(Fig. 6-14a).** When glucose binds to hexokinase, the lobes swing together, engulfing the sugar (Fig. 6-14b). The result of hinge bending is that the substrate glucose is positioned near the substrate ATP such that a phosphoryl group can be easily transferred from the ATP to a hydroxyl group of the sugar. Not even a water molecule can enter the closed active site. This is beneficial, since water in the active site could lead to wasteful hydrolysis of ATP:

$$ATP + H_2O \rightarrow ADP + P_i$$

A glucose molecule in solution is surrounded by ordered water molecules in a hydration shell (see Section 2-1). These water molecules must be shed in order for glucose to fit into the active site of an enzyme, such as hexokinase. However, once the desolvated substrate is in the enzyme active site, the reaction can proceed quickly because there are no solvent molecules to interfere. In solution, rearranging the hydrogen bonds of surrounding water molecules as the reactants approach each other and pass through the transition state is energetically costly. *By sequestering substrates in the active site, an enzyme can eliminate the energy barrier imposed by the ordered solvent molecules, thereby accelerating the reaction.*

This phenomenon is sometimes described as **electrostatic catalysis** since the nonaqueous active site allows more powerful electrostatic interactions between the enzyme and substrate than could occur in aqueous solution (for example, a low-barrier hydrogen bond can form in an active site but not in the presence of solvent molecules that would form ordinary hydrogen bonds).

CONCEPT REVIEW
- What are the roles of proximity and orientation effects in enzyme catalysis?
- Why does the exclusion of water from an active site promote catalysis?
- Why is the lock-and-key model of enzyme action only partially valid?
- What is the role of the oxyanion hole in chymotrypsin?
- How does a low-barrier hydrogen bond promote catalysis?

6-4 Some Additional Features of Enzymes

KEY CONCEPTS
- Evolution has produced a number of serine proteases that differ in overall structure and substrate specificity.
- Inactive zymogens are activated by proteolysis.
- Active enzymes may be regulated by inhibitors.

Chymotrypsin serves as a model for the structures and functions of a large family of serine proteases. And as with myoglobin and hemoglobin (Section 5-1), a close look at chymotrypsin reveals some general features of enzyme function, including evolution, substrate specificity, and inhibition.

Not all serine proteases are related by evolution

The first three proteases to be examined in detail were the digestive enzymes chymotrypsin, trypsin, and elastase, which have strikingly similar three-dimensional structures **(Fig. 6-15)**. This was not expected on the basis of their limited sequence similarity (Table 6-3). However, careful examination revealed that most of the sequence variation is on the enzyme surface, and the positions of the catalytic residues in the three actives sites are virtually identical. It is believed that *these proteins diverged from a common ancestor and have retained their overall structure and catalytic mechanism.*

Some bacterial proteases with a catalytically essential Ser are structurally related to the mammalian digestive serine proteases. However, the bacterial serine protease subtilisin (**Fig. 6-16**) shows no sequence similarity to chymotrypsin and no overall structural similarity, although it has the same Asp–His–Ser catalytic triad and an oxyanion hole in its active site. Subtilisin is an example of **convergent evolution,** a phenomenon whereby unrelated proteins evolve similar characteristics.

As many as five groups of serine proteases, each with a different overall backbone conformation, have undergone convergent evolution to arrive at the same Asp, His, and Ser catalytic groups. In some other hydrolases, the substrate is attacked by a nucleophilic Ser or Thr residue that is located in a catalytic triad such as His–His–Ser or Asp–Lys–Thr. It would appear that natural selection favors this sort of arrangement of catalytic residues.

Enzymes with similar mechanisms exhibit different substrate specificity

Despite similarities in their catalytic mechanisms, chymotrypsin, trypsin, and elastase differ significantly from one another in their substrate specificity. Chymotrypsin preferentially cleaves peptide bonds following large hydrophobic residues. Trypsin prefers the basic residues Arg and Lys, and elastase cleaves the peptide bonds following small hydrophobic residues such as Ala, Gly, and Val (these residues predominate in elastin, an animal protein responsible for the elasticity of some tissues). *The varying specificities of these enzymes are largely explained by the chemical character of the so-called **specificity pocket,** a cavity on the enzyme surface at the active site that accommodates the residue on the N-terminal side of the scissile peptide bond* (**Fig. 6-17**). In chymotrypsin, the specificity pocket is about 10 Å deep and 5 Å wide, which offers a snug fit for an aromatic ring (whose dimensions are 6 Å × 3.5 Å). The specificity pocket in trypsin is similarly sized but has an Asp residue rather than Ser at the bottom. Consequently, the trypsin specificity pocket readily binds the side chain of Arg or Lys, which has a diameter of about 4 Å and a cationic group at the

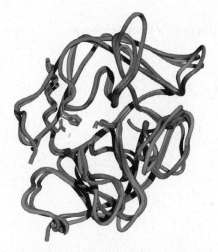

Figure 6-15 Structures of chymotrypsin, trypsin, and elastase. The superimposed backbone traces of bovine chymotrypsin (blue), bovine trypsin (green), and porcine elastase (red) are shown along with the side chains of the active site Asp, His, and Ser residues. [Chymotrypsin structure (pdb 4CHA) determined by H. Tsukada and D. M. Blow; trypsin structure (pdb 3PTN) determined by J. Walker, W. Steigemann, T. P. Singh, H. Bartunik, W. Bode, and R. Huber; elastase structure (3EST) determined by E. F. Meyer, G. Cole, R. Radhakrishnan, and O. Epp.]

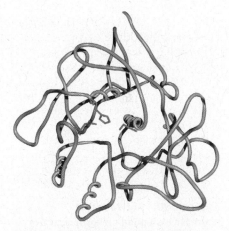

Figure 6-16 Structure of subtilisin from *Bacillus amyloliquefaciens*. The residues of the catalytic triad are highlighted in red. Compare the structure of this enzyme with those of the three serine proteases shown in Figure 6-15. [Structure (pdb 1CSE) determined by W. Bode.]

[**TABLE 6-3**]

Percent Sequence Identity among Three Serine Proteases

Bovine trypsin	100%
Bovine chymotrypsinogen	53%
Porcine elastase	48%

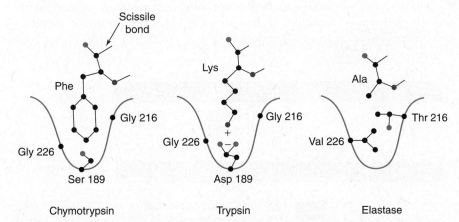

Figure 6-17 Specificity pockets of three serine proteases. The side chains of key residues that determine the size and nature of the specificity pocket are shown along with a representative substrate for each enzyme. Chymotrypsin prefers large hydrophobic side chains; trypsin prefers Lys or Arg; and elastase prefers Ala, Gly, or Val. For convenience, the residues of all three enzymes are numbered to correspond to the sequence of residues in chymotrypsin.

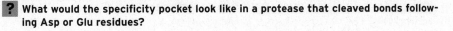

? **What would the specificity pocket look like in a protease that cleaved bonds following Asp or Glu residues?**

end. In elastase, the specificity pocket is only a small depression due to the replacement of two Gly residues on the walls of the specificity pocket (residues 216 and 226 in chymotrypsin) with the bulkier Val and Thr. Elastase therefore preferentially binds small nonpolar side chains. Although these same side chains could easily enter the chymotrypsin and trypsin specificity pockets, they do not fit well enough to immobilize the substrate at the active site, as required for efficient catalysis. Other serine proteases, such as those that participate in blood coagulation, exhibit exquisite substrate specificity, in keeping with their highly specialized physiological functions.

Chymotrypsin is activated by proteolysis

Relatively nonspecific proteases could do considerable damage to the cells where they are synthesized unless their activity is carefully controlled. In many organisms, *the activity of proteases is limited by the action of protease inhibitors* (some of which are discussed further below) *and by synthesizing the proteases as inactive precursors (called zymogens) that are later activated when and where they are needed.*

The inactive precursor of chymotrypsin is called chymotrypsinogen, and it is synthesized by the pancreas along with the zymogens of trypsin (trypsinogen), elastase (proelastase), and other hydrolytic enzymes. All these zymogens are activated by proteolysis after they are secreted into the small intestine. An intestinal protease called enteropeptidase activates trypsinogen by catalyzing the hydrolysis of its Lys 6—Ile 7 bond. Enteropeptidase catalyzes a highly specific reaction; it appears to recognize a string of Asp residues near the N-terminus of its substrate:

$$H_3\overset{+}{N}—Val—Asp—Asp—Asp—Asp—Lys—Ile—\cdots$$

$$H_2O \downarrow \text{enteropeptidase}$$

$$H_3\overset{+}{N}—Val—Asp—Asp—Asp—Asp—Lys—COO^- + H_3\overset{+}{N}—Ile—\cdots$$

Trypsin, now itself active, cleaves the N-terminal peptide of the other pancreatic zymogens, including trypsinogen. The activation of trypsinogen by trypsin is an example of **autoactivation.**

The Arg 15—Ile 16 bond of chymotrypsinogen is susceptible to trypsin-catalyzed hydrolysis. Cleavage of this bond generates a species of active chymotrypsin (called π chymotrypsin), which then undergoes two autoactivation steps to generate fully active chymotrypsin (also called α chymotrypsin; **Fig. 6-18**). A similar process, in which zymogens are sequentially activated through proteolysis, occurs during blood coagulation (Box 6-B).

Figure 6-18 Activation of chymotrypsinogen. Trypsin activates chymotrypsinogen by catalyzing hydrolysis of the Arg 15—Ile 16 bond of the zymogen. The resulting active chymotrypsin then excises the Ser 14–Arg 15 dipeptide (by cleaving the Leu 13—Ser 14 bond) and the Thr 147–Asn 148 dipeptide (by cleaving the Tyr 146—Thr 147 and Asn 148—Ala 149 bonds). All three species of chymotrypsin (π, δ, and α) have proteolytic activity.

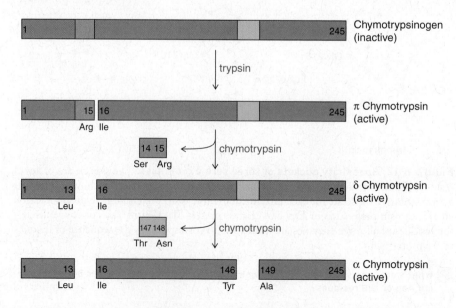

BOX 6-B ⬢ CLINICAL CONNECTION

Blood Coagulation Requires a Cascade of Proteases

When a blood vessel is injured by mechanical force, infection, or some other pathological process, red and white blood cells and the plasma (fluid) that surrounds them can leak out. Except in the most severe trauma, the loss of blood can be halted through formation of a clot at the site of injury. The clot consists of aggregated platelets (tiny enucleated cells that rapidly adhere to the damaged vessel wall and to each other) and a mesh of the protein fibrin, which reinforces the platelet plug and traps larger particles such as red blood cells.

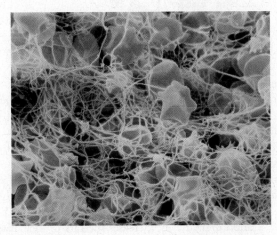

[P. Motta/Dept. of Anatomy, University La Sapienza, Rome/Science Photo Library/Photo Researchers.]

Fibrin polymers can form rapidly because they are generated at the site of injury from the soluble protein fibrinogen, which circulates in the blood plasma. Fibrinogen is an elongated molecule with a molecular mass of 340,000 and consists of three pairs of polypeptide chains. The proteolytic removal of short (14- or 16-residue) peptides from the N-termini of four of the six chains causes the protein to polymerize in end-to-end and side-to-side fashion to produce a thick fiber.

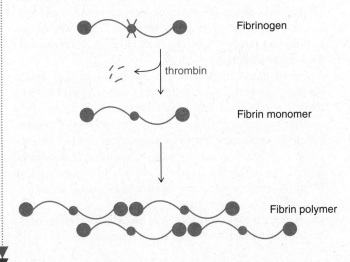

The conversion of fibrinogen to fibrin is the final step of **coagulation,** a series of proteolytic reactions involving a number of proteins and additional factors from platelets and damaged tissue. The enzyme responsible for cleaving fibrinogen to fibrin is known as thrombin. It is similar to trypsin in its sequence (38% of their residues are identical) and in its structure and catalytic mechanism.

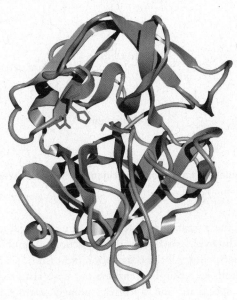

[Thrombin structure (pdb 1PPB) determined by W. Bode.]

Compare the structure of thrombin shown here with that of chymotrypsin (Fig. 6-1). Thrombin's catalytic residues (Asp, His, and Ser) are highlighted in red.

Like trypsin, thrombin cleaves peptide bonds following Arg residues, but it is highly specific for the two cleavage sites in the fibrinogen sequence. Thrombin, like fibrinogen, circulates as an inactive precursor. Its zymogen, called prothrombin, contains a serine protease domain along with several other structural motifs. These elements interact with other coagulation factors to help ensure that thrombin—and therefore fibrin—is produced only when needed.

A serine protease known as factor Xa catalyzes the specific hydrolysis of prothrombin to generate thrombin. Factor Xa (the *a* stands for *active*) is the protease form of the zymogen factor X. To initiate coagulation, factor X is activated by a protease known as factor VIIa, working in association with an accessory protein called tissue factor, which is exposed when a blood vessel is broken. During the later stages of coagulation, factor Xa is generated by the activity of factor IXa, which is generated from its zymogen by the activity of factor XIa or factor VIIa–tissue factor. Factor XIa is in turn generated by the proteolysis of its zymogen by trace amounts of thrombin produced earlier in the coagulation process. The cascade of activation reactions is depicted on the next page. Many of these enzymatic reactions require accessory factors that are not shown in this simplified diagram.

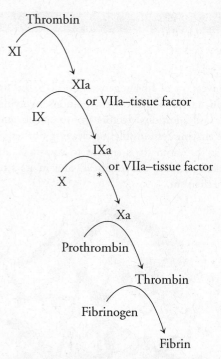

The coagulation cascade. The triggering step is marked with an asterisk.

Note that the coagulation proteases are named according to their order of discovery, not their order of action (thrombin is also know as factor II). All the coagulation proteases appear to have evolved from a trypsin-like enzyme but have acquired narrow substrate specificity and a correspondingly narrow range of physiological activities.

The coagulation reactions have an amplifying effect because each protease is a catalyst for the activation of another catalyst. Thus, a very small amount of factor IXa can activate a larger amount of factor Xa, which can then activate an even larger amount of thrombin. This amplification effect is reflected in the plasma concentrations of the coagulation factors.

Plasma concentrations of some human coagulation factors

Factor	Concentration (μM)[a]
XI	0.06
IX	0.09
VII	0.01
X	0.18
Prothrombin	1.39
Fibrinogen	8.82

[a]Concentrations calculated from data in High, K. A., and Roberts, H. R., eds., *Molecular Basis of Thrombosis and Hemostasis, Marcel Dekker* (1995).

A complex process such as coagulation is subject to regulation at a variety of points, including the activation and inhibition of the various proteases. Protease inhibitors account for about 10% of the protein that circulates in plasma. One inhibitor, known as antithrombin, blocks the proteolytic activities of factor IXa, factor Xa, and thrombin, thereby limiting the extent and duration of clot formation. Residues 377–400 of the 432-residue antithrombin form a loop (yellow in the figure above right) that extends away from the rest of the protein, and an arginine residue (red) at position 393 serves as a "bait" for the Arg-specific coagulation proteases. The protease recognizes the inhibitor as a substrate but is unable to complete the hydrolysis reaction. The protease and inhibitor form a stable acyl–enzyme intermediate that is removed from the circulation within a few minutes.

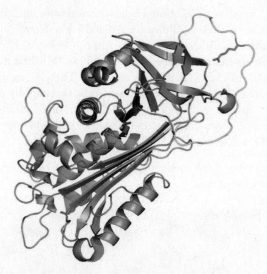

[Antithrombin structure (pdb 2ANT) determined by R. Skinner, J. P. Abrahams, J. C. Whisstock, A. M. Lesk, R. W. Carrell, and M. R. Wardell.]

Heparin, a sulfated polysaccharide (Section 11-3) first purified from liver, enhances the activity of antithrombin by two mechanisms: A short segment (5 monosaccharide residues) acts as an allosteric activator of antithrombin, and a longer heparin polymer (containing at least 18 residues) can bind simultaneously to both antithrombin and its target protease such that their co-localization dramatically increases the rate of their reaction. Heparin or a synthetic version of it is used clinically as an anticoagulant following surgery.

Defects in many of the proteins involved in blood coagulation or its regulation have been linked to bleeding (or clotting) disorders. For example, one form of hemophilia, a tendency to bleed following minor trauma, results from a genetic deficiency of factor IX. A deficiency of antithrombin results in an increased risk of clot formation in the veins. If dislodged, the clots may end up blocking an artery in the lungs or brain, with dire consequences.

● **Questions:**

1. Hemophiliacs who are deficient in factor IX can be treated with injections of pure factor IX. Occasionally, a patient develops antibodies to this material, and it becomes ineffective. In such cases, the patient can be given factor VII instead. Explain why factor VII injections would be a useful treatment for a factor IX deficiency.

2. A genetic defect in factor IX causes hemophilia, a serious disease. However, a genetic defect in factor XI may have no clinical symptoms. Explain this discrepancy in terms of the cascade mechanism for activation of coagulation proteases.

3. Factor IX deficiency has been successfully treated by gene therapy, in which a normal copy of the gene is introduced into the body (see Section 3-4). Explain why restoration of factor IX levels to just a few percent of normal is enough to cure the abnormal bleeding.

4. Of all the proteases included in the cascade diagram of coagulation, which one is the oldest in evolutionary terms?

5. Following severe trauma, surgery, or infection, a patient may develop disseminated intravascular coagulation (DIC), in which numerous small clots form throughout the circulatory system. Explain why patients with DIC subsequently exhibit excessive bleeding.

6. In one thrombin variant, a point mutation produces a protease with normal activity toward fibrinogen but decreased activity toward antithrombin. Would this genetic defect increase the risk of bleeding or of clotting?

7. Why can't heparin be administered orally?

8. Researchers have developed drugs based on hirudin, a thrombin inhibitor from the medicinal leech *Hirudo medicinalis*. Why would the leech be a good source for an anticoagulant?

The two dipeptides that are excised during chymotrypsinogen activation are far removed from the active site (Fig. 6-19). How does their removal boost catalytic activity? A comparison of the X-ray structures of chymotrypsin and chymotrypsinogen reveals that the conformations of their active site Asp, His, and Ser residues are virtually identical (in fact, the zymogen can catalyze hydrolysis extremely slowly). However, the substrate specificity pocket and the oxyanion hole are incompletely formed in the zymogen. Proteolysis of the zymogen elicits small conformational changes that open up the substrate specificity pocket and oxyanion hole. *Thus, the enzyme becomes maximally active only when it can efficiently bind its substrates and stabilize the transition state.*

Protease inhibitors limit protease activity

The pancreas, in addition to synthesizing the zymogens of digestive proteases, synthesizes small proteins that act as **protease inhibitors.** The liver also produces a variety of protease inhibitor proteins that circulate in the bloodstream. If the pancreatic enzymes were prematurely activated or escaped from the pancreas through trauma, they would be rapidly inactivated by protease inhibitors. *The inhibitors pose as protease substrates but are not completely hydrolyzed.* For example, when trypsin attacks a Lys residue of bovine pancreatic trypsin inhibitor, the reaction halts during formation of the tetrahedral intermediate. The inhibitor remains in the active site, preventing any further catalytic activity (Fig. 6-20). The noncovalent complex between trypsin and bovine pancreatic trypsin inhibitor is one of the strongest protein–protein interactions known, with a dissociation constant of 10^{-14} M. An imbalance between the activities of proteases and the activities of protease inhibitors may contribute to disease (see Box 6-B).

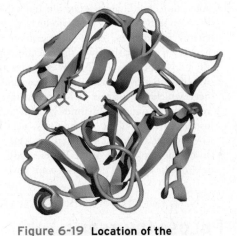

Figure 6-19 Location of the dipeptides removed during the activation of chymotrypsinogen. The Ser 14–Arg 15 dipeptide (lower *right*, green) and the Thr 147–Asn 148 dipeptide (*right*, blue) are located at some distance from the active site residues (red) in chymotrypsinogen. [Chymotrypsinogen structure (pdb 2CGA) determined by D. Wang, W. Bode, and R. Huber.]

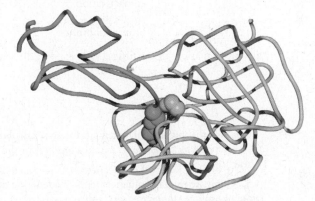

Figure 6-20 The complex of trypsin with bovine pancreatic trypsin inhibitor. Ser 195 of trypsin (gold) attacks the peptide bond of Lys 15 (green) in the inhibitor, but the reaction is arrested on the way to the tetrahedral intermediate stage. [Structure (pdb 2PTC) determined by R. Huber and J. Deisenhofer.]

> **CONCEPT REVIEW**
> - How are chymotrypsin, trypsin, and elastase similar?
> - What is the chemical basis for their different substrate specificities?
> - How is chymotrypsin activated and why is this step necessary?
> - Why are protease inhibitors necessary?

[SUMMARY]

6-1 What Is an Enzyme?

- Enzymes accelerate chemical reactions with high specificity under mild conditions.

6-2 The Chemistry of Catalysis

- A reaction coordinate diagram illustrates the change in free energy between the reactants and products as well as the activation energy required to reach the transition state. The higher the activation energy, the fewer the reactant molecules that can reach the transition state and the slower the reaction.
- An enzyme provides a route from reactants (substrates) to products, which has a lower activation energy than the uncatalyzed reaction. Enzymes, sometimes with the assistance of a cofactor, use chemical catalytic mechanisms such as acid–base catalysis, covalent catalysis, and metal ion catalysis.
- In chymotrypsin, an Asp–His–Ser triad catalyzes peptide bond hydrolysis through acid–base and covalent catalysis and by stabilizing the transition state via the oxyanion hole and low-barrier hydrogen bonds.

6-3 The Unique Properties of Enzyme Catalysts

- In addition to transition state stabilization, enzymes use proximity and orientation effects, induced fit, and electrostatic catalysis to facilitate reactions.

6-4 Some Additional Features of Enzymes

- Serine proteases that have evolved from a common ancestor share their overall structure and catalytic mechanism but differ in their substrate specificity.
- The activities of some proteases are limited by their synthesis as zymogens that are later activated and by their interaction with protease inhibitors.

[GLOSSARY TERMS]

hydrolysis	coenzyme	catalytic triad
reactant	acid–base catalysis	scissile bond
catalyst	acid catalysis	lock-and-key model
enzyme	base catalysis	oxyanion hole
ribozyme	tautomer	low-barrier hydrogen bond
active site	carbanion	proximity and orientation effects
reaction specificity	covalent catalysis	induced fit
substrate	imine	electrostatic catalysis
isozymes	Schiff base	convergent evolution
reaction coordinate	nucleophile	specificity pocket
$\Delta G^{\ddagger}$	electrophile	zymogen
transition state	metal ion catalysis	autoactivation
$\Delta G_{reaction}$	chemical labeling	coagulation
cofactor	serine protease	protease inhibitor

[PROBLEMS]

6-1 What Is an Enzyme?

1. Why are most enzymes globular rather than fibrous proteins?

2. Explain why the motor proteins myosin and kinesin (described in Section 5-3) are enzymes and write the reaction that each catalyzes.

3. The uncatalyzed rate of amide bond hydrolysis in hippuryl-phenylalanine is 1.3×10^{-10} s^{-1}, whereas the rate of hydrolysis of the amide bond in the same substrate catalyzed by carboxypeptidase is 61 s^{-1}. What is the rate enhancement for amide bond hydrolysis by this enzyme for this particular substrate?

4. The half-life for the hydrolysis of the glycosidic bond in the sugar trehalose is 6.6×10^6 years.
   (a) What is the rate constant for the uncatalyzed hydrolysis of this bond? (*Hint:* For a first-order reaction, the rate constant, k, is equal to $0.693/t_{1/2}$.)

(b) What is the rate enhancement for glycosidic bond hydrolysis catalyzed by trehalase if the rate constant for the catalyzed reaction is 2.6×10^3 s^{-1}?

5. Compare the rate enhancements of adenosine deaminase and triose phosphate isomerase (see Table 6-1).

6. What is the relationship between the rate of an enzyme-catalyzed reaction and the rate of the corresponding uncatalyzed reaction? Do enzymes enhance the rates of slow uncatalyzed reactions as much as they enhance the rates of fast uncatalyzed reactions?

7. Draw an arrow to identify the bonds in the peptide below that would be hydrolyzed in the presence of the appropriate peptidase.

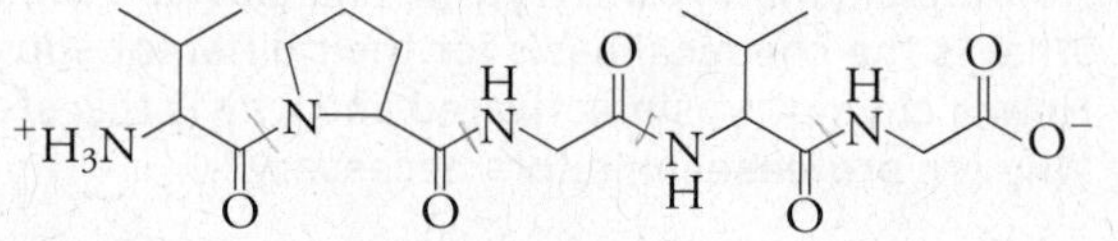

8. (a) The molecule shown here is a substrate for trypsin. Draw the reaction products.

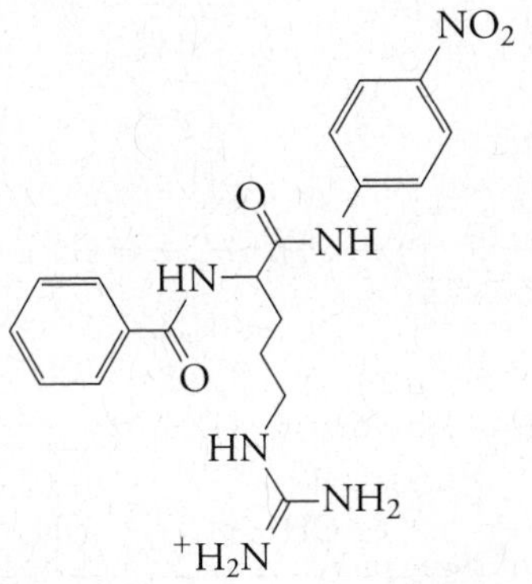

(b) Design an assay that would allow you to measure the rate of trypsin-catalyzed hydrolysis of the molecule shown in part (a).

9. The reactions catalyzed by the enzymes listed below are presented in this chapter. To which class does each enzyme belong? Explain your answers.
(a) pyruvate decarboxylase
(b) alanine aminotransferase
(c) alcohol dehydrogenase
(d) hexokinase
(e) chymotrypsin

10. To which class do the enzymes that catalyze the following reactions belong?

(a)

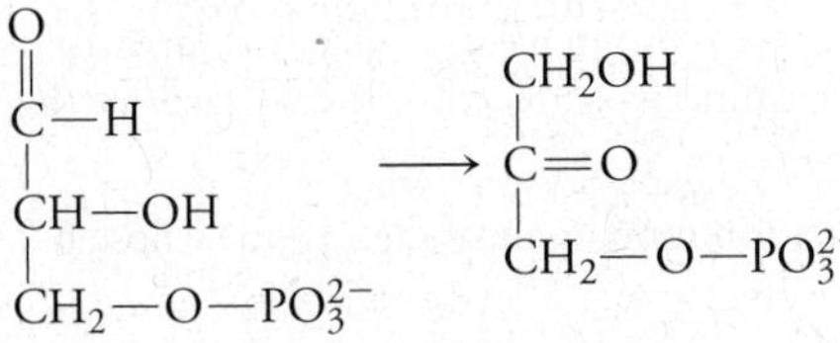

$$\begin{array}{c} O \\ \| \\ C-H \\ | \\ CH-OH \\ | \\ CH_2-O-PO_3^{2-} \end{array} \longrightarrow \begin{array}{c} CH_2OH \\ | \\ C=O \\ | \\ CH_2-O-PO_3^{2-} \end{array}$$

(b)

$$\begin{array}{c} COO^- \\ | \\ CH-O-PO_3^{2-} \\ | \\ CH_2OH \end{array} \longrightarrow \begin{array}{c} COO^- \\ | \\ C-O-PO_3^{2-} \\ \| \\ CH_2 \end{array}$$

(c)

$$\begin{array}{c} COO^- \\ | \\ C=O \\ | \\ CH_3 \end{array} \longrightarrow \begin{array}{c} COO^- \\ | \\ HO-C-H \\ | \\ CH_3 \end{array}$$

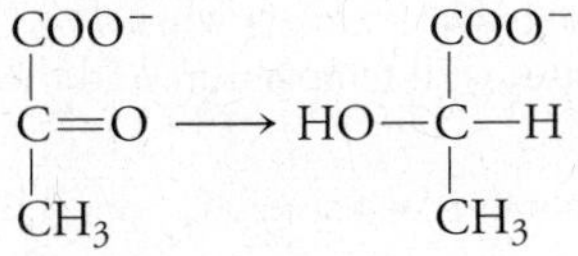

(d)

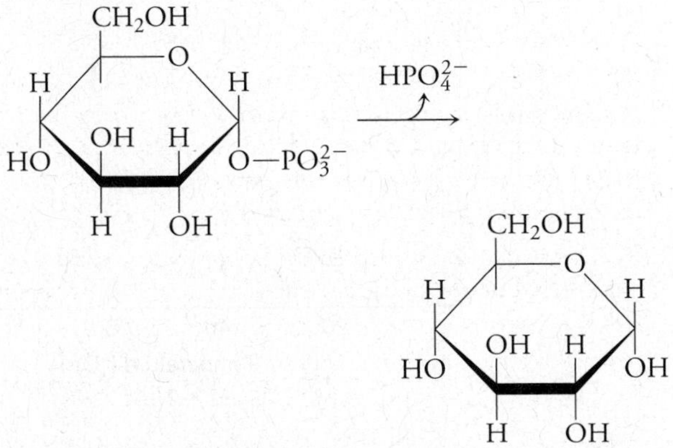

11. Draw the oxidized product of the reaction catalyzed by succinate dehydrogenase. To which class of enzymes does succinate dehydrogenase belong?

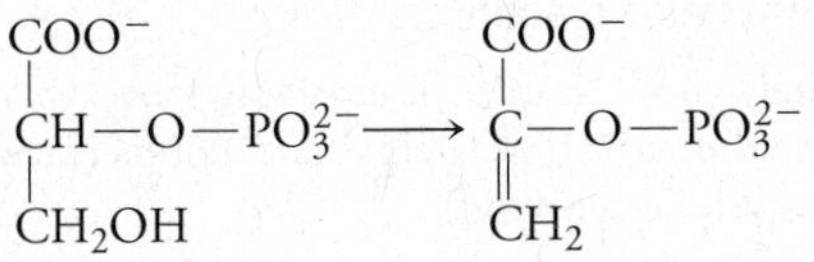

$$\begin{array}{c} COO^- \\ | \\ H-C-H \\ | \\ H-C-H \\ | \\ COO^- \end{array} \xrightarrow{\text{succinate dehydrogenase}}$$

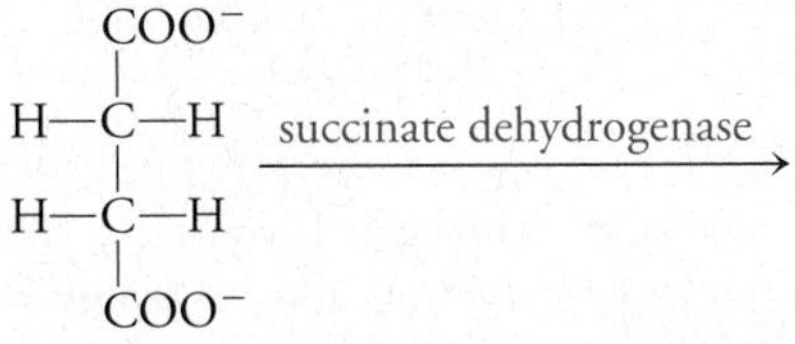

Succinate

12. Malate dehydrogenase catalyzes a reaction in which C2 of malate is oxidized. Draw the structure of the product. To which class of enzymes does malate dehydrogenase belong?

$$\begin{array}{c} CH_2-COO^- \\ | \\ CH-OH \\ | \\ COO^- \end{array} \xrightarrow{\text{malate dehydrogenase}}$$

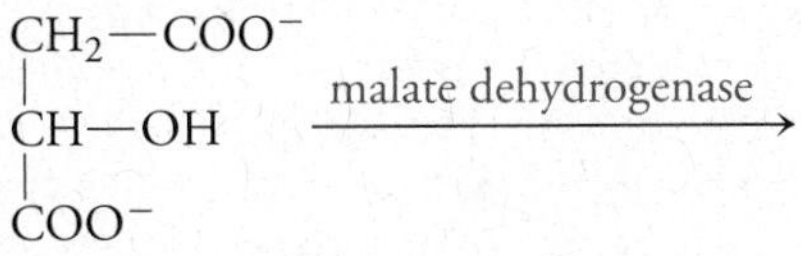

Malate

13. Examine the reaction catalyzed by hexokinase (Section 6-3). Draw the product of the reaction catalyzed by creatine kinase, which acts on creatine in a similar manner. What would you predict to be the usual function of a kinase?

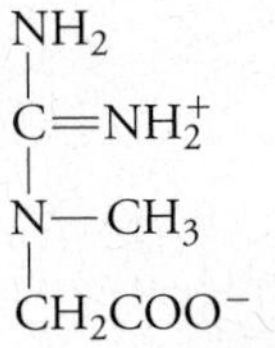

$$\begin{array}{c} NH_2 \\ | \\ C=NH_2^+ \\ | \\ N-CH_3 \\ | \\ CH_2COO^- \end{array}$$

Creatine

14. Propose a name for the enzymes that catalyze the following reactions (reactions may not be balanced):

(a)

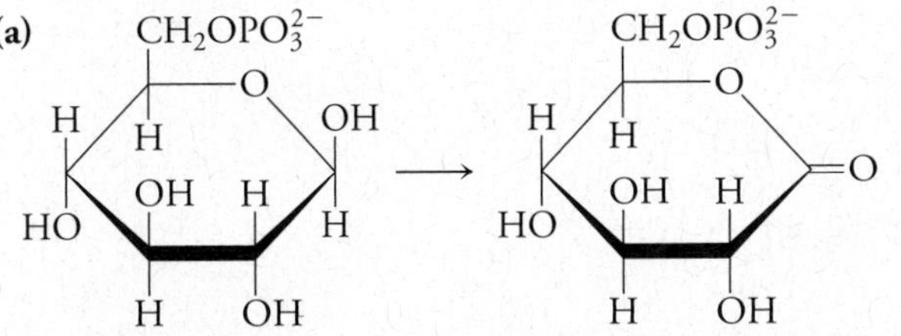

Glucose-6-phosphate → 6-Phosphoglucono-δ-lactone

(b)

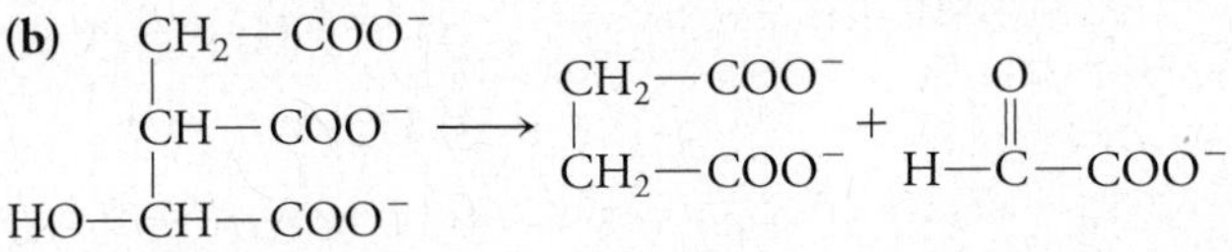

$$\begin{array}{c} CH_2-COO^- \\ | \\ CH-COO^- \\ | \\ HO-CH-COO^- \end{array} \longrightarrow \begin{array}{c} CH_2-COO^- \\ | \\ CH_2-COO^- \end{array} + \begin{array}{c} O \\ \| \\ H-C-COO^- \end{array}$$

Isocitrate · · · · · Succinate · · · · · Glyoxylate

(c)

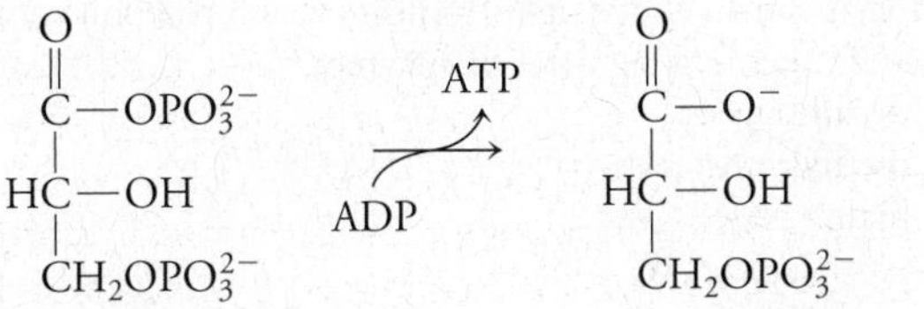

$$\begin{array}{c} O \\ \| \\ C-OPO_3^{2-} \\ | \\ HC-OH \\ | \\ CH_2OPO_3^{2-} \end{array} \xrightarrow[\text{ADP}]{\text{ATP}} \begin{array}{c} O \\ \| \\ C-O^- \\ | \\ HC-OH \\ | \\ CH_2OPO_3^{2-} \end{array}$$

1,3-Bisphosphoglycerate · · · · · 3-Phosphoglycerate

(d)

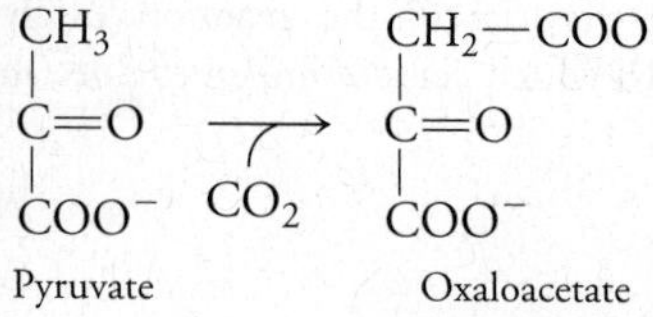

$$
\begin{array}{c}
CH_3 \\
| \\
C\!=\!O \\
| \\
COO^-
\end{array}
\xrightarrow[\;CO_2\;]{}
\begin{array}{c}
CH_2\!-\!COO^- \\
| \\
C\!=\!O \\
| \\
COO^-
\end{array}
$$

Pyruvate   Oxaloacetate

15. The amino acid tryptophan is converted to the hormone melatonin as shown here. Indicate which reaction is catalyzed by each of the following types of enzymes.

(a) methyltransferase
(b) hydroxylase
(c) acetyltransferase
(d) decarboxylase

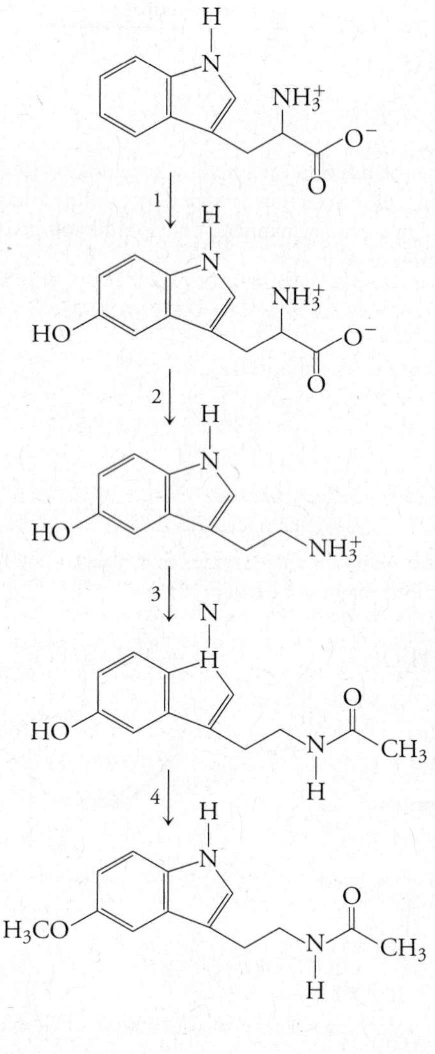

16. The amino acid tyrosine is converted to the hormone epinephrine as shown above right. Indicate which reaction is catalyzed by each of the following types of enzymes.

(a) decarboxylase
(b) methyltransferase
(c) hydroxylase

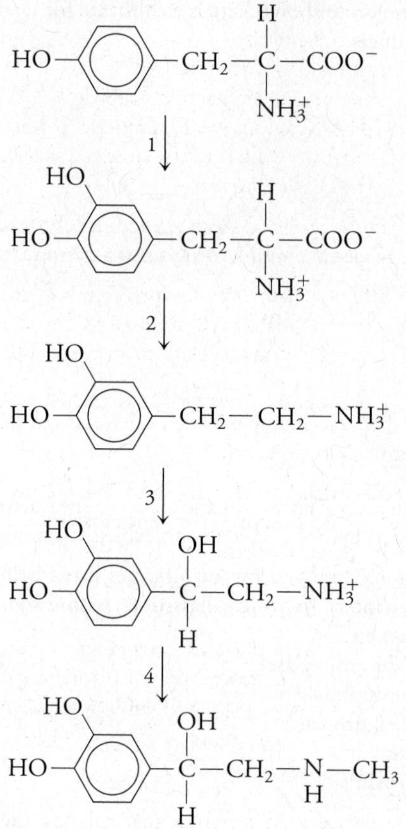

17. Use the enzyme nomenclature database at http://enzyme.expasy.org to determine what reaction is catalyzed by the enzyme called catalase.

18. What is the common name of the enzyme whose EC number is 4.3.2.1?

6-2 The Chemistry of Catalysis

19. Approximately how much does staphylococcal nuclease (Table 6-1) decrease the activation energy ($\Delta G^{\ddagger}$) of its reaction (the hydrolysis of a phosphodiester bond)?

20. The rate of an enzyme-catalyzed reaction is measured at several temperatures, generating the curve shown below. Explain why the enzyme activity increases with temperature and then drops off sharply.

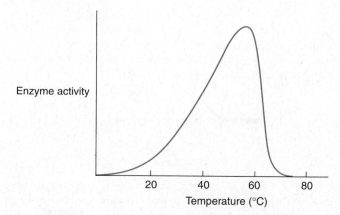

21. Draw pairs of free energy diagrams that show the differences between the following reactions:
(a) a fast reaction vs. a slow reaction
(b) a one-step reaction vs. a two-step reaction
(c) an endergonic reaction vs. an exergonic reaction
(d) a two-step reaction with an initial slow step vs. a two-step reaction with an initial fast step

22. Consult Figure 6-10 and draw a reaction coordinate diagram for the chymotrypsin-catalyzed hydrolysis of a peptide bond.

23. Under certain conditions, peptide bond formation is more thermodynamically favorable than peptide bond hydrolysis. Would you expect chymotrypsin to catalyze peptide bond formation?

24. What is the relationship between the nucleophilicity and the acidity of an amino acid side chain?

25. Amino acids such as Gly, Ala, and Val are not known to participate directly in acid–base or covalent catalysis.
(a) Explain why this is the case.
(b) Mutating a Gly, Ala, or Val residue in an enzyme's active site can still have dramatic effects on catalysis. Why?

26. Using what you know about the structures of the amino acid side chains and the mechanisms presented in this chapter, choose an amino acid side chain to play the following roles in an enzymatic mechanism:
(a) participate in proton transfer,
(b) act as a nucleophile.

27. Ribozymes are RNA molecules that catalyze chemical reactions.
(a) What features of a nucleic acid would be important for it to act as an enzyme?
(b) What part of the RNA structure might fulfill the roles listed in Problem 26?
(c) Why can RNA but not DNA act as a catalyst?

28. RNA is susceptible to base catalysis in which the hydroxide ion abstracts a proton from the 2′ OH group and then the resulting 2′ O⁻ nucleophilically attacks the 5′ phosphate.
(a) Is DNA susceptible to base hydrolysis?
(b) Does this observation help you explain why DNA rather than RNA evolved as the genetic material?

29. Use what you know about the mechanism of chymotrypsin to explain why DIPF inactivates the enzyme.

30. Solutions of chymotrypsin prepared for experimental purposes are stored in a dilute solution of hydrochloric acid as a "preservative." Explain.

31. Sarin is an organophosphorous compound similar to DIPF. In 1995, terrorists released it on a Japanese subway. Injuries resulted from the reaction of sarin with a serine esterase involved in nerve transmission called acetylcholinesterase. Draw the structure of the enzyme's catalytic residue modified by sarin.

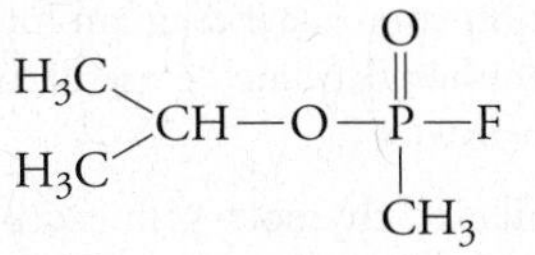

Sarin

32. Parathion is an insecticide that kills pests by reacting irreversibly with acetylcholinesterase, a serine esterase involved in nerve transmission. Parathion doesn't react with the enzyme directly; it is first converted to paraoxon, which then reacts with acetylcholinesterase in a manner similar to the reaction of DIPF with chymotrypsin. Draw the structure of the enzyme's catalytic serine reside modified by paraoxon.

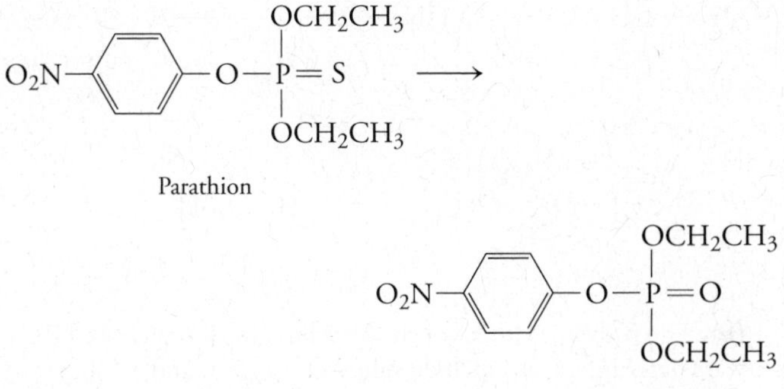

Parathion

Paraoxon

33. Angiogenin, an enzyme involved in the formation of blood vessels, was treated with bromoacetate ($BrCH_2COO^-$). An essential histidine was modified, and the activity of the enzyme was reduced by 95%. Show the chemical reaction for the modification of the histidine side chain with bromoacetate. Why would such a modification inactivate an enzyme that required a His side chain for activity?

34. (a) Diagram the hydrogen-bonding interactions of the catalytic triad His–Lys–Ser during catalysis in a hypothetical hydrolytic enzyme.
(b) In a chemical labeling study of the enzyme, a reagent that covalently modifies Lys residues also abolishes enzyme activity. Why doesn't this observation prove that the active site includes a Lys residue?

35. Treatment of the enzyme D-amino acid oxidase with 1-fluoro-2,4-dinitrophenol (FDNP, below) inactivates the enzyme. Analysis of the derivatized enzyme revealed that one of the tyrosines in the enzyme was unusually reactive. In a second experiment, when benzoate (a compound that structurally resembles the enzyme's substrate) was added prior to the addition of the FDNP, the tyrosine did not react.
(a) Show the chemical reaction between FDNP and the tyrosine residue.
(b) Why does benzoate prevent the reaction? What does this tell you about the role of the tyrosine in this enzyme?

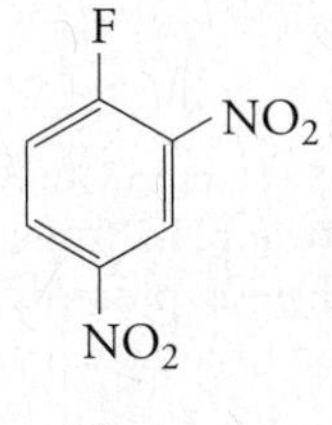

36. Propose two variants of a protease catalytic triad that take the form X–His–Y.

37. How would chymotrypsin's catalytic triad be affected by extremely low and extremely high pH values (assuming that the rest of the protein structure remained intact)?

38. *E. coli* ribonuclease HI is an enzyme that catalyzes the hydrolysis of phosphodiester bonds in RNA. Its proposed mechanism involves a "carboxylate relay."

(a) Using the structures below as a guide, draw arrows that indicate how the hydrolysis reaction might be initiated.

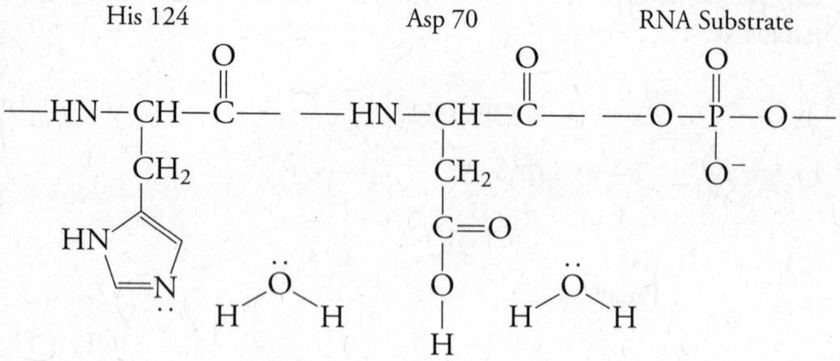

His 124   Asp 70   RNA Substrate

(b) The pK values for all of the histidines in ribonuclease HI were determined and include values of 7.1, 5.5, and <5.0. Which value is most likely to correspond to His 124? Explain.
(c) Substituting an alanine residue for His 124 resulted in a dramatic decrease in enzyme activity. Explain.

39. Lysozyme catalyzes the hydrolysis of a polysaccharide component of bacterial cell walls. The damaged bacteria subsequently lyse (rupture). Part of lysozyme's mechanism is shown below. The enzyme catalyzes cleavage of a bond between two sugar residues (represented by hexagons). Catalysis involves the side chains of Glu 35 and Asp 52. One of the residues has a pK of 4.5; the other has a pK of 5.9.

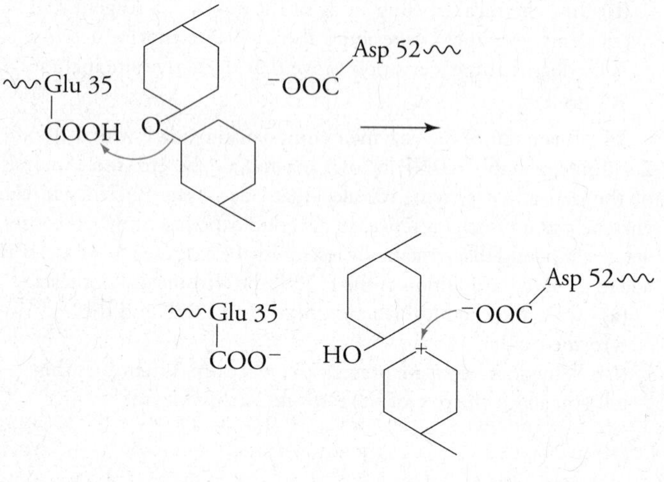

(a) Assign the given pK values to Glu 35 and Asp 52.
(b) Lysozyme is inactive at pH 2.0 and at pH 8.0. Explain.
(c) The lysozyme mechanism proceeds via a covalent intermediate. Use this information to complete the drawing of the lysozyme mechanism.

40. RNase A is a digestive enzyme secreted by the pancreas into the small intestine, where it hydrolyzes RNA into its component nucleotides. The optimum pH for RNase A is about 6, and the pK values of the two histidines that serve as catalytic residues are 5.4 and 6.4. The first step of the mechanism is shown.
(a) Does ribonuclease proceed via acid–base catalysis, covalent catalysis, metal ion catalysis, or some combination of these strategies? Explain.
(b) Assign the appropriate pK values to His 12 and His 119.
(c) Explain why the pH optimum of ribonuclease is pH 6.

(d) Ribonuclease catalyzes the hydrolysis of RNA but not DNA. Explain why.

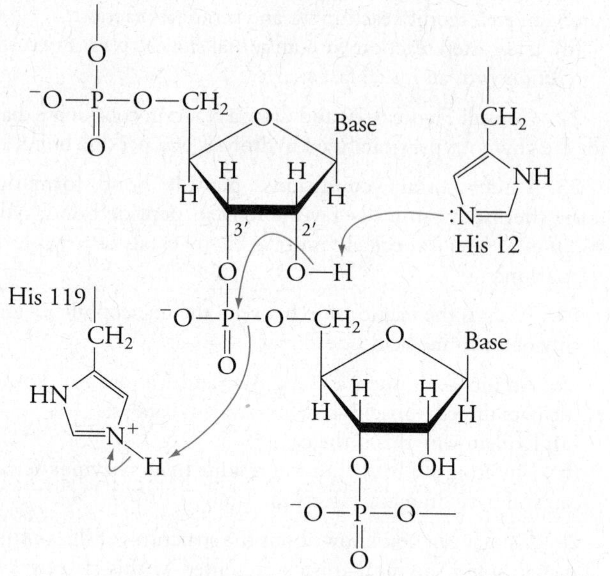

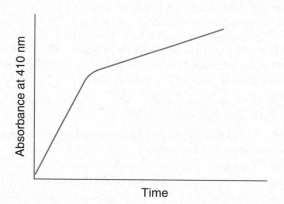

41. Chymotrypsin hydrolyzes the artificial substrate *p*-nitrophenylacetate using a similar mechanism to that of peptide bond hydrolysis. The product *p*-nitrophenolate is bright yellow and absorbs light at 410 nm; thus the reaction can be monitored spectrophotometrically. When *p*-nitrophenylacetate is mixed with chymotrypsin, *p*-nitrophenolate initially forms extremely rapidly. This is followed by a steady-state phase in which the product forms at a uniform rate.

(a) Using what you know about the mechanism of chymotrypsin, explain why these results were observed.
(b) Draw a reaction coordinate diagram for the reaction.
(c) Would *p*-nitrophenylacetate be useful for monitoring the activity of trypsin *in vitro*?

42. Renin is similar to chymotrypsin except that it belongs to a class of enzymes called aspartyl proteases in which aspartate replaces serine in the enzyme active site. Renin inhibitors are potentially valuable in the treatment of high blood pressure. Renin catalyzes the hydrolysis of a peptide bond in the protein angiotensinogen, converting it to angiotensin I, a precursor of a hormone involved in regulating blood pressure. Two aspartate residues in the renin active site, Asp 32 and Asp 215, constitute a catalytic dyad, rather than the catalytic triad found in chymotrypsin.

(a) Begin with the figure provided and draw the reaction mechanism for peptide hydrolysis catalyzed by renin.

(b) What type of catalytic strategy is utilized by renin?

(c) Compare the pK values of the Asp 32 and Asp 215 residues.

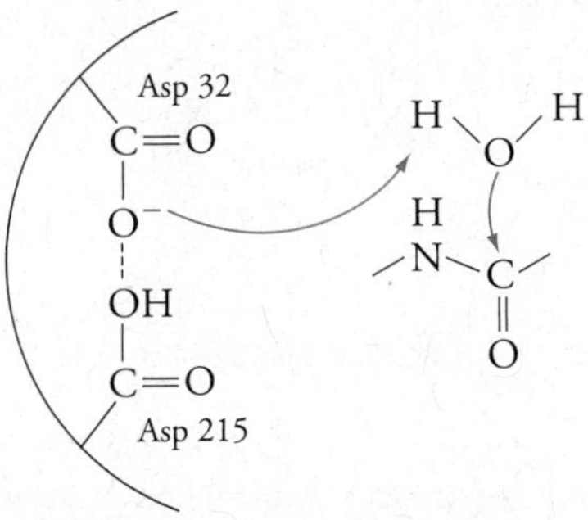

43. The protease from the human immunodeficiency virus (HIV) is a target of the "cocktail" of drugs now on the market to treat HIV/AIDS. The protease has a mechanism similar to that of renin, described in Problem 42, except that renin consists of a single protein with two catalytic Asp residues, whereas the HIV protease consists of two identical subunits, each of which contributes an Asp residue. Is this an example of convergent or divergent evolution? Explain.

44. The enzyme bromelain (found in pineapple) is also a protease but is a member of a family of enzymes, referred to as cysteine proteases, in which a cysteine residue plays a catalytic role similar to serine in serine proteases. Bromelain has been found to relieve inflammation, and some studies have shown that it has antitumor activity. Like the serine proteases, bromelain contains a histidine, but unlike serine proteases, bromelain lacks an aspartate in its active site.

(a) Using what you know about chymotrypsin's mechanism, draw a reaction mechanism for the hydrolysis of a peptide bond by bromelain.

(b) Does the mechanism you have drawn employ acid–base catalysis, covalent catalysis, or both?

(c) Draw a reaction coordinate diagram that is consistent with the mechanism of bromelain.

(d) Why did natives of tropical countries use bromelain as a meat tenderizer?

(e) The pK values for the active site residues in bromelain are ~3 and ~8. The pH optimum for the reaction is 6.0. Assign pK values to the appropriate amino acid side chains and explain your reasoning.

45. Sulfhydryl groups can react with the alkylating reagent *N*-ethylmaleimide (NEM). When NEM is added to a solution of creatine kinase, Cys 278 is alkylated, but no other Cys residues in the protein are modified. What can you infer about the role of the Cys 278 residue based on this information?

46. Cysteine proteases (see Problem 44) can often be inactivated by iodoacetate (ICH_2COO^-). Show the chemical reaction for the modification of the Cys side chain by iodoacetate. Why would such a modification inactivate an enzyme that required a Cys side chain for activity?

47. The botulinum neurotoxin secreted by *Clostridium botulinum* consists of three polypeptide chains, one of which is a protease. The enzyme belongs to a family of enzymes called metalloproteases. The active site includes a Zn^{2+} ion coordinated by two His residues and a Glu residue. The active site contains a second Glu residue (Glu 224). Use the figure provided as a starting point to draw the mechanism of peptide bond hydrolysis catalyzed by this enzyme.

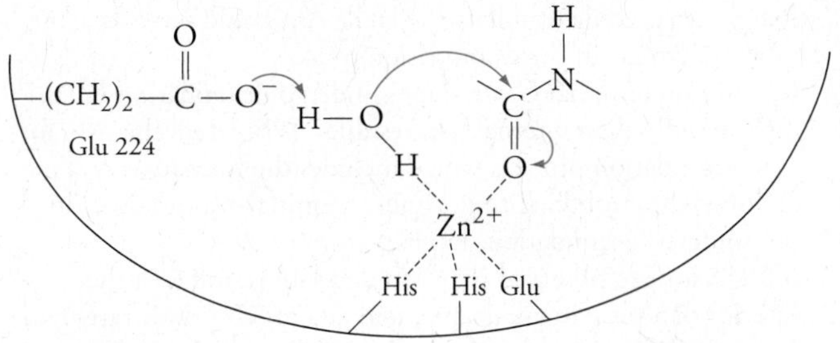

48. The enzyme carbonic anhydrase is one of the fastest enzymes known and, like the enzyme in Problem 47, is a metalloenzyme. It catalyzes the hydration of carbon dioxide to form bicarbonate and a hydrogen ion:

$$CO_2 + H_2O \rightleftharpoons H^+ + HCO_3^-$$

The enzyme's active site contains a zinc ion that is coordinated to the imidazole rings of three histidine residues. (A fourth histidine residue is located nearby and participates in catalysis). A fourth coordination position is occupied by a water molecule. Draw the reaction mechanism (the first step is shown) of the hydration of carbon dioxide and show the regeneration of the enzyme.

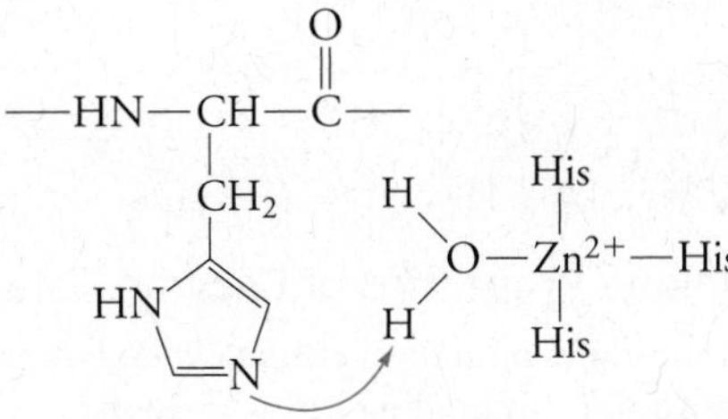

49. Asn and Gln residues in proteins are sometimes nonenzymatically hydrolytically deamidated to Asp and Glu residues, respectively. It has been suggested that deamidation might function as a molecular timer for protein turnover, since deamidation of some proteins increases their susceptibility to degradation by cellular proteases. An exhaustive study of proteins known to undergo deamidation has revealed that deamidated Asn residues are most likely to be preceded by Ser, Thr, or Lys and to be followed by Gly, Ser, or Thr.

(a) Write the balanced chemical equation for the hydrolytic deamidation of Asn to Asp.

(b) The mechanism of deamidation of the amide side chain involves an acid catalyst (HA). The first step of the reaction is shown below. Draw the rest of the reaction mechanism.

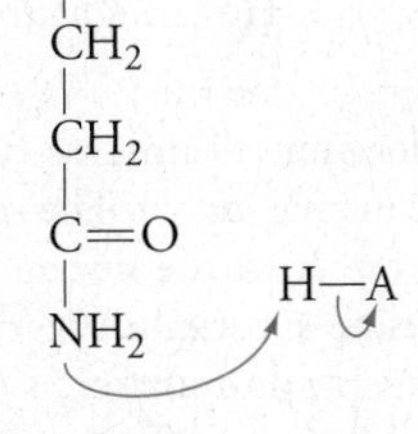

(c) Since the deamidation of Asn residues is known to be nonenzymatic, the HA acid catalyst cannot be provided by an enzyme. Instead, the *catalytic* groups are believed to be provided by neighboring amino acids in the protein undergoing deamidation. Refer to your answer in part (b) and examine the mechanism you have written. Describe how amino acid side chains that either precede or follow the labile Asn could serve as catalytic groups in the deamidation process.

(d) Amino terminal Gln residues undergo deamidation much more rapidly than internal Gln residues. Write a mechanism for this deamidation process, which includes the formation of a five-membered pyrrolidone ring. Amino terminal Asn residues do not undergo deamidation. Explain why.

(e) It has been observed that Asn and Gln residues in the interior of a protein are deamidated at a much slower rate than Asn and Gln residues on the surface of the protein. Explain why.

50. Enzymes in the amidase family catalyze the hydrolysis of amide bonds using a mechanism similar to that of the serine proteases. But the catalytic triad contains the residues Ser–Ser–Lys instead of the Asp–His–Ser triad found in the serine proteases. The Ser–Ser–Lys triad is shown below. Propose a mechanism for the amidase enzyme.

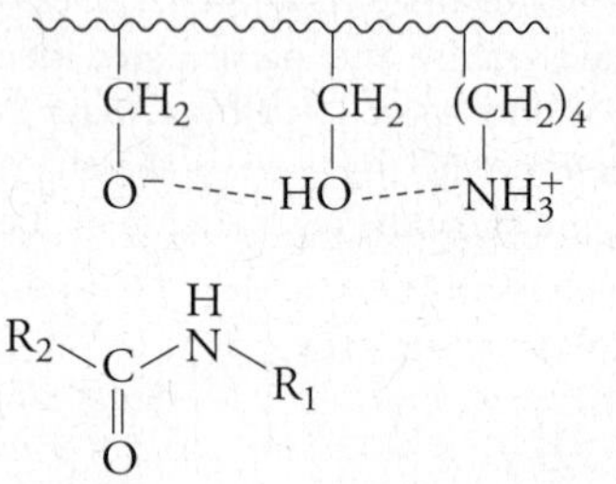

6-3 The Unique Properties of Enzyme Catalysts

51. When substrates bind to an enzyme, their free energy may be lowered. Why doesn't this binding defeat catalysis?

52. Substrates and reactive groups in an enzyme's active site must be precisely aligned in order for a productive reaction to occur. Why, then, is some conformational flexibility also a requirement for catalysis?

53. Daniel Koshland has noted that hexokinase undergoes a large conformational change on substrate binding, which apparently prevents water from entering the active site and participating in hydrolysis. However, the serine proteases do not undergo large conformational changes on substrate binding. Explain.

54. When ATP and the sugar xylose (which resembles glucose but has only five carbons) are added to hexokinase, the enzyme produces xylose-5-phosphate along with some free phosphate. Does this observation support Koshland's induced fit hypothesis?

55. Refer to Problem 47. What is the role of the zinc ion in catalysis? In transition state stabilization?

56. Refer to Problem 50. How might the amidase stabilize the transition state?

57. The enzyme adenosine deaminase catalyzes the conversion of adenosine to inosine (above, right). The compound 1,6-dihydropurine ribonucleoside binds to the enzyme with a much greater affinity than the adenosine substrate. What does this tell us about the mechanism of adenosine deaminase?

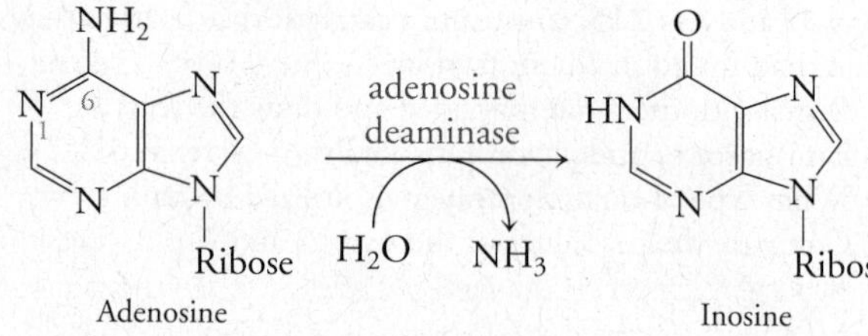

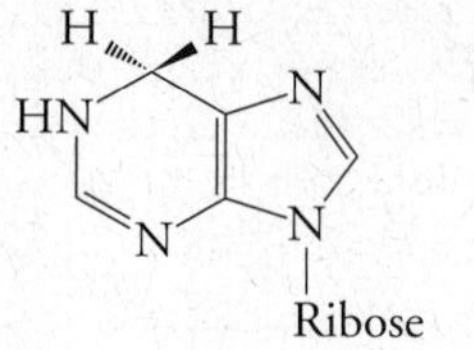

Adenosine

Inosine

1,6-Dihydropurine nucleoside

58. Imidazole reacts with *p*-nitrophenylacetate to produce the *N*-acetylimidazolium and *p*-nitrophenolate ions.

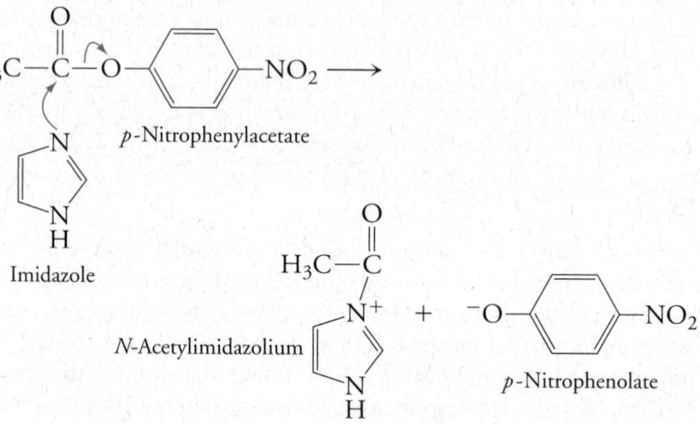

p-Nitrophenylacetate

Imidazole

N-Acetylimidazolium

p-Nitrophenolate

(a) The compound shown below reacts to form *p*-nitrophenolate. Draw the reaction mechanism for this reaction.

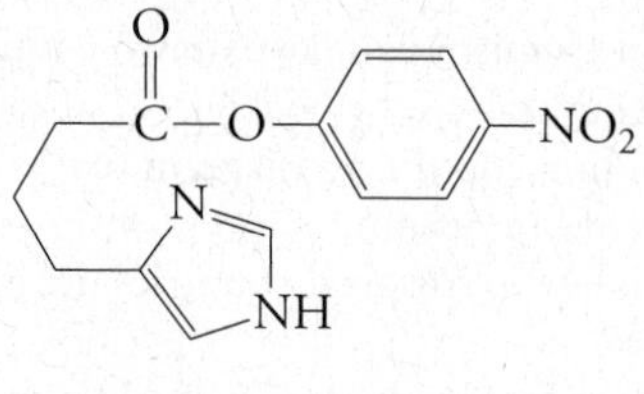

(b) The compound described in part (a) reacts 24 times faster than imidazole. Explain why.

(c) What can the results of this experiment tell us about the way enzymes speed up reaction rates?

6-4 Some Additional Features of Enzymes

59. Many genetic mutations prevent the synthesis of a protein or give rise to an enzyme with diminished catalytic activity. Is it possible for a mutation to increase the catalytic activity of an enzyme?

60. If the active site of chymotrypsin excludes water when its substrate binds, how is it possible for a water molecule to participate in the second stage of catalysis (as shown in Fig. 6-10)?

61. The amino acid Asp 189 lies at the base of the substrate specificity pocket in the enzyme trypsin.

(a) How is this related to trypsin's substrate specificity? What kinds of interactions take place between the Asp 189 and the side chain of the residue preceding the scissile bond?

(b) In site-directed mutagenesis studies, Asp 189 was replaced with lysine. How do you think this would affect substrate specificity?

(c) The investigators who carried out the experiment described in part (b) analyzed the three-dimensional structure of the mutant enzyme and found that Lys 189 is actually not located in the substrate specificity pocket. Instead, the Lys side chain reaches out of the base of the pocket, rendering the specificity pocket nonpolar. With this additional information, determine how the substrate specificity would differ in the Lys 189 mutant enzyme.

62. An enzyme involved in the regulation of blood pressure was recently purified and characterized. It is an aspartyl aminopeptidase, which hydrolyzes peptide bonds on the carboxyl side of aspartate residues. The investigators who purified the enzyme were interested in learning about the substrate binding site, so they synthesized a series of artificial peptides and tested the ability of the enzyme to hydrolyze them. The peptides and their reaction rates are listed in the table above right. The residues are labeled P1–P1′–P2′–P3′, and the hydrolyzed bond is between P1 and P1′.

(a) It is likely that both the P1′ residue and the P2′ residue fit in adjacent "pockets" on the aspartyl aminopeptidase enzyme. Describe the characteristics of these pockets on the enzyme, using the kinetic data in the table.

(b) The investigators have proposed that the "natural" or endogenous substrate for aspartyl aminopeptidase is angiotensin II, which is involved in the regulation of blood pressure. However, the authors also mention that a potential exogenous substrate might be the artificial sweetener aspartame (Nutrasweet®), which is the methylated dipeptide aspartylphenylalanine (Asp–Phe–OCH₃). Based on the results presented here, do you think that aspartame would be a good substrate for the enzyme?

(c) Draw the structure of aspartame and the products of its hydrolysis by aspartyl aminopeptidase.

(d) Under nondenaturing conditions, the enzyme appears to be the same size as the protein ferritin, which has a molecular mass of 440 kD. Under denaturing conditions, the purified enzyme behaves like a single polypeptide of 55 kD. What can you determine about the enzyme's structure from these data?

Peptide (P1–P1′–P2′–P3′)	Catalyzed rate (s^{-1})
Asp–Ala–Ala–Leu	5.3
Asp–Phe–Ala–Leu	9.9
Asp–Lys–Ala–Leu	2.8
Asp–Asp–Ala–Leu	9.8
Asp–Ala–Phe–Leu	17.2
Asp–Ala–Lys–Leu	5.0
Asp–Ala–Asp–Leu	2.3

63. Chymotrypsin is usually described as an enzyme that catalyzes hydrolysis of peptide bonds following Phe, Trp, or Tyr residues. Is this information consistent with the description of the bonds cleaved during chymotrypsin activation (Figure 6-18)? What does this tell you about chymotrypsin's substrate specificity?

64. Describe how the activation of the digestive enzyme chymotrypsin follows a cascade mechanism. Draw a diagram of your cascade.

65. Would chymotrypsin catalyze hydrolysis of the substrate shown in Problem 8? Explain why or why not.

66. Does DIPF inactivate trypsin or elastase (see Problem 29)? Explain.

67. Hereditary pancreatitis is caused by a mutation in the autocatalytic domain of trypsinogen that results in persistent activity of the enzyme.

(a) What are the physiological consequences of this disease?

(b) Describe a strategy to treat this disease.

68. Some plants contain compounds that inhibit serine proteases. It has been hypothesized that these compounds protect the plant from proteolytic enzymes of insects and microorganisms that would damage the plant. Tofu, or bean curd, possesses these compounds. Manufacturers of tofu treat it to eliminate serine protease inhibitors. Why is this treatment necessary?

69. Not all proteases are synthesized as zymogens or have inhibitors that block their activity. What limits the potentially destructive power of these proteases?

70. A chymotrypsin inhibitor isolated from snake venom resembles the bovine pancreatic trypsin inhibitor (Fig. 6-20) but has an Asn residue in place of Lys 15. Why is this finding unexpected?

[SELECTED READINGS]

Di Cera, E., Serine proteases, *IUBMB Life* **61**, 510–515 (2009). [Summarizes some of the biological roles and mechanistic features of proteases.]

Fersht, A., *Structure and Mechanism in Protein Science: A Guide to Enzyme Catalysis and Protein Folding*, W. H. Freeman (1999). [Includes detailed reaction mechanisms for chymotrypsin and other enzymes.]

Gutteridge, A., and Thornton, J. M., Understanding nature's catalytic toolkit, *Trends Biochem. Sci.* **30**, 622–629 (2005).

[Describes how certain sets of amino acid side chains form catalytic units that appear in many different enzymes.]

Radisky, E. S., Lee, J. M., Lu, C.-J. K., and Koshland, D. E., Jr., Insights into the serine protease mechanism from atomic resolution structures of trypsin reaction intermediates, *Proc. Nat. Acad. Sci.* **103**, 6835–6840 (2006).

Ringe, D., and Petsko, G. A., How enzymes work, *Science* **320**, 1428–1429 (2008). [Briefly summarizes some general features of enzyme function.]

chapter 7

ENZYME KINETICS AND INHIBITION

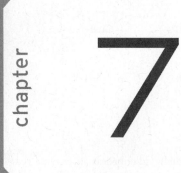

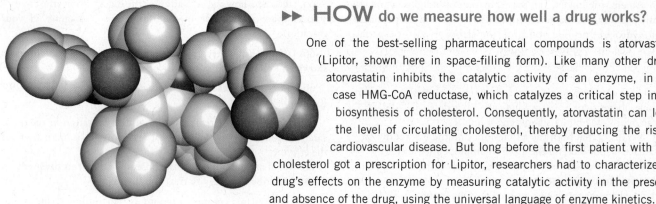

▸▸ HOW do we measure how well a drug works?

One of the best-selling pharmaceutical compounds is atorvastatin (Lipitor, shown here in space-filling form). Like many other drugs, atorvastatin inhibits the catalytic activity of an enzyme, in this case HMG-CoA reductase, which catalyzes a critical step in the biosynthesis of cholesterol. Consequently, atorvastatin can lower the level of circulating cholesterol, thereby reducing the risk of cardiovascular disease. But long before the first patient with high cholesterol got a prescription for Lipitor, researchers had to characterize the drug's effects on the enzyme by measuring catalytic activity in the presence and absence of the drug, using the universal language of enzyme kinetics.

THIS CHAPTER IN CONTEXT

Part 1 Foundations

Part 2 Molecular Structure and Function

7 Enzyme Kinetics and Inhibition

Part 3 Metabolism

Part 4 Genetic Information

Do You Remember?

- **Living organisms obey the laws of thermodynamics (Section 1-3).**
- **Noncovalent forces, including hydrogen bonds, ionic interactions, and van der Waals forces, act on biological molecules (Section 2-1).**
- **An acid's p*K* value describes its tendency to ionize (Section 2-3).**
- **The 20 amino acids differ in the chemical characteristics of their R groups (Section 4-1).**
- **Some proteins can adopt more than one stable conformation (Section 4-3).**

In the preceding chapter, we examined the basic features of enzyme catalytic activity, primarily by exploring the various catalytic mechanisms used by chymotrypsin. This chapter extends the discussion of enzymes by introducing enzyme kinetics, the mathematical analysis of enzyme activity. Here we describe how an enzyme's reaction speed and specificity can be quantified and how this information can be used to evaluate the enzyme's physiological function. We also look at the regulation of enzyme activity by inhibitors that bind to the enzyme and alter its activity. The discussion also focuses on allosteric regulation, a mechanism for inhibiting as well as activating enzymes.

7-1 Introduction to Enzyme Kinetics

The structure and chemical mechanism of an enzyme (for example, chymotrypsin, as discussed in Chapter 6) often reveal a great deal about how that enzyme functions *in vivo*. However, structural information alone does not provide a full accounting of an enzyme's physiological role. For example, one might need to know exactly how fast the enzyme catalyzes a reaction or how well it recognizes different substrates or how its activity is affected by other substances. These questions are not trivial—consider that a single cell contains thousands of different enzymes, all operating simultaneously and in the presence of one anothers' substrates and products. To fully describe enzyme activity, enzymologists apply mathematical tools to quantify an enzyme's catalytic power and its substrate affinity as well as its response to inhibitors. This analysis is part of the area of study known as enzyme **kinetics** (from the Greek *kinetos,* which means "moving").

Some of the earliest biochemical studies examined the reactions of crude preparations of yeast cells and other organisms. Even without isolating any enzymes, researchers could mathematically analyze their activity by measuring the concentrations of substrates and reaction products and observing how these quantities changed over time. For example, consider the simple reaction catalyzed by triose phosphate isomerase, which interconverts two three-carbon sugars (trioses):

O═C—H
|
H—C—OH
|
CH$_2$OPO$_3^{2-}$

Glyceraldehyde-3-phosphate

→ triose phosphate isomerase →

H
|
H—C—OH
|
C═O
|
CH$_2$OPO$_3^{2-}$

Dihydroxyacetone phosphate

Over the course of the reaction, the concentration of the substrate glyceraldehyde-3 phosphate falls as the concentration of the product dihydroxyacetone phosphate rises (**Fig. 7-1**). *The progress of this or any reaction can be expressed as a **velocity***

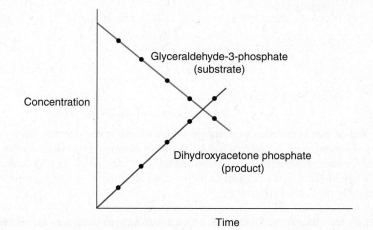

Figure 7-1 Progress of the triose phosphate isomerase reaction. Over time, the concentration of the substrate glyceraldehyde-3-phosphate decreases and the concentration of the product dihydroxyacetone phosphate increases.

? Extend the lines in the graph to show that the reaction eventually reaches equilibrium, when the ratio of product concentration to reactant concentration is about 22.

(v), *either the rate of disappearance of the substrate (S) or the rate of appearance of the product (P):*

$$v = -\frac{d[S]}{dt} = \frac{d[P]}{dt}$$ **[7-1]**

where [S] and [P] represent the concentrations of the substrate and product, respectively. Not surprisingly, *the more catalyst (enzyme) present, the faster the reaction* (**Fig. 7-2**).

When the enzyme concentration is held constant, the reaction velocity varies with the substrate concentration, but in a nonlinear fashion (**Fig. 7-3**). The shape of this velocity versus substrate curve is an important key to understanding how enzymes interact with their substrates. *The hyperbolic, rather than linear, shape of the curve suggests that an enzyme physically combines with its substrate to form an **enzyme–substrate (ES) complex.*** Therefore, the enzyme-catalyzed conversion of S to P

$$S \xrightarrow{\text{E}} P$$

can be more accurately written as

$$E + S \rightarrow ES \rightarrow E + P$$

As small amounts of substrate are added to the enzyme preparation, enzyme activity (measured as the reaction velocity) appears to increase almost linearly. However, the enzyme's activity increases less dramatically as more substrate is added. At very high substrate concentrations, enzyme activity appears to level off as it approaches a maximum value. This behavior shows that at low substrate concentrations, the enzyme quickly converts all the substrate to product, but as more substrate is added, the enzyme becomes **saturated** with substrate—that is, there are many more substrate molecules than enzyme molecules, so not all the substrate can be converted to product in a given time. These so-called saturation kinetics are a feature of many binding phenomena, including the binding of O_2 to myoglobin (see Section 5-1).

The curve shown in Figure 7-3 reveals considerable information about a given enzyme and substrate under a chosen set of reaction conditions. All simple enzyme-catalyzed reactions yield a hyperbolic velocity versus substrate curve, but the exact shape of the curve depends on the enzyme, its concentration, the concentrations of

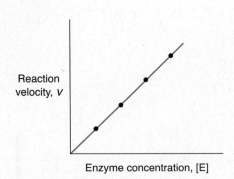

Figure 7-2 Progress of the triose phosphate isomerase reaction. The more enzyme present, the faster the reaction.

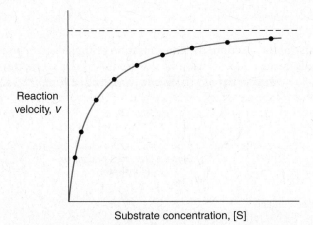

Figure 7-3 A plot of reaction velocity versus substrate concentration. Varying amounts of substrate are added to a fixed amount of enzyme. The reaction velocity is measured for each substrate concentration and plotted. The resulting curve takes the form of a hyperbola, a mathematical function in which the values first increase steeply but eventually approach a maximum.

? **Compare this diagram to Figure 5-3, which shows oxygen binding to myoglobin.**

enzyme inhibitors, the pH, the temperature, and so on. By analyzing such curves, it is possible to address some basic questions, for example,

- How fast does the enzyme operate?
- How efficiently does the enzyme convert different substrates to products?
- How susceptible is the enzyme to various inhibitors and how do these inhibitors affect enzyme activity?

The answers to these questions, in turn, may reveal

- Whether an enzyme is likely to catalyze a particular reaction *in vivo*
- What substances are likely to serve as physiological regulators of the enzyme's activity
- Which enzyme inhibitors might be effective drugs

CONCEPT REVIEW
- Describe two ways to express the velocity of an enzymatic reaction.
- Why does a plot of velocity versus substrate concentration yield a curve?

7-2 Derivation and Meaning of the Michaelis–Menten Equation

The mathematical analysis of enzyme behavior centers on the equation that describes the hyperbolic shape of the velocity versus substrate plot (see Fig. 7-3). We can analyze an enzyme-catalyzed reaction, such as the one catalyzed by triose phosphate isomerase, by conceptually breaking it down into smaller steps and using the terms that apply to simple chemical processes.

Rate equations describe chemical processes

Consider a **unimolecular reaction** (one that involves a single reactant) such as the conversion of compound A to compound B:

$$A \rightarrow B$$

The progress of this reaction can be mathematically described by a **rate equation** in which the reaction rate (the velocity) is expressed in terms of a constant (the **rate constant**) and the reactant concentration [A]:

$$v = -\frac{d[A]}{dt} = k[A] \qquad \text{[7-2]}$$

Here, **k** is the rate constant and has units of reciprocal seconds (s^{-1}). This equation shows that the reaction velocity is directly proportional to the concentration of reactant A. Such a reaction is said to be **first-order** because its rate depends on the concentration of one substance.

A **bimolecular** or **second-order reaction,** which involves two reactants, can be written

$$A + B \rightarrow C$$

Its rate equation is

$$v = -\frac{d[A]}{dt} = -\frac{d[B]}{dt} = k[A][B] \qquad \text{[7-3]}$$

Here, k is a second-order rate constant and has units of $M^{-1} \cdot s^{-1}$. The velocity of a second-order reaction is therefore proportional to the product of the two reactant concentrations (see Sample Calculation 7-1).

KEY CONCEPTS
- Simple chemical reactions are described in terms of rate constants.
- The Michaelis–Menten equation describes enzyme-catalyzed reactions in terms of K_M and V_{max}.
- The kinetic parameters K_M, k_{cat}, and k_{cat}/K_M are experimentally determined.
- K_M and V_{max} values can be derived for enzymes that do not follow the Michaelis–Menten model.

PROBLEM

Determine the velocity of the reaction $X + Y \rightarrow Z$ when the sample contains 3 μM X and 5 μM Y and k for the reaction is 400 $M^{-1} \cdot s^{-1}$.

SOLUTION

Use Equation 7-3 and make sure that all units are consistent:

$$
\begin{aligned}
v &= k[X][Y] \\
&= (400\ M^{-1} \cdot s^{-1})(3\ \mu M)(5\ \mu M) \\
&= (400\ M^{-1} \cdot s^{-1})(3 \times 10^{-6}\ M)(5 \times 10^{-6}\ M) \\
&= 6 \times 10^{-9}\ M \cdot s^{-1} = 6\ nM \cdot s^{-1}
\end{aligned}
$$

●●● PRACTICE PROBLEMS

1. Calculate k for the reaction $X + Y \rightarrow Z$ when the sample contains 5 μM A and 5 μM Y and the velocity is 5 μM $\cdot s^{-1}$.
2. Determine the velocity of the reaction in Problem 1 when $[X] = 20$ μM and $[Y] = 10$ μM.
3. If k for the reaction $X + Y \rightarrow Z$ is 0.5 $mM^{-1} \cdot s^{-1}$ and $[X] = [Y]$, determine the substrate concentration at which $v = 8\ mM \cdot s^{-1}$.

The Michaelis–Menten equation is a rate equation for an enzyme-catalyzed reaction

In the simplest case, an enzyme binds its substrate (in an enzyme–substrate complex) before converting it to product, so *the overall reaction actually consists of first-order and second-order processes, each with a characteristic rate constant:*

$$
E + S \underset{k_{-1}}{\overset{k_1}{\rightleftharpoons}} ES \overset{k_2}{\longrightarrow} E + P \qquad \textbf{[7-4]}
$$

The initial collision of E and S is a bimolecular reaction with the second-order rate constant k_1. The ES complex can then undergo one of two possible unimolecular reactions: k_2 is the first-order rate constant for the conversion of ES to E and P, and k_{-1} is the first-order rate constant for the conversion of ES back to E and S. The bimolecular reaction that would represent the formation of ES from E and P is not shown because we assume that the formation of product from ES (the step described by k_2) does not occur in reverse.

The rate equation for product formation is

$$
v = \frac{d[P]}{dt} = k_2[ES] \qquad \textbf{[7-5]}
$$

To calculate the rate constant, k_2, for the reaction, we would need to know the reaction velocity and the concentration of ES. The velocity can be measured relatively easily, for example, by using a synthetic substrate that is converted to a light-absorbing or fluorescent product. (The velocity of the reaction is just the rate of appearance of the product as monitored by a spectrophotometer or fluorimeter.) However, measuring [ES] is more difficult because the concentration of the enzyme–substrate complex depends on its rate of formation from E and S and its rate of decomposition to E + S and E + P:

$$
\frac{d[ES]}{dt} = k_1[E][S] - k_{-1}[ES] - k_2[ES] \qquad \textbf{[7-6]}
$$

To simplify our analysis, we choose experimental conditions such that the substrate concentration is much greater than the enzyme concentration ($[S] \gg [E]$). Under these conditions, after E and S have been mixed together, *the concentration of*

Figure 7-4 Changes in concentration for a simple enzyme-catalyzed reaction. For most of the duration of the reaction, [ES] remains constant while S is converted to P. In this idealized reaction, all the substrate is converted to product. **See Animated Figure. Progress curves for an enzyme-catalyzed reaction.**

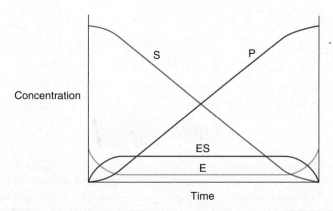

ES remains constant until nearly all the substrate has been converted to product. This is shown graphically in Figure 7-4. [ES] is said to maintain a **steady state** (it has a constant value) and

$$\frac{d[\mathrm{ES}]}{dt} = 0 \qquad [7\text{-}7]$$

According to the steady-state assumption, the rate of ES formation must therefore balance the rate of ES consumption:

$$k_1[\mathrm{E}][\mathrm{S}] = k_{-1}[\mathrm{ES}] + k_2[\mathrm{ES}] \qquad [7\text{-}8]$$

At any point during the reaction, [E]—like [ES]—is difficult to determine, but the total enzyme concentration, $[\mathrm{E}]_\mathrm{T}$, is usually known:

$$[\mathrm{E}]_\mathrm{T} = [\mathrm{E}] + [\mathrm{ES}] \qquad [7\text{-}9]$$

Thus, $[\mathrm{E}] = [\mathrm{E}]_\mathrm{T} - [\mathrm{ES}]$. This expression for [E] can be substituted into Equation 7-8 to give

$$k_1([\mathrm{E}]_\mathrm{T} - [\mathrm{ES}])[\mathrm{S}] = k_{-1}[\mathrm{ES}] + k_2[\mathrm{ES}] \qquad [7\text{-}10]$$

Rearranging (by dividing both sides by [ES] and k_1) gives an expression in which all three rate constants are together:

$$\frac{([\mathrm{E}]_\mathrm{T} - [\mathrm{ES}])[\mathrm{S}]}{[\mathrm{ES}]} = \frac{k_{-1} + k_2}{k_1} \qquad [7\text{-}11]$$

At this point, we can define the **Michaelis constant, K_M,** as a collection of rate constants:

$$K_\mathrm{M} = \frac{k_{-1} + k_2}{k_1} \qquad [7\text{-}12]$$

Consequently, Equation 7-11 becomes

$$\frac{([\mathrm{E}]_\mathrm{T} - [\mathrm{ES}])[\mathrm{S}]}{[\mathrm{ES}]} = K_\mathrm{M} \qquad [7\text{-}13]$$

or

$$K_\mathrm{M}[\mathrm{ES}] = ([\mathrm{E}]_\mathrm{T} - [\mathrm{ES}])[\mathrm{S}] \qquad [7\text{-}14]$$

Dividing both sides by [ES] gives

$$K_\mathrm{M} = \frac{[\mathrm{E}]_\mathrm{T}[\mathrm{S}]}{[\mathrm{ES}]} - [\mathrm{S}] \qquad [7\text{-}15]$$

or

$$\frac{[\mathrm{E}]_\mathrm{T}[\mathrm{S}]}{[\mathrm{ES}]} = K_\mathrm{M} + [\mathrm{S}] \qquad [7\text{-}16]$$

Solving for [ES] yields

$$[ES] = \frac{[E]_T [S]}{K_M + [S]}$$ [7-17]

The rate equation for the formation of product (Equation 7-5) is $v = k_2[ES]$, so we can express the reaction velocity as

$$v = k_2[ES] = \frac{k_2 [E]_T [S]}{K_M + [S]}$$ [7-18]

Now we have an equation containing known quantities: $[E]_T$ and [S]. Although some S is consumed in forming the ES complex, we can ignore it because $[S]_T \gg [E]_T$.

Typically, kinetic measurements are made soon after the enzyme and substrate are mixed together, before more than about 10% of the substrate molecules have been converted to product molecules (this is also the reason why we can ignore the reverse reaction, $E + P \rightarrow ES$). Therefore, the velocity at the start of the reaction (at time zero) is expressed as v_0 (the **initial velocity**):

$$v_0 = \frac{k_2 [E]_T [S]}{K_M + [S]}$$ [7-19]

We can make one additional simplification: When [S] is very high, virtually all the enzyme is in its ES form (it is saturated with substrate) and therefore approaches its point of maximum activity (see Fig. 7-3). The maximum reaction velocity, designated V_{max}, can be expressed as

$$V_{max} = k_2[E]_T$$ [7-20]

which is similar to Equation 7-5. By substituting Equation 7-20 into Equation 7-19, we obtain

$$v_0 = \frac{V_{max} [S]}{K_M + [S]}$$ [7-21]

This relationship is called the **Michaelis–Menten equation** after Leonor Michaelis and Maude Menten, who derived it in 1913. *It is the rate equation for an enzyme-catalyzed reaction and is the mathematical description of the hyperbolic curve shown in Figure 7-3.* (See Sample Calculation 7-2.)

●●● SAMPLE CALCULATION 7-2

PROBLEM An enzyme-catalyzed reaction has a K_M of 1 mM and a V_{max} of 5 nM · s^{-1}. What is the reaction velocity when the substrate concentration is 0.25 mM?

SOLUTION Use the Michaelis–Menten equation (Equation 7-21):

$$v_0 = \frac{(5 \text{ nM} \cdot \text{s}^{-1})(0.25 \text{ mM})}{(1 \text{ mM}) + (0.25 \text{ mM})}$$

$$= \frac{1.25}{1.25} \text{ nM} \cdot \text{s}^{-1}$$

$$= 1 \text{ nM} \cdot \text{s}^{-1}$$

●●● PRACTICE PROBLEMS

4. Using the above information, determine the reaction velocity when the substrate concentration is 1.5 mM.

5. Using the above information, determine the reaction velocity when the substrate concentration is 10 mM.

6. If V_{max} for a reaction is 7.5 μM · s^{-1} and the reaction velocity is 5 μM · s^{-1} when the substrate concentration is 1 μM, what is the K_M?

7. Using the same information as in Problem 6, determine the substrate concentration at which $v = 2.5$ μM · s^{-1}.

K_M is the substrate concentration at which velocity is half-maximal

We have seen that the Michaelis constant, K_M, is a combination of three rate constants (Equation 7-12), but it can be fairly easily determined from experimental data. Kinetic measurements are usually made over a range of substrate concentrations. When $[S] = K_M$, the reaction velocity (v_0) is equal to half its maximum value ($v_0 = V_{max}/2$), as shown in **Figure 7-5**. You can prove that this is true by substituting K_M for $[S]$ in the Michaelis–Menten equation (Equation 7-21). *Since K_M is the substrate concentration at which the reaction velocity is half-maximal, it indicates how efficiently an enzyme selects its substrate and converts it to product.* The lower the value of K_M, the more effective the enzyme is at low substrate concentrations; the higher the value of K_M, the less effective the enzyme is. The K_M is unique for each enzyme–substrate pair. Consequently, K_M values are useful for comparing the activities of two enzymes that act on the same substance or for assessing the ability of different substrates to be recognized by a single enzyme.

In practice, *K_M is often used as a measure of an enzyme's affinity for a substrate.* In other words, it approximates the dissociation constant of the ES complex:

$$K_M \approx \frac{[E][S]}{[ES]} \qquad \text{[7-22]}$$

Note that this relationship is strictly true only when the rate of the $ES \rightarrow E + P$ reaction is much slower than the rate of the $ES \rightarrow E + S$ reaction (that is, when $k_2 \ll k_{-1}$).

The catalytic constant describes how quickly an enzyme can act

It is also useful to know how fast an enzyme operates after it has selected and bound its substrate. In other words, how fast does the ES complex proceed to $E + P$? This parameter is termed the **catalytic constant** and is symbolized k_{cat}. For any enzyme-catalyzed reaction,

$$k_{cat} = \frac{V_{max}}{[E]_T} \qquad \text{[7-23]}$$

For a simple reaction, such as the one diagrammed in Equation 7-4,

$$k_{cat} = k_2 \qquad \text{[7-24]}$$

Thus, *k_{cat} is the rate constant of the reaction when the enzyme is saturated with substrate* (when $[ES] \approx [E]_T$ and $v_0 \approx V_{max}$). We have already seen this relationship in Equation 7-20. k_{cat} is also known as the enzyme's **turnover number** because it is the number of catalytic cycles that each active site undergoes per unit time, or the number of substrate molecules transformed to product molecules by a single enzyme in a given period of time. The turnover number is a first-order rate constant and therefore has units of s^{-1}. As shown in Table 7-1, the catalytic constants of enzymes vary over many orders of magnitude.

Keep in mind that the rate of an enzymatic reaction is a function of the number of reactant molecules that can achieve the high-energy transition state per unit time (as explained in Section 6-2). While an enzyme can accelerate a chemical reaction by providing a mechanistic pathway with a lower activation energy barrier, the enzyme cannot alter the free energies of the reactants and products. This means that the enzyme can make a reaction happen faster, but only when the overall change in free energy is less than zero (that is, the products have lower free energy than the reactants).

k_{cat}/K_M indicates catalytic efficiency

An enzyme's effectiveness as a catalyst depends on how avidly it binds its substrates and how rapidly it converts them to products. Thus, a measure of catalytic efficiency must reflect both binding and catalytic events. The quantity **k_{cat}/K_M** satisfies this

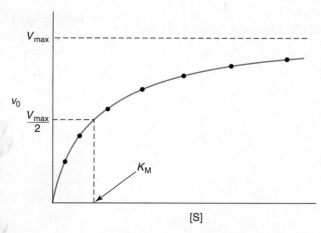

Figure 7-5 Graphical determination of K_M. K_M corresponds to the substrate concentration at which the reaction velocity is half-maximal. It can be visually estimated from a plot of v_0 versus $[S]$. ⊕ See Animated Figure. **Plot of initial velocity versus substrate concentration.**

....

? **Explain why doubling the substrate concentration does not necessarily double the reaction rate.**

[TABLE **7-1**]

Catalytic Constants of Some Enzymes

Enzyme	k_{cat} (s^{-1})
Staphylococcal nuclease	95
Cytidine deaminase	299
Triose phosphate isomerase	4300
Cyclophilin	13,000
Ketosteroid isomerase	66,000
Carbonic anhydrase	1,000,000

[Data from Radzicka, A., and Wolfenden, R., *Science* **267**, 90–93 (1995).]

requirement. At low concentrations of substrate ($[S] < K_M$), very little ES forms and $[E] \approx [E]_T$. Equation 7-18 can then be simplified (the [S] term in the denominator becomes insignificant):

$$v_0 = \frac{k_2 [E]_T [S]}{K_M + [S]} \qquad [7\text{-}25]$$

$$v_0 \approx \frac{k_2}{K_M} [E][S] \qquad [7\text{-}26]$$

Equation 7-26 is the rate equation for the second-order reaction of E and S. k_{cat}/K_M, which has units of $M^{-1} \cdot s^{-1}$, is the apparent second-order rate constant. As such, it indicates how the reaction velocity varies according to how often the enzyme and substrate combine with each other. *The value of k_{cat}/K_M, more than either K_M or k_{cat} alone, represents the enzyme's overall ability to convert substrate to product.*

What limits the catalytic power of enzymes? Electronic rearrangements during formation of the transition state occur on the order of 10^{-13} s, about the lifetime of a bond vibration. But enzyme turnover numbers are much slower than this (see Table 7-1). An enzyme's overall speed is further limited by how often it collides productively with its substrate. The upper limit for the rate of this second-order reaction (a bimolecular reaction) is about 10^8 to 10^9 $M^{-1} \cdot s^{-1}$, which is the maximum rate at which two freely diffusing molecules can collide with each other in aqueous solution.

This so-called **diffusion-controlled limit** for the second-order reaction between an enzyme and a substrate is achieved by several enzymes, including triose phosphate isomerase, whose value of k_{cat}/K_M is 2.4×10^8 $M^{-1} \cdot s^{-1}$. This enzyme is therefore said to have reached **catalytic perfection** because *its overall rate is diffusion-controlled: It catalyzes a reaction as rapidly as it encounters its substrate.* However, many enzymes perform their physiological roles with more modest k_{cat}/K_M values.

K_M and V_{max} are experimentally determined

Kinetic data are usually collected by adding a small amount of an enzyme to varying amounts of substrate and then monitoring the reaction mixture for the appearance of product over a period of time. In order to meet the assumptions of the Michaelis–Menten model, the concentration of the substrate must be much greater than the concentration of the enzyme (so that the concentration of the ES complex will be constant and the formation of the ES complex will be limited by the affinity of E for S, not the amount of S available), and measurements must be taken of the initial velocity, before product begins to accumulate and the reverse reaction becomes significant.

Velocity versus substrate plots such as Figure 7-5 can be useful for visually estimating the kinetic parameters K_M and V_{max} (from which k_{cat} can be derived by Equation 7-23). However, in practice, hyperbolic curves are prone to misinterpretation because it is difficult to estimate the upper limit of the curve (V_{max}). In order to more accurately determine V_{max} and K_M (the substrate concentration at $V_{max}/2$), it is necessary to perform one of the following steps:

1. Analyze the data by a curve-fitting computer program that mathematically calculates the upper limit for the reaction velocity.
2. Transform the data to a form that can be plotted as a line. The best-known linear transformation of the velocity versus substrate curve is known as a **Lineweaver–Burk plot,** whose equation is

$$\frac{1}{v_0} = \left(\frac{K_M}{V_{max}}\right) \frac{1}{[S]} + \frac{1}{V_{max}} \qquad [7\text{-}27]$$

Equation 7-27 has the familiar form $y = mx + b$. A plot of $1/v_0$ versus $1/[S]$ gives a straight line whose slope is K_M/V_{max} and whose intercept on the $1/v_0$ axis is $1/V_{max}$. The extrapolated intercept on the $1/[S]$ axis is $-1/K_M$ (Fig. 7-6). A

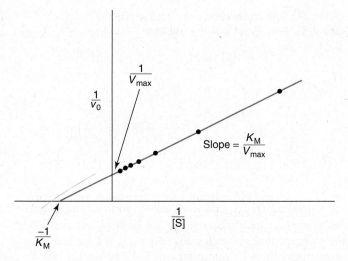

Figure 7-6 A Lineweaver–Burk plot. Plotting the reciprocals of [S] and v_0 yields a line whose slope and intercepts yield values of K_M and V_{max}. The plotted points correspond to the points in Figure 7-5. ⊕ **See Animated Figure. Lineweaver–Burk plot.**

⋯⋯⋯⋯⋯⋯⋯⋯⋯⋯⋯⋯⋯⋯⋯⋯⋯⋯⋯

? **Sketch a line to represent a reaction with a larger value of K_M and the same value of V_{max}.**

comparison of Figures 7-5 and 7-6, made from the same data, illustrates how evenly spaced points on a velocity versus substrate plot become compressed in a Lineweaver–Burk plot (see Sample Calculation 7-3).

SAMPLE CALCULATION 7-3 ●●●

PROBLEM

The velocity of an enzyme-catalyzed reaction was measured at several substrate concentrations. Calculate K_M and V_{max} for the reaction.

[S] (μM)	v_0 (mM · s^{-1})
0.25	0.75
0.5	1.20
1.0	1.71
2.0	2.18
4.0	2.53

SOLUTION

Calculate the reciprocals of the substrate concentration and velocity, then make a plot of $1/v_0$ versus $1/[S]$ (a Lineweaver–Burk plot).

1/[S] (μM^{-1})	$1/v_0$ (mM^{-1} · s^{-1})
4.0	1.33
2.0	0.83
1.0	0.58
0.5	0.46
0.25	0.40

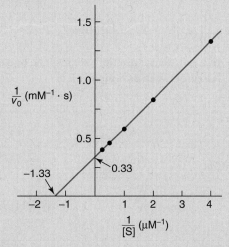

The intercept on the 1/[S] axis (which is equal to $-1/K_M$) is -1.33 μM^{-1}. Therefore,

$$K_M = -\left(\frac{1}{-1.33 \ \mu M^{-1}}\right) = 0.75 \ \mu M$$

The intercept on the $1/v_0$ axis (which is equal to $1/V_{max}$) is 0.33 mM^{-1} · s. Therefore,

$$V_{max} = \frac{1}{0.33 \ mM^{-1} \cdot s} = 3.0 \ mM \cdot s^{-1}$$

(*continued on next page*)

8. Calculate K_M and V_{max} from the following data.

[S] (mM)	v_0 (mM · s^{-1})
1	1.82
2	3.33
4	5.71
8	8.89
18	12.31

Ideally, experimental conditions are chosen so that velocity measurements can be made for substrate concentrations that are both higher and lower than K_M. This yields the most accurate values for K_M and V_{max}. A Lineweaver–Burk plot, whether constructed manually or by computer, offers the advantage that K_M and V_{max} can be quickly estimated by eye. Linear plots are also more convenient than curves for comparing multiple data sets, such as different enzyme preparations, or a single enzyme in the presence of different concentrations of an inhibitor.

Not all enzymes fit the simple Michaelis–Menten model

So far, the discussion has focused on the very simplest of enzyme-catalyzed reactions, namely, a reaction with one substrate and one product. Such reactions represent only a small portion of known enzymatic reactions, many of which involve multiple substrates and products, proceed via multiple steps, or do not meet the assumptions of the Michaelis–Menten kinetic model for other reasons. Nevertheless, the kinetics of these reactions can still be evaluated.

1. Multisubstrate Reactions

More than half of all known biochemical reactions involve two substrates. Most of these **bisubstrate reactions** are either oxidation–reduction reactions or transferase reactions. In an oxidation–reduction reaction, electrons are transferred between substrates:

$$X_{oxidized} + Y_{reduced} \rightarrow X_{reduced} + Y_{oxidized}$$

In a transferase reaction, such as the one catalyzed by transketolase, a group is transferred between two molecules:

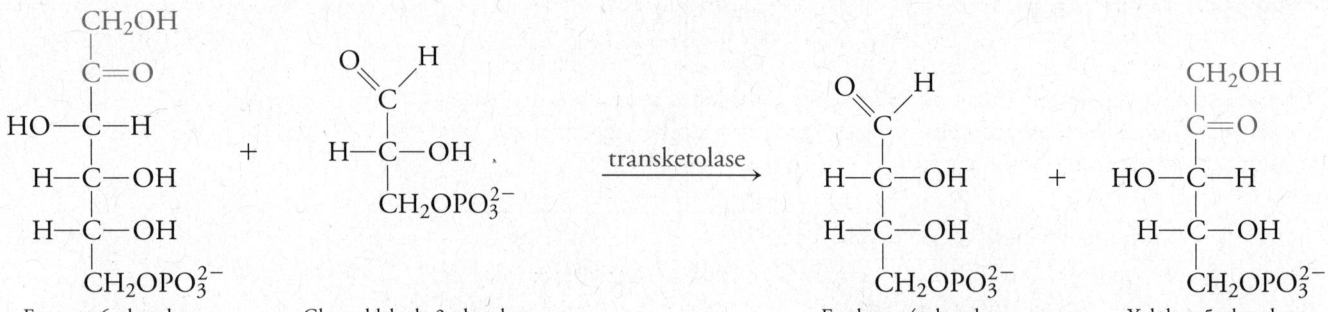

Fructose-6-phosphate   Glyceraldehyde-3-phosphate   Erythrose-4-phosphate   Xylulose-5-phosphate

The transketolase reaction is ubiquitous in nature; it functions in the synthesis and degradation of carbohydrates. As written here, it transforms a six-carbon sugar and a three-carbon sugar to a four-carbon sugar and a five-carbon sugar. *Each of the substrates interacts with the enzyme with a characteristic K_M.* To experimentally determine each K_M, the reaction velocity is measured at different concentrations of one substrate while the other substrate is present at a saturating concentration. V_{max} *is the maximum reaction velocity when both substrates are present at concentrations that saturate their binding sites on the enzyme.*

In some bisubstrate reactions, the substrates can bind in any order, as long as they both end up in the active site at the same time. These reactions are said to follow a **random mechanism.** Enzymes in which one substrate must bind before the other follow an **ordered mechanism.** In a **ping pong mechanism,** one substrate binds and one product is released before the other substrate binds and the second product is released. Transketolase catalyzes a ping pong reaction: Fructose-6-phosphate binds first and surrenders a two-carbon fragment to the enzyme, and the first product (erythrose-4-phosphate) leaves the active site before the second substrate (glyceraldehyde-3-phosphate) binds and receives the two-carbon fragment to yield the second product (xylulose-5-phosphate).

2. Multistep Reactions

As the transketolase reaction illustrates, an enzyme-catalyzed reaction may contain many steps. In this example, the reaction includes an intermediate in which the two-carbon fragment removed from fructose-6-phosphate remains bound to the enzyme while awaiting the arrival of the second substrate. (The chymotrypsin reaction mechanism outlined in Fig. 6-10 similarly requires several steps.) The multistep transketolase reaction can be broken down into a series of simple mechanistic steps and diagrammed as follows:

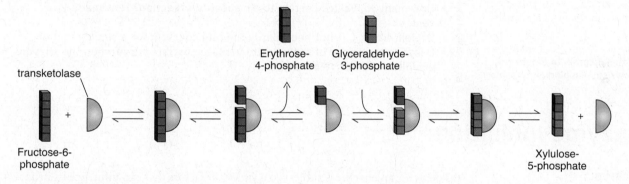

Each step of this process has characteristic forward and reverse rate constants. Consequently, k_{cat} for the overall reaction

$$\text{fructose-6-phosphate} + \text{glyceraldehyde-3-phosphate} \rightleftharpoons$$
$$\text{erythrose-4-phosphate} + \text{xylulose-5-phosphate}$$

is a complicated function of many individual rate constants (only for a very simple reaction, for example, Equation 7-4, does $k_{cat} = k_2$). Nevertheless, *the meaning of* k_{cat}—*the enzyme's turnover number—is the same as for a single-step reaction.*

The rate constants of individual steps in a multistep reaction can sometimes be measured during the initial stages of the reaction, that is, *before* a steady state is established. This requires instruments that can rapidly mix the reactants and then monitor the mixture on a time scale from 1 s to 10^{-7} s.

3. Nonhyperbolic Reactions

Many enzymes, particularly oligomeric enzymes with multiple active sites, do not obey the Michaelis–Menten rate equation and therefore do not yield hyperbolic velocity versus substrate curves. In these **allosteric enzymes,** *the presence of a substrate at one active site can affect the catalytic activity of the other active sites.* This **cooperative** behavior occurs when the enzyme subunits are structurally linked to each other so that a substrate-induced conformational change in one subunit elicits conformational changes in the remaining subunits. (Cooperative behavior also occurs in hemoglobin, when O_2 binding to the heme group in one subunit alters the O_2 affinity of the other subunits; see Section 5-1.) Like hemoglobin, an allosteric enzyme has two possible quaternary structures. The T (or "tense") state has lower catalytic activity, and the R (or "relaxed") state has higher activity. Because of the interactions between

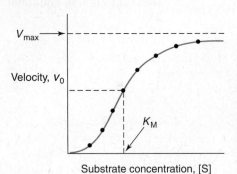

Figure 7-7 Effect of cooperative substrate binding. The velocity versus substrate curve is sigmoidal rather than hyperbolic when substrate binding to one active site in an oligomeric enzyme alters the catalytic activity of the other active sites. The maximum reaction velocity is V_{max}, and K_M is the substrate concentration when the velocity is half-maximal.

? Compare this diagram to Figure 5-7, which shows oxygen binding to hemo-globin.

individual subunits, the entire enzyme can switch between the T and R conformations. The result of allosteric behavior is a sigmoidal (S-shaped) velocity versus substrate curve (**Fig.** 7-7). Although the standard Michaelis–Menten equation does not apply here, K_M and V_{max} can be estimated and used to characterize enzyme activity.

CONCEPT REVIEW
- What are the rate equations for first-order and second-order reactions?
- Describe the three possible reactions of an ES complex.
- Describe the changes in the concentrations of S, P, E_T, and ES during the course of an enzyme-catalyzed reaction.
- Write the rate equation for an enzyme-catalyzed reaction.
- What is the formal definition of K_M?
- What other meaning is it often given?
- What is the meaning of k_{cat}?
- How can V_{max} be used to calculate k_{cat}?
- Why is k_{cat}/K_M a better indicator of enzyme efficiency than either k_{cat} or K_M alone?
- Why is triose phosphate isomerase considered to be a catalytically perfect enzyme?
- Why is a Lineweaver–Burk plot, rather than a velocity versus substrate plot, often used to determine K_M and V_{max}?
- How many K_M values pertain to a bisubstrate reaction? How many V_{max} values?
- Explain why k_{cat} is not necessarily the rate constant for a single reaction.
- Describe the shape of the velocity versus substrate curve when the enzyme exhibits cooperative behavior.

7-3 Enzyme Inhibition

KEY CONCEPTS
- Substances that bind irreversibly to an enzyme can inhibit its activity.
- A competitive inhibitor appears to increase K_M without affecting V_{max}.
- Transition state analogs can act as competitive inhibitors.
- Noncompetitive, mixed, and uncompetitive inhibitors decrease k_{cat}.
- Allosteric regulators can inhibit or activate enzymes.

Inside a cell, an enzyme is subject to a variety of factors that can influence its behavior. Substances that interact with the enzyme can interfere with substrate binding and/or catalysis. Many naturally occurring antibiotics, pesticides, and other poisons are substances that inhibit the activity of essential enzymes. From a strictly scientific point of view, these inhibitors are useful probes of an enzyme's active-site structure and catalytic mechanism. Enzyme inhibitors are also used therapeutically as drugs. The ongoing pursuit of more effective drugs requires knowledge of how enzyme inhibitors work and how they can be altered to better inhibit their target enzymes. Box 7-A describes some of the steps that must be taken to turn an enzyme inhibitor into an effective drug (also see Bioinformatics Project 5).

BOX 7-A ⬡ **CLINICAL CONNECTION**

Drug Development

The path of a drug from a biochemist's laboratory to a patient's medicine cabinet is typically long, arduous, and expensive. Whether a new drug results from a surprise discovery or dedicated efforts by pharmaceutical scientists, a newly identified drug candidate is almost never the exact same compound that makes it to the pharmacist's shelves. Drug development is a process of refinement and testing that can take years and cost billions of dollars, but it usually yields a substance that is therapeutically useful and safe.

The majority of drugs currently in use block the activity of proteins that participate in cell-signaling pathways, but they have much in common with drugs that act as enzyme inhibitors. Every potential drug starts out as a synthetic compound or a

natural product that is then altered and tested for the desired biological effect. For enzyme inhibitors, modern approaches take advantage of knowledge about an enzyme's active-site structure and mechanism to design a compound that will precisely block catalysis. This process is sometimes called **rational drug design.** A drug candidate, or lead compound, may be systematically altered by adding or deleting various chemical groups (including fluorine, Box 2-A) and then retested for inhibitory activity. Robotic procedures for chemical synthesis and testing allow thousands of substances to be analyzed. Alternatively, computer simulations, based on the enzyme's structure, can predict whether a modified structure might be a better inhibitor.

The goal throughout the drug development process is to create a compound with the following properties: The drug must be a tight-binding inhibitor, it must be highly selective for its target enzyme (so that it doesn't interfere with the activity of other enzymes), and it must not be toxic. In addition, the drug's **pharmacokinetics,** or behavior in the body, must be assessed. For example, the drug must be water-soluble so that it can be transported via the bloodstream; at the same time, it must be lipid-soluble so that it can pass through the walls of the intestine to be absorbed and through the walls of blood vessels so that it can reach the tissues. In many cases, the drug is inactive until this point; once inside a cell, it is converted to its bioactive form by cellular enzymes.

Acylcovir (Zovirax), which is used to treat herpes virus infections, provides an interesting example of this phenomenon. The drug mimics guanosine with an incomplete ribose group. Infected cells contain a viral kinase that converts the drug to a nucleotide analog that then interferes with DNA synthesis.

Acyclovir

Uninfected cells lack the kinase and are therefore unaffected by acyclovir.

A large part of the research relevant to drug development focuses on how the body metabolizes foreign compounds. The cytochrome P450 class of enzymes is responsible for detoxifying a variety of naturally occurring compounds that make their way into the body; drugs are also potential substrates. Cytochrome P450 contains a heme prosthetic group (see Section 5-1) that participates in an oxidation–reduction reaction that adds a hydroxyl group to the substrate to make it more water-soluble and therefore more easily excreted. Consequently, the activity of a cytochrome P450 enzyme can decrease the effectiveness of a drug. In some cases, the hydroxylation reaction can convert a drug to a toxin. This occurs when acetaminophen, a commonly used fever-reducer (also known as paracetamol), is consumed in large doses:

Acetaminophen

Acetimidoquinone
(toxic)

Individuals vary in the amount and type of cytochrome P450 enzymes they express, so it is difficult to predict how a drug might be metabolized.

Consequently, the best laboratory-based predictions—or even testing in animals—cannot guarantee a successful outcome in the human body. Ultimately, a drug's effectiveness and safety must be assessed in **clinical trials.** This type of testing is organized into three consecutive phases, each involving a larger number of subjects. In Phase I clinical trials, the drug candidate is administered to a small group of healthy volunteers at levels up to the expected therapeutic dose. The goal of Phase I trials is to ensure that the drug is safe and is well tolerated by humans. Researchers can also examine the drug's pharmacokinetics and select a version of the drug and a dosing regimen suitable for larger-scale testing.

After the safety of the drug has been verified, its effectiveness is tested in Phase II trials, which typically involve several hundred subjects who have the disease targeted by the drug. Assessments of safety and optimum dosing continue during Phase II. Patients are usually randomly assigned to test or control groups. Since many clinical trials test whether a new drug works better than an existing drug, the control group receives the old drug rather than no drug at all. To avoid the placebo effect, in which a patient's or physician's expectations can alter the outcome, the trial is "blinded" so that the patients do not know which treatment they are receiving (a single-blind trial) or, better yet, neither the patients nor their physicians know (a double-blind trial). Drug trials are usually easy to blind, since the test and control substances can be made to appear identical, but it is difficult if not impossible to conduct blind trials in other situations, such as comparing a drug to a surgical intervention. In these cases, objectivity can be maximized by not telling the statisticians who analyze the results which patient group received which treatment.

Phase III clinical trials are conducted with large numbers of patients (hundreds to thousands). Large numbers are needed to provide robust statistical evidence that the new drug works as hoped and is safe for widespread use. Like Phase II trials, Phase III trials are randomized and blinded if possible. Successful completion of Phase III, the longest phase of testing, usually leads to regulatory approval of the drug by the Food and Drug Administration (in the United States), although the drug may be marketed to a limited extent before full approval has been granted.

Even after a drug has been approved, marketed, and widely adopted, the patient population is continually scrutinized for rare side effects or side effects that might not have developed during the short time period of Phase II or Phase III trials. This surveillance period is sometimes called Phase IV. Its importance is highlighted by the case of the widely used pain reliever rofecoxib (Vioxx), which had been approved but was withdrawn when it was discovered that it increased the risk of heart attacks. Similarly, the use of the antidiabetic drug rosiglitazone (Avandia) has been limited since analysis of large numbers of patients revealed an increase in heart attacks.

(*continued on next page*)

1. Medicinal chemists apply certain rules to assess the suitability of drug candidates. For example, they prefer to work with compounds that have a molecular mass less than 500 and contain fewer than five hydrogen bond donor groups and fewer than ten hydrogen bond acceptor groups. Explain why these properties would be desirable in a drug.

2. The structure of rosiglitazone is shown here. Does it meet the "rules" described in Question 1?

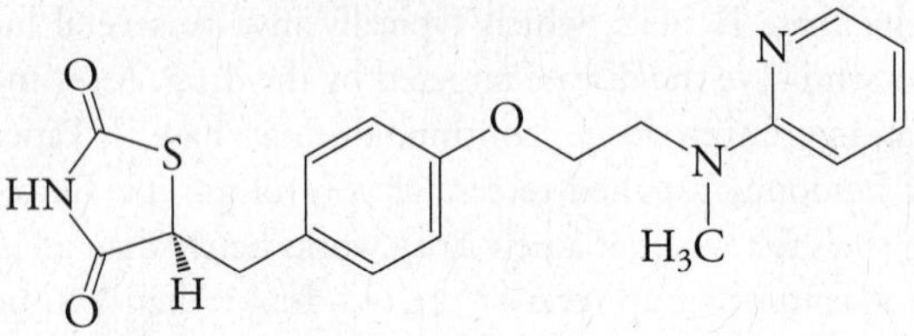

3. Explain why some drugs must be administered intravenously rather than swallowed in the form of a pill.

4. The toxicity of the acetaminophen derivative acetamidoquinone results from its ability to react with the Cys groups of proteins. Explain why the toxic effects of acetamidoquinone are localized to the liver.

5. The drug warfarin is widely prescribed as a "blood thinner" because it inhibits the post-translational modification of some of the proteins involved in blood coagulation (Box 6-B). In order to choose an effective dose of warfarin, physicians may test patients to identify which cytochrome P450 enzymes they express. Explain the purpose of this genetic testing.

6. Enalapril (Vasotec) inhibits an enzyme that converts the peptide angiotensin I to angiotensin II, a potent vasoconstrictor.

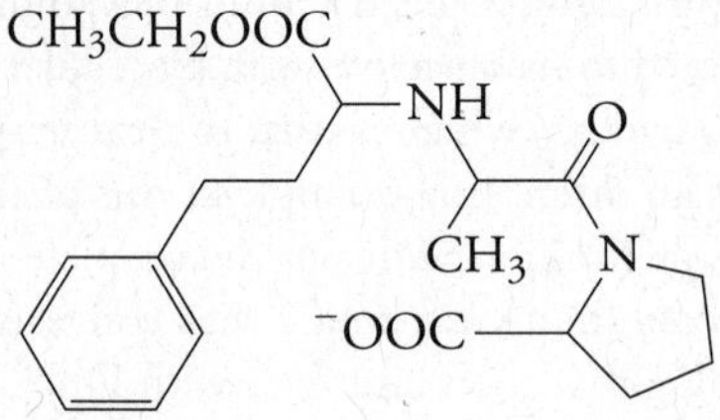

(a) What type of disorders might be treated by enalapril?
(b) Enalapril is inactive until acted on by an esterase. Draw the structure of the resulting bioactive derivative.

Some inhibitors act irreversibly

➕ **See Guided Exploration. Michaelis-Menten kinetics, Lineweaver-Burk plots, and enzyme inhibition.**

Certain compounds interact with enzymes so tightly that their effects are essentially irreversible. For example, diisopropylphosphofluoridate (DIPF), the reagent used to identify the active-site Ser residue of chymotrypsin (see Section 6-2), is an **irreversible inhibitor** of the enzyme. When DIPF reacts with chymotrypsin, leaving the DIP group covalently attached to the Ser hydroxyl group, the enzyme becomes catalytically inactive. In general, *any reagent that covalently modifies an amino acid side chain in a protein can potentially act as an irreversible enzyme inhibitor.*

Some irreversible enzyme inhibitors are called **suicide substrates** because they enter the enzyme's active site and begin to react, just as a normal substrate would. However, they are unable to undergo the complete reaction and hence become "stuck" in the active site. For example, thymidylate synthase is the enzyme that converts the nucleotide deoxyuridylate (dUMP) to deoxythymidylate (dTMP) by adding a methyl group to C5:

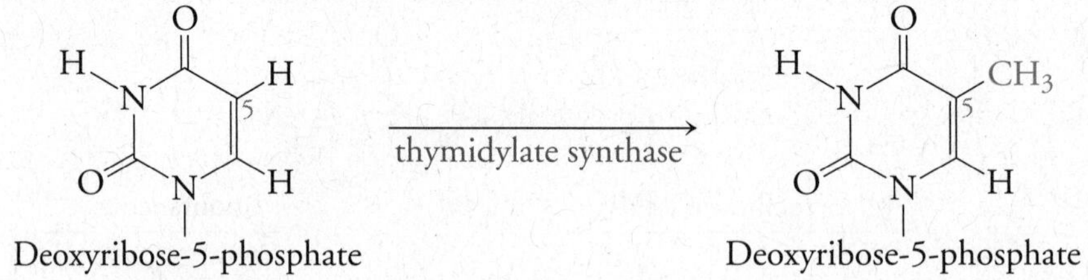

Deoxyribose-5-phosphate
dUMP

thymidylate synthase →

Deoxyribose-5-phosphate
dTMP

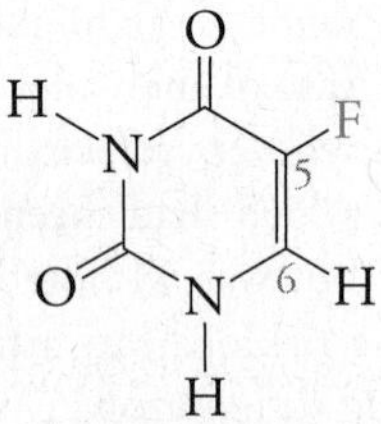

When the synthetic compound 5-fluorouracil (*left*) is taken up by cells, it is readily converted to the nucleotide 5-fluorodeoxyuridylate. This compound, like dUMP, enters the active site of thymidylate synthase, where a Cys—SH group adds to C6. Normally, this enhances the nucleophilicity (electron richness) of C5 so that it can accept an electron-poor methyl group. However, the presence of the electron-withdrawing F atom prevents methylation. The inhibitor therefore remains in the active site, bound to the Cys side chain, rendering thymidylate synthase inactive. For this reason, 5-fluorouracil is used to disrupt DNA synthesis in rapidly dividing cancer cells.

5-Fluorouracil

Competitive inhibition is the most common form of reversible enzyme inhibition

As its name implies, reversible enzyme inhibition results when a substance binds reversibly (that is, noncovalently) to an enzyme so as to alter its catalytic properties. A reversible inhibitor may affect the enzyme's K_M, k_{cat}, or both. The most common form of reversible enzyme inhibition is known as **competitive inhibition.** In this situation, *the inhibitor is a substance that directly competes with a substrate for binding to the enzyme's active site* (**Fig. 7-8**). As expected, the inhibitor usually resembles the substrate in overall size and chemical properties so that it can bind to the enzyme, but it lacks the exact electronic structure that allows it to react.

One well-known competitive inhibitor affects the activity of succinate dehydrogenase, which catalyzes the oxidation (dehydrogenation) of succinate to produce fumarate:

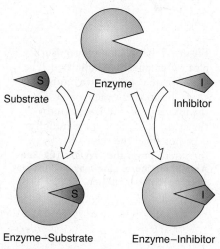

Figure 7-8 Competitive enzyme inhibition. In its simplest form, competitive inhibition of an enzyme occurs when the inhibitor and substrate compete for binding in the enzyme active site. A competitive inhibitor often resembles the substrate in size and shape but cannot undergo a reaction. In all cases, binding of the inhibitor and the substrate is mutually exclusive.

The compound malonate

inhibits the reaction because it binds to the dehydrogenase active site but cannot be dehydrogenated. Apparently, the enzyme active site can accommodate either the substrate succinate or the competitive inhibitor malonate, although they differ slightly in size.

A plot of an enzyme's reaction velocity in the presence of a competitive inhibitor, as a function of the substrate concentration, is shown in **Figure 7-9**. Because the inhibitor prevents some of the substrate from reaching the active site, the K_M appears to increase (the enzyme's affinity for the substrate appears to decrease). But because the inhibitor binds reversibly, it constantly dissociates from and reassociates with the enzyme, which allows a substrate molecule to occasionally enter the active site. High concentrations of substrate can overcome the effect of the inhibitor because when [S] $\gg$ [I], the enzyme is more likely to bind S than I. The presence of a competitive inhibitor does not affect the enzyme's k_{cat}, so as [S] approaches infinity, v_0 approaches V_{max}. To summarize, *a competitive inhibitor increases the apparent K_M of the enzyme but does not affect k_{cat} or V_{max}.*

The Michaelis–Menten equation for a competitively inhibited reaction has the form

$$v_0 = \frac{V_{max}[S]}{\alpha K_M + [S]} \qquad \textbf{[7-28]}$$

where α is a factor that makes K_M appear larger. The value of α—the degree of inhibition—depends on the inhibitor's concentration and its affinity for the enzyme:

$$\alpha = 1 + \frac{[I]}{K_I} \qquad \textbf{[7-29]}$$

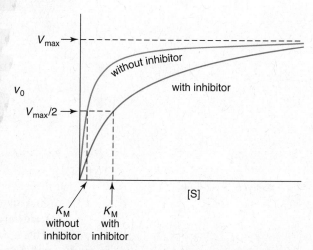

Figure 7-9 Effect of a competitive inhibitor on reaction velocity. In a plot of velocity versus substrate concentration, the inhibitor increases the apparent K_M because it competes with the substrate for binding to the enzyme. The inhibitor does not affect k_{cat}, so at high [S], v_0 approaches V_{max}.

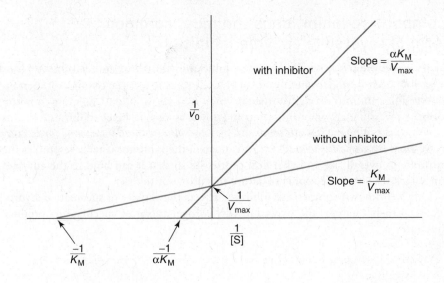

Figure 7-10 A Lineweaver–Burk plot for competitive inhibition. The presence of a competitive inhibitor alters the apparent value of $-1/K_M$, the intercept on the 1/[S] axis, by the factor α. Note that the inhibitor does not affect the value of $1/V_{max}$, the intercept on the $1/v_0$ axis. ⊕ See Animated Figure. Lineweaver–Burk plot of competitive inhibition.

K_I is the **inhibition constant;** it is the dissociation constant for the **enzyme–inhibitor (EI) complex:**

$$K_I = \frac{[E][I]}{[EI]} \qquad \text{[7-30]}$$

The lower the value of K_I, the tighter the inhibitor binds to the enzyme. It is possible to derive α (and therefore K_I) by plotting the reaction velocity as a function of substrate concentration in the presence of a known concentration of inhibitor. When the data are replotted in Lineweaver–Burk form, the intercept on the 1/[S] axis is $-1/\alpha K_M$ (**Fig. 7-10**; also see Sample Calculation 7-4).

●●● SAMPLE CALCULATION 7-4

PROBLEM An enzyme has a K_M of 8 μM in the absence of a competitive inhibitor and an apparent K_M of 12 μM in the presence of 3 μM of the inhibitor. Calculate K_I.

SOLUTION The inhibitor increases K_M by a factor α (Equation 7-28). Since the value of K_M with the inhibitor is 1.5 times greater than the value of K_M without the inhibitor (12 μM ÷ 8 μM), $\alpha = 1.5$. Equation 7-29, which gives the relationship between α, [I], and K_I, can be rearranged to solve for K_I:

$$K_I = \frac{[I]}{\alpha - 1}$$
$$= \frac{3\ \mu M}{1.5 - 1}$$
$$= \frac{3\ \mu M}{0.5} = 6\ \mu M$$

●●● PRACTICE PROBLEMS

9. An enzyme has a K_M of 1 mM in the absence of a competitive inhibitor and an apparent K_M of 3 mM in the presence of 10 μM of the inhibitor. Calculate K_I.

10. The K_I value for a certain inhibitor is 2 μM. When no inhibitor is present, the K_M value is 10 μM. Calculate the apparent K_M when 4 μM inhibitor is present.

11. Inhibitor A at a concentration of 2 μM doubles the apparent K_M for an enzymatic reaction, whereas inhibitor B at a concentration of 9 μM quadruples the apparent K_M. What is the ratio of the K_I for inhibitor B to the K_I for inhibitor A?

K_I values are useful for assessing the inhibitory power of different substances, such as a series of compounds being tested for usefulness as drugs. For example, atorvastatin, the drug introduced at the start of the chapter, binds to the enzyme HMG-CoA reductase with a K_I value of about 8 nM (the enzyme's substrate has a K_M of about 4 μM). Keep in mind that an effective drug is not necessarily the compound with the lowest K_I (the tightest binding), since other factors, such as the drug's solubility or its stability, must also be considered.

Product inhibition occurs when the product of a reaction occupies the enzyme's active site, thereby preventing the binding of additional substrate molecules. This is one reason why measurements of enzyme activity are made early in the reaction, before product has significantly accumulated.

▶▶ HOW do we measure how well a drug works?

Transition state analogs inhibit enzymes

Studies of enzyme inhibitors can reveal information about the chemistry of the reaction and the enzyme's active site. For example, the inhibition of succinate dehydrogenase by malonate, shown earlier, suggests that the dehydrogenase active site recognizes and binds substances with two carboxylate groups. Similarly, the ability of an inhibitor to bind to an enzyme's active site may confirm a proposed reaction mechanism. The triose phosphate isomerase reaction (introduced in Section 7-1) is believed to proceed through an enediolate transition state (shown inside brackets):

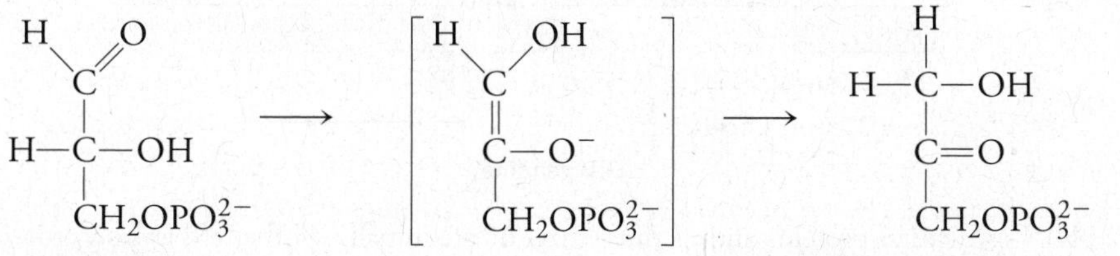

Glyceraldehyde-3-phosphate · Enediolate · Dihydroxyacetone phosphate

Recall from Section 6-2 that the transition state corresponds to a high-energy structure in which bonds are in the process of breaking and forming. The compound phosphoglycohydroxamate

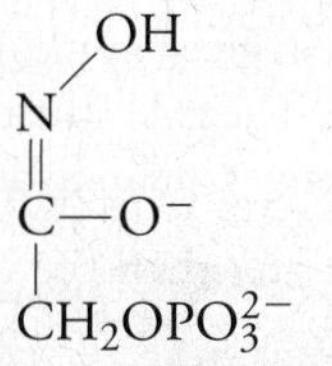

Phosphoglycohydroxamate

resembles the proposed transition state and, in fact, binds to triose phosphate isomerase about 300 times more tightly than glyceraldehyde-3-phosphate or dihydroxyacetone phosphate binds to the enzyme.

Numerous studies demonstrate that *whereas substrate analogs make good competitive inhibitors,* **transition state analogs** *make even better inhibitors.* This is because in order to catalyze a reaction, the enzyme must bind to (stabilize, or lower the energy of) the reaction's transition state. A compound that mimics the transition state can take advantage of features in the active site in a way that a substrate analog cannot. For example, the nucleoside adenosine is converted to inosine as follows:

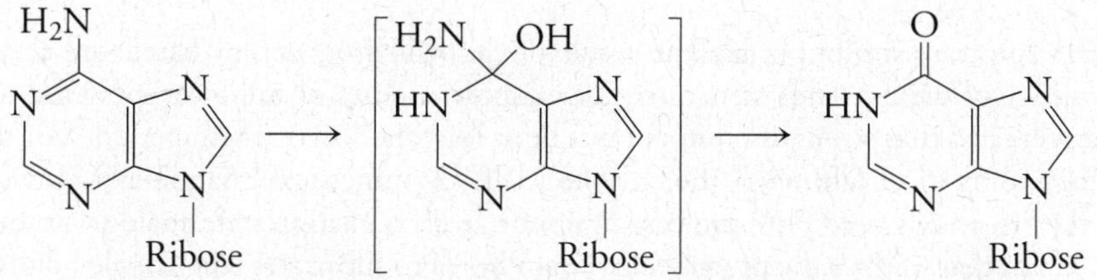

Adenosine · Inosine

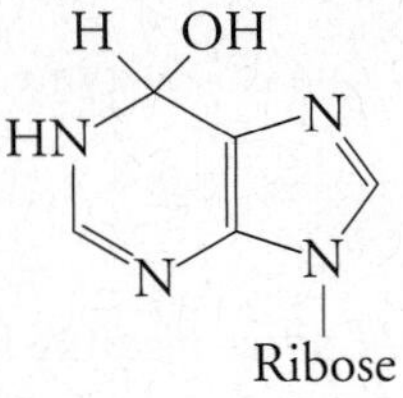

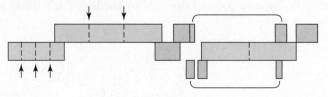

H   OH

HN

N

N

N

Ribose

1,6-Dihydroinosine

The K_M of the enzyme for the substrate adenosine is 3×10^{-5} M. The product inosine acts as an inhibitor of the reaction, with a K_I of 3×10^{-4} M. The transition state analog 1,6-dihydroinosine (*left*) inhibits the reaction with a K_I of 1.5×10^{-13} M.

Not only do such inhibitors shed light on the probable structure of the reaction's transition state, they may provide a starting point for the design of even better inhibitors. Some of the drugs used to treat infection by HIV (the human immunodeficiency virus) were first designed by considering how transition state analogs inhibit viral enzymes (Box 7-B).

BOX 7-B   BIOCHEMISTRY NOTE

Inhibitors of HIV Protease

Human immunodeficiency virus (HIV), the causative agent of acquired immune deficiency syndrome (AIDS), has an RNA genome that codes for 15 different proteins. Six of these are structural proteins (green in the diagram below), three are enzymes (purple), and six are accessory proteins (blue) that are required for viral gene expression and assembly of new viral particles. Several genes overlap, and two genes are composed of noncontiguous segments of RNA.

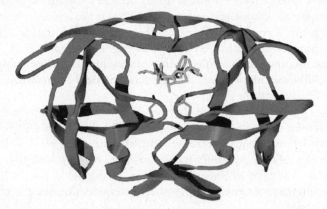

HIV genome

HIV's structural proteins and its three enzymes are initially synthesized as **polyproteins** whose individual members are later separated by proteolysis (at the sites indicated by arrows). The enzyme responsible for this activity is HIV protease, one of the three viral enzymes. A small amount of the protease is present in the virus particle when it first infects a cell; more is generated as the viral genome is transcribed and translated. HIV protease catalyzes the hydrolysis of Tyr—Pro or Phe—Pro peptide bonds in the viral polyproteins. Catalytic activity is centered on two Asp residues (shown in green in the model below), each contributed by one subunit of the homodimeric enzyme. The gold structure represents a peptide substrate analog. The side chains of peptide substrates bind in hydrophobic pockets near the active site.

HIV protease. [Structure (pdb 1HXW) determined by C. H. Park, V. Nienaber, and X. P. Kong.]

⊕ **See Interactive Exercise. HIV protease.**

HIV protease inhibitors are the result of rational drug design based on detailed knowledge of the enzyme's structure. For example, studies of inhibitor–protease complexes revealed that strong inhibitors must be at least the size of a tetrapeptide but need not be symmetrical (although the enzyme itself is symmetrical). Saquinavir (Invirase) was the first widely used HIV protease inhibitor. It is a transition state analog with bulky side chains that mimics the protease's natural Phe—Pro substrates (the scissile bonds are shown in red).

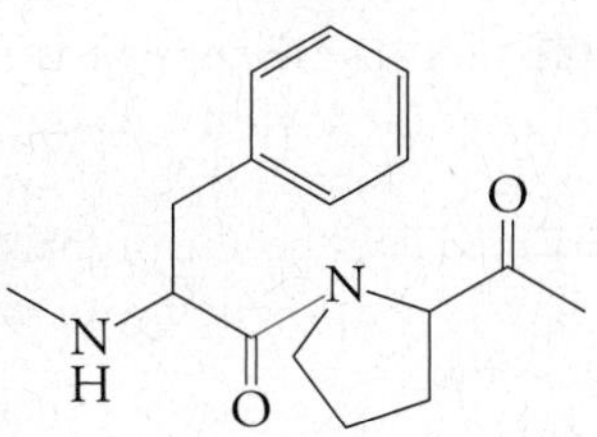

—Phe—Pro—

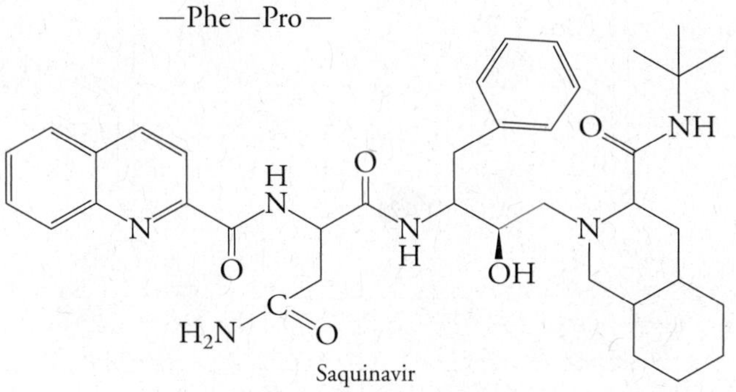

Saquinavir

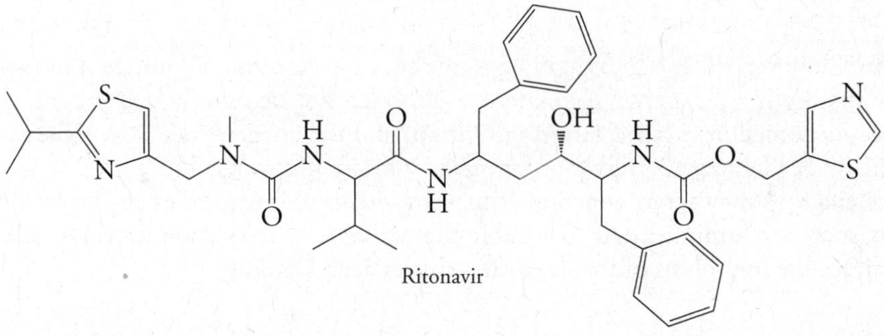

Ritonavir

Saquinavir acts as a competitive inhibitor with a K_I of 0.15 nM (for comparison, synthetic peptide substrates have K_M values of about 35 μM). Efforts to develop other drugs have focused on improving the solubility (and therefore the bioavailability) of protease inhibitors by adding polar groups without diminishing binding to the protease. Such efforts have yielded, for example, ritonavir (Norvir), with a K_I of 0.17 nM.

One of the challenges of developing antiviral drugs is to *select a target that is unique to the virus so that the drugs will not disrupt the host's normal metabolic reactions.* The HIV protease inhibitors are effective antiviral agents because mammalian proteases do not recognize compounds containing amide bonds to Pro or Pro analogs. Nevertheless, the drugs that target HIV protease do have side effects. In addition, the high rate of mutation in HIV increases the risk of the virus developing resistance to a drug. For this reason, HIV infection is typically treated with a combination of several drugs, including protease inhibitors and inhibitors of reverse transcriptase and integrase, the virus's other two enzymes (see Box 20-A).

● **Question:** In HIV patients with high cholesterol, administering both HIV protease inhibitors and a statin drug such as atorvastatin can lead to dangerously high levels of the statin. Based on what you know about drug metabolism, propose an explanation for this observation.

Other types of inhibitors affect V_{max}

Some reversible enzyme inhibitors diminish an enzyme's activity not only by interfering with substrate binding (as approximated by K_M) but also by directly affecting k_{cat}. This situation usually occurs when the inhibitor binds to a site on the enzyme other than the active site and elicits a conformational change that affects the structure or chemical properties of the active site. As a result, k_{cat} *and the apparent* V_{max} *decrease but* K_M *does not change.* This situation is called **noncompetitive inhibition** (Fig. 7-11).

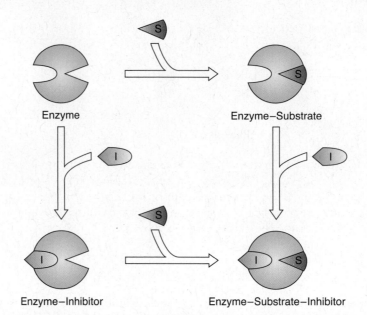

Figure 7-11 Noncompetitive enzyme inhibition. The inhibitor and substrate bind to separate sites. Inhibitor binding does not prevent substrate binding but alters the catalytic activity of the enzyme (thereby decreasing the apparent V_{max}). In mixed inhibition, the inhibitor also affects substrate binding so that the K_M appears to increase or decrease.

Often, however, the inhibitor binding to the enzyme alters its conformation in such a way that both V_{max} and K_M are affected, although not necessarily in the same way. This phenomenon is called **mixed inhibition,** and the apparent K_M may increase or decrease. A Lineweaver–Burk plot for mixed inhibition is shown in **Figure 7-12**.

Metal ions may act as noncompetitive enzyme inhibitors. For example, trivalent ions such as aluminum (Al^{3+}) inhibit the activity of acetylcholinesterase, which catalyzes the hydrolysis of the neurotransmitter acetylcholine:

$$H_3C-\overset{\overset{\displaystyle O}{\|}}{C}-O-CH_2-CH_2-\overset{+}{N}(CH_3)_3 \xrightarrow[\text{acetylcholinesterase}]{H_2O \qquad H^+} H_3C-\overset{\overset{\displaystyle O}{\|}}{C}-O^- \ + \ HO-CH_2-CH_2-\overset{+}{N}(CH_3)_3$$

Acetylcholine   Acetate   Choline

This reaction limits the duration of certain nerve impulses (see Section 9-4). Al^{3+} inhibits acetylcholinesterase noncompetitively by binding to the enzyme at a site distinct from the active site. Consequently, Al^{3+} can bind to the free enzyme or to the enzyme–substrate complex.

In a multisubstrate reaction, an inhibitor can bind to the enzyme *after* one substrate has bound, in a way that prevents the reaction from continuing and yielding product (**Fig. 7-13**). In the presences of such a substance, called an **uncompetitive inhibitor,** k_{cat} is affected so that the apparent V_{max} is lowered, and the apparent K_M is lowered to the same degree.

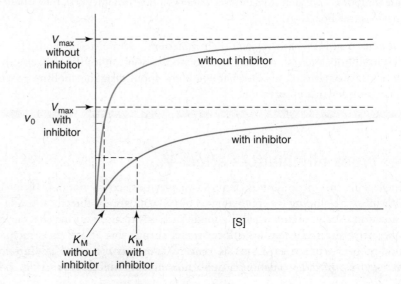

Figure 7-12 Effect of a mixed inhibitor on reaction velocity. As shown here, the inhibitor affects both substrate binding (represented as K_M) and k_{cat} so that the apparent K_M increases and the apparent V_{max} decreases. In some cases, the apparent K_M may decrease or remain unchanged (noncompetitive inhibition).

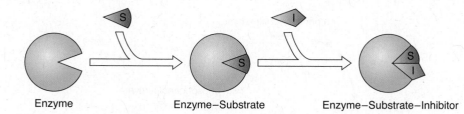

Enzyme   Enzyme–Substrate   Enzyme–Substrate–Inhibitor

Figure 7-13 Uncompetitive enzyme inhibition. The inhibitor binds to the enzyme after the substrate binds. As a result, V_{max} and K_M appear to be reduced by the same amount.

Competitive inhibition can be distinguished from mixed, noncompetitive, and uncompetitive inhibition because increasing the concentration of the substrate reverses the effects of a competitive inhibitor. Increasing the substrate concentration does not alleviate mixed, noncompetitive, and uncompetitive inhibition because the inhibitor does not prevent substrate binding to the active site.

Allosteric enzyme regulation includes inhibition and activation

Oligomeric enzymes—those with multiple active sites in one multisubunit protein—are commonly subject to **allosteric regulation,** which includes inhibition as well as activation. Just as ligand binding to one subunit of an oligomeric enzyme may alter the activity of the other active sites (as occurs in hemoglobin; see Section 5-1), inhibitor (or activator) binding to one subunit of an enzyme may decrease (or increase) the catalytic activity of all the subunits.

Allosteric effects are part of the physiological regulation of the enzyme phosphofructokinase, which catalyzes the reaction

$$^{-2}O_3POCH_2 \overset{6}{}\quad O \quad \overset{1}{}CH_2OH \quad + \text{ ATP} \xrightarrow{\text{phosphofructokinase}} \quad ^{-2}O_3POCH_2 \overset{6}{}\quad O \quad \overset{1}{}CH_2OPO_3^{2-} \quad + \text{ ADP}$$

Fructose-6-phosphate   Fructose-1,6-bisphosphate

This phosphorylation reaction is step 3 of glycolysis, the glucose degradation pathway that is an important source of ATP in virtually all cells (see Section 13-1). The phosphofructokinase reaction is inhibited by phosphoenolpyruvate, the product of Reaction 9 of glycolysis:

$$
\begin{array}{c}
\text{O} \quad \text{O}^- \\
\backslash\!\!\diagup\!\!\text{C} \\
| \\
\text{C}-\text{OPO}_3^{2-} \\
\| \\
\text{CH}_2
\end{array}
$$

Phosphoenolpyruvate

Phosphoenolpyruvate is an example of a **feedback inhibitor:** When its concentration in the cell is sufficiently high, *it shuts down its own synthesis by blocking an earlier step in its biosynthetic pathway:*

Glucose → → ——— X ———→ → → → → → Phosphoenolpyruvate
                phosphofructokinase

Phosphofructokinase from the bacterium *Bacillus stearothermophilus* is a tetramer with four active sites. The subunits are arranged as a dimer of dimers (**Fig. 7-14**). Each of the four fructose-6-phosphate binding sites is made up of residues from both dimers. *B. stearothermophilus* phosphofructokinase binds fructose-6-phosphate

Figure 7-14 Structure of phosphofructokinase from *B. stearothermophilus.* The four identical subunits are arranged as a dimer of dimers (one dimer is shown with blue subunits, the other with purple subunits). [Structure (pdb 6PFK) determined by P. R. Evans, T. Schirmer, and M. Auer.]

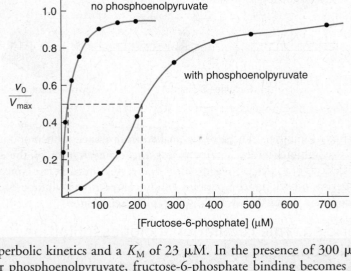

Figure 7-15 Effect of phosphoenolpyruvate on phosphofructokinase activity. In the absence of the inhibitor (green line), *B. stearothermophilus* phosphofructokinase binds the substrate fructose-6-phosphate with a K_M of 23 μM. In the presence of 300 μM phosphoenolpyruvate (red line), the K_M increase to about 200 μM. [Data from Zhu, X., Byrnes, N., Nelson, J. W., and Chang, S. H., *Biochemistry* 34, 2560–2565 (1995).]

with hyperbolic kinetics and a K_M of 23 μM. In the presence of 300 μM of the inhibitor phosphoenolpyruvate, fructose-6-phosphate binding becomes sigmoidal and the K_M increases to about 200 μM (Fig. 7-15). The inhibitor does not affect V_{max}, but phosphofructokinase becomes less active because its apparent affinity for fructose-6-phosphate decreases.

How does phosphoenolpyruvate exert its inhibitory effects? The sigmoidal velocity versus substrate curve (see Fig. 7-15) indicates that the phosphofructokinase active sites behave cooperatively in the presence of phosphoenolpyruvate. In each subunit, the inhibitor binds in a pocket that is separated from the fructose-6-phosphate binding site of the neighboring dimer by a loop of protein. When phosphoenolpyruvate occupies its binding site, the protein closes in around it. This causes a conformational change in which two residues in the loop switch positions: Arg 162 moves away from the fructose-6-phosphate binding site of the neighboring subunit and is replaced by Glu 161 (Fig. 7-16). This conformational switch diminishes fructose-6-phosphate binding because the positively charged side chain of Arg 162, which helps stabilize the negatively charged phosphate group of fructose-6-phosphate, is replaced by the negatively charged side chain of Glu 161, which repels the phosphate group. *The effect of phosphoenolpyruvate is communicated to the entire protein (thereby explaining the cooperative effect) because phosphoenolpyruvate binding to one subunit of phosphofructokinase affects fructose-6-phosphate binding to the neighboring subunit in the other dimer.* Using the terminology for allosteric proteins, phosphoenolpyruvate binding causes the entire tetramer to switch to the T (low-activity) conformation, as measured by fructose-6-phosphate binding affinity. Phosphoenolpyruvate is therefore known as a **negative effector** of the enzyme.

Figure 7-16 Conformational change upon phosphoenolpyruvate binding to phosphofructokinase. The green structure represents the conformation of the enzyme that readily binds the substrate fructose-6-phosphate (labeled F6P). The red structure represents the enzyme with a bound allosteric inhibitor (a phosphoenolpyruvate analog labeled PEP). Phosphoenolpyruvate binding to the enzyme causes a conformational change in which Arg 162 (which forms part of the fructose-6-phosphate binding site of the neighboring subunit) changes places with Glu 161. Because the subunits act cooperatively, the inhibitor diminishes substrate binding to the entire enzyme, causing the K_M to increase. [After Schirmer, T., and Evans, P. R., *Nature* 343, 142 (1990).]

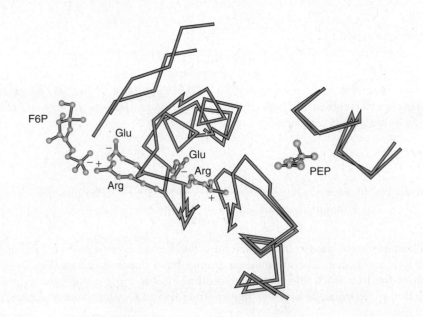

Phosphofructokinase is allosterically inhibited by phosphoenolpyruvate, but it can also be allosterically activated by ADP, a **positive effector** of the enzyme. Although ADP is a product of the phosphofructokinase reaction, it is also a general signal of the cell's need for more ATP since the metabolic consumption of ATP yields ADP:

$$ATP + H_2O \rightarrow ADP + P_i$$

Because phosphofructokinase catalyzes step 3 of the 10-step glycolytic pathway (one of whose ultimate products is ATP), increasing phosphofructokinase activity can increase the rate of ATP produced by the pathway as a whole.

Interestingly, the activator ADP binds not to the active site (which accommodates the substrate ATP and the reaction product ADP) but to the same site where the inhibitor phosphoenolpyruvate binds. But because ADP is much larger than phosphoenolpyruvate, the enzyme cannot close around it, and the conformational change to the low-activity T state cannot occur. Instead, ADP binding forces Arg 162 to remain where it can stabilize fructose-6-phosphate binding (in other words, it helps keep the enzyme in the high-activity R state). The overall result is that ADP counteracts the inhibitory effect of phosphoenolpyruvate and boosts phosphofructokinase activity.

Several factors may influence enzyme activity

So far, we have examined how small molecules that bind to an enzyme inhibit (or sometimes activate) that enzyme. These relatively simple phenomena are not the only means for regulating enzyme activity *in vivo*. Listed below and in **Figure 7-17** are some additional mechanisms. Keep in mind that several of these mechanisms, along with enzyme inhibition or activation, may operate in concert to precisely adjust the activity of a given enzyme.

1. A change in the rate of an enzyme's synthesis or degradation can alter the amount of enzyme available to catalyze a reaction (since $V_{max} = k_{cat}[E]_T$; Equation 7-23).
2. A change in subcellular location, for example, from an intracellular membrane to the cell surface, can bring an enzyme into proximity to its substrate and thereby increase reaction velocity. The opposite effect—sequestering an enzyme away from its substrate—dampens the reaction velocity.
3. An ionic "signal" such as a change in pH or the release of stored Ca^{2+} ions can activate or deactivate an enzyme by altering its conformation.
4. Covalent modification of an enzyme can affect the enzyme's K_M or k_{cat}, just as with an allosteric activator or inhibitor. Most commonly, a phosphoryl ($-PO_3^{2-}$) group or a fatty acyl (lipid) group is added to an enzyme so as to alter its catalytic activity. The effects of covalent modification are actually considered to be reversible, since cells contain enzymes that catalyze the removal of the modifying group as well as its addition. As we will see in Chapter 10 on signaling, phosphorylation and dephosphorylation can dramatically alter the activities of certain proteins.

> **CONCEPT REVIEW**
> - Why does an agent that chemically modifies an amino acid residue sometimes act as an irreversible enzyme inhibitor?
> - What is the meaning of K_I and how is it determined?
> - How does a competitive inhibitor affect K_M and V_{max}?
> - Why do some reaction products and transition state analogs act as competitive enzyme inhibitors?
> - How do noncompetitive, mixed, and uncompetitive inhibitors affect K_M and V_{max}?
> - How can noncompetitive, mixed, and uncompetitive inhibition be distinguished from competitive inhibition?
> - How is phosphofructokinase allosterically inhibited and activated?
> - List all the ways a cell could regulate the activity of an enzyme.

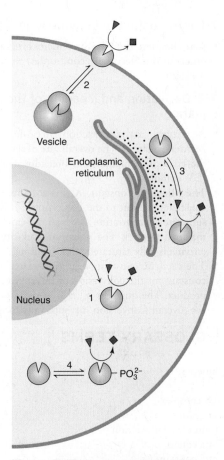

Figure 7-17 Some mechanisms for regulating enzyme activity. In this diagram, enzymes are shown as circular shapes, substrates as small triangles, and products as small squares. The amount of enzyme may depend on its rates of synthesis and degradation (1). The reaction velocity may depend on the enzyme's location (2). A signal such as a burst of Ca^{2+} ions released from the endoplasmic reticulum may affect the enzyme's activity (3). Covalent modification, such as phosphorylation, may activate an enzyme; the enzyme may be inactive when dephosphorylated (4).

? Which of the mechanisms shown in the diagram would be fastest (or slowest) to alter an enzyme's activity?

[SUMMARY]

7-1 Introduction to Enzyme Kinetics

- Rate equations describe the velocity of simple unimolecular (first-order) or bimolecular (second-order) reactions in terms of a rate constant.

7-2 Derivation and Meaning of the Michaelis–Menten Equation

- An enzyme-catalyzed reaction can be described by the Michaelis–Menten equation. The overall rate of the reaction is a function of the rates of formation and breakdown of an enzyme–substrate (ES) complex.
- The Michaelis constant, K_M, is a combination of the three rate constants relevant to the ES complex. It is also equivalent to the substrate concentration at which the enzyme is operating at half-maximal velocity. The maximum velocity is achieved when the enzyme is fully saturated with substrate.
- The catalytic constant, k_{cat}, for a reaction is the first-order rate constant for the conversion of the enzyme–substrate complex to product. The quotient k_{cat}/K_M, a second-order rate constant for the overall conversion of substrate to product, indicates an enzyme's catalytic efficiency because it accounts for both the binding and catalytic activities of the enzyme.
- Values for K_M and V_{max} (from which k_{cat} can be calculated) are often derived from Lineweaver–Burk, or double-reciprocal, plots. Not all enzymatic reactions obey the simple Michaelis–Menten model, but their kinetic parameters can still be estimated.

7-3 Enzyme Inhibition

- Some substances react irreversibly with enzymes to permanently block catalytic activity.
- The most common reversible enzyme inhibitors, which may be transition state analogs, compete with substrate for binding to the active site, thereby increasing the apparent K_M.
- Reversible enzyme inhibitors that affect an enzyme's V_{max} may be noncompetitive, mixed, or uncompetitive inhibitors.
- Oligomeric enzymes such as bacterial phosphofructokinase are regulated by allosteric inhibitors and activators.
- The activities of enzymes may also be regulated by changes in enzyme concentration, location, ion concentrations, and covalent modification.

[GLOSSARY TERMS]

kinetics	catalytic constant (k_{cat})	irreversible inhibitor
v	turnover number	suicide substrate
ES complex	k_{cat}/K_M	competitive inhibition
saturation	diffusion-controlled limit	inhibition constant (KI)
unimolecular reaction	catalytic perfection	EI complex
rate equation	Lineweaver–Burk plot	product inhibition
rate constant (k)	bisubstrate reaction	transition state analog
first-order reaction	random mechanism	polyprotein
bimolecular reaction	ordered mechanism	noncompetitive inhibition
second-order reaction	ping pong mechanism	mixed inhibition
steady state	allosteric enzyme	uncompetitive inhibition
Michaelis constant (K_M)	cooperativity	allosteric regulation
v_0	rational drug design	feedback inhibitor
V_{max}	pharmacokinetics	negative effector
Michaelis–Menten equation	clinical trial	positive effector

BIOINFORMATICS | **PROJECT 5** | Examine the interactions between enzymes and drug molecules and explore the significance of single nucleotide polymorphisms in cytochrome P 450.

ENZYME INHIBITORS AND RATIONAL DRUG DESIGN

[PROBLEMS]

7-1 Introduction to Enzyme Kinetics

1. At about the time scientists began analyzing the hyperbolic velocity versus substrate curve (Fig. 7-3), Emil Fischer was formulating his lock-and-key hypothesis of enzyme action (Section 6-3). Explain why kinetic data are consistent with Fischer's model of enzyme action.

2. Explain why it is usually easier to calculate an enzyme's reaction velocity from the rate of appearance of the product rather than the rate of disappearance of the substrate.

3. The rate of hydrolysis of sucrose to glucose and fructose is quite slow in the absence of a catalyst.

$$sucrose + H_2O \rightarrow glucose + fructose$$

If the initial concentration of sucrose is 0.050 M, it takes 440 years for the concentration of the sucrose to decrease by half to 0.025 M. What is the rate of disappearance of sucrose in the absence of a catalyst?

4. If a catalyst is present, the hydrolysis of sucrose (see Problem 3) is much more rapid. If the initial concentration of sucrose is 0.050 M, it takes 6.9×10^{-5} s for the concentration to decrease by half to 0.025 M. What is the rate of disappearance of sucrose in the presence of a catalyst?

5. The uncatalyzed rate of amide bond hydrolysis in hippuryl-phenylalanine was measured by monitoring the formation of the product, phenylalanine. A 30-mM solution of hippurylphenylala-nine was monitored for amide bond hydrolysis for 50 days. At the end of this time, 25 μM phenylalanine was detected in the solution. What is the rate of formation (in units of $M \cdot s^{-1}$) of phenylalanine in this reaction?

6. Draw a graph of the reaction described in Problem 5 in which the velocity of the reaction (measured as the rate of the appearance of phenylalanine per unit time) is plotted as a function of the reactant concentration.

7. The amide bond in hippurylphenylalanine (see Problem 5) is hydrolyzed 4.7×10^{11} times faster in the presence of an enzyme. What is the rate of formation of phenylalanine in the presence of an enzyme?

8. Draw a graph of the reaction described in Problem 7 in which the appearance of phenylalanine is plotted as a function of substrate concentration. Compare this graph to the one you drew in Problem 6. How is this graph different, and why?

9. A bacterial enzyme catalyzes the hydrolysis of maltose as shown in the equation below.

$$\text{maltose} + H_2O \rightarrow 2 \text{ glucose}$$

During an interval of one minute, the concentration of maltose decreases by 65 mM. What is the rate of disappearance of maltose in the enzyme-catalyzed reaction?

10. In the reaction shown in Problem 9, what is the rate of appearance of glucose?

7-2 Derivation and Meaning of the Michaelis–Menten Equation

11. Complete the table for the following one-step reactions:

Reaction	Molecularity	Rate equation	Units of k
$A \rightarrow B + C$			
$A + B \rightarrow C$			
$2 A \rightarrow B$			
$2 A \rightarrow B + C$			

Reaction	Reaction velocity proportional to . . .	Order
$A \rightarrow B + C$		
$A + B \rightarrow C$		
$2 A \rightarrow B$		
$2 A \rightarrow B + C$		

12. The rate of the reaction described in Problem 3 doubles when the concentration of sucrose doubles. The reaction rate is not affected by the concentration of water. Write a rate law for the reaction that is consistent with these observations. What is the order of the reaction?

13. Refer to your rate law written for Problem 12 and Sample Calculation 7-1 to determine the velocity of the uncatalyzed reaction when the concentration of sucrose is 0.050 M. The rate constant k for the uncatalyzed reaction is $5.0 \times 10^{-11} \text{ s}^{-1}$.

14. Repeat the calculation you performed in Problem 13 for the enzyme-catalyzed hydrolysis of sucrose. The rate constant k for the catalyzed reaction is $1.0 \times 10^4 \text{ s}^{-1}$.

15. An unprotonated primary amino group in a blood protein can react with carbon dioxide to form a carbamate as shown here:

$$R-NH_2 + CO_2 \rightarrow R-NH-COO^- + H^+$$
$$\text{Carbamate}$$

The rate constant k for this reaction is $4950 \text{ M}^{-1} \cdot s^{-1}$.
   (a) What is the order of this reaction?
   (b) Calculate the velocity of the reaction of an α-amino group in a blood protein at 37°C if its concentration is 0.6 mM and the partial pressure of carbon dioxide is 40 torr (*Hint*: Convert the units of partial pressure to molar concentration using the ideal gas law. The value of R is 0.0821 L $\cdot$ atm $\cdot$ K^{-1} $\cdot$ mol^{-1}.)
   (c) How would the rate constant for this reaction vary with pH? Explain.

16. Refer to the carbamate reaction in Problem 15. What CO_2 partial pressure is required to yield a velocity of 0.045 M $\cdot$ s^{-1} for the reaction?

17. What portions of the velocity versus substrate curve (Fig. 7-3) correspond to zero-order and first-order processes?

18. Draw curves that show the appropriate relationship between the variables of each plot.

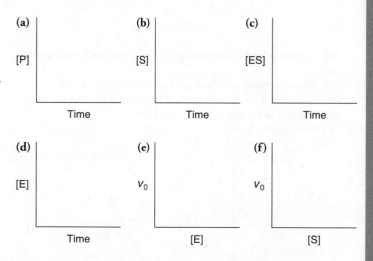

19. You are attempting to determine K_M by measuring the reaction velocity at different concentrations, but you do not realize that the substrate tends to precipitate under the experimental conditions you have chosen. How would this affect your measurement of K_M?

20. You are constructing a velocity versus substrate curve for an enzyme whose K_M is believed to be about 2 μM. The enzyme concentration is 200 nM and the substrate concentrations range from 0.1 μM to 10 μM. What is wrong with this experimental setup and how could you fix it?

21. **(a)** Is it necessary for measurements of reaction velocity to be expressed in units of concentration per unit time (M/s, for example) in order to calculate an enzyme's K_M?
   (b) Is it necessary to know $[E]_T$ in order to determine K_M, V_{max}, or k_{cat}?

22. The enzyme-catalyzed reaction described in Problem 9 has a K_M of 0.135 μM and a V_{max} of 65 μmol $\cdot$ min^{-1}. What is the reaction velocity when the concentration of maltose is 1.0 μM?

23. The enzyme phenylalanine hydroxylase (PAH) catalyzes the hydroxylation of phenylalanine to tyrosine and is deficient in patients with the disease PKU (see Box 18-B). The K_M for PAH is 0.5 mM and the V_{max} is 7.5 μmol $\cdot$ min^{-1} $\cdot$ mg^{-1}. What is the velocity of the reaction when the phenylalanine concentration is 0.15 mM?

24. In the absence of allosteric effectors, the enzyme phospho-fructokinase displays Michaelis–Menten kinetics (see Fig. 7-15). The v_0/V_{max} ratio is 0.9 when the concentration of its substrate, fructose-6-phosphate, is 0.10 mM. Use Equation 7-21 to calculate the K_M for phosphofructokinase under these conditions.

25. Use the plot provided to estimate values of K_M and V_{max} for an enzyme-catalyzed reaction.

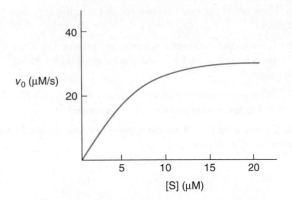

26. Estimate the K_M and the V_{max} for each enzyme from the plot shown.
(a) Which enzyme generates product more rapidly when [S] = 1 mM? What is the relationship between [S] and K_M for each enzyme at this substrate concentration?
(b) Answer part (a) for a [S] of 10 mM.

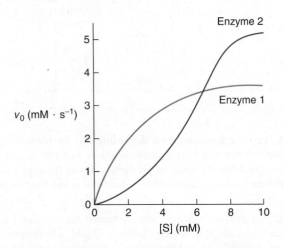

27. What relationship exists between K_M and [S] when an enzyme-catalyzed reaction proceeds at **(a)** 75% V_{max} and **(b)** 90% V_{max}?

28. When [S] = 5 K_M, how close is v_0 to V_{max}? When [S] = 20 K_M, how close is v_0 to V_{max}? What do these results tell you about the accuracy of estimating V_{max} from a plot of v_0 versus [S]?

29. An insect aminopeptidase was purified and its catalytic activity was investigated using an artificial peptide substrate. The V_{max} was 4.0×10^{-7} M·s^{-1} and the K_M was 1.4×10^{-4} M. The enzyme concentration used in the assay was 1.0×10^{-7} M. What is the value of k_{cat}? What is the meaning of k_{cat}?

30. Calculate the catalytic efficiency of the aminopeptidase described in Problem 29.

31. Three enzymatic reactions carried out under different conditions are shown in the figure above right. Estimate the K_M and V_{max} values from the plot. Which reaction has the lowest K_M? Which reaction has the highest V_{max}?

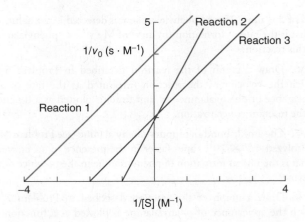

32. The formation of an enzyme–substrate complex for a one-substrate reaction can be expressed as E + S ⇌ ES. For an enzymatic reaction with two substrates, A and B, the expression would be E + A + B ⇌ EAB. Explain why the second process is approached as two bimolecular reactions rather than as a single termolecular (three-reactant) reaction.

33. The K_M values for the reaction of chymotrypsin with two different substrates are given.

Substrate	K_M (M)
N-Acetylvaline ethyl ester	8.8×10^{-2}
N-Acetyltyrosine ethyl ester	6.6×10^{-4}

(a) Which substrate has the higher apparent affinity for the enzyme? Use what you learned about chymotrypsin in Chapter 6 to explain why.
(b) Which substrate is likely to give a higher value for V_{max}?

34. The enzyme hexokinase acts on both glucose and fructose. The K_M and V_{max} values are given in the table. Using these data, compare and contrast the interaction of hexokinase with both of these substrates.

Substrate	K_M (M)	V_{max} (relative)
Glucose	1.0×10^{-4}	1.0
Fructose	7.0×10^{-4}	1.8

35. Use the data provided to determine whether the reactions catalyzed by enzymes A, B, and C are diffusion-controlled.

Enzyme	Reaction	K_M	K_{cat}
A	S → P	0.3 mM	5000 s^{-1}
B	S → Q	1 nM	2 s^{-1}
C	S → R	2 μM	850 s^{-1}

36. Refer to the data presented in Problem 35. A reaction is carried out in which 5 nM substrate S is added to a reaction mixture containing equivalent amounts of enzymes A, B, and C. After 30 seconds, which product will be more abundant: P, Q, or R?

37. The enzyme γ-glutamylcysteine synthetase (γ-GCS) from the protozoan *Trypanosoma brucei*, the parasite that causes African sleeping sickness, catalyzes the first step of the biosynthesis of trypanothione, a compound the parasite requires to maintain proper redox balance.

$$\text{aminobutyric acid + glutamate + ATP}$$

$$\xrightarrow{\gamma\text{-GCS}} \rightarrow \rightarrow \text{trypanothione}$$

The K_M values for each of the substrates have been measured.

Substrate	K_M (mM)
Glutamate	5.9
Aminobutyric acid	6.1
ATP	1.4

(a) Does this reaction obey Michaelis–Menten kinetics?
(b) Describe how the K_M values for each of the three substrates were determined.
(c) How would V_{max} be achieved for the γ-GCS reaction?

38. A galactosyl transferase, β3-GalT, has recently been purified. β3-GalT catalyzes the reaction shown below, an early step in the production of glycoproteins important in immune recognition.

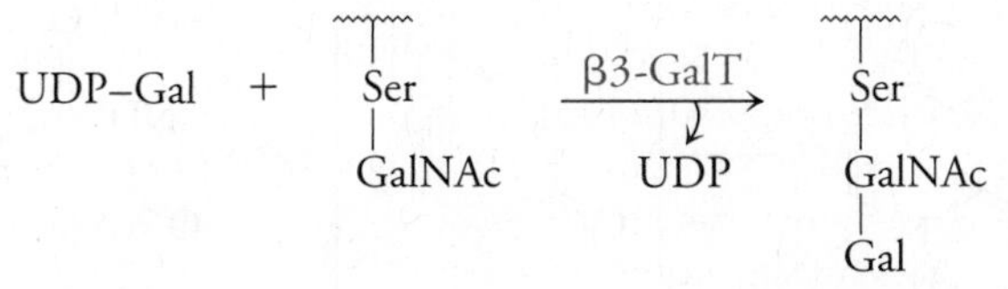

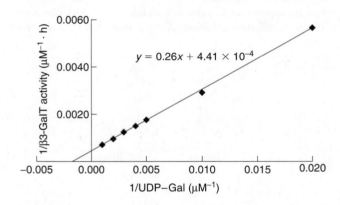

The reaction velocity was measured in the presence of UDP-[³H]Gal. Use the Lineweaver–Burk plot provided to determine the K_M and the V_{max} for β3-GalT.

$y = 0.26x + 4.41 \times 10^{-4}$

39. Subtilisin, an alkaline serine protease produced by *Bacillus* bacteria, is used in the detergent and food-processing industries. Protein engineers constructed a mutant subtilisin enzyme in an attempt to improve its catalytic activity. The isoleucine at position 31 (adjacent to the Asp residue in the Asp–His–Ser catalytic triad) was replaced with leucine. Both the mutant and wild-type enzymes were assayed for catalytic activity using the artificial substrate *N*-succinyl-Ala-Ala-Pro-Phe-*p*-nitroanilide (AAPF). The results are shown in the table.

Enzyme	K_M (mM)	k_{cat} (s⁻¹)	k_{cat}/K_M (mM⁻¹ · s⁻¹)
Ile 31 wild-type subtilisin E	1.9 ± 0.2	21 ± 4	11
Leu 31 mutant subtilisin E	2.0 ± 0.3	120 ± 15	60

(a) Explain how reaction velocity is measured using the AAPF substrate.
(b) What effect did the replacement of Leu for Ile at position 31 have on the catalytic activity of subtilisin E for the hydrolysis of AAPF? Comment on the K_M and k_{cat} values.
(c) Both enzymes were tested for their ability to cleave peptide bonds in a "natural" substrate, milk casein. The results are

shown in the table. Compare the activities of the two enzymes for the casein substrate.

Enzyme	Specific activity (units/mg)
Ile 31 wild-type subtilisin E	109 ± 9
Leu 31 mutant subtilisin E	297 ± 30

(d) Why did the Ile → Leu substitution alter the enzyme's catalytic efficiency?
(e) Would subtilisin as a detergent additive be effective in removing protein-based stains (such as milk or blood) from clothing?

40. Aspartate aminotransferase (AspAT) catalyzes the following reaction:

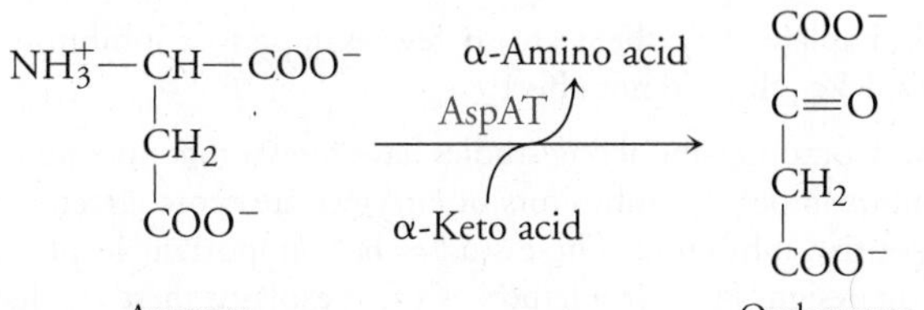

The AspAT enzyme has two active-site arginines, Arg 386 and Arg 292, which interact with the α-carboxylate and β-carboxylate groups on the aspartate substrate, respectively. Mutant AspAT enzymes were constructed in which either or both of the essential arginines were replaced with a lysine residue. The kinetic parameters for the wild-type enzyme and mutant enzymes are shown in the table.

Enzyme	K_M Aspartate (mM)	k_{cat} (s⁻¹)
Wild-type AspAT (Arg 292 Arg 386)	4	530
Mutant AspAT (Lys 292 Arg 386)	326	4.5
Mutant AspAT (Arg 292 Lys 386)	72	9.6
Mutant AspAT (Lys 292 Lys 386)	300	0.055

(a) Compare the ability of aspartate to bind to the wild-type and mutant enzymes.
(b) Why does replacing arginine with lysine affect substrate binding to AspAT?
(c) Evaluate the catalytic efficiency of the wild-type and mutant enzymes.
(d) Why does the replacement of arginine with lysine affect the catalytic activity of the enzyme?

7-3 Enzyme Inhibition

41. Based on some preliminary measurements, you suspect that a sample of enzyme contains an irreversible inhibitor. You decide to dilute the sample 100-fold and remeasure the enzyme's activity. What would your results show if the inhibitor in the sample is **(a)** irreversible or **(b)** reversible?

42. How would diisopropylphosphofluoridate (DIPF) affect the apparent K_M and V_{max} of a sample of chymotrypsin?

43. Inhibitors of acetylcholinesterase, such as edrophonium, are used to treat Alzheimer's disease. The substrate for acetylcholinesterase is acetylcholine. Structures of both molecules are shown.

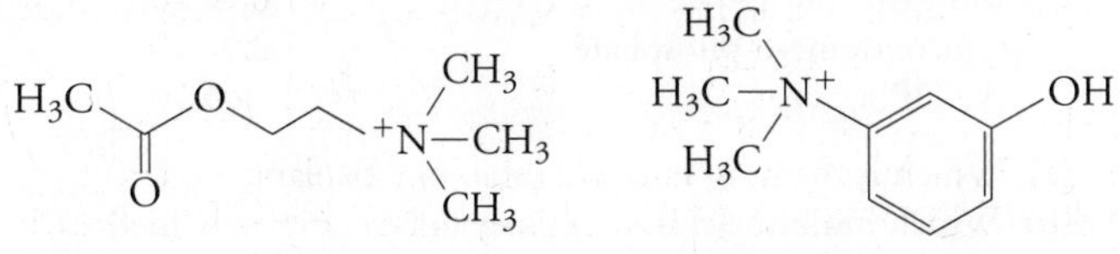

Acetylcholine

Edrophonium

(a) What kind of inhibitor is edrophonium? Explain.

(b) Can inhibition by edrophonium be overcome *in vitro* by increasing the substrate concentration? Explain.

(c) Does this inhibitor bind reversibly or irreversibly to the enzyme? Explain.

44. Would indole be a more effective competitive inhibitor of chymotrypsin, trypsin, or elastase? Explain.

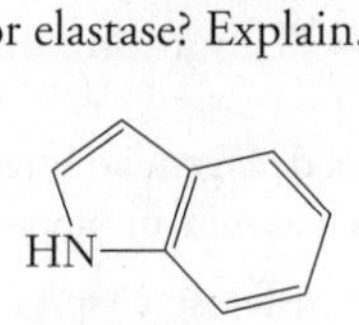

Indole

45. Explain why there are so few examples of inhibitors that decrease V_{max} but do not affect K_M.

46. Computer-modeling studies have shown that uncompetitive and noncompetitive inhibitors of enzymes are more effective than competitive inhibitors. These studies have important implications in drug design. Propose a hypothesis that explains these results.

47. Glucose-6-phosphate dehydrogenase catalyzes the following reaction (see Section 13-4):

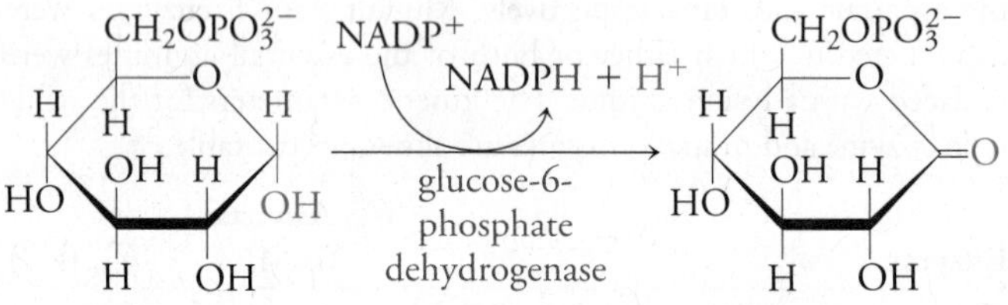

Glucose-6-phosphate   6-Phosphoglucono-δ-lactone

Kinetic data for the enzyme isolated from the thermophilic bacterium *T. maritima* are presented in the following table.

Substrate	K_M (mM)	V_{max} (U · mg^{-1})
Glucose-6-phosphate	0.15	20
NADP$^+$	0.03	20
NAD$^+$	12.0	6

(a) NADPH inhibits glucose-6-phosphate dehydrogenase. What kind of inhibitor is NADPH?

(b) How are the values of K_M and V_{max} likely to differ from those listed in the table if NADPH is present?

(c) Does glucose-6-phosphate dehydrogenase prefer NAD$^+$ or NADP$^+$ as a cofactor?

48. Glucose-6-phosphate dehydrogenase is also found in yeast, and it catalyzes the same reaction shown in Problem 47. The K_M for glucose-6-phosphate in yeast is 2.0×10^{-5} M. The K_M for NADP$^+$ is 2.0×10^{-6} M. Yeast glucose-6-phosphate dehydrogenase can be inhibited by a number of cellular agents whose K_I values are listed in the following table.

Inhibitor	K_I (M)
Inorganic phosphate	1.0×10^{-1}
Glucosamine-6-phosphate	7.2×10^{-4}
NADPH	2.7×10^{-5}

(a) Which is the most effective inhibitor? Explain.

(b) Which inhibitor(s) is (are) likely to be completely ineffective under normal cellular conditions? Explain.

49. The bacterial enzyme proline racemase catalyzes the interconversion of two isomers of proline:

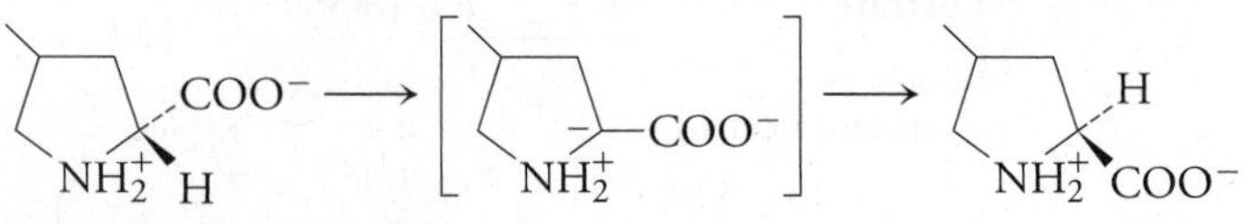

The compound shown below is an inhibitor of proline racemase. Explain why.

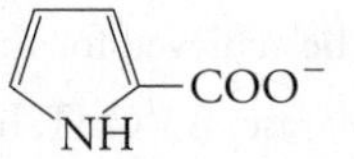

50. Cytidine deaminase catalyzes the following reaction:

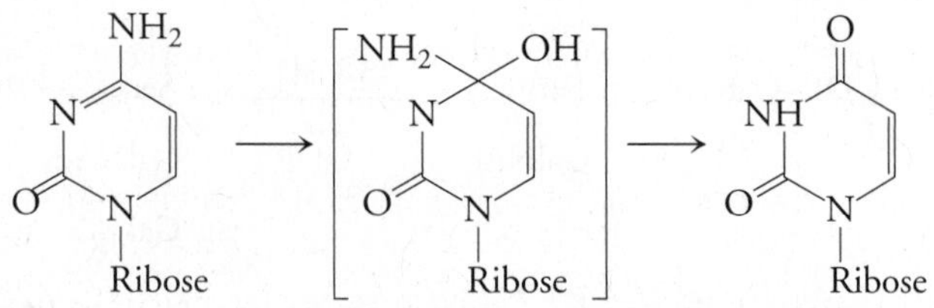

Cytidine   Uridine

Both of the compounds shown below inhibit the reaction. The compounds have K_I values of 3×10^{-5} M and 1.2×10^{-2} M. Assign the appropriate K_I value to each inhibitor. Which compound is the more effective inhibitor? Give a structural basis for your answer.

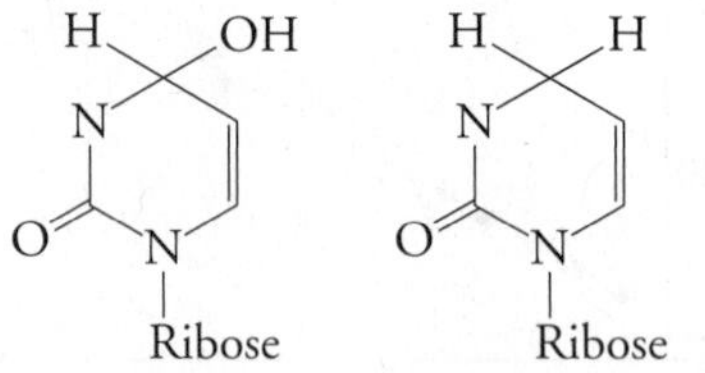

A   B

51. The observation that adenosine deaminase is inhibited by 1,6-dihydroinosine allowed scientists to propose a structure for the transition state of this enzyme. The compound coformycin has also been shown to inhibit adenosine deaminase; its K_I value is about 0.25 μM. Does this observation support or refute the proposed transition state for adenosine deaminase?

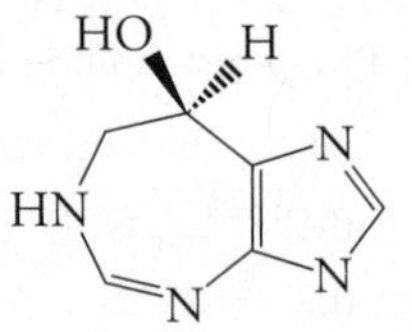

Coformycin

52. The flu virus enzyme neuraminidase hydrolyzes sialic acid residues from cell-surface glycoproteins. This activity helps newly made viruses to escape from the host cell. The drugs oseltamivir (Tamiflu) and zanamivir (Relenza) are transition state analogs that inhibit neuraminidase to block viral infection. Wild-type and mutant strains of the flu virus, in which neuraminidase residue 274 is changed from His to Tyr, exhibit the following kinetic parameters.

	K_M for sialic acid (μM)	V_{max} (relative units)
Wild-type enzyme	6.3	1.0
His274Tyr mutant enzyme	27.0	0.8

	K_I for oseltamivir (nM)	K_I for zanamivir (nM)
Wild-type enzyme	0.32	0.1
His274Tyr mutant enzyme	85	0.19

(a) For the wild-type virus, which drug would work better?
(b) Does the mutation appear to affect substrate binding or turnover?
(c) How does the mutation affect inhibition by oseltamivir and zanamivir? Which drug would be more effective against the mutant flu strain?

53. A phospholipase hydrolyzes its lipid substrate with a K_M of 10 μM and a V_{max} of 7 $\mu mol \cdot mg^{-1} \cdot min^{-1}$. In the presence of 30 μM palmitoylcarnitine inhibitor, the K_M increases to 40 μM and the V_{max} remains unchanged. What is the K_I of the inhibitor?

54. A cysteine protease from a malaria parasite was recently isolated. A compound named P56 was found to inhibit the enzyme. Compounds of this type have the potential to be used as drugs to treat malaria. Data were plotted as shown below.

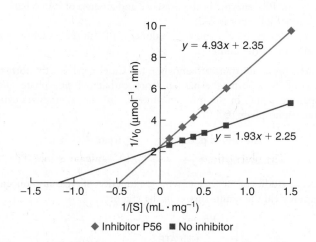

◆ Inhibitor P56 ■ No inhibitor

(a) What kind of inhibitor is P56?
(b) Use the equations of the lines provided to calculate K_M and V_{max} for the enzyme with and without the inhibitor.
(c) Calculate the value of K_I in the presence of 0.22 mM inhibitor.

55. The tyrosinase enzyme catalyzes reactions that produce brown-colored products. In mammals, it is responsible for the production of melanin in the skin; in plants, it is responsible for the browning that occurs when foods such as apples or mushrooms are sliced open and exposed to air. Food scientists interested in preventing the browning process in plant-based foods have shown that dodecyl gallate is an inhibitor of tyrosinase. Using the Lineweaver–Burk plot provided, calculate the V_{max} and K_M for the enzyme in the presence and absence of the inhibitor. What type of inhibitor is dodecyl gallate?

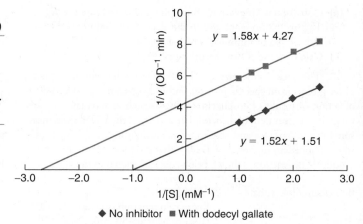

◆ No inhibitor ■ With dodecyl gallate

56. The HIV-1 protease, an important enzyme in the life cycle of the human immunodeficiency virus-1 (HIV-1), is a good drug target for the treatment of HIV and AIDS. A protein produced by the virus, p6*, is an HIV-1 protease inhibitor. The activity of the HIV-1 protease was measured in the presence and absence of p6* using an assay involving an artificial substrate, as shown below.

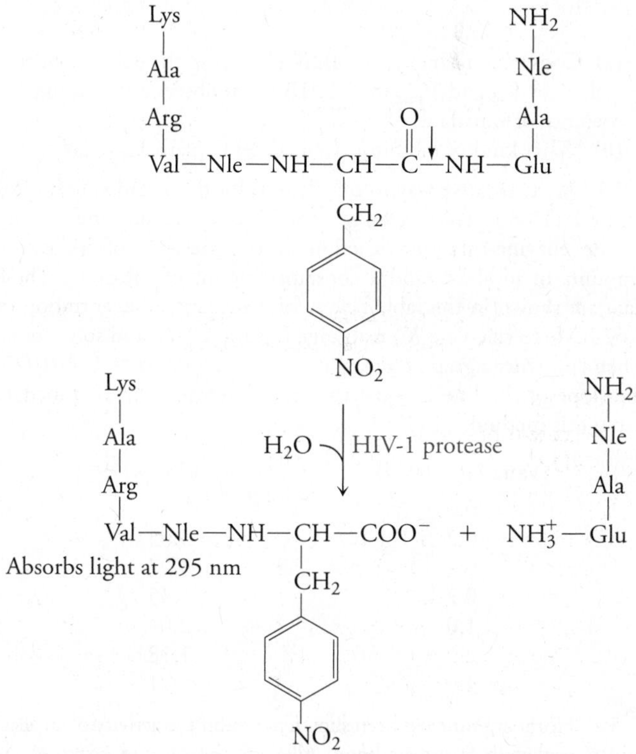

Kinetic assays were carried out in the presence (10 μM) and in the absence of p6*. The data are shown in the table.

[S] (μM)	v_0 without p6* (nmol $\cdot$ min^{-1})	v_0 with p6* (nmol $\cdot$ min^{-1})
10	4.63	2.70
15	5.88	3.46
20	6.94	4.74
25	9.26	6.06
30	10.78	6.49
40	12.14	8.06
50	14.93	9.71

(a) Construct a Lineweaver–Burk plot and determine the K_M and V_{max} in the presence and in the absence of inhibitor.

(b) What type of inhibitor is p6*? Explain.

(c) Calculate the K_I for the inhibitor.

57. The phosphatase enzyme PTP1B catalyzes the removal of phosphate groups from specific proteins and is involved in the mechanism of action of insulin. Phosphatase inhibitors such as vanadate may be useful in the treatment of diabetes. The activity of the PTP1B was measured in the presence and absence of vanadate using an artificial substrate, fluorescein diphosphate (FDP), that produces a light-absorbing product. The reaction is shown below; data are shown in the table.

$$\text{fluorescein diphosphate} \xrightarrow{\text{PTP1B}}$$
$$\text{fluorescein monophosphate} + HPO_4^{2-}$$
$$\text{(absorbs light at 450 nm)}$$

[FDP] (μM)	v_0 without vanadate (nM · s^{-1})	v_0 with vanadate (nM · s^{-1})
6.67	5.7	0.71
10	8.3	1.06
20	12.5	2.04
40	16.7	3.70
100	22.2	8.00
200	25.4	12.5

(a) Construct a Lineweaver–Burk plot using the data provided. Calculate K_M and V_{max} for PTP1B in the absence and in the presence of vanadate.

(b) What kind of inhibitor is vanadate? Explain.

58. An alternative way to calculate K_I for the vanadate inhibitor in the PTP1B reaction (see Problem 57) is to measure the velocity of the enzyme-catalyzed reaction in the presence of increasing amounts of inhibitor and a constant amount of substrate. These data are shown in the table below for a substrate concentration of 6.67 μM. To calculate K_I, rearrange Equation 7-28 and solve for α. Then construct a graph plotting α versus [I]. Since $\alpha = 1 + [I]/K_I$, the slope of the line is equal to $1/K_I$. Determine K_I for vanadate using this method.

[Vanadate] (μM)	v_0 (nM · s^{-1})
0.0	5.70
0.2	3.83
0.4	3.07
0.7	2.35
1.0	2.04
2.0	1.18
4.0	0.71

59. Homoarginine has been shown to inhibit the activity of an alkaline phosphatase found in bone. Alkaline phosphatase enzymes are widely distributed in tissues, and serum levels of these enzymes are often used as a diagnostic tool for various diseases. The artificial substrate phenyl phosphate was incubated with bone alkaline phosphatase in the presence of 4 mM homoarginine. The data are shown in the table.

[Phenyl phosphate] (mM)	v_0, no inhibitor (nM · min^{-1})	v_0, with inhibitor (nM · min^{-1})
4.00	1.176	0.476
2.00	0.909	0.436
1.00	0.667	0.385
0.67	0.556	0.333
0.50	0.455	0.286
0.33	0.345	0.244

(a) Construct a Lineweaver–Burk plot using the data provided. Calculate K_M and V_{max} for the alkaline phosphatase enzyme in the absence and in the presence of homoarginine.

(b) What kind of inhibitor is homoarginine? Explain.

(c) Homoarginine does not inhibit intestinal alkaline phosphatase. Why might homoarginine inhibit bone alkaline phosphatase but not the enzyme found in the intestine?

60. Protein phosphatase 1 (PP1) catalyzes a reaction which yields products that regulate cell division; consequently, PP1 is a possible drug target to treat certain types of cancers. The PP1 enzyme catalyzes the hydrolysis of a phosphate group from myelin basic protein (MBP). The reaction is shown below:

$$\text{MBP-phosphate} \xrightarrow{\text{PP1}} \text{MBP} + P_i$$

The activity of PP1 was measured in the presence and absence of the inhibitor phosphatidic acid (PA). The concentration of PA was 300 nM.

[MBP] (mg/mL)	v_0 without PA (nmol P_i released · mL^{-1} · min^{-1})	v_0 with PA (nmol P_i released · mL^{-1} · min^{-1})
0.010	0.0209	0.00381
0.015	0.0335	0.00620
0.025	0.0419	0.00931
0.050	0.0838	0.0140

(a) Use the data provided to construct a Lineweaver–Burk plot for the PP1 enzyme in the presence and absence of PA. What kind of inhibitor is PA?

(b) Report the K_M and V_{max} values for PP1 in the presence and absence of the inhibitor.

61. Aspartate transcarbamoylase (ATCase) catalyzes the formation of *N*-carbamoyl aspartate from carbamoyl phosphate and aspartate, a step in the multienzyme process that synthesizes cytidine triphosphate (CTP).

$$\text{carbamoyl phosphate} + \text{aspartate} \xrightarrow{\text{ATCase}}$$
$$\textit{N}\text{-carbamoylaspartate} \to \to \to UMP \to \to UTP \to CTP$$

Kinetic studies of ATCase activity as a function of aspartate concentration yield the results shown in the graph.

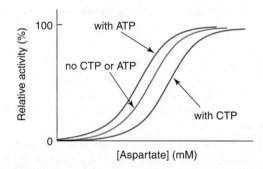

(a) Is ATCase an allosteric enzyme? How do you know?

(b) What kind of an effector is CTP? Explain. What is the biological significance of CTP's effect on ATCase?

(c) What kind of an effector is ATP? Explain. What is the biological significance of ATP's effect on ATCase?

62. The activity of ATCase (see Problem 61) was measured in the presence of several nucleotide combinations. The results are shown in the table below. A value greater than 1 indicates the enzyme was activated, a value less than 1 indicates inhibition.

(a) What combination gives the most effective inhibition?

(b) What is the physiological significance of this combination?

(c) Redraw the graph shown in Problem 61 to include this new information. How does the K_M of the nucleotide combination compare with the values for the nucleotides alone?

Nucleotide effectors	Activity in the presence of 5 mM aspartate
ATP	1.35
CTP	0.43
UTP	0.95
ATP + CTP	0.85
ATP + UTP	1.52
CTP + UTP	0.06

63. The redox state of the cell (the likelihood that certain groups will be oxidized or reduced) is believed to regulate the activities of some enzymes. Explain how the reversible formation of an intramolecular disulfide bond (—S—S—) from two Cys —SH groups could affect the activity of an enzyme.

64. Pyruvate kinase catalyzes the last reaction in the glycolytic pathway (see Fig. 13-2) as shown here.

$$\text{phosphoenolpyruvate (PEP)} + \text{ADP} \rightarrow \text{pyruvate} + \text{ATP}$$

There are four different mammalian forms of pyruvate kinase. All catalyze the same reaction, but they differ in their response to the glycolytic metabolite fructose-1,6-bisphosphate (F16BP).

The activity of the M_1 form of pyruvate kinase was measured at various concentrations of PEP, both in the presence (blue circles) and absence (red circles) of F16BP. The results are shown in the graph above right.

(a) Is the M_1 form of pyruvate kinase an allosteric enzyme? Does F16BP affect the enzyme activity, and, if so, how?

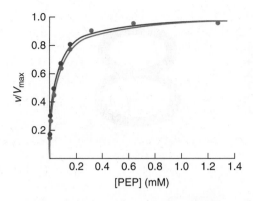

(b) The investigators carried out a site-directed mutagenesis experiment in which they mutated Ala 398 to Arg. The mutated residue was at one of the intersubunit contact positions. The activity of the mutant enzyme, in the presence and absence of F16BP, was measured as described in part (a). The results are shown in the graph below. What effect did the mutation have on the enzyme?

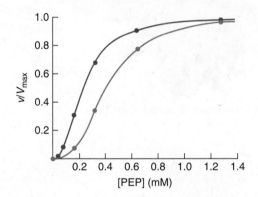

[SELECTED READINGS]

Corey, E. J., Czakó, B., and Kürti, L., *Molecules and Medicine*, Wiley (2007). [Describes the discovery, structure, and action of over a hundred drugs affecting a variety of medical conditions.]

Schramm, V. L., Enzymatic transition state theory and transition state analogue design, *J. Biol. Chem.* **282**, 28297–28300 (2007). [Describes how transition state analog inhibitors work, with examples.]

Segel, I. H., *Enzyme Kinetics*, Wiley (1993). [Focuses on the equations that describe enzyme activity and various forms of inhibition.]

Wlodawer, A., and Vondrasek, J., Inhibitors of HIV-1 protease: A major success of structure-assisted drug design, *Annu. Rev. Biophys. Biomol. Struct.* **27**, 249–284 (1998). [Reviews the development of some HIV-1 protease inhibitors.]

LIPIDS AND MEMBRANES

▶▶ **HOW** can a cell membrane be both flexible and impermeable?

The aquatic organism *Amoeba proteus* (left) is a single cell enclosed by a membrane consisting of lipid and protein. As the cell changes shape, so does the membrane. Yet the membrane continues to function as a barrier, keeping extracellular materials out and preventing the loss of intracellular materials. In this chapter, we'll explore the properties that allow lipids to spontaneously form a bilayer that is both a dynamic structure and a powerful separator of what's inside and what's outside.

[Gerd Guenther/Photo Researchers, Inc.]

THIS CHAPTER IN CONTEXT

Part 1   Foundations

Part 2 Molecular Structure and Function

   8 Lipids and Membranes

Part 3   Metabolism

Part 4   Genetic Information

Do You Remember?

- Cells contain four major types of biological molecules and three major types of polymers (Section 1-2).
- Noncovalent forces, including hydrogen bonds, ionic interactions, and van der Waals forces, act on biological molecules (Section 2-1).
- The hydrophobic effect, which is driven by entropy, excludes nonpolar substances from water (Section 2-2).
- Amphiphilic molecules form micelles or bilayers (Section 2-2).
- A folded polypeptide assumes a shape with a hydrophilic surface and a hydrophobic core (Section 4-3).

All cells—and the various compartments inside eukaryotic cells—are surrounded by membranes. In fact, the formation of a membrane is believed to be a defining event in the evolutionary history of the cell (Section 1-4); without membranes, a cell would be unable to retain essential resources. To begin to understand how membranes work, we will examine them as composite structures containing both lipids and proteins.

The **lipids** that make up the physical structure of the membrane—the **lipid bilayer**—aggregate to form sheets that are impermeable to ions and other solutes. The key to this behavior is the hydrophobicity of the lipid molecules. Hydrophobicity is also a useful feature of lipids that perform other roles, such as energy storage. Although lipids exhibit enormous variety in shape and size and carry out all sorts of biological tasks, they are united by their hydrophobicity.

8-1 Lipids

The molecules that fit the label of *lipid* do not follow a single structural template or share a common set of functional groups, as nucleotides and amino acids do. In fact, *lipids are defined primarily by the absence of functional groups*. Because they consist mainly of C and H atoms and have few if any N- or O-containing functional groups, they lack the ability to form hydrogen bonds and are therefore largely insoluble in water (most lipids are soluble in nonpolar organic solvents). Although some lipids do contain polar or charged groups, the bulk of their structure is hydrocarbon-like.

KEY CONCEPTS
- Lipids are predominantly hydrophobic molecules that can be esterified but cannot form polymers.
- Glycerophospholipids and sphingolipids are amphipathic molecules.
- Cholesterol and other lipids that do not form bilayers have a variety of other functions.

Fatty acids contain long hydrocarbon chains

The simplest lipids are the **fatty acids,** which are long-chain carboxylic acids (at physiological pH, they are ionized to the carboxylate form). These molecules may contain up to 24 carbon atoms, but the most common fatty acids in plants and animals are the even-numbered C_{16} and C_{18} species such as palmitate and stearate:

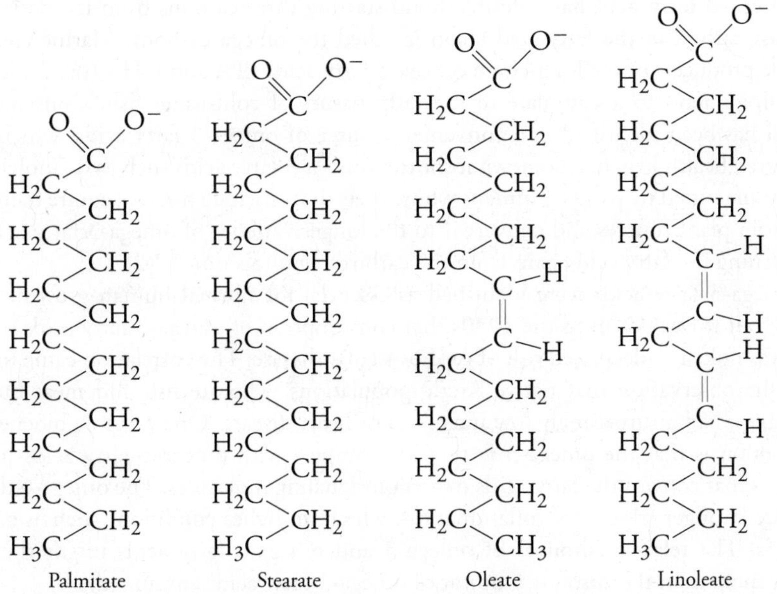

Palmitate   Stearate   Oleate   Linoleate

Such molecules are called **saturated fatty acids** because all their tail carbons are "saturated" with hydrogen. **Unsaturated fatty acids** (which contain one or more double bonds) such as oleate and linoleate are also common in biological systems. In these molecules, the double bond usually has the *cis* configuration (in which the two

[TABLE 8-1] Some Common Fatty Acids

Number of Carbon Atoms	Common Name	Systematic Name[a]	Structure
Saturated fatty acids			
12	Lauric acid	Dodecanoic acid	$CH_3(CH_2)_{10}COOH$
14	Myristic acid	Tetradecanoic acid	$CH_3(CH_2)_{12}COOH$
16	Palmitic acid	Hexadecanoic acid	$CH_3(CH_2)_{14}COOH$
18	Stearic acid	Octadecanoic acid	$CH_3(CH_2)_{16}COOH$
20	Arachidic acid	Eicosanoic acid	$CH_3(CH_2)_{18}COOH$
22	Behenic acid	Docosanoic acid	$CH_3(CH_2)_{20}COOH$
24	Lignoceric acid	Tetracosanoic acid	$CH_3(CH_2)_{22}COOH$
Unsaturated fatty acids			
16	Palmitoleic acid	9-Hexadecenoic acid	$CH_3(CH_2)_5CH{=}CH(CH_2)_7COOH$
18	Oleic acid	9-Octadecenoic acid	$CH_3(CH_2)_7CH{=}CH(CH_2)_7COOH$
18	Linoleic acid	9,12-Octadecadienoic acid	$CH_3(CH_2)_4(CH{=}CHCH_2)_2(CH_2)_6COOH$
18	α-Linolenic acid	9,12,15-Octadecatrienoic acid	$CH_3CH_2(CH{=}CHCH_2)_3(CH_2)_6COOH$
18	γ-Linolenic acid	6,9,12-Octadecatrienoic acid	$CH_3(CH_2)_4(CH{=}CHCH_2)_3(CH_2)_3COOH$
20	Arachidonic acid	5,8,11,14-Eicosatetraenoic acid	$CH_3(CH_2)_4(CH{=}CHCH_2)_4(CH_2)_2COOH$
20	EPA	5,8,11,14,17-Eicosapentaenoic acid	$CH_3CH_2(CH{=}CHCH_2)_5(CH_2)_2COOH$
22	DHA	4,7,10,13,16,19-Docosahexaenoic acid	$CH_3CH_2(CH{=}CHCH_2)_6CH_2COOH$

[a]Numbers indicate the starting position of the double bond; the carboxylate carbon is in position 1.

? **Which of the unsaturated fatty acids listed here are omega-3 fatty acids?**

hydrogens are on the same side). Some common saturated and unsaturated fatty acids are listed in Table 8-1. Although human cells can synthesize a variety of unsaturated fatty acids, they are unable to make any with double bonds past carbon 9 (counting from the carboxylate end). Some organisms can do so, however, producing what are known as omega-3 fatty acids (Box 8-A).

BOX 8-A ⬡ **BIOCHEMISTRY NOTE**

Omega-3 Fatty Acids

An **omega-3 fatty acid** has a double bond starting three carbons from its methyl end (the last carbon in the fatty acid chain is called the omega carbon). Marine algae are notable producers of the long-chain omega-3 fatty acids EPA and DHA (see Table 8-1); these lipids tend to accumulate in the fatty tissues of cold-water fish. Consequently, fish oil has been identified as a convenient source of omega-3 fatty acids, which have purported health benefits. Somewhat shorter omega-3 fatty acids such as α-linoleic acid are manufactured by plants. Humans whose diets don't include fish oil acquire α-linoleic acid from plant sources and convert it to the longer varieties of omega-3 fatty acids by lengthening the fatty acid chain from the carboxyl end (Section 17-2).

Omega-3 fatty acids were identified as essential for normal human growth in the 1930s, but it was not until the 1970s that consumption of omega-3 fatty acids such as EPA was linked to decreased risk of cardiovascular disease. The correlation came to light from the observation that native Arctic populations, who ate fish and meat but few vegetables, had a surprisingly low incidence of heart disease. One possible biochemical explanation is that the omega-3 fatty acids compete with omega-6 fatty acids for the enzymes that convert the fatty acids to certain signaling molecules. The omega-6 derivatives are stronger triggers of inflammation, which underlies conditions such as atherosclerosis. The relative amounts of omega-3 and omega-6 fatty acids might therefore matter more than the absolute amount of omega-3 fatty acids consumed.

The 22-carbon DHA is abundant in the brain and retina, and its concentration decreases with age. DHA is converted *in vivo* to substances that are believed to protect neural tissues from damage following a stroke. However, DHA supplements do not appear to reverse the cognitive decline associated with disorders such as Alzheimer's disease (see Box 4-C).

Numerous studies have attempted to demonstrate the ability of omega-3 fatty acids to prevent or treat other conditions such as arthritis and cancer, but the findings have been mixed, with the omega-3 fatty acid supplements in some cases exacerbating the disease. For this reason, the true role of omega-3 fatty acids in human health demands additional investigation.

● **Question:** Explain why vegans exhibit DHA levels only slightly lower than the levels in individuals who consume large amounts of fish. What does this information reveal about the usefulness of consuming DHA supplements?

Free fatty acids are relatively scarce in biological systems. Instead, they are usually esterified, for example, to glycerol (*right*). The fats and oils found in animals and plants are **triacylglycerols** (sometimes called **triglycerides**) in which the **acyl groups** (the R—CO— groups) of three fatty acids are esterified to the three hydroxyl groups of glycerol (*right*). The ester bond linking each acyl group is the result of a condensation reaction. This is as close as lipids come to forming polymers: They cannot be linked end-to-end to form long chains, as the other types of biological molecules can.

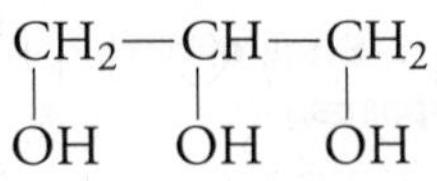

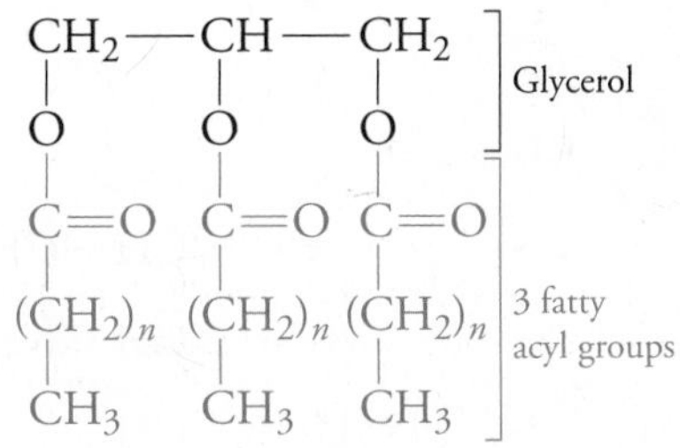

The three fatty acids of a given triacylglycerol may be the same or different. For reasons that are outlined below, triacylglycerols do not form bilayers and so are not important components of biological membranes. However, they do aggregate in large globules, serving as a storage depot for fatty acids that can be broken down to release metabolic energy (these reactions are described in Section 17-1). The hydrophobic nature of triacylglycerols and their tendency to aggregate means that cells can store a large amount of this material without it interfering with other activities that take place in an aqueous environment.

Some lipids contain polar head groups

Among the major lipids of biological membranes are the **glycerophospholipids,** which contain a glycerol backbone with fatty acyl groups esterified at positions 1 and 2 and a phosphate derivative, called a head group, esterified at position 3. As in triacylglycerols, the fatty acyl components of glycerophospholipids vary from molecule to molecule. These lipids are usually named according to their head group, for example,

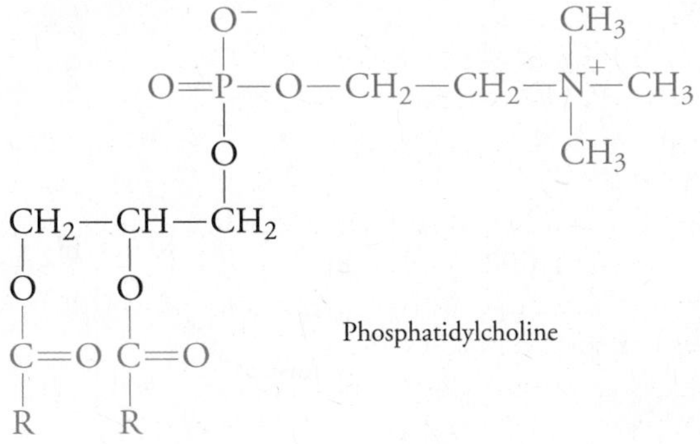

Glycerol

Phosphatidylcholine

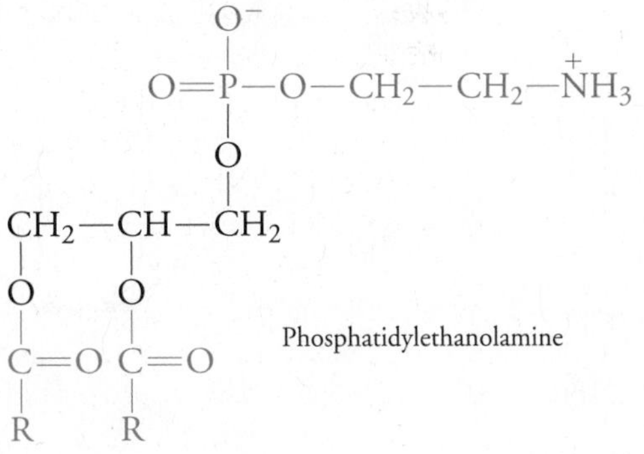

Phosphatidylethanolamine

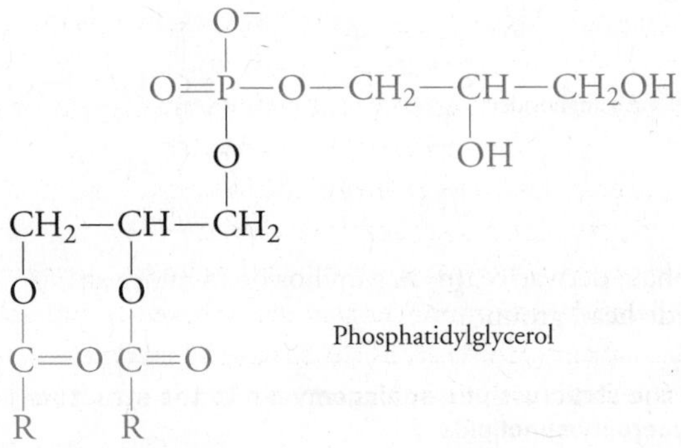

Phosphatidylglycerol

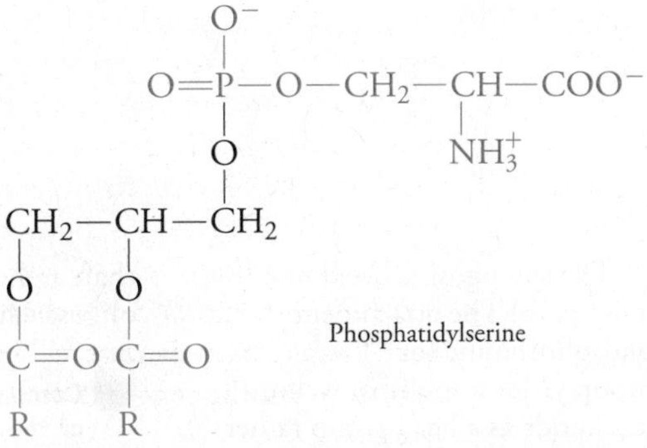

Phosphatidylserine

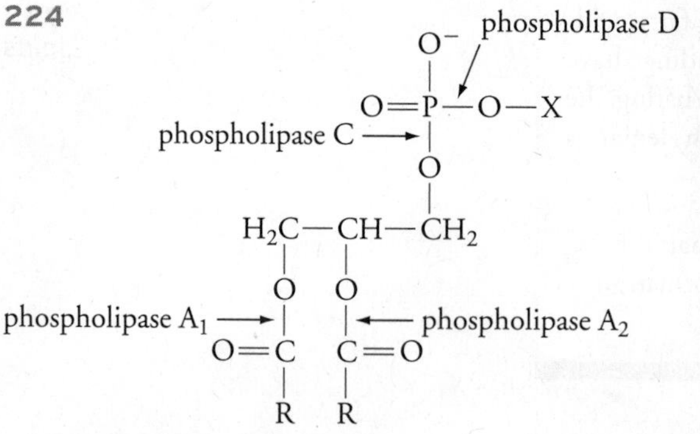

Figure 8-1 Sites of action of phospholipases.

? Which phospholipase-catalyzed reactions release charged portions of the phospholipid?

Note that the glycerophospholipids are not completely hydrophobic: *They are* **amphipathic,** *with hydrophobic tails attached to polar or charged head groups.* As we will see, their structure is ideal for forming bilayers.

The bonds that link the various components of a glycerophospholipid can be hydrolyzed by phospholipases to release the acyl chains or portions of the head group (**Fig. 8-1**). These enzymatic reactions are not just for degrading lipids. Some products derived from membrane lipids act as signaling molecules inside cells or between neighboring cells.

Membranes also contain amphipathic lipids known as **sphingolipids.** The **sphingomyelins,** with phosphocholine or phosphoethanolamine head groups, are sterically similar to their glycerophospholipid counterparts. The major difference is that sphingomyelins are not built on a glycerol backbone. Instead, their basic component is sphingosine, a derivative of serine and the fatty acid palmitate. In a sphingolipid, a second fatty acyl group is attached via an amide bond to the serine nitrogen (**Fig. 8-2**). Some sphingolipids include head groups consisting of one or more carbohydrate groups. These **glycolipids** are known as cerebrosides and gangliosides.

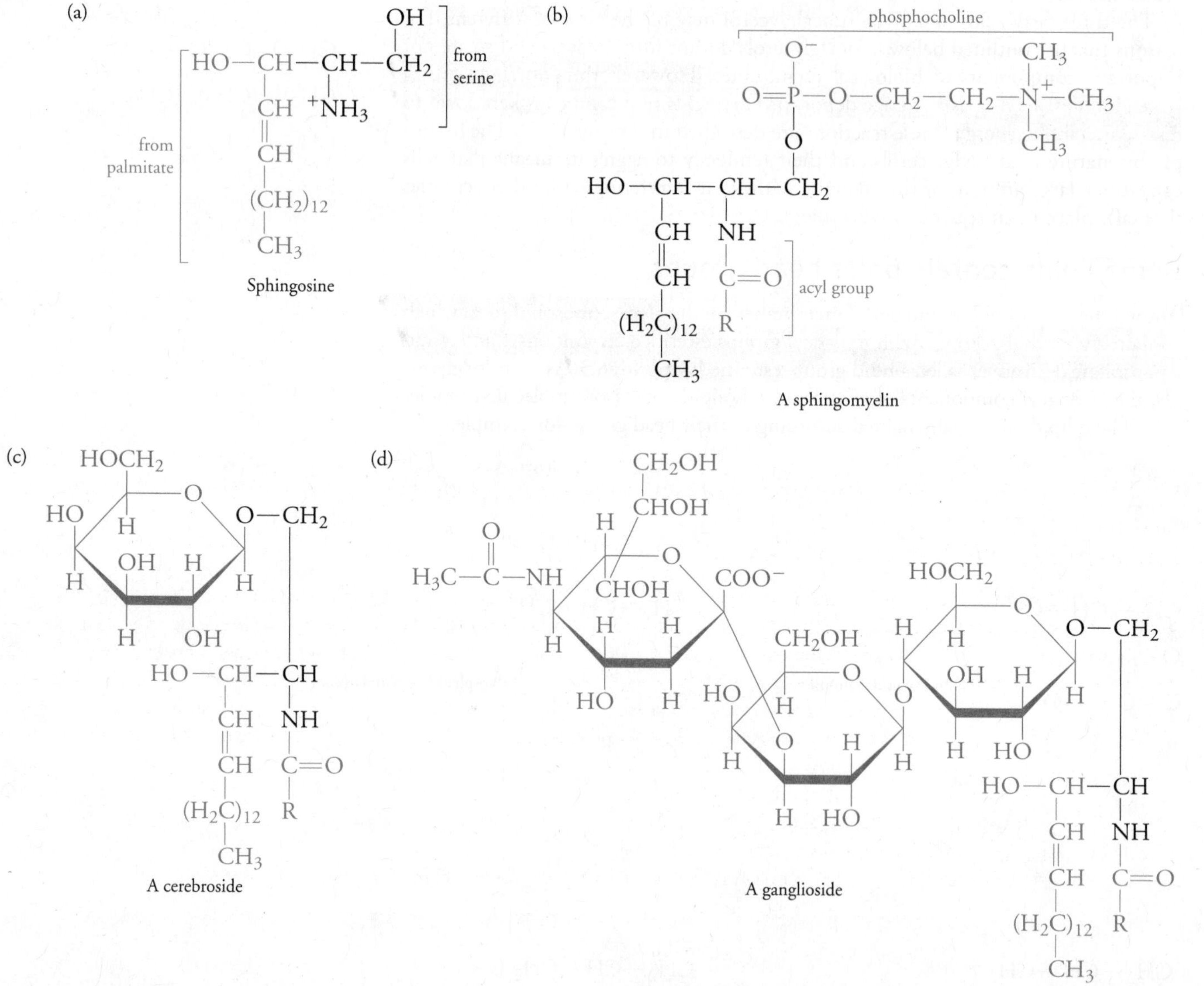

Figure 8-2 Sphingolipids. (a) The sphingosine backbone is derived from serine and palmitate. (b) The attachment of a second acyl group and a phosphocholine (or phosphoethanolamine) head group yields a sphingomyelin. (c) A cerebroside has a monosaccharide as a head group rather than a phosphate derivative. (d) A ganglioside includes an oligosaccharide head group.

? Compare the structure of a sphingomyelin to the structure of the glycerophospholipids.

Lipids perform a variety of physiological functions

In addition to glycerophospholipids and sphingolipids, many other types of lipids occur in membranes and elsewhere in the cell. One of these is cholesterol, a 27-carbon, four-ring molecule:

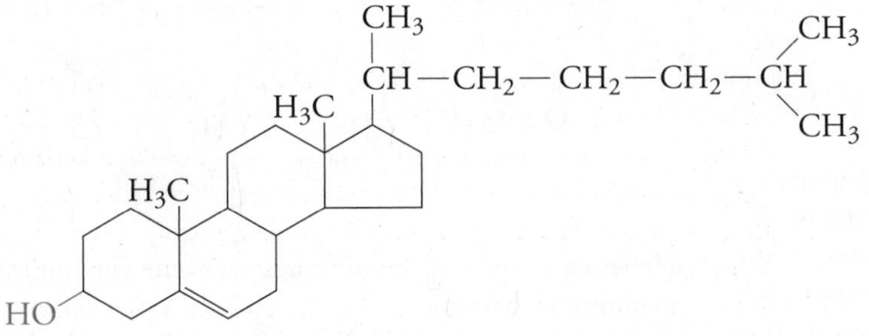

Cholesterol

Cholesterol is an important component of membranes and is also a metabolic precursor of steroid hormones such as estrogen and testosterone.

Cholesterol is one of many types of terpenoids, or **isoprenoids,** lipids that are constructed from 5-carbon units with the same carbon skeleton as isoprene (*right*). For example, the isoprenoid ubiquinone is a compound that is reversibly reduced and oxidized in the mitochondrial membrane (it is described further in Section 12-2):

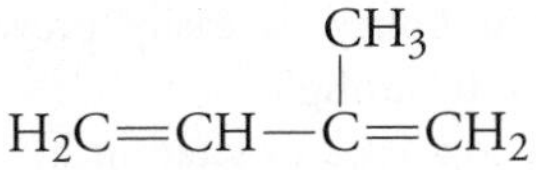

Isoprene

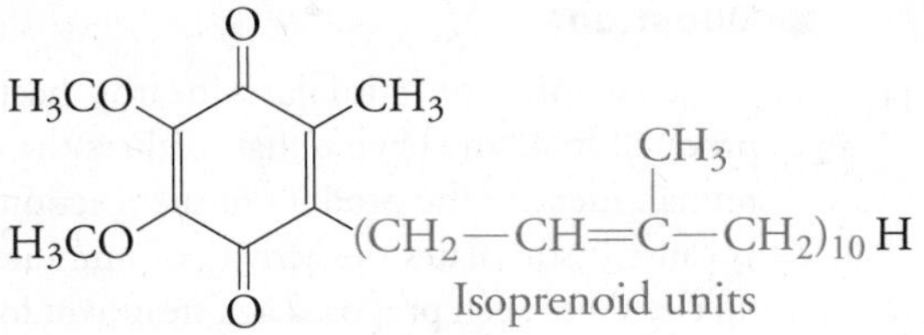

Ubiquinone

The molecules known as vitamins A, D, E, and K are all isoprenoids that perform a variety of physiological roles not related to membrane structure (Box 8-B).

BOX 8-B ⬡ **CLINICAL CONNECTION**

The Lipid Vitamins A, D, E, and K

The plant kingdom is rich in isoprenoid compounds, which serve as pigments, molecular signals (hormones and pheromones), and defensive agents. During the course of evolution, vertebrate metabolism has co-opted several of these compounds for other purposes. The compounds have become **vitamins,** which are substances that an animal cannot synthesize but must obtain from its food. Vitamins A, D, E, and K are lipids, but many other vitamins are water-soluble. Other than being lipids, vitamins A, D, E, and K have little in common.

The first vitamin to be discovered was vitamin A, or retinol.

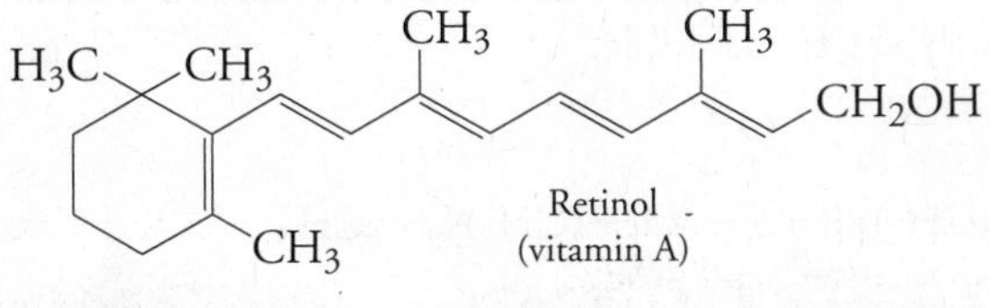

and tomatoes). Retinol is oxidized to retinal, an aldehyde, which functions as a light receptor in the eye. Light causes the retinal to isomerize, triggering an impulse through the optic nerve. A severe deficiency of vitamin A can lead to blindness. The retinol derivative retinoic acid behaves like a hormone by stimulating tissue repair. It is sometimes used to treat severe acne and skin ulcers.

The steroid derivative vitamin D is actually two similar compounds—one (vitamin D_2) derived from plants and the other (vitamin D_3) from endogenously produced cholesterol.

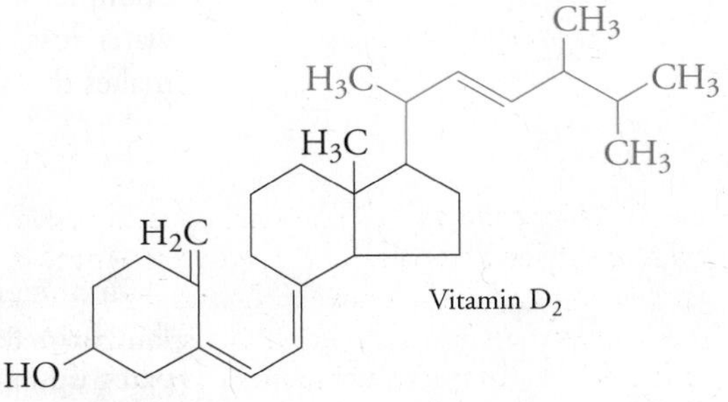

Retinol
(vitamin A)

Vitamin D_2

It is derived mainly from plant pigments such as β-carotene (an orange pigment that is present in green vegetables as well as carrots

Vitamin D₃

Ultraviolet light is required for the formation of vitamins D_2 and D_3, giving rise to the saying that sunlight makes vitamin D. Two hydroxylation reactions carried out by enzymes in the liver and kidney convert vitamin D to its active form, which stimulates calcium absorption in the intestine. The resulting high concentration of Ca^{2+} in the bloodstream promotes Ca^{2+} deposition in the bones and teeth. Rickets, a vitamin D–deficiency disease characterized by stunted growth and deformed bones, is easily prevented by good nutrition and exposure to sunlight.

α-Tocopherol, or vitamin E,

α-Tocopherol
(vitamin E)

is a highly hydrophobic molecule that is incorporated into cell membranes. The traditional view is that it reacts with free radicals generated during oxidative reactions. Vitamin E activity would thereby help prevent the peroxidation of polyunsaturated fatty acids in membrane lipids. However, compounds that are closely related to α-tocopherol in structure do not exhibit this free radical–scavenging activity, and it has been proposed that the observed antioxidant effect of vitamin E instead stems from its activity as a regulatory molecule that may suppress free radical formation by inhibiting the production or activation of oxidative enzymes. In this respect, vitamin E resembles other lipids that function as signaling molecules.

Vitamin K is named for the Danish word *koagulation*. It participates in the enzymatic carboxylation of Glu residues in some of the proteins involved in blood coagulation (see Box 6-B). A vitamin K deficiency prevents Glu carboxylation, which inhibits

the normal function of the proteins, leading to excessive bleeding. Vitamin K can be obtained from green plants, as phylloquinone,

Phylloquinone
(vitamin K)

However, about half the daily uptake of the vitamin is supplied by intestinal bacteria.

Because vitamins A, D, E, and K are water-insoluble, they can accumulate in fatty tissues over time. Excessive vitamin D accumulation can lead to kidney stones and abnormal calcification of soft tissues. High levels of vitamin K have few adverse effects, but extremely high levels of vitamin A can produce a host of nonspecific symptoms as well as birth defects. In general, vitamin toxicities are rare and usually result from overconsumption of commercial vitamin supplements rather than from natural causes.

◆ Questions

1. Bacterial infection stimulates nearby host cells to increase production of an enzyme that oxidizes the aldehyde group of retinal. Identify the product of the reaction.
2. Because it stimulates the activity of immune system cells, retinoic acid has been proposed as a treatment for infections. However, administering retinoic acid can also exacerbate conditions such as arthritis and other inflammatory diseases. Explain.
3. Does vitamin D fit the definition of a vitamin as a substance that an organism requires but cannot synthesize?
4. Vitamin D deficiency is hypothesized to contribute to the development of multiple sclerosis because the prevalence of this autoimmune disease increases with increasing latitude. Explain.
5. Explain why long-term use of antibiotics may lead to a vitamin K deficiency.
6. Explain why obese individuals require larger amounts of vitamins A, D, E, and K in their diets.
7. Relatively few large-scale clinical trials (see Box 7-A) have been conducted to test the purported health benefits of vitamins. Explain why pharmaceutical companies are reluctant to test vitamins the same way that they test drug candidates.

What are some other functions of lipids that are not used to construct bilayers? Due to their hydrophobicity, some lipids function as waterproofing agents. For example, waxes produced by plants protect the surfaces of leaves and fruits against water loss. Beeswax contains an ester of palmitate and a 30-carbon alcohol that makes this substance extremely water-insoluble.

$$H_3C-(CH_2)_{14}-\overset{\overset{\displaystyle O}{\|}}{C}-O-(CH_2)_{29}-CH_3$$

In humans, derivatives of the C_{20} fatty acid arachidonate are signaling molecules that help regulate blood pressure, pain, and inflammation (Section 10-4). Many plant lipids function as attractants for pollinators or repellants for herbivores. For example, geraniol is produced by many flowering plants (it has a roselike smell). Limonene gives citrus fruits their characteristic odor.

CH3 (Geraniol structure)

Geraniol

Limonene

Capsaicin, the compound that gives chili peppers their "hot" taste, is an irritant to the digestive tracts of many animals, but it is consumed by humans worldwide.

Capsaicin

Its hydrophobicity explains why it cannot be washed away with water. Capsaicin has been used therapeutically as a pain reliever. It appears to activate receptors on neurons that sense both pain and heat; by overwhelming the receptors with a "hot" signal, capsaicin prevents the neurons from receiving pain signals.

> **CONCEPT REVIEW**
> - What are the structural features that distinguish fatty acids, triacylglycerols, glycerophospholipids, sphingomyelins, and cholesterol?
> - How are lipids combined to make larger molecules?
> - List some functions of lipid molecules.

8-2 The Lipid Bilayer

The fundamental component of a biological membrane is the lipid bilayer, *a two-dimensional array of amphipathic molecules whose tails associate with each other, out of contact with water, and whose head groups interact with the aqueous solvent* (Fig. 8-3). The beauty of the bilayer as a barrier for biological systems is that it forms spontaneously (without the input of free energy) due to the influence of the hydrophobic effect, which favors the aggregation of nonpolar groups that are energetically costly to individually hydrate (Section 2-2). In addition, a bilayer is self-sealing and, despite its thinness, it can enclose a relatively vast compartment or an entire cell. Once it has formed, a bilayer is quite stable.

Glycerophospholipids and sphingolipids have the appropriate geometry to form bilayers, whereas fatty acids and triacylglycerols do not (Fig. 8-4). Pure cholesterol cannot form a bilayer on its own because it is entirely hydrophobic except for a single polar hydroxyl group. It is found mostly buried in the hydrophobic region of a membrane, where its planar ring structure inserts among the acyl chains of other lipids. Similarly, other lipids are soluble in membranes although they do not contribute to the overall structure of the bilayer.

The bilayer is a fluid structure

Naturally occurring bilayers are mixtures of many different lipids. This is one reason why *a lipid bilayer has no clearly defined geometry*. Most lipid bilayers have a total thickness of between 30 and 40 Å, with a hydrophobic core about 25 to 30 Å thick. The exact thickness varies according to the lengths of the acyl chains and how they

KEY CONCEPTS
- Bilayer fluidity depends on the length and saturation of its lipids and on the presence of cholesterol.
- Lipid asymmetry is maintained by the slow rate of diffusion between leaflets.

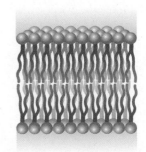

Figure 8-3 **A lipid bilayer.**

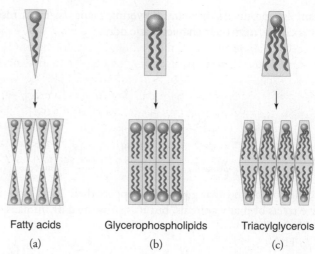

Fatty acids

(a)

Glycerophospholipids

(b)

Triacylglycerols

(c)

Figure 8-4 Bilayer-forming abilities of some lipids. (a) Free fatty acids have a relatively large head group attached to their single hydrophobic tail and therefore cannot align side-by-side to form a bilayer. (b) In most glycerophospholipids, the width of the head group is comparable to the width of the two acyl chains. The lipids can therefore form a bilayer with no voids between lipid molecules. (c) Triacylglycerols, with a small and weakly polar head group, do not form a bilayer.

▶▶ HOW can a cell membrane be both flexible and impermeable?

bend and interdigitate. In addition, the head groups of membrane lipids are not of uniform size, and their distance from the membrane center depends on how they nestle in with neighboring head groups. Finally, the lipid bilayer is impossible to describe in precise terms because it is not a static structure. Rather, it is a dynamic assembly: The head groups bob up and down, and the hydrocarbon tails of the lipids are in constant rapid motion. At its very center, the bilayer is as fluid as a sample of pure hydrocarbon (**Fig. 8-5**). This structure provides an answer to the question posed at the start of the chapter: The bilayer is flexible enough to accommodate changes in cell shape, but its hydrocarbon interior remains intact, as an oily layer between two aqueous compartments.

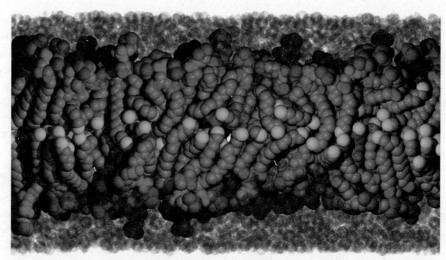

Figure 8-5 Simulation of a lipid bilayer. In this model of a dipalmitoyl phosphatidylcholine bilayer, C atoms are gray (except for the terminal carbon of each lipid tail, which is yellow), ester O atoms are red, phosphate groups are green, and choline head groups are magenta. Water molecules on each side of the bilayer are shown as blue spheres. [Courtesy Richard Pastor and Richard Venable, National Institutes of Health.]

⋯⋯

? Indicate where the molecules shown in Box 8-B would be located.

It is useful to describe the fluidity of a given membrane lipid in terms of its **melting point,** the temperature of transition from an ordered crystalline state to a more fluid state. In the crystalline phase, all the acyl chains in the sample pack together tightly in van der Waals contact. In the fluid phase, the methylene ($-CH_2-$) groups of the acyl chains in the sample can rotate freely. The melting point of a particular acyl chain depends on its length and degree of saturation. For a saturated acyl chain, the melting point increases with increasing chain length. This is because more free energy (a higher temperature) is required to disrupt the more extensive van der Waals interactions between longer chains. A shorter acyl chain melts at a lower temperature because it has less surface area to make van der Waals contacts. A double bond introduces a kink into the acyl chain, so an unsaturated acyl chain is less able to pack efficiently against its neighbors:

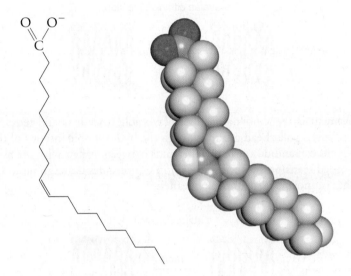

Consequently, the melting point of an acyl chain decreases as the degree of unsaturation increases.

What does all this mean for a mixed bilayer at a constant temperature? In general, *longer acyl chains tend to be less mobile (more crystalline) than shorter acyl chains, and saturated acyl chains are less mobile than unsaturated chains.* Because a fluid membrane is essential for many metabolic processes, organisms endeavor to maintain constant membrane fluidity by adjusting the lipid composition of the bilayer. For example, during adaptation to lower temperatures, an organism may increase its production of lipids with shorter and less saturated acyl chains.

The membranes of most organisms remain fluid over a range of temperatures. This is partly because biological membranes include a variety of different lipids (with different melting points) and do not undergo a sharp transition between liquid and crystalline phases, as a sample of pure lipid would. In addition, cholesterol helps maintain constant membrane fluidity over a range of temperatures through two opposing mechanisms:

1. In a bilayer of mixed lipids, cholesterol's rigid and planar ring system restricts the movements of nearby acyl chains, thereby decreasing membrane fluidity.
2. By inserting between membrane lipids, cholesterol prevents their close packing (that is, their crystallization), which tends to increase membrane fluidity.

Different areas of a naturally occurring membrane may be characterized by different degrees of fluidity. Regions known as membrane **rafts** are thought to contain tightly packed cholesterol and sphingolipids and have a near-crystalline consistency. Certain proteins appear to associate with rafts, so these structures may have functional importance for processes such as transport and signaling. However, the physical characteristics of lipid rafts have been difficult to pin down, and it is possible that such structures may have only a fleeting existence.

Natural bilayers are asymmetric

The two leaflets of the bilayer in a biological membrane seldom have identical compositions. For example, sphingolipids with carbohydrate head groups occur almost exclusively on the outer leaflet of the plasma membrane, facing the extracellular space. The polar head groups of phosphatidylcholine also usually face the cell exterior, whereas phosphatidylserine is usually found in the inner leaflet.

Lipid asymmetry is mostly a consequence of the orientation of lipid-synthesizing enzymes in the endoplasmic reticulum and Golgi apparatus, but the distinct compositions of the inner and outer leaflets are preserved by the extremely slow rate at which most membrane lipids undergo **transverse diffusion,** or **flip-flop** (the movement from one leaflet to the other):

Transverse diffusion (flip-flop)

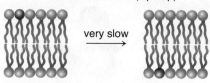

very slow

This movement is thermodynamically unfavorable since it would require the passage of a solvated polar head group through the hydrophobic interior of the bilayer. However, cells can and do move certain lipids between leaflets with the assistance of enzymes called **translocases** or **flippases.** Lipid molecules undergo rapid **lateral diffusion,** that is, movement within one leaflet:

Lateral diffusion

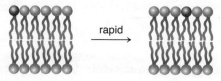

rapid

In a membrane bilayer, a lipid changes places with its neighbors as often as 10^7 times per second. Thus, the image of the bilayer in Figure 8-5 represents a bilayer frozen for an instant in time.

> **CONCEPT REVIEW**
> - Describe the overall structure of a lipid bilayer.
> - How do chain length and saturation influence the melting point of a fatty acid?
> - What is the role of cholesterol in maintaining bilayer fluidity?
> - What are the differences between transverse and lateral diffusion?

8-3 Membrane Proteins

> **KEY CONCEPTS**
> - Integral membrane proteins completely span the bilayer by forming one or more α helices or a β barrel.
> - Covalently attached lipids anchor some proteins in the bilayer.

Biological membranes consist of proteins as well as lipids. On average, a membrane is about 50% protein by weight, but this value varies widely, depending on the source of the membrane. Some bacterial plasma membranes and organelle membranes are as much as three-quarters protein. By itself, a lipid bilayer serves mainly as a barrier to the diffusion of polar substances, and virtually all the additional functions of a biological membrane depend on membrane proteins. For example, some membrane proteins sense exterior conditions and communicate them to the cell interior. Other membrane proteins carry out specific metabolic reactions or function as transporters to move substances from one side of the membrane to the other. Membrane proteins fall into different groups, depending

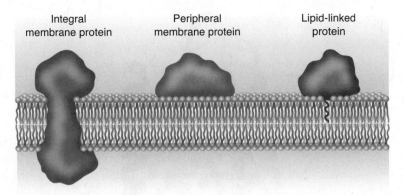

Integral membrane protein   Peripheral membrane protein   Lipid-linked protein

Figure 8-6 Types of membrane proteins. This schematic cross-section of a membrane shows an integral protein spanning the width of the membrane, a peripheral membrane protein associated with the membrane surface, and a lipid-linked protein whose attached hydrophobic tail is incorporated into the lipid bilayer.

? Which type of membrane protein would be easiest to separate from the lipid bilayer?

on how they are specialized for interaction with the hydrophobic interior of the lipid bilayer (**Fig. 8-6**).

Integral membrane proteins span the bilayer

In the membrane proteins known as **integral** or **intrinsic membrane proteins,** a portion of the structure is fully buried in the lipid bilayer. These proteins are customarily contrasted with **peripheral** or **extrinsic membrane proteins,** which are more loosely associated with the membrane via interactions with lipid head groups or integral membrane proteins (see Fig. 8-6). Except for their weak affiliation with the membrane, peripheral proteins are not notably different from ordinary water-soluble proteins.

All but a few integral membrane proteins completely span the lipid bilayer, so they are exposed to the hydrophobic interior as well as the aqueous environment on each side of the membrane. The solvent-exposed portions of an integral membrane protein are typical of other proteins: a polar surface surrounding a hydrophobic core. However, the portion of the protein that penetrates the lipid bilayer must have a hydrophobic surface, since the energetic cost of burying a polar protein group (and its solvating water molecules) is too great.

An α helix can cross the bilayer

One way for a polypeptide chain to cross a lipid bilayer is by forming an α helix whose side chains are all hydrophobic. The hydrogen-bonding tendencies of the amino and carboxyl groups of the backbone are satisfied through hydrogen bonding in the α helix (see Fig. 4-4). The hydrophobic side chains project outward from the helix to mingle with the acyl chains of the lipids.

To span a 30-Å hydrophobic bilayer core, an α helix must contain at least 20 amino acids. A transmembrane helix is often easy to spot by its sequence: It is rich in highly hydrophobic amino acids such as Ile, Leu, Val, and Phe. Polar aromatic groups (Trp and Tyr) and Asn and Gln often occur where the helix approaches the more polar lipid head groups. Highly polar residues such as Asp, Glu, Lys, and Arg often mark the point where the polypeptide leaves the membrane and is exposed to the solvent (**Fig. 8-7**).

Many integral membrane proteins contain several membrane-spanning α helices bundled together (**Fig. 8-8**). These α helices interact much like the left-handed coiled coils in keratin (Section 5-2). Some of the helix–helix interactions involve

(a) Pro–Glu–Trp–Ile–Trp–Leu–Ala–
Leu–Gly–Thr–Ala–Leu–Met–
Gly–Leu–Gly–Thr–Leu–Tyr–Phe–
Leu–Val–Lys–Gly

(b)

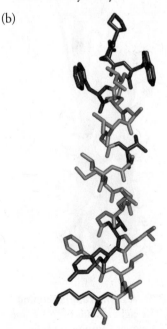

Figure 8-7 A membrane-spanning α helix. (a) A portion of the amino acid sequence from the protein bacteriorhodopsin. (b) Three-dimensional structure of the same sequence. Polar residues are purple and nonpolar residues are green.

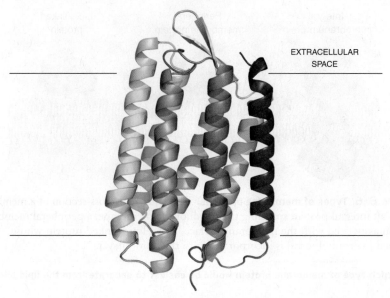

EXTRACELLULAR
SPACE

Figure 8-8 Bacteriorhodopsin. This integral membrane protein consists of a bundle of seven membrane-spanning α helices connected by loops that project into the solution on each side of the membrane. The helices are colored in rainbow order from blue (N-terminus) to red (C-terminus). The horizontal lines approximate the outer surfaces of the membrane. [Structure (pdb 1QHJ) determined by H. Belrhali, P. Nollert, A. Royant, C. Menzel, J. P. Rosenbusch, E. M. Landau, and E. Pebay-Peyroula.]

the electrostatic pairing of polar residues, but the surface of the helix bundle— where it contacts the lipid tails—is predominantly hydrophobic.

A transmembrane β sheet forms a barrel

A polypeptide that crossed the membrane as a β strand would leave its hydrogen-bonding backbone groups unsatisfied. However, if several β strands together form a fully hydrogen-bonded β sheet, they can cross the membrane in an energetically favorable way. In order to maximize hydrogen bonding, the β sheet must close up on itself to form a **β barrel.**

The smallest possible β barrel contains eight strands. The exterior surface of the barrel includes a band, about 22 Å wide, of hydrophobic side chains. This band is flanked on each side by aromatic side chains, which are more polar and form an interface with the lipid head groups **(Fig. 8-9)**. Larger β barrels, containing up to 22 strands, sometimes include a central water-filled passageway that allows small molecules to diffuse from one side of the membrane to the other (Section 9-2).

Because the side chains in a β sheet point alternately to each face, some side chains in a β barrel point into the barrel interior, and others face the lipid bilayer. The absence of a discrete stretch of hydrophobic residues, as in a membrane-spanning α helix, makes it difficult to detect membrane-spanning β strands by examining a protein's sequence.

Lipid-linked proteins are anchored in the membrane

A second group of membrane proteins consists of **lipid-linked proteins.** Many of these are otherwise soluble proteins that are anchored in the lipid bilayer by a covalently attached lipid group. A few lipid-linked proteins also contain membrane-spanning polypeptide segments. In some lipid-linked proteins, a fatty acyl group such as a myristoyl residue (from the 14-carbon saturated fatty acid myristate) is attached via an amide bond to the N-terminal Gly residue of a protein **(Fig. 8-10a)**. Other proteins contain a palmitoyl group (from the 16-carbon palmitate) attached to the sulfur of a Cys side chain via a thioester bond (Fig. 8-10b). Palmitoylation, in contrast to myristoylation, is reversible *in vivo*. Consequently,

Figure 8-9 A membrane-spanning β barrel. The eight strands of this *E. coli* protein, known as OmpX, are fully hydrogen-bonded where they span the width of the bilayer. Hydrophobic side chains (green) on the barrel exterior face the bilayer core. Aromatic residues (gold) are located mostly near the lipid head groups. [Structure (pdb 1QJ9) determined by J. Vogt and G. E. Schulz.]

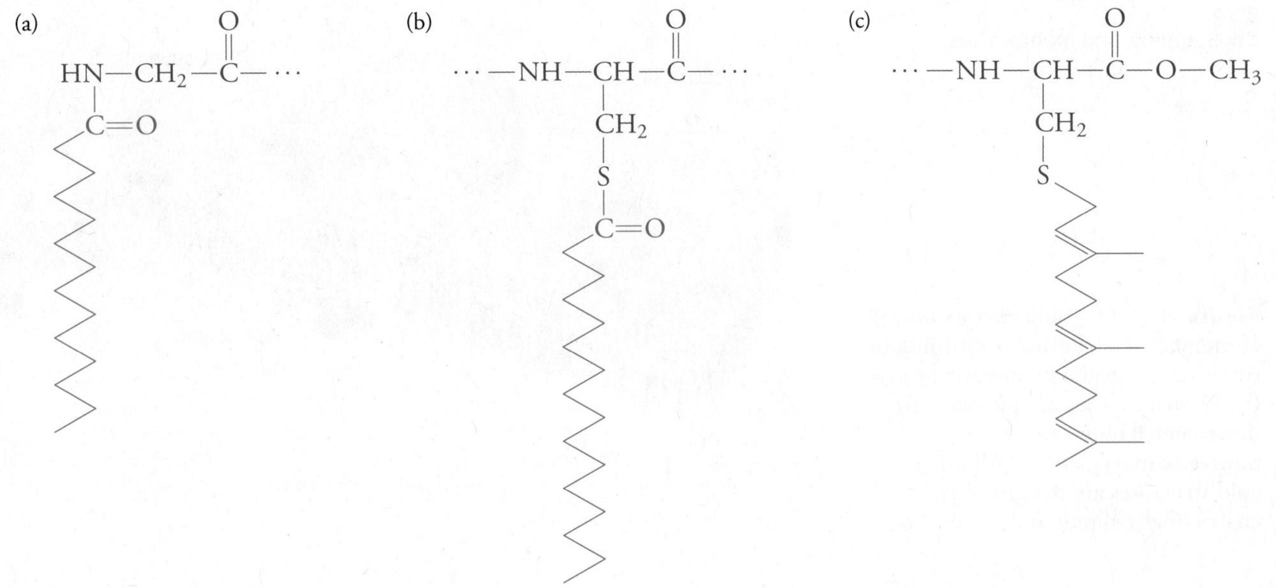

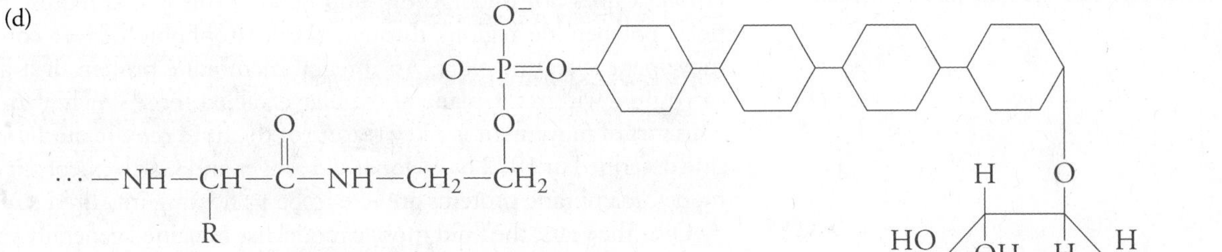

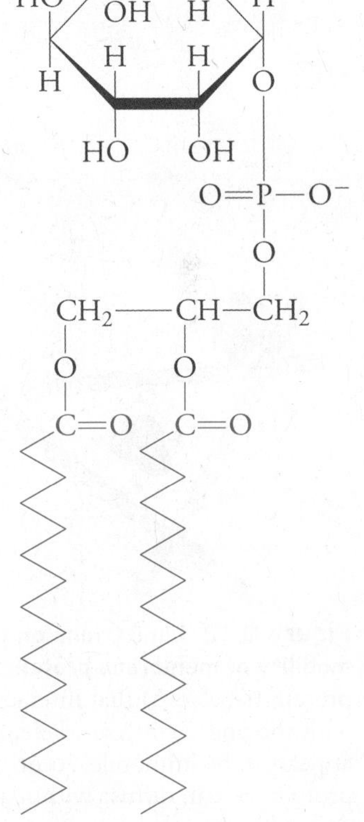

Figure 8-10 Attachment of lipids in some lipid-linked proteins. Proteins are shown in blue, and the lipid anchors are in green. Other groups are purple. (a) Myristoylation. (b) Palmitoylation. (c) Prenylation. The lipid anchor is the 15-carbon farnesyl group. (d) Linkage to a glycosylphosphatidylinositol group. The hexagons represent different monosaccharide residues.

proteins with a myristoyl group are permanently anchored to the membrane, but proteins with a palmitoyl group may become soluble if the acyl group is removed.

Other lipid-linked proteins in eukaryotes are prenylated; that is, they contain a 15- or 20-carbon isoprenoid group linked to a C-terminal Cys residue via a thioether bond (Fig. 8-10c). The C-terminus is also usually carboxymethylated.

Finally, many eukaryotes, particularly protozoans, contain proteins linked to a lipid–carbohydrate group, known as glycosylphosphatidylinositol, at the C-terminus (Fig. 8-10d). These lipid-linked proteins almost always face the external surface of the cell and are often found in sphingolipid–cholesterol rafts.

> **CONCEPT REVIEW**
> * What are the sequence requirements for a membrane-spanning α helix and β barrel?
> * What are the similarities and differences between integral membrane proteins, peripheral membrane proteins, and lipid-linked proteins?

8-4 The Fluid Mosaic Model

Biological membranes consist of both proteins and lipids, although the mixture may not be entirely random. For example, *certain lipids appear to associate specifically with certain proteins,* possibly to stabilize the protein's structure or modulate its function. A given membrane protein has a characteristic orientation; that is, it faces one side of

KEY CONCEPT
* A membrane's structure can be described as proteins diffusing laterally within a lipid bilayer.

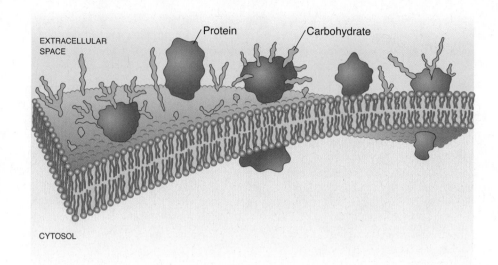

Figure 8-11 The fluid mosaic model of membrane structure. According to this model, integral membrane proteins (blue) float in a sea of lipid and can move laterally but cannot undergo transverse movement (flip-flop). The gold structures are the carbohydrate chains of glycolipids and glycoproteins.

⊕ **See Guided Exploration. Membrane structure and the fluid mosaic model.**

the membrane or the other. After it has assumed its mature conformation and orientation, it does not undergo flip-flop because this would require the passage of large polar polypeptide regions through the hydrophobic bilayer core. However, lateral movement is still possible. An integral membrane protein or a lipid-linked protein can diffuse within the plane of the bilayer, albeit more slowly than a membrane lipid. This sort of movement is a key feature of the **fluid mosaic model** of membrane structure described in 1972 by S. Jonathan Singer and Garth Nicolson. According to their model, membrane proteins are like icebergs floating in a lipid sea (**Fig. 8-11**).

Over the years, the fluid mosaic model has remained generally valid, although it has been refined. For example, *many membrane proteins do not diffuse as freely as first imagined*. Their movements are hindered to some degree according to whether they interact with other membrane proteins or with cytoskeletal elements that lie just beneath the membrane. Thus, a given membrane protein may be virtually immobile (if it is firmly attached to the cytoskeleton), mobile within a small area (if it is confined within a space defined by other membrane and cytoskeletal proteins), or fully free to diffuse (**Fig. 8-12**). The presence of lipid rafts, another feature not described in the original fluid mosaic model, may further define the boundaries for a membrane protein.

Membrane glycoproteins face the cell exterior

Like membrane lipids, membrane proteins are distributed asymmetrically between the two leaflets. For example, most lipid-linked proteins face the cell interior (glycosylphosphatidylinositol-linked proteins are an exception). The exterior face of the membrane in vertebrate cells is rich in carbohydrate-bearing glycolipids (such as cerebrosides and gangliosides) and **glycoproteins.** The oligosaccharide chains (polymers of monosaccharide residues) that are covalently attached to membrane lipids and proteins shroud the cell in a fuzzy coat (see Fig. 8-11). When fully solvated, the highly hydrophilic carbohydrates tend to occupy a large volume.

As we will see in Chapter 11, monosaccharide residues can be linked to each other in different ways and in potentially unlimited sequences. This diversity, present in both glycolipids and glycoproteins, is a form of biological information. For example, the well-known ABO blood group system is based on differences in the composition of the carbohydrate components of glycolipids and glycoproteins on red blood cells (discussed in Box 11-B). Many other cells appear to recognize each other through mutual interactions between membrane proteins.

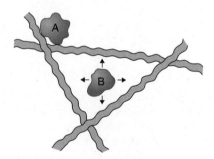

Figure 8-12 Limitations on the mobility of membrane proteins. A protein (labeled A) that interacts tightly with the underlying cytoskeleton appears to be immobile. Another protein (B) can diffuse within a space defined by cytoskeletal proteins. Some proteins appear to diffuse throughout the membrane with no constraints.

CONCEPT REVIEW
- Describe the fluid mosaic model of membrane structure.
- What factors limit membrane protein mobility?
- Why do glycoproteins and glycolipids face the cell exterior?

[SUMMARY]

8-1 Lipids

- Lipids are largely hydrophobic molecules. Fatty acids can be esterified to form triacylglycerols.
- Glycerophospholipids contain two fatty acyl groups attached to a glycerol backbone that bears a phosphate derivative head group. Sphingomyelins are functionally similar but lack a glycerol backbone. Cholesterol and some other lipids are isoprenoids.

8-2 The Lipid Bilayer

- Lipid bilayers are dynamic structures. Their fluidity depends on the length and degree of saturation of their fatty acyl groups: Shorter and less saturated chains are more fluid. Cholesterol helps maintain membrane fluidity over a range of temperatures.

- Membrane lipids can freely diffuse laterally but undergo transverse diffusion very slowly. Membranes may contain crystalline rafts composed of cholesterol and sphingolipids.

8-3 Membrane Proteins

- An integral membrane protein spans the lipid bilayer as one or a bundle of α helices or as a β barrel. Some membrane proteins are anchored in the bilayer by a covalently linked lipid group.

8-4 The Fluid Mosaic Model

- According to the fluid mosaic model, membrane proteins diffuse within the plane of the bilayer. The mobility of proteins may be limited by their interaction with cytoskeletal proteins. Membrane glycolipids and glycoproteins face the cell exterior.

[GLOSSARY TERMS]

lipid	amphipathic	translocase (flippase)
lipid bilayer	sphingolipid	lateral diffusion
fatty acid	sphingomyelin	integral (intrinsic) protein
saturated fatty acid	glycolipid	peripheral (extrinsic) protein
unsaturated fatty acid	isoprenoid	β barrel
omega-3 fatty acid	vitamin	lipid-linked protein
triacylglycerol (triglyceride)	melting point	fluid mosaic model
acyl group	raft	glycoprotein
glycerophospholipid	transverse diffusion (flip-flop)	

[PROBLEMS]

8-1 Lipids

1. The structures of fatty acids can be represented in a shorthand form consisting of two numbers separated by a colon. The first number is the number of carbons; the second number is the number of double bonds. For example, palmitate would be represented by the shorthand 16:0. For unsaturated fatty acids, the quantity n-x is used, where n is the total number of carbons and x is the last double-bonded carbon counting from the methyl end. Unless indicated otherwise, it is assumed that the double bonds are *cis* and that one methylene group separates each double bond. For example, oleate would be represented by the shorthand 18:1n-9. Using the shorthand form as a guide, draw the structures of the following fatty acids:

(a) myristate, 14:0

(b) palmitoleate, 16:1n-7

(c) α-linolenate, 18:3 n-3

(d) nervonate, 24:1 n-9

2. Fish oils contain the fatty acids EPA and DHA. People who eat at least two fish meals a week have a lower incidence of cardiovascular disease because of the positive physiological effects of these lipids. Use the shorthand form described in Problem 1 to draw the structures of (a) EPA (eicosapentaenoate, 20:5 n-3) and (b) DHA (docosahexaenoate 22:6 n-3).

3. Some species of plants contain desaturase enzymes capable of producing fatty acids not found in the animal kingdom. An example of an unusual fatty acid is sciadonate, which is designated as 20:3 $\Delta^{5,11,14}$. This is a different shorthand nomenclature than that described in Problem 1. The superscripts refer to the position of the double bonds beginning at the carboxyl end. This form of shorthand nomenclature is less common but is sometimes used when the positions of the double bonds do not conform to the pattern described in Problem 1. Using this form of shorthand as a guide, draw the structure of sciadonate.

4. Several hundred unusual fatty acids have been found in plants. Some of these fatty acids are unusually short or long; some have double bonds in unexpected positions (see Problem 3); others have additional functional groups. These fatty acids are important as starting materials in the industrial synthesis of lubricants and polymers. Using what you have learned about shorthand notation in Problem 3, draw the structures of the following unusual plant fatty acids:

(a) erucic acid (22:1 Δ^{13})

(b) calendic acid (18:3 $\Delta^{8 \text{ trans},10 \text{ trans}, 12 \text{ cis}}$)

(c) ricinoleic acid (12-hydroxy-18:1 Δ^9)

5. A class of fatty acids called demospongic fatty acids was so named because of their occurrence in Demospongia sponges; however, it has since been discovered that these fatty acids have a wider distribution. Use the shorthand method described in Problem 3 as a guide to draw the structures of the following lipids found in marine mollusks:

(a) 24:2 $\Delta^{5,9}$

(b) 24:4 $\Delta^{5,9,15,18}$

6. Minerval (2-hydroxyoleate) is a fatty acid that has been shown to induce apoptosis (a form of programmed cell death) when added to leukemia cells in culture. Draw the structure of minerval.

7. *Trans* fatty acids occur naturally in beef and milk products but are also produced when oils undergo partial hydrogenation to convert the liquid oil into a semi-solid fat. Draw the structure of elaidic acid ($18:1 \, \Delta^{9 \, trans}$). How does the melting point of elaidic acid compare with the melting point of oleic acid ($18:1 \, \Delta^9$) in which the double bond is in the *cis* configuration?

8. Marine organisms are also good sources of unusual fatty acids, which have potential therapeutic properties. A monoacylglycerol isolated from a sponge contained 10-methyl-9-*cis*-octadecenoic acid as its component fatty acid, esterified to C1 of glycerol. Draw the structure of the monoacylglycerol.

9. When certain nutrients are limiting, some marine phytoplankton can change their membrane lipid composition, producing substitute lipids such as sulfoquinovosyldiacylglycerol (SQDG).

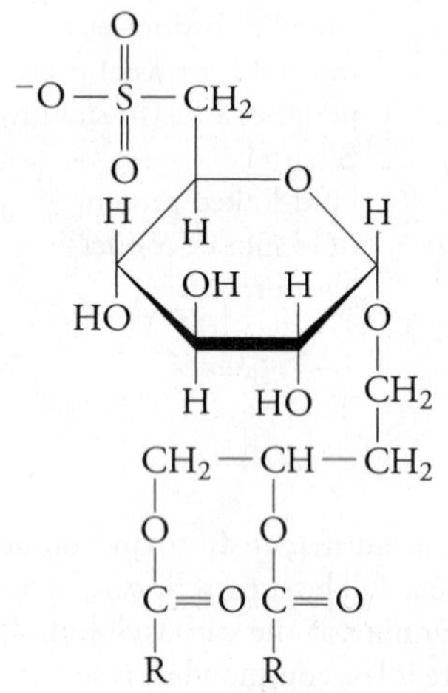

Sulfoquinovosyldiacylglycerol
(SQDG)

(a) Is SQDG more likely to substitute for phosphatidylethanolamine or phosphatidylglycerol?
(b) What element must be in short supply to induce the organism to increase its synthesis of SQDG?

10. Draw the structure of tripalmitin, a triacylglycerol containing three palmitate groups.

11. Phosphatidylinositols (PI) are glycerophospholipids important in cell signaling. Draw the structure of a PI, given the structure of *myo*-inositol. The hydroxyl group involved in the formation of the bond between the inositol and the phosphate group is circled.

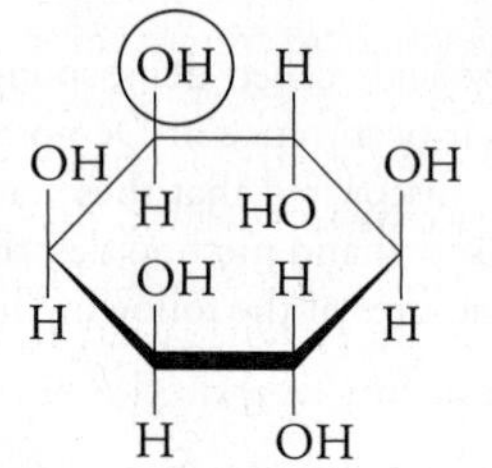

myo-Inositol

12. Some signaling pathways generate signal molecules derived from phosphatidylinositol that has been phosphorylated at multiple sites. How many additional phosphate groups can potentially be attached to phosphatidylinositol?

13. Dipalmitoylphosphatidylcholine (DPPC) is the major lipid of lung surfactant, a protein–lipid mixture essential for pulmonary function. Surfactant production in the developing fetus is low until just before birth, so infants may develop respiratory difficulties if born prematurely. Draw the structure of DPPC.

14. An unusual sphingosine variant has recently been isolated from the nerve of the squid *Loligo pealeii*. Its chemical name is 2-amino-9-methyl-4,8,10-octadecatriene-1,3-diol. Draw the structure of this sphingosine variant.

15. Complex lipids in mammalian skin serve as a waterproof layer. One of these lipids is a glucocerebroside in which the amide-linked acyl group has 28 carbons and an omega hydroxyl group to which linoleate is esterified. Draw the structure of this lipid.

16. The lipid in Problem 15 undergoes hydrolysis to remove the glucose and linolenate groups, followed by linkage of the omega hydroxyl group to the side chain of a protein Glu residue. Draw the structure of the protein–lipid complex.

17. Which of the glycerophospholipids shown in Section 8-1 have hydrogen-bonding head groups?

18. Which of the glycerophospholipids shown in Section 8-1 are charged? Which are neutral?

19. In some autoimmune diseases, an individual develops antibodies that recognize cell constituents such as DNA and phospholipids. Some of the antibodies actually react with both DNA and phospholipids. What is the biochemical basis for this cross-reactivity?

20. The points of attack of several phospholipases are indicated in Figure 8-1. Draw the products of the following reactions:
(a) phosphatidylserine + phospholipase A_1
(b) phosphatidylcholine + phospholipase C
(c) phosphatidylglycerol + phospholipase D

21. Spicy Indian dishes flavored with hot peppers are often served with a side dish made from whole-milk yogurt. Why is a spoonful of yogurt preferable to a drink of water after a mouthful of spicy food?

22. Olestra® is a synthetic lipid that passes through the intestinal tract undigested and is used to make "low-calorie" chips and snacks. Prior to its approval, the FDA required Procter & Gamble, which markets Olestra®, to add vitamins A, D, and K to products containing the synthetic lipid. Explain.

23. A nutrition study was carried out in volunteers who consumed salads with and without added avocado (a rich source of monounsaturated lipids). Blood samples drawn from the volunteers showed a dramatic increase in β-carotene (an orange pigment found in plants that is processed by cells to yield vitamin A) following the consumption of salad containing avocado. Explain these findings.

24. Why does the consumption of excessive amounts of vitamins D and A cause adverse health effects, whereas consumption of vitamin C (structure shown in Box 5-D) well in excess of the recommended daily allowance generally does not lead to toxicity?

8-2 The Lipid Bilayer

25. Classify the following molecules as polar, nonpolar, or amphipathic:

(a) $CH_3CH_2(CH{=}CHCH_2)_3(CH_2)_6COO^-$

(b)
$$
\begin{array}{l}
CH_2{-}OH \\
HC{-}OH \\
CH_2{-}OH
\end{array}
$$

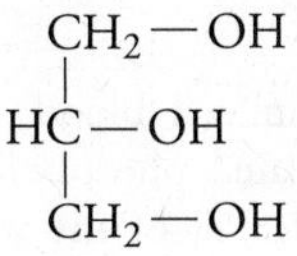

(c)

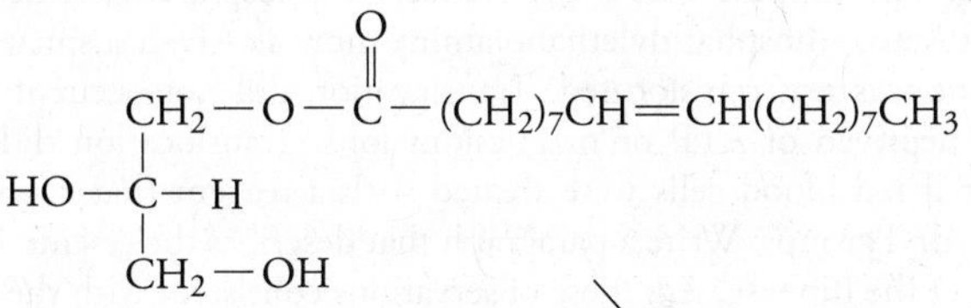

(d)

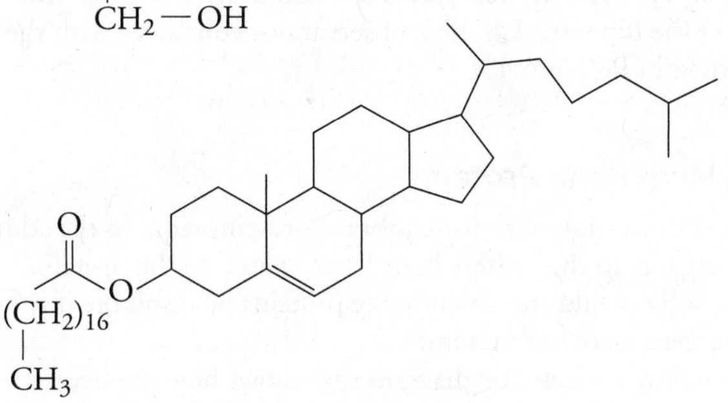

(e)

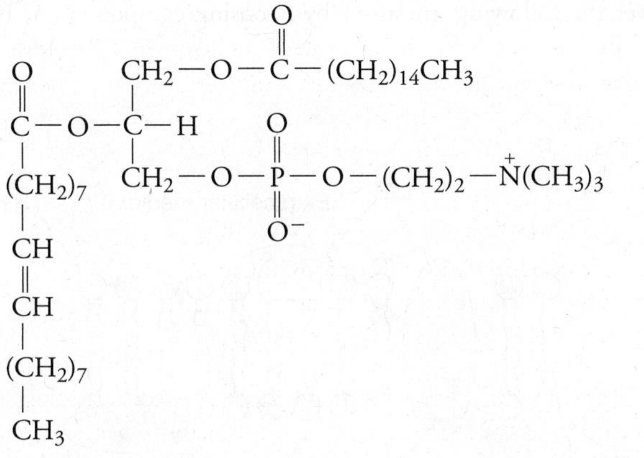

26. Which molecules in Problem 25 can form bilayers? For the molecules that do not form bilayers, explain why not.

27. The lipid shown below is a plasmalogen.

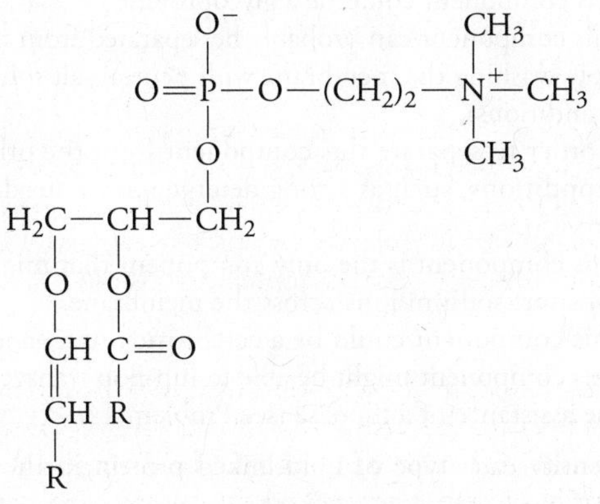

A plasmalogen

(a) How does it differ from a glycerophospholipid?

(b) Would the presence of this lipid have a dramatic effect on a bilayer that contained only phosphatidylcholine?

28. Use a simple diagram such as the one in Figure 8-4 to show why bilayer curvature would be affected by replacing a glycerophospholipid bearing two saturated acyl chains with one bearing two highly unsaturated acyl chains.

29. Why can't triacylglycerols form a lipid bilayer?

30. Red blood cells lyse (break apart) when treated with phospholipase A_1, an enzyme found in the venom of many poisonous insects. Why does treatment with the enzyme result in the destruction of the red blood cell membrane?

31. The melting points of some common saturated and unsaturated fatty acids are shown in the table. What important factors influence a fatty acid's melting point?

Fatty acid	Melting point (°C)
Laurate (12:0)	44.2
Linoleate (18:2)	−9
Linolenate (18:3)	−17
Myristate (14:0)	52
Oleate (18:1)	13.2
Palmitate (16:0)	63.1
Stearate (18:0)	69.1

32. Rank the melting points of the following fatty acids:

(a)

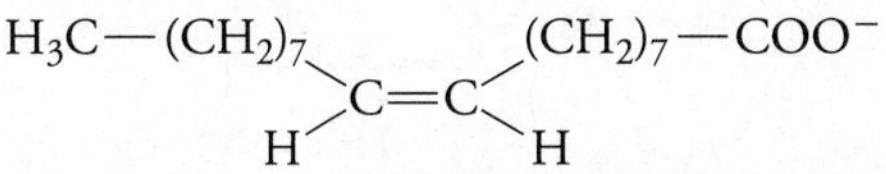

cis-Oleate

(b)

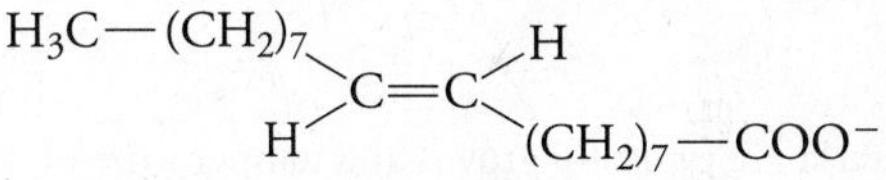

trans-Oleate

(c) linoleate

33. The triacylglycerols of animals tend to be solids (fats), whereas the triacylglycerols of plants tend to be liquids (oils) at room temperature. What can you conclude about the nature of the fatty acyl chains in animal and plant triacylglycerols?

34. Peanut oil contains a high percentage of monounsaturated triacylglycerols (having acyl chains with only one double bond), whereas vegetable oil contains a higher percentage of polyunsaturated triacylglycerols (having acyl chains with more than one double bond). A bottle of peanut oil and a bottle of vegetable oil are stored in a pantry with an outside wall. During a cold spell, the peanut oil freezes but the vegetable oil remains liquid. Explain why.

35. Reindeer meat is an important food source in northern Europe. A study was undertaken to compare meat from reindeer slaughtered in October (in relatively good health) with meat from reindeer slaughtered in February (in relatively poor health after a harsh winter). The investigators found that the percentage of lipids containing oleic acid, linoleic acid, and α-linolenic acid was decreased in the legs of the reindeer slaughtered in February. How would this affect the ability of the animals to survive the winter?

36. Phytol is an alcohol produced from chlorophyll that becomes part of the diet of mammals consuming the plant. Phytol is converted to phytanic acid in a three-step process, then oxidized to obtain metabolic energy. Inborn errors of metabolism exist in which one of the enzymes in the oxidation pathway is defective. In these individuals, the phytanic acid accumulates in the membranes of nerve cells and causes neurological disorders. What effect would the presence of phytanic acid have on nerve cell membrane fluidity?

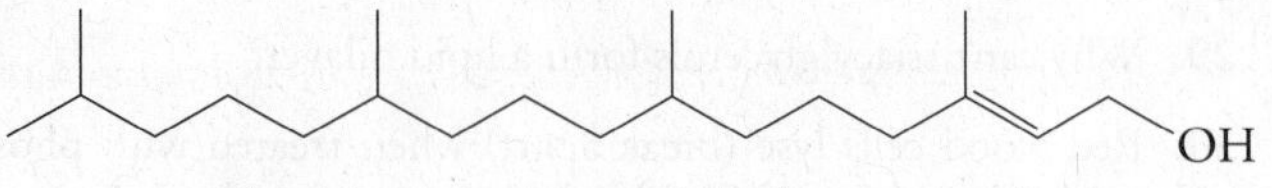

Phytol (3,7,11,15-Tetramethyl-2-hexadecen-1-ol)

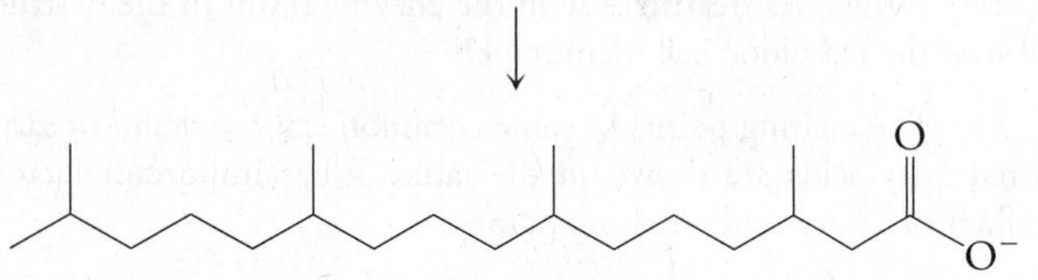

Phytanic acid (3,7,11,15-Tetramethylhexadecanoic acid)

37. Bacteria of the genus *Lactobacillus* colonize the human digestive tract and are considered "friendly" bacteria that are often used to treat digestive disorders. These bacteria produce lactobacillic acid, a 19-carbon fatty acid containing a cyclopropane ring. Is the melting point of this fatty acid closer to the melting point of stearate (18:0) or oleate (18:1)? Rank the melting points of these three fatty acids.

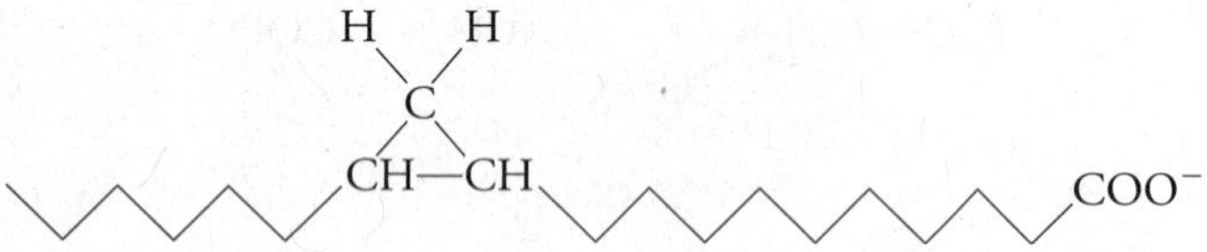

38. Bacteria are typically grown at a temperature of 37°C. What happens to the membrane lipid composition if the temperature is increased to 42°C?

39. A membrane consisting only of phospholipids undergoes a sharp transition from the crystalline form to the fluid form as it is heated. However, a membrane containing 80% phospholipid and 20% cholesterol undergoes a more gradual change from crystalline to fluid form when heated over the same temperature range. Explain why.

40. Why is fluidity greatest at the center of a lipid bilayer?

41. Plants can synthesize trienoic acids (fatty acids with three double bonds) by introducing another double bond into a dienoic acid. Would you expect plants growing at higher temperatures to convert more of their dienoic acids into trienoic acids?

42. *Chorispora bungeana* is a plant that is well adapted to growth at freezing temperatures. Plants grown at −4°C showed a greater percentage of 18:3 fatty acids compared to control plants grown at 25°C. The increase in 18:3 fatty acids was accompanied by a decrease in 18:0, 18:1, and 18:2 fatty acids. Propose a hypothesis consistent with these data.

43. The lipid distribution in membranes is asymmetric. Phosphatidylserine (PS) is exclusively found in the cytosolic-facing leaflet of the membrane bilayer. Phosphatidylethanolamine (PE) is also more likely to be found in this leaflet. In contrast, phosphatidylcholine (PC) and sphingomyelin (SM) are more likely to be found in the extracellular leaflet of the membrane bilayer.
   (a) What functional group do PS and PE have in common?
   (b) What functional group do PC and SM have in common?
   (c) Is one side of a membrane more likely to carry a charge than the other side, or do both sides of the membrane have the same charge?

44. A phospholipid flippase was studied in red blood cells. Experiments showed that the flippase translocated phospholipids from the extracellular leaflet to the cytosolic leaflet of the membrane. The flippase had a preference for phosphatidylserine and translocated phosphatidylethanolamine more slowly. Phosphatidylcholine was not translocated. Translocation did not occur if cells were deprived of ATP or magnesium ions. Translocation did not occur if red blood cells were treated with a reagent that alkylated sulfhydryl groups. Write a paragraph that describes the essential features of the flippase. Are these observations consistent with the data presented in Problem 43?

8-3 Membrane Proteins

45. Purification of transmembrane proteins requires the addition of detergents to the buffers in order to solubilize the proteins.
   (a) Why would transmembrane proteins be insoluble if the detergent were not present?
   (b) Draw a schematic diagram that shows how the detergent sodium dodecyl sulfate interacts with a transmembrane protein.

46. Use the simplified diagram of the plasma membrane to answer the following questions by choosing component A, B, C, D, or E.

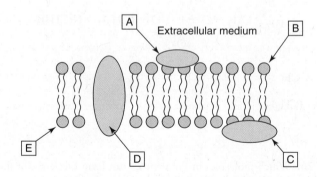

   (a) This component could be a glycoprotein.
   (b) This component can probably be separated from the others by simply washing the membrane with neutral salt solutions (mild conditions).
   (c) In order to separate this component from the others, harsh conditions, such as strong detergents, are needed (see Problem 45).
   (d) This component is the only component that might bind and transport sodium ions across the membrane.
   (e) This component could be a cerebroside or a ganglioside.
   (f) This component might be able to flip-flop transversely with the assistance of a flippase (see Problem 44).

47. Identify each type of lipid-linked protein in the following drawing.

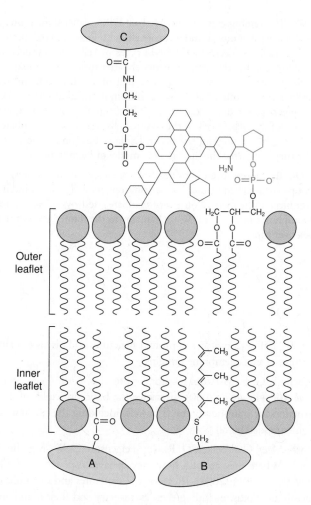

structure assumed by melittin when it is membrane-associated. Artificial membranes were prepared using phosphatidylcholine (PC) esterified with palmitate (16:0) at position 1. Lipids at position 2 were varied. It was found that melittin was conformationally restricted when associated with PC containing oleate (18:1) at position 2. But when the lipids contained arachidonate (20:4) at position 2, the melittin peptide was less conformationally restricted. Propose a hypothesis consistent with the observed results.

53. There is some evidence that membranes contain structured domains called lipid rafts. These structures are thought to be composed of loosely packed glycosphingolipids, with the gaps filled in with cholesterol. The fatty acyl chains of the phospholipids associated with lipid rafts tend to be saturated.
(a) Why do glycosphingolipids pack together loosely?
(b) Given the description provided here, do you expect a lipid raft to be more or less fluid than the surrounding membrane?

54. Cholesterol is transported through the blood in association with phospholipids and proteins that make up complexes called lipoproteins. A schematic drawing of low-density lipoprotein (LDL) is shown below. High levels of LDL are associated with an increased risk of cardiovascular disease.

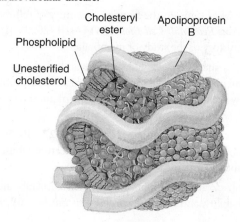

Cholesteryl ester
Apolipoprotein B
Phospholipid
Unesterified cholesterol

(a) Why is it necessary for cholesterol and cholesteryl esters (see Problem 25d) to be packaged into LDL for transport through the blood?
(b) How is the structure of LDL similar to the structure of a membrane? How is it different?
(c) The protein apolipoprotein B was purified in the mid-1980s. Why was this protein so difficult to purify?

48. Cytochrome *c*, a protein of the electron transport chain in the inner mitochondrial membrane, can be removed by relatively mild means, such as extraction with salt solution. In contrast, cytochrome oxidase from the same source can be removed only by extraction into detergent solutions or organic solvents. What kind of membrane proteins are cytochrome *c* and cytochrome oxidase? Explain. Draw a schematic diagram of what each protein looks like in the membrane.

49. Glycophorin A is a 131-residue integral membrane protein that includes one bilayer-spanning segment. Identify that segment in the glycophorin A amino acid sequence (which uses one-letter abbreviations):

A F I L M P V W

LSTTEVAMHTTTSSSVSKSYISSQTNDTHKRDTYAATPRA-
HEVSEISVRTVYPPEEETGERVQLAHHFSEPEITLIIFGVMA-
GVIGTILLISYGIRRLIKKSPSDVKPLPSPDTDVPLSSVEIEN-
PETSDQ

50. Proteins that form a transmembrane β barrel always have an even number of β strands.
(a) Explain why.
(b) Why are the strands antiparallel?
(c) Could some of them possibly be parallel?

51. Peptide hormones must bind to receptors on the extracellular surface of their target cells before their effects are communicated to the cell interior. In contrast, receptors for steroid hormones such as estrogen are intracellular proteins. Why is this possible?

52. Melittin, a 26–amino acid peptide, is known to associate with membranes. The presence of a tryptophan residue in the peptide allows fluorescence spectroscopy to be used to monitor the

8-4 The Fluid Mosaic Model

55. Around the turn of the twentieth century, Charles Overton noted that low-molecular-weight aliphatic alcohols, ether, chloroform, and acetone could pass through membranes easily, while sugars, amino acids, and salts could not. This was a radical notion at the time, since most scientists believed that membranes were impermeable to all compounds but water.
(a) Using what you know about membrane structure, explain Charles Overton's results.
(b) Propose a hypothesis to explain how the polar water molecule could be transported across a membrane.

56. In 1935, Davson and Danielli described a "sandwich model" for membrane structure which proposed that the membrane consisted of outer and inner layers of protein (the sandwich bread) with a "filling" of lipid. This model is no longer accepted because of inconsistencies between the model and experimental data. Using what we now know about membrane structure, explain some of the shortcomings of the sandwich model.

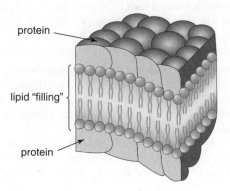

57. In a famous experiment, Michael Edidin labeled the proteins on the surface of mouse and human cells with green and red fluorescent markers, respectively. The two types of cells were induced to fuse, forming hybrid cells. Immediately after fusion, green markers could be seen on the surface of one half of a hybrid cell and red markers on the other half. After a 40-minute incubation at 37°C, the green and red markers became intermingled over the entire surface of the hybrid cell. If the hybrid cells were instead incubated at 15°C, this mixing did not occur. Explain these observations and why they supported the fluid mosaic model of membrane structure.

58. In fluorescence photobleaching recovery studies, fluorescent groups are attached to membrane components in a cell. An intense laser beam pulse focused on a very small area destroys (bleaches) the fluorophores in that area. What happens to the fluorescence in that area over time?

[SELECTED READINGS]

Edidin, M., Lipids on the frontier: A century of cell-membrane bilayers, *Nat. Rev. Mol. Cell Biol.* **4,** 414–418 (2003). [Briefly reviews the history of the study of membranes.]

Engel, A., and Gaub, H. E., Structure and mechanics of membrane proteins, *Annu. Rev. Biochem.* **77,** 127–148 (2008). [Discusses new techniques for examining the structures of membrane proteins.]

Lingwood, D., and Simons, K., *Science* **327,** 46–50 (2010). [Discusses lipid rafts as well as some general features of membranes and membrane proteins.]

Popot, J.-L., and Engelman, D. M., Helical membrane protein folding, stability, and evolution, *Annu. Rev. Biochem.* **69,** 881–922

(2000). [Shows a number of protein structures and discusses many features of transmembrane proteins.]

Schulz, G. E., β-Barrel membrane proteins, *Curr. Opin. Struct. Biol.* **10,** 443–447 (2000). [Reviews the basic principles of construction for transmembrane β barrels, including the smallest, an eight-stranded barrel.]

van Meer, G., Voelker, D. R., and Feigenson, G. W., Membrane lipids: Where they are and how they behave, *Nat. Rev. Mol. Cell Biol.* **9,** 112–124 (2008). [Reviews the structures and functions of membrane lipids, including their asymmetry and their liquid and gel phases.]

MEMBRANE TRANSPORT

▶▶ **HOW** does a small hole kill a cell?

The fungus *Cryptococcus neoformans,* which infects lung and skin tissue, is difficult to eliminate because it is a eukaryotic pathogen whose metabolism is similar to its host's. Moreover, it is surrounded by a thick cell wall that resists attack by the immune system. The cell's membrane, however, is vulnerable to small compounds that preferentially bind to ergosterol, a fungal counterpart of the cholesterol found in mammalian cell membranes. A small hole forms where the molecules insert themselves into the *Cryptococcus* membrane, killing the fungus without harming the host.

[E. Gueho/CNR/Photo Researchers, Inc.]

THIS CHAPTER IN CONTEXT

Do You Remember?

- Living organisms obey the laws of thermodynamics (Section 1-3).
- Amphiphilic molecules form micelles or bilayers (Section 2-2).
- An enzyme provides a lower-energy pathway from reactants to products (Section 6-2).
- Integral membrane proteins completely span the bilayer by forming one or more α helices or a β barrel (Section 8-3).

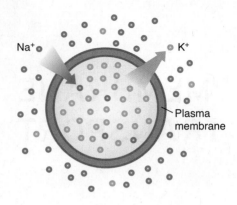

Na⁺

K⁺

Plasma membrane

Figure 9-1 Distribution of Na⁺ and K⁺ ions in an animal cell. The extracellular Na⁺ concentration (about 150 mM) is much greater than the intracellular concentration (about 12 mM), whereas the extracellular K⁺ concentration (about 4 mM) is much less than the intracellular concentration (about 140 mM). If the plasma membrane were completely permeable to ions, Na⁺ would flow into the cell down its concentration gradient (purple arrow), and K⁺ would flow out of the cell down its concentration gradient (orange arrow).

Some of the best-understood membrane-related events occur during operation of the nervous system. The ability of neurons to transmit signals from cell to cell depends on electrical changes that result from the regulated flow of charged particles across the cells' plasma membranes. As noted in Section 2-2, all animal cells—including neurons—maintain intracellular ion concentrations that differ from those outside the cell (see Fig. 2-13). For example, intracellular sodium ion concentrations are much lower than extracellular sodium ion concentrations, and the opposite is true for potassium ions. Neither ion is at equilibrium.

To reach equilibrium, Na⁺ would have to enter the cell, by spontaneously moving down its concentration gradient. Likewise, K⁺ would have to exit the cell, also moving down its concentration gradient (Fig. 9-1). However, the distributions of the ions do *not* change, because the plasma membrane presents a barrier to their diffusion. Proteins that establish and maintain the gradients and proteins that allow ions to move down their gradients are both essential for neuronal signaling.

9-1 The Thermodynamics of Membrane Transport

KEY CONCEPTS
- During a nerve impulse, ion movements alter membrane potential, producing an action potential that travels along the axon.
- Transporters obey the laws of thermodynamics, providing a way for solutes to move down their concentration gradients or using ATP to move substances against their gradients.

Although mammalian membranes are largely impermeable to ions, a small percentage of K⁺ ions do leak out of the cell. The movement of K⁺ and other ions places relatively more positive charges outside the cell and leaves relatively more negative charges inside the cell. The resulting charge imbalance, though small, generates a voltage across the membrane, which is called the **membrane potential** and is symbolized **Δψ.** In the simplest case, Δψ is a function of the ion concentration on each side of a membrane:

$$\Delta\psi = \frac{RT}{Z\mathcal{F}} \ln \frac{[\text{ion}]_{in}}{[\text{ion}]_{out}} \qquad \textbf{[9-1]}$$

where R is the **gas constant** ($8.3145 \text{ J} \cdot \text{K}^{-1} \cdot \text{mol}^{-1}$), T is temperature in Kelvin ($20°C = 293 \text{ K}$), Z is the net charge per ion, and $\mathcal{F}$ is the **Faraday constant,** the charge of one mole of electrons ($96,485 \text{ coulombs} \cdot \text{mol}^{-1}$ or $96,485 \text{ J} \cdot \text{V}^{-1} \cdot \text{mol}^{-1}$). Δψ is expressed in units of volts (V) or millivolts (mV). For a monovalent ion ($Z = 1$) at 20°C, the equation reduces to

$$\Delta\psi = 0.058 \text{ V} \log_{10} \frac{[\text{ion}]_{in}}{[\text{ion}]_{out}} \qquad \textbf{[9-2]}$$

See Sample Calculation 9-1. In a neuron, the membrane potential is actually a more complicated function of the concentrations and membrane permeabilities of several different ions, although K⁺ is the most important.

●●● SAMPLE CALCULATION 9-1

PROBLEM

Calculate the intracellular concentration of Na⁺ when the extracellular concentration is 160 mM. Assume that the membrane potential, −50 mV at 20°C, is due entirely to Na⁺.

SOLUTION

Use Equation 9-2 and solve for $[\text{Na}^+]_{in}$:

$$\Delta\psi = 0.058 \log \frac{[\text{Na}^+]_{in}}{[\text{Na}^+]_{out}}$$

$$\frac{\Delta\psi}{0.058} = \log[Na^+]_{in} - \log[Na^+]_{out}$$

$$\log[Na^+]_{in} = \frac{\Delta\psi}{0.058} + \log[Na^+]_{out}$$

$$\log[Na^+]_{in} = \frac{-0.050}{0.058} + \log(0.160)$$

$$\log[Na^+]_{in} = -0.862 - 0.796$$

$$\log[Na^+]_{in} = -1.66$$

$$[Na^+]_{in} = 0.022\ M = 22\ mM$$

PRACTICE PROBLEMS ●●

1. Calculate the intracellular concentration of Na^+ when the membrane potential is -100 mV at 20°C.
2. Calculate the membrane potential at 20°C when $[Na^+]_{in} = 10$ mM and $[Na^+]_{out} = 100$ mM.
3. Calculate the membrane potential at 20°C when $[Na^+]_{in} = 40$ mM and $[Na^+]_{out} = 25$ mM.

Ion movements alter membrane potential

Most animal cells maintain a membrane potential of about -70 mV. The negative sign indicates that the inside (the cytosol) is more negative than the outside (the extracellular fluid). A sudden flux of ions across the cell membrane can dramatically alter the membrane potential, and this is exactly what happens when a neuron fires.

When a nerve is stimulated, either mechanically or by a signal ultimately derived from one of the sensory organs, Na^+ channels in the plasma membrane open. Sodium ions immediately move into the cell, since their concentration inside is much less than outside. The inward movement of Na^+ makes the membrane potential more positive, increasing it from its resting value of -70 mV to as much as $+50$ mV. This reversal of membrane potential, or depolarization, is called the **action potential.**

The Na^+ channels remain open for less than a millisecond. However, the action potential has already been generated, and it has two effects. First, *it triggers the opening of nearby voltage-gated K^+ channels* (these channels open only in response to the change in membrane potential). The open K^+ channels allow K^+ ions to diffuse out of the cell, following their concentration gradient. This action restores the membrane potential to about -70 mV (**Fig. 9-2**).

The action potential also stimulates the opening of additional Na^+ channels farther along the **axon** (the elongated portion of the cell). This induces another round of depolarization and repolarization, and then another. In this way, the action potential travels down the axon. The signal cannot travel backward because once the ion channels have shut, they remain closed for a few milliseconds. These events are summarized in **Figure 9-3**.

In mammals, action potentials propagate extremely rapidly because the axons are insulated by a so-called **myelin sheath.** This structure consists of several layers of membrane, derived from another cell, coiled around the axon (**Fig. 9-4**). The myelin sheath is rich in sphingomyelins and contains little protein (about 18%; a typical membrane contains about 50% protein). Because the myelin sheath prevents ion movements except at the points, or nodes, in between myelinated segments of the axon, the action potential appears to jump from node to node, propagating about 20 times faster than it would in an unwrapped axon. Deterioration of the myelin sheath in diseases such as multiple sclerosis results in the progressive loss of motor control.

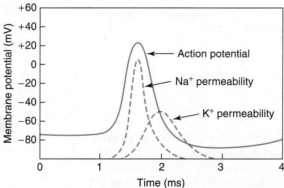

Figure 9-2 An action potential.
The neuron membrane undergoes depolarization as Na^+ channels are opened (green dashed line), then repolarizes as K^+ channels are opened (red dashed line). Following the action potential, the membrane may be hyperpolarized ($\Delta\psi < -70$ mV) but returns to normal within a few milliseconds.

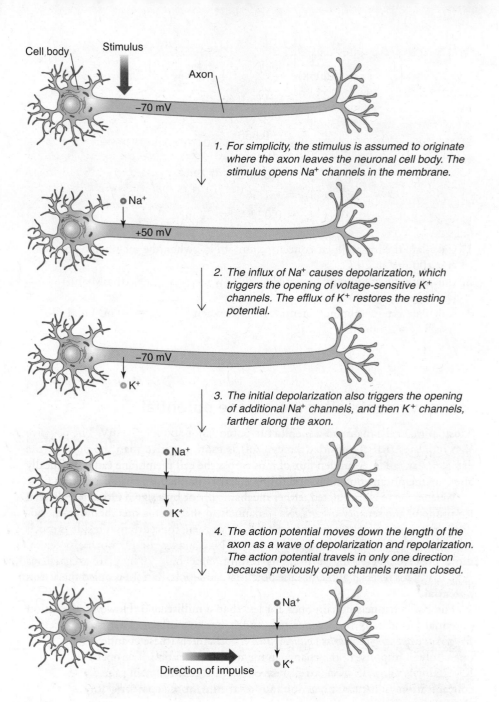

Cell body

Stimulus

Axon

−70 mV

1. *For simplicity, the stimulus is assumed to originate where the axon leaves the neuronal cell body. The stimulus opens Na⁺ channels in the membrane.*

Na^+

+50 mV

2. *The influx of Na⁺ causes depolarization, which triggers the opening of voltage-sensitive K⁺ channels. The efflux of K⁺ restores the resting potential.*

−70 mV

K^+

3. *The initial depolarization also triggers the opening of additional Na⁺ channels, and then K⁺ channels, farther along the axon.*

Na^+

K^+

4. *The action potential moves down the length of the axon as a wave of depolarization and repolarization. The action potential travels in only one direction because previously open channels remain closed.*

Na^+

K^+

Direction of impulse

Figure 9-3 Propagation of a nerve impulse.

Figure 9-4 Myelination of an axon.
(a) Cross-sectional diagram showing how an accessory cell coils around the axon so that multiple layers of its plasma membrane coat the axon. (b) Electron micrograph of myelinated axons. The myelin sheath may be 10 to 15 layers thick. [Courtesy Cedric S. Raine, Albert Einstein College of Medicine.]

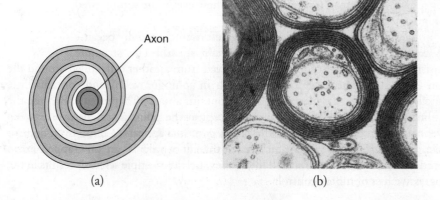

Axon

(a)

(b)

Transporters mediate transmembrane ion movement

The Na^+ and K^+ channels that participate in the propagation of an action potential are just two members of a large group of transport proteins that occur in the plasma membranes of all cells and in the internal membranes of eukaryotes. Transport proteins go by many different names, depending somewhat arbitrarily on their mode of action: transporters, translocases, permeases, pores, channels, and pumps, to list a few. These proteins can also be classified by the type of substance they transport across the membrane and by whether they are always open or gated (open only when stimulated). Nearly 6000 transporters have been grouped into five mechanistic classes in the Transporter Classification Database (www.tcdb.org). The most important distinction among transport proteins, however, is whether they require a source of free energy to operate. The neuronal Na^+ and K^+ channels are considered **passive transporters** because they provide a means for ions to move down a concentration gradient, a thermodynamically favorable event.

For any transport protein operating independently of the effects of membrane potential, the free energy change for the transmembrane movement of a substance X from the outside to the inside is

$$\Delta G = RT \ln \frac{[X]_{in}}{[X]_{out}} \qquad \text{[9-3]}$$

Consequently, *the free energy change is negative (the process is spontaneous) only when X moves from an area of high concentration on the outer side of the membrane to an area of low concentration on the inner side of the membrane* (see Sample Calculation 9-2).

SAMPLE CALCULATION 9-2 ●●●

PROBLEM

Show that $\Delta G < 0$ when glucose moves from outside the cell (where its concentration is 10 mM) to the cytosol (where its concentration is 0.1 mM).

SOLUTION

The cytosol is *in* and the endoplasmic reticulum is *out*.

$$\Delta G = RT \ln \frac{[glucose]_{in}}{[glucose]_{out}}$$

$$= RT \ln \frac{[10^{-4}]}{[10^{-2}]} = RT(-4.6)$$

Because the logarithm of $(10^{-4}/10^{-2})$ is a negative quantity, ΔG is also negative.

PRACTICE PROBLEMS ●●●

4. Calculate the value of ΔG for the process described above. Assume that $T = 20°C$.
5. Calculate the value of ΔG for the movement of glucose from outside to inside at 20°C when the extracellular concentration is 5 mM and the cytosolic concentration is 0.5 mM.
6. What is the free energy cost of moving glucose from the outside of the cell (where its concentration is 0.5 mM) to the cytosol (where its concentration is 5 mM) when $T = 20°C$?

If the transported substance is an ion, there will be a charge difference across the membrane, so a term containing the membrane potential must be added to Equation 9-3:

$$\Delta G = RT \ln \frac{[X]_{in}}{[X]_{out}} + Z\mathcal{F}\Delta\psi \qquad \text{[9-4]}$$

Equation 9-4 can be used to determine the free energy change for transporting an ion when *out* is the ion's initial location and *in* is the final location (see Sample Calculation 9-3). Note that for an anionic substance with charge Z, transport may not be thermodynamically favored, depending on the membrane potential $\Delta\psi$, even if the concentration gradient alone favors transport.

PROBLEM

Calculate the free energy change for the movement of Na$^+$ into a cell when its concentration outside is 150 mM and its cytosolic concentration is 10 mM. Assume that $T = 20°C$ and $\Delta\psi = -50$ mV (inside negative).

SOLUTION

Use Equation 9-4:

$$\Delta G = RT \ln\frac{[X]_{in}}{[X]_{out}} + Z\mathcal{F}\Delta\psi$$

$$= (8.3145 \text{ J} \cdot \text{K}^{-1} \cdot \text{mol}^{-1})(293 \text{ K}) \ln\frac{(0.010)}{(0.150)}$$

$$+ (1)(96,485 \text{ J} \cdot \text{V}^{-1} \cdot \text{mol}^{-1})(-0.05 \text{ V})$$

$$= -6600 \text{ J} \cdot \text{mol}^{-1} - 4820 \text{ J} \cdot \text{mol}^{-1}$$

$$= -11,600 \text{ J} \cdot \text{mol}^{-1} = -11.6 \text{ kJ} \cdot \text{mol}^{-1}$$

●●● **PRACTICE PROBLEMS**

7. Calculate the free energy change for the movement of K$^+$ into a cell when the K$^+$ concentration outside is 15 mM and the cytosolic K$^+$ concentration is 50 mM. Assume that $T = 20°C$ and $\Delta\psi = -50$ mM (inside negative). Is this process spontaneous?

8. Calculate the free energy cost of moving Na$^+$ ions across a membrane from a compartment (*outside*) where [Na$^+$] = 100 mM to a compartment (*inside*) where [Na$^+$] = 25 mM. Assume that $T = 20°C$ and $\Delta\psi = +50$ mV. Is this process spontaneous?

In contrast to the passive ion channels in neurons, the protein that initially establishes and maintains the cell's Na$^+$ and K$^+$ gradients is an **active transporter** that needs the free energy of ATP to move ions against their concentration gradients. In the following sections we will examine various types of transport proteins. Keep in mind that small nonpolar substances can cross a membrane without the aid of any transport protein; they simply diffuse through the lipid bilayer.

> **CONCEPT REVIEW**
> - What is the membrane potential?
> - What is the role of the cell membrane in maintaining membrane potential?
> - Describe how an action potential is generated and propagated.
> - When is the transmembrane movement of a substance thermodynamically favorable? What if the substance is an ion?
> - Summarize the difference between active and passive transport.

9-2 Passive Transport

KEY CONCEPTS
- Porins are β barrel channels with some solute selectivity.
- Ion channels include a selectivity filter and may be gated.
- Aquaporins allow only water molecules to pass through.
- Transport proteins alternate between conformations to expose binding sites on each side of the membrane.

Transporters of one kind or another have been described for virtually every substance that cannot easily diffuse through a bilayer on its own. Some transporters create a straightforward opening through the membrane, while more complicated transporters function much like enzymes. We begin our discussion with porins, the simplest transporters.

Porins are β barrel proteins

Porins are located in the outer membranes of bacteria, mitochondria, and chloroplasts (some bacteria and the organelles descended from them have a second outer membrane in addition to the membrane that encloses the cytosol). *All known porins are trimers in which each subunit forms a 16- or 18-stranded membrane-spanning β barrel* (Fig. 9-5). A β barrel of this size has a water-filled core lined with hydrophilic side chains, which forms a passageway for the transmembrane movement of ions or

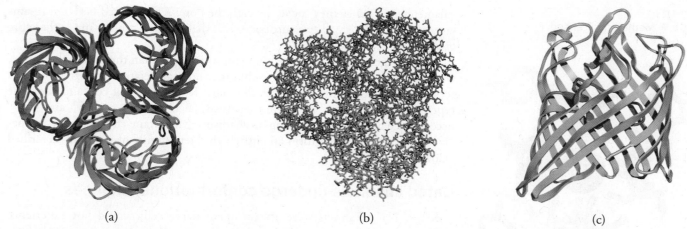

(a)   (b)   (c)

Figure 9-5 The *E. coli* OmpF porin. Each subunit of the trimeric protein forms a transmembrane β barrel that permits the passage of ions or small molecules. (a) Ribbon model, viewed from the extracellular side of the membrane. (b) Stick model. (c) In each subunit, the 16 β strands are connected by loops, one of which (blue) constricts the barrel core and makes the porin specific for small cationic solutes. [Structure (pdb 1OPF) determined by S. W. Cowan, T. Schirmer, R. A Pauptit, and J. N. Jansonius.]

molecules with a molecular mass up to about 1000 D. In the eight-stranded β barrel shown in Fig. 8-9, the protein core is too tightly packed with amino acid side chains for it to function as a pore.

In the 16-stranded OmpF barrel, long loops connect the β strands (Fig. 9-5c). One of these loops in each monomer folds down into the β barrel and constricts its diameter to about 7 Å at one point, thereby preventing the passage of substances larger than 600 D. The loop bears several carboxylate side chains, which make this porin weakly selective for cationic substances. Other porins exhibit a greater degree of solute selectivity, depending on the geometry of the barrel interior and the nature of the side chains that project into it. For example, some porins are specific for anions or small carbohydrates. A porin is considered to be always open, and a solute can travel through it in either direction.

Ion channels are highly selective

The ion channels in neurons and in other eukaryotic and prokaryotic cells are more complicated proteins than the porins. Many are multimers of identical or similar subunits with α-helical membrane-spanning segments. The ion passageway itself lies along the central axis of the protein, where the subunits meet. One of the best known of these proteins is the K⁺ channel from the bacterium *Streptomyces lividans*. Each subunit of this tetrameric protein includes two long α helices. One helix forms part of the wall of the transmembrane pore, and the other helix faces the hydrophobic membrane interior **(Fig. 9-6)**. A third, smaller helix is located on the extracellular side of the protein.

The K⁺ channel is about 10,000 times more permeant to K⁺ than to Na⁺, even though Na⁺ is smaller and should easily pass through the central pore. *The high selectivity for K⁺ reflects the geometry of the selectivity filter,* an arrangement of protein groups that define the extracellular mouth of the pore. At one point, the pore narrows to ~3 Å, and the four polypeptide backbones fold so that their carbonyl groups project into the pore. The carbonyl oxygen atoms are arranged with a geometry suitable for coordinating desolvated K⁺ ions

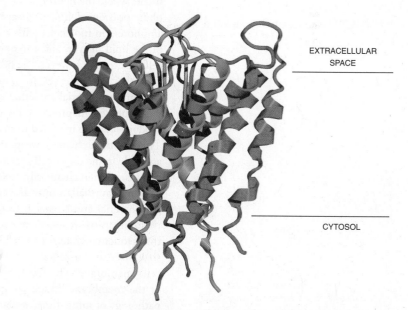

EXTRACELLULAR SPACE

CYTOSOL

Figure 9-6 Structure of the K⁺ channel from *S. lividans*. The four subunits are shown in different colors. Each subunit consists mostly of an inner helix that forms part of the central pore and an outer helix that contacts the membrane interior. [Structure (pdb 1BL8) determined by D. A. Doyle, J. M. Cabral, R. A. Pfuetzner, A. Kuo, J. M. Gulbis, S. L. Cohen, B. T. Chait, and R. MacKinnon.] ⊕ See Interactive Exercise. **The K⁺ channel selectivity filter.**

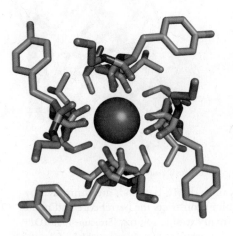

Figure 9-7 The K⁺ channel selectivity filter. This model shows a portion of the pore looking from the extracellular space into the selectivity filter. The pore is lined by backbone carbonyl groups with a geometry suitable for coordinating a K⁺ ion (purple sphere). A rigid protein network that includes Tyr residues prevents the pore from contracting to accommodate the smaller Na⁺ ion. Atoms are color-coded: C green, N blue, and O red.

? **Explain why a small anion would not pass through the channel.**

▶▶ **HOW** does a small hole kill a cell?

(diameter 2.67 Å) as they move through the pore. A desolvated Na⁺ ion (diameter 1.90 Å) is too small to coordinate with the carbonyl groups and is therefore excluded from the pore (Fig. 9-7).

The voltage-gated K⁺ channel in neurons is larger than the bacterial channel, with six helices in each of its four subunits, and it associates with other proteins to form a large complex. However, like most K⁺ channels, it contains the same type of selectivity filter. Other ion channels, such as those for Na⁺ and Ca²⁺, necessarily have different filtering mechanisms. Membrane channels that are specific for water molecules form an entirely different family of proteins (described below).

Gated channels undergo conformational changes

If K⁺ or Na⁺ channels were always open, nerve cells would not experience action potentials, and the intracellular and extracellular concentrations of ions would quickly reach equilibrium, thereby killing the cell (Box 9-A). Consequently, these channels—and many others—are **gated;** that is, they open or

---

BOX 9-A ⬡ **BIOCHEMISTRY NOTE**

Pores Can Kill

The antifungal agent amphotericin B kills *Cryptococcus* and other pathogenic fungi, as described on the first page of this chapter. Amphotericin B is a relatively small cyclic compound with decidedly hydrophobic and hydrophilic faces. Atoms are color-coded: C green, N blue, O red, and H white.

An estimated four to six amphotericin molecules insert into the fungal cell membrane, where the hydrophobic portions interact strongly with ergosterol, and the hydrophilic portions define a passageway from one side of the membrane to the other. The amphotericin molecule, with a length of only 23 Å, barely spans the hydrophobic core of the lipid bilayer, and the opening it forms is too small to allow the mass exit of the cell's contents. However, the pore is apparently sufficient to permit to flow of Na⁺, K⁺, and other ions. The resulting disruption of ion concentration gradients and the loss of membrane potential are lethal to the cell, which otherwise remains intact.

Amphotericin, produced by soil bacteria, is just one example of a wide array of microbial products that are intended to disrupt the membrane integrity of competitors or predators. Typically, these chemical weapons are excreted individually and, in pairs or small multimers, assemble in the target cell's membrane to form a passageway for ions. Some of these compounds, often in chemically modified form, have been adopted for use as antibiotics.

The mammalian immune system also relies on pore-forming mechanisms to combat bacterial and fungal infections. The presence of certain cell-wall components in these organisms triggers the activation of **complement,** a set of circulating proteins that sequentially activate each other and lead to the formation of a doughnut-shaped structure, the so-called membrane attack complex, that creates a pore in the target cell's membrane. The resulting loss of ions kills the cells. The inappropriate assembly of the membrane attack complex on the surface of human cells contributes to the pathology of some diseases that are caused by the immune system mistakenly responding to the body's own components.

⬡ **Questions:** What prevents pore formation in the membrane of the cell that produces amphotericin or a similar compound? Explain why bacteria with elaborate outer membranes are more resistant to pore-forming antibiotics.

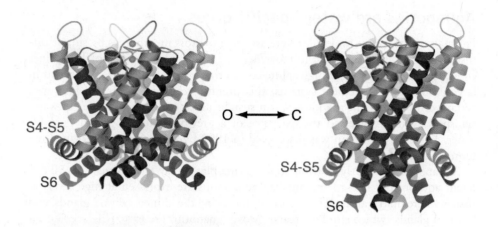

Figure 9-8 Operation of the voltage-gated K⁺ channel. In the closed conformation (right), the so-called S4-S5 linker helix (shown in red) pushes down on the S6 helix, pinching off the intracellular end of the pore. When depolarization occurs, the linker helix swings upward, and the S6 helix bends, opening up the pore (left). [Courtesy Roderick MacKinnon, Rockefeller University and Howard Hughes Medical Institute.]

close in response to a specific signal. Some ion channels respond to changes in pH or to the binding of a specific ligand such as Ca^{2+} or a small molecule. The cystic fibrosis Cl^- channel (the CFTR protein, described in Box 3-A) opens when it is phosphorylated.

Analysis of gated channels reveals a variety of mechanisms for opening and closing off a pore. The neuronal K^+ channel is voltage-gated; it opens in response to depolarization. The gating mechanism involves the motion of helices near the intracellular side of the membrane, which move enough to expose the entrance of the pore without significantly disrupting the structure of the rest of the protein (Fig. 9-8).

In addition to voltage gating, the K^+ channel in neurons is subject to inactivation by a process in which an N-terminal segment of the protein (not shown in Fig. 9-8) is repositioned to block the cytoplasmic opening of the pore. This inactivation occurs a few milliseconds after the K^+ channel first opens and explains why the channel cannot immediately reopen. As a result, the action potential can only travel forward.

In bacterial mechanosensitive channels, which open in response to membrane tension, a set of α helices slide past each other to alter their packing arrangement (Fig. 9-9). Interestingly, in the closed state, the pore is not 100% occluded. However, neither water nor ions can pass through because the opening is lined with bulky hydrophobic residues. Although a single water molecule or desolvated ion might fit geometrically, the high energetic cost of passing a polar solute past this hydrophobic barrier effectively closes off the pore.

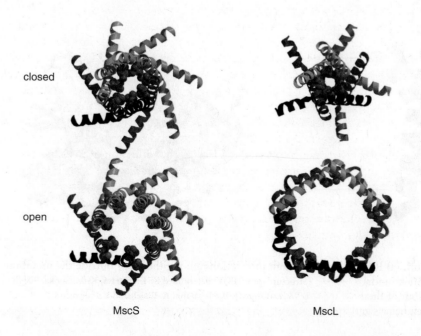

Figure 9-9 Closed and open conformations of mechanosensitive channels. In the bacterial proteins MscS and MscL, a set of α helices surround the pore (the rest of the proteins are not shown). The helices slide past each other, opening and closing the pore, much like the iris of the eye. Hydrophobic residues that block the pore in the closed state are shown in magenta. [Courtesy Douglas C. Rees, California Institute of Technology.]

Aquaporins are water-specific pores

For many years, water molecules were assumed to cross membranes by simple diffusion (technically **osmosis,** the movement of water from regions of low solute concentration to regions of high solute concentration). Because water is present in large amounts in biological systems, this premise seemed reasonable. However, certain cells, such as in the kidney, can sustain unexpectedly rapid rates of water transport, which suggested the existence of a previously unrecognized pore for water. The elusive protein was discovered in 1992 by Peter Agre, who coined the term aquaporin.

Aquaporins are widely distributed in nature; plants may have as many as 50 different aquaporins. The 13 mammalian aquaporins are expressed at high levels in tissues where fluid transport is important, including the kidney, salivary glands, and lacrimal glands (which produce tears). Most aquaporins are extremely specific for water molecules and do not permit the transmembrane passage of other small polar molecules such as glycerol or urea.

$$H_2N - \overset{\overset{\displaystyle O}{\|}}{C} - NH_2$$
$$\text{Urea}$$

The best-defined member of the aquaporin family (aquaporin 1, or AQP1) is a homotetramer with carbohydrate chains on its extracellular surface. Each subunit consists mostly of six membrane-spanning α helices plus two shorter helices that lie within the dimensions of the bilayer (Fig. 9-10).

Unlike the K^+ channel, whose pore lies in the center of the four protein subunits, each aquaporin subunit contains a pore. At its narrowest, the pore is about 3 Å in diameter (the diameter of a water molecule is 2.8 Å). The dimensions of the pore clearly restrict the passage of larger molecules. The pore is lined with hydrophobic residues except for two Asn side chains, which have an important function (Fig. 9-11).

If water were to pass through aquaporin as a chain of hydrogen-bonded molecules, then protons could also easily pass through (recall from Section 2-3 that a proton is equivalent to H_3O^+ and that a proton can appear to jump rapidly through a network of hydrogen-bonded water molecules). However, aquaporin does not transport protons (other proteins that do transport protons play important roles in energy metabolism). To prevent proton transport, aquaporin interrupts the hydrogen-bonded chain of water molecules in its pores, which occurs when the Asn side chains transiently form hydrogen bonds to a water molecule passing by.

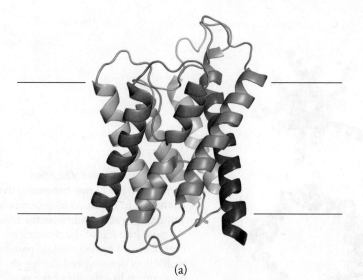

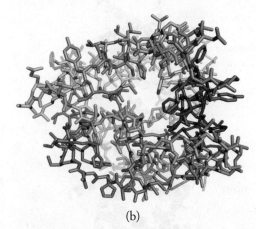

(a)

(b)

Figure 9-10 Structure of an aquaporin subunit. (a) Ribbon model viewed from within the membrane. (b) Stick model viewed from one end. In the intact aquaporin, four of these subunits associate via hydrogen bonding between helices and through interactions among the loops outside the membrane. [Structure (pdb 1FQY) determined by K. Murata, K. Mitsuoka, T. Hirai, T. Walz, P. Agre, J. B. Heymann, A. Engel, and Y. Fujiyoshi.]

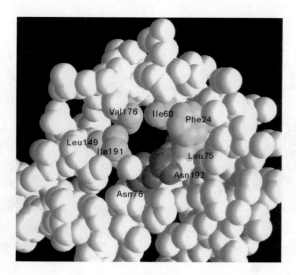

Figure 9-11 View of the aquaporin pore. Hydrophobic residues are colored yellow and the two Asn residues are red. [Courtesy Yoshinori Fujiyoshi, Kyoto University.]

Some transport proteins alternate between conformations

Not all proteins that mediate transmembrane traffic have an obvious membrane-spanning pore, as in porins and ion channels. A protein such as the glucose transporter from red blood cells undergoes a conformational change in order to move a solute from one side of the membrane to the other. Experimental evidence indicates that the protein has a glucose-binding site that alternately faces the cell interior and exterior. When glucose binds to the protein on one face of the membrane, it triggers a conformational change that exposes the bound glucose to the other face (Fig. 9-12). Because the two conformational states of this passive transporter are in equilibrium, it can move glucose in either direction across the cell membrane, depending on the relative concentrations of glucose inside and outside the cell.

Other transport proteins resemble the glucose transporter. *They are all transmembrane proteins that alternate between conformations in order to bind and release a ligand on opposite sides of the membrane.* They function like enzymes in that they accelerate the rate at which a substance crosses the membrane. And like enzymes, they can be saturated by high concentrations of their "substrate" and they are susceptible to competitive and other types of inhibition. For obvious reasons, transport proteins tend to be more solute-selective than porins or ion channels. Their great variety reflects the need to transport many different kinds of metabolic fuels and building blocks into and out of cells and organelles. An estimated 10% of the genes in microorganisms encode transport proteins.

Some transport proteins can bind more than one type of ligand, so it is useful to classify them according to how they operate (Fig. 9-13):

1. A **uniporter** such as the glucose transporter moves a single substance at a time.
2. A **symporter** transports two different substances across the membrane.
3. An **antiporter** moves two different substances in opposite directions across the membrane.

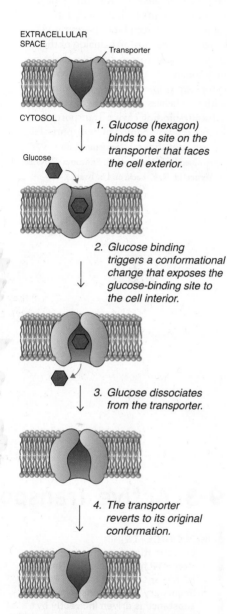

EXTRACELLULAR SPACE

Transporter

CYTOSOL

1. *Glucose (hexagon) binds to a site on the transporter that faces the cell exterior.*

Glucose

2. *Glucose binding triggers a conformational change that exposes the glucose-binding site to the cell interior.*

3. *Glucose dissociates from the transporter.*

4. *The transporter reverts to its original conformation.*

Figure 9-12 Operation of the red blood cell glucose transporter. ⊕ See Animated Figure. Model for glucose transport.

? Would this transport protein allow water or ions to move across the membrane?

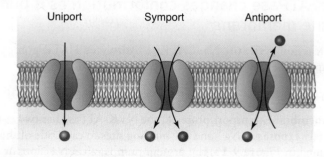

Uniport   Symport   Antiport

Figure 9-13 Some types of membrane transport systems.

Figure 9-14 **Structure of *E. coli* lactose permease.** (a) Ribbon model with the protein's helices colored in rainbow order from blue (N-terminus) to red (C-terminus). An analog of the disaccharide lactose is shown as dark gray spheres. The binding site faces the cytoplasm (top). (b) Space-filling model with two helices removed to reveal the substrate-binding cavity. Transport of lactose is accompanied by transport of a proton (not shown). [Structure (pdb 1PV7) determined by J. Abramson, I. Smirnova, V. Kasho, G. Verner, H. R. Kaback, and S. Iwata.]

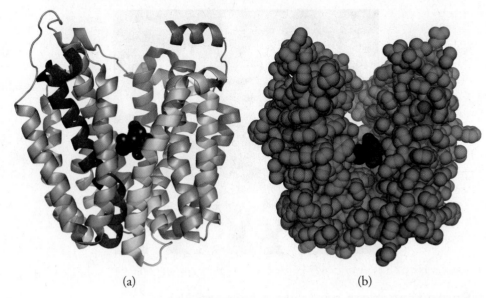

(a)　　　　　　　　　　　　(b)

Although the structure of the glucose transporter is not fully known, the structures of some other saccharide transporters have been described in detail. These catalyze symport processes and, like the glucose transporter, appear to work by a rocking mechanism. The structure of *E. coli* lactose permease shows how the protein's 12 α helices define a binding cleft open to the cytoplasm. It is easy to envision how a slight conformational change could tilt the two halves of the protein in order to expose the bound sugar to the other side of the membrane (Fig. 9-14).

> **CONCEPT REVIEW**
> • Compare the overall structure, solute selectivity, and general mechanism of porins, ion channels, aquaporins, and the red blood cell glucose transporter.
> • Which of these transporters is continually open?
> • Explain why these transport systems allow solute movement in either direction.

9-3 Active Transport

> **KEY CONCEPTS**
> • Conformational changes resulting from ATP hydrolysis drive Na$^+$ and K$^+$ transport in the Na,K-ATPase.
> • Secondary active transport of a substance is driven indirectly by the ATP-dependent formation of a gradient of a second substance.

The differing Na$^+$ and K$^+$ concentrations inside and outside of eukaryotic cells are maintained largely by an antiport protein known as the Na,K-ATPase. This active transporter pumps Na$^+$ out of and K$^+$ into the cell, working against the ion concentration gradients. As its name implies, ATP is its source of free energy. Other ATP-requiring transport proteins pump a variety of substances against their concentration gradients.

The Na,K-ATPase changes conformation as it pumps ions across the membrane

With each reaction cycle, the Na,K-ATPase hydrolyzes 1 ATP, pumps 3 Na$^+$ ions out, and pumps 2 K$^+$ ions in:

$$3\,Na_{in}^+ + 2\,K_{out}^+ + ATP + H_2O \rightarrow 3\,Na_{out}^+ + 2\,K_{in}^+ + ADP + P_i$$

Like other membrane transport proteins, the Na,K-ATPase has two conformations that alternately expose the Na$^+$ and K$^+$ binding sites to each side of the membrane. As diagrammed in **Figure 9-15**, the protein pumps out 3 Na$^+$ ions at a time, then transports in 2 K$^+$ ions at a time as it hydrolyzes ATP. *The energetically favorable reaction of converting ATP to ADP + P$_i$ drives the energetically unfavorable transport of Na$^+$ and K$^+$.* The ATP hydrolysis reaction is coupled to ion transport so that phosphoryl-group transfer from ATP to the protein triggers one conformational change (steps 3 and 4) and the subsequent release of the phosphoryl group as P$_i$ triggers another conformational change (steps 5 and 6). This multistep process, which involves a phosphorylated protein intermediate, ensures that the transporter operates in only one direction and prevents Na$^+$ and K$^+$ from diffusing back down their concentration gradients. A similar mechanism operates in motor proteins (Section 5-3), where ADP and P$_i$ are released in separate steps and so cannot recombine to re-form ATP and drive the reaction cycle in reverse.

The Na,K-ATPase consists of a large α subunit with 10 transmembrane helices plus smaller β and γ subunits containing one transmembrane helix each. The structure of the pump in its outward-facing form is shown in **Figure 9-16**. The ATP binding site and the Asp residue that becomes phosphorylated during the reaction cycle are located in cytoplasmic domains, indicating that ATP-binding and phosphate-transfer events must be communicated over a considerable distance to the membrane-spanning region where cations are bound and released.

The Na,K-ATPase is known as a P-type ATPase (P stands for phosphorylation). Other types of ATP-dependent pumps are the V-type ATPases, which operate in plant vacuoles and other organelles, and the F-type ATPases, which actually operate in reverse to *synthesize* ATP in mitochondria (Section 15-4) and chloroplasts (Section 16-2).

ABC transporters mediate drug resistance

All cells have some ability to protect themselves from toxic substances that insert themselves into the lipid bilayer and alter membrane structure and function. This defense depends on the action of membrane proteins known as **ABC transporters** (ABC refers to ATP-binding cassette, a common structural motif in these proteins). Unfortunately, many antibiotics and other drugs are lipid-soluble and so are substrates for these same transporters. Drug resistance in cancer chemotherapy and antibiotic resistance in bacteria have been linked to the expression or overexpression of ABC transporters. In humans, this transporter is also known as P-glycoprotein or the multidrug-resistance transporter.

ABC transporters function much like other transport proteins and ATPase pumps: *ATP-dependent conformational changes in cytoplasmic portions of the protein are coupled to conformational changes in the membrane-embedded portion of the protein.* As expected, the transporter consists of two halves that presumably reorient relative to each other to expose the ligand-binding site to each side of the membrane in turn. Each half of the transporter includes a bundle of membrane-spanning α helices linked to a globular nucleotide-binding domain where the ATP reaction takes place (**Fig. 9-17**).

Some ABC transporters are specific for ions, sugars, amino acids, or other polar substances. P-glycoprotein and other drug-resistance transporters prefer nonpolar substrates. In this case, the transmembrane domain of the protein allows the entry

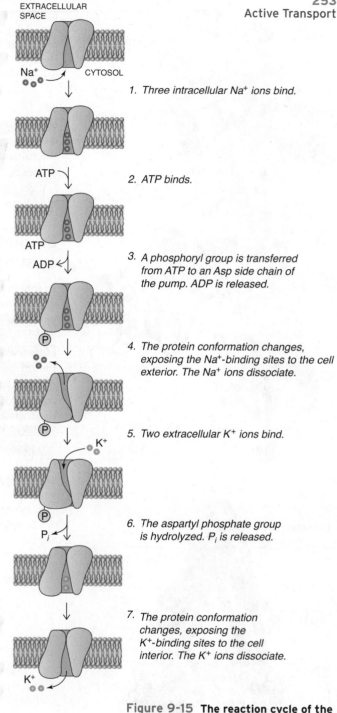

EXTRACELLULAR SPACE

Na$^+$ CYTOSOL

ATP

ATP

ADP

P

P

K$^+$

P

P$_i$

K$^+$

1. *Three intracellular Na$^+$ ions bind.*

2. *ATP binds.*

3. *A phosphoryl group is transferred from ATP to an Asp side chain of the pump. ADP is released.*

4. *The protein conformation changes, exposing the Na$^+$-binding sites to the cell exterior. The Na$^+$ ions dissociate.*

5. *Two extracellular K$^+$ ions bind.*

6. *The aspartyl phosphate group is hydrolyzed. P$_i$ is released.*

7. *The protein conformation changes, exposing the K$^+$-binding sites to the cell interior. The K$^+$ ions dissociate.*

Figure 9-15 The reaction cycle of the Na,K-ATPase.

? What does the ion transport mechanism have in common with the mechanism of motor proteins (see Figs. 5-35 and 5-37)?

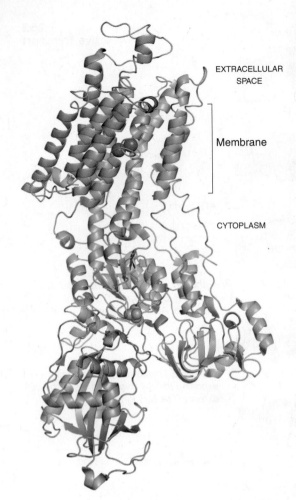

EXTRACELLULAR SPACE

Membrane

CYTOPLASM

Figure 9-16 Structure of the Na,K-ATPase. The α subunit is shown in ribbon form (the β and γ subunits are not included). Two Rb⁺ ions (purple) mark the location where two K⁺ ions would bind. An analog of the phosphate group is shown in orange, marking the site of phosphorylation. [Structure (pdb 3B8E) determined by J. P. Morth, P. B. Pedersen, M. S. Toustrup-Jensen, T. L. M. Soerensen, J. Petersen, J. P. Andersen, B. Vilsen, and P. Nissen.]

of substances from within the lipid bilayer. The substance may then be entirely expelled from the cell, as occurs in drug resistance. Alternatively, the substance may simply move from one leaflet to the other. Some lipid flippases (Section 8-2) are ABC transporters that transport lipids between leaflets, generating nonequilibrium distributions of certain lipids in membranes.

Secondary active transport exploits existing gradients

In some cases, the "uphill" transmembrane movement of a substance is not directly coupled to the conversion of ATP to ADP + P$_i$. Instead, *the transporter takes advantage of a gradient already established by another pump,* which is often an ATPase. This indirect use of the free energy of ATP is known as **secondary active transport.** For example, the high Na⁺ concentration outside of intestinal cells (a gradient established by the Na,K-ATPase) helps drive glucose into the cells via a symport protein (**Fig. 9-18**). The free energy released by the movement of Na⁺ into the cells (down its concentration gradient) drives the inward movement of glucose (against its gradient). This mechanism allows the intestine to collect glucose from digested food and then release it into the bloodstream. Lactose permease (see Fig. 9-14) is another secondary transporter. It transports lactose into the cell along with a proton, using the free energy of a proton gradient.

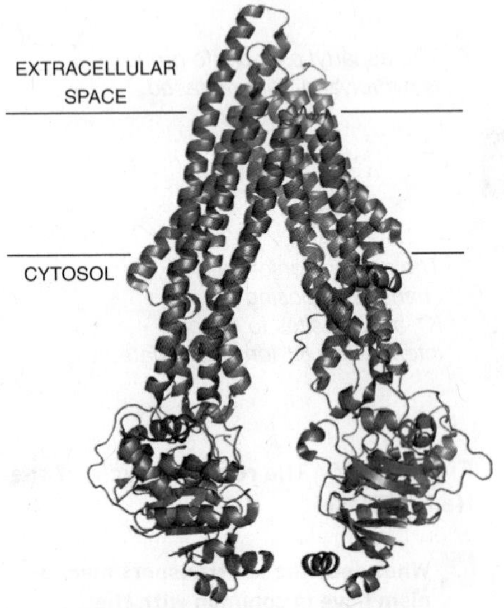

EXTRACELLULAR SPACE

CYTOSOL

Figure 9-17 Structure of mouse P-glycoprotein. The transporter, which is built from a single polypeptide chain, is shown as a ribbon model. Note that the internal cavity is open to both the cytoplasm and the inner leaflet of the membrane. [Structure (pdb 3G5U) determined by S. G. Aller, J. Yu, A. Ward, Y. Weng, S. Chittaboina, R. Zhao, P. M. Harrell, Y. T. Trinh, Q. Zhang, I. L. Urbatsch, and G. Chang.]

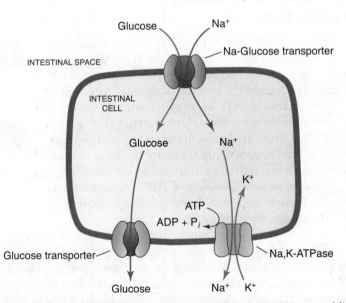

Glucose   Na⁺

Na-Glucose transporter

INTESTINAL SPACE

INTESTINAL CELL

Glucose   Na⁺

K⁺

ATP

ADP + P$_i$

Glucose transporter

Na,K-ATPase

Glucose

Na⁺ K⁺

Figure 9-18 Glucose transport into intestinal cells. The Na,K-ATPase establishes a concentration gradient in which [Na⁺]$_{out}$ > [Na⁺]$_{in}$. Sodium ions move into the cell, down their concentration gradient, along with glucose molecules via a symport protein that transports Na⁺ and glucose simultaneously. Glucose thereby becomes more concentrated inside the cell, which it then exits, down its concentration gradient, via a passive uniport glucose transporter similar to the red blood cell transporter described in Figure 9-12. Energetically favorable movements are indicated by green arrows; energy-requiring movements are indicated by red arrows.

- How does the Na,K-ATPase compare to other transport systems in terms of mechanism and energy requirement?
- Summarize the role of ATP in P-type ATPases and in ABC transporters.
- How do secondary active transporters compare to other transporters in terms of mechanism and energy requirement?

9-4 Membrane Fusion

The final steps in the transmission of a signal from one neuron to the next, or to a gland or muscle cell, culminate in the release of substances known as **neurotransmitters** from the end of the axon. Common neurotransmitters include amino acids and compounds derived from them. In the case of a synapse linking a motor neuron and its target muscle cell, the neurotransmitter is acetylcholine:

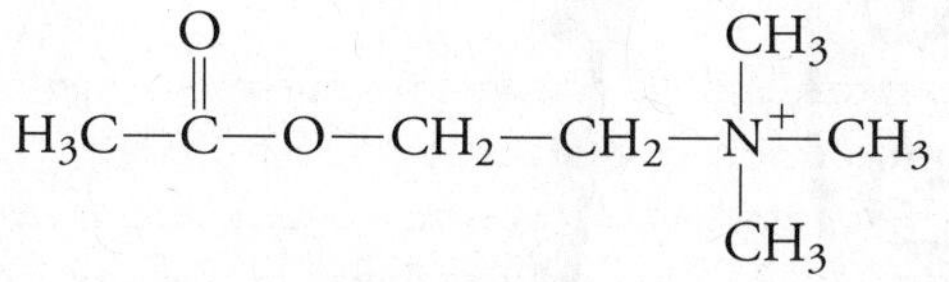

$$H_3C-\overset{\overset{O}{\|}}{C}-O-CH_2-CH_2-\overset{\overset{CH_3}{|}}{\underset{\underset{CH_3}{|}}{N^+}}-CH_3$$

Acetylcholine

Acetylcholine is stored in membrane-bounded compartments, called **synaptic vesicles,** about 400 Å in diameter. When the action potential reaches the axon terminus, voltage-gated Ca^{2+} channels open and allow the influx of extracellular Ca^{2+} ions. The increase in the local intracellular Ca^{2+} concentration, from less than 1 μM to as much as 100 μM, triggers **exocytosis** of the vesicles (exocytosis is the fusion of the vesicle with the plasma membrane such that the vesicle contents are released into the extracellular space). The acetylcholine diffuses across the synaptic cleft, the narrow space between the axon terminus and the muscle cell, and binds to integral membrane protein receptors on the muscle cell surface. This binding initiates a sequence of events that result in muscle contraction (Fig. 9-19). Blocking neurotransmission prevents muscle contraction (Box 9-B). In general, the response of the cell that receives a neurotransmitter depends on the nature of the neurotransmitter and the cellular proteins that are activated when the neurotransmitter binds to its receptor. We will explore other receptor systems in Chapter 10.

The events at the nerve–muscle synapse occur rapidly, within about one millisecond, but the effects of a single action potential are limited. First, the Ca^{2+} that triggers neurotransmitter release is quickly pumped back out of the cell by a Ca^{2+}-ATPase. Second, acetylcholine in the synaptic cleft is degraded within a few milliseconds by a lipid-linked or soluble acetylcholinesterase, which catalyzes the reaction

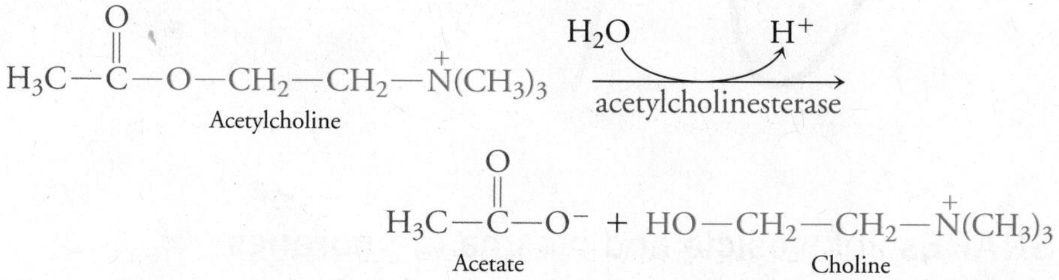

$$H_3C-\overset{\overset{O}{\|}}{C}-O-CH_2-CH_2-\overset{+}{N}(CH_3)_3 \xrightarrow[\text{acetylcholinesterase}]{H_2O \qquad H^+}$$

Acetylcholine

$$H_3C-\overset{\overset{O}{\|}}{C}-O^- + HO-CH_2-CH_2-\overset{+}{N}(CH_3)_3$$

Acetate   Choline

Other types of neurotransmitters are recycled rather than destroyed. They are transported back into the cell that released them by the action of secondary active transporters (Box 9-C). A neuron may contain hundreds of synaptic vesicles, only a small percentage of which undergo exocytosis at a time, so the cell can release neurotransmitters repeatedly (as often as 50 times per second).

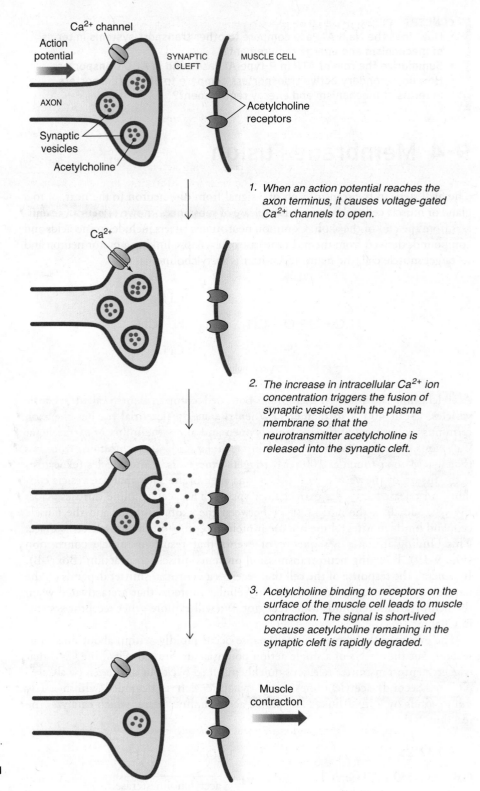

1. When an action potential reaches the axon terminus, it causes voltage-gated Ca^{2+} channels to open.

2. The increase in intracellular Ca^{2+} ion concentration triggers the fusion of synaptic vesicles with the plasma membrane so that the neurotransmitter acetylcholine is released into the synaptic cleft.

3. Acetylcholine binding to receptors on the surface of the muscle cell leads to muscle contraction. The signal is short-lived because acetylcholine remaining in the synaptic cleft is rapidly degraded.

Figure 9-19 Events at the nerve-muscle synapse.

? Describe the events that must occur to restore the neuron and muscle cell to their original states.

SNAREs link vesicle and plasma membranes

Membrane fusion is a multistep process that begins with the targeting of one membrane (for example, the vesicle) to another (for example, the plasma membrane). A number of proteins participate in tethering the two membranes and readying them for fusion. However, many of these proteins may be only accessory factors for the SNAREs, the proteins that physically pair the two membranes and induce them to fuse.

BOX 9-B ⬡ BIOCHEMISTRY NOTE

257
Membrane Fusion

Some Drugs Interfere with Neuronal Signaling

A variety of drugs interfere with different aspects of neuronal function, thus affecting muscle activity, perception of pain, or consciousness. Some of these substances have been used for medical purposes for centuries, such as opium (from poppies); others are the result of more recent drug development efforts.

Succinylcholine, which resembles acetylcholine, is used as a muscle relaxant. It binds to acetylcholine receptors on muscle cells, which activates them. However, because succinylcholine is only slowly degraded (it is not a substrate for acetylcholinesterase, which rapidly breaks down the normal acetylcholine neurotransmitter), its effects persist. The muscle cell cannot return to its initial state or respond to additional stimulation, so the net result is relaxation of the muscle. Succinylcholine's effects last for only a few minutes, and it does not interfere with pain sensation or consciousness. It is used mostly in emergency situations, often to allow insertion of an endotracheal (breathing) tube.

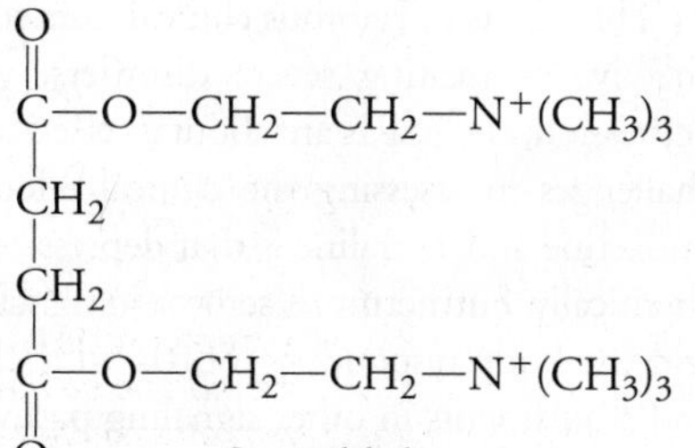

$$
\begin{array}{l}
O \\
\| \\
C-O-CH_2-CH_2-N^+(CH_3)_3 \\
| \\
CH_2 \\
| \\
CH_2 \\
| \\
C-O-CH_2-CH_2-N^+(CH_3)_3 \\
\| \\
O
\end{array}
$$

Succinylcholine

Some local anesthetics interfere with neuronal signaling in a limited portion of the body by blocking the pores of Na^+ channels involved in generating action potentials. One widely used drug is lidocaine, which is used to numb tissues for dental procedures and minor surgery. It is also used topically to relieve pain and severe itching. Lidocaine is broken down slowly by the liver, so its effects can last up to two hours.

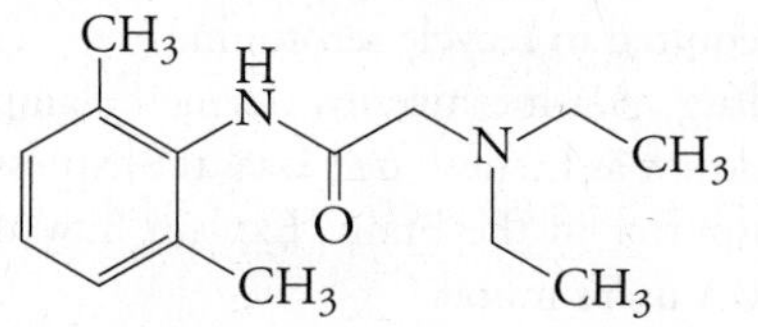

Lidocaine

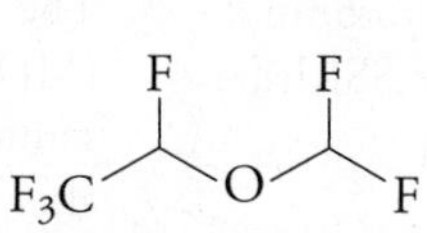

Desflurane

A variety of substances are used as general anesthetics, including several volatile substances that are inhaled, such as desflurane. At one time, such compounds were believed to exert their anesthetic effects by nonspecifically inserting into and altering the fluidity of membranes. However, X-ray structures indicate that they act more specifically by binding to certain neurotransmitter receptors that function as ligand-gated ion channels. The anesthetic molecules slip into a cavity in the transmembrane region of the receptor, preventing ion movement and thereby blocking neuronal signaling.

⬡ **Question:** Explain why general anesthetics tend to be highly hydrophobic substances.

SNAREs are integral membrane proteins (their name is coined from the term "soluble N-ethylmaleimide-sensitive-factor attachment protein receptor"). Two SNAREs from the plasma membrane and one from the synaptic vesicle form a complex that includes a 120-Å-long coiled-coil structure containing four helices (two of the SNAREs contribute one helix each, and one SNARE contributes two helices).

BOX 9-C ⬡ CLINICAL CONNECTION

Antidepressants Block Serotonin Transport

The neurotransmitter serotonin, a derivative of tryptophan, is released by cells in the central nervous system.

Serotonin

Sertraline

Serotonin signaling leads to feelings of well-being, suppression of appetite, and wakefulness, among other things. Seven different families of receptor proteins respond to serotonin signals, sometimes in opposing ways, so the pathways by which this neurotransmitter affects mood and behavior have not been completely defined.

Unlike acetylcholine, serotonin is not broken down in the synapse, but instead about 90% of it is transported back into the cell that released it and is reused. Because the extracellular concentration of serotonin is relatively low, it reenters cells via symport with Na^+, whose extracellular concentration is greater than its intracellular concentration. The rate at which the serotonin transporter takes up the neurotransmitter helps regulate the extent of signaling. Research suggests that genetic variation in the transporter protein may explain an individual's susceptibility to conditions such as depression and post-traumatic stress disorder, but such correlations are difficult to prove, since the level of expression of the transporter appears to vary between individuals and even within an individual.

Drugs known as selective serotonin reuptake inhibitors (SSRIs) block the transporter and thereby enhance serotonin signaling. Some of the most widely prescribed drugs are SSRIs, including fluoxetine (Prozac) and sertraline (Zoloft).

Fluoxetine

These drugs are used primarily as antidepressants, although they are also prescribed for anxiety disorders and obsessive-compulsive disorder. Despite decades of research, the interactions between the serotonin transporter and the drugs are not completely understood at the molecular level, and some studies indicate that different inhibitors may bind to different sites on the transporter.

The results of rigorous clinical tests suggest that SSRIs are most effective at treating severe disorders, whereas in mild cases of depression, the SSRIs are about as effective as a placebo. One of the challenges in assessing the clinical effectiveness of drugs such as fluoxetine and sertraline is that depression is difficult to define biochemically. Furthermore, serotonin signaling pathways are complex, and the body responds to SSRIs with changes in gene expression and adjustments in other signaling pathways so that antidepressive effects may not be apparent for several weeks. The list of SSRI side effects is long and highly variable among individuals, and, disturbingly, includes a small increase in risk of suicidal behavior.

● Questions

1. What types of food could contribute to serotonin production in the body?
2. The Na,K-ATPase is essential for establishing the ion gradients that make neuronal signaling possible. Explain why the pump is also required to recycle serotonin.
3. The illicit drug 3,4-methylenedioxymethylamphetamine (MDMA, also know as Ecstasy) decreases the expression of the serotonin transporter in the brain. Explain how this would alter the MDMA user's mood.
4. In addition to being an SSRI, fluoxetine may also bind to and block signaling by one type of serotonin receptor. Would you expect this activity to be consistent with fluoxetine's ability to treat the symptoms of depression?
5. Serotonin and other so-called monoamine signaling molecules are degraded in the liver, beginning with a reaction catalyzed by a monoamine oxidase (MAO). Explain why doctors avoid prescribing an MAO inhibitor along with an SSRI.

The four helices, each with about 70 residues, line up in parallel fashion (Fig. 9-20). Unlike other coiled-coil proteins such as keratin (Section 5-2), the four-helix bundle is not a perfectly geometric structure but varies irregularly in diameter.

The mutual interactions between SNAREs in the vesicle and plasma membranes serve as an addressing system so that the proper membranes fuse with each other. Initially, the individual SNARE proteins are unfolded, and they spontaneously zip up to form the four-helix complex. This action necessarily brings the two membranes close together (Fig. 9-21). *Because the formation of the SNARE complex is thermodynamically favorable, membrane fusion is also spontaneous.* The rapid rate of acetylcholine release *in vivo* indicates that at least some synaptic vesicles are already docked at the plasma membrane, awaiting the Ca^{2+} signal for fusion to proceed.

Membrane fusion requires changes in bilayer curvature

Experiments with pure lipid vesicles demonstrate that SNAREs are not essential for membrane fusion to occur *in vitro*, but the rate of fusion does depend on the membranes' lipid compositions. The explanation for this observation is that the lipid bilayers of the fusing membranes undergo rearrangement: The lipids in the contacting leaflets must mix before a pore forms **(Fig. 9-22)**. Certain types of lipids appear to promote the required changes in membrane curvature.

In living cells, bilayer shape changes could be facilitated by the tension exerted by the SNARE complex. In addition, membrane lipids may undergo active remodeling. For example, the enzymatic removal of an acyl chain would convert a cylindrical lipid to a cone-shaped lipid:

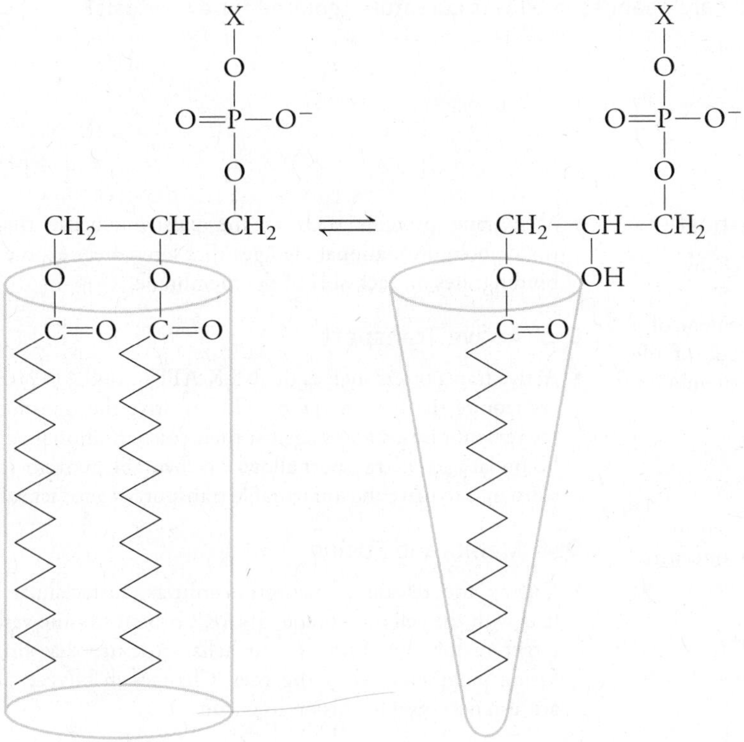

Figure 9-20 Structure of the four-helix bundle of the SNARE complex. The three proteins (one includes two helices) are in different colors. Portions of the SNAREs that do not form the helix bundle were cleaved off before X-ray crystallography. [Structure (pdb 1SFC) determined by R. B. Sutton and A. T. Brunger.]

Clustering of such lipids would cause the bilayer to bow outward. Conversely, removing large lipid head groups would cause the bilayer to bow inward.

As a result of exocytosis, the plasma membrane becomes augmented with material from the membranes of the fused synaptic vesicles. The neuron recycles some

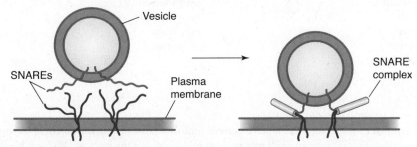

Figure 9-21 Model for SNARE-mediated membrane fusion. Formation of the complex of SNAREs from the vesicle and plasma membranes brings the membranes close together so that they can fuse.

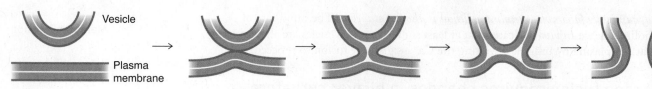

Vesicle

Plasma membrane

Figure 9-22 **Schematic view of membrane fusion.** For simplicity, the vesicle and plasma membranes are depicted as bilayers.

of its membrane material by forming new synaptic vesicles. A new vesicle forms by budding off of an existing membrane, a process known as **endocytosis.** It is the opposite of exocytosis and follows, in reverse, the scheme diagrammed in Figure 9-22.

CONCEPT REVIEW
- How does acetylcholine inside a synaptic vesicle reach a target muscle cell?
- How is the structure of the SNARE complex related to its function?
- Why are changes in bilayer curvature required for exocytosis?

[SUMMARY]

9-1 The Thermodynamics of Membrane Transport
- The transmembrane movements of ions generate changes in membrane potential, $\Delta\psi$, during neuronal signaling.
- The free energy change for the transmembrane movement of a substance depends on the concentrations on each side of the membrane and, if the substance is charged, on the membrane potential.

9-2 Passive Transport
- Passive transport proteins such as porins allow the transmembrane movement of substances according to their concentration gradients. Aquaporins mediate the transport of water molecules.
- Ion channels have a selectivity filter that allows passage of one type of ion. Gated channels open and close in response to some other event.

- Membrane proteins such as the passive glucose transporter undergo conformational changes that alternately expose ligand-binding sites on each side of the membrane.

9-3 Active Transport
- Active transporters such as the Na,K-ATPase and ABC transporters require the free energy of ATP to drive the transmembrane movement of substances against their concentration gradients.
- Secondary active transport allows the favorable movement of one substance to drive the unfavorable transport of another substance.

9-4 Membrane Fusion
- During the release of neurotransmitters, intracellular vesicles fuse with the cell membrane. SNARE proteins in the vesicle and target membranes form a four-helix structure that brings the fusing membranes close together. Changes in bilayer curvature are also necessary for fusion to occur.

[GLOSSARY TERMS]

membrane potential ($\Delta\psi$)	active transport	antiporter
gas constant (R)	porin	ABC transporter
Z	complement	secondary active transport
Faraday constant ($\mathcal{F}$)	gated channel	neurotransmitter
action potential	osmosis	synaptic vesicle
axon	uniporter	exocytosis
myelin sheath	symporter	endocytosis
passive transport		

[PROBLEMS]

9-1 The Thermodynamics of Membrane Transport

1. The resting membrane potential maintained in most nerve cells is about -70 mV. Use Equation 9-2 to calculate the ratio of $[Na^+]_{in}/[Na^+]_{out}$ at the resting potential.

2. When a nerve is stimulated, the membrane potential increases from -70 mV to $+50$ mV. Calculate the ratio of $[Na^+]_{in}/[Na^+]_{out}$ in the depolarized nerve cell. How does this ratio compare to the ratio you calculated in Problem 1 and what is the significance of the change?

3. Use Equation 9-4 to calculate the free energy change for the movement of Na^+ into a cell at the resting potential (described in Problem 1). Assume the temperature is 37°C. Is this process favorable?

4. Use Equation 9-4 to calculate the free energy change for the movement of Na^+ into the depolarized nerve cell described in Problem 2. Assume the temperature is 37°C. How does this compare with the value you calculated in Problem 3, and what is the significance of the difference?

5. In typical marine organisms, the intracellular concentrations of Na^+ and Ca^{2+} are 10 mM and 0.1 μM, respectively. Extracellular concentrations of Na^+ and Ca^{2+} are 450 mM and 4 mM, respectively. Use Equation 9-4 to calculate the free energy changes at 20°C for the transmembrane movement of these ions. In which direction do the ions move? Assume the membrane potential is −70 mV.

6. Calculate the free energy changes at 20°C for the transmembrane movement of Na^+ and K^+ ions using the conditions presented in Figure 9-1. Assume the membrane potential is −70 mV. In which direction do the ions move?

7. The concentration of Ca^{2+} in the endoplasmic reticulum (*outside*) is 1 mM, and the concentration of Ca^{2+} in the cytosol (*inside*) is 0.1 μM. Calculate ΔG at 37°C when the membrane potential is (a) −50 mV (cytosol negative) and (b) +50 mV. In which case is Ca^{2+} movement more thermodynamically favorable?

8. The concentration of Ca^{2+} in the cytosol (*inside*) is 0.1 μM and the concentration of Ca^{2+} in the extracellular medium (*outside*) is 2 mM. Calculate ΔG at 37°C, assuming the membrane potential is −50 mV. In which direction is Ca^{2+} movement thermodynamically favored?

9. A high fever can interfere with normal neuronal activity. Since temperature is one of the terms in Equation 9-1, which defines membrane potential, a fever could potentially alter a neuron's resting membrane potential.
(a) Calculate the effect of a change in temperature from 98°F to 104°F (37°C to 40°C) on a neuron's membrane potential. Assume that the normal resting potential is −70 mV and that the distribution of ions does not change.
(b) How else might an elevated temperature affect neuronal activity?

10. During mitochondrial electron transport (Chapter 15), protons are transported across the inner mitochondrial membrane from inside the mitochondrial matrix to the intermembrane space. The pH of the mitochondrial matrix is 7.78. The pH of the intermembrane space is 6.88.
(a) What is the membrane potential under these conditions?
(b) Use Equation 9-4 to determine the free energy change for proton transport at 37°C.

11. Glucose absorbed by the epithelial cells lining the intestine leaves these cells and travels to the liver via the portal vein. After a high-carbohydrate meal, the concentration of glucose in the portal vein can reach 15 mM.
(a) What is the ΔG for transport of glucose from portal vein blood to the interior of the liver cell, where the concentration of glucose is 0.5 mM?
(b) What is the ΔG for transport under fasting conditions when the blood glucose level falls to 4 mM? Assume the temperature is 37°C.

12. What is the ΔG for transport of glutamate from the outside of the cell, where the concentration is 0.1 mM, to the inside of the cell, where the concentration is 10 mM? Assume the cell potential is −70 mV.

13. Rank the rate of transmembrane diffusion of the following compounds:

$$H_3C-\overset{\displaystyle O}{\overset{\|}{C}}-NH_2 \qquad H_3C-CH_2-CH_2-\overset{\displaystyle O}{\overset{\|}{C}}-NH_2$$

A. Acetamide   B. Butyramide

$$H_2N-\overset{\displaystyle O}{\overset{\|}{C}}-NH_2$$

C. Urea

14. The permeability coefficient indicates a solute's tendency to move from an aqueous solvent to a nonpolar lipid membrane. The permeability coefficients for the compounds shown in Problem 13 are listed in the table below. How do the permeability coefficients assist you in ranking the rate of the transmembrane diffusion of these compounds?

	Permeability coefficient ($cm \cdot s^{-1}$)
Acetamide	9×10^{-6}
Butyramide	1×10^{-5}
Urea	1×10^{-7}

15. The permeability coefficients (see Problem 14) of glucose and mannitol for both natural and artificial membranes are shown in the table below.
(a) Compare the permeability coefficients for the two solutes (structures shown below). Which solute moves more easily through the synthetic bilayer and why?
(b) Compare the permeability coefficients of each solute for the two types of membrane. For which solute is the difference more dramatic and why?

	Permeability coefficient ($cm \cdot s^{-1}$)	
	Glucose	**Mannitol**
Synthetic lipid bilayer	2.4×10^{-10}	4.4×10^{-11}
Red blood cell membrane	2.0×10^{-4}	5.0×10^{-9}

```
        CHO              CH2OH
   H ——— OH         HO ——— H
  HO ——— H          HO ——— H
   H ——— OH          H ——— OH
   H ——— OH          H ——— OH
        CH2OH            CH2OH
     Glucose           Mannitol
```

16. Explain why carbon dioxide can cross a cell membrane without the aid of a transport protein.

9-2 Passive Transport

17. The bacterium *Pseudomonas aeruginosa* expresses a phosphate-specific porin when phosphate in its growth medium is limiting. Noting that there were three lysines clustered in the surface-exposed amino

terminal region of the protein, investigators constructed mutants in which the lysine residues were replaced with glutamate residues.

(a) Why did the investigators hypothesize that lysine residues might play an important role in phosphate transport in the bacterium?

(b) Predict the effect of the Lys → Glu substitution on the transport activity of the porin.

18. As noted in the text, the OmpF porin in *E. coli* has a con-stricted diameter that prevents the passage of large substances. The protein loops in this constricted site contain a D-E-K-A sequence, which makes the porin weakly selective for cations. If you wanted to construct a mutant porin that was highly selective for calcium ions, what changes would you make to the amino acid sequence at the constricted site?

19. In addition to neurons, muscle cells undergo depolarization, although smaller and slower than in the neuron, as a result of the activity of the acetylcholine receptor.

(a) The acetylcholine receptor is also a gated ion channel. What triggers the gate to open?

(b) The acetylcholine receptor/ion channel is specific for Na^+ ions. Do Na^+ ions flow in or out?

(c) How does the Na^+ flow through the ion channel change the membrane potential?

20. Explain why acid-gated channel proteins include a set of Asp or Glu residues in their acid sensor. How would these groups par-ticipate in gating?

21. When a bacterial cell is transferred from a solute-rich envi-ronment to pure water, mechanosensitive channels open and allow the efflux (outflow) of cytoplasmic contents. Use osmotic effects to explain how this prevents cell death.

22. Ammonia, like water, was long believed to diffuse across membranes without the aid of a channel protein. Recently, researchers deleted the gene *Rhcg* from mice and examined the NH_3 permeability of their kidney cells. Cells from wild-type mice exhibited an NH_3 flux about three times higher then the mutant cells.

(a) What do these results suggest about the role of the protein encoded by *Rhcg*?

(b) In light of what you know about aquaporin, does this surprise you?

23. The selectivity filter in the bacterial chloride channel ClC is formed in part by the hydroxyl groups of Ser and Tyr side chains and main-chain NH groups. Explain how these groups would allow Cl^- but not cations to pass through.

24. The selectivity filter of the KcsA channel has the sequence TVGYG. The corresponding sequences of some other ion channels, and their ion specificities, are given below. What do you expect would be the ion selectivity of channels with the following sequences: (a) SVGFG, (b) ITMEG?

Channel	Sequence	Ion specificity
KcsA	T V G Y G	K^+
NaK	T V G D G	Na^+ and K^+
Ca^{2+} II	L T G E D	Ca^{2+}
Na^+ IV	T T S A G	Na^+

25. A plot of the glucose transport rate versus glucose concentration for the passive glucose transporter of red blood cells is shown here.

(a) Explain why the plot yields a hyperbolic curve.

(b) Use what you have learned about Michaelis–Menten kinetics to estimate the V_{max} and the K_M values for this glucose transporter.

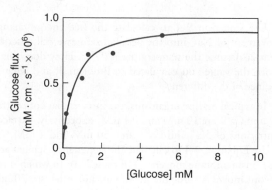

26. The compound 6-*O*-benzyl-D-galactose competes with glu-cose for binding to the glucose transporter. Sketch a curve similar to the one shown in Problem 25b that illustrates the kinetics of glucose transport in the presence of this inhibitor.

27. Experiments with erythrocyte "ghosts" were carried out to learn more about the glucose transporter in red blood cells. "Ghosts" are prepared by lysing red blood cells in a hypotonic medium and washing away the cytoplasmic contents. Suspension in an isotonic buffer allows the ghost membranes to reseal. If ghosts are prepared so that the enzyme trypsin is incorporated into the ghost interior, glucose transport does not occur. But glucose transport is not affected if trypsin is located in the extracellular ghost medium. What can you conclude about the glucose trans-porter, given these observations?

28. If a propyl group is added to the hydroxyl group on C1 of glucose, the modified glucose is unable to bind to its transporter on the extracellular surface. In another experiment, it was shown that if a propyl group is added to the hydroxyl group on C6 of glucose, the modified glucose is unable to bind to its transporter on the cyto-solic surface of the membrane. What do these observations tell us about the mechanism of passive glucose transport?

29. Glutamate acts as a neurotransmitter in the brain, and it is reabsorbed and recycled by glial cells associated with the neurons. The glutamate transporter also transports 3 Na^+ and 1 H^+ along with glutamate, and it transports 1 K^+ in the opposite direction. What is the net charge imbalance across the cell membrane for each glutamate transported?

30. The bacterium *Oxalobacter formigenes* is found in the intes-tine, where it plays a role in digesting the oxalic acid found in some fruits and vegetables (spinach is a particularly rich source). The bacterium takes up oxalic acid from the extracellular medium in the form of oxalate. Once inside the cell, the oxalate undergoes decar-boxylation to produce formate, which leaves the cell in antiport with the oxalate.

(a) What is the net charge imbalance generated by this system?

(b) Why do you suppose the investigators who elucidated this mechanism referred to the process as a "hydrogen pump"?

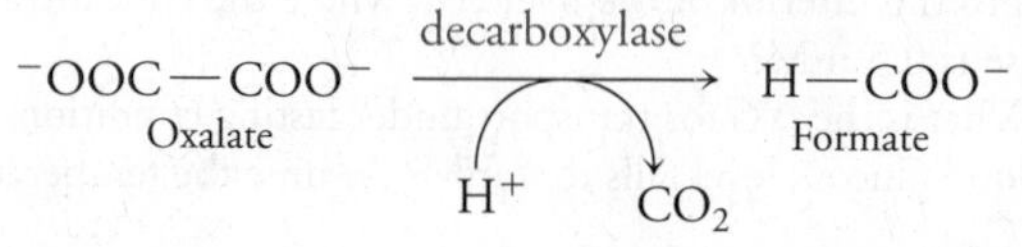

$$^-OOC-COO^- \xrightarrow[\text{H}^+ \quad \text{CO}_2]{\text{decarboxylase}} H-COO^-$$

Oxalate → Formate

31. As discussed in Section 5-1, tissues produce CO_2 as a waste product during respiration. The CO_2 enters the red blood cell and

combines with water to form carbonic acid in a reaction catalyzed by carbonic anhydrase. A red blood cell protein called Band 3 transports HCO_3^- ions in exchange for Cl^- ions. What role does Band 3 play in transporting CO_2 to the lungs, where it can be exhaled?

32. Band 3 is unusual in that it can transport HCO_3^- in exchange for Cl^- ions, in either direction, via passive transport. Why is it important that Band 3 operate in both directions?

9-3 Active Transport

33. The Na,K-ATPase first binds sodium ions, then reacts with ATP to form a "high-energy" aspartyl phosphate intermediate. Draw the structure of the phosphorylated Asp residue.

34. Ouabain, an extract of the East African ouabio tree, has been used as an arrow poison. Ouabain binds to an extracellular site on the Na,K-ATPase and prevents the hydrolysis of the "high-energy" phosphorylated intermediate. Why is ouabain a lethal poison?

35. In eukaryotes, ribosomes (approximate mass 2.5×10^6 D) are synthesized inside the nucleus, which is enclosed by a double membrane. Protein synthesis occurs in the cytosol.
(a) Would you expect that a protein similar to a porin or the glucose transporter would be responsible for transporting ribosomes into the cytoplasm? Explain.
(b) Do you think that free energy would be required to move ribosomes from the nucleus to the cytoplasm? Why or why not?

36. Osteoclasts in bone tissue are particularly rich in one of the carbonic anhydrase isozymes. Proper enzyme function is critical to the development of healthy tissue.
(a) Carbonic anhydrase catalyzes the reaction between water and carbon dioxide to yield carbonic acid. The carbonic acid then undergoes dissociation. Write the two equations that describe these processes.
(b) Proper bone development requires that the osteoclast acidify its extracellular environment (the bone-resorbing compartment). Several transporters are involved in the acidification process: a Na^+/H^+ exchanger, a Cl^-/HCO_3^- exchanger, and the Na,K-ATPase, which exchanges Na^+ and K^+ ions. A partial diagram of the osteoclast is shown below. Fill in the blanks in the diagram indicating the roles of carbonic anhydrase and the exchangers in the acidification of the bone-resorbing compartment. Include the reactants and products of the appropriate intracellular reaction(s) and note in which direction each ion is transported in the osteoclast.

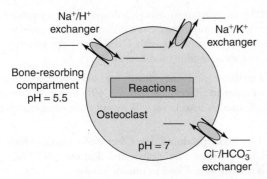

37. The retina of the eye contains equal amounts of endothelial and pericyte cells. Basement membrane thickening in pericytes occurs during the early stages of diabetic retinopathy, eventually leading to blindness. Glucose uptake was measured in both types of cells in culture in the presence of increasing amounts of sodium. The results are shown in the figure.

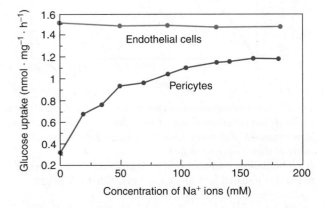

(a) What is your interpretation of these results?
(b) What information is conveyed by the shapes of the curves?
(c) By what mechanism might the pericytes use sodium ions to assist with glucose import?

38. The kinetics of transport through protein transporters can be described using the language of Michaelis and Menten. The transported substance is analogous to the substrate, and the protein transporter is analogous to the enzyme. K_M and V_{max} values can be determined to describe the binding and transport process where K_M is a measure of the affinity of the transported substance for its transporter and V_{max} is a measure of the rate of the transport process. Use the information provided in the figure to estimate the K_M and V_{max} for glucose uptake by the pericyte transporter described in Problem 37.

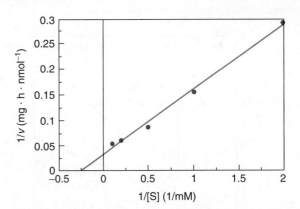

39. Liver cells use a choline transport protein to import choline from the portal circulation. Choline transport was measured in mouse cells transfected with the gene for the hepatic transporter. The uptake of radioactively labeled choline by the transfected cells was measured at increasing choline concentrations.

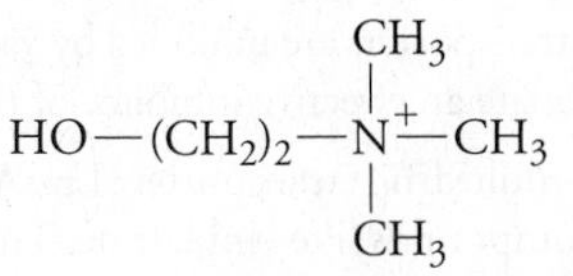

$$HO-(CH_2)_2-\overset{CH_3}{\underset{CH_3}{\overset{|}{\underset{|}{N^+}}}}-CH_3$$

Choline

(a) Use the language of Michaelis and Menten as described in Problem 38 to estimate K_M and V_{max} from the curve shown.

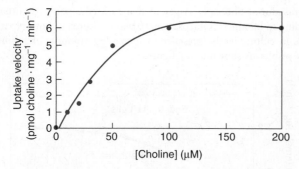

(b) Plasma concentrations of choline range from 10 to 80 μM, although the concentration may be higher in portal circulation after ingestion of choline. Does the transporter operate effectively at these concentrations?

(c) The choline transport protein is inhibited by low external pH and stimulated by high external pH. What role might protons play in the transport of choline?

(d) Propose a mechanism that explains how tetraethylammonium (TEA) ions inhibit choline transport.

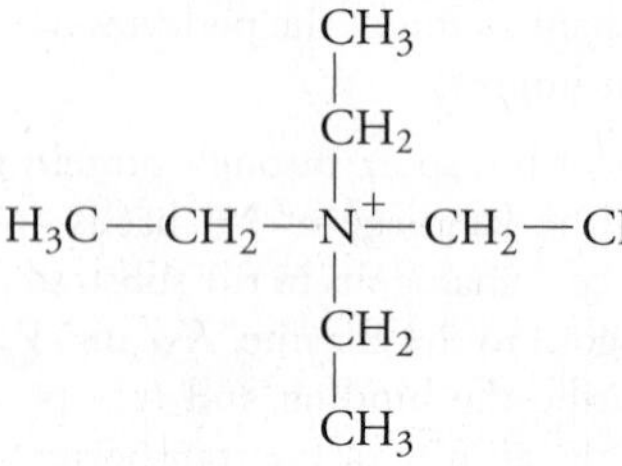

$$CH_3$$
$$|$$
$$CH_2$$
$$|$$
$$H_3C—CH_2—N^+—CH_2—CH_3$$
$$|$$
$$CH_2$$
$$|$$
$$CH_3$$

Tetraethylammonium (TEA)

40. Unidirectional glucose transport into the brain was measured in the presence and absence of phlorizin. The velocity of transport was determined at various glucose concentrations and is displayed on the Lineweaver–Burk plot below.

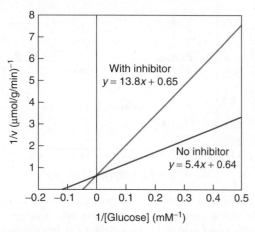

(a) Calculate the K_M and the V_{max} in the presence and absence of phlorizin.

(b) What kind of inhibitor is phlorizin? Explain.

41. Many ABC transporters are inhibited by vanadate, a phosphate analog. Why is vanadate an effective inhibitor of these transporters?

42. The ABC multidrug transporter LmrA from *Lactococcus* exports cytotoxic compounds like vinblastine. There are two binding sites for vinblastine on the transporter; one vinblastine binds with an association constant (equivalent to $1/K_d$) of 150 nM; the association constant of the second vinblastine is 30 nM. What do these values tell you about the mode of vinblastine binding to LmrA?

43. Kidney cells include two antiport proteins, a H^+/Na^+ exchanger and a Cl^-/HCO_3^- exchanger (see Box 2-D). What is the source of free energy that drives the transmembrane movement of all these ions?

44. Many cells have a mechanism for exporting ammonium ions. Describe how this could occur through secondary active transport.

45. The PEPT1 transporter aids in digestion by transporting di- and tripeptides into the cells lining the small intestine. There are three components of this system: **(a)** a symport transporter that ferries di- and tripeptides across the membrane, along with an H^+ ion, **(b)** a $Na^+–H^+$ antiporter, and **(c)** a Na,K-ATPase. Draw a diagram that illustrates how these three transporters work together to transport peptides into the cell.

46. The X-ray structure of a Ca^{2+}-ATPase was determined by crystallizing the enzyme along with adenosine-5'-(βγ-methylene) triphosphate (AMPPCP), an ATP analog (structure shown below). How did the co-crystallization strategy assist the crystallographers in capturing the image of this protein?

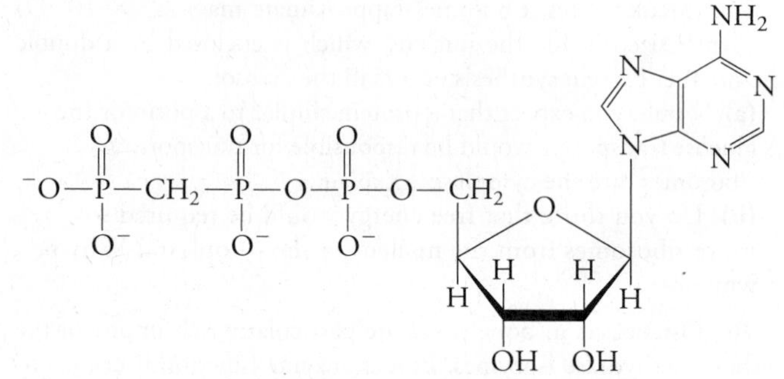

Adenosine-5'-(β,γ-methylene) triphosphate (AMPPCP)

9-4 Membrane Fusion

47. Myasthenia gravis is an autoimmune disorder characterized by muscle weakness and fatigue. Patients with the disease generate antibodies against the acetylcholine receptor in the postsynaptic cell; this results in a decrease in the number of receptors. The disease can be treated by administering drugs that inhibit acetylcholinesterase. Why is this an effective strategy for treating the disease?

48. Lambert–Eaton syndrome is another autoimmune muscle disorder, but in this disease, antibodies against the voltage-gated calcium channels in the presynaptic cell prevent the channels from opening. Patients with this disease suffer from muscle weakness. Explain why.

49. Like chymotrypsin, acetylcholinesterase is a member of the serine protease family because it contains an active site serine, and like chymotrypsin, it reacts with DIPF. Draw the structure of the enzyme's catalytic residue modified by DIPF.

50. Parathion and malathion are organophosphorus compounds similar to DIPF (see Problem 49). These compounds are sometimes used as insecticides. Why are these compounds deadly poisons?

51. The toxin produced by *Clostridium tetani*, which causes tetanus, is a protease that cleaves and destroys SNAREs. Explain why this activity would lead to muscle paralysis.

52. The drug known as Botox is a preparation of botulinum toxin, which is similar to the tetanus toxin (Problem 51). Describe the biochemical basis for its use by plastic surgeons, who inject small amounts of it to alleviate wrinkling in areas such as around the eyes.

53. Phosphatidylinositol is a membrane glycerophospholipid whose head group includes a monosaccharide (inositol) group. A certain kinase adds another phosphate group to a phosphorylated phosphatidylinositol:

54. Some studies show that prior to membrane fusion, the proportion of diacylglycerol in the bilayer increases. Explain how the presence of this lipid would aid the fusion process.

55. In autophagy (literally, "self-eating"), a damaged or unneeded cellular organelle becomes enclosed in a compartment called an autophagosome. Autophagosome formation begins with an assembly of lipids and proteins that grows by acquiring additional lipids until it forms a small bilayer-enclosed compartment called a phagophore. This structure continues to expand, encircling the damaged organelle. A membrane fusion event closes off the autophagosome. Complete the diagram of autophagosome formation shown here. How many membranes now separate the damaged organelle from the rest of the cell?

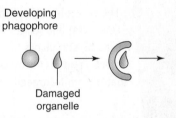

56. After an autophagosome has formed (Problem 55), a lysosome fuses with it to deliver hydrolytic enzymes that will eventually degrade the damaged organelle inside the autophagosome. Using the diagram you made for Problem 55 as a starting point, show that the lysosomal enzymes must degrade a lipid bilayer before they can begin to degrade the organelle.

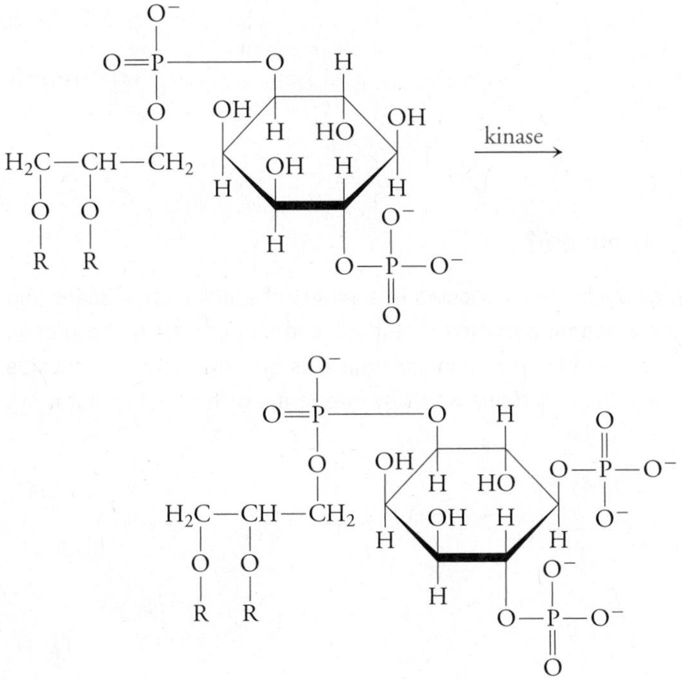

Why might this activity be required during the production of a new vesicle, which forms by budding from an existing membrane?

[SELECTED READINGS]

Gouaux, E., and MacKinnon, R., Principles of selective ion transport in channels and pumps, *Science* **310,** 1461–1465 (2005). [Compares several transport proteins of known structure and discusses the selectivity of Na⁺, K⁺, Ca²⁺, and Cl⁻ transport.]

Higgins, C. F., Multiple molecular mechanisms for multidrug resistance transporters, *Nature* **446,** 749–757 (2007). [Describes some general features of ABC transporters and others that function in drug resistance.]

King, L. S., Kozono, D., and Agre, P., From structure to disease: the evolving tale of aquaporin biology, *Nat. Rev. Mol. Cell Biol.* **5,** 687–698 (2004). [Reviews aquaporin structure and function.]

Morth, J. P., Pederson, B. P., Buch-Pederson, M. J., Andersen, J. P., Vilsen, B., Palmgren, M. G., and Nissen, P., A structural overview of the plasma membrane Na⁺, K⁺-ATPase and H⁺-ATPase ion pumps, *Nat. Rev. Mol. Cell Biol.* **12,** 60–70 (2011). [Summarizes the structures and physiological importance of two active transporters.]

Südhof, T. C., and Rothman, J. E., Membrane fusion: grappling with SNARE and SM proteins, *Science* **323,** 474–477 (2009). [Describes some of the proteins involved in membrane fusion.]

10

SIGNALING

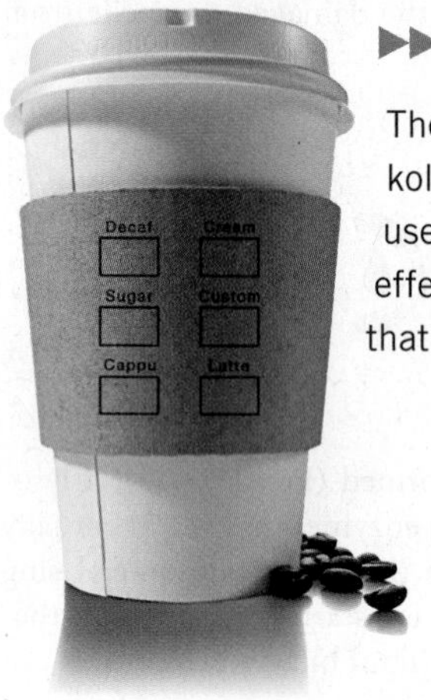

▶▶ **WHY** does coffee keep you awake?

The active ingredient in coffee is caffeine, a substance that occurs in a variety of seeds (coffee beans and kola nuts) and leaves (tea), where it acts as a natural pesticide. Solutions containing caffeine have been used as stimulants for thousands of years. The world's most popular drug acts quickly, produces few side effects, and is only mildly addictive. Like many drugs, caffeine works by interfering with signaling pathways that transmit extracellular signals to the cell's interior.

[Paul Eekhoff/Getty Images, Inc.]

THIS CHAPTER IN CONTEXT

Part 1 Foundations

Part 2 Molecular Structure and Function

10 Signaling

Part 3 Metabolism

Part 4 Genetic Information

Do You Remember?

- Some proteins can adopt more than one stable conformation (Section 4-3).
- Allosteric regulators can inhibit or activate enzymes (Section 7-3).
- Cholesterol and other lipids that do not form bilayers have a variety of other functions (Section 8-1).
- Integral membrane proteins completely span the bilayer by forming one or more α helices or a β barrel (Section 8-3).
- Conformational changes resulting from ATP hydrolysis drive Na^+ and K^+ transport in the Na,K-ATPase (Section 9-3).

All cells, including prokaryotes, must have mechanisms for sensing external conditions and responding to them. Because the cell membrane creates a barrier between outside and inside, communication typically involves an extracellular molecule binding to a cell-surface receptor. The receptor then changes its conformation to transmit information to the cell interior. **Signal transduction** may require many proteins, from the receptor itself to the intracellular proteins that ultimately respond to the signal by changing their behavior. We begin this chapter by describing some characteristics of signal transduction pathways and then examine some well-known signaling systems that involve G proteins and receptor tyrosine kinases.

10-1 General Features of Signaling Pathways

Every signal transduction pathway requires a **receptor,** most commonly an integral membrane protein, that specifically binds a small molecule called a **ligand.** A receptor does not merely bind its ligand in the way that hemoglobin binds oxygen; rather, *a receptor interacts with its ligand in such a way that some kind of response occurs inside the cell.*

A ligand binds to a receptor with a characteristic affinity

Extracellular signals can take many forms, including amino acids and their derivatives, peptides, lipids, and other small molecules (Table 10-1). Some are formally called **hormones,** which are substances produced in one tissue that affect the functions of other tissues, but many signals go by other names. Keep in mind that other stimuli—such as light, mechanical stress, odorants, and tastants—also serve as signals for cells, although we will not discuss them here. Some bacterial signal molecules are described in Box 10-A.

Signaling molecules behave much like enzyme substrates: *They bind to their receptors with high affinity, reflecting the structural and electronic complementarity between each ligand and its binding site.* Receptor–ligand binding can be written as a reversible reaction, where R represents the receptor and L the ligand:

$$R + L \rightleftharpoons R \cdot L$$

Biochemists express the strength of receptor–ligand binding as a dissociation constant, K_d, which is the reciprocal of the association constant. For this reaction,

$$K_d = \frac{[R][L]}{[R \cdot L]} \qquad \text{[10-1]}$$

(See Sample Calculation 10-1.) In keeping with other binding phenomena, such as oxygen binding to myoglobin (Section 5-1) or substrate binding to an enzyme (Section 7-2), K_d is the ligand concentration at which the receptor is half-saturated with ligand; in other words, half the receptor molecules have bound ligand (Fig. 10-1).

> **KEY CONCEPTS**
> - Receptor-ligand binding is described in terms of a dissociation constant.
> - G protein-coupled receptors and receptor tyrosine kinases are the two major types of receptors that transduce extracellular signals to the cell interior.
> - Regulatory mechanisms limit the extent of signaling.

[**TABLE 10-1**] Examples of Extracellular Signals

Hormone	Chemical Class	Source	Physiological Function
Auxin	Amino acid derivative	Most plant tissues	Promotes cell elongation and flowering in plants
Cortisol	Steroid	Adrenal gland	Suppresses inflammation
Epinephrine	Amino acid derivative	Adrenal gland	Prepares the body for action
Erythropoietin	Polypeptide (165 residues)	Kidneys	Stimulates red blood cell production
Growth hormone	Polypeptide (19 residues)	Pituitary gland	Stimulates growth and metabolism
Nitric oxide	Gas	Vascular endothelial cells	Triggers vasodilation
Thromboxane	Eicosanoid	Platelets	Activates platelets and triggers vasoconstriction

| BOX 10-A | BIOCHEMISTRY NOTE |

Bacterial Quorum Sensing

Even free-living single-celled organisms occasionally need to communicate with others of their kind. In bacteria, one form of intercellular communication is known as **quorum sensing,** which allows the cells to monitor population density and adjust gene expression accordingly. As a result, groups of cells can undertake communal endeavors such as producing the polysaccharides and other materials that form a protective matrix called a biofilm (discussed in Section 11-2). Quorum sensing also prepares cells to take up or exchange DNA, an occasional necessity for asexually reproducing organisms. Pathogenic bacteria may use quorum sensing to wait until their numbers are sufficiently high before synthesizing toxins and other proteins needed to attack a host organism.

The essence of quorum sensing is that *cells respond to some signal molecule that increases in concentration as cell density increases.* Only a few types of these molecules have been identified. One type consists of acyl homoserine lactones. The acyl chains of these molecules, which may include from 4 to 18 carbons, are derived either from fatty acids within the cell or from exogenous lipids.

An acyl homoserine lactone

A number of different acyl chains can be attached to the homoserine lactone group, making this class of compounds highly diverse. These lipids are poorly soluble in water, so they may be released from cells in the form of lipid vesicles (Section 2-2) that also contain other types of molecules. Because they are hydrophobic, the molecules can diffuse across the membranes of recipient cells and combine with a receptor protein in the cytosol. The receptor–ligand complex then binds to DNA to turn on the expression of certain genes.

A given bacterial species may produce dozens of different molecules that could be used for quorum sensing. Some of these molecules appear to do double duty as toxins for other species of bacteria, thus coordinating the growth of one species at the expense of others. In retaliation, some bacteria have evolved mechanisms to degrade the signals produced by other species or to synthesize molecules that competitively inhibit receptor binding by the other signals.

● **Question:** Microbiologists have proposed that drugs that interfere with quorum sensing would be useful as antibiotics, particularly since there would be little selective pressure for individual bacteria to evolve resistance to the drug. Explain.

Figure 10-1 Receptor–ligand binding. As the ligand concentration [L] increases, more receptor molecules bind ligand. Consequently, the fraction of receptors that have bound ligand [R · L] approaches 1.0. [R]$_T$ is the total concentration of receptors. The dissociation constant K_d is the ligand concentration at which half of the receptor molecules have bound ligand.

? Compare this graph to Figures 5-3 and 7-5.

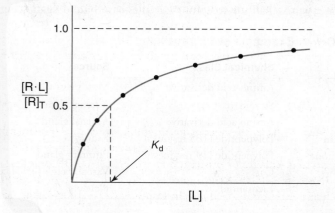

SAMPLE CALCULATION 10-1

PROBLEM

A sample of cells has a total receptor concentration of 10 mM. Twenty-five percent of the receptors are occupied with ligand, and the concentration of free ligand is 15 mM. Calculate K_d for the receptor–ligand interaction.

SOLUTION

Because 25% of the receptors are occupied, $[\text{R} \cdot \text{L}] = 2.5$ mM and $[\text{R}] = 7.5$ mM. Use Equation 10-1 to calculate K_d:

$$K_d = \frac{[\text{R}][\text{L}]}{[\text{R} \cdot \text{L}]}$$
$$= \frac{(0.0075)(0.015)}{(0.0025)}$$
$$= 0.045 \text{ M} = 45 \text{ mM}$$

PRACTICE PROBLEMS

1. A sample of cells has a total receptor concentration of 24 μM, and 40% of the receptors are occupied with ligand. The concentration of free ligand is 10 μM. Calculate K_d.
2. K_d for a receptor–ligand interaction is 3 mM. When the concentration of free ligand is 18 mM and the concentration of free receptor is 5 mM, what is the concentration of receptor that is occupied by ligand?
3. The total concentration of receptors in a sample is 20 mM. The concentration of free ligand is 5 mM, and K_d is 10 mM. Calculate the percentage of receptors that are occupied by ligand.

A ligand that binds to a receptor and elicits a biological effect is known as an **agonist.** For example, adenosine is the natural agonist of the adenosine receptor. Adenosine signaling in cardiac muscle slows the heart, and adenosine signaling in the brain leads to a decrease in neurotransmitter release, producing a sedative effect.

▶▶ WHY does coffee keep you awake?

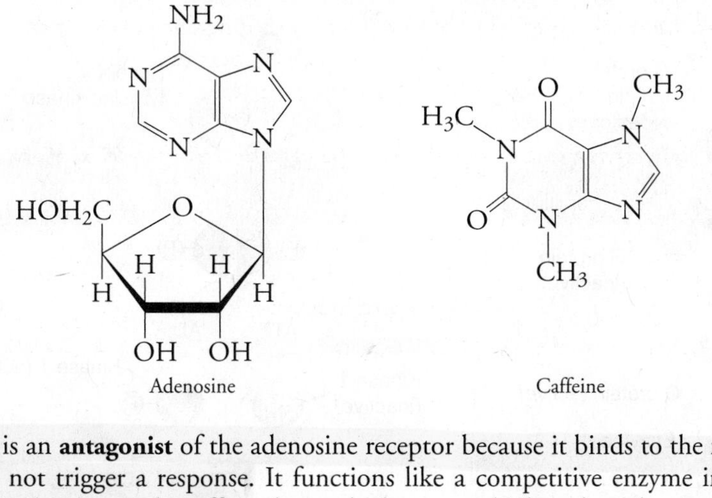

Adenosine   Caffeine

Caffeine is an **antagonist** of the adenosine receptor because it binds to the receptor but does not trigger a response. It functions like a competitive enzyme inhibitor (Section 7-3). As a result, caffeine keeps the heart rate high and produces a sense of wakefulness. Like caffeine, the majority of drugs currently in clinical use act as agonists or antagonists for various receptors involved in regulating such things as blood pressure, reproduction, and inflammation.

Most signaling occurs through two types of receptors

When an agonist binds to the adenosine receptor, which is a transmembrane protein, the receptor undergoes a conformational change so that it can interact with an intracellular protein called a **G protein.** Such receptors are therefore called **G protein–coupled receptors (GPCRs).** G proteins are named for their ability to bind guanine nucleotides (GTP and GDP). In response to receptor–ligand binding, the

G protein becomes activated and in turn interacts with and thereby activates additional intracellular proteins. Often, one of these is an enzyme that generates a small molecule product that diffuses throughout the cell. These small molecules are called **second messengers** because they represent the intracellular response to the extracellular, or first, message that binds to the GPCR. A variety of substances serve as second messengers in cells, including nucleotides, nucleotide derivatives, and the polar and nonpolar portions of membrane lipids. The presence of a second messenger can alter the activities of cellular proteins, leading ultimately to changes in metabolic activity and gene expression. These events are summarized in **Figure 10-2a.**

A second type of receptor, also a transmembrane protein, becomes activated as a kinase as a result of ligand binding. A **kinase** is an enzyme that transfers a phosphoryl group from ATP to another molecule. In this case, the phosphoryl group is condensed with the hydroxyl group of a Tyr side chain on a target protein, so these receptors are termed **receptor tyrosine kinases.** In some signal transduction pathways involving receptor tyrosine kinases, the target protein is also a kinase that becomes catalytically active when phosphorylated. The result may be a series of kinase-activation events that eventually lead to changes in metabolism and gene expression (Fig. 10-2b).

Some receptor systems include both G proteins and tyrosine kinases, and others operate by entirely different mechanisms. For example, the acetylcholine receptor on muscle cells (Fig. 9-19) is a ligand-gated ion channel. When acetylcholine is released into the neuromuscular synapse and binds to the receptor, Na^+ ions flow into the muscle cell, causing depolarization that leads to an influx of Ca^{2+} ions to trigger muscle contraction.

The effects of signaling are limited

The multistep nature of signaling pathways and the participation of enzyme catalysts ensure that the signal represented by an extracellular ligand will be amplified as it is transduced inside the cell (see Fig. 10-2). Consequently, *a relatively small extracellular*

Figure 10-2 Overview of signal transduction pathways. Ligand binding to a cell-surface receptor causes a signal to be transduced to the cell interior, leading eventually to changes in the cell's behavior. (a) Ligand binding to a G protein–coupled receptor triggers the activation of a G protein, which then activates an enzyme that produces a second messenger. Second messenger molecules diffuse away to activate or inhibit the activity of target proteins in the cell. (b) Ligand binding to a receptor tyrosine kinase activates the kinase activity of the receptor so that intracellular proteins become phosphorylated. A series of kinase reactions activates or inhibits target proteins by adding phosphoryl groups to them.

? **Which pathway components are enzymes?**

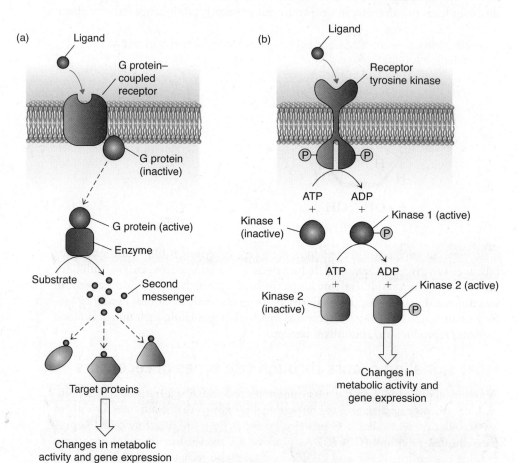

signal can have a dramatic effect on a cell's behavior. The cell's responses to the signal, however, are regulated in various ways.

The speed, strength, and duration of a signaling event may depend on the cellular location of the components of the signaling pathway. There is evidence that components for some pathways are preassembled in multiprotein complexes in or near the plasma membrane so that they can be quickly activated when the ligand docks with its receptor. Components that must diffuse long distances to reach their targets, or that move from the cytoplasm to the nucleus, may need more time to trigger cellular responses.

Because signaling pathways tend to be branched rather than completely linear, the same intracellular components may participate in more than one signal transduction pathway, so two different extracellular signals could ultimately achieve the same intracellular results. Alternatively, two signals could cancel each other's effects. The response of a given cell, which expresses many different types of receptors, therefore depends on how various signals are integrated. Similarly, different types of cells may include different intracellular components and therefore respond to the same ligand in different ways.

In a biological system that obeys the law of homeostasis, *any process that is turned on must eventually be turned off.* Such control applies to signaling pathways. For example, shortly after a G protein has been activated by its associated receptor, it becomes inactive again. The action of kinases is undone by the action of enzymes that remove phosphoryl groups from target proteins. These and other reactions restore the signaling components to their resting state so that they can be ready to respond again when another ligand binds to its receptor.

Finally, most people are aware that a strong odor loses its pungency after a few minutes. This occurs because olfactory receptors, like other types of receptors, become **desensitized.** In other words, the receptors becomes less able to transmit a signal even when continually exposed to ligand. Desensitization may allow the signaling machinery to reset itself at a certain level of stimulation so that it can better respond to subsequent changes in ligand concentration.

> **CONCEPT REVIEW**
> - What do hormone receptors have in common with enzymes and simple binding proteins?
> - What is the purpose of second messengers?
> - How are extracellular signals amplified inside the cell?
> - Explain why a receptor would need to be turned off or desensitized.

10-2 G Protein Signaling Pathways

At least 800 genes in the human genome encode G protein–coupled receptors, and these proteins are responsible for transducing the majority of extracellular signals. In this section we will describe some features of these receptors, their associated G proteins, and various second messengers and their intracellular targets.

G protein-coupled receptors include seven transmembrane helices

The GPCRs are known as 7-transmembrane (7TM) receptors because they include seven α helices, which are arranged much like those of the membrane protein bacteriorhodopsin (Fig. 8-8). Many G protein–coupled receptors are palmitoylated at a Cys residue, so they are also lipid-linked proteins (Section 8-3). In the GPCR family, the helical segments are more strongly conserved than the loops that join them on the intracellular and extracellular sides of the membrane.

KEY CONCEPTS
- Ligand binding to a G protein-coupled receptor alters its conformation so that an intracellular G protein becomes activated.
- The G protein stimulates adenylate cyclase to produce the second messenger cAMP, which activates protein kinase A.
- G protein-dependent signaling is limited by several mechanisms.
- The phosphoinositide signaling system activates a G protein, which leads to the production of lipid-derived second messengers and the activation of protein kinase C.
- Cross-talk results when signaling pathways share components.

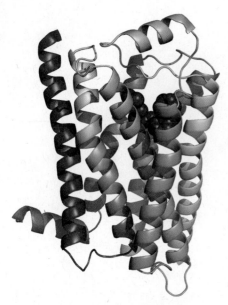

Figure 10-3 The β₂-adrenergic receptor. The backbone structure of the protein is colored in rainbow order from the N-terminus (blue) to the C-terminus (red). A ligand is shown in space-filling form in blue. [Structure (pdb 2RH1) determined by V. Cherezov, D. M. Rosenbaum, M. A. Hanson, S. G. F. Rasmussen, F. S. Thian, T. S. Kobilka, H. J. Choi, P. Kuhn, W. I. Weis, B. K. Kobilka, and R. C. Stevens.]

? Explain why the receptor for a polar hormone must be a transmembrane protein.

Figure 10-4 A GPCR-G protein complex. (a) Side view. The GPCR is purple, a bound agonist is red, and the G protein is yellow, green, and blue. [Structure (pdb 3SN6) determined by S. G. F. Rasmussen, B. T. DeVree, Y. Zou, A. C. Kruse, K. Y. Chung, T. S. Kobilka, F. S. Thian, P. S. Chae, E. Pardon, D. Calinski, J. M. Mathiesen, S. T. A. Shah, J. A. Lyons, M. Caffrey, S.H. Gellman, J. Steyaert, G. Skiniotis, W. I. Weis, R. K. Sunahara, and B. K. Kobilka.] (b) The β subunit (green) has a propeller-like structure. The small γ subunit (yellow) associates tightly with the β subunit. The α subunit (blue) binds a guanine nucleotide (GDP, orange) in a cleft between two domains. The α and β subunits are covalently attached to lipids, so they are anchored to the cytosolic leaflet of the plasma membrane near the receptor. [Structure (pdb 1GP2) determined by M. A. Wall and S. R. Sprang.] ⊕ **See Interactive Exercise. A heterotrimeric G protein.**

The structure of one of these proteins, the β₂-adrenergic receptor, is shown in **Figure 10-3**. The ligand-binding site of a GPCR is defined by a portion of the helical core of the protein as well as its extracellular loops. Aside from this general location, there are few similarities in the structures of the binding sites in different GPCRs, which is consistent with the observation that each receptor is specific for only one or a few of many possible ligands, which may be large or small, polar or nonpolar substances.

The physiological ligands for the β₂-adrenergic receptor are the hormones epinephrine and norepinephrine, which are synthesized by the adrenal glands from the amino acid tyrosine.

Epinephrine   Norepinephrine

These same substances, sometimes called adrenaline and noradrenaline, also function as neurotransmitters. They are responsible for the fight-or-flight response, which is characterized by fuel mobilization, dilation of blood vessels and bronchi (airways), and increased cardiac action. Antagonists that prevent signaling via the β₂-adrenergic receptor, known as β-blockers, are used to treat high blood pressure.

How does the receptor transmit the extracellular hormonal signal to the cell interior? *Signal transduction depends on conformational changes involving the receptor's transmembrane helices.* Two of the helices shift slightly to accommodate the ligand, which repositions one of the cytoplasmic protein loops. Studies with a variety of different ligands indicate that the receptor can actually adopt a range of conformations, suggesting that the receptor is not merely an on–off switch but can mediate the effects of strong as well as weak agonists.

The receptor activates a G protein

Ligand-induced conformational changes in a G protein–coupled receptor open up a pocket on its cytoplasmic side to create a binding site for a specific G protein (**Fig. 10-4a**). The G protein is presumably already close to the receptor, since it is lipid-linked. The trimeric G proteins associated with GPCRs consist of three subunits, designated α, β, and γ (**Fig. 10-4b**; other types of G proteins do not have

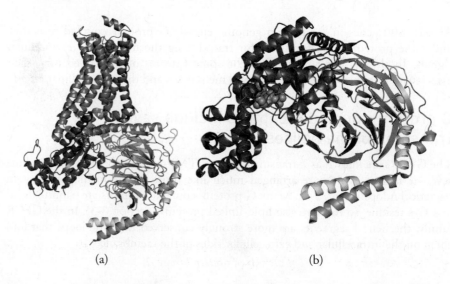

(a)   (b)

this three-part structure). In the resting state, GDP is bound to the α subunit, but association with the receptor causes the G protein to release GDP and bind GTP in its place. The third phosphate group of GTP is not easily accommodated in the αβγ trimer, so the α subunit dissociates from the β and γ subunits, which remain together. Once they dissociate, the α subunit and the βγ dimer are both active; that is, *they interact with additional cellular components in the signal transduction pathway.* However, since both the α and β subunits include lipid anchors, the G protein subunits do not diffuse far from the receptor that activated them.

The signaling activity of the G protein is limited by the intrinsic GTPase activity of the α subunit, which slowly converts the bound GTP to GDP:

$$GTP + H_2O \rightarrow GDP + P_i$$

Hydrolysis of the GTP allows the α and βγ units to reassociate as an inactive trimer (**Fig. 10-5**). The cost for the cell to switch a G protein on and then off again is the free energy of the GTP hydrolysis reaction (GTP is energetically equivalent to ATP).

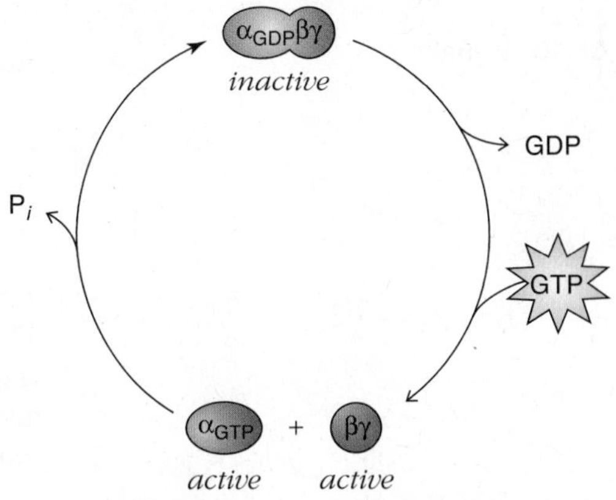

Adenylate cyclase generates the second messenger cyclic AMP

Cells contain a number of different types of G proteins that interact with various targets in the cell and activate or inhibit them. A single receptor may interact with more than one G protein, so the effects of ligand binding are amplified at this point. One of the major targets of the activated G protein is an integral membrane enzyme called adenylate cyclase. When the α subunit of the G protein binds, the enzyme's catalytic domains convert ATP to a molecule known as cyclic AMP (**cAMP**). cAMP is a second messenger that can freely diffuse in the cell.

Figure 10-5 **The G protein cycle.** The αβγ trimer, with GDP bound to the α subunit, is inactive. Ligand binding to a receptor associated with the G protein triggers a conformational change that causes GTP to replace GDP and the α subunit to dissociate from the βγ dimer. Both portions of the G protein are active in the signaling pathway. The GTPase activity of the α subunit returns the G protein to its inactive trimeric state.

⊕ **See Guided Exploration. Hormone signaling by the adenylate cyclase system.**

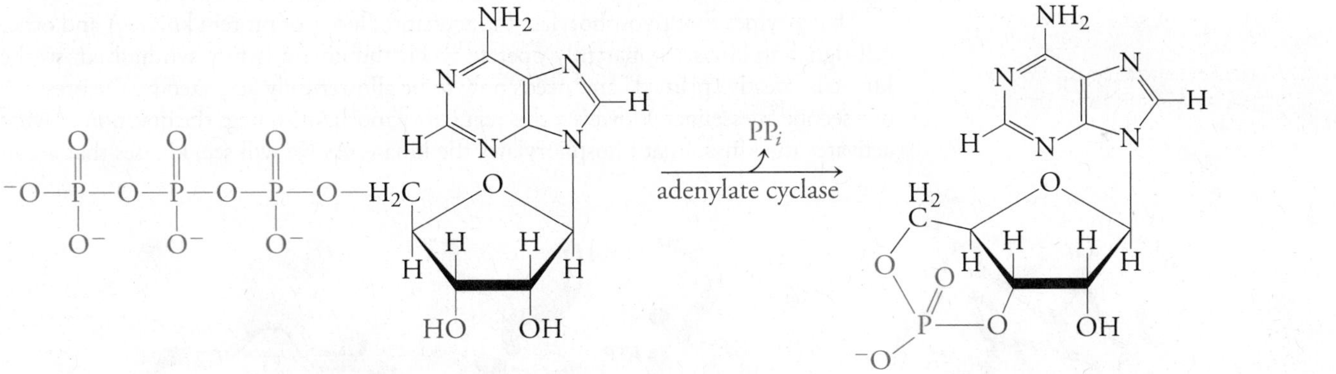

ATP → (adenylate cyclase, PP_i) → Cyclic AMP (cAMP)

Cyclic AMP activates protein kinase A

Among the targets of cAMP is an enzyme called protein kinase A or PKA. In the absence of cAMP, this kinase is an inactive tetramer of two regulatory (R) and two catalytic (C) subunits. A segment of each R subunit occupies the active site in a C subunit so that the kinase is unable to phosphorylate any substrates. cAMP binding to the regulatory subunits relieves the inhibition, causing the tetramer to release the two active catalytic subunits.

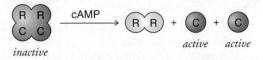

Consequently, cAMP functions as an allosteric activator of the kinase, and *the level of cAMP determines the level of activity of protein kinase A.*

Protein kinase A is known as a Ser/Thr kinase because it transfers a phosphoryl group from ATP to the Ser or Thr side chain of a target protein.

$$—CH_2—O—PO_3^{2-} \qquad —CH—O—PO_3^{2-}$$
$$\underset{\text{Phospho-Ser}}{} \qquad \underset{\text{Phospho-Thr}}{\overset{\displaystyle |}{CH_3}}$$

The substrates for the reaction bind in a cleft defined by the two lobes of the protein (**Fig. 10-6a**). Other kinases share this same core structure but often have additional domains that determine their subcellular location or provide additional regulatory functions.

In addition to regulation by cAMP binding to the R subunit, *protein kinase A itself is regulated by phosphorylation.* The protein's so-called activation loop, a segment of polypeptide near the entrance to the active site, includes a phosphorylatable Thr residue. When the loop is not phosphorylated, the kinase's active site is blocked. When phosphorylated, the loop swings aside and the kinase's catalytic activity increases; for some protein kinases, activity increases by several orders of magnitude. This activation effect is not merely a matter of improving substrate access to the active site but also appears to involve conformational changes that affect catalysis. For example, the negatively charged phospho-Thr interacts with a positively charged Arg residue in the active site. Efficient catalysis requires that this Arg residue and an adjacent Asp residue be repositioned for phosphoryl-group transfer from ATP to a protein substrate (Fig 10-6b).

Among the many targets of protein kinase A are enzymes involved in glycogen metabolism (Section 19-2). One result of signaling via the β_2-adrenergic receptor, which leads to protein kinase A activation by cAMP, is the phosphorylation and activation of an enzyme called glycogen phosphorylase, which catalyzes the removal of glucose residues, the cell's primary metabolic fuel, from glycogen, the cell's glucose-storage depot. Consequently, a signal such as epinephrine can mobilize the metabolic fuel needed to power the body's fight-or-flight response.

The enzymes that phosphorylate the activation loops of protein kinase A and other cell-signaling kinases apparently operate when the kinase is first synthesized, so the kinase is already "primed" and needs only to be allosterically activated by the presence of a second messenger. However, this regulatory mechanism begs the question of what activates the kinase that phosphorylates the kinase. As we will see, kinases that act in

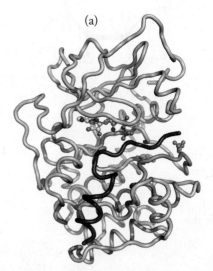

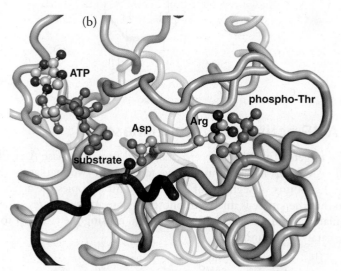

Figure 10-6 Protein kinase A. (a) The backbone of the catalytic subunit is light green, with its activation loop dark green. The phospho-Thr residue (right side) and ATP (left side) are shown in stick form. A peptide that mimics a target protein is blue. (b) Close-up view of the active site region. When the activation loop is phosphorylated, the phospho-Thr residue interacts with an Arg residue, and the adjacent Asp residue is positioned near the third phosphate group of ATP and the peptide substrate. Atoms are color-coded: C gray or green, O red, N blue, and P gold. [Structure (pdb 1ATP) determined by J. Zheng, E. A. Trafny, D. R. Knighton, N.-H. Xuong, S. S. Taylor, L. F. Teneyck, and J. M. Sowadski.]
➕ **See Interactive Exercise. C subunit of protein kinase A.**

series are common in biological signaling pathways, and many of these pathways are interconnected, making it difficult to trace simple cause-and-effect relationships.

Signaling pathways are also switched off

What happens after a ligand binds to a receptor, a G protein responds, a second messenger is produced, an effector enzyme such as protein kinase A is activated, and target proteins are phosphorylated? *To restore the cell to a resting state, any or all of the events of the signal transduction pathway can be blocked or reversed.* We have already seen that the intrinsic GTPase activity of G proteins limits their activity. And second messengers often have short lifetimes due to their rapid degradation in the cell. For example, cAMP is hydrolyzed by the enzyme cAMP phosphodiesterase:

$$\text{cAMP} \xrightarrow[\text{phosphodiesterase}]{H_2O} \text{AMP}$$

cAMP   AMP

Caffeine, in addition to being an adenosine receptor antagonist, can diffuse inside cells and inhibit cAMP phosphodiesterase. As a result, the cAMP concentration remains high, the action of protein kinase A is sustained, and stored metabolic fuels are mobilized, readying the body for action rather than sleep.

Some of the cell's G proteins may inhibit rather than activate adenylate cyclase and therefore decrease the cellular cAMP level. Some G proteins activate cAMP phosphodiesterase, with similar effects on cAMP-dependent processes. *A cell's response to a hormone signal depends in part on which G proteins respond.* Because a single type of hormone may stimulate several types of G proteins, the signaling system may be active for only a brief time before it is turned off.

The phosphorylations catalyzed by protein kinase A (and other kinases) can be reversed by the work of protein **phosphatases,** which catalyze a hydrolytic reaction to remove phosphoryl groups from protein side chains. Like kinases, phosphatases are generally specific for Ser or Thr, or Tyr, although some "dual specificity" phosphatases remove phosphoryl groups from all three side chains. The active-site pocket of the Tyr phosphatases is deeper than the pocket of Ser/Thr phosphatases in order to accommodate the larger phospho-Tyr side chain. Some protein phosphatases are transmembrane proteins; others are entirely intracellular. They tend to have multiple domains or multiple subunits, consistent with their ability to form numerous protein–protein interactions and participate in complex regulatory networks.

Ultimately, the dissociation of an extracellular ligand from its receptor can halt a signal transduction process, or the receptor can become desensitized. Desensitization of a G protein–coupled receptor begins with phosphorylation of the ligand-bound receptor by a GPCR kinase. The phosphorylated receptor is then recognized by a protein known as arrestin (**Fig. 10-7**), which includes Lys and Arg residues that bind the phosphoryl group. This binding halts signaling, presumably by blocking interactions with a G protein (hence the name *arrestin*), and promotes the movement of the receptor from the cell surface to an intracellular compartment by the process of endocytosis. Interestingly, arrestins also serve as scaffold proteins for organizing components of other signaling pathways. Experimental evidence suggests that some GPCRs, including the β2-adrenergic receptor, interact with arrestin so that they can initiate an intracellular response without ever recruiting a G protein.

▶▶ **WHY** does coffee keep you awake?

Figure 10-7 Arrestin. This arrestin, bovine β-arrestin 1, includes two cuplike domains that are believed to move relative to each other in order to cradle the phosphorylated G protein–coupled receptor and reduce its ability to activate a G protein. [Structure (pdb 1G4R) determined by C. Schubert and M. Han.]

The phosphoinositide signaling pathway generates two second messengers

The diversity of G protein–coupled receptors, together with the diversity of G proteins, creates almost unlimited possibilities for adjusting the levels of second messengers and altering the activities of cellular enzymes. Epinephrine, the hormone that activates the β_2-adrenergic receptor, also binds to a receptor known as the α-adrenergic receptor, which is part of the **phosphoinositide signaling system.** The α- and β-adrenergic receptors are situated in different tissues and mediate different physiological effects, even though they bind the same hormone. The G protein associated with the α-adrenergic receptor activates the cellular enzyme phospholipase C, which acts on the membrane lipid phosphatidylinositol bisphosphate. Phosphatidylinositol is a minor component of the plasma membrane (4–5% of the total phospholipids), and the bisphosphorylated form (with a total of three phosphate groups) is rarer still. Phospholipase C converts this lipid to inositol trisphosphate and diacylglycerol.

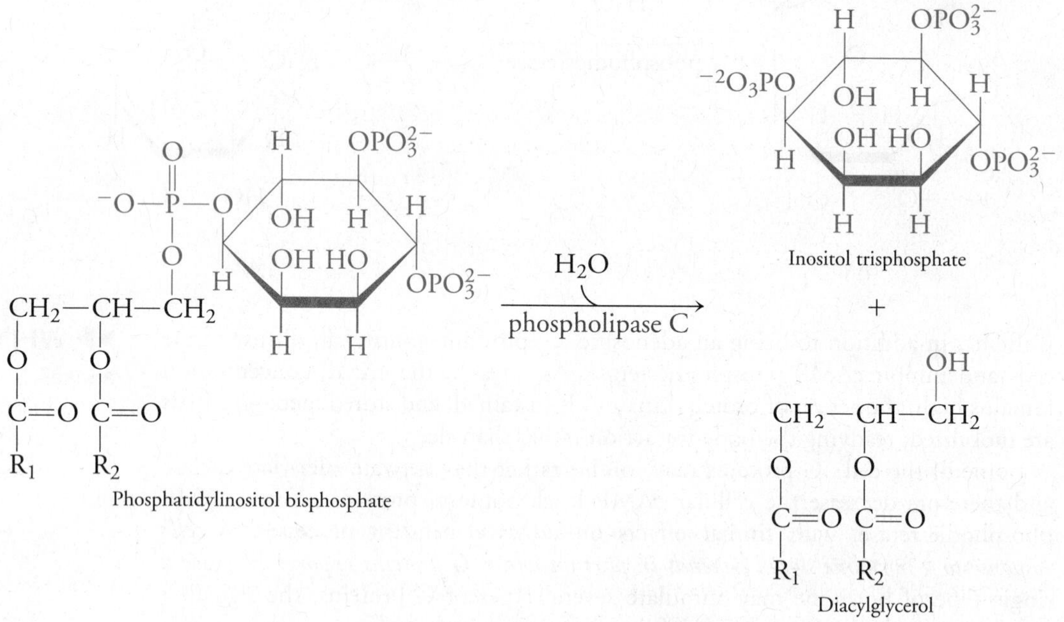

Phosphatidylinositol bisphosphate

H_2O → phospholipase C

Inositol trisphosphate

+

Diacylglycerol

⊕ **See Animated Figure. The phosphoinositide signaling system.**

The highly polar inositol trisphosphate is a second messenger that triggers the opening of calcium channels in the endoplasmic reticulum membrane, allowing Ca^{2+} ions to flow into the cytosol. The flux of Ca^{2+} initiates numerous events in the cell, including the activation of a Ser/Thr kinase known as protein kinase B or Akt. Inositol trisphosphate also directly activates kinases and can undergo additional phosphorylations to generate a series of second messengers containing up to eight phosphoryl groups.

The hydrophobic diacylglycerol product of the phospholipase C reaction is also a second messenger. Although it remains in the cell membrane, it can diffuse laterally to activate protein kinase C, which phosphorylates its target proteins at Ser or Thr residues. In its resting state, protein kinase C is a cytosolic protein with an activation loop blocking its active site. Noncovalent binding to diacylglycerol docks the enzyme at the membrane surface so that it changes its conformation, repositions the activation loop, and becomes catalytically active. As in protein kinase A (which shares about 40% sequence identity), phosphorylation of a Thr residue in the activation loop, a requirement for catalytic activity, has already occurred. Full activation of some forms of protein kinase C also requires Ca^{2+}, which is presumably available as a result of the activity of the inositol trisphosphate second messenger. Among the targets of protein kinase C are proteins involved in the regulation of gene expression and cell division. Certain compounds that mimic diacylglycerol can activate protein kinase C, leading eventually to the uncontrolled growth characteristic of cancer.

Phospholipase C can be activated not only by a G protein–coupled receptor such as the α-adrenergic receptor but also by other signaling systems involving receptor tyrosine kinases. This is an example of **cross-talk,** the interconnections between signaling pathways that share some intracellular components. The phosphoinositide signaling pathway is regulated in part by the action of lipid phosphatases that remove phosphoryl groups from the phosphatidylinositol bisphosphate substrate that gives rise to the second messengers.

Another example of the overlap between signaling pathways involves sphingolipids such as sphingomyelin (Fig. 8-2), which is a normal component of membranes. Ligand binding to certain receptor tyrosine kinases leads to activation of sphingomyelinases that release sphingosine and ceramide (ceramide is sphingomyelin without its phosphocholine head group). Ceramide is a second messenger that activates kinases, phosphatases, and other cellular enzymes. Sphingosine, which undergoes phosphorylation (by a receptor tyrosine kinase–dependent mechanism) to sphingosine-1-phosphate, is both an intracellular and extracellular signaling molecule. It inhibits phospholipase C inside the cell, and it apparently exits the cell via an ABC transport protein (Section 9-3), then binds to a G protein–coupled receptor to trigger additional cellular responses.

Calmodulin mediates some Ca^{2+} signals

In some cases where Ca^{2+} ions elicit a change in an enzyme's activity, the change is mediated by a Ca^{2+}-binding protein known as calmodulin. This small (148-residue) protein binds two Ca^{2+} ions in each of its two globular domains, which are separated by a long α helix **(Fig. 10-8a)**. Free calmodulin has an extended shape, but in the presence of Ca^{2+} and a target protein, the helix partially unwinds and calmodulin bends in half to grasp the target protein to activate or inhibit it (Fig. 10-8b).

(a)

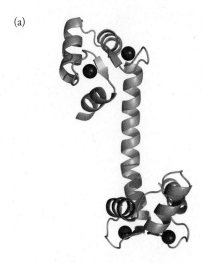

(b)

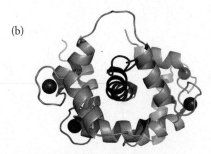

Figure 10-8 Calmodulin. (a) Isolated calmodulin has an extended shape. The four bound Ca^{2+} ions are shown as blue spheres. (b) When bound to a target protein (blue helix), calmodulin's long central helix unwinds and bends so that the protein can wrap around its target. [Structure of calmodulin (pdb 3CLN) determined by Y. S. Babu, C.E. Bugg, and W. J. Cook. Structure of calmodulin bound to a 26-residue target (pdb 2BBM) determined by G. M. Clore, A. Bax, M. Ikura, and A. M. Gronenborn.]

CONCEPT REVIEW
- Draw a simple diagram that includes all the components of the β$_2$-adrenergic receptor signaling pathway. Draw a similar diagram for the α-adrenergic receptor signaling pathway.
- For each of these pathways, indicate points where signaling activity can be switched off.
- What is the relationship between an extracellular hormone and a second messenger in terms of concentration and specificity?
- What is the relationship between kinases and phosphatases?
- Explain how the same hormone can elicit different responses in different cells.
- Explain how different hormones can elicit the same response in a cell.

10-3 Receptor Tyrosine Kinases

A number of hormones and other signaling molecules that regulate cell growth and division bind to cell-surface receptors that operate as tyrosine kinases. Most of these receptors are monomeric, with a single membrane-spanning segment. Ligand binding allows them to form dimers, and in this state, their cytoplasmic domains become catalytically active kinases. The insulin receptor serves as a model for other receptor tyrosine kinases, although it exists as a dimer in its resting state.

The insulin receptor has two ligand-binding sites

Insulin, a 51-residue polypeptide hormone that regulates many aspects of fuel metabolism in mammals, binds to receptors that are present in most of the body's tissues. The receptor is constructed from two long polypeptides that are cleaved after their synthesis, so the mature receptor has an α$_2$β$_2$ structure in which all four polypeptide segments are held together by disulfide bonds **(Fig. 10-9a)**.

KEY CONCEPTS
- Ligands such as insulin activate the tyrosine kinase activity of their receptors.
- Receptor tyrosine kinases trigger cellular responses by phosphorylating target proteins and by activating Ras.

⊕ **See Guided Exploration. Hormone signaling by the receptor tyrosine kinase system.**

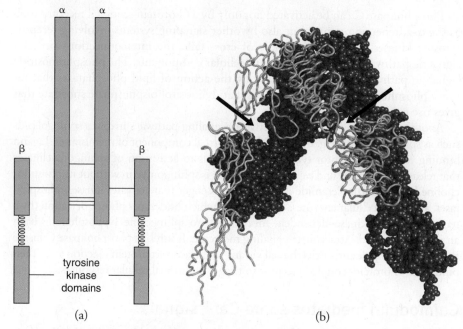

Figure 10-9 The insulin receptor. (a) Schematic diagram. The receptor consists of two α and two β subunits joined by disulfide bonds (horizontal lines). The α subunits bind insulin, and the β subunits each include a membrane-spanning segment (coil) and a cytoplasmic tyrosine kinase domain. (b) The extracellular portion of the insulin receptor. The cell surface is at the bottom. One αβ pair of subunits is shown in space-filling form, and the other is shown as a backbone trace. Insulin binds to one of the two binding sites indicated by arrows. [Structure (pdb 2DTG) determined by M. C. Lawrence and V. A. Streltsov.]

The extracellular portion of the insulin receptor has an inverted V shape with multiple structural domains (Fig. 10-9b). Segments of the α subunits define the two insulin-binding sites, but the sites are too far apart (~65 Å) to simultaneously bind a single hormone molecule. Instead, biochemical evidence indicates that binding to just one site pulls the two α subunits together in such a way that the second binding site cannot bind insulin. Interdomain interactions, along with the disulfide linkages, suggest that the receptor is rigid, and this feature is believed to be important in transducing the message (insulin binding to the extracellular α subunits) to the intracellular signaling apparatus (the tyrosine kinase domains of the β subunits). Presumably, other receptor tyrosine kinases that exist as inactive monomers come together in a similar fashion after binding their ligands and reposition and thereby activate their tyrosine kinase domains.

The receptor undergoes autophosphorylation

The ligand-induced conformational change in a receptor tyrosine kinase brings its two tyrosine kinase domains close enough that they can phosphorylate each other. Because the kinases appear to phosphorylate themselves, this process is termed **autophosphorylation.** Each tyrosine kinase domain has the typical kinase core structure, including an activation loop that lies across the active site to prevent substrate binding. Phosphorylation of three Tyr residues in the activation loop causes a conformational change that allows the enzyme to bind ATP and protein substrates and to catalyze phosphoryl group transfer **(Fig. 10-10).**

To initiate the responses that promote cell growth and division, growth factor receptors and other receptor tyrosine kinases phosphorylate various intracellular target proteins. They also switch on other pathways that involve small monomeric G proteins (which are also GTPases) such as Ras. The tyrosine kinase domain of the receptor does not directly interact with Ras but instead relies on one or more adapter proteins that form a bridge between Ras and the phospho-Tyr residues of the receptor **(Fig. 10-11).** These proteins also stimulate Ras to release GDP and bind GTP.

Like other G proteins, *Ras is active as long as it has GTP bound to it.* The Ras · GTP complex allosterically activates a Ser/Thr kinase, which becomes active and phos-

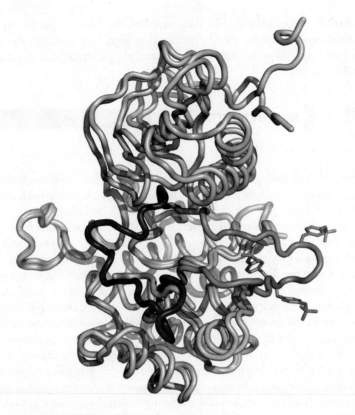

Figure 10-10 Activation of the insulin receptor tyrosine kinase. The backbone structure of the inactive tyrosine kinase domain of the insulin receptor is shown in light blue, with the activation loop in dark blue. The structure of the active tyrosine kinase domain is shown in light green, with the activation loop in dark green. Note that three Tyr side chains in the activation loop have been phosphorylated (a result of each tyrosine kinase domain phosphorylating the other) so that the activation loop has swung aside to better expose the active site. The phospho-Tyr side chains are shown in stick form with atoms color coded: C green, O red, and P orange. [Structure of the inactive kinase domain (pdb 1IRK) determined by S. R. Hubbard, L. Wei, L. Ellis, and W. A. Hendrickson. Structure of the active kinase domain (pdb 1IR3) determined by S. R. Hubbard.] ⊕ **See Interactive Exercise. Tyrosine kinase domain of the insulin receptor.**

? **How would a phosphatase affect the activity of the insulin receptor?**

phorylates another kinase, activating it, and so on. Cascades of several kinases can therefore amplify the initial growth factor signal.

The ultimate targets of Ras-dependent signaling cascades are nuclear proteins, which, when phosphorylated, bind to specific sequences of DNA to induce (turn on) or repress (turn off) gene expression. The altered activity of these **transcription factors** means that the original hormonal signal not only alters the activities of cellular enzymes on a short time scale (seconds to minutes) via phosphorylation but also affects protein synthesis, a process that may require several hours or more.

Ras signaling activity is shut down by the action of proteins that enhance the GTPase activity of Ras so that it returns to its inactive GDP-bound form. In addition, phosphatases reverse the effects of the various kinases. Like the other signaling pathways we have examined, the receptor tyrosine kinase pathways are not linear and they are capable of cross-talk. For example, some receptor tyrosine

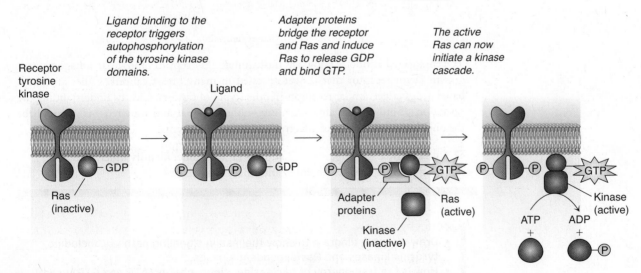

Figure 10-11 The Ras pathway. Ras links receptor–hormone binding to an intracellular kinase cascade.

? **Describe the role of protein conformational changes in each step of the pathway.**

kinases directly or indirectly (via Ras) activate the kinase that phosphorylates phosphatidylinositol lipids, thereby promoting signaling through the phosphoinositide pathway. Abnormalities in these signaling pathways can promote tumor growth (Box 10-B).

BOX 10-B ⬢ BIOCHEMISTRY NOTE

Cell Signaling and Cancer

The progress of a cell through the cell cycle, from DNA replication through the phases of mitosis, depends on the orderly activity of signaling pathways. Cancer, which is the uncontrolled growth of cells, can result from a variety of factors, including overactivation of the signaling pathways that stimulate cell growth. In fact, *the majority of cancers include mutations in the genes for proteins that participate in signaling via Ras and the phosphoinositide pathways.* These altered genes are termed **oncogenes,** from the Greek *onkos,* meaning "tumor."

Oncogenes were first discovered in certain cancer-causing viruses. The viruses presumably picked up the normal gene from a host cell, then mutated. In some cases, an oncogene encodes a growth factor receptor that has lost its ligand-binding domain but retains its tyrosine kinase domain. As a result, the kinase is constitutively (constantly) active, promoting cell growth and division even in the absence of the growth factor. Some *RAS* oncogenes generate mutant forms of Ras that hydrolyze GTP extremely slowly, thus maintaining the signaling pathway in the "on" state. Note that oncogenic mutations can strengthen an activating event or weaken an inhibitory event; in either case, the outcome is excessive signaling activity.

The importance of various kinases in triggering or sustaining tumor growth has made these enzymes attractive targets for anticancer drugs. Some forms of leukemia (a cancer of white blood cells) are triggered by a chromosomal rearrangement that generates a kinase with constitutive signaling activity, called Bcr-Abl. The drug imatinib (Gleevec) specifically inhibits this kinase without affecting any of the cell's numerous other kinases. The result is an effective anticancer treatment with few side effects.

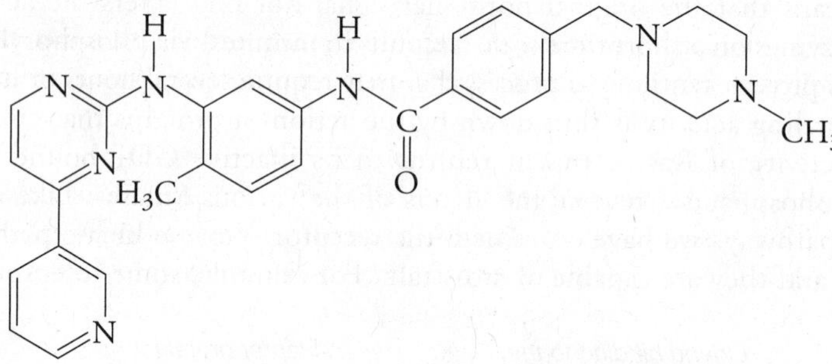

Gleevec (imatinib)

The engineered antibody known as trastuzumab (Herceptin) binds as an antagonist to a growth factor receptor that is overexpressed in many breast cancers. Other antibody-based drugs target similar receptors in other types of cancers. Clearly, understanding the operation of growth-signaling pathways—both normal and mutated—is essential for the ongoing development of effective anticancer treatments.

⬢ **Question:** Using Figures 10-2 and 10-11 as a guide, identify other types of signaling proteins that are potential anticancer drug targets.

CONCEPT REVIEW
- Draw a simple diagram to show the insulin signaling pathway, including tyrosine kinases and Ras-dependent kinases.
- How is the free energy of nucleoside triphosphates (ATP and GTP) used to turn on cellular responses to extracellular signals?
- Explain how kinases and transcription factors mediate cellular responses over different time scales.

10-4 Lipid Hormone Signaling

KEY CONCEPTS
- Lipid hormones bind to intracellular receptors that regulate gene expression.
- Eicosanoids are local mediators that act through G protein-coupled receptors.

Some hormones do not need to bind to cell-surface receptors because they are lipids and can cross the membrane to interact with intracellular receptors. For example, retinoic acid and the thyroid hormones thyroxine (T_4) and triiodothyronine (T_3) belong to this class of hormones **(Fig. 10-12)**. Retinoic acid (retinoate), a compound that regulates cell growth and differentiation, particularly in the immune system, is synthesized from retinol, a derivative of β-carotene (Box 8-B). The thyroid hormones, which generally stimulate metabolism, are derived from a large precursor protein called thyroglobulin: Tyr side chains are enzymatically iodinated, then two of these residues undergo condensation, and the hormones are liberated from thyroglobulin by proteolysis.

The 27-carbon cholesterol, introduced in Section 8-1, is the precursor of a large number of hormones that regulate metabolism, salt and water balance, and reproductive functions. Androgens (which are primarily male hormones) have 19 carbons, and estrogens (which are primarily female hormones) contain 18 carbons. Cortisol, a C_{21} glucocorticoid hormone, affects the metabolic activities of a wide variety of tissues.

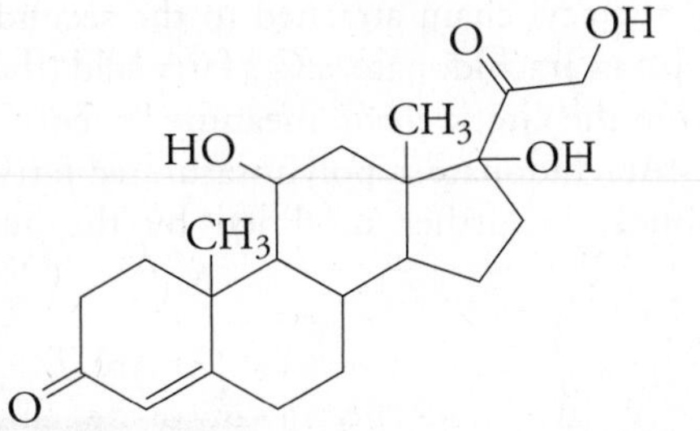

Cortisol

Retinoic acid, thyroid hormones, and steroids are all hydrophobic molecules that are carried in the bloodstream either by specific carrier proteins or by albumin, a sort of all-purpose binding protein.

The receptors to which the lipid hormones bind are located inside the appropriate target cell, either in the cytoplasm or the nucleus. Ligand binding often—but not always—causes the receptors to form dimers. Each receptor subunit is constructed from several modules, which include a ligand-binding domain and a DNA-binding domain. The ligand-binding domains are as varied as their hormone ligands, but the DNA-binding domains exhibit a common structure that includes two zinc fingers, which are cross-links formed by the interaction of four Cys side chains with a Zn^{2+} ion (Section 4-3). In the absence of a ligand, the receptor cannot bind to DNA.

Following ligand binding and dimerization, the receptor moves to the nucleus (if it is not already there) and binds to specific DNA sequences called **hormone response elements.** Although the hormone response elements vary for each receptor–ligand complex, they are all composed of two identical 6-bp sequences separated by a few base pairs. Simultaneous binding of the two hormone response element sequences explains why many of the lipid hormone receptors are dimers **(Fig. 10-13)**.

The receptors function as transcription factors so that the genes near the hormone response elements may experience higher or lower levels of expression. For example, glucocorticoids such as cortisol stimulate the production of phosphatases, which dampen the stimulating effects of kinases. This property makes cortisol and its derivatives useful as drugs to treat conditions such as chronic inflammation or asthma. However, because so many tissues respond to glucocorticoids, the side effects of these drugs can be significant and tend to limit their long-term use.

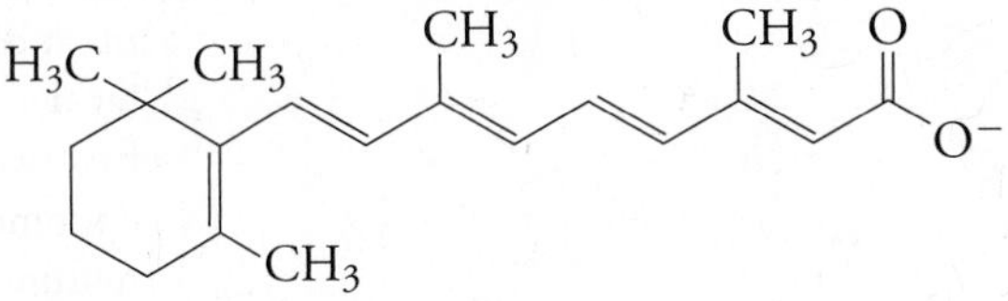

Retinoate

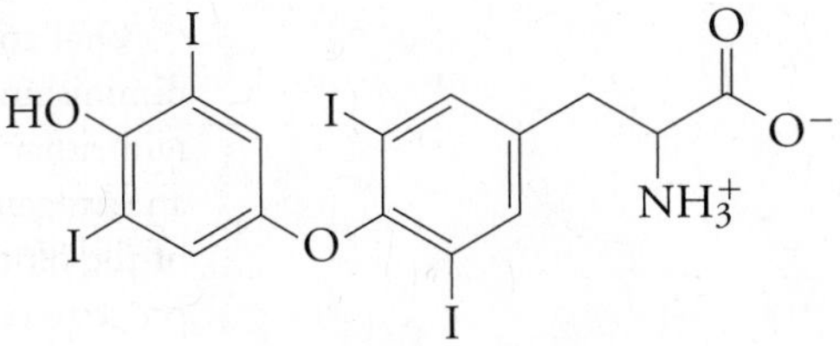

Thyroxine (T_4)

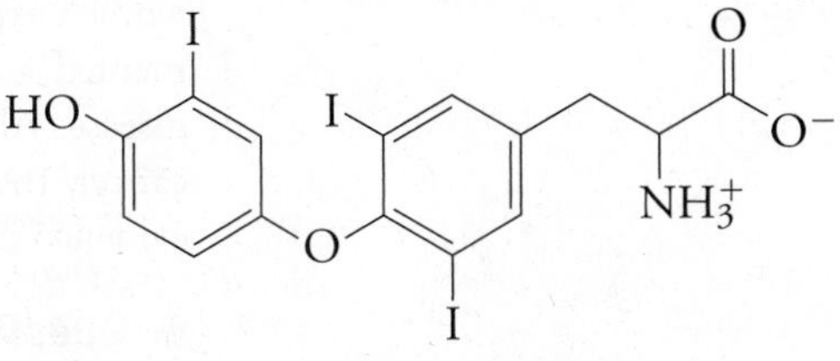

Triiodothyronine (T_3)

Figure 10-12 Some lipid hormones.

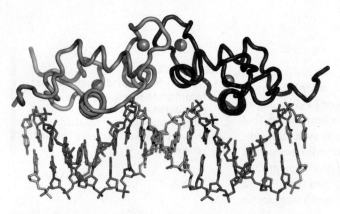

Figure 10-13 The glucocorticoid receptor–DNA complex. The two DNA-binding zinc finger domains of the glucocorticoid receptor are blue and green. The Zn^{2+} ions are shown as gray spheres. The hormone response element sequences of the DNA (bottom) are colored red. Two protein helices make sequence-specific contacts with nucleotides. [Structure (pdb 1GLU) determined by B. F. Luisi, W. X. Xu, Z. Otwinowski, L. P. Freedman, K. R. Yamamoto, and P. B. Sigler.] ⊕ **See Interactive Exercise. Glucocorticoid receptor DNA-binding domain in complex with DNA.**

? **What is the surface charge of the receptor's DNA-binding domain?**

The steroid hormone analogs used as contraceptives, however, are well tolerated (Box 10-C).

The changes in gene expression triggered by steroids and other lipid hormones require many hours to take effect. However, cellular responses to some lipid hormones are evident within seconds or minutes, suggesting that the hormones also participate in signaling pathways with shorter time courses, such as those centered on G proteins and/or kinases. In these instances, the receptors must be located on the cell surface.

Eicosanoids are short-range signals

Many of the hormones discussed in this chapter are synthesized and stockpiled to some extent before they are released, but some lipid hormones are synthesized as a response to other signaling events (sphingosine-1-phosphate is one example; Section 10-2). The lipid hormones called **eicosanoids** are produced when the enzyme phospholipase A_2 is activated by phosphorylation and by the presence of Ca^{2+}. One substrate of the phospholipase is the membrane lipid phosphatidylinositol. In this lipid, cleavage of the acyl chain attached to the second glycerol carbon often releases arachidonate, a C_{20} fatty acid (the term *eicosanoid* comes from the Greek *eikosi*, meaning "twenty").

Arachidonate, a polyunsaturated fatty acid with four double bonds, is further modified by the action of enzymes that

---

BOX 10-C ⬡ BIOCHEMISTRY NOTE

Oral Contraceptives

The female reproductive cycle depends on a set of pituitary and ovarian hormones, including estrogen and progesterone. Estrogen, which is actually a mix of six different substances, all derived from cholesterol (Section 8-1), is responsible for the development of female reproductive structures and female characteristics such as the pattern of fat deposition and hair growth. Production of estrogen by the ovaries rises during the first two weeks of a typical 28-day reproductive cycle, promoting the thickening of the endometrium (the lining of the uterus) and helping trigger ovulation, the release of a mature ovum (egg). The ovary continues to produce some estrogen, along with progesterone, whose concentration peaks at around the third week of the cycle. Unless the egg is fertilized, the production of hormones then declines, and menstruation commences.

Oral contraceptives, or birth-control pills, are designed to prevent ovulation, thereby diminishing the chance of fertilization. The most popular contraceptive formulations contain either a progesterone analog, called a progestin, or a combination of a progestin and an estrogen analog. These hormones, usually taken in regular daily doses, inhibit the release of the pituitary hormones necessary for ovulation and prevent the surge of estrogen and progesterone that normally occur during the monthly cycle. Progestin by itself thickens cervical mucus to help prevent sperm entry, further decreasing the risk of conception.

The widespread use of oral contraceptives over half a century reveals that most women experience few side effects. Among the serious side effects that have been documented are venous thrombosis (blood clots) and an increased risk of cardiovascular disease. Although oral contraceptives can decrease the risk of ovarian and colorectal cancer, they appear to increase the rate of some other cancers. There is no scientific evidence that contraceptive use contributes to weight gain or depression.

● **Question:** Why does it makes sense for ovarian cells to synthesize hormones such as estrogen and progesterone as well as receptors for these hormones?

catalyze cyclization and oxidation reactions (Fig. 10-14). A wide variety of eicosanoids can be produced in a tissue-dependent fashion, and their functions are similarly varied. Eicosanoids regulate such things as blood pressure, blood coagulation, inflammation, pain, and fever. The eicosanoid thromboxane, for example, helps activate platelets (cell fragments that participate in blood coagulation) and induces vasoconstriction. Other eicosanoids have the opposite effects: They prevent platelet activation and promote vasodilation. The use of aspirin as a "blood thinner" stems from its ability to inhibit the enzyme that initiates the conversion of arachidonate to thromboxane (see Fig. 10-14). A number of other drugs interfere with the production of eicosanoids by blocking the same enzymatic step (Box 10-D).

The receptors for eicosanoids are G protein–coupled receptors that trigger cAMP-dependent and phosphoinositide-dependent responses. However, eicosanoids are degraded relatively quickly. This instability, along with their hydrophobicity, means that *their effects are relatively limited in time and space.* Eicosanoids tend to elicit responses only in the cells that produce them and in nearby cells. In contrast, many other hormones travel throughout the body, eliciting effects in any tissue that exhibits the appropriate receptors. For this reason, eicosanoids are sometimes called local mediators rather than hormones.

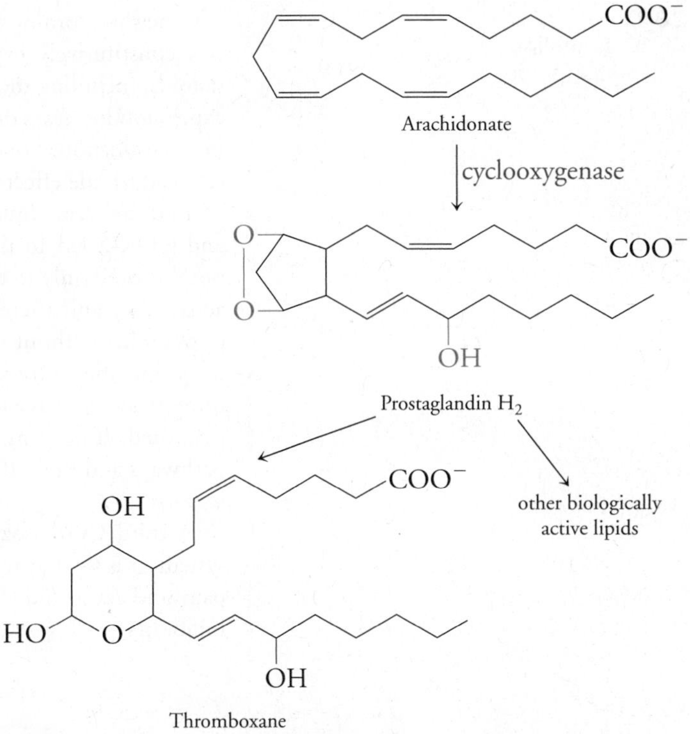

Figure 10-14 Arachidonate conversion to eicosanoid signal molecules. The first step is catalyzed by cyclooxygenase. Only two of the many dozens of eicosanoids are shown here.

BOX 10-D ⬡ BIOCHEMISTRY NOTE

Aspirin and Other Inhibitors of Cyclooxygenase

The bark of the willow *Salix alba* has been used since ancient times to relieve pain and fever. The active ingredient is acetylsalicylate, or aspirin.

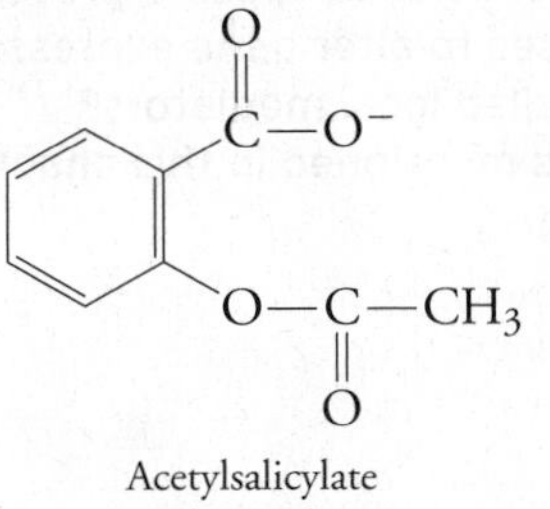

Acetylsalicylate
(aspirin)

Aspirin was first prepared in 1853, but it was not used clinically for another 50 years or so. Effective promotion of aspirin by the Bayer chemical company at the start of the twentieth century marked the beginning of the modern pharmaceutical industry.

Despite its universal popularity, aspirin's mode of action was not discovered until 1971. It inhibits the production of prostaglandins (which induce pain and fever, among other things) by inhibiting the activity of cyclooxygenase (also known as COX), the enzyme that acts on arachidonate (see Fig. 10-14). COX inhibition results from acetylation of a Ser residue located near the active site in a cavity that accommodates the arachidonate substrate. Other pain-relieving substances such as ibuprofen also bind to COX to prevent the synthesis of prostaglandins, although this drug does not acetylate the enzyme.

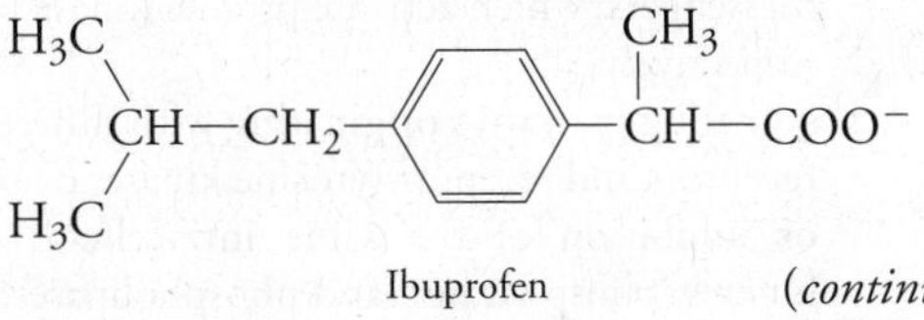

Ibuprofen   (*continued on the next page*)

One shortcoming of aspirin is that it inhibits more than one COX isozyme. COX-1 is a constitutively expressed enzyme that is responsible for generating various eicosanoids, including those that maintain the stomach's protective layer of mucus. COX-2 expression increases during tissue injury or infection and generates eicosanoids involved in inflammation. Long-term aspirin use suppresses the activity of both isozymes, which can lead to side effects such as gastric ulcers.

Rational drug design (Box 7-A) based on the slightly different structures of COX-1 and COX-2 led to the development of the drugs Celebrex and Vioxx. These compounds bind only to the active site of COX-2 (they are too large to fit into the COX-1 active site) and therefore can selectively block the production of pro-inflammatory eicosanoids without damaging gastric tissue. Unfortunately, the side effects of these drugs include an increased risk of heart attacks, through a mechanism that is not fully understood, and as a result Vioxx has been taken off the market and the use of Celebrex is limited. If nothing else, this story illustrates the complexity of biological signaling pathways and the difficulty of understanding how to manipulate them for therapeutic reasons.

A third COX isozyme, COX-3, is expressed at high levels in the central nervous system. It is the target of the widely used drug acetaminophen (Box 7-A), which reduces pain and fever and does not appear to incur the side effects of the COX-2–specific inhibitors.

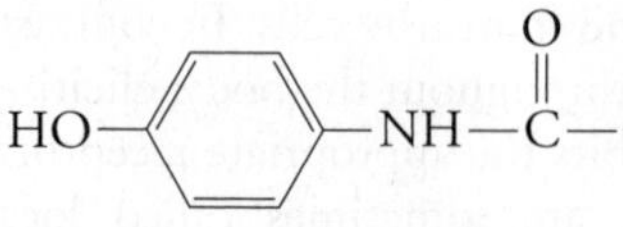

Acetaminophen

⬢ **Question:** Which of the drugs shown here is chiral (see Section 4-1), with two different configurations?

CONCEPT REVIEW
- Explain how lipid hormones can bypass G protein–coupled receptors and receptor tyrosine kinases to alter gene expression.
- Why are eicosanoids called local mediators?
- Make a list of the drugs mentioned in this chapter, and indicate how they interfere with signaling.

[SUMMARY]

10-1 General Features of Signaling Pathways

- Agonist or antagonist binding to a receptor can be quantified by a dissociation constant.
- G protein–coupled receptors and receptor tyrosine kinases are the most common types of receptors.
- While signaling systems amplify extracellular signals, they are also regulated so that signaling can be turned off, and the receptor may become desensitized.

10-2 G Protein Signaling Pathways

- A ligand such as epinephrine binds to a G protein–coupled receptor. A G protein responds to the receptor–ligand complex by releasing GDP, binding GTP, and splitting into an α subunit and a βγ dimer.
- The α subunit of the G protein activates adenylate cyclase, which converts ATP to cAMP. cAMP is a second messenger that triggers

a conformational change in protein kinase A that repositions its activation loop to achieve full catalytic activity.
- cAMP-dependent signaling activity is limited by the reduction of second messenger production through the GTPase activity of G proteins and the action of phosphodiesterases and by the activity of phosphatases that reverse the effects of protein kinase A. Ligand dissociation and receptor desensitization through phosphorylation and arrestin binding also limit signaling via G protein–coupled receptors.
- G protein–coupled receptors that lead to activation of phospholipase C generate inositol trisphosphate and diacylglycerol second messengers, which activate protein kinase B and protein kinase C, respectively.
- Signaling pathways originating with different G protein–coupled receptors and receptor tyrosine kinases overlap through activation or inhibition of the same intracellular components, such as kinases, phosphatases, and phospholipases.

10-3 Receptor Tyrosine Kinases

• Receptor tyrosine kinases are dimeric molecules with a single ligand-binding site. Ligand binding brings the monomers together such that the cytoplasmic tyrosine kinase domains can phosphorylate each other.

• In addition to acting as kinases, the receptor tyrosine kinases initiate other kinase cascades by activating the small monomeric G protein Ras.

10-4 Lipid Hormone Signaling

• Steroids and other lipid hormones bind primarily to intracellular receptors that dimerize and bind to hormone response elements on DNA to induce or repress the expression of nearby genes.

• Eicosanoids, which are synthesized from membrane lipids, function as signals over short ranges and for a limited time.

[GLOSSARY TERMS]

signal transduction	GPCR	phosphoinositide signaling system
receptor	second messenger	cross-talk
ligand	kinase	autophosphorylation
hormone	receptor tyrosine kinase	transcription factor
quorum sensing	desensitization	oncogene
agonist	cAMP	hormone response element
antagonist	phosphatase	eicosanoid
G protein		

[PROBLEMS]

10-1 General Features of Signaling Pathways

1. Which of the signal molecules listed in Table 10-1 would not require a cell-surface receptor?

2. The structure of the drug prednisone is shown below. What kind of molecule is this and by what pathway is it likely to exert its effects?

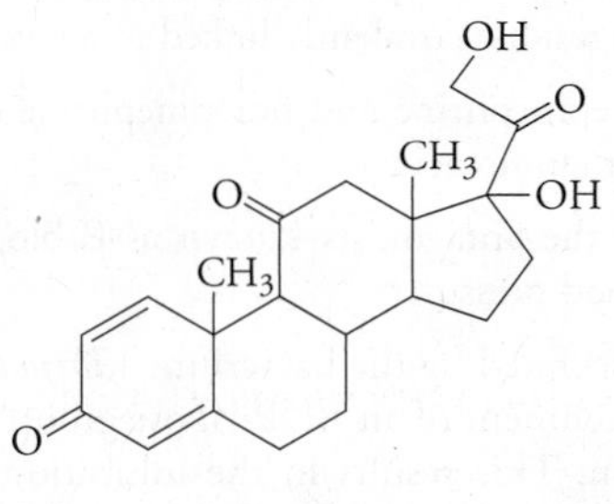

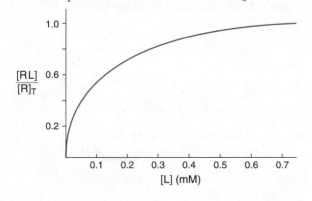

3. The total concentration of receptors in a sample is 10 mM. The concentration of free ligand is 2.5 mM, and K_d is 1.5 mM. Calculate the percentage of receptors that are occupied by ligand.

4. The total concentration of receptors in a sample is 10 mM. The concentration of free ligand is 2.5 mM, and K_d is 0.3 mM. Calculate the percentage of receptors that are occupied by ligand. Compare the answer to this problem with the answer you obtained in Problem 3 and explain the difference.

5. Use the plot below to estimate a value for K_d.

6. In an experiment, the ligand adenosine is added to heart cells in culture. The number of receptors with ligand bound is measured and the data, when plotted, yield a curve like the one shown in Figure 10-1. What would the results look like if the experiment was repeated in the presence of caffeine?

7. The definition of K_d is given in Equation 10-1, which shows the relationship between [R], the free receptor concentration; [L], the ligand concentration; and [RL], the concentration of receptor–ligand complexes. The value of [R], like [RL], is difficult to evaluate, but various experimental techniques can determine $[R]_T$, the total number of receptors. $[R]_T$ is the sum of [R] and [RL]. Using this information, begin with Equation 10-1 and derive an expression for the $[RL]/[R]_T$ ratio. (Note that your derived expression will be similar to the Michaelis–Menten equation and that Equations 7-9 through 7-17 may give you an idea how to proceed.)

8. The $[RL]:[R]_T$ ratio gives the fraction of receptors that have bound ligand. Use the expression you derived in Problem 7 to express [RL] as a fraction of $[R]_T$ for the following situations:
(a) $K_d = 5[L]$
(b) $K_d = [L]$
(c) $5K_d = [L]$

9. If a cell has 1000 surface receptors for erythropoietin, and if only 10% of those receptors need to bind ligand to achieve a maximal response, what ligand concentration is required to achieve a maximal response? Use the equation you derived in Problem 7. The K_d for erythropoietin is 1.0×10^{-10} M.

10. Suppose the number of surface receptors on the cell described in Problem 9 decreases to 150. What ligand concentration is required to achieve a maximal response?

11. ADP binds to platelets in order to initiate the activation process. Two binding sites were identified on platelets, one with a K_d of 0.35 μM and one with a K_d of 7.9 μM.
(a) Which of these is a low-affinity binding site and which is a high-affinity binding site?

(b) The ADP concentration required to activate a platelet is in the range of 0.1–0.5 μM. Which receptor will be more effective at activating the platelet?

(c) Two ADP agonists were also found to bind to platelets: 2-methylthio-ADP bound with a K_d of 7 μM and 2-(3-aminopropylthio)-ADP bound with a K_d of 200 μM. Can these agonists effectively compete with ADP for binding to platelets?

12. In the study described in Problem 11, 160,000 high-affinity binding sites were identified on each platelet. What is the concentration of ADP required to achieve 85% binding?

13. Like the Michaelis–Menten equation, the equation derived in Problem 7 can be converted to an equation for a straight line. A double-reciprocal plot for a ligand binding to its receptor is shown below. Use the information in the plot to estimate a value for K_d.

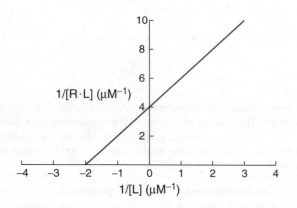

14. A Scatchard plot is another method of representing ligand binding data using a straight line. In a Scatchard plot, $[R \cdot L]/[L]$ is plotted versus $[R \cdot L]$. The slope is equal to $-1/K_d$. Use the Scatchard plot provided to estimate a value for K_d for calmodulin binding to calcineurin.

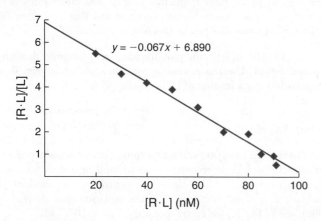

$$y = -0.067x + 6.890$$

15. Why might it be difficult to purify cell-surface receptors using the techniques described in Section 4-5?

16. Affinity chromatography is often used as a technique to purify cell-surface receptors. Describe the steps you would take to purify a cell-surface receptor using this technique.

17. Epinephrine can bind to several different types of G protein–linked receptors. Each of these receptors triggers a different cellular response. Explain how this is possible.

18. In the liver, both glucagon and epinephrine bind to different members of the G protein–coupled receptor family, yet binding of each of these ligands results in the same response—glycogen breakdown. How is it possible that two different ligands can trigger the same cellular response?

19. Many receptors become desensitized in the presence of high concentrations of signaling ligand. This can occur in a variety of ways. Sometimes the receptors are removed from the cell surface by endocytosis; in other cases the receptor may be phosphorylated. Why are these effective desensitization strategies?

20. What is the advantage of activating D using the strategy shown in the figure? Why is the strategy shown more effective than a simple activation of D that occurs in one step?

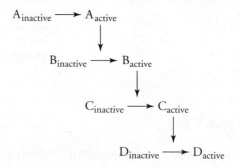

10-2 G Protein Signaling Pathways

21. As described in the text, G protein–coupled receptors are often palmitoylated at a Cys residue. Draw the structure of the palmitate (16:0) residue covalently linked to a Cys residue.

22. How do epinephrine and norepinephrine differ from tyrosine, their parent amino acid?

23. Why are the antagonists known as β-blockers effective at treating high blood pressure?

24. A toxin secreted by the bacterium *Vibrio cholerae* catalyzes the covalent attachment of an ADP-ribose group to the α subunit of the G protein. This results in the inhibition of the intrinsic GTPase activity of the G protein. How does this affect the activity of adenylate cyclase? How are intracellular levels of cAMP affected?

25. Some G protein–linked receptors are associated with a protein called RGS (regulator of G protein signaling). RGS stimulates the GTPase activity of the G protein associated with the receptor. What effect does RGS have on the signaling process?

26. Addition of the nonhydrolyzable analog GTPγS to cultured cells is a common practice in signal transduction experiments. What effect does GTPγS have on cellular cAMP levels?

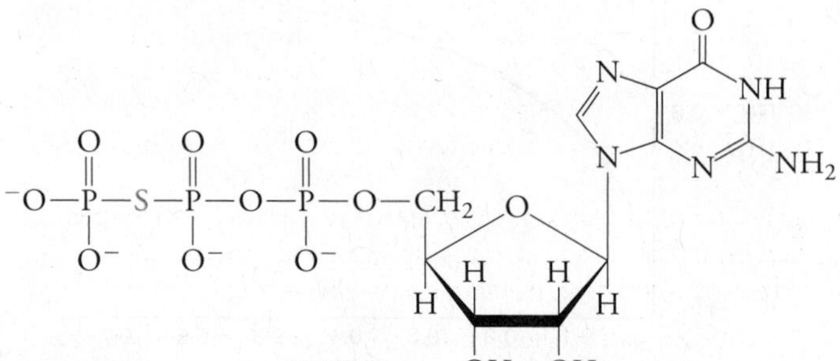

27. (a) Draw the reaction that shows the protein kinase A–catalyzed phosphorylation of a threonine residue on a target protein.

(b) Draw the reaction that shows the phosphatase-catalyzed hydrolysis of the phosphorylated threonine.

28. Some bacterial signaling systems involve kinases that transfer a phosphoryl group to a His side chain. Draw the structure of the phospho-His side chain.

29. Phorbol esters, which are compounds isolated from plants, are structurally similar to diacylglycerol. How does the addition of phorbol esters affect the cellular signaling pathways of cells in culture?

30. As described in the text, ligand binding to certain receptor tyrosine kinases results in the activation of a sphingomyelinase enzyme. Draw the reaction that shows the sphingomyelinase-catalyzed hydrolysis of sphingomyelin to ceramide.

31. In unstimulated T cells, a transcription factor called NFAT (nuclear factor of activated T cells) resides in the cytosol in a phosphorylated form. When the cell is stimulated, the cytosolic Ca^{2+} concentration increases and activates a phosphatase called calcineurin. The activated calcineurin catalyzes the hydrolysis of the phosphate group from NFAT, exposing a nuclear localization signal that allows the NFAT to enter the nucleus and stimulate the expression of genes essential for T-cell activation. Describe the cell-signaling events that resulted in the activation of NFAT.

32. The immunosuppressive drug cyclosporine A is an inhibitor of calcineurin (see Problem 31). Why is cyclosporine A an effective immunosuppressant?

33. Pathways that lead to the activation of protein kinase B (Akt) are considered to be anti-apoptotic (apoptosis is programmed cell death). In other words, protein kinase B stimulates a cell to grow and proliferate. Like all biological events, signaling pathways that are turned on must also be turned off. A phosphatase called PTEN plays a role in removing phosphate groups from proteins, but it is highly specific for removing a phosphate group from inositol trisphosphate. If PTEN is overexpressed in mammalian cells, do these cells grow or do they undergo apoptosis?

34. Would you expect to find mutations in the gene for PTEN (see Problem 33) in human cancers? Explain why or why not.

35. Nitric oxide (NO) is a naturally occurring signaling molecule (see Table 10-1) that is produced from the decomposition of arginine to NO and citrulline in endothelial cells. The enzyme that catalyzes this reaction, NO synthase, is stimulated by cytosolic Ca^{2+}, which increases when acetylcholine binds to endothelial cells.

(a) What is the source of the acetylcholine ligand?
(b) Propose a mechanism that describes how acetylcholine binding leads to the activation of NO synthase.
(c) NO formed in endothelial cells quickly diffuses into neighboring smooth muscle cells and binds to a cytosolic receptor that catalyzes the formation of the second messenger cyclic GMP. Draw the reaction that shows the formation of cGMP, and propose a name for the enzyme that catalyzes the reaction.
(d) Cyclic GMP next activates protein kinase G, which then acts on muscle proteins, resulting in smooth muscle cell relaxation. Propose a mechanism for this process.

36. As discussed in the text, any signal transduction event that is turned on must subsequently be turned off. Refer to your answer to

Problem 35 and describe the events that would lead to the cessation of each step of the signaling pathway you described.

37. NO synthase knockout mice (animals missing the NO synthase enzyme) have elevated blood pressure, an increased heart rate, and enlarged left ventricle chambers. Explain the reasons for these symptoms.

38. Clotrimazole is a calmodulin antagonist (see Solution 35b). What effect does the addition of clotrimazole have on endothelial cells in culture?

39. Nitroglycerin placed under the tongue has been used since the late nineteenth century to treat angina pectoris (chest pains resulting from reduced blood flow to the heart). But only recently have scientists elucidated its mechanism of action. Propose a hypothesis that explains why nitroglycerin placed under the tongue relieves the pain of angina.

Nitroglycerin

40. Viagra, a drug used to treat erectile dysfunction, is a cGMP phosphodiesterase inhibitor. Propose a mechanism that explains why the drug is effective in treating this condition.

41. *Bacillus anthracis*, the cause of anthrax, produces a three-part toxin. One part facilitates the entry of the two other toxins into the cytoplasm of a mammalian cell. The toxin known as edema factor (EF) is an adenylate cyclase.

(a) Explain how EF could disrupt normal cell signaling.
(b) EF must first be activated by $Ca^{2+} \cdot$ calmodulin binding to it. Explain how this requirement could also disrupt cell signaling.

42. *B. anthracis* also makes a toxin known as lethal factor (LF; see Problem 41). LF is a protease that specifically cleaves and inactivates a protein kinase that is part of a pathway for stimulating cell proliferation. Explain why the entry of LF into white blood cells promotes the spread of *B. anthracis* in the body.

43. Ligand binding to some growth factor receptors triggers kinase cascades and also leads to activation of enzymes that convert O_2 to hydrogen peroxide (H_2O_2), which acts as a second messenger. Describe the likely effect of H_2O_2 on the activity of cellular phosphatases.

44. Hydrogen peroxide has been shown to act as a second messenger, as described in Problem 43, and affects PTEN (see Problem 33) as well as other cellular phosphatases. Does H_2O_2 activate or inhibit PTEN?

10-3 Receptor Tyrosine Kinases

45. The activity of Ras is regulated in part by two proteins, a guanine nucleotide exchange factor (GEF) and a GTPase activating protein (GAP). The GEF protein binds to Ras $\cdot$ GDP and promotes dissociation of bound GDP. The GAP protein binds to Ras $\cdot$ GTP and stimulates the intrinsic GTPase activity of Ras. How is downstream activity of a signaling pathway affected by the presence of GEF? By the presence of GAP?

46. Mutant Ras proteins have been found to be associated with various types of cancer. What is the effect on a cell if the mutant Ras is able to bind GTP but is unable to hydrolyze it?

47. Stimulation of the insulin receptor by ligand binding and autophosphorylation eventually leads to the activation of both protein kinase B (Akt) and protein kinase C. Protein kinase B phosphorylates glycogen synthase kinase 3 (GSK3) and inactivates it. (Active GSK3 inactivates glycogen synthase by phosphorylating it.) Glycogen synthase catalyzes synthesis of glycogen from glucose. In the presence of insulin, GSK3 is inactivated, so glycogen synthase is not phosphorylated and is active. Protein kinase C stimulates the translocation of glucose transporters to the plasma membrane by a mechanism not currently understood. One strategy for treating diabetes is to develop drugs that act as inhibitors of the phosphatase enzymes that remove phosphate groups from the phosphorylated tyrosines on the insulin receptor. Why might this be an effective treatment for diabetes?

48. When insulin binds to its receptor, a conformational change occurs that results in autophosphorylation of the receptor on specific Tyr residues. In the next step of the signaling pathway, an adaptor protein called IRS-1 (insulin receptor substrate-1) docks on the phosphorylated receptor (the involvement of adaptor proteins in cell signaling is shown in Fig. 10-11). This step is essential for the downstream activation of protein kinases B and C (see Problem 47). If IRS-1 is overexpressed in muscle cells in culture, what effects, if any, would you expect to see on glucose transporter translocation and glycogen synthesis?

49. As shown in Figure 10-11, Ras can activate a kinase cascade. The most common cascade is the MAP kinase pathway, which is activated when growth factors bind to cell-surface receptors and activate Ras. This leads to the eventual activation of transcription factors and other gene regulatory proteins and results in growth, proliferation, and differentiation. Use this information to explain why phorbol esters (see Problem 29) promote tumor development.

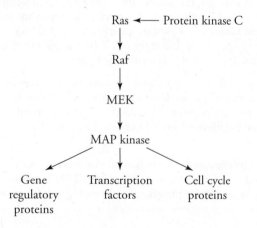

50. How might a signaling molecule activate the MAP kinase cascade (see Problem 49) via a G protein–coupled receptor rather than a receptor tyrosine kinase?

51. PKR is a protein kinase that recognizes double-stranded RNA molecules such as those that form during the intracellular growth of certain viruses. The structure of PKR includes the standard kinase domains as well as an RNA-binding module. In the presence of viral RNA, PKR undergoes autophosphorylation and is then able to phosphorylate cellular target proteins that initiate antiviral responses. Short (<30 bp) RNAs inhibit activation of PKR, but RNAs longer than 33 bp are strong activators of PKR. Explain the role of RNA in PKR activation.

52. The bacterium *Yersinia pestis*, the pathogen responsible for bubonic plague, caused the deaths of about a third of the population of Europe in the fourteenth century. The bacterium produces a phosphatase called YopH, which hydrolyzes phosphorylated tyrosines and is much more catalytically active than mammalian phosphatases.

(a) What happens when the *Yersinia* bacterium injects YopH into a mammalian cell?

(b) Why is the bacterium itself not affected by YopH?

(c) Scientists are interested in developing YopH inhibitors in order to treat *Yersinia* infection, a reemerging disease. What are some important considerations in the development of a YopH inhibitor?

10-4 Lipid Hormone Signaling

53. Steroid hormone receptors have different cellular locations. The progesterone receptor is located in the nucleus and interacts with DNA once progesterone has bound. But the glucocorticoid receptor is located in the cytosol and does not move into the nucleus until its ligand has bound. What structural feature must be different in these two receptor molecules?

54. Abnormal changes in steroid hormone levels in the breast, uterus, ovaries, prostate, and testes are observed in cancers of these steroid-responsive tissues.

(a) Using what you know about the mechanism of steroid hormone–induced stimulation in these tissues, what strategies could you use to design drugs to treat cancers in these tissues?

(b) Given what you know about cell signaling and cancer (see Box 10-B), is it reasonable to suspect that a "nonclassical" pathway may be involved in the development of cancers in these tissues?

55. The production of sphingosine-1-phosphate and ceramide-1-phosphate is shown in the diagram below. These signaling molecules can act in the cell where they are produced (intracellularly) or they can exit the cell and act on neighboring cells, as described in Section 10-2. There is quite a bit of cross-talk between the pathway shown in the diagram and other pathways discussed in this chapter.

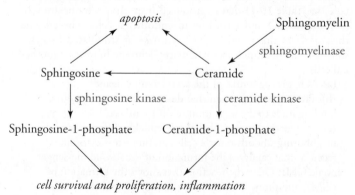

Ceramide-1-phosphate (C1P) has been shown to promote the release of arachidonate from the membrane. Sphingosine-1-phosphate (S1P) has been shown to stimulate the activity of COX-2. Are these observations consistent with the inflammatory properties attributed to C1P and S1P?

56. Sphingosine-1-phosphate (see Problem 55) has been shown to stimulate the activation of protein kinase B (Akt) (see Problem 33). What effect would this have on the cell?

57. Sphingosine-1-phosphate's pro-survival effects probably result from the interaction of S1P with multiple cellular pathways. In addition to S1P's ability to activate Akt (see Problem 56), how else might S1P act to promote cell survival?

58. There is a strong association between inflammation and cancer. Using the information presented in Problem 55, propose pharmacologic products that might be useful as anticancer drugs.

59. Aspirin inhibits COX by acetylating a Ser residue on the enzyme (see Box 10-D). Draw the reaction that shows how aspirin acetylates the serine side chain. Propose a hypothesis that explains why acetylation of the Ser residue inhibits the enzyme.

60. Cyclooxygenase uses arachidonate as a substrate for the synthesis of prostaglandins (see Box 10-D). In platelets, a similar pathway yields thromboxanes, compounds that stimulate vasoconstriction and platelet aggregation (see Fig. 10-14). Why do some people take a daily aspirin as protection against heart attacks?

61. Analogs of cortisol, such as prednisone, are used as anti-inflammatory drugs, although their mechanism of action is not entirely understood. Explain how inhibition of phospholipase A_2 by prednisone could decrease inflammation.

62. Tetrahydrocannabinol (THC), the active ingredient in marijuana, binds to a receptor in the brain. The natural ligand for the receptor is anandamide.

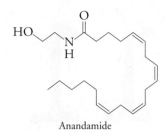

Anandamide

Anandamide has a short half-life because it is rapidly broken down by a hydrolase. One product is ethanolamine. Name the other product of anandamide breakdown.

63. A complex signaling pathway in yeast allows the cells to accumulate high concentrations of glycerol if they are exposed to a high extracellular concentration of salt or glucose. The increased osmolarity of the extracellular medium activates Ras, which in turn activates adenylate cyclase. A second pathway, the HOG (high osmolarity glycerol) pathway, activates the MAP kinase pathway (see Problem 49). The target protein is the enzyme PFK2, which is activated by phosphorylation. (PFK2 produces an allosteric regulator that activates glycolysis, which ultimately produces glycerol.) Draw a diagram that shows how the Ras and MAP kinase pathway converge to result in the phosphorylation and activation of PFK2.

64. Yeast mutants lacking components of the HOG pathway (see Problem 63) were exposed to high concentrations of glucose and then the PFK2 activity was measured. How does the PFK2 activity in the mutants compare with the PFK2 activity in mutants exposed to isotonic conditions?

[SELECTED READINGS]

De Meyts, P., The insulin receptor: A prototype of dimeric, allosteric membrane receptors? *Trends Biochem. Sci.* **33**, 376–384 (2008).

Eyster, K. M., The membrane and lipids as integral participants in signal transduction: Lipid signal transduction for the non-lipid biochemist, *Adv. Physiol. Educ.* **31**, 5–16 (2007). [Summarizes the roles of phospholipases, lipid kinases and phosphatases, and lipid hormones and second messengers.]

Milligan, G., and Kostenis, E., Heterotrimeric G-proteins: A short history, *Br. J. Pharmacol.* **147**, S46–55 (2006).

Newton, A. C., Lipid activation of protein kinases, *J. Lipid Res.* **50**, S266–S271 (2009). [Describes how lipid second messengers affect kinases such as protein kinase C.]

Rasmussen, S. G. F., DeVree, B. T., Zou, Y., Kruse, A. C., Chung, K. Y., Kobilka, T. S., Thian, F. S., Chae, P. S., Pardon, E., Calinski, D., Mathiesen, J. M., Shah, S. T. A., Lyons, J. A., Caffrey, M., Gellman, S. H., Steyaert, J., Skiniotis, G., Weis, W. I., Sunahara, R. K., and Kobilka, B. K., Crystal structure of the β₂ adrenergic receptor–Gs protein complex, *Nature* **477**, 549–555 (2011). [Provides detailed information about the interaction between a G protein and its associated receptor.]

Taylor, S. S., and Kornev, A. P., Protein kinases: Evolution of dynamic regulatory proteins, *Trends Biochem. Sci.* **36**, 65–77 (2011). [Reviews the general structure of protein kinases and their mechanism of activation by phosphorylation of an activation loop.]

11

CARBOHYDRATES

▶▶ **WHY** are some carbohydrates indigestible?

Of the major molecular building blocks of cells—nucleotides, amino acids, lipids, and carbohydrates—carbohydrates are the most abundant. Virtually all the foods we eat contain carbohydrates. Yet there are subtle chemical differences in the way small sugars are connected to form large molecules, which means that not all carbohydrates occur in forms that can be readily broken down. As a result, we cannot digest and use all the carbohydrates we eat.

[Dean Turner/iStockphoto]

THIS CHAPTER IN CONTEXT

Part 1 Foundations

Part 2 Molecular Structure and Function

   11 Carbohydrates

Part 3 Metabolism

Part 4 Genetic Information

Do You Remember?

- Cells contain four major types of biological molecules and three major types of polymers (Section 1-2).
- The polar water molecule forms hydrogen bonds with other molecules (Section 2-1).

Although their atomic composition is largely limited to C, H, and O, carbohydrates take on a variety of biological functions from energy metabolism to cellular structure. **Carbohydrates,** also known as sugars or saccharides, occur as **monosaccharides** (simple sugars), small polymers (**disaccharides, trisaccharides,** and so on), and larger **polysaccharides** (sometimes called complex carbohydrates). Monosaccharides follow the molecular formula $(CH_2O)_n$, where $n \geq 3$ (hence the name *carbohydrate*). But even saccharide derivatives—many of which include groups containing nitrogen, phosphorus, and other elements—are easy to recognize by their large number of hydroxyl (—OH) groups. This chapter surveys monosaccharides and their derivatives, some common disaccharides and polysaccharides, and carbohydrates linked to proteins.

11-1 Monosaccharides

KEY CONCEPTS

- Monosaccharides contain different numbers of carbon atoms and may be aldoses or ketoses with multiple enantiomeric and epimeric forms.
- α and β anomers can freely interconvert unless the anomeric carbon participates in a glycosidic bond.
- The functional groups of a monosaccharide can be modified to form a variety of derivatives.

The simplest sugars are the three-carbon compounds glyceraldehyde and dihydroxyacetone:

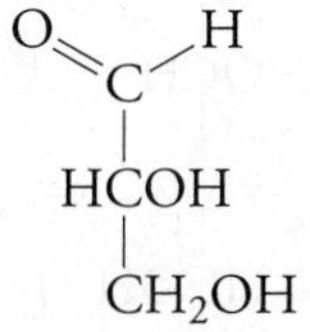

Glyceraldehyde

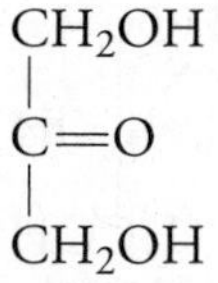

Dihydroxyacetone

A sugar such as glyceraldehyde, in which the carbonyl group is an aldehyde, is known as an **aldose,** and a carbohydrate such as dihydroxyacetone, in which the carbonyl group is a **ketone,** is known as a ketose. In most ketoses, the carbonyl group occurs at the second carbon (C2).

Monosaccharides can also be described according to the number of carbon atoms they contain; for example, the three-carbon compounds shown above are **trioses. Tetroses** contain four carbons, **pentoses** five, **hexoses** six, and so on. The aldopentose ribose (below) is a component of ribonucleic acid (RNA; its derivative 2′-deoxyribose occurs in deoxyribonucleic acid, DNA). By far the most abundant monosaccharide is glucose, an aldohexose. A common ketohexose is fructose:

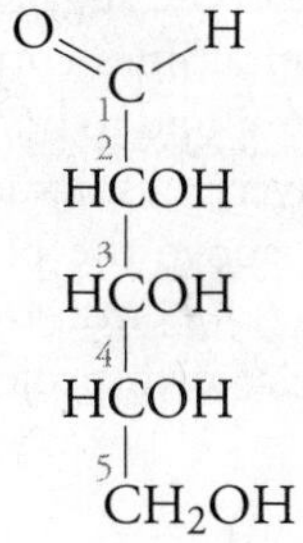

Ribose

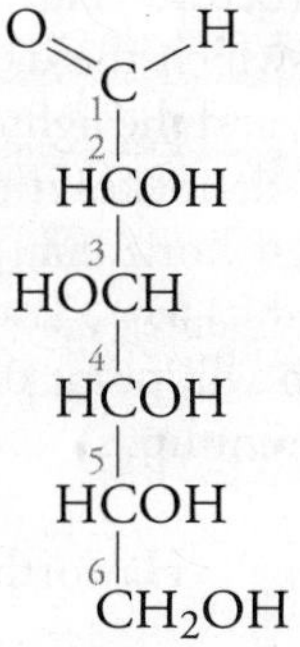

Glucose

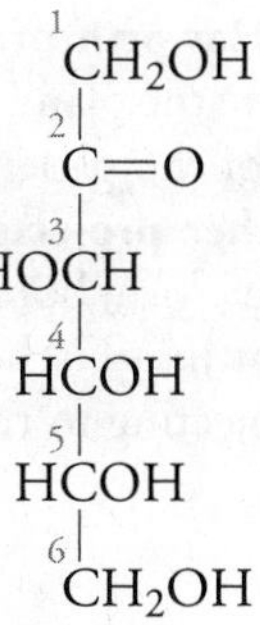

Fructose

Most carbohydrates are chiral compounds

Note that glucose, shown above, is a **chiral** compound because several of its carbon atoms (all except C1 and C6) bear four different substituents (see Section 4-1 for a discussion of chirality). As a result, *glucose has a number of stereoisomers, as do nearly all monosaccharides* (the symmetric dihydroxyacetone is one exception). Several types of stereoisomerism apply to carbohydrates.

Like the amino acids (Section 4-1), glyceraldehyde has two different structures that exhibit mirror symmetry. *Such pairs of structures, known as* ***enantiomers,*** *cannot be superimposed by rotation.* By convention, these structures are given the designations L and D, derived from the Greek *levo,* "left," and *dextro,* "right." The enantiomeric forms of larger monosaccharides are given the D or L designation by comparing their structures to D- and L-glyceraldehyde (*left*). In a **D sugar,** the asymmetric carbon farthest from the carbonyl group (that would be C5 in glucose) has the same spatial arrangement as the chiral carbon of D-glyceraldehyde. In an **L sugar,** that carbon has the same arrangement as in L-glyceraldehyde. Thus, every D sugar is the mirror image of an L sugar.

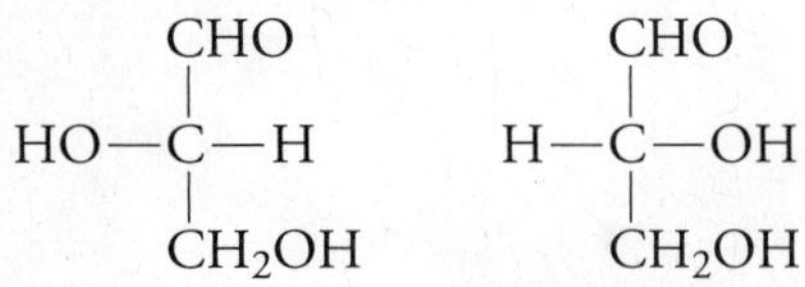

Although enantiomers behave identically in a strictly chemical sense, they are not biologically equivalent. This is because biological systems, which are built of other chiral compounds, such as L-amino acids, can distinguish D and L sugars. Most naturally occurring sugars have the D configuration, so the D and L prefixes are often omitted from their names.

Glucose, in addition to its enantiomeric carbon (C1), has four other asymmetric carbons, so there are stereoisomers for the configuration at each of these positions. *Carbohydrates that differ in configuration at one of these carbons are known as **epimers.*** For example, the common monosaccharide galactose is an epimer of glucose, at position C4:

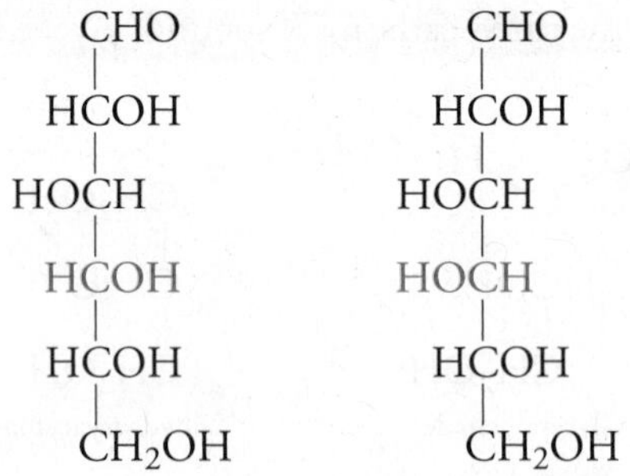

```
      CHO                    CHO
       |                      |
     HCOH                   HCOH
       |                      |
     HOCH                   HOCH
       |                      |
     HCOH                   HOCH
       |                      |
     HCOH                   HCOH
       |                      |
     CH₂OH                  CH₂OH
   D-Glucose              D-Galactose
```

Both ketoses and aldoses have epimeric forms. And like enantiomers, epimers are not biologically interchangeable: An enzyme whose active site accommodates glucose may not recognize galactose at all.

Cyclization generates α and β anomers

The numerous hydroxyl groups that characterize carbohydrate structures also provide multiple points for chemical reactions to occur. One such reaction is an intramolecular rearrangement in which the sugar's carbonyl group reacts with one of its —OH groups to form a cyclic structure (**Fig. 11-1**). The cyclic sugars are represented as **Haworth projections** in which the darker horizontal lines correspond to bonds above the plane of the paper, and the lighter lines correspond to bonds behind the plane of the paper. A simple rule makes it easy to convert a structure from its linear **Fischer projection** (in which horizontal bonds are above the plane of the paper and vertical bonds are behind it) to a Haworth projection: Groups projecting to the right in a Fischer projection will point down in a Haworth projection, and groups projecting to the left will point up.

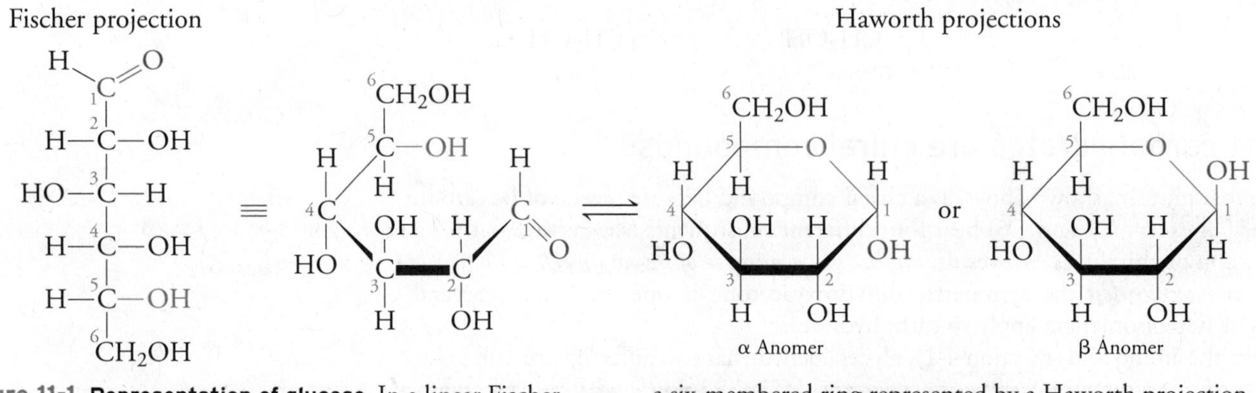

Fischer projection   Haworth projections

Figure 11-1 Representation of glucose. In a linear Fischer projection, the horizontal bonds point out of the page and the vertical bonds point below the page. Glucose cyclizes to form a six-membered ring represented by a Haworth projection, in which the heaviest bonds point out from the plane of the page. The α and β anomers freely interconvert.

As a result of the cyclization reaction, the hydroxyl group attached to what was the carbonyl carbon (C1 in the case of glucose) may point either up or down. In the **α anomer,** this hydroxyl group lies on the opposite side of the ring from the CH_2OH group of the chiral carbon that determines the D or L configuration (in the α anomer of glucose, the hydroxyl group points down; see Fig. 11-1). In the **β anomer,** the hydroxyl group lies on the same side of the ring as the CH_2OH group of the chiral carbon that determines the D or L configuration (in the β anomer of glucose, the hydroxyl group points up; see Fig. 11-1).

Unlike enantiomers and epimers, which are not interchangeable, anomers in an aqueous solution freely interconvert between the α and β forms, unless the hydroxyl group attached to the anomeric carbon is linked to another molecule. In fact, a solution of glucose molecules consists of about 64% β anomer, about 36% α anomer, and only trace amounts of the linear or open-chain form.

Hexoses and pentoses, which also undergo cyclization, do not form planar structures, as a Haworth projection might suggest. Instead, the sugar ring puckers so that each C atom can retain its tetrahedral bonding geometry. The substituents of each carbon may point either above the ring (axial positions) or outward (equatorial positions). Glucose can adopt a chair conformation in which all its bulky ring substituents (the —OH and —CH_2OH groups) occupy equatorial positions.

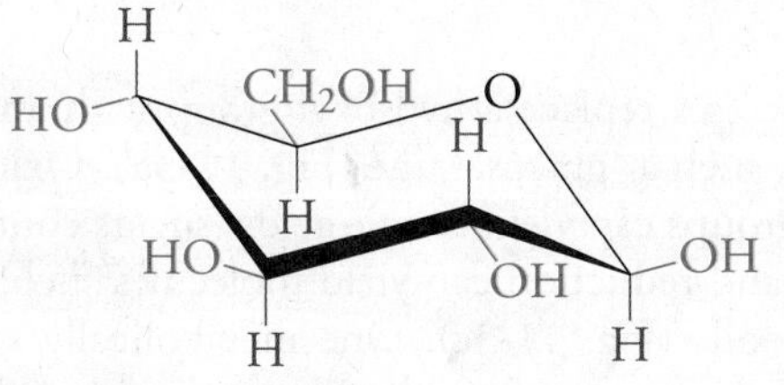

Glucose

In all other hexoses, some of these groups must occupy the more crowded—and therefore less stable—axial positions. The greater stability of glucose may be one reason for its abundance among monosaccharides.

Monosaccharides can be derivatized in many different ways

The anomeric carbon of a monosaccharide is easy to recognize: It is the carbonyl carbon in the straight-chain form of the sugar, and it is the carbon bonded to both the ring oxygen and a hydroxyl group in the cyclic form of the sugar. The anomeric carbon can undergo oxidation, so it can reduce substances such as Cu(II) to Cu(I). This chemical reactivity, often assayed using a copper-containing solution known as Benedict's reagent, can distinguish a free monosaccharide, called a **reducing sugar,** from a monosaccharide in which the anomeric carbon has already reacted with another molecule. For example, when a glucose molecule (a reducing sugar) reacts with methanol (CH_3OH), the result is a **nonreducing sugar** (Fig. 11-2). Because the

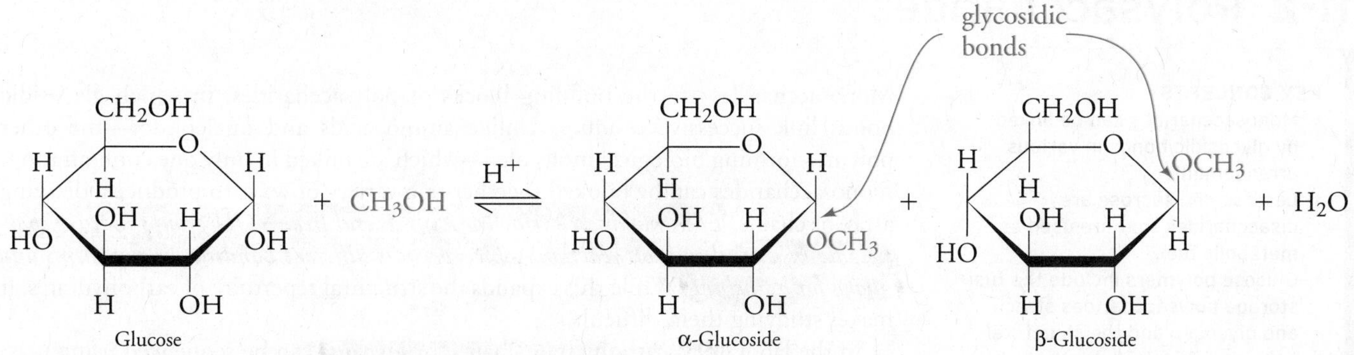

Figure 11-2 Reaction of glucose with methanol. The addition of methanol to the anomeric carbon blocks the ability of glucose to function as a reducing sugar. The glycosidic bond that forms between the anomeric carbon and the oxygen of methanol may have the α or β configuration.

anomeric carbon is involved in the reaction, the methyl group can end up in either the α or β position. The bond that links the anomeric carbon to the other group is called a **glycosidic bond,** and a molecule consisting of a sugar linked to another molecule is called a **glycoside.** Glycosidic bonds link the monomers in oligo- and polysaccharides (Section 11-2) and also link the ribose groups to the purine and pyrimidine bases of nucleotides (Section 3-1).

Phosphorylated sugars, including glyceraldehyde-3-phosphate and fructose-6-phosphate, appear as intermediates in the metabolic pathways for breaking down glucose (glycolysis; Section 13-1) and synthesizing it (photosynthesis; Section 16-3).

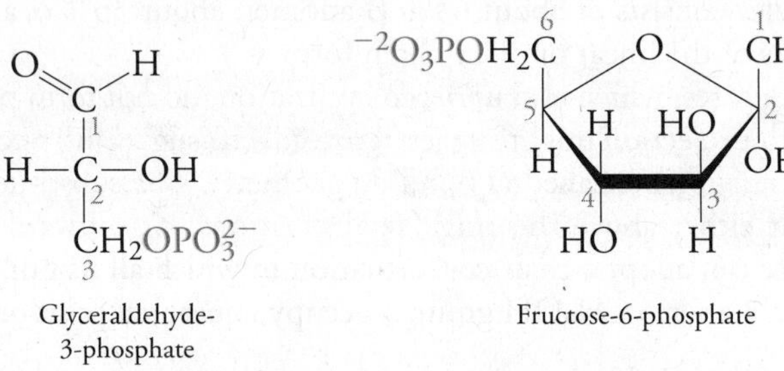

Glyceraldehyde-3-phosphate

Fructose-6-phosphate

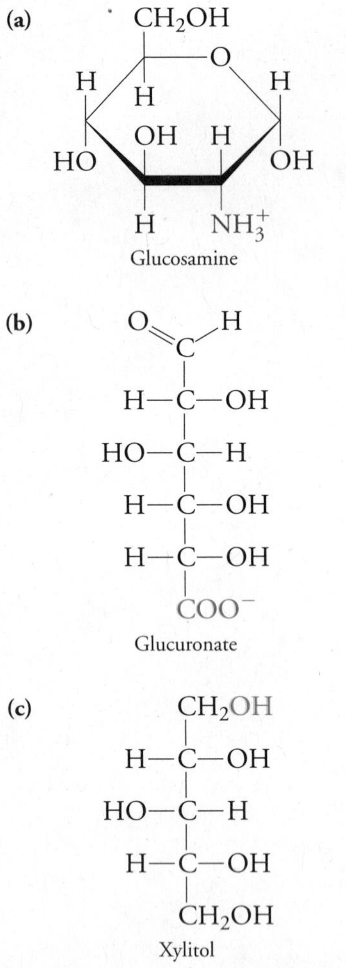

Other metabolic processes replace a hydroxyl group with an amino group to produce an amino sugar, such as glucosamine (**Fig. 11-3a**). Oxidation of a sugar's carbonyl and hydroxyl groups can yield uronic acids (sugars containing carboxylic acid groups; Fig. 11-3b), and reduction can yield molecules such as xylitol, a sweetener used in "sugarless" foods (Fig. 11-3c). One metabolically essential carbohydrate-modifying reaction is the one catalyzed by ribonucleotide reductase, which reduces the 2'-OH group of ribose to convert a ribonucleotide to a deoxyribonucleotide for DNA synthesis (Section 18-3):

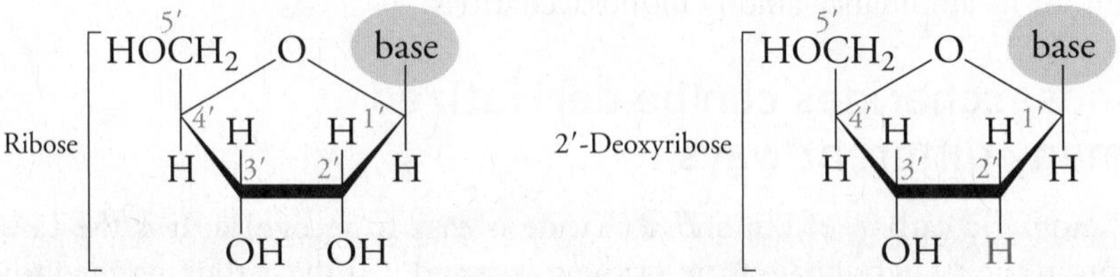

Ribose

2'-Deoxyribose

Figure 11-3 Some monosaccharide derivatives. In an amino sugar (a), —NH$_3^+$ replaces an —OH group. Oxidation and reduction reactions yield sugars with carboxylate groups (b) or additional hydroxyl groups (c).

? Identify the net charge of each sugar.

CONCEPT REVIEW

- Draw the straight-chain form of D-glucose, a ketose isomer of glucose, the L enantiomer of glucose, and one of its epimers.
- Explain why the α and β anomers of a monosaccharide can interconvert.
- List some types of monosaccharide derivatives.

11-2 Polysaccharides

KEY CONCEPTS

- Monosaccharides can be linked by glycosidic bonds in various arrangements.
- Lactose and sucrose are disaccharides that are used as metabolic fuel.
- Glucose polymers include the fuel-storage polysaccharides starch and glycogen and the structural polysaccharide cellulose.
- Other structural polysaccharides include chitin and components of bacterial biofilms.

Monosaccharides are the building blocks of polysaccharides, in which glycosidic bonds link successive residues. Unlike amino acids and nucleotides—the other polymer-forming biological molecules—which are linked in only one configuration, monosaccharides can be hooked together in a variety of ways to produce a dizzying array of chains. *Each monosaccharide contains several free —OH groups that can participate in a condensation reaction, which permits different bonding arrangements and allows for branching.* While this expands the structural repertoire of carbohydrates, it makes studying them difficult.

In the laboratory, carbohydrate chains, or **glycans,** can be sequenced using mass spectrometry (Section 4-5), although the results are sometimes ambiguous due to the inability to distinguish isomers, which have the same mass. Glycan three-dimensional structures are typically studied using NMR techniques (Section 4-5), since these yield an average conformation for the molecules, which tend to be highly flexible in solution. Due to the challenges of defining carbohydrate sequences and structures, **glycomics,** the systematic study of carbohydrates, is not as fully developed as genomics or proteomics.

The most complex glycans are the oligosaccharides that are commonly linked to other molecules, for example, in glycoproteins. Polysaccharides, some of which are truly enormous molecules, generally do not exhibit the heterogeneity and complexity of oligosaccharides. Instead, they tend to consist of one or a pair of monosaccharides that are linked over and over in the same fashion. This sort of structural homogeneity is well suited to the function of polysaccharides as fuel-storage molecules and architectural elements. We begin our survey with the simplest of the polysaccharides, the disaccharides.

Lactose and sucrose are the most common disaccharides

A glycosidic bond links two monosaccharides to generate a disaccharide. In nature, disaccharides occur as intermediates in the digestion of polysaccharides and as a source of metabolic fuel. For example, lactose, secreted into the milk of lactating mammals, consists of galactose and glucose:

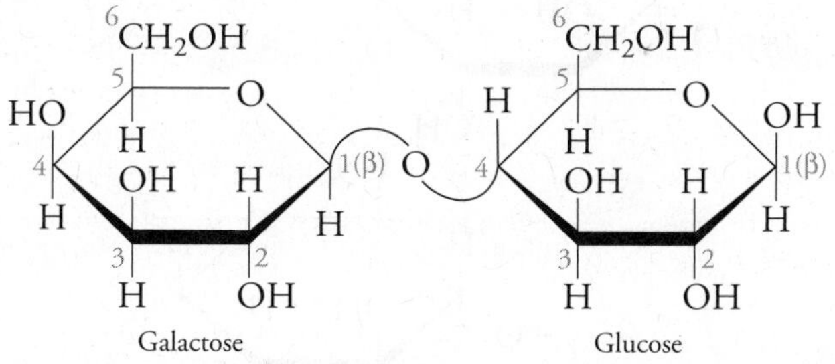

Lactose

Note that the anomeric carbon (C1) of galactose is linked to C4 of glucose via a β-glycosidic bond. If the two sugars were linked by an α-glycosidic bond, or if the galactose anomeric carbon were linked to a different glucose carbon, the result would be an entirely different disaccharide. Lactose serves as a major food for newborn mammals. Most adult mammals, including humans, produce very little lactase (also called β-galactosidase), the enzyme that breaks the glycosidic bond of lactose, and therefore cannot efficiently digest this disaccharide.

Sucrose, or table sugar, is the most abundant disaccharide in nature:

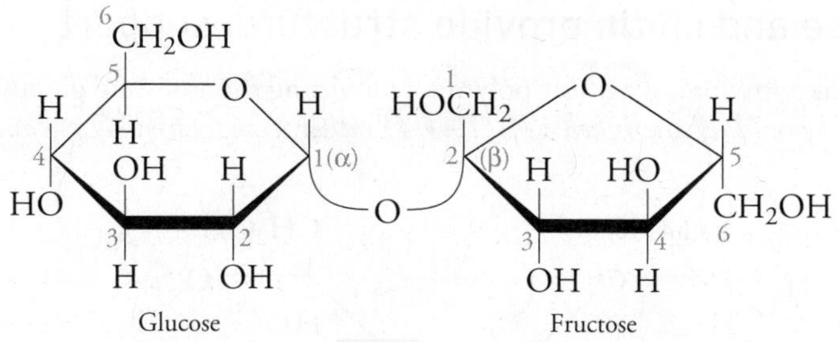

Sucrose

In this molecule, the anomeric carbon of glucose (in the α configuration) is linked to the anomeric carbon of fructose (in the β configuration). Sucrose is the major form in which newly synthesized carbohydrates are transported from a plant's leaves, where most photosynthesis occurs, to other plant tissues to be used as a fuel or stored as starch for later use.

Starch and glycogen are fuel-storage molecules

Starch and glycogen are polymers of glucose residues linked by glycosidic bonds designated α(1 → 4); in other words, the anomeric carbon (carbon 1) of one residue is linked by an α-glycosidic bond to carbon 4 of the next residue:

Plants manufacture a linear form of starch, called amylose, which can consist of several thousand glucose residues. Amylopectin, an even larger molecule, includes α(1 → 6) glycosidic linkages every 24 to 30 glucose residues to generate a branched polymer:

Compiling many monosaccharide residues in a single polysaccharide is an efficient way to store glucose, the plant's primary metabolic fuel. The α-linked chains curve into helices so that the entire molecule forms a relatively compact particle (**Fig. 11-4**).

Animals store glucose in the form of glycogen, a polymer that resembles amylopectin but with branches every 12 residues or so. Due to its highly branched structure, a glycogen molecule can be quickly assembled or disassembled according to the metabolic needs of the cell because the enzymes that add or remove glucose residues work from the ends of the branches.

Cellulose and chitin provide structural support

Cellulose, like amylose, is a linear polymer containing thousands of glucose residues. However, *the residues are linked by β(1 → 4) rather than α(1 → 4) glycosidic bonds:*

Figure 11-4 Structure of amylose. This unbranched polysaccharide, consisting of α(1 → 4)-linked glucose residues (six are shown here), forms a large left-handed helix.

? **In what part of a cell would amylose be located?**

This simple difference in bonding has profound structural consequences: Whereas starch molecules form compact granules inside the cell, cellulose forms extended

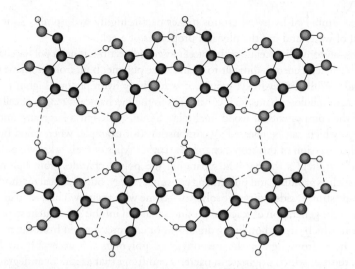

Figure 11-5 Cellulose structure.
Glucose residues are represented as hexagons, with C atoms gray and O atoms red. Not all H atoms (small circles) are shown. Hydrogen bonds (dashed lines) link residues in the same and adjacent chains so that a bundle of cellulose polymers forms an extended rigid fiber.

fibers that lend rigidity and strength to plant cell walls. Individual cellulose polymers form bundles with extensive hydrogen bonding within and between adjacent chains (Fig. 11-5). Plant cell walls include other polymers that, together with cellulose, yield a strong but resilient substance. Recovering the carbohydrates from materials such as wood remains a challenge for the biofuels industry (Box 11-A).

BOX 11-A ⬢ BIOCHEMISTRY NOTE

Cellulosic Biofuel

Cellulose is by far the most abundant polysaccharide in nature, with chains containing thousands of monosaccharide residues. For this reason, cellulose-rich materials, including wood and agricultural waste, are sources of sugars that can be converted to biofuels such as ethanol, potentially replacing petroleum-derived fuels. Unfortunately, cellulose does not exist in pure form in nature; plant cell walls typically contain other polymers, including hemicellulose, pectin, and lignin.

Hemicellulose is the name given to a class of polysaccharides whose chains are shorter than cellulose (500–3000 residues) and may be branched. Hemicellulose is a heteropolymer, indicating that it contains a variety of monomeric units, in this case 5- and 6-carbon sugars. Xylose is the most abundant of these:

$$
\begin{array}{c}
\text{CHO} \\
|\\
\text{HCOH} \\
|\\
\text{HOCH} \\
|\\
\text{HCOH} \\
|\\
\text{CH}_2\text{OH}
\end{array}
$$

Xylose

While cellulose forms rigid fibers and hemicellulose forms a network, the spaces in between these polymers are occupied by pectin, a heteropolymer containing galacturonate and rhamnose residues, among others:

Galacturonate   Rhamnose

(*continued on the next page*)

The large number of hydroxyl groups makes pectin highly hydrophilic, so it "holds" a great deal of water and has the physical properties of a gel.

Lignin—in contrast to cellulose, hemicellulose, and pectin—is not a polysaccharide at all. It is a highly heterogeneous, difficult-to-characterize polymer built from aromatic (phenolic) compounds. With few hydroxyl groups, it is relatively hydrophobic. Lignin is covalently linked to hemicellulose chains, so it contributes to the mechanical strength of cell walls.

All of the components of wood, including lignin, represent a large amount of stored free energy, which can be released by combustion (for example, when wood burns). The industrial conversion of this stored energy into other types of fuels is known as bioconversion. The first step is the hardest: separating the polysaccharides from lignin. Physical methods such as grinding and pulverizing consume energy, but chemical methods—which may include strong acids or organic solvents—come with their own hazards. An additional drawback is the generation of reaction products that can inhibit subsequent steps in biofuel production, which depend on living organisms or enzymes derived from them.

Once freed from lignin, the carbohydrate polymers are accessible to hydrolytic enzymes, most of which originate in bacteria and fungi that are adept at degrading plant materials. The result is a mixture of monosaccharides. Fungi such as yeast efficiently ferment the glucose to ethanol (Section 13-1), which can be distilled and used as a fuel. Other organisms can convert monosaccharides such as xylose to ethanol, but these pathways are not always efficient, and they may yield other substances, such as lactate and acetate, as end products. The most promising approach appears to be bioengineering microorganisms to carry out polysaccharide hydrolysis and then convert the resulting monosaccharides to ethanol, which is relatively stable and easy to transport and store. Alternative strategies use microorganisms to convert the sugar mixture into hydrocarbons that can be used in place of diesel fuel.

⬣ **Question:** How do xylose, galacturonate, and rhamnose differ from glucose?

▶▶ **WHY** are some carbohydrates indigestible?

Animals do not synthesize cellulose, and most cannot digest it in order to use its glucose residues as an energy source. Organisms such as termites and ruminants (grazing mammals), who do derive energy from cellulose-rich foods, harbor microorganisms that produce cellulases capable of hydrolyzing the $\beta(1 \rightarrow 4)$ bonds between glucose residues. Humans lack these microorganisms, so although up to 80% of the dry weight of plants consists of glucose, much of this is not available for metabolism in humans. However, its bulk, called fiber, is required for the normal function of the digestive system.

The exoskeletons of insects and crustaceans and the cell walls of many fungi contain a cellulose-like polymer called chitin, in which the $\beta(1 \rightarrow 4)$-linked residues are the glucose derivative N-acetylglucosamine (glucosamine with an acetyl group linked to its amino group):

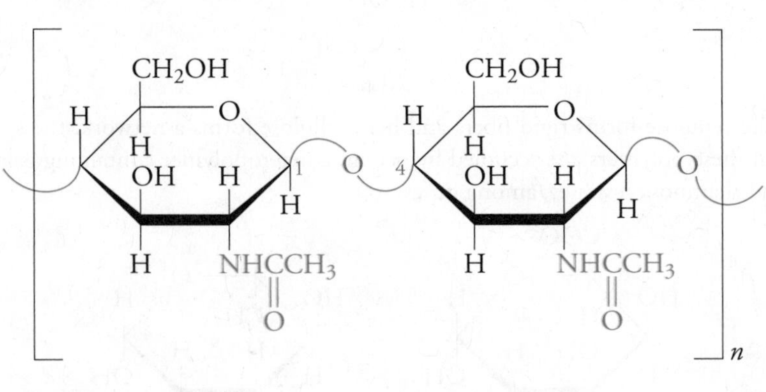

Bacterial polysaccharides form a biofilm

Prokaryotes do not synthesize cellulosic cell walls (see Section 11-3) or store fuel as starch or glycogen, but they do produce extracellular polysaccharides that provide a protective matrix for their growth. *A **biofilm** is attached to a surface and harbors a community of embedded bacteria that contribute to biofilm production and maintenance* (Fig. 11-6). The extracellular material of the biofilm includes an assortment of highly hydrated polysaccharides containing glucuronate and *N*-acetylglucosamine. A biofilm can be difficult to characterize because it typically houses a mixture of species and the proportions of its component polysaccharides depend on many environmental factors.

The gel-like consistency of a biofilm, such as the plaque that forms on teeth, prevents bacterial cells from being washed away and protects them from desiccation. Biofilms that develop on medical apparatus, such as catheters, are problematic because they offer a foothold for pathogenic organisms and create a barrier for antibiotics and cells of the immune system.

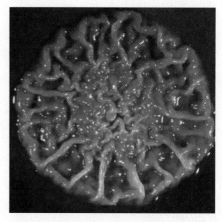

Figure 11-6 A *Pseudomonas aeruginosa* biofilm. These pathogenic bacteria growing on the surface of an agar plate form a biofilm with a complex three-dimensional shape. [Courtesy Roberto Kolter, Harvard Medical School.]

CONCEPT REVIEW

- Explain why it is possible for two monosaccharides to form more than one type of disaccharide.
- Summarize the physiological roles of lactose, sucrose, starch, glycogen, cellulose, and chitin.
- How do the physical properties of polysaccharides—such as overall size, shape, branching, and composition—relate to their biological functions?

11-3 Glycoproteins

Because there are so many different monosaccharides and so many ways to link them, the number of possible structures for oligosaccharides with even just a few residues is enormous. Organisms take advantage of this complexity to mark various structures—mainly proteins and lipids—with unique oligosaccharides. Most of the proteins that are secreted from eukaryotic cells or remain on their surface are glycoproteins in which one or more oligosaccharide chains are covalently attached to the polypeptide chain shortly after its synthesis.

KEY CONCEPTS

- Carbohydrates are attached to proteins as *N*-linked or *O*-linked oligosaccharides.
- The long glycosaminoglycan chains of proteoglycans are highly hydrated.
- Bacteria build cell walls from peptidoglycan, a three-dimensional network of carbohydrate chains and short peptides.

N-linked oligosaccharides undergo processing

In eukaryotes, the oligosaccharides attached to glycoproteins are usually linked either to an Asn side chain (**N-linked oligosaccharides**) or to a Ser or Thr side chain (**O-linked oligosaccharides**).

N-Linked oligosaccharide

O-Linked oligosaccharide

N-Glycosylation begins while a protein is being synthesized by a ribosome associated with the rough endoplasmic reticulum (ER). As the protein is translocated into the ER lumen (the internal space), an oligosaccharide chain of 14 residues is attached

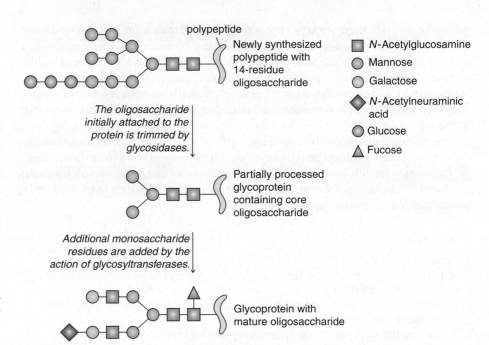

Figure 11-7 Processing of an N-linked oligosaccharide. Glycosidases and glycosyltransferases in the Golgi apparatus process the 14-residue oligosaccharide that is attached to the newly synthesized protein in the ER. The 5-residue core oligosaccharide (with 3 mannose and 2 *N*-acetylglucosamine residues, center) is common to all *N*-linked oligosaccharides. Only one of many possible mature oligosaccharides is shown.

? **How many different glycosyltransferases would be required to synthesize the oligosaccharide shown here?**

to an Asn residue (**Fig. 11-7**). When the newly synthesized protein leaves the ER and traverses the Golgi apparatus (a series of membrane-bounded compartments), *enzymes known as* **glycosidases** *remove various monosaccharide residues, and other enzymes, called* **glycosyltransferases,** *add new monosaccharides.* These processing enzymes are highly specific for the identities of the monosaccharides and the positions of the glycosidic bonds.

Apparently, the amino acid sequence or local structure of the protein, as well as the set of processing enzymes present in the cell, roughly determines which sugars are added and deleted. The net result is a great deal of heterogeneity in the oligosaccharide chains attached to different glycoproteins or even to different molecules of the same glycoprotein. An example of an *N*-linked oligosaccharide is shown in **Figure 11-8**.

O-linked oligosaccharides tend to be large

O-Linked oligosaccharides are built, one residue at a time, primarily in the Golgi apparatus, through the action of glycosyltransferases. Unlike *N*-linked oligosaccharides, the *O*-linked oligosaccharides do not undergo processing by glycosidases. Glycoproteins with *O*-linked saccharide chains tend to have many such groups, and the glycan chains tend to be longer than those of *N*-linked oligosaccharides. Such glycoproteins, which may be 80% carbohydrate, are major components of the mucus that provides a protective layer for the respiratory and digestive tracts.

What is the purpose of the oligosaccharide groups?

Because oligosaccharides are highly hydrophilic and are conformationally flexible, they occupy a large effective volume above the protein's surface. This may serve a protective function or help stabilize the protein's structure. In fact, certain chaperones recognize partially glycosylated proteins and help them fold to their native conformations. In some cases, oligosaccharide groups constitute a sort of intracellular addressing system so that newly synthesized proteins can be delivered to their proper cellular location, such as a lysosome. In other cases, the oligosaccharide groups act as recognition and attachment points for interactions between different types of cells. For example, the familiar A, B, and O blood types are determined by the presence of different oligosaccharides on the surface of red blood cells (Box 11-B). Circulating white blood cells latch onto glycoproteins on the cells lining the blood vessels in order to leave the bloodstream and migrate to sites of injury or infection. Unfortunately, many viruses and pathogenic bacteria also recognize specific carbohydrate groups on cell surfaces and attach themselves to these sites before invading the host cell.

Figure 11-8 Structure of an N-linked oligosaccharide. The 11 monosaccharide residues, linked by glycosidic bonds, experience considerable conformational flexibility, so the structure shown here is just one of many possible. The arrow indicates the N atom of the Asn side chain to which the glycan is attached. [Structure of the carbohydrate chain of soybean agglutinin determined by A. Darvill and H. Halbeek.]

BOX 11-B **BIOCHEMISTRY NOTE**

301
Glycoproteins

The ABO Blood Group System

The carbohydrates on the surface of red blood cells and other human cells form 15 different blood group systems. The best known and one of the clinically important carbohydrate-classification schemes is the ABO blood group system, which has been known for about a century. Biochemically, the ABO system involves the oligosaccharides attached to sphingolipids and proteins on red blood cells and other cells.

In individuals with type A blood, the oligosaccharide has a terminal *N*-acetylated galactose group. In type B individuals, the terminal sugar is galactose. Neither of these groups appears in the oligosaccharides of type O individuals.

Blood groups are genetically determined: Type A and B individuals have slightly different versions of the gene for a glycosyltransferase that adds the final monosaccharide residue to the oligosaccharide. Type O individuals have a mutation such that they lack the enzyme entirely and therefore produce an oligosaccharide without the final residue.

Type A individuals develop antibodies that recognize and cross-link red blood cells bearing the type B oligosaccharide. Type B individuals develop antibodies to the type A oligosaccharide. Therefore, a transfusion of type B blood cannot be given to a type A individual, and vice versa. Individuals with type AB blood bear both types of oligosaccharides and therefore do not develop antibodies to either type. They can receive transfusions of either type A or type B blood. Type O individuals develop both anti-A and anti-B antibodies. If they receive a transfusion of type A, type B, or type AB blood, their antibodies will react with the transfused cells, which causes them to lyse or to clump together and block blood vessels. On the other hand, type O individuals are universal donors: Type A, B, or AB individuals can safely receive type O blood (these individuals do not develop antibodies to the O-type oligosaccharide because it occurs naturally in these individuals as a precursor of the A-type and B-type oligosaccharides).

● **Question:** In rare cases, individuals do not produce an A-, B-, or O-type oligosaccharide at all. Can these individuals receive type A, B, or O blood transfusions? Can they donate blood to others?

Proteoglycans contain long glycosaminoglycan chains

Proteoglycans are glycoproteins in which the protein chain serves mainly as an attachment site for enormous linear *O*-linked polysaccharides called *glycosaminoglycans.* Most glycosaminoglycan chains consist of a repeating disaccharide of an amino sugar (often *N*-acetylated) and a uronic acid (a sugar with a carboxylate group).

Figure 11-9 The repeating disaccharide of chondroitin sulfate. A chondroitin sulfate chain may include hundreds of these disaccharide units, and the degree of sulfation may vary along its length.

? What ions are likely to be associated with chondroitin sulfate?

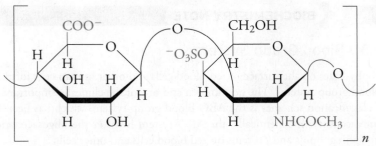

Chondroitin sulfate

After synthesis, various hydroxyl groups may be enzymatically sulfated (an $-OSO_3^-$ group added). The repeating disaccharide of the proteoglycan known as chondroitin sulfate is shown in **Figure 11-9.** Proteoglycans may be transmembrane proteins or lipid-linked (Section 8-3), but *the glycosaminoglycan chains are invariably on the extracellular side of the plasma membrane.* Extracellular proteoglycans and glycosaminoglycan chains that are not attached to a protein scaffold play an important structural role in connective tissue.

The many hydrophilic groups on the glycosaminoglycans attract water molecules, so glycosaminoglycans are highly hydrated and occupy the spaces between cells and other components of the extracellular matrix, such as collagen fibrils (Section 5-2). Under mechanical pressure, some of the water can be squeezed out of the glycosaminoglycans, which allows connective tissue and other structures to accommodate the body's movements. Pressure also brings the negatively charged sulfate and carboxylate groups of the polysaccharides close together. When the pressure abates, the glycosaminoglycans quickly spring back to their original shape as the repulsion between anionic groups is relieved and water is drawn back into the molecule. This spongelike action of glycosaminoglycans in the spaces of the joints provides shock absorption.

Bacterial cell walls are made of peptidoglycan

A network of cross-linked carbohydrate chains and peptides constitutes the cell walls of bacteria. This material, called **peptidoglycan,** *surrounds the plasma membrane of the cell and determines its overall shape.* The carbohydrate component in many species is a repeating $\beta(1 \rightarrow 4)$-linked disaccharide:

Peptides of four or five amino acids covalently cross-link the saccharide chains in three dimensions to form a structure as thick as 250 Å in some species. Antibiotics of the penicillin family block the formation of the peptide cross-links, thereby killing bacteria (Box 11-C).

A model of the cell wall from the bacterium *Staphylococcus aureus*, based on NMR data, shows that the glycan chains are perpendicular to the cell surface. The peptide cross-links form a honeycomb-like structure with spaces that could accommodate proteins (**Fig. 11-10**).

BOX 11-C ⬡ **BIOCHEMISTRY NOTE**

303
Glycoproteins

Antibiotics and Bacterial Cell Walls

The dozens of antibiotics currently prescribed to treat bacterial infections interfere with bacterial growth in several ways, for example, by inhibiting the synthesis of DNA, RNA, or proteins. However, the most common mechanism of antibiotic action is disruption of cell wall synthesis. Penicillin,

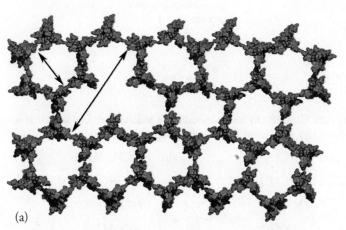

Penicillin G

the first antibiotic to be used clinically, kills bacteria by inhibiting an enzyme that cross-links peptidoglycan. The resulting structural defect weakens the cell wall, which normally resists osmotic pressure from the cell. Without a strong wall to contain it, the cell quickly swells and bursts, accounting for the rapid action of penicillin. Other β-lactam antibiotics, such as methicillin, amoxicillin, and the cephalosporins, have a similar mechanism of action (the four-atom ring in penicillin and related compounds is known as a β-lactam).

Bacteria that are resistant to β-lactam antibiotics produce an enzyme that cleaves the amide bond of the β-lactam ring. Methicillin-resistant *Staphylococcus aureus* (MRSA) is therefore difficult to treat, since none of the β-lactam antibiotics are effective against it (methicillin itself is no longer used clinically). Although many healthy individuals harbor MRSA, it can cause severe infections, particularly in immunocompromised individuals. In such cases, a different class of antibiotic may be effective. However, the emergence of multiple-drug-resistant strains of MRSA (and other pathogens) demands constant vigilance by clinicians and dedication by the scientists who must continually develop new antibiotics.

● **Question:** Explain why human cells are not sensitive to antibiotics such as penicillin.

100 Å

Figure 11-10 Model of a bacterial cell wall. The carbohydrate chains are colored orange and contain eight disaccharide repeats. The peptides are shown in green. (a) Top view, looking down onto the cell surface. Arrows indicate spaces that could be occupied by cell wall–spanning proteins. (b) Side view. [From PNAS 103: 4404-4409, Fig 4 (2006). Copyright 2006 National Academy of Sciences, U.S.A. Photo courtesy Shahriar Mobashery, University of Notre Dame.]

CONCEPT REVIEW
- Summarize the processing of an *N*-linked oligosaccharide.
- List some functions of the carbohydrate groups of glycoproteins.
- What are some differences between *N*-linked and *O*-linked oligosaccharides?
- How do proteoglycans function as shock absorbers?
- Briefly describe the structure and function of peptidoglycans.

[SUMMARY]

11-1 Monosaccharides

- Carbohydrates, which have the general formula $(CH_2O)_n$, exist as monosaccharides and polysaccharides of different sizes. The monosaccharides may be aldoses or ketoses and exist as enantiomers, which are mirror images, and as epimers, which differ in configuration at individual carbon atoms.
- Cyclization of a monosaccharide produces α and β anomers. Formation of a glycosidic bond prevents the interconversion of the α and β forms.
- Monosaccharide derivatives include phosphorylated sugars; sugars with amino, carboxylate, and extra hydroxyl groups; and deoxy sugars.

11-2 Polysaccharides

- Lactose consists of galactose linked $\beta(1 \rightarrow 4)$ to glucose. Sucrose consists of glucose linked $\alpha(1 \rightarrow 2)$ to fructose.

- Starch is a linear polymer of $\alpha(1 \rightarrow 4)$-linked glucose residues; glycogen also contains $\alpha(1 \rightarrow 6)$ branch points. Cellulose consists of $\beta(1 \rightarrow 4)$-linked glucose residues; in chitin, the residues are *N*-acetylglucosamine.
- A bacterial biofilm is a community of cells embedded in an extracellular polysaccharide matrix.

11-3 Glycoproteins

- Oligosaccharides are attached to proteins as *N*- and *O*-linked oligosaccharides. *N*-Linked oligosaccharides undergo processing by glycosidases and glycosyltransferases. The carbohydrate chains of glycoproteins function as protection and as recognition markers.
- Proteoglycans consist mainly of long glycosaminoglycan chains that can be compressed and regain their shape.
- Peptidoglycan in bacterial cell walls is constructed of cross-linked oligosaccharides and peptides.

[GLOSSARY TERMS]

carbohydrate	enantiomers	glycoside
monosaccharide	D sugar	glycan
disaccharide	L sugar	glycomics
trisaccharide	epimer	biofilm
polysaccharide	Haworth projection	*N*-linked oligosaccharide
aldose	Fischer projection	*O*-linked oligosaccharide
ketose	α anomer	glycosidase
triose	β anomer	glycosyltransferase
tetrose	reducing sugar	proteoglycan
pentose	nonreducing sugar	glycosaminoglycan
hexose	glycosidic bond	peptidoglycan
chirality		

[PROBLEMS]

11-1 Monosaccharides

1. Classify the following sugars as aldoses or ketoses:

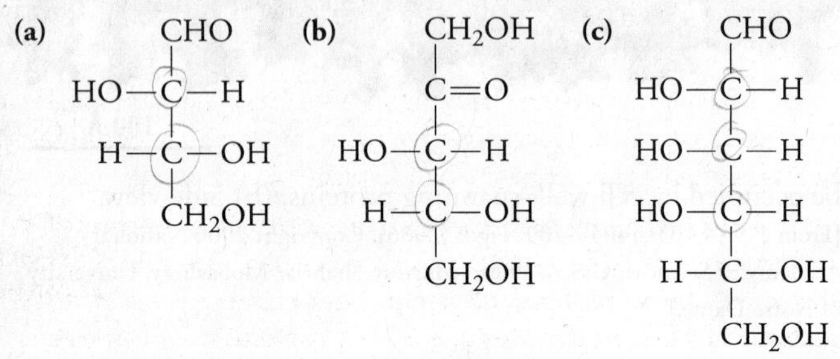

(a)
```
        CHO
        |
  HO—C—H
        |
   H—C—OH
        |
      CH2OH
```

(b)
```
      CH2OH
        |
      C=O
        |
  HO—C—H
        |
   H—C—OH
        |
      CH2OH
```

(c)
```
        CHO
        |
  HO—C—H
        |
  HO—C—H
        |
  HO—C—H
        |
   H—C—OH
        |
      CH2OH
```

2. Glucose can be described as an aldohexose. Use similar terminology to describe the following sugars:

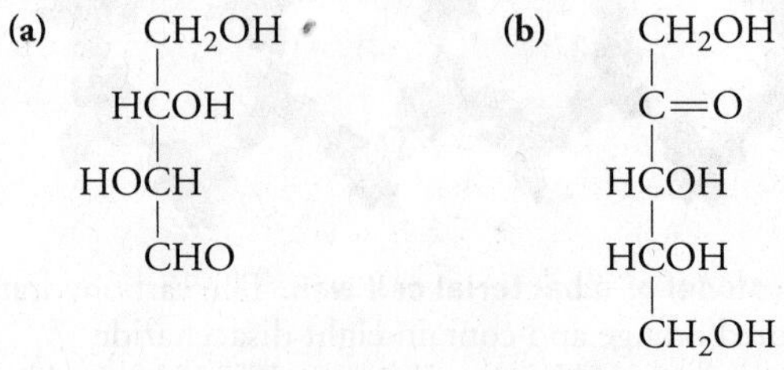

(a)
```
      CH2OH
        |
      HCOH
        |
      HOCH
        |
      CHO
```

(b)
```
      CH2OH
        |
      C=O
        |
      HCOH
        |
      HCOH
        |
      CH2OH
```

3. Which of the carbons in the sugars shown in Problem 1 are chiral?

4. Identify the monosaccharide(s) present in coenzyme A, NAD, and FAD (see Fig. 3-3).

5. Identify the following sugars as D or L:

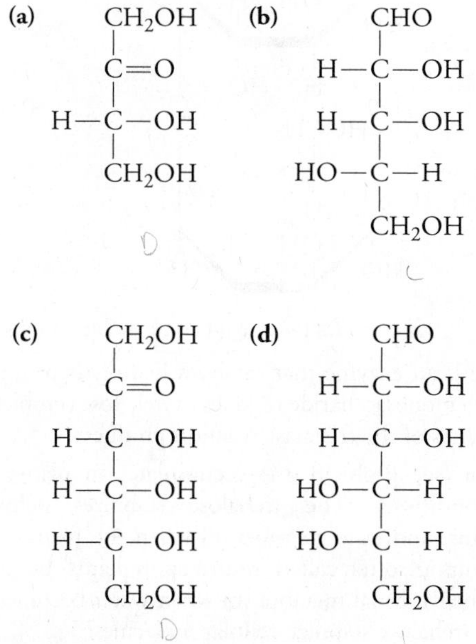

(a)
```
    CH2OH
    |
    C=O
    |
H—C—OH
    |
   CH2OH
```

(b)
```
    CHO
    |
H—C—OH
    |
H—C—OH
    |
HO—C—H
    |
   CH2OH
```

(c)
```
    CH2OH
    |
    C=O
    |
H—C—OH
    |
H—C—OH
    |
H—C—OH
    |
   CH2OH
```

(d)
```
    CHO
    |
H—C—OH
    |
H—C—OH
    |
HO—C—H
    |
HO—C—H
    |
   CH2OH
```

6. How many stereoisomers are possible for a **(a)** ketopentose, **(b)** ketohexose, and **(c)** ketoheptose if the number of stereoisomers is equal to 2^n where n is the number of chiral carbons?

7. Which type of isomer is represented by each pair of sugars? **(a)** D-sorbose and D-psicose

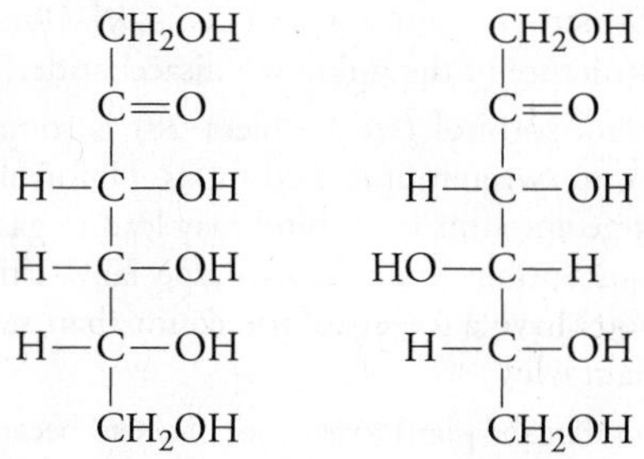

```
    CH2OH              CH2OH
    |                  |
    C=O                C=O
    |                  |
H—C—OH             H—C—OH
    |                  |
H—C—OH             HO—C—H
    |                  |
H—C—OH             H—C—OH
    |                  |
   CH2OH              CH2OH
  D-Psicose          D-Sorbose
```

(b) D-sorbose and D-fructose
(c) D-fructose and L-fructose
(d) D-ribose and D-ribulose (structure shown in Problem 2b)

8. What type of isomers are α-D-glucose and β-D-glucose?

9. Which of the following are isomers of glucose? **(a)** glucose-6-phosphate, **(b)** fructose, **(c)** galactose, and **(d)** ribose.

10. Mannose is the C2 epimer of glucose. Draw its structure.

11. Tagatose is an "artificial" sweetener that is similar to sucrose in sweetness and replaces sucrose in Pepsi Slurpees®. Tagatose is a C4 epimer of fructose.

(a) Draw the structure of D-tagatose.

(b) L-Tagatose is also about as sweet as sugar, but it costs more to manufacture than its D isomer; thus, an early decision to market L-tagatose was abandoned. Draw the structure of L-tagatose.

(c) Only about 30% of tagatose is absorbed in the small intestine; effectively, tagatose has 30% of the calories of sucrose. Why is D-tagatose absorbed less efficiently in the small intestine?

12. D-Psicose, the C3 epimer of fructose, has no calories and has also been found to have hypoglycemic effects, which might be useful in treating diabetes. Draw the structure of D-psicose.

13. Which is more stable: the α or the β anomer of mannose?

14. Carry out a cyclization reaction with galactose and draw the Haworth structures of the two possible reaction products.

15. Like glucose, ribose can undergo a cyclization reaction with its aldehyde group and the C5 hydroxyl group to form a six-membered ring. Draw the structures of the two possible reaction products.

16. Ribose can undergo a cyclization reaction different from the one presented in Problem 15. In this case, the aldehyde group reacts with the C4 hydroxyl group. Draw the structures of the two possible reaction products. What is the size of the ring?

17. (a) Like glucose, fructose can undergo a cyclization reaction. The most common reaction involves the ketone group and the C5 hydroxyl group. Draw the structures of the two possible reaction products. What is the size of the ring? **(b)** Repeat the exercise described in part (a) but use the C6 hydroxyl group instead. What is the size of the resulting ring?

18. The linear structure of *N*-acetylneuraminic acid (NANA) is shown below. It is synthesized from *N*-acetylmannosamine and pyruvic acid. This monosaccharide can undergo a cyclization reaction to form a six-membered ring. Draw the structure of the α anomer of NANA.

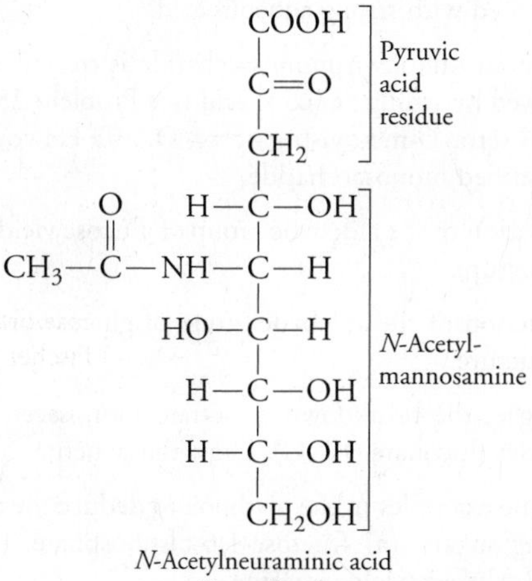

```
              COOH   ⎤
              |      ⎥  Pyruvic
              C=O    ⎥  acid
              |      ⎥  residue
              CH2    ⎦
       O    H—C—OH   ⎤
       ‖      |       
CH3—C—NH—C—H          
           HO—C—H    ⎥  N-Acetyl-
              |      ⎥  mannosamine
           H—C—OH    ⎥
              |      ⎥
           H—C—OH    ⎥
              |      ⎥
             CH2OH   ⎦
```
N-Acetylneuraminic acid
(linear form)

19. An enzyme recognizes only the α anomer of glucose as a substrate and converts it to product. If the enzyme is added to a mixture of the α and β anomers, explain why all the sugar molecules in the sample will eventually be converted to product.

20. As described in the text, a solution of glucose molecules consists of about 64% β anomer and about 36% α anomer. Why isn't the mixture 50% β anomer and 50% α anomer? In other words, why is the β anomer favored?

21. When glucose from the blood enters a cell, intracellular enzymes convert it to glucose-6-phosphate. This strategy results in the "trapping" of glucose in the cell. Explain.

22. During baking, reducing sugars react with proteins in food to generate adducts with brown color and different flavors (this is why bread becomes darker and has a different taste when toasted). The first step in this process is a condensation between a carbonyl group and an amino group. For example, the carbonyl carbon of glucose can condense with the ε-amino group of a lysine side chain to form a Schiff base. Draw the product of this reaction.

23. Benedict's solution is an alkaline copper sulfate solution which is used to detect the presence of aldehyde groups. In the presence of Benedict's solution, the aldehyde group is oxidized and the aqueous blue Cu^{2+} ion is reduced to a red Cu_2O precipitate. Sugars such as glucose, which produces a red precipitate when Benedict's solution is added, are called reducing sugars because they can reduce Cu^{2+} to Cu^+. Why is glucose a reducing sugar while fructose is not?

24. Which of the following carbohydrates would you expect to give a positive reaction with Benedict's reagent (see Problem 23)?
(a) galactose
(b) xylitol
(c) β-ethylglucoside
(d) fructose-6-phosphate

25. (a) If methanol is added to a sugar in the presence of an acid catalyst, only the anomeric hydroxyl group becomes methylated, as shown in Figure 11-2. If a stronger methylating agent is used, such as methyl iodide, CH_3I, all hydroxyl groups become methylated. Draw the product that results when methyl iodide is added to a solution of α-D-glucose. **(b)** If the product that formed in part (a) is treated with a strong aqueous acid solution, the glycosidic methyl group is readily hydrolyzed, but the methyl ethers are not. Draw the structure of the product that results when the compound formed in part (a) is treated with strong aqueous acid.

26. When an unknown monosaccharide is treated with methyl iodide followed by strong aqueous acid (see Problem 25), the product is 2,3,5,6-tetra-O-methyl-D-glucose. Draw a Haworth structure of the unmodified monosaccharide.

27. Oxidation of the aldehyde group of glucose yields gluconate. Draw its structure.

28. Reduction of the aldehyde group of glucose yields sorbitol. Draw its structure.

29. In algae, the breakdown of certain monosaccharides yields 2-keto-3-deoxygluconate (KDG). Draw the structure of KDG.

30. Use the examples in the textbook to deduce the structures of the following sugars: **(a)** fructose-1,6-bisphosphate, **(b)** galactosamine, and **(c)** N-acetylglucosamine.

11-2 Polysaccharides

31. Explain why lactose is a reducing sugar, whereas sucrose is not.

32. As described in the text, lactase is the enzyme that catalyzes the hydrolysis of lactose. The enzyme is expressed in newborns, but its activity typically decreases as the organism matures, and adults become *lactose-intolerant*. However, adult members of populations whose ancestors domesticated cows (northern Europe) or goats (Africa) are lactose-tolerant. Explain why.

33. Cellobiose is a disaccharide composed of two glucose monomers linked by a β(1 → 4) glycosidic bond. Draw the structure of cellobiose. Is cellobiose a reducing sugar?

34. The structure of trehalose is shown. Is trehalose a reducing sugar?

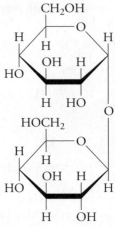

35. Trehalase is an enzyme that catalyzes hydrolysis of the bond that links the two monosaccharide residues of trehalose (Problem 34). Draw the structures of the trehalase reaction products.

36. Trehalose (see Problem 34) accumulates in plants under dehydrating conditions. The trehalose stabilizes dehydrated enzymes, proteins, and lipid membranes when the plant is desiccated. These plants are often called "resurrection plants" because the plants resume their normal metabolism when water becomes available. How does trehalose stabilize cellular molecules?

37. Propose a structure for the disaccharide maltose that is consistent with the following information: Complete hydrolysis yields only D-glucose; it reduces copper(II) to Cu_2O, and it is hydrolyzed by α-glucosidase but not β-glucosidase.

38. An unknown disaccharide is subjected to methylation with CH_3I followed by acid hydrolysis (see Problem 25). The reaction yields 2,3,4,6-tetra-O-methyl-D-glucose and 2,3,4-tri-O-methyl-D-glucose. Draw the structure of the unknown disaccharide.

39. The sugar alcohol sorbitol (see Problem 28) is sometimes used instead of sucrose to sweeten processed foods. Unfortunately, the consumption of large amounts of sorbitol may lead to gastrointestinal distress in susceptible individuals, especially children. Sorbitol-sweetened foods have a lower calorie count than sucrose-sweetened foods. Explain why.

40. A friend tells you that she plans to eat a lot of celery because she heard that the digestion of celery consumes more calories than the food contains, and this will help her to lose weight. How do you respond?

41. The presence of the oligosaccharide raffinose (below) in beans is blamed for causing flatulence when beans are consumed. Undigested raffinose is acted upon by bacteria in the large intestine, which produce gas as a metabolic by-product. Why are humans unable to digest raffinose?

Raffinose

42. Scientists in Singapore injected soybeans with an enzyme extract from the fungus *R. oligosporus* before allowing the soybeans to germinate. After three days, the concentration of raffinose (see Problem 41) decreased dramatically. (As an added bonus, the concentration of cancer-fighting isoflavones in the soybeans increased.) What enzymes must have been present in the fungus?

43. In a homopolymer, all the monomers are the same; in a heteropolymer, the monomers are different. Which of the polysaccharides discussed in this chapter are homopolymers and which are heteropolymers?

44. How many reducing ends are in a molecule of potato amylopectin that contains 500,000 residues with a branch every 250 residues?

45. Instead of starch, some plants produce inulin, which is a polymer of $\beta(2 \rightarrow 1)$-linked fructose residues. Chicory root is a particularly rich source of inulin.
(a) Draw the structure of an inulin disaccharide.
(b) Why do some food manufacturers add inulin to their products (yogurt, ice cream, and even drinks) to boost their fiber content?

46. When amylose is suspended in water in the presence of I_2, a blue color results due to the ability of iodine to occupy the interior of the helix. A drop of yellow iodine solution on a potato slice turns blue, but when a drop of iodine is placed on an apple slice, the color of the solution remains yellow. Explain why.

47. Explain why pectin (see Box 11-A) is sometimes added to fruit extracts to make jams and jellies.

48. Hemicellulose is another polysaccharide produced in plants. It is not a cellulose derivative but is a random heteropolymer of various monosaccharides. D-Xylose residues (see Box 11-A) are present in the largest amount and are linked with $\beta(1 \rightarrow 4)$ glycosidic bonds. Draw the structure of a hemicellulose disaccharide consisting of D-xylose.

49. Explain why a growing plant cell, which produces cellulose, must also produce cellulase.

50. Stonewashed cotton clothing, which appears faded and weathered, can be prepared by tumbling a garment with water and stones. Alternatively, the garment can be briefly soaked in a solution containing cellulase.
(a) Explain what cellulase does.
(b) What would happen if the cellulase treatment was prolonged?

51. Brown algae are an attractive source of biofuels, since these organisms do not require land, fertilizer, or fresh water and lack lignin. One major polysaccharide in brown algae is laminarin, which consists of glucose residues linked by $\beta(1 \rightarrow 3)$ glycosidic bonds. Draw the structure of the disaccharide unit of laminarin.

52. Mycobacteria contain several unusual polymethylated polysaccharides. One consists of repeating units of 3-*O*-methylmannose linked with $\alpha(1 \rightarrow 4)$ glycosidic bonds. Draw the structure of the disaccharide unit of this polysaccharide.

11-3 Glycoproteins

53. Identify the *N*- and *O*-linked saccharides shown at the start of Section 11-3. Are the linkages α- or β-glycosidic bonds?

54. A common *O*-glycosidic attachment of an oligosaccharide to a glycoprotein is β-galactosyl-$(1 \rightarrow 3)$-α-*N*-acetylgalactosyl-Ser. Draw the structure of the oligosaccharide and its linkage to the glycoprotein.

55. Identify the parent monosaccharides of the chondroitin sulfate disaccharide (see Fig. 11-9) and identify the linkages between them.

56. Calculate the net charge of a chondroitin sulfate molecule containing 100 disaccharide units.

57. Collagen isolated from a deep-sea hydrothermal vent worm contains a glycosylated threonine residue in the "Y" position of the (Gly-X-Y)$_n$ repeating triplet. A galactose residue is covalently attached to the threonine via a β-glycosidic bond. Draw the structure of the galactosylated threonine residue.

58. In the past, it was believed that glycosylated proteins were found only in eukaryotic cells, but recently glycoproteins have been found in prokaryotic cells. The flagellin protein from the grampositive bacterium *L. monocytogenes* has been shown to be glycosylated with a single *N*-acetylglucosamine at up to six different serine or threonine residues in the protein. Draw the structure of this linkage.

59. Identify the monosaccharide precursor of the repeating disaccharide in peptidoglycan.

60. Lysozyme, an enzyme present in tears and mucus secretions, is a $\beta(1 \rightarrow 4)$ glycosidase. Explain how lysozyme helps prevent bacterial infections.

61. In one type of peptide cross-link in peptidoglycan, an Ala residue forms an amide bond with the repeating disaccharide. Identify the location of this linkage.

62. Transmembrane proteins destined for the plasma membrane are post-translationally processed: Oligosaccharide chains are added in the ER to selected Asn residues, then the oligosaccharide chains are processed in the Golgi apparatus (Fig. 11-7). One of the sugars added to the core oligosaccharide is *N*-acetylneuraminic acid, also called sialic acid (Problem 18). It has been discovered that tumor cells frequently overexpress sialic acid on their surface, which might contribute to a tumor cell's ability to detach from the tumor and travel through the bloodstream to form additional tumors. The sialic acid residues on tumor cells are typically not recognized by the immune system, which compounds the problem. (a) How does the presence of sialic acid on the cell surface facilitate detachment? (b) What strategies would you use to design therapeutic agents to kill these types of tumor cells?

[SELECTED READINGS]

Frank, M., and Schloissnig, S., Bioinformatics and molecular modeling in glycobiology, *Cell. Mol. Life Sci.* **67,** 2749–2772 (2010). [Describes the databases and software used to study carbohydrates.]

Hart, G. W., and Copeland, R. J., Glycomics hits the big time, *Cell* **143,** 672–676 (2010). [Describes the biological roles of carbohydrates and some approaches used to study them.]

Kolter, R., and Greenberg, P., The superficial life of microbes, *Nature* **441,** 300–302 (2006). [A brief review of bacterial biofilms.]

Schwarz, F., and Aebi, M., Mechanism and principles of N-linked protein glycosylation, *Curr. Opin. Struct. Biol.* **21,** 576–582 (2011). [Reviews the synthesis and biological importance of oligosaccharides.]

Spiro, R. G., Protein glycosylation: nature, distribution, enzymatic formation, and disease implications of glycopeptide bonds, *Glycobiology* **12,** 43R–56R (2002). [Describes the different ways carbohydrates are linked to proteins.]

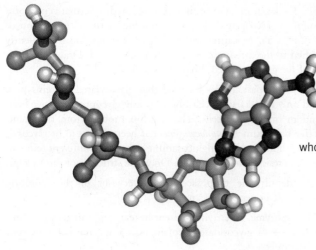

▶▶ **WHAT'S** so special about ATP?

ATP is a relatively abundant nucleotide, serving as a building block for RNA and, in its deoxy form, for DNA. Yet ATP is also known as the cell's energy currency, and we can speak of the energetic cost of a metabolic process in terms of the ATP that a cell must spend. In this chapter we'll see that ATP is not some kind of magic coin with a special chemical structure. Rather, it is an ordinary nucleotide whose *reactions* play a vital part in the metabolism of all cells.

THIS CHAPTER IN CONTEXT

Do You Remember?

- Living organisms obey the laws of thermodynamics (Section 1-3).
- Amino acids are linked by peptide bonds to form a polypeptide (Section 4-1).
- Allosteric regulators can inhibit or activate enzymes (Section 7-3).
- Lipids are predominantly hydrophobic molecules that can be esterified but cannot form polymers (Section 8-1).
- Monosaccharides can be linked by glycosidic bonds in various arrangements (Section 11-2).

Some organisms, known as **chemoautotrophs** (from the Greek *trophe,* "nourishment"), obtain virtually all their metabolic building materials and free energy from the simple inorganic compounds CO_2, N_2, H_2, and S_2. **Photoautotrophs,** such as the familiar green plants, need little more than CO_2, H_2O, a source of nitrogen, and sunlight. In contrast, **heterotrophs,** a group that includes animals, directly or indirectly obtain all their building materials and free energy from organic compounds produced by chemo- or photoautotrophs. Despite their different trophic strategies, all organisms have remarkably similar cellular structures, make the same types of biomolecules, and use similar enzymes to build and break down those molecules.

Cells break down or **catabolize** large molecules to release free energy and small molecules. The cells then use the free energy and small molecules to rebuild larger molecules, a process called **anabolism** (Fig. 12-1). The set of all catabolic and anabolic activities constitutes an organism's **metabolism.** A catalog of all the metabolic reactions undertaken by plants, animals, and bacteria is far beyond the scope of this book. Instead, we will examine a few common metabolic processes, focusing primarily on mammalian systems. In the next few chapters, we will examine some catabolic processes that release free energy and some anabolic processes that consume free energy. But first we will introduce a few of the major molecular players in metabolism, including their precursors and degradation products, and further explore the meaning of free energy in biological systems.

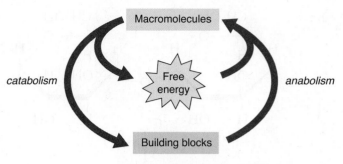

Figure 12-1 Catabolism and anabolism. Catabolic (degradative) reactions yield free energy and small molecules that can be used for anabolic (synthetic) reactions. Metabolism is the sum of all catabolic and anabolic processes.

12-1 Food and Fuel

As heterotrophs, mammals rely on food produced by other organisms. After food is digested and absorbed, it becomes a source of metabolic energy and materials to support the animal's growth and other activities. The human diet includes the four types of biological molecules introduced in Section 1-2 and described in more detail in subsequent chapters. These molecules are often present as macromolecular polymers, namely proteins, nucleic acids, polysaccharides, and triacylglycerols (technically, fats are not polymers since the monomeric units are not linked to each other but to glycerol). *Digestion reduces the polymers to their monomeric components: amino acids, nucleotides, monosaccharides, and fatty acids.* The breakdown of nucleotides does not yield significant amounts of metabolic free energy, so we will devote more attention to the catabolism of other types of biomolecules.

KEY CONCEPTS
- The macromolecules in food are hydrolyzed, and the monomeric products are absorbed by the intestine.
- Cells store fatty acids, glucose, and amino acids in the form of polymers.
- Metabolic fuels can be mobilized by breaking down glycogen, triacylglycerols, and proteins.

Cells take up the products of digestion

Digestion takes place extracellularly in the mouth, stomach, and small intestine and is catalyzed by hydrolytic enzymes (Fig. 12-2). For example, salivary amylase begins to break down starch, which consists of linear polymers of glucose residues (amylose) and branched polymers (amylopectin; Section 11-2). Gastric and pancreatic proteases (including trypsin, chymotrypsin, and elastase) degrade proteins to small peptides and amino acids. Lipases synthesized by the pancreas and secreted into the small intestine catalyze the release of fatty acids from triacylglycerols. Water-insoluble lipids do not freely mix with the other digested molecules but instead form micelles (Fig. 2-10).

The products of digestion are absorbed by the cells lining the intestine. Monosaccharides enter the cells via active transporters such as the Na^+-glucose system diagrammed in Figure 9-18. Similar symport systems bring amino acids and di- and tripeptides into the cells. Some highly hydrophobic lipids diffuse through the cell membrane; others require transporters. Inside the cell, the triacylglycerol digestion

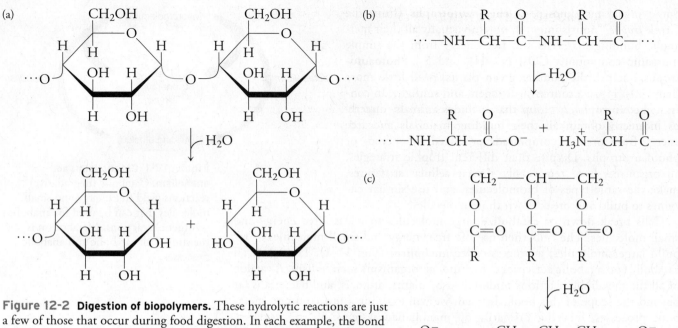

Figure 12-2 Digestion of biopolymers. These hydrolytic reactions are just a few of those that occur during food digestion. In each example, the bond to be cleaved is colored red. (a) The chains of glucose residues in starch are hydrolyzed by amylases. (b) Proteases catalyze the hydrolysis of peptide bonds in proteins. (c) Lipases hydrolyze the ester bonds linking fatty acids to the glycerol backbone of triacylglycerols.

? Explain why the reactions shown here are thermodynamically favorable.

products re-form triacylglycerols, and some fatty acids are linked to cholesterol to form cholesteryl esters, for example,

Cholesteryl stearate

Triacylglycerols and cholesteryl esters are packaged, together with specific proteins, to form **lipoproteins.** These particles, known specifically as chylomicrons, are released into the lymphatic circulation before entering the bloodstream for delivery to tissues.

Water-soluble substances such as amino acids and monosaccharides leave the intestinal cells and enter the portal vein, which drains the intestine and other visceral organs and leads directly to the liver. *The liver therefore receives the bulk of a meal's nutrients and catabolizes them, stores them, or releases them back into the bloodstream.* The liver also takes up chylomicrons and repackages the lipids with different proteins to form other lipoproteins, which circulate throughout the body, carrying cholesterol, triacylglycerols, and other lipids (lipoproteins are discussed in greater detail in Chapter 17). The allocation of resources following a meal varies with the individual's needs at that time and with the type of nutrients consumed. Fortunately, the body does this efficiently, regardless of what food was eaten (Box 12-A).

BOX 12-A ◆ BIOCHEMISTRY NOTE

311
Food and Fuel

Dietary Guidelines

Nutritionists have yet to come up with the ideal diet; the best they can do is identify the body's overall needs and roughly outline dietary requirements. For example, scientists have compiled lists of recommended daily intakes for various substances in terms of grams of the substance or the proportion of total energy intake contributed by that substance:

Distribution of Macronutrients for Adults

Carbohydrate	45–65%
Fat	20–35%
Protein	10–35%

However, few foods are composed of pure substances, so more practical guidelines focus on types of foods, with units that are more familiar to most consumers, such as ounces or cups. One source of information is the U.S. Department of Agriculture, which has published the following guidelines:

Food Group Choices

	Moderately active female, age 21–25	Moderately active male, age 21–25
Total calories	2200	2800
Fruits	2 cups	2.5 cups
Grains	7 oz	10 oz
Milk products	3 cups	3 cups
Meat, beans, nuts	6 oz	7 oz
Oils	6 tsp	8 tsp
Vegetables	3 cups	3.5 cups

[From www.cnpp.usda.gov/]

Even these recommendations are somewhat clumsy, since most individuals do not determine the volume or mass of what they place on their plates. Nutrition educators strive to translate some formal quantities into yet more familiar units: A cup of rice or a medium apple is about the size of a baseball, and three ounces of meat is about the size of a deck of cards.

An additional drawback of dietary guidelines formulated like those above is that in the United States, recommendations are based on a traditional Western diet that includes meat and dairy products. Vegetarians (who do not consume meat), vegans (who avoid consuming any animal products), and those who do not drink milk must be more diligent in assessing whether the foods they consume meet the basic requirements for carbohydrates, proteins, and so on.

Finally, a serious challenge for anyone interested in tracking their nutrient consumption is that many foods are processed; that is, raw ingredients are combined, sometimes in unknown proportions, to generate a product that can be sold as a convenience item (think: instant soup). Such foods are typically accompanied by a nutrition facts label that lists, among other things, the serving size; calories per serving; and the quantities of carbohydrates, fats, and proteins (in grams) and their percentage of the recommended daily value.

The availability of different types of dietary guidelines, along with a plethora of advice (which may or may not be grounded in the scientific method), suggests that there is significant leeway regarding what humans can or should consume. Indeed, consideration of how eating patterns have varied across centuries and across continents indicates that the human body must be remarkably versatile in converting a variety of raw materials into the molecular building blocks and metabolic energy required to sustain life.

● **Question:** How would the recommended intake of protein vary from infancy to old age? Should intakes be adjusted according to body mass?

Monomers are stored as polymers

Immediately following a meal, the circulating concentrations of monomeric compounds are relatively high. All cells can take up these materials to some extent to fulfill their immediate needs, but *some tissues are specialized for the long-term storage of nutrients*. For example, fatty acids are used to build triacylglycerols, many of which travel in the form of lipoproteins to adipose tissue. Here, adipocytes take up the triacylglycerols and store them as intracellular fat globules. Because the mass of lipid is hydrophobic and does not interfere with activities in the aqueous cytoplasm, the fat globule can be enormous, occupying most of the volume of the adipocyte (**Fig. 12-3**).

Virtually all cells can take up monosaccharides and immediately catabolize them to produce free energy. Some tissues, primarily liver and muscle (which makes up a significant portion of the human body), use monosaccharides to synthesize glycogen, the storage polymer of glucose. Glycogen is a highly branched polymer with a compact shape. Several glycogen molecules may clump together to form granules that are visible by electron microscopy (**Fig. 12-4**). Glycogen's branched structure means that a single molecule can be expanded quickly, by adding glucose residues to its many branches, and degraded quickly, by simultaneously removing glucose from the ends of many branches. Glucose that does not become part of glycogen can be catabolized to two-carbon acetyl units and converted into fatty acids for storage as triacylglycerols.

Amino acids can be used to build polypeptides. A protein is not a dedicated storage molecule for amino acids, as glycogen is for glucose and triacylglycerols are for fatty acids, so excess amino acids cannot be saved for later. However, in certain cases, such as during starvation, proteins are catabolized to supply the body's energy needs. If the intake of amino acids exceeds the body's immediate protein-building needs, the excess amino acids can be broken down and converted to carbohydrate (which can be stored as glycogen) or converted to acetyl units (which can then be converted to fat).

Amino acids and glucose are both required to synthesize nucleotides. Asp, Gln, and Gly supply some of the carbon and nitrogen atoms used to build the purine and pyrimidine bases (Section 18-3). The ribose-5-phosphate component of nucleotides is derived from glucose by a pathway that converts the six-carbon sugar to a five-carbon sugar (Section 13-4). In sum, the allocation of resources within a cell depends on the type of tissue and its need to build cellular structures, provide free energy, or stockpile resources in anticipation of future needs.

Fuels are mobilized as needed

Amino acids, monosaccharides, and fatty acids are known as **metabolic fuels** *because they can be broken down by processes that make free energy available for the cell's activities.* After a meal, free glucose and amino acids are catabolized to release their free energy. When these fuel supplies are exhausted, the body **mobilizes** its stored resources; that is, it converts its polysaccharide and triacylglycerol storage molecules (and sometimes proteins) to their respective monomeric units. Most of the body's tissues prefer to use glucose as their primary metabolic fuel, and the central nervous system can run on almost nothing else. In response to this demand, the liver mobilizes glucose by breaking down glycogen.

In general, depolymerization reactions are hydrolytic, but in the case of glycogen, the molecule that breaks the bonds between glucose residues is not water but phosphate. Thus, the degradation of glycogen is called **phosphorolysis**. This reaction is catalyzed by glycogen phosphorylase, which releases residues from the ends of branches in the glycogen polymer.

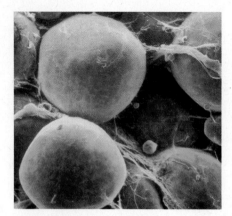

Figure 12-3 Adipocytes. These cells, which make up adipose tissue, contain a small amount of cytoplasm surrounding a large globule of triacylglycerols (fat). [© CNRI/ Phototake.]

(a)

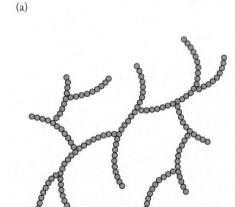

(b)

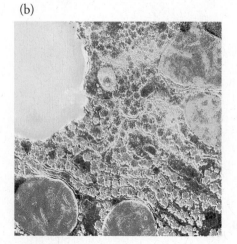

Figure 12-4 Glycogen structure. (a) Schematic diagram of a glycogen molecule. Each circle represents a glucose monomer, and branches occur every 8 to 14 residues. (b) Electron micrograph of a liver cell showing glycogen granules (colored pink). Mitochondria are green, and a fat globule is yellow. [© CNRI/Science Photo Library/Photo Researchers.]

Glycogen

Glucose-1-phosphate

The phosphate group of glucose-1-phosphate is removed before glucose is released from the liver into the circulation. Other tissues absorb glucose from the blood. In the disease **diabetes mellitus,** this does not occur, and the concentration of circulating glucose may become elevated.

Only when the supply of glucose runs low does adipose tissue mobilize its fat stores. A lipase hydrolyzes triacylglycerols so that fatty acids can be released into the bloodstream. These free fatty acids are not water-soluble and therefore bind to circulating proteins. Except for the heart, which uses fatty acids as its primary fuel, the body does not have a budget for burning fatty acids. In general, as long as dietary carbohydrates and amino acids can meet the body's energy needs, stored fat will not be mobilized, even if the diet includes almost no fat. This feature of mammalian fuel metabolism is a source of misery for many dieters!

Amino acids are not mobilized to generate energy except during a fast, when glycogen stores are depleted (in this situation, the liver can also convert some amino acids into glucose). However, cellular proteins are continuously degraded and rebuilt with the changing demand for particular enzymes, transporters, cytoskeletal elements, and so on. There are two major mechanisms for degrading unneeded proteins. In the first, the **lysosome,** an organelle containing proteases and other hydrolytic enzymes, breaks down proteins that are enclosed in a membranous vesicle. Membrane proteins and extracellular proteins taken up by endocytosis are degraded by this pathway, but intracellular proteins that become enclosed in vesicles can also be broken down by lysosomal enzymes.

A second pathway for degrading intracellular proteins requires a barrel-shaped structure known as a **proteasome.** The 700-kD core of this multiprotein complex encloses an inner chamber with multiple active sites that carry out peptide bond hydrolysis (**Fig. 12-5**). A protein can enter the proteasome only after it has been covalently tagged with a small protein called ubiquitin. This 76-residue protein is ubiquitous (hence its name) and highly conserved in eukaryotes (**Fig. 12-6**). Ubiquitin is attached to a protein by the action of a set of enzymes that link the C-terminus of ubiquitin to a Lys side chain. Additional ubiquitin molecules are then added to the first, each one linked via its C-terminus to a Lys side chain of the preceding ubiquitin. A chain of at least four ubiquitins is required to mark a protein for destruction by a proteasome.

The structural features that allow a protein to be ubiquitinated are not completely understood, but the system is sophisticated enough to allow unneeded or defective proteins to be destroyed while sparing essential proteins. A cap at the end of the proteasome barrel (not shown in Fig. 12-5) regulates the entry of ubiquitinated proteins into the inner chamber. The free energy of ATP drives conformational

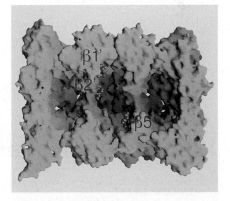

Figure 12-5 Structure of the yeast proteasome core. This cutaway view shows the inner chamber, where proteolysis occurs. Additional protein complexes (not shown) assist the entry of proteins into the proteasome. The red structures mark the locations of three protease active sites. [Courtesy Robert Huber, Max-Planck-Institut fur Biochemie, Germany.]

Figure 12-6 Ubiquitin. Several copies of this 76-residue protein are linked to Lys residues in proteins that are to be degraded by a proteasome. Atoms are color-coded: C green, O red, N blue, and H white. [Structure (pdb 1UBQ) determined by S. Vijay-Kumar, C. E. Bugg, and W. J. Cook.]

? Draw the linkage between a protein's C-terminus and a ubiquitin Lys residue.

changes that apparently help the condemned protein to unfold so that it can be more easily hydrolyzed. The ubiquitin molecules are not degraded; instead they are detached and reused. The three protease active sites inside the proteasome cleave the unfolded polypeptide substrate, releasing peptides of about eight residues that can diffuse out of the proteasome (**Fig. 12-7**). These peptides are further broken down by cytosolic peptidases so that the amino acids can be catabolized or recycled.

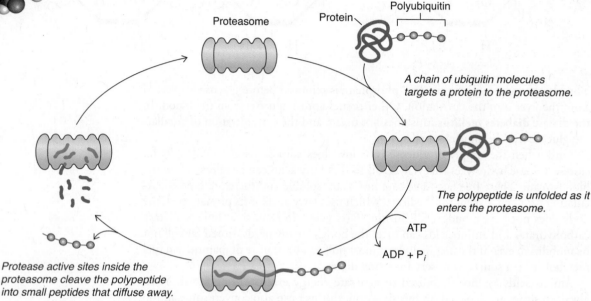

Proteasome

Protein

Polyubiquitin

A chain of ubiquitin molecules targets a protein to the proteasome.

The polypeptide is unfolded as it enters the proteasome.

ATP

ADP + P$_i$

Protease active sites inside the proteasome cleave the polypeptide into small peptides that diffuse away.

Figure 12-7 Protein degradation by the proteasome.

CONCEPT REVIEW
• Review the steps by which nutrients from food molecules reach the body's tissues.
• What are metabolic fuels and how are they stored?
• How are metabolic fuels mobilized?
• Describe the pathways for intracellular protein degradation.

12-2 Metabolic Pathways

KEY CONCEPTS
• A few metabolites appear in several metabolic pathways.
• Coenzymes such as NAD$^+$ and ubiquinone collect electrons from compounds that become oxidized.
• Metabolic pathways in cells are connected and are regulated.
• Many vitamins, substances that humans cannot synthesize, are components of coenzymes.

The interconversion of a biopolymer and its monomeric units is usually accomplished in just one or a few enzyme-catalyzed steps. In contrast, many steps are required to break down the monomeric compounds or build them up from smaller precursors. These series of reactions are known as **metabolic pathways.** A metabolic pathway can be considered from many viewpoints: as a series of intermediates or **metabolites,** as a set of enzymes that catalyze the reactions by which metabolites are interconverted, as an energy-producing or energy-requiring phenomenon, or as a dynamic process whose activity can be turned up or down. As we explore metabolic pathways in the coming chapters, we will take on each of these issues.

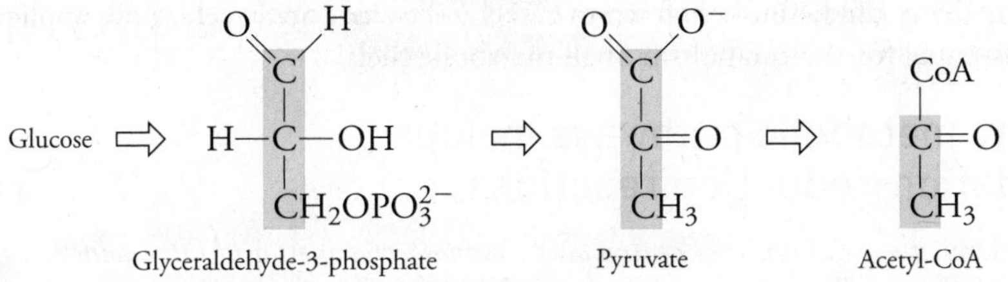

Glucose ⟹ [Glyceraldehyde-3-phosphate structure with CHO, H—C—OH, $CH_2OPO_3^{2-}$] ⟹ [Pyruvate structure with COO^-, C=O, CH_3] ⟹ [Acetyl-CoA structure with CoA, C=O, CH_3]

Glyceraldehyde-3-phosphate   Pyruvate   Acetyl-CoA

Figure 12-8 Some intermediates resulting from glucose catabolism.

? Compare the oxidation states of the carbons in glyceraldehyde-3-phosphate and in pyruvate.

Some major metabolic pathways share a few common intermediates

One of the challenges of studying metabolism is dealing with the large number of reactions that occur in a cell—involving thousands of different intermediates. However, *a handful of metabolites appear as precursors or products in the pathways that lead to or from virtually all other types of biomolecules.* These intermediates are worth examining at this point, since they will reappear several times in the coming chapters.

In **glycolysis,** the pathway that degrades the monosaccharide glucose, the six-carbon sugar is phosphorylated and split in half, yielding two molecules of glyceraldehyde-3-phosphate (Fig. 12-8). This compound is then converted in several more steps to another three-carbon molecule, pyruvate. The decarboxylation of pyruvate (removal of a carbon atom as CO_2) yields acetyl-CoA, in which a two-carbon acetyl group is linked to the carrier molecule coenzyme A (CoA).

Glyceraldehyde-3-phosphate, pyruvate, and acetyl-CoA are key players in other metabolic pathways. For example, glyceraldehyde-3-phosphate is the metabolic precursor of the three-carbon glycerol backbone of triacylglycerols. In plants, it is also the entry point for the carbon "fixed" by photosynthesis; in this case, two molecules of glyceraldehyde-3-phosphate combine to form a six-carbon monosaccharide. Pyruvate can undergo a reversible amino-group transfer reaction to yield alanine (at right). This makes pyruvate both a precursor of an amino acid and the degradation product of one. Pyruvate can also be carboxylated to yield oxaloacetate, a four-carbon precursor of several other amino acids:

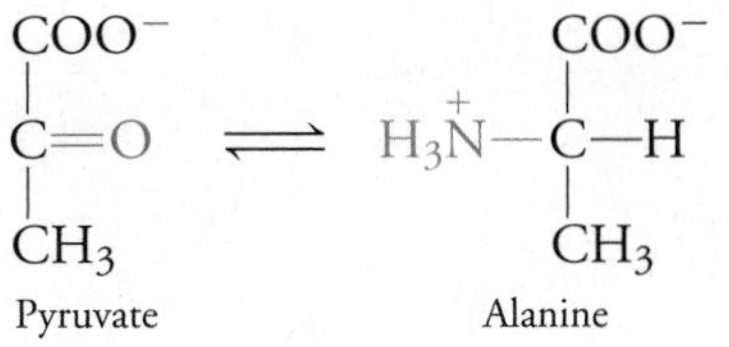

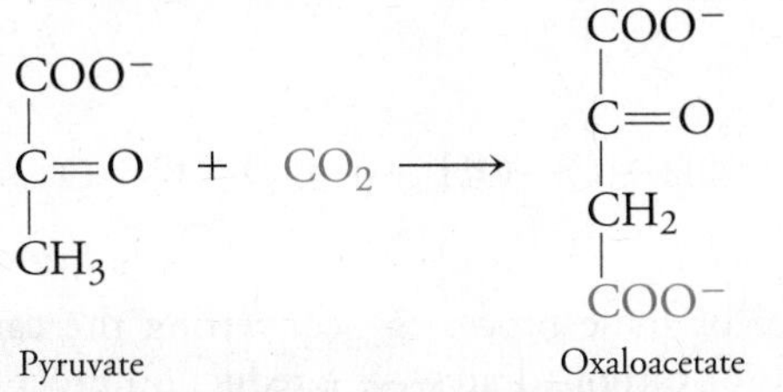

[Pyruvate structure: COO^-, C=O, CH_3] + CO_2 ⟶ [Oxaloacetate structure: COO^-, C=O, CH_2, COO^-]

Pyruvate   Oxaloacetate

Fatty acids are built by the sequential addition of two-carbon units derived from acetyl-CoA; fatty acid breakdown yields acetyl-CoA. These relationships are summarized in Figure 12-9. If not used to synthesize other compounds, two-carbon

[Pyruvate ⇌ Alanine structures: Pyruvate: COO^-, C=O, CH_3 ⇌ Alanine: COO^-, $H_3\overset{+}{N}$—C—H, CH_3]

Pyruvate   Alanine

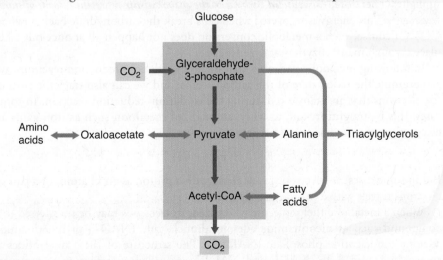

Glucose → Glyceraldehyde-3-phosphate ← CO_2; Amino acids ⇌ Oxaloacetate ⇌ Pyruvate ⇌ Alanine → Triacylglycerols; Pyruvate → Acetyl-CoA ← Fatty acids; Acetyl-CoA → CO_2

Figure 12-9 Some of the metabolic roles of the common intermediates.

? Without looking at the text, draw the structures of glyceraldehyde-3-phosphate, pyruvate, and oxaloacetate.

intermediates can be broken down to CO_2 by the **citric acid cycle,** a metabolic pathway essential for the catabolism of all metabolic fuels.

Many metabolic pathways include oxidation-reduction reactions

In general, *the catabolism of amino acids, monosaccharides, and fatty acids is a process of oxidizing carbon atoms, and the synthesis of these compounds involves carbon reduction.* Recall from Section 1-3 that **oxidation** is defined as the loss of electrons and **reduction** is the gain of electrons. Oxidation–reduction, or **redox,** reactions occur in pairs so that as one compound becomes more oxidized (gives up electrons or loosens its hold on them), another compound becomes reduced (receives the electrons or tightens its grip on them).

For the metabolic reactions that we are concerned with, the oxidation of carbon atoms frequently appears as the replacement of C—H bonds (in which the C and H atoms share the bonding electrons equally) with C—O bonds (in which the more electronegative O atom "pulls" the electrons away from the carbon atom). Carbon has given up some of its electrons, even though the electrons are still participating in a covalent bond.

The transformation of methane to carbon dioxide represents the conversion of carbon from its most reduced state to its most oxidized state:

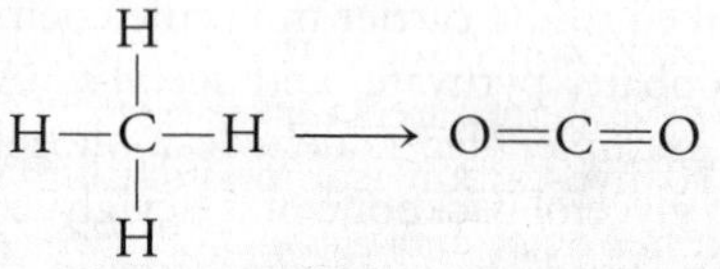

$$\begin{array}{c} \text{H} \\ | \\ \text{H}-\text{C}-\text{H} \\ | \\ \text{H} \end{array} \longrightarrow \text{O}=\text{C}=\text{O}$$

Similarly, oxidation occurs during the catabolism of a fatty acid, when saturated methylene (—CH_2—) groups are converted to CO_2 and when the carbons of a carbohydrate (represented as CH_2O) are converted to CO_2:

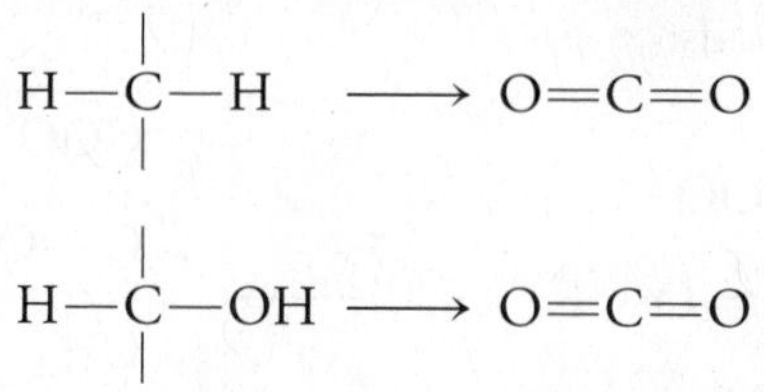

$$\begin{array}{c} | \\ \text{H}-\text{C}-\text{H} \\ | \end{array} \longrightarrow \text{O}=\text{C}=\text{O}$$

$$\begin{array}{c} | \\ \text{H}-\text{C}-\text{OH} \\ | \end{array} \longrightarrow \text{O}=\text{C}=\text{O}$$

The reverse of either of these processes—converting the carbons of CO_2 to the carbons of fatty acids or carbohydrates—is a reduction process (this is what occurs during photosynthesis, for example). In reduction processes, the carbon atoms regain electrons as C—O bonds are replaced by C—H bonds.

Turning CO_2 into carbohydrate (CH_2O) requires the input of free energy (think: sunlight). Therefore, *the reduced carbons of the carbohydrate represent a form of stored free energy.* This energy is recovered when cells break the carbohydrate back down to CO_2. Of course, such a metabolic conversion does not happen all at once but takes place through many enzyme-catalyzed steps.

In following metabolic pathways that include oxidation–reduction reactions, we can examine the redox state of the carbon atoms, and we can also trace the path of the electrons that are transferred during the oxidation–reduction reaction. In some cases, this is straightforward, as when an oxidized metal ion such as iron gains an electron (represented as e^-) to become reduced.

$$Fe^{3+} + e^- \rightarrow Fe^{2+}$$

But in some cases, an electron travels along with a proton as an H atom, or a pair of electrons travels with a proton as a hydride ion (H^-).

When a metabolic fuel molecule is oxidized, its electrons may be transferred to a compound such as nicotinamide adenine dinucleotide (NAD^+) or nicotinamide adenine dinucleotide phosphate ($NADP^+$). The structure of these nucleotides is shown in Figure 3-3b. NAD^+ and $NADP^+$ are called **cofactors** or **coenzymes,** organic compounds that allow an enzyme to carry out a particular chemical reaction (Section 6-2). The redox-active portion of NAD^+ and $NADP^+$ is the nicotinamide group, which accepts a hydride ion to form NADH or NADPH.

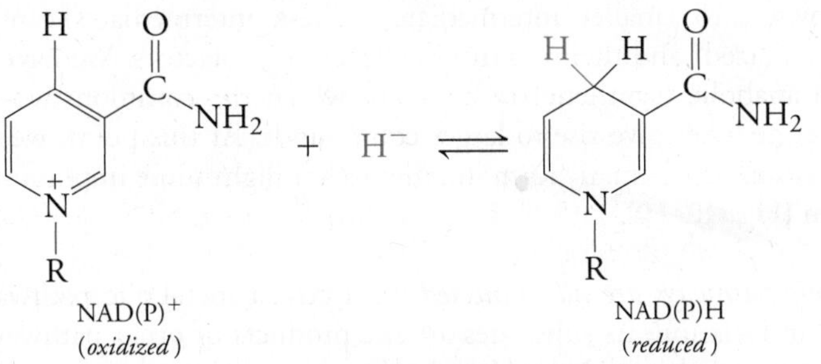

NAD(P)$^+$
(*oxidized*)   NAD(P)H
(*reduced*)

This reaction is reversible, so the reduced cofactors can become oxidized by giving up a hydride ion. In general, NAD^+ participates in catabolic reactions and $NADP^+$ in anabolic reactions. Because these electron carriers are soluble in aqueous solution, they can travel throughout the cell, shuttling electrons from reduced compounds to oxidized compounds.

Many cellular oxidation–reduction reactions take place at membrane surfaces, for example, in the inner membranes of mitochondria and chloroplasts in eukaryotes and in the plasma membrane of prokaryotes. In these cases, a membrane-associated enzyme may transfer electrons from a substrate to a lipid-soluble electron carrier such as ubiquinone (coenzyme Q, abbreviated Q; see Section 8-1). Ubiquinone's hydrophobic tail, containing 10 five-carbon isoprenoid units in mammals, allows it to diffuse within the membrane. *Ubiquinone can take up one or two electrons* (in contrast to NAD^+, which is strictly a two-electron carrier). A one-electron reduction of ubiquinone (addition of an H atom) produces a semiquinone, a stable free radical (shown as QH·). A two-electron reduction (two H atoms) yields ubiquinol (QH_2):

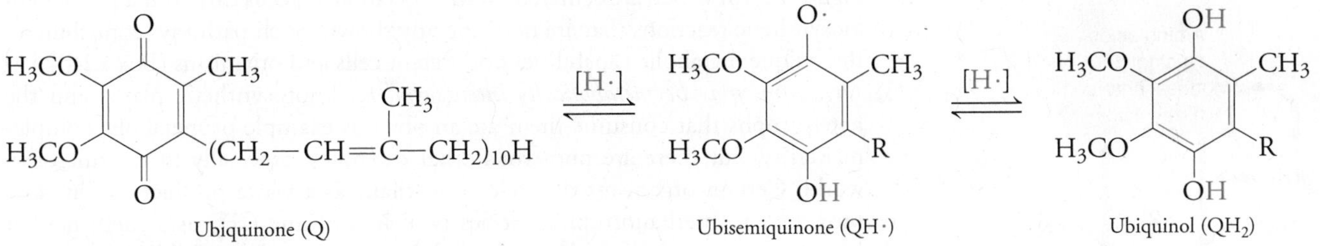

Ubiquinone (Q)   Ubisemiquinone (QH·)   Ubiquinol (QH_2)

The reduced ubiquinol can then diffuse through the membrane to donate its electrons in another oxidation–reduction reaction.

Catabolic pathways, such as the citric acid cycle, generate considerable amounts of reduced cofactors. Some of them are reoxidized in anabolic reactions. The rest are reoxidized by a process that is accompanied by the synthesis of ATP from ADP and P_i. In mammals, the reoxidation of NADH and QH_2 and the concomitant production of ATP require the reduction of O_2 to H_2O. This pathway is known as **oxidative phosphorylation.**

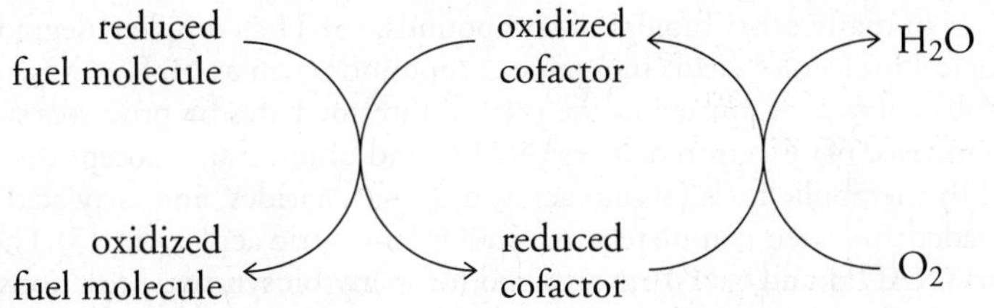

In effect, NAD^+ and ubiquinone collect electrons (and hence free energy) from reduced fuel molecules. When the electrons are ultimately transferred to O_2, this free energy is harvested in the form of ATP.

Metabolic pathways are complex

So far we have sketched the outlines of mammalian fuel metabolism, in which macromolecules are stored and mobilized so that their monomeric units can be broken down into smaller intermediates. These intermediates can be further degraded (oxidized) and their electrons collected by cofactors. We have also briefly mentioned anabolic (synthetic) reactions in which the common two- and three-carbon intermediates give rise to larger compounds. At this point, we can present this information in schematic form in order to highlight some important features of metabolism (**Fig. 12-10**).

1. *Metabolic pathways are all connected.* In a cell, a metabolic pathway does not operate in isolation; its substrates are the products of other pathways, and vice versa. For example, the NADH and QH_2 generated by the citric acid cycle are the starting materials for oxidative phosphorylation.

2. *Pathway activity is regulated.* Cells do not synthesize polymers when monomers are in short supply. Conversely, they do not catabolize fuels when the need for ATP is low. The **flux,** or flow of intermediates, through a metabolic pathway is regulated in various ways according to substrate availability and the cell's need for the pathway's products. The activity of one or more enzymes in a pathway may be controlled by allosteric effectors (Sections 5-1 and 7-3). These changes in turn may reflect extracellular signals that activate intracellular kinases, phosphatases, and second messengers (Section 10-1). Regulation of pathways is especially critical when the simultaneous operation of two opposing processes, such as fatty acid synthesis and degradation, would be wasteful.

3. *Not every cell carries out every pathway.* Figure 12-10 is a composite of a number of metabolic processes, and a given cell or organism may undertake only a subset of these. Mammals do not perform photosynthesis, and only the liver and kidney can synthesize glucose from noncarbohydrate precursors.

4. *Each cell has a unique metabolic repertoire.* In addition to the pathways outlined in Figure 12-10, which are centered on fuel metabolism, cells carry out a plethora of biosynthetic reactions that are not explicitly shown. Such pathways contribute to the unique metabolic capabilities of different cells and organisms (Box 12-B).

5. *Organisms may be metabolically interdependent.* Photosynthetic plants and the heterotrophs that consume them are an obvious example of metabolic complementarity, but there are numerous other examples, especially in the microbial world. Certain organisms that release methane as a waste product live in close proximity to methanotrophic species (which consume CH_4 as a fuel); neither organism can survive without the other. Humans also exhibit interspecific

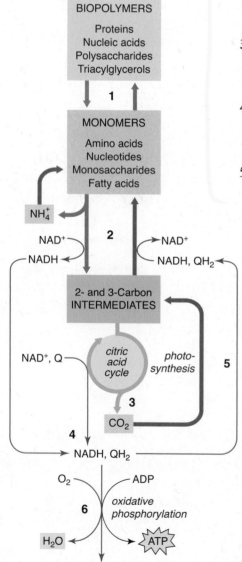

Figure 12-10 Outline of metabolism. In this composite diagram, downward arrows represent catabolic processes, and upward arrows represent anabolic processes. Red arrows indicate some major oxidation–reduction reactions. The major metabolic processes are highlighted: (1) Biological polymers (proteins, nucleic acids, polysaccharides, and triacylglycerols) are built from and are degraded to monomers (amino acids, nucleotides, monosaccharides, and fatty acids). (2) The monomers are broken down into two- and three-carbon intermediates such as glyceraldehyde-3-phosphate, pyruvate, and acetyl-CoA, which are also the precursors of many other biological compounds. (3) The complete degradation of biological molecules yields inorganic compounds such as NH_4^+, CO_2, and H_2O. These substances are returned to the pool of intermediates by processes such as photosynthesis. (4) Electron carriers (NAD^+ and ubiquinone) accept the electrons released by metabolic fuels (amino acids, monosaccharides, and fatty acids) as they are degraded and then completely oxidized by the citric acid cycle. (5) The reduced cofactors (NADH and QH_2) are required for many biosynthetic reactions. (6) The reoxidation of reduced cofactors drives the production of ATP from ADP + P_i (oxidative phosphorylation).

? **Is the citric acid cycle a process of carbon oxidation or reduction? Is photosynthesis a process of carbon oxidation or reduction?**

cooperativity: Thousands of different microbial species, amounting to some 100 trillion cells, can live in or on the human body. Collectively, these organisms express millions of different genes and carry out a correspondingly rich set of metabolic activities.

BOX 12-B BIOCHEMISTRY NOTE

The Transcriptome, the Proteome, and the Metabolome

Modern biologists have developed research tools that use the power of computers to collect enormous data sets and analyze them. Such endeavors provide great insights but also have limitations. As we saw in Section 3-3, genomics, the study of an organism's complete set of genes, yields a glimpse of that organism's overall metabolic repertoire. But what the organism, or a single cell, is actually doing at a particular moment depends in part on which genes are active.

A cell's population of mRNA molecules represents genes that are turned on, or transcribed. The study of these mRNAs is known as **transcriptomics.** Identifying and quantifying all the mRNA transcripts (the **transcriptome**) from a single cell type can be done by assembling short strands of DNA with known sequences on a solid support, then allowing them to hybridize, or form double-stranded structures, with fluorescent-labeled mRNAs from a cell preparation. The strength of fluorescence indicates how much mRNA binds to a particular complementary DNA sequence. The collection of DNA sequences is called a **microarray** or **DNA chip** because thousands of sequences fit in a few square centimeters. The microarray may represent an entire genome or just a few selected genes. Each bright spot in the DNA chip shown here represents a DNA sequence to which a fluorescent mRNA molecule has bound.

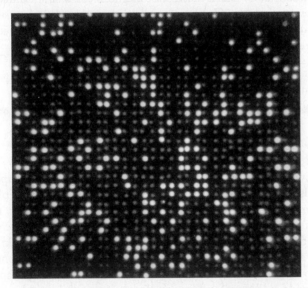

[Voker Steger/Science Photo Library/Photo Researchers.]

Biologists use DNA chips to identify genes whose expression changes under certain conditions or at different developmental stages.

Unfortunately, the correlation between the amount of a particular mRNA and the amount of its protein product is not perfect; some mRNAs are rapidly degraded, whereas others are translated many times, yielding large quantities of the corresponding protein. Hence, a more reliable way to assess gene expression is through **proteomics**—by examining a cell's **proteome,** the complete set of proteins that are synthesized by the cell at a particular point in its life cycle. However, this approach is limited by the technical problems of detecting minute quantities of thousands of different proteins. Nucleic acids can be amplified by the polymerase chain reaction (see Section 3-4), but there is no comparable procedure for amplifying proteins.

(*continued on the next page*)

Where genomics, transcriptomics, and proteomics fall short, **metabolomics** steps in, attempting to pin down the actual metabolic activity in a cell or tissue by identifying and quantifying all its metabolites, that is, its **metabolome.** This is no trivial task, as a cell may contain tens of thousands of different types of compounds, whose concentrations may range over many orders of magnitude. These substances include nonfood molecules such as toxins, preservatives, drugs, and their degradation products. Metabolites are typically detected through column chromatography, nuclear magnetic resonance (NMR) spectroscopy, or mass spectrometry (Section 4-5). In the example shown below, approximately 20 metabolites are visible in an ^1H NMR spectrum of a 10-μL sample of rat brain.

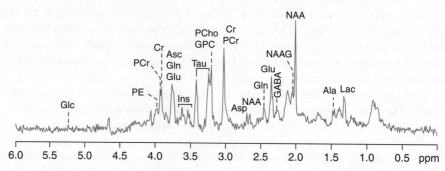

Metabolite profile of rat brain. [Courtesy Raghavendra Rao, University of Minnesota, Minneapolis.]

As has been done for genomics and proteomics and other areas of bioinformatics, metabolomic data are deposited in publicly accessible databases for retrieval and analysis. One hope for metabolomics is that disease diagnosis could be streamlined by obtaining a complete metabolic profile of a patient's urine or blood. Industrial applications include monitoring biological processes such as winemaking and bioremediation (using microorganisms to detoxify contaminated environments).

⬣ **Question:** Compare the metabolomic complexity of a single-celled prokaryote and that of a multicellular eukaryote.

An overview such as Figure 12-10 does not convey the true complexity of cellular metabolism, which takes place in a milieu crowded with multiple substrates, competing enzymes, and layers of regulatory mechanisms. Moreover, Figure 12-10 does not include any of the reactions involved in transmitting and decoding genetic information (these topics are covered in the final section of this book). However, a diagram such as Figure 12-10 is a useful tool for mapping the relationships among metabolic processes, and we will refer back to it in the coming chapters. Online databases provide additional information about metabolic pathways, enzymes, intermediates, and metabolic diseases (see Bioinformatics Project 6, Metabolic Enzymes, Microarrays, and Proteomics).

Human metabolism depends on vitamins

Humans lack many of the biosynthetic pathways that occur in plants and microorganisms and so rely on other species to provide certain raw materials. Some amino acids and unsaturated fatty acids are considered **essential** because the human body cannot synthesize them and must obtain them from food (Table 12-1; see also Box 8-B). **Vitamins** likewise are compounds that humans need but cannot make. Presumably, the pathways for synthesizing these substances, which require many specialized enzymes, are not necessary for heterotrophic organisms and have been lost through evolution.

[TABLE 12-1] Some Essential Substances for Humans

Amino Acids		Fatty Acids		Other	
Isoleucine		Linoleate	$CH_3(CH_2)_4(CH{=}CHCH_2)_2(CH_2)_6COO^-$	Choline	$(CH_3)_3N^+CH_2CH_2OH$
Leucine		Linolenate	$CH_3CH_2(CH{=}CHCH_2)_3(CH_2)_6COO^-$		
Lysine					
Methionine					
Phenylalanine					
Threonine					
Tryptophan					
Valine					

The word *vitamin* comes from *vital amine,* a term coined by Casimir Funk in 1912 to describe organic compounds that are required in small amounts for normal health. It turns out that most vitamins are not amines, but the name has stuck. Table 12-2 lists the vitamins and their metabolic roles. Vitamins A, D, E, and K are lipids; their functions were described in Box 8-B. Many of the water-soluble vitamins are the precursors of coenzymes, which we will describe as we encounter them in the context of their particular metabolic reactions. Vitamins are a diverse group of compounds,

[TABLE 12-2] Vitamins and Their Roles

Vitamin	Coenzyme Product	Biochemical Function	Human Deficiency Disease	Text Reference
Water-Soluble				
Ascorbic acid (C)	Ascorbate	Cofactor for hydroxylation of collagen	Scurvy	Box 5-D
Biotin (B_7)	Biocytin	Cofactor for carboxylation reactions	*	Section 13-1
Cobalamin (B_{12})	Cobalamin coenzymes	Cofactor for alkylation reactions	Anemia	Section 17-1
Folic acid	Tetrahydrofolate	Cofactor for one-carbon transfer reactions	Anemia	Section 18-2
Lipoic acid	Lipoamide	Cofactor for acyl transfer reactions	*	Section 14-1
Nicotinamide (niacin, B_3)	Nicotinamide coenzymes (NAD^+, $NADP^+$)	Cofactor for oxidation–reduction reactions	Pellagra	Fig. 3-3, Section 12-2
Pantothenic acid (B_5)	Coenzyme A	Cofactor for acyl transfer reactions	*	Fig. 3-3, Section 12-3
Pyridoxine (B_6)	Pyridoxal phosphate	Cofactor for amino-group transfer reactions	*	Section 18-1
Riboflavin (B_2)	Flavin coenzymes (FAD, FMN)	Cofactor for oxidation–reduction reactions	*	Fig. 3-3
Thiamine (B_1)	Thiamine pyrophosphate	Cofactor for aldehyde transfer reactions	Beriberi	Sections 12-2, 14-1
Fat-Soluble				
Vitamin A (retinol)		Light-absorbing pigment	Blindness	Box 8-B
Vitamin D		Hormone that promotes Ca^{2+} absorption	Rickets	Box 8-B
Vitamin E (tocopherol)		Antioxidant	*	Box 8-B
Vitamin K (phylloquinone)		Cofactor for carboxylation of blood coagulation proteins	Bleeding	Box 8-B

*Deficiency in humans is rare or unobserved.

whose discoveries and functional characterization have provided some of the more colorful stories in the history of biochemistry.

Many vitamins were discovered through studies of nutritional deficiencies. One of the earliest links between nutrition and disease was observed centuries ago in sailors suffering from scurvy, an illness caused by vitamin C deficiency (Box 5-D). A study of the disease beriberi led to the discovery of the first B vitamin. Beriberi, characterized by leg weakness and swelling, is caused by a deficiency of thiamine (vitamin B_1).

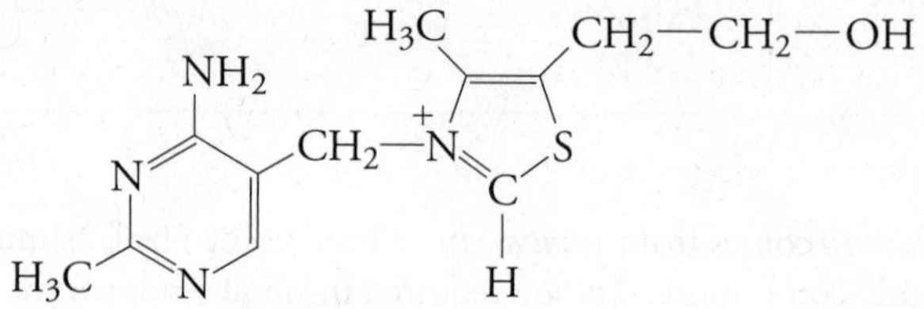

Thiamine (vitamin B_1)

Thiamine acts as a prosthetic group in some essential enzymes, including the one that converts pyruvate to acetyl-CoA. Rice husks are rich in thiamine, and individuals whose diet consists largely of polished (huskless) rice can develop beriberi. The disease was originally thought to be infectious, until the same symptoms were observed in chickens and prisoners fed a diet of polished rice. Thiamine deficiency can occur in chronic alcoholics and others with a limited diet and impaired nutrient absorption.

Niacin, a component of NAD^+ and $NADP^+$, was first identified as the factor missing in the vitamin-deficiency disease pellagra.

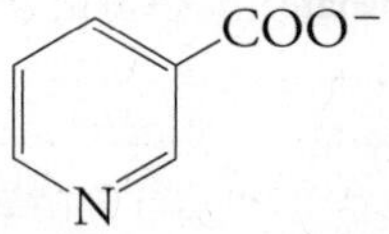

Niacin

The symptoms of pellagra, including diarrhea and dermatitis, can be alleviated by boosting the intake of the essential amino acid tryptophan, which humans can convert to niacin. Niacin deficiency was once common in certain populations whose diet consisted largely of maize (corn). This grain is low in tryptophan and its niacin is covalently bound to other molecules; hence, it is not easily absorbed during digestion. In South America, where maize originated, the kernels are traditionally prepared by soaking or boiling them in an alkaline solution, a treatment that releases niacin and prevents pellagra. Unfortunately, this food-preparation custom did not spread to other parts of the world that adopted maize farming.

Most vitamins are readily obtained from a balanced diet, although poor nutrition, particularly in impoverished parts of the world, still causes vitamin-deficiency diseases. Intestinal bacteria, as well as plant- and animal-derived foods, are the natural sources of vitamins. However, plants do not contain cobalamin, so individuals who follow a strict vegetarian diet are at higher risk for developing a cobalamin deficiency.

CONCEPT REVIEW
- Why are compounds such as glyceraldehyde-3-phosphate, pyruvate, and acetyl-CoA so important in metabolism?
- What role do cofactors such as NAD^+ and ubiquinone play in metabolic reactions?
- What is the importance of reoxidizing NADH and QH_2 by molecular oxygen?
- Summarize the main features of metabolic pathways.
- Explain the relationship between vitamins and coenzymes.

12-3 Free Energy Changes in Metabolic Reactions

KEY CONCEPTS

- The free energy change for a reaction depends on the equilibrium constant for the reaction and on the actual concentrations of the reacting species.
- A reaction with a large negative change in free energy can be coupled to another, unfavorable reaction.
- A reaction that breaks a phosphoanhydride bond in ATP occurs with a large change in free energy.
- Cells also use the free energy of other phosphorylated compounds, thioesters, reduced cofactors, and electrochemical gradients.
- Nonequilibrium reactions often serve as metabolic control points.

We have introduced the idea that catabolic reactions tend to release free energy and anabolic reactions tend to consume it (see Fig. 12-1), but, in fact, *all reactions in vivo occur with a net decrease in free energy; that is, ΔG is always less than zero* (free energy is discussed in Section 1-3). In a cell, metabolic reactions are not isolated but are linked, so the free energy of a thermodynamically favorable reaction can be used to pull a second, unfavorable reaction forward. How can free energy be transferred from one reaction to another? Free energy is not a substance or the property of a single molecule, so it is misleading to refer to a molecule or a bond within that molecule as having a large amount of free energy. Rather, *free energy is a property of a system, and it changes when the system undergoes a chemical reaction.*

The free energy change depends on reactant concentrations

The change in free energy of a system is related to the concentrations of the reacting substances. When a reaction such as $A + B \rightleftharpoons C + D$ is at equilibrium, the concentrations of the four reactants define the **equilibrium constant, K_{eq},** for the reaction:

$$K_{eq} = \frac{[C]_{eq}[D]_{eq}}{[A]_{eq}[B]_{eq}} \qquad \textbf{[12-1]}$$

(the brackets indicate the molar concentration of each substance). Recall that at equilibrium, the rates of the forward and reverse reactions are balanced, so there is no net change in the concentration of any reactant. Equilibrium does *not* mean that the concentrations of the reactants and products are equal.

When the system is not at equilibrium, the reactants experience a driving force to reach their equilibrium values. This force is the ***standard free energy change for the reaction, $\Delta G^{\circ\prime}$****, which is defined as*

$$\Delta G^{\circ\prime} = -RT \ln K_{eq} \qquad \textbf{[12-2]}$$

R is the gas constant ($8.3145\ J \cdot K^{-1} \cdot mol^{-1}$) and T is the temperature in Kelvin. Recall from Section 1-3 that free energy has units of joules per mole (Box 12-C). Equation 12-2 can be used to calculate $\Delta G^{\circ\prime}$ from K_{eq} and vice versa (see Sample Calculation 12-1).

SAMPLE CALCULATION 12-1

PROBLEM

Calculate $\Delta G^{\circ\prime}$ for a reaction at 25°C when $K_{eq} = 5.0$.

SOLUTION

Use Equation 12-2:

$$\begin{aligned}\Delta G^{\circ\prime} &= -RT \ln K_{eq} \\ &= -(8.3145\ J \cdot K^{-1} \cdot mol^{-1})(298\ K)\ln 5.0 \\ &= -4000\ J \cdot mol^{-1} = -4.0\ kJ \cdot mol^{-1}\end{aligned}$$

PRACTICE PROBLEMS

1. Calculate $\Delta G^{\circ\prime}$ for a reaction at 25°C when $K_{eq} = 0.25$.
2. If the temperature for the reaction in Practice Problem 1 was raised to 37°C, how would $\Delta G^{\circ\prime}$ change?
3. If $\Delta G^{\circ\prime}$ for a reaction at 37°C is $-10\ kJ \cdot mol^{-1}$, what is K_{eq}?

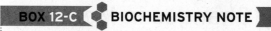

BOX 12-C BIOCHEMISTRY NOTE

What Is a Calorie?

Most biochemists express quantities using the International System of units (Box 1-A), which includes the joule, named after physicist James Prescott Joule. The joule is actually a derived unit that can be defined in terms of different kinds of work (energy is the capacity to do work). For example, a joule is equivalent to the work done by applying a force of one newton through a distance of one meter; that is, $1 \text{ J} = 1 \text{ N} \cdot \text{m}$.

The joule has largely replaced the calorie, which is the amount of heat required to increase the temperature of 1 g of water by 1°C. In practice, a calorie is a difficult thing to measure, but the term remains popular for referring to the energy content of food. However, a calorie is actually a fairly small quantity, so kilocalories (kcal), also known as large calories (Cal), are typically used. Thus, a nutrition label indicating that a tablespoon of peanut butter contains 95 calories should really say 95 Cal, 95 kcal, or 95,000 cal. To avoid confusion, calories can always be converted to joules: $1 \text{ cal} = 4.1484 \text{ J}$ and $1 \text{ J} = 0.239 \text{ cal}$.

● **Question:** How many joules are in one tablespoon of peanut butter?

By convention, measurements of standard free energy are valid under **standard conditions,** where the temperature is 25°C (298 K) and the pressure is 1 atm (these conditions are indicated by the degree symbol after ΔG). For a chemist, standard conditions specify an initial activity of 1 for each reactant (activity is the reactant's concentration corrected for its nonideal behavior). However, these conditions are impractical for biochemists since most biochemical reactions occur near neutral pH (where $[H^+] = 10^{-7}$ M rather than 1 M) and in aqueous solution (where $[H_2O] = 55.5$ M). The biochemical standard conditions are summarized in Table 12-3. Biochemists use a prime symbol to indicate the standard free energy change for a reaction under biochemical standard conditions. In most equilibrium expressions, $[H^+]$ and $[H_2O]$ are set to 1 so that these terms can be ignored. And because biochemical reactions typically involve dilute solutions of reactants, molar concentrations can be used instead of activities.

Like K_{eq}, $\Delta G^{\circ\prime}$ is a constant for a particular reaction. It may be a positive or negative value, and it indicates whether the reaction can proceed spontaneously ($\Delta G^{\circ\prime} < 0$) or not ($\Delta G^{\circ\prime} > 0$) under standard conditions. In a living cell, reactants and products are almost never present at standard-state concentrations and the temperature may not be 25°C, yet reactions do occur with some change in free energy. Thus, it is important to distinguish the standard free energy change of a reaction from its actual free energy change, ΔG. ΔG is a function of the actual concentrations of the reactants and the temperature (37°C or 310 K in humans). ΔG is related to the standard free energy change for the reaction:

$$\Delta G = \Delta G^{\circ\prime} + RT \ln \frac{[C][D]}{[A][B]}$$ [12-3]

Here, the bracketed quantities represent the actual, nonequilibrium concentrations of the reactants. The concentration term in Equation 12-3 is sometimes called the **mass action ratio.**

When the reaction is at equilibrium, $\Delta G = 0$ and

$$\Delta G^{\circ\prime} = -RT \ln \frac{[C]_{eq}[D]_{eq}}{[A]_{eq}[B]_{eq}}$$ [12-4]

which is equivalent to Equation 12-2. Note that Equation 12-3 shows that *the criterion for spontaneity for a reaction is ΔG, a property of the actual concentrations of the reactants, not the constant $\Delta G^{\circ\prime}$*. Thus, a reaction with a positive standard free energy change (a reaction that cannot occur when the reactants are present at standard

[**TABLE 12-3**]

Biochemical Standard State

Temperature	25°C (298 K)
Pressure	1 atm
Reactant concentration	1 M
pH	7.0
	($[H^+] = 10^{-7}$ M)
Water concentration	55.5 M

concentrations) may proceed *in vivo*, depending on the concentrations of reactants in the cell (see Sample Calculation 12-2). Keep in mind that thermodynamic spontaneity does not imply a rapid reaction. Even a substance with a strong tendency to undergo reaction ($\Delta G \ll 0$) will usually not react until acted upon by an enzyme that catalyzes the reaction.

SAMPLE CALCULATION 12-2

PROBLEM

The standard free energy change for the reaction catalyzed by phosphoglucomutase is $-7.1\ \text{kJ} \cdot \text{mol}^{-1}$. Calculate the equilibrium constant for the reaction. Calculate ΔG at 37°C when the concentration of glucose-1-phosphate is 1 mM and the concentration of glucose-6-phosphate is 25 mM. Is the reaction spontaneous under these conditions?

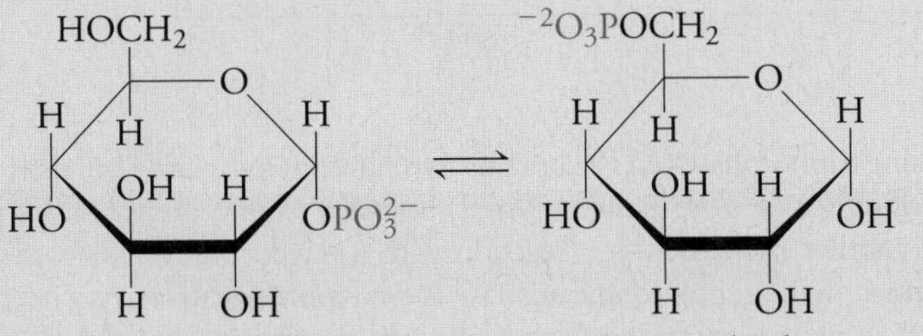

Glucose-1-phosphate Glucose-6-phosphate

SOLUTION

The equilibrium constant K_{eq} can be derived by rearranging Equation 12-2.

$$
\begin{aligned}
K_{eq} &= e^{-\Delta G^{\circ\prime}/RT} \\
&= e^{-(-7100\ \text{J}\cdot\text{mol}^{-1})/(8.3145\ \text{J}\cdot\text{K}^{-1}\cdot\text{mol}^{-1})(298\ \text{K})} \\
&= e^{2.87} = 17.6
\end{aligned}
$$

At 37°C, $T = 310$ K.

$$
\begin{aligned}
\Delta G &= \Delta G^{\circ\prime} + RT \ln \frac{[\text{glucose-6-phosphate}]}{[\text{glucose-1-phosphate}]} \\
&= -7100\ \text{J} \cdot \text{mol}^{-1} + (8.3145\ \text{J} \cdot \text{K}^{-1} \cdot \text{mol}^{-1})(310\ \text{K})\ln(0.025/0.001) \\
&= -7100\ \text{J} \cdot \text{mol}^{-1} + 8300\ \text{J} \cdot \text{mol}^{-1} \\
&= +1200\ \text{J} \cdot \text{mol}^{-1} = +1.2\ \text{kJ} \cdot \text{mol}^{-1}
\end{aligned}
$$

The reaction is not spontaneous because ΔG is greater than zero.

PRACTICE PROBLEMS

4. Calculate ΔG for the reaction shown here when the initial concentration of glucose-1-phosphate is 5 mM and the initial concentration of glucose-6-phosphate is 20 mM. Is the reaction spontaneous under these conditions?
5. At equilibrium, the concentration of glucose-6-phosphate is 35 mM. What is the concentration of glucose-1-phosphate?
6. Calculate the ratio of the concentration of glucose-6-phosphate to the concentration of glucose-1-phosphate that gives a free energy change of $-2.0\ \text{kJ} \cdot \text{mol}^{-1}$.

Unfavorable reactions are coupled to favorable reactions

A biochemical reaction may at first seem to be thermodynamically forbidden because its free energy change is greater than zero. Yet the reaction can proceed *in vivo* when it is coupled to a second reaction whose value of ΔG is very large and negative so that the *net* change in free energy for the combined reactions is less than zero. *ATP is often involved in such coupled processes because its reactions occur with a relatively large negative change in free energy.*

Figure 12-11 Adenosine triphosphate.
The three phosphate groups are
sometimes described by the Greek
letters α, β, and γ. The linkage between
the first (α) and second (β) phosphoryl
groups, and between the second (β)
and third (γ), is a phosphoanhydride
bond. A reaction in which one or two
phosphoryl groups are transferred to
another compound (a reaction in which
a phosphoanhydride bond is cleaved)
has a large negative value of $\Delta G°'$.

? **How does hydrolysis of a phosphoanhydride bond affect the net charge
of a nucleotide?**

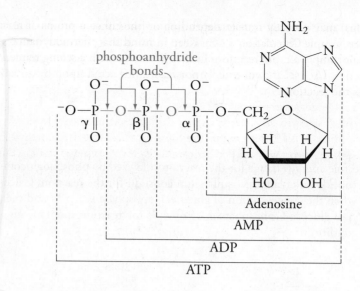

Adenosine triphosphate (ATP) contains two phosphoanhydride bonds (**Fig. 12-11**).
Cleavage of either of these bonds—that is, transfer of one or two of its phosphoryl
groups to another molecule—is a reaction with a large negative standard free energy
change (under physiological conditions, ΔG is even more negative). As a reference point,
biochemists use the reaction in which a phosphoryl group is transferred to water—in
other words, hydrolysis of the phosphoanhydride bond, such as

$$ATP + H_2O \rightarrow ADP + P_i$$

This is a spontaneous reaction with a $\Delta G°'$ value of -30 kJ $\cdot$ mol^{-1}.

The following example illustrates the role of ATP in a coupled reaction. Consider
the phosphorylation of glucose by inorganic phosphate (HPO_4^{2-} or P_i), a thermodynamically unfavorable reaction ($\Delta G°' = +13.8$ kJ $\cdot$ mol^{-1}):

When this reaction is combined with the ATP hydrolysis reaction, the values of
$\Delta G°'$ for each reaction are added:

	$\Delta G°'$
glucose + $P_i \rightarrow$ glucose-6-phosphate + H_2O	$+13.8$ kJ $\cdot$ mol^{-1}
ATP + $H_2O \rightarrow$ ADP + P_i	-30.5 kJ $\cdot$ mol^{-1}
glucose + ATP $\rightarrow$ glucose-6-phosphate + ADP	-16.7 kJ $\cdot$ mol^{-1}

The net chemical reaction, the phosphorylation of glucose, is thermodynamically
favorable ($\Delta G < 0$). *In vivo*, this reaction is catalyzed by hexokinase (introduced in
Section 6-3), and a phosphoryl group is transferred from ATP directly to glucose.
The ATP is not actually hydrolyzed, and there is no free phosphoryl group floating
around the enzyme. However, writing out the two coupled reactions, as shown
above, makes it easier to see what's going on thermodynamically.

Some biochemical processes appear to occur with the concomitant hydrolysis of
ATP to ADP + P_i, for example, the operation of myosin and kinesin (Section 5-3)
or the Na,K-ATPase ion pump (Section 9-3). But a closer look reveals that in all

these processes, ATP actually transfers a phosphoryl group to a protein. Later, the phosphoryl group is transferred to water, so the net reaction takes the form of ATP hydrolysis. The same ATP "hydrolysis" effect applies to some reactions in which the AMP moiety of ATP (rather than a phosphoryl group) is transferred to a substance, leaving inorganic pyrophosphate (PP_i). Cleavage of the phosphoanhydride bond of PP_i also has a large negative value of $\Delta G°'$.

Because ATP appears to drive many thermodynamically unfavorable reactions, it is tempting to think of ATP as an agent that transfers packets of free energy around the cell. This is one reason why ATP is commonly called the energy currency of the cell. The general role of ATP in linking exergonic ATP-producing processes to endergonic ATP-consuming processes can be diagrammed as

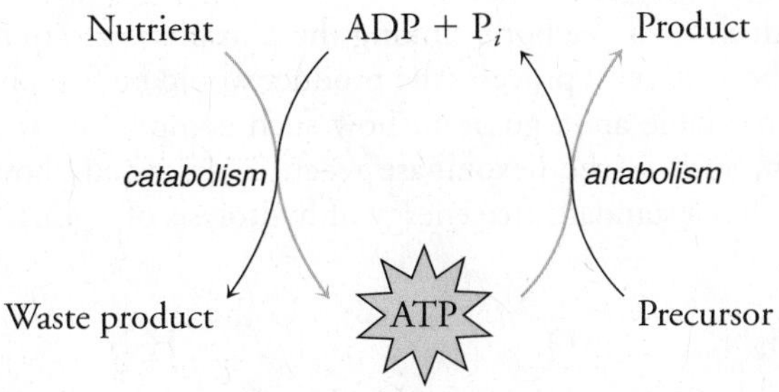

▶▶ **WHAT'S** so special about ATP?

In this scheme, it appears that the "energy" of the catabolized nutrient is transferred to ATP, then the "energy" of ATP is transferred to another product in a biosynthetic reaction. However, free energy is not a tangible item, and there is nothing magic about ATP, as the question at the start of the chapter indicates. The two phosphoanhydride bonds of ATP are sometimes called "high-energy" bonds, but they are no different from other covalent bonds. All that matters is that *breaking these bonds is a process with a large negative free energy change.* Using the simple example of ATP hydrolysis, we can state that a large amount of free energy is released when ATP is hydrolyzed because the products of the reaction have less free energy than the reactants. It is worth examining two reasons why this is so.

1. *The ATP hydrolysis products are more stable than the reactants.* At physiological pH, ATP has three to four negative charges (its pK is close to 7), and the anionic groups repel each other. In the products ADP and P_i, separation of the charges relieves some of this unfavorable electrostatic repulsion.
2. *A compound with a phosphoanhydride bond experiences less resonance stabilization than its hydrolysis products.* **Resonance stabilization** reflects the degree of electron delocalization in a molecule and can be roughly assessed by the number of different ways of depicting the molecule's structure. There are fewer equivalent ways of arranging the bonds of the terminal phosphoryl group of ATP than there are in free P_i.

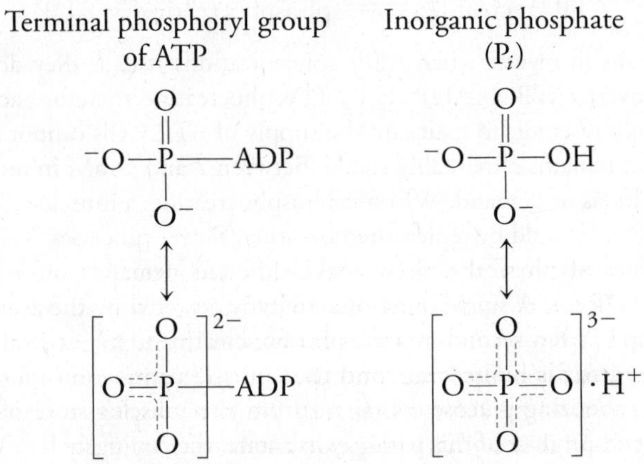

To summarize, *ATP functions as an energy currency because its reaction is highly exergonic ($\Delta G \ll 0$). The favorable ATP reaction (ATP → ADP) can therefore pull*

[TABLE 12-4]

Standard Free Energy Change for Phosphate Hydrolysis

Compound	$\Delta G^{\circ\prime}$ (kJ · mol^{-1})
Phosphoenolpyruvate	−61.9
1,3-Bisphosphoglycerate	−49.4
ATP → AMP + PP$_i$	−45.6
Phosphocreatine	−43.1
ATP → ADP + P$_i$	−30.5
Glucose-1-phosphate	−20.9
PP$_i$ → 2 P$_i$	−19.2
Glucose-6-phosphate	−13.8
Glycerol-3-phosphate	−9.2

another, unfavorable reaction with it, provided that the sum of the free energy changes for both reactions is less than zero. In effect, the cell "spends" ATP to make another process happen.

Free energy can take different forms

ATP is not the only substance that functions as energy currency in the cell. Other compounds that participate in reactions with large negative changes in free energy can serve the same purpose. For example, a number of phosphorylated compounds other than ATP can give up their phosphoryl group to another molecule. Table 12-4 lists the standard free energy changes for some of these reactions in which the phosphoryl group is transferred to water.

Although hydrolysis of the bond linking the phosphate group to the rest of the molecule could be a wasteful process (the product would be free phosphate, P$_i$), the values listed in the table are a guide to how such compounds would behave in a coupled reaction, such as the hexokinase reaction described above. For example, phosphocreatine has a standard free energy of hydrolysis of -43.1 kJ · mol^{-1}:

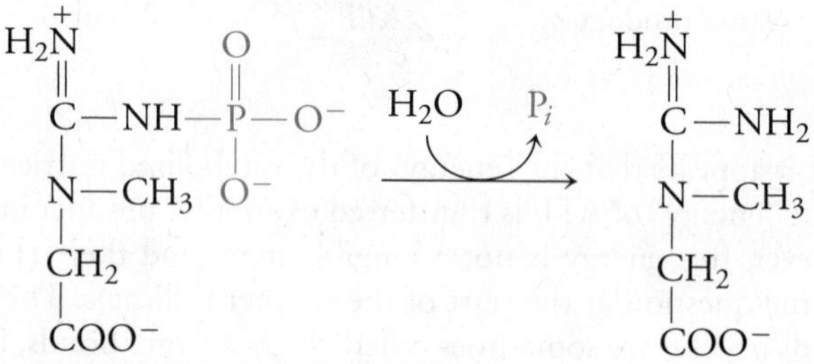

Creatine has lower free energy than phosphocreatine since it has two, rather than one, resonance forms; this resonance stabilization contributes to the large negative free energy change when phosphocreatine transfers its phosphoryl group to another compound. In muscles, phosphocreatine transfers a phosphoryl group to ADP to produce ATP (Box 12-D).

BOX 12-D   BIOCHEMISTRY NOTE

Powering Human Muscles

In resting muscles, when the demand for ATP is low, creatine kinase catalyzes the transfer of a phosphoryl group from ATP to creatine to produce phosphocreatine:

$$\text{ATP} + \text{creatine} \rightleftharpoons \text{phosphocreatine} + \text{ADP}$$

This reaction runs in reverse when ADP concentrations rise, as they do when muscle contraction converts ATP to ADP + P$_i$. Phosphocreatine therefore acts as a sort of phosphoryl-group reservoir to maintain the supply of ATP. Cells cannot stockpile ATP; its concentration remains remarkably stable (between 2 and 5 mM in most cells) under widely varying levels of demand. Without phosphocreatine, a muscle would exhaust its ATP supply before it could be replenished by other, slower processes.

Different types of physical activity make different demands on a muscle's ATP-generating mechanisms. A single burst of activity is powered by the available ATP. Activities lasting up to a few seconds require phosphocreatine to maintain the ATP supply. Phosphocreatine itself is limited, so continued muscle contraction must rely on ATP produced by catabolizing glucose (obtained from the muscle's store of glycogen) via glycolysis. The end product of this pathway is lactate, the conjugate base of a weak acid, and muscle pain sets in as the acid accumulates and the pH begins to drop. Up to this point, the muscle functions anaerobically (without the participation of O$_2$). To continue its activity, it must switch to aerobic (O$_2$-dependent) metabolism and further oxidize glucose via the citric acid cycle. The muscle also catabolizes fatty acids, whose products also enter the citric acid cycle. Recall that the citric acid cycle generates reduced cofactors that must be reoxidized by molecular oxygen. Aerobic metabolism of glucose and fatty acids is slower than anaerobic glycolysis, but it generates considerably more ATP. Some forms of physical activity and the systems that power them are diagrammed here.

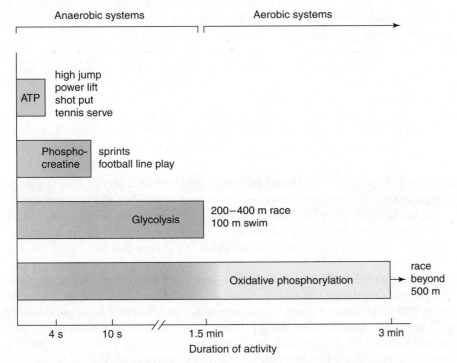

[Figure adapted from McArdle, W. D., Katch, F. I., and Katch, V. L., *Exercise Physiology* (2nd ed.), p. 348, Lea & Febiger (1986).]

A casual athlete can detect the shift from anaerobic to aerobic metabolism after about a minute and a half. In world-class athletes, the breakpoint occurs at about 150 to 170 seconds, which corresponds roughly to the finish line in a 1000-meter race.

The muscles of sprinters have a high capacity for anaerobic ATP generation, whereas the muscles of distance runners are better adapted to produce ATP aerobically. Such differences in energy metabolism are visibly manifest in the flight muscles of birds. Migratory birds such as geese, which power their long flights primarily with fatty acids, have large numbers of mitochondria to carry out oxidative phosphorylation. The reddish-brown color of the mitochondria gives the flight muscles a dark color. Birds that rarely fly, such as chickens, have fewer mitochondria and lighter-colored muscles. When these birds do fly, it is usually only a short burst of activity that is powered by anaerobic mechanisms.

● **Question:** Why do some athletes believe that creatine supplements boost their performance?

Like ATP, other nucleoside triphosphates have large negative standard free energies of hydrolysis. GTP rather than ATP serves as the energy currency for reactions that occur during cellular signaling (Section 10-2) and protein synthesis (Section 22-3). In the cell, nucleoside triphosphates are freely interconverted by reactions such as the one catalyzed by nucleoside diphosphate kinase, which transfers a phosphoryl group from ATP to a nucleoside diphosphate (NDP):

$$ATP + NDP \rightleftharpoons ADP + NTP$$

Because the reactants and products are energetically equivalent, $\Delta G^{\circ\prime}$ values for these reactions are near zero.

Another class of compounds that can release a large amount of free energy upon hydrolysis are **thioesters,** such as acetyl-CoA. Coenzyme A is a nucleotide derivative with a side chain ending in a sulfhydryl (SH) group (see Fig. 3-3a). An acyl or acetyl group (the "A" for which coenzyme A was named) is linked to the sulfhydryl group by a thioester bond. Hydrolysis of this bond has a $\Delta G^{\circ\prime}$ value of $-31.5 \text{ kJ} \cdot \text{mol}^{-1}$, comparable to that of ATP hydrolysis:

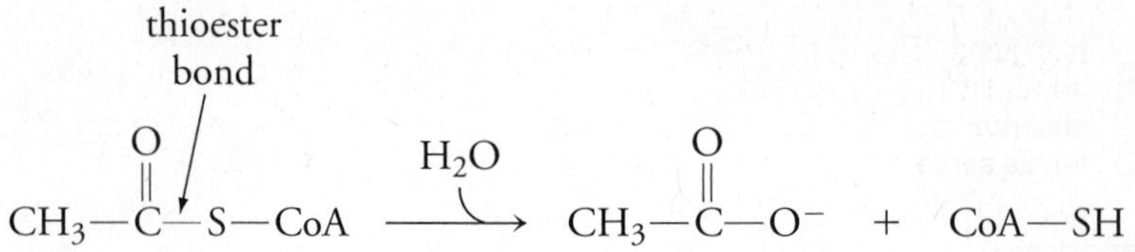

Hydrolysis of a thioester is more exergonic than the hydrolysis of an ordinary (oxygen) ester because thioesters have less resonance stability than oxygen esters, owing to the larger size of an S atom relative to an O atom. An acetyl group linked to coenzyme A can be readily transferred to another molecule because formation of the new linkage is powered by the favorable free energy change of breaking the thioester bond.

We have already seen that in oxidation–reduction reactions, cofactors such as NAD^+ and ubiquinone can collect electrons. The reduced cofactors are a form of energy currency because their subsequent reoxidation by another compound occurs with a negative change in free energy. Ultimately, the transfer of electrons from one reduced cofactor to another and finally to oxygen, the final electron acceptor in many cells, releases enough free energy to drive the synthesis of ATP.

Keep in mind that free energy changes occur not just as the result of chemical changes such as phosphoryl-group transfer or electron transfer. As decreed by the first law of thermodynamics (Section 1-3), *energy can take many forms.* We will see that ATP production in cells depends on the energy of an electrochemical gradient, that is, an imbalance in the concentration of a substance (in this case, protons) on the two sides of a membrane. The free energy change of dissipating this gradient (allowing the system to move toward equilibrium) is converted to the mechanical energy of an enzyme that synthesizes ATP. In photosynthetic cells, the chemical reactions required to generate carbohydrates are ultimately driven by the free energy changes of reactions in which light-excited molecules return to a lower-energy state.

Regulation occurs at the steps with the largest free energy changes

In a series of reactions that make up a metabolic pathway, some reactions have ΔG values near zero. These near-equilibrium reactions are not subject to a strong driving force to proceed in either direction. Rather, flux can go forward or backward, according to slight fluctuations in the concentrations of reactants and products. When the concentrations of metabolites change, the enzymes that catalyze these near-equilibrium reactions tend to act quickly to restore the near-equilibrium state.

Reactions with large changes in free energy have a longer way to go to reach equilibrium; these are the reactions that experience the greatest "urge" to proceed forward. However, the enzymes that catalyze these reactions do not allow the reaction to reach equilibrium because they work too slowly. Often the enzymes are already saturated with substrate, so the reactions cannot go any faster (when $[S] \gg K_M$, $v \approx V_{max}$; Section 7-2). The rates of these far-from-equilibrium reactions limit flux through the entire pathway because the reactions function like dams.

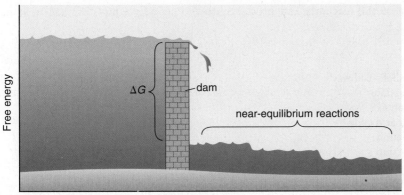

Cells can regulate flux through a pathway by adjusting the rate of a reaction with a large free energy change. This can be done by increasing the amount of enzyme that catalyzes that step or by altering the intrinsic activity of the enzyme through allosteric mechanisms (see Fig. 7-17). As soon as more metabolite has gotten past the dam, the near-equilibrium reactions go with the flow, allowing the pathway intermediates to move toward the final product. Most metabolic pathways do not have a single flow-control point, as the dam analogy might suggest. Instead, flux is typically controlled at several points to ensure that the pathway can work efficiently as part of the cell's entire metabolic network.

CONCEPT REVIEW

- Why must free energy changes be negative for reactions *in vivo*?
- What is the standard free energy change for a reaction and how is it related to the reaction's equilibrium constant?
- Distinguish ΔG and $\Delta G^{\circ\prime}$. How are they related?
- Why is it misleading to refer to ATP as a high-energy molecule?
- Explain why cleavage of one of ATP's phosphoanhydride bonds releases large amounts of free energy.
- How do phosphorylated compounds, thioesters, and reduced cofactors appear to transfer free energy? What other forms of energy do cells use?
- Why do cells control the metabolic reactions that have large free energy changes?

[SUMMARY]

12-1 Food and Fuel

- Polymeric food molecules such as starch, proteins, and triacylglycerols are broken down to their monomeric components (glucose, amino acids, and fatty acids), which are absorbed. These materials are stored as polymers in a tissue-specific manner.
- Metabolic fuels are mobilized from glycogen, fat, and proteins as needed.

12-2 Metabolic Pathways

- Series of reactions known as metabolic pathways break down and synthesize biological molecules. Several pathways make use of the same small molecule intermediates.
- During the oxidation of amino acids, monosaccharides, and fatty acids, electrons are transferred to carriers such as NAD^+ and ubiquinone. Reoxidation of the reduced cofactors drives the synthesis of ATP by oxidative phosphorylation.

- Metabolic pathways form a complex network, but not all cells or organisms carry out all possible metabolic processes. Humans rely on other organisms to supply vitamins and other essential materials.

12-3 Free Energy Changes in Metabolic Reactions

- The standard free energy change for a reaction is related to the equilibrium constant, but the actual free energy change is related to the actual cellular concentrations of reactants and products.
- A thermodynamically unfavorable reaction may proceed when it is coupled to a favorable process involving ATP, whose phosphoanhydride bonds release a large amount of free energy when cleaved.
- Other forms of cellular energy currency include phosphorylated compounds, thioesters, and reduced cofactors.
- Cells regulate metabolic pathways at the steps that are farthest from equilibrium.

[GLOSSARY TERMS]

chemoautotroph	metabolic pathway	microarray (DNA chip)
photoautotroph	metabolite	proteomics
heterotroph	glycolysis	proteome
catabolism	citric acid cycle	metabolomics
anabolism	oxidation	metabolome
metabolism	reduction	essential compound
lipoprotein	redox reaction	vitamin
metabolic fuel	cofactor	equilibrium constant (K_{eq})
mobilization	coenzyme	standard free energy change ($\Delta G^{\circ\prime}$)
phosphorolysis	oxidative phosphorylation	standard conditions
diabetes mellitus	flux	mass action ratio
lysosome	transcriptomics	resonance stabilization
proteasome	transcriptome	thioester

BIOINFORMATICS ▶ **PROJECT 6**   Learn to use the KEGG database and explore the technology behind microarrays and two-dimensional gel electrophoresis.

METABOLIC ENZYMES, MICROARRAYS, AND PROTEOMICS

[PROBLEMS]

12-1 Food and Fuel

1. Classify the following organisms as chemoautotrophs, photoautotrophs, or heterotrophs:

(a) *Hydrogenobacter*, which converts molecular hydrogen and oxygen to water

(b) *Arabidopsis thaliana*, a green plant

(c) The nitrosifying bacteria, which oxidize NH_3 to nitrite

(d) *Saccharomyces cerevisiae*, yeast

(e) *Caenorhabditis elegans*, a nematode worm

(f) The *Thiothrix* bacteria, which oxidize hydrogen sulfide

(g) Cyanobacteria (erroneously termed "blue-green algae" in the past)

2. The purple nonsulfur bacteria obtain their cellular energy from a photosynthetic process that does not produce oxygen. These bacteria also require an organic carbon source. Using the terms in this chapter, coin a new term that describes the trophic strategy of this organism.

3. Digestion of carbohydrates begins in the mouth, where salivary amylases act on dietary starch. When the food is swallowed and enters the stomach, carbohydrate digestion ceases (it resumes in the small intestine). Why does carbohydrate digestion not occur in the stomach?

4. Pancreatic amylase, which is similar to salivary amylase, is secreted by the pancreas into the small intestine. The active site of pancreatic amylase accommodates five glucosyl residues and cleaves the glycosidic bond between the second and third residues. The enzyme cannot accommodate branched chains.

(a) What are the main products of amylose digestion?

(b) What are the products of amylopectin digestion?

5. Starch digestion is completed by the enzymes isomaltase (or α-dextrinase), which catalyzes the hydrolysis of α(1 → 6) glycosidic bonds, and maltase, which hydrolyzes α(1 → 4) bonds. Why are these enzymes needed in addition to α-amylase?

6. Monosaccharides, the products of polysaccharide and disaccharide digestion, enter the cells lining the intestine via a specialized transport system. What is the source of free energy for this transport process?

7. Unlike the monosaccharides described in Problem 6, sugar alcohols such as sorbitol (see Solution 11-28) are absorbed via passive diffusion. Why? What process occurs more rapidly, passive diffusion or passive transport?

8. Use what you know about the properties of alcohol (ethanol) to describe how it is absorbed in both the stomach and the small intestine. What effect does the presence of food have on the absorption of ethanol?

9. Nucleic acids that are present in food are hydrolyzed by digestive enzymes. What type of mechanism most likely mediates the entry of the reaction products into intestinal cells?

10. Hydrolysis of proteins begins in the stomach, catalyzed by the hydrochloric acid secreted into the stomach by parietal cells. Draw the reaction that shows the hydrolysis of a peptide bond.

11. How does the low pH of the stomach affect protein structure in such a way that the proteins are prepared for hydrolytic digestion?

12. Like the serine proteases (see Section 6-4), pepsin is made as a zymogen and is inactive at its site of synthesis, where the pH is 7. Pepsin becomes activated when secreted into the stomach, where it encounters a pH of ~2. Pepsinogen contains a "basic peptide" that blocks its active site at pH 7. The basic peptide dissociates from the active site at pH 2 and is cleaved, resulting in the formation of the active form of the enzyme. What amino acids residues are found in the active site of pepsin? Why does the basic peptide bind tightly to the active site at pH 7 and why does it dissociate at the lower pH?

13. The cleavage of peptide bonds in the stomach is catalyzed both by hydrochloric acid (see Problem 10) and by the stomach enzyme pepsin. Peptide bond cleavage continues in the small intestine, catalyzed by the pancreatic enzymes trypsin and chymotrypsin. At what pH does pepsin function optimally; that is, at what pH is the V_{max} for pepsin greatest? Is the pH optimum for pepsin different from that for trypsin and chymotrypsin? Explain.

14. Scientists have recently discovered why the botulinum toxin survives the acidic environment of the stomach. The toxin forms a complex with a second nontoxic protein that acts as a shield to protect the botulinum toxin from being digested by stomach enzymes. Upon entry into the small intestine, the two proteins dissociate and the botulinum toxin is released. What is the likely interaction between the botulinum toxin and the nontoxic protein, and why does the complex form readily in the stomach but not in the small intestine?

15. Free amino acid transport from the intestinal lumen into intestinal cells requires Na^+ ions. Draw a diagram that illustrates amino acid transport into these cells.

16. In oral rehydration therapy (ORT), patients suffering from diarrhea are given a solution consisting of a mixture of glucose and electrolytes. Some formulations also contain amino acids. Why are electrolytes added to the mixture?

17. Triacylglycerol digestion begins in the stomach. Gastric lipase catalyzes hydrolysis of the fatty acid from the third glycerol carbon.
(a) Draw the reactants and products of this reaction.
(b) Conversion of the triacylglycerol to a diacylglycerol and a fatty acid promotes emulsification of fats in the stomach; that is, the products are more easily incorporated into micelles. Explain why.

18. Most of the fatty acids produced in the reaction described in Problem 17 form micelles and are absorbed as such, but a small percentage of fatty acids are free and are transported into the intestinal epithelial cells without the need for a transport protein. Explain why a transport protein is not required.

19. The cells lining the small intestine absorb cholesterol but not cholesteryl esters. Draw the reaction catalyzed by cholesteryl esterase that produces cholesterol from cholesteryl stearate.

20. Some cholesterol is converted back to cholesteryl esters in the epithelial cells lining the small intestine (the reverse of the reaction described in Problem 19). Both cholesterol and cholesteryl esters are packaged into particles called chylomicrons, which consist of lipid and protein. Use what you know about the physical properties of cholesterol and cholesteryl esters to describe their locations in the chylomicron particle.

21. (a) Consider the physical properties of a polar glycogen molecule and an aggregation of hydrophobic triacylglycerols. On a per-weight basis, why is fat a more efficient form of energy storage than glycogen?
(b) Explain why there is an upper limit to the size of a glycogen molecule but there is no upper limit to the amount of triacylglycerols that an adipocyte can store.

22. Glycogen can be expanded quickly, by adding glucose residues to its many branches, and degraded quickly, by simultaneously removing glucose from the ends of these branches. Are the enzymes that catalyze these processes specific for the reducing or nonreducing ends of the glycogen polymer? Explain.

23. The phosphorolysis reaction that removes glucose residues from glycogen yields as its product glucose-1-phosphate. Glucose-1-phosphate is isomerized to glucose-6-phosphate; then the phosphate group is removed in a hydrolysis reaction. Why is it necessary to remove the phosphate group before the glucose exits the cell to enter the circulation?

24. Hydrolytic enzymes encased within the membrane-bound lysosomes all work optimally at pH ~5. This feature serves as a cellular "insurance policy" in the event of lysosomal enzyme leakage into the cytosol. Explain.

12-2 Metabolic Pathways

25. The common intermediates listed in the table below appear as reactants or products in several pathways. Place a checkmark in the box that indicates the appropriate pathway for each reactant.

	Glycolysis	Citric acid cycle	Fatty acid metabolism
Acetyl-CoA			
Glyceraldehyde-3-phosphate			
Pyruvate			

	Triacylglycerol synthesis	Photosynthesis	Transamination
Acetyl-CoA			
Glyceraldehyde-3-phosphate			
Pyruvate			

26. For each of the (unbalanced) reactions shown below, tell whether the reactant is being oxidized or reduced.
(a) A reaction from the catabolic glycolytic pathway

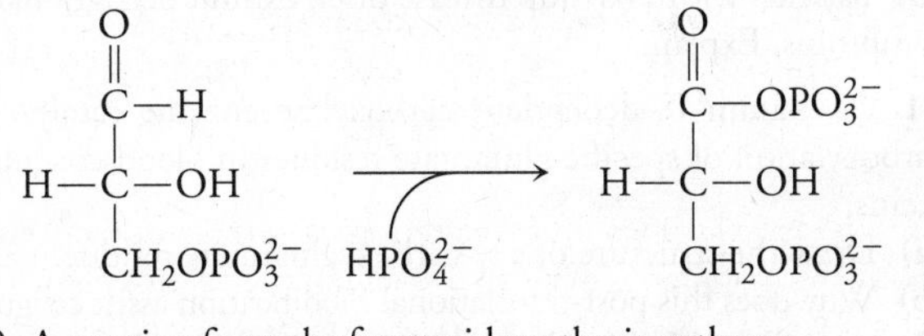

(b) A reaction from the fatty acid synthesis pathway

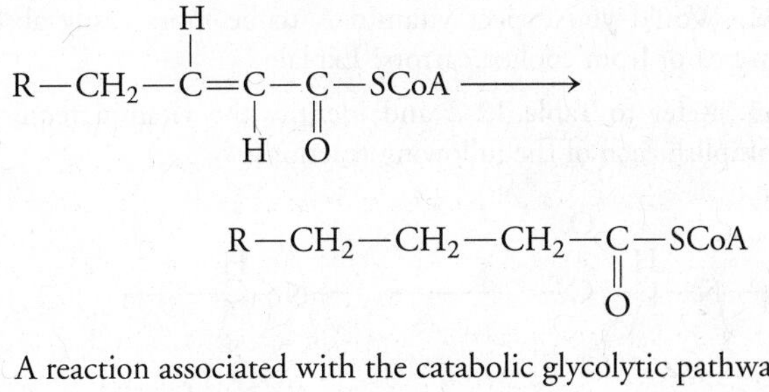

(c) A reaction associated with the catabolic glycolytic pathway

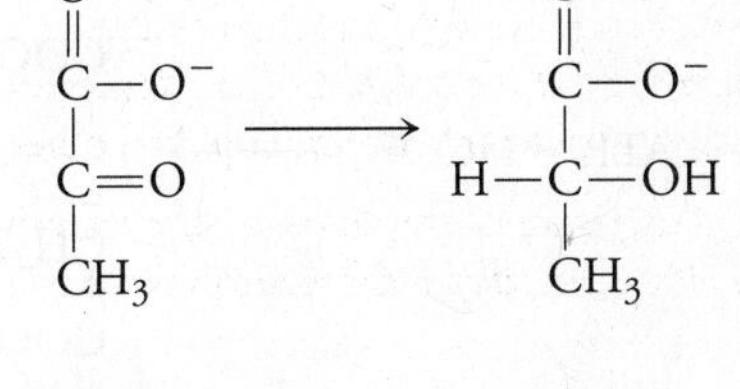

334

(d) A reaction associated with the anabolic pentose phosphate pathway

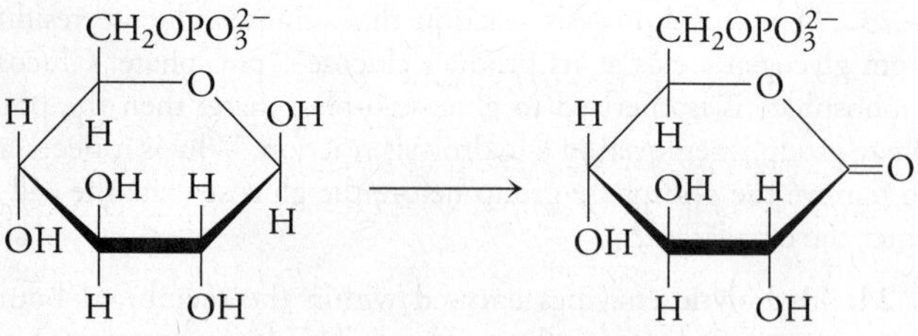

27. For each of the reactions shown in Problem 26, identify the cofactor as NAD^+, $NADP^+$, NADH, or NADPH.

28. A potential way to reduce the concentration of methane, a greenhouse gas, is to take advantage of sulfate-reducing bacteria. **(a)** Complete the chemical equation for methane consumption by these organisms:

$$CH_4 + SO_4^{2-} \rightarrow \underline{\hspace{1cm}} + HS^- + H_2O$$

Identify the reaction component that undergoes **(b)** oxidation and **(c)** reduction.

29. Vitamin B_{12} is synthesized by certain gastrointestinal bacteria and is also found in foods of animal origin such as meat, milk, eggs, and fish. When vitamin B_{12}–containing foods are consumed, the vitamin is released from the food and binds to a salivary vitamin B_{12}–binding protein called haptocorrin. The haptocorrin–vitamin B_{12} complex passes from the stomach to the small intestine, where the vitamin is released from the haptocorrin and then binds to intrinsic factor (IF). The IF–vitamin B_{12} complex then enters the cells lining the intestine by receptor-mediated endocytosis. Using this information, make a list of individuals most at risk for vitamin B_{12} deficiency.

30. Hartnup disease is a hereditary disorder caused by a defective transporter for nonpolar amino acids.
(a) The symptoms of the disease (photosensitivity and neurological abnormalities) can be prevented through dietary adjustments. What sort of diet would be effective?
(b) Patients with Hartnup disease often exhibit pellagra-like symptoms. Explain.

31. A vitamin K–dependent carboxylase enzyme catalyzes the γ-carboxylation of specific glutamate residues in blood coagulation proteins.
(a) Draw the structure of a γ-carboxyglutamate residue.
(b) Why does this post-translational modification assist coagulation proteins in binding the Ca^{2+} ions required for blood clotting?

32. Would you expect vitamin A to be more easily absorbed from raw or from cooked carrots? Explain.

33. Refer to Table 12-2 and identify the vitamin required to accomplish each of the following reactions:

(a)

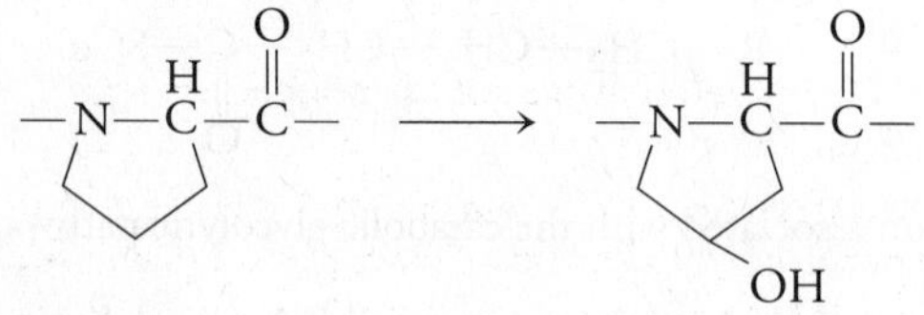

(b)

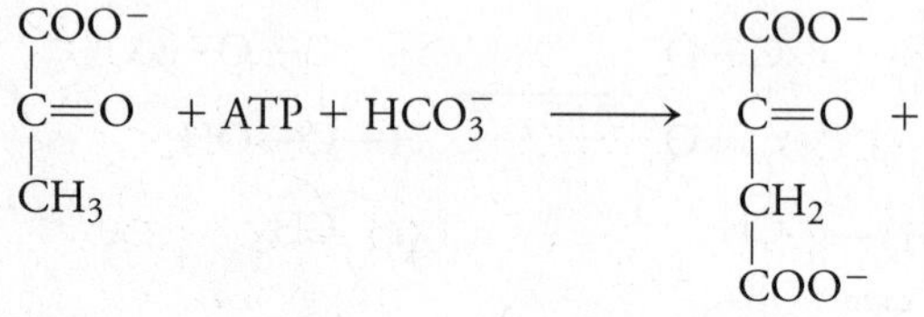

(c)

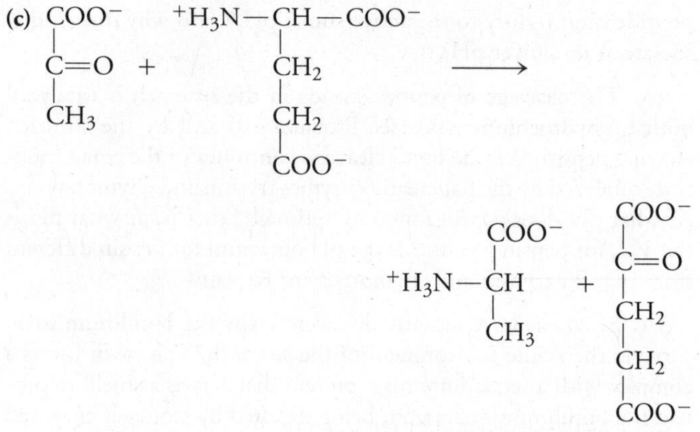

(d)

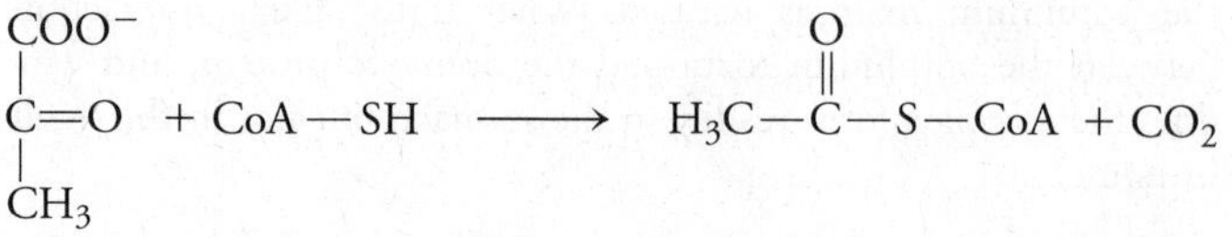

34. Why is niacin technically not a vitamin?

12-3 Free Energy Changes in Metabolic Reactions

35. Consider two reactions: A $\rightleftharpoons$ B and C $\rightleftharpoons$ D. K_{eq} for the A $\rightleftharpoons$ B reaction is 10, and K_{eq} for the C $\rightleftharpoons$ D reaction is 0.1. You place 1 mM A in tube 1 and 1 mM C in tube 2 and allow the reactions to reach equilibrium. Without doing any calculations, determine whether the concentration of B in tube 1 will be greater than or less than the concentration of D in tube 2.

36. Calculate the $\Delta G^{\circ\prime}$ values for the reactions described in Problem 35. Assume a temperature of 37°C.

37. For the reaction E $\rightleftharpoons$ F, $K_{eq} = 1$.
(a) Without doing any calculations, what can you conclude about the $\Delta G^{\circ\prime}$ value for the reaction?
(b) You place 1 mM F in a tube and allow the reaction to reach equilibrium. Determine the final concentrations of E and F.

38. Refer to the hypothetical reaction described in Problem 37. Determine the direction the reaction will proceed if you place 5 mM E and 2 mM F in a test tube. What are the final concentrations of E and F?

39. Calculate the ΔG value for the A $\rightleftharpoons$ B reaction described in Problem 35 when the concentrations of A and B are 0.9 mM and 0.1 mM, respectively. In which direction will the reaction proceed?

40. Calculate the ΔG value for the C $\rightleftharpoons$ D reaction described in Problem 35 when the concentrations of C and D are 0.9 mM and 0.1 mM, respectively. In which direction will the reaction proceed?

41. **(a)** The $\Delta G^{\circ\prime}$ value for a hypothetical reaction is 10 kJ · mol^{-1}. Compare the K_{eq} for this reaction with the K_{eq} for a reaction whose $\Delta G^{\circ\prime}$ value is twice as large.
(b) Carry out the same exercise for a hypothetical reaction whose $\Delta G^{\circ\prime}$ value is -10 kJ · mol^{-1}.

42. Use the standard free energies provided in Table 12-4 to calculate the $\Delta G^{\circ\prime}$ for the isomerization of glucose-1-phosphate to glucose-6-phosphate.
(a) Is this reaction spontaneous under standard conditions?
(b) Is the reaction spontaneous when the concentration of glucose-6-phosphate is 5 mM and the concentration of glucose-1-phosphate is 0.1 mM?

43. Calculate ΔG for the hydrolysis of ATP under cellular conditions, where [ATP] = 3 mM, [ADP] = 1 mM, and [P_i] = 5 mM.

44. The standard free energy change for the reaction catalyzed by triose phosphate isomerase is 7.9 kJ · mol^{-1}.

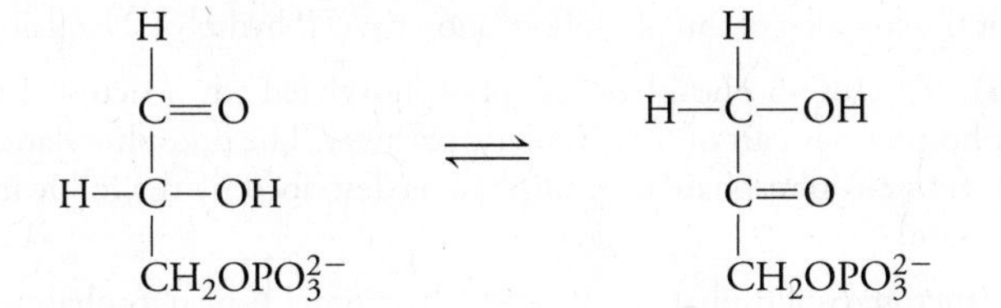

Glyceraldehyde-3-phosphate   Dihydroxyacetone phosphate

(a) Calculate the equilibrium constant for the reaction.
(b) Calculate ΔG at 37°C when the concentration of glyceraldehyde-3-phosphate is 0.1 mM and the concentration of dihydroxyacetone phosphate is 0.5 mM.
(c) Is the reaction spontaneous under these conditions? Would the reverse reaction be spontaneous?

45. An apple contains about 72 Calories. Express this quantity in terms of ATP equivalents (that is, how many ATP → ADP + P_i reactions?).

46. A large hot chocolate with whipped cream purchased at a national coffee chain contains 760 calories. Express this quantity in terms of ATP equivalents (see Problem 45).

47. Use the graph below to sketch the free energy changes for **(a)** the glucose + P_i → glucose-6-phosphate reaction, **(b)** the ATP + H_2O → ADP + P_i reaction, and **(c)** the coupled reaction (see Section 12-3).

G

Reaction coordinate

48. Some studies (but not all) show that creatine supplementation increases performance in high-intensity exercises lasting less than 30 seconds. Would you expect creatine supplements to affect endurance exercise?

49. The $\Delta G°'$ for the hydrolysis of ATP under standard conditions at pH 7 and in the presence of magnesium ions is −30.5 kJ · mol^{-1}.
(a) How would this value change if ATP hydrolysis was carried out at a pH of less than 7? Explain.
(b) How would this value change if magnesium ions were not present?

50. The $\Delta G°'$ for the formation of UDP–glucose from glucose-1-phosphate and UTP is about zero. Yet the production of UDP–glucose is highly favorable. What is the driving force for this reaction?

$$\text{glucose-1-phosphate} + \text{UTP} \rightleftharpoons \text{UDP–glucose} + PP_i$$

51. (a) The complete oxidation of glucose releases a considerable amount of energy. The $\Delta G°'$ for the reaction shown below is −2850 kJ · mol^{-1}.

$$C_6H_{12}O_6 + 6\,O_2 \rightarrow 6\,CO_2 + 6\,H_2O$$

How many moles of ATP could be produced under standard conditions from the oxidation of one mole of glucose, assuming about 33% efficiency?
(b) The oxidation of palmitate, a 16-carbon saturated fatty acid, releases 9781 kJ · mol^{-1}.

$$C_{16}H_{32}O_2 + 23\,O_2 \rightarrow 16\,CO_2 + 16\,H_2O$$

How many moles of ATP could be produced under standard conditions from the oxidation of one mole of palmitate, assuming 33% efficiency?
(c) Calculate the number of ATP molecules produced per carbon for glucose and palmitate. Explain the reason for the difference.

52. A moderately active adult female weighing 125 pounds must consume 2200 Calories of food daily.
(a) If this energy is used to synthesize ATP, calculate the number of moles of ATP that would be synthesized each day under standard conditions (assuming 33% efficiency).
(b) Calculate the number of grams of ATP that would be synthesized each day. The molar mass of ATP is 505 g · mol^{-1}. What is the mass of ATP in pounds? (2.2 kg = 1 lb)
(c) There is approximately 40 g of ATP in the adult 125-lb female. Considering this fact and your answer to part (b), suggest an explanation that is consistent with these findings.

53. Calculate how many apples (see Solution 45) would be required to provide the amount of ATP calculated in Problem 52.

54. Calculate how many large hot chocolate drinks (see Solution 46) would be required to provide the amount of ATP calculated in Problem 52.

55. Which of the compounds listed in Table 12-4 could be involved in a reaction coupled to the synthesis of ATP from ADP + P_i?

56. Which of the compounds listed in Table 12-4 could be involved in a reaction coupled to the hydrolysis of ATP to ADP + P_i?

57. Citrate is isomerized to isocitrate in the citric acid cycle (Chapter 14). The reaction is catalyzed by the enzyme aconitase. The $\Delta G°'$ of the reaction is 5 kJ · mol^{-1}. The kinetics of the reaction are studied *in vitro*, where 1 M citrate and 1 M isocitrate are added to an aqueous solution of the enzyme at 25°C.
(a) What is the K_{eq} for the reaction?
(b) What are the equilibrium concentrations of the reactant and product?
(c) What is the preferred direction of the reaction under standard conditions?
(d) The aconitase reaction is the second step of an eight-step pathway and occurs in the direction shown in the figure. How can you reconcile these facts with your answer to part (c)?

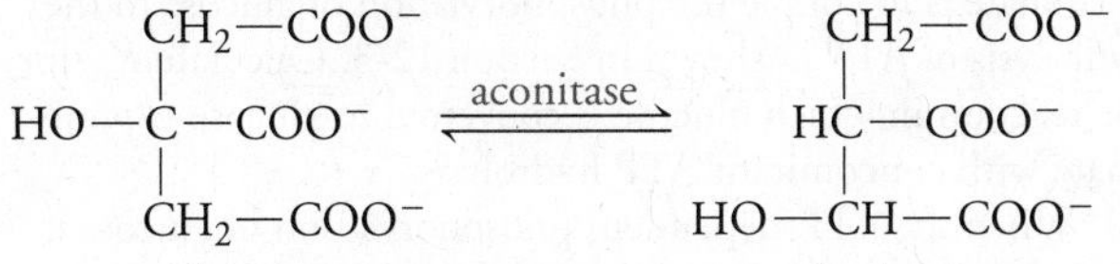

Citrate   aconitase   Isocitrate

58. The equilibrium constant for the conversion of glucose-6-phosphate to fructose-6-phosphate is 0.41. The reaction is reversible and is catalyzed by the enzyme phosphoglucose isomerase.

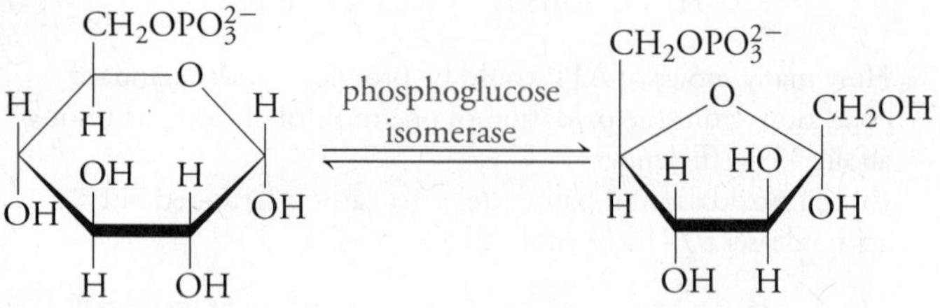

Glucose-6-phosphate   Fructose-6-phosphate

(a) What is the $\Delta G^{\circ\prime}$ for this reaction? Would this reaction proceed in the direction written under standard conditions?
(b) What is the ΔG for this reaction at 37°C when the concentration of glucose-6-phosphate is 2.0 mM and the concentration of the fructose-6-phosphate is 0.5 mM? Would the reaction proceed in the direction written under these cellular conditions?

59. The conversion of glutamate to glutamine is unfavorable. In order for this transformation to occur in the cell, it must be coupled to the hydrolysis of ATP. Consider two possible mechanisms:

Mechanism 1: glutamate + NH_3 $\rightleftharpoons$ glutamine
ATP + H_2O $\rightleftharpoons$ ADP + P_i

Mechanism 2: glutamate + ATP $\rightleftharpoons$ γ-glutamylphosphate + ADP

γ-glutamylphosphate + H_2O + NH_3 $\rightleftharpoons$ glutamine + P_i

Write the overall equation for the reaction for each mechanism. Is one mechanism more likely than the other? Or are both mechanisms equally feasible for the conversion of glutamate to glutamine? Explain.

60. The phosphorylation of glucose to glucose-6-phosphate is the first step of glycolysis (Chapter 13). The phosphorylation of glucose by phosphate is described by the following equation:

glucose + P_i $\rightleftharpoons$ glucose-6-phosphate + H_2O
$\Delta G^{\circ\prime} = +13.8$ kJ · mol^{-1}.

(a) Calculate the equilibrium constant for the above reaction.
(b) What would the equilibrium concentration of glucose-6-phosphate be under cellular conditions of [glucose] = [P_i] = 5 mM if glucose was phosphorylated according to the reaction above? Does this reaction provide a feasible route for the production of glucose-6-phosphate for the glycolytic pathway?
(c) One way to increase the amount of product is to increase the concentrations of the reactants. This would decrease the mass action ratio (see Equation 12-3) and would theoretically make the reaction as written more favorable. If the cellular concentration of phosphate is 5 mM, what concentration of glucose would be required to achieve a glucose-6-phosphate concentration of 250 μM? Is this strategy physiologically feasible, given that the solubility of glucose in aqueous medium is less than 1 M?
(d) Another way to promote the formation of glucose-6-phosphate is to couple the phosphorylation of glucose to the hydrolysis of ATP as shown in Section 12-3. Calculate K_{eq} for the reaction in which glucose is converted to glucose-6-phosphate with concomitant ATP hydrolysis.
(e) When the ATP-dependent phosphorylation of glucose is carried out, what concentration of glucose is needed to

achieve a 250-μM intracellular concentration of glucose-6-phosphate when the concentrations of ATP and ADP are 5.0 mM and 1.25 mM, respectively?
(f) Which route is more feasible to accomplish the phosphorylation of glucose to glucose-6-phosphate: the direct phosphorylation by P_i or the coupling of this phosphorylation to ATP hydrolysis? Explain.

61. Fructose-6-phosphate is phosphorylated to fructose-1,6-bisphosphate as part of the glycolytic pathway. The phosphorylation of fructose-6-phosphate by phosphate is described by the following equation:

fructose-6-phosphate + P_i $\rightleftharpoons$ fructose-1,6-bisphosphate
$\Delta G^{\circ\prime} = 47.7$ kJ · mol^{-1}.

(a) What is the ratio of fructose-1,6-bisphosphate to fructose-6-phosphate at equilibrium if the concentration of phosphate in the cell is 5 mM?
(b) Suppose that the phosphorylation of fructose-6-phosphate is coupled to the hydrolysis of ATP.

ATP + H_2O $\rightleftharpoons$ ADP + P_i $\Delta G^{\circ\prime} = -30.5$ kJ · mol^{-1}

Write the new equation that describes the phosphorylation of fructose-6-phosphate coupled with ATP hydrolysis. Calculate the $\Delta G^{\circ\prime}$ for the reaction.
(c) What is the ratio of fructose-1,6-bisphosphate to fructose-6-phosphate at equilibrium for the reaction you wrote in part (b) if the equilibrium concentration of ATP = 3 mM and [ADP] = 1 mM?
(d) Write a concise paragraph that summarizes your findings above.
(e) One can envision two mechanisms for coupling ATP hydrolysis to the phosphorylation of fructose-6-phosphate, yielding the same overall reaction:

Mechanism 1: ATP is hydrolyzed as fructose-6-phosphate is transformed to fructose-1,6-bisphosphate:

fructose-6-phosphate + P_i $\rightleftharpoons$ fructose-1,6-bisphosphate
ATP + H_2O $\rightleftharpoons$ ADP + P_i

Mechanism 2: ATP transfers its γ-phosphate directly to fructose-6-phosphate in one step, producing fructose-1,6-bisphosphate.

fructose-6-phosphate + ATP + H_2O $\rightleftharpoons$
fructose-1,6-bisphosphate + ADP

Choose one of the above mechanisms as the more biochemically feasible and provide a rationale for your choice.

62. Glyceraldehyde-3-phosphate (GAP) is eventually converted to 3-phosphoglycerate (3PG) in the glycolytic pathway.

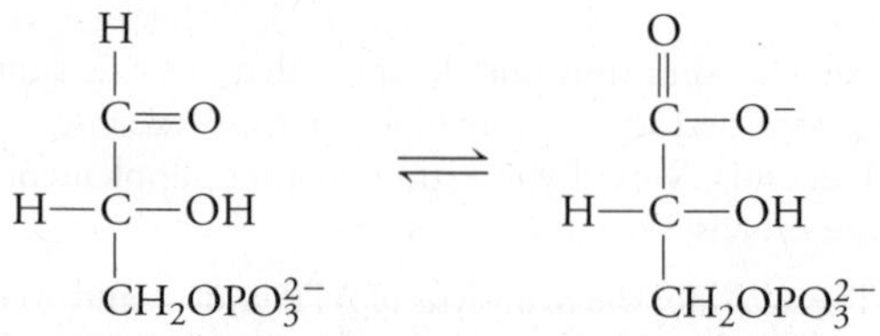

Glyceraldehyde-3-phosphate   3-Phosphoglycerate

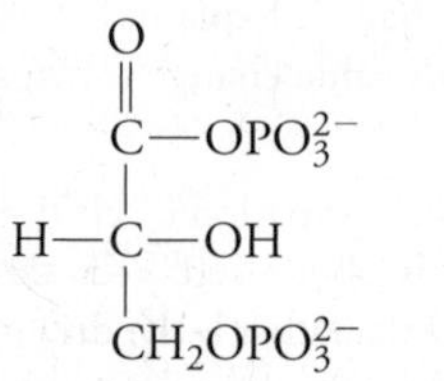

1,3-Bisphosphoglycerate (1,3-BPG)

Consider these two scenarios:

I. GAP is oxidized to 1,3-BPG ($\Delta G^{\circ\prime} = 6.7$ kJ $\cdot$ mol^{-1}), which is subsequently hydrolyzed to yield 3PG ($\Delta G^{\circ\prime} = -49.3$ kJ $\cdot$ mol^{-1})

II. GAP is oxidized to 1,3-BPG, which then transfers its phosphate to ADP, yielding ATP ($\Delta G^{\circ\prime} = -18.8$ kJ $\cdot$ mol^{-1}).

Write the overall equations for the two scenarios. Which is more likely to occur in the cell, and why?

63. Palmitate is activated in the cell by forming a thioester bond to coenzyme A. The $\Delta G^{\circ\prime}$ for the synthesis of palmitoyl-CoA from palmitate and coenzyme A is 31.5 kJ $\cdot$ mol^{-1}.

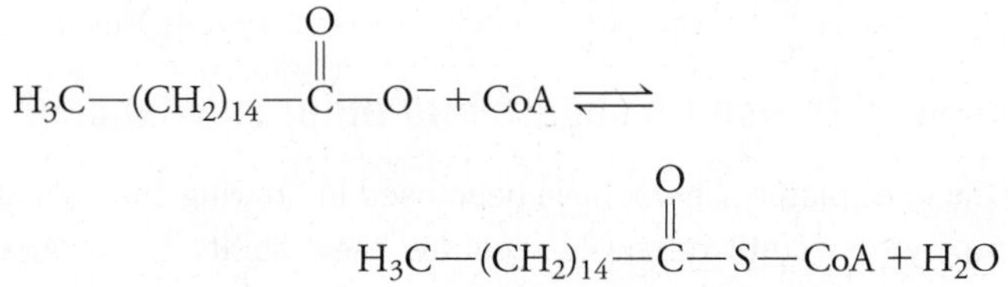

$$H_3C-(CH_2)_{14}-\overset{\overset{\textstyle O}{\|}}{C}-O^- + CoA \rightleftharpoons$$

$$H_3C-(CH_2)_{14}-\overset{\overset{\textstyle O}{\|}}{C}-S-CoA + H_2O$$

(a) What is the ratio of products to reactants at equilibrium for the reaction? Is the reaction favorable? Explain.

(b) Suppose the synthesis of palmitoyl-CoA were coupled with ATP hydrolysis. The standard free energy for the hydrolysis of the ATP to ADP is listed in Table 12-4. Write the new equation for the activation of palmitate when coupled with ATP hydrolysis to ADP. Calculate $\Delta G^{\circ\prime}$ for the reaction. What is the ratio of products to reactants at equilibrium for the reaction? Is the reaction favorable? Compare your answer to the answer you obtained in part (a).

(c) Suppose the reaction described in part (a) were coupled with ATP hydrolysis to AMP; the standard free energy for the hydrolysis of ATP to AMP is listed in Table 12-4. Write the new equation for the activation of palmitate when coupled with ATP hydrolysis to AMP. Calculate $\Delta G^{\circ\prime}$ for the reaction. What is the ratio of products to reactants at equilibrium for the reaction? Is the reaction favorable? Compare your answer to the answer you obtained in part (b).

(d) Pyrophosphate, PP$_i$, is hydrolyzed to 2 P$_i$, as shown in Table 12-4. The activation of palmitate, as described in part (c), is coupled to the hydrolysis of pyrophosphate. Write the equation for this coupled reaction and calculate the $\Delta G^{\circ\prime}$. What is the ratio of products to reactants at equilibrium for the reaction? Is the reaction favorable? Compare your answer to the answers you obtained in parts (b) and (c).

64. DNA containing broken phosphodiester bonds ("nicks") can be repaired by the action of a ligase enzyme. Formation of a new phosphodiester bond in DNA requires the free energy of ATP phosphoanhydride bond cleavage. In the ligase-catalyzed reaction, ATP is hydrolyzed to AMP:

$$\text{ATP} + \text{nicked bond} \underset{\text{ligase}}{\overset{\text{ligase}}{\rightleftharpoons}} \text{AMP} + \text{PP}_i + \text{phosphodiester bond}$$

The equilibrium constant expression for this reaction can be rearranged to define a constant, C, as follows:

$$K_{eq} = \frac{[\text{phosphodiester bond}][\text{AMP}][\text{PP}_i]}{[\text{nick}][\text{ATP}]}$$

$$\frac{[\text{nick}]}{[\text{phosphodiester bond}]} = \frac{[\text{AMP}][\text{PP}_i]}{K_{eq}[\text{ATP}]}$$

$$C = \frac{[\text{PP}_i]}{K_{eq}[\text{ATP}]}$$

$$\frac{[\text{nick}]}{[\text{phosphodiester bond}]} = C[\text{AMP}]$$

Researchers have determined the ratio of nicked bonds to phosphodiester bonds at various concentrations of AMP.

(a) Using the data provided, construct a plot of [nick]/[phosphodiester bond] versus [AMP] and determine the value of C from the plot.

[AMP] (mM)	[nick]/[phosphodiester bond]
10	4.0×10^{-5}
15	4.3×10^{-5}
20	5.47×10^{-5}
25	6.67×10^{-5}
30	8.67×10^{-5}
35	9.47×10^{-5}
40	9.30×10^{-5}
45	1.0×10^{-4}
50	1.13×10^{-4}

(b) Determine the value of K_{eq} for the reaction, given that the concentrations of PP$_i$ and ATP were held constant at 1.0 mM and 14 μM, respectively.

(c) What is the value of $\Delta G^{\circ\prime}$ for the reaction?

(d) What is the value of $\Delta G^{\circ\prime}$ for the following reaction?

$$\text{nicked bond} \rightleftharpoons \text{phosphodiester bond}$$

Note that the $\Delta G^{\circ\prime}$ for the hydrolysis of ATP to AMP and PP$_i$ is -48.5 kJ $\cdot$ mol^{-1} in the presence of 10 mM Mg^{2+}, the conditions used in these experiments.

(e) The $\Delta G^{\circ\prime}$ for the hydrolysis of a typical phosphomonoester to yield P$_i$ and an alcohol is -13.8 kJ $\cdot$ mol^{-1}. Compare the stability of the phosphodiester bond in DNA to the stability of a typical phosphomonoester bond.

[SELECTED READINGS]

Falkowski, P. G., Fenchel, T., and Delong, E. F., The microbial engines that drive Earth's biogeochemical cycles, *Science* **320,** 1034–1038 (2008). [Discusses the diversity and interconnectedness of metabolic processes.]

Hanson, R. W., The role of ATP in metabolism, *Biochem. Ed.* **17,** 86–92 (1989). [Provides an excellent explanation of why ATP is an energy transducer rather than an energy store.]

Wishart, D. S., Knox, C., Guo, A. C., *et al.,* HMDB: a knowledgebase for the human metabolome, *Nuc. Acids Res.* **37,** D603–D610 (2009). [Describes the human metabolome database, with approximately 7000 entries. Available at http://www.hmdb.ca.]

chapter 13

GLUCOSE METABOLISM

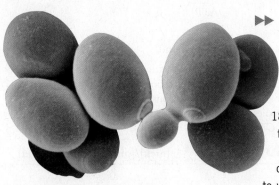

[SciMAT/Photo Researchers, Inc.]

▸▸ HOW do yeast transform sugars into other substances?

Yeast, such as the ones pictured here, have been used in brewing and baking for thousands of years. Until relatively recently, their ability to produce bubbles (CO_2 gas) and intoxicants (ethanol) was believed to be a unique property of living things that possessed a "vital force." However, in the mid-1800s, scientists began developing techniques for preparing cell extracts and then for isolating individual enzymes, and it became clear that the conversion of glucose to CO_2, ethanol, and other substances was the result of a series of enzyme-catalyzed chemical reactions. Modern biochemists, who continue to work with model organisms such as yeast, strive to describe each chemical process in detail, revealing a great deal about how yeast—and all organisms—carry out essential metabolic activities.

THIS CHAPTER IN CONTEXT

Part 1 Foundations

Part 2 Molecular Structure and Function

Part 3 Metabolism

13 Glucose Metabolism

Part 4 Genetic Information

Do You Remember?

- **Enzymes accelerate chemical reactions using acid–base catalysis, covalent catalysis, and metal ion catalysis (Section 6-2).**
- **Glucose polymers include the fuel-storage polysaccharides starch and glycogen and the structural polysaccharide cellulose (Section 11-2).**
- **Coenzymes such as NAD^+ and ubiquinone collect electrons from compounds that become oxidized (Section 12-2).**
- **A reaction with a large negative change in free energy can be coupled to another unfavorable reaction (Section 12-3).**
- **A reaction that breaks a phosphoanhydride bond in ATP occurs with a large change in free energy (Section 12-3).**
- **Nonequilibrium reactions often serve as metabolic control points (Section 12-3).**

Glucose occupies a central position in the metabolism of most cells. It is a major source of metabolic energy (in some cells, it is the only source), and it provides the precursors for the synthesis of other biomolecules. Recall that glucose is stored in polymeric form as starch in plants and as glycogen in animals (Section 11-2). The breakdown of these polymers provides glucose monomers that can be catabolized to release energy. The conversion of the six-carbon glucose to the three-carbon pyruvate, a pathway we now call **glycolysis,** occurs in ten steps. As a result of many years of research, we know a great deal about the pathway's nine intermediates and the enzymes that mediate their chemical transformations. We have also learned that glycolysis, along with other metabolic pathways, exhibits the following properties:

1. Each step of the pathway is catalyzed by a distinct enzyme.
2. The free energy consumed or released in certain reactions is transferred by molecules such as ATP and NADH.
3. The rate of the pathway can be controlled by altering the activity of individual enzymes.

If metabolic processes did not occur via multiple enzyme-catalyzed steps, cells would have little control over the amount and type of reaction products and no way to manage free energy. For example, the combustion of glucose and O_2 to CO_2 and H_2O—if allowed to occur in one grand explosion—would release about 2850 kJ · mol^{-1} of free energy all at once. In the cell, *glucose combustion requires many steps so that the cell can recover its free energy in smaller, more useful quantities.*

In this chapter, we will examine the major metabolic pathways involving glucose. **Figure 13-1** shows how these pathways relate to the general metabolic scheme outlined in Figure 12-10. The highlighted pathways include the interconversion of the monosaccharide glucose with its polymeric form glycogen, the degradation of glucose to the three-carbon intermediate pyruvate (the glycolytic pathway), the synthesis of glucose from smaller compounds (**gluconeogenesis**), and the conversion of glucose to the five-carbon monosaccharide ribose. For all the pathways, we will present the intermediates and some of the relevant enzymes. We will also examine the thermodynamics of reactions that release or consume large amounts of free energy and discuss how some of these reactions are regulated.

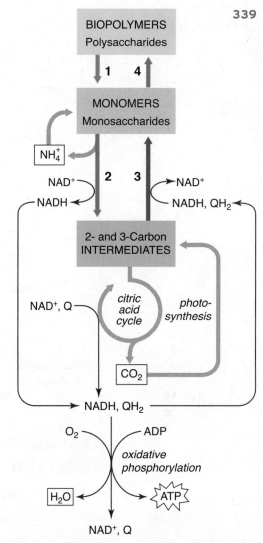

Figure 13-1 Glucose metabolism in context. (1) The polysaccharide glycogen is degraded to glucose, which is then catabolized by the glycolytic pathway (2) to the three-carbon intermediate pyruvate. Gluconeogenesis (3) is the pathway for the synthesis of glucose from smaller precursors. Glucose can then be reincorporated into glycogen (4). The conversion of glucose to ribose, a component of nucleotides, is not shown in this diagram.

13-1 Glycolysis

Glycolysis appears to be an ancient metabolic pathway. The fact that it does not require molecular oxygen suggests that it evolved before photosynthesis increased the level of atmospheric O_2. Overall, glycolysis is a series of 10 enzyme-catalyzed steps in which a six-carbon glucose molecule is broken down into two three-carbon pyruvate molecules. This catabolic pathway is accompanied by the phosphorylation of two molecules of ADP (to produce 2 ATP) and the reduction of two molecules of NAD$^+$. The net equation for the pathway (ignoring water and protons) is

$$glucose + 2\,NAD^+ + 2\,ADP + 2\,P_i \rightarrow 2\,pyruvate + 2\,NADH + 2\,ATP$$

It is convenient to divide the 10 reactions of glycolysis into two phases. In the first (Reactions 1–5), the hexose is phosphorylated and cleaved in half. In the second (Reactions 6–10), the three-carbon molecules are converted to pyruvate (**Fig. 13-2**).

KEY CONCEPTS
- Glycolysis is a 10-step pathway in which glucose is converted to two molecules of pyruvate.
- Energy is invested in the first half of the pathway, and the second half of the pathway generates 2 ATP and 2 NADH.
- Flux through the pathway is controlled primarily at the phosphofructokinase step.
- Pyruvate can be converted to lactate, acetyl-CoA, or oxaloacetate.

➕ **See Guided Exploration. Glycolysis overview.**

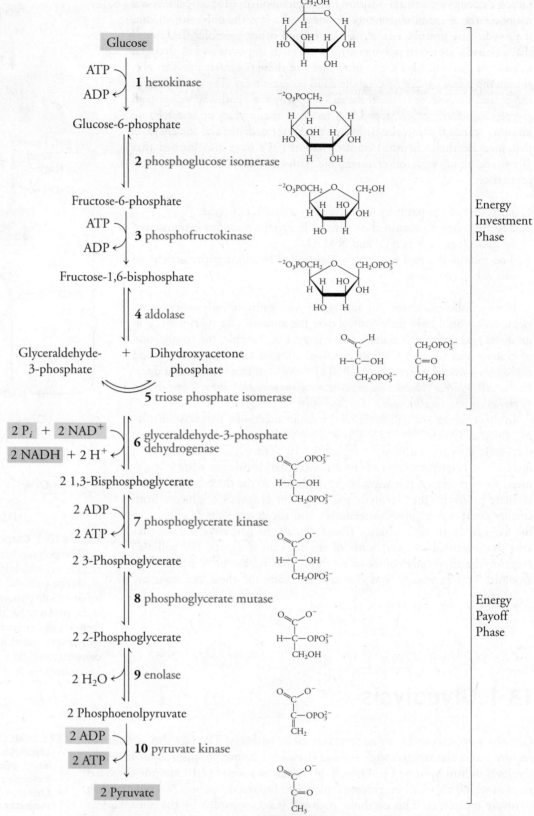

Figure 13-2 The reactions of glycolysis. The substrates, products, and enzymes corresponding to the 10 steps of the pathway are shown. Shading indicates the substrates (purple) and products (green) of the pathway as a whole. ⊕ **See Animated Figure. Overview of glycolysis.**

? Next to each reaction, write the term that describes the type of chemical change that occurs.

As you examine each of the reactions of glycolysis described in the following pages, note how the reaction substrates are converted to products by the action of an enzyme (and note how the enzyme's name often reveals its purpose). Pay attention also to the free energy change of each reaction.

Reactions 1–5 are the energy-investment phase of glycolysis

The first five reactions of glycolysis can be considered a preparatory phase for the second, energy-producing phase. In fact, the first phase requires the *investment* of free energy in the form of two ATP molecules.

1. Hexokinase

In the first step of glycolysis, the enzyme hexokinase transfers a phosphoryl group from ATP to the C6 OH group of glucose to form glucose-6-phosphate:

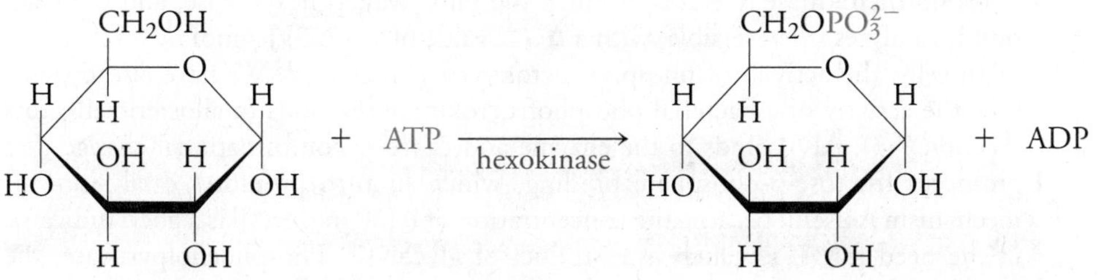

Glucose + ATP $\xrightarrow{\text{hexokinase}}$ Glucose-6-phosphate + ADP

A **kinase** is an enzyme that transfers a phosphoryl group from ATP (or another nucleoside triphosphate) to another substance.

Recall from Section 6-3 that the hexokinase active site closes around its substrates so that a phosphoryl group is efficiently transferred from ATP to glucose. The standard free energy change for this reaction, which cleaves one of ATP's phosphoanhydride bonds, is $-16.7 \text{ kJ} \cdot \text{mol}^{-1}$ (ΔG, the actual free energy change for the reaction inside a cell, has a similar value). The magnitude of this free energy change means that the reaction proceeds in only one direction; the reverse reaction is extremely unlikely since its standard free energy change would be $+16.7 \text{ kJ} \cdot \text{mol}^{-1}$. Consequently, hexokinase is said to catalyze a **metabolically irreversible reaction** that prevents glucose from backing out of glycolysis. *Many metabolic pathways have a similar irreversible step near the start that commits a metabolite to proceed through the pathway.*

2. Phosphoglucose Isomerase

The second reaction of glycolysis is an isomerization reaction in which glucose-6-phosphate is converted to fructose-6-phosphate:

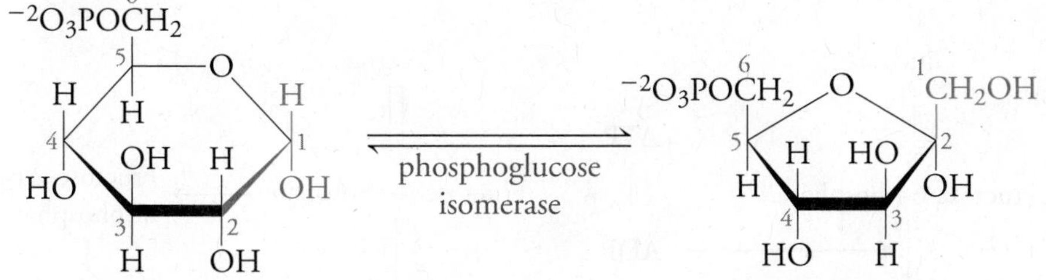

Glucose-6-phosphate $\xrightleftharpoons{\substack{\text{phosphoglucose}\\\text{isomerase}}}$ Fructose-6-phosphate

Because fructose is a six-carbon ketose (Section 11-1), it forms a five-membered ring.

The standard free energy change for the phosphoglucose isomerase reaction is $+2.2 \text{ kJ} \cdot \text{mol}^{-1}$, but the reactant concentrations *in vivo* yield a ΔG value of about $-1.4 \text{ kJ} \cdot \text{mol}^{-1}$. A value of ΔG near zero indicates that the reaction operates close to equilibrium (at equilibrium, $\Delta G = 0$). *Such **near-equilibrium** reactions are considered to be freely reversible, since a slight excess of products can easily drive the* *reaction in reverse by mass action effects.* In a metabolically irreversible reaction, such as the hexokinase reaction, the concentration of product could never increase enough to compensate for the reaction's large value of ΔG.

3. Phosphofructokinase

The third reaction of glycolysis consumes a second ATP molecule in the phosphorylation of fructose-6-phosphate to yield fructose-1,6-bisphosphate.

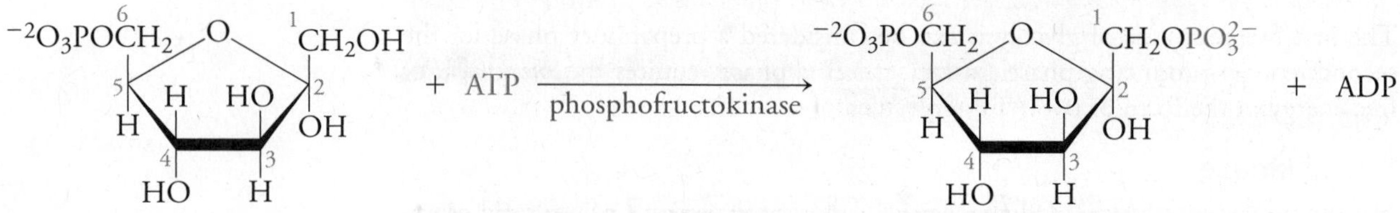

Fructose-6-phosphate   Fructose-1,6-bisphosphate

Phosphofructokinase operates in much the same way as hexokinase, and the reaction it catalyzes is irreversible, with a $\Delta G°'$ value of $-17.2\ \text{kJ} \cdot \text{mol}^{-1}$.

In cells, the activity of phosphofructokinase is regulated. We have already seen how the activity of a bacterial phosphofructokinase responds to allosteric effectors (Section 7-3). ADP binds to the enzyme and causes a conformational change that promotes fructose-6-phosphate binding, which in turn promotes catalysis. This mechanism is useful because the concentration of ADP in the cell is a good indicator of the need for ATP, which is a product of glycolysis. Phosphoenolpyruvate, the product of step 9 of glycolysis, binds to bacterial phosphofructokinase and causes it to assume a conformation that destabilizes fructose-6-phosphate binding, thereby diminishing catalytic activity. Thus, when the glycolytic pathway is producing plenty of phosphoenolpyruvate and ATP, the phosphoenolpyruvate can act as a feedback inhibitor to slow the pathway by decreasing the rate of the reaction catalyzed by phosphofructokinase (**Fig. 13-3a**).

The most potent activator of phosphofructokinase in mammals, however, is the compound fructose-2,6-bisphosphate, which is synthesized from fructose-6-phosphate by an enzyme known as phosphofructokinase-2. (The glycolytic enzyme is therefore sometimes called phosphofructokinase-1).

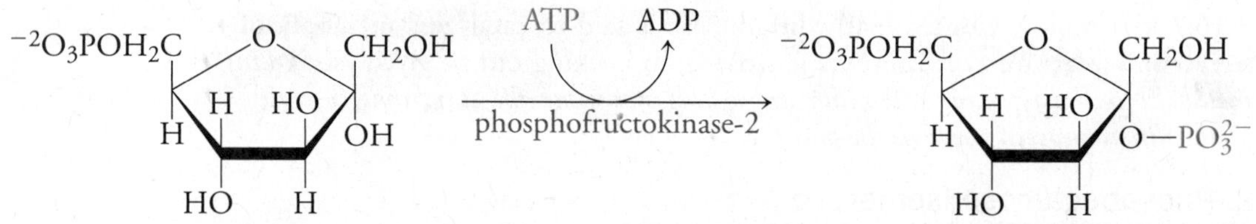

Fructose-6-phosphate   Fructose-2,6-bisphosphate

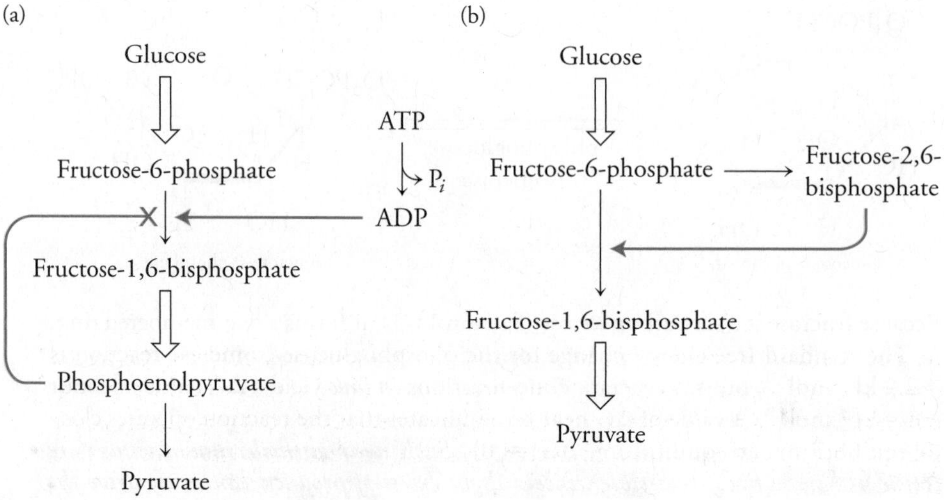

Figure 13-3 Regulation of phosphofructokinase. (a) Regulation in bacteria. ADP, produced when ATP is consumed elsewhere in the cell, stimulates the activity of phosphofructokinase (green arrow). Phosphoenolpyruvate, a late intermediate of glycolysis, inhibits phosphofructokinase (red symbol), thereby decreasing the rate of the entire pathway. (b) Regulation in mammals.

The activity of phosphofructokinase-2 is hormonally stimulated when the concentration of glucose in the blood is high. The resulting increase in fructose-2,6-bisphosphate concentration activates phosphofructokinase to increase the flux of glucose through the glycolytic pathway (Fig. 13-3b).

The phosphofructokinase reaction is the primary control point for glycolysis. It is the slowest reaction of glycolysis, so the rate of this reaction largely determines the flux (rate of flow) of glucose through the entire pathway. In general, a **rate-determining reaction**—such as the phosphofructokinase reaction—operates far from equilibrium; that is, it has a large negative free energy change and is irreversible under metabolic conditions. The rate of the reaction can be altered by allosteric effectors but not by fluctuations in the concentrations of its substrates or products. Thus, it acts as a one-way valve. In contrast, a near-equilibrium reaction—such as the phosphoglucose isomerase reaction—cannot serve as a rate-determining step for a pathway because it can respond to small changes in reactant concentrations by operating in reverse.

4. Aldolase

Reaction 4 converts the hexose fructose-1,6-bisphosphate to two three-carbon molecules, each of which bears a phosphate group.

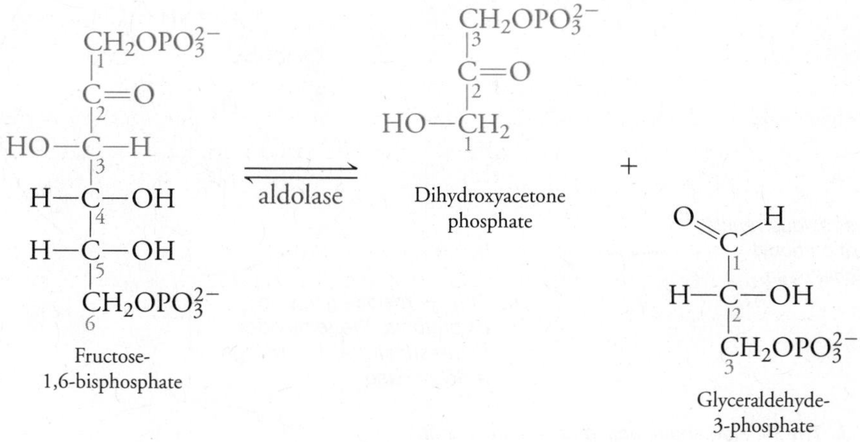

Fructose-1,6-bisphosphate ⇌ (aldolase) Dihydroxyacetone phosphate + Glyceraldehyde-3-phosphate

This reaction is the reverse of an aldol (aldehyde–alcohol) condensation, so the enzyme that catalyzes the reaction is called aldolase. It is worth examining its mechanism. The active site of mammalian aldolase contains two catalytically important residues: a Lys residue that forms a Schiff base (imine) with the substrate and an ionized Tyr residue that acts as a base catalyst (Fig. 13-4).

Early studies of aldolase implicated a Cys residue in catalysis because iodoacetate, a reagent that reacts with the Cys side chain, also inactivates the enzyme:

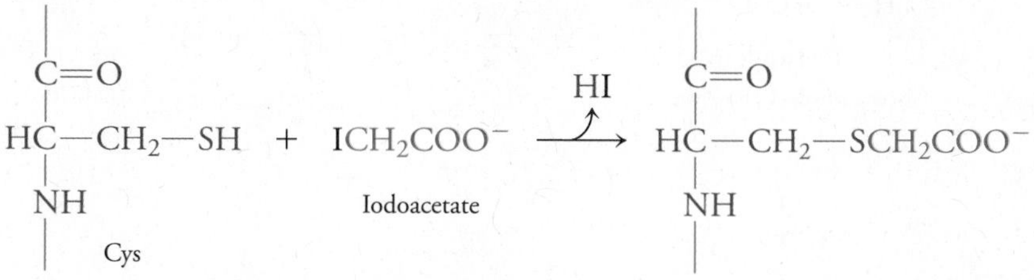

Researchers used iodoacetate to help identify the intermediates of glycolysis: In the presence of iodoacetate, fructose-1,6-bisphosphate accumulates because the next step is blocked. Acetylation of the Cys residue, which turned out not to be part of the active site, probably interferes with a conformational change that is necessary for aldolase activity.

NH$_2$—(CH$_2$)$_4$—
Lys

$\overline{O}$

Tyr

1. *Fructose-1,6-bisphosphate, shown in its linear form, binds to the enzyme.*

CH$_2$OPO$_3^{2-}$
C=O   H$_2$N—(CH$_2$)$_4$—
HO—C—H
H—C—O—H
H—C—OH
CH$_2$OPO$_3^{2-}$

Fructose-1,6-bisphosphate

2. *The active-site Lys residue reacts with the substrate's carbonyl group to form a Schiff base.*

H$_2$O

CH$_2$OPO$_3^{2-}$
C=NH—(CH$_2$)$_4$—
HO—C—H
H—C—O—H
H—C—OH
CH$_2$OPO$_3^{2-}$

3. *The Tyr side chain, acting as a base, accepts a proton from the substrate, and the bond between C3 and C4 breaks, releasing the first product, an aldehyde. The reaction is facilitated by the protonated Schiff base, which is a better electron-withdrawing group than the carbonyl group that was originally at C2.*

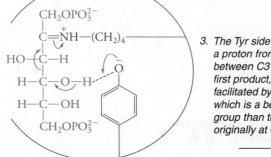

H   O
  \ //
   C
H—C—OH
CH$_2$OPO$_3^{2-}$

Glyceraldehyde-3-phosphate

Figure 13-4   The aldolase reaction.
⊕ See Animated Figure. Mechanism of aldolase.

....

❓ **Does this reaction follow an ordered or ping pong mechanism (see Section 7-2)?**

5. *The Schiff base is hydrolyzed, releasing the second product and regenerating the enzyme.*

NH$_2$—(CH$_2$)$_4$—

$\overline{O}$

CH$_2$OPO$_3^{2-}$
C=O
CH$_2$OH

Dihydroxyacetone phosphate

H$_2$O

CH$_2$OPO$_3^{2-}$
C=NH—(CH$_2$)$_4$—
HO—C—H
     H
     $\overline{O}$

4. *The Tyr residue gives up its proton to the remainder of the substrate, re-forming a Schiff base.*

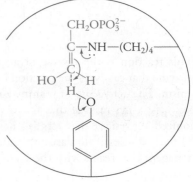

CH$_2$OPO$_3^{2-}$
C—NH—(CH$_2$)$_4$—
HO—C—H
      H—O

The $\Delta G°'$ value for the aldolase reaction is $+22.8$ kJ · mol^{-1}, indicating that the reaction is unfavorable under standard conditions. However, the reaction proceeds *in vivo* (ΔG is actually less than zero) because the products of the reaction are quickly whisked away by subsequent reactions. In essence, the rapid consumption of glyceraldehyde-3-phosphate and dihydroxyacetone phosphate "pulls" the aldolase reaction forward.

5. Triose Phosphate Isomerase

The products of the aldolase reaction are both phosphorylated three-carbon compounds, but only one of them—glyceraldehyde-3-phosphate—proceeds through the remainder of the pathway. Dihydroxyacetone phosphate is converted to glyceraldehyde-3-phosphate by triose phosphate isomerase:

$$
\begin{array}{c}
\text{H} \\
| \\
\text{H}-\text{C}-\text{OH} \\
| \\
\text{C}{=}\text{O} \\
| \\
\text{CH}_2\text{OPO}_3^{2-}
\end{array}
\quad \xrightleftharpoons[\text{triose phosphate isomerase}]{} \quad
\begin{array}{c}
\text{O}{=}\text{C}-\text{H} \\
| \\
\text{H}-\text{C}-\text{OH} \\
| \\
\text{CH}_2\text{OPO}_3^{2-}
\end{array}
$$

Dihydroxyacetone phosphate   Glyceraldehyde-3-phosphate

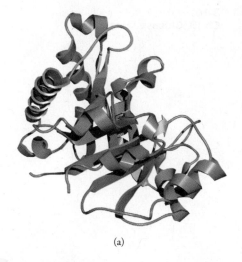

(a)

Triose phosphate isomerase was introduced in Section 7-2 as an example of a catalytically perfect enzyme, one whose rate is limited only by the rate at which its substrates can diffuse to its active site. The catalytic mechanism of triose phosphate isomerase may involve low-barrier hydrogen bonds (which also help stabilize the transition state in serine proteases; see Section 6-3). In addition, the catalytic power of triose phosphate isomerase depends on a protein loop that closes over the active site (Fig. 13-5).

The standard free energy change for the triose phosphate isomerase reaction is slightly positive, even under physiological conditions ($\Delta G^{\circ\prime} = +7.9 \cdot \text{kJ} \cdot \text{mol}^{-1}$ and $\Delta G = +4.4 \text{ kJ} \cdot \text{mol}^{-1}$), but the reaction proceeds because glyceraldehyde-3-phosphate is quickly consumed in the next reaction, so more dihydroxyacetone phosphate is constantly being converted to glyceraldehyde-3-phosphate.

(b)

Reactions 6-10 are the energy-payoff phase of glycolysis

So far, the reactions of glycolysis have consumed 2 ATP, but this investment pays off in the second phase of glycolysis when 4 ATP are produced, for a net gain of 2 ATP. All of the reactions of the second phase involve three-carbon intermediates, but keep in mind that *each glucose molecule that enters the pathway yields two of these three-carbon units.*

Some species convert glucose to glyceraldehyde-3-phosphate by different pathways than the one presented above. However, the second phase of glycolysis, which converts glyceraldehyde-3-phosphate to pyruvate, is the same in all organisms. This suggests that glycolysis may have evolved from the "bottom up"; that is, it first evolved as a pathway for extracting free energy from abiotically produced small molecules, before cells developed the ability to synthesize larger molecules such as hexoses.

6. Glyceraldehyde-3-Phosphate Dehydrogenase

In the sixth reaction of glycolysis, glyceraldehyde-3-phosphate is both oxidized and phosphorylated:

$$
\begin{array}{c}
\text{O}{=}\overset{1}{\text{C}}-\text{H} \\
| \\
\text{H}-\overset{2}{\text{C}}-\text{OH} \\
| \\
\overset{3}{\text{CH}_2}\text{OPO}_3^{2-}
\end{array}
+ \text{NAD}^+ + \text{P}_i
\quad \xrightleftharpoons[\substack{\text{glyceraldehyde-3-phosphate} \\ \text{dehydrogenase}}]{} \quad
\begin{array}{c}
\text{O}{=}\overset{1}{\text{C}}-\text{OPO}_3^{2-} \\
| \\
\text{H}-\overset{2}{\text{C}}-\text{OH} \\
| \\
\overset{3}{\text{CH}_2}\text{OPO}_3^{2-}
\end{array}
+ \text{NADH} + \text{H}^+
$$

Glyceraldehyde-3-phosphate   1,3-Bisphosphoglycerate

Figure 13-5 Conformational changes in yeast triose phosphate isomerase.
(a) One loop of the protein, comprising residues 166–176, is highlighted in green. (b) When a substrate binds to the enzyme, the loop closes over the active site to stabilize the reaction's transition state. In this model, the transition state analog 2-phosphoglycolate (orange) occupies the active site. Triose phosphate isomerase is actually a homodimer; only one subunit is pictured here. [Structure of the enzyme alone (pdb 1YPI) determined by T. Alber, E. Lolis, and G. A. Petsko; structure of the enzyme with the analog (pdb 2YPI) determined by E. Lolis and G. A. Petsko.] ⊕ **See Interactive Exercise. Triose phosphate isomerase.**

Unlike the kinases that catalyze Reactions 1 and 3, glyceraldehyde-3-phosphate dehydrogenase does not use ATP as a phosphoryl-group donor; it adds inorganic phosphate to the substrate. This reaction is also an oxidation–reduction reaction in which the aldehyde group of glyceraldehyde-3-phosphate is oxidized and the cofactor NAD^+ is reduced to NADH. In effect, glyceraldehyde-3-phosphate dehydrogenase catalyzes the removal of an H atom (actually, a hydride ion); hence the name "dehydrogenase." Note that the reaction product NADH must eventually be reoxidized to NAD^+, or else glycolysis will come to a halt. In fact, the reoxidation of NADH, which is a form of "energy currency," can generate ATP (Chapter 15).

An active-site Cys residue participates in the glyceraldehyde-3-phosphate dehydrogenase reaction **(Fig. 13-6)**. The enzyme is inhibited by arsenate (AsO_4^{3-}), which competes with P_i (PO_4^{3-}) for binding in the enzyme active site.

7. Phosphoglycerate Kinase

The product of Reaction 6, 1,3-bisphosphoglycerate, is an acyl phosphate.

$$R-\overset{\overset{\displaystyle O}{\|}}{C}-O-\overset{\overset{\displaystyle O}{\|}}{\underset{\underset{\displaystyle O^-}{|}}{P}}-O^-$$

An acyl phosphate

The subsequent removal of its phosphoryl group releases a large amount of free energy in part because the reaction products are more stable (the same principle con-

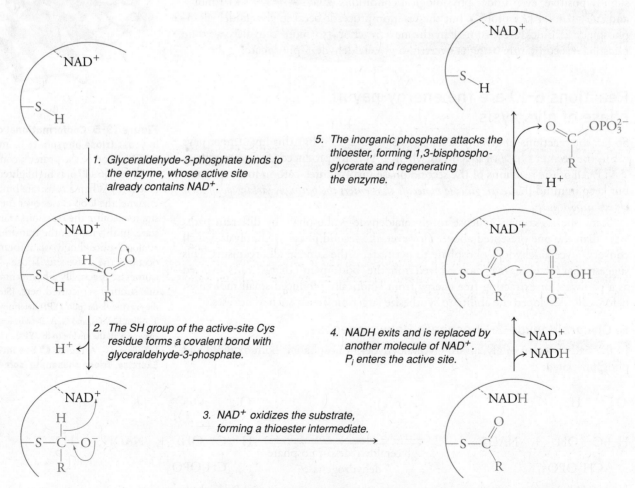

Figure 13-6 The glyceraldehyde-3-phosphate dehydrogenase reaction. ⊕ **See Animated Figure. Mechanism of glyceraldehyde-3-phosphate dehydrogenase.**

? **Identify the reactant that undergoes oxidation and the reactant that undergoes reduction.**

tributes to the large negative value of ΔG for reactions involving cleavage of ATP's phosphoanhydride bonds; see Section 12-3). *The free energy released in this reaction is used to drive the formation of ATP, as 1,3-bisphosphoglycerate donates its phosphoryl group to ADP:*

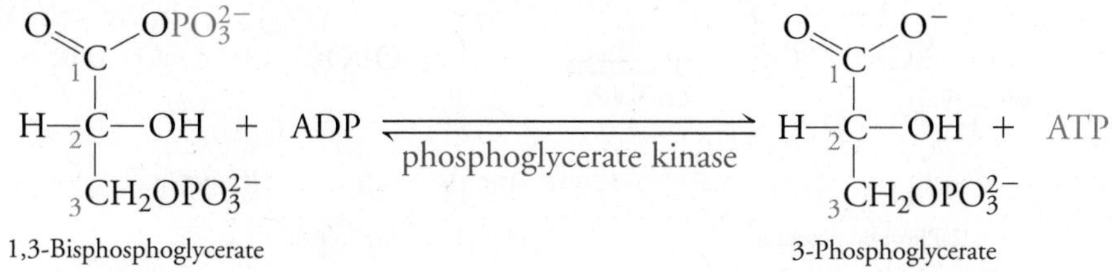

1,3-Bisphosphoglycerate   3-Phosphoglycerate

Note that the enzyme that catalyzes this reaction is called a kinase since it transfers a phosphoryl group between ATP and another molecule.

The standard free energy change for the phosphoglycerate kinase reaction is -18.8 kJ $\cdot$ mol^{-1}. This strongly exergonic reaction helps pull the glyceraldehyde-3-phosphate dehydrogenase reaction forward, since its standard free energy change is greater than zero ($\Delta G^{\circ\prime} = +6.7$ kJ $\cdot$ mol^{-1}). This pair of reactions provides a good example of the coupling of a thermodynamically favorable and unfavorable reaction so that both proceed with a net decrease in free energy: -18.8 kJ $\cdot$ mol^{-1} + 6.7 kJ $\cdot$ mol^{-1} = -12.1 kJ $\cdot$ mol^{-1}. Under physiological conditions, ΔG for the paired reactions is close to zero.

8. Phosphoglycerate Mutase

In the next reaction, 3-phosphoglycerate is converted to 2-phosphoglycerate:

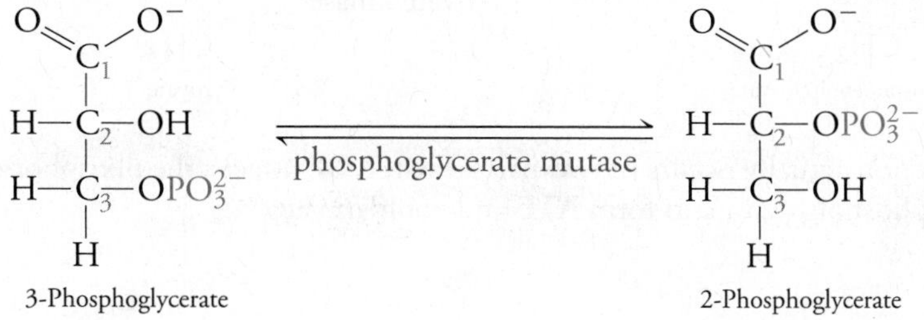

3-Phosphoglycerate   2-Phosphoglycerate

Although the reaction appears to involve the simple intramolecular transfer of a phosphoryl group, the reaction mechanism is a bit more complicated and requires an enzyme active site that contains a phosphorylated His residue. The phospho-His transfers its phosphoryl group to 3-phosphoglycerate to generate 2,3-bisphosphoglycerate, which then gives a phosphoryl group back to the enzyme, leaving 2-phosphoglycerate and phospho-His:

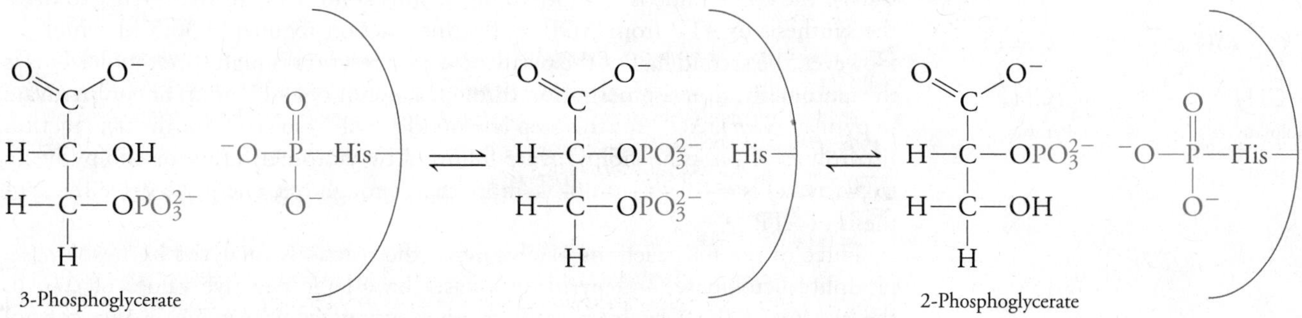

3-Phosphoglycerate   2-Phosphoglycerate

As can be guessed from its mechanism, the phosphoglycerate mutase reaction is freely reversible *in vivo*.

9. Enolase

Enolase catalyzes a dehydration reaction, in which water is eliminated:

$$
\begin{array}{c}
\text{O}=\overset{\text{O}^-}{\underset{\text{C}_1}{\text{C}}} \\
\text{H}-\overset{\text{OPO}_3^{2-}}{\underset{\text{C}_2}{\text{C}}} \\
\text{H}-\overset{\text{OH}}{\underset{\text{C}_3}{\text{C}}} \\
\text{H}
\end{array}
\underset{\text{enolase}}{\rightleftharpoons}
\begin{array}{c}
\text{O}=\overset{\text{O}^-}{\text{C}} \\
\text{C}-\text{OPO}_3^{2-} \\
\text{H}-\text{C} \\
\text{H}
\end{array}
+ \text{H}_2\text{O}
$$

2-Phosphoglycerate Phosphoenolpyruvate

The enzyme active site includes an Mg^{2+} ion that apparently coordinates with the OH group at C3 and makes it a better leaving group. Fluoride ion and P_i can form a complex with the Mg^{2+} and thereby inhibit the enzyme. In early studies demonstrating the inhibition of glycolysis by F^-, 2-phosphoglycerate, the substrate of enolase, accumulated. The concentration of 3-phosphoglycerate also increased in the presence of F^- since phosphoglycerate mutase readily converted the excess 2-phosphoglycerate back to 3-phosphoglycerate.

10. Pyruvate Kinase

The tenth reaction of glycolysis is catalyzed by pyruvate kinase, which converts phosphoenolpyruvate to pyruvate and transfers a phosphoryl group to ADP to produce ATP:

$$
\begin{array}{c}
\text{O}=\overset{\text{O}^-}{\text{C}} \\
\text{C}-\text{OPO}_3^{2-} \\
\text{CH}_2
\end{array}
+ \text{ADP}
\xrightarrow{\text{pyruvate kinase}}
\begin{array}{c}
\text{O}=\overset{\text{O}^-}{\text{C}} \\
\text{C}=\text{O} \\
\text{CH}_3
\end{array}
+ \text{ATP}
$$

Phosphoenolpyruvate Pyruvate

The reaction actually occurs in two parts. First, ADP attacks the phosphoryl group of phosphoenolpyruvate to form ATP and enolpyruvate:

$$
\begin{array}{c}
\text{O}=\overset{\text{O}^-}{\text{C}} \\
\text{C}-\text{O}-\overset{\text{O}}{\underset{\text{O}^-}{\text{P}}}-\text{O}^- \\
\text{CH}_2
\end{array}
+
\begin{array}{c}
\text{O} \\
{}^-\text{O}-\overset{\parallel}{\underset{\text{O}^-}{\text{P}}}-\text{O}-\text{AMP}
\end{array}
\overset{\text{ATP}}{\underset{\text{H}^+}{\longrightarrow}}
\begin{array}{c}
\text{O}=\overset{\text{O}^-}{\text{C}} \\
\text{C}-\text{OH} \\
\text{CH}_2
\end{array}
$$

Phosphoenolpyruvate ADP Enolpyruvate

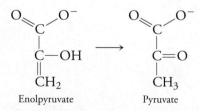

Enolpyruvate Pyruvate

Removal of phosphoenolpyruvate's phosphoryl group is not a particularly exergonic reaction: When written as a hydrolytic reaction (transfer of the phosphoryl group to water), the $\Delta G^{\circ\prime}$ value is $-16\ \text{kJ} \cdot \text{mol}^{-1}$. This is not enough free energy to drive the synthesis of ATP from ADP + P_i (this reaction requires $+30.5\ \text{kJ} \cdot \text{mol}^{-1}$). However, the second half of the pyruvate kinase reaction is highly exergonic. This is the **tautomerization** (isomerization through the shift of an H atom) of enolpyruvate to pyruvate (*left*). $\Delta G^{\circ\prime}$ for this step is $-46\ \text{kJ} \cdot \text{mol}^{-1}$, so $\Delta G^{\circ\prime}$ for the net reaction (hydrolysis of phosphoenolpyruvate followed by tautomerization of enolpyruvate to pyruvate) is $-61.9\ \text{kJ} \cdot \text{mol}^{-1}$, more than enough free energy to drive the synthesis of ATP.

Three of the ten reactions of glycolysis (the reactions catalyzed by hexokinase, phosphofructokinase, and pyruvate kinase) have large negative values of ΔG. In theory, any of these far-from-equilibrium reactions could serve as a flux-control point for the pathway. The other seven reactions function near equilibrium ($\Delta G \approx 0$) and can therefore accommodate flux in either direction. The free energy changes for the ten reactions of glycolysis are shown graphically in **Figure 13-7**.

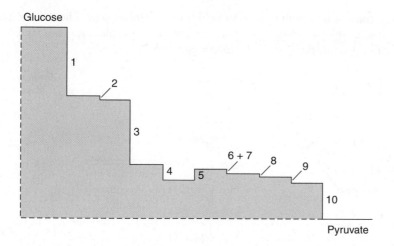

Figure 13-7 Graphical representation of the free energy changes of glycolysis. Three steps have large negative values of ΔG; the remaining steps are near equilibrium ($\Delta G \approx 0$). The height of each step corresponds to its ΔG value in heart muscle, and the numbers correspond to glycolytic enzymes. Keep in mind that ΔG values vary slightly among tissues. [Data from Newsholme, E. A., and Start, C., *Regulation in Metabolism*, p. 97, Wiley (1973).]

We have already discussed the mechanisms for regulating phosphofructokinase activity, the major control point for glycolysis. Hexokinase also catalyzes an irreversible reaction and is subject to inhibition by its product, glucose-6-phosphate. However, hexokinase cannot be the sole control point for glycolysis because glucose can also enter the pathway as glucose-6-phosphate, bypassing the hexokinase reaction. Finally, it would not be efficient for the pyruvate kinase reaction to be the primary regulatory step for glycolysis because it occurs at the very end of the 10-step pathway. Even so, pyruvate kinase activity can be adjusted. In some organisms, fructose-1,6-bisphosphate activates pyruvate kinase at an allosteric site. This is an example of **feed-forward activation:** Once a monosaccharide has entered glycolysis, fructose-1,6-bisphosphate helps ensure rapid flux through the pathway.

To sum up the second phase of glycolysis: Glyceraldehyde-3-phosphate is converted to pyruvate with the synthesis of 2 ATP (in Reactions 7 and 10). Since each molecule of glucose yields two molecules of glyceraldehyde-3-phosphate, the reactions of the second phase of glycolysis must be doubled, so the yield is 4 ATP. Two molecules of ATP are invested in phase 1, bringing the net yield to 2 ATP produced per glucose molecule. Two NADH are also generated for each glucose molecule. Monosaccharides other than glucose are metabolized in a similar fashion to yield ATP (Box 13-A).

BOX 13-A ⬡ BIOCHEMISTRY NOTE

Catabolism of Other Sugars

A typical human diet contains many carbohydrates other than glucose and its polymers. For example, lactose, a disaccharide of glucose and galactose, is present in milk and food derived from it (Section 11-2). Lactose is cleaved in the intestine by the enzyme lactase, and the two monosaccharides are absorbed, transported to the liver, and metabolized. Galactose undergoes phosphorylation and isomerization and enters the glycolytic pathway as glucose-6-phosphate, so its energy yield is the same as that of glucose.

Sucrose, the other major dietary disaccharide, is composed of glucose and fructose (Section 11-2); it is present in a variety of foods of plant origin. Like lactose, sucrose is hydrolyzed in the small intestine, and its components glucose and fructose are absorbed. The monosaccharide fructose is also present in many foods, particularly fruits and honey. It tastes sweeter than sucrose, is more soluble, and is inexpensive to produce in the form of high-fructose corn syrup—all of which make fructose attractive to the manufacturers of soft drinks and other processed foods. This is the primary reason why the consumption of fructose in the United States has increased by about 61% over the past 30 years.

The overconsumption of fructose may be contributing to the obesity epidemic. One possible explanation relates to the catabolism of fructose, which differs somewhat from the catabolism of glucose. Fructose is metabolized primarily by the liver, but the form of hexokinase present in the liver (called glucokinase) has very low affinity for fructose. Fructose therefore enters glycolysis by a different route.

(continued on next page)

First, fructose is phosphorylated to yield fructose-1-phosphate. The enzyme fructose-1-phosphate aldolase then splits the six-carbon molecule into two three-carbon molecules: glyceraldehyde and dihydroxyacetone phosphate:

Dihydroxyacetone phosphate is converted to glyceraldehyde-3-phosphate by triose phosphate isomerase and can proceed through the second phase of glycolysis. The glyceraldehyde can be phosphorylated to glyceraldehyde-3-phosphate, but it can also be converted to glycerol-3-phosphate, a precursor of the backbone of triacylglycerols. This may contribute to an increase in fat deposition. A second potential hazard of the fructose catabolic pathway is that fructose catabolism bypasses the phosphofructokinase-catalyzed step of glycolysis and thus avoids a major regulatory point. This may disrupt fuel metabolism in such a way that fructose catabolism leads to greater production of lipid than does glucose catabolism. The metabolic consequences of consuming high-fructose foods may therefore go beyond their caloric content.

● **Question:** When its concentration is extremely high, fructose is converted to fructose-1-phosphate much faster than it can be cleaved by the aldolase. How would this affect the cell's ATP supply?

Pyruvate is converted to other substances

What happens to the pyruvate generated by the catabolism of glucose? It can be further broken down to acetyl-CoA or used to synthesize other compounds such as oxaloacetate. The fate of pyruvate depends on the cell type and the need for metabolic free energy and molecular building blocks. Some of the options are diagrammed in **Figure 13-8**.

During exercise, pyruvate may be temporarily converted to lactate. In a highly active muscle cell, glycolysis rapidly provides ATP to power muscle contraction, but the pathway also consumes NAD^+ at the glyceraldehyde-3-phosphate dehydrogenase step.

Figure 13-8 Fates of pyruvate. Pyruvate may be converted to a two-carbon acetyl group linked to the carrier coenzyme A. Acetyl-CoA may be further broken down by the citric acid cycle or used to synthesized fatty acids. In muscle, pyruvate is reduced to lactate to regenerate NAD^+ for glycolysis. Yeast degrade pyruvate to ethanol and CO_2. Pyruvate can also be carboxylated to produce the four-carbon oxaloacetate.

? **Beside each arrow, write the names of the enzymes that catalyze the process.**

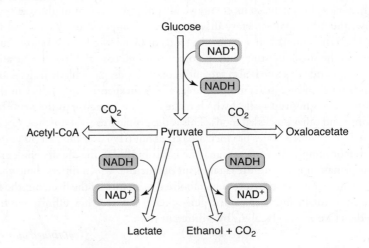

The two NADH molecules generated for each glucose molecule catabolized can be reoxidized in the presence of oxygen. However, this process is too slow to replenish the NAD^+ needed for the rapid production of ATP by glycolysis. *To regenerate NAD^+, the enzyme lactate dehydrogenase reduces pyruvate to lactate:*

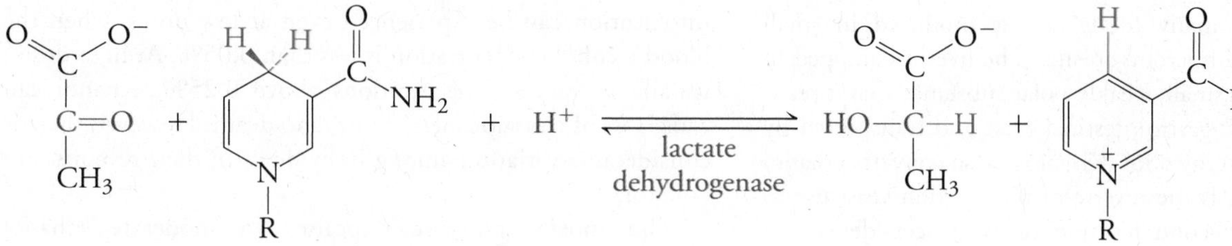

Pyruvate   NADH   Lactate NAD^+

This reaction, sometimes called the eleventh step of glycolysis, allows the muscle to function anaerobically for a minute or so (see Box 12-D). The net reaction for anaerobic glucose catabolism is

$$\text{glucose} + 2\,\text{ADP} + 2\,P_i \rightarrow 2\,\text{lactate} + 2\,\text{ATP}$$

Lactate represents a sort of metabolic dead end: Its only options are to be eventually converted back to pyruvate (the lactate dehydrogenase reaction is reversible) or to be exported from the cell. The liver takes up lactate, oxidizes it back to pyruvate, and then uses it for gluconeogenesis. The glucose produced in this manner may eventually make its way back to the muscle to help fuel continued muscle contraction. When the muscle is functioning aerobically, NADH produced by the glyceraldehyde-3-phosphate dehydrogenase reaction is reoxidized by oxygen and the lactate dehydrogenase reaction is not needed.

Organisms such as yeast growing under anaerobic conditions can regenerate NAD^+ by producing alcohol. In the mid-1800s, Louis Pasteur called this process **fermentation,** meaning life without air, although yeast also ferment sugars in the presence of O_2. And, to answer the question posed at the start of the chapter, yeast transform sugars to pyruvate by glycolysis, then carry out a two-step fermentation process. First, pyruvate decarboxylase (an enzyme not present in animals) catabolizes the removal of pyruvate's carboxylate group to produce acetaldehyde. Next, alcohol dehydrogenase reduces acetaldehyde to ethanol.

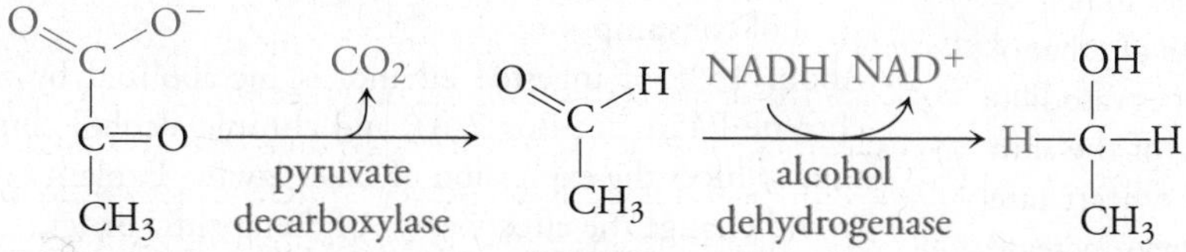

►► **HOW** do yeast transform sugars into other substances?

Pyruvate   Acetaldehyde   Ethanol

Ethanol is considered to be a waste product of sugar metabolism; its accumulation is toxic to other organisms (Box 13-B), including the yeast that produce it. This is one reason why the alcohol content of yeast-fermented beverages such as wine is limited to about 13%. "Hard" liquor must be distilled to increase its ethanol content.

Although glycolysis is an oxidative pathway, its end product pyruvate is still a relatively reduced molecule. The further catabolism of pyruvate begins with its decarboxylation to form a two-carbon acetyl group linked to coenzyme A.

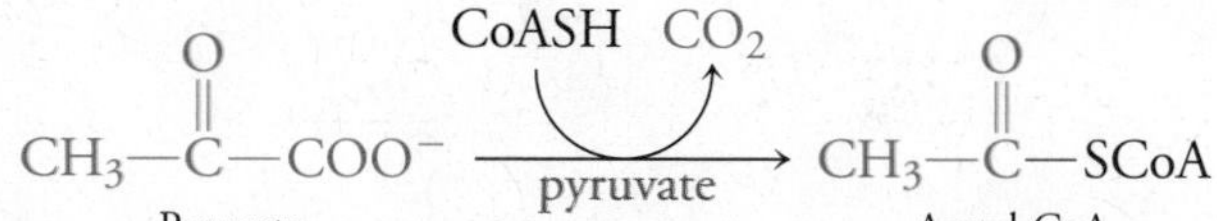

Pyruvate   Acetyl-CoA

BOX 13-B ⬡ **CLINICAL CONNECTION**

Alcohol Metabolism

Unlike yeast, mammals do not produce ethanol, although it is naturally present in many foods and is produced in small amounts by intestinal microorganisms. The liver is equipped to metabolize ethanol, a small, weakly polar substance that is readily absorbed from the gastrointestinal tract and transported by the bloodstream. First, alcohol dehydrogenase converts ethanol to acetaldehyde. This is the reverse of the reaction yeast use to produce ethanol. A second reaction converts acetaldehyde to acetate:

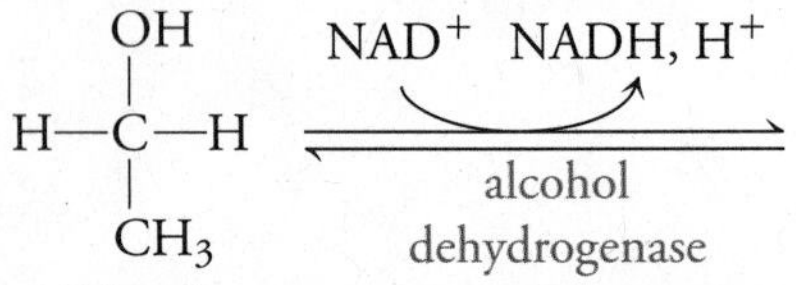

Note that both of these reactions require NAD^+, a cofactor used in many other oxidative cellular processes, including glycolysis. The liver uses the same two-enzyme pathway to metabolize the excess ethanol obtained from alcoholic beverages. Ethanol itself is mildly toxic, and the physiological effects of alcohol also reflect the toxicity of acetaldehyde and acetate in tissues such as the liver and brain.

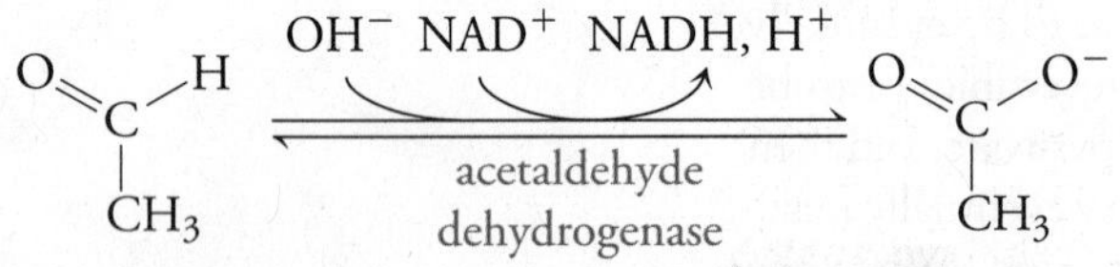

Over the short term and at low doses, alcohol triggers relaxation, often leading to animated movements and talkativeness. Some of these responses may be psychological (resulting from social cues rather than chemical effects), since changes in behavior sometimes occur even before significant amounts of ethanol have been absorbed. Once in the body, ethanol induces vasodilation, apparent as flushing (warming and reddening of the skin due to increased blood flow). At the same time, the heart rate and respiration rate become slightly lower. The kidneys increase the excretion of water as ethanol interferes with the ability of the hypothalamus (a region of the brain) to properly sense osmotic pressure.

Ethanol is considered to be a psychoactive drug because of its effects on the central nervous system. It stimulates signaling from certain neurotransmitter receptors that function as ligand-gated ion channels (Section 9-2) to inhibit neuronal signaling, producing a sedative effect. Sensory, motor, and cognitive functions are impaired, leading to delayed reaction time, loss of balance, and blurred vision. Some of the symptoms of ethanol intoxication can be experienced even at low doses, when the blood alcohol concentration is less than 0.05%. At high doses, usually at blood concentrations above 0.25%, ethanol can cause loss of consciousness, coma, and death. However, there is considerable variation among individuals in their responses to ethanol.

The mostly pleasant responses to moderate ethanol consumption are followed by a period of recovery, when the concentrations of ethanol metabolites are relatively high. The unpleasant symptoms of a hangover in part reflect the chemistry of producing acetaldehyde and acetate. As shown at left, their production in the liver consumes NAD^+, thereby lowering the cell's NAD^+:NADH ratio. Without sufficient NAD^+, the liver's ability to produce ATP by glycolysis is diminished (since NAD^+ is required for the glyceraldehyde-3-phosphate dehydrogenase reaction). Acetaldehyde itself can react with liver proteins, inactivating them. Acetate (acetic acid) production lowers blood pH.

Long-term, excessive alcohol consumption exacerbates the toxic effects of ethanol and its metabolites. For example, a shortage of liver NAD^+ slows fatty acid breakdown (like glycolysis, a process that requires NAD^+) and promotes fatty acid synthesis, leading to fat accumulation in the liver. Over time, cell death causes permanent loss of function in the central nervous system. The death of liver cells and their replacement by fibrous scar tissue causes liver cirrhosis.

⬤ **Questions:**

1. Explain why alcohol consumption is associated with increased risk of developing hypothermia.
2. Drinking a glass of water for each alcoholic drink is a popular hangover-prevention strategy. Explain how increased water consumption might relieve some of the negative effects of alcohol consumption.
3. About 15% of ingested ethanol is metabolized by a cytochrome P450 (see Box 7-A), and chronic alcohol consumption induces the expression of this enzyme. Explain how this would change the effectiveness of therapeutic drugs.
4. Normally, the liver converts lactate, produced mainly by muscles, back to pyruvate so that the pyruvate can be converted to glucose by gluconeogenesis (Section 13-2). How do the activities of alcohol dehydrogenase and acetaldehyde dehydrogenase contribute to hypoglycemia?
5. Acetate can be broken down further but only if it is first converted to acetyl-CoA in a reaction that requires ATP. Explain why this metabolic process is inhibited when the concentration of acetate is high.

| **TABLE 13-1** | Standard Free Energy Changes for Glucose Catabolism |

Catabolic Process	$\Delta G^{\circ\prime}$ (kJ · mol^{-1})
$C_6H_{12}O_6 \rightarrow 2\ C_3H_5O_3^- + 2\ H^+$ (glucose) (lactate)	-196
$C_6H_{12}O_6 + 6\ O_2 \rightarrow 6\ CO_2 + 6\ H_2O$ (glucose)	-2850

The resulting acetyl-CoA is a substrate for the citric acid cycle (Chapter 14). The complete oxidation of the six glucose carbons to CO_2 releases much more free energy than the conversion of glucose to lactate (Table 13-1). Much of this energy is recovered in the synthesis of ATP by the reactions of the citric acid cycle and oxidative phosphorylation (Chapter 15), pathways that require the presence of molecular oxygen.

Pyruvate is not always destined for catabolism. *Its carbon atoms provide the raw material for synthesizing a variety of molecules,* including, in the liver, more glucose (discussed in the following section). Fatty acids, the precursors of triacylglycerols and many membrane lipids, can be synthesized from the two-carbon units of acetyl-CoA derived from pyruvate. This is how fat is produced from excess carbohydrate.

Pyruvate is also the precursor of oxaloacetate, a four-carbon molecule that is an intermediate in the synthesis of several amino acids. It is also one of the intermediates of the citric acid cycle. Oxaloacetate is synthesized by the action of pyruvate carboxylase:

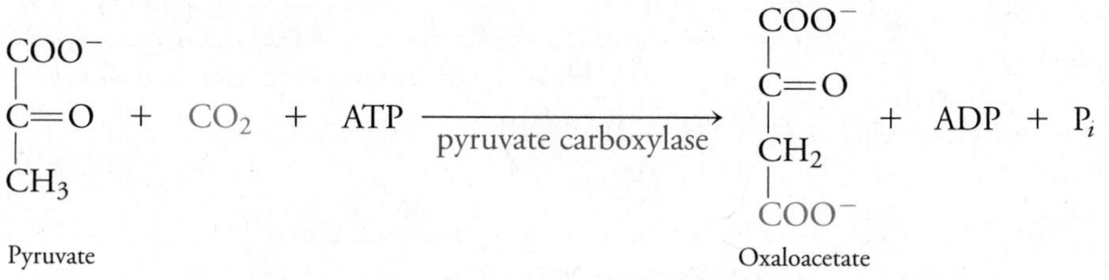

The pyruvate carboxylase reaction is interesting because of its unusual chemistry. The enzyme has a biotin prosthetic group that acts as a carrier of CO_2. Biotin is considered a vitamin, but a deficiency is rare because it is present in many foods and is synthesized by intestinal bacteria. The biotin group is covalently linked to an enzyme Lys residue:

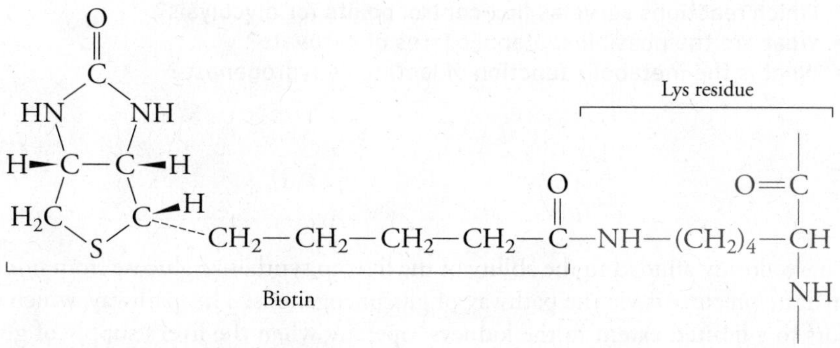

The Lys side chain and its attached biotin group form a 14-Å-long flexible arm that swings between two active sites in the enzyme. In one active site, a CO_2 molecule is first "activated" by its reaction with ATP, then transferred to the biotin. The second active site transfers the carboxyl group to pyruvate to produce oxaloacetate (Fig. 13-9).

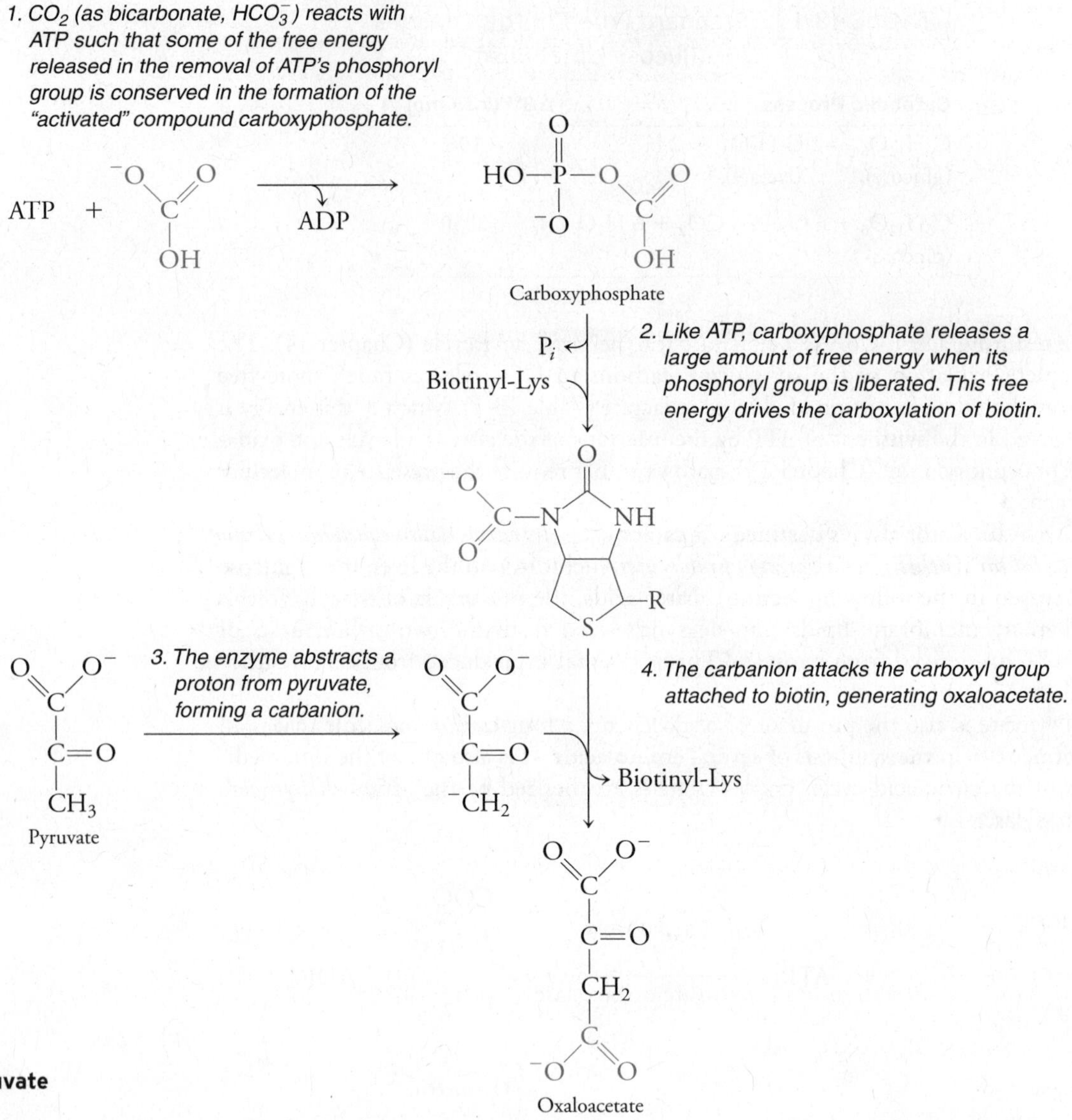

1. CO_2 (as bicarbonate, HCO_3^-) reacts with ATP such that some of the free energy released in the removal of ATP's phosphoryl group is conserved in the formation of the "activated" compound carboxyphosphate.

ATP + [carboxylate structure] $\xrightarrow{\text{ADP}}$ Carboxyphosphate

2. Like ATP, carboxyphosphate releases a large amount of free energy when its phosphoryl group is liberated. This free energy drives the carboxylation of biotin.

P_i

Biotinyl-Lys

3. The enzyme abstracts a proton from pyruvate, forming a carbanion.

Pyruvate

4. The carbanion attacks the carboxyl group attached to biotin, generating oxaloacetate.

Biotinyl-Lys

Oxaloacetate

Figure 13-9 The pyruvate carboxylase reaction.

CONCEPT REVIEW
- Draw the structures of the substrates and products of the ten glycolytic reactions, name the enzyme that catalyzes each step, and indicate whether ATP or NADH is involved.
- Which glycolytic reactions consume ATP? Which generate ATP?
- What is the net yield of ATP and NADH per glucose molecule?
- Which reactions serve as flux-control points for glycolysis?
- What are the possible metabolic fates of pyruvate?
- What is the metabolic function of lactate dehydrogenase?

13-2 Gluconeogenesis

KEY CONCEPTS
- Pyruvate is converted to glucose by glycolytic enzymes operating in reverse and by enzymes that bypass the irreversible steps of glycolysis.
- Gluconeogenic flux is regulated primarily by fructose-2,6-bisphosphate.

We have already alluded to the ability of the liver to synthesize glucose from noncarbohydrate precursors via the pathway of gluconeogenesis. This pathway, which also occurs to a limited extent in the kidneys, operates when the liver's supply of glycogen is exhausted. Certain tissues, such as the central nervous system and red blood cells, which burn glucose as their primary metabolic fuel, rely on the liver to supply them with newly synthesized glucose.

Gluconeogenesis is considered to be the reversal of glycolysis, that is, the conversion of two molecules of pyruvate to one molecule of glucose. Although some of the steps of gluconeogenesis are catalyzed by glycolytic enzymes operating in reverse, the gluconeogenic pathway contains several unique enzymes that bypass the three irreversible steps of glycolysis—the steps catalyzed by pyruvate kinase, phosphofructokinase, and hexokinase (Fig. 13-10). This principle applies to all pairs of opposing metabolic pathways: *The pathways may share some near-equilibrium reactions but cannot use the same enzymes to catalyze thermodynamically favorable irreversible reactions.* The three irreversible reactions of glycolysis are clearly visible in the "waterfall" diagram (see Fig. 13-7).

Four gluconeogenic enzymes plus some glycolytic enzymes convert pyruvate to glucose

Pyruvate cannot be converted directly back to phosphoenolpyruvate because pyruvate kinase catalyzes an irreversible reaction (Reaction 10 of glycolysis). To get around this thermodynamic barrier, pyruvate is carboxylated by pyruvate carboxylase to yield

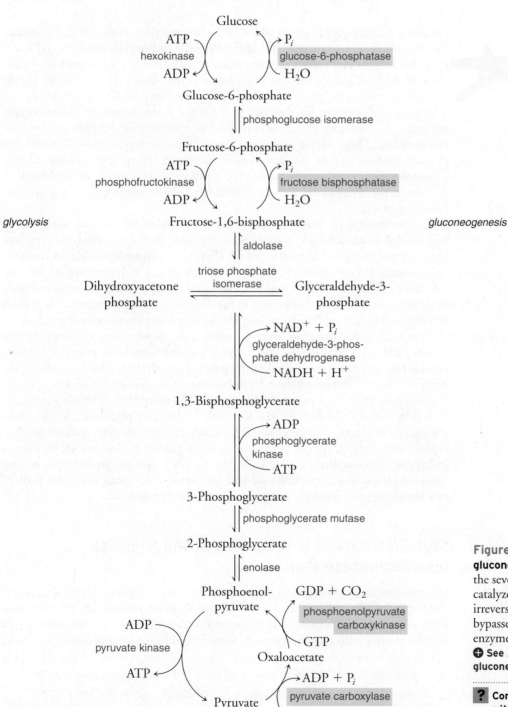

Figure 13-10 The reactions of gluconeogenesis. The pathway uses the seven glycolytic enzymes that catalyze reversible reactions. The three irreversible reactions of glycolysis are bypassed in gluconeogenesis by the four enzymes that are highlighted in blue.
⊕ See Animated Figure. Comparison of gluconeogenesis and glycolysis.

? Compare the ATP yield of glycolysis with the ATP consumption of gluconeogenesis.

the four-carbon compound oxaloacetate (the same reaction shown in Fig. 13-9). Next, phosphoenolpyruvate carboxykinase catalyzes the decarboxylation of oxalo-acetate to form phosphoenolpyruvate:

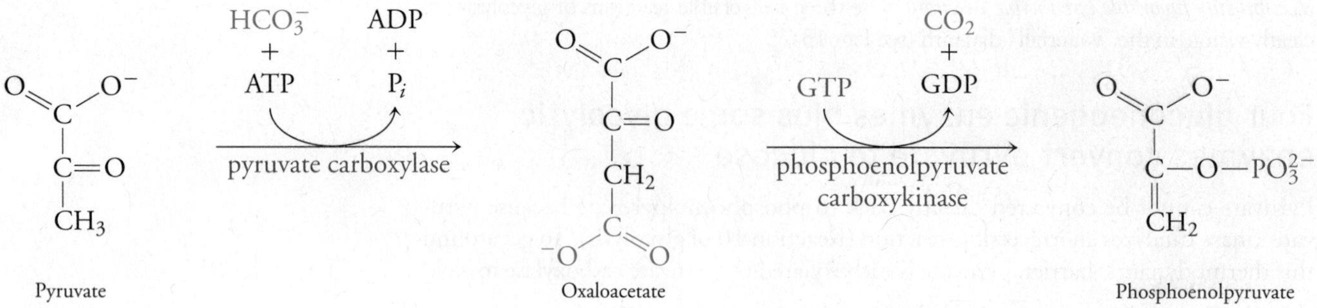

$$\text{Pyruvate} \xrightarrow[\text{pyruvate carboxylase}]{\text{HCO}_3^- + \text{ATP} \rightarrow \text{ADP} + \text{P}_i} \text{Oxaloacetate} \xrightarrow[\substack{\text{phosphoenolpyruvate} \\ \text{carboxykinase}}]{\text{GTP} \rightarrow \text{CO}_2 + \text{GDP}} \text{Phosphoenolpyruvate}$$

Note that the carboxylate group added in the first reaction is released in the second. The two reactions are energetically costly: Pyruvate carboxylase consumes ATP, and phosphoenolpyruvate carboxykinase consumes GTP (which is energetically equivalent to ATP). Cleavage of two phosphoanhydride bonds is required to supply enough free energy to "undo" the highly exergonic pyruvate kinase reaction.

Amino acids (except for Leu and Lys) are the main sources of gluconeogenic precursors because they can all be converted to oxaloacetate and then to phosphoenolpyruvate. Thus, during starvation, proteins can be broken down and used to produce glucose to fuel the central nervous system. Fatty acids cannot serve as gluconeogenic precursors because they cannot be converted to oxaloacetate. (However, the three-carbon glycerol backbone of triacylglycerols is a gluconeogenic precursor.)

Two molecules of phosphoenolpyruvate are converted to one molecule of fructose-1,6-bisphosphate in a series of six reactions that are all catalyzed by glycolytic enzymes (steps 4–9 in reverse order). These reactions are reversible because they are near equilibrium ($\Delta G \approx 0$), and the direction of flux is determined by the concentrations of substrates and products. Note that the phosphoglycerate kinase reaction consumes ATP when it operates in the direction of gluconeogenesis. NADH is also required to reverse the glyceraldehyde-3-phosphate dehydrogenase reaction.

The final three reactions of gluconeogenesis require two enzymes unique to this pathway. The first step undoes the phosphofructokinase reaction, the irreversible reaction that is the major control point of glycolysis. In gluconeogenesis, the enzyme fructose bisphosphatase hydrolyzes the C1 phosphate of fructose-1,6-bisphosphate to yield fructose-6-phosphate. This reaction is thermodynamically favorable, with a ΔG value of $-8.6 \text{ kJ} \cdot \text{mol}^{-1}$. Next, the glycolytic enzyme phosphoglucose isomerase catalyzes the reverse of step 2 of glycolysis to produce glucose-6-phosphate. Finally, the gluconeogenic enzyme glucose-6-phosphatase catalyzes a hydrolytic reaction that yields glucose and P_i. Note that the hydrolytic reactions catalyzed by fructose bisphosphatase and glucose-6-phosphatase undo the work of two kinases in glycolysis (phosphofructokinase and hexokinase).

Gluconeogenesis is regulated at the fructose bisphosphatase step

Gluconeogenesis is energetically expensive. Producing 1 glucose from 2 pyruvate consumes 6 ATP, 2 each at the steps catalyzed by pyruvate carboxylase, phosphoenolpyruvate carboxykinase, and phosphoglycerate kinase. If glycolysis occurred simultaneously with gluconeogenesis, there would be a net consumption of ATP:

glycolysis	glucose + 2 ADP + 2 P$_i$ → 2 pyruvate + 2 ATP
gluconeogenesis	2 pyruvate + 6 ATP → glucose + 6 ADP + 6 P$_i$
net	4 ATP → 4 ADP + 4 P$_i$

To avoid this waste of metabolic free energy, gluconeogenic cells (mainly liver cells) carefully regulate the opposing pathways of glycolysis and gluconeogenesis according to the cell's energy needs. *The major regulatory point is centered on the interconversion of fructose-6-phosphate and fructose-1,6-bisphosphate.* We have already seen that fructose-2,6-bisphosphate is a potent allosteric activator of phosphofructokinase, which catalyzes step 3 of glycolysis. Not surprisingly, fructose-2,6-bisphosphate is a potent *inhibitor* of fructose bisphosphatase, which catalyzes the opposing gluconeogenic reaction.

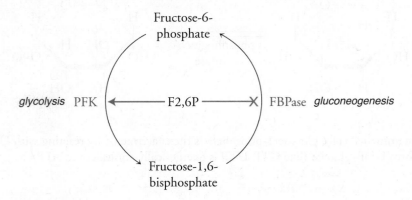

This mode of allosteric regulation is efficient because *a single compound can control flux through two opposing pathways in a reciprocal fashion.* Thus, when the concentration of fructose-2,6-bisphosphate is high, glycolysis is stimulated and gluconeogenesis is inhibited, and vice versa.

Many cells that do not carry out gluconeogenesis do contain the gluconeogenic enzyme fructose bisphosphatase. What is the reason for this? When both fructose bisphosphatase (FBPase) and phosphofructokinase (PFK) are active, the net result is the hydrolysis of ATP:

PFK	fructose-6-phosphate + ATP → fructose-1,6-bisphosphate + ADP
FBPase	fructose-1,6-bisphosphate + H_2O → fructose-6-phosphate + P_i
net	ATP + H_2O → ADP + P_i

This combination of metabolic reactions is called a **futile cycle** since it seems to have no useful result. However, Eric Newsholme realized that such futile cycles could actually provide a means for fine-tuning the output of a metabolic pathway. For example, flux through the phosphofructokinase step of glycolysis is diminished by the activity of fructose bisphosphatase. An allosteric compound such as fructose-2,6-bisphosphate modulates the activity of both enzymes so that as the activity of one enzyme increases, the activity of the other one decreases. This dual regulatory effect results in a greater possible range of net flux than if the regulator merely activated or inhibited a single enzyme.

> **CONCEPT REVIEW**
> - Which reactions of gluconeogenesis are catalyzed by glycolytic enzymes?
> - Why are some enzymes unique to gluconeogenesis?
> - What is a futile cycle and what is its purpose?

13-3 Glycogen Synthesis and Degradation

Dietary glucose and the glucose produced by gluconeogenesis are stored in the liver and other tissues as glycogen. Later, glucose units can be removed from the glycogen polymer by phosphorolysis (see Section 12-1). Because glycogen degradation is thermodynamically spontaneous, glycogen synthesis requires the input of free energy. The two opposing pathways use different sets of enzymes so that each process can be thermodynamically favorable under cellular conditions.

KEY CONCEPTS
- The substrate for glycogen synthase is UDP-glucose, whose production costs the free energy of one phosphoanhydride bond.
- Glycogen is phosphorolyzed to produce glucose that can exit the cell or be catabolized by glycolysis.

Glycogen synthesis consumes the free energy of UTP

The monosaccharide unit that is incorporated into glycogen is glucose-1-phosphate, which is produced from glucose-6-phosphate (the penultimate product of gluconeogenesis) by the action of the enzyme phosphoglucomutase:

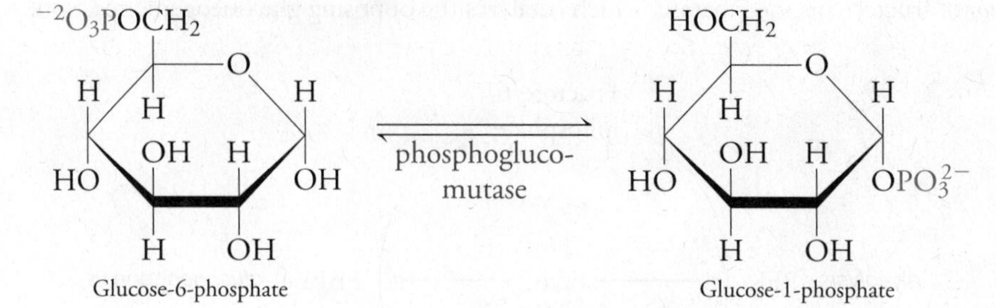

Glucose-6-phosphate   phosphogluco-mutase   Glucose-1-phosphate

In mammalian cells, glucose-1-phosphate is then "activated" by reacting with UTP to form UDP–glucose (like GTP, UTP is energetically equivalent to ATP).

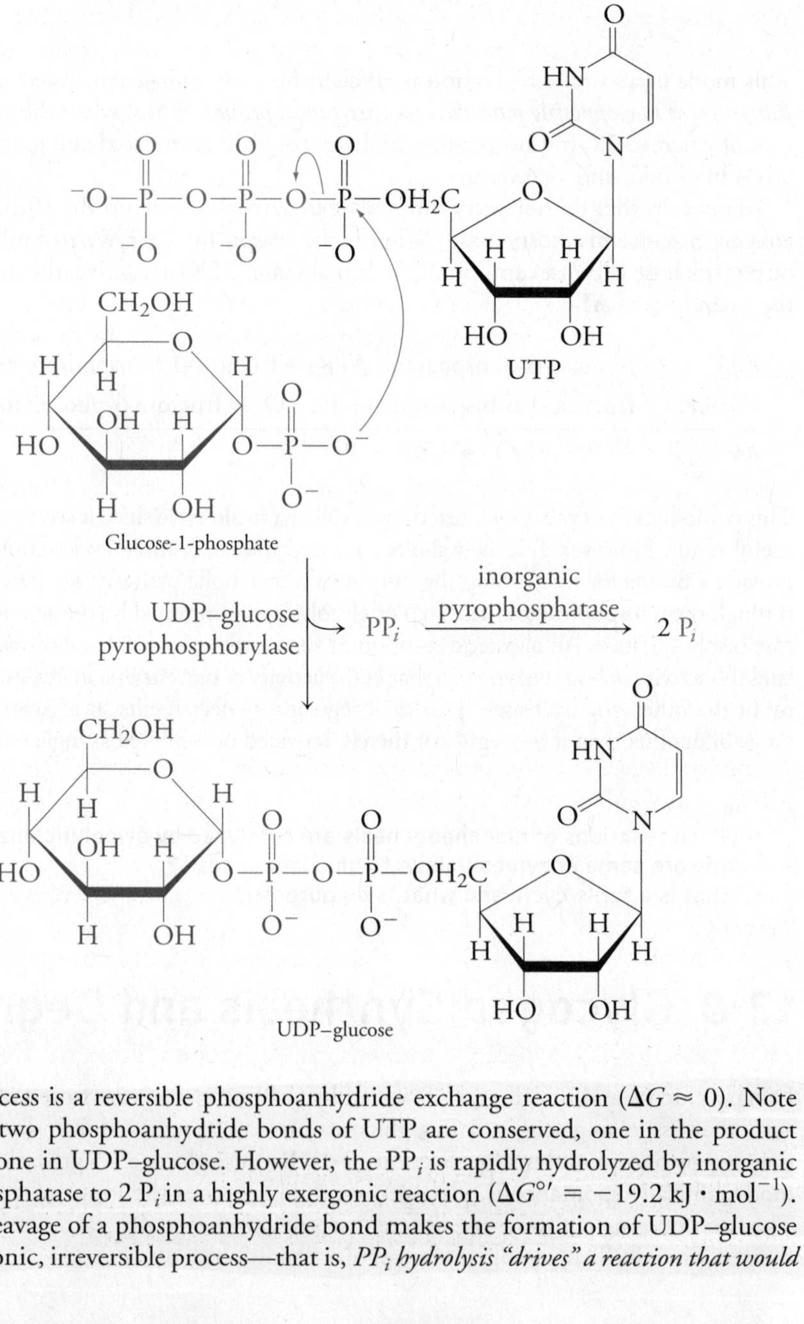

UTP

Glucose-1-phosphate

UDP–glucose pyrophosphorylase $\longrightarrow$ PP_i $\xrightarrow{\text{inorganic pyrophosphatase}}$ 2 P_i

UDP–glucose

This process is a reversible phosphoanhydride exchange reaction ($\Delta G \approx 0$). Note that the two phosphoanhydride bonds of UTP are conserved, one in the product PP_i and one in UDP–glucose. However, the PP_i is rapidly hydrolyzed by inorganic pyrophosphatase to 2 P_i in a highly exergonic reaction ($\Delta G^{\circ\prime} = -19.2\ \text{kJ} \cdot \text{mol}^{-1}$). Thus, cleavage of a phosphoanhydride bond makes the formation of UDP–glucose an exergonic, irreversible process—that is, *PP_i hydrolysis "drives" a reaction that would* *otherwise be near equilibrium.* The hydrolysis of PP_i by inorganic pyrophosphatase is a common strategy in biosynthetic reactions; we will see this reaction again in the synthesis of other polymers, namely DNA, RNA, and polypeptides.

Finally, glycogen synthase transfers the glucose unit to the C4 OH group at the end of one of glycogen's branches to extend the linear polymer of $\alpha(1 \rightarrow 4)$-linked residues (*right*).

A separate enzyme, called a transglycosylase or branching enzyme, cleaves off a seven-residue segment and reattaches it to a glucose C6 OH group to create an $\alpha(1 \rightarrow 6)$ branch point.

The steps of glycogen synthesis can be summarized as follows:

UDP–glucose pyrophosphorylase	glucose-1-phosphate + UTP $\rightleftharpoons$ UDP–glucose + PP_i
pyrophosphatase	$PP_i + H_2O \rightarrow 2\,P_i$
glycogen synthase	UDP–glucose + glycogen (*n* residues) $\rightarrow$ glycogen (*n* +1 residues) + UDP
net	glucose-1-phosphate + glycogen + UTP + $H_2O \rightarrow$ glycogen + UDP + $2\,P_i$

The energetic price for adding one glucose unit to glycogen is the cleavage of one phosphoanhydride bond of UTP. Nucleotides are also required for the synthesis of other saccharides. For example, lactose is synthesized from glucose and UDP–galactose. In plants, starch is synthesized using ADP–glucose, and cellulose is synthesized using CDP–glucose as starting materials.

Glycogen phosphorylase catalyzes glycogenolysis

Glycogen breakdown follows a different set of steps than glycogen synthesis. *In* **glycogenolysis,** *glycogen is phosphorolyzed, not hydrolyzed, to yield glucose-1-phosphate.* However, a debranching enzyme can remove $\alpha(1 \rightarrow 6)$-linked residues by hydrolysis. In the liver, phosphoglucomutase converts glucose-1-phosphate to glucose-6-phosphate, which is then hydrolyzed by glucose-6-phosphatase to release free glucose.

This glucose leaves the cell and enters the bloodstream. Only gluconeogenic tissues such as the liver can make glucose available to the body at large. Other tissues that store glycogen, such as muscle, lack glucose-6-phosphatase and so break down glycogen only for their own needs. In these tissues, the glucose-1-phosphate liberated by phosphorolysis of glycogen is converted to glucose-6-phosphate, which then enters glycolysis at the phosphoglucose isomerase reaction (step 2). The hexokinase reaction (step 1) is skipped, thereby sparing the consumption of ATP. Consequently, *glycolysis using glycogen-derived glucose has a higher net yield of ATP than glycolysis using glucose supplied by the bloodstream.*

Because the mobilization of glucose must be tailored to meet the energy demands of a particular tissue or the entire body, the activity of glycogen phosphorylase is carefully regulated by a variety of mechanisms linked to hormonal signaling. Likewise, the activity of glycogen synthase is subject to hormonal control. In Chapter 19 we will examine some of the mechanisms for regulating different aspects of fuel metabolism, including glycogen synthesis and degradation. Box 13-C discusses the metabolic disorders known as **glycogen storage diseases.**

CONCEPT REVIEW
- What is the role of UTP in glycogen synthesis?
- What is the advantage of breaking down glycogen by phosphorolysis?
- Why do only some tissues contain glucose-6-phosphatase?

BOX 13-C ❖ **CLINICAL CONNECTION**

Glycogen Storage Diseases

The glycogen storage diseases are a set of inherited disorders of glycogen metabolism, not all of which result in glycogen accumulation, as the name might suggest. The symptoms of the glycogen storage diseases vary, depending on whether the affected tissue is liver or muscle or both. In general, the disorders that affect the liver cause hypoglycemia (too little glucose in the blood) and an enlarged liver. Glycogen storage diseases that affect primarily muscle are characterized by muscle weakness and cramps. The incidence of glycogen storage diseases is estimated to be as high as 1 in 20,000 births, although some disorders are not apparent until adulthood. Twelve types of glycogen storage diseases have been described, and the defect in each is listed in the table on the next page. The following discussion focuses on the most common of these conditions.

A defect of glucose-6-phosphatase (type I glycogen storage disease, also called von Gierke's disease) affects both gluconeogenesis and glycogenolysis, since glucose-6-phosphatase catalyzes the final step of gluconeogenesis and makes free glucose available from glycogenolysis. The enlarged liver and hypoglycemia can lead to a host of other symptoms, including irritability, lethargy, and, in severe cases, death. A related defect is the deficiency of the transport protein that imports glucose-6-phosphate into the endoplasmic reticulum, where the phosphatase is located.

Type III glycogen storage disease, or Cori's disease, results from a deficiency of the glycogen debranching enzyme. This condition accounts for about one-quarter of all cases of glycogen storage disease and usually affects both liver and muscle. The symptoms include muscle weakness and liver enlargement due to the accumulation of glycogen that cannot be efficiently broken down. The symptoms of type III glycogen storage disease often improve with age and disappear by early adulthood.

The most common type of glycogen storage disease is type IX. In this disorder, the kinase that activates glycogen phosphorylase is defective. Symptoms range from severe to mild and may fade with time. The complexity of this disease reflects the fact that the phosphorylase kinase consists of four subunits, with isoforms that are differentially expressed in the liver and other tissues. Genes for the α chain (the kinase catalytic subunit) are located on the X chromosome, so one form of disease (type VIII glycogen storage disease) is inherited in a sex-linked manner (more males than females are affected). Genes for the β, γ, and δ subunits of the kinase, which have regulatory functions, are on other chromosomes, so defects in these genes affect males and females equally.

Type II glycogen storage disease, the deficiency of a muscle glucosidase, is not common, but it causes death within the first year. The missing enzyme is a lysosomal hydrolase that does not participate in the main pathways of glycogen degradation but,

like many lysosomal enzymes, apparently plays a role in recycling cellular materials. Glycogen accumulates in the lysosomes, eventually killing the cell.

In the past, glycogen storage diseases were diagnosed on the basis of symptoms, blood tests, and painful biopsies of liver or muscle to assess its glycogen content. Current diagnostic methods are centered on analyzing the relevant genes for mutations, a noninvasive approach. Treatment of glycogen storage diseases typically includes a regimen of frequent, small, carbohydrate-rich meals to alleviate hypoglycemia. However, because dietary therapy does not completely eliminate the symptoms of some glycogen storage diseases, and because the metabolic abnormalities, such as chronic hypoglycemia and liver damage, can severely impair physical growth as well as cognitive development, liver transplant has proved to be an effective treatment. At lease one disorder, the type II glycogen storage disease, has been treated with infusions of the missing enzyme. The glycogen storage diseases are single-gene defects, which makes them attractive targets for gene therapy (see Section 3-4).

Type	Enzyme Deficiency
I	Glucose-6-phosphatase
II	α-1,4-Glucosidase
III	Amylo-1,6-glucosidase (debranching enzyme)
IV	Amylo-(1,4 → 1,6)-transglycosylase (branching enzyme)
V	Muscle glycogen phosphorylase
VI	Liver glycogen phosphorylase
VII	Phosphofructokinase
VIII, IX, X	Phosphorylase kinase
XI	GLUT2 transporter
0	Glycogen synthase

● Questions:

1. Most patients with moderate to severe glycogen storage disease experience some growth retardation. What feature of the glycogen storage diseases would account for this?

2. Patients with type I glycogen storage disease sometimes have enlarged kidneys. Explain.

3. Would frequent small feedings of cornstarch relieve the symptoms of type 0 glycogen storage disease?

4. Would a liver transplant cure all the symptoms of type III glycogen storage disease? Explain.

5. The symptoms of type XI glycogen storage disease include hypoglycemia and hypergalactosemia. What does this tell you about the function of the GLUT2 glucose transport protein?

13-4 The Pentose Phosphate Pathway

We have already seen that glucose catabolism can lead to pyruvate, which can be further oxidized to generate more ATP or used to synthesize amino acids and fatty acids. Glucose is also a precursor of the ribose groups used for nucleotide synthesis. The **pentose phosphate pathway,** which converts glucose-6-phosphate to ribose-5-phosphate, is an oxidative pathway that occurs in all cells. But unlike glycolysis, the pentose phosphate pathway generates NADPH rather than NADH. The two cofactors are not interchangeable and are easily distinguished by degradative enzymes (which generally use NAD^+) and biosynthetic enzymes (which generally use $NADP^+$). The pentose phosphate pathway is by no means a minor feature of glucose metabolism. As much as 30% of glucose in the liver may be catabolized by the pentose phosphate pathway. This pathway can be divided into two phases: a series of oxidative reactions followed by a series of reversible interconversion reactions.

KEY CONCEPTS
- The pentose phosphate pathway is an oxidative pathway for producing NADPH and converting glucose to ribose.
- The reversible reactions of the pathway allow the interconversion of ribose and intermediates of glycolysis and gluconeogenesis.

The oxidative reactions of the pentose phosphate pathway produce NADPH

The starting point of the pentose phosphate pathway is glucose-6-phosphate, which can be derived from free glucose, from the glucose-1-phosphate produced by glycogen phosphorolysis, or from gluconeogenesis. In the first step of the pathway, glucose-6-phosphate dehydrogenase catalyzes the metabolically irreversible transfer of a hydride ion from glucose-6-phosphate to $NADP^+$, forming a lactone and NADPH:

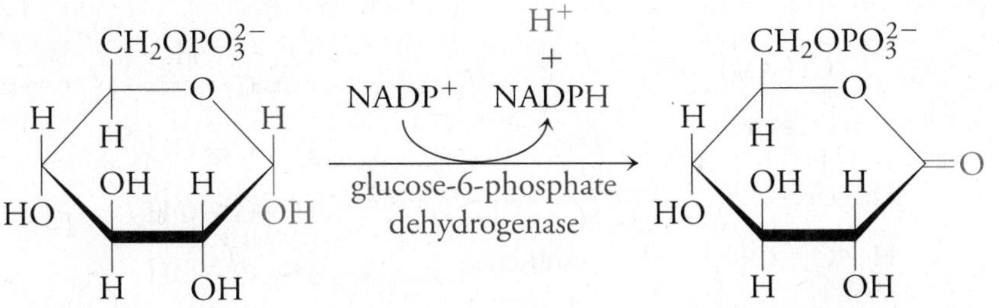

Glucose-6-phosphate → 6-Phosphoglucono-δ-lactone

A deficiency of glucose-6-phosphate dehydrogenase is the most common human enzyme deficiency. This defect, which decreases the cellular production of NADPH, interferes with the normal function of certain oxidation–reduction processes and makes the cells more susceptible to oxidative damage. However, individuals with glucose-6-phosphate dehydrogenase deficiency are also more resistant to malaria. Thus, the gene for the defective enzyme (like the gene for sickle cell hemoglobin described in Box 5-C) persists because it confers a selective advantage.

The lactone intermediate is hydrolyzed to 6-phosphogluconate by the action of 6-phosphogluconolactonase, although this reaction can also occur in the absence of the enzyme:

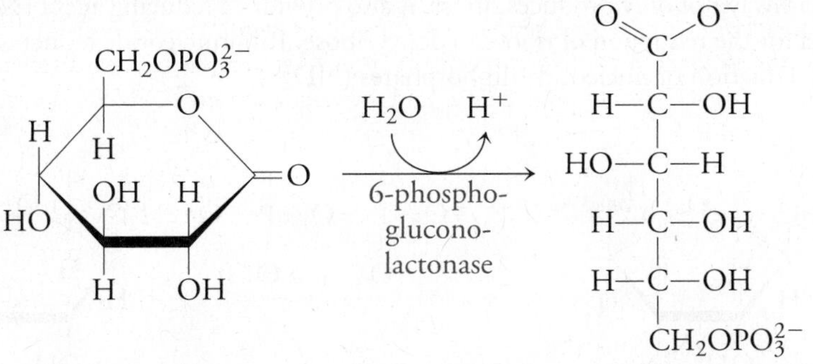

6-Phosphoglucono-δ-lactone → 6-Phosphogluconate

In the third step of the pentose phosphate pathway, 6-phosphogluconate is oxidatively decarboxylated in a reaction that converts the six-carbon sugar to a five-carbon sugar and reduces a second $NADP^+$ to NADPH:

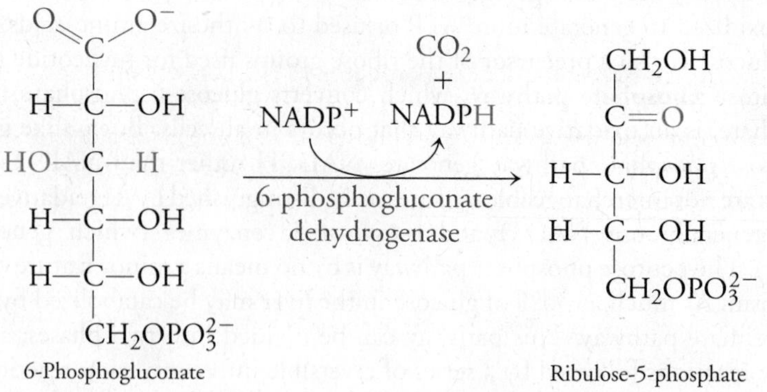

6-Phosphogluconate → Ribulose-5-phosphate

The two molecules of NADPH produced for each glucose molecule that enters the pathway are used primarily for biosynthetic reactions, such as fatty acid synthesis and the synthesis of deoxynucleotides.

Isomerization and interconversion reactions generate a variety of monosaccharides

The ribulose-5-phosphate product of the oxidative phase of the pentose phosphate pathway can isomerize to ribose-5-phosphate:

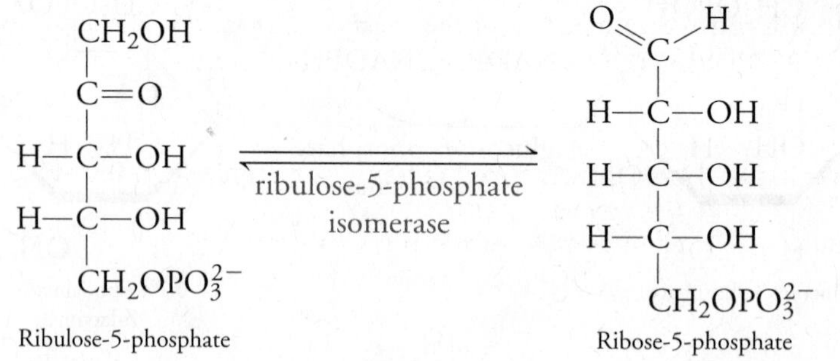

Ribulose-5-phosphate ⇌ Ribose-5-phosphate

Ribose-5-phosphate is the precursor of the ribose unit of nucleotides. In many cells, this marks the end of the pentose phosphate pathway, which has the net equation

glucose-6-phosphate + 2 $NADP^+$ + H_2O →
$\qquad$ ribose-5-phosphate + 2 NADPH + CO_2 + 2 H^+

Not surprisingly, the activity of the pentose phosphate pathway is high in rapidly dividing cells that must synthesize large amounts of DNA. In fact, the pentose phosphate pathway not only produces ribose, it also provides a reducing agent (NADPH) required for the reduction of ribose to deoxyribose. Ribonucleotide reductase carries out the reduction of nucleotide diphosphates (NDPs):

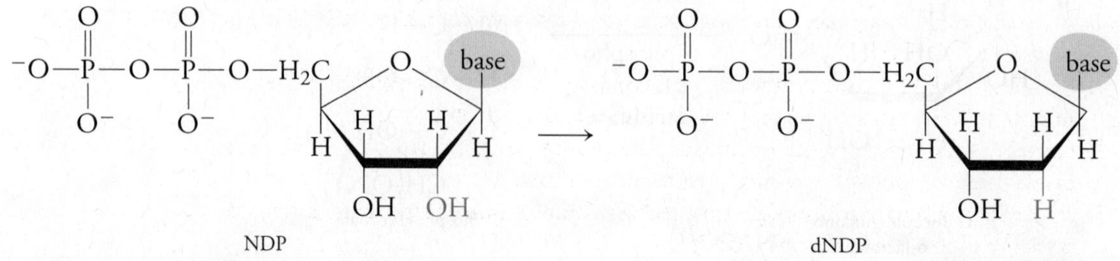

NDP → dNDP

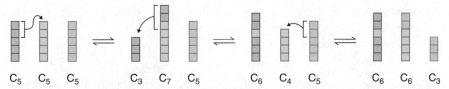

$C_5 \quad C_5 \quad C_5 \qquad C_3 \quad C_7 \quad C_5 \qquad C_6 \quad C_4 \quad C_5 \qquad C_6 \quad C_6 \quad C_3$

Figure 13-11 Rearrangements of the products of the pentose phosphate pathway.
Three of the five-carbon products of the oxidative phase of the pentose phosphate pathway are converted to two fructose-6-phosphate and one glyceraldehyde-3-phosphate by reversible reactions involving the transfer of two- and three-carbon units. Each square represents a carbon atom in a monosaccharide. This pathway also allows ribose carbons to be used in glycolysis and gluconeogenesis.

The enzyme, which is oxidized in the process, is restored to its original state by a series of reactions in which NADPH is reduced.

In some cells, however, the need for NADPH for other biosynthetic reactions is greater than the need for ribose-5-phosphate. In this case, *the excess carbons of the pentose are recycled into intermediates of the glycolytic pathway so that they can be degraded to pyruvate or used in gluconeogenesis,* depending on the cell type and its metabolic needs.

A set of reversible reactions transform five-carbon ribulose units into six-carbon units (fructose-6-phosphate) and three-carbon units (glyceraldehyde-3-phosphate). These transformations are accomplished mainly by the enzymes transketolase and transaldolase, which transfer two- and three-carbon units among various intermediates to produce a set of sugars containing three, four, five, six, or seven carbons (the reaction catalyzed by transketolase was introduced in Section 7-2). **Figure 13-11** is a schematic view of this process. Because all these interconversions are reversible, *glycolytic intermediates can also be siphoned from glycolysis or gluconeogenesis to synthesize ribose-5-phosphate.* Thus, the cell can use some or all of the steps of the pentose phosphate pathway to generate NADPH, to produce ribose, and to interconvert other monosaccharides.

CONCEPT REVIEW
- What are the main products of the pentose phosphate pathway and how does the cell use them?
- How does the cell catabolize excess ribose groups?

A summary of glucose metabolism

The central position of glucose metabolism in all cells warrants its close study. Indeed, the enzymes of glycogen metabolism, glycolysis, gluconeogenesis, and the pentose phosphate pathway are among the best-studied proteins. In nearly all cases, detailed knowledge of their molecular structures has provided insight into their catalytic mechanisms and mode of regulation.

Although our coverage of glucose metabolism is far from exhaustive, this chapter describes quite a few enzymes and reactions, which are compiled in **Figure 13-12**. As you examine this diagram, keep in mind the following points, which also apply to the metabolic pathways we will encounter in subsequent chapters:

1. A metabolic pathway is a series of enzyme-catalyzed reactions, so the pathway's substrate is converted to its product in discrete steps.
2. A monomeric compound such as glucose is interconverted with its polymeric form (glycogen), with other monosaccharides (fructose-6-phosphate and ribose-5-phosphate, for example), and with smaller metabolites such as the three-carbon pyruvate.
3. Although anabolic and catabolic pathways may share some steps, their irreversible steps are catalyzed by enzymes unique to each pathway.
4. Certain reactions consume or produce free energy in the form of ATP. In most cases, these are phosphoryl-group transfer reactions.
5. Some steps are oxidation–reduction reactions that require or generate a reduced cofactor such as NADH or NADPH.

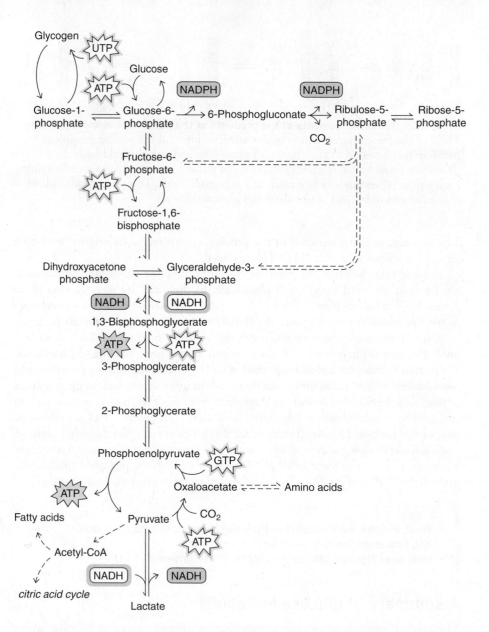

Figure 13-12 Summary of glucose metabolism. This diagram includes the pathways of glycogen synthesis and degradation, glycolysis, gluconeogenesis, and the pentose phosphate pathway. Dotted lines are used where the individual reactions are not shown. Filled gold symbols indicate ATP production; shadowed gold symbols indicate ATP consumption. Filled and shadowed red symbols represent the production and consumption of the reduced cofactors NADH and NADPH. ⊕ **See Animated Figure. Overview of glucose metabolism.**

[SUMMARY]

13-1 Glycolysis

- The pathway of glucose catabolism, or glycolysis, is a series of enzyme-catalyzed steps in which free energy is conserved as ATP or NADH.
- The 10 reactions of glycolysis convert the six-carbon glucose to two molecules of pyruvate and produce two molecules of NADH and two molecules of ATP. The first phase (reactions catalyzed by hexokinase, phosphoglucose isomerase, phosphofructokinase, aldolase, and triose phosphate isomerase) requires the investment of two ATP. The irreversible reaction catalyzed by phosphofructokinase is the rate-determining step and the major control point for glycolysis. The second phase of the pathway (reactions catalyzed by glyceraldehyde-3-phosphate dehydrogenase, phosphoglycerate kinase, phosphoglycerate mutase, enolase, and pyruvate kinase) generates four ATP per glucose.
- Pyruvate may be reduced to lactate or ethanol, further oxidized by the citric acid cycle, or converted to other compounds.

13-2 Gluconeogenesis

- The pathway of gluconeogenesis converts two molecules of pyruvate to one molecule of glucose at a cost of six ATP. The pathway

uses seven glycolytic enzymes, and the activities of pyruvate carboxylase, phosphoenolpyruvate carboxykinase, fructose bisphosphatase, and glucose-6-phosphatase bypass the three irreversible steps of glycolysis.
- A futile cycle involving phosphofructokinase and fructose bisphosphatase helps regulate the flux through glycolysis and gluconeogenesis.

13-3 Glycogen Synthesis and Degradation

- Glucose residues are incorporated into glycogen after first being activated by attachment to UDP.
- Phosphorolysis of glycogen produces phosphorylated glucose that can enter glycolysis. In the liver, this glucose is dephosphorylated and exported.

13-4 The Pentose Phosphate Pathway

- The pentose phosphate catabolic pathway for glucose yields NADPH and ribose groups. The five-carbon sugar intermediates can be converted to glycolytic intermediates.

[GLOSSARY TERMS]

glycolysis

gluconeogenesis

kinase

metabolically irreversible reaction

near-equilibrium reaction

rate-determining reaction

tautomerization

feed-forward activation

fermentation

futile cycle

glycogenolysis

glycogen storage disease

pentose phosphate pathway

[PROBLEMS]

13-1 Glycolysis

1. Which of the 10 reactions of glycolysis are **(a)** phosphorylations, **(b)** isomerizations, **(c)** oxidation–reductions, **(d)** dehydrations, and **(e)** carbon–carbon bond cleavages?

2. Which reactions of glycolysis can be reversed? Which reactions are irreversible? What is the significance of the metabolically irreversible reactions?

3. The $\Delta G°'$ value for the hexokinase reaction is -16.7 kJ $\cdot$ mol^{-1}, while the ΔG value under cellular conditions is similar.
(a) What is the ratio of glucose-6-phosphate to glucose under standard conditions if the ratio of [ATP] to [ADP] is 10:1?
(b) How high would the ratio of glucose-6-phosphate to glucose have to be in order to reverse the hexokinase reaction by mass action?

4. What is the ratio of fructose-6-phosphate to glucose-6-phosphate under **(a)** standard conditions and **(b)** cellular conditions? In which direction does the reaction proceed under cellular conditions?

5. Except during starvation, the brain burns glucose as its sole metabolic fuel and consumes up to 40% of the body's circulating glucose.
(a) Why would hexokinase be the primary rate-determining step of glycolysis in the brain? (In tissues such as muscle, phosphofructokinase rather than hexokinase catalyzes the rate-determining step.)
(b) Brain hexokinase has a K_M for glucose that is 100 times lower than the concentration of circulating glucose (5 mM). What is the advantage of this low K_M?

6. Glucose is frequently administered intravenously (injected directly into the bloodstream) to patients as a food source. A new resident at a hospital where you are doing one of your rotations suggests administering glucose-6-phosphate instead. You recall from biochemistry class that the transformation of glucose to glucose-6-phosphate requires ATP and you consider the possibility that administering glucose-6-phosphate might save the patient energy. Should you use the resident's suggestion?

7. ADP stimulates the activity of phosphofructokinase (PFK), yet it is a product of the reaction, not a reactant. Explain this apparently contradictory regulatory strategy.

8. We can apply the T and R nomenclature used to describe the low- and high-affinity forms of hemoglobin (see Section 5-1) to allosteric enzymes like PFK. Allosteric inhibitors stabilize the T form, which has a low affinity for its substrate, while activators stabilize the high-affinity R form. Do the following allosteric effectors stabilize the T form of PFK or the R form?
(a) ADP (bacteria)
(b) PEP (bacteria)
(c) fructose-2,6-bisphosphate (mammals)

9. PFK isolated from the bacterium *Bacillus stearothermophilus* is a tetramer that binds fructose-6-phosphate with hyperbolic kinetics and a K_M of 23 μM. What happens to the K_M in the presence of phosphoenolpyruvate (PEP; see Figure 7-15)? Use the T $\rightleftharpoons$ R terminology to explain what happens.

10. Refer to Figure 7-16. Why does the conformational change that results when Arg 162 changes places with Glu 161 result in a form of the enzyme that has a low affinity for its substrate?

11. Researchers isolated a yeast mutant that was deficient in the enzyme phosphofructokinase. The mutant yeast was able to grow on glycerol as an energy source, but not glucose. Explain why.

12. Researchers isolated a yeast PFK mutant in which a serine at the fructose-2,6-bisphosphate binding site was replaced with an aspartate residue. The amino acid substitution completely abolished the binding of fructose-2,6-bisphosphate (F26BP) to PFK. There was a dramatic decline in glucose consumption and ethanol production in the mutant compared to control yeast.
(a) Propose a hypothesis that explains why the mutant PFK cannot bind fructose-2,6-bisphosphate.
(b) What does the decline of glucose consumption and ethanol production in the yeast reveal about the role of fructose-2,6-bisphosphate in glycolysis?

13. Explain why iodoacetate was useful for determining the order of intermediates in glycolysis but provided misleading information about an enzyme's active site.

14. Biochemists use transition state analogs to determine the structure of a short-lived intermediate in an enzyme-catalyzed reaction. Because an enzyme binds tightly to the transition state, a compound that resembles the transition state should be a potent competitive inhibitor. Phosphoglycohydroxamate binds 150 times more tightly than dihydroxyacetone phosphate to triose phosphate isomerase. Based on this information, propose a structure for the intermediate of the triose phosphate isomerase reaction.

Phosphoglycohydroxamate

15. What is the ratio of glyceraldehyde-3-phosphate (GAP) to dihydroxyacetone phosphate (DHAP) in cells at 37°C under nonequilibrium conditions? Considering your answer to this question, how do you account for the fact that the conversion of DHAP to GAP occurs readily in cells?

16. Cancer cells have elevated levels of glyceraldehyde-3-phosphate dehydrogenase (GAPDH), which may account for the high rate of glycolysis seen in cancer cells. The compound methylglyoxal has been shown to inhibit GAPDH in cancer cells but not in normal cells. This observation may lead to the development of rapid screening assays for cancer cells and to the development of drugs for treatment of cancerous tumors.

(a) What mechanisms might be responsible for the elevated levels of GAPDH in cancer cells?

(b) Why might methylglyoxal inhibit GAPDH in cancer cells but not in normal cells?

17. Arsenate, AsO_4^{3-}, acts as a phosphate analog and can replace phosphate in the GAPDH reaction. The product of this reaction is 1-arseno-3-phosphoglycerate. It is unstable and spontaneously hydrolyzes to form 3-phosphoglycerate, as shown below. What is the effect of arsenate on cells undergoing glycolysis?

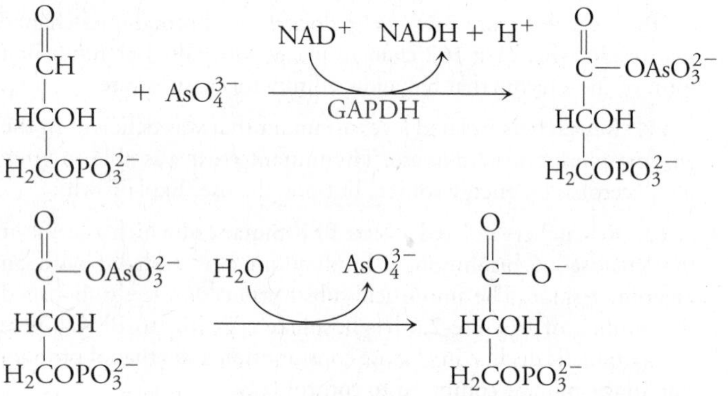

18. In several species of bacteria, activity of GADPH is controlled by the NADH/NAD$^+$ ratio. Does the activity of GAPDH increase or decrease when the NADH/NAD$^+$ ratio increases? Explain. Assume that only the forward direction of the reaction is relevant.

19. Phosphoglycerate kinase in red blood cells is bound to the plasma membrane. This allows the kinase reaction to be coupled to the Na,K-ATPase pump. How does the proximity of the enzyme to the membrane facilitate the action of the pump?

20. Red blood cells synthesize and degrade 2,3-bisphosphoglycerate (2,3-BPG) as a detour from the glycolytic pathway, as shown in the figure.

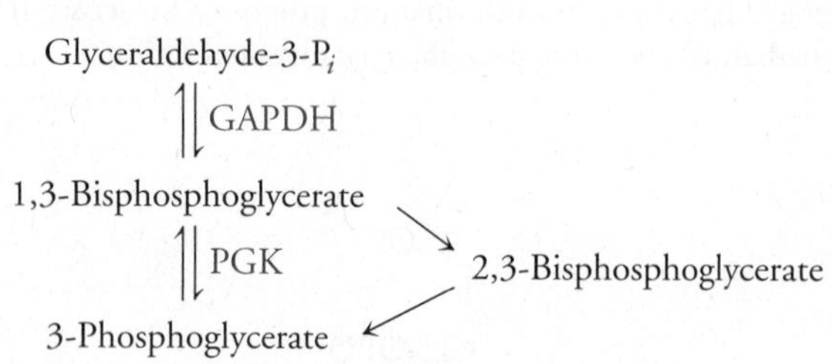

Glyceraldehyde-3-P$_i$

‖ GAPDH

1,3-Bisphosphoglycerate

‖ PGK   2,3-Bisphosphoglycerate

3-Phosphoglycerate

2,3-BPG decreases the oxygen affinity of hemoglobin by binding in the central cavity of the deoxygenated form of hemoglobin. This encourages delivery of oxygen to tissues. A defect in one of the glycolytic enzymes may affect levels of 2,3-BPG. The plot above right shows oxygen-binding curves for normal erythrocytes and for hexokinase- and pyruvate kinase–deficient erythrocytes. Identify which curve corresponds to which enzyme deficiency.

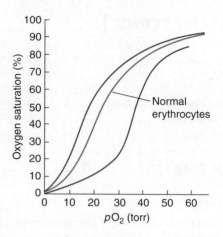

21. Vanadate, VO_4^{3-}, inhibits GAPDH, not by acting as a phosphate analog, but by interacting with essential —SH groups on the enzyme. What happens to cellular levels of phosphate, ATP, and 2,3-bisphosphoglycerate (see Problem 20) when red blood cells are incubated with vanadate?

22. The mechanism of plant phosphoglycerate mutase is different from the mechanism of mammalian phosphoglycerate mutase presented in the text. 3-Phosphoglycerate (3PG) binds to the plant enzyme, transfers its phosphate to the enzyme, and then the enzyme transfers the phosphate group back to the substrate to form 2-phosphoglycerate (2PG). When [^{32}P]-labeled 3PG is added to cultured **(a)** hepatocytes or **(b)** plant cells, what is the fate of the [^{32}P] label?

23. Which intermediates of glycolysis accumulate if fluoride ions are present?

24. Assuming a standard free energy change of 30.5 kJ · mol^{-1} for the synthesis of ATP from ADP and P$_i$, how many molecules of ATP can be theoretically produced by the catabolism of glucose to **(a)** lactate or **(b)** CO$_2$ (see Table 13-1)?

25. What happens to the [ADP]/[ATP] and [NAD$^+$]/[NADH] ratios in red blood cells with a pyruvate kinase deficiency (see Problem 20)?

26. One of the symptoms of a pyruvate kinase deficiency (see Problem 25) is hemolytic anemia, in which red blood cells swell and eventually lyse. Explain why the deficiency of the enzyme brings about this symptom.

27. Organisms such as yeast growing under anaerobic conditions can convert pyruvate to alcohol in a process called fermentation, as described in the text. Instead of being converted to lactate, pyruvate is converted to ethanol in a two-step reaction. Why is the second step of this process essential to the yeast cell?

28. Several studies have shown that aluminum inhibits PFK in liver cells.

(a) Compare the production of pyruvate by perfused livers in control and aluminum-treated rats using fructose as an energy source.

(b) What would the experimental results be if glucose was used instead of fructose?

29. Drinking methanol can cause blindness and death, depending on the dosage. The causative agent is formaldehyde derived from methanol.

(a) Draw the balanced chemical reaction for the conversion of methanol to formaldehyde.

(b) Why would administering whiskey (ethanol) to a person poisoned with methanol be a good antidote?

30. The term *turbo design* has been used to describe pathways such as glycolysis that have one or more ATP-consuming steps followed by one or more ATP-producing steps with a net yield of ATP production for the pathway overall. Mathematical models have shown that "turbo" pathways have the risk of substrate-accelerated death unless there is a "guard at the gate," that is, a mechanism for inhibiting an early step of the pathway. In yeast, hexokinase is inhibited by a complex mechanism mediated by trehalose-6-phosphate synthase (TPSI). Mutant yeast in which TPSI is defective (there is no "guard at the gate") die if grown under conditions of high glucose concentration. Explain why.

31. Recent studies have shown that the halophilic organism *Halococcus saccharolyticus* degrades glucose via the Entner–Doudoroff pathway rather than by the glycolytic pathway presented in this chapter. A modified scheme of the Entner–Doudoroff pathway is shown here.

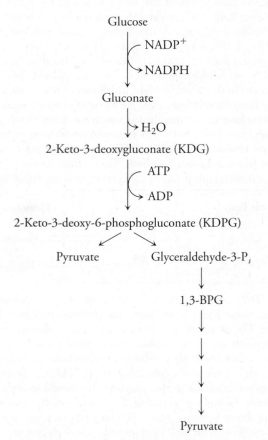

(a) What is the ATP yield per mole of glucose for this pathway?
(b) Describe (in general) what kinds of reactions would need to follow the Entner–Doudoroff pathway in this organism.

32. Trypanosomes living in the bloodstream obtain all their free energy from glycolysis. They take up glucose from the host's blood and excrete pyruvate as a waste product. In this part of their life cycle, trypanosomes do not carry out any oxidative phosphorylation, but they do use another oxygen-dependent pathway, which is absent in mammals, to oxidize NADH.
(a) Why is this other pathway necessary?
(b) Would the pathway be necessary if the trypanosome excreted lactate rather than pyruvate?
(c) Why would this pathway be a good target for antiparasitic drugs?

13-2 Gluconeogenesis

33. Flux through the opposing pathways of glycolysis and gluconeogenesis is controlled in several ways.

(a) Explain how the activation of pyruvate carboxylase by acetyl-CoA affects glucose metabolism.
(b) Pyruvate can undergo a reversible amino-group transfer reaction to yield alanine (see Section 12-2). Alanine is an allosteric effector of pyruvate kinase. Would you expect alanine to stimulate or inhibit pyruvate kinase? Explain.

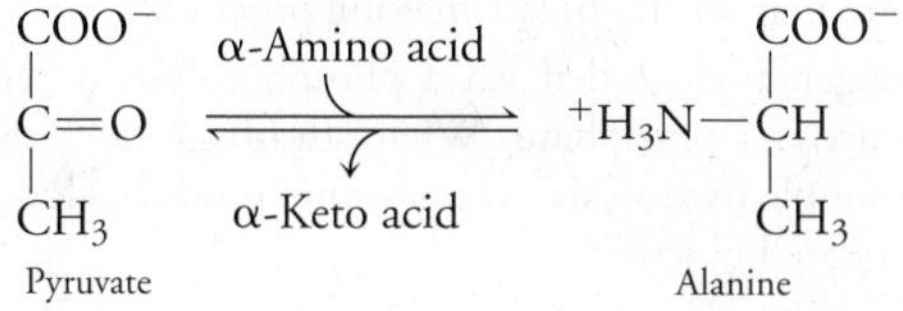

$$
\begin{array}{ccc}
\text{COO}^- & & \text{COO}^- \\
| & \alpha\text{-Amino acid} & | \\
\text{C}=\text{O} & \rightleftharpoons & {}^+\text{H}_3\text{N}-\text{CH} \\
| & \alpha\text{-Keto acid} & | \\
\text{CH}_3 & & \text{CH}_3 \\
\text{Pyruvate} & & \text{Alanine}
\end{array}
$$

34. A liver biopsy of a four-year-old boy indicated that the fructose-1,6-bisphosphatase enzyme activity was 20% of normal. The patient's blood glucose levels were normal at the beginning of a fast but then decreased suddenly. Pyruvate and alanine concentrations were also elevated, as was the glyceraldehyde-3-phosphate/dihydroxyacetone phosphate ratio. Explain the reason for these symptoms.

35. Insulin is one of the major hormones that regulates gluconeogenesis. Insulin acts in part by decreasing the transcription of genes coding for certain gluconeogenic enzymes. For which genes would you expect insulin to suppress transcription?

36. Type 2 diabetes is characterized by insulin resistance, in which insulin is unable to perform its many functions. What symptom would you expect in a type 2 diabetic patient if insulin is unable to perform the function described in Problem 35?

37. The concentration of fructose-2,6-bisphosphate (F26BP) is regulated in the cell by a homodimeric enzyme with two catalytic activities: a kinase that phosphorylates fructose-6-phosphate on the C2 hydroxyl group to form fructose-2,6-bisphosphate and a phosphatase that removes the phosphate group.
(a) Which enzyme activity, the kinase or the phosphatase, would you expect to be active under fasting conditions? Explain.
(b) Which hormone is likely to be responsible for inducing this activity?
(c) Consult Section 10-2 and propose a mechanism for this induction.

38. The "carbon skeletons" of most amino acids can be converted to glucose, a process that may require many enzymatic steps. Which amino acids can enter the gluconeogenic pathway directly after undergoing deamination (a reaction in which the carbon with the amino group becomes a ketone)?

39. Brazilin, a compound found in aqueous extracts of sappan wood, has been used to treat diabetics in Korea. Brazilin increases the activity of the enzyme that produces fructose-2,6-bisphosphate, and the compound also stimulates the activity of pyruvate kinase.
(a) What is the effect of adding brazilin to hepatocytes (liver cells) in culture?
(b) Why would brazilin be an effective treatment for diabetes?

40. Metformin is a drug that decreases the expression of phosphoenolpyruvate carboxykinase. Explain why metformin would be helpful in treating diabetes.

41. Draw a diagram that illustrates how lactate released from the muscle is converted back to glucose in the liver. What is the cost (in ATP) of running this cycle?

42. Draw a diagram that illustrates how alanine (see Problem 33b) released from the muscle is converted back to glucose in the liver. What is the physiological cost if this cycle runs for a prolonged period of time?

13-3 Glycogen Synthesis and Degradation

43. Beer is produced from raw materials such as wheat and barley. Explain why the grains are allowed to sprout, a process in which their starch is broken down to glucose, before fermentation begins.

44. Some bread manufacturers add amylase to bread dough prior to the fermentation process. What role does this enzyme (see Section 12-1) play in the bread-making process?

45. Glycogen is degraded via a phosphorolysis process, which produces glucose-1-phosphate. What advantage does this process have over a simple hydrolysis, which would produce glucose instead of phosphorylated glucose?

46. The equation for the degradation of glycogen is shown below.
(a) What is the ratio of $[P_i]/[G1P]$ under standard conditions?
(b) What is the value of ΔG under cellular conditions when the $[P_i]/[G1P]$ ratio is 50:1?

$$\text{glycogen } (n \text{ residues}) + P_i \xrightleftharpoons{\text{phosphorylase}}$$

$$\text{glycogen } (n - 1 \text{ residues}) + G1P \quad \Delta G^{\circ\prime} = +3.1 \text{ kJ} \cdot \text{mol}^{-1}$$

47. During even mild exertion, individuals with McArdle's disease experience painful muscle cramps due to a genetic defect in glycogen phosphorylase, the enzyme that breaks down glycogen. Yet the muscles in these individuals contain normal amounts of glycogen. What did this observation tell researchers about the pathways for glycogen degradation and glycogen synthesis?

48. Patients with McArdle's disease have normal liver glycogen content and structure. Identify the type of glycogen storage disease as listed in the table in Box 13-C.

49. A patient with McArdle's disease performs ischemic (anaerobic) exercise for as long as he is able to do so. Blood is withdrawn from the patient every few minutes during the exercise period and tested for lactate. The patient's samples are compared with control samples from a patient who does not suffer from a glycogen storage disease. The results are shown in the figure. Why does the lactate concentration increase in the normal patient? Why is there no corresponding increase in the patient's lactate concentration?

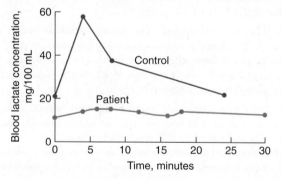

50. Does a patient with McArdle's disease (see Problems 47–49) suffer from hypoglycemia, hyperglycemia, or neither?

51. Patients with von Gierke's disease (type I glycogen storage disease) have a deficiency of glucose-6-phosphatase. One of the most prominent symptoms of the disease is a protruding abdomen due to an enlarged liver. Explain why the liver is enlarged in patients with von Gierke's disease.

52. Does a patient with von Gierke's disease (see Problem 51) suffer from hypoglycemia, hyperglycemia, or neither?

53. The mechanism of the phosphoglucomutase enzyme is similar to that of the plant mutase described in Problem 22 and is shown here. On occasion, the glucose-1,6-bisphosphate dissociates from the enzyme. Why does the dissociation of glucose-1,6-bisphosphate inhibit the enzyme?

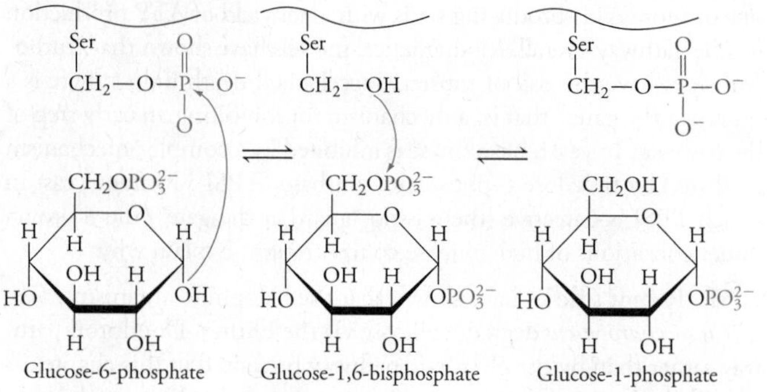

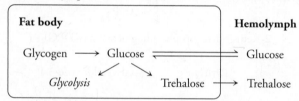

Glucose-6-phosphate   Glucose-1,6-bisphosphate   Glucose-1-phosphate

54. Trehalose is one of the major sugars in the insect hemolymph (the fluid that circulates through the insect's body). It is a disaccharide consisting of two linked glucose residues. In the hemolymph, trehalose serves as a storage form of glucose and also helps protect the insect from desiccation and freezing. Its concentration in the hemolymph must be closely regulated. Trehalose is synthesized in the insect fat body, which plays a role in metabolism analogous to the vertebrate liver. Recent studies on the insect *Manduca sexta* have shown that during starvation, hemolymph glucose concentration decreases, which results in an increase in fat body glycogen phosphorylase activity and a decrease in the concentration of fructose-2,6-bisphosphate. What effect do these changes have on hemolymph trehalose concentration in the fasted insect?

Fat body		**Hemolymph**
Glycogen ⟶ Glucose ⇌		Glucose
Glycolysis ↙ ↘ Trehalose ⟶		Trehalose

55. The glycolytic pathway in the thermophilic archaebacterium *Thermoproteus tenax* differs from the pathway presented in this chapter. The phosphofructokinase reaction in *T. tenax* is reversible and depends on pyrophosphate rather than ATP. In addition, *T. tenax* has two glyceraldehyde-3-phosphate dehydrogenase (GADPH) isozymes. The "phosphorylating GAPDH" is similar to the enzyme described in this chapter. The second isozyme is the irreversible "nonphosphorylating GAPDH," which catalyzes the reaction shown below. *T. tenax* relies on glycogen stores as a source of energy. What is the ATP yield for one mole of glucose oxidized by the pathway that uses the nonphosphorylating GAPDH enzyme?

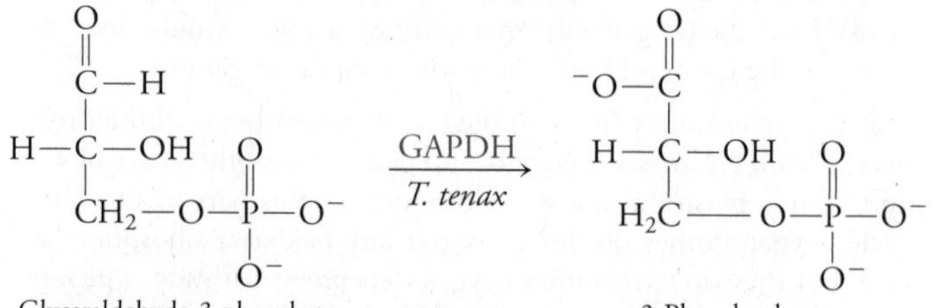

Glyceraldehyde-3-phosphate
+
NAD^+

$\xrightarrow{\text{GAPDH} \atop T. \, tenax}$

3-Phosphoglycerate
+
$NADH + H^+$

56. Individuals with fructose intolerance lack fructose-1-phosphate aldolase, a liver enzyme essential for catabolizing fructose. In the absence of fructose-1-phosphate aldolase, fructose-1-phosphate accumulates in the liver and inhibits glycogen phosphorylase and fructose-1,6-bisphosphatase.

(a) Explain why individuals with fructose intolerance exhibit hypoglycemia (low blood sugar).

(b) Administering glycerol and dihydroxyacetone phosphate does not alleviate the hypoglycemia, but administering galactose does relieve the hypoglycemia. Explain.

13-4 The Pentose Phosphate Pathway

57. Most metabolic pathways include an enzyme-catalyzed reaction that commits a metabolite to continue through the pathway.

(a) Identify the first committed step of the pentose phosphate pathway. Explain your reasoning.

(b) Hexokinase catalyzes an irreversible reaction at the start of glycolysis. Does this step commit glucose to continue through glycolysis?

58. A given metabolite may follow more than one metabolic pathway. List all the possible fates of glucose-6-phosphate in **(a)** a liver cell and **(b)** a muscle cell.

59. Reduced glutathione, a tripeptide containing a Cys residue, is found in red blood cells, where it reduces organic peroxides formed in cellular structures exposed to high concentrations of reactive oxygen.

$$2 \ \gamma\text{-Glu}-\text{Cys}-\text{Gly} \ + \ \text{R}-\text{O}-\text{OH} \longrightarrow$$
$$\quad\quad\quad\quad | $$
$$\quad\quad\quad\quad \text{SH}$$

Reduced glutathione   Organic peroxide

$$\gamma\text{-Glu}-\text{Cys}-\text{Gly}$$
$$\quad\quad | $$
$$\quad\quad \text{S} $$
$$\quad\quad | \quad\quad\quad + \ \text{R}-\text{OH} \ + \ \text{H}_2\text{O}$$
$$\quad\quad \text{S} $$
$$\quad\quad | $$
$$\gamma\text{-Glu}-\text{Cys}-\text{Gly}$$

Oxidized glutathione

Reduced glutathione also plays a role in maintaining normal red blood cell structure and keeping the iron ion of hemoglobin in the +2 oxidation state. Glutathione is regenerated as shown in the following reaction:

$$\gamma\text{-Glu}-\text{Cys}-\text{Gly}$$
$$\quad\quad | $$
$$\quad\quad \text{S} $$
$$\quad\quad | \quad\quad\quad + \ \text{NADPH} \ + \ \text{H}^+ \longrightarrow$$
$$\quad\quad \text{S} $$
$$\quad\quad | $$
$$\gamma\text{-Glu}-\text{Cys}-\text{Gly}$$

$$2 \ \gamma\text{-Glu}-\text{Cys}-\text{Gly} \ + \ \text{NADP}^+$$
$$\quad\quad\quad\quad | $$
$$\quad\quad\quad\quad \text{SH}$$

Use this information to predict the physiological effects of a glucose-6-phosphate dehydrogenase deficiency.

60. Experiments were carried out in cultured cells to determine the relationship between glucose-6-phosphate dehydrogenase (G6PDH) activity and rates of cell growth. Cells were cultured in a medium supplemented with serum, which contains growth factors that stimulate G6PDH activity. Predict how the cellular NADPH/NADP$^+$ ratio would change under the following circumstances:

(a) Serum is withdrawn from the medium.

(b) DHEA, an inhibitor of glucose-6-phosphate dehydrogenase, is added.

(c) The oxidant H_2O_2 is added.

(d) Serum is withdrawn and H_2O_2 is added.

61. Write a mechanism for the nonenzymatic hydrolysis of 6-phosphogluconolactone to 6-phosphogluconate.

62. Enzymes in the soil fungus *Aspergillus nidulans* use NADPH as a coenzyme when converting nitrate to the ammonium ion. When the fungus was cultured in a growth medium containing nitrate, it was discovered that the activities of several enzymes involved in glucose metabolism increased. What enzymes are good candidates for regulation under these conditions? Explain.

63. Several studies have shown that the metabolite glucose-1,6-bisphosphate (G16BP) regulates several pathways of carbohydrate metabolism by inhibiting or activating key enzymes. The effect of G16BP on several important enzymes is summarized in the table below. What pathways are active when G16BP is present? What pathways are inactive? What is the overall effect? Explain.

Enzyme	Effect of G16BP
Hexokinase	Inhibits
Phosphofructokinase (PFK)	Activates
Pyruvate kinase (PK)	Activates
Phosphoglucomutase	Activates
6-Phosphogluconate dehydrogenase	Inhibits

64. Xylulose-5-phosphate acts as an intracellular signaling molecule that activates kinases and phosphatases in liver cells. As a result of this signaling, there is an increase in the activity of the enzyme that produces fructose-2,6-bisphosphate, and the expression of genes for lipid synthesis is increased. What is the net effect of these responses?

[SELECTED READINGS]

Brosnan, J. T., Comments on metabolic needs for glucose and the role of gluconeogenesis, *Eur. J. Clin. Nutr.* **53,** S107–S111 (1999). [A very readable review that discusses possible reasons why carbohydrates are used universally as metabolic fuels, why glucose is stored as glycogen, and why the pentose phosphate pathway is important.]

Greenberg, C. C., Jurczak, M. J., Danos, A. M., and Brady, M. J., Glycogen branches out: new perspectives on the role of glycogen metabolism in the integration of metabolic pathways, *Am. J. Physiol. Endocrinol. Metab.* **291,** E1–E8 (2006). [Describes the roles of glycogen in the liver and muscles.]

Özen, H., Glycogen storage diseases: New perspectives, *World J. Gastroenterology* **13,** 2541–2553 (2007). [Describes the symptoms, biochemistry, and treatment of the major forms of these diseases.]

Roach, P. J., Depaoli-Roach, A. A., Hurley, T. D., and Tagliabracci, V. S., Glycogen and its metabolism: Some new developments and old themes. *Biochem. J.* **441,** 763–787 (2012). [Includes discussions of the hormone-mediated regulation of key enzymes of glycogen synthesis and breakdown.]

14

THE CITRIC ACID CYCLE

▶▶ **WHERE** does exhaled CO_2 come from?

Inflating a toy balloon is one way to capture exhaled breath, which contains CO_2. It is tempting to believe that in air-breathing animals, the oxygen that is inhaled is transformed into carbon dioxide that is exhaled. In fact, the two types of molecules never directly interact inside the body. In this chapter we will see that exhaled CO_2, a waste product of cellular metabolism, is generated mostly by operation of the citric acid cycle. This metabolic pathway converts the carbons of metabolic fuels into CO_2, saving their energy for ATP synthesis.

[Tom Merton/OJO Images/Getty Images, Inc.]

THIS CHAPTER IN CONTEXT

Part 1   Foundations

Part 2   Molecular Structure and Function

Part 3 Metabolism
  14 The Citric Acid Cycle

Part 4   Genetic Information

Do You Remember?

- Enzymes accelerate chemical reactions using acid–base catalysis, covalent catalysis, and metal ion catalysis (Section 6-2).
- Coenzymes such as NAD^+ and ubiquinone collect electrons from compounds that become oxidized (Section 12-2).
- Metabolic pathways in cells are connected and are regulated (Section 12-2).
- Many vitamins, substances that humans cannot synthesize, are components of coenzymes (Section 12-2).
- Pyruvate can be converted to lactate, acetyl-CoA, or oxaloacetate (Section 13-1).

The **citric acid cycle** is a pathway that occupies a central place in the metabolism of most cells. It converts two-carbon groups, in the form of acetyl-CoA, into CO_2 and therefore represents the final stage in the oxidation of metabolic fuels—not just carbohydrates but also fatty acids and amino acids (Fig. 14-1). As the carbons become fully oxidized to CO_2, their energy is conserved and subsequently used to produce ATP. The eight reactions of the citric acid cycle take place in the cytosol of prokaryotes and in the mitochondria of eukaryotes.

Unlike a linear pathway such as glycolysis (see Fig. 13-2) or gluconeogenesis (see Fig. 13-10), the citric acid cycle always returns to its starting position, essentially behaving as a multistep catalyst. However, it is still possible to follow the chemical transformations that occur at each step.

An examination of the citric acid cycle also illustrates an important feature of metabolic pathways in general, namely, that a pathway is less like an element of plumbing and more like a web or network. In other words, the pathway does not function like a simple pipeline, where one substance enters one end and another emerges from the other end. Instead, the pathway's intermediates can participate in many reactions, serving as both precursors and products of a large variety of biological molecules.

Although the carbon atoms that enter the citric acid cycle may be derived from amino acids, fatty acids, or carbohydrates, we will use pyruvate, the end product of glycolysis, as the starting point for our study of the citric acid cycle. We will examine the eight reactions of the citric acid cycle and discuss how this sequence of reactions might have evolved. Finally, we will consider the citric acid cycle as a multifunctional pathway with links to other metabolic processes.

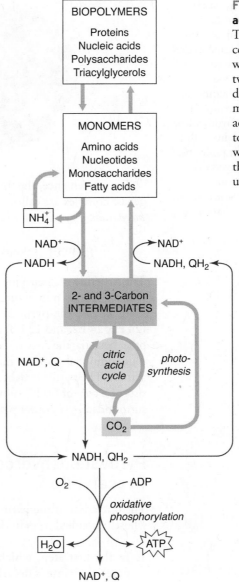

Figure 14-1 The citric acid cycle in context. The citric acid cycle is a central metabolic pathway whose starting material is two-carbon acetyl units derived from amino acids, monosaccharides, and fatty acids. These are oxidized to the waste product CO_2, with the reduction of the cofactors NAD^+ and ubiquinone (Q).

14-1 The Pyruvate Dehydrogenase Reaction

The end product of glycolysis is the three-carbon compound pyruvate. In aerobic organisms, these carbons are ultimately oxidized to 3 CO_2 (although the oxygen atoms come not from molecular oxygen but from water and phosphate). The first molecule of CO_2 is released when pyruvate is decarboxylated to an acetyl unit. The second and third CO_2 molecules are products of the citric acid cycle.

> **KEY CONCEPT**
> • The pyruvate dehydrogenase complex includes three types of enzymes that collectively remove a carboxylate group from pyruvate and produce acetyl-CoA and NADH.

The pyruvate dehydrogenase complex contains multiple copies of three different enzymes

The decarboxylation of pyruvate is catalyzed by the pyruvate dehydrogenase complex. In eukaryotes, this enzyme complex, and the enzymes of the citric acid cycle itself, are located inside the mitochondrion (an organelle surrounded by a double membrane and whose interior is called the **mitochondrial matrix**). Accordingly, pyruvate produced by glycolysis in the cytosol must first be transported into the mitochondria.

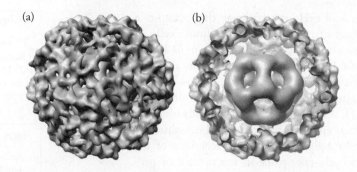

(a) (b)

Figure 14-2 Model of the pyruvate dehydrogenase complex from *B. stearothermophilus*. These images are based on cryoelectron microscope studies of the pyruvate dehydrogenase complex. (a) Surface view. (b) Cutaway view showing the core of 60 E2 subunits. In this model the outer shell contains only E3; in the native pyruvate dehydrogenase complex, the outer shell contains both E1 and E3, which occupy similar positions. The space between the two layers of protein is about 75–90 Å. [Courtesy Jacqueline L. S. Milne and Sriram Subramaniam, National Cancer Institute, National Institutes of Health.]

For convenience, the three kinds of enzymes that make up the pyruvate dehydrogenase complex are called E1, E2, and E3. Together *they catalyze the oxidative decarboxylation of pyruvate and the transfer of the acetyl unit to coenzyme A*:

$$\text{pyruvate} + \text{CoA} + \text{NAD}^+ \rightarrow \text{acetyl-CoA} + \text{CO}_2 + \text{NADH}$$

The structure of coenzyme A, a nucleotide derivative containing the vitamin pantothenate, is shown in Figure 3-3a.

In *E. coli*, the pyruvate dehydrogenase complex contains 60 protein subunits (24 E1, 24 E2, and 12 E3) and has a mass of about 4600 kD. In mammals and some other bacteria, the enzyme complex is even larger, with 42–48 E1, 60 E2, and 6–12 E3 plus additional proteins that hold the complex together and regulate its enzymatic activity. The pyruvate dehydrogenase complex from *Bacillus stearothermophilus* consists of a core of 60 E2 subunits that form a dodecahedron (a 12-sided polyhedron) surrounded by an outer protein shell (**Fig. 14-2**).

Pyruvate dehydrogenase converts pyruvate to acetyl-CoA

The operation of the pyruvate dehydrogenase complex requires several coenzymes, whose functional roles in the five-step reaction are described below.

1. In the first step, which is catalyzed by E1 (also called pyruvate dehydrogenase), pyruvate is decarboxylated. This reaction requires the cofactor thiamine pyrophosphate (TPP; **Fig. 14-3**). TPP attacks the carbonyl carbon of pyruvate, and the departure of CO_2 leaves a hydroxyethyl group attached to TPP. This carbanion is stabilized by the positively charged thiazolium ring group of TPP:

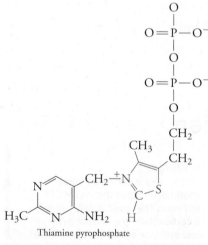

Thiamine pyrophosphate

Figure 14-3 Thiamine pyrophosphate (TPP). This cofactor is the phosphorylated form of thiamine, also known as vitamin B_1 (see Section 12-2). The central thiazolium ring (blue) is the active portion. An acidic proton (red) dissociates, and the resulting carbanion is stabilized by the nearby positively charged nitrogen. TPP is a cofactor for several different decarboxylases.

2. The hydroxyethyl group is then transferred to E2 of the pyruvate dehydrogenase complex. The hydroxyethyl acceptor is a lipoamide prosthetic group (**Fig. 14-4**). The transfer reaction regenerates the TPP cofactor of E1 and oxidizes the hydroxyethyl group to an acetyl group:

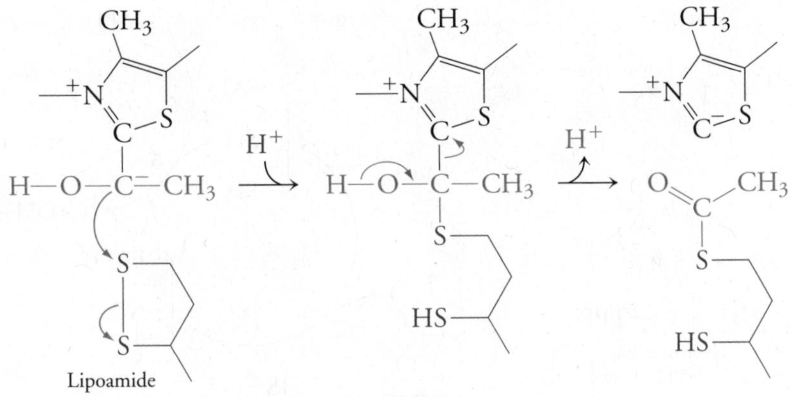

Lipoamide

3. Next, E2 transfers the acetyl group to coenzyme A, producing acetyl-CoA and leaving a reduced lipoamide group.

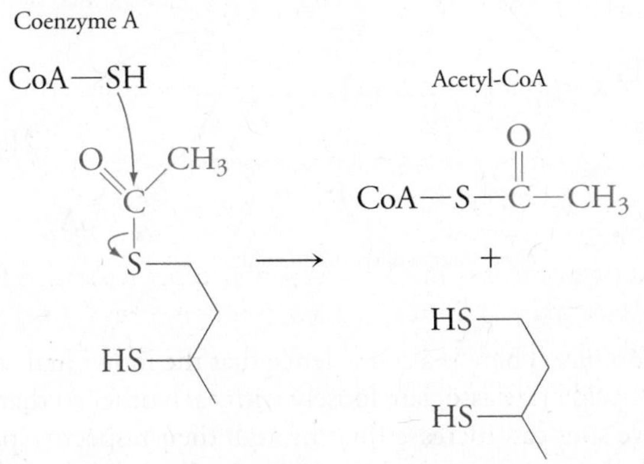

Coenzyme A

CoA—SH

Acetyl-CoA

CoA—S—C—CH₃

Lipoamide

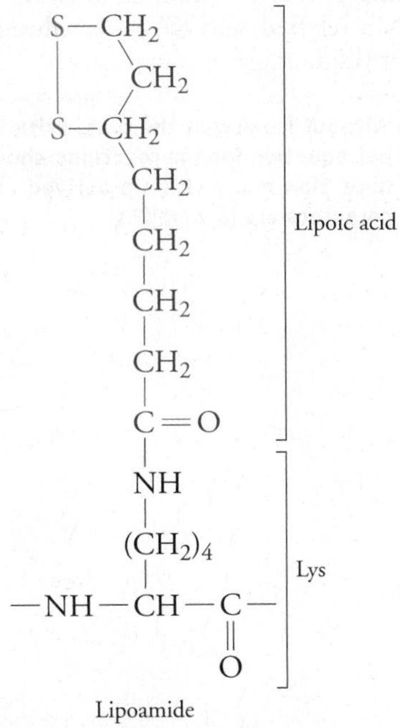

Figure 14-4 **Lipoamide.** This prosthetic group consists of lipoic acid (a vitamin) linked via an amide bond to the ε-amino group of a protein Lys residue. The active portion of the 14-Å-long lipoamide is the disulfide bond (red), which can be reversibly reduced.

Recall that acetyl-CoA is a thioester, a form of energy currency (see Section 12-3). Some of the free energy released in the oxidation of the hydroxyethyl group to an acetyl group is conserved in the formation of acetyl-CoA.

4. The final two steps of the reaction restore the pyruvate dehydrogenase complex to its original state. E3 reoxidizes the lipoamide group of E2 by transferring electrons to a Cys–Cys disulfide group in the enzyme.

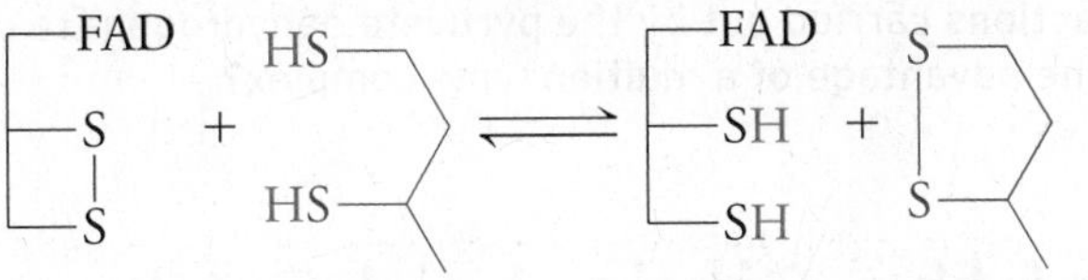

5. Finally, NAD⁺ reoxidizes the reduced Cys sulfhydryl groups. This electron-transfer reaction is facilitated by an FAD prosthetic group (the structure of FAD, a nucleotide derivative, is shown in Fig. 3-3c).

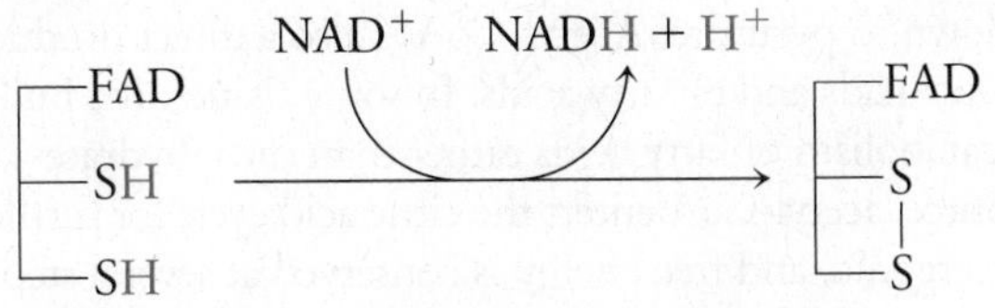

$NAD^+ \quad NADH + H^+$

During the five-step reaction (summarized in **Fig. 14-5**), the long lipoamide group of E2 acts as a swinging arm that visits the active sites of E1, E2, and E3 within the multienzyme complex. The arm picks up an acetyl group from an E1 subunit and transfers it to coenzyme A in an E2 active site. The arm then swings to an E3 active site, where it is reoxidized. Some other multienzyme complexes also include swinging arms, often attached to hinged protein domains to maximize their mobility.

A **multienzyme complex** such as the pyruvate dehydrogenase complex can carry out a multistep reaction sequence efficiently because *the product of one reaction can quickly become the substrate for the next reaction without diffusing away or reacting*

Figure 14-5 **Reactions of the pyruvate dehydrogenase complex.**
In these five reactions, an acetyl group from pyruvate is transferred to CoA, CO_2 is released, and NAD^+ is reduced to NADH.

? **Without looking at the text, write the net equation for the reactions shown here. How many vitamin-derived cofactors are involved?**

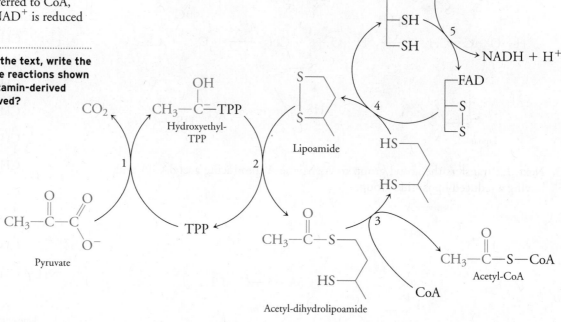

with another substance. There is also evidence that the individual enzymes of glycolysis and the citric acid cycle associate loosely with each other so that the close proximity of their active sites can increase flux through their respective pathways.

Flux through the pyruvate dehydrogenase complex is regulated by product inhibition: Both NADH and acetyl-CoA act as inhibitors. The activity of the complex is also regulated by hormone-controlled phosphorylation and dephosphorylation, which suits its function as the gatekeeper for the entry of a metabolic fuel into the citric acid cycle.

> **CONCEPT REVIEW**
> • Describe the functional importance of the coenzymes that participate in the reactions carried out by the pyruvate dehydrogenase complex.
> • What is the advantage of a multienzyme complex?

14-2 The Eight Reactions of the Citric Acid Cycle

KEY CONCEPTS
• The citric acid cycle is a set of eight reactions in which an acetyl group is condensed with oxaloacetate, two CO_2 are lost, and oxaloacetate is regenerated.
• Each round of the citric acid cycle generates three NADH, one QH_2, and one GTP or ATP.
• Flux through the citric acid cycle is regulated primarily by feedback inhibition at three steps.
• The citric acid cycle likely evolved by the combination of oxidative and reductive pathways.

⊕ **See Guided Exploration. Citric acid cycle overview.**

An acetyl-CoA molecule derived from pyruvate is a product of carbohydrate catabolism as well as a product of amino acid catabolism, since the carbon skeletons of many amino acids are broken down to pyruvate. Acetyl-CoA is also a direct product of the degradation of certain amino acids and of fatty acids. In some tissues, the bulk of acetyl-CoA is derived from the catabolism of fatty acids rather than carbohydrates or amino acids.

Whatever its source, acetyl-CoA enters the citric acid cycle for further oxidation. This process is highly exergonic, and free energy is conserved at several steps in the form of a nucleotide triphosphate (GTP) and reduced cofactors. *For each acetyl group that enters the citric acid cycle, two molecules of fully oxidized CO_2 are produced, representing a loss of four pairs of electrons.* These electrons are transferred to 3 NAD^+ and 1 ubiquinone (Q) to produce 3 NADH and 1 QH_2. The net equation for the citric acid cycle is therefore

$$\text{acetyl-CoA} + \text{GDP} + P_i + 3\,NAD^+ + Q \rightarrow$$
$$2\,CO_2 + \text{CoA} + \text{GTP} + 3\,\text{NADH} + QH_2$$

In this section we examine the sequence of eight enzyme-catalyzed reactions of the citric acid cycle, focusing on a few interesting reactions. The entire pathway is summarized in **Figure 14-6.**

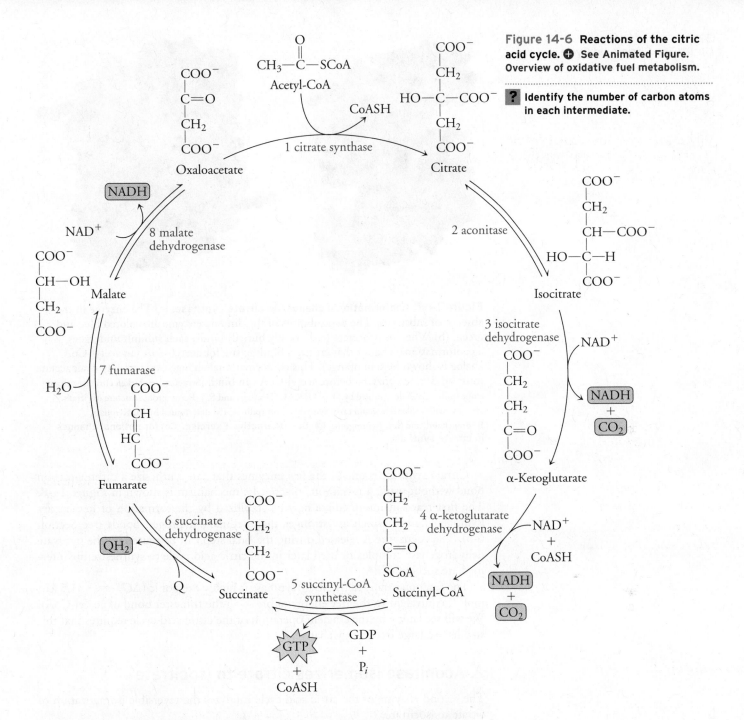

Figure 14-6 Reactions of the citric acid cycle. ⊕ See Animated Figure. Overview of oxidative fuel metabolism.

⸩ Identify the number of carbon atoms in each intermediate.

1. Citrate synthase adds an acetyl group to oxaloacetate

In the first reaction of the citric acid cycle, the acetyl group of acetyl-CoA condenses with the four-carbon compound oxaloacetate to produce the six-carbon compound citrate:

Citrate synthase is a dimer that undergoes a large conformational change upon substrate binding (**Fig. 14-7**).

Figure 14-7 Conformational changes in citrate synthase. (a) The enzyme in the absence of substrates. The two subunits of the dimeric enzyme are colored blue and green. (b) When oxaloacetate (red, mostly buried) binds, each subunit undergoes a conformational change that creates a binding site for acetyl-CoA (an acetyl-CoA analog is shown here in orange). This conformational change explains why oxaloacetate must bind to the enzyme before acetyl-CoA can bind. [Structure of chicken citrate synthase alone (pdb 5CSC) determined by D.-I. Liao, M. Karpusas, and S. J. Remington; structure of citrate synthase with oxaloacetate and carboxymethyl-CoA (pdb 5CTS) determined by M. Karpusas, B. Branchaud, and S. J. Remington.] ⊕ **See Interactive Exercise. Conformational changes in citrate synthase.**

Citrate synthase is one of the few enzymes that can synthesize a carbon–carbon bond without using a metal ion cofactor. Its mechanism is shown in **Figure 14-8**. The first reaction intermediate may be stabilized by the formation of low-barrier hydrogen bonds, which are stronger than ordinary hydrogen bonds (see Section 6-3). The coenzyme A released during the final step can be reused by the pyruvate dehydrogenase complex or used later in the citric acid cycle to synthesize the intermediate succinyl-CoA.

The reaction catalyzed by citrate synthase is highly exergonic ($\Delta G^{\circ\prime} = -31.5$ kJ · mol^{-1}, equivalent to the free energy of breaking the thioester bond of acetyl-CoA). We will see later why the efficient operation of the citric acid cycle requires that this step have a large free energy change.

2. Aconitase isomerizes citrate to isocitrate

The second enzyme of the citric acid cycle catalyzes the reversible isomerization of citrate to isocitrate:

$$
\begin{array}{ccccc}
\text{COO}^- & & \left[\begin{array}{c}\text{COO}^-\end{array}\right. & & \text{COO}^- \\
| & & | & & | \\
\text{CH}_2 & \xrightarrow{\text{H}_2\text{O}} & \text{CH}_2 & \xrightarrow{\text{H}_2\text{O}} & \text{CH}_2 \\
| & \rightleftharpoons & | & \rightleftharpoons & | \\
\text{HO}-\text{C}-\text{COO}^- & & \text{C}-\text{COO}^- & & \text{H}-\text{C}-\text{COO}^- \\
| & & \parallel & & | \\
\text{CH}_2 & & \text{CH} & & \text{HO}-\text{C}-\text{H} \\
| & & | & & | \\
\text{COO}^- & & \left.\text{COO}^-\right] & & \text{COO}^- \\
\text{Citrate} & & \text{Aconitate} & & \text{Isocitrate}
\end{array}
$$

The enzyme is named after the reaction intermediate.

Citrate is a symmetrical molecule, yet only one of its two carboxymethyl arms ($-\text{CH}_2-\text{COO}^-$) undergoes dehydration and rehydration during the aconitase reaction. This stereochemical specificity long puzzled biochemists, including Hans Krebs, who first described the citric acid cycle (also known as the Krebs cycle).

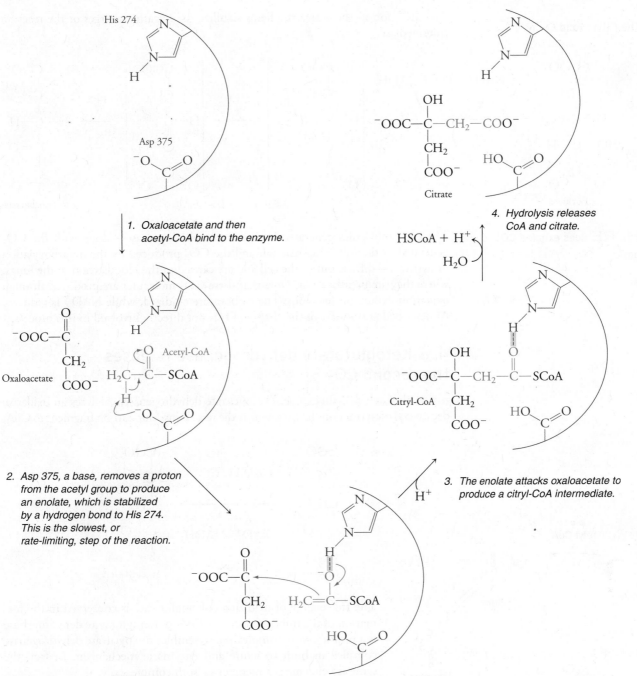

Figure 14-8 The citrate synthase reaction.

Eventually, Alexander Ogston pointed out that although citrate is symmetrical, its two carboxymethyl groups are no longer identical when it is bound to an asymmetrical enzyme (Fig. 14-9). In fact, a three-point attachment is not even necessary for an enzyme to distinguish two groups in a molecule such as citrate, which are related by mirror symmetry. You can prove this yourself with a simple organic chemistry model kit. By now you should appreciate that biological systems, including enzyme, are inherently chiral (also see Section 4-1).

3. Isocitrate dehydrogenase releases the first CO$_2$

The third reaction of the citric acid cycle is the oxidative decarboxylation of isocitrate to α-ketoglutarate. The substrate is first oxidized in a reaction accompanied by the reduction of NAD$^+$ to NADH. Then the carboxylate group β to the ketone function (that is, two carbon atoms away from the ketone) is eliminated as CO$_2$.

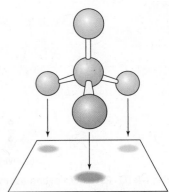

Figure 14-9 Stereochemistry of citrate synthase. The three-point attachment of citrate to the enzyme allows only one carboxymethyl group (shown in green) to react.

An Mn^{2+} ion in the active site helps stabilize the negative charges of the reaction intermediate.

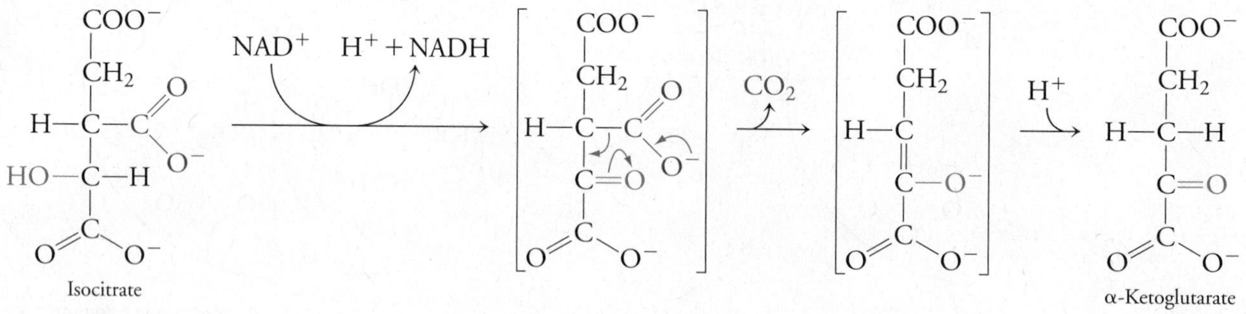

▶▶ **WHERE** does exhaled CO_2 come from?

The CO_2 molecules generated by isocitrate dehydrogenase—along with the CO_2 generated in the following reaction and the CO_2 produced by the decarboxylation of pyruvate—diffuse out of the cell and are carried in the bloodstream to the lungs, where they are breathed out. Note that these CO_2 molecules are produced through oxidation–reduction reactions: The carbons are oxidized, while NAD^+ is reduced. As described at the start of the chapter, O_2 is not directly involved in this process.

4. α-Ketoglutarate dehydrogenase releases the second CO_2

α-Ketoglutarate dehydrogenase, like isocitrate dehydrogenase, catalyzes an oxidative decarboxylation reaction. It also transfers the remaining four-carbon fragment to CoA:

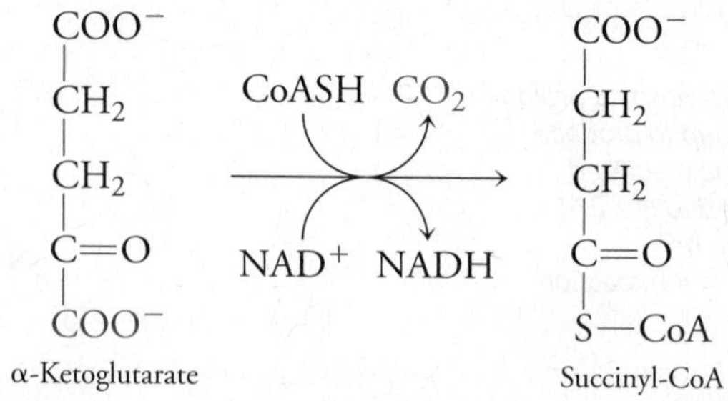

The free energy of oxidizing α-ketoglutarate is conserved in the formation of the thioester succinyl-CoA. α-Ketoglutarate dehydrogenase is a multienzyme complex that resembles the pyruvate dehydrogenase complex in both structure and enzymatic mechanism. In fact, the same E3 enzyme is a member of both complexes.

The isocitrate dehydrogenase and α-ketoglutarate dehydrogenase reactions both release CO_2. These two carbons are not the ones that entered the citric acid cycle as acetyl-CoA; those acetyl carbons are released in subsequent rounds of the cycle (Fig. 14-10). *However, the net result of each round of the citric acid cycle is the loss of two carbons as CO_2 for each acetyl-CoA that enters the cycle.*

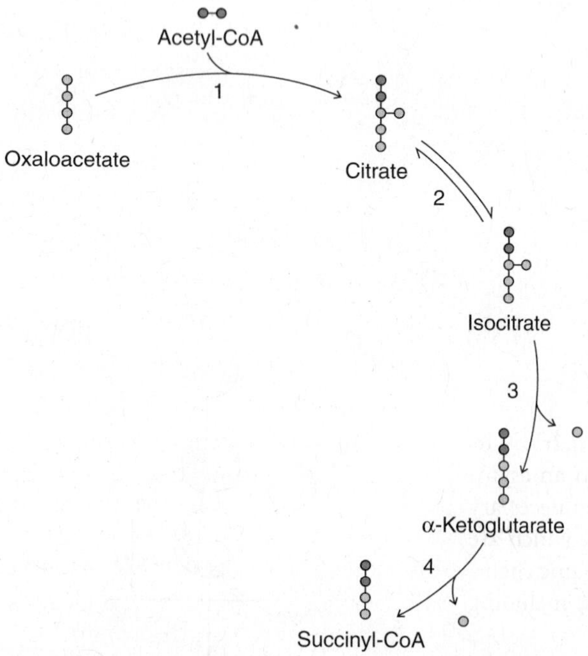

5. Succinyl-CoA synthetase catalyzes substrate-level phosphorylation

The thioester succinyl-CoA releases a large amount of free energy when it is cleaved ($\Delta G^{\circ\prime} = -32.6$ kJ · mol^{-1}). This is enough free energy to drive the synthesis of a nucleoside triphosphate from a nucleoside diphosphate and P_i ($\Delta G^{\circ\prime} = 30.5$ kJ · mol^{-1}). The change in free energy for the net reaction is near zero, so the reaction is reversible. In fact, the enzyme is named for the reverse reaction. Succinyl-CoA synthetase in the mammalian citric acid cycle generates GTP,

Figure 14-10 Fates of carbon atoms in the citric acid cycle. The two carbon atoms that are lost as CO_2 in the reactions catalyzed by isocitrate dehydrogenase (step 3) and α-ketoglutarate dehydrogenase (step 4) are not the same carbons that entered the cycle as acetyl-CoA (red). The acetyl carbons become part of oxaloacetate and are lost in subsequent rounds of the cycle.

1. A phosphate group displaces CoA in succinyl-CoA. The product, succinyl phosphate, is an acyl phosphate, which releases a large amount of free energy when cleaved.

Figure 14-11 **The succinyl-CoA synthetase reaction.**

? What is the fate of the free CoA molecule?

Succinyl-CoA

Succinyl phosphate

2. Succinyl phosphate donates its phosphoryl group to a His residue on the enzyme, producing a phospho-His intermediate and releasing succinate.

His residue

Succinate

GTP

GDP + P$_i$

3. The phosphoryl group is then transferred to GDP to form GTP.

Phospho-His

whereas the plant and bacterial enzymes generate ATP (recall that GTP is energetically equivalent to ATP). An exergonic reaction coupled to the transfer of a phosphoryl group to a nucleoside diphosphate is termed **substrate-level phosphorylation** to distinguish it from oxidative phosphorylation (Section 15-4) and photophosphorylation (Section 16-2), which are more indirect ways of synthesizing ATP.

How does succinyl-CoA synthetase couple thioester cleavage to the synthesis of a nucleoside triphosphate? The reaction is a series of phosphoryl-group transfers that involve an active-site His residue (**Fig. 14-11**). The phospho-His reaction intermediate must move a large distance to shuttle the phosphoryl group between the succinyl group and the nucleoside diphosphate (**Fig. 14-12**).

6. Succinate dehydrogenase generates ubiquinol

The final three reactions of the citric acid cycle convert succinate back to the cycle's starting substrate, oxaloacetate. Succinate dehydrogenase catalyzes the reversible dehydrogenation of succinate to fumarate. This oxidation–reduction reaction requires an FAD prosthetic group, which is reduced to FADH$_2$ during the reaction:

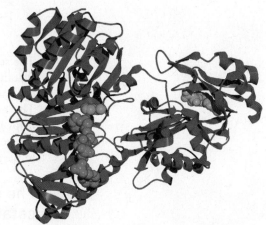

Figure 14-12 Substrate binding in succinyl-CoA synthetase. Succinyl-CoA (represented by coenzyme A, red) binds to the enzyme, and its succinyl group is phosphorylated. The succinyl phosphate then transfers its phosphoryl group to the His 246 side chain (green). A protein loop containing the phospho-His side chain must undergo a large movement because the nucleoside diphosphate awaiting phosphorylation (ADP, orange) binds to a site about 35 Å away. [Structure of *E. coli* succinyl-CoA synthetase (pdb 1CQI) determined by M. A. Joyce, M. E. Fraser, M. N. G. James, W. A. Bridger, and W. T. Wolodko.]

Enz-FAD   Enz-FADH$_2$

succinate dehydrogenase

Succinate

Fumarate

To regenerate the enzyme, the FADH$_2$ group must be reoxidized. Since succinate dehydrogenase is embedded in the inner mitochondrial membrane (it is the only one of the eight citric acid cycle enzymes that is not soluble in the mitochondrial matrix), it can be reoxidized by the lipid-soluble electron carrier ubiquinone (see Section 12-2) rather than by the soluble cofactor NAD^+. Ubiquinone (abbreviated Q) acquires two electrons to become ubiquinol (QH_2).

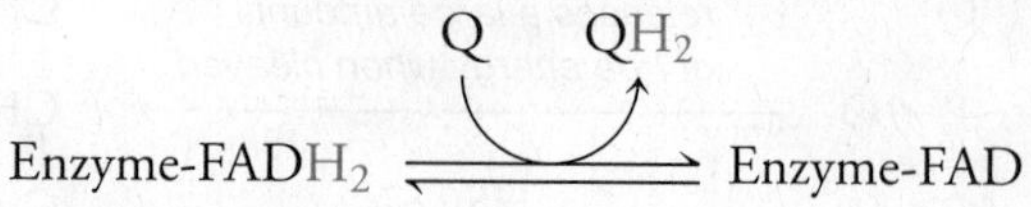

$$\text{Enzyme-FADH}_2 \xrightleftharpoons{\quad Q \quad QH_2 \quad} \text{Enzyme-FAD}$$

7. Fumarase catalyzes a hydration reaction

In the seventh reaction, fumarase (also known as fumarate hydratase) catalyzes the reversible hydration of a double bond to convert fumarate to malate:

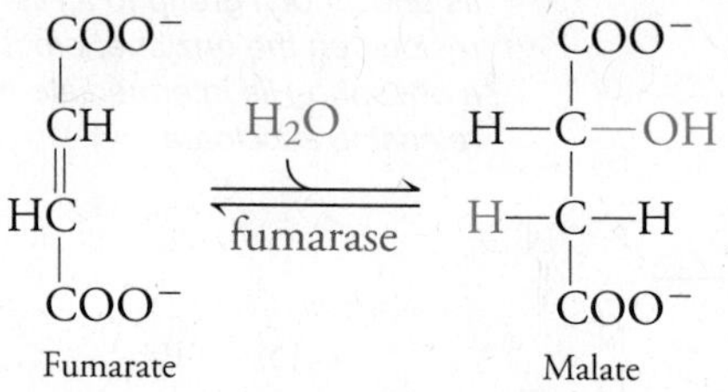

Fumarate   Malate

8. Malate dehydrogenase regenerates oxaloacetate

The citric acid cycle concludes with the regeneration of oxaloacetate from malate in an NAD^+-dependent oxidation reaction:

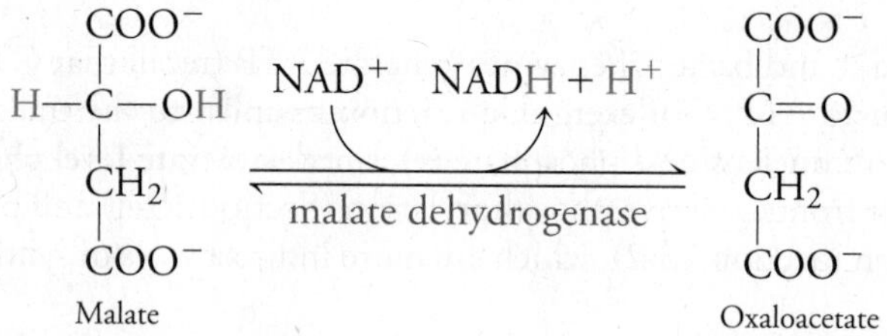

Malate   Oxaloacetate

The standard free energy change for this reaction is $+29.7$ kJ $\cdot$ mol^{-1}, indicating that the reaction has a low probability of occurring as written. However, the product oxaloacetate is a substrate for the next reaction (Reaction 1 of the citric acid cycle). The highly exergonic—and therefore highly favorable—citrate synthase reaction helps pull the malate dehydrogenase reaction forward. This is the reason for the apparent waste of free energy released by cleaving the thioester bond of acetyl-CoA in the first reaction of the citric acid cycle.

The citric acid cycle is an energy-generating catalytic cycle

Because the eighth reaction of the citric acid cycle returns the system to its original state, *the entire pathway acts in a catalytic fashion to dispose of carbon atoms derived from amino acids, carbohydrates, and fatty acids*. Albert Szent-Györgyi discovered the catalytic nature of the pathway by observing that small additions of organic compounds such as succinate, fumarate, and malate stimulated O_2 uptake in a tissue preparation. Because the O_2 consumption was much greater than would be required for the direct oxidation of the added substances, he inferred that the compounds acted catalytically.

We now know that oxygen is consumed during oxidative phosphorylation, the process that reoxidizes the reduced cofactors (NADH and QH_2) that are produced by the citric acid cycle. Although the citric acid cycle generates one molecule of GTP (or ATP), considerably more ATP is generated when the reduced cofactors are re-oxidized by O_2. Each NADH yields approximately 2.5 ATP, and each QH_2 yields approximately 1.5 ATP (we will see in Section 15-4 why these values are not whole numbers). Every acetyl unit that enters the citric acid cycle can therefore generate a total of 10 ATP equivalents. The energy yield of a molecule of glucose, which generates two acetyl units, can be calculated:

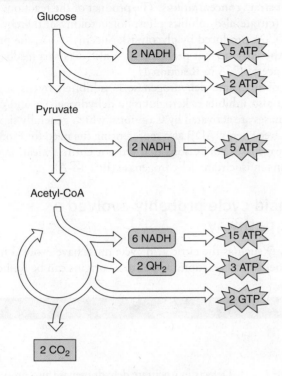

A muscle operating anaerobically produces only 2 ATP per glucose, but under aerobic conditions when the citric acid cycle is fully functional, each glucose molecule generates about 32 ATP equivalents. This general phenomenon is called the **Pasteur effect,** after Louis Pasteur, who first observed that the rate of glucose consumption by yeast cells decreased dramatically when the cells were shifted from anaerobic to aerobic growth conditions.

The citric acid cycle is regulated at three steps

Flux through the citric acid cycle is regulated primarily at the cycle's three metabolically irreversible steps: those catalyzed by citrate synthase (Reaction 1), isocitrate dehydrogenase (Reaction 3), and α-ketoglutarate dehydrogenase (Reaction 4). The major regulators are shown in **Figure 14-13.**

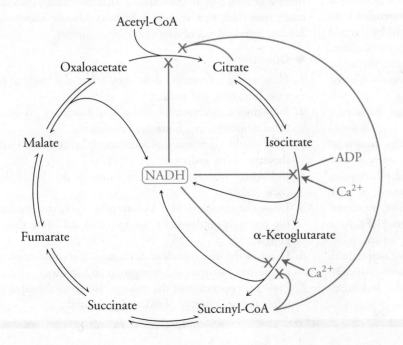

Figure 14-13 Regulation of the citric acid cycle. Inhibition is represented by red symbols, activation by green symbols.

Neither acetyl-CoA nor oxaloacetate is present at concentrations high enough to saturate citrate synthase, so flux through the first step of the citric acid cycle depends largely on the substrate concentrations. The product of the reaction, citrate, inhibits citrate synthase (citrate also inhibits phosphofructokinase, thereby decreasing the supply of acetyl-CoA produced by glycolysis). Succinyl-CoA, the product of Reaction 4, inhibits the enzyme that produces it. It also acts as a feedback inhibitor by competing with acetyl-CoA in Reaction 1.

The activity of isocitrate dehydrogenase is inhibited by its reaction product, NADH. NADH also inhibits α-ketoglutarate dehydrogenase and citrate synthase. Both dehydrogenases are activated by Ca^{2+} ions, which generally signify the need to generate cellular free energy. ADP, also representing the need for more ATP, activates isocitrate dehydrogenase. For reasons that are not entirely clear, many cancer cells develop mutations in isocitrate dehydrogenase (Box 14-A).

The citric acid cycle probably evolved as a synthetic pathway

A circular pathway such as the citric acid cycle must have evolved from a linear set of preexisting biochemical reactions. Clues to its origins can be found by examining

---

BOX 14-A   CLINICAL CONNECTION

Mutations in Citric Acid Cycle Enzymes

Possibly because the citric acid cycle is a central metabolic pathway, severe defects in any of its components are expected to be incompatible with life. However, researchers have documented mutations in the genes for several of the cycle's enzymes, including α-ketoglutarate dehydrogenase, succinyl-CoA synthetase, and succinate dehydrogenase. These defects, which are all rare, typically affect the central nervous system, causing symptoms such as movement disorders and neurodegeneration. A rare form of fumarase deficiency results in brain malformation and developmental disabilities. Individuals who have a different fumarase defect show a higher risk of developing leiomyomas, noncancerous tumors such as uterine fibroids. A small percentage of these do become malignant. The other citric acid cycle enzyme mutations are also associated with cancer. One possible explanation is that a defective enzyme contributes to **carcinogenesis** (the development of cancer, or uncontrolled cell growth) by directly interfering with the cell's vital pathways for energy metabolism. Another possibility is that a defective enzyme causes the accumulation of particular metabolites, which are responsible for altering the cell's developmental fate.

Fumarase appears to be linked to cancer through the second mechanism. Normal cells respond to a drop in oxygen availability (hypoxia) by activating transcription factors known as hypoxia-inducible factors (HIFs). These proteins interact with DNA to turn on the expression of genes for glycolytic enzymes and a growth factor that promotes the development of new blood vessels. When the fumarase gene is defective, fumarate accumulates and inhibits a protein that destabilizes HIFs. As a result, the fumarase deficiency promotes glycolysis (an anaerobic pathway) and the growth of blood vessels. These two adaptations would favor tumors, whose growth, although characteristically rapid, may be limited by the availability of oxygen and other nutrients delivered by the bloodstream.

Defects in isocitrate dehydrogenase also promote cancer in an indirect fashion. Many cancerous cells exhibit a mutation in one of the two genes for the enzyme, suggesting that the unaltered copy is necessary for maintaining the normal activity of the citric acid cycle, while the mutated copy plays a role in carcinogenesis. Interestingly, the mutation usually converts an active-site Arg residue to His, indicating some strong selective pressure for a gain-of-function mutation (most mutations in proteins lead to loss of function). The mutated isocitrate dehydrogenase no longer carries out the usual reaction (converting isocitrate to α-ketoglutarate) but instead converts α-ketoglutarate to 2-hydroxyglutarate in an NADPH-dependent manner. The mechanism whereby 2-hydroxyglutarate contributes to carcinogenesis is not clear, but its involvement is bolstered by the observation that individuals who harbor other mutations that lead to 2-hydroxyglutarate accumulation have an increased risk of developing brain tumors.

Questions:

1. How would a fumarase deficiency affect the levels of pyruvate, fumarate, and malate?
2. Why does a succinate dehydrogenase deficiency produce the same symptoms as a fumarase deficiency?
3. Why would a deficiency of succinate dehydrogenase lead to a shortage of free coenzyme A?
4. Individuals who are deficient in fumarase develop lactic acidosis. Explain.
5. Draw the structure of 2-hydroxyglutarate, the product of the reaction catalyzed by the mutated isocitrate dehydrogenase.
6. Describe the single-nucleotide changes that could convert the arginine in isocitrate dehydrogenase to histidine.
7. How does operation of the mutated isocitrate dehydrogenase affect the cell's supply of reduced cofactors?

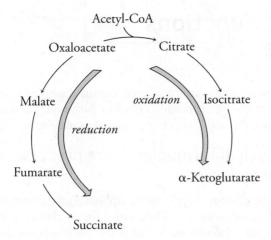

Figure 14-14 Pathways that might have given rise to the citric acid cycle. The pathway starting from oxaloacetate and proceeding to the right is an oxidative biosynthetic pathway, whereas the pathway that proceeds to the left is a reductive pathway. The modern citric acid cycle may have evolved by connecting these pathways.

the metabolism of organisms that resemble earlier life-forms. Such organisms emerged before atmospheric oxygen was available and may have used sulfur as their ultimate oxidizing agent, reducing it to H_2S. Their modern-day counterparts are anaerobic autotrophs that harvest free energy by pathways that are independent of the pathways of carbon metabolism. These organisms therefore do not use the citric acid cycle to generate reduced cofactors that are subsequently oxidized by molecular oxygen. However, all organisms must synthesize small molecules that can be used to build proteins, nucleic acids, carbohydrates, and so on.

Even organisms that do not use the citric acid cycle contain genes for some citric acid cycle enzymes. For example, the cells may condense acetyl-CoA with oxaloacetate, leading to α-ketoglutarate, which is a precursor of several amino acids. They may also convert oxaloacetate to malate, proceeding to fumarate and then to succinate. Together, these two pathways resemble the citric acid cycle, with the right arm following the usual oxidative sequence of the cycle and the left arm following a reversed, reductive sequence (**Fig. 14-14**). The reductive sequence of reactions might have evolved as a way to regenerate the cofactors reduced during other catabolic reactions (for example, the NADH produced by the glyceraldehyde-3-phosphate dehydrogenase reaction of glycolysis; see Section 13-1).

It is easy to theorize that the evolution of an enzyme to interconvert α-ketoglutarate and succinate could have created a cyclic pathway similar to the modern citric acid cycle. Interestingly, *E. coli*, which uses the citric acid cycle under aerobic growth conditions, uses an interrupted citric acid cycle like the one diagrammed in Figure 14-14 when it is growing anaerobically.

Since the final four reactions of the modern citric acid cycle are metabolically reversible, the primitive citric acid cycle might easily have accommodated one-way flux in the clockwise direction, forming an oxidative cycle. If the complete cycle proceeded in the counterclockwise direction, the result would have been a reductive biosynthetic pathway (**Fig. 14-15**). This pathway, which would incorporate, or "fix," atmospheric CO_2 into biological molecules, may have preceded the modern CO_2-fixing pathway found in green plants and some photosynthetic bacteria (described in Section 16-3).

CONCEPT REVIEW

- Describe the sources of the acetyl groups that enter the citric acid cycle.
- List the substrates and products for each of the cycle's eight reactions.
- Which products of the citric acid cycle represent forms of energy currency for the cell?
- Which substrates and products of the citric acid cycle regulate flux through the pathway?
- Describe how primitive oxidative and reductive biosynthetic pathways might have been combined to generate a circular metabolic pathway.

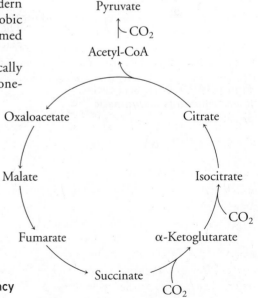

Figure 14-15 A proposed reductive biosynthetic pathway based on the citric acid cycle. This pathway might have operated to incorporate CO_2 into biological molecules.

14-3 Anabolic and Catabolic Functions of the Citric Acid Cycle

KEY CONCEPTS
- The citric acid cycle supplies precursors for the synthesis of other compounds.
- Citric acid cycle intermediates can be replenished.

In mammals, six of the eight citric acid cycle intermediates (all except isocitrate and succinate) are the precursors or products of other pathways. For this reason, it is impossible to designate the citric acid cycle as a purely catabolic or anabolic pathway.

Citric acid cycle intermediates are precursors of other molecules

Intermediates of the citric acid cycle can be siphoned off to form other compounds (Fig. 14-16). For example, succinyl-CoA is used for the synthesis of heme. The five-carbon α-ketoglutarate (sometimes called 2-oxoglutarate) can undergo reductive amination by glutamate dehydrogenase to produce the amino acid glutamate:

$$
\begin{array}{c}
\text{COO}^- \\
| \\
\text{CH}_2 \\
| \\
\text{CH}_2 \\
| \\
\text{C}=\text{O} \\
| \\
\text{COO}^-
\end{array}
\quad + \quad \text{NH}_4^+
\quad \xrightleftharpoons[\text{glutamate dehydrogenase}]{\text{NADH} + \text{H}^+ \quad \text{NAD}^+}
\quad
\begin{array}{c}
\text{COO}^- \\
| \\
\text{CH}_2 \\
| \\
\text{CH}_2 \\
| \\
\text{H}-\text{C}-\text{NH}_3^+ \\
| \\
\text{COO}^-
\end{array}
\quad + \quad \text{H}_2\text{O}
$$

α-Ketoglutarate$\qquad\qquad$Glutamate

Glutamate is a precursor of the amino acids glutamine, arginine, and proline. Glutamine in turn is a precursor for the synthesis of purine and pyrimidine nucleotides. We have already seen that oxaloacetate is a precursor of monosaccharides (Section 13-2). Consequently, *any of the citric acid cycle intermediates, which can be converted by the cycle to oxaloacetate, can ultimately serve as gluconeogenic precursors.*

Citrate produced by the condensation of acetyl-CoA with oxaloacetate can be transported out of the mitochondria to the cytosol. ATP-citrate lyase then catalyzes the reaction

$$\text{ATP} + \text{citrate} + \text{CoA} \rightarrow \text{ADP} + \text{P}_i + \text{oxaloacetate} + \text{acetyl-CoA}$$

The resulting acetyl-CoA is used for fatty acid and cholesterol synthesis, which take place in the cytosol. Note that the ATP-citrate lyase reaction undoes the work of the

Figure 14-16 Citric acid cycle intermediates as biosynthetic precursors.

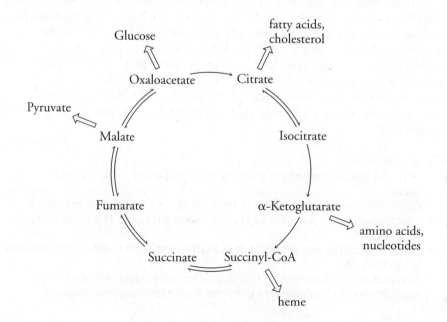

exergonic citrate synthase reaction. This seems wasteful, but cytosolic ATP-citrate lyase is essential because acetyl-CoA, which is produced in the mitochondria, cannot cross the mitochondrial membrane to reach the cytosol, whereas citrate can. The oxaloacetate product of the ATP-citrate lyase reaction can be converted to malate by a cytosolic malate dehydrogenase operating in reverse. Malate is then decarboxylated by the action of malic enzyme to produce pyruvate:

$$
\begin{array}{c}
\text{COO}^- \\
| \\
\text{CH—OH} \\
| \\
\text{CH}_2 \\
| \\
\text{COO}^-
\end{array}
\quad
\xrightleftharpoons[\text{malic enzyme}]{\text{NADP}^+ \quad \text{NADPH}}
\quad
\begin{array}{c}
\text{COO}^- \\
| \\
\text{C=O} \\
| \\
\text{CH}_3
\end{array}
\quad + \quad \text{CO}_2
$$

Malate   Pyruvate

Pyruvate can reenter the mitochondria and be converted back to oxaloacetate to complete the cycle shown in **Figure 14-17**. In plants, isocitrate is diverted from the citric acid cycle in a biosynthetic pathway known as the **glyoxylate pathway** (Box 14-B).

Anaplerotic reactions replenish citric acid cycle intermediates

Intermediates that are diverted from the citric acid cycle for other purposes can be replenished through **anaplerotic reactions** (from the Greek *ana*, "up," and *plerotikos*, "to fill"; **Fig. 14-18**). One of the most important of these reactions is catalyzed by pyruvate carboxylase (this is also the first step of gluconeogenesis; Section 13-2):

pyruvate + CO_2 + ATP + H_2O → oxaloacetate + ADP + P_i

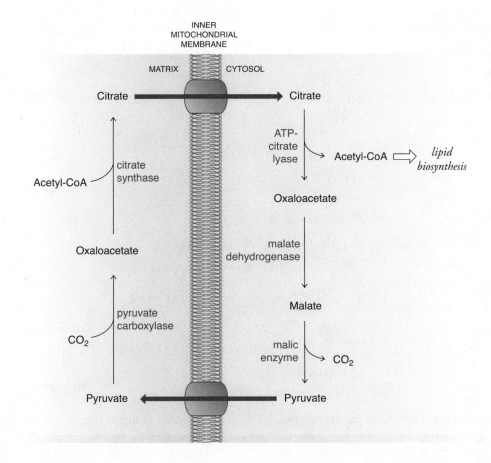

Figure 14-17 The citrate transport system. Both citrate and pyruvate cross the inner mitochondrial membrane via specific transport proteins. This system allows carbon atoms from mitochondrial acetyl-CoA to be transferred to the cytosol for the synthesis of fatty acids and cholesterol.

BOX 14-B ⬡ BIOCHEMISTRY NOTE

The Glyoxylate Pathway

Plants and some bacterial cells contain certain enzymes that act together with some citric acid cycle enzymes to convert acetyl-CoA to oxaloacetate, a gluconeogenic precursor. Animals lack the enzymes to do this and therefore cannot undertake the net synthesis of carbohydrates from two-carbon precursors. In plants, the glyoxylate pathway includes reactions that take place in the mitochondria and the **glyoxysome,** an organelle that, like the peroxisome, contains enzymes that carry out some essential metabolic processes.

In the glyoxysome, acetyl-CoA condenses with oxaloacetate to form citrate, which is then isomerized to isocitrate, as in the citric acid cycle. However, the next step is not the isocitrate dehydrogenase reaction but a reaction catalyzed by the glyoxysome enzyme isocitrate lyase, which converts isocitrate to succinate and the two-carbon compound glyoxylate. Succinate continues as usual through the mitochondrial citric acid cycle to regenerate oxaloacetate.

In the glyoxysome, the glyoxylate condenses with a second molecule of acetyl-CoA in a reaction catalyzed by the glyoxysome enzyme malate synthase to form the four-carbon compound malate. Malate can then be converted to oxaloacetate for gluconeogenesis. The two reactions that are unique to the glyoxylate pathway are shown in green in the figure below; reactions that are identical to those of the citric acid cycle are shown in blue.

In essence, the glyoxylate pathway bypasses the two CO_2-generating steps of the citric acid cycle (catalyzed by isocitrate dehydrogenase and α-ketoglutarate dehydrogenase) and incorporates a second acetyl unit (at the malate synthase step). The net result of the glyoxylate pathway is the production of a four-carbon compound that can be used to synthesize glucose. This pathway is highly active in germinating seeds, where stored oils (triacylglycerols) are broken down to acetyl-CoA. The glyoxylate pathway thus provides a route for synthesizing glucose from fatty acids. Because animals lack isocitrate lyase and malate synthase, they cannot undertake the net synthesis of carbohydrates from fat.

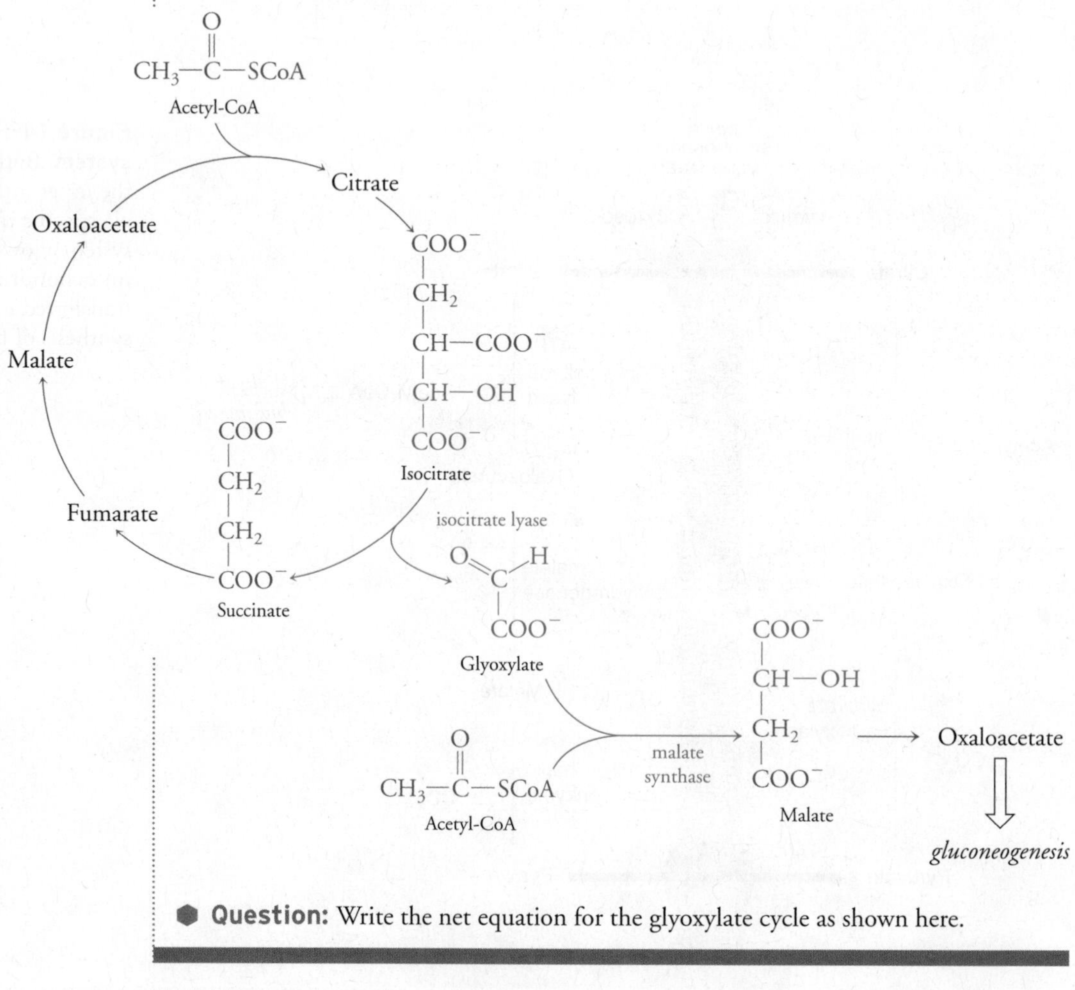

⬡ **Question:** Write the net equation for the glyoxylate cycle as shown here.

Acetyl-CoA activates pyruvate carboxylase, so when the activity of the citric acid cycle is low and acetyl-CoA accumulates, more oxaloacetate is produced. The concentration of oxaloacetate is normally low since the malate dehydrogenase reaction is thermodynamically unfavorable and the citrate synthase reaction is highly favorable. The replenished oxaloacetate is converted to citrate, isocitrate, α-ketoglutarate, and so on, so the concentrations of all the citric acid cycle intermediates increase and the cycle can proceed more quickly. *Since the citric acid cycle acts as a catalyst, increasing the concentrations of its components increases flux through the pathway.*

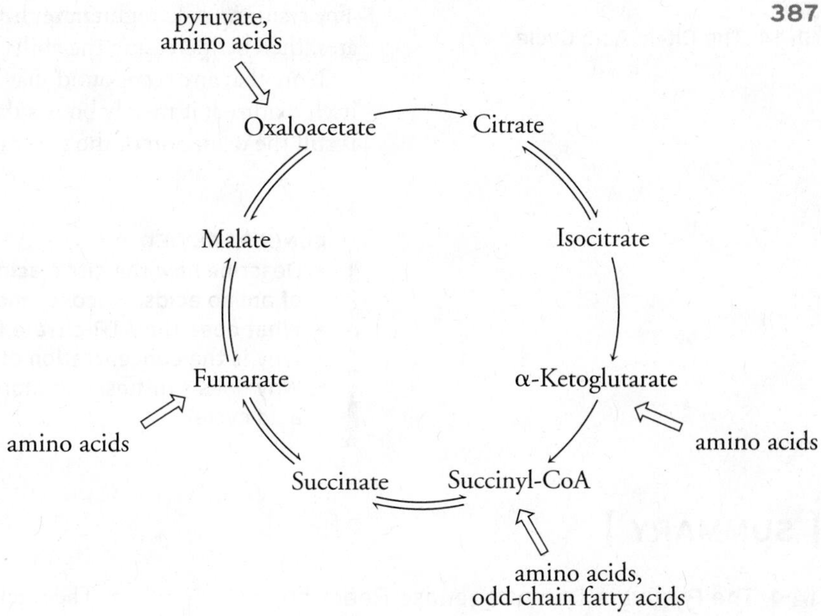

The degradation of fatty acids with an odd number of carbon atoms yields the citric acid cycle intermediate succinyl-CoA. Other anaplerotic reactions include the pathways for the degradation of some amino acids, which produce α-ketoglutarate, succinyl-CoA, fumarate, and oxaloacetate. Some of these reactions are transaminations, such as

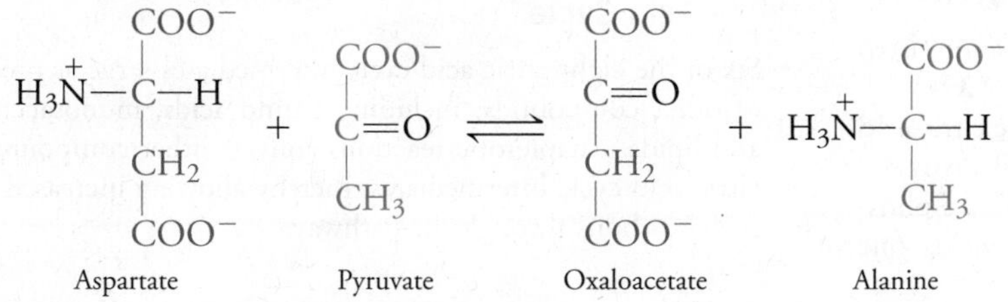

Figure 14-18 Anaplerotic reactions of the citric acid cycle.

Because transamination reactions have ΔG values near zero, the direction of flux into or out of the pool of citric acid cycle intermediates depends on the relative concentrations of the reactants.

In vigorously exercising muscle, the concentrations of citric acid cycle intermediates increase about three- to fourfold within a few minutes. This may help boost the energy-generating activity of the citric acid cycle, but it cannot be the sole mechanism, since flux through the citric acid cycle actually increases as much as 100-fold due to the increased activity of the three enzymes at the control points: citrate synthase, isocitrate dehydrogenase, and α-ketoglutarate dehydrogenase. The increase in citric acid cycle intermediates may actually be a mechanism for accommodating the large increase in pyruvate that results from rapid glycolysis at the start of exercise. Not all the pyruvate is converted to lactate (Section 13-1); some is shunted into the pool of citric acid cycle intermediates via the pyruvate carboxylase reaction. Some pyruvate also undergoes a reversible reaction catalyzed by alanine aminotransferase

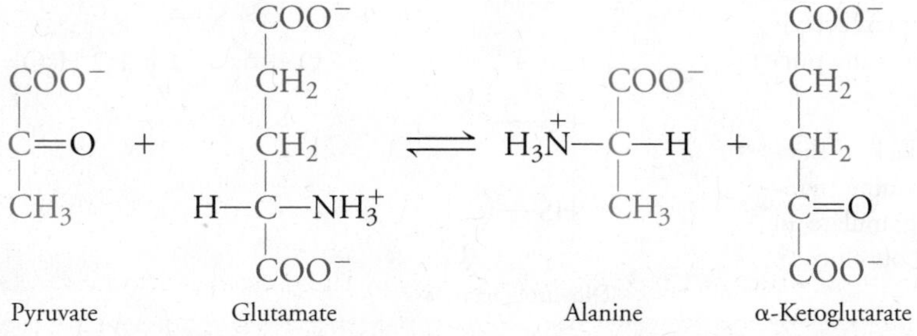

The resulting α-ketoglutarate then augments the pool of citric acid cycle intermediates, thereby increasing the ability of the cycle to oxidize the extra pyruvate.

Note that any compound that enters the citric acid cycle as an intermediate is not itself oxidized; it merely boosts the catalytic activity of the cycle, whose net reaction is still the oxidation of the two carbons of acetyl-CoA.

CONCEPT REVIEW
- Describe how the citric acid cycle supplies the precursors for the synthesis of amino acids, glucose, and fatty acids.
- What does the ATP-citrate lyase reaction accomplish?
- Why is the concentration of oxaloacetate low?
- Why does synthesizing more oxaloacetate increase flux through the citric acid cycle?

[SUMMARY]

14-1 The Pyruvate Dehydrogenase Reaction

- In order for pyruvate, the product of glycolysis, to enter the citric acid cycle, it must undergo oxidative decarboxylation catalyzed by the multienzyme pyruvate dehydrogenase complex, which yields acetyl-CoA, CO_2, and NADH.

14-2 The Eight Reactions of the Citric Acid Cycle

- The eight reactions of the citric acid cycle function as a multistep catalyst to convert the two carbons of acetyl-CoA to 2 CO_2.
- The electrons released in the oxidative reactions of the citric acid cycle are transferred to 3 NAD^+ and to ubiquinone. The reoxidation of the reduced cofactors generates ATP by oxidative phosphorylation. In addition, succinyl-CoA synthetase yields one molecule of GTP or ATP.

- The regulated reactions of the citric acid cycle are its irreversible steps, catalyzed by citrate synthase, isocitrate dehydrogenase, and α-ketoglutarate dehydrogenase.
- The citric acid cycle most likely evolved from biosynthetic pathways leading to α-ketoglutarate or succinate.

14-3 Anabolic and Catabolic Functions of the Citric Acid Cycle

- Six of the eight citric acid cycle intermediates serve as precursors of other compounds, including amino acids, monosaccharides, and lipids. Anaplerotic reactions convert other compounds into citric acid cycle intermediates, thereby allowing increased flux of acetyl carbons through the pathway.

[GLOSSARY TERMS]

citric acid cycle	substrate-level phosphorylation	glyoxysome
mitochondrial matrix	Pasteur effect	glyoxylate pathway
multienzyme complex	carcinogenesis	anaplerotic reaction

[PROBLEMS]

14-1 The Pyruvate Dehydrogenase Reaction

1. What are four possible transformations of pyruvate in mammalian cells?

2. Determine which one of the five steps of the pyruvate dehydrogenase complex reaction is metabolically irreversible and explain why.

3. The product of the pyruvate dehydrogenase complex, acetyl-CoA, is released in step 3 of the overall reaction. What is the purpose of steps 4 and 5?

4. Beriberi is a disease that results from a dietary lack of thiamine, the vitamin that serves as the precursor for thiamine pyrophosphate (TPP). There are two metabolites that accumulate in individuals with beriberi, especially after ingestion of glucose. Which metabolites accumulate and why?

5. Arsenite is toxic in part because it binds to sulfhydryl compounds such as lipoamide, as shown in the figure. What effect would the presence of arsenite have on the citric acid cycle?

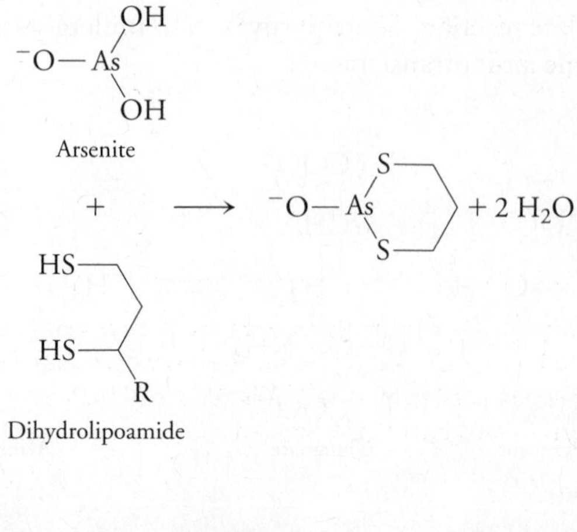

Arsenite

Dihydrolipoamide

6. Using the pyruvate dehydrogenase complex reaction as a model, reconstruct the TPP-dependent yeast pyruvate decarboxylase reaction in alcoholic fermentation (see Section 13-1).

7. How is the activity of the pyruvate dehydrogenase complex affected by **(a)** a high [NADH]/[NAD$^+$] ratio or **(b)** a high [acetyl-CoA]/[CoASH] ratio?

8. The activity of the pyruvate dehydrogenase complex is also controlled by phosphorylation. Pyruvate dehydrogenase kinase catalyzes the phosphorylation of a specific Ser residue on the E1 subunit of the enzyme, rendering it inactive. The pyruvate dehydrogenase phosphatase enzyme reverses the inhibition by catalyzing the removal of this phosphate group. The kinase and the phosphatase enzymes themselves are controlled by cytosolic Ca^{2+} levels. In the muscle, Ca^{2+} levels rise when the muscle contracts. Which of these two enzymes is inhibited by Ca^{2+} and which is activated by Ca^{2+}?

9. Most cases of pyruvate dehydrogenase deficiency disease that have been studied to date involve a mutation in the E1 subunit of the enzyme. The disease is extremely difficult to treat successfully, but physicians who identify patients with a pyruvate dehydrogenase deficiency will administer thiamine as a first course of treatment. Explain why.

10. A second strategy to treat a pyruvate dehydrogenase deficiency disease (see Problem 9) involves administering dichloroacetate, a compound that inhibits pyruvate dehydrogenase kinase (see Problem 8). How might this strategy be effective?

14-2 The Eight Reactions of the Citric Acid Cycle

11. Why is it advantageous for citrate, the product of Reaction 1 of the citric acid cycle, to inhibit phosphofructokinase, which catalyzes the third reaction of glycolysis (see Section 13-1)?

12. Animals that have ingested the leaves of the poisonous South African plant *Dichapetalum cymosum* exhibit a 10-fold increase in levels of cellular citrate. The plant contains fluoroacetate, which is converted to fluoroacetyl-CoA. Describe the mechanism that leads to increased levels of citrate in animals that have ingested this poisonous plant. (*Note:* Fluoroacetyl-CoA is not an inhibitor of citrate synthase.)

13. Site-directed mutagenesis techniques were used to synthesize a mutant citrate synthase enzyme in which the active-site histidine was converted to an alanine. Why did the mutant citrate synthase enzyme exhibit decreased catalytic activity?

14. The compound *S*-acetonyl-CoA can be synthesized from 1-bromoacetone and coenzyme A.

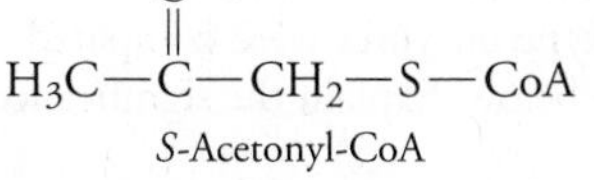

$$H_3C-\overset{\overset{\textstyle O}{\|}}{C}-CH_2-S-CoA$$
S-Acetonyl-CoA

(a) Write the reaction for the formation of *S*-acetonyl-CoA.
(b) The Lineweaver–Burk plot for the citrate synthase reaction with and without *S*-acetonyl-CoA is shown. What type of inhibitor is *S*-acetonyl-CoA? Explain.

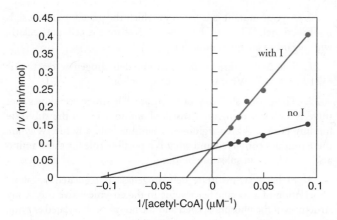

(c) Acetyl-CoA acts as an allosteric activator of pyruvate carboxylase. *S*-acetonyl CoA does not activate pyruvate carboxylase, and it cannot compete with acetyl-CoA for binding to the enzyme. What does this tell you about the binding requirements for an allosteric activator of pyruvate carboxylase?

15. The compound carboxymethyl-CoA (shown below) is a competitive inhibitor of citrate synthase and is a proposed transition state analog. Use this information to propose a structure for the reaction intermediate derived from acetyl-CoA in the rate-limiting step of the reaction, just prior to its reaction with oxaloacetate.

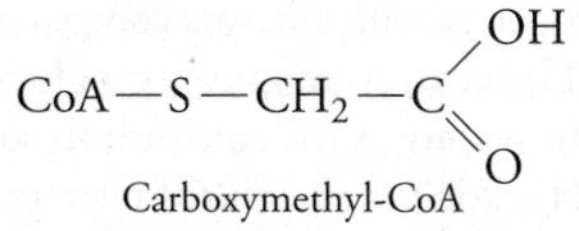

$$CoA-S-CH_2-\overset{\overset{\textstyle OH}{\diagup}}{\underset{\underset{\textstyle O}{\diagdown}}{C}}$$
Carboxymethyl-CoA
(transition state analog)

16. Citrate competes with oxaloacetate for binding to citrate synthase. Isocitrate dehydrogenase is activated by Ca^{2+} ions, which are released when muscle contracts. How do these two regulatory strategies assist the cell in making the transition from the rested state (low citric acid cycle activity) to the exercise state (high citric acid cycle activity)?

17. Administration of high concentrations of oxygen (hyperoxia) is effective in the treatment of lung injuries but at the same time can also be quite damaging.
(a) It has been shown that lung aconitase activity is dramatically decreased during hyperoxia. How would the concentration of citric acid cycle intermediates be affected?
(b) The decreased aconitase activity and decreased mitochondrial respiration in hyperoxia are accompanied by elevated levels of glycolysis and the pentose phosphate pathway. Explain why.

18. The scientists who carried out the hyperoxia experiments described in Problem 17 noted that they could mimic this effect by administering either fluoroacetate or fluorocitrate to cells in culture. Explain. (*Hint:* See Solution 12.)

19. Kinetic studies with aconitase in the 1970s revealed that *trans*-aconitate is a competitive inhibitor of the enzyme if *cis*-aconitate is used as the substrate. But if citrate is used as the substrate, *trans*-aconitate is a noncompetitive inhibitor. Propose a hypothesis that explains this observation.

20. A yeast mutant is isolated in which the gene for aconitase is nonfunctional. What are the consequences for the cell, particularly with regard to energy production?

21. The $\Delta G^{\circ\prime}$ value for the isocitrate dehydrogenase reaction is $-21 \text{ kJ} \cdot \text{mol}^{-1}$. What is K_{eq} for this reaction?

22. The crystal structure of isocitrate dehydrogenase shows that there is a cluster of highly conserved amino acids in the substrate binding pocket—three arginines, a tyrosine, and a lysine. Why are these residues conserved and what is a possible role for these amino acid side chains in substrate binding?

23. In bacteria, isocitrate dehydrogenase is regulated by phosphorylation of a specific Ser residue in the enzyme active site. X-ray structures of the phosphorylated and the nonphosphorylated enzyme show no significant conformational differences.
(a) How does phosphorylation regulate isocitrate dehydrogenase activity?
(b) To confirm their hypothesis, the investigators constructed a mutant enzyme in which the Ser residue was replaced with an Asp residue. The mutant was unable to bind isocitrate. Are these results consistent with the hypothesis you proposed in part (a)?

24. The expression of several enzymes changes when yeast grown on glucose are abruptly shifted to a two-carbon food source such as acetate.
(a) Why does the level of expression of isocitrate dehydrogenase increase when the yeast are shifted from glucose to acetate?
(b) The metabolism of a yeast mutant with a nonfunctional isocitrate dehydrogenase enzyme was compared to that of a wild-type yeast. The yeast were grown on glucose and then abruptly shifted to acetate as the sole carbon source. The [NAD$^+$]/[NADH] ratio was measured over a period of 48 hours. The results are shown below. Why does the ratio increase slightly at 36 hours for the wild-type yeast? Why is there a more dramatic increase in the ratio for the mutant?

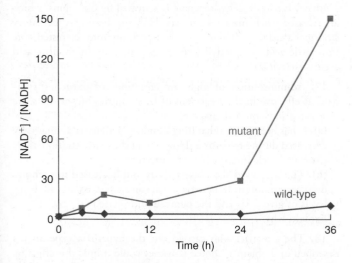

25. Using the pyruvate dehydrogenase complex reaction as a model, draw the intermediates of the α-ketoglutarate dehydrogenase reaction. Describe what happens in each of the five reaction steps.

26. Using the mechanism you drew for Problem 25, explain how succinyl phosphonate (above right) inhibits α-ketoglutarate dehydrogenase.

$$\begin{array}{c} O^- \\ | \\ {}^-O-P=O \\ | \\ C=O \\ | \\ CH_2 \\ | \\ CH_2 \\ | \\ COO^- \end{array}$$

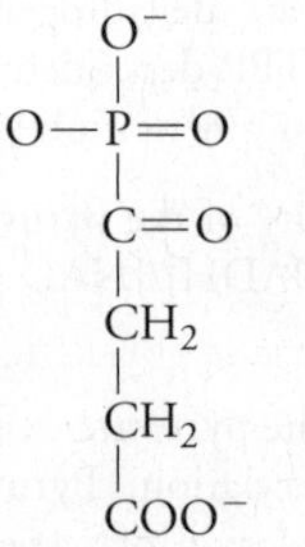

Succinyl phosphonate

27. Succinyl-CoA inhibits both citrate synthase and α-ketoglutarate dehydrogenase. How is succinyl-CoA able to inhibit both enzymes?

28. A patient with an α-ketoglutarate deficiency exhibits a small increase in blood pyruvate level and a large increase in blood lactate level, resulting in a [lactate]/[pyruvate] ratio that is many times greater than normal. Explain the reason for these symptoms.

29. Succinyl-CoA synthetase is also called succinate thiokinase. Why is the enzyme considered to be a kinase?

30. Succinyl-CoA synthetase is a dimer composed of an α subunit and a β subunit. A single gene codes for the α subunit protein. Two genes code for two different β subunit proteins—one subunit is specific for GDP, and one is specific for ADP.
(a) The β subunit specific for ADP is expressed in "catabolic tissues" such as brain and muscles, whereas the β subunit specific for GDP is expressed in "anabolic tissues" such as liver and kidneys. Propose a hypothesis to explain this observation.
(b) Individuals who are born with a mutation in the gene coding for the α subunit of the enzyme experience severe lactic acidosis and usually die within a few days of birth. Why is this mutation so deleterious?
(c) Individuals who are born with a mutation in the gene coding for the ADP-specific β subunit of the enzyme experience normal to moderately elevated concentrations of lactate and usually survive to their early 20s. Why is the prognosis for these patients better than for patients with a mutation in the gene for the α subunit?

31. Malonate is a competitive inhibitor of succinate dehydrogenase. What citric acid cycle intermediates accumulate if malonate is present in a preparation of isolated mitochondria?

32. Succinate dehydrogenase is not considered to be part of the glyoxylate pathway, yet it is vital to the proper functioning of the pathway. Why?

33. The $\Delta G^{\circ\prime}$ for the fumarase reaction is $23.4 \text{ kJ} \cdot \text{mol}^{-1}$, but the ΔG value is close to zero. What is the ratio of fumarate to malate under cellular conditions at 37°C? Is this reaction likely to be a control point for the citric acid cycle?

34. A mutant bacterial fumarase was constructed by replacing the Glu (E) at position 315 with Gln (Q). The kinetic parameters of the mutant and wild-type enzymes were compared, and the results are shown in the table below. Explain the significance of the changes.

	Wild-type enzyme	E315Q mutant enzyme
V_{max} (μmol · min^{-1} · mg^{-1})	345	32
K_M (mM)	0.21	0.25
k_{cat} (s^{-1})	1150	107
k_{cat}/K_M (M^{-1} · s^{-1})	5.6×10^6	4.3×10^5

35. Reaction 8 and Reaction 1 of the citric acid cycle can be considered to be coupled because the exergonic cleavage of the thioester bond of acetyl-CoA in Reaction 1 drives the regeneration of oxaloacetate in Reaction 8.

(a) Write the equation for the overall coupled reaction and calculate its $\Delta G^{\circ\prime}$.

(b) What is the equilibrium constant for the coupled reaction? Compare this equilibrium constant with the equilibrium constant of Reaction 8 alone.

36. Malate dehydrogenase is more active in cells oxidizing glucose aerobically than in cells oxidizing glucose anaerobically. Explain why.

37. (a) Oxaloacetate labeled at C4 with ^{14}C is added to a suspension of respiring mitochondria. What is the fate of the labeled carbon?

(b) Acetyl-CoA labeled at C1 with ^{14}C is added to a suspension of respiring mitochondria. What is the fate of the labeled carbon?

38. When leavened bread is made, the bread dough is "punched" down and then put in a warm place to "rise" to increase its volume. Give a biochemical explanation for this observation.

39. Flux through the citric acid cycle is regulated by the simple mechanisms of **(a)** substrate availability, **(b)** product inhibition, and **(c)** feedback inhibition. Give examples of each.

40. Certain microorganisms with an incomplete citric acid cycle decarboxylate α-ketoglutarate to produce succinate semialdehyde. A dehydrogenase then converts succinate semialdehyde to succinate. These reactions can be combined with other standard citric acid cycle reactions to create a pathway from citrate to oxaloacetate. How does this alternative pathway compare to the standard citric acid cycle in its ability to make free energy available to the cell?

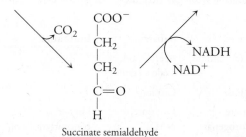

14-3 Anabolic and Catabolic Functions of the Citric Acid Cycle

41. Why is the reaction catalyzed by pyruvate carboxylase the most important anaplerotic reaction of the citric acid cycle?

42. Why is the activation of pyruvate carboxylase by acetyl-CoA a good regulatory strategy?

43. Many amino acids are broken down to intermediates of the citric acid cycle.

(a) Why can't these amino acid "remnants" be completely oxidized to CO_2 by the citric acid cycle?

(b) Explain why amino acids that are broken down to pyruvate can be completely oxidized by the citric acid cycle.

44. Describe how the transamination reaction below could function as an anaplerotic reaction for the citric acid cycle.

45. Is net synthesis of glucose in mammals possible from the following compounds?

(a) The fatty acid palmitate (16:0), which is degraded to eight acetyl-CoA

(b) The fatty acid pentadecanoate (15:0), which is degraded to six acetyl-CoA and one propionyl-CoA

(c) Glyceraldehyde-3-phosphate

(d) Leucine, which is degraded to acetyl-CoA and acetoacetate (a compound that is metabolically equivalent to two acetyl-CoA groups)

(e) Tryptophan, which is degraded to alanine and acetoacetate

(f) Phenylalanine, which is degraded to acetoacetate and fumarate

46. Pancreatic islet cells cultured in the presence of 1–20 mM glucose showed increased activities of pyruvate carboxylase and the E1 subunit of the pyruvate dehydrogenase complex proportional to the increase in glucose concentration. Explain why.

47. A physician is attempting to diagnose a neonate with a pyruvate carboxylase deficiency. An injection of alanine normally leads to a gluconeogenic response, but in the patient no such response occurs. Explain.

48. The physician treats the patient described in Problem 47 by administering glutamine. Explain why glutamine supplements are effective in treating the disease.

49. Physicians often attempt to treat a pyruvate carboxylase deficiency by administering biotin. Explain why this strategy might be effective.

50. Patients with a pyruvate dehydrogenase deficiency and patients with a pyruvate carboxylase deficiency (see Problems 47–49) both have high blood levels of pyruvate and lactate. Explain why.

51. Metabolites in rat muscle were measured before and after exercising. After exercise, the rat muscle showed an increase in oxaloacetate concentration, a decrease in phosphoenolpyruvate concentration, and no change in pyruvate concentration. Explain.

52. Oxygen does not appear as a reactant in any of the citric acid cycle reactions, yet it is essential for the proper functioning of the cycle. Explain why.

53. The activity of isocitrate dehydrogenase in *E. coli* is regulated by the covalent attachment of a phosphate group to the enzyme. Phosphorylated isocitrate dehydrogenase is inactive. When acetate is the food source for a culture of *E. coli,* isocitrate dehydrogenase is phosphorylated.

(a) Draw a diagram showing how acetate is metabolized in *E. coli.*

(b) When glucose is added to the culture, the phosphate group is removed from isocitrate dehydrogenase. How does flux through the metabolic pathways change in *E. coli* when glucose is the food source instead of acetate?

54. Yeast are unusual in that they are able to use ethanol as a gluconeogenic substrate. Ethanol is converted to glucose using the assistance of the glyoxylate pathway. Describe how the ethanol → glucose conversion takes place.

55. Animals lack a glyoxylate pathway and cannot convert fats to carbohydrates. If an animal is fed a fatty acid with all its carbons replaced by the isotope ^{14}C, some of the labeled carbons later appear in glucose. How is this possible?

56. A bacterial mutant with low levels of isocitrate dehydrogenase is able to grow normally when the culture medium is supplemented with glutamate. Explain why.

57. The purine nucleotide cycle (shown below) is an important pathway in muscle cells. The activity of the cycle increases during periods of high muscle activity. Explain how the purine nucleotide cycle contributes to the ability of the muscle cell to generate energy during intense exercise. IMP is inosine monophosphate.

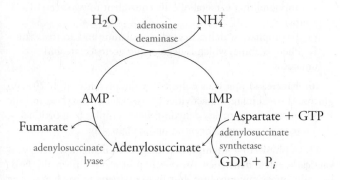

58. The plant metabolite hydroxycitrate is advertised as an agent that prevents fat buildup.

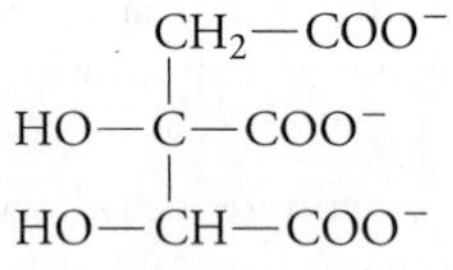

$$
\begin{array}{c}
CH_2-COO^- \\
| \\
HO-C-COO^- \\
| \\
HO-CH-COO^-
\end{array}
$$
Hydroxycitrate

(a) How does this compound differ from citrate?
(b) Hydroxycitrate inhibits the activity of ATP-citrate lyase. What kind of inhibition is likely to occur?
(c) Why might inhibition of ATP-citrate lyase block the conversion of carbohydrates to fats?
(d) The synthesis of what other compounds would be inhibited by hydroxycitrate?

59. *Helicobacter pylori* is a bacterium that colonizes the upper gastrointestinal tract in humans and is the causative agent of chronic gastritis, ulcers, and possibly gastric cancer. Knowledge of the intermediary metabolism of this organism will be helpful in the development of effective drug therapies to treat these diseases. The citric acid "cycle" in *H. pylori* is a noncyclic, branched pathway that is used to produce biosynthetic intermediates instead of metabolic energy. Succinate is produced in the "reductive branch," whereas

α-ketoglutarate is produced in the "oxidative branch." The two branches are linked by the α-ketoglutarate oxidase reaction. The pathway is shown in the diagram below.

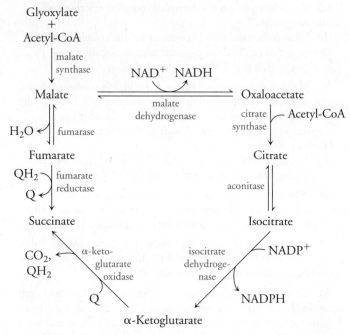

(a) Compare and contrast the citric acid cycle in *H. pylori* with the citric acid cycle in mammals.
(b) The K_M values for the enzymes listed in the table below are higher than the K_M values for the corresponding enzymes in other species of bacteria. What does this tell you about the conditions under which the citric acid cycle operates in *H. pylori*?
(c) Compare the properties of *H. pylori* citrate synthase with mammalian citrate synthase.
(d) What enzymes might serve to regulate the citric acid cycle in *H. pylori*?
(e) What enzymes might be used as drug targets for persons suffering from gastritis, ulcers, or gastric cancer?

Enzyme	Substrate	Inhibitors	Activators
Citrate synthase	Acetyl-CoA, oxaloacetate	ATP	
Aconitase	Citrate		
Isocitrate dehydrogenase (NADP$^+$-dependent)	Isocitrate, NADP$^+$	Higher concentrations of NADP$^+$, isocitrate	AMP (slight)
α-ketoglutarate oxidase	α-ketoglutarate, FAD		CoASH
Malate dehydrogenase	Oxaloacetate, NADH		
Fumarase	Malate		
Fumarate reductase	Fumarate, QH$_2$		
Malate synthase	Glyoxylate, acetyl-CoA		

60. *H. pylori,* whose citric acid cycle has an oxidative branch and a reductive branch (see Problem 59), uses amino acids and fatty acids present in the gastrointestinal tract as a source of biosynthetic intermediates.

(a) Describe how *H. pylori* uses the acetyl-CoA derived from fatty acid breakdown to synthesize glucose and glutamate.

(b) Describe how *H. pylori* converts aspartate to glutamate.

61. Yeast cells that are grown on nonfermentable substrates and then abruptly switched to glucose exhibit substrate-induced inactivation of several enzymes. Which enzymes would glucose cause to be inactivated and why?

62. Phagocytes such as macrophages and neutrophils are cellular components of the immune system that protect the host against damage caused by invading microorganisms. Phagocytes engulf and internalize a foreign microbe, forming a membrane-bound structure called a phagosome. The phagosome then fuses with the lysosome, a cellular organelle that contains a wide variety of proteolytic enzymes that destroy the pathogen if the host is fortunate. But some microbes can survive the harsh conditions of the phagolysosome. An example is *Mycobacterium tuberculosis,* which can reside in the macrophage in a dormant state for a prolonged period of time. Investigators noted that following engulfment by the macrophage, the levels of bacterial isocitrate lyase, malate synthase, citrate synthase, and malate dehydrogenase increased to levels as much as 20 times above normal.

(a) What pathway(s) does *M. tuberculosis* employ while in the phagosome and why are these pathways essential to its survival?

(b) What might be good drug targets for treating a patient infected with *M. tuberculosis*?

63. Bacteria and plants (but not animals) possess the enzyme phosphoenolpyruvate carboxylase (PPC), which catalyzes the reaction shown here.

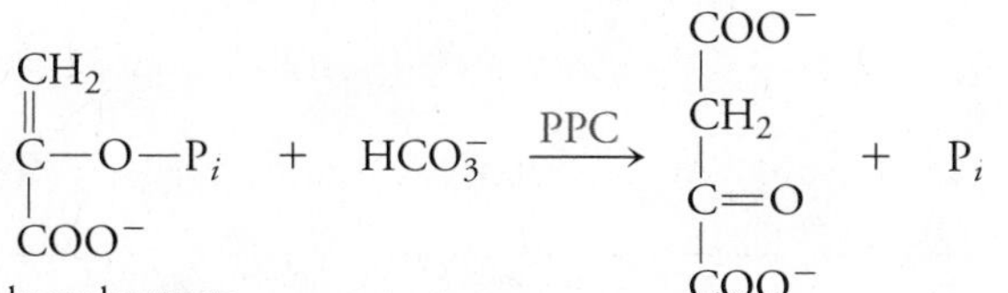

Phosphoenolpyruvate + HCO_3^- →(PPC) Oxaloacetate + P_i

(a) What is the importance of this reaction to the organism?

(b) PPC is allosterically activated by both acetyl-CoA and fructose-1,6-bisphosphate. Explain these regulatory strategies.

64. Succinic acid is an important compound used by the pharmaceutical, cosmetics, and food industries. "Green" production of succinic acid by fermenting bacteria is a more environmentally responsible way to produce the compound, which has traditionally been synthesized from petrochemicals. Investigators interested in optimizing bacterial succinic acid production noted that malate dehydrogenase activity increased under anaerobic conditions.

(a) Draw a reaction scheme outlining how succinic acid can be produced from phosphoenolpyruvate. Include the names of all reactants, products, and enzymes.

(b) Why is it essential that the production of succinic acid take place under anaerobic conditions?

65. Experiments with cancer cells grown in culture show that glutamine is consumed at a high rate and used for biosynthetic reactions, aside from protein synthesis. One possible pathway involves the conversion of glutamine to glutamate and then to α-ketoglutarate. The α-ketoglutarate can then be used to produce pyruvate for gluconeogenesis.

(a) Describe the types of reactions that convert glutamine to α-ketoglutarate.

(b) Give the sequence of enzymes that can convert α-ketoglutarate to pyruvate.

66. Many cancer cells carry out glycolysis at a high rate but convert most of the resulting pyruvate to lactate rather than to acetyl-CoA. Acetyl-CoA, however, is required for the synthesis of fatty acids, which are needed in large amounts by rapidly growing cancer cells. In these cells, the isocitrate dehydrogenase reaction apparently operates in reverse. Explain why this reaction could facilitate the conversion of amino acids such as glutamate into fatty acids.

[SELECTED READINGS]

Barry, M. J., Enzymes and symmetrical molecules, *Trends Biochem. Sci.* **22,** 228–230 (1997). [Recounts how experiments and insight revealed that the symmetrical citrate molecule can react asymmetrically.]

Brière, J.-J., Favier, J., Giminez-Roqueplo, A.-P., and Rustin, P., Tricarboxylic acid cycle dysfunction as a cause of human disease and tumor formation, *Am. J. Physiol. Cell Physiol.* **291,** C1114–C1120 (2006). [Includes discussions of the metabolic role of the citric acid cycle and its location.]

Milne, J. L. S., Wu, X., Borgnia, M. J., Lengyel, J. S., Brooks, B. R., Shi, D., Perham, R. N., and Subramaniam, S., Molecular structure of a 9-MDa icosahedral pyruvate dehydrogenase subcomplex containing the E2 and E3 enzymes using cryoelectron microscopy, *J. Biol. Chem.* **281,** 4364–4370 (2006).

Owen, O. E., Kalhan, S. C., and Hanson, R. W., The key role of anaplerosis and cataplerosis for citric acid cycle function, *J. Biol. Chem.* **277,** 30409–30412 (2002). [Describes the addition (anaplerosis) and removal (cataplerosis) of citric acid cycle intermediates in different organ systems.]

chapter 15

OXIDATIVE PHOSPHORYLATION

[Steve Byland/iStockphoto]

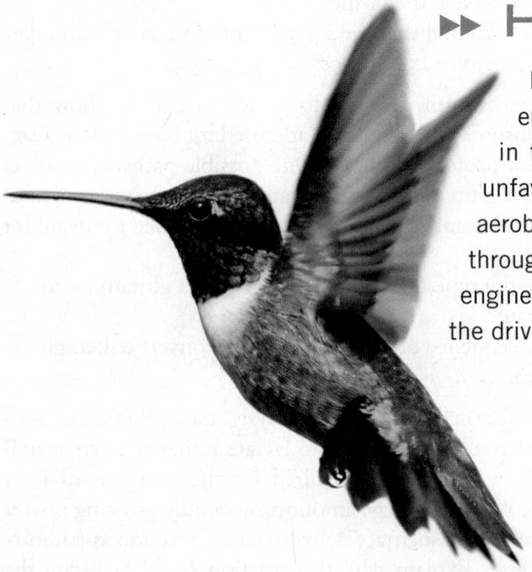

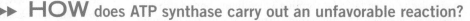

▸▸ HOW does ATP synthase carry out an unfavorable reaction?

In previous chapters, we have seen numerous examples of cells using the free energy of the ATP reaction to carry out cellular work. But in order to produce ATP in the first place, cells must add a third phosphate to ADP, a thermodynamically unfavorable reaction. Glycolysis and the citric acid cycle provide some ATP, but in aerobic organisms, such as the hummingbird shown here, most of the ATP is made through the action of a mitochondrial ATP synthase. This enzyme functions like a rotary engine, attaching a phosphoryl group to ADP as it spins. In this chapter, we'll examine the driving force for this unusual molecular machine.

THIS CHAPTER IN CONTEXT

Part 1 Foundations

Part 2 Molecular Structure and Function

Part 3 Metabolism

15 Oxidative Phosphorylation

Part 4 Genetic Information

Do You Remember?

- **Living organisms obey the laws of thermodynamics (Section 1-3).**
- **Transporters obey the laws of thermodynamics, providing a way for solutes to move down their concentration gradients or using ATP to move substances against their gradients (Section 9-1).**
- **Coenzymes such as NAD^+ and ubiquinone collect electrons from compounds that become oxidized (Section 12-2).**
- **A reaction that breaks a phosphoanhydride bond in ATP occurs with a large change in free energy (Section 12-3).**

The oxidation of metabolic fuels such as glucose, fatty acids, and amino acids as well as the oxidation of acetyl carbons to CO_2 via the citric acid cycle yields the reduced cofactors NADH and ubiquinol (QH_2). These compounds are forms of energy currency (see Section 12-3) not because they are chemically special but because their reoxidation—ultimately by molecular oxygen in aerobic organisms—is an exergonic reaction. The free energy released thereby is harvested to synthesize ATP, a phenomenon called **oxidative phosphorylation.** In the scheme introduced in Figure 12-10, oxidative phosphorylation represents the final phase of the catabolism of metabolic fuels and the major source of the cell's ATP (**Fig. 15-1**).

Oxidative phosphorylation differs from the conventional biochemical reactions we have focused on in the last two chapters. In particular, ATP synthesis is not directly coupled to a single discrete chemical reaction, such as a kinase-catalyzed reaction. Rather, *oxidative phosphorylation is a more indirect process in which free energy is converted to, or conserved as, a transmembrane gradient of protons that is then used to drive ATP synthesis.*

To understand oxidative phosphorylation, we must first consider how the reduced cofactors produced in other metabolic reactions are reoxidized by molecular oxygen. The flow of electrons from reduced compounds such as NADH and QH_2 to an oxidized compound such as O_2 is a thermodynamically favorable process. We will see that the free energy changes for electron-transfer reactions can be quantified by considering the reduction potentials of the chemical species involved. Next, we will track the movements of the electrons through a series of electron carriers that include small molecules as well as prosthetic groups of large integral membrane proteins. As electrons are shuttled from NADH and QH_2 to molecular oxygen, the membrane proteins translocate protons from one side of the membrane to the other. This process is the first step of chemiosmosis, the formation and dissipation of a transmembrane chemical gradient. Finally, we will examine the structure of ATP synthase, the enzyme complex that taps the free energy of the proton gradient to synthesize ATP from ADP + P_i.

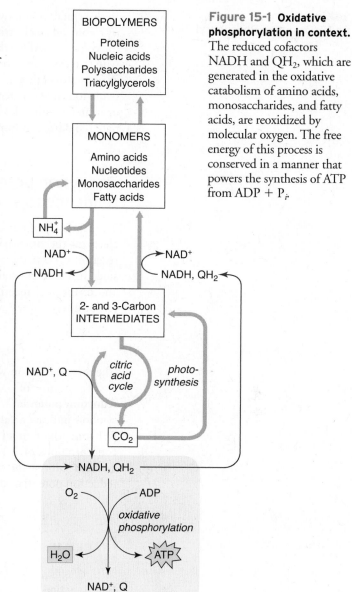

Figure 15-1 Oxidative phosphorylation in context. The reduced cofactors NADH and QH_2, which are generated in the oxidative catabolism of amino acids, monosaccharides, and fatty acids, are reoxidized by molecular oxygen. The free energy of this process is conserved in a manner that powers the synthesis of ATP from ADP + P_i.

15-1 The Thermodynamics of Oxidation-Reduction Reactions

Oxidation–reduction reactions (or redox reactions, introduced in Section 12-2) are similar to other chemical reactions in which a portion of a molecule—electrons in this case—is transferred. In any oxidation–reduction reaction, one reactant (called the **oxidizing agent** or **oxidant**) is reduced as it gains electrons. The other reactant (called the **reducing agent** or **reductant**) is oxidized as it gives up electrons:

$$A_{oxidized} + B_{reduced} \rightleftharpoons A_{reduced} + B_{oxidized}$$

For example, in the succinate dehydrogenase reaction (step 6 of the citric acid cycle; see Section 14-2), the two electrons of the reduced $FADH_2$ prosthetic group of the enzyme are transferred to ubiquinone (Q) so that $FADH_2$ is oxidized and ubiquinone is reduced:

$$\underset{(reduced)}{FADH_2} + \underset{(oxidized)}{Q} \rightleftharpoons \underset{(oxidized)}{FAD} + \underset{(reduced)}{QH_2}$$

KEY CONCEPTS
- The standard reduction potential indicates a substance's tendency to become reduced; the actual reduction potential depends on the concentrations of reactants.
- Electrons are transferred from a substance with a lower reduction potential to a substance with a higher reduction potential.
- The free energy change for an oxidation–reduction reaction depends on the change in reduction potential.

In this reaction, the two electrons are transferred as H atoms (an H atom consists of a proton and an electron, or H^+ and e^-). In oxidation–reduction reactions involving the cofactor NAD^+, the electron pair takes the form of a hydride ion (H^-, a proton with two electrons). In biological systems, electrons usually travel in pairs, although, as we will see, they may also be transferred one at a time. Note that the change in oxidation state of a reactant may be obvious, such as when Fe^{3+} is reduced to Fe^{2+}, or it may require closer inspection of the molecule's structure, such as when succinate is oxidized to fumarate (Section 14-2).

Reduction potential indicates a substance's tendency to accept electrons

The tendency of a substance to accept electrons (to become reduced) or to donate electrons (become oxidized) can be quantified. Although an oxidation–reduction reaction necessarily requires both an oxidant and a reductant, it is helpful to consider just one substance at a time, that is, a **half-reaction.** Using the example above, the half-reaction for ubiquinone (by convention, written as a reduction reaction) is

$$Q + 2\,H^+ + 2e^- \rightleftharpoons QH_2$$

(the reverse reaction would describe an oxidation half-reaction).

The affinity of a substance such as ubiquinone for electrons is its **standard reduction potential ($\mathcal{E}^{\circ\prime}$),** which has units of volts (note that the degree and prime symbols indicate a value under standard biochemical conditions where the pressure is 1 atm, the temperature is 25°C, the pH is 7.0, and all species are present at concentrations of 1 M). *The greater the value of $\mathcal{E}^{\circ\prime}$, the greater the tendency of the oxidized form of the substance to accept electrons and become reduced.* The standard reduction potentials of some biological substances are given in Table 15-1.

[TABLE 15-1] Standard Reduction Potentials of Some Biological Substances

Half-Reaction	$\mathcal{E}^{\circ\prime}$(V)
$\frac{1}{2}O_2 + 2\,H^+ + 2\,e^- \rightleftharpoons H_2O$	0.815
$SO_4^{2-} + 2\,H^+ + 2\,e^- \rightleftharpoons SO_3^{2-} + H_2O$	0.48
$NO_3^- + 2\,H^+ + 2e^- \rightleftharpoons NO_2^- + H_2O$	0.42
Cytochrome a_3 (Fe^{3+}) + $e^- \rightleftharpoons$ cytochrome a_3 (Fe^{2+})	0.385
Cytochrome a (Fe^{3+}) + $e^- \rightleftharpoons$ cytochrome a (Fe^{2+})	0.29
Cytochrome c (Fe^{3+}) + $e^- \rightleftharpoons$ cytochrome c (Fe^{2+})	0.235
Cytochrome c_1 (Fe^{3+}) + $e^- \rightleftharpoons$ cytochrome c_1 (Fe^{2+})	0.22
Cytochrome b (Fe^{3+}) + $e^- \rightleftharpoons$ cytochrome b (Fe^{2+})(*mitochondrial*)	0.077
Ubiquinone + 2 H^+ + 2 $e^- \rightleftharpoons$ ubiquinol	0.045
Fumarate$^-$ + 2 H^+ + 2 $e^- \rightleftharpoons$ succinate$^-$	0.031
FAD + 2 H^+ + 2 $e^- \rightleftharpoons$ FADH$_2$ (*in flavoproteins*)	~ 0.
Oxaloacetate$^-$ + 2 H^+ + 2 $e^- \rightleftharpoons$ malate$^-$	− 0.166
Pyruvate$^-$ + 2 H^+ + 2 $e^- \rightleftharpoons$ lactate$^-$	− 0.185
Acetaldehyde + 2 H^+ + 2 $e^- \rightleftharpoons$ ethanol	− 0.197
S + 2 H^+ + 2 $e^- \rightleftharpoons$ H$_2$S	− 0.23
Lipoic acid + 2 H^+ + 2 $e^- \rightleftharpoons$ dihydrolipoic acid	− 0.29
NAD^+ + H^+ + 2 $e^- \rightleftharpoons$ NADH	− 0.315
$NADP^+$ + H^+ + 2 $e^- \rightleftharpoons$ NADPH	− 0.320
Acetoacetate$^-$ + 2 H^+ + 2 $e^- \rightleftharpoons$ 3-hydroxybutyrate$^-$	− 0.346
Acetate$^-$ + 3 H^+ + 2 $e^- \rightleftharpoons$ acetaldehyde + H$_2$O	− 0.581

Source: Mostly from Loach, P. A., in Fasman, G. D. (ed.), *Handbook of Biochemistry and Molecular Biology* (3rd ed.), Physical and Chemical Data, Vol. I, pp. 123–130, CRC Press (1976).

Like a ΔG value, *the actual reduction potential depends on the actual concentrations of the oxidized and reduced species.* The actual reduction potential ($\mathcal{E}$) is related to the standard reduction potential ($\mathcal{E}^{\circ\prime}$) by the **Nernst equation:**

$$\mathcal{E} = \mathcal{E}^{\circ\prime} - \frac{RT}{n\mathcal{F}} \ln \frac{[A_{reduced}]}{[A_{oxidized}]} \qquad \text{[15-1]}$$

R (the gas constant) has a value of 8.3145 J $\cdot$ K^{-1} $\cdot$ mol^{-1}, T is the temperature in Kelvin, n is the number of electrons transferred (one or two in most of the reactions we will encounter), and $\mathcal{F}$ is the **Faraday constant** (96,485 J $\cdot$ V^{-1} $\cdot$ mol^{-1}; it is equivalent to the electrical charge of one mole of electrons). At 25°C (298 K), the Nernst equation reduces to

$$\mathcal{E} = \mathcal{E}^{\circ\prime} - \frac{0.026 \text{ V}}{n} \ln \frac{[A_{reduced}]}{[A_{oxidized}]} \qquad \text{[15-2]}$$

In fact, for many substances in biological systems, the concentrations of the oxidized and reduced species are similar, so the logarithmic term is small (recall that $\ln 1 = 0$) and $\mathcal{E}$ is close to $\mathcal{E}^{\circ\prime}$ (Sample Calculation 15-1).

SAMPLE CALCULATION 15-1 ●●●

PROBLEM

Calculate the reduction potential of fumarate ($\mathcal{E}^{\circ\prime} = 0.031$ V) at 25°C when [fumarate] = 40 μM and [succinate] = 200 μM.

SOLUTION

Use Equation 15-2. Fumarate is the oxidized compound and succinate is the reduced compound.

$$\mathcal{E} = \mathcal{E}^{\circ\prime} - \frac{0.026 \text{ V}}{n} \ln \frac{[A_{reduced}]}{[A_{oxidized}]}$$

$$= 0.031 \text{ V} - \frac{0.026 \text{ V}}{2} \ln \frac{(2 \times 10^{-4})}{(4 \times 10^{-5})}$$

$$= 0.031 \text{ V} - 0.021 \text{ V} = 0.010 \text{ V}$$

PRACTICE PROBLEMS ●●●

1. Calculate the reduction potential of fumarate at 37°C when [fumarate] = 80 μM and [succinate] = 100 μM.
2. Calculate the standard reduction potential of substance A when $\mathcal{E} = 0.5$ V at 25°C, [A$_{reduced}$] = 5 μM, and [A$_{oxidized}$] = 200 μM. Assume that $n = 2$.

The free energy change can be calculated from the change in reduction potential

Knowing the reduction potentials of different substances is useful for predicting the movement of electrons between the two substances. When the substances are together in solution or connected by wire in an electrical circuit, *electrons flow spontaneously from the substance with the lower reduction potential to the substance with the higher reduction potential.* For example, in a system containing Q/QH$_2$ and NAD$^+$/ NADH, we can predict whether electrons will flow from QH$_2$ to NAD$^+$ or from NADH to Q. Using the standard reduction potentials given in Table 15-1, we note that $\mathcal{E}^{\circ\prime}$ for NAD$^+$ (-0.315 V) is lower than $\mathcal{E}^{\circ\prime}$ for ubiquinone (0.045 V). Therefore, NADH will tend to transfer its electrons to ubiquinone; that is, NADH will be oxidized and Q will be reduced.

A complete oxidation–reduction reaction is just a combination of two half-reactions. For the NADH–ubiquinone reaction, the net reaction is the ubiquinone reduction half-reaction (the half-reaction as listed in Table 15-1) combined with the NADH oxidation half-reaction (the reverse of the half-reaction listed in Table 15-1).

Note that because the NAD^+ half-reaction has been reversed to indicate oxidation, we have also reversed the sign of its $\mathcal{E}^{\circ\prime}$ value:

$$NADH \rightleftharpoons NAD^+ + H^+ + 2\,e^- \qquad\qquad \mathcal{E}^{\circ\prime} = +0.315\ V$$
$$Q + 2\,H^+ + 2\,e^- \rightleftharpoons QH_2 \qquad\qquad \mathcal{E}^{\circ\prime} = 0.045\ V$$

net: $NADH + Q + H^+ \rightleftharpoons NAD^+ + QH_2 \qquad \Delta\mathcal{E}^{\circ\prime} = +0.360\ V$

When the two half-reactions are added, their reduction potentials are also added, yielding a $\Delta\mathcal{E}^{\circ\prime}$ value. Keep in mind that the reduction potential is a property of the half-reaction and is independent of the direction in which the reaction occurs. Reversing the sign of $\mathcal{E}^{\circ\prime}$, as shown above, is just a shortcut to simplify the task of calculating $\Delta\mathcal{E}^{\circ\prime}$. Another method for calculating $\Delta\mathcal{E}^{\circ\prime}$ uses the following equation:

$$\Delta\mathcal{E}^{\circ\prime} = \mathcal{E}^{\circ\prime}_{(e^-\ acceptor)} - \mathcal{E}^{\circ\prime}_{(e^-\ donor)} \qquad \textbf{[15-3]}$$

Not surprisingly, *the larger the difference in $\mathcal{E}$ values (the greater the $\Delta\mathcal{E}$ value), the greater the tendency of electrons to flow from one substance to the other, and the greater the change in free energy of the system.* ΔG is related to $\Delta\mathcal{E}$ as follows:

$$\Delta G^{\circ\prime} = -n\mathcal{F}\mathcal{E}^{\circ\prime} \quad or \quad \Delta G = -n\mathcal{F}\Delta\mathcal{E} \qquad \textbf{[15-4]}$$

Accordingly, an oxidation–reduction reaction with a large positive $\Delta\mathcal{E}$ value has a large negative value of ΔG (see Sample Calculation 15-2). Depending on the relevant reduction potentials, an oxidation–reduction reaction can release considerable amounts of free energy. This is what happens in the mitochondria, where the reduced cofactors generated by the oxidation of metabolic fuels are reoxidized. The free energy released in this process powers ATP synthesis by oxidative phosphorylation. **Figure 15-2** shows the major mitochondrial electron transport components arranged by their reduction potentials. Each stage of electron transfer, from NADH to O_2, the final electron acceptor, occurs with a negative change in free energy.

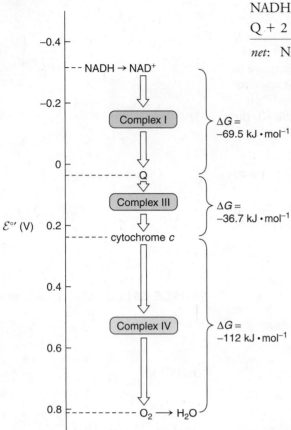

Figure 15-2 Overview of mitochondrial electron transport. The reduction potentials of the key electron carriers are indicated. The oxidation–reduction reactions mediated by Complexes I, III, and IV release free energy.

? Determine the total free energy change for the oxidation of NADH by O_2.

CONCEPT REVIEW
- Explain why an oxidation-reduction reaction must include both an oxidant and a reductant.
- When two reactants are mixed together, how can you predict which one will become reduced and which one will become oxidized?
- Explain how adding the $\mathcal{E}^{\circ\prime}$ values for two half-reactions yields a value of $\Delta\mathcal{E}^{\circ\prime}$ and $\Delta G^{\circ\prime}$ for an oxidation-reduction reaction.

●●● SAMPLE CALCULATION 15-2

PROBLEM

Calculate the standard free energy change for the oxidation of malate by NAD^+. Is the reaction spontaneous under standard conditions?

SOLUTION

Method 1

Write the relevant half-reactions, reversing the malate half-reaction (so that it becomes an oxidation reaction) and reversing the sign of its $\mathcal{E}^{\circ\prime}$:

$$malate \rightarrow oxaloacetate + 2\,H^+ + e^- \qquad\qquad \mathcal{E}^{\circ\prime} = +0.166\ V$$
$$NAD^+ + H^+ + 2e^- \rightarrow NADH \qquad\qquad \mathcal{E}^{\circ\prime} = -0.315\ V$$

net: $malate + NAD^+ \rightarrow oxaloacetate + NADH + H^+ \qquad \Delta\mathcal{E}^{\circ\prime} = -0.149\ V$

Method 2

Identify the electron acceptor (NAD^+) and electron donor (malate). Substitute their standard reduction potentials into Equation 15-3:

$$\Delta \mathcal{E}^{\circ\prime} = \mathcal{E}^{\circ\prime}_{(e^- \text{ acceptor})} - \mathcal{E}^{\circ\prime}_{(e^- \text{ donor})}$$
$$= -0.315 \text{ V} - (-0.166 \text{ V})$$
$$= -0.149 \text{ V}$$

Both Methods

The $\Delta \mathcal{E}^{\circ\prime}$ for the net reaction is -0.149 V. Use Equation 15-4 to calculate $\Delta G^{\circ\prime}$:

$$\Delta G^{\circ\prime} = -n\mathcal{F}\Delta\mathcal{E}^{\circ\prime}$$
$$= -(2)(96,485 \text{ J} \cdot \text{V}^{-1} \cdot \text{mol}^{-1})(-0.149 \text{ V})$$
$$= +28,750 \text{ J} \cdot \text{mol}^{-1} = +28.8 \text{ kJ} \cdot \text{mol}^{-1}$$

The reaction has a positive value of $\Delta G^{\circ\prime}$ and so is not spontaneous. (*In vivo*, this endergonic reaction occurs as step 8 of the citric acid cycle and is coupled to step 1, which is exergonic.)

PRACTICE PROBLEMS ●●●

3. Calculate the standard free energy change for the oxidation of malate by ubiquinone. Is the reaction spontaneous under standard conditions?
4. In yeast, alcohol dehydrogenase reduces acetaldehyde to ethanol (Section 13-1). Calculate the free energy change for this reaction under standard conditions.
5. In cells, cytochrome c oxidizes cytochrome c_1. Calculate the change in free energy for the reverse reaction.

15-2 Mitochondrial Electron Transport

In aerobic organisms, the NADH and ubiquinol produced by glycolysis, the citric acid cycle, fatty acid oxidation, and other metabolic pathways is ultimately reoxidized by molecular oxygen, a process called **respiration.** The standard reduction potential of $+0.815$ V for the reduction of O_2 to H_2O indicates that O_2 is a more effective oxidizing agent than any other biological compound (see Table 15-1). The oxidation of NADH by O_2, that is, the transfer of electrons from NADH directly to O_2, would release a large amount of free energy, but this reaction does not occur in a single step. Instead, *electrons are shuttled from NADH to O_2 in a multistep process that offers several opportunities to conserve the free energy of oxidation.* In eukaryotes, all the steps of oxidative phosphorylation are carried out by a series of membrane-bound protein complexes in the mitochondria (in prokaryotes, plasma membrane proteins perform similar functions). The following sections describe how electrons flow through this "respiratory chain" from reduced cofactors to oxygen.

Mitochondrial membranes define two compartments

In accordance with its origin as a bacterial symbiont, the **mitochondrion** (plural, *mitochondria*) has two membranes. The outer membrane, analogous to the outer membrane of some bacteria, is relatively porous due to the presence of porin-like proteins that permit the transmembrane diffusion of substances with masses up to about 10 kD (see Section 9-2 for an example of porin structure and function). The inner membrane has a convoluted architecture that encloses a space called the **mitochondrial matrix.** Because the inner mitochondrial membrane prevents the transmembrane movements of ions and small molecules (except via specific transport proteins), the composition of the matrix differs from that of the space between the inner and outer membranes. In fact, the ionic composition of the **intermembrane space** is considered to be equivalent to that of the cytosol due to the presence of the porins in the outer mitochondrial membrane (**Fig. 15-3**).

Mitochondria are customarily shown as kidney-shaped organelles with the inner mitochondrial membrane forming a system of baffles called **cristae** (**Fig. 15-4a**).

KEY CONCEPTS

- The inner mitochondrial membrane encloses the matrix and includes specific transport proteins.
- Complex I transfers electrons from NADH to ubiquinone.
- The citric acid cycle, fatty acid oxidation, and other processes also generate mitochondrial ubiquinol.
- The Q cycle mediated by Complex III reduces cytochrome c.
- Complex IV uses electrons from cytochrome c to reduce O_2 to H_2O.

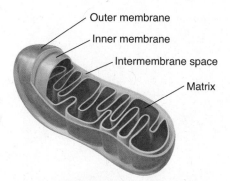

Outer membrane

Inner membrane

Intermembrane space

Matrix

Figure 15-3 Model of mitochondrial structure. The relatively impermeable inner mitochondrial membrane encloses the protein-rich matrix. The intermembrane space has an ionic composition similar to that of the cytosol because the outer mitochondrial membrane is permeable to substances with masses of less than about 10 kD.

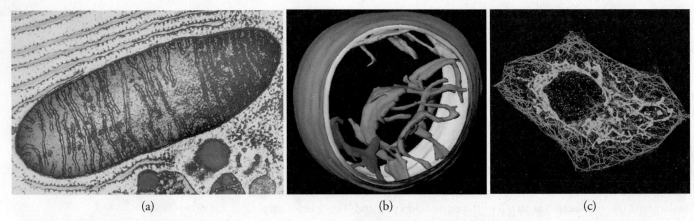

(a)　　　　　　　　　　　　(b)　　　　　　　　　　　　(c)

Figure 15-4 Images of mitochondria. (a) Conventional electron micrograph showing cristae as a system of planar baffles. [K. Porter/Photo Researchers.] (b) Three-dimensional reconstruction of a mitochondrion by electron tomography, showing irregular tubular cristae. [Courtesy Carmen Mannella, Wadsworth Center, Albany, New York.] (c) Electron micrograph of a mammalian fibroblast, showing a network of tubular mitochondria (labeled with a green fluorescent dye). The remainder of the cytosol is delineated by microtubules (labeled with a red fluorescent dye). [Courtesy Michael P. Yaffe. From *Science* 283, 1493–1497 (1991).]

⊕ **See Guided Exploration. Electron transport and oxidative phosphorylation overview.**

However, **electron tomography,** a technique for visualizing cellular structures in three dimensions by analyzing micrographs of sequential cell slices, reveals that mitochondria are highly variable structures. For example, the cristae may be irregular and bulbous rather than planar and may make several tubular connections with the rest of the inner mitochondrial membrane (Fig. 15-4b). Moreover, a cell may contain hundreds to thousands of discrete bacteria-shaped mitochondria, or a single tubular organelle may take the form of an extended network with many branches and interconnections (Fig. 15-4c). Individual mitochondria can move around the cell and undergo fusion (joining) and fission (separating).

Reflecting its ancient origin as a free-living organism, the mitochondrion has its own genome and protein-synthesizing machinery consisting of mitochondrially encoded rRNA and tRNA. The mitochondrial genome encodes 13 proteins, all of which are components of the respiratory chain complexes. This is only a small subset of the approximately 1500 proteins required for mitochondrial function; the other respiratory chain proteins, matrix enzymes, transporters, and so on are encoded by the cell's nuclear genome, synthesized in the cytosol, and imported into the mitochondria (across one or both membranes) by special mechanisms.

Much of the cell's NADH and QH_2 is generated by the citric acid cycle in the mitochondrial matrix. Fatty acid oxidation also takes place largely in the matrix and yields NADH and QH_2. These reduced cofactors transfer their electrons to the protein complexes of the respiratory electron transport chain, which are tightly associated with the inner mitochondrial membrane. However, NADH produced by glycolysis and other oxidative processes in the cytosol cannot directly reach the respiratory chain. There is no transport protein that can ferry NADH across the inner mitochondrial membrane. Instead, "reducing equivalents" are imported into the matrix by the chemical reactions of systems such as the malate–aspartate shuttle system (Fig. 15-5).

Mitochondria also need a mechanism to export ATP and to import ADP and P_i, since most of the cell's ATP is generated in the matrix by oxidative phosphorylation and is consumed in the cytosol. A transport protein called the adenine nucleotide translocase exports ATP and imports ADP, binding one or the other and changing its conformation to release the bound nucleotide on the other side of the membrane (Fig. 15-6a). Inorganic phosphate, a substrate for oxidative phosphorylation, is imported from the cytosol in symport with H^+ (Fig. 15-6b).

The protein complexes that carry out electron transport and ATP synthesis are oriented in the inner mitochondrial membrane so that they can bind the NADH, ADP, and P_i present in the matrix. Electron microscopy studies strongly suggest that

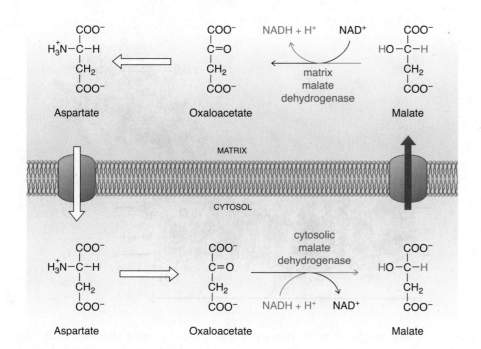

Figure 15-5 The malate-aspartate shuttle system. Cytosolic oxaloacetate is reduced to malate for transport into mitochondria. Malate is then reoxidized in the matrix. The net result is the transfer of "reducing equivalents" from the cytosol to the matrix. Mitochondrial oxaloacetate can be exported back to the cytosol after being converted to aspartate by an aminotransferase.

the complexes associate with each other, which would increase the efficiency of electron transfer between them.

Complex I transfers electrons from NADH to ubiquinone

The path electrons travel through the respiratory chain begins with Complex I, also called NADH:ubiquinone oxidoreductase or NADH dehydrogenase. This enzyme catalyzes the transfer of a pair of electrons from NADH to ubiquinone:

$$NADH + H^+ + Q \rightleftharpoons NAD^+ + QH_2$$

Complex I is the largest of the electron transport proteins in the mitochondrial respiratory chain, with 45 different subunits and a total mass of about 980 kD in mammals. The crystal structure of the smaller (550-kD) bacterial Complex I reveals an L-shaped protein with numerous transmembrane helices and a peripheral arm (**Fig. 15-7**). Electron transport takes place in the peripheral arm, which includes several prosthetic groups that undergo reduction as they receive electrons and become oxidized as they give up their electrons to the next group. All these groups, or **redox centers,** appear to have reduction potentials approximately between the reduction potentials of NAD^+ ($\mathcal{E}°' = -0.315$ V) and ubiquinone ($\mathcal{E}°' = +0.045$ V). This allows them to form a chain where the electrons travel a path of increasing reduction potential. The redox centers do not need to be in intimate contact with each

Figure 15-6 Mitochondrial transport systems. (a) The adenine nucleotide translocase binds either ATP or ADP and changes its conformation to release the nucleotide on the opposite side of the inner mitochondrial membrane. This transporter can therefore export ATP and import ADP. (b) A P_i–H^+ symport protein permits the simultaneous movement of inorganic phosphate and a proton into the mitochondrial matrix.

? **Does the activity of either of these transporters contribute to the mitochondrial membrane potential?**

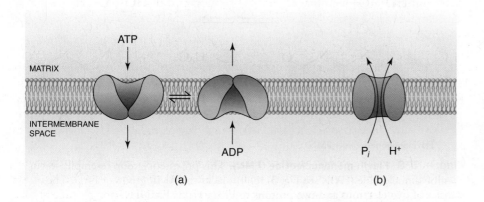

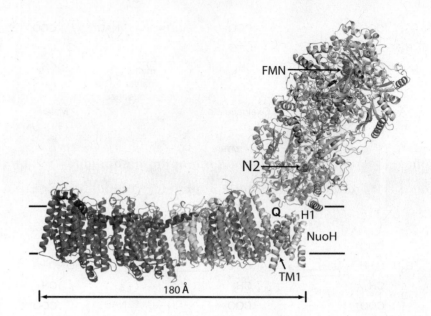

other, as they would be if the transferred group were a larger chemical entity. An electron can move between redox centers up to 14 Å apart by "tunneling" through the covalent bonds of the protein.

The two electrons donated by NADH are first picked up by flavin mononucleotide (FMN; Fig. 15-8). This noncovalently bound prosthetic group, which is similar to FAD, then transfers the electrons, one at a time, to a second type of redox center, an iron–sulfur (Fe–S) cluster. Complex I bears nine of these prosthetic groups, which contain equal numbers of iron and sulfide ions (Fig. 15-9). Unlike the electron carriers we have introduced so far, Fe–S clusters are one-electron carriers. They have an oxidation state of either +3 (oxidized) or +2 (reduced), regardless of the number of Fe atoms in the cluster (each cluster is a conjugated structure that functions as a single unit). Electrons travel between several Fe–S clusters before reaching ubiquinone. Like FMN, ubiquinone is a two-electron carrier (see Section 12-2), but it accepts one electron at a time from an Fe–S donor. Iron–sulfur clusters may be among the most ancient of electron carriers, dating from a time when the earth's abundant iron and sulfur were major players in prebiotic chemical reactions.

As electrons are transferred from NADH to ubiquinone, Complex I transfers four protons from the matrix to the intermembrane space. Comparisons with other transport proteins and detailed analysis of the crystal structure indicate the presence of

Figure 15-8 Flavin mononucleotide (FMN). This prosthetic group resembles flavin adenine dinucleotide (FAD; see Fig. 3-3c) but lacks the AMP group of FAD. The transfer of two electrons and two protons to FMN yields $FMNH_2$.

four proton-translocating "channels" in the membrane-embedded arm of Complex I. When redox groups in the peripheral arm are transiently reduced and reoxidized, the protein undergoes conformational changes that are transmitted from the peripheral arm to the membrane arm in part by a horizontally oriented helix that lies within the membrane portion of the complex (see Fig. 15-7). These conformational changes do not open passageways, as occurs in Na^+ and K^+ transporters (Section 9-2). Instead, each proton passes from one side of the membrane to the other via a **proton wire,** a series of hydrogen-bonded protein groups plus water molecules that form a chain through which a proton can be rapidly relayed. (Recall from Fig. 2-14 that protons readily jump between water molecules.) Note that in this relay mechanism, the protons taken up from the matrix are not the same ones that are released into the intermembrane space. The reactions of Complex I are summarized in **Figure 15-10.**

Other oxidation reactions contribute to the ubiquinol pool

The reduced quinone product of the Complex I reaction joins a pool of quinones that are soluble in the inner mitochondrial membrane by virtue of their long hydrophobic isoprenoid tails (see Section 12-2). *The pool of reduced quinones is augmented by the activity of other oxidation–reduction reactions.* One of these is catalyzed by succinate dehydrogenase, which carries out step 6 of the citric acid cycle (see Section 14-2).

$$\text{succinate} + Q \rightleftharpoons \text{fumarate} + QH_2$$

Succinate dehydrogenase is the only one of the citric acid cycle enzymes that is not soluble in the mitochondrial matrix; it is embedded in the inner membrane. Like the other respiratory complexes, it contains several redox centers, including an FAD group. Succinate dehydrogenase is also called Complex II of the mitochondrial respiratory chain. However, it is more like a tributary because it does not undertake proton translocation and therefore does not directly contribute the free energy of its oxidation–reduction reaction toward ATP synthesis. Nevertheless, it does feed reducing equivalents as ubiquinol into the electron transport chain (**Fig. 15-11a**).

A major source of ubiquinol is fatty acid oxidation, another energy-generating catabolic pathway that takes place in the mitochondrial matrix. A membrane-bound fatty acyl-CoA dehydrogenase catalyzes the oxidation of a C—C bond in a fatty acid attached to coenzyme A. The electrons removed in this dehydrogenation reaction are transferred to ubiquinone (Fig. 15-11b). As we will see in Section 17-1, the complete oxidation of a fatty acid also produces NADH that is reoxidized by the mitochondrial electron transport chain, starting with Complex I.

Electrons from cytosolic NADH can also enter the mitochondrial ubiquinol pool through the actions of a cytosolic and a mitochondrial glycerol-3-phosphate dehydrogenase (Fig. 15-11c). This system, which shuttles electrons from NADH to ubiquinol, bypasses Complex I.

Complex III transfers electrons from ubiquinol to cytochrome *c*

Ubiquinol is reoxidized by Complex III, an integral membrane protein with 11 subunits in each of its two monomeric units. Complex III, also called ubiquinol:cytochrome *c* oxidoreductase or cytochrome bc_1, transfers electrons to the peripheral membrane protein cytochrome *c*. **Cytochromes** are proteins with heme prosthetic groups. The name *cytochrome* literally means "cell color"; cytochromes are largely responsible for the purplish-brown color of mitochondria. Cytochromes are commonly named with a letter (*a*, *b*, or *c*) indicating the exact

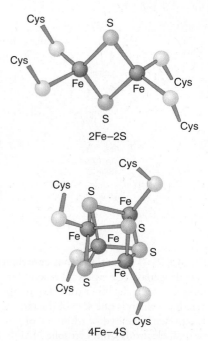

Figure 15-9 Iron–sulfur clusters. Although some Fe–S clusters contain up to eight Fe atoms, the most common are the 2Fe–2S and 4Fe–4S clusters. In all cases, the iron–sulfur clusters are coordinated by the S atoms of Cys side chains. These prosthetic groups undergo one-electron redox reactions.

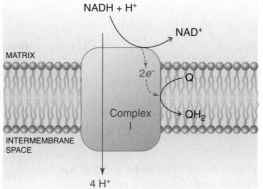

Figure 15-10 Complex I function. As two electrons from the water-soluble NADH are transferred to the lipid-soluble ubiquinone, four protons are translocated from the matrix into the intermembrane space.

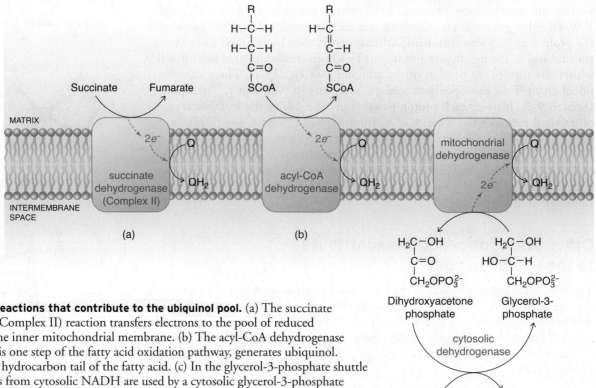

Figure 15-11 Reactions that contribute to the ubiquinol pool. (a) The succinate dehydrogenase (Complex II) reaction transfers electrons to the pool of reduced ubiquinone in the inner mitochondrial membrane. (b) The acyl-CoA dehydrogenase reaction, which is one step of the fatty acid oxidation pathway, generates ubiquinol. R represents the hydrocarbon tail of the fatty acid. (c) In the glycerol-3-phosphate shuttle system, electrons from cytosolic NADH are used by a cytosolic glycerol-3-phosphate dehydrogenase to reduce dihydroxyacetone phosphate to glycerol-3-phosphate. The mitochondrial enzyme, embedded in the inner membrane, then reoxidizes the glycerol-3-phosphate, ultimately transferring the two electrons to the membrane ubiquinone pool.

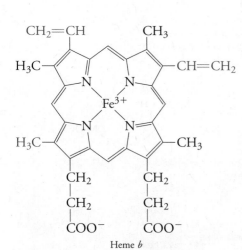

Figure 15-12 The heme group of a *b* cytochrome. The planar porphyrin ring surrounds a central Fe atom, shown here in its oxidized (Fe^{3+}) state. The heme substituent groups that are colored blue differ in the *a* and *c* cytochromes (the heme group of hemoglobin and myoglobin has the *b* structure; see Section 5-1).

structure of the porphyrin ring of their heme group (**Fig. 15-12**). The structure of the heme group and the surrounding protein microenvironment influence the protein's absorption spectrum. They also determine the reduction potentials of cytochromes, which range from about -0.080 V to about $+0.385$ V.

Unlike the heme prosthetic groups of hemoglobin and myoglobin, the heme groups of cytochromes undergo reversible one-electron reduction, with the central Fe atom cycling between the Fe^{3+} (oxidized) and Fe^{2+} (reduced) states. Consequently, the net reaction for Complex III, in which two electrons are transferred, includes two cytochrome *c* proteins:

$$QH_2 + 2 \text{ cytochrome } c \text{ (Fe}^{3+}) \rightleftharpoons Q + 2 \text{ cytochrome } c \text{ (Fe}^{2+}) + 2\,H^+$$

Complex III itself contains two cytochromes (cytochrome *b* and cytochrome c_1) that are integral membrane proteins. These two proteins, along with an iron–sulfur protein (also called the Rieske protein), form the functional core of Complex III (these same three subunits are the only ones that have homologs in the corresponding bacterial respiratory complex). Altogether, each monomer of Complex III is anchored in the membrane by 14 transmembrane α helices (**Fig. 15-13**).

The flow of electrons through Complex III is complicated, in part because the two electrons donated by ubiquinol must split up in order to travel through a series of one-electron carriers that includes the 2Fe–2S cluster of the iron–sulfur protein, cytochrome c_1, and cytochrome *b* (which actually contains two heme groups with slightly different reduction potentials). Except for the 2Fe–2S cluster, all the redox centers are arranged in such a way that electrons can tunnel from one to another. The iron–sulfur protein must change its conformation by rotating and moving about 22 Å in order to pick up and deliver an electron. Further complicating the picture is the fact that each monomeric unit of Complex III has two active sites where quinone cofactors undergo reduction and oxidation.

The circuitous route of electrons from ubiquinol to cytochrome *c* is described by the two-round **Q cycle**, diagrammed in **Figure 15-14**. *The net result of the Q cycle is*

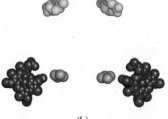

(a)

(b)

Figure 15-13 Structure of mammalian Complex III.
(a) Backbone model. Eight transmembrane helices in each monomer of the dimeric complex are contributed by cytochrome *b* (yellow). The iron–sulfur protein (orange) and cytochrome *c*$_1$ (red) project into the intermembrane space. The approximate position of the membrane is indicated.
(b) Arrangement of prosthetic groups. The two heme groups of each cytochrome *b* (yellow) and the heme group of cytochrome *c*$_1$ (red), along with the iron–sulfur clusters (orange), provide a pathway for electrons between ubiquinol (in the membrane) and cytochrome *c* (in the intermembrane space). [Structure (pdb 1BE3) determined by S. Iwata, J. W. Lee, K. Okada, J. K. Lee, M. Iwata, S. Ramaswamy, and B. K. Jap.] ⊕ **See Interactive Exercise. Complex III.**

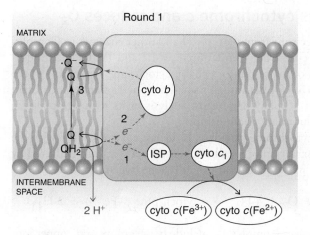

Round 1

1. In the first round, QH$_2$ donates one electron to the iron–sulfur protein (ISP). The electron then travels to cytochrome *c*$_1$ and then to cytochrome *c*.

2. QH$_2$ donates its other electron to cytochrome *b*. The two protons from QH$_2$ are released into the intermembrane space.

3. The oxidized ubiquinone diffuses to another quinone-binding site, where it accepts the electron from cytochrome *b*, becoming a half-reduced semiquinone (·Q$^-$).

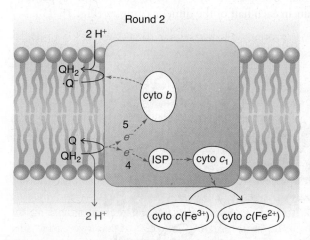

Round 2

4. In the second round, a second QH$_2$ surrenders its two electrons to Complex III and its two protons to the intermembrane space. One electron goes to reduce cytochrome *c*.

5. The other electron goes to cytochrome *b* and then to the waiting semiquinone produced in the first part of the cycle. This step regenerates QH$_2$, using protons from the matrix.

Figure 15-14 The Q cycle.

? Write an equation for round 1 and for round 2.

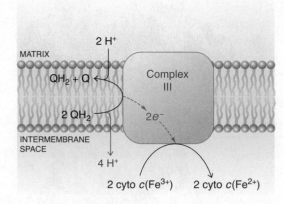

Figure 15-15 Complex III function. For every two electrons that pass from ubiquinol to cytochrome *c*, four protons are translocated to the intermembrane space.

- -

? **How does the proton-translocating mechanism of Complex III differ from the one in Complex I?**

that two electrons from QH₂ reduce two molecules of cytochrome c. In addition, four protons are translocated to the intermembrane space, two from QH₂ in the first round of the Q cycle and two from QH₂ in the second round. This proton movement contributes to the transmembrane proton gradient. The reactions of Complex III are summarized in **Figure 15-15**.

Complex IV oxidizes cytochrome c and reduces O₂

Just as ubiquinone ferries electrons from Complex I and other enzymes to Complex III, cytochrome *c* ferries electrons between Complexes III and IV. Unlike ubiquinone and the other proteins of the respiratory chain, cytochrome *c* is soluble in the intermembrane space (**Fig. 15-16**). Because this small peripheral membrane protein is central to the metabolism of many organisms, analysis of its sequence played a large role in elucidating evolutionary relationships.

Complex IV, also called cytochrome *c* oxidase, is the last enzyme to deal with the electrons derived from the oxidation of metabolic fuels. Four electrons delivered by cytochrome *c* are consumed in the reduction of molecular oxygen to water:

$$4 \text{ cytochrome } c \text{ (Fe}^{2+}) + O_2 + 4 \text{ H}^+ \rightarrow 4 \text{ cytochrome } c \text{ (Fe}^{3+}) + 2 \text{ H}_2\text{O}$$

The redox centers of mammalian Complex IV include heme groups and copper ions situated among the 13 subunits in each half of the dimeric complex (**Fig. 15-17**).

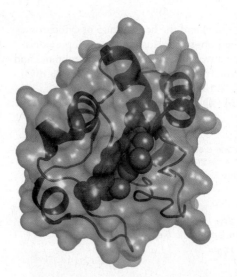

Figure 15-16 Cytochrome c. The protein is shown as a gray transparent surface over its ribbon backbone. The heme group (pink) lies in a deep pocket. Cytochrome *c* transfers one electron at a time from Complex III to Complex IV. [Structure of tuna cytochrome *c* (pdb 5CYT) determined by T. Takano.]

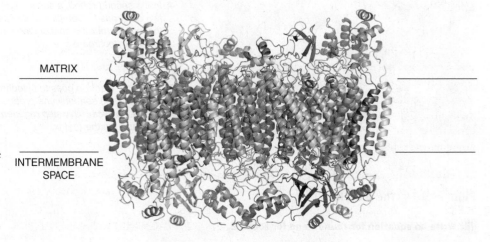

Figure 15-17 Structure of cytochrome c oxidase. The 13 subunits in each monomeric half of the mammalian complex comprise 28 transmembrane α helices. [Structure (pdb 2OCC) determined by T. Tsukihara and M. Yao.] ⊕ See Interactive Exercise. Cytochrome *c* oxidase.

Each electron travels from cytochrome c to the Cu_A redox center, which has two copper ions, and then to a heme a group. From there it travels to a binuclear center consisting of the iron atom of heme a_3 and a copper ion (Cu_B). The four-electron reduction of O_2 occurs at the Fe–Cu binuclear center. Note that the chemical reduction of O_2 to H_2O consumes four protons from the mitochondrial matrix. One possible sequence of reaction intermediates is shown in **Figure 15-18**. The incomplete reduction of O_2 to H_2O is believed to generate free radicals that can damage mitochondria (Box 15-A).

Cytochrome c oxidase also relays four additional protons from the matrix to the intermembrane space (two protons for every pair of electrons). The protein complex appears to harbor two proton wires. One delivers H^+ ions from the matrix to the oxygen-reducing active site. The other one spans the 50-Å distance between the matrix and intermembrane faces of the protein. Protons are relayed through the proton wires when the protein changes its conformation in response to changes in its oxidation

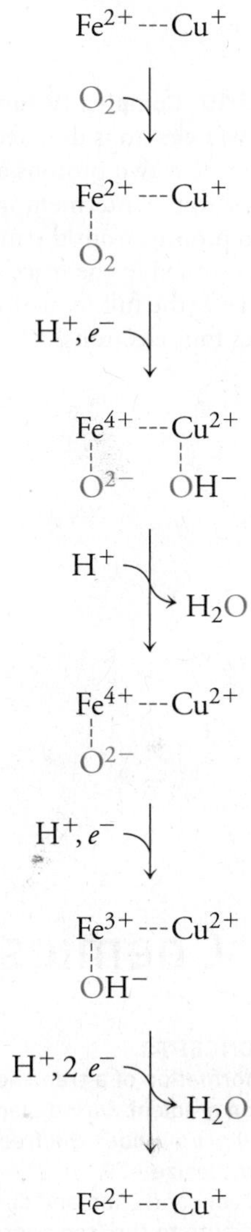

Figure 15-18 A proposed model for the cytochrome *c* oxidase reaction. Although the exact sequence of proton and electron transfers is not known, the reaction intermediates shown here are inferred from spectroscopic and other evidence. An enzyme tyrosine radical (not shown) plays a role in electron transfer.

BOX 15-A ⬡ BIOCHEMISTRY NOTE

Free Radicals and Aging

The oxidizing power of molecular oxygen allows aerobic metabolism—a far more efficient strategy than anaerobic metabolism—but comes at a cost. Partial reduction of O_2 by Complex IV, or possibly via side reactions carried out in Complexes I and III, can produce the superoxide free radical, $\cdot O_2^-$.

$$O_2 + e^- \rightarrow \cdot O_2^-$$

A **free radical** is an atom or molecule with a single unpaired electron and is highly reactive as it seeks another electron to form a pair. Such reactivity means that a free radical, although extremely short-lived (the half-life of $\cdot O_2^-$ is 1×10^{-6} seconds), can chemically alter nearby molecules. Presumably, the most damage is felt by mitochondria, whose proteins, lipids, and DNA are all susceptible to oxidation as superoxide steals an electron.

As the damage accumulates, the mitochondria become less efficient and eventually nonfunctional, at which point the cell self-destructs. According to the free radical theory of aging, oxidative damage mediated by $\cdot O_2^-$ and other free radicals is responsible for the degeneration of tissues that occurs with aging. Oxidative damage has also been implicated in the pathogenesis of disorders such as Parkinson's disease and Alzheimer's disease (see Box 4-C).

Several lines of evidence support the link between free radicals and aging. First, the tissues of some individuals with progeria, a form of accelerated aging, appear to produce higher than normal levels of oxygen free radicals. Second, cells of all kinds are equipped with antioxidant mechanisms, suggesting that these components perform an essential function. For example, the enzyme superoxide dismutase converts superoxide to a less toxic product, peroxide:

$$2 \cdot O_2^- + 2 H^+ \rightarrow H_2O_2 + O_2$$

Other cellular components, such as ascorbate (see Box 5-D) and α-tocopherol (see Box 8-B) may protect cells from oxidative damage by scavenging free radicals. Finally, animal experiments suggest that caloric restriction, which extends life spans, generates fewer free radicals by decreasing the availability of fuel molecules that undergo oxidative metabolism. Unfortunately, studies in humans have not yielded conclusive evidence that consuming particular antioxidants or decreasing fuel consumption diminishes the degeneration that normally accompanies aging.

● **Question:** Free radicals have been identified as hormone-like signaling molecules in both animals and plants. How might this information alter the free radical theory of aging?

Figure 15-19 Complex IV function.
For every two electrons donated
by cytochrome *c*, two protons are
translocated to the intermembrane
space. Two protons from the matrix
are also consumed in the reaction
$\frac{1}{2} O_2 \rightarrow H_2O$ (the full reduction of
O_2 requires four electrons).

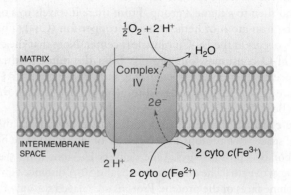

state. *The production of water and the proton relays both deplete the matrix proton concentration and thereby contribute to the formation of a proton gradient across the inner mitochondrial membrane* (**Fig. 15-19**).

> **CONCEPT REVIEW**
> • Describe the compartments of a mitochondrion.
> • What transport proteins occur in the inner mitochondrial membrane?
> • List the different types of redox centers in the respiratory electron transport chain. Which are one-electron carriers and which are two-electron carriers?
> • How are electrons delivered to Complexes I, III, and IV?
> • Why is oxygen the final electron acceptor of the respiratory chain?
> • What is the function of a proton wire?

15-3 Chemiosmosis

> **KEY CONCEPTS**
> • The formation of a transmembrane proton gradient during electron transport provides the free energy to synthesize ATP.
> • Both concentration and charge contribute to the free energy of the proton gradient.

The electrons collected from metabolic fuels during their oxidation are now fully disposed of in the reduction of O_2 to H_2O. However, their free energy has been conserved. How much free energy is potentially available? Using the ΔG values calculated from the standard reduction potentials of the substrates and products of Complexes I, III, and IV (presented graphically in Fig. 15-2), we can see that each of the three respiratory complexes theoretically releases enough free energy to drive the endergonic phosphorylation of ADP to form ATP ($\Delta G^{\circ\prime} = +30.5$ kJ · mol^{-1}).

Complex I: NADH $\rightarrow$ QH$_2$	$\Delta G^{\circ\prime} = -69.5$ kJ · mol^{-1}
Complex III: QH$_2$ $\rightarrow$ cytochrome c	$\Delta G^{\circ\prime} = -36.7$ kJ · mol^{-1}
Complex IV: cytochrome c $\rightarrow$ O$_2$	$\Delta G^{\circ\prime} = -112.0$ kJ · mol^{-1}
NADH $\rightarrow$ O$_2$	$\Delta G^{\circ\prime} = -218.2$ kJ · mol^{-1}

Chemiosmosis links electron transport and oxidative phosphorylation

Until the 1960s, the connection between respiratory electron transport (measured as O_2 consumption) and ATP synthesis was a mystery. Credit for discovering the connection belongs primarily to Peter Mitchell, who was inspired by his work on mitochondrial phosphate transport and recognized the importance of compartmentation in biological systems. Mitchell's **chemiosmotic theory** proposed that the proton-translocating activity of the electron transport complexes in the inner mitochondrial membrane generates a proton gradient across the membrane. The protons cannot diffuse back into the matrix because the membrane is impermeable to ions. *The imbalance of protons represents a source of free energy, also called a **protonmotive force**, that can drive the activity of an ATP synthase.*

We now know that for each pair of electrons that flow through Complexes I, III, and IV, 10 protons are translocated from the matrix to the intermembrane space (which is ionically equivalent to the cytosol). In bacteria, electron transport complexes in the plasma membrane translocate protons from the cytosol to the cell exterior. Mitchell's theory of chemiosmosis actually explains more than just aerobic respiration. It also applies to systems where the energy from sunlight is used to generate a transmembrane proton gradient (this aspect of photosynthesis is described in Section 16-2).

The proton gradient is an electrochemical gradient

When the mitochondrial complexes translocate protons across the inner mitochondrial membrane, the concentration of H^+ outside increases and the concentration of H^+ inside decreases (**Fig. 15-20**). *This imbalance of protons, a nonequilibrium state, has an associated free energy (the force that would restore the system to equilibrium).* The free energy of the proton gradient has two components, reflecting the difference in the concentration of the chemical species and the difference in electrical charge of the positively charged protons (for this reason, the mitochondrial proton gradient is referred to as an electrochemical gradient rather than a simple concentration gradient). The free energy change for generating the chemical imbalance of protons is

$$\Delta G = RT \ln \frac{[H^+_{out}]}{[H^+_{in}]} \qquad \textbf{[15-5]}$$

The pH ($-\log [H^+]$) of the intermembrane space (*out*) is typically about 0.75 units less than the pH of the matrix (*in*).

The free energy change for generating the electrical imbalance of protons is

$$\Delta G = ZF\Delta\psi \qquad \textbf{[15-6]}$$

where Z is the ion's charge ($+1$ in this case) and $\Delta\psi$ is the membrane potential caused by the imbalance in positive charges (see Section 9-1). For mitochondria, $\Delta\psi$ is positive, usually 150 to 200 mV. This value indicates that the intermembrane space or cytosol is more positive than the matrix (recall from Section 9-1 that for a whole cell, the cytosol is more negative than the extracellular space and $\Delta\psi$ is negative).

Combining the chemical and electrical effects gives an overall free energy change for transporting protons from the matrix (*in*) to the intermembrane space (*out*):

$$\Delta G = RT \ln \frac{[H^+_{out}]}{[H^+_{in}]} + ZF\Delta\psi \qquad \textbf{[15-7]}$$

Typically, the free energy change for translocating one proton out of the matrix is about $+20$ kJ $\cdot$ mol^{-1} (see Sample Calculation 15-3 for a detailed application of Equation 15-7). This is a thermodynamically costly event. Passage of the proton back *into* the matrix, following its electrochemical gradient, would have a free energy change of about -20 kJ $\cdot$ mol^{-1}. This event is thermodynamically favorable, but it does not provide enough free energy to drive the synthesis of ATP. However, the 10 protons translocated for each pair of electrons transferred from NADH to O_2 have an associated proton motive force of over 200 kJ $\cdot$ mol^{-1}, enough to drive the phosphorylation of several molecules of ADP.

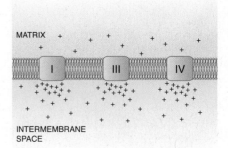

Figure 15-20 Generation of a proton gradient. During the oxidation–reduction reactions catalyzed by mitochondrial Complexes I, III, and IV, protons (represented by positive charges) are translocated out of the matrix into the intermembrane space. This creates an imbalance in both proton concentration and electrical charge.

SAMPLE CALCULATION 15-3 ●●●

PROBLEM

Calculate the free energy change for translocating a proton out of the mitochondrial matrix, where pH$_{matrix}$ = 7.8, pH$_{cytosol}$ = 7.15, $\Delta\psi$ = 170 mV, and T = 25°C.

SOLUTION

Since pH = $-\log [H^+]$ (Equation 2-4), the logarithmic term of Equation 15-7 can be rewritten. Equation 15-7 then becomes

$$\Delta G = 2.303\, RT\, (\text{pH}_{in} - \text{pH}_{out}) + ZF\Delta\psi$$

(*continued on next page*)

Substituting known values gives

$$\Delta G = 2.303\ (8.3145\ \text{J} \cdot \text{K}^{-1} \cdot \text{mol}^{-1})(298\ \text{K})(7.8 - 7.15)$$
$$+ (1)(96,485\ \text{J} \cdot \text{V}^{-1} \cdot \text{mol}^{-1})(0.170\ \text{V})$$
$$= 3700\ \text{J} \cdot \text{mol}^{-1} + 16,400\ \text{J} \cdot \text{mol}^{-1}$$
$$= +20.1\ \text{kJ} \cdot \text{mol}^{-1}$$

● ● ● ● **PRACTICE PROBLEMS**

6. Calculate the free energy change for transporting a proton out of the mitochondrial matrix, where $\text{pH}_{matrix} = 7.6$, $\text{pH}_{cytosol} = 7.35$, $\Delta\psi = 170$ mV, and $T = 37°\text{C}$.
7. What size pH gradient (the difference between pH_{matrix} and $\text{pH}_{cytosol}$) would correspond to a free energy change of 30.5 kJ $\cdot$ mol^{-1}? Assume that $\Delta\psi = 170$ mV and $T = 25°\text{C}$.

> **CONCEPT REVIEW**
> • What is the source of the protons that form the gradient across the inner mitochondrial membrane?
> • Why does the free energy of the proton gradient have a chemical and an electrical component?

15-4 ATP Synthase

> **KEY CONCEPTS**
> • Proton translocation drives the rotation of a portion of ATP synthase.
> • Rotation-induced conformational changes allow ATP synthase to bind ADP and P_i, to phosphorylate ADP, and to release ATP.
> • Because ATP synthesis is indirectly linked to electron transport, the P:O ratio is not a whole number.
> • The supply of reduced cofactors determines the rate of oxidative phosphorylation.

⊕ **See Guided Exploration. ATP synthase.**

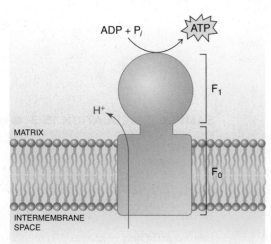

Figure 15-21 **ATP synthase function.** As protons flow through the F_0 component from the intermembrane space to the matrix, the F_1 component catalyzes the synthesis of ATP from ADP + P_i.

The protein that taps the electrochemical proton gradient to phosphorylate ADP is known as the F-ATP synthase (or Complex V). One part of the protein, called F_0, functions as a transmembrane channel that permits H^+ to flow back into the matrix, following its gradient. The F_1 component catalyzes the reaction ADP + P_i → ATP + H_2O (Fig. 15-21). This section describes the structures of the two components of ATP synthase and shows how their activities are linked so that exergonic H^+ transport can be coupled to endergonic ATP synthesis.

ATP synthase rotates as it translocates protons

Not surprisingly, the overall structure of ATP synthase is conserved among different species. The F_1 component consists of three α and three β subunits surrounding a central shaft. The membrane-embedded portion of ATP synthase includes an *a* subunit, two *b* subunits that extend upward to interact with the F_1 component, and a ring of *c* subunits (Fig. 15-22). The exact number of *c* subunits varies with the source; bovine mitochondrial ATP synthase, for example, has 8 *c* subunits, while some bacterial enzymes have 15 *c* subunits.

In all species, proton transport through ATP synthase involves the rotation of the *c* ring past the stationary *a* subunit. The carboxylate side chain of a highly conserved Asp or Glu residue on each *c* subunit serves as a proton binding site (Fig. 15-23). When properly positioned at the *a* subunit, a *c* subunit can take up a proton from the intermembrane space. A slight rotation of the *c* ring brings another *c* subunit into position so that it can release its bound proton into the matrix. The favorable thermodynamics of proton translocation force the *c* ring to keep moving in one direction. Experiments show that depending on the relative concentrations of protons on the two sides of the membrane, the *c* ring can actually spin in either direction. Related proteins, known as P- and V-ATPases, in fact function as

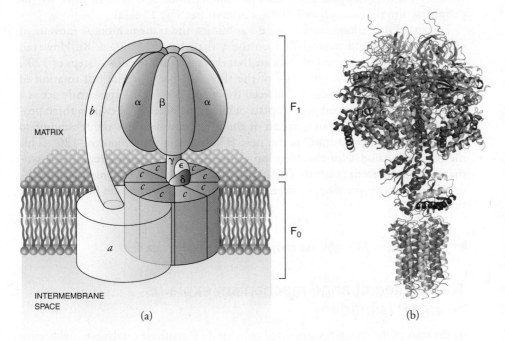

Figure 15-22 **Structure of ATP synthase.** (a) Model of the mammalian enzyme with individual subunits labeled. A spherical structure consisting of three α and three β subunits is connected via a central stalk (subunits γ, δ, and ε) to the membrane-embedded c ring. The a subunit is closely associated with the c ring, and a peripheral stalk containing several subunits (including b) links subunit a to the catalytic domain. (b) X-Ray structure of bovine ATP synthase at 3.5-Å resolution. Some subunits are not visible in the crystal structure. [Structure (pdb 2XND) determined by I. N. Watt, M. G. Montgomery, M. J. Runswick, A. G. W. Leslie, and J. E. Walker.]

active transporters that use the free energy of the ATP hydrolysis reaction to drive ion movement across the membrane.

Attached to the c ring and rotating along with it are the γ, δ, and ε subunits (Fig. 15-22). The δ and ε subunits are relatively small, but the γ subunit consists of two long α helices arranged as a bent coiled coil that protrudes into the center of the globular F_1 structure. The three α and three β subunits of F_1 have similar tertiary structures and are arranged like the sections of an orange around the γ subunit (**Fig. 15-24**). Although all six subunits can bind adenine nucleotides, only the β subunits have catalytic activity (nucleotide binding to the α subunits may play a regulatory role).

A close examination of the F_1 assembly reveals that the γ subunit interacts asymmetrically with the three pairs of αβ units. In fact, each αβ unit has a slightly different conformation, and model-building indicates that for steric reasons, the three units cannot simultaneously adopt the same conformation. *The three αβ pairs change their conformations as the γ subunit rotates* (it is like a shaft driven by the c ring

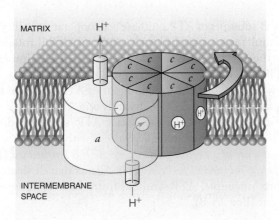

Figure 15-23 **Mechanism of proton transport by ATP synthase.** When a c subunit (purple) binds a proton from one side of the membrane, it moves away from the a subunit (blue). Because the c subunits form a ring, rotation brings another c subunit toward the a subunit, where it releases its bound proton to the opposite side of the membrane. In mammalian ATP synthase (shown here), one complete rotation of the c ring corresponds to the translocation of eight protons.

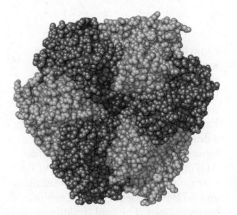

Figure 15-24 **Structure of the F_1 component of ATP synthase.** The alternating α (blue) and β (green) subunits form a hexamer around the end of the γ shaft (purple). This view is looking from the matrix down onto the top of the ATP synthase structure shown in Figure 15-22b. [Structure (pdb 1E79) determined by C. Gibbons, M. G. Montgomery, A. G. W. Leslie, and J. E. Walker.]

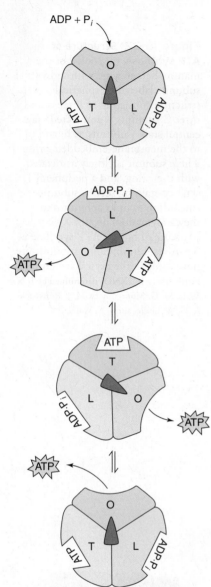

Figure 15-25 The binding change mechanism. The diagram shows the catalytic (β) subunits of the F_1 component of ATP synthase from the same perspective as in Fig. 15-24. Each of the three β subunits adopts a different conformation: open (O), loose (L), or tight (T). The substrates ADP and P_i bind to a loose site, ATP is synthesized when the site becomes tight, and ATP is released when the subunit becomes open. The conformational shifts are triggered by the 120° rotation of the γ subunit, arbitrarily represented by the purple shape. Because each of the three catalytic sites cycles through the three conformational states, ATP is released from one of the three β subunits with each 120° rotation of the γ subunit. ⊕ **See Animated Figure. The binding change mechanism of ATP synthesis.**

"rotor"). The αβ hexamer itself does not rotate, since it is held in place by the peripheral arm that is anchored to the *a* subunit (see Fig. 15-22a).

For an ATP synthase containing 8 *c* subunits, the transmembrane movement of each proton could potentially turn the γ shaft by 45° (360° ÷ 8). However, videomicroscopy experiments indicate that the γ subunit rotates in steps of 120°, interacting successively with each of the three αβ pairs in one full rotation of 360°. Electrostatic interactions between the γ and β subunits apparently act as a catch that holds the γ subunit in place while translocation of two to three protons builds up strain. Translocation of the next proton causes the γ subunit to suddenly snap into position at the next β subunit, a movement of 120°. This mechanism accounts for the variation in the number of *c* subunits in ATP synthases from different sources. The *c* ring spins in small increments (24° to 45°, depending on the number of *c* subunits), but the γ subunit makes just three large shifts of 120°.

▶▶ **HOW** does ATP synthase carry out an unfavorable reaction?

The binding change mechanism explains how ATP is made

At the start of the chapter, we pointed out that ATP synthase catalyzes a highly endergonic reaction ($\Delta G^{\circ\prime} = +30.5$ kJ · mol^{-1}) in order to produce the bulk of a cell's ATP supply. This enzyme operates in an unusual fashion, using mechanical energy (rotation) to form a chemical bond (the attachment of a phosphoryl group to ADP). In other words, the enzyme converts mechanical energy to the chemical energy of ATP. The interaction between the γ subunit and the αβ hexamer explains this energy transduction.

According to the **binding change mechanism** described by Paul Boyer, *rotation-driven conformational changes alter the affinity of each catalytic β subunit for an adenine nucleotide.* At any moment, each catalytic site has a different conformation (and binding affinity), referred to as the open, loose, or tight state. ATP synthesis occurs as follows (**Fig. 15-25**):

1. The substrates ADP and P_i bind to a β subunit in the loose state.
2. The substrates are converted to ATP as rotation of the γ subunit causes the β subunit to shift to the tight conformation.
3. The product ATP is released after the next rotation, when the β subunit shifts to the open conformation.

Because the three β subunits of ATP synthase act cooperatively, they all change their conformations simultaneously as the γ subunit turns. A full rotation of 360° is required to restore the enzyme to its initial state, but each rotation of 120° results in the release of ATP from one of the three active sites.

Experiments with the isolated F_1 component of ATP synthase show that in the absence of F_0, F_1 functions as an ATPase, hydrolyzing ATP to ADP + P_i (a thermodynamically favorable reaction). In the intact ATP synthase, dissipation of the proton gradient is tightly coupled to ATP synthesis with near 100% efficiency. Consequently, *in the absence of a proton gradient, no ATP is synthesized because there is no free energy to drive the rotation of the γ subunit.* Agents that dissipate the proton gradient can therefore "uncouple" ATP synthesis from electron transport, the source of the proton gradient (Box 15-B).

The P:O ratio describes the stoichiometry of oxidative phosphorylation

Since the γ shaft of ATP synthase is attached to the *c*-subunit rotor, 3 ATP molecules are synthesized for every complete *c*-ring rotation. However, the number

BOX 15-B ⬢ **BIOCHEMISTRY NOTE** ▶

413
ATP Synthase

Uncoupling Agents Prevent ATP Synthesis

When the metabolic need for ATP is low, the oxidation of reduced cofactors proceeds until the transmembrane proton gradient builds up enough to halt further electron transport. When the protons reenter the matrix via the F_0 component of ATP synthase, electron transport resumes. However, if the protons leak back into the matrix by a route other than ATP synthase, then electron transport will continue without any ATP being synthesized. ATP synthesis is said to be "uncoupled" from electron transport, and the agent that allows the proton gradient to dissipate in this way is called an **uncoupler.**

One well-known uncoupler is the dinitrophenolate ion, which can take up a proton (its pK is near neutral):

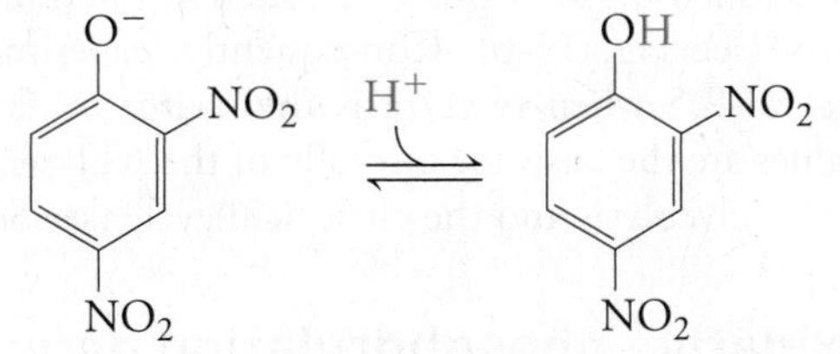

In its neutral state, dinitrophenol can diffuse through a lipid bilayer and release the proton on the other side. Dinitrophenol can therefore dissipate a proton gradient by providing a way for protons to cross the membrane.

Dinitrophenol, of course, is not usually present in mitochondria; however, physiological uncoupling does occur. Dissipating a proton gradient prevents ATP synthesis, but it allows oxidative metabolism to continue at a high rate. The by-product of this metabolic activity is heat.

Uncoupling for **thermogenesis** (heat production) occurs in specialized adipose tissue known as brown fat (its dark color is due to the relatively high concentration of cytochrome-containing mitochondria; ordinary adipose tissue is lighter). The inner membrane of the mitochondria in brown fat contains a transmembrane proton channel called a UCP (uncoupling protein). Protons translocated to the intermembrane space during respiration can reenter the mitochondrial matrix via the uncoupling protein, bypassing ATP synthase. The free energy of respiration is therefore given off as heat rather than used to synthesize ATP. Brown fat is abundant in hibernating mammals and newborn humans, and the activity of the UCP is under the control of hormones that also mobilize the stored fatty acids to be oxidized in the brown fat mitochondria.

⬢ **Question:** Why would increasing the activity of UCP promote weight loss?

of protons translocated per ATP depends on the number of c subunits. For mammalian ATP synthase, which has 8 c subunits, the stoichiometry is 8 H^+ per 3 ATP, or 2.7 H^+ per ATP. Such nonintegral values would be difficult to reconcile with most biochemical reactions, but they are consistent with the chemiosmotic theory: *Chemical energy (from the respiratory oxidation–reduction reactions) is transduced to a proton motive force, then to the mechanical movement of a rotary engine (the c ring and its attached γ shaft), and finally back to chemical energy in the form of ATP.*

The relationship between respiration (the activity of the electron transport complexes) and ATP synthesis is traditionally expressed as a **P:O ratio,** that is, the number of phosphorylations of ADP relative to the number of oxygen atoms reduced. For example, the oxidation of NADH by O_2 (carried out by the sequential activities of Complexes I, III, and IV) translocates 10 protons into the intermembrane space. The movement of these 10 protons back into the matrix via the F_0 component would theoretically drive the synthesis of about 3.7 ATP since 1 ATP can be made for every 2.7 protons translocated, at least in mammalian mitochondria:

$$\frac{1 \text{ ATP}}{2.7 \text{ H}^+} \times 10 \text{ H}^+ = 3.7$$

Thus, the P:O ratio would be about 3.7 (3.7 ATP per $\frac{1}{2} O_2$ reduced). For an electron pair originating as QH_2, only 6 protons would be translocated (by the activities of Complexes III and IV), and the P:O ratio would be approximately 2.2:

$$\frac{1 \text{ ATP}}{2.7 \text{ H}^+} \times 6 \text{ H}^+ = 2.2$$

In vivo, the P:O ratios are actually a bit lower than the theoretical values, because some of the protons translocated during electron transport do leak across the membrane or are consumed in other processes, such as the transport of P_i into the mitochondrial matrix (see Fig. 15-6). Consequently, experimentally determined P:O ratios are closer to 2.5 when NADH is the source of electrons and 1.5 for ubiquinol. These values are the basis for our tally of the ATP yield for the complete oxidation of glucose by glycolysis and the citric acid cycle (see Section 14-2).

The rate of oxidative phosphorylation depends on the rate of fuel catabolism

In most metabolic pathways, control is exerted at highly exergonic (irreversible) steps. In oxidative phosphorylation, this step would be the reaction catalyzed by cytochrome *c* oxidase (Complex IV; see Fig. 15-2). However, there are no known effectors of cytochrome *c* oxidase activity. Apparently, *the close coupling between generation of the proton gradient and ATP synthesis allows oxidative phosphorylation to be regulated primarily by the availability of reduced cofactors (NADH and QH_2) produced by other metabolic processes.*

Less important regulatory mechanisms may involve the availability of the substrates ADP and P_i (which depend on the activity of their respective transport proteins). Experiments with ATP synthase show that when ADP and P_i are absent, the β subunits cannot undergo the conformational changes required by the binding change mechanism. The γ subunit therefore remains immobile, and no protons are translocated through the *c* ring. This tight coupling between ATP synthesis and proton translocation prevents the waste of the free energy of the proton gradient.

There is also evidence that mitochondria contain a regulatory protein that binds to ATP synthase to inhibit its rate of ATP hydrolysis. The inhibitor is sensitive to pH, so it does not bind to ATP synthase when the matrix pH is high (as it is when electron transport is occurring). However, if the matrix pH drops as a result of a momentary disruption of the proton gradient, the inhibitor binds to ATP synthase. This regulatory mechanism prevents ATP synthase from operating in reverse as an ATPase.

CONCEPT REVIEW

- How does ATP synthase dissipate the proton gradient? How is this activity related to the activity of the ATP-synthesizing catalytic sites?
- Describe how the three conformational states of the β subunits of ATP synthase are involved in ATP synthesis.
- How does the binding change mechanism account for ATP hydrolysis by ATP synthase?
- Why does the number of protons translocated per ATP synthesized vary among species?
- What is a P:O ratio and why is it nonintegral?
- Explain why the availability of reduced substrates is the primary mechanism for regulating oxidative phosphorylation.

[SUMMARY]

15-1 The Thermodynamics of Oxidation–Reduction Reactions

- The electron affinity of a substance participating in an oxidation–reduction reaction, which involves the transfer of electrons, is indicated by its reduction potential, $\mathcal{E}^{\circ\prime}$.
- The difference in reduction potential between species undergoing oxidation and reduction is related to the free energy change for the reaction.

15-2 Mitochondrial Electron Transport

- Oxidation of reduced cofactors generated by metabolic reactions takes place in the mitochondrion. Shuttle systems and transport proteins allow the transmembrane movement of reducing equivalents, ATP, ADP, and P_i.
- The electron transport chain consists of a series of integral membrane protein complexes that contain multiple redox groups, including iron–sulfur clusters, flavins, cytochromes, and copper ions, and that are linked by mobile electron carriers. Starting from NADH, electrons travel a path of increasing reduction potential through Complex I, ubiquinone, Complex III, cytochrome c, and then to Complex IV, where O_2 is reduced to H_2O.
- As electrons are transferred, protons are translocated to the intermembrane space via proton wires in Complexes I and IV and by the action of the Q cycle associated with Complex III.

15-3 Chemiosmosis

- The chemiosmotic theory describes how proton translocation during mitochondrial electron transport generates an electrochemical gradient whose free energy drives ATP synthesis.

15-4 ATP Synthase

- The energy of the proton gradient is tapped as protons spontaneously flow through ATP synthase. Proton transport allows rotation of a ring of integral membrane c subunits. The linked γ subunit thereby rotates, triggering conformational changes in the F_1 portion of ATP synthase.
- According to the binding change mechanism, the three functional units of the F_1 portion cycle through three conformational states to sequentially bind ADP and P_i, convert the substrates to ATP, and release ATP.
- The P:O ratio quantifies the link between electron transport and oxidative phosphorylation in terms of the ATP synthesized and the O_2 reduced. Because these processes are coupled, the rate of oxidative phosphorylation is controlled primarily by the availability of reduced cofactors.

[GLOSSARY TERMS]

oxidative phosphorylation
oxidizing agent (oxidant)
reducing agent (reductant)
half-reaction
$\mathcal{E}^{\circ\prime}$
$\mathcal{E}$
Nernst equation
$\mathcal{F}$
respiration
mitochondrion

mitochondrial matrix
intermembrane space
cristae
electron tomography
redox center
proton wire
cytochrome
Q cycle
free radical
chemiosmotic theory

protonmotive force
Z
binding change mechanism
P:O ratio
uncoupler
thermogenesis

[PROBLEMS]

15-1 The Thermodynamics of Oxidation–Reduction Reactions

1. Calculate the standard free energy change for the reduction of pyruvate by NADH. Consult Table 15-1 for the relevant half-reactions. Is this reaction spontaneous under standard conditions?

2. Calculate the standard free energy change for the reduction of oxygen by cytochrome a_3. Consult Table 15-1 for the relevant half-reactions. Is this reaction spontaneous under standard conditions?

3. In one of the final steps of the pyruvate dehydrogenase reaction (see Section 14-1), E3 reoxidizes the lipoamide group of E2, then NAD^+ reoxidizes E3. Calculate $\Delta G^{\circ\prime}$ for the electron transfer from dihydrolipoic acid to NAD^+.

4. Each electron from cytochrome c is donated to a Cu_A redox center in Complex IV. The $\mathcal{E}^{\circ\prime}$ value for the Cu_A redox center is 0.245 V. Calculate $\Delta G^{\circ\prime}$ for this electron transfer.

5. Acetaldehyde may be oxidized to acetate. Would NAD^+ be an effective oxidizing agent? Explain.

6. Acetoacetate may be reduced to 3-hydroxybutyrate. What serves as a better reducing agent, NADH or $FADH_2$? Explain.

7. For every two QH_2 that enter the Q cycle, one is regenerated and the other passes its two electrons to two cytochrome c_1 centers. The overall equation is

$$QH_2 + 2 \text{ cyt } c_1 \text{ (Fe}^{3+}) + 2 \text{ H}^+ \rightarrow Q + 2 \text{ cyt } c_1 \text{ (Fe}^{2+}) + 4 \text{ H}^+$$

Calculate the free energy change associated with the Q cycle.

8. Why is succinate oxidized by FAD instead of by NAD^+?

9. (a) What is the $\Delta\mathcal{E}$ value for the oxidation of ubiquinol by cytochrome c when the ratio of QH_2/Q is 10 and the ratio of cyt c $(Fe^{3+})/$cyt c (Fe^{2+}) is 5?

(b) Calculate the ΔG for the reaction in part (a).

10. An iron–sulfur protein in Complex III donates an electron to cytochrome c_1. The reduction half-reactions and $\mathcal{E}°'$ values are shown below. Write the balanced equation for the reaction and calculate the standard free energy change. How can you account for the fact that this reaction occurs spontaneously in the cell?

$$FeS\ (ox) + e^- \rightarrow FeS\ (red) \qquad \mathcal{E}°' = 0.280\ V$$

$$cyt\ c_1\ (Fe^{3+}) + e^- \rightarrow cyt\ c_1\ (Fe^{2+}) \qquad \mathcal{E}°' = 0.215\ V$$

11. Calculate the overall efficiency of oxidative phosphorylation, assuming standard conditions, by comparing the free energy potentially available from the oxidation of NADH by O_2 and the free energy required to synthesize 2.5 ATP from 2.5 ADP.

12. Using the percent efficiency calculated in Problem 11, calculate the number of ATP generated by **(a)** Complex I (where NADH is oxidized by ubiquinone), **(b)** Complex III (where ubiquinol is oxidized by cytochrome c), and **(c)** Complex IV (where cytochrome c is oxidized by molecular oxygen).

13. Calculate $\Delta\mathcal{E}°'$ and $\Delta G°'$ for the succinate dehydrogenase (Complex II) reaction.

14. Refer to the solution for Problem 13. Does this reaction provide sufficient free energy to drive ATP synthesis under standard conditions? Explain.

15-2 Mitochondrial Electron Transport

15. The sequence of events in electron transport was elucidated in part by the use of inhibitors that block electron transfer at specific points along the chain. For example, adding rotenone (a plant toxin) or amytal (a barbiturate) blocks electron transport in Complex I; antimycin A (an antibiotic) blocks electron transport in Complex III; and cyanide (CN^-) blocks electron transport in Complex IV by binding to the Fe^{2+} in the Fe–Cu binuclear center.

(a) What happens to oxygen consumption when these inhibitors are added to a suspension of respiring mitochondria?

(b) What is the redox state of the electron carriers in the electron transport chain when each of the inhibitors is added separately to the mitochondrial suspension?

16. What is the effect of added succinate to rotenone-blocked, amytal-blocked, or cyanide-blocked mitochondria (see Problem 15)? In other words, can succinate help "bypass" the block? Explain.

17. The compound tetramethyl-p-phenylenediamine (TMPD) donates a pair of electrons directly to Complex IV. What is the P:O ratio of this compound?

18. Ascorbate (vitamin C) can donate a pair of electrons to cytochrome c. What is the P:O ratio for ascorbate?

19. Can tetramethyl-p-phenylenediamine (see Problem 17) act as a bypass for the rotenone-blocked, amytal-blocked, or cyanide-blocked mitochondria described in Problem 15? Can ascorbate (see Problem 18) act as a bypass? Explain.

20. When the antifungal agent myxothiazol is added to a suspension of respiring mitochondria, the QH_2/Q ratio increases. Where in the electron transport chain does myxothiazol inhibit electron transfer?

21. If cyanide poisoning (see Problem 15) is diagnosed immediately, it can be treated by administering nitrites that can oxidize the Fe^{2+} in hemoglobin to Fe^{3+}. Why is this treatment effective?

22. The effect of the drug fluoxetine (Prozac) on isolated rat brain mitochondria was examined by measuring the rate of electron transport (units not given) in the presence of various combinations of substrates and inhibitors (see Problems 15–17).

(a) How do pyruvate and malate serve as substrates for electron transport?

(b) What is the effect of fluoxetine on electron transport? Explain.

(c) Fluoxetine can also inhibit ATP synthase. Why might long-term use of fluoxetine be a concern?

Rate of Electron Transport

[Fluoxetine], (mM)	Pyruvate + malate	Succinate + rotenone	Ascorbate + TMPD
0	163 ± 15.1	145 ± 14.2	184 ± 22.2
0.15	77 ± 7.3	131 ± 13.5	116 ± 13.9

23. Complex I, succinate dehydrogenase, acyl-CoA dehydrogenase, and glycerol-3-phosphate dehydrogenase (see Fig. 15-11) are all flavoproteins; that is, they contain an FMN or FAD prosthetic group. Explain the function of the flavin group in these enzymes.

24. What side chains would you expect to find as part of a proton wire in a proton-translocating membrane protein?

25. Ubiquinone is not anchored in the mitochondrial membrane but is free to diffuse laterally throughout the membrane among the electron transport chain components. What aspects of its structure account for this behavior?

26. Explain why the ubiquinone-binding site of Complex I (Fig. 15-7) is located at the end of the peripheral arm closest to the membrane.

27. Cytochrome c is easily dissociated from isolated mitochondrial membrane preparations, but the isolation of cytochrome c_1 requires the use of strong detergents. Explain why.

28. Release of cytochrome c from the mitochondrion to the cytosol is one of the signals that induces apoptosis, a form of programmed cell death. What structural features of cytochrome c allow it to play this role?

29. In coastal marine environments, high concentrations of nutrients from terrestrial runoff may lead to algal blooms. When the nutrients are depleted, the algae die and sink and are degraded by other microorganisms. The algal die-off may be followed by a sharp drop in oxygen in the depths, which can kill fish and bottom-dwelling invertebrates. Why do these "dead zones" form?

30. Chromium is most toxic and highly soluble in its oxidized Cr(VI) state but is less toxic and less soluble in its more reduced Cr(III) state. Efforts to detoxify Cr-contaminated groundwater have involved injecting chemical reducing agents underground. Another approach is bioremediation, which involves injecting molasses or cooking oil into the contaminated groundwater. Explain how these substances would promote the reduction of Cr(VI) to Cr(III).

31. At one time, it was believed that myoglobin functioned simply as an oxygen-storage protein. New evidence suggests that myoglobin plays a much more active role in the muscle cell. The phrase *myoglobin-facilitated oxygen diffusion* describes myoglobin's role in transporting oxygen from the muscle cell sarcolemma to the mitochondrial membrane surface. Mice in which the myoglobin gene was knocked out had higher tissue capillary density, elevated red blood cell counts, and increased coronary blood flow. Explain the reasons for these compensatory mechanisms in the knockout mice.

32. The myoglobin and cytochrome c oxidase content were determined in several animals, as shown in the table. What is the relationship between the two proteins? Explain.

	Myoglobin content, $mmol \cdot kg^{-1}$	Cytochrome c oxidase activity
Hare	0.1	900
Sheep	0.19	950
Ox	0.31	1200
Horse	0.38	1800

33. It has recently been found that myoglobin distribution is not confined to muscle cells. Tumor cells, which generally exist in hypoxic (low-oxygen) conditions because of limited blood flow, have been found to express myoglobin. How does this adaptation increase the chances of tumor cell survival?

34. Cancer cells, even when sufficient oxygen is available, produce large amounts of lactate. It has been observed that the concentration of fructose-2,6-bisphosphate is much higher in cancer cells than in normal cells. Why would this result in anaerobic metabolism being favored, even when oxygen is available?

35. A group of elderly patients who did not exercise regularly were asked to participate in a 12-week exercise program. Data collected from the patients are shown in the table. What was the result of the exercise intervention and why did this occur?

	Prior to exercise intervention	Post 12-week exercise intervention
Total mitochondrial DNA (copies per diploid genome)	1300	1900
Complex II activity	0.13	0.20
Complex I–IV activity	0.51	1.00

36. Amyotrophic lateral sclerosis (ALS) is a neurodegenerative disease that causes muscle paralysis and eventually death. Researchers measured the activity of the electron transport chain complexes in various regions of the nervous system in patients with ALS. In a certain region of the spinal cord, Complex I showed decreased activity but not decreased concentration. How does this contribute to progression of the disease?

15-3 Chemiosmosis

37. What is the free energy change for generating the electrical imbalance of protons in neuroblastoma cells, where $\Delta\psi$ is 81 mV?

38. Calculate the free energy change for translocating a proton out of the mitochondrial matrix, where $pH_{matrix} = 7.6$, $pH_{cytosol} = 7.2$, $\Delta\psi = 200$ mV, and $T = 37°C$.

39. Several key experimental observations were important in the development of the chemiosmotic theory. Explain how each observation listed below is consistent with the chemiosmotic theory as described by Peter Mitchell.

(a) The pH of the intermembrane space is lower than the pH of the mitochondrial matrix.

(b) Oxidative phosphorylation does not occur in mitochondrial preparations to which detergents have been added.

(c) Lipid-soluble compounds such as DNP (see Box 15-B) inhibit oxidative phosphorylation while allowing electron transport to continue.

40. Mitchell's original chemiosmotic hypothesis relies on the impermeability of the inner mitochondrial membrane to ions other than H^+, such as Na^+ and Cl^-.

(a) Why was this thought to be important?

(b) Could ATP still be synthesized if the membrane were permeable to other ions?

41. Nigericin is an antibiotic that integrates into membranes and functions as a K^+/H^+ antiporter. Another antibiotic, valinomycin, is similar, but it allows the passage of K^+ ions. When both antibiotics are added simultaneously to suspensions of respiring mitochondria, the electrochemical gradient completely collapses.

(a) Draw a diagram of a mitochondrion in which nigericin and valinomycin have integrated into the inner mitochondrial membrane, in a manner that is consistent with the experimental results.

(b) Explain why the electrochemical gradient dissipates. What happens to ATP synthesis?

42. How does transport of inorganic phosphate from the intermembrane space to the mitochondrial matrix affect the pH difference across the inner mitochondrial membrane?

15-4 ATP Synthase

43. How much ATP can be obtained by the cell from the complete oxidation of one mole of glucose? Compare this value with the amount of ATP obtained when glucose is anaerobically converted to lactate or ethanol.

44. The glycerol-3-phosphate shuttle can transport NADH generated in the cytosol into the mitochondrial matrix (see Figure 15-11c). In this shuttle, the protons and electrons are donated to FAD, which is reduced to $FADH_2$. These protons and electrons are subsequently donated to coenzyme Q in the electron transport chain. How much ATP is generated per mole of glucose when the glycerol-3-phosphate shuttle is used?

45. The complete oxidation of glucose yields $-2850 \text{ kJ} \cdot mol^{-1}$ of free energy. Incomplete oxidation by conversion to lactate yields $-196 \text{ kJ} \cdot mol^{-1}$ and by alcoholic fermentation yields $-235 \text{ kJ} \cdot mol^{-1}$. Calculate the overall efficiencies of glucose oxidation by these three processes.

46. Do organisms that can completely oxidize glucose have an advantage over organisms that cannot? (*Hint:* See Problem 45).

47. A culture of yeast grown under anaerobic conditions is exposed to oxygen, resulting in a dramatic decrease in glucose consumption by the cells. This phenomenon is referred to as the Pasteur effect.

(a) Explain the Pasteur effect.

(b) The $[NADH]/[NAD^+]$ and $[ATP]/[ADP]$ ratios also change when an anaerobic culture is exposed to oxygen. Explain how these ratios change and what effect this has on glycolysis and the citric acid cycle in the yeast.

48. Experiments in the late 1970s attributed the Pasteur effect (see Problem 47) to the stimulation of hexokinase and phosphofructokinase under anaerobic conditions. Upon exposure to oxygen, the stimulation of these enzymes ceases. Why are these enzymes more active in the absence of oxygen?

49. Consider the adenine nucleotide translocase and the P_i–H^+ symport protein that import ADP and P_i, the substrates for oxidative phosphorylation, into the mitochondrion (see Fig. 15-6).

(a) How does the activity of the adenine nucleotide translocase affect the electrochemical gradient across the mitochondrial membrane?

(b) How does the activity of the P_i–H^+ symport protein affect the gradient?

(c) What can you conclude about the thermodynamic force that drives the two transport systems?

50. The compounds atractyloside and bongkrekic acid both bind tightly to and inhibit the ATP/ADP translocase. What is the effect of these compounds on ATP synthesis? On electron transport?

51. Dicyclohexylcarbodiimide (DCCD) is a reagent that reacts with Asp or Glu residues. Explain why the reaction of DCCD with just one c subunit completely blocks both the ATP-synthesizing and ATP-hydrolyzing activity of ATP synthase.

52. Oligomycin is an antibiotic that blocks proton transfer through the F_0 proton channel of ATP synthase. What is the effect on (a) ATP synthesis, (b) electron transport, and (c) oxygen consumption when oligomycin is added to a suspension of respiring mitochondria? (d) What changes occur when dinitrophenol is then added to the suspension?

53. The compound DNP was introduced as a "diet pill" in the 1920s. Its use was discontinued because the side effects were fatal in some cases. What was the rationale for believing that DNP would be an effective diet aid?

54. The compound carbonylcyanide-p-trifluoromethoxy phenylhydrazone (FCCP) is an uncoupler similar to DNP. Describe how FCCP acts as an uncoupler.

FCCP

55. In the 1950s, experiments with isolated mitochondria showed that organic compounds are oxidized and O_2 is consumed only when ADP is included in the preparation. When the ADP supply runs out, oxygen consumption halts. Explain these results.

56. A patient seeks treatment because her metabolic rate is twice normal and her temperature is elevated. A biopsy reveals that her muscle mitochondria are structurally unusual and not subject to normal respiratory controls. Electron transport takes place regardless of the concentration of ADP.

(a) What is the P:O ratio (compared to normal) of NADH that enters the electron transport chain in the mitochondria of this patient?

(b) Why are the patient's metabolic rate and temperature elevated?

(c) Will this patient be able to carry out strenuous exercise?

57. In experimental systems, the F_0 component of ATP synthase can be reconstituted into a membrane. F_0 can then act as a proton channel that is blocked when the F_1 component is added to the system. What molecule must be added to the system in order to restore the proton-translocating activity of F_0? Explain.

58. Calculate the ratio of protons translocated to ATP synthesized for yeast ATP synthase, which has 10 c subunits, and for spinach chloroplast ATP synthase, which has 14 c subunits.

59. A bacterial ATP synthase has 10 c subunits, and a chloroplast ATP synthase has 14 c subunits. Would you expect the bacterium or the chloroplast to have a higher P:O ratio?

60. Experiments indicate that the c ring of ATP synthase spins at a rate of 6000 rpm. How many ATP molecules are generated each second?

61. Mutations that impair ATP synthase function are rare. Laboratory studies indicate that adding α-ketoglutarate boosts ATP production in ATP synthase–deficient cells, but only when aspartate is also added to the cells. Explain.

62. In yeast, pyruvate may be converted to ethanol in a two-step pathway catalyzed by pyruvate decarboxylase and alcohol dehydrogenase (see Section 13-1). Pyruvate may also be converted to acetyl-CoA by pyruvate dehydrogenase. Yeast mutants in which the pyruvate decarboxylase gene is missing ($pdc–$) are useful for studying the regulation of the pyruvate dehydrogenase enzyme. When wild-type yeast were pulsed with glucose, glycolytic flux increased dramatically and the rate of respiration increased. But when the same experiments were performed with $pdc–$ mutants, only a small increase in glycolysis was observed, and pyruvate was the main product excreted by the yeast cells. Explain these results.

63. During anaerobic fermentation in yeast, the majority of the available glucose is oxidized via the glycolytic pathway and the rest enters the pentose phosphate pathway to generate NADPH and ribose. This occurs during aerobic respiration as well, except that the percentage of glucose entering the pentose phosphate pathway is much greater in aerobic respiration than during anaerobic fermentation. Explain why.

64. UCP1 is an uncoupling protein in brown fat (see Box 15-B). Experiments using UCP1-knockout mice (animals missing the gene for UCP1) resulted in the discovery of a second uncoupling protein named UCP2.

(a) Oxygen consumption increased over twofold when a β_3 adrenergic agonist that stimulates UCP1 was injected into normal mice. This was not observed when the agonist was injected into the knockout mice. Explain these results.

(b) The UCP1-knockout mice were essentially normal except for increased lipid deposition in their adipose tissue. Explain why.

(c) In one experiment, normal mice and UCP1-knockout mice were placed in a cold (5°C) room overnight. The normal mice were able to maintain their body temperature at 37°C even after 24 hours in the cold. But the body temperatures of the cold-exposed knockout mice decreased 10°C or more. Explain.

(d) UCP1-knockout mice did not become obese when fed a high-fat diet. Propose a hypothesis that explains this observation.

65. The Eastern skunk cabbage can maintain its temperature 15–35°C higher than ambient temperature during the months of February and March, when ambient temperatures range from −15 to +15°C. Thermogenesis in the skunk cabbage is critical to the survival of the plant since the spadix (a flower component) is not frost-resistant. An uncoupling protein was found to be responsible for the observed thermogenesis.

(a) The spadix relies on the skunk cabbage's massive root system, which stores appreciable quantities of starch. Why is a large quantity of starch required for the skunk cabbage to carry out sustained thermogenesis for weeks rather than hours?

(b) Oxygen consumption by the skunk cabbage increases as the temperature decreases, nearly doubling with every 10°C drop in ambient temperature. Oxygen consumption was observed to decrease during the day, when temperatures were close to 30°C, and increase at night. What is the biochemical explanation for these observations?

66. The gene that codes for the uncoupling protein (see Problem 65) in potatoes was recently isolated. The results of a Northern blot analysis (which detects mRNA) are shown below. What is your interpretation of these results? How does the mRNA level affect thermogenesis in the potato?

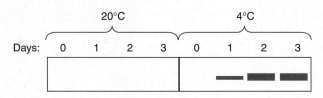

67. Glutamate can be used as an artificial substrate for mitochondrial respiration, as shown in the diagram below. When ceramide is added to a mitochondrial suspension respiring in the presence of glutamate, respiration decreases, leading scientists to hypothesize that ceramide might regulate mitochondrial function *in vivo*.

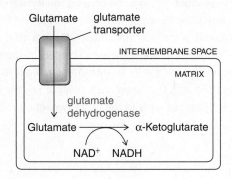

(a) How does glutamate act as a substrate for mitochondrial respiration?

(b) Ceramide-induced inhibition of respiration could be due to several different factors. List several possibilities.

(c) Mitochondria treated with ceramide were exposed to an uncoupler, but the respiration rate did not increase. What site(s) of inhibition can be ruled out?

(d) In another experiment, mitochondria were subjected to a freeze–thaw cycle that rendered the inner mitochondrial membrane permeable to NADH. NADH could then be added to a mitochondrial suspension as a substrate for electron transport. When NADH was used as a substrate, ceramide decreased the respiration rate to the same extent as when glutamate was the substrate. What site(s) of inhibition can be ruled out?

68. When cells cannot carry out oxidative phosphorylation, they can synthesize ATP through substrate-level phosphorylation.

(a) Which enzymes of glycolysis and the citric acid cycle catalyze substrate-level phosphorylation?

(b) The O_2 that we breathe in is not directly converted to the CO_2 that we breathe out. Write a balanced equation for the complete combustion of glucose and oxygen.

[SELECTED READINGS]

Boekema, E. J., and Braunm H.-P., Supramolecular structure of the mitochondrial oxidative phosphorylation system, *J. Biol. Chem.* **282,** 1– 4 (2007). [Presents evidence that Complexes I through IV form "supercomplexes."]

Boyer, P. D., Catalytic site forms and controls in ATP synthase catalysis, *Biochim. Biophys. Acta* **1458,** 252–262 (2000). [Reviews the steps of ATP synthesis and hydrolysis, along with experimental evidence and alternative explanations; by the originator of the binding change mechanism.]

Hosler, J. P., Ferguson-Miller, S., and Mills, D. A., Energy transduction: proton transfer through the respiratory complexes, *Annu. Rev. Biochem.* **75,** 165–187 (2006). [Reviews respiration, with a focus on Complex IV.]

Watt, I. N., Montgomery, M. G., Runswick, M. J., Leslie, A. G. W., and Walker, J. E., Bioenergetic cost of making an adenosine triphosphate molecule in animal mitochondria, *Proc. Nat. Acad. Sci.* **107,** 16823–16827 (2010). [Includes details of ATP synthase structure.]

▶▶ **WHY** do plants produce O_2?

Plants take up water and carbon dioxide and produce sugar during photosynthesis. A by-product of this process is molecular oxygen— O_2 gas—visible as tiny bubbles emanating from a plant placed under water. In the context of cellular respiration, O_2 is a vital reactant, providing the oxidizing power that is ultimately responsible for generating most of the cell's ATP. In this chapter, we'll explore the photosynthetic machinery and see why O_2, rather than being a key player at the end of the electron transport chain, is produced early in the photosynthetic process.

[Elena Elisseeva/Alamy Limited]

THIS CHAPTER IN CONTEXT

Part 1 Foundations

Part 2 Molecular Structure and Function

Part 3 Metabolism
   16 Photosynthesis

Part 4 Genetic Information

Do You Remember?

- Glucose polymers include the fuel-storage polysaccharides starch and glycogen and the structural polysaccharide cellulose (Section 11-2).
- Coenzymes such as NAD^+ and ubiquinone collect electrons from compounds that become oxidized (Section 12-2).
- Electrons are transferred from a substance with a lower reduction potential to a substance with a higher reduction potential (Section 15-1).
- The formation of a transmembrane proton gradient during electron transport provides the free energy to synthesize ATP (Section 15-3).

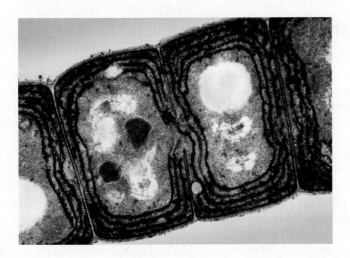

Figure 16-1 Cyanobacteria. The first photosynthetic organisms were probably similar to these bacterial cells. [Biophoto Associates/Photo Researchers, Inc.]

Every year, plants and bacteria convert an estimated 6×10^{16} grams of carbon to organic compounds by **photosynthesis.** About half of this activity occurs in forests and savannas, and the rest occurs in the ocean and under ice—wherever water, carbon dioxide, and light are available. The organic materials produced by photosynthetic organisms sustain them as well as the organisms that feed on them.

The ability to use sunlight as an energy source evolved about 3.5 billion years ago. Before that, cellular metabolism probably centered around the inorganic reductive reactions associated with hydrothermal vents. The first photosynthetic organisms produced various pigments (light-absorbing molecules) to capture solar energy and thereby drive the reduction of metabolites. The descendants of some of these organisms are known today as purple bacteria and green sulfur bacteria. By about 2.5 billion years ago, the cyanobacteria had evolved (**Fig. 16-1**; these organisms are sometimes misleadingly called blue-green algae, although they are prokaryotes and true algae are eukaryotes). Cyanobacteria absorb enough solar energy to undertake the energetically costly oxidation of water to molecular oxygen. In fact, the dramatic increase in the level of atmospheric oxygen (from an estimated 1% to the current level of about 20%) around 2.4 billion years ago is attributed to the rise of cyanobacteria. Modern plants are the result of the symbiosis of early eukaryotic cells with cyanobacteria.

Photosynthesis as a biochemical phenomenon is worth studying not just because of its importance in global carbon and oxygen cycles. It also affords an opportunity to compare the biochemistry of plants and mammals. To begin, we will examine light-absorbing chloroplasts, which are the relics of bacterial symbionts, as are mitochondria. We will then look at the electron-transporting complexes that convert solar energy to biologically useful forms of free energy such as ATP and the reduced cofactor NADPH. Finally, we will see how plants use these energy currencies to carry out biosynthetic reactions—the **fixation** of carbon dioxide into organic compounds.

Although the apparatus and reactions of photosynthesis are not found in all organisms, we can place them in the context of the metabolic scheme outlined in Chapter 12 (**Fig. 16-2**). As we describe the harvest of solar energy and its use in the incorporation of CO_2 into three-carbon compounds, we will emphasize how these processes differ from and resemble the metabolic processes that occur in animal cells.

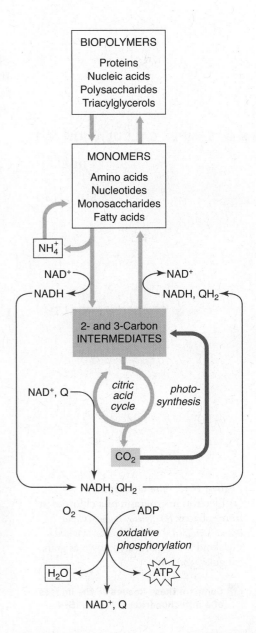

Figure 16-2 Photosynthesis in context. Photosynthetic organisms incorporate atmospheric CO_2 into three-carbon compounds that are the precursors of biological molecules such as carbohydrates and amino acids. Photosynthesis requires light energy to drive the production of the ATP and NADPH consumed in biosynthetic reactions.

16-1 Chloroplasts and Solar Energy

KEY CONCEPTS
- Photosynthetic pigments absorb different wavelengths of light to become excited.
- Light-harvesting complexes direct light energy to the reaction center.

Photosynthesis in green plants takes place in **chloroplasts,** discrete organelles that are descended from cyanobacteria. Like mitochondria, chloroplasts contain their own DNA, in this case coding for 100 to 200 chloroplast proteins. DNA in the cell's nucleus contains close to a thousand more genes whose products are essential for photosynthesis.

The chloroplast is enclosed by a porous outer membrane and an ion-impermeable inner membrane **(Fig. 16-3).** The inner compartment, called the **stroma,** is analogous to the mitochondrial matrix and is rich in enzymes, including those required for carbohydrate synthesis. Within the stroma is a membranous structure called the **thylakoid.** Unlike the planar or tubular mitochondrial cristae (see Fig. 15-4), the thylakoid membrane folds into stacks of flattened vesicles and encloses a compartment called the thylakoid lumen. The energy-transducing reactions of photosynthesis take place in the thylakoid membrane. The analogous reactions in photosynthetic bacteria typically take place in folded regions of the plasma membrane.

Pigments absorb light of different wavelengths

Light can be considered as both a wave and a particle, the **photon.** The energy (E) of a photon depends on its wavelength, as expressed by **Planck's law:**

$$E = \frac{hc}{\lambda} \qquad \text{[16-1]}$$

where h is Planck's constant (6.626×10^{-34} J $\cdot$ s), c is the speed of light (2.998×10^8 m $\cdot$ s^{-1}), and λ is the wavelength (about 400 to 700 nm for visible light; see Sample Calculation 16-1). *This energy is absorbed by the photosynthetic apparatus of the chloroplast and transduced to chemical energy.*

●●● SAMPLE CALCULATION 16-1

PROBLEM

Calculate the energy of a photon with a wavelength of 550 nm.

SOLUTION

$$E = \frac{hc}{\lambda}$$

$$E = \frac{(6.626 \times 10^{-34} \text{ J} \cdot \text{s})(2.998 \times 10^8 \text{ m} \cdot \text{s}^{-1})}{550 \times 10^{-9} \text{ m}}$$

$$E = 3.6 \times 10^{-19} \text{ J}$$

●●● PRACTICE PROBLEMS

1. Calculate the energy of one mole of photons (6.022×10^{23} mol^{-1}) with a wavelength of 550 nm.
2. At what wavelength would a mole of photons have an energy of 250 kJ?

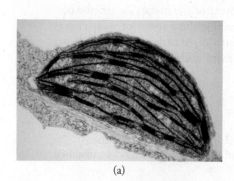

Figure 16-3 The chloroplast.
(a) Electron micrograph of a chloroplast from tobacco. [Dr. Jeremy Burgess/Photo Researchers, Inc.] (b) Model. The stacked thylakoid membranes are known as grana (singular, *granum*).

? **Compare these images to the images of a mitochondrion in Figure 15-4.**

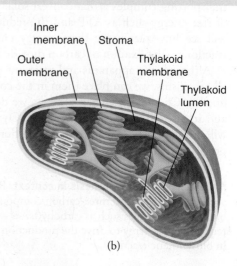

Inner membrane   Stroma
Outer membrane
Thylakoid membrane
Thylakoid lumen

(a)   (b)

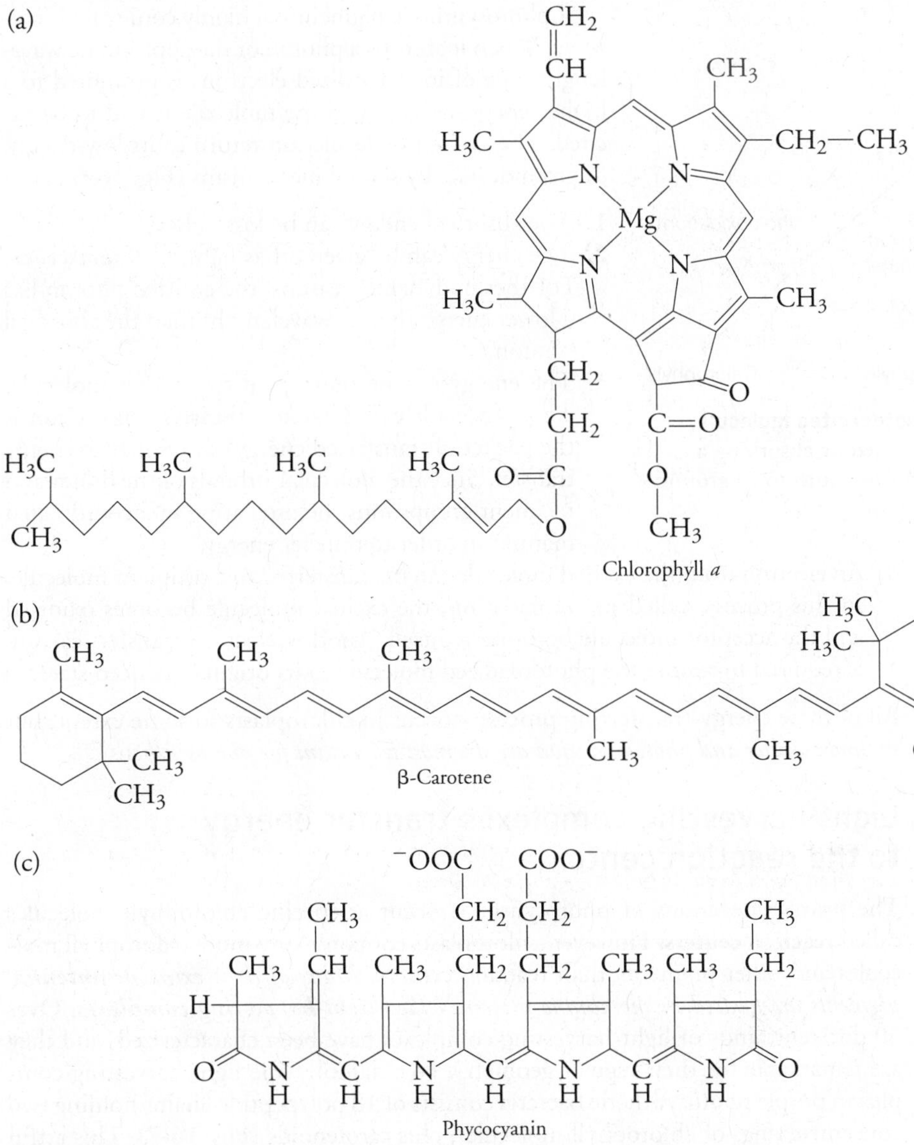

(a)

Chlorophyll *a*

(b)

β-Carotene

(c)

Phycocyanin

Figure 16-4 Some common chloroplast photoreceptors.
(a) Chlorophyll *a*. In chlorophyll *b*, a methyl group (blue) is replaced by an aldehyde group. Chlorophyll resembles the heme groups of hemoglobin and cytochromes (see Fig. 15-12), but it has a central Mg^{2+} rather than an Fe^{2+} ion; it includes a fused cyclopentane ring, and it has a long lipid side chain. (b) The carotenoid β-carotene, a precursor of vitamin A (see Box 8-B). (c) Phycocyanin, a linear tetrapyrrole. It resembles an unfolded chlorophyll molecule.

Chloroplasts contain a variety of light-absorbing groups called pigments or **photoreceptors** (**Fig. 16-4**). Chlorophyll is the principal photoreceptor. It appears green because it absorbs both blue and red light. The second most common pigments are the red carotenoids, which absorb blue light. Pigments such as phycocyanin, which absorb longer-wavelength red light, are common in aquatic systems because water absorbs blue light. Together, these types of pigments absorb all the wavelengths of visible light (**Fig. 16-5**).

Figure 16-5 Visible light absorption by some photosynthetic pigments.
The wavelengths of absorbed light correspond to the peak of the solar energy that reaches the earth.

? **Use this diagram to explain the color of each type of pigment molecule.**

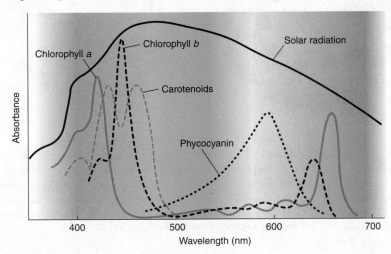

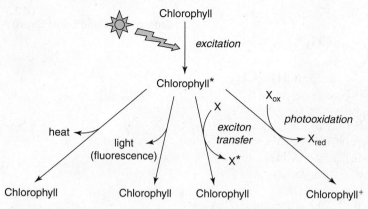

Figure 16-6 Dissipation of energy in a photoexcited molecule.
A pigment molecule such as chlorophyll is excited by absorbing a
photon. The excited molecule (chlorophyll*) can return to its ground
state by one of several mechanisms.

A photosynthetic pigment is a highly conjugated molecule. When it absorbs a photon of the appropriate wavelength, one of its delocalized electrons is promoted to a higher-energy orbital, and the molecule is said to be excited. The excited molecule can return to its low-energy, or ground, state by several mechanisms (**Fig. 16-6**):

1. The absorbed energy can be lost as heat.
2. The energy can be given off as light, or **fluorescence.** For thermodynamic reasons, the emitted photon has a lower energy (longer wavelength) than the absorbed photon.
3. The energy can be transferred to another molecule. This process is called **exciton transfer** (an exciton is the packet of transferred energy) or resonance energy transfer, since the molecular orbitals of the donor and recipient groups must be oscillating in a coordinated manner in order to transfer energy.
4. An electron from the excited molecule can be transferred to a recipient molecule. In this process, called **photooxidation,** the excited molecule becomes oxidized and the acceptor molecule becomes reduced. Another electron-transfer reaction is required to restore the photooxidized molecule to its original reduced state.

All of these energy-transferring processes occur in chloroplasts to some extent, but *exciton transfer and photooxidation are the most important for photosynthesis.*

Light-harvesting complexes transfer energy to the reaction center

The primary reactions of photosynthesis occur at specific chlorophyll molecules called **reaction centers.** However, chloroplasts contain many more chlorophyll molecules and other pigments than reaction centers. *Many of these extra, or **antenna,** pigments are located in membrane proteins called **light-harvesting complexes.*** Over 30 different kinds of light-harvesting complexes have been characterized, and they are remarkable for their regular geometry. For example, one light-harvesting complex in purple photosynthetic bacteria consists of 18 polypeptide chains holding two concentric rings of chlorophyll molecules, plus carotenoids (**Fig. 16-7**). This artful

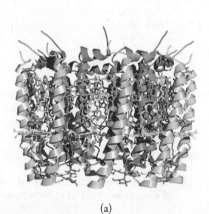

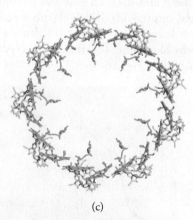

(a) (b) (c)

Figure 16-7 A light-harvesting complex from
Rhodopseudomonas acidophila. The nine pairs of subunits
(light and dark gray) are mostly buried in the membrane and
form a scaffold for two rings of chlorophyll molecules (yellow
and green) and carotenoids (red). The pigments are all within
a few angstroms of each other. (a) Side view. The extracellular
side is at the top. (b) Top view. (c) Top view showing only the
chlorophyll molecules. The 18 chlorophyll molecules in the
inner ring (green) overlap so that excitation energy may be
delocalized over the entire ring. [Structure (pdb 1KZU) determined
by R. J. Cogdell, A. A. Freer, N. W. Isaacs, A. M. Hawthornthwaite-Lawless,
G. McDermott, M. Z. Papiz, and S. M. Prince.] ⊕ **See Interactive
Exercise. Light-harvesting complex LH-2.**

arrangement of light-absorbing groups is essential for the function of the light-harvesting complex.

The protein microenvironment of each photoreceptor influences the wavelength (and therefore the energy) of the photon it can absorb (just as cytochrome protein structure influences the reduction potential of its heme group; see Section 15-2). Consequently, *the various light-harvesting complexes with their multiple pigments can absorb light of many different wavelengths.* Within a light-harvesting complex, the precisely aligned pigment molecules can quickly transfer their energy to other pigments. *Exciton transfer eventually brings the energy to the chlorophyll at the reaction center* (Fig. 16-8). Without light-harvesting complexes to collect and concentrate light, the reaction center chlorophyll could collect only a small fraction of the incoming solar radiation. Even so, a leaf captures only about 1% of the available solar energy.

During periods of high light intensity, some accessory pigments may function to dissipate excess solar energy as heat so that it does not damage the photosynthetic apparatus by inappropriate photooxidation. Various pigment molecules may also act as photosensors to regulate the plant's growth rate and shape and to coordinate the plant's activities—such as germination, flowering, and dormancy—according to daily or seasonal light levels.

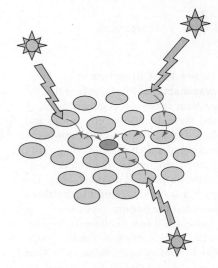

Figure 16-8 Function of light-harvesting complexes. A typical photosynthetic system consists of a reaction center (dark green) surrounded by light-harvesting complexes (light green), whose multiple pigments absorb light of different wavelengths. Exciton transfer funnels this captured solar energy to the chlorophyll at the reaction center. Because the exciton must move from higher-energy to lower-energy states, the antenna pigments farthest from the reaction center have the highest-energy excited states.

> **CONCEPT REVIEW**
> - How is a photon's energy related to its wavelength?
> - Why is it advantageous for photosynthetic pigments to absorb different colors of light?
> - Describe the four mechanisms by which a photoexcited molecule can return to its ground state.
> - What is the function of a light-harvesting complex?

16-2 The Light Reactions

In plants and cyanobacteria, the energy captured by the antenna pigments of the light-harvesting complexes is funneled to two photosynthetic reaction centers. *Excitation of the reaction centers drives a series of oxidation–reduction reactions whose net results are the oxidation of water, the reduction of $NADP^+$, and the generation of a transmembrane proton gradient that powers ATP synthesis.* These events are known as the **light reactions** of photosynthesis. (Most photosynthetic bacteria undertake similar reactions but have a single reaction center and do not produce oxygen.) The two photosynthetic reaction centers that mediate light energy transduction are part of protein complexes called Photosystem I and Photosystem II. These, along with other integral and peripheral proteins of the thylakoid membrane, operate in a series, much like the mitochondrial electron transport chain.

Photosystem II is a light-activated oxidation–reduction enzyme

In plants and cyanobacteria, the light reactions begin with Photosystem II (the number indicates that it was the second to be discovered). This integral membrane protein complex is dimeric, with more bulk on the lumenal side of the thylakoid membrane than on the stromal side. The cyanobacterial Photosystem II contains at least 19 subunits (14 of them integral membrane proteins). Its numerous prosthetic groups include light-absorbing pigments and redox-active cofactors (Fig. 16-9).

Several dozen chlorophyll molecules in Photosystem II function as internal antennas, funneling energy to the two reaction centers, each of which includes a pair of chlorophyll molecules known as P680 (680 nm is the wavelength of one of their absorption peaks). The reaction center chlorophylls overlap so that they are electronically coupled and function as a single unit. When P680 is excited, as indicated by the notation P680*, it quickly gives up an electron, dropping to a lower-energy

> **KEY CONCEPTS**
> - The P680 reaction center of Photosystem II undergoes photooxidation.
> - Photosystem II splits water to replace the lost P680 electron and generate O_2.
> - Electrons from Photosystem II travel via plastoquinone, cytochrome $b_6 f$, and plastocyanin to Photosystem I.
> - Photooxidation of P700 in Photosystem I drives cyclic and noncyclic electron flow.
> - The proton gradient across the thylakoid membrane drives ATP synthesis.

⊕ **See Guided Exploration. Two-center photosynthesis overview.**

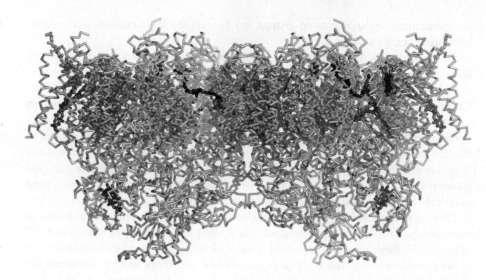

Figure 16-9 Structure of cyanobacterial Photosystem II. The proteins are shown as gray ribbons, and the various prosthetic groups and cofactors are shown as stick models and color-coded: chlorophyll, green; pheophytin, orange; β-carotene, red; heme, purple; and quinone, blue. The stroma is at the top and the thylakoid lumen at the bottom. [Structure of Photosystem II from *Synechococcus elongatus* (1S5L) determined by K. N. Ferreira, T. M. Iverson, K. Maghlaoui, J. Barber, and S. Iwata.]

state, $P680^+$. In other words, light has oxidized P680. The photooxidized chlorophyll molecule must be reduced in order to return to its original state.

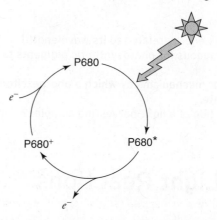

The two P680 groups are located near the lumenal side of Photosystem II. The electron given up by each photooxidized P680 travels through several redox groups (**Fig. 16-10**). Although the prosthetic groups in Photosystem II are arranged more or less symmetrically, they do not all directly participate in electron transfer. The electron eventually reaches a plastoquinone molecule on the stromal side of Photosystem II. Plastoquinone (PQ) is similar to mammalian mitochondrial ubiquinone (see Section 12-2).

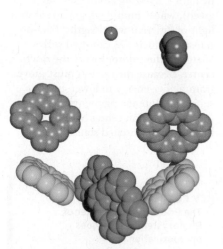

Figure 16-10 Arrangement of prosthetic groups in Photosystem II. The green chlorophyll groups constitute the photooxidizable P680. The two "accessory" chlorophyll groups (yellow) do not undergo oxidation or reduction. An electron from P680 travels to one of the pheophytin groups (orange), which are essentially chlorophyll molecules without the central Mg^{2+} ion. Next, the electron is transferred to a tightly bound plastoquinone molecule (blue) and then to a loosely bound plastoquinone (not shown). An iron atom (red) may assist the final electron transfer. The lipid tails of the prosthetic groups are not shown.

Plastoquinone

It functions in the same way as a two-electron carrier. The fully reduced plastoquinol (PQH_2) joins a pool of plastoquinones that are soluble in the thylakoid membrane. Two electrons (two photooxidations of P680) are required to fully reduce plastoquinone to PQH_2. This reaction also consumes two protons, which are taken from the stroma.

The oxygen-evolving complex of Photosystem II oxidizes water

▶▶ **WHY** do plants produce O_2?

At the start of the chapter, we pointed out that O_2 is a waste product of the photosynthetic process. O_2 is generated from H_2O by a lumenal portion of Photosystem II called the oxygen-evolving center. This reaction can be written as

$$2\,H_2O \rightarrow O_2 + 4\,H^+ + 4\,e^-$$

The electrons derived from H_2O are used to restore photooxidized P680 to its reduced state.

The catalyst for the water-splitting reaction is a cofactor with the composition Mn_4CaO_5 (Fig. 16-11). This unusual inorganic cofactor occurs in all Photosystem II complexes, which suggests a unique chemistry that has remained unaltered for about 2.5 billion years. No synthetic catalyst can match the manganese cluster in its ability to extract electrons from water to form O_2. The water-splitting reaction is rapid, with about 50 O_2 produced per second per Photosystem II, and generates most of the earth's atmospheric O_2.

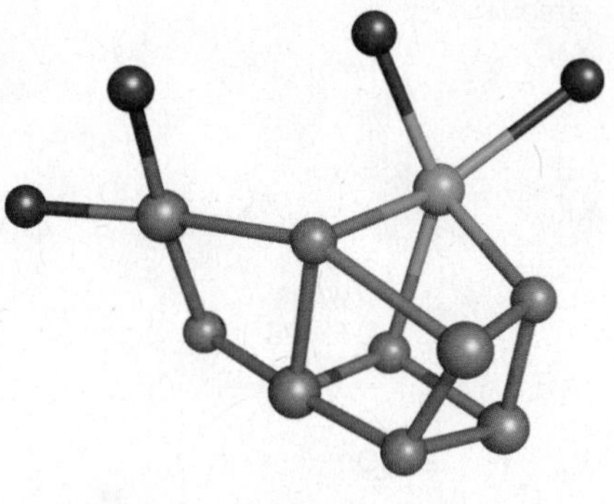

During water oxidation, the manganese cluster undergoes multiple changes in its oxidation state, somewhat reminiscent of the changes in the Fe–Cu binuclear center of cytochrome c oxidase (mitochondrial Complex IV; see Fig. 15-18), which carries out the reverse reaction. The four water-derived protons are released into the thylakoid lumen, contributing to a drop in pH relative to the stroma. A tyrosine radical (Y·) in Photosystem II transfers each of the four water-derived electrons to P680$^+$ (a tyrosine radical also plays a role in electron transfer in cytochrome c oxidase; see Section 15-2).

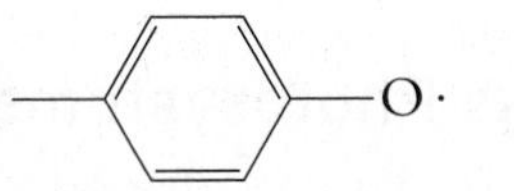

Tyrosine radical

The oxidation of water is a thermodynamically demanding reaction because O_2 has an extremely high reduction potential (+0.815 V) and electrons spontaneously flow from a group with a lower reduction potential to a group with a higher reduction potential (see Section 15-1). In fact, P680 is the most powerful biological oxidant, with a reduction potential of about +1.15 V. Upon photoexcitation, the reduction potential of P680 (now P680*) is dramatically diminished, to about −0.8 V. *This low reduction potential allows P680* to surrender an electron to a series of groups with increasingly positive reduction potentials* (Fig.16-12). The overall result is that the input of solar energy allows an electron to travel a thermodynamically favorable path from water to plastoquinone. *Four photooxidation events in Photosystem II are required to oxidize two H_2O molecules and produce one O_2 molecule.* Figure 16-13 summarizes the functions of Photosystem II.

Figure 16-11 Structure of the Mn_4CaO_5 cluster. Atoms are color-coded: Mn purple, Ca green, and O red. One or more of the four H_2O molecules associated with the cluster (oxygen atoms in blue) may be substrates for the water-splitting reaction. The Asp, Glu, and His side chains that hold the cluster in place are not shown. [Structure of PSII (pdb 3ARC) determined by Y. Umena, K. Kawakami, J.-R. Shen, and N. Kamiya.]

Figure 16-12 Reduction potential and electron flow in Photosystem II. Electrons flow spontaneously from a group with a low reduction potential to a group with a high reduction potential. The transfer of electrons from H_2O to plastoquinone (dashed red lines) is made possible by the excitation of P680 (wavy arrow), which dramatically lowers its reduction potential.

? Will pheophytin be in the oxidized or reduced state when it is dark?

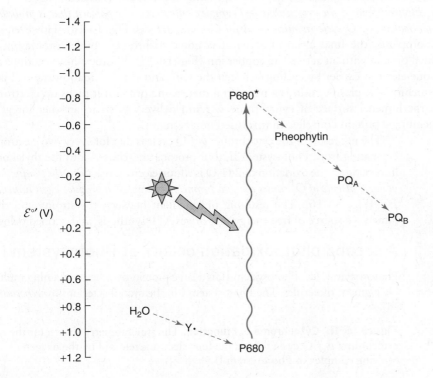

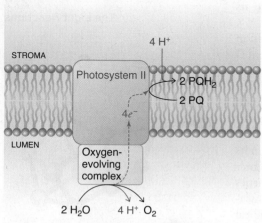

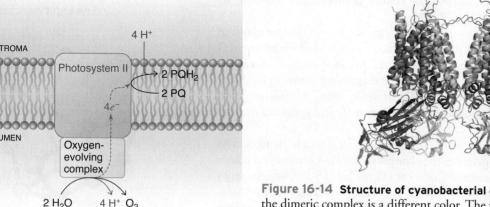

Figure 16-14 Structure of cyanobacterial cytochrome b_6f. Each subunit of the dimeric complex is a different color. The prosthetic groups are not shown. [Structure (pdb 1VF5) determined by G. Kurisu, H. Zhang, J. L. Smith, and W. A. Cramer.]

> **?** Compare this structure to the functionally similar mitochondrial cytochrome bc_1 (Complex III) in Figure 15-13.

Figure 16-13 Photosystem II function. For every oxygen molecule evolved, two plastoquinone molecules are reduced.

Cytochrome b_6f links Photosystems I and II

After they leave Photosystem II as plastoquinol, electrons reach a second membrane-bound protein complex known as cytochrome b_6f. This complex resembles mitochondrial Complex III (also called cytochrome bc_1)—from the entry of electrons in the form of a reduced quinone, through the circular flow of electrons among its redox groups, to the final transfer of electrons to a mobile electron carrier.

The cytochrome b_6f complex contains eight subunits in each of its monomeric halves (Fig. 16-14). Three subunits bear electron-transporting prosthetic groups. One of these subunits is cytochrome b_6, which is homologous to mitochondrial cytochrome b. The second is cytochrome f, whose heme group is actually of the c type. Although it shares no sequence homology with mitochondrial cytochrome c_1, it functions similarly. The chloroplast complex also contains a Rieske iron–sulfur protein with a 2Fe–2S group that behaves like its mitochondrial counterpart. However, the cytochrome b_6f complex also contains subunits with prosthetic groups that are absent in the mitochondrial complex: a chlorophyll molecule and a β-carotene. These light-absorbing molecules do not appear to participate in electron transfer and may instead help regulate the activity of cytochrome b_6f by registering the amount of available light.

Electron flow in the cytochrome b_6f complex follows a cyclic pattern that is probably identical to the Q cycle in mitochondrial Complex III (see Fig. 15-14). However, in chloroplasts, the final electron acceptor is not cytochrome c but plastocyanin, a small protein with an active-site copper ion (Fig. 16-15). Plastocyanin functions as a one-electron carrier by cycling between the Cu^+ and Cu^{2+} oxidation states. Like cytochrome c, plastocyanin is a peripheral membrane protein; it picks up electrons at the lumenal surface of cytochrome b_6f and delivers them to another integral membrane protein complex, in this case Photosystem I.

The net result of the cytochrome b_6f Q cycle is that for every two electrons emanating from Photosystem II, four protons are released into the thylakoid lumen. Since the oxidation of 2 H_2O is a four-electron reaction, *the production of one molecule of O_2 causes the cytochrome b_6f complex to produce eight lumenal H^+* (Fig. 16-16). The resulting pH gradient between the stroma and the lumen is a source of free energy that drives ATP synthesis, as described below.

A second photooxidation occurs at Photosystem I

Photosystem I, like Photosystem II, is a large protein complex containing multiple pigment molecules. The Photosystem I in the cyanobacterium *Synechococcus*

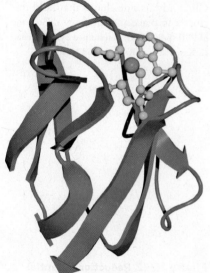

Figure 16-15 Plastocyanin. The redox active copper ion (green) is coordinated by a Cys, a Met, and two His residues (yellow). [Structure of plastocyanin from poplar leaves (pdb 1PLC) determined by J. M. Guss and H. C. Freeman.]

Figure 16-16 Cytochrome b_6f function. The stoichiometry shown for the cytochrome b_6f Q cycle reflects the four electrons released by the oxygen-evolving complex of Photosystem II.

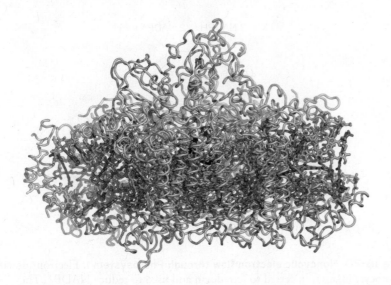

Figure 16-17 Structure of cyanobacterial Photosystem I. The protein is shown as a gray ribbon, and the various prosthetic groups are color-coded: chlorophyll, green; β-carotene, red; phylloquinone, blue; and Fe–S clusters, orange. Only one monomer of the trimeric complex is shown. The stroma is at the top. [Structure of the *Synechococcus* Photosystem I (pdb IJB0) determined by P. Jordan, P. Fromme, H. T. Witt, O. Klukas, W. Saenger, and N. Krauss.]

is a symmetric trimer with 12 proteins in each monomer **(Fig. 16-17)**. Ninety-six chlorophyll molecules and 22 carotenoids operate as a built-in light-harvesting complex for Photosystem I.

In the core of each monomer, a pair of chlorophyll molecules constitute the photoactive group known as P700 (it has a slightly longer-wavelength absorbance maximum than P680). Like P680, P700 undergoes exciton transfer from an antenna pigment. P700* gives up an electron to achieve a low-energy oxidized state, P700$^+$. The group is then reduced by accepting an electron donated by plastocyanin.

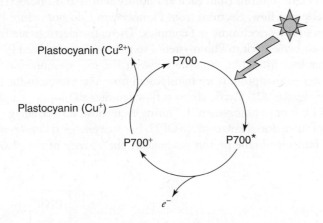

P700 is not a particularly good reducing agent (its reduction potential is relatively high, about +0.45 V). However, excited P700 (P700*) has an extremely low $\mathcal{E}^{\circ\prime}$ value (about −1.3 V), so electrons can spontaneously flow from P700* to the other redox groups of Photosystem I. These groups include four additional chlorophyll molecules, quinones, and iron–sulfur clusters of the 4Fe–4S type **(Fig. 16-18)**. As in Photosystem II, these prosthetic groups are arranged with approximate symmetry. However, in Photosystem I, all the redox groups appear to undergo oxidation and reduction.

Each electron given up by photooxidized P700 eventually reaches ferredoxin, a small peripheral protein on the stromal side of the thylakoid membrane. Ferredoxin undergoes a one-electron reduction at a 2Fe–2S cluster **(Fig. 16-19)**. Reduced ferredoxin participates in two different electron transport pathways in the chloroplast, which are known as noncyclic and cyclic electron flow.

In **noncyclic electron flow,** ferredoxin serves as a substrate for ferredoxin–NADP$^+$ reductase. This stromal enzyme uses two electrons (from two separate ferredoxin molecules) to reduce NADP$^+$ to NADPH **(Fig. 16-20)**. *The net result of noncyclic electron flow is therefore the transfer of electrons from water, through Photosystem II, cytochrome b$_6$f, Photosystem I, and then on to NADP$^+$.* Photosystem I does not contribute to the transmembrane proton gradient except by consuming stromal protons in the reduction of NADP$^+$ to NADPH.

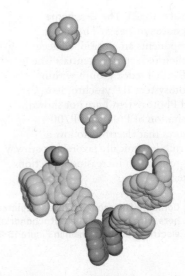

Figure 16-18 Prosthetic groups in Photosystem I. The groups include P700 (the green chlorophyll molecules), "accessory" chlorophylls (yellow), quinones (marked by blue spheres), and 4Fe–4S clusters (orange).

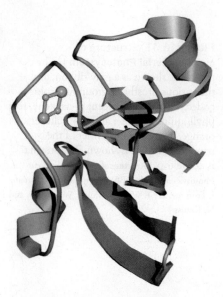

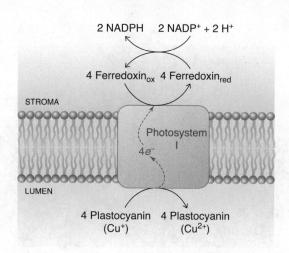

Figure 16-20 Noncyclic electron flow through Photosystem I. Electrons donated by plastocyanin are transferred to ferredoxin and used to reduce $NADP^+$. The stoichiometry reflects the four electrons released by the oxidation of 2 H_2O in Photosystem II. Therefore, 2 NADPH are produced for every molecule of O_2.

Figure 16-19 Ferredoxin. The 2Fe–2S cluster is shown in orange. [Structure of ferredoxin from the cyanobacterium *Anabaena* (pdb 1CZP) determined by R. Morales, M. H. Charon, and M. Frey.]

When plotted according to reduction potential, the electron-carrying groups of the pathway from water to $NADP^+$ form a diagram called the **Z-scheme** of photosynthesis (**Fig. 16-21**). The zigzag pattern is due to the two photooxidation events, which markedly decrease the reduction potentials of P680 and P700. Note that the four-electron process of producing one O_2 and two NADPH is accompanied by the absorption of eight photons (four each at Photosystem II and Photosystem I).

In **cyclic electron flow,** electrons from Photosystem I do not reduce $NADP^+$ but instead return to the cytochrome $b_6 f$ complex. There, the electrons are transferred to plastocyanin and flow back to Photosystem I to reduce photooxidized $P700^+$. Meanwhile, plastoquinol molecules circulate between the two quinone-binding sites of cytochrome $b_6 f$ so that protons are translocated from the stroma to the lumen, as in the Q cycle (**Fig. 16-22**). Cyclic electron flow requires the input of light energy at Photosystem I but not Photosystem II. During cyclic flow, no free energy is recovered in the form of the reduced cofactor NADPH, but free energy is conserved in the formation of a transmembrane proton gradient by the activity of the cytochrome $b_6 f$

Figure 16-21 The Z-scheme of photosynthesis. The major components are positioned according to their reduction potentials (the individual redox groups within Photosystem II, cytochrome $b_6 f$, and Photosystem I are not shown). Excitation of P680 and P700 ensures that electrons follow a thermodynamically favorable pathway to groups with increasing reduction potential.

? Compare the redox changes depicted here with those of the mitochondrial electron transport chain in Figure 15-2.

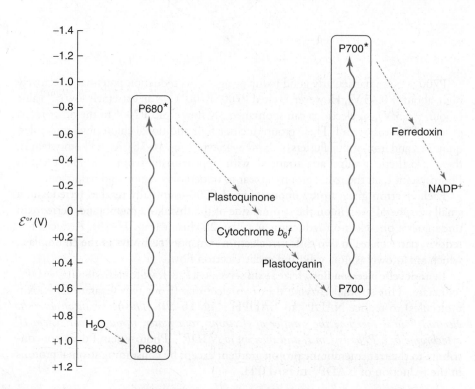

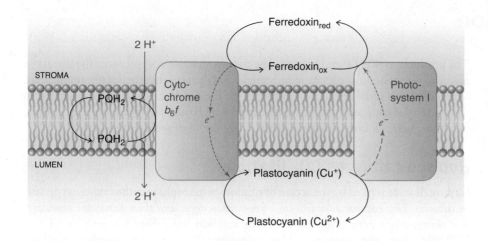

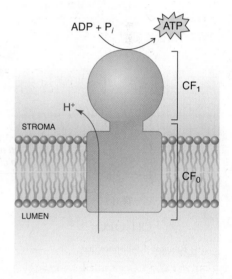

Figure 16-22 Cyclic electron flow. Electrons circulate between Photosystem I and the cytochrome b_6f complex. No NADPH or O_2 is produced, but the activity of cytochrome b_6f builds up a proton gradient that drives ATP synthesis.

complex. Consequently, *cyclic electron flow augments ATP generation by chemiosmosis* (in some bacteria with just a single reaction center, electrons flow through a similar pathway that does not produce O_2 or NADPH). By varying the proportion of electrons that follow the noncyclic and cyclic pathways through Photosystem I, a photosynthetic cell can vary the proportions of ATP and NADPH produced by the light reactions.

Chemiosmosis provides the free energy for ATP synthesis

Chloroplasts and mitochondria use the same mechanism to synthesize ATP: *They couple the dissipation of a transmembrane proton gradient to the phosphorylation of ADP.* In photosynthetic organisms, this process is called **photophosphorylation.** Chloroplast ATP synthase is highly homologous to mitochondrial and bacterial ATP synthases. The CF_1CF_0 complex ("C" indicates chloroplast) consists of a proton-translocating integral membrane component (CF_0) mechanically linked to a soluble CF_1 component where ATP synthesis occurs by a binding change mechanism (as described in Fig. 15-25). The movement of protons from the thylakoid lumen to the stroma provides the free energy to drive ATP synthesis **(Fig. 16-23)**.

As in mitochondria, the proton gradient has both chemical and electrical components. In chloroplasts, the pH gradient (about 3.5 pH unit) is much larger than in mitochondria (about 0.75 units). However, in chloroplasts, the electrical component is less than in mitochondria because of the permeability of the thylakoid membrane to ions such as Mg^{2+} and Cl^-. Diffusion of these ions tends to minimize the difference in charge due to protons.

Assuming noncyclic electron flow, 8 photons are absorbed (4 by Photosystem II and 4 by Photosystem I) to generate 4 lumenal protons from the oxygen-evolving complex and 8 protons from the cytochrome b_6f complex. Theoretically, these 12 protons can drive the synthesis of about 3 ATP, which is consistent with experimental results showing approximately 3 ATP generated for each molecule of O_2.

Figure 16-23 Photophosphorylation. As protons traverse the CF_0 component of chloroplast ATP synthase (following their concentration gradient from the lumen to the stroma), the CF_1 component carries out ATP synthesis.

> **CONCEPT REVIEW**
> * Summarize the functions of Photosystem II, the oxygen-evolving complex, plastoquinone, the cytochrome b_6f complex, plastocyanin, Photosystem I, and ferredoxin.
> * How does absorption of a photon drive electron transfer from water to plastoquinone?
> * Describe the Z-scheme of photosynthesis and explain its zigzag shape.
> * Discuss the yields of O_2, NADPH, and ATP in cyclic and noncyclic electron flow.
> * Compare and contrast the chloroplast light reactions and mitochondrial electron transport.
> * How does photophosphorylation resemble oxidative phosphorylation?
> * Which reactions contribute to the pH gradient that drives the activity of chloroplast ATP synthase?

16-3 Carbon Fixation

KEY CONCEPTS
- Rubisco catalyzes carbon fixation by adding CO_2 to a five-carbon acceptor molecule.
- The Calvin cycle shuffles sugars for the net conversion of three CO_2 to one glyceraldehyde-3-phosphate.
- Light-dependent mechanisms regulate the activity of the Calvin cycle.
- Newly synthesized sugars are incorporated into sucrose and polysaccharides.

The production of ATP and NADPH by the photoactive complexes of the thylakoid membrane (or bacterial plasma membrane) is only part of the story of photosynthesis. The rest of this chapter focuses on the use of the products of the light reactions in the so-called **dark reactions.** These reactions, which occur in the chloroplast stroma, fix atmospheric carbon dioxide in biologically useful organic molecules.

Rubisco catalyzes CO_2 fixation

Carbon dioxide is fixed by the action of ribulose bisphosphate carboxylase/oxygenase, or rubisco. *This enzyme adds CO_2 to a five-carbon sugar and then cleaves the product to two three-carbon units* (**Fig. 16-24**). This reaction itself does not require ATP or NADPH, but the reactions that transform the rubisco reaction product, 3-phosphoglycerate, to the three-carbon sugar glyceraldehyde-3-phosphate require both ATP and NADPH, as we will see below.

Three-carbon compounds are the biosynthetic precursors of monosaccharides, amino acids, and—indirectly—nucleotides. They also give rise to the two-carbon acetyl units used to build fatty acids. The metabolic importance of these small molecular building blocks is one reason why the scheme shown in Figure 16-2 presents photosynthesis as a process in which CO_2 is converted to two- and three-carbon intermediates.

Rubisco is a notable enzyme, in part because its activity directly or indirectly sustains most of the earth's biomass. Plant chloroplasts are packed with the enzyme, which accounts for about half of the chloroplast's protein content. Rubisco is easily the most abundant biological catalyst. One reason for its high concentration is that it is not a particularly efficient enzyme. Its catalytic output is only about three CO_2 fixed per second.

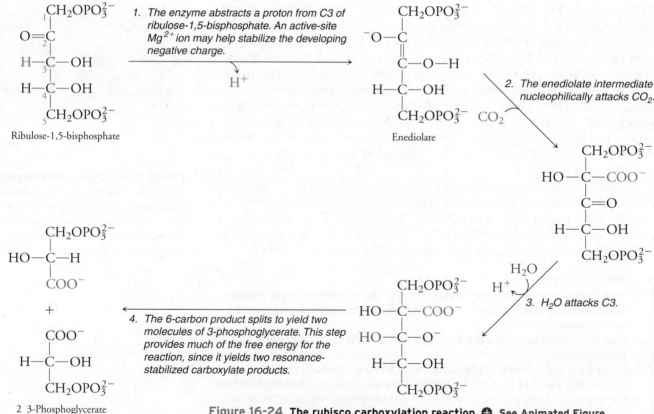

Figure 16-24 The rubisco carboxylation reaction. ⊕ See Animated Figure. Mechanism of RUBP carboxylase.

? Before this reaction was understood, scientists believed that carbon fixation involved the reaction of CO_2 with a two-carbon molecule. Explain.

Bacterial rubisco is usually a small dimeric enzyme, whereas the plant enzyme is a large multimer of eight large and eight small subunits (**Fig. 16-25**). In some archaebacteria, rubisco has ten identical subunits. Enzymes with multiple catalytic sites typically exhibit cooperative behavior and are regulated allosterically, but this does not seem to be true for plant rubisco, whose eight active sites operate independently. Multimerization may simply be an efficient way to pack more active sites into the limited space of the chloroplast.

Despite its metabolic importance, rubisco is not a highly specific enzyme. It also acts as an oxygenase (as reflected in its name) by reacting with O_2, which chemically resembles CO_2. The products of the oxygenase reaction are a three-carbon and a two-carbon compound:

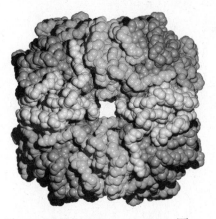

Figure 16-25 Spinach rubisco. The complex has a mass of approximately 550 kD. The eight catalytic sites are located in the large subunits (dark colors). Only four of eight small subunits (light colors) are visible in this image. [Structure (pdb 1RCX) determined by T. C. Taylor and I. Anderson.]

$$
\begin{array}{c}
\text{CH}_2\text{OPO}_3^{2-} \\
| \\
\text{O}=\text{C} \\
| \\
\text{H}-\text{C}-\text{OH} \quad + \quad \text{O}_2 \\
| \\
\text{H}-\text{C}-\text{OH} \\
| \\
\text{CH}_2\text{OPO}_3^{2-} \\
\text{Ribulose bisphosphate}
\end{array}
\xrightarrow[\text{rubisco}]{\text{H}_2\text{O}}
\begin{array}{c}
\text{CH}_2\text{OPO}_3^{2-} \\
| \\
^-\text{O}-\text{C}=\text{O} \\
\text{2-Phosphoglycolate} \\
+ \\
\text{COO}^- \\
| \\
\text{H}-\text{C}-\text{OH} \\
| \\
\text{CH}_2\text{OPO}_3^{2-} \\
\text{3-Phosphoglycerate}
\end{array}
$$

The 2-phosphoglycolate product of the rubisco oxygenation reaction is subsequently metabolized by a pathway that consumes ATP and NADPH and produces CO_2. This process, called **photorespiration,** uses the products of the light reactions and therefore wastes some of the free energy of captured photons.

Oxygenase activity is a feature of all known rubisco enzymes and must play an essential role that has been conserved throughout plant evolution. *Photorespiration apparently provides a mechanism for plants to dissipate excess free energy under conditions where the CO_2 supply is insufficient for carbon fixation.* Photorespiration may consume significant amounts of ATP and NADPH at high temperatures, which favor oxygenase activity. Some plants have evolved a mechanism, called the **C_4 pathway,** to minimize photorespiration (Box 16-A).

BOX 16-A　BIOCHEMISTRY NOTE

The C_4 Pathway

On hot, bright days, high temperatures favor photorespiration, and CO_2 supplies are low as plants close their stomata (pores in the leaf surface) to avoid evaporative water loss. This combination of events can bring photosynthesis to a halt. Some plants avoid this possibility by stockpiling CO_2 in four-carbon molecules so that photosynthesis can proceed even while stomata are closed (the closed stomata also limit the availability of O_2 for photorespiration).

The mechanism for storing carbon begins with the condensation of bicarbonate (HCO_3^-) with phosphoenolpyruvate to yield oxaloacetate, which is then reduced to malate. These four-carbon acids give the C_4 pathway its name. The subsequent oxidative decarboxylation of malate regenerates CO_2 and NADPH to be used in the Calvin cycle. The three-carbon remnant, pyruvate, is recycled back to phosphoenolpyruvate.

Because the C_4 pathway and the rubisco reaction compete for CO_2, they take place in different types of cells or at different times of day. For example, in some plants, carbon accumulates in mesophyll cells, which are near the leaf surface and lack rubisco. The C_4 compounds then enter bundle sheath cells in the leaf interior, which contain abundant rubisco. In other plants, the C_4 pathway occurs at night, when the stomata are open and water loss is minimal, and carbon is fixed by rubisco during the day.

The C_4 pathway is energetically expensive, so it requires lots of sunlight. Consequently, C_4 plants grow more slowly than conventional, or C_3, plants when light is limited, but they have the advantage in hot, dry climates. About 5% of the earth's plants, including the economically important maize (corn), sugarcane, and sorghum, use the C_4 pathway.

(continued on next page)

The recognition of global warming has led to predictions that C_4 "weeds" may overtake economically important C_3 plants as temperatures increase. In fact, the increase in atmospheric CO_2 that is driving the warming trend appears to promote the growth of C_3 plants, which can obtain CO_2 more easily without losing too much water via their stomata. However, if water is limited, C_4 plants may still have a competitive edge, as they are adapted not just for hot environments but for arid ones.

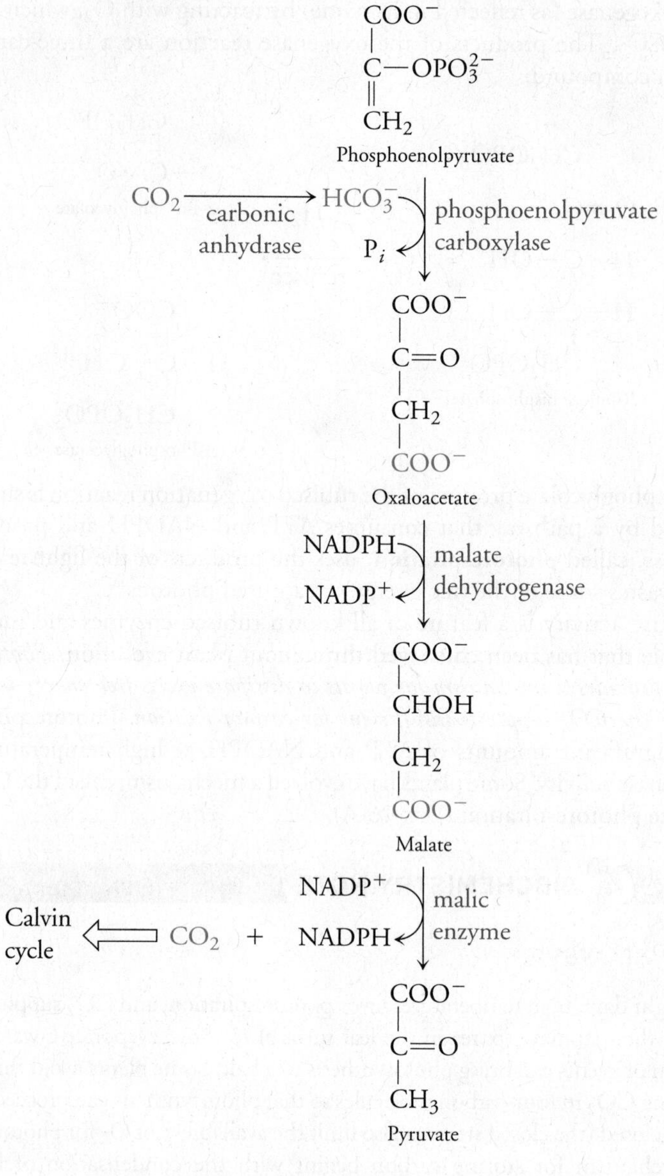

● **Question:** Operation of the C_4 pathway consumes ATP. Using the pathway shown here, indicate where this occurs.

The Calvin cycle rearranges sugar molecules

If rubisco is responsible for fixing CO_2, what is the origin of its other substrate, ribulose bisphosphate? The answer—elucidated over many years by Melvin Calvin, James Bassham, and Andrew Benson—is a metabolic pathway known as the **Calvin cycle.** Early experiments to study the fate of ^{14}C-labeled CO_2 in algae showed that within a few minutes, the cells had synthesized a complex mixture of sugars, all containing the radioactive label. Rearrangements among these sugar molecules generate the five-carbon substrate for rubisco.

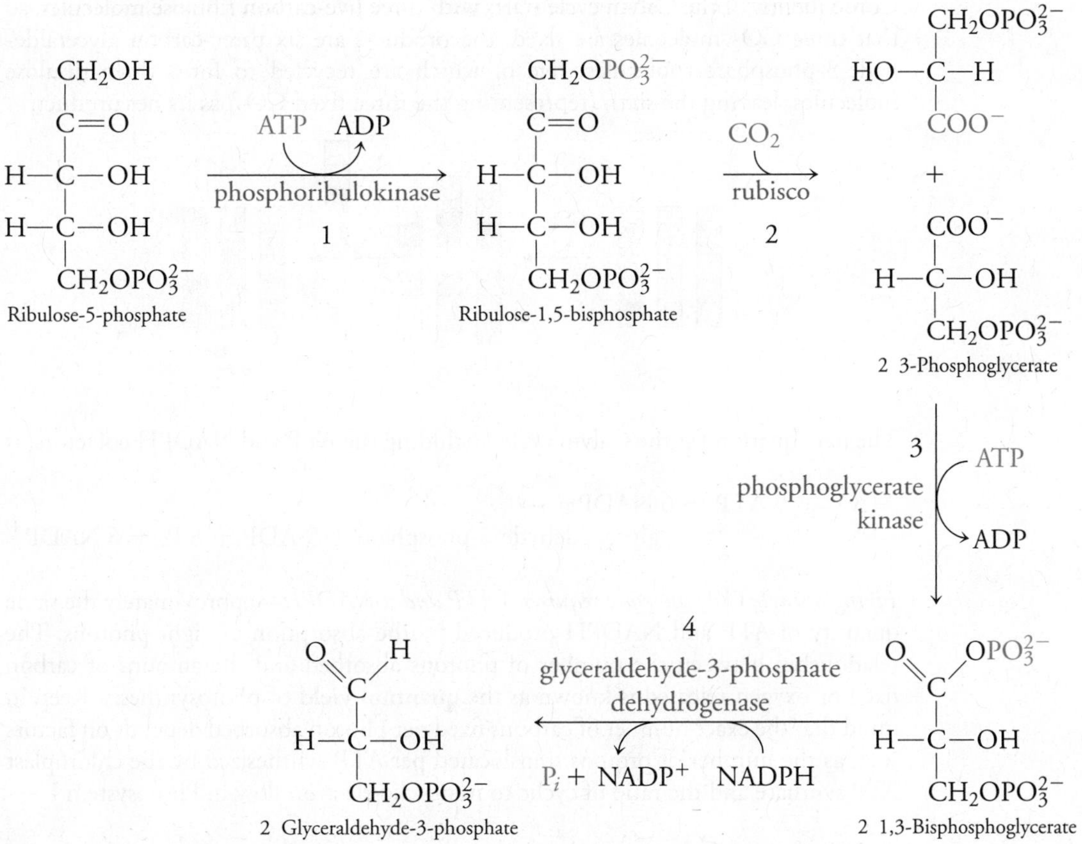

Figure 16-26 Initial reactions of the Calvin cycle. Note that ATP and NADPH, products of the light-dependent reactions, are consumed in the process of converting CO_2 to glyceraldehyde-3-phosphate.

The Calvin cycle actually begins with a sugar monophosphate, ribulose-5-phosphate, which is phosphorylated in an ATP-dependent reaction (**Fig. 16-26**). The resulting bisphosphate is a substrate for rubisco, as we have already seen. Each 3-phosphoglycerate product of the rubisco reaction is then phosphorylated, again at the expense of ATP. This phosphorylation reaction (step 3 of Fig. 16-26) is identical to the phosphoglycerate kinase reaction of glycolysis (see Section 13-1). Next, bisphosphoglycerate is reduced by the chloroplast enzyme glyceraldehyde-3-phosphate dehydrogenase, which resembles the glycolytic enzyme but uses NADPH rather than NADH. The NADPH is the product of the light reactions of photosynthesis.

Some glyceraldehyde-3-phosphate is siphoned from the Calvin cycle for metabolic fates such as glucose or amino acid synthesis. Recall from Section 13-2 that the pathway from glyceraldehyde-3-phosphate to glucose consists of reactions that require no further input of ATP. Glyceraldehyde-3-phosphate can also be converted to pyruvate and then to oxaloacetate, both of which can undergo transamination to generate amino acids. Additional reactions lead to other metabolites.

The glyceraldehyde-3-phosphate that is not used for biosynthetic pathways remains part of the Calvin cycle and enters a series of isomerization and group-transfer reactions that regenerate ribulose-5-phosphate. These interconversion reactions are similar to those of the pentose phosphate pathway (Section 13-4). We can represent them simply by showing how carbon atoms are shuffled among three- to seven-carbon sugars:

$$C_3 + C_3 \rightarrow C_6$$
$$C_3 + C_6 \rightarrow C_4 + C_5$$
$$C_3 + C_4 \rightarrow C_7$$
$$C_3 + C_7 \rightarrow C_5 + C_5$$

net reaction: $5 \, C_3 \rightarrow 3 \, C_5$

Consequently, if the Calvin cycle starts with three five-carbon ribulose molecules, so that three CO_2 molecules are fixed, the products are six three-carbon glyceraldehyde-3-phosphate molecules, five of which are recycled to form three ribulose molecules, leaving the sixth (representing the three fixed CO_2) as its net product.

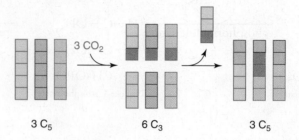

The net equation for the Calvin cycle, including the ATP and NADPH cofactors, is

$$3\ CO_2 + 9\ ATP + 6\ NADPH \rightarrow$$
$$\text{glyceraldehyde-3-phosphate} + 9\ ADP + 8\ P_i + 6\ NADP^+$$

Fixing a single CO_2 therefore requires 3 ATP and 2 NADPH—approximately the same quantity of ATP and NADPH produced by the absorption of eight photons. The relationship between the number of photons absorbed and the amount of carbon fixed or oxygen released is known as the **quantum yield** of photosynthesis. Keep in mind that the exact number of carbons fixed per photon absorbed depends on factors such as the number of protons translocated per ATP synthesized by the chloroplast ATP synthase and the ratio of cyclic to noncyclic electron flow in Photosystem I.

The availability of light regulates carbon fixation

Plants must coordinate light availability with carbon fixation. During the day, both processes occur. At night, when the photosystems are inactive, the plant turns off the "dark" reactions to conserve ATP and NADPH while it turns on pathways to regenerate these cofactors by metabolic pathways such as glycolysis and the pentose phosphate pathway. It would be wasteful for these catabolic processes to proceed simultaneously with the Calvin cycle. Thus, the "dark" reactions do not actually occur in the dark!

All the mechanisms for regulating the Calvin cycle are directly or indirectly linked to the availability of light energy. Some of the regulatory mechanisms are highlighted here. For example, a catalytically essential Mg^{2+} ion in the rubisco active site is coordinated in part by a carboxylated Lys side chain that is produced by the reaction of CO_2 with the ε-amino group:

$$-(CH_2)_4-NH_2 + CO_2 \rightleftharpoons -(CH_2)_4-NH-COO^- + H^+$$
$$\text{Lys}$$

By forming the Mg^{2+}-binding site, this "activating" CO_2 molecule promotes the ability of rubisco to fix additional substrate CO_2 molecules. The carboxylation reaction is favored at high pH, a signal that the light reactions are working (depleting the stroma of protons) and that ATP and NADPH are available for the Calvin cycle.

Magnesium ions also directly activate rubisco and several of the Calvin cycle enzymes. During the light reactions, the rise in stromal pH triggers the flux of Mg^{2+} ions from the lumen to the stroma (this ion movement helps balance the charge of the protons that are translocated in the opposite direction). Some of the Calvin cycle enzymes are also activated when the ratio of reduced ferredoxin to oxidized ferredoxin is high, another signal that the photosystems are active.

Calvin cycle products are used to synthesize sucrose and starch

Many of the three-carbon sugars produced by the Calvin cycle are converted to sucrose or starch. The polysaccharide starch is synthesized in the chloroplast stroma as a temporary storage depot for glucose. It is also synthesized as a long-term storage molecule elsewhere in the plant, including leaves, seeds, and roots. In the first stage of starch synthesis, two molecules of glyceraldehyde-3-phosphate are converted to glucose-6-phosphate by reactions analogous to those of mammalian gluconeogenesis (see Fig. 13-10). Phosphoglucomutase then carries out an isomerization reaction to produce glucose-1-phosphate. Next, this sugar is "activated" by its reaction with ATP to form ADP–glucose:

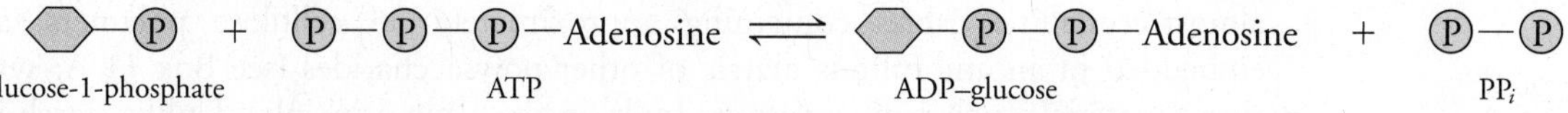

Glucose-1-phosphate   ATP   ADP–glucose   PP$_i$

(Recall from Section 13-3 that glycogen synthesis uses the chemically related nucleotide sugar UDP–glucose.) Starch synthase then transfers the glucose residue to the end of a starch polymer, forming a new glycosidic linkage (*right*). The overall reaction is driven by the exergonic hydrolysis of the PP$_i$ released in the formation of ADP–glucose. Thus, one phosphoanhydride bond is consumed in lengthening a starch molecule by one glucose residue.

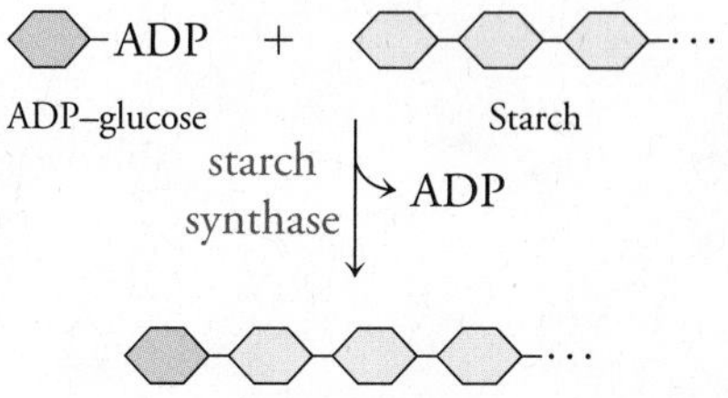

Sucrose, a disaccharide of glucose and fructose, is synthesized in the cytosol. Glyceraldehyde-3-phosphate or its isomer dihydroxyacetone phosphate is transported out of the chloroplast by an antiport protein that exchanges phosphate for a phosphorylated three-carbon sugar. Two of these sugars combine to form fructose 6-phosphate, and two others combine to form glucose-1-phosphate, which is subsequently activated by UTP. Next, fructose-6-phosphate reacts with UDP–glucose to produce sucrose-6-phosphate. Finally, a phosphatase converts the phosphorylated sugar to sucrose:

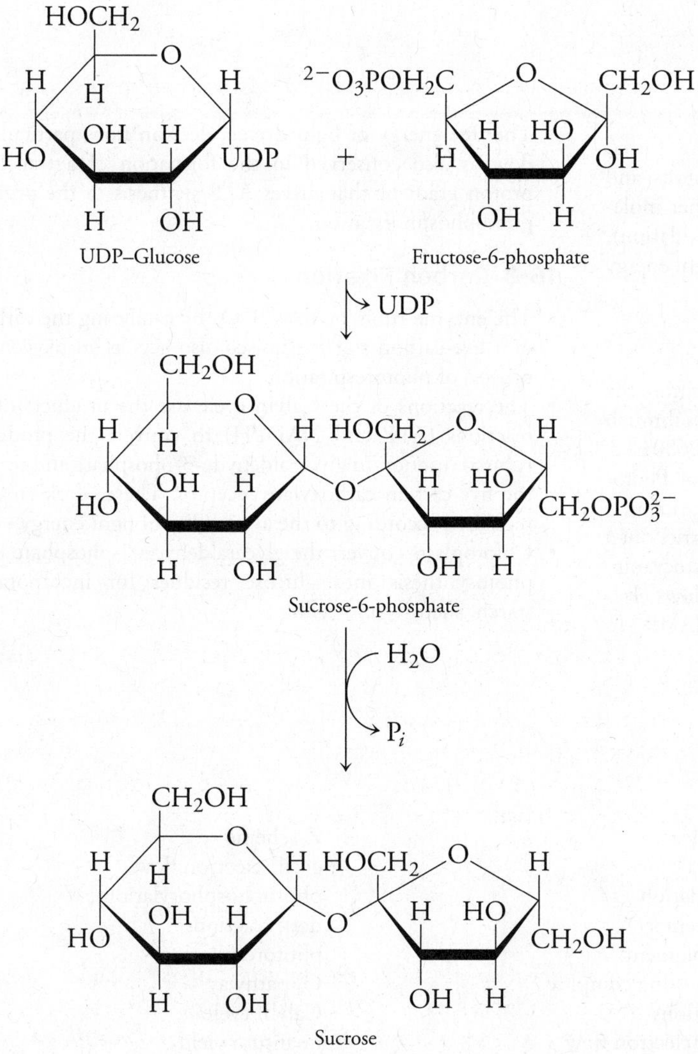

ADP–glucose   Starch

starch synthase → ADP

UDP–Glucose   Fructose-6-phosphate

↘ UDP

Sucrose-6-phosphate

H$_2$O

↘ P$_i$

Sucrose

Sucrose can then be exported to other plant tissues. This disaccharide probably became the preferred transport form of carbon in plants because its glycosidic linkage is insensitive to amylases (starch-digesting enzymes) and other common hydrolases. Also, its two anomeric carbons are tied up in the glycosidic bond and therefore cannot react nonenzymatically with other substances.

Cellulose, the other major polysaccharide of plants, is also synthesized from UDP–glucose (cellulose is described in Section 11-2). Plant cell walls consist of almost-crystalline cables containing approximately 36 cellulose polymers, all embedded in an amorphous matrix of other polysaccharides (see Box 11-A; synthetic materials such as fiberglass are built on the same principle). Unlike starch in plants or glycogen in mammals, cellulose is synthesized by enzyme complexes in the plant plasma membrane and is extruded into the extracellular space.

CONCEPT REVIEW
- What are the products of the two reactions catalyzed by rubisco?
- Compare the physiological implications of carbon fixation and photorespiration.
- How does the carbon from a molecule of fixed CO_2 become incorporated into other compounds such as monosaccharides?
- What is the source of the ribulose-1,5-bisphosphate used for carbon fixation by rubisco?
- Describe some of the mechanisms for regulating the activity of the "dark" reactions.
- What is the role of nucleotides in the synthesis of starch and sucrose?

[SUMMARY]

16-1 Chloroplasts and Solar Energy

- Plant chloroplasts contain pigments that absorb photons and release the energy, primarily by transferring it to another molecule (exciton transfer) or giving up an electron (photooxidation). Light-harvesting complexes act to capture and funnel light energy to the photosynthetic reaction centers.

16-2 The Light Reactions

- In the so-called light reactions of photosynthesis, electrons from the photooxidized P680 reaction center of Photosystem II pass through several prosthetic groups and then to plastoquinone. The P680 electrons are replaced when the oxygen-evolving complex of Photosystem II converts water to O_2, a four-electron oxidation reaction.
- Electrons flow next to a cytochrome $b_6 f$ complex that carries out a proton-translocating Q cycle, and then to the protein plastocyanin.
- A second photooxidation at P700 of Photosystem I allows electrons to flow to the protein ferredoxin and finally to $NADP^+$ to produce NADPH.

- The free energy of light-driven electron flow, particularly cyclic flow, is also conserved in the formation of a transmembrane proton gradient that drives ATP synthesis in the process called photophosphorylation.

16-3 Carbon Fixation

- The enzyme rubisco "fixes" CO_2 by catalyzing the carboxylation of a five-carbon sugar. Rubisco also acts as an oxygenase in the process of photorespiration.
- The reactions of the Calvin cycle use the products of the light reactions (ATP and NADPH) to convert the product of the rubisco reaction to glyceraldehyde-3-phosphate and to regenerate the five-carbon carboxylate receptor. These "dark" reactions are regulated according to the availability of light energy.
- Chloroplasts convert the glyceraldehyde-3-phosphate product of photosynthesis into glucose residues for incorporation into starch, sucrose, and cellulose.

[GLOSSARY TERMS]

photosynthesis	fluorescence	Z-scheme
carbon fixation	exciton transfer	cyclic electron flow
chloroplast	photooxidation	photophosphorylation
stroma	reaction center	dark reactions
thylakoid	antenna pigment	photorespiration
photon	light-harvesting complex	C_4 pathway
Planck's law	light reactions	Calvin cycle
photoreceptor	noncyclic electron flow	quantum yield

[PROBLEMS]

16-1 Chloroplasts and Solar Energy

1. Indicate with a C or an M whether the following occur in chloroplasts, mitochondria, or both:

_____ proton translocation
_____ photophosphorylation
_____ photooxidation
_____ quinones
_____ oxygen reduction
_____ water oxidation
_____ electron transport
_____ oxidative phosphorylation
_____ carbon fixation
_____ NADH oxidation
_____ Mn cofactor
_____ heme groups
_____ binding change mechanism
_____ iron–sulfur clusters
_____ $NADP^+$ reduction

2. Compare and contrast the structures of chloroplasts and mitochondria.

3. The thylakoid membrane contains some unusual lipids. One of these is galactosyl diacylglycerol, which consists of a glycerol with fatty acids esterified at the second and third positions. C1 of a β-D-galactose residue is attached to the first glycerol carbon. Draw the structure of galactosyl diacylglycerol.

4. Thylakoid membranes contain lipids with a high degree of unsaturation. What does this tell you about the character of the thylakoid membrane?

5. Calculate the energy per mole of photons with wavelengths of **(a)** 400 nm and **(b)** 700 nm.

6. Assuming 100% efficiency, calculate the number of moles of ATP that could be generated per mole of photons with the energies calculated in Problem 5.

7. Use Equation 15-4 to calculate the free energy change for transforming one mole of P680 to P680*.

8. Determine the wavelength of the photons whose absorption would supply the free energy to transform one mole of P680 to P680* (see Problem 7).

9. Red tides are caused by algal blooms that cause seawater to become visibly red. In the photosynthetic process, red algae take advantage of wavelengths not absorbed by other organisms. Describe the photosynthetic pigments of the red algae.

10. Some photosynthetic bacteria live in murky ponds where visible light does not penetrate easily. What wavelengths might the photosynthetic pigments in these organisms absorb?

11. Compare the structures of chlorophyll *a* (see Fig. 16-4) and the reduced form of heme *b* (see Fig. 15-12).

12. You are investigating the functional similarities of chloroplast cytochrome *f* and mitochondrial cytochrome c_1.
(a) Which would provide more useful information: the amino acid sequences of the proteins or models of their three-dimensional shapes? Explain.
(b) Would it be better to examine high-resolution models of the two apoproteins (the polypeptides without their heme groups) or low-resolution models of the holoproteins (polypeptides plus heme groups)?

13. Under conditions of very high light intensity, excess absorbed solar energy is dissipated by the action of "photoprotective" proteins in the thylakoid membrane. Explain why it is advantageous for these proteins to be activated by a buildup of a proton gradient across the membrane.

14. Of the four mechanisms for dissipating light energy shown in Figure 16-6, which would be best for "protecting" the photosystems from excess light energy?

16-2 The Light Reactions

15. The three electron-transporting complexes of the thylakoid membrane can be called plastocyanin–ferredoxin oxidoreductase, plastoquinone–plastocyanin oxidoreductase, and water–plastoquinone oxidoreductase. What are the common names of these enzymes and in what order do they act?

16. Photosystem II is located mostly in the tightly stacked regions of the thylakoid membrane, whereas Photosystem I is located mostly in the unstacked regions (see Fig. 16-3). Why might it be important for the two photosystems to be separated?

17. The herbicide 3-(3,4-dichlorophenyl)-1,1-dimethylurea (DCMU) blocks electron flow from Photosystem II to Photosystem I. What is the effect on oxygen production and photophosphorylation when DCMU is added to plants?

18. When the antifungal agent myxothiazol is added to a suspension of chloroplasts, the PQH_2/PQ ratio increases. Where does myxothiazol inhibit electron transfer?

19. Plastoquinone is not firmly anchored to any thylakoid membrane component but is free to diffuse laterally throughout the membrane among the photosynthetic components. What aspects of its structure account for this behavior?

20. The Photosystem II reaction center contains a protein called D1 that contains the PQ_B binding site. The D1 protein in the single-celled alga *Chlamydomonas reinhardtii* is predicted to have five hydrophobic membrane-spanning helical regions. A loop between the fourth and fifth segments is located in the stroma. This loop is believed to lie along the membrane surface, and it contains several highly conserved amino acid residues. D1 proteins with mutations at the Ala 251 position were evaluated for photosynthetic activity and herbicide susceptibility. The results are shown in the table. What are the essential properties of the amino acid at position 251 in the D1 protein?

Amino acid at position 251	Characteristics
Ala	Wild-type
Cys	Similar to wild-type
Gly, Pro, Ser	Impaired in photosynthesis
Ile, Leu, Val	Impaired in photoautotrophic growth
	Impaired in photosynthesis Resistant to herbicides
Arg, Gln, Glu, His, and Asp	Not photosynthetically competent

21. Calculate the free energy of translocating a proton out of the stroma when the lumenal pH is 3.5 units lower than the stromal pH and $\Delta\psi$ is −50 mV.

22. Compare the free energy of proton translocation you calculated in Problem 21 to the free energy of translocating a proton out of a mitochondrion where the pH difference is 0.75 units and $\Delta\psi$ is 200 mV. Compare both types of translocation. Are the processes exergonic or endergonic? Which contributes a larger component of the free energy for each process, the pH difference or the membrane potential?

23. Calculate the standard free energy change for the oxidation of one molecule of water by $NADP^+$.

24. Calculate the energy available in two photons of wavelength 600 nm. Compare this value with the standard free energy changes you calculated in Problem 23. Do two photons supply enough energy to drive the oxidation of one molecule of water by $NADP^+$?

25. Photophosphorylation in chloroplasts is similar to oxidative phosphorylation in mitochondria. What is the final electron acceptor in photosynthesis? What is the final electron acceptor in mitochondrial electron transport?

26. If radioactively labeled water ($H_2^{18}O$) is provided to a plant, where does the label appear?

27. Predict the effect of an uncoupler such as dinitrophenol (see Box 15-B) on production of **(a)** ATP and **(b)** NADPH by a chloroplast.

28. Antimycin A (an antibiotic) blocks electron transport in Complex III of the electron transport chain in mitochondria. How would the addition of antimycin A to chloroplasts affect chloroplast ATP synthesis and NADPH production?

29. Does the quantum yield of photosynthesis increase or decrease for systems where **(a)** the CF_0 component of ATP synthase contains more c subunits and **(b)** the proportion of cyclic electron flow through Photosystem I increases?

30. Oligomycin inhibits the proton channel (F_0) of the ATP synthase enzyme in mitochondria but does not inhibit CF_0. When oligomycin is added to plant cells undergoing photosynthesis, the cytosolic ATP/ADP ratio decreases, whereas the chloroplastic ATP/ADP ratio is unchanged or even increases. Explain these results.

16-3 Carbon Fixation

31. Defend or refute this statement: The "dark" reactions are so named because these reactions occur only at night.

32. Examine the net equation for the light and "dark" reactions of photosynthesis—that is, the incorporation of one molecule of CO_2 into carbohydrate, which has the chemical formula $(CH_2O)_n$.

$$CO_2 + 2\,H_2O \rightarrow (CH_2O) + O_2 + H_2O$$

How would this equation differ for a bacterial photosynthetic system in which H_2S rather than H_2O serves as a source of electrons?

33. Melvin Calvin and his colleagues noted that when $^{14}CO_2$ was added to algal cells, a single compound was radiolabeled within 5 seconds of exposure. What is the compound, and where does the radioactive label appear?

34. As described in the text, rubisco is not a particularly specific enzyme. Scientists have wondered why millions of years of evolution failed to produce a more specific enzyme. What advantage would be conferred upon a plant that evolved a rubisco enzyme that was able to react with CO_2 but not oxygen?

35. A tiny acorn grows into a massive oak tree. Using what you know about photosynthesis, what accounts for the increase in mass?

36. The $\Delta G^{\circ\prime}$ value for the rubisco reaction is -35.1 kJ $\cdot$ mol^{-1}, and the ΔG is -41.0 kJ $\cdot$ mol^{-1}. What is the ratio of products to reactants under normal cellular conditions?

37. Efforts to engineer a more efficient rubisco, one that could fix CO_2 more quickly than three per second, could improve farming by allowing plants to grow larger and/or faster. Explain why the engineered rubisco might also decrease the need for nitrogen-containing fertilizers.

38. Some plants synthesize the sugar 2-carboxyarabinitol-1-phosphate. This compound inhibits the activity of rubisco.

$$
\begin{array}{c}
CH_2OPO_3^{2-} \\
| \\
HO-C-CO_2^- \\
| \\
H-C-OH \\
| \\
H-C-OH \\
| \\
CH_2OH
\end{array}
$$

2-Carboxyarabinitol-
1-phosphate

(a) What is the probable mechanism of action of the inhibitor?
(b) Why do plants synthesize the inhibitor at night and break it down during the day?

39. An "activating" CO_2 reacts with a Lys side chain on rubisco to carboxylate it. The carboxylation reaction is favored at high pH. Explain why.

40. Chloroplast phosphofructokinase (PFK) is inhibited by ATP and NADPH. How does this observation link the light reactions to the regulation of glycolysis?

41. Crab grass, a C_4 plant, remains green during a long spell of hot, dry weather when C_3 grasses turn brown. Explain this observation.

42. Chloroplasts contain thioredoxin, a small protein with two Cys residues that can form an intramolecular disulfide bond. The sulfhydryl/disulfide interconversion in thioredoxin is catalyzed by an enzyme known as ferredoxin–thioredoxin reductase. This enzyme, along with some of the Calvin cycle enzymes, also includes two Cys residues that undergo sulfhydryl/disulfide transitions. Show how disulfide interchange reactions involving thioredoxin could coordinate the activity of Photosystem I with the activity of the Calvin cycle.

43. The inner chloroplast membrane is impermeable to large polar and ionic compounds such as NADH and ATP. However, the membrane has an antiport protein that facilitates the passage of dihydroxyacetone phosphate or 3-phosphoglycerate in exchange for P_i. This system permits the entry of P_i for photophosphorylation and the exit of the products of carbon fixation. Show how the same antiport could "transport" ATP and reduced cofactors from the chloroplast to the cytosol.

44. The sedoheptulose bisphosphatase (SBPase) enzyme in the Calvin cycle catalyzes the removal of a phosphate group from C1 of sedoheptulose-1,7-bisphosphate (SBP) to produce sedoheptulose-7-phosphate (S7P). The $\Delta G^{\circ\prime}$ for this reaction is -14.2 kJ $\cdot$ mol^{-1} and the ΔG is -29.7 kJ $\cdot$ mol^{-1}. What is the ratio of products to reactants under normal cellular conditions? Is this enzyme likely to be regulated in the Calvin cycle?

45. Phosphoenolpyruvate carboxylase (PEPC) catalyzes the carboxylation of phosphoenolpyruvate (PEP) to oxaloacetate (OAA).

The enzyme is commonly found in plants but is absent in animals. The reaction is shown below:

(a) Why is PEPC referred to as an anaplerotic enzyme?

(b) Acetyl-CoA is an allosteric regulator of PEPC. Does acetyl-CoA activate or inhibit PEPC? Explain.

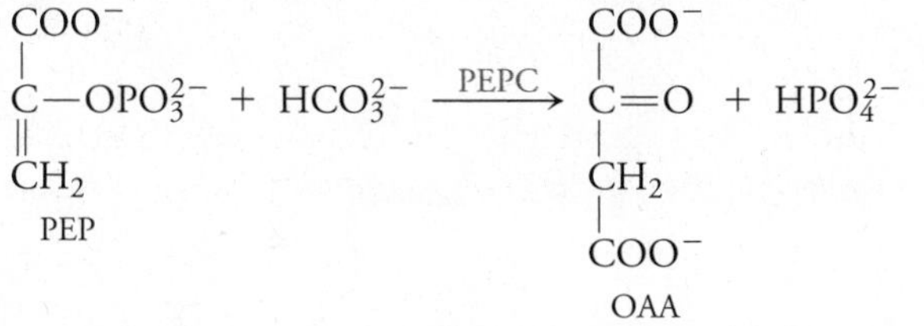

$$
\begin{array}{c}
COO^- \\
| \\
C-OPO_3^{2-} \\
\| \\
CH_2 \\
\text{PEP}
\end{array}
+ HCO_3^{2-}
\xrightarrow{\text{PEPC}}
\begin{array}{c}
COO^- \\
| \\
C=O \\
| \\
CH_2 \\
| \\
COO^- \\
\text{OAA}
\end{array}
+ HPO_4^{2-}
$$

46. In germinating oil seeds, triacylglycerols are rapidly converted to sucrose and protein. What is the role of PEPC (see Problem 45) in this process?

47. The use of transgenic plants has been increasing over the past decade. These plants are constructed by inserting a new gene into the plant genome to give the plant a desirable characteristic, such as resistance to pesticides or frost, or to confer a nutritional benefit. Transformed plants contain the gene of interest, a promoter (to induce expression of the gene), and a terminator. The cauliflower mosaic virus (CaMV) is often used as a promoter in transgenic plants. A partial sequence of DNA is shown below. Design a set of 18-bp PCR primers that you could use to detect the presence of this promoter in a transgenic plant.

GTAGTGGGATTGTGCGTCATCCCTTACGTCAGT···
(112 bases) ··· TCAACGATGGCCTTTCCTTTATCGCAATGA-
TGGCATTTGTAGGAGC

48. Novartis has constructed a corn cultivar that contains a gene from the bacterium *Bacillus thuringiensis* (Bt) that codes for an endotoxin protein. When insects of the order Lepidoptera (which includes the European corn borer but also unfortunately includes the Monarch butterfly) eat the corn, the ingested endotoxin enters the high-pH environment of the insect's midgut. Under these conditions, the endotoxin forms a pore in the membrane of the cells lining the midgut, causing ions to flow into the cell and ultimately resulting in the death of the organism. Is the Bt endotoxin toxic to humans? Explain why or why not.

49. The Monsanto Company has constructed a transgenic "Roundup Ready" soybean cultivar in which the bacterial gene for the enzyme EPSPS has been inserted into the plant genome. EPSPS catalyzes an important step in the synthesis of aromatic amino acids, as shown in the diagram below. The herbicide Roundup® contains glyphosate, a compound that competitively inhibits plant EPSPS but not the bacterial form of the enzyme. Explain the strategy for using glyphosate-containing herbicide on a soybean crop to kill weeds.

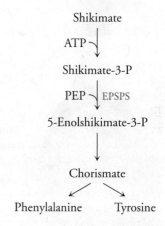

50. Is Roundup® (see Problem 49) toxic to humans? Explain why or why not.

[SELECTED READINGS]

Allen, J. F., and Martin, W., Out of thin air, *Nature* **445**, 610–612 (2007). [A short discussion of the cyanobacterial shift to oxygenic photosynthesis billions of years ago.]

Andersson, I., Catalysis and regulation in rubisco, *J. Exp. Botany* **59**, 1555–1568 (2008). [Discusses the structure, chemistry, and regulation of rubisco.]

Heathcote, P., Fyfe, P. K., and Jones, M. R., Reaction centres: the structure and evolution of biological solar power, *Trends Biochem. Sci.* **27**, 79–87 (2002).

Nelson, N., and Yocum, C. F., Structure and function of photosystems I and II, *Annu. Rev. Plant Biol.* **57**, 521–565 (2006).

Umena, Y., Kawakami, K., Shen, J.-R., and Kamiya, N., Crystal structure of oxygen-evolving photosystem II at a resolution of 1.9 Å. *Nature* **473**, 55–60 (2011). [Includes the structure of the Mn_4CaO_5 cluster.]

chapter 17

LIPID METABOLISM

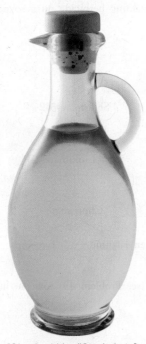

[Olga Lyubkina/iStockphoto]

▸▸ WHY are unsaturated fatty acids less fattening?

Cells can oxidize a variety of metabolic fuels, producing CO_2 as a waste product and conserving free energy in the form of ATP. On a per-mass basis, lipids provide more energy than carbohydrates, which helps account for the weight loss of dieters consuming low-fat diets. However, the type of lipid also matters. Triacylglycerols containing unsaturated fatty acids (as in the oil shown here) yield less free energy (fewer calories) than triacylglycerols containing saturated fatty acids (solid fats). We'll see why this is the case when we explore the energy-producing reactions of fatty acid metabolism.

THIS CHAPTER IN CONTEXT

Part 1 Foundations

Part 2 Molecular Structure and Function

Part 3 Metabolism

17 Lipid Metabolism

Part 4 Genetic Information

Do You Remember?

- **Lipids are predominantly hydrophobic molecules that can be esterified but cannot form polymers (Section 8-1).**
- **Metabolic fuels can be mobilized by breaking down glycogen, triacylglycerols, and proteins (Section 12-1).**
- **A few metabolites appear in several metabolic pathways (Section 12-2).**
- **Cells also use the free energy of other phosphorylated compounds, thioesters, reduced cofactors, and electrochemical gradients (Section 12-3).**

Approximately half of all deaths in the United States are linked to the vascular disease **atherosclerosis** (a term derived from the Greek *athero*, "paste," and *sclerosis*, "hardness"). Atherosclerosis is a slow progressive disease that begins with the accumulation of lipids in the walls of large blood vessels. The trapped lipids initiate inflammation by triggering the production of chemical signals that attract white blood cells, particularly macrophages. These cells engorge themselves by taking up the accumulated lipids and continue to recruit more macrophages, thereby perpetuating the inflammation.

The damaged vessel wall forms a plaque with a core of cholesterol, cholesteryl esters, and remnants of dead macrophages, surrounded by proliferating smooth muscle cells that may undergo calcification, as occurs in bone formation. This accounts for the "hardening" of the arteries. Although a very large plaque can occlude the lumen of the artery (Fig. 17-1), blood flow is usually not completely blocked unless the plaque ruptures, triggering formation of a blood clot that can prevent circulation to the heart (a heart attack) or brain (a stroke).

What is the source of the lipids that accumulate in vessel walls? They are deposited by lipoproteins known as LDL (for low-density lipoproteins). **Lipoproteins** (particles consisting of lipids and specialized proteins) are the primary form of circulating lipid (Fig. 17-2). Recall from Section 12-1 that dietary lipids travel from the intestine to other tissues as chylomicrons. These lipoproteins are relatively large (1000 to 5000 Å in diameter) with a protein content of only 1% to 2%. Their primary function is to transport dietary triacylglycerols to adipose tissue and cholesterol to the liver. The liver repackages the cholesterol and other lipids—including triacylglycerols, phospholipids, and cholesteryl esters—into other lipoproteins known as VLDL (very-low-density lipoproteins). VLDL have a triacylglycerol content of about 50% and a diameter of about 500 Å. As they circulate in the bloodstream, VLDL give up triacylglycerols to the tissues, becoming smaller, denser, and richer in cholesterol and cholesteryl esters. After passing through an intermediate state (IDL, or intermediate-density lipoproteins), they become LDL, about 200 Å in diameter and about 45% cholesteryl ester (Table 17-1).

High concentrations of circulating LDL, measured as serum cholesterol (popularly called "bad cholesterol"), are a major factor in atherosclerosis. Some high-fat diets (especially those rich in saturated fats) may contribute to atherosclerosis by boosting LDL levels, but genetic factors, smoking, and infection also increase the risk of atherosclerosis. The disease is less likely to occur in individuals who consume low-cholesterol diets and who have high levels of HDL (high-density lipoproteins, sometimes called "good" cholesterol). HDL particles are even smaller and denser than LDL (see Table 17-1), and their primary function is to transport the body's excess cholesterol back to the liver. HDL therefore counter the atherogenic tendencies of LDL. The roles of the various lipoproteins are summarized in Figure 17-3.

The opposing actions of LDL and HDL are just one part of the body's efforts to regulate lipid metabolism, which consists of multiple pathways. For example, lipids are obtained by digesting food; they are synthesized from smaller precursors; they

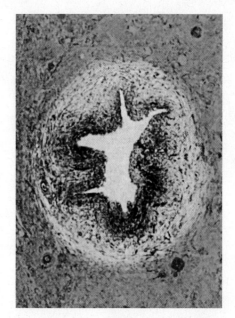

Figure 17-1 An atherosclerotic plaque in an artery. Note the thickening of the vessel wall. [James Cavallini/BSIP/Phototake.]

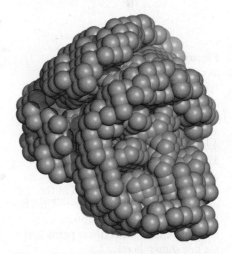

Figure 17-2 Structure of a lipoprotein. This image is based on small angle neutron scattering of HDL particles. Three copies of apolipoprotein A1 (orange) wrap around a core containing phospholipids, cholesterol, and cholesteryl esters. The particle has dimensions of about 110 Å × 96 Å. The proteins help target the particle to cell surfaces and modulate the activities of enzymes that act on the component lipids. The various types of lipoproteins differ in size, lipid composition, protein composition, and density (a function of the relative proportions of lipid and protein). [From Wu et al., *J. Biol. Chem.* 286, 12495–12508 (2011). Reproduced with permission.]

? In what part of the lipoprotein would you expect to find cholesteryl esters? Phospholipids?

[**TABLE 17-1**] Characteristics of Lipoproteins

Lipoprotein	Diameter (Å)	Density (g·cm⁻³)	% Protein	% Triacylglycerol	% Cholesterol and Cholesteryl Ester
Chylomicrons	1000–5000	<0.95	1–2	85–90	4–8
VLDL	300–800	0.95–1.006	5–10	50–65	15–25
IDL	250–350	1.006–1.019	10–20	20–30	40–45
LDL	180–250	1.019–1.063	20–25	7–15	45–50
HDL	50–120	1.063–1.210	40–55	3–10	15–20

Figure 17-3 Lipoprotein function.
Large chylomicrons, which are mostly lipid, transport dietary lipids to the liver and other tissues. The liver produces triacylglycerol-rich very-low-density lipoproteins (VLDL). As they circulate in the tissues, VLDL give up their triacylglycerols, becoming cholesterol-rich low-density lipoproteins (LDL), which are taken up by tissues. High-density lipoproteins (HDL), the smallest and densest of the lipoproteins, transport cholesterol from the tissues back to the liver.

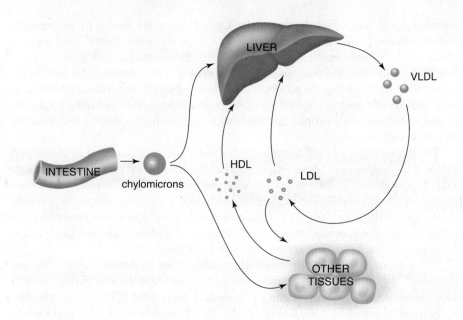

are used by cells as a source of free energy, as building materials, and as signaling molecules; they are stored in adipose tissue; and they are transported between tissues via lipoproteins. Much of this chapter focuses on the opposing pathways of lipid synthesis and degradation. For the most part, these reactions fall within the shaded portions of the metabolic map shown in **Figure 17-4.**

Figure 17-4 Lipid metabolism in context. Triacylglycerols, the "polymeric" form of fatty acids, are hydrolyzed to release fatty acids (1) that are oxidatively degraded to the two-carbon intermediate acetyl-CoA (2). Acetyl-CoA is also the starting material for the reductive biosynthesis of fatty acids (3), which can then be stored as triacylglycerols (4) or used in the synthesis of other lipids. Acetyl-CoA is also the precursor of lipids that are not built from fatty acids (these pathways are not shown here).

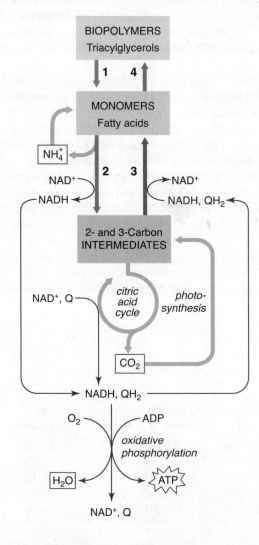

17-1 Fatty Acid Oxidation

KEY CONCEPTS

- Fatty acids to be degraded are linked to coenzyme A and then transported into mitochondria.
- The four reactions of each cycle of β oxidation produce acetyl-CoA, QH_2, and NADH.
- Additional enzymes are required to break down unsaturated fatty acids.
- Fatty acids with an odd number of carbons yield propionyl-CoA that is ultimately converted to acetyl-CoA.
- Peroxisomes oxidize long-chain and branched fatty acids, producing H_2O_2.

The degradation (oxidation) of fatty acids is a source of metabolic free energy. In this section we describe how cells obtain, activate, and oxidize fatty acids. In humans, *dietary triacylglycerols are the primary source of fatty acids used as metabolic fuel.* The triacylglycerols are carried by lipoproteins to tissues, where hydrolysis releases their fatty acids from the glycerol backbone.

Triacylglycerol Glycerol Fatty acyl groups

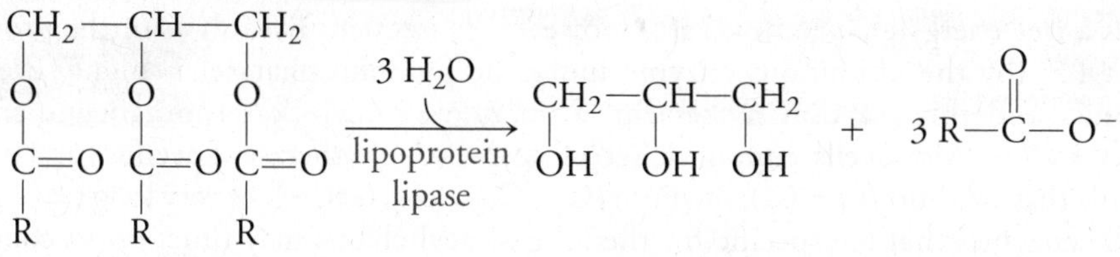

Hydrolysis occurs extracellularly, catalyzed by lipoprotein lipase, an enzyme associated with the outer surface of cells.

Triacylglycerols that are stored in adipose tissue are mobilized (their fatty acids are released to be used as fuel) by an intracellular hormone-sensitive lipase. The mobilized fatty acids travel through the bloodstream, not as part of lipoproteins, but bound to albumin, a 66-kD protein that accounts for about half of the serum protein (it also binds metal ions and hormones, serving as an all-purpose transport protein).

The concentration of free fatty acids in the body is very low because these molecules are detergents (which form micelles; see Section 2-2) and can disrupt cell membranes. After they enter cells, probably with the assistance of proteins, the fatty acids are either broken down for energy or re-esterified to form triacylglycerols or other complex lipids (as described in Section 17-3). Many free fatty acids are deployed to the liver and muscle cells, especially heart muscle, which prefers to burn fatty acids even when carbohydrate fuels are available.

Fatty acids are activated before they are degraded

To be oxidatively degraded, a fatty acid must first be activated. Activation is a two-step reaction catalyzed by acyl-CoA synthetase. First, the fatty acid displaces the diphosphate group of ATP, then coenzyme A (HSCoA) displaces the AMP group to form an acyl-CoA:

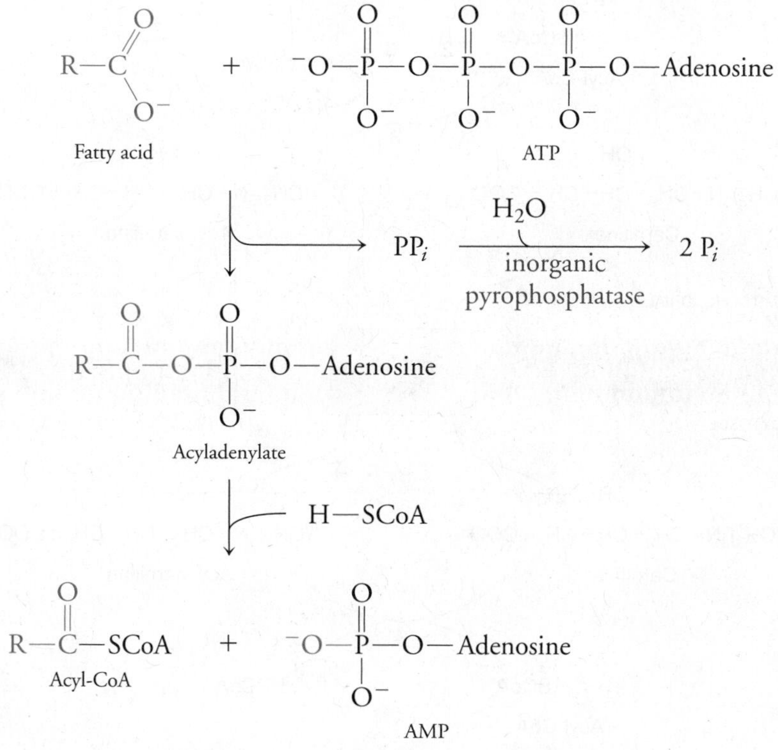

The acyladenylate product of the first step has a large free energy of hydrolysis (in other words, its cleavage would release a large amount of free energy), so it conserves the free energy of the cleaved phosphoanhydride bond in ATP. The second step, transfer of the acyl group to CoA (the same molecule that carries acetyl groups as acetyl-CoA), likewise conserves free energy in the formation of a thioester bond (see Section 12-3). Consequently, the overall reaction

$$\text{fatty acid} + \text{CoA} + \text{ATP} \rightleftharpoons \text{acyl-CoA} + \text{AMP} + \text{PP}_i$$

has a free energy change near zero. However, subsequent hydrolysis of the product PP_i (by the ubiquitous enzyme inorganic pyrophosphatase) is highly exergonic, and this reaction makes the formation of acyl-CoA spontaneous and irreversible. Most cells contain a set of acyl-CoA synthetases specific for fatty acids that are short (C_2-C_3), medium (C_4-C_{12}), long ($\geq C_{12}$), or very long ($\geq C_{22}$). The enzymes that are specific for the longest acyl chains may function in cooperation with a membrane transport protein so that the fatty acid is activated as it enters the cell. Once the large and polar coenzyme A is attached, the fatty acid is unable to diffuse back across the membrane and remains inside the cell to be metabolized.

Fatty acids are activated in the cytosol, but the rest of the oxidation pathway occurs in the mitochondria. Because there is no transport protein for CoA adducts, acyl groups must enter the mitochondria via a shuttle system involving the small molecule carnitine (Fig. 17-5). The acyl group is now ready to be oxidized.

Each round of β oxidation has four reactions

The pathway known as **β oxidation** degrades an acyl-CoA in a way that produces acetyl-CoA molecules for further oxidation and energy production by the citric acid cycle. In fact, in some tissues or under certain conditions, β oxidation supplies far more acetyl groups to the citric acid cycle than does glycolysis. β Oxidation also feeds electrons directly into the mitochondrial electron transport chain, which generates ATP by oxidative phosphorylation.

Figure 17-5 The carnitine shuttle system. (1) A cytosolic carnitine acyltransferase transfers an acyl group from CoA to carnitine. (2) The carnitine transporter allows the acyl-carnitine to enter the mitochondrial matrix. (3) A mitochondrial carnitine acyltransferase transfers the acyl group to a mitochondrial CoA molecule. (4) Free carnitine returns to the cytosol via the transport protein.

? **Why do acyl groups move into the mitochondrion, not out of it?**

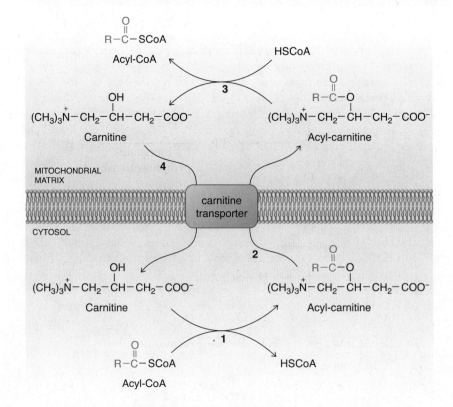

β Oxidation is a spiral pathway. *Each round consists of four enzyme-catalyzed steps that yield one molecule of acetyl-CoA and an acyl-CoA shortened by two carbons, which becomes the starting substrate for the next round.* Seven rounds of β oxidation degrade a C$_{16}$ fatty acid to eight molecules of acetyl-CoA:

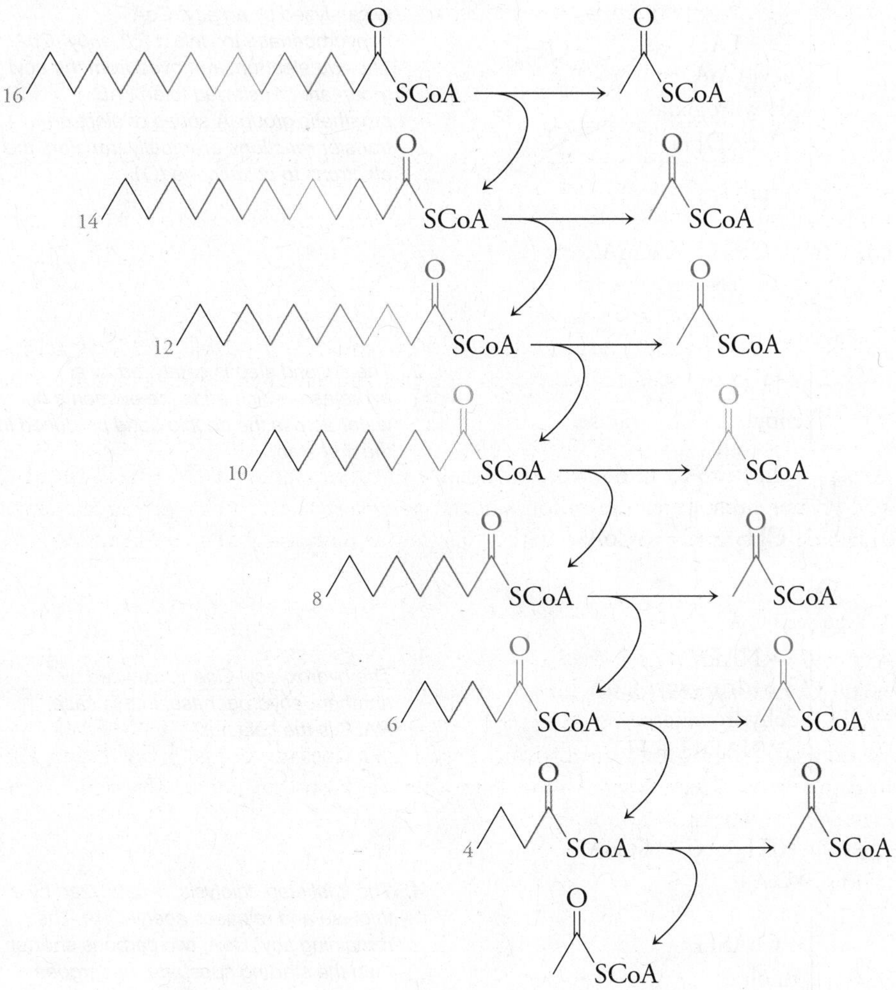

Figure 17-6 shows the reactions of β oxidation. The β indicates that oxidation occurs at the β position, which is the carbon atom that is two away from the carbonyl carbon (C3 is the β carbon). Note that acetyl units are not lost from the methyl end of the fatty acid but from the activated, CoA end.

The oxidation of fatty acids by the successive removal of two-carbon units was discovered over 100 years ago, and the enzymatic steps were elucidated about 40 years ago. But β oxidation still offers surprises in the details. For example, many enzymes are required to fully degrade an acyl-CoA to acetyl-CoA. Each of the four steps shown in Figure 17-6 appears to be catalyzed by two to five different enzymes with different chain-length specificities, as in the acyl-CoA synthetase reaction. The existence of some of these isozymes was inferred from studies of patients with disorders of fatty acid oxidation. One of these often-fatal diseases is due to a deficiency of medium-chain acyl-CoA dehydrogenase; affected individuals cannot degrade acyl-CoAs having 4 to 12 carbons, and derivatives of these molecules accumulate in the liver and are excreted in the urine.

β Oxidation is a major source of cellular free energy, especially during a fast, when carbohydrates are not available. Each round of β oxidation produces one QH$_2$, one NADH, and one acetyl-CoA. The citric acid cycle oxidizes the acetyl-CoA to produce an additional three NADH, one QH$_2$, and one GTP. Oxidation of all the reduced cofactors yields approximately 13 ATP: 3 from the two QH$_2$ and 10 from the four NADH (recall from the discussion of P:O ratios that oxidative phosphorylation is an indirect process, so the amount of

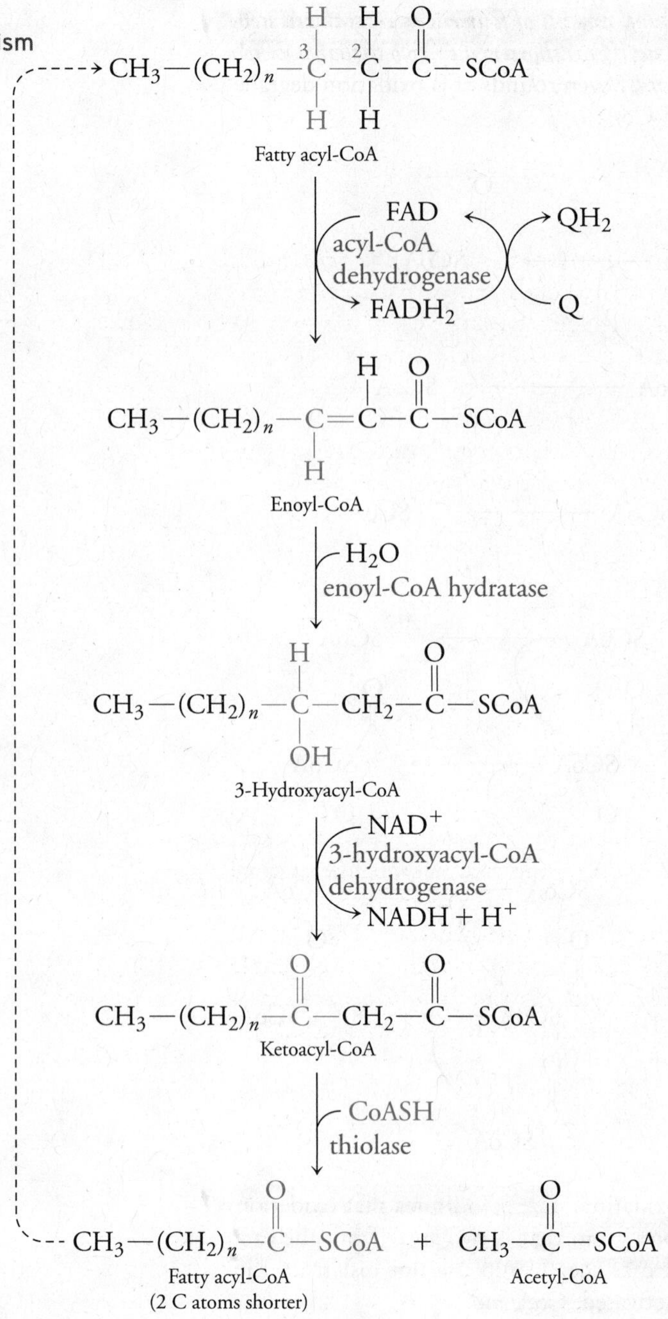

Figure 17-6 **The reactions of β oxidation.** ⊕ **See Animated Figure. β Oxidation pathway.**

? Which steps of the pathway are redox reactions?

1. Oxidation of acyl-CoA at the 2,3 position is catalyzed by an acyl-CoA dehydrogenase to yield a 2,3-enoyl-CoA. The two electrons removed from the acyl group are transferred to an FAD prosthetic group. A series of electron-transfer reactions eventually transfers the electrons to ubiquinone (Q).

2. The second step is catalyzed by a hydratase, which adds the elements of water across the double bond produced in the first step.

3. The hydroxyacyl-CoA is oxidized by another dehydrogenase. In this case, NAD^+ is the cofactor.

4. The final step, thiolysis, is catalyzed by a thiolase and releases acetyl-CoA. The remaining acyl-CoA, two carbons shorter than the starting substrate, undergoes another round of the four reactions (dotted line).

ATP produced per pair of electrons entering the electron transport chain is not a whole number; see Section 15-4). A total of 14 ATP are generated from each round of β oxidation:

One round of β oxidation	Citric acid cycle	Oxidative phosphorylation
1 QH_2		→ 1.5 ATP
1 NADH		→ 2.5 ATP
1 Acetyl-CoA →	3 NADH	→ 7.5 ATP
	1 QH_2	→ 1.5 ATP
	1 GTP	→ 1 ATP

Total 14 ATP

β Oxidation is regulated primarily by the availability of free CoA (to make acyl-CoA) and by the ratios of $NAD^+/NADH$ and Q/QH_2 (these reflect the state of the oxidative phosphorylation system). Some individual enzymes are also regulated by product inhibition.

Degradation of unsaturated fatty acids requires isomerization and reduction

Common fatty acids such as oleate and linoleate (*right*) contain *cis* double bonds that present obstacles to the enzymes that catalyze β oxidation. For linoleate, the first three rounds of β oxidation proceed as usual. But the acyl-CoA that begins the fourth round has a 3,4 double bond (originally the 9,10 double bond). Furthermore, this molecule is a *cis* enoyl-CoA, but enoyl-CoA hydratase (the enzyme that catalyzes step 2 of β oxidation) recognizes only the *trans* configuration. This metabolic obstacle is removed by the enzyme enoyl-CoA isomerase, which converts the *cis* 3,4 double bond to a *trans* 2,3 double bond so that β oxidation can continue.

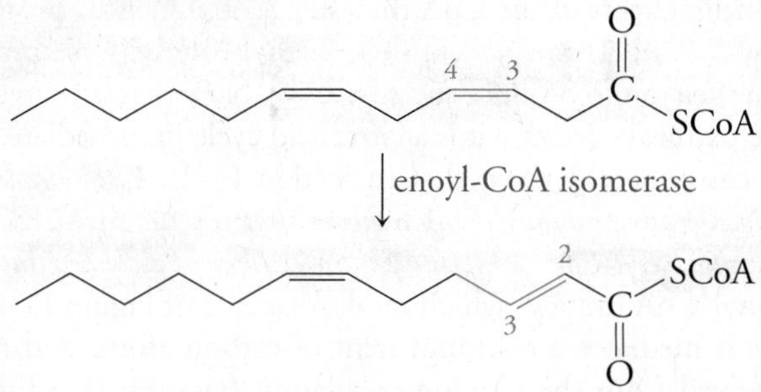

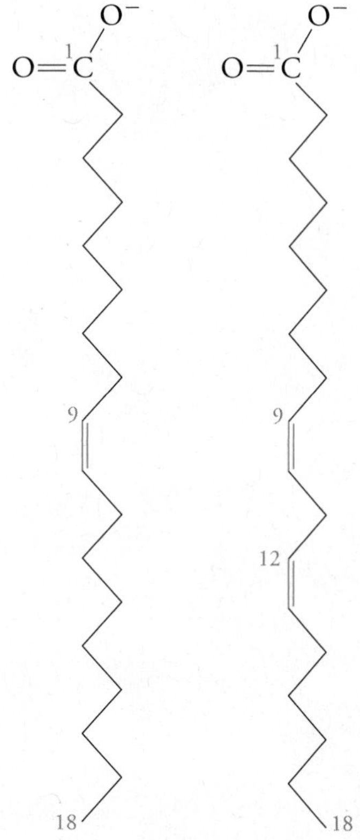

A second obstacle arises after the first reaction of the fifth round of β oxidation. Acyl-CoA dehydrogenase introduces a 2,3 double bond as usual, but the original 12,13 double bond of linoleate is now at the 4,5 position. The resulting dienoyl-CoA is not a good substrate for the next enzyme, enoyl-CoA hydratase. The dienoyl-CoA must undergo an NADPH-dependent reduction to convert its two double bonds to a single *trans* 3,4 double bond. This product must then be isomerized to produce the *trans* 2,3 double bond that is recognized by enoyl-CoA hydratase.

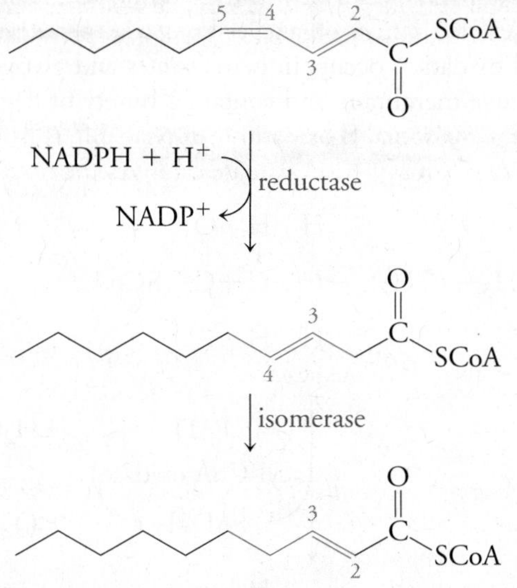

Oleate   Linoleate

Carbon compounds with double bonds are slightly more oxidized than saturated compounds (see Table 1-3), so less energy is released in converting them to CO_2. Accordingly, a diet rich in unsaturated fatty acids contains fewer calories than a diet rich in saturated fatty acids. The bypass reactions described above provide the molecular explanation why *unsaturated fatty acids yield less free energy than saturated fatty acids*. First, the enoyl-CoA isomerase reaction bypasses the QH_2-producing acyl-CoA dehydrogenase step, so 1.5 fewer ATP are produced. Second, the NADPH-dependent reductase consumes 2.5 ATP equivalents because NADPH is energetically equivalent to NADH.

▶▶ **WHY** are unsaturated fatty acids less fattening?

Oxidation of odd-chain fatty acids yields propionyl-CoA

Most fatty acids have an even number of carbon atoms (this is because they are synthesized by the addition of two-carbon acetyl units, as we will see later in this chapter). However, some plant and bacterial fatty acids that make their way into the human system have an odd number of carbon atoms. *The final round of β oxidation of these molecules leaves a three-carbon fragment, propionyl-CoA, rather than the usual acetyl-CoA.*

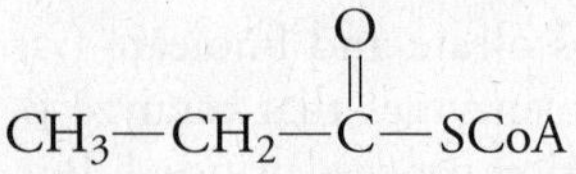

$$CH_3—CH_2—\overset{\overset{\textstyle O}{\|}}{C}—SCoA$$

Propionyl-CoA

This intermediate can be further metabolized by the sequence of steps outlined in **Figure 17-7**. At first, this pathway seems longer than necessary. For example, adding a carbon to C3 of the propionyl group would immediately generate succinyl-CoA. However, such a reaction is not chemically favored, because C3 is too far from the electron-delocalizing effects of the CoA thioester. Consequently, propionyl-CoA carboxylase must add a carbon to C2, and then methylmalonyl-CoA mutase must rearrange the carbon skeleton to produce succinyl-CoA. Note that succinyl-CoA is not the end point of the pathway. Because it is a citric acid cycle intermediate, it acts catalytically and is not consumed by the cycle (see Section 14-2). *The complete catabolism of the carbons derived from propionyl-CoA requires that the succinyl-CoA be converted to pyruvate and then to acetyl-CoA, which enters the citric acid cycle as a substrate.*

Methylmalonyl-CoA mutase, which catalyzes step 3 of Figure 17-7, is an unusual enzyme because it mediates a rearrangement of carbon atoms and requires a prosthetic group derived from the vitamin cobalamin (vitamin B_{12}; **Fig. 17-8**). Only about a dozen enzymes are known to use cobalamin cofactors. The small amounts of cobalamin required for human health are usually easily obtained from a diet that contains animal products. Vegans, however, are advised to consume B_{12} supplements. A disorder of vitamin B_{12} absorption causes the disease pernicious anemia.

Some fatty acid oxidation occurs in peroxisomes

The majority of a mammalian cell's fatty acid oxidation occurs in mitochondria, but a small percentage is carried out in organelles known as **peroxisomes** (**Fig. 17-9**). In plants, all fatty acid oxidation occurs in peroxisomes and glyoxysomes. Peroxisomes are enclosed by a single membrane and contain a variety of degradative and biosynthetic enzymes. The peroxisomal β oxidation pathway differs from the mitochondrial pathway in the first step. An acyl-CoA oxidase catalyzes the reaction

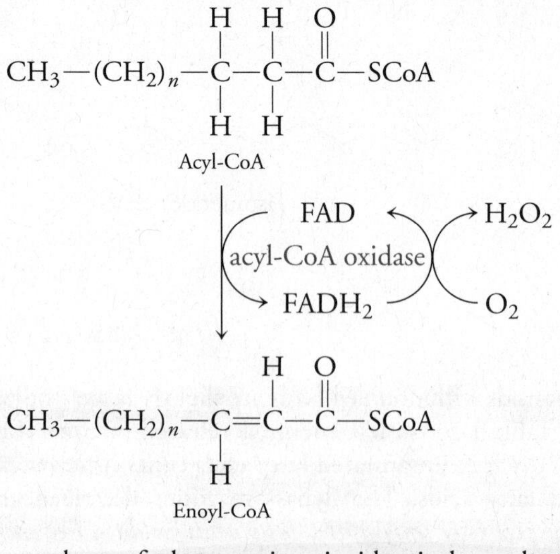

$$CH_3—(CH_2)_n—\overset{\overset{\textstyle H}{|}}{\underset{\underset{\textstyle H}{|}}{C}}—\overset{\overset{\textstyle H}{|}}{\underset{\underset{\textstyle H}{|}}{C}}—\overset{\overset{\textstyle O}{\|}}{C}—SCoA$$

Acyl-CoA

FAD ← → H_2O_2

acyl-CoA oxidase

→ $FADH_2$ ← O_2

$$CH_3—(CH_2)_n—\overset{\overset{\textstyle H}{|}}{C}=\overset{\overset{\textstyle H}{|}}{\underset{\underset{\textstyle H}{|}}{C}}—\overset{\overset{\textstyle O}{\|}}{C}—SCoA$$

Enoyl-CoA

The enoyl-CoA product of the reaction is identical to the product of the mitochondrial acyl-CoA dehydrogenase reaction (see Fig. 17-6), but the electrons removed from the acyl-CoA are transferred not to ubiquinone but directly to molecular oxygen to produce hydrogen peroxide, H_2O_2. This reaction product,

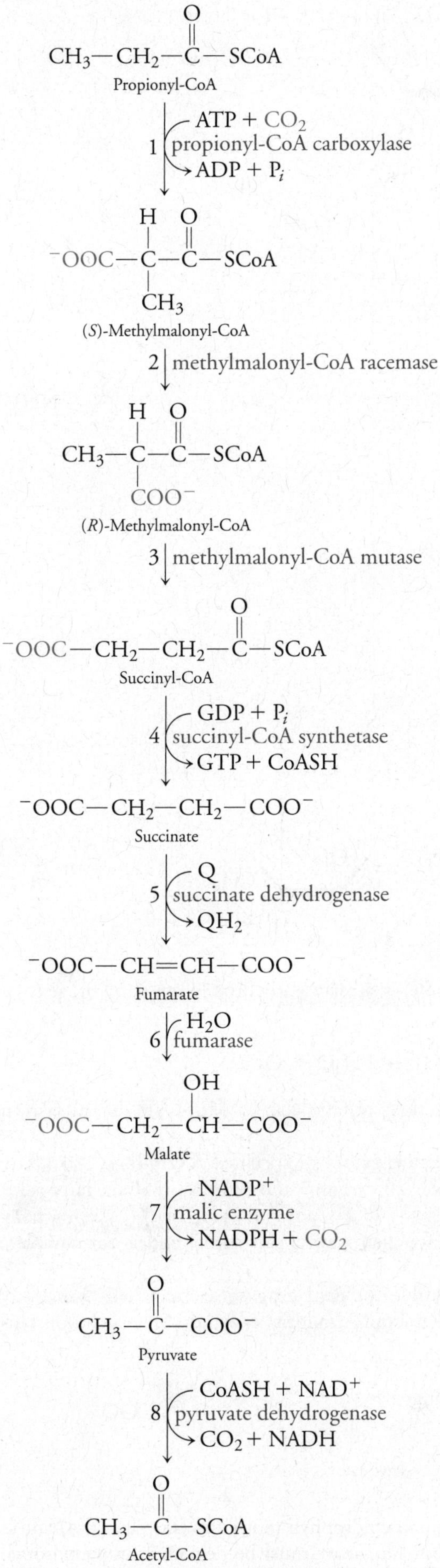

$$CH_3-CH_2-\overset{\overset{\displaystyle O}{\|}}{C}-SCoA$$

Propionyl-CoA

1 | propionyl-CoA carboxylase
ATP + CO_2 ⟶ ADP + P_i

1. Propionyl-CoA carboxylase adds a carboxyl group at C2 of the propionyl group to form a four-carbon methylmalonyl group.

$$^-OOC-\overset{\overset{\displaystyle H}{|}}{\underset{\underset{\displaystyle CH_3}{|}}{C}}-\overset{\overset{\displaystyle O}{\|}}{C}-SCoA$$

(S)-Methylmalonyl-CoA

2 | methylmalonyl-CoA racemase

2. A racemase interconverts the two different methylmalonyl-CoA stereoisomers (the two configurations are indicated by the R and S symbols).

$$CH_3-\overset{\overset{\displaystyle H}{|}}{\underset{\underset{\displaystyle COO^-}{|}}{C}}-\overset{\overset{\displaystyle O}{\|}}{C}-SCoA$$

(R)-Methylmalonyl-CoA

3 | methylmalonyl-CoA mutase

3. Methylmalonyl-CoA mutase rearranges the carbon skeleton to generate succinyl-CoA.

$$^-OOC-CH_2-CH_2-\overset{\overset{\displaystyle O}{\|}}{C}-SCoA$$

Succinyl-CoA

4 | succinyl-CoA synthetase
GDP + P_i ⟶ GTP + CoASH

4–6. Succinyl-CoA, a citric acid cycle intermediate, is converted to malate by reactions 5–7 of the citric acid cycle (see Fig. 14-6).

$$^-OOC-CH_2-CH_2-COO^-$$

Succinate

5 | succinate dehydrogenase
Q ⟶ QH_2

$$^-OOC-CH=CH-COO^-$$

Fumarate

6 | fumarase
H_2O

$$^-OOC-CH_2-\overset{\overset{\displaystyle OH}{|}}{C}H-COO^-$$

Malate

7 | malic enzyme
$NADP^+$ ⟶ NADPH + CO_2

7. After being exported from the mitochondria, malate is decarboxylated by malic enzyme to produce pyruvate in the cytosol.

$$CH_3-\overset{\overset{\displaystyle O}{\|}}{C}-COO^-$$

Pyruvate

8 | pyruvate dehydrogenase
CoASH + NAD^+ ⟶ CO_2 + NADH

8. Pyruvate, imported back into the mitochondria, can then be converted to acetyl-CoA by the pyruvate dehydrogenase complex.

$$CH_3-\overset{\overset{\displaystyle O}{\|}}{C}-SCoA$$

Acetyl-CoA

Figure 17-7 Catabolism of propionyl-CoA.

? **How many ATP equivalents can be produced from propionyl-CoA?**

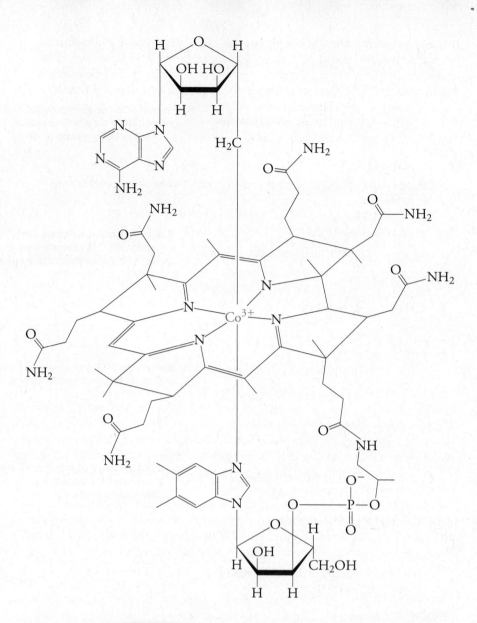

Figure 17-8 The cobalamin-derived cofactor. The prosthetic group of methylmalonyl-CoA mutase is a derivative of the vitamin cobalamin. The structure includes a hemelike ring structure with a central cobalt ion. Note that one of the Co ligands is a carbon atom, an extremely rare instance of a carbon–metal bond in a biological system.

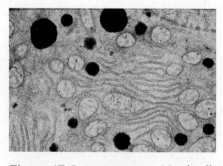

Figure 17-9 Peroxisomes. Nearly all eukaryotic cells contain these single-membrane-bound organelles (dark structures), which are similar to plant glyoxysomes (see Box 14-B). [Don W. Fawcett/Photo Researchers, Inc.]

which gives the peroxisome its name, is subsequently broken down by the peroxisomal enzyme catalase:

$$2\ H_2O_2 \rightarrow 2\ H_2O + O_2$$

The second, third, and fourth reactions of fatty acid oxidation are the same as in mitochondria.

Because the peroxisomal oxidation enzymes are specific for very-long-chain fatty acids (such as those containing over 20 carbons) and bind short-chain fatty acids with low affinity, *the peroxisome serves as a chain-shortening system.* The partially degraded fatty acyl-CoAs then make their way to the mitochondria for complete oxidation.

The peroxisome is also responsible for degrading some branched-chain fatty acids, which are not recognized by the mitochondrial enzymes. One such nonstandard fatty acid is phytanate,

Phytanate

which is derived from the side chain of chlorophyll molecules (see Fig. 16-4) and is present in all plant-containing diets. Phytanate must be degraded by peroxisomal enzymes because the methyl group at C3 prevents dehydrogenation by 3-hydroxyacyl-CoA dehydrogenase (step 3 of the standard β oxidation pathway). A deficiency of any of the phytanate-degrading enzymes results in Refsum's disease, a degenerative neuronal disorder characterized by an accumulation of phytanate in the tissues. The importance of peroxisomal enzymes in lipid metabolism (both catabolic and anabolic) is confirmed by the fatal outcome of most diseases stemming from deficient peroxisomal enzymes or improper synthesis of the peroxisomes themselves.

> **CONCEPT REVIEW**
> - Summarize the role of lipoproteins in making fatty acids available for oxidation.
> - How is a fatty acid activated? What is the cost in free energy?
> - Summarize the chemical reactions that occur in β oxidation.
> - How do these reactions lead to ATP production in the cell?
> - Summarize the enzyme activities that are required to fully metabolize unsaturated and odd-chain fatty acids.
> - What is the role of peroxisomes in fatty acid catabolism?

17-2 Fatty Acid Synthesis

At first glance, fatty acid synthesis appears to be the exact reverse of fatty acid oxidation. For example, fatty acyl groups are built and degraded two carbons at a time, and several of the reaction intermediates in the two pathways are similar or identical. However, *the pathways for fatty acid synthesis and degradation must differ for thermodynamic reasons,* as we saw for glycolysis and gluconeogenesis. Since fatty acid oxidation is a thermodynamically favorable process, simply reversing the steps of this pathway would be energetically unfavorable.

In mammalian cells, the opposing metabolic pathways of fatty acid synthesis and degradation are entirely separate. β Oxidation takes place in the mitochondrial matrix, and synthesis occurs in the cytosol. Furthermore, the two pathways use different cofactors. In β oxidation, the acyl group is attached to coenzyme A, but a growing fatty acyl chain is bound by an acyl-carrier protein (ACP; **Fig. 17-10**). β Oxidation funnels electrons to ubiquinone and NAD^+, but in fatty acid synthesis, NADPH is the source of reducing power. Finally, β oxidation requires two ATP equivalents (two phosphoanhydride bonds) to "activate" the acyl group, but the biosynthetic pathway consumes one ATP for every two carbons incorporated into a fatty acid. In this section, we focus on the reactions of fatty acid synthesis, comparing and contrasting them to β oxidation.

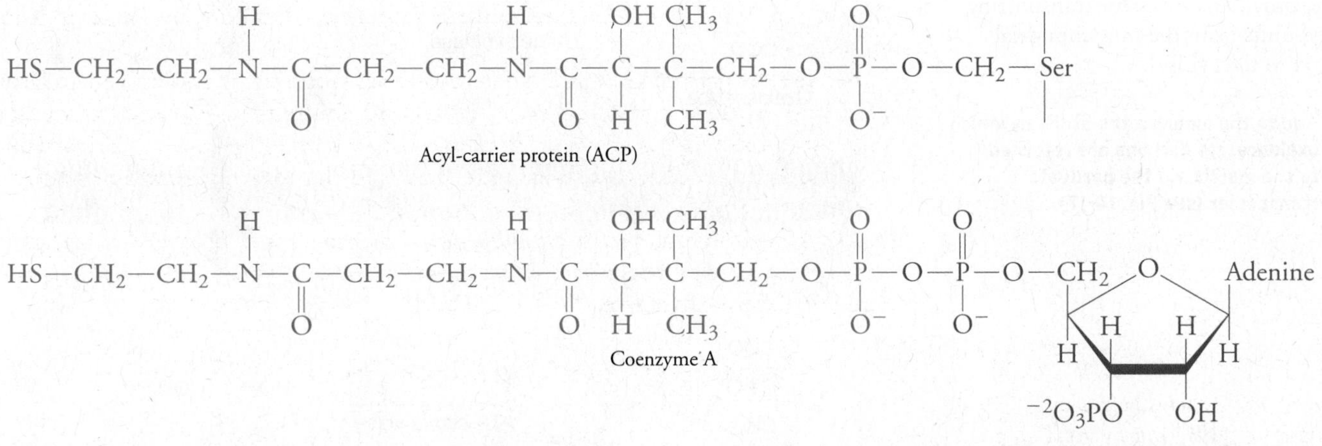

KEY CONCEPTS
- Fatty acid synthesis begins with the carboxylation of acetyl-CoA in the cytosol.
- Fatty acid synthase catalyzes seven separate reactions to extend a fatty acid by two carbons.
- Elongases and desaturases modify newly synthesized fatty acids.
- Various metabolites contribute to the regulation of fatty acid synthesis.
- Ketogenesis converts acetyl-CoA to small soluble ketone bodies.

Figure 17-10 Acyl-carrier protein and coenzyme A. Both acyl-carrier protein (ACP) and coenzyme A (CoA) include a pantothenate (vitamin B_5) derivative ending with a sulfhydryl group that forms a thioester with an acyl or acetyl group. In CoA, the pantothenate derivative is esterified to an adenine nucleotide; in ACP, the group is esterified to a Ser OH group of a polypeptide (in mammals, ACP is part of a larger multifunctional protein, fatty acid synthase).

Acetyl-CoA carboxylase catalyzes the first step of fatty acid synthesis

The starting material for fatty acid synthesis is acetyl-CoA, which may be generated in the mitochondria by the action of the pyruvate dehydrogenases complex (Section 14-1). But just as cytosolic acyl-CoA cannot directly enter the mitochondria to be oxidized, mitochondrial acetyl-CoA cannot exit to the cytosol for biosynthetic reactions. *The transport of acetyl groups to the cytosol involves citrate, which has a transport protein.* Citrate synthase (the enzyme that catalyzes the first step of the citric acid cycle; see Fig. 14-6) combines acetyl-CoA with oxaloacetate to produce citrate, which then leaves the mitochondria. ATP-citrate lyase "undoes" the citrate synthase reaction to produce acetyl-CoA and oxaloacetate in the cytosol (**Fig. 17-11**). Note that ATP is consumed in the ATP-citrate lyase reaction to drive the formation of a thioester bond.

The first step of fatty acid synthesis is the carboxylation of acetyl-CoA, an ATP-dependent reaction carried out by acetyl-CoA carboxylase. This enzyme catalyzes the rate-controlling step for the fatty acid synthesis pathway. The acetyl-CoA carboxylase mechanism is similar to that of propionyl-CoA carboxylase (step 1 in Fig. 17-7) and pyruvate carboxylase (see Fig. 13-9). First, CO_2 (as bicarbonate, HCO_3^-) is "activated" by its attachment to a biotin prosthetic group in a reaction that converts ATP to ADP + P_i:

$$\text{biotin} + HCO_3^- + ATP \rightarrow \text{biotin}-COO^- + ADP + P_i$$

Next, the carboxybiotin prosthetic group transfers the carboxylate group to acetyl-CoA to form the three-carbon malonyl-CoA and regenerate the enzyme:

$$\text{biotin}-COO^- + CH_3-\overset{\displaystyle O}{\overset{\|}{C}}-SCoA \longrightarrow {}^-OOC-CH_2-\overset{\displaystyle O}{\overset{\|}{C}}-SCoA + \text{biotin}$$

Acetyl-CoA   Malonyl-CoA

Malonyl-CoA is the donor of the two-carbon acetyl units that are used to build a fatty acid. The carboxylate group added by the carboxylation reaction is lost in a

Figure 17-11 The citrate transport system. The citrate transport protein, along with mitochondrial citrate synthase and cytoplasmic ATP-citrate lyase, provides a route for transporting acetyl units from the mitochondrial matrix to the cytosol.

? Add to the diagram the steps by which oxaloacetate carbons are returned to the matrix via the pyruvate transporter (see Fig. 14-17).

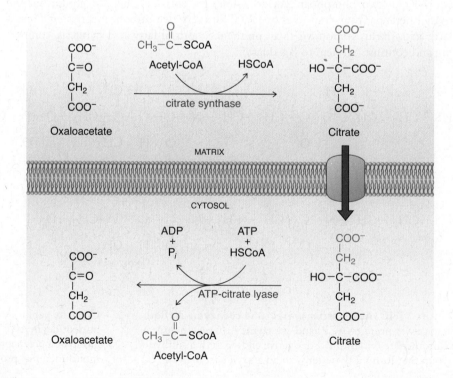

subsequent decarboxylation reaction. This sequence of carboxylation followed by decarboxylation also occurs in the conversion of pyruvate to phosphoenolpyruvate in gluconeogenesis (see Section 13-2). Note that fatty acid synthesis requires a C_3 intermediate, whereas β oxidation involves only two-carbon acetyl units.

Fatty acid synthase catalyzes seven reactions

The protein that carries out the main reactions of fatty acid synthesis in animals is a 540-kD **multifunctional enzyme** made of two identical polypeptides (**Fig. 17-12**). Each polypeptide of this fatty acid synthase has six active sites to carry out seven discrete reactions, which are summarized in **Figure 17-13**. In plants and bacteria, the reactions are catalyzed by separate polypeptides, but the chemistry is the same.

Reactions 1 and 2 are transacylation reactions that serve to prime or load the enzyme with the reactants for the condensation reaction (step 3). In the condensation reaction, decarboxylation of the malonyl group allows C2 to attack the acetyl thioester group to form acetoacetyl-ACP:

ACP   Cys   ACP   Cys

$| \qquad | \qquad\qquad\qquad | \qquad\qquad |$

S   S   S   SH

C=O   C=O CO_2 C=O

CH_2 CH_3 ⟶ CH_2

C=O   C=O

O^- CH_3

Malonyl-ACP   Acetyl-Cys   Acetoacetyl-ACP

This chemistry explains the necessity for carboxylating an acetyl group to a malonyl group: C2 of an acetyl group would not be sufficiently reactive.

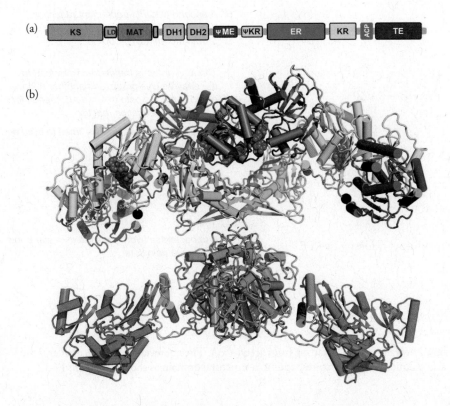

Figure 17-12 Mammalian fatty acid synthase. (a) Domain organization of the fatty acid synthase polypeptide. The domains labeled KS, MAT, DH1/DH2, ER, KR, and TE are the six enzymes. ACP is the acyl-carrier protein, whose pantothenate arm swings between the active sites. The domains labeled ψME and ψKR have no enzymatic activity. (b) Three-dimensional structure of the fatty acid synthase dimer, with the domains of each monomer colored as in part (a). The ACP and TE (thioesterase) modules are missing in this model. The $NADP^+$ molecules are shown in blue, and black spheres represent the anchor points for the acyl-carrier protein domains. [From T. Maier, M. Leibundgut, N. Ban, *Science* 321: 5894, 2008. Reprinted with permission from AAAS.]

? **Provide an example of another multifunctional enzyme.**

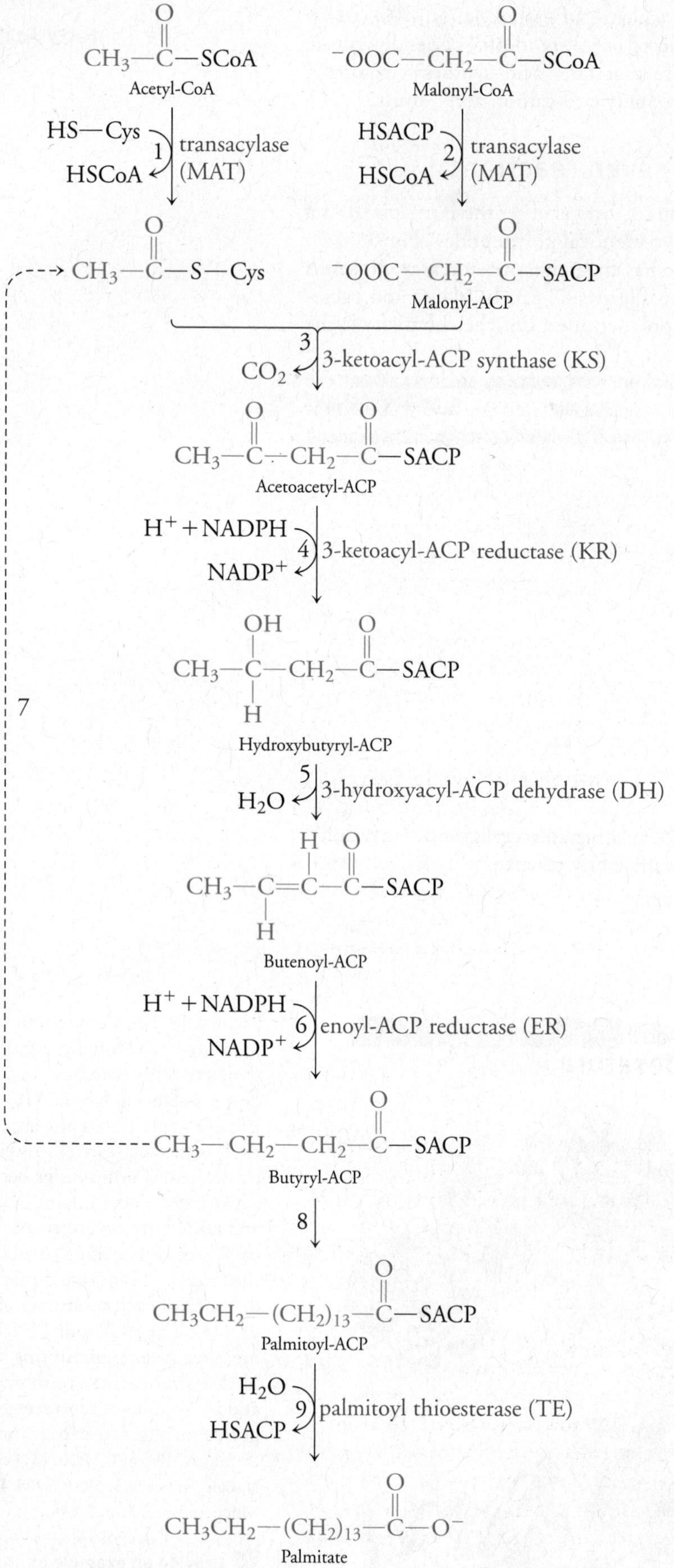

$$CH_3-\overset{\overset{\displaystyle O}{\|}}{C}-SCoA$$
Acetyl-CoA

$$^-OOC-CH_2-\overset{\overset{\displaystyle O}{\|}}{C}-SCoA$$
Malonyl-CoA

HS—Cys ⟍
1 transacylase (MAT)
HSCoA ↙

HSACP ⟍
2 transacylase (MAT)
HSCoA ↙

1. The two-carbon acetyl group that will be lengthened is transferred from CoA to a Cys side chain of fatty acid synthase.

2. The malonyl group that will donate an acetyl group to the growing fatty acyl chain is transferred from CoA to the ACP domain of the enzyme.

$$CH_3-\overset{\overset{\displaystyle O}{\|}}{C}-S-Cys$$

$$^-OOC-CH_2-\overset{\overset{\displaystyle O}{\|}}{C}-SACP$$
Malonyl-ACP

3 ⟍ 3-ketoacyl-ACP synthase (KS)
CO_2 ↙

3. In this condensation reaction, the malonyl group is decarboxylated and the resulting two-carbon fragment attacks the acetyl group to form a four-carbon product.

$$CH_3-\overset{\overset{\displaystyle O}{\|}}{C}-CH_2-\overset{\overset{\displaystyle O}{\|}}{C}-SACP$$
Acetoacetyl-ACP

$H^+ + NADPH$ ⟍
4 3-ketoacyl-ACP reductase (KR)
$NADP^+$ ↙

4. The 3-ketoacyl product of step 3 is reduced.

$$CH_3-\overset{\overset{\displaystyle OH}{\|}}{\underset{\underset{\displaystyle H}{\|}}{C}}-CH_2-\overset{\overset{\displaystyle O}{\|}}{C}-SACP$$
Hydroxybutyryl-ACP

5 ⟍ 3-hydroxyacyl-ACP dehydrase (DH)
H_2O ↙

5. A dehydration introduces a 2,3 double bond.

$$CH_3-\overset{\overset{\displaystyle H}{\|}}{C}=\overset{\underset{\displaystyle H}{\|}}{C}-\overset{\overset{\displaystyle O}{\|}}{C}-SACP$$
Butenoyl-ACP

$H^+ + NADPH$ ⟍
6 enoyl-ACP reductase (ER)
$NADP^+$ ↙

6. A second NADPH-dependent reduction completes the conversion of the condensation product to an acyl group.

7

$$CH_3-CH_2-CH_2-\overset{\overset{\displaystyle O}{\|}}{C}-SACP$$
Butyryl-ACP

7. The acyl group is transferred from ACP to the enzyme Cys group, and another malonyl group is loaded onto the free ACP, ready for another condensation reaction.

8 ↓

8. Steps 3–6 are repeated six times to build a C_{16} fatty acid.

$$CH_3CH_2-(CH_2)_{13}-\overset{\overset{\displaystyle O}{\|}}{C}-SACP$$
Palmitoyl-ACP

H_2O ⟍
9 palmitoyl thioesterase (TE)
HSACP ↙

9. A thioesterase hydrolyzes the thioester bond to release palmitate.

$$CH_3CH_2-(CH_2)_{13}-\overset{\overset{\displaystyle O}{\|}}{C}-O^-$$
Palmitate

Figure 17-13 Fatty acid synthesis. The steps show how fatty acid synthase carries out the synthesis of the C_{16} fatty acid palmitate, starting from acetyl-CoA. The abbreviation next to each enzyme corresponds to a structural domain as shown in Figure 17-12.

The hydroxyacyl product of Reaction 4 is chemically similar to the hydroxyacyl product of step 2 of β oxidation, but the intermediates of the two pathways have opposite configurations (see Section 4-1):

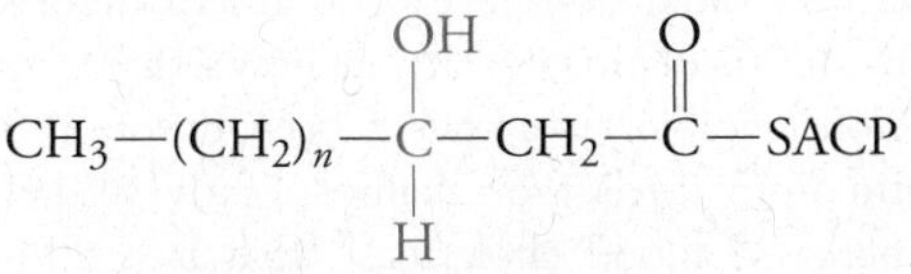

$$CH_3-(CH_2)_n-\underset{\underset{H}{|}}{\overset{\overset{OH}{|}}{C}}-CH_2-\overset{\overset{O}{\|}}{C}-SACP \qquad CH_3-(CH_2)_n-\underset{\underset{OH}{|}}{\overset{\overset{H}{|}}{C}}-CH_2-\overset{\overset{O}{\|}}{C}-SCoA$$

3-Hydroxyacyl-ACP intermediate of
fatty acid synthesis
(D configuration)

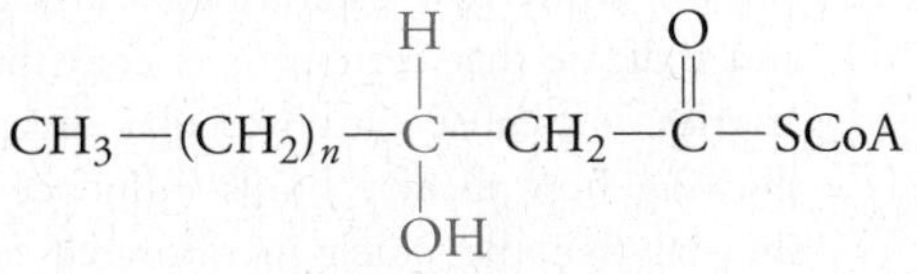

3-Hydroxyacyl-ACP intermediate of
β oxidation
(L configuration)

Note also that growth of the acyl chain—like chain shortening in β oxidation—occurs at the thioester end of the molecule.

The NADPH required for the two reduction steps of fatty acid synthesis (steps 4 and 6) is supplied mostly by the pentose phosphate pathway (see Section 13-4). The synthesis of one molecule of palmitate (the usual product of fatty acid synthase) requires the production of 7 malonyl-CoA, at a cost of 7 ATP. The seven rounds of fatty acid synthesis consume 14 NADPH, which is equivalent to 14×2.5, or 35, ATP, bringing the total cost to 42 ATP. Still, this energy investment is much less than the free energy yield of oxidizing palmitate.

During fatty acid synthesis, the long flexible arm of the pantothenate derivative in ACP (see Fig. 17-10) shuttles intermediates between the active sites of fatty acid synthase (the lipoamide group in the pyruvate dehydrogenase complex functions similarly; see Section 14-1). In the fatty acid synthase dimer, two fatty acids can be built simultaneously.

Packaging several enzyme activities into one multifunctional protein like mammalian fatty acid synthase allows the enzymes to be synthesized and controlled in a coordinated fashion. Also, the product of one reaction can quickly diffuse to the next active site. Bacterial and plant fatty acid synthase systems may lack the efficiency of a multifunctional protein, but because the enzymes are not locked together, a wider variety of fatty acid products can be more easily made. In mammals, fatty acid synthase produces mostly the 16-carbon saturated fatty acid palmitate.

Other enzymes elongate and desaturate newly synthesized fatty acids

Some sphingolipids contain C_{22} and C_{24} fatty acyl groups. *These and other long-chain fatty acids are generated by enzymes known as elongases,* which extend the C_{16} fatty acid produced by fatty acid synthase. Elongation can occur in either the endoplasmic reticulum or mitochondria. The endoplasmic reticulum reactions use malonyl-CoA as the acetyl-group donor and are chemically similar to those of fatty acid synthase. In the mitochondria, fatty acids are elongated by reactions that more closely resemble the reversal of β oxidation but use NADPH.

Desaturases introduce double bonds into saturated fatty acids. These reactions take place in the endoplasmic reticulum, catalyzed by membrane-bound enzymes. The electrons removed in the dehydrogenation of the fatty acid are eventually transferred to molecular oxygen to produce H_2O. The most common unsaturated fatty acids in animals are palmitoleate (a C_{16} molecule) and oleate (a C_{18} fatty acid; see Section 8-1), both with one *cis* double bond at the 9,10 position. *Trans* fatty acids are relatively rare in plants and animals, but they are abundant in some prepared foods, which has produced confusion among individuals concerned with eating the "right" kinds of fats (Box 17-A).

Elongation can follow desaturation (and vice versa), so animals can synthesize a variety of fatty acids with different chain lengths and degrees of unsaturation. However, mammals cannot introduce double bonds at positions beyond C9 and

BOX 17-A ⬡ BIOCHEMISTRY NOTE

Fats, Diet, and Heart Disease

Years of study have established a link between elevated LDL levels and atherosclerosis and indicate that certain diets contribute to the formation of fatty deposits that clog arteries and cause cardiovascular disease. Considerable research has been devoted to showing how dietary lipids influence serum lipid levels. For example, early studies showed that diets rich in saturated fats increased blood cholesterol (that is, LDL), whereas diets in which unsaturated vegetable oils replaced the saturated fats had the opposite effect. These and other findings led to recommendations that individuals at risk for atherosclerosis avoid butter, which is rich in saturated fat as well as cholesterol, and instead use margarine, which is prepared from cholesterol-free vegetable oils.

The production of semisolid margarine from liquid plant oils (triacylglycerols containing unsaturated fatty acids) often includes a hydrogenation step to chemically saturate the carbons of the fatty acyl chains. In this process, some of the original *cis* double bonds are converted to *trans* double bonds. In clinical studies, *trans* fatty acids are comparable to saturated fatty acids in their tendency to increase LDL levels and decrease HDL levels. Dietary guidelines now warn against the excessive intake of *trans* fatty acids in the form of hydrogenated vegetable oils (small amounts of *trans* fatty acids also occur naturally in some animal fats). This would mean avoiding processed foods whose list of ingredients includes "partially hydrogenated vegetable oil."

So should you consume butter or margarine? Linking specific types of dietary fats to human health and disease has always been a risky venture because quantitative information comes mainly from epidemiological and clinical studies, which are typically time-consuming and often inconclusive or downright contradictory. Scientists still do not fully understand *how* the consumption of specific fatty acids—saturated or unsaturated, *cis* or *trans*—influences lipoprotein metabolism. Other dietary factors also play a role. For example, one consequence of low-fat diets is that individuals consume relatively more carbohydrates. And when they reduce their meat intake (an obvious source of fat), people eat more fruits and vegetables, which may have health-enhancing effects of their own.

⬡ **Question:** Rank the "healthiness" of the following sources of fatty acids: animal fat, olive oil, and hydrogenated soybean oil.

therefore cannot synthesize fatty acids such as linoleate and linolenate. These molecules are precursors of the C_{20} fatty acid arachidonate and other lipids with specialized biological activities (**Fig. 17-14**). *Mammals must therefore obtain linoleate and linolenate from their diet.* These **essential fatty acids** are abundant in fish and plant oils. Unsaturated fatty acids with a double bond three carbons from the

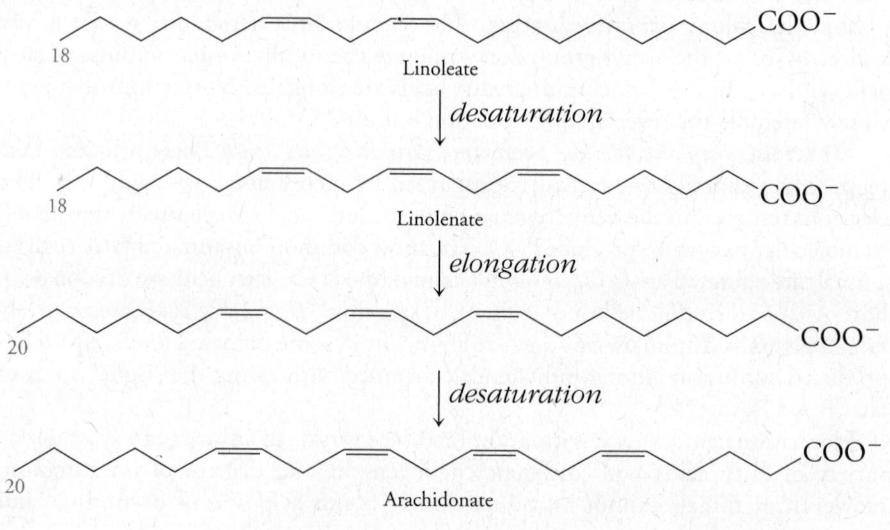

Figure 17-14 Synthesis of arachidonate. Linoleate (or linolenate) is elongated and desaturated to produce arachidonate, a C_{20} fatty acid with four double bonds.

end, omega-3 fatty acids, may have health benefits (Box 8-A). A deficiency of essential fatty acids resulting from a very-low-fat diet may elicit symptoms such as slow growth and poor wound healing.

Fatty acid synthesis can be activated and inhibited

Under conditions of abundant metabolic fuel, the products of carbohydrate and amino acid catabolism are directed toward fatty acid synthesis, and the resulting fatty acids are stored as triacylglycerols. The rate of fatty acid synthesis is controlled by acetyl-CoA carboxylase, which catalyzes the first step of the pathway. This enzyme is inhibited by a pathway product (palmitoyl-CoA) and is allosterically activated by citrate (which signals abundant acetyl-CoA). The enzyme is also subject to allosteric regulation by hormone-stimulated phosphorylation and dephosphorylation.

The concentration of malonyl-CoA is also critical for preventing the wasteful simultaneous activity of fatty acid synthesis and fatty acid oxidation. Malonyl-CoA is the source of acetyl groups that are incorporated into fatty acids, and it also blocks β oxidation by inhibiting carnitine acyltransferase, the enzyme involved in shuttling acyl groups into the mitochondria (see Fig. 17-5). Consequently, *when fatty acid synthesis is under way, no acyl groups are transported into the mitochondria for oxidation.* Some of the mechanisms that regulate fatty acid metabolism are summarized in **Figure 17-15**.

There are both natural and synthetic inhibitors of fatty acid synthase, such as the widely used antibacterial agent triclosan and drugs that target pathogens more specifically (Box 17-B). Fatty acid synthase inhibitors are of great scientific and popular interest, given that excess body weight (due to fat) is a major health problem, affecting about two-thirds of the population of the United States. And because many tumors sustain high levels of fatty acid synthesis, fatty acid synthase inhibitors may also be useful for treating cancer.

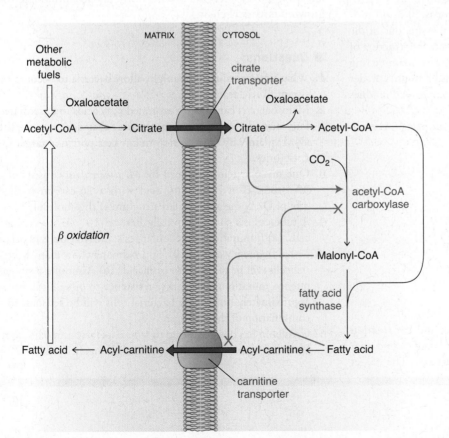

Figure 17-15 Some control mechanisms in fatty acid metabolism. Red symbols indicate inhibition, and the green symbol indicates activation.

BOX 17-B ⬡ CLINICAL CONNECTION

Inhibitors of Fatty Acid Synthesis

Because fatty acid synthesis is an essential metabolic activity, inhibiting the process in pathogenic organisms—but not in their mammalian hosts—is a useful strategy for preventing or curing certain infectious diseases. For example, many cosmetics, toothpastes, antiseptic soaps, and even plastic toys and kitchenware contain the compound 5-chloro-2-(2,4-dichlorophenoxy)-phenol, better known as triclosan:

Triclosan

This compound has been used since the 1970s as an antibacterial agent, although its mechanism of action was not understood until 1998.

Triclosan was long believed to act as a general microbicide, an agent that kills nonspecifically, much like household bleach or ultraviolet light. Such microbicides are effective because it is difficult for bacteria to evolve specific resistance mechanisms. However, triclosan actually operates more like an antibiotic with a specific biochemical target, in this case, enoyl-ACP reductase, which catalyzes step 6 of fatty acid synthesis (see Fig. 17-13).

The enzyme's natural substrate has a K_M of about 22 μM, but the dissociation constant for the inhibitor is 20 to 40 pM, indicating extremely tight binding. In the active site, one of the phenyl rings of triclosan, whose structure mimics the structure of the reaction intermediate, stacks on top of the nicotinamide ring of the NADH cofactor. Triclosan also binds through van der Waals interactions and hydrogen bonds with amino acid residues in the active site.

The antibiotic isoniazid has been used for the past 60 years to treat *Mycobacterium tuberculosis* infections.

Isoniazid

Inside the bacterial cell, isoniazid is oxidized, and the reaction product combines with NAD^+ to generate a compound that inhibits one of the cell's enoyl-ACP reductases. The target enzyme is specific for extremely long-chain fatty acids, which are incorporated into mycolic acids, the waxlike components of the mycobacterial cell wall.

Isoniazid is typically administered along with other antibiotics that target different metabolic activities. The combination therapy reduces the likelihood of the bacteria becoming drug-resistant. Even so, the drugs must be taken for many months, since mycobacteria replicate very slowly and can remain dormant inside host cells, protected from drugs as well as the host's immune system.

Some fungal species are susceptible to cerulenin, which inhibits 3-ketoacyl-ACP synthase (step 3 of fatty acid synthesis; see Fig. 17-13) by blocking the reaction of malonyl-ACP, that is, the condensation step.

Cerulenin

Cerulenin is also effective against *M. tuberculosis,* inhibiting the production of long-chain fatty acids required for cell-wall synthesis. The drug, which contains a reactive epoxide group, reacts irreversibly with the enzyme's active-site Cys residue, forming a C2—S covalent bond. Cerulenin's hydrocarbon tail occupies the site that would normally accommodate the growing fatty acyl chain.

● **Questions:**

1. What types of mutations might allow bacteria to become resistant to triclosan?
2. Many bacteria can convert saturated fatty acids to unsaturated fatty acids by a dehydration reaction that does not require O_2. Explain why this reaction might be a potential target for antibiotics.
3. One mycolic acid produced by *M. tuberculosis* consists of a 50-carbon β-hydroxy fatty acid with a 26-carbon α-alkyl group. Draw the abbreviated structure of this molecule.
4. A tuberculosis patient typically harbors 10^{12} *M. tuberculosis* cells. (a) If mutations that confer resistance to isoniazid occur with a frequency of 1 in 10^8, approximately how many bacterial cells will be resistant to isoniazid? (b) Assuming a similar rate for mutations that confer resistance to other antibiotics, approximately how many bacterial cells will be resistant to a combination of three drugs?
5. Cerulenin has been reported to stimulate fatty acid oxidation. Does this activity contribute to its antifungal effects?

Acetyl-CoA can be converted to ketone bodies

During a prolonged fast, when glucose is unavailable from the diet and liver glycogen has been depleted, many tissues depend on fatty acids released from stored triacylglycerols to meet their energy needs. However, the brain does not burn fatty acids because they pass poorly through the blood–brain barrier. Gluconeogenesis helps supply the brain's energy needs, but the liver also produces **ketone bodies** to supplement gluconeogenesis. The ketone bodies—acetoacetate and 3-hydroxybutyrate (also called β-hydroxybutyrate)—are synthesized from acetyl-CoA in liver mitochondria by a process called **ketogenesis.** Because ketogenesis uses fatty acid–derived acetyl groups, it helps spare amino acids that would otherwise be diverted to gluconeogenesis.

The assembly of ketone bodies is somewhat reminiscent of the synthesis of fatty acids or the oxidation of fatty acids in two-carbon steps (Fig. 17-16). In fact, the hydroxymethylglutaryl-CoA intermediate is chemically similar to the 3-hydroxyacyl intermediates of β oxidation and fatty acid synthesis.

Because they are small and water-soluble, ketone bodies are transported in the bloodstream without specialized lipoproteins, and they can easily pass into the

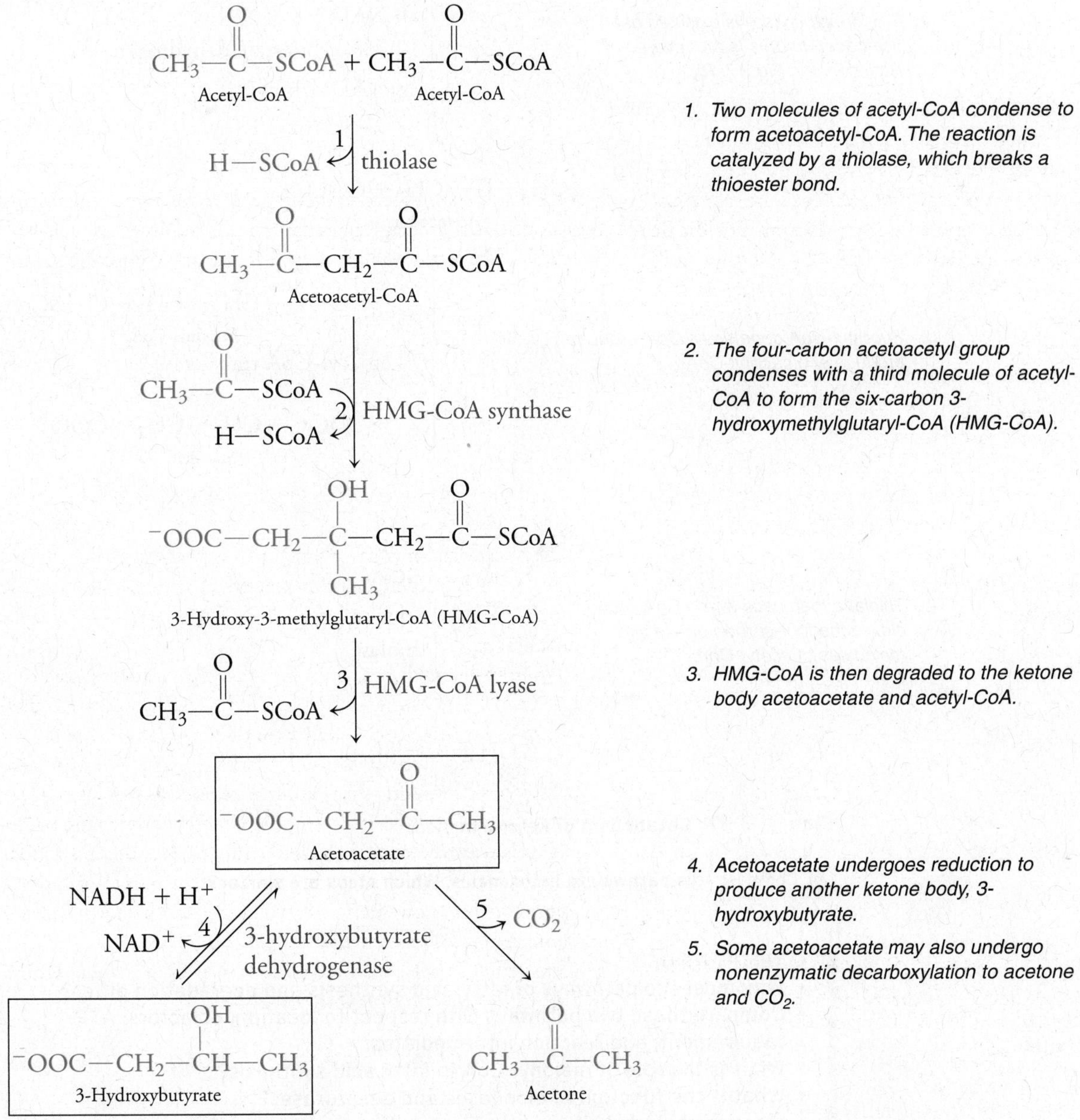

1. Two molecules of acetyl-CoA condense to form acetoacetyl-CoA. The reaction is catalyzed by a thiolase, which breaks a thioester bond.

2. The four-carbon acetoacetyl group condenses with a third molecule of acetyl-CoA to form the six-carbon 3-hydroxymethylglutaryl-CoA (HMG-CoA).

3. HMG-CoA is then degraded to the ketone body acetoacetate and acetyl-CoA.

4. Acetoacetate undergoes reduction to produce another ketone body, 3-hydroxybutyrate.

5. Some acetoacetate may also undergo nonenzymatic decarboxylation to acetone and CO_2.

Figure 17-16 Ketogenesis. The ketone bodies are boxed.

central nervous system. During periods of high ketogenic activity, such as in diabetes, ketone bodies may be produced faster than they are consumed. Some of the excess acetoacetate breaks down to acetone, which gives the breath a characteristic sweet smell. Ketone bodies are also acids, with a pK of about 3.5. Their overproduction can lead to a drop in the pH of the blood, a condition called ketoacidosis. Mild symptoms may also develop in some individuals following a high-protein, low-carbohydrate diet, when ketogenesis increases to offset the shortage of dietary carbohydrates.

Ketone bodies produced by the liver are used by other tissues as metabolic fuels after being converted back to acetyl-CoA (**Fig. 17-17**). The liver itself cannot catabolize ketone bodies because it lacks one of the required enzymes, 3-ketoacyl-CoA transferase.

1. 3-Hydroxybutyrate is oxidized back to acetoacetate (this is the reverse of Reaction 4 in Fig. 17-16).

2. Succinyl-CoA donates its CoA group to produce acetoacetyl-CoA.

3. Thiolase then uses a free CoA group to cleave the four-carbon unit to two molecules of acetyl-CoA.

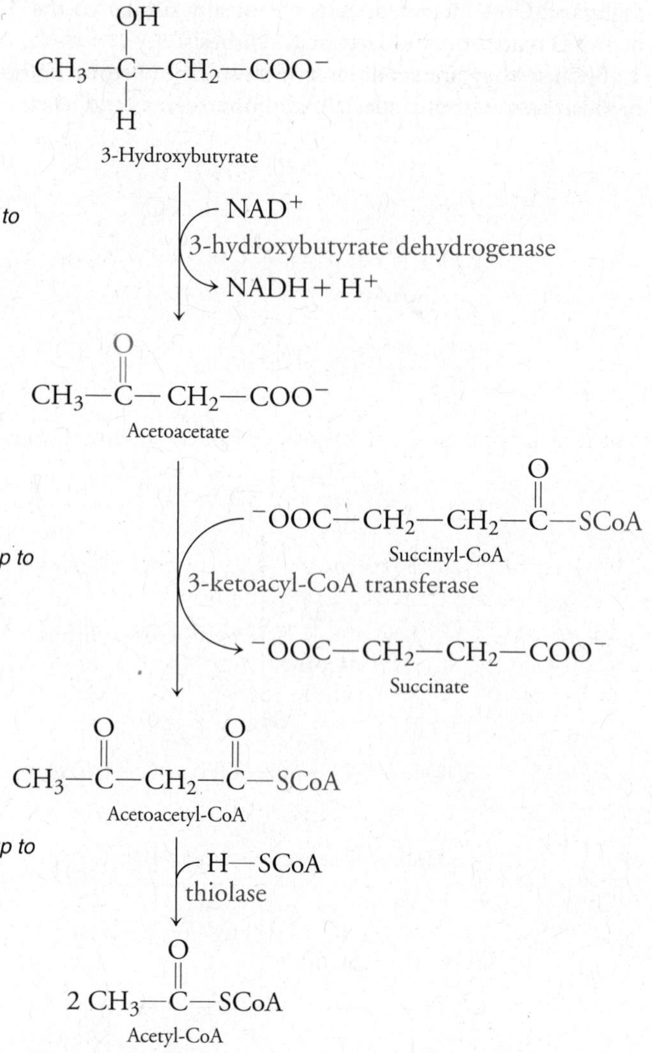

Figure 17-17 **Catabolism of ketone bodies.**

? **Compare this pathway to ketogenesis. Which steps are similar?**

CONCEPT REVIEW
• Why must the pathways of fatty acid synthesis and degradation differ?
• Compare these two pathways with respect to location, cofactors, ATP requirement, and reaction intermediates.
• What is the role of malonyl-CoA in fatty acid synthesis?
• What is the function of elongases and desaturases?
• What are ketone bodies and when are they synthesized?

17-3 Synthesis of Other Lipids

Lipid metabolism encompasses many chemical reactions involving fatty acids, which are structural components of other lipids such as triacylglycerols, glycerophospholipids, and sphingolipids. Fatty acids such as arachidonate are also the precursors of eicosanoids that function as signaling molecules (Section 10-4). This section covers the biosynthesis of some of the major types of lipids, including the synthesis of cholesterol from acetyl-CoA.

KEY CONCEPTS
- Acyl groups are transferred from CoA to a glycerol backbone to generate triacylglycerols and phospholipids.
- Cholesterol is synthesized from acetyl-CoA.
- Cholesterol can be used both inside and outside the cell.

Triacylglycerols and phospholipids are built from acyl-CoA groups

Cells have a virtually unlimited capacity for storing fatty acids in the form of triacylglycerols, which aggregate in the cytoplasm to form droplets surrounded by a layer of amphipathic phospholipids. *Triacylglycerols are synthesized by attaching fatty acyl groups to a glycerol backbone derived from phosphorylated glycerol or from glycolytic intermediates,* for example, dihydroxyacetone phosphate:

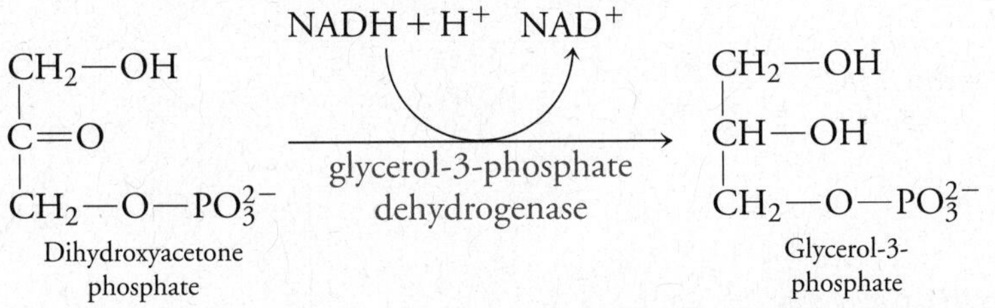

The fatty acyl groups are first activated to CoA thioesters in an ATP-dependent manner:

$$\text{fatty acid} + \text{CoA} + \text{ATP} \rightleftharpoons \text{acyl-CoA} + \text{AMP} + \text{PP}_i$$

This reaction is catalyzed by acyl-CoA synthetase, the same enzyme that activates fatty acids for oxidation. Triacylglycerols are assembled as shown in **Figure 17-18**. The acyl transferases that add fatty acids to the glycerol backbone are not highly specific with respect to chain length or degree of unsaturation of the fatty acyl group, but human triacylglycerols usually contain palmitate at C1 and unsaturated oleate at C2.

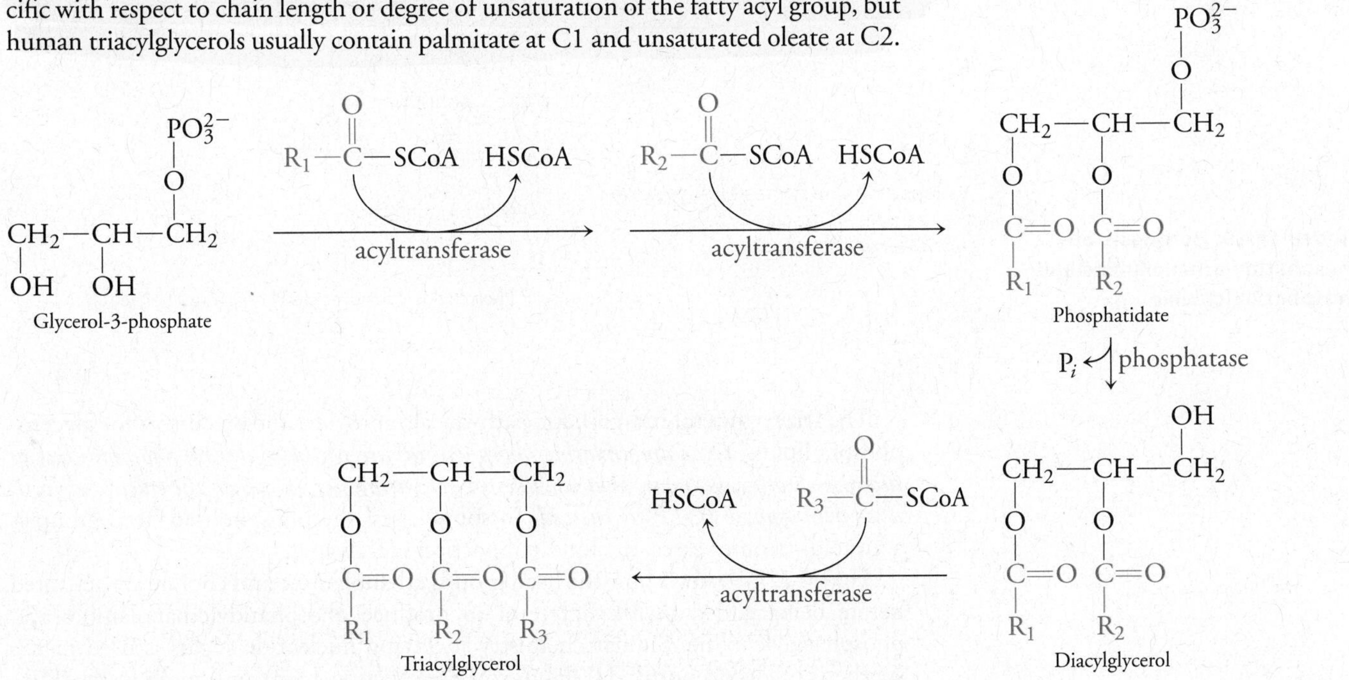

Figure 17-18 Triacylglycerol synthesis. An acyltransferase appends a fatty acyl group to C1 of glycerol-3-phosphate. A second acyltransferase reaction adds an acyl group to C2, yielding phosphatidate. A phosphatase removes P_i to produce diacylglycerol. The addition of a third acyl group yields a triacylglycerol.

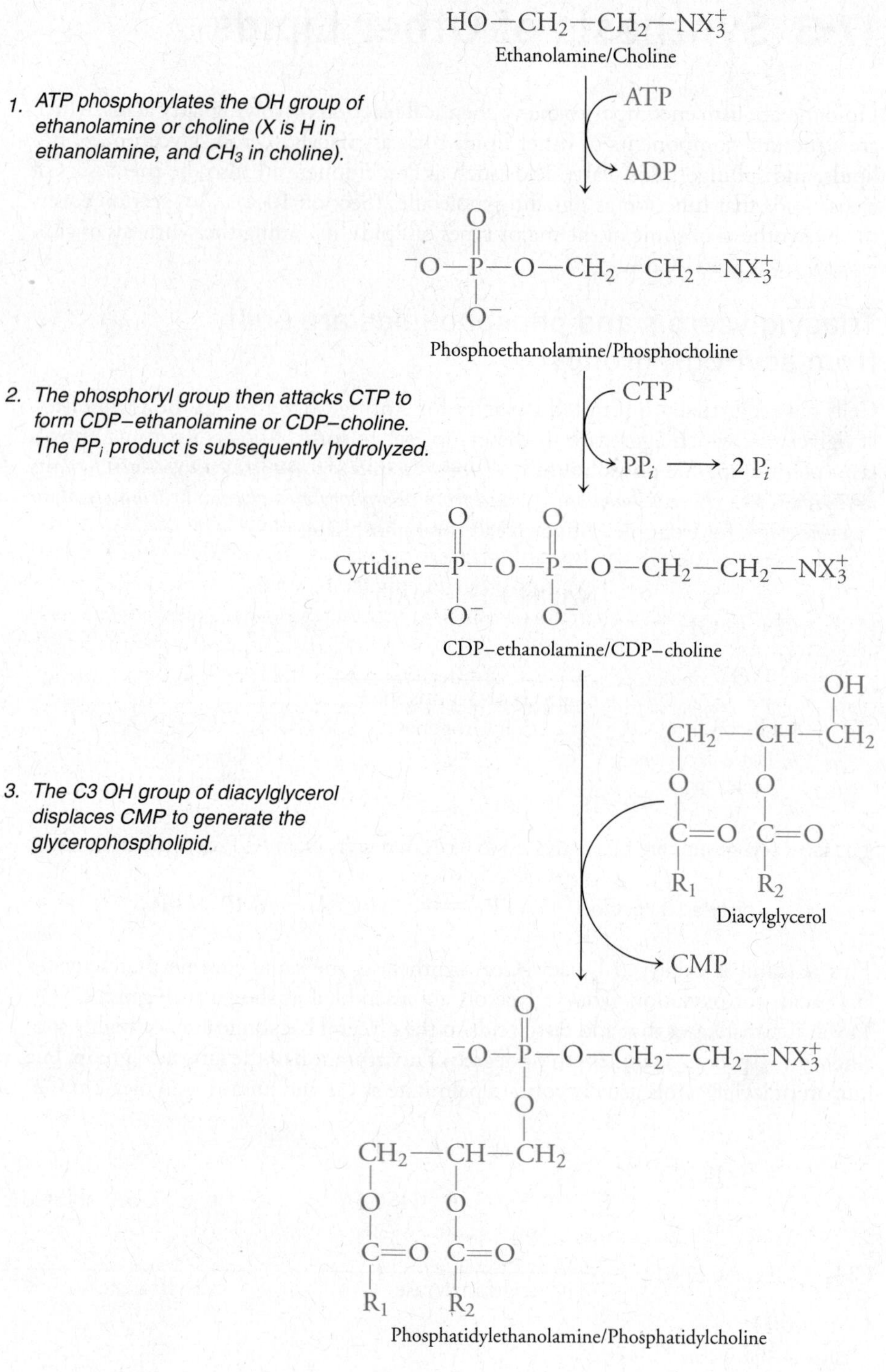

$$HO-CH_2-CH_2-NX_3^+$$

Ethanolamine/Choline

1. *ATP phosphorylates the OH group of ethanolamine or choline (X is H in ethanolamine, and CH3 in choline).*

ATP → ADP

$$^-O-\overset{\displaystyle O}{\underset{\displaystyle O^-}{P}}-O-CH_2-CH_2-NX_3^+$$

Phosphoethanolamine/Phosphocholine

2. *The phosphoryl group then attacks CTP to form CDP–ethanolamine or CDP–choline. The PP_i product is subsequently hydrolyzed.*

CTP → $PP_i \longrightarrow 2\,P_i$

$$Cytidine-\overset{\displaystyle O}{\underset{\displaystyle O^-}{P}}-O-\overset{\displaystyle O}{\underset{\displaystyle O^-}{P}}-O-CH_2-CH_2-NX_3^+$$

CDP–ethanolamine/CDP–choline

3. *The C3 OH group of diacylglycerol displaces CMP to generate the glycerophospholipid.*

Diacylglycerol

→ CMP

Phosphatidylethanolamine/Phosphatidylcholine

Figure 17-19 Synthesis of phosphatidylethanolamine and phosphatidylcholine.

The triacylglycerol biosynthetic pathway also provides the precursors for glycerophospholipids. *These amphipathic phospholipids are synthesized from phosphatidate or diacylglycerol by pathways that include an activating step in which the nucleotide cytidine triphosphate (CTP) is cleaved.* In some cases, the phospholipid head group is activated; in other cases, the lipid tail portion is activated.

Figure 17-19 shows how the head groups ethanolamine and choline are activated before being added to diacylglycerol to produce phosphatidylethanolamine and phosphatidylcholine. Similar chemistry involving nucleotide sugars is used in the synthesis of glycogen from UDP–glucose (see Section 13-3) and starch from ADP–glucose (see Section 16-3).

Phosphatidylserine is synthesized from phosphatidylethanolamine by a head-group exchange reaction in which serine displaces the ethanolamine head group:

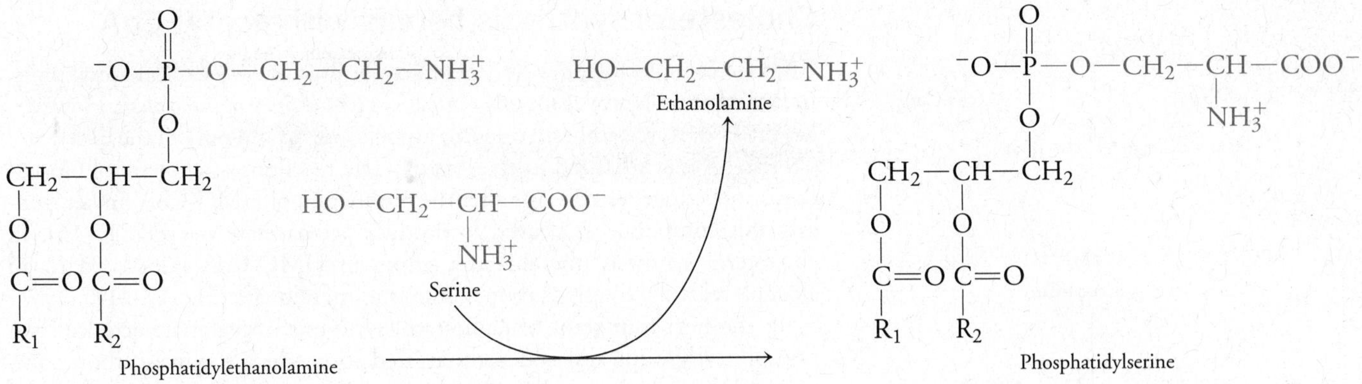

Phosphatidylethanolamine → (via Serine, releasing Ethanolamine) → Phosphatidylserine

In the synthesis of phosphatidylinositol, the diacylglycerol component is activated, rather than the head group, so that the inositol head group adds to CDP–diacylglycerol (Fig. 17-20).

Glycerophospholipids (and sphingolipids) are components of cellular membranes. New membranes are formed by inserting proteins and lipids into preexisting membranes, mainly in the endoplasmic reticulum. The newly synthesized membrane components reach their final cellular destinations primarily via vesicles that bud off the endoplasmic reticulum and, in some cases, by diffusing at points where two membranes make physical contact. Glycerophospholipids may undergo remodeling through the action of phospholipases and acyltransferases that remove and reattach different fatty acyl groups.

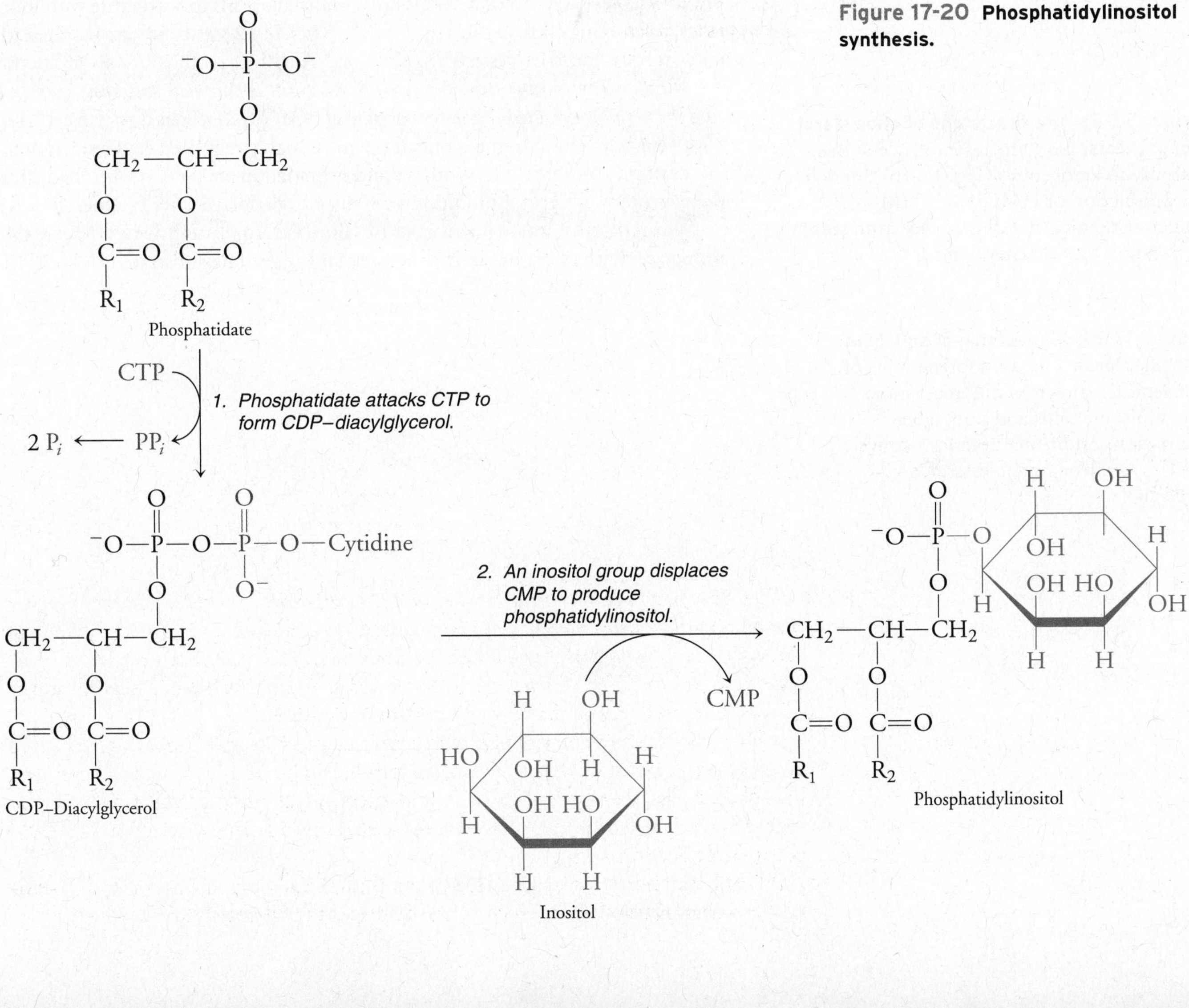

Figure 17-20 Phosphatidylinositol synthesis.

Phosphatidate

1. Phosphatidate attacks CTP to form CDP–diacylglycerol.

CTP → PP$_i$ → 2 P$_i$

CDP–Diacylglycerol

2. An inositol group displaces CMP to produce phosphatidylinositol.

Inositol → CMP

Phosphatidylinositol

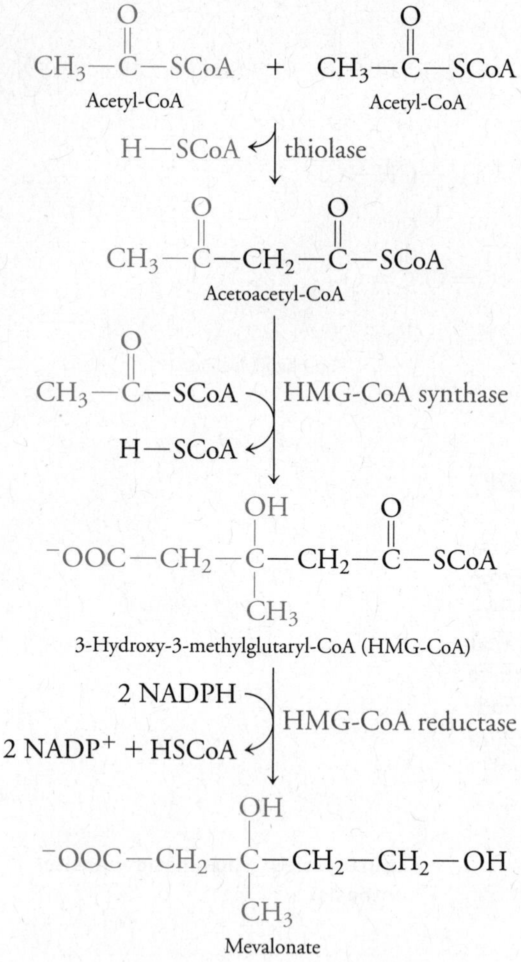

Cholesterol synthesis begins with acetyl-CoA

Cholesterol molecules, like fatty acids, are built from two-carbon acetyl units. In fact, *the first steps of cholesterol synthesis resemble those of ketogenesis.* However, ketone bodies are synthesized in the mitochondria (and only in the liver), and cholesterol is synthesized in the cytosol. The reactions of cholesterol biosynthesis and ketogenesis diverge after the production of HMG-CoA. In ketogenesis, this compound is cleaved to produce acetoacetate (see Fig. 17-16). In cholesterol synthesis, the thioester group of HMG-CoA is reduced to an alcohol, releasing the six-carbon compound mevalonate (**Fig. 17-21**).

In the next four steps of cholesterol synthesis, mevalonate acquires two phosphoryl groups and is decarboxylated to produce the five-carbon compound isopentenyl pyrophosphate:

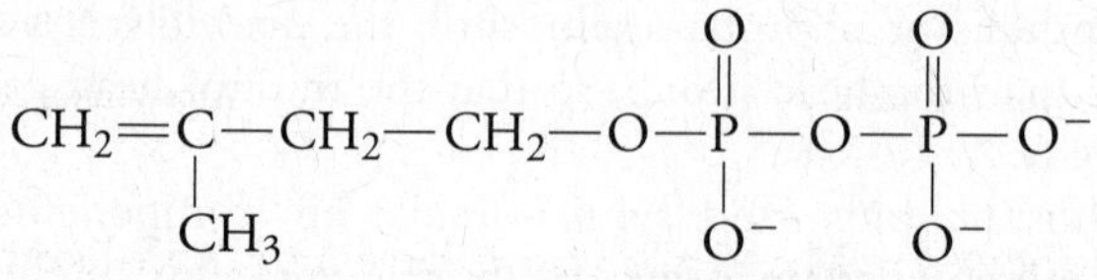

Isopentenyl pyrophosphate

This isoprene derivative is the precursor of cholesterol as well as other isoprenoids, such as ubiquinone, the C_{15} farnesyl group that is attached to some lipid-linked membrane proteins, and pigments such as β-carotene. Isoprenoids are an extremely diverse group of compounds, particularly in plants, with about 25,000 characterized to date.

In cholesterol synthesis, six isoprene units condense to form the C_{30} compound squalene. Cyclization of this linear molecule leads to a structure with four rings, resembling cholesterol (**Fig. 17-22**). A total of 21 reactions are required to convert squalene to cholesterol. NADH or NADPH is required for several steps.

The rate-determining step of cholesterol synthesis (a pathway with over 30 steps) and the major control point is the conversion of HMG-CoA to mevalonate by HMG-CoA reductase. This enzyme is one of the most highly regulated enzymes known. For example, the rates of its synthesis and degradation are tightly controlled, and the enzyme is subject to inhibition by phosphorylation of a Ser residue.

Synthetic inhibitors known as statins bind extremely tightly to HMG-CoA reductase, with K_I values in the nanomolar range. The substrate HMG-CoA

Figure 17-21 The first steps of cholesterol biosynthesis. Note the resemblance of this pathway to ketogenesis (Fig. 17-16) through the production of HMG-CoA. HMG-CoA reductase then catalyzes a four-electron reductive deacylation to yield mevalonate.

Figure 17-22 Conversion of squalene to cholesterol. The six isoprene units of squalene are shown in different colors. The molecule folds and undergoes cyclization. Additional reactions convert the C_{30} squalene to cholesterol, a C_{27} molecule.

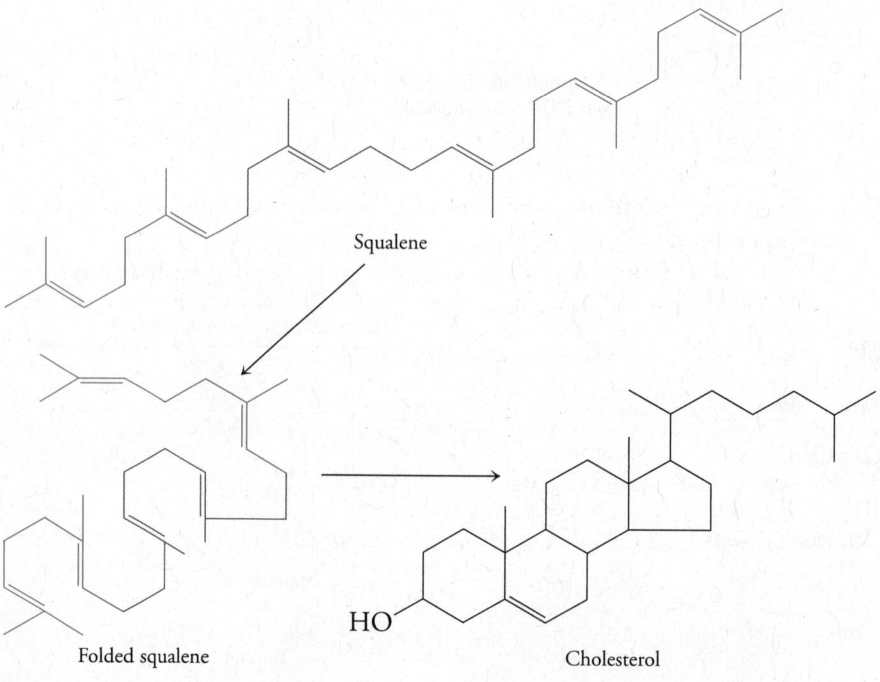

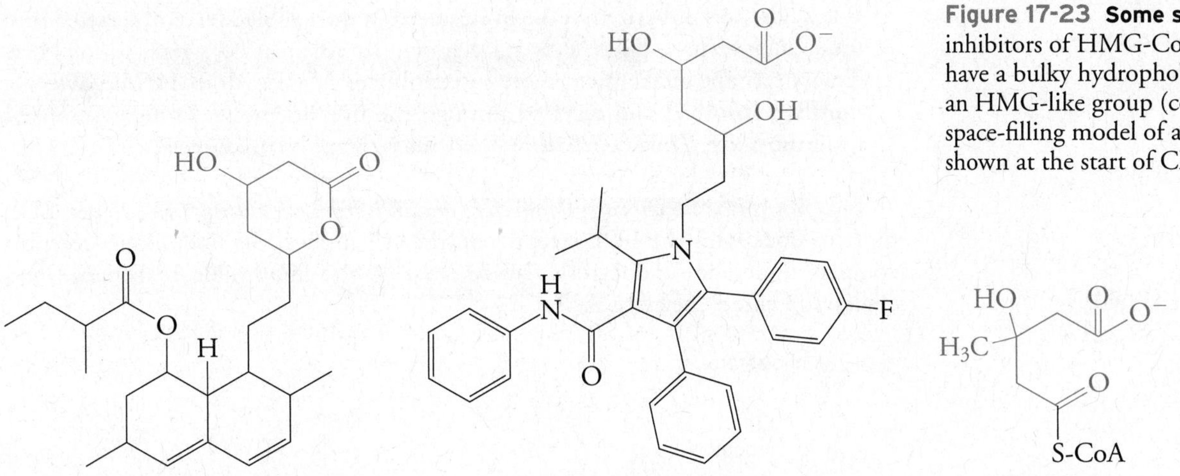

Lovastatin (Mevacor)   Atorvastatin (Lipitor)   HMG-CoA

Figure 17-23 Some statins. These inhibitors of HMG-CoA reductase have a bulky hydrophobic group plus an HMG-like group (colored red). A space-filling model of atorvastatin is shown at the start of Chapter 7.

has a K_M of about 4 μM. All the statins have an HMG-like group that acts as a competitive inhibitor of HMG-CoA binding to the enzyme (Fig. 17-23). Their rigid hydrophobic groups also prevent the enzyme from forming a structure that would accommodate the pantothenate moiety of CoA. The physiological effect of the statins is to lower serum cholesterol levels by blocking mevalonate synthesis. Cells must then obtain cholesterol from circulating lipoproteins. But since mevalonate is also the precursor of other isoprenoids such as ubiquinone, the long-term use of statins can have negative side effects.

Cholesterol can be used in several ways

Newly synthesized cholesterol has several fates:

1. It can be incorporated into a cell membrane.
2. It may be acylated to form a cholesteryl ester for storage or, in liver, for packaging in VLDL.

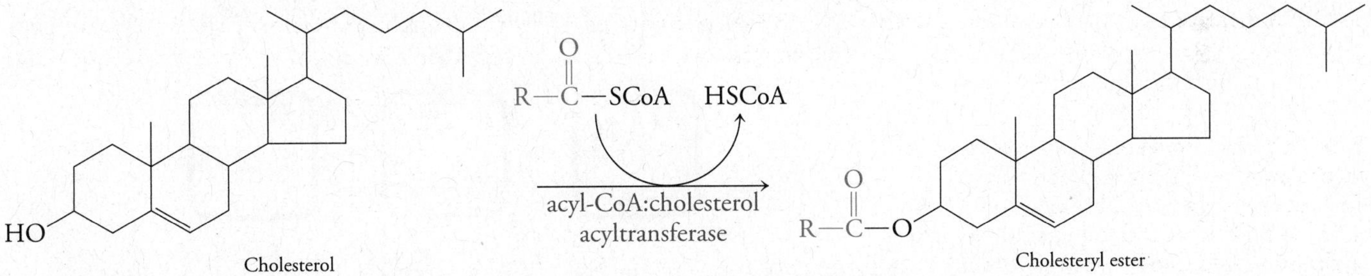
R—C(=O)—SCoA
HSCoA
acyl-CoA:cholesterol
acyltransferase

Cholesterol   Cholesteryl ester

3. It is a precursor of steroid hormones such as testosterone and estrogen in the appropriate tissues.
4. It is a precursor of bile acids such as cholate:

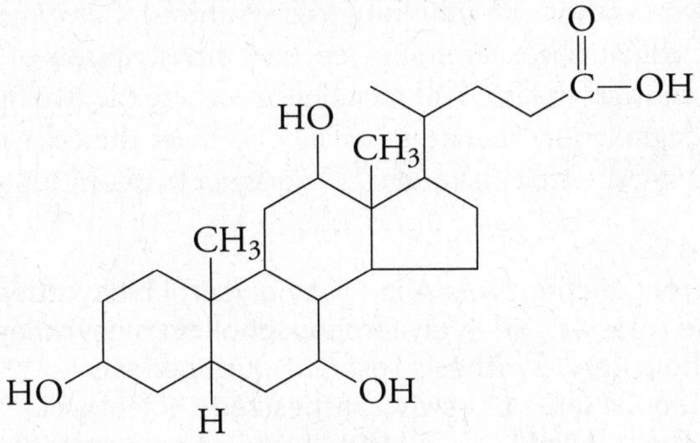

Cholate

Bile acids are synthesized in the liver, stored in the gallbladder, and secreted into the small intestine. There, they aid digestion by acting as detergents to solubilize dietary fats and make them more susceptible to lipases. Although bile acids are mostly reabsorbed and recycled through the liver for reuse, some are excreted from the body. *This is virtually the only route for cholesterol disposal.*

Cells can synthesize cholesterol as well as obtain it from circulating LDL. When LDL proteins dock with the LDL receptor on the cell surface, the lipoprotein–receptor complex undergoes endocytosis. Inside the cell, the lipoprotein is degraded and cholesterol enters the cytosol.

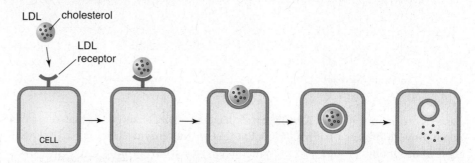

The role of LDL in delivering cholesterol to cells is dramatically illustrated by the disease familial hypercholesterolemia, which is due to a genetic defect in the LDL receptor. The cells of homozygotes are unable to take up LDL, so the concentration of serum cholesterol is about three times higher than normal. This contributes to atherosclerosis, and many individuals die of the disease before age 30.

High-density lipoproteins (HDL) are essential for removing excess cholesterol from cells. The efflux of cholesterol requires the close juxtaposition of the cell membrane and an HDL particle as well as specific cell-surface proteins. One of these is an ABC transporter (Section 9-3) that acts as a flippase to move cholesterol from the cytosolic leaflet to the extracellular leaflet, from which it can diffuse into the HDL particle.

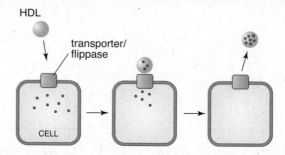

Defects in the gene for the transporter cause Tangier disease, which is characterized by accumulations of cholesterol in tissues and a high risk of heart attack.

Because cells do not break down cholesterol and because the accumulation of cholesterol is potentially toxic (it could disrupt membrane structure), the body must coordinate cholesterol synthesis and transport among tissues. For example, cholesterol shuts down its own synthesis by inhibiting the synthesis of enzymes such as HMG-CoA reductase. Cellular cholesterol also represses transcription of the gene for the LDL receptor. In contrast to fatty acid metabolism, where the two opposing pathways of synthesis and degradation operate in balance to meet the cell's needs, cholesterol metabolism in many cells is characterized by a balance between influx and efflux.

CONCEPT REVIEW
- What is the role of coenzyme A in triacylglycerol biosynthesis?
- What are the roles of CTP in glycerophospholipid biosynthesis?
- How does cholesterol synthesis resemble ketogenesis?
- List the metabolic fates of newly synthesized cholesterol.
- Summarize the roles of LDL and HDL in cholesterol metabolism.

A summary of lipid metabolism

The processes of breaking down and synthesizing lipids illustrate some general principles related to how the cell carries out opposing metabolic pathways. The diagram in **Figure 17-24** includes the major lipid metabolic pathways covered in this chapter. Several features are worth noting:

1. The pathways for fatty acid catabolism and synthesis as well as the synthesis of some other compounds all converge at the common intermediate acetyl-CoA, which is also a product of carbohydrate metabolism (see Section 14-1) and a key player in amino acid metabolism (which will be covered in Chapter 18).
2. The pathways for fatty acid degradation and fatty acid synthesis have a certain degree of symmetry, with similar intermediates and a role for thioesters, but the

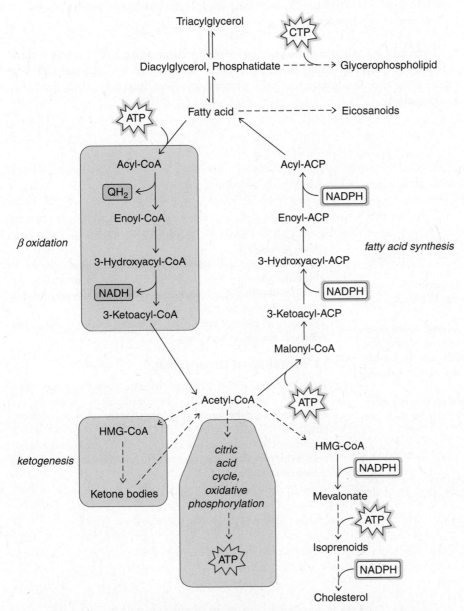

Figure 17-24 Summary of lipid metabolism. Only the major pathways covered in this chapter are included. Open gold symbols indicate ATP consumption; filled gold symbols indicate ATP production. Open and filled red symbols represent the consumption and production of reduced cofactors (NADH, NADPH, and QH$_2$). The shaded portions of the diagram indicate reactions that occur in mitochondria.

pathways have very different free energy considerations. β Oxidation produces reduced cofactors and requires only two ATP equivalents; fatty acid synthesis consumes NADPH and requires the input of ATP in each round. Other metabolic pathways, including cholesterol synthesis, consume reduced cofactors generated by catabolic reactions.

3. The catabolic pathway of β oxidation, the conversion of acetyl-CoA to ketone bodies, the oxidation of acetyl-CoA via the citric acid cycle, and the reoxidation of reduced cofactors occur in mitochondria (although some lipid metabolic reactions also occur in peroxisomes). In contrast, many lipid biosynthetic reactions take place in the cytosol or in association with the endoplasmic reticulum. Various pathways therefore require transmembrane transport systems and/or separate pools of substrates and cofactors.

4. Although the central pathways of lipid metabolism, as outlined in Figure 17-24, comprise just a few different reactions, complexity is introduced in the form of isozymes with different acyl-chain-length specificities; in additional enzymes to deal with odd-chain, branched, and unsaturated fatty acids; and in tissue-specific reactions leading to particular products such as eicosanoids or isoprenoids.

Understanding lipid metabolism is essential for diagnosing and treating certain human diseases, including those that reflect deficiencies of lipid-metabolizing enzymes or abnormalities among the proteins involved in transporting lipids via lipoproteins.

[SUMMARY]

17-1 Fatty Acid Oxidation

- Lipoproteins transport lipids, including cholesterol, in the bloodstream. High levels of LDL are associated with the development of atherosclerosis.
- Fatty acids released from triacylglycerols by the action of lipases are activated by their attachment to CoA in an ATP-dependent reaction.
- In the process of β oxidation, a series of four enzymatic reactions degrades a fatty acyl-CoA two carbons at a time, producing one QH_2, one NADH, and one acetyl-CoA, which can be further oxidized by the citric acid cycle. Reoxidation of the reduced cofactors generates considerable ATP.
- Oxidation of unsaturated and odd-chain fatty acids requires additional enzymes. Very-long-chain and branched fatty acids are oxidized in peroxisomes.

17-2 Fatty Acid Synthesis

- Fatty acids are synthesized by a pathway that resembles the reverse of β oxidation. In the first step of fatty acid synthesis, acetyl-CoA

carboxylase catalyzes an ATP-dependent reaction that converts acetyl-CoA to malonyl-CoA, which becomes the donor of two-carbon groups.
- Mammalian fatty acid synthase is a multifunctional enzyme in which the growing fatty acyl chain is attached to acyl-carrier protein rather than CoA. Elongases and desaturases may modify newly synthesized fatty acids.
- The liver can convert acetyl-CoA to ketone bodies to be used as metabolic fuels in other tissues.

17-3 Synthesis of Other Lipids

- Triacylglycerols are synthesized by attaching three fatty acyl groups to a glycerol backbone. Intermediates of the triacylglycerol pathway are the starting materials for the synthesis of phospholipids.
- Cholesterol is synthesized from acetyl-CoA. The rate-determining step of this pathway is the target of drugs known as statins. Excess cholesterol circulates via HDL.

[GLOSSARY TERMS]

atherosclerosis
lipoprotein
β oxidation

peroxisome
multifunctional enzyme
essential fatty acid

ketone bodies
ketogenesis
bile acids

[PROBLEMS]

17-1 Fatty Acid Oxidation

1. Use the information in Table 17-1 to explain the density rankings of the lipoproteins.

2. What amino acid side chains of apolipoprotien A1 would be expected to be in contact with the lipid core of the lipoprotein shown in Figure 17-2?

3. Biodiesel, a fuel typically derived from plant oils, can be manufactured by treating the oil with a methanol/KOH mixture to produce fatty acid methyl esters. Indicate the structures of the products generated by treating the triacylglycerol 1-palmitoyl-2,3-dioleoyl-glycerol with methanol/KOH.

4. The methanol/KOH reaction described in Problem 3 must be carried out in "dry" methanol. If any H_2O were present, what would happen?

5. The overall reaction for the activation of a fatty acid to fatty acyl-CoA, with concomitant hydrolysis of ATP to AMP, has a free energy change of about zero. The reaction is favorable because of subsequent hydrolysis of pyrophosphate to orthophosphate (the reaction has a $\Delta G^{\circ\prime}$ value of -19.2 kJ $\cdot$ mol^{-1}). Write the equation for the coupled reaction and calculate **(a)** $\Delta G^{\circ\prime}$ and **(b)** the equilibrium constant for the reaction.

6. Fatty acid activation catalyzed by acyl-CoA synthetase begins with nucleophilic attack by the negatively charged carboxylate oxygen of the fatty acid on the α-phosphate (the innermost phosphate) of ATP. An acyladenylate mixed anhydride is formed. Write the mechanism of the reaction.

7. A deficiency of carnitine results in muscle cramps, which are exacerbated by fasting or exercise. Give a biochemical explanation for the muscle cramping, and explain why cramping increases during fasting and exercise.

8. Muscle biopsy and enzyme assays of a carnitine-deficient individual show that medium-chain (C_8–C_{10}) fatty acids can be metabolized normally, despite the carnitine deficiency. What does this tell you about the role of carnitine in fatty acid transport across the inner mitochondrial membrane?

9. Some individuals suffer from medium-chain acyl-CoA dehydrogenase (MCAD) deficiency. Which intermediates accumulate in individuals with MCAD deficiency?

10. How should a patient with a medium-chain acyl-CoA dehydrogenase deficiency be treated?

11. The first three reactions of the β oxidation pathway are similar to three reactions of the citric acid cycle. Which reactions are these, and why are they similar?

12. During β oxidation, methylene ($-CH_2-$) groups in a fatty acid are oxidized to carbonyl ($C=O$) groups, yet no oxygen is consumed by the reactions of β oxidation. How is this possible?

13. The β oxidation pathway was elucidated in part by Franz Knoop in 1904. He fed dogs fatty acid phenyl derivatives and then analyzed their urine for the resulting metabolites. What metabolite was produced when the dogs were fed **(a)** phenylpropionate and **(b)** phenylbutyrate?

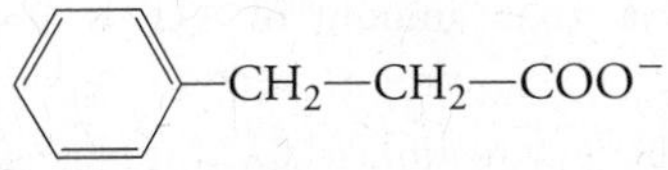

Phenylpropionate

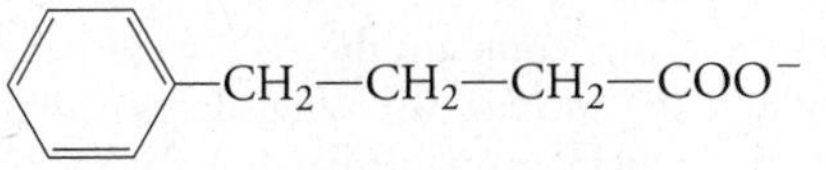

Phenylbutyrate

14. A deficiency of phytanate-degrading enzymes in peroxisomes results in Refsum's disease, a neuronal disorder caused by phytanate accumulation. Patients with Refsum's disease cannot convert phytanate to pristanate because they lack the enzymes involved in this α oxidation reaction. Pristanate normally enters the peroxisomal β oxidation pathway.

The α oxidation of phytanate to pristanate is shown below. Show how pristanate is oxidized via β oxidation, and list the products of pristanate oxidation.

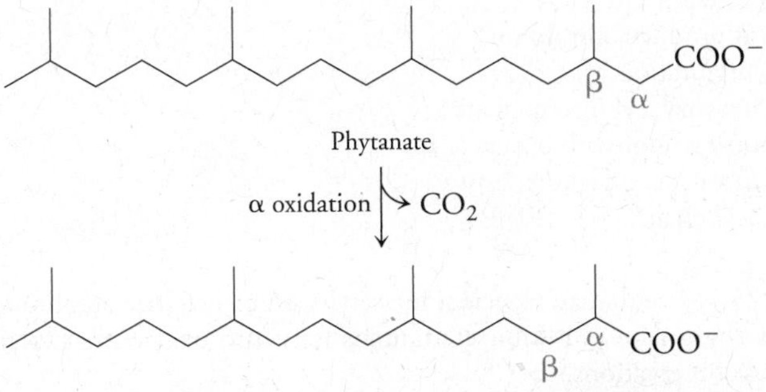

Phytanate

α oxidation $\rightarrow CO_2$

Pristanate

15. How many molecules of ATP are generated when **(a)** palmitate and **(b)** stearate are completely oxidized via the β oxidation pathway in mitochondria?

16. How many molecules of ATP are generated when **(a)** oleate and **(b)** linoleate are completely oxidized via the β oxidation pathway?

17. How many molecules of ATP are generated when a fully saturated 17-carbon fatty acid is oxidized via the β oxidation pathway?

18. How many molecules of ATP are generated when oxidation of a fully saturated C_{24} fatty acid begins in the peroxisome and is completed by the mitochondrion when 12 carbons remain?

19. A vitamin B_{12} deficiency causes the disease pernicious anemia. Often the disease is caused not by the lack of the vitamin itself but by the lack of a protein called intrinsic factor, which is secreted by gastric parietal cells. Intrinsic factor binds to vitamin B_{12} and facilitates its absorption into the small intestine. Use this information to devise a treatment for a patient diagnosed with pernicious anemia.

20. If you were a physician and you wanted to test a patient for pernicious anemia (see Problem 19), what metabolite would you measure in the patient's blood or urine, and why?

21. Both fatty acid oxidation and glucose oxidation by glycolysis lead to the formation of large amounts of ATP. Explain why a cell preparation containing all the enzymes required for either pathway cannot generate ATP when a fatty acid or glucose is added, unless a small amount of ATP is also added to the preparation.

22. The complete oxidation to CO_2 of glucose and palmitate releases considerable free energy: $\Delta G^{\circ\prime} = -2850$ kJ $\cdot$ mol^{-1} for glucose oxidation and $\Delta G^{\circ\prime} = -9781$ kJ $\cdot$ mol^{-1} for palmitate. For each fuel molecule, compare the ATP yield per carbon atom **(a)** in theory and **(b)** *in vivo*. **(c)** What do these results tell you about the relative efficiency of oxidizing carbohydrates and fatty acids?

17-2 Fatty Acid Synthesis

23. Compare fatty acid degradation and fatty acid synthesis by filling in the table below.

	Fatty acid degradation	Fatty acid synthesis
Cellular location		
Acyl-group carrier		
Electron carrier(s)		
ATP requirement		
Unit product/unit donor		
Configuration of hydroxyacyl intermediate		
Shortening/growth occurs at which end of the fatty acyl chain?		

24. Mice that are deficient in acetyl-CoA carboxylase are thinner than normal and exhibit continuous fatty acid oxidation. Explain these observations.

25. Write the mechanism for the carboxylation of acetyl-CoA to malonyl-CoA catalyzed by acetyl-CoA carboxylase.

26. During fatty acid synthesis, why is the condensation of an acetyl group and a malonyl group energetically favorable, whereas the condensation of two acetyl groups would be unfavorable?

27. What do acetyl-CoA carboxylase, pyruvate carboxylase, and propionyl-CoA carboxylase have in common?

28. The activity of acetyl-CoA carboxylase is regulated by hormone-controlled phosphorylation and dephosphorylation. Based on what you know about signaling via epinephrine (Section 10-2), describe the effect of epinephrine on acetyl-CoA carboxylase and fatty acid metabolism. Is this consistent with epinephrine's effect on glycogen metabolism?

29. Is the K_I (see Equation 7-30) for inhibition of acetyl-CoA carboxylase by palmitoyl-CoA higher or lower when the enzyme is phosphorylated (see Problem 28)?

30. Would higher or lower concentrations of citrate be required to activate acetyl-CoA carboxylase when the enzyme is phosphorylated (see Problem 28)?

31. What is the cost of synthesizing palmitate from acetyl-CoA?

32. On what carbon atoms does the $^{14}CO_2$ used to synthesize malonyl-CoA from acetyl-CoA appear in palmitate?

33. Why does triclosan inhibit bacterial fatty acid synthase but not mammalian fatty acid synthase?

34. Cancer cells have a greater-than-normal level of fatty acid synthesis, which has been attributed to enhanced expression of fatty acid synthase. Rapid fatty acid synthesis is required for tumor growth. This observation has led cancer researchers to investigate the use of fatty acid synthesis inhibitors as antitumor agents.

(a) A series of potential inhibitors of fatty acid synthase were synthesized. All the inhibitors had the structure shown in the figure; only the alkyl chain (R) varied in length. Why were these compounds effective as inhibitors of fatty acid synthase?

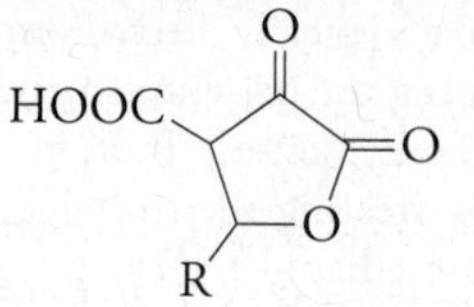

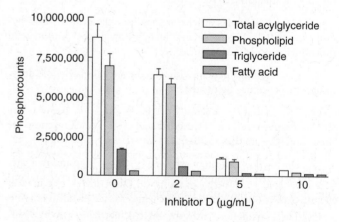

(b) Each compound was tested for its ability to inhibit fatty acid synthase activity in both normal cells and breast cancer cells. ID_{50} values were determined (this is the concentration of inhibitor required to inhibit cell growth by 50%). The results are shown in the table. Which inhibitors are most effective? What are the characteristics of the effective inhibitors? (*Hint:* Calculate the ratio of the ID_{50} values for normal and breast cancer cells.) An effective inhibitor must also be soluble in aqueous solution, so consider solubility as an additional factor in your answer.

Compound	Alkyl side chain (R)	Breast cancer cells ID_{50} (mg/mL)	Normal cells ID_{50} (mg/mL)
A	—$C_{13}H_{27}$	3.9	10.6
B	—$C_{11}H_{23}$	4.8	29.0
C	—C_9H_{19}	5.2	12.8
D	—C_8H_{17}	5.0	21.3
E	—C_7H_{15}	4.8	21.7
F	—C_6H_{13}	8.4	12.4

(c) Inhibitor D was tested for its ability to inhibit triacylglycerol synthesis in leukemia cells. The cells were incubated with ^{14}C-labeled acetate, and the amount of radioactivity in various types of cellular lipids was measured. What is your interpretation of the results shown in the graph?

35. Draw the structures of the fatty acids listed below. Which are essential fatty acids for humans?
(a) Oleic acid (18:1 *n*-9)
(b) Linoleic acid (18:2 *n*-6)
(c) α-linolenic acid (18:3 *n*-3)
(d) Palmitoleic acid (16:1 *n*-6)

36. Why is docosahexaenoic acid (DHA, 22:6n-3), a fatty acid commonly found in fish, added to baby formula?

37. Compare the carboxylation/decarboxylation sequence of reactions in gluconeogenesis (Fig. 13-10) and fatty acid synthesis. Discuss the source of free energy for these steps in each pathway.

38. Does fatty acid synthase activity increase or decrease under the following conditions? Explain.
(a) High-carbohydrate diet (liver fatty acid synthase)
(b) High-fat diet (liver fatty acid synthase)
(c) Mid to late pregnancy (mammary gland fatty acid synthase)

39. Isolated heart cells undergo contraction even in the absence of glucose and fatty acids if they are supplied with acetoacetate.
(a) How does this compound act as a metabolic fuel?
(b) Even with plentiful acetoacetate, the rate of flux through the citric acid cycle gradually drops off unless pyruvate is added to the cells. Explain.

40. When glucose is unavailable, the liver begins to break down fatty acids to supply the rest of the body with metabolic fuel. Explain why fatty acid–derived acetyl-CoA is not catabolized by the citric acid cycle but is instead diverted to ketogenesis when no glucose is available.

41. Discuss the energetic costs of converting two acetyl-CoA to the ketone body 3-hydroxybutyrate in the liver and then converting the 3-hydroxybutyrate back to two acetyl-CoA in the muscle.

42. It has been said that "fats burn in the flame of carbohydrates." Give a biochemical explanation for this statement.

43. Two siblings born two years apart were separately diagnosed with a pyruvate carboxylase deficiency. Both neonates died less than a month after they were born. Blood samples taken before the death of the infants showed a high concentration of ketone bodies in the blood. Explain why the ketone body concentration was elevated.

44. A symptom of a pyruvate carboxylase deficiency (see Problem 43) is a decreased β-hydroxybutryate:acetoacetate ratio. Explain.

17-3 Synthesis of Other Lipids

45. The glycerol-3-phosphate required for the first step of triacylglycerol synthesis can be obtained from either glucose or pyruvate. Explain how this occurs. Can all cells obtain glycerol-3-phosphate from either precursor?

46. In a site-directed mutagenesis experiment, an essential serine of HMG-CoA reductase was replaced with an alanine residue. In normal cells, HMG-CoA reductase levels decreased when the cells were incubated with media containing LDL particles, but in cells expressing the mutant enzyme, no change was observed in enzyme activity. What do these results indicate regarding the regulation of HMG-CoA reductase?

47. Fumonisins are mycotoxins isolated from fungi commonly found on corn and other grains. They are both toxic and carcinogenic and can cause disease in animals grazing on fungus-contaminated grain. The structure of fumonisin B$_1$ is shown. Note the structural similarity to sphingosine.

Fumonisin B$_1$

Fumonisins inhibit one of the enzymes in the ceramide synthetic pathway, which is outlined below. Ceramide is an important cell-signaling molecule, and its regulation is critical to cell survival.

(a) Deduce which enzyme in the signaling pathway is inhibited by fumonisin given the following clues: (1) The addition of fumonisin B$_1$ to rat hepatocytes almost completely inhibited ceramide synthesis. The synthesis of other phospholipids was not affected. (2) Addition of fumonisin B$_1$ to the cultured cells did not significantly change the rate of formation of 3-ketosphinganine. (3) There was no accumulation of 3-ketosphinganine. (4) When radioactively labeled serine was added to culture medium containing fumonisin B$_1$, there was an increase in the amount of label in sphinganine as compared to controls.
(b) How does fumonisin inhibit the target enzyme identified in part (a)?

48. How many phosphoanhydride bonds must be cleaved to synthesize phosphatidylcholine from choline and diacylglycerol?

49. Manufacturers of cooking oil can chemically convert triglycerides to diglycerides. What kind of chemical reaction occurs and what is its purpose?

50. Cancer cells appear to increase the expression of choline kinase. Why is this enzyme necessary for phospholipid synthesis?

51. The malaria parasite can synthesize large amounts of phosphatidylcholine even in the absence of choline. To do this, it relies on an enzyme called phosphoethanolamine methyltransferase.
(a) Describe the reaction catalyzed by this enzyme.
(b) What information would you need in order to assess whether this methyltransferase would be a suitable antimalarial drug target?

52. Fungi produce ergosterol rather than cholesterol. List the ways that ergosterol differs from cholesterol.

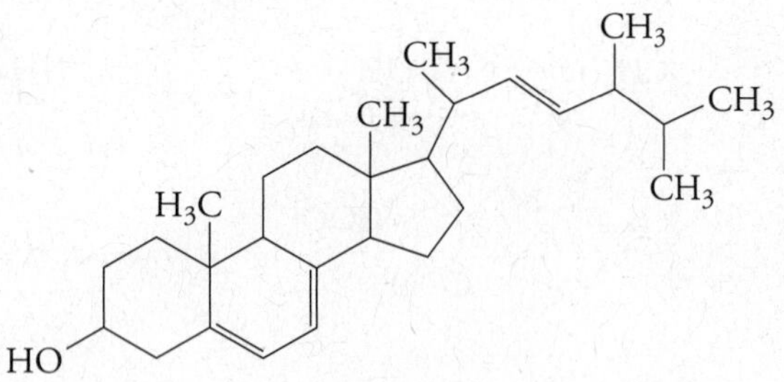

Ergosterol

53. Cholesterol is poorly soluble in aqueous solution, yet cells must be able to sense the cholesterol level in order to regulate uptake and biosynthesis, in part by altering the expression of genes for HMG-CoA reductase and the LDL receptor. The cellular cholesterol sensors are proteins called SREBPs (sterol regulatory element binding proteins). In the absence of cholesterol, an SREBP residing in the endoplasmic reticulum is proteolytically cleaved to release a large soluble N-terminal domain that includes a structural motif found in many DNA-binding proteins.

(a) Why is it important that the SREBP be an integral membrane protein?
(b) Why is proteolysis of the SREBP required?
(c) How might the SREBP regulate the transcription of enzymes related to cholesterol metabolism?

54. Premenopausal women typically have higher HDL levels than men.
(a) Why would this tend to decrease the risk of heart disease in these women?
(b) Why is HDL level alone not a good indicator of the risk of developing heart disease?

55. Individuals with a mutation in the gene for apolipoprotein B-100 produce very low levels of this protein, which is a component of LDL.
(a) Explain why these individuals exhibit accumulation of fat in the liver.
(b) Would such individuals exhibit hypercholesterolemia or hypocholesterolemia?

56. Individuals who are unable to produce chylomicrons display symptoms consistent with vitamin A deficiency. Explain.

[SELECTED READINGS]

Houten, S. M., and Wanders, R. J. A., A general introduction to the biochemistry of mitochondrial fatty acid β-oxidation, *J. Inherit. Metab. Dis.* **33,** 469–477 (2010). [Includes a review of the relevant chemical reaction and diseases resulting from enzyme deficiencies.]

Maier, T., Leibundgut, M., and Ban, N., The crystal structure of a mammalian fatty acid synthase, *Science* **321,** 1315–1322 (2008). [Presents the structure of all but two domains of the multifunctional protein.]

Mizuno, Y., Jacob, R. F., and Mason, R. P., Inflammation and the Development of Atherosclerosis Effects of Lipid-Lowering Therapy, *J. Atheroscler. Thromb.* **18,** 351–358 (2011). [Describes the pathology of atherosclerosis, the role of lipoproteins, and drug treatments.]

Zhang, Y.-M., White, S. W., and Rock, C. O., Inhibiting bacterial fatty acid synthesis, *J. Biol. Chem.* **281,** 17541–17544 (2006). [Describes each enzyme of bacterial pathway along with inhibitors.]

chapter 18

NITROGEN METABOLISM

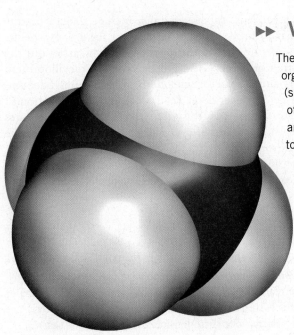

▸▸ WHY don't humans excrete ammonia?

The acquisition and allocation of nitrogen accounts for a large portion of every organism's metabolic activities. Humans can incorporate the ammonium ion (shown at left) into biological molecules as well as transfer it—in the form of an amino group—to other molecules to make amino acids, nucleotides, and other nitrogen-containing substances. But the reverse process is not tolerated because the catabolism of nitrogenous compounds would release free ammonia, which is toxic. Consequently, humans and many other organisms rely on elaborate pathways to convert waste nitrogen into safer forms for disposal.

THIS CHAPTER IN CONTEXT

Do You Remember?

- **DNA and RNA are polymers of nucleotides, each of which consists of a purine or pyrimidine base, deoxyribose or ribose, and phosphate (Section 3-1).**
- **The 20 amino acids differ in the chemical characteristics of their R groups (Section 4-1).**
- **A few metabolites appear in several metabolic pathways (Section 12-2).**
- **Many vitamins, substances that humans cannot synthesize, are components of coenzymes (Section 12-2).**
- **The citric acid cycle supplies precursors for the synthesis of other compounds (Section 14-3).**

18-1 Nitrogen Fixation and Assimilation

KEY CONCEPTS
- Nitrogen fixation by the activity of nitrogenase is part of the nitrogen cycle.
- Other enzymes incorporate amino groups into glutamine and glutamate.
- Transaminases transfer amino groups to interconvert amino acids and α-keto acids.

Approximately 80% of the air we breathe is nitrogen (N_2), but we cannot use this form of nitrogen for the synthesis of amino acids, nucleotides, and other nitrogen-containing biomolecules. Instead, we—along with most macroscopic and many microscopic life-forms—depend on the activity of a few types of microorganisms that can "fix" gaseous N_2 by transforming it into biologically useful forms. The availability of fixed nitrogen—as nitrite, nitrate, and ammonia—is believed to limit the biological productivity in much of the world's oceans. It also limits the growth of terrestrial organisms, which is why farmers use fertilizer (a source of fixed nitrogen, among other things) to promote crop growth.

Nitrogenase converts N_2 to NH_3

The known **nitrogen-fixing** organisms, or **diazotrophs,** include certain marine cyanobacteria and bacteria that colonize the root nodules of leguminous plants (**Fig. 18-1**). These bacteria make the enzyme nitrogenase, which carries out the energetically expensive reduction of N_2 to NH_3. Nitrogenase is a metalloprotein containing iron–sulfur centers and a cofactor with both iron and molybdenum, which resembles an elaborate Fe–S cluster (**Fig. 18-2**). The industrial fixation of nitrogen also involves metal catalysts, but this nonbiological process requires temperatures of 300 to 500°C and pressures of over 300 atm in order to break the triple bond between the two nitrogen atoms.

Biological N_2 reduction consumes large amounts of ATP and requires a strong reducing agent such as ferredoxin (see Section 16-2) to donate electrons. The net reaction is

$$N_2 + 8\,H^+ + 8\,e^- + 16\,ATP + 16\,H_2O \rightarrow 2\,NH_3 + H_2 + 16\,ADP + 16\,P_i$$

Note that eight electrons are required for the nitrogenase reaction, although N_2 reduction formally requires only six electrons; the two extra electrons are used to produce H_2. *In vivo,* the inefficiency of the reaction boosts the ATP toll to about 20 or 30 per N_2 reduced. Oxygen inactivates nitrogenase, so many nitrogen-fixing bacteria are confined to anaerobic habitats or carry out nitrogen fixation when O_2 is scarce.

Biologically useful nitrogen also originates from nitrate (NO_3^-), which is naturally present in water and soils. *Nitrate is reduced to NH_3 by plants, fungi, and many bacteria.* First, nitrate reductase catalyzes the two-electron reduction of nitrate to nitrite (NO_2^-):

$$NO_3^- + 2\,H^+ + 2\,e^- \rightarrow NO_2^- + H_2O$$

Next, nitrite reductase converts nitrite to ammonia:

$$NO_2^- + 8\,H^+ + 6\,e^- \rightarrow NH_4^+ + 2\,H_2O$$

Under physiological conditions, ammonia exists primarily in the protonated form, NH_4^+ (the ammonium ion), which has a pK of 9.25.

Nitrate is also produced by certain bacteria that oxidize NH_4^+ to NO_2^- and then NO_3^-, a process called **nitrification.** Still other organisms convert nitrate back to N_2, which is called **denitrification.** All the reactions we have discussed so far constitute the earth's **nitrogen cycle** (**Fig. 18-3**).

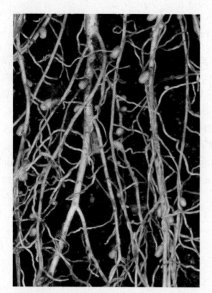

Figure 18-1 Root nodules from clover. Legumes (such as beans, clover, and alfalfa) and some other plants harbor nitrogen-fixing bacteria in root nodules. The symbiotic relationship revolves around the ability of the bacteria to fix nitrogen and the ability of the plant to make other nutrients available to the bacteria. [Dr. Jeremy Burgess/Science Photo Library/Photo Researchers, Inc.]

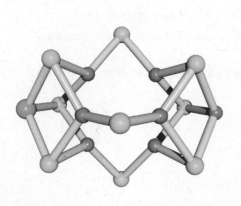

Figure 18-2 Model of the FeMo cofactor of nitrogenase. This prosthetic group in the enzyme nitrogenase consists of iron atoms (orange), sulfur atoms (yellow), and a molybdenum atom (green). The central cavity includes a carbon atom coordinated with the six iron atoms. The manner in which N_2 interacts with the FeMo cofactor is not understood. [Structure of the FeMo cofactor in nitrogenase (pdb 1QGU) determined by S. M. Mayer, M. Lawson, C. A. Gormal, S. M. Roe, and B. E. Smith.]

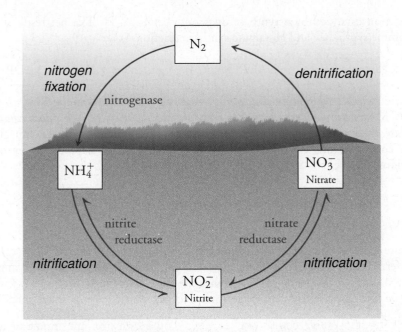

Figure 18-3 **The nitrogen cycle.**
Nitrogen fixation converts N_2 to the biologically useful NH_4^+. Nitrate can also be converted to NH_4^+. Ammonia is transformed back to N_2 by nitrification followed by denitrification.

? **Indicate which processes are oxidations and which are reductions.**

Ammonia is assimilated by glutamine synthetase and glutamate synthase

The enzyme glutamine synthetase is found in all organisms. In microorganisms, it is a metabolic entry point for fixed nitrogen. In animals, it helps mop up excess ammonia, which is toxic. In the first step of the reaction, ATP donates a phosphoryl group to glutamate. Then ammonia reacts with the reaction intermediate, displacing P_i to produce glutamine:

The name *synthetase* indicates that ATP is consumed in the reaction.

Glutamine, along with glutamate, is usually present in organisms at much higher concentrations than the other amino acids, which is consistent with its role as a carrier of amino groups. Not surprisingly, *the activity of glutamine synthetase is tightly regulated to maintain a supply of accessible amino groups.* For example, the dodecameric glutamine synthetase from *E. coli* is regulated allosterically and by covalent modification (**Fig. 18-4**).

The glutamine synthetase reaction that introduces fixed nitrogen (ammonia) into biological compounds requires a nitrogen-containing compound (glutamate) as a substrate. So what is the source of the nitrogen in glutamate? In bacteria and plants, the enzyme glutamate synthase catalyzes the reaction (a reaction catalyzed by a synthase does not require ATP). The net result of the glutamine synthetase and glutamate synthase reactions is

$$\alpha\text{-ketoglutarate} + NH_4^+ + NADPH + ATP \rightarrow glutamate + NADP^+ + ADP + P_i$$

In other words, *the combined action of these two enzymes assimilates fixed nitrogen (NH_4^+) into an organic compound (α-ketoglutarate, a citric acid cycle intermediate) to produce an amino acid (glutamate).* Mammals lack glutamate synthase, but glutamate concentrations are relatively high because glutamate is produced by other reactions.

Transamination moves amino groups between compounds

Because reduced nitrogen is so precious but free ammonia is toxic, amino groups are transferred from molecule to molecule, with glutamate often serving as an amino-group donor. We saw some of these **transamination** reactions in Section 14-3 when we examined how citric acid cycle intermediates participate in other metabolic pathways.

A transaminase (also called an aminotransferase) catalyzes the transfer of an amino group to an α-keto acid. For example,

Figure 18-4 *E. coli* glutamine synthetase. The 12 identical subunits of this enzyme are arranged in two stacked rings of 6 subunits (only the upper ring is visible here). The symmetrical arrangement of subunits is a general feature of enzymes that are regulated by allosteric effectors: Changes in activity at one of the active sites can be efficiently communicated to the other active sites. [Structure (pdb 2GLS) determined by D. Eisenberg, R. J. Almassy, and M. M. Yamashita.]

During such an amino-group transfer reaction, the amino group is transiently attached to a prosthetic group of the enzyme. This group is pyridoxal-5′-phosphate (PLP), a derivative of pyridoxine (an essential nutrient also known as vitamin B_6):

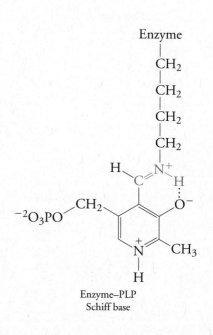

Enzyme–PLP
Schiff base

PLP is covalently attached to the enzyme via a Schiff base (imine) linkage to the ε-amino group of a Lys residue (*left*). The amino acid substrate of the transaminase displaces this Lys amino group, which then acts as an acid–base catalyst. The steps of the reaction are diagrammed in **Figure 18-5**.

The transamination reaction is freely reversible, so transaminases participate in pathways for amino acid synthesis as well as degradation. Note that if the α-keto acid produced in step 4 reenters the active site, then the amino group that was removed from the starting amino acid is restored. However, most transaminases accept only α-ketoglutarate or oxaloacetate as the α-keto acid substrate for the second

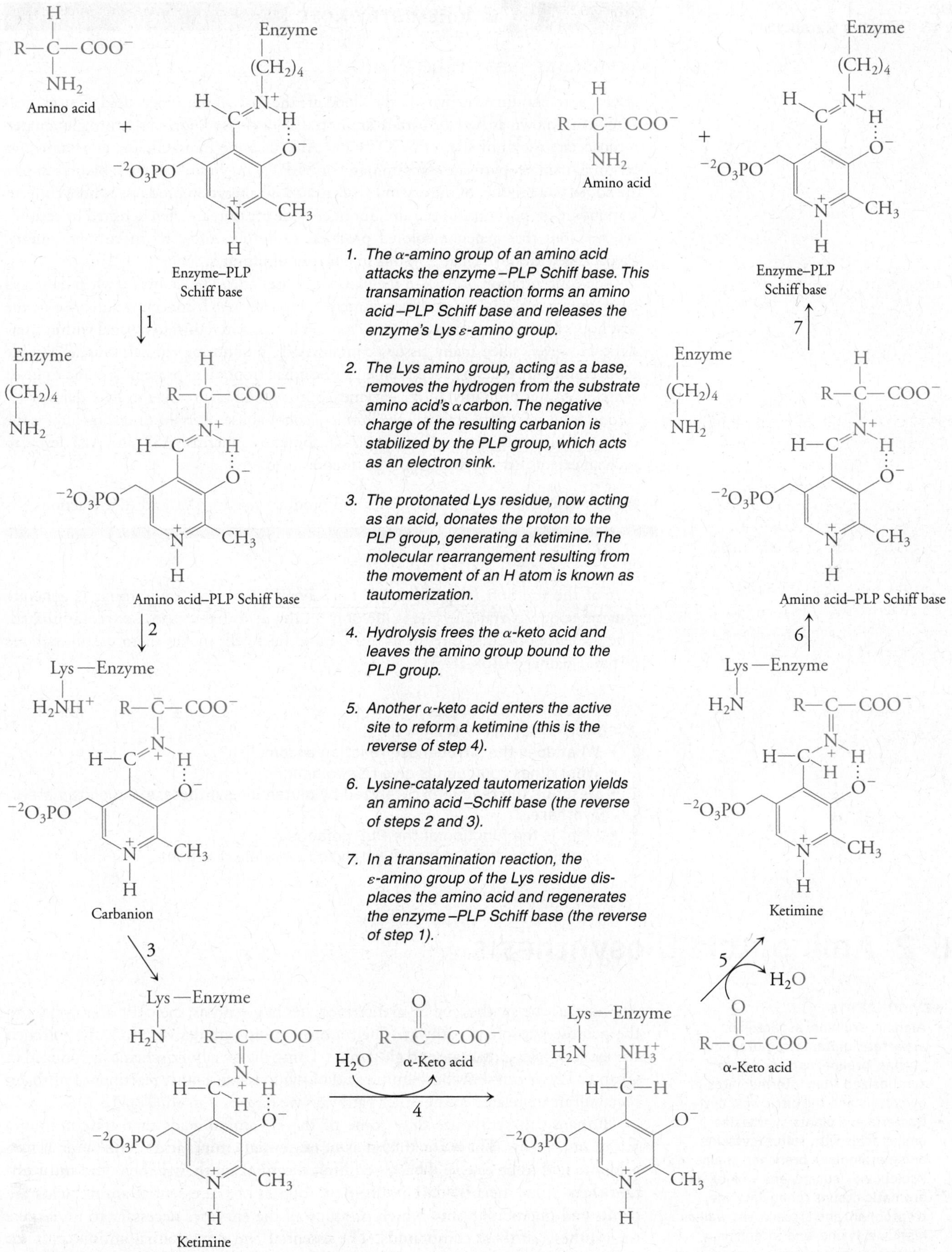

1. The α-amino group of an amino acid attacks the enzyme–PLP Schiff base. This transamination reaction forms an amino acid–PLP Schiff base and releases the enzyme's Lys ε-amino group.

2. The Lys amino group, acting as a base, removes the hydrogen from the substrate amino acid's α carbon. The negative charge of the resulting carbanion is stabilized by the PLP group, which acts as an electron sink.

3. The protonated Lys residue, now acting as an acid, donates the proton to the PLP group, generating a ketimine. The molecular rearrangement resulting from the movement of an H atom is known as tautomerization.

4. Hydrolysis frees the α-keto acid and leaves the amino group bound to the PLP group.

5. Another α-keto acid enters the active site to reform a ketimine (this is the reverse of step 4).

6. Lysine-catalyzed tautomerization yields an amino acid–Schiff base (the reverse of steps 2 and 3).

7. In a transamination reaction, the ε-amino group of the Lys residue displaces the amino acid and regenerates the enzyme–PLP Schiff base (the reverse of step 1).

Figure 18-5 PLP-catalyzed transamination. ⊕ See Animated Figure. Mechanism of PLP-dependent transamination.

BOX 18-A BIOCHEMISTRY NOTE

Transaminases in the Clinic

Assays of transaminase activity in the blood are the basis of the widely used clinical measurements known as AST (aspartate aminotransferase; also known as serum glutamate–oxaloacetate transaminase, or SGOT) and ALT (alanine transaminase; also known as serum glutamate–pyruvate transaminase, or SGPT). In clinical lab tests, blood samples are added to a mixture of the enzymes' substrates. The reaction products, whose concentrations are proportional to the amount of enzyme present, are then detected by secondary reactions that generate colored products easily quantified by spectrophotometry. Prepackaged kits give reliable results in a matter of minutes.

The concentration of AST in the blood increases after a heart attack, when damaged heart muscle leaks its intracellular contents. Typically, AST concentrations rise in the first hours after a heart attack, peak in 24 to 36 hours, and return to normal within a few days. However, since many tissues contain AST, monitoring cardiac muscle damage more commonly relies on measurements of cardiac troponins (proteins specific to heart muscle). ALT is primarily a liver enzyme, so it is useful as a marker of liver damage resulting from infection, trauma, or chronic alcohol abuse. Certain drugs, including the cholesterol-lowering statins (Section 17-3), sometimes increase AST and ALT levels to such an extent that the drugs must be discontinued.

● **Question:** Identify the substrates and products for the AST and ALT reactions.

part of the reaction (steps 5 to 7). This means that most transaminases generate glutamate or aspartate. Lysine is the only amino acid that cannot be transaminated. The presence of transaminases in muscle and liver cells makes them useful markers of tissue damage (Box 18-A).

> **CONCEPT REVIEW**
> • What does the nitrogenase reaction accomplish?
> • What other compounds give rise to ammonia?
> • Describe the reactions catalyzed by glutamine synthetase and glutamate synthase.
> • What is the function of the PLP cofactor?
> • Explain why transaminases catalyze reversible reactions.

18-2 Amino Acid Biosynthesis

KEY CONCEPTS
• Alanine, arginine, asparagine, aspartate, glutamate, glutamine, glycine, proline, and serine are synthesized from intermediates of glycolysis and the citric acid cycle.
• Bacteria and plants synthesize amino acids with sulfur (cysteine and methionine), branched chains (isoleucine, leucine, and valine), and aromatic groups (phenylalanine, tryptophan, and tyrosine) as well as histidine, lysine, and threonine.
• Glutamate and tyrosine are modified to generate neurotransmitters and hormones.

Amino acids are synthesized from intermediates of glycolysis, the citric acid cycle, and the pentose phosphate pathway. Their amino groups are derived from the nitrogen carrier molecules glutamate and glutamine. Using the metabolic scheme introduced in Chapter 12, we can show how amino acid biosynthesis and other reactions of nitrogen metabolism are related to the other pathways we have examined (Fig. 18-6).

Humans can synthesize only some of the 20 amino acids that are commonly found in proteins. These are known as **nonessential** amino acids. The other amino acids are said to be **essential** because humans cannot synthesize them and must obtain them from their food. The ultimate sources of the essential amino acids are plants and microorganisms, which produce all the enzymes necessary to undertake the synthesis of these compounds. The essential and nonessential amino acids for humans are listed in Table 18-1. This classification scheme can be somewhat confusing. For example, some nonessential amino acids, such as arginine, may be essential for young children; that is, dietary sources must supplement what the body can

Figure 18-6 Nitrogen metabolism in context. Amino acids are synthesized mostly from three-carbon intermediates of glycolysis and from intermediates of the citric acid cycle. Amino acid catabolism yields some of the same intermediates, as well as the two-carbon acetyl-CoA. Amino acids are also the precursors of nucleotides. Both types of molecules contain nitrogen, so a discussion of amino acid metabolism includes pathways for obtaining, using, and disposing of amino groups.

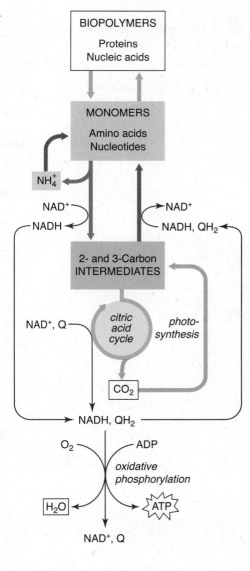

produce on its own. Human cells cannot synthesize histidine, so it is classified as an essential amino acid, even though a dietary requirement has never been defined (probably because sufficient quantities are naturally supplied by intestinal microorganisms). Tyrosine can be considered essential in that it is synthesized directly from the essential amino acid phenylalanine. Likewise, cysteine synthesis depends on the availability of sulfur provided by the essential amino acid methionine.

Several amino acids are easily synthesized from common metabolites

We have already seen that *some amino acids can be produced by transamination reactions*. In this way, alanine is produced from pyruvate, aspartate from oxaloacetate, and glutamate from α-ketoglutarate. We have already seen that glutamine synthetase catalyzes the amidation of glutamate to produce glutamine. Asparagine synthetase, which uses glutamine as an amino-group donor rather than ammonia, converts aspartate to asparagine:

So far, we have seen that three common metabolic intermediates (pyruvate, oxaloacetate, and α-ketoglutarate) give rise to five nonessential amino acids by simple transamination and amidation reactions.

Slightly longer pathways convert glutamate to proline and arginine, which each have the same five-carbon core:

[TABLE 18-1]

Essential and Nonessential Amino Acids

Essential	Nonessential
Histidine	Alanine
Isoleucine	Arginine
Leucine	Asparagine
Lysine	Aspartate
Methionine	Cysteine
Phenylalanine	Glutamate
Threonine	Glutamine
Tryptophan	Glycine
Valine	Proline
	Serine
	Tyrosine

Serine is derived from the glycolytic intermediate 3-phosphoglycerate in three steps:

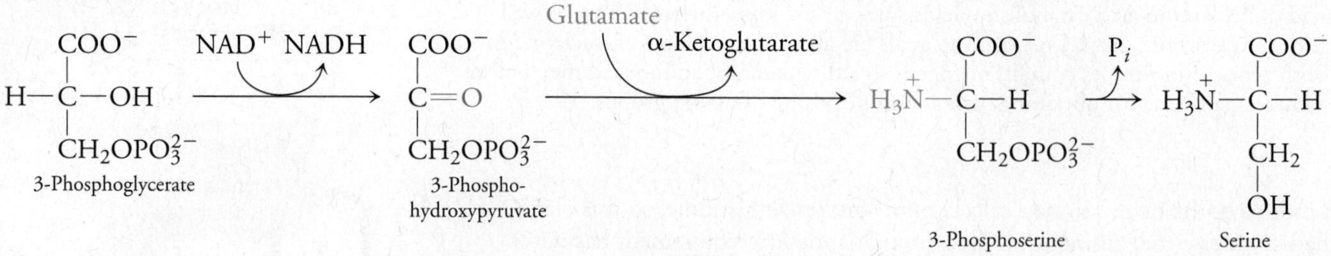

$$\begin{array}{c}
\text{COO}^- \\
| \\
\text{H}-\text{C}-\text{OH} \\
| \\
\text{CH}_2\text{OPO}_3^{2-} \\
\text{3-Phosphoglycerate}
\end{array}
\xrightarrow{\text{NAD}^+ \quad \text{NADH}}
\begin{array}{c}
\text{COO}^- \\
| \\
\text{C}=\text{O} \\
| \\
\text{CH}_2\text{OPO}_3^{2-} \\
\text{3-Phospho-} \\
\text{hydroxypyruvate}
\end{array}
\xrightarrow{\text{Glutamate} \quad \alpha\text{-Ketoglutarate}}
\begin{array}{c}
\text{COO}^- \\
| \\
\text{H}_3\overset{+}{\text{N}}-\text{C}-\text{H} \\
| \\
\text{CH}_2\text{OPO}_3^{2-} \\
\text{3-Phosphoserine}
\end{array}
\xrightarrow{P_i}
\begin{array}{c}
\text{COO}^- \\
| \\
\text{H}_3\overset{+}{\text{N}}-\text{C}-\text{H} \\
| \\
\text{CH}_2 \\
| \\
\text{OH} \\
\text{Serine}
\end{array}$$

Serine, a three-carbon amino acid, gives rise to the two-carbon glycine in a reaction catalyzed by serine hydroxymethyltransferase (the reverse reaction converts glycine to serine). This enzyme uses a PLP-dependent mechanism to remove the hydroxymethyl ($-$CH$_2$OH) group attached to the α carbon of serine; this one-carbon fragment is then transferred to the cofactor tetrahydrofolate:

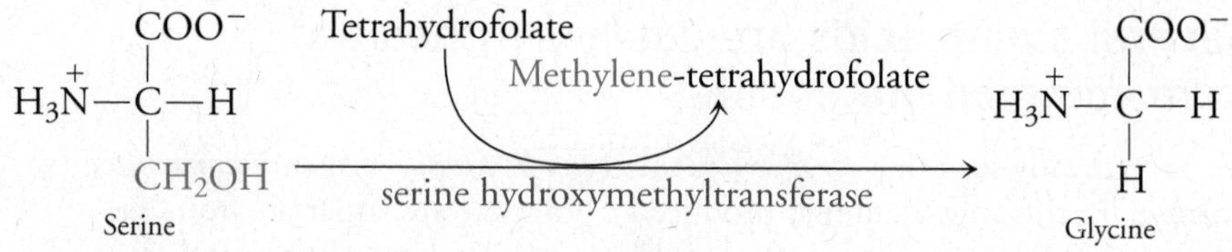

$$\begin{array}{c}
\text{COO}^- \\
| \\
\text{H}_3\overset{+}{\text{N}}-\text{C}-\text{H} \\
| \\
\text{CH}_2\text{OH} \\
\text{Serine}
\end{array}
\xrightarrow[\text{serine hydroxymethyltransferase}]{\text{Tetrahydrofolate} \quad \text{Methylene-tetrahydrofolate}}
\begin{array}{c}
\text{COO}^- \\
| \\
\text{H}_3\overset{+}{\text{N}}-\text{C}-\text{H} \\
| \\
\text{H} \\
\text{Glycine}
\end{array}$$

Tetrahydrofolate functions as a carrier of one-carbon units in several reactions of amino acid and nucleotide metabolism (Fig. 18-7). Mammals cannot synthesize folate (the oxidized form of tetrahydrofolate) and must therefore obtain it as a vitamin from their diet. Folate is abundant in foods such as fortified cereal, fruits, and vegetables. The requirement for folate increases during the first few weeks of pregnancy, when the fetal nervous system begins to develop. Supplemental folate appears to prevent certain neural tube defects such as spina bifida, in which the spinal cord remains exposed.

Amino acids with sulfur, branched chains, or aromatic groups are more difficult to synthesize

We have just described how a few metabolites—pyruvate, 3-phosphoglycerate, oxaloacetate, and α-ketoglutarate—are converted in a few enzyme-catalyzed steps to nine different amino acids. Synthesis of the other amino acids (the essential amino acids and those derived directly from them) also begins with common metabolites. However, these biosynthetic pathways tend to be more complicated. At some point in their evolution, animals lost the ability to synthesize these amino acids, probably because the pathways were energetically expensive and the compounds were already available in food. In general, humans cannot synthesize branched-chain amino acids

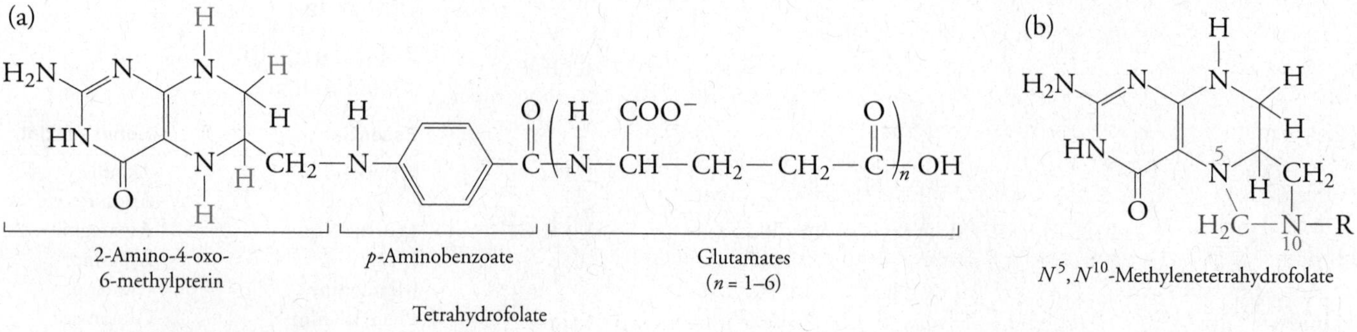

(a)

2-Amino-4-oxo-
6-methylpterin   p-Aminobenzoate   Glutamates
                                                  (n = 1–6)

Tetrahydrofolate

(b)

N^5, N^{10}-Methylenetetrahydrofolate

Figure 18-7 Tetrahydrofolate. (a) This cofactor consists of a pterin derivative, a *p*-aminobenzoate residue, and up to six glutamate residues. It is a reduced form of the vitamin folate. The four H atoms of the tetrahydro form are colored red. (b) In the conversion of serine to glycine, a methylene group (blue) becomes attached to both N5 and N10 of tetrahydrofolate. Tetrahydrofolate can carry carbon units of different oxidation states. For example, a methyl group can attach to N5, and a formyl group ($-$HCO) can attach at N5 or N10.

or aromatic amino acids and cannot incorporate sulfur into compounds such as methionine. In this section, we will focus on a few interesting points related to the synthesis of essential amino acids.

The bacterial pathway for producing sulfur-containing amino acids begins with serine and uses sulfur that comes from inorganic sulfide:

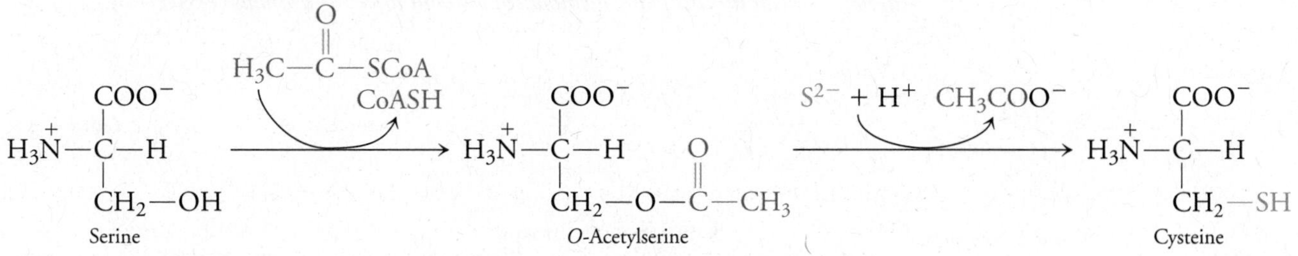

$$
\underset{\text{Serine}}{
\begin{array}{c}
COO^- \\
| \\
H_3\overset{+}{N}-C-H \\
| \\
CH_2-OH
\end{array}}
\xrightarrow{\;\;H_3C-C(=O)-SCoA,\;\;CoASH\;\;}
\underset{\text{O-Acetylserine}}{
\begin{array}{c}
COO^- \\
| \\
H_3\overset{+}{N}-C-H \\
| \;\;\;\;\;\;O \\
CH_2-O-C-CH_3
\end{array}}
\xrightarrow{\;\;S^{2-}+H^+,\;\;CH_3COO^-\;\;}
\underset{\text{Cysteine}}{
\begin{array}{c}
COO^- \\
| \\
H_3\overset{+}{N}-C-H \\
| \\
CH_2-SH
\end{array}}
$$

Cysteine can then donate its sulfur atom to a four-carbon compound derived from aspartate, forming the nonstandard amino acid homocysteine. The final step of methionine synthesis is catalyzed by methionine synthase, which adds to homocysteine a methyl group carried by tetrahydrofolate:

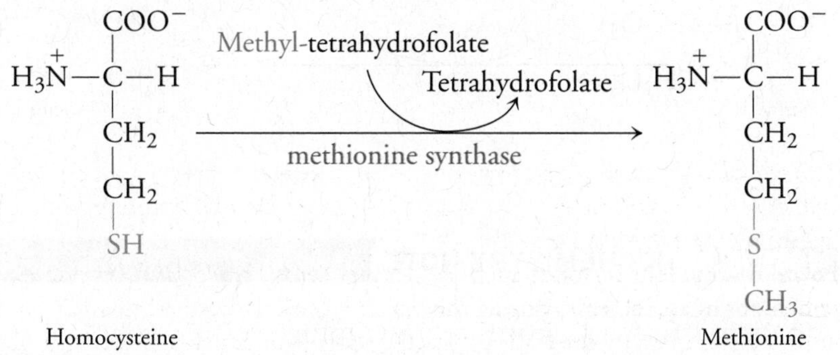

$$
\underset{\text{Homocysteine}}{
\begin{array}{c}
COO^- \\
| \\
H_3\overset{+}{N}-C-H \\
| \\
CH_2 \\
| \\
CH_2 \\
| \\
SH
\end{array}}
\xrightarrow[\text{methionine synthase}]{\text{Methyl-tetrahydrofolate} \;\to\; \text{Tetrahydrofolate}}
\underset{\text{Methionine}}{
\begin{array}{c}
COO^- \\
| \\
H_3\overset{+}{N}-C-H \\
| \\
CH_2 \\
| \\
CH_2 \\
| \\
S \\
| \\
CH_3
\end{array}}
$$

In humans, serine reacts with homocysteine to yield cysteine:

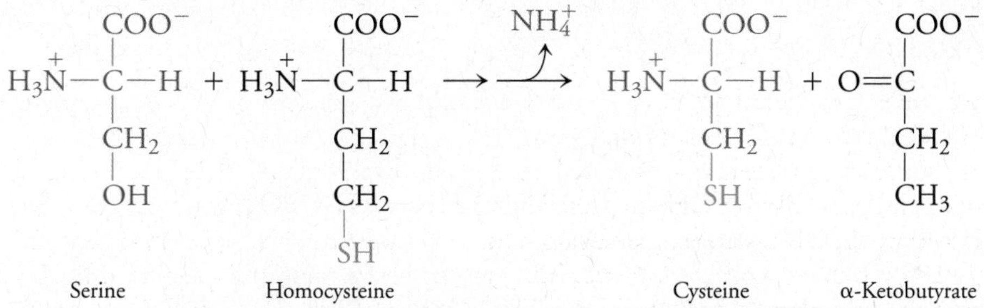

$$
\underset{\text{Serine}}{
\begin{array}{c}
COO^- \\
| \\
H_3\overset{+}{N}-C-H \\
| \\
CH_2 \\
| \\
OH
\end{array}}
+
\underset{\text{Homocysteine}}{
\begin{array}{c}
COO^- \\
| \\
H_3\overset{+}{N}-C-H \\
| \\
CH_2 \\
| \\
CH_2 \\
| \\
SH
\end{array}}
\xrightarrow{\;\;NH_4^+\;\;}
\underset{\text{Cysteine}}{
\begin{array}{c}
COO^- \\
| \\
H_3\overset{+}{N}-C-H \\
| \\
CH_2 \\
| \\
SH
\end{array}}
+
\underset{\text{α-Ketobutyrate}}{
\begin{array}{c}
COO^- \\
| \\
O=C \\
| \\
CH_2 \\
| \\
CH_3
\end{array}}
$$

This pathway is the reason why cysteine is considered a nonessential amino acid, although its sulfur atom must come from another amino acid.

High levels of homocysteine in the blood are associated with cardiovascular disease. The link was first discovered in individuals with homocystinuria, a disorder in which excess homocysteine is excreted in the urine. These individuals develop atherosclerosis as children, probably because the homocysteine directly damages the walls of blood vessels even in the absence of elevated LDL levels (see Chapter 17). Increasing the intake of folate, the vitamin precursor of tetrahydrofolate, helps decrease the level of homocysteine by promoting its conversion to methionine.

Aspartate, the precursor of methionine, is also the precursor of the essential amino acids threonine and lysine. Since these amino acids are derived from another amino acid, they already have an amino group. The branched-chain amino acids (valine, leucine, and isoleucine) are synthesized by pathways that use pyruvate as the starting substrate. These amino acids require a step catalyzed by a transaminase (with glutamate as a substrate) to introduce an amino group.

In plants and bacteria, the pathway for synthesizing the aromatic amino acids (phenylalanine, tyrosine, and tryptophan) begins with the condensation of the C_3 compound phosphoenolpyruvate (a glycolytic intermediate) and erythrose-4-phosphate (a four-carbon intermediate of the pentose phosphate pathway). The seven-carbon reaction product then cyclizes and undergoes additional modifications, including the addition of three more carbons from phosphoenolpyruvate, before becoming chorismate, the last common intermediate in the synthesis of the three aromatic amino acids. *Because animals do not synthesize chorismate, this pathway is an obvious target for agents that can inhibit plant metabolism without affecting animals* (Box 18-B).

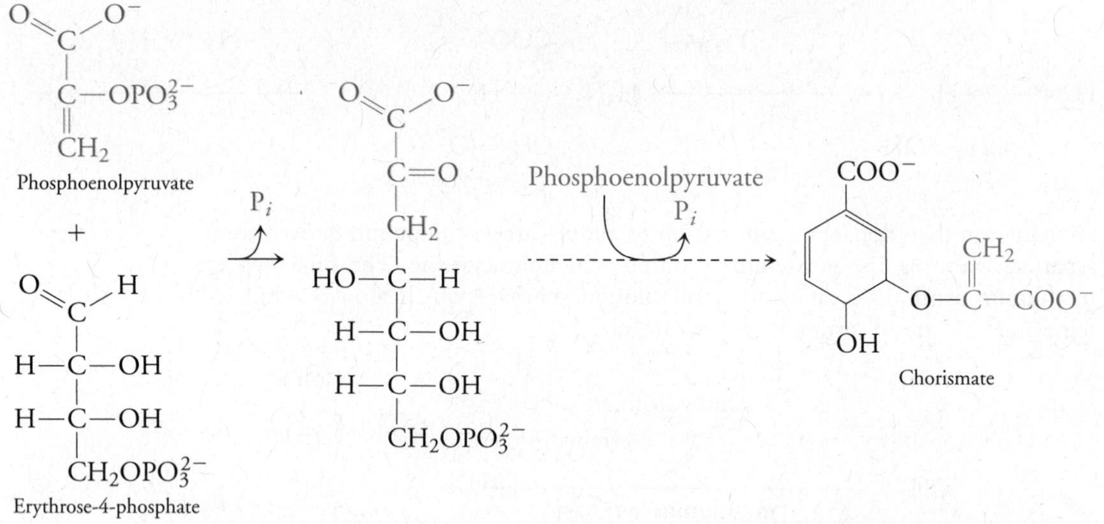

Phosphoenolpyruvate

+

Erythrose-4-phosphate

P_i

Phosphoenolpyruvate
P_i

Chorismate

BOX 18-B BIOCHEMISTRY NOTE

Glyphosate, the most popular herbicide

Glycine phosphonate, also known as glyphosate or Roundup (its trade name), competes with the second phosphoenolpyruvate in the pathway leading to chorismate:

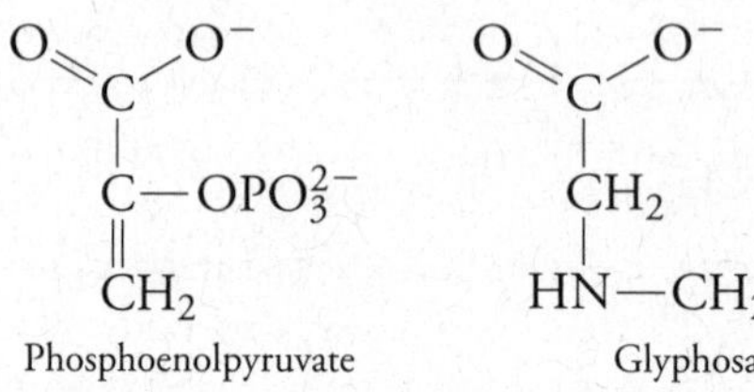

Phosphoenolpyruvate

Glyphosate

Because plants cannot manufacture aromatic amino acids without chorismate, glyphosate acts as an herbicide. Used widely in agriculture as well as home gardens, it has become the most popular herbicide in the United States, replacing other, more toxic compounds. Glyphosate that is not directly absorbed by the plant appears to bind tightly to soil particles and then is rapidly broken down by bacteria. Consequently, glyphosate has less potential to contaminate water supplies than do more stable compounds. In order to be an effective herbicide, glyphosate must enter plant tissues, so it is often packaged along with a surfactant (an amphiphilic compound) that helps it penetrate the waxy coatings on leaves.

Farmers can take advantage of glyphosate's weed-killing properties by planting glyphosate-resistant crops and then spraying the field with glyphosate when weeds emerge and begin to compete with the crop plants. Such "Roundup-Ready" species include soybeans, corn (maize), and cotton. These plants have been genetically engineered to express a bacterial version of the enzyme that uses phosphoenolpyruvate but is not inhibited by glyphosate. Predictably, the use of glyphosate creates selective pressure for herbicide resistance, so many types of weeds have already evolved resistance to glyphosate.

● **Questions:** In addition to plants, what other types of organisms synthesize chorismate as a precursor of aromatic amino acids? How would glyphosate affect them?

Phenylalanine and tyrosine are derived from chorismate by diverging pathways. In humans, tyrosine is generated by hydroxylating phenylalanine, which is why tyrosine is not considered an essential amino acid.

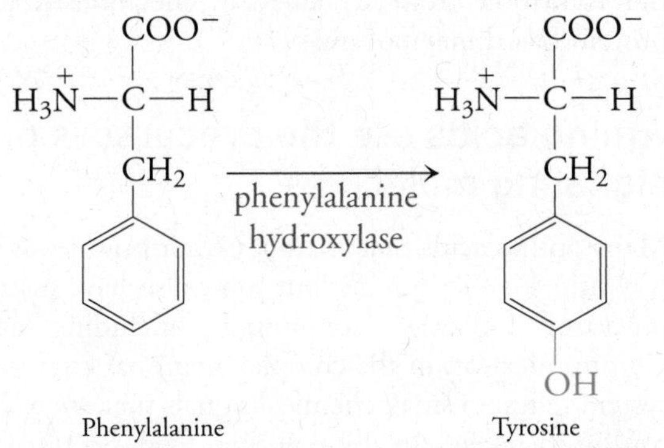

Phenylalanine → Tyrosine (phenylalanine hydroxylase)

The final two reactions of the tryptophan biosynthetic pathway (which has 13 steps altogether) are catalyzed by tryptophan synthase, a bifunctional enzyme with an $\alpha_2\beta_2$ quaternary structure. The α subunit cleaves indole-3-glycerol phosphate to indole and glyceraldehyde-3-phosphate, then the β subunit adds serine to indole to produce tryptophan:

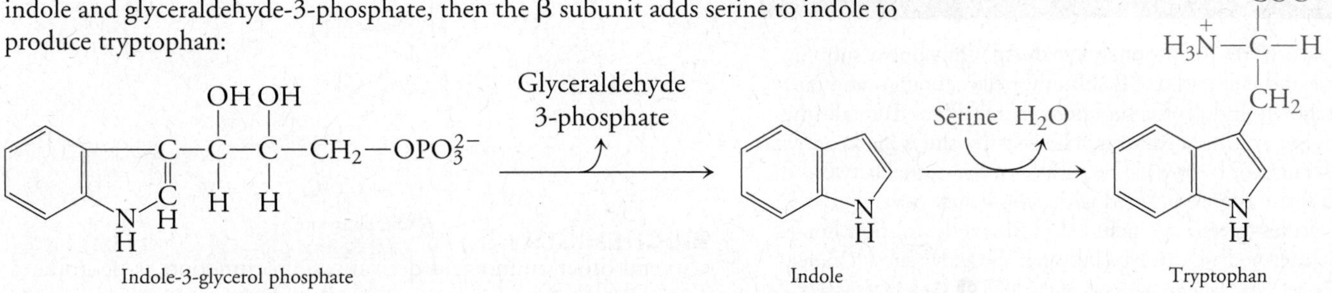

Indole-3-glycerol phosphate → Indole → Tryptophan

Indole, the product of the α-subunit reaction and the substrate for the β-subunit reaction, never leaves the enzyme. Instead, it diffuses directly from one active site to the other without entering the surrounding solvent. The X-ray structure of the enzyme reveals that the active sites in adjacent α and β subunits are 25 Å apart but are connected by a tunnel through the protein that is large enough to accommodate indole (**Fig. 18-8**). *The movement of a reactant between two active sites is called **channeling**, and it increases the rate of a metabolic process by preventing the loss of intermediates.* Channeling is known to occur in a few other multifunctional enzymes.

All but one of the 20 standard amino acids are synthesized entirely from precursors produced by the main carbohydrate-metabolizing pathways. The exception is histidine, to which ATP provides one nitrogen and one carbon atom. Glutamate and glutamine donate the other two nitrogen atoms, and the remaining five carbons are derived from a phosphorylated monosaccharide, 5-phosphoribosyl pyrophosphate (PRPP):

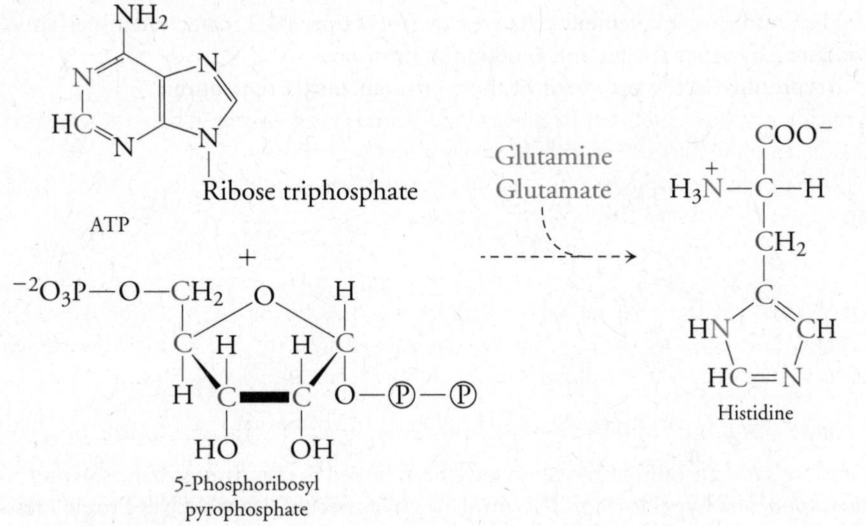

ATP + 5-Phosphoribosyl pyrophosphate → Histidine (Glutamine, Glutamate)

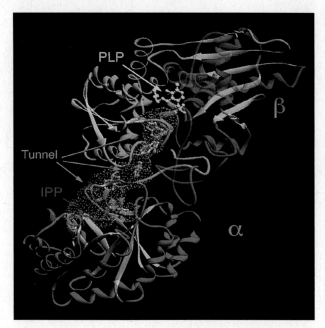

Figure 18-8 Tryptophan synthase. Only one α subunit (blue and tan) and one β subunit (yellow, orange, and tan) are shown. Indolepropanol phosphate (IPP; red) marks the active site of the α subunit. The β active site is marked by its PLP cofactor (yellow). The surface of the tunnel between the two active sites is outlined with yellow dots. Several indole molecules (green) are included in the model to show how this intermediate can pass between the active sites. [Courtesy of Craig Hyde, National Institutes of Health.] ⊕ **See Interactive Exercise. The bifunctional enzyme tryptophan synthase.**

5-Phosphoribosyl-pyrophosphate is also the source of the ribose group of nucleotides. This suggests that histidine might have been one of the first amino acids synthesized by an early life-form making the transition from an all-RNA metabolism to an RNA-and-protein-based metabolism.

Amino acids are the precursors of some signaling molecules

Many amino acids that are ingested or built from scratch find their way into a cell's proteins, but some also have essential functions as precursors of other compounds, including **neurotransmitters.** Communication in the complex neuronal circuitry of the nervous system relies on small chemical signals that are released by one neuron and taken up by another (see Section 9-4). Common neurotransmitters include the amino acids glycine and glutamate and a glutamate derivative (its carboxylate group has been removed) known as γ-aminobutyric acid (GABA) or γ-aminobutyrate.

$$\overset{+}{H_3}N-CH_2$$
$$|$$
$$CH_2$$
$$|$$
$$CH_2$$
$$|$$
$$COO^-$$

γ-Aminobutyrate

Several other amino acid derivatives also function as neurotransmitters. For example, tyrosine gives rise to dopamine, norepinephrine, and epinephrine. These compounds are called catecholamines, reflecting their resemblance to catechol.

Tyrosine ⟹ Dopamine   Norepinephrine   Epinephrine   Catechol

A deficiency of dopamine produces the symptoms of Parkinson's disease: tremor, rigidity, and slow movements. As we saw in Section 10-2, catecholamines are also produced by other tissues and function as hormones.

Tryptophan is the precursor of the neurotransmitter serotonin:

Tryptophan ⟹ Serotonin

Low levels of serotonin in the brain have been linked to conditions such as depression, aggression, and hyperactivity. The antidepressive effect of drugs such as Prozac® results from their ability to increase serotonin levels by blocking the reabsorption of the released neurotransmitter (see Box 9-C). Serotonin is the precursor of melatonin (*right*). This tryptophan derivative is synthesized in the pineal gland and retina. Its concentration is low during the day, rising during darkness. Because melatonin appears to govern the synthesis of some other neurotransmitters that control circadian (daily) rhythms, it has been touted as a cure for sleep disorders and jet lag.

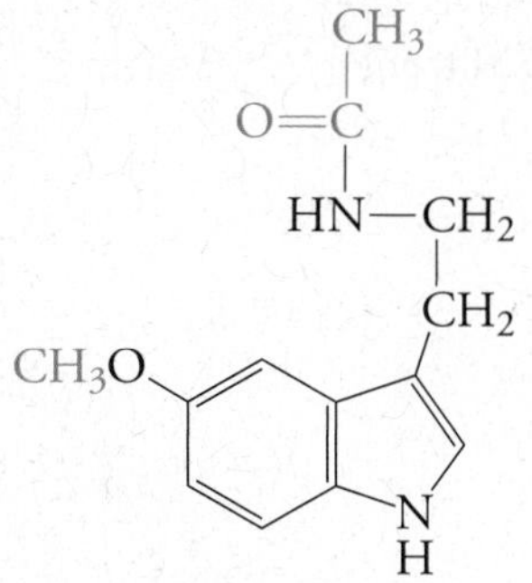

Arginine is also the precursor of a signaling molecule that was discovered only a few years ago to be the free radical gas nitric oxide (NO; Box 18-C).

Melatonin

BOX 18-C ⬡ BIOCHEMISTRY NOTE

Nitric Oxide

In the 1980s, vascular biologists were investigating the nature of an endothelial cell–derived "relaxation factor" that caused blood vessels to dilate. This substance diffused quickly, acted locally, and disappeared within seconds. To the surprise of many, the mysterious factor turned out to be the free radical nitric oxide ($\cdot$NO). Although NO was known to elicit vasodilation, it had not been considered a good candidate for a biological signaling molecule because its unpaired electron makes it extremely reactive and it breaks down to yield the corrosive nitric acid.

NO is a signaling molecule in a wide array of tissues. At low concentrations it induces blood vessel dilation; at high concentrations (along with oxygen radicals) it kills pathogens. NO is synthesized from arginine by nitric oxide synthase, an enzyme whose cofactors include FMN, FAD, tetrahydrobiopterin (discussed in Section 18-4), and a heme group. The first step of NO production is a hydroxylation reaction. In the second step, one electron oxidizes *N*-hydroxyarginine.

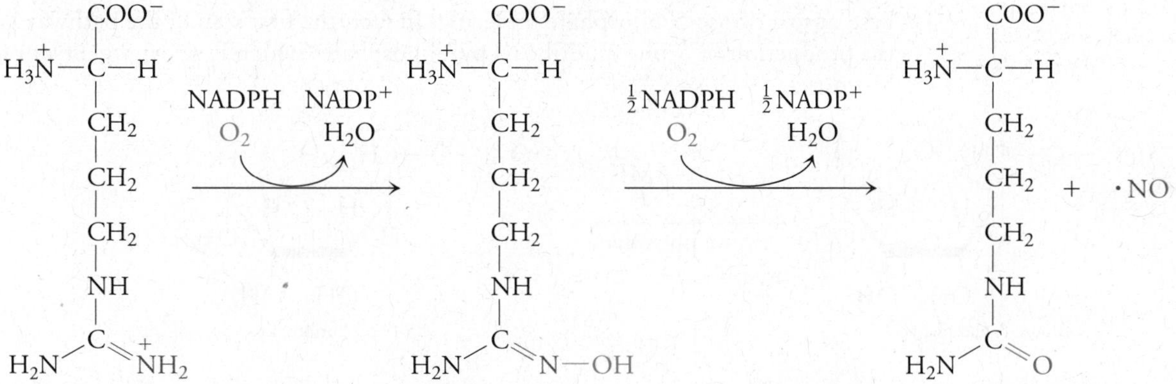

NO is unusual among signaling molecules for several reasons: It cannot be stockpiled for later release; it diffuses into cells, so it does not need a cell-surface receptor; and it needs no degradative enzyme because it breaks down on its own. NO is produced only when and where it is needed. A free radical gas such as NO cannot be directly introduced into the body, but an indirect source of NO has been clinically used for over a century. Individuals who suffer from angina pectoris, a painful condition caused by obstruction of the coronary blood vessels, can relieve their symptoms by taking nitroglycerin:

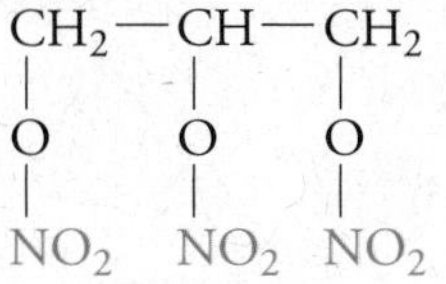

Nitroglycerin

In vivo, nitroglycerin yields NO, which rapidly stimulates vasodilation, temporarily relieving the symptoms of angina.

⬡ **Question:** Explain why blood vessels express constant amounts of nitric oxide synthase whereas white blood cells must be induced to produce the enzyme.

CONCEPT REVIEW
- List the metabolites used as precursors for the nonessential amino acids.
- Which amino acids are synthesized by simple transamination reactions?
- Describe the role of tetrahydrofolate in amino acid biosynthesis.
- Why do herbicides target the pathway for synthesizing aromatic amino acids?
- How does His differ from other amino acids in its synthesis?
- Which amino acids are neurotransmitters?
- List some amino acid derivatives that act as signaling molecules.

18-3 Nucleotide Biosynthesis

KEY CONCEPTS
- AMP and GMP are derived from the purine nucleotide IMP.
- Pyrimidine nucleotide synthesis produces UTP and then CTP.
- Ribonucleotide reductase converts NDPs to dNDPs using a free radical mechanism.
- dUMP is methylated to produce dTMP.
- Nucleotides are degraded for excretion and to supply materials for salvage or other pathways.

Nucleotides are synthesized from precursors that include several amino acids. The human body can also recycle nucleotides from nucleic acids and nucleotide cofactors that are broken down. Although food supplies nucleotides, the biosynthetic and recycling pathways are so efficient that there is no true dietary requirement for purines and pyrimidines. In this section we will take a brief look at the biosynthetic pathways for purine and pyrimidine nucleotides in mammals.

Purine nucleotide synthesis yields IMP and then AMP and GMP

Purine nucleotides (AMP and GMP) are synthesized by building the purine base onto a ribose-5-phosphate molecule. In fact, the first step of the pathway is the production of 5-phosphoribosyl pyrophosphate (which is also a precursor of histidine):

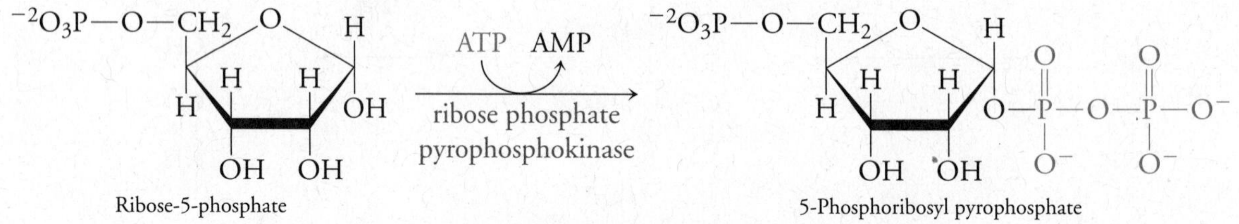

Ribose-5-phosphate   5-Phosphoribosyl pyrophosphate

The subsequent ten steps of the pathway require as substrates glutamine, glycine, aspartate, bicarbonate, plus one-carbon formyl (—HC=O) groups donated by tetrahydrofolate. The product is inosine monophosphate (IMP), a nucleotide whose base is the purine hypoxanthine:

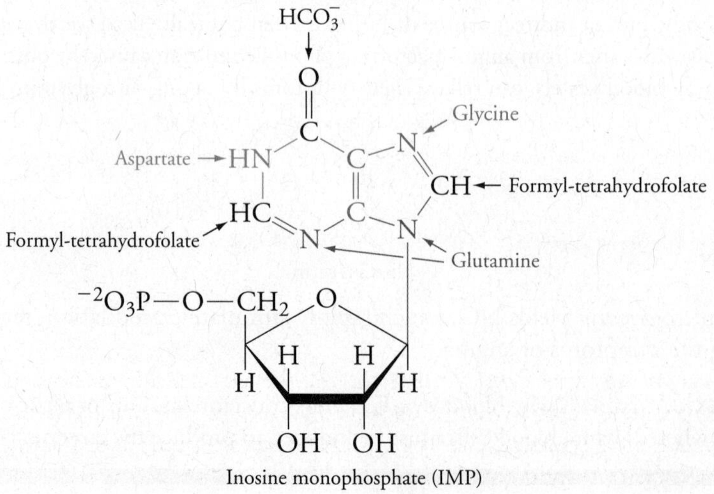

Inosine monophosphate (IMP)

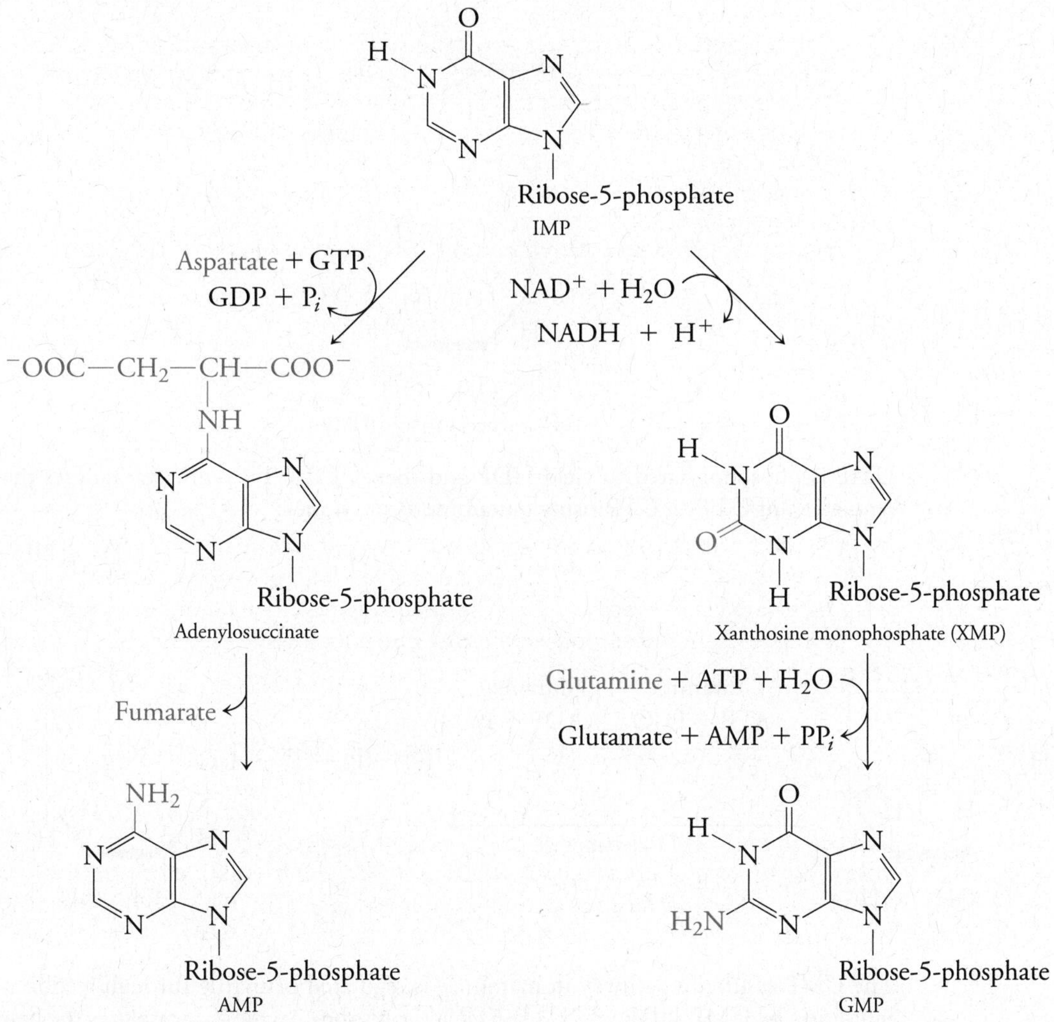

Figure 18-9 AMP and GMP synthesis from IMP.

IMP is the substrate for two short pathways that yield AMP and GMP. In AMP synthesis, an amino group from aspartate is transferred to the purine; in GMP synthesis, glutamate is the source of the amino group (Fig. 18-9). Kinases then catalyze phosphoryl-group transfer reactions to convert the nucleoside monophosphates to diphosphates and then triphosphates (ATP and GTP).

Figure 18-9 indicates that GTP participates in AMP synthesis and ATP participates in GMP synthesis. High concentrations of ATP therefore promote GMP production, and high concentrations of GTP promote AMP production. *This reciprocal relationship is one mechanism for balancing the production of adenine and guanine nucleotides.* (Because most nucleotides are destined for DNA or RNA synthesis, they are required in roughly equal amounts.) The pathway leading to AMP and GMP is also regulated by feedback inhibition at several points, including the first step, the production of 5-phosphoribosyl pyrophosphate from ribose-5-phosphate, which is inhibited by both ADP and GDP.

Pyrimidine nucleotide synthesis yields UTP and CTP

In contrast to purine nucleotides, pyrimidine nucleotides are synthesized as a base that is subsequently attached to 5-phosphoribosyl pyrophosphate to form a nucleotide. The six-step pathway that yields uridine monophosphate (UMP) requires glutamine, aspartate, and bicarbonate.

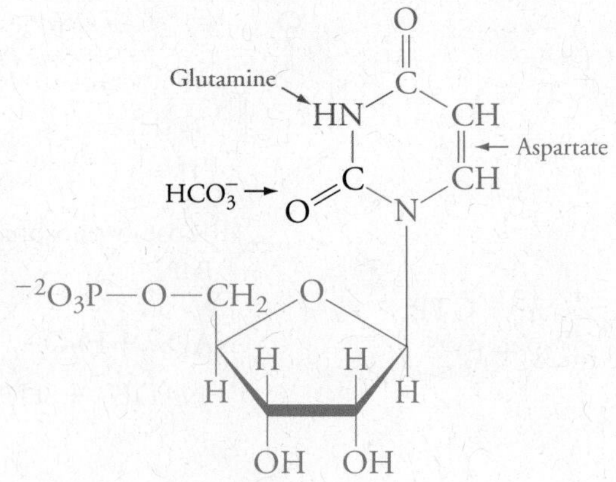

Uridine monophosphate (UMP)

UMP is phosphorylated to yield UDP and then UTP. CTP synthase catalyzes the amination of UTP to CTP, using glutamine as the donor:

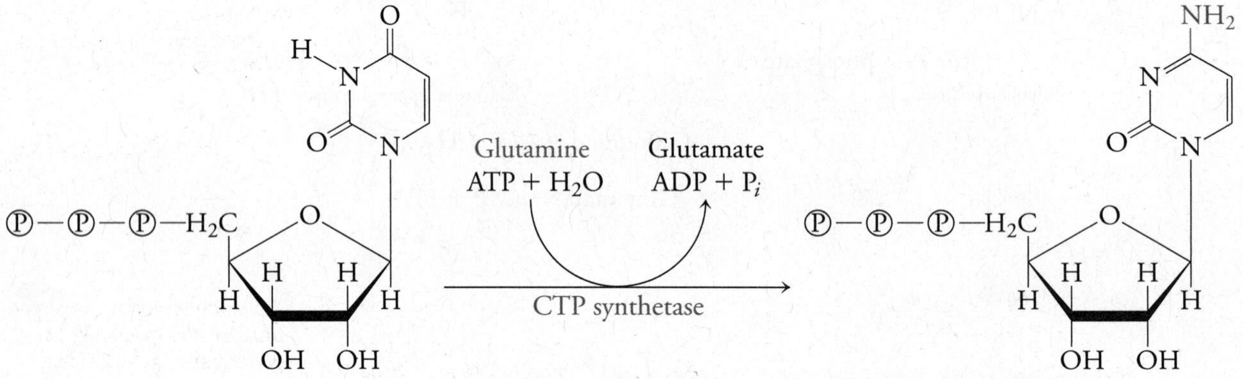

The UMP synthetic pathway in mammals is regulated primarily through feedback inhibition by UMP, UDP, and UTP. ATP activates the enzyme that catalyzes the first step; this helps balance the production of purine and pyrimidine nucleotides.

Ribonucleotide reductase converts ribonucleotides to deoxyribonucleotides

So far, we have accounted for the synthesis of ATP, GTP, CTP, and UTP, which are substrates for the synthesis of RNA. DNA, of course, is built from deoxynucleotides. In deoxynucleotide synthesis, each of the four nucleoside triphosphates (NTPs) is converted to its diphosphate (NDP) form, ribonucleotide reductase replaces the 2′ OH group with H, then the resulting deoxynucleoside diphosphate (dNDP) is phosphorylated to produce the corresponding triphosphate (dNTP).

Ribonucleotide reductase is an essential enzyme that carries out a chemically difficult reaction using a mechanism that involves free radicals. Three types of ribonucleotide reductases, which differ in their catalytic groups, have been described. Class I enzymes (the type that occurs in mammals and most bacteria) have two Fe^{3+} ions and an unusually stable tyrosine radical (most free radicals, which have one unpaired electron, are highly reactive and short-lived).

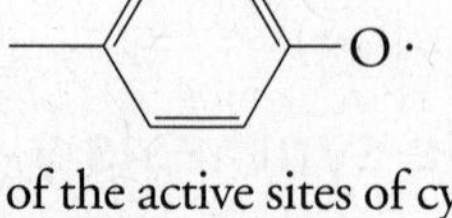

Tyrosine radicals are also features of the active sites of cytochrome c oxidase (mitochondrial Complex IV) and Photosystem II in plants. Class II ribonucleotide reductases use adenosylcobalamin (the cofactor used in the isomerization of methylmalonyl-CoA; see Section 17-1), and class III enzymes use a glycyl radical. The job of all these groups is to interact with a Cys side chain to generate a thiyl radical that attacks the ribonucleotide substrate. A possible reaction mechanism is shown in **Figure 18-10**.

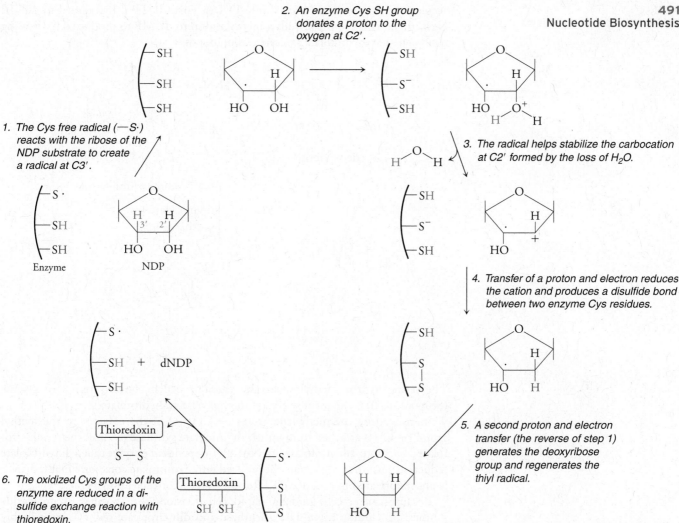

2. An enzyme Cys SH group donates a proton to the oxygen at C2'.

1. The Cys free radical (—S·) reacts with the ribose of the NDP substrate to create a radical at C3'.

Enzyme   NDP

3. The radical helps stabilize the carbocation at C2' formed by the loss of H₂O.

4. Transfer of a proton and electron reduces the cation and produces a disulfide bond between two enzyme Cys residues.

—SH + dNDP

Thioredoxin

6. The oxidized Cys groups of the enzyme are reduced in a disulfide exchange reaction with thioredoxin.

Thioredoxin

5. A second proton and electron transfer (the reverse of step 1) generates the deoxyribose group and regenerates the thiyl radical.

Figure 18-10 Proposed mechanism for ribonucleotide reductase. Only part of the nucleotide's ribose ring is shown. ⊕ **See Interactive Exercise. _E. coli_ ribonucleotide reductase.**

The final step of the reaction, which regenerates the enzyme, requires the small protein thioredoxin. The oxidized thioredoxin must then undergo reduction to return to its original state. _This reaction uses NADPH, which is therefore the ultimate source of reducing power for the synthesis of deoxyribonucleotides._ Recall that the pentose phosphate pathway, which provides the ribose-5-phosphate for nucleotide synthesis, also generates NADPH (Section 13-4).

Not surprisingly, the activity of ribonucleotide reductase is tightly regulated so that the cell can balance the levels of ribo- and deoxyribonucleotides as well as the proportions of each of the four deoxyribonucleotides. Control of the enzyme involves two regulatory sites that are distinct from the substrate-binding site. For example, ATP binding to the so-called activity site activates the enzyme. Binding of the deoxyribonucleotide dATP decreases enzyme activity. Several nucleotides bind to the so-called substrate specificity site. Here, ATP binding induces the enzyme to act on pyrimidine nucleotides, and dTTP binding causes the enzyme to prefer GDP as a substrate. These mechanisms, in concert with other mechanisms for balancing the amounts of the various nucleotides, help make all four deoxynucleotides available for DNA synthesis.

Thymidine nucleotides are produced by methylation

The ribonucleotide reductase reaction, followed by kinase-catalyzed phosphorylation, generates dATP, dCTP, dGTP, and dUTP. However, dUTP is not used for DNA synthesis. Instead, _it is rapidly converted to thymine nucleotides_ (which helps prevent the accidental incorporation of uracil into DNA). First, dUTP is hydrolyzed to dUMP. Next, thymidylate synthase adds a methyl group to dUMP to produce dTMP, using methylene-tetrahydrofolate as a one-carbon donor.

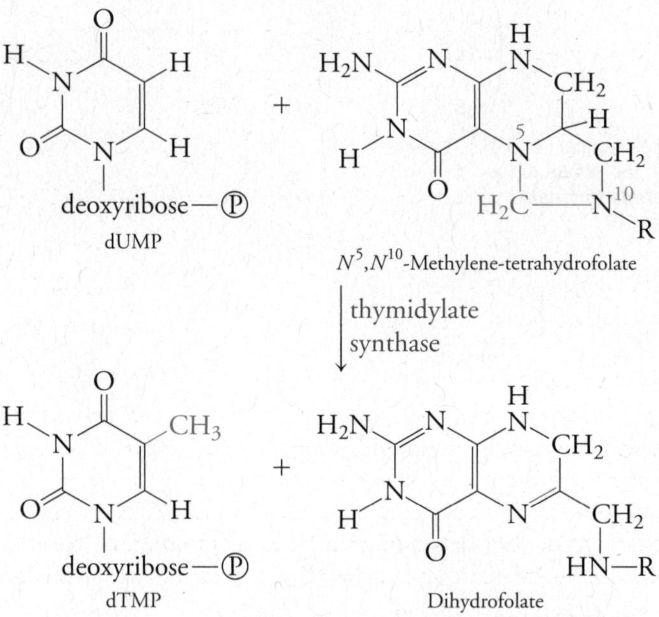

The serine hydroxymethyltransferase reaction, which converts serine to glycine (Section 18-2), is the main source of methylene-tetrahydrofolate.

In converting the methylene group ($-CH_2-$) of the cofactor to the methyl group ($-CH_3$) attached to thymine, thymidylate synthase oxidizes the tetrahydrofolate cofactor to dihydrofolate. An NADPH-dependent enzyme called dihydrofolate reductase must then regenerate the reduced tetrahydrofolate cofactor. Finally, dTMP is phosphorylated to produce dTTP, the substrate for DNA polymerase.

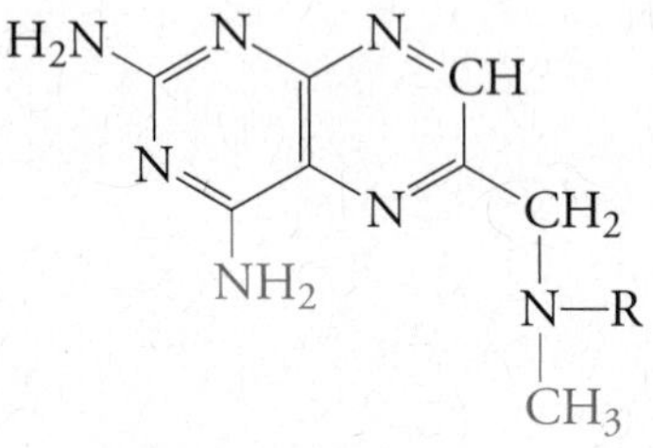

Because cancer cells undergo rapid cell division, the enzymes of nucleotide synthesis, including thymidylate synthase and dihydrofolate reductase, are highly active. Compounds that inhibit either of these reactions can therefore act as anti-cancer agents. For example, the dUMP analog 5-fluorodeoxyuridylate, introduced in Section 7-3, inactivates thymidylate synthase. "Antifolates" such as methotrexate (*left*) are competitive inhibitors of dihydrofolate reductase because they compete with dihydrofolate for binding to the enzyme. In the presence of methotrexate, a cancer cell cannot regenerate the tetrahydrofolate required for dTMP production, and the cell dies. Most noncancerous cells, which grow much more slowly, are not as sensitive to the effect of the drug.

Methotrexate

Nucleotide degradation produces uric acid or amino acids

Nucleotides that are obtained from food or synthesized by cells can be broken down, releasing ribose groups and a purine or pyrimidine that can be further catabolized and excreted (purines) or used as a metabolic fuel (pyrimidines). At several points in the degradation pathways, intermediates may be redirected toward the synthesis of new nucleotides by so-called **salvage pathways.** For example, a free adenine base can be reattached to ribose by the reaction

$$\text{adenine} + \text{5-phosphoribosyl pyrophosphate} \rightleftharpoons \text{AMP} + \text{PP}_i$$

Degradation of a nucleoside monophosphate begins with dephosphorylation to produce a nucleoside. In a subsequent step, a phosphorylase breaks the glycosidic bond between the base and the ribose by adding phosphate (a similar phosphorolysis reaction occurs during glycogenolysis; Section 13-3).

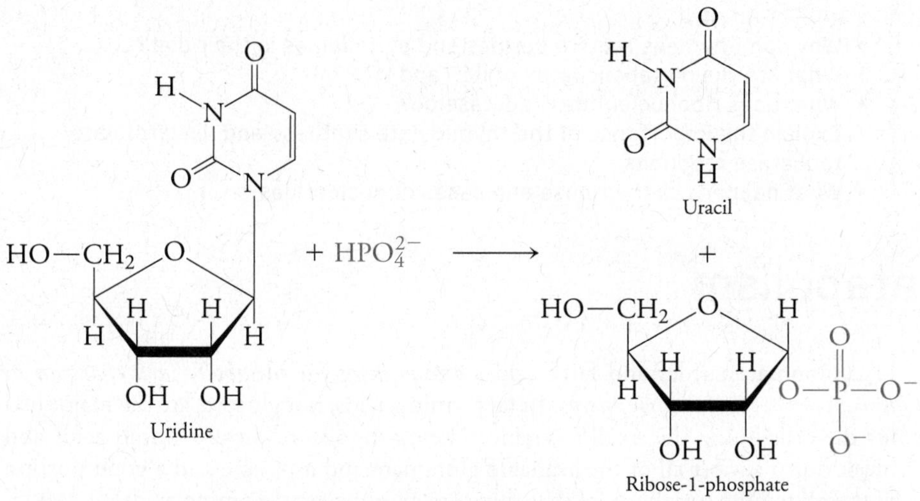

Uracil

Uridine $+ HPO_4^{2-} \longrightarrow$ + Ribose-1-phosphate

The phosphorylated ribose can be catabolized or salvaged and converted to 5-phosphoribosyl pyrophosphate for synthesis of another nucleotide. The fate of the base depends on whether it is a purine or a pyrimidine.

The purine bases are eventually converted to uric acid in a process that may require deamination and oxidation, depending on whether the original base was adenine, guanine, or hypoxanthine. Uric acid has a pK of 5.4, so it exists mainly as urate.

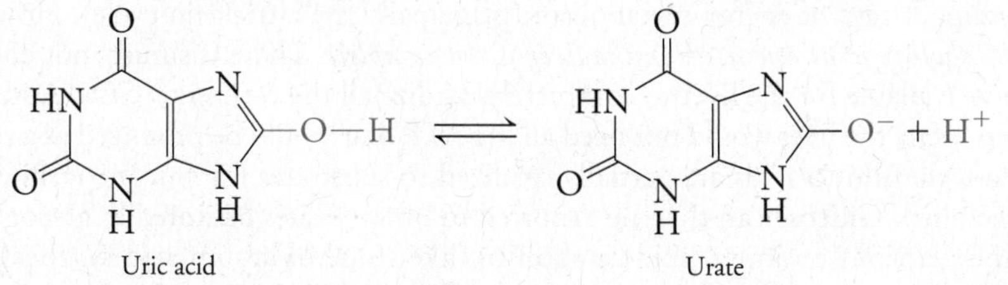

Uric acid $\rightleftharpoons$ Urate $+ H^+$

In humans, urate, a poorly soluble compound, is excreted in the urine. Excess urate may precipitate as crystals of sodium urate in the kidneys (kidney "stones"). Deposits of urate in the joints, primarily the knees and toes, cause a painful condition called gout. Other organisms may further catabolize urate to generate more soluble waste products such as urea and ammonia.

The pyrimidines cytosine, thymine, and uracil undergo deamination and reduction, after which the pyrimidine ring is opened. Further catabolism produces the nonstandard amino acid β-alanine (from cytosine and uracil) or β-aminoisobutyrate (from thymine), both of which feed into other metabolic pathways.

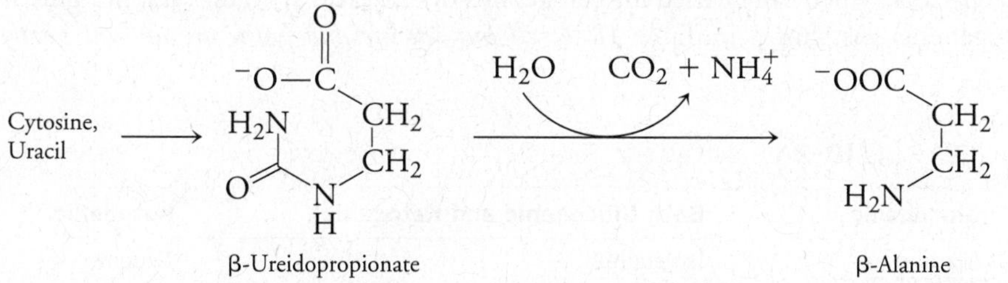

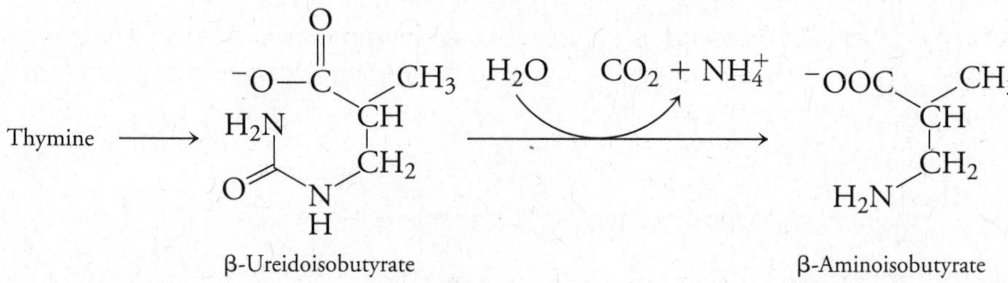

Cytosine, Uracil $\longrightarrow$ β-Ureidopropionate $\xrightarrow{\quad H_2O \quad CO_2 + NH_4^+ \quad}$ β-Alanine

Thymine $\longrightarrow$ β-Ureidoisobutyrate $\xrightarrow{\quad H_2O \quad CO_2 + NH_4^+ \quad}$ β-Aminoisobutyrate

Consequently, pyrimidine catabolism contributes to the pool of cellular metabolites for both anabolic and catabolic processes. In contrast, purine catabolism generates a waste product that is excreted from the body.

CONCEPT REVIEW
- Why don't humans require purines and pyrimidines in their diet?
- What are the metabolic fates of IMP and UTP?
- What does ribonucleotide reductase do?
- Explain the importance of the thymidylate synthase and dihydrofolate reductase reactions.
- What happens to the ribose and bases of nucleotides?

18-4 Amino Acid Catabolism

KEY CONCEPT
- Degradation of the carbon skeletons of amino acids produces acetyl-CoA and precursors for gluconeogenesis.

Like monosaccharides and fatty acids, *amino acids are metabolic fuels that can be broken down to release free energy.* In fact, amino acids, not glucose, are the major fuel for the cells lining the small intestine. These cells absorb dietary amino acids and break down almost all of the available glutamate and aspartate and a good portion of the glutamine supply (note that these are all nonessential amino acids).

Other tissues, mainly the liver, also catabolize amino acids originating from the diet and from the normal turnover of intracellular proteins. During periods when dietary amino acids are not available, such as during a prolonged fast, amino acids are mobilized through the breakdown of muscle tissue, which accounts for about 40% of the total protein in the body. The amino acids undergo transamination reactions to remove their α-amino groups, and their carbon skeletons then enter the central pathways of energy metabolism (principally the citric acid cycle). However, *the catabolism of amino acids in the liver is not complete.* There is simply not enough oxygen available for the liver to completely oxidize all the carbon to CO_2. And even if there were, the liver would not need all the ATP that would be produced as a result. Instead, the amino acids are partially oxidized to substrates for gluconeogenesis (or ketogenesis). Glucose can then be exported to other tissues or stored as glycogen.

The reactions of amino acid catabolism, like those of amino acid synthesis, are too numerous to describe in full here, and the catabolic pathways do not necessarily mirror the anabolic pathways, as they do in carbohydrate and fatty acid metabolism. In this section, we will focus on some general principles and a few interesting chemical aspects of amino acid catabolism. In the following section we will see how organisms dispose of the nitrogen component of catabolized amino acids.

Amino acids are glucogenic, ketogenic, or both

It is useful to classify amino acids in humans as **glucogenic** (giving rise to gluconeogenic precursors such as citric acid cycle intermediates) or **ketogenic** (giving rise to acetyl-CoA, which can be used for ketogenesis or fatty acid synthesis, but not gluconeogenesis). As shown in Table 18-2, *all but leucine and lysine are at least partly*

[**TABLE 18-2**] Catabolic Fates of Amino Acids

Glucogenic	Both Glucogenic and Ketogenic	Ketogenic
Alanine	Isoleucine	Leucine
Arginine	Phenylalanine	Lysine
Asparagine	Threonine	
Aspartate	Tryptophan	
Cysteine	Tyrosine	
Glutamate		
Glutamine		
Glycine		
Histidine		
Methionine		
Proline		
Serine		
Valine		

glucogenic, most of the nonessential amino acids are glucogenic, and the large skeletons of the aromatic amino acids are both glucogenic and ketogenic.

Three amino acids are converted to gluconeogenic substrates by simple transamination (the reverse of their biosynthetic reactions): alanine to pyruvate, aspartate to oxaloacetate, and glutamate to α-ketoglutarate. Glutamate can also be deaminated in an oxidation reaction that we will examine in the following section. Asparagine undergoes a simple hydrolytic deamidation to aspartate, which is then transaminated to oxaloacetate:

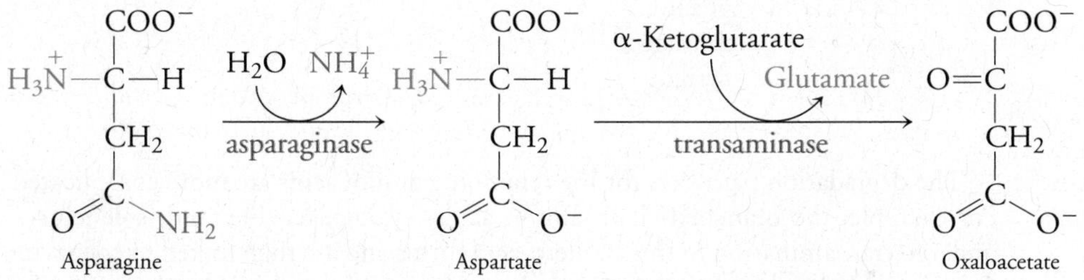

Similarly, glutamine is deamidated by a glutaminase to glutamate, and the glutamate dehydrogenase reaction yields α-ketoglutarate. Serine is converted to pyruvate:

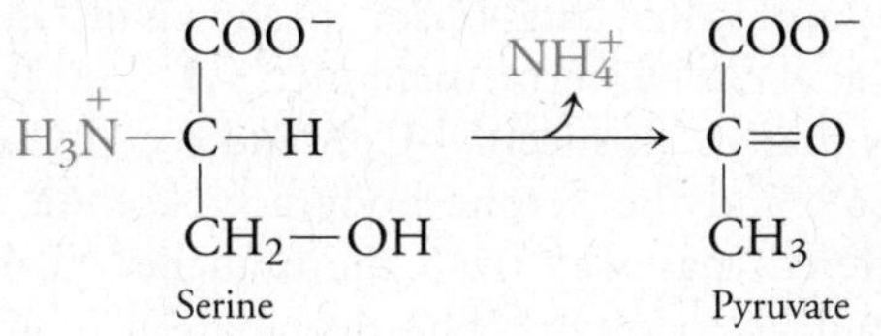

Note that in this reaction and in the conversion of asparagine and glutamine to their acid counterparts, the amino group is released as NH_4^+ rather than being transferred to another compound.

Arginine and proline (which are synthesized from glutamate) as well as histidine are catabolized to glutamate, which is then converted to α-ketoglutarate. Amino acids of the glutamate "family," namely arginine, glutamine, histidine, and proline, constitute about 25% of dietary amino acids, so their potential contribution to energy metabolism is significant.

Cysteine is converted to pyruvate by a process that releases ammonia as well as sulfur:

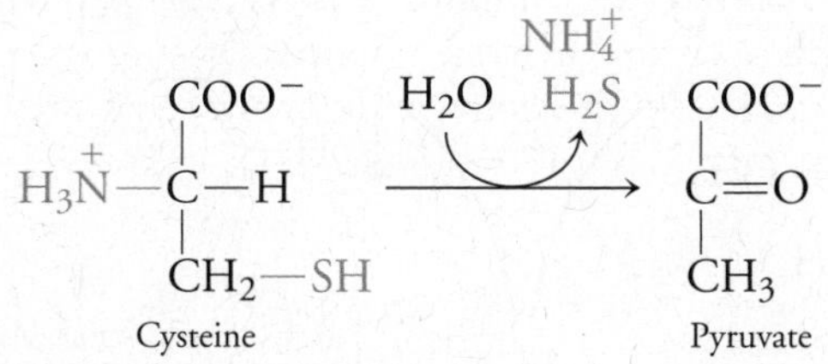

The products of the reactions listed so far—pyruvate, oxaloacetate, and α-ketoglutarate—are all gluconeogenic precursors. Threonine is both glucogenic and ketogenic because it is broken down to acetyl-CoA and glycine:

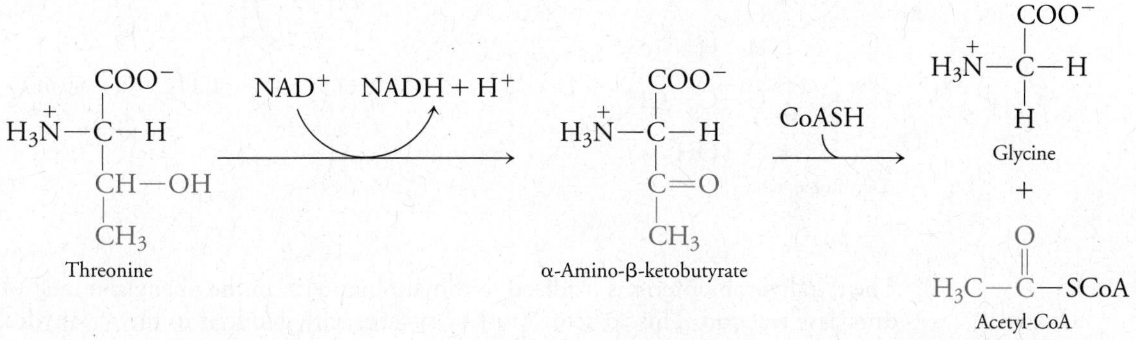

The acetyl-CoA is a precursor of ketone bodies (see Section 17-2), and the glycine is potentially glucogenic—if it is first converted to serine by the action of serine hydroxymethyltransferase. The major route for glycine disposal, however, is catalyzed by a multiprotein complex known as the glycine cleavage system:

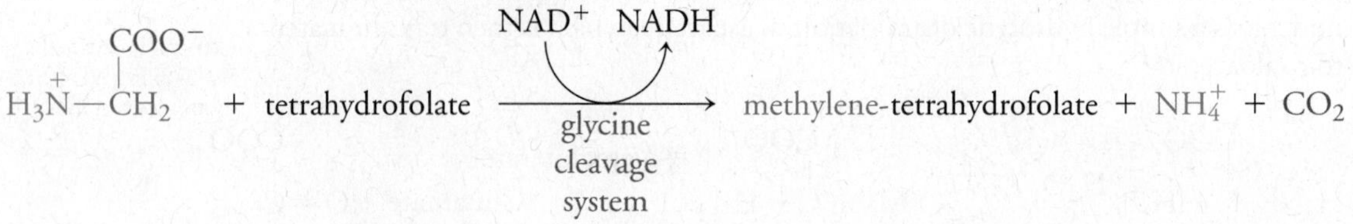

The degradation pathways for the remaining amino acids are more complicated. For example, the branched-chain amino acids—valine, leucine, and isoleucine—undergo transamination to their α-keto-acid forms and are then linked to coenzyme A in an oxidative decarboxylation reaction. This step is catalyzed by the branched-chain α-keto-acid dehydrogenase complex, a multienzyme complex that resembles the pyruvate dehydrogenase complex (see Section 14-1) and even shares some of the same subunits.

The initial reactions of valine catabolism are shown in **Fig. 18-11**. Subsequent steps yield the citric acid cycle intermediate succinyl-CoA. Isoleucine is degraded by a similar pathway that yields succinyl-CoA and acetyl-CoA. Leucine degradation yields acetyl-CoA and the ketone body acetoacetate. Lysine degradation, which follows a different pathway from the branched-chain amino acids, also yields acetyl-CoA and acetoacetate. The degradation of methionine produces succinyl-CoA.

Finally, the cleavage of the aromatic amino acids—phenylalanine, tyrosine, and tryptophan—yields the ketone body acetoacetate as well as a glucogenic compound (alanine or fumarate). The first step of phenylalanine degradation is a hydroxylation reaction that produces tyrosine, as we have already seen. This reaction is worth noting because it uses the cofactor tetrahydrobiopterin (which, like folate, contains a pterin moiety).

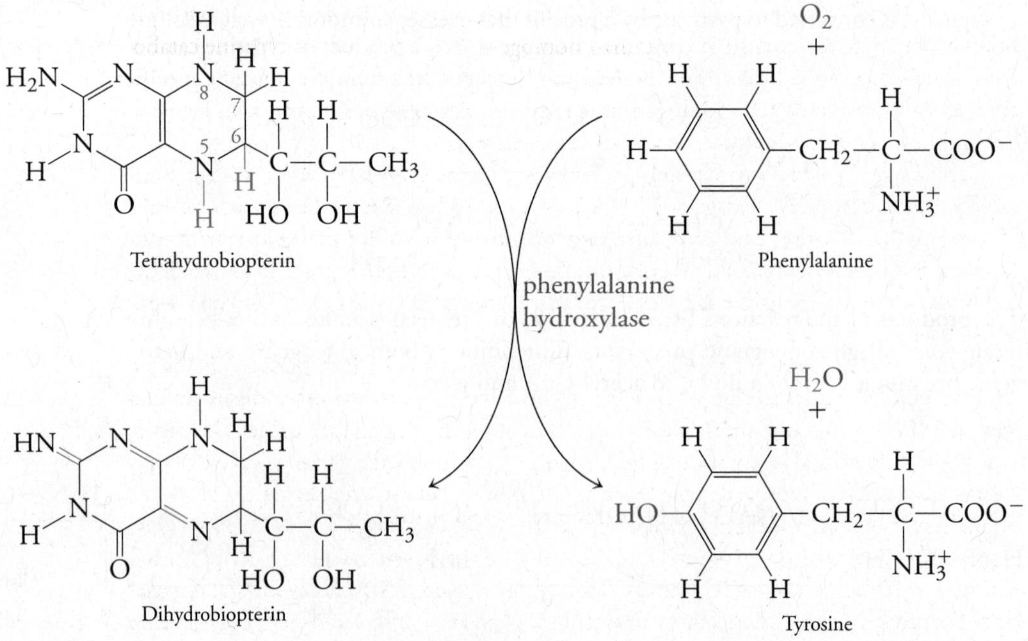

The tetrahydrobiopterin is oxidized to dihydrobiopterin in the phenylalanine hydroxylase reaction. This cofactor must be subsequently reduced to the tetrahydro

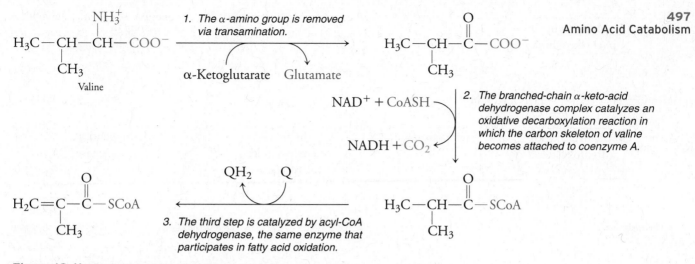

1. The α-amino group is removed via transamination.

2. The branched-chain α-keto-acid dehydrogenase complex catalyzes an oxidative decarboxylation reaction in which the carbon skeleton of valine becomes attached to coenzyme A.

3. The third step is catalyzed by acyl-CoA dehydrogenase, the same enzyme that participates in fatty acid oxidation.

Figure 18-11 **The initial steps of valine degradation.**

? List all the cofactors involved in this process. (*Hint:* Step 2 is carried out by a multienzyme complex.)

form by a separate NADH-dependent enzyme. Another step of the phenylalanine (and tyrosine) degradation pathway is also notable because a deficiency of the enzyme was one of the first-characterized "inborn errors of metabolism" (Box 18-D).

BOX 18-D BIOCHEMISTRY NOTE

Inborn Errors of Metabolism

Today we understand that inherited diseases result from defective genes. We are also beginning to understand that malfunctioning genes underlie many noninherited diseases as well. The link between genes and disease was first recognized by the physician Archibald Garrod, who coined the term *inborn error of metabolism* in 1902. Garrod's insights came from his studies of individuals with alcaptonuria. Their urine turned black upon exposure to air because it contained homogentisate, a product of tyrosine catabolism. Garrod concluded that this inherited condition resulted from the lack of a specific enzyme. We now know that homogentisate is excreted because the enzyme that breaks it down, homogentisate dioxygenase, is missing or defective.

Garrod's findings built on Mendel's discovery of the laws of inheritance but went largely unappreciated for about half a century. In the 1950s, George Beadle and Edward Tatum popularized the "one gene, one enzyme" theory based on their work with mutants of the mold *Neurospora*. Around the same time, Vernon Ingram discovered the molecular defect responsible for sickle cell hemoglobin. At this point, Garrod's work finally seemed relevant.

Garrod also described a number of other inborn errors of metabolism, including albinism, cystinuria (excretion of cysteine in the urine), and several other disorders that were not life-threatening and left easily detected clues in the patient's urine. Of course, many "inborn errors" are catastrophic. For example, phenylketonuria (PKU) results from a deficiency of phenylalanine hydroxylase, the first enzyme in the pathway shown on the next page. Phenylalanine cannot be broken down, although it can undergo transamination. The resulting α-keto-acid derivative phenylpyruvate accumulates and is excreted in the urine, giving it a mousy odor. If not treated, PKU causes mental retardation. Fortunately, the disease can be detected in newborns. Afflicted individuals develop normally if they consume a diet that is low in phenylalanine. Sadly, the biochemical defects behind many other diseases are not as well understood, making them difficult to identify and treat.

(*continued on next page*)

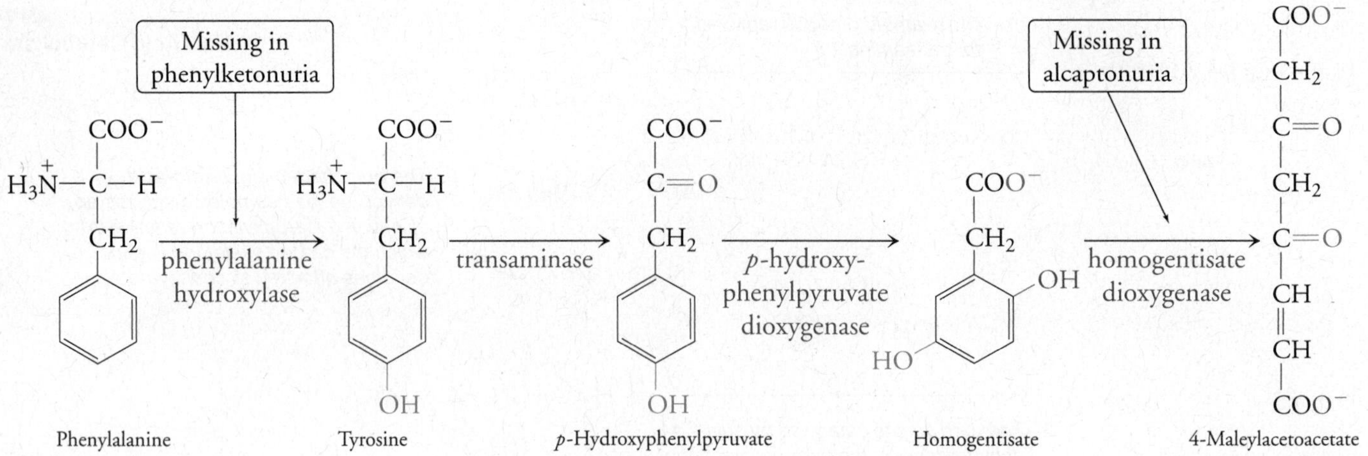

Missing in
phenylketonuria

Missing in
alcaptonuria

$$COO^-$$
$$H_3N^+-C-H$$
$$CH_2$$

phenylalanine
hydroxylase

$$COO^-$$
$$H_3N^+-C-H$$
$$CH_2$$

OH

transaminase

$$COO^-$$
$$C=O$$
$$CH_2$$

OH

p-hydroxy-
phenylpyruvate
dioxygenase

$$COO^-$$
$$CH_2$$

OH
HO

homogentisate
dioxygenase

$$COO^-$$
$$CH_2$$
$$C=O$$
$$CH_2$$
$$C=O$$
$$CH$$
$$CH$$
$$COO^-$$

Phenylalanine   Tyrosine   p-Hydroxyphenylpyruvate   Homogentisate   4-Maleylacetoacetate

⬢ **Question:** Individuals with PKU must consume a certain amount of phenylalanine. Describe the three metabolic fates of this amino acid.

CONCEPT REVIEW
• What is the role of the liver in catabolizing amino acids?
• Distinguish the glucogenic and ketogenic amino acids.
• How does coenzyme A participate in amino acid catabolism?

18-5 Nitrogen Disposal: The Urea Cycle

KEY CONCEPTS
• Ammonia released by the glutamate dehydrogenase reaction is incorporated into carbamoyl phosphate.
• The four reactions of the urea cycle incorporate two amino groups into urea, a highly water-soluble waste product.

▶▶ **WHY** don't humans excrete ammonia?

When the supply of amino acids exceeds the cell's immediate needs for protein synthesis or other amino acid–consuming pathways, the carbon skeletons are broken down and the nitrogen disposed of. All amino acids except lysine can be deaminated by the action of transaminases, but this merely transfers the amino group to another molecule; it does not eliminate it from the body.

Some catabolic reactions do release free ammonia, which can be excreted as a waste product in the urine. In fact, the kidney is a major site of glutamine catabolism, and the resulting NH_4^+ facilitates the excretion of metabolic acids such as H_2SO_4 that arise from the catabolism of methionine and cysteine. However, ammonia production is not feasible for disposing of large amounts of excess nitrogen. First, high concentrations of NH_4^+ in the blood cause alkalosis. Second, ammonia is highly toxic. It easily enters the brain, where it activates the NMDA receptor, whose normal agonist is the neurotransmitter glutamate. The activated receptor is an ion channel that normally opens to allow Ca^{2+} and Na^+ ions to enter the cell and K^+ ions to exit the cell. However, the large Ca^{2+} influx triggered by ammonia binding to the receptor results in neuronal cell death, a phenomenon called excitotoxicity. Humans and many other organisms have therefore evolved safer ways to deal with excess amino groups.

Approximately 80% of the body's excess nitrogen is excreted in the form of urea,

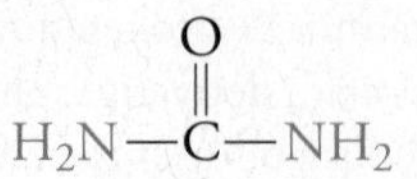

$$O$$
$$H_2N-C-NH_2$$
Urea

*which is produced in the liver by the reactions of the **urea cycle**.* This catabolic cycle was elucidated in 1932 by Hans Krebs and Kurt Henseleit; Krebs went on to outline another circular pathway—the citric acid cycle—in 1937.

Glutamate supplies nitrogen to the urea cycle

Because many transaminases use α-ketoglutarate as the amino-group acceptor, glutamate is one of the most abundant amino acids inside cells. Glutamate can be deaminated to regenerate α-ketoglutarate and release NH_4^+ in an oxidation–reduction reaction catalyzed by glutamate dehydrogenase:

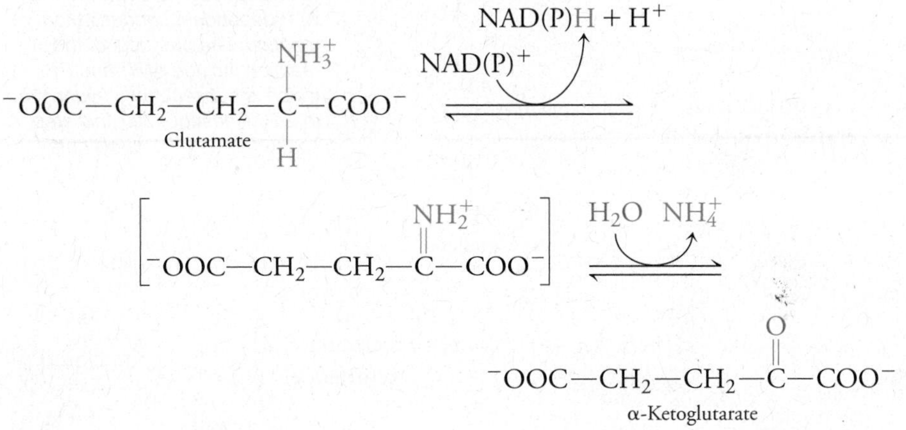

This mitochondrial enzyme is unusual: It is the only known enzyme that can use either NAD^+ or $NADP^+$ as a cofactor. The glutamate dehydrogenase reaction is a major route for feeding amino acid–derived amino groups into the urea cycle and, not surprisingly, is subject to allosteric activation and inhibition.

The starting substrate for the urea cycle is an "activated" molecule produced by the condensation of bicarbonate and ammonia, as catalyzed by carbamoyl phosphate synthetase (**Fig. 18-12**). The NH_4^+ may be contributed by the glutamate dehydrogenase reaction or another process that releases ammonia. The bicarbonate is the source of the urea carbon. Note that the phosphoanhydride bonds of two ATP molecules are consumed in the energetically costly production of carbamoyl phosphate.

The urea cycle consists of four reactions

The four enzyme-catalyzed reactions of the urea cycle proper are shown in **Figure 18-13**. The cycle also provides a means for synthesizing arginine: The five-carbon ornithine is derived from glutamate, and the urea cycle converts it to arginine. However, the arginine needs of children exceed the biosynthetic capacity of the urea cycle, so arginine is classified as an essential amino acid.

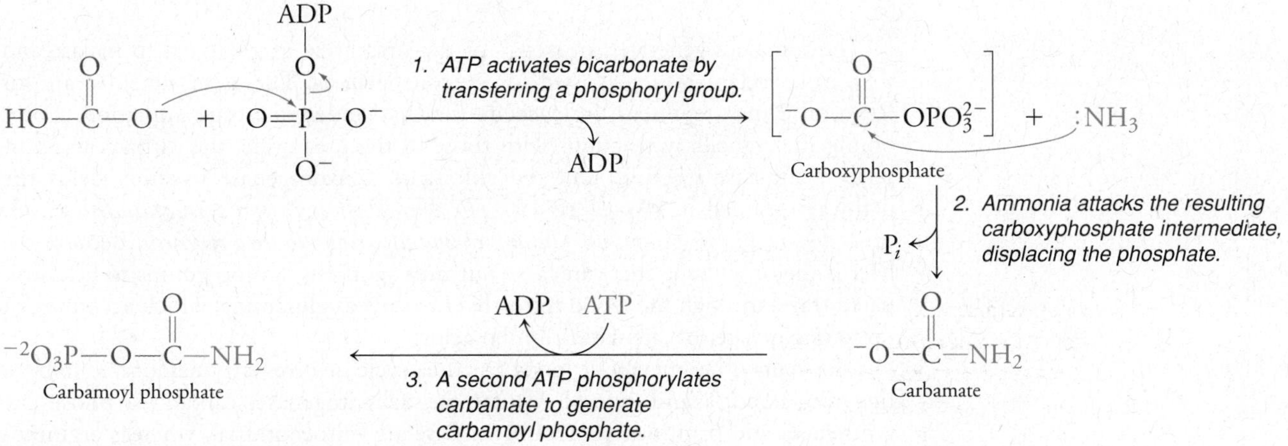

Figure 18-12 The carbamoyl phosphate synthetase reaction.

? **Bicarbonate (as CO_2) and ammonia can easily diffuse into mitochondria. Why doesn't the reaction product diffuse out?**

500

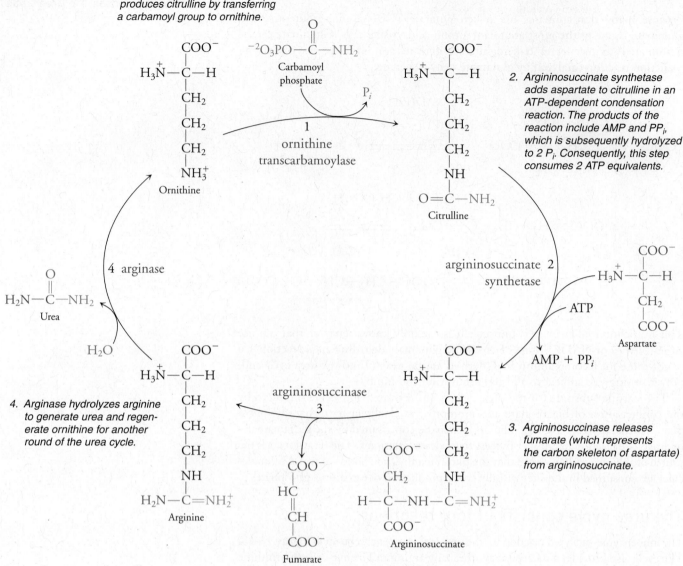

Figure 18-13 The four reactions of the urea cycle.

❓ **Which steps are likely to be irreversible *in vivo*?**

The fumarate generated in step 3 of the urea cycle is converted to malate and then oxaloacetate, which is used for gluconeogenesis. The aspartate substrate for Reaction 2 may represent oxaloacetate that has undergone transamination. Combining these ancillary reactions with those of the urea cycle, the carbamoyl phosphate synthetase reaction, and the glutamate dehydrogenase reaction yields the pathway outlined in **Figure 18-14**. *The overall effect is that transaminated amino acids donate amino groups, via glutamate and aspartate, to urea synthesis.* Because the liver is the only tissue that can carry out urea synthesis, amino groups to be eliminated travel through the blood to the liver mainly as glutamine, which accounts for up to one-quarter of circulating amino acids.

Like many other metabolic loops, the urea cycle involves enzymes located in both the mitochondria and cytosol. Glutamate dehydrogenase, carbamoyl phosphate synthetase, and ornithine transcarbamoylase are mitochondrial, whereas argininosuccinate synthetase, argininosuccinase, and arginase are cytosolic. Consequently, citrulline is produced in the mitochondria but must be transported to the cytosol for the next step, and ornithine produced in the cytosol must be imported into the mitochondria to begin a new round of the cycle.

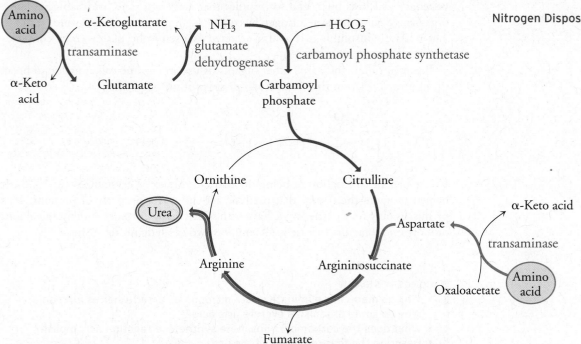

Figure 18-14 **The urea cycle and related reactions.** Two routes for the disposal of amino groups are highlighted. The blue pathway shows how an amino group from an amino acid enters the urea cycle via glutamate and carbamoyl phosphate. The red pathway shows how an amino group from an amino acid enters via aspartate.

? **Identify all the products of this metabolic scheme. Which ones are recycled?**

The carbamoyl phosphate synthetase reaction and the argininosuccinate synthetase reactions each consume 2 ATP equivalents, so the cost of the urea cycle is 4 ATP per urea. However, when considered in context, operation of the urea cycle is often accompanied by ATP synthesis. The glutamate dehydrogenase reaction produces NADH (or NADPH), whose free energy is conserved in the synthesis of 2.5 ATP by oxidative phosphorylation. Catabolism of the carbon skeletons of the amino acids that donated their amino groups via transamination also yields ATP.

The rate of urea production is controlled largely by the activity of carbamoyl phosphate synthetase. This enzyme is allosterically activated by *N*-acetylglutamate, which is synthesized from glutamate and acetyl-CoA:

$$\underset{\text{Glutamate}}{\begin{array}{c} \text{COO}^- \\ | \\ \overset{+}{\text{H}_3\text{N}}-\text{C}-\text{H} \\ | \\ \text{CH}_2 \\ | \\ \text{CH}_2 \\ | \\ \text{COO}^- \end{array}} + \underset{\text{Acetyl-CoA}}{\begin{array}{c} \text{O} \\ || \\ \text{H}_3\text{C}-\text{C}-\text{SCoA} \end{array}} \xrightarrow{\text{HSCoA}} \underset{N\text{-Acetylglutamate}}{\begin{array}{c} \text{O} \qquad \text{COO}^- \\ || \qquad\quad | \\ \text{H}_3\text{C}-\text{C}-\text{NH}-\text{C}-\text{H} \\ | \\ \text{CH}_2 \\ | \\ \text{CH}_2 \\ | \\ \text{COO}^- \end{array}}$$

When amino acids are undergoing transamination and being catabolized, the resulting increases in the cellular glutamate and acetyl-CoA concentrations boost production of *N*-acetylglutamate. This stimulates carbamoyl phosphate synthetase activity, and flux through the urea cycle increases. Such a regulatory system allows the cell to efficiently dispose of the nitrogen released from amino acid degradation.

Urea is relatively nontoxic and easily transported through the bloodstream to the kidneys for excretion in the urine. However, the polar urea molecule requires large amounts of water for its efficient excretion. This presents a problem for flying vertebrates such as birds and for reptiles that are adapted to arid habitats. These organisms deal with waste nitrogen by converting it to uric acid via purine synthesis. The relatively insoluble uric acid is excreted as a semisolid paste, which conserves water.

Bacteria, fungi, and some other organisms use an enzyme called urease to break down urea, a reaction that completes our story of nitrogen disposal.

$$H_2N-\overset{\overset{O}{\|}}{C}-NH_2 \ + \ H_2O \ \xrightarrow{\text{urease}} \ 2\,NH_3 \ + \ CO_2$$

Urease has the distinction of being the first enzyme to be crystallized (in 1926). It helped promote the theory that catalytic activity was a property of proteins. This premise is only partly true, as we have seen, since many enzymes contain metal ions or inorganic cofactors (urease itself contains two catalytic nickel atoms).

> **CONCEPT REVIEW**
> - Why do mammals eliminate waste nitrogen as urea rather as ammonia? Why do some organisms excrete uric acid?
> - What does the carbamoyl phosphate synthetase reaction accomplish?
> - Describe the four reactions of the urea cycle.
> - How are amino groups from amino acids incorporated into urea?

[SUMMARY]

18-1 Nitrogen Fixation and Assimilation

- Nitrogen-fixing organisms convert N_2 to NH_3 in the ATP-consuming nitrogenase reaction. Nitrate and nitrite can also be reduced to NH_3.
- Ammonia is incorporated into glutamine by the action of glutamine synthetase.
- Transaminases use a PLP prosthetic group to catalyze the reversible interconversion of α-amino acids and α-keto acids.

18-2 Amino Acid Biosynthesis

- In general, the nonessential amino acids are synthesized from common metabolic intermediates such as pyruvate, oxaloacetate, and α-ketoglutarate.
- The essential amino acids, which include the sulfur-containing, branched-chain, and aromatic amino acids, are synthesized by more elaborate pathways in bacteria and plants.
- Amino acids are the precursors of some neurotransmitters and hormones.

18-3 Nucleotide Biosynthesis

- The synthesis of nucleotides requires glutamate, glycine, and aspartate as well as ribose-5-phosphate. The pathways for purine and pyrimidine biosynthesis are regulated to balance the production of the various nucleotides.
- A ribonucleotide reductase uses a free radical mechanism to convert nucleotides to deoxynucleotides.
- Thymidine production requires a methyl group donated by the cofactor tetrahydrofolate.
- In humans, purines are degraded to uric acid for excretion, and pyrimidines are converted to β-amino acids.

18-4 Amino Acid Catabolism

- Following removal of their amino groups by transamination, amino acids are broken down to intermediates that can be converted to glucose or acetyl-CoA for use in the citric acid cycle, fatty acid synthesis, or ketogenesis.

18-5 Nitrogen Disposal: The Urea Cycle

- In mammals, excess amino groups are converted to urea for disposal. The urea cycle is regulated at the carbamoyl phosphate synthetase step, an entry point for ammonia. Other organisms convert excess nitrogen to compounds such as uric acid.

[GLOSSARY TERMS]

nitrogen fixation	transamination	salvage pathway
diazotroph	nonessential amino acid	glucogenic amino acid
nitrification	essential amino acid	ketogenic amino acid
denitrification	channeling	urea cycle
nitrogen cycle	neurotransmitter	

[PROBLEMS]

18-1 Nitrogen Fixation and Assimilation

1. The $\mathcal{E}°'$ for the half-reaction

$$N_2 + 6\,H^+ + 6\,e^- \rightleftharpoons 2\,NH_3$$

is -0.34 V. The reduction potential of the nitrogenase component that donates electrons for nitrogen reduction is about -0.29 V. ATP hydrolysis apparently induces a conformational change in the protein that alters its reduction potential (a change with a magnitude of about 0.11 V). Does this change increase or decrease the $\mathcal{E}°'$ of the electron donor and why is this change necessary?

2. Why is it a good agricultural practice to plant a field with alfalfa every few years?

3. Plants whose root nodules contain nitrogen-fixing bacterial symbionts synthesize a heme-containing protein, called leghemoglobin, which resembles myoglobin in its structure. What is the function of this protein in the root nodules?

4. Photosynthetic cyanobacteria carry out nitrogen fixation in specialized cells that have Photosystem I but lack Photosystem II. Explain the reason for this strategy.

5. The highly versatile prokaryotic cell can incorporate ammonia into amino acids using two different mechanisms, depending on the concentration of available ammonia.
(a) One method involves coupling glutamine synthetase, glutamate synthase, and transamination reactions. Write the overall balanced equation for this process.
(b) A second method involves coupling the freely reversible glutamate dehydrogenate reaction and a transamination reaction. Write the overall balanced equation for this process.

6. Refer to your answer to Problem 5.
(a) Which process is used when the concentration of available ammonia is low? When the concentration of ammonia is high? (*Hint:* The K_M of glutamine synthetase for ammonia is lower than the K_M of glutamate dehydrogenase for the ammonium ion.)
(b) Why is the prokaryotic cell at a disadvantage when the concentration of ammonia is low? Explain.

7. Cancer cells show a much higher rate of glutamine utilization than normal cells. What are two strategies that cancer cells could employ to increase the cellular glutamine content? How could you use this information to devise a therapeutic agent for treating cancer?

8. *E. coli* glutamine synthetase is regulated by adenylylation; that is, an AMP group is covalently attached to a tyrosine side chain. The enzyme is less active in the adenylylated form. The reaction is catalyzed by an adenylyltransferase (ATase) enzyme.
(a) Write a balanced equation for this reaction and show the structure of the adenylylated tyrosine side chain.
(b) Would you expect α-ketoglutarate to stimulate or inhibit the adenylylation of the enzyme?

9. Draw the products of the following transamination reactions:
(a) glycine + α-ketoglutarate → glutamate + _____
(b) arginine + α-ketoglutarate → glutamate + _____
(c) serine + α-ketoglutarate → glutamate + _____
(d) phenylalanine + α-ketoglutarate → glutamate + _____

10. Draw the products of the following transamination reactions. What do all of the products have in common?
(a) aspartate + α-ketoglutarate → glutamate + _____
(b) alanine + α-ketoglutarate → glutamate + _____
(c) glutamate + oxaloacetate → aspartate + _____

11. Which amino acid generates each of the following products in a transamination reaction with α-ketoglutarate?

12. Serine hydroxymethyltransferase catalyzes the conversion of threonine to glycine in a PLP-dependent reaction. The mechanism is slightly different from that shown in the transamination reaction in Figure 18-5. The degradation of threonine to glycine begins with a threonine C_α—C_β bond cleavage. Draw the structure of the threonine–Schiff base intermediate that forms in this reaction and show how C_α—C_β bond cleavage occurs.

18-2 Amino Acid Biosynthesis

13. Glutamine synthetase and asparagine synthetase catalyze reactions to produce glutamine and asparagine, respectively. It might be reasonable to expect that these enzymes are similar, but they catalyze the formation of their amino acid products very differently. Compare and contrast the two enzymes.

14. Asparagine synthetase can use a similar mechanism to catalyze the synthesis of asparagine from aspartate using ammonium ions rather than glutamate as the nitrogen donor (see Problem 13).
(a) Write a balanced equation for this reaction.
(b) Compare this reaction to the glutamine synthetase reaction.

15. The expression of phosphoglycerate dehydrogenase, which catalyzes the first step of serine biosynthesis, is dramatically elevated in some cancers, which are highly metabolically active. Experiments suggest that in addition to supplying extra serine, the elevated dehydrogenase activity increases flux through the citric acid cycle. Explain.

16. Researchers have identified several metabolites whose concentrations in urine may be elevated in prostate cancer. One of these metabolites is sarcosine (*N*-methylglycine).
(a) Draw its structure.
(b) Folate deficiency causes sarcosine levels in the blood to increase. Propose an explanation for this observation.

17. All of the nonessential amino acids, with the exception of tyrosine, can be synthesized from four metabolites: pyruvate, oxaloacetate, α-ketoglutarate, and 3-phosphoglycerate. Draw a diagram that shows how the 10 amino acids are obtained from these metabolites.

18. In many bacteria, the pathway for synthesizing cysteine and methionine consists of enzymes that contain relatively few Cys and Met residues. Explain why this is an advantage.

19. Taurine, a naturally occurring compound that is added to some energy drinks, is used in the synthesis of bile salts (Section 17-3); it may also help regulate cardiovascular function and lipoprotein metabolism.

From what amino acid is taurine derived, and what types of reactions are required to convert this amino acid to taurine?

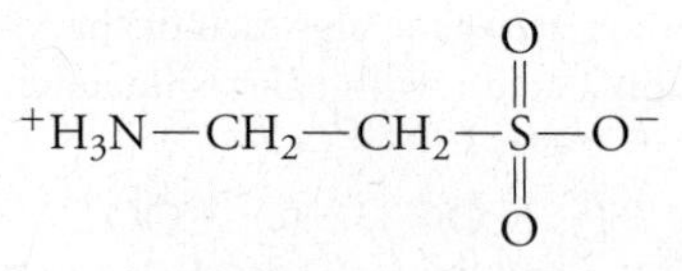

$$^+H_3N-CH_2-CH_2-\overset{\overset{\displaystyle O}{\|}}{\underset{\underset{\displaystyle O}{\|}}{S}}-O^-$$

Taurine

20. Sulfonamides (sulfa drugs) act as antibiotics by inhibiting the synthesis of folate in bacteria.

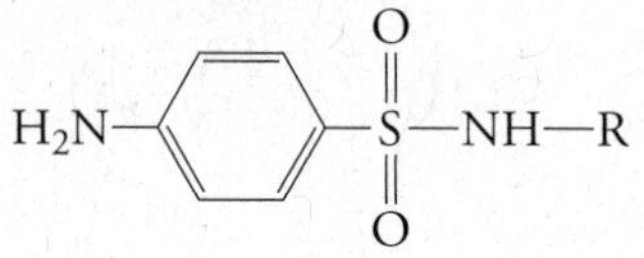

A sulfonamide

(a) Which portion of the folate molecule does the sulfonamide resemble?
(b) Why do sulfonamides kill bacteria without harming their mammalian host?

21. A person whose diet is poor in just one of the essential amino acids may enter a state of negative nitrogen balance, in which nitrogen excretion is greater than nitrogen intake. Explain why this occurs, even when the supply of other amino acids is high.

22. A dilute solution of gelatin, which is derived from the protein collagen, is sometimes given to ill children who have not been able to consume solid food for several days.
(a) Explain why gelatin is not a good source of essential amino acids.
(b) What is the advantage of giving gelatin, rather than a sugar solution, to someone who has not eaten for several days?

23. Threonine deaminase catalyzes the committed step of the biosynthetic pathway leading to the branched-chain amino acid isoleucine. The enzyme catalyzes the dehydration and deamination of threonine to α-ketobutyrate.

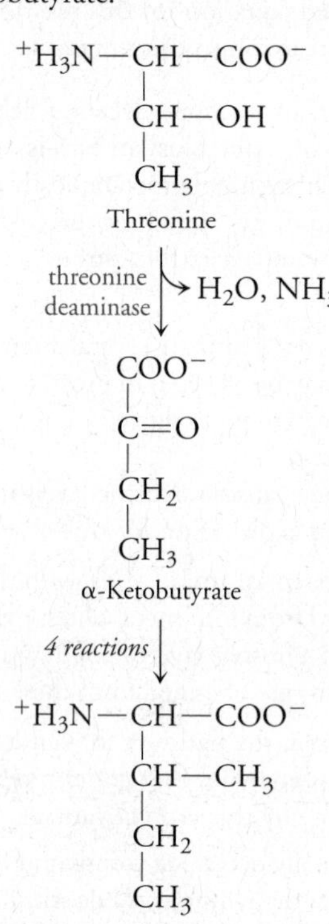

$$^+H_3N-CH-COO^-$$
$$|$$
$$CH-OH$$
$$|$$
$$CH_3$$

Threonine

threonine deaminase ↘ H_2O, NH_3

$$COO^-$$
$$|$$
$$C=O$$
$$|$$
$$CH_2$$
$$|$$
$$CH_3$$

α-Ketobutyrate

4 reactions ↓

$$^+H_3N-CH-COO^-$$
$$|$$
$$CH-CH_3$$
$$|$$
$$CH_2$$
$$|$$
$$CH_3$$

Isoleucine

Threonine deaminase is a tetramer and has a low affinity, or T, form and a high affinity, or R, form. The enzyme is allosterically regulated. The activity of threonine deaminase was measured at increasing concentrations of its substrate, threonine (curve A). Measurements were also made in the presence of isoleucine (curve B) and valine (curve C). The kinetic constants obtained from the graphs are shown in the table.

	No modulators present (Thr only)	Isoleucine added	Valine added
V_{max} ($\mu mol \cdot mg^{-1} \cdot min^{-1}$)	214	180	225
K_M (mM)	8.0	74	5.7

(a) What can you determine about threonine deaminase from these data?
(b) What is the effect of isoleucine on threonine deaminase activity? To which form of the enzyme does isoleucine bind?
(c) What is the effect of valine (a product of a parallel pathway) on threonine deaminase activity? To which form of the enzyme does valine bind?

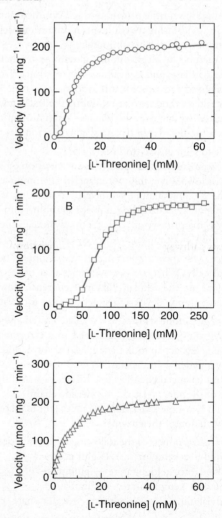

24. Bacteria synthesize the essential amino acids lysine, methionine, and threonine using aspartate as a substrate. The pathway is summarized on the next page. Certain enzymes in this pathway serve as regulatory points so that the cell can maintain appropriate concentrations of each type of amino acid. The amino acids themselves serve as allosteric regulators of enzyme activity. Which enzyme(s) is(are) good candidates for regulation of the synthesis of each of these amino acids: **(a)** lysine, **(b)** methionine, and **(c)** threonine?

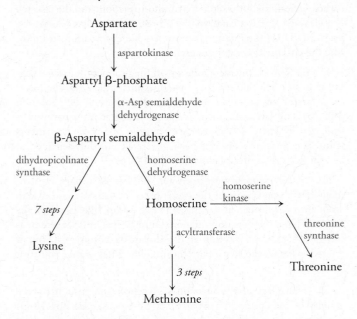

18-3 Nucleotide Biosynthesis

25. Although methotrexate is still used for cancer chemotherapy, it is also prescribed to treat some autoimmune diseases such as rheumatoid arthritis. Explain.

26. Multifunctional enzymes are common in eukaryotic metabolism. Explain why it would be an advantage for dihydrofolate reductase (DHFR) and thymidylate synthase (TS) activities to be part of the same protein.

27. Purine nucleotide synthesis is a highly regulated process. The main objective is to provide the cell with approximately equal concentrations of ATP and GTP for DNA synthesis. The purine nucleotide synthesis pathway is outlined in the figure below.

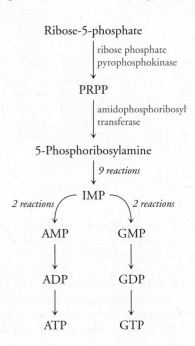

(a) How do ADP and GDP regulate ribose phosphate pyrophosphokinase?
(b) Amidophosphoribosyltransferase catalyzes the committed step of the IMP synthetic pathway. How might 5-phosphoribosyl pyrophosphate (PRPP), AMP, ADP, ATP, GMP, GDP, and GTP affect the activity of this enzyme? Explain your reasoning.

28. The purine nucleotide synthesis pathway shown in Problem 27 is the *de novo* synthetic pathway and is virtually identical in all organisms. Most organisms have additional salvage pathways in which purines released from degradative processes are recycled to form their respective nucleotides. (Some organisms that do not have the *de novo* pathway rely exclusively on a salvage pathway to synthesize purine nucleotides.) One salvage pathway involves the conversion of adenine to AMP, catalyzed by the enzyme adenine phosphoribosyltransferase (APRT). If adenine is present in large amounts, it is converted to dihydroxyadenine, which can form kidney stones.

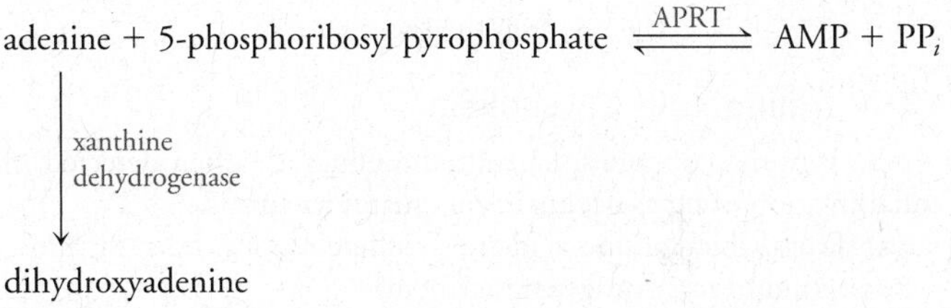

$$adenine + \text{5-phosphoribosyl pyrophosphate} \xrightleftharpoons{APRT} AMP + PP_i$$

$$\downarrow \text{xanthine dehydrogenase}$$

dihydroxyadenine

(a) A mutation in the APRT gene results in a 10-fold increase in the K_M value for one of the APRT substrates. What is the consequence of this mutation?
(b) How would you treat the condition resulting from this mutation?

29. A second purine salvage pathway (see Problem 28) is catalyzed by hypoxanthine-guanine phosphoribosyl transferase (HGPRT), which catalyzes the following reactions:

$$hypoxanthine + \text{5-phosphoribosyl pyrophosphate} \xrightleftharpoons{HGPRT} IMP + PP_i$$

$$guanine + \text{5-phosphoribosyl pyrophosphate} \xrightleftharpoons{HGPRT} GMP + PP_i$$

Some intracellular protozoan parasites have high levels of HGPRT. Inhibitors of this enzyme are being studied for their effectiveness in blocking parasite growth. What is the metabolic effect of inhibiting the parasite's HGPRT, and what does this tell you about the parasite's metabolic capabilities?

30. Lesch–Nyhan syndrome is a disease caused by a severe deficiency in HGPRT activity (see Problem 29). The disease is characterized by the accumulation of excessive amounts of uric acid, a product of nucleotide degradation, which causes neurological abnormalities and destructive behavior, including self-mutilation. Explain why the absence of HGPRT causes uric acid to accumulate.

31. Antibody-producing B-lymphocytes use both the *de novo* (see Problem 27) and salvage pathways (see Problem 28) to synthesize nucleotides, but their survival rate in culture is only 7 to 10 days. Myeloma cells lack the HGPRT enzyme (see Problem 29) and survive indefinitely in culture. The preparation of long-lived antibody-producing hybridoma cells involves fusion of a lymphocyte and a myeloma cell to produce a hybridoma, then selecting for hybridomas that can grow in HAT medium. The medium contains hypoxanthine, aminopterin (an antibiotic that inhibits enzymes of the *de novo* nucleotide synthetic pathway), and thymidine. How does the HAT medium select for hybridoma cells?

32. The compound 5-fluorouracil (shown below) is often used topically to treat minor skin cancers. When 5-fluorouracil is added to cells in culture, the concentration of dUTP increases, while dTTP is depleted. How do you account for these observations, and how does 5-fluorouracil kill cancer cells?

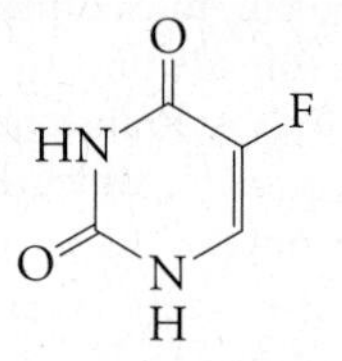

5-Fluorouracil

18-4 Amino Acid Catabolism

33. Peptides containing citrulline residues are often detected in inflammation, such as occurs in rheumatoid arthritis.
(a) From which amino acid are citrulline residues derived, and what kind of reaction generates them?
(b) Why are citrullinated peptides likely to trigger an autoimmune response?

34. Urinary 3-methylhistidine (a modified amino acid found primarily in actin) is used as an indicator of the rate of muscle degradation. When actin is degraded, 3-methylhistidine is excreted because it cannot be reused for protein synthesis. Why? Explain why monitoring 3-methylhistidine offers only an approximation of the rate of muscle degradation.

35. The catabolic pathways for the 20 amino acids vary considerably, but all amino acids are degraded to one of seven metabolites: pyruvate, α-ketoglutarate, succinyl-CoA, fumarate, oxaloacetate, acetyl-CoA, or acetoacetate. What is the fate of each of these metabolites?

36. Although amino acids are classified as glucogenic, ketogenic, or both, it is possible for all their carbon skeletons to be broken down to acetyl-CoA. Explain.

37. Mouse embryonic stem cells are small but divide extremely rapidly. To maintain high metabolic flux, these cells require a high concentration of threonine and express high levels of threonine dehydrogenase, which catalyzes the first step of threonine breakdown. Explain how threonine catabolism contributes to citric acid cycle activity and nucleotide biosynthesis.

38. Isoleucine is degraded to acetyl-CoA and propionyl-CoA by a pathway in which the first steps are identical to those of valine degradation (see Fig. 18-11) and the last steps are identical to those of fatty acid oxidation.
(a) Draw the intermediates of isoleucine degradation and indicate the enzyme that catalyzes each step.
(b) Which reaction in the degradation scheme is analogous to the reaction catalyzed by pyruvate dehydrogenase?
(c) What reaction is analogous to the reaction catalyzed by acyl-CoA dehydrogenase in fatty acid oxidation?

39. What is the fate of the propionyl-CoA that is produced upon degradation of isoleucine?

40. Leucine is degraded to acetyl-CoA and acetoacetate by a pathway whose first two steps are identical to those of valine degradation (see Fig. 18-11). The third step is the same as the first step in fatty acid oxidation. The fourth step involves an ATP-dependent carboxylation, the fifth step is a hydration reaction, and the last step is a cleavage reaction catalyzed by a lyase enzyme that releases the products. Draw the intermediates of leucine degradation and indicate the enzyme that catalyzes each step.

41. (a) What is the metabolic fate of the products of leucine degradation (see Solution 40) and how does this differ from the metabolic fate of the products of isoleucine degradation? (see Solution 39)? **(b)** The last reaction in the leucine degradation pathway is catalyzed by HMG-CoA lyase (see Solution 40). Why do patients who are missing this enzyme need to restrict not only leucine but also fat from their diets? (*Hint:* See Fig. 17-16.)

42. Maple syrup urine disease (MSUD) is an inborn error of metabolism that results in the excretion of branched-chain α-keto acids into the urine, imparting a maple syrup–like smell to the urine. What enzyme is nonfunctional in these patients? Severe neurological symptoms develop if the disease is not treated, but if treated, patients can live a fairly normal life. How would you treat the disease?

43. The effect of the hormones glucagon and insulin on the activity of the enzyme phenylalanine hydroxylase (see Box 18-D) was investigated, with and without preincubation with phenylalanine (Phe has been shown to convert the enzyme from the dimeric to the tetrameric form.) The results are shown below.
(a) What hormone stimulates the enzyme?
(b) What is the role of Phe in the regulation of enzyme activity?
(c) In a separate experiment, the amount of radioactively labeled phosphate incorporated into the enzyme with glucagon treatment was found to be nearly sevenfold greater than in controls. How do you explain this observation?
(d) Is phenylalanine hydroxylase more active in the fed or fasting state?

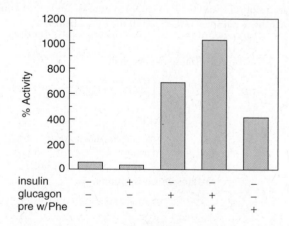

44. Phenylketonuria (PKU) is an inherited disease that results from the lack of the enzyme phenylalanine hydroxylase (see Problem 43). Phenylalanine hydroxylase catalyzes the first step in the degradation of phenylalanine (see Box 18-D). Individuals with phenylketonuria cannot break down phenylalanine, which accumulates in the blood and is eventually transaminated to phenylpyruvate, a phenylketone compound. The accumulation of phenylpyruvate causes irreversible brain damage if the disease is not treated.

(a) Draw the structure of phenylpyruvate, the product of phenylalanine transamination.

(b) Why do children with a deficiency of tetrahydrobiopterin excrete large quantities of the phenylketone?

(c) Individuals diagnosed with PKU are placed on a low phenylalanine diet. Why would a phenylalanine-free diet be undesirable?

(d) Why should patients with PKU avoid the artificial sweetener aspartame (see Problem 4-15)?

(e) Explain why individuals on a low-phenylalanine diet may need to increase their tyrosine intake.

45. Nonketotic hyperglycinemia (NKH) is an inborn error of metabolism characterized by high levels of glycine in the blood, urine, and cerebrospinal fluid. Babies with this disease suffer from hypotonia, seizures, and intellectual disability. What enzyme is most likely to be nonfunctional in patients with NKH?

46. In mammals, metabolic fuels can be stored: glucose as glycogen and fatty acids as triacylglycerols. What type of molecule could be considered as a sort of storage depot for amino acids? How does it differ from other fuel-storage molecules?

18-5 Nitrogen Disposal: The Urea Cycle

47. At one time, ammonia's toxicity was believed to result from its participation in the glutamate dehydrogenase reaction, which is reversible. Explain how this reaction could affect the brain's energy metabolism.

48. Glutamate dehydrogenase is allosterically regulated by a variety of metabolites. Predict the effect of each of the following on glutamate dehydrogenase activity.

(a) GTP

(b) ADP

(c) NADH

49. Recall that ketone bodies are produced during starvation (see Section 17-2). Under these conditions, the kidneys increase their uptake of glutamine. Explain how this helps counteract the effects of ketosis.

50. Identify the source of the two nitrogen atoms in urea.

51. List all the reactions shown in this chapter that generate free ammonia.

52. Which three mammalian enzymes can potentially "mop up" excess NH_4^+?

53. A complete deficiency of a urea cycle enzyme usually causes death soon after birth, but a partial deficiency may be tolerated.

(a) Explain why hyperammonemia (high levels of ammonia in the blood) accompanies a urea cycle enzyme deficiency.

(b) What dietary adjustments might minimize ammonia toxicity?

54. The drug Ucephan, a mixture of sodium salts of phenylacetate and benzoate, is used to treat the hyperammonemia associated with a urea cycle enzyme deficiency (see Problem 53). Phenylacetate reacts with glutamine, and benzoate reacts with glycine; the reaction products are excreted in the urine. Draw the structures of the products. What is the biochemical rationale for this treatment?

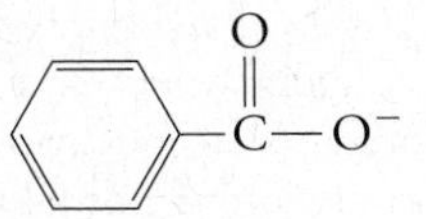

55. An inborn error of metabolism results in the deficiency of argininosuccinase. What could be added to the diet to boost urea production?

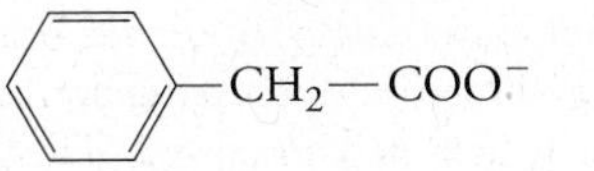

56. Production of the enzymes that catalyze the reactions of the urea cycle can increase or decrease according to the metabolic needs of the organism. High levels of these enzymes are associated with high-protein diets as well as starvation. Explain this paradox.

57. Describe how the reversible glutamate dehydrogenase reaction **(a)** contributes to amino acid biosynthesis and **(b)** functions as an anaplerotic reaction for the citric acid cycle.

58. The kidneys play a role in regulating acid–base balance in humans by releasing amino groups from glutamine so that the resulting ammonium ions can neutralize metabolic acids (see Box 2-D). Which two kidney enzymes are responsible for removing glutamine's amino groups? Write the reaction catalyzed by each enzyme, and write the net reaction.

59. Vigorous exercise is known to break down muscle proteins. What is the probable metabolic fate of the resulting free amino acids?

60. The brain NMDA receptor is named for one of its agonists, N-methyl-D-aspartate. Draw the structure of this compound.

61. Bacterial glutamine synthetase has an elaborate control system, which includes both allosteric and covalent control. The bacterial enzyme consists of 12 identical subunits arranged at corners of a hexagonal prism (see Fig. 18-4). Its activity is inhibited by adenylylation (covalent attachment of an AMP group, see Problem 8). The enzyme has 12 adenylylation sites, one on each subunit. The level of adenylylation is controlled by the activity of a uridylyltransferase (UTase), an enzyme that attaches a UMP group on a regulatory protein termed P_{II}. [A uridylyl-removing enzyme (UR) removes the UMP group from P_{II}.] The P_{II} protein is associated with the adenylyltransferase (ATase), which catalyzes both adenylylation and deadenylylation of glutamine synthetase. When P_{II} is uridylylated, the ATase enzyme catalyzes deadenylylation of glutamine synthetase; when the UMP group is removed from P_{II}, adenylylation of glutamine synthetase takes place. This complex metabolic strategy is illustrated in the following figure.

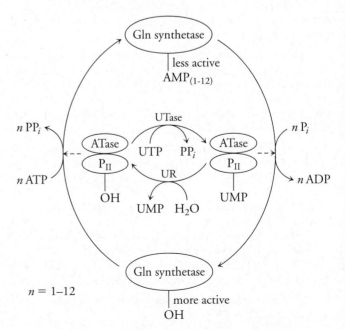

$n = 1–12$

Benzoate   Phenylacetate

(a) The activity of the UTase is affected by cellular concentrations of α-ketoglutarate, ATP, glutamine, and inorganic phosphate. The activity of UR is affected by cellular concentrations of glutamine. Classify each of these substances as inhibitors or activators of UTase and UR.

(b) Increased levels of adenylylation of glutamine synthetase render the enzyme more susceptible to allosteric inhibition by the following metabolites: histidine, tryptophan, carbamoyl phosphate, glucosamine-6-phosphate, AMP, CTP, NAD$^+$, alanine, serine, and glycine. Why do these substances inhibit glutamine synthetase?

(c) When glutamine synthetase was first purified from *E. coli*, the enzyme was isolated from cells grown in medium containing glycerol as the only carbon source and glutamate as the only nitrogen source. The enzyme prepared from these cells was sensitive to the allosteric inhibitors listed above. A second batch of enzyme was prepared from *E. coli* cells grown in medium containing glucose and ammonium chloride. The enzyme purified from these cells was insensitive to the inhibitors listed above. Provide an explanation for these observations.

62. Gastric ulcers result from infection by *Helicobacter pylori*. To survive in the extreme acidity of the stomach, the bacteria express high levels of the enzyme urease.

(a) Why is urease activity essential for *H. pylori* survival?

(b) Why is it important for at least some urease to be associated with the bacterial cell surface?

[SELECTED READINGS]

Brosnan, J. T., Glutamate, at the interface between amino acid and carbohydrate metabolism, *J. Nutr.* **130,** 988S–990S (2000). [Summarizes the metabolic roles of glutamate as a fuel and as a participant in deamination and transamination reactions.]

Reichard, P., Ribonucleotide reductases: Substrate specificity by allostery, *Biochem. Biophys. Res. Commun.* **396,** 19–23 (2010). [Includes a discussion of how these enzymes are regulated.]

Toney, M. D., Controlling reaction specificity in pyridoxal phosphate enzymes, *Biochim. Biophys Acta.* **1814,** 1407–1418 (2011). [Discusses how these enzymes use the same cofactor for a variety of reactions.]

Williams, R. A., Mamotte, C. D. S., and Burnett, J. R., Phenylketonuria: An inborn error of phenylalanine metabolism, *Clin., Biochem. Rev.* **29,** 31–41 (2008). [Includes the history, biochemistry, diagnosis, and treatment of the disease.]

Withers, P. C., Urea: Diverse functions of a 'waste product,' *Clin. Exp. Pharm. Physiol.* **25,** 722–727 (1998). [Discusses the various roles of urea, contrasting it with ammonia with respect to toxicity, acid–base balance, and nitrogen transport.]

19

REGULATION OF MAMMALIAN FUEL METABOLISM

▶▶ **WHAT** goes wrong in diabetes?

Humans are adapted for long-term stability: Over many decades, the adult body does not usually undergo much change in size or shape. Even the dramatic variations in food consumption and activity levels that occur on a daily basis do not appear to perturb the system. Yet when the regulatory mechanisms are disrupted or overwhelmed, the results can be deadly. Type 2 diabetes, a disorder that is increasing worldwide at an alarming rate, represents a failure of the body to control fuel consumption, expenditure, and storage. And even injections of insulin, a key hormone involved in regulating fuel metabolism, may not be effective.

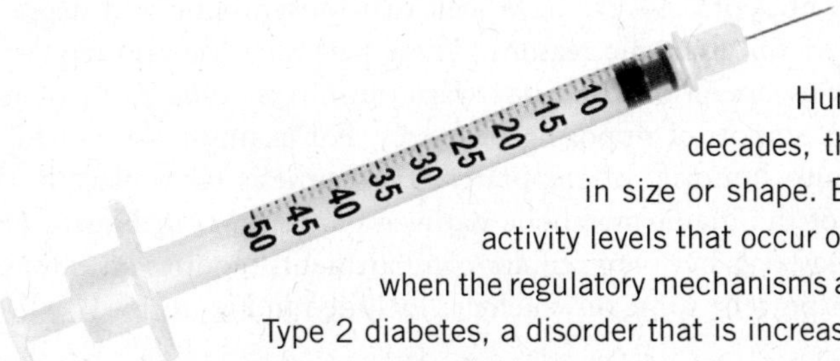

[spxChrome/iStockphoto]

THIS CHAPTER IN CONTEXT

Part 1   Foundations

Part 2   Molecular Structure and Function

Part 3 Metabolism

19 Regulation of Mammalian Fuel Metabolism

Part 4   Genetic Information

Do You Remember?

- Allosteric regulators can inhibit or activate enzymes (Section 7-3).
- G protein–coupled receptors and receptor tyrosine kinases are the two major types of receptors that transduce extracellular signals to the cell interior (Section 10-1).
- Metabolic fuels can be mobilized by breaking down glycogen, triacylglycerols, and proteins (Section 12-1).
- Metabolic pathways in cells are connected and are regulated (Section 12-2).

As a machine obeying the laws of thermodynamics, the human body is remarkably flexible in managing its resources. Humans and other mammals rely on different organs that are specialized for using, storing, and interconverting metabolic fuels. The exchange of materials between organs and communication between organs allows the body to operate as a unified whole. But for the same reasons, disorders in one aspect of fuel metabolism can have body-wide consequences. We begin this chapter by reviewing the roles of different organs. Then we can explore how metabolic processes are coordinated and how their disruption leads to disease.

19-1 Integration of Fuel Metabolism

KEY CONCEPTS
- Various organs are specialized for fuel storage, mobilization, and other functions.
- Metabolites travel between tissues in interorgan metabolic pathways.

In examining the various pathways for the catabolism and anabolism of the major metabolic fuels and building blocks in mammals—carbohydrates, fatty acids, amino acids, and nucleotides—we have seen that biosynthetic and degradative pathways differ for thermodynamic reasons. These pathways are also regulated so that their simultaneous operation does not waste resources. One form of regulation is the compartmentation of opposing processes. For example, fatty acid oxidation takes place in mitochondria, whereas fatty acid synthesis takes place in the cytosol. The locations of the major metabolic pathways are shown in **Figure 19-1**. The movement of materials between cellular compartments requires an extensive set of membrane transporters, some of which are included in Figure 19-1.

Organs are specialized for different functions

Compartmentation also takes the form of organ specialization: *Different tissues have different roles in energy storage and use.* For example, the liver carries out most metabolic processes as well as liver-specific functions such as gluconeogenesis, ketogenesis, and urea production. Adipose tissue is specialized to store about 95% of the body's triacylglycerols. Some tissues, such as red blood cells, do not store glycogen or fat and rely primarily on glucose supplied by the liver.

The functions of some organs as fuel depositories or fuel sources depend on whether the body is experiencing abundance (for example, immediately following a meal) or deprivation (after many hours of fasting). The major metabolic functions of some organs, including their reciprocating roles in storing and mobilizing fuels, are diagrammed in **Figure 19-2**.

Figure 19-1 Cellular locations of major metabolic pathways. In mammalian cells, most metabolic reactions occur in either the cytosol or the mitochondrial matrix. The urea cycle requires enzymes located in the matrix and cytosol. Amino acid degradation also occurs in both compartments. Other reactions, not pictured here, occur in the peroxisome, endoplasmic reticulum, Golgi apparatus, and lysosome. The diagram includes some transport proteins that transfer substrates and products between the mitochondria and cytosol.

? **For each transporter, indicate the primary direction of solute movement and identify its metabolic purpose.**

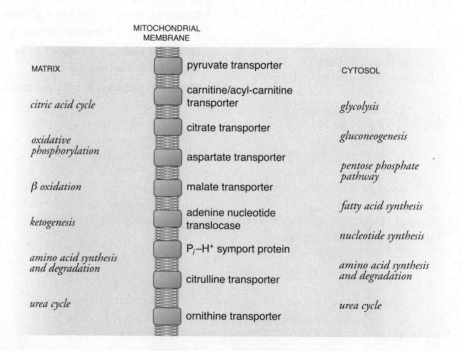

MITOCHONDRIAL MEMBRANE

MATRIX		CYTOSOL
	pyruvate transporter	
citric acid cycle	carnitine/acyl-carnitine transporter	glycolysis
oxidative phosphorylation	citrate transporter	gluconeogenesis
	aspartate transporter	pentose phosphate pathway
β oxidation	malate transporter	fatty acid synthesis
ketogenesis	adenine nucleotide translocase	nucleotide synthesis
amino acid synthesis and degradation	P$_i$–H$^+$ symport protein	amino acid synthesis and degradation
	citrulline transporter	
urea cycle	ornithine transporter	urea cycle

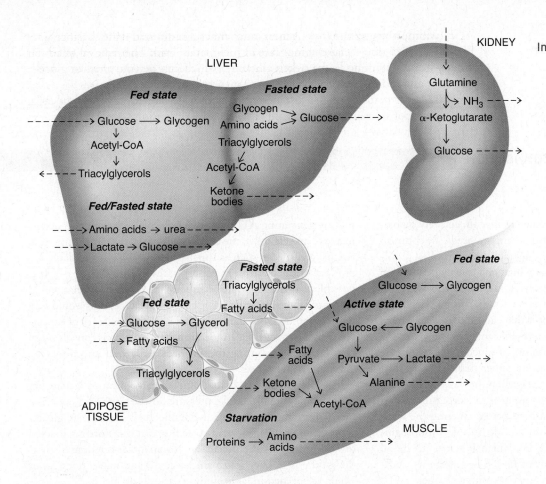

Figure 19-2 The major metabolic roles of the liver, kidney, muscle, and adipose tissue.

? Identify the processes that are mainly catabolic and mainly anabolic.

Following a meal, the liver takes up glucose and converts it to glycogen for storage. Excess glucose and amino acids are catabolized to acetyl-CoA, which is used to synthesize fatty acids. The fatty acids are esterified to glycerol, and the resulting triacylglycerols, along with dietary triacylglycerols, are exported to other tissues. During a fast, the liver mobilizes glucose from glycogen stores and releases it into the circulation for other tissues to use. Triacylglycerols are broken down to acetyl-CoA, which can be converted to ketone bodies to power the brain and heart when glucose is in short supply. Amino acids derived from proteins can be converted to glucose by gluconeogenesis (the nonglucogenic amino acids can be converted to ketone bodies). The liver also deals with lactate and alanine produced by muscle activity, converting these molecules to glucose and disposing of amino groups through urea synthesis.

Muscle cells take up glucose when it is available and store it as glycogen, although there is a limit to how much glycogen can be stockpiled. During exercise, the glycogen is quickly broken down for glycolysis, a pathway for the rapid—if inefficient—production of ATP. Muscle cells can also burn fatty acids and ketone bodies. Heart muscle, which maintains a near-constant level of activity, is specialized to burn fatty acids as its primary fuel and is rich in mitochondria to carry out this aerobic activity. During intense prolonged activity, muscle cells export lactate and alanine (see below). Muscle protein can be tapped as a source of metabolic fuel during starvation, when amino acids are needed to generate glucose.

Adipocytes take up glucose and convert it to glycerol; this plus fatty acids taken up from the circulation are the raw materials for triacylglycerols that are stored as a fat globule inside the cell. Fatty acids are mobilized in times of need and released from adipose tissue into the circulation.

In addition to eliminating wastes and maintaining acid–base balance (see Box 2-D), the kidneys play a minor role in fuel metabolism. The removal of amino groups from glutamine leaves α-ketoglutarate, which can be converted to glucose (the liver and kidney are the only organs that carry out gluconeogenesis).

The typical patterns of fuel use described above are altered in cancerous cells, whose metabolism must support rapid growth and cell division (Box 19-A).

BOX 19-A ⬡ CLINICAL CONNECTION

Cancer Metabolism

Normal differentiated cells grow slowly, if at all, and rely on oxidative phosphorylation to meet their energy needs. In contrast, most cancer cells are characterized by rapid, uncontrolled proliferation and carry out glycolysis at a high rate, even when oxygen is plentiful. This anaerobic glycolysis, known as the **Warburg effect,** has puzzled biochemists since Otto Warburg described it in the 1920s. One would expect that cancer cells would burn fuel and generate ATP using more efficient pathways, yet the cells instead consume large amounts of glucose and eliminate waste carbons in the form of lactate rather than CO_2. In fact, heightened glucose uptake by cancer cells is the basis for the PET (positron emission tomography) scan used to locate tumors and monitor their growth. About an hour before entering the scanner, a patient is given an injection of 2-deoxy-2-[^{18}F]fluoroglucose (fluorodeoxyglucose, or FDG), which is taken up by all glucose-metabolizing cells. Decay of the ^{18}F isotope emits a positron (which is like an electron with a positive charge) that is ultimately detected as a flash of light. The PET scanner generates a two- or three-dimensional map showing the location of the tracer, that is, tissues with a high rate of glucose uptake.

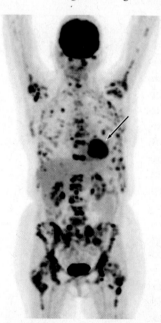

The dark areas of the PET scan indicate tissues that rely heavily on glucose metabolism. The arrow points to a tumor.
[Living Art Enterprises/Photo Researchers.]

So why do cancer cells use glucose in an apparently inefficient manner? One possible advantage of this metabolic strategy is that by generating ATP mainly by glycolysis (which also occurs in active muscle cells), the cancer cells can avoid generating reactive oxygen species such as superoxide, $\cdot O_2^-$, an unavoidable by-product of the electron transport pathway (see Box 15-A). It is possible that the high rate of superoxide production that would accompany accelerated aerobic activity in cancer cells would lead to catastrophic DNA damage. However, one hallmark of cancerous cells is damaged DNA (Section 20-4), which is considered to be a factor contributing to, not necessarily resulting from, the development of cancer.

Another possibility is that glucose catabolism to lactate can supply enough ATP to take pressure off the citric acid cycle as a mechanism for converting metabolic fuels to ATP. A greater proportion of the citric acid cycle flux can therefore be directed toward producing the precursors for anabolic reactions such as the synthesis of fatty acids, amino acids, and nucleotides, all of which are needed in large amounts as cells divide.

The need for biosynthetic precursors also seems to explain why—despite the high rate of glucose catabolism—many cancer cells express a variant pyruvate kinase that has *lower* enzymatic activity. The resulting bottleneck in glycolytic flux may help divert some glucose carbons through amino acid biosynthetic pathways or through the pentose phosphate pathway (Section 13-4) in order to generate NADPH and ribose groups for nucleotide synthesis.

⬡ Questions:

1. Draw the structure of 2-deoxy-2-fluoroglucose (FDG).
2. FDG is taken up by cancer cells and phosphorylated by hexokinase. **(a)** What prevents the reaction product from leaving the cells? **(b)** The reaction product accumulates in the cancer cells. Explain why it cannot be further metabolized.
3. Whole-body PET scans show high levels of FDG in the brain and bladder. Explain.
4. Which glycolytic intermediates are used for amino acid biosynthesis in humans?
5. In addition to glucose, cancer cells consume relatively large amounts of glutamine. Explain.
6. Which citric acid cycle intermediates can be used for nucleotide biosynthesis?
7. In cancer cells with low pyruvate kinase activity, phosphoenolpyruvate accumulates and can donate its phosphoryl group to the active-site His in phosphoglycerate mutase. The phospho-His then spontaneously hydrolyzes. What is accomplished by this alternative pathway for phosphoenolpyruvate?

Metabolites travel between organs

The body's organs are connected to each other by the circulatory system so that metabolites synthesized by one organ, such as glucose produced by the liver, can easily reach other tissues. Amino acids released from various tissues travel to the liver or kidney for disposal of their amino groups. Materials are also exchanged between the body's organs and the microorganisms that inhabit the intestines (Box 19-B).

Some metabolic pathways are circuits that include interorgan transport. For example, the **Cori cycle** (named after Carl and Gerty Cori, who first described it) is a metabolic pathway involving the muscles and liver. During periods of high activity, muscle glycogen is broken down to glucose, which undergoes glycolysis to produce

BOX 19-B ⬡ BIOCHEMISTRY NOTE

The Intestinal Microbiome Contributes to Metabolism

The human body is estimated to contain about 10 trillion (10^{13}) cells, and there are probably 10 times that number of microorganisms, primarily living in the intestine. These organisms, mostly bacteria, form an integrated community called the **microbiome.** At one time, the existence of microbes inside the human host was believed to be a form of commensalism, a relationship in which neither party has much to gain or lose in the partnership. It is now clear, however, that the microbiome plays an active role in providing nutrients (including some vitamins), regulating fuel use and storage, and preventing disease.

DNA sequencing studies have revealed that an individual may host up to 10,000 different species. Establishing this community of microbes begins at birth and is mostly complete after about a year. The mix of species remains fairly constant throughout the person's lifetime but can vary markedly among individuals, even in the same household. However, the overall metabolic capabilities of the microbial community seem to matter more than which species are actually present. Contrary to expectations, there is no common, or core, microbiome. The Human Microbiome Project (http://commonfund.nih.gov/hmp/index.aspx) aims to better characterize the species that inhabit the human body and assess their contribution to human health and disease.

Bacteria and fungi in the small intestine ferment undigested carbohydrates, mainly polysaccharides that cannot be broken down by human digestive enzymes, and produce acetate, propionate, and butyrate. These short-chain fatty acids are absorbed by the host and are transformed into triacylglycerols for long-term storage. Intestinal bacteria also produce vitamin K, biotin, and folate, some of which can be taken up and used by the host. The importance of microbial digestion is illustrated by mice grown in a germ-free environment. Without the normal bacterial partners, the mice need to consume about 30% more food than normal animals whose digestive systems have been colonized by microorganisms.

There is evidence that thin and obese individuals harbor different proportions of two major types of bacteria, the Bacteroidetes and the Firmicutes. It is not the case that microbes from obese individuals are more efficient at extracting nutrients from food so that the host absorbs and stores the excess as fat. Rather, different classes of microbes appear to play a role in promoting or inhibiting immune responses in the gut, and the resulting degree of inflammation influences fuel-use patterns throughout the host.

A growing body of evidence suggests that the microbiome plays a role in the development of disorders such as diabetes and autoimmunity. For that reason, so-called probiotic microbes have been proposed as a treatment or cure for such ailments, and a variety of capsules and foods containing live organisms are commercially available. Clinical trials have shown that probiotic yogurt can alter the metabolic properties of the intestinal microbiome in humans. But whether such changes are large enough or last long enough to truly impact human health has not yet been proved.

● **Question:** Formulate a hypothesis to explain the link between increased antibiotic use over the past 60 years and the increasing incidence of obesity in humans.

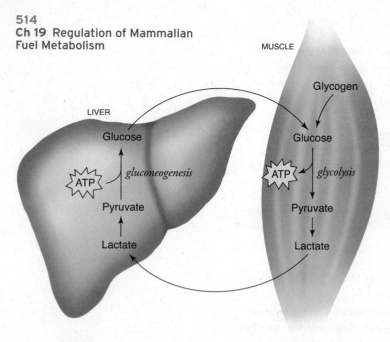

the ATP required for muscle contraction. The rapid catabolism of muscle glucose exceeds the capacity of the mitochondria to reoxidize the resulting NADH and so generates lactate as an end product. This three-carbon molecule is excreted from the muscle cells and travels via the bloodstream to the liver, where it serves as a substrate for gluconeogenesis. The newly synthesized glucose can then return to the muscle cells to sustain ATP production even after the muscle glycogen has been depleted (**Fig. 19-3**). The free energy to drive gluconeogenesis in the liver is derived from ATP produced by the oxidation of fatty acids. In effect, *the Cori cycle transfers free energy from the liver to the muscles.*

A second interorgan pathway, the **glucose–alanine cycle,** also links the muscles and liver. During vigorous exercise, muscle proteins break down, and the resulting amino acids undergo transamination to generate intermediates to boost the activity of the citric acid cycle (Section 14-3). Transamination reactions convert pyruvate, a product of glycolysis, to alanine, which travels by the blood to the liver. There, the amino group is used for urea synthesis (Section 18-5) and the resulting pyruvate is converted back to glucose by the reactions of gluconeogenesis. As in the Cori cycle, the glucose returns to the muscle cells to complete the metabolic loop (**Fig. 19-4**). *The net effect of the glucose–alanine cycle is to transport nitrogen from muscles to the liver.*

Figure 19-3 The Cori cycle. The product of muscle glycolysis is lactate, which travels to the liver. Lactate dehydrogenase converts the lactate back to pyruvate, which can then be used to synthesize glucose by gluconeogenesis. The input of free energy in the liver (in the form of ATP) is recovered when glucose returns to the muscles to be catabolized. ⊕ **See Animated Figure. The Cori cycle.**

Figure 19-4 The glucose-alanine cycle. The pyruvate produced by muscle glycolysis undergoes transamination to alanine, which delivers amino groups to the liver. The carbon skeleton of alanine is converted back to glucose to be used by the muscles, and the nitrogen is converted to urea for disposal. ⊕ **See Animated Figure. The glucose-alanine cycle.**

? What is the fate of the α-keto acids in muscle?

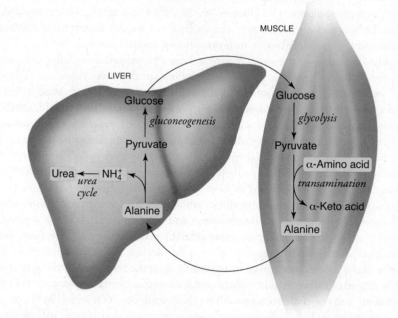

CONCEPT REVIEW
- Why is it efficient for metabolic pathways to take place in different cellular compartments or different organs?
- Summarize the role of the liver, muscle, adipose tissue, and kidney in storing and mobilizing metabolic fuel.
- Which metabolic pathways are unique to the liver?
- Describe the individual steps and the net effect of the Cori cycle and the glucose-alanine cycle.

19-2 Hormonal Control of Fuel Metabolism

Individual cells or organs must regulate the activities of their respective pathways according to their metabolic needs and the availability of fuel and building materials, which are supplied intermittently. The body buffers itself against fluctuations in the fuel supply by storing metabolic fuels, mobilizing them as needed, and replenishing them after the next meal. Maintaining a steady supply of glucose is especially critical for the brain, which exerts a large and relatively constant demand for glucose, regardless of how the dietary intake of carbohydrates varies or how much carbohydrate is oxidized to support other activities.

How does the body control the level of glucose and other fuels from hour to hour or day to day? The activities of organs that store and release fuels are coordinated by **hormones,** which are substances produced by one tissue that affect the functions of other tissues throughout the body. The most important hormones involved in fuel metabolism are insulin, glucagon, and the catecholamines epinephrine and norepinephrine, but a host of substances produced by other organs participate in a network that regulates appetite, fuel allocation, and body weight.

The ability of a cell to respond to an extracellular signal depends on cell-surface receptors that recognize the hormone and transmit a signal to the cell interior. Intracellular responses to the hormone include changes in enzyme activity and gene expression. The major signal transduction pathways are described in Chapter 10.

Insulin is released in response to glucose

Insulin plays a large role in regulating fuel metabolism by stimulating activities such as glucose uptake and inhibiting processes such as glycogen breakdown. A lack of insulin or an inability to respond to it results in the disease **diabetes mellitus** (Section 19-3). Immediately following a meal, blood glucose concentrations may rise to about 8 mM, from a normal concentration of about 3.6 to 5.8 mM. The increase in circulating glucose triggers the release of the hormone insulin, a 51–amino acid polypeptide (**Fig. 19-5**). Insulin is synthesized in the β cells of pancreatic islets, which are small clumps of cells that produce hormones rather than digestive enzymes (**Fig. 19-6**). The hormone is named after the Latin *insula,* "island."

The mechanism that triggers the release of insulin from the β cells is not well understood. The pancreatic cells do not express a glucose receptor on their surface, as might be expected. Instead, *the cellular metabolism of glucose itself seems to generate the signal to release insulin.* In liver and pancreatic β cells, the glycolytic degradation of glucose begins with a reaction catalyzed by glucokinase (an isozyme of hexokinase; see Section 13-1):

$$\text{glucose} + \text{ATP} \rightarrow \text{glucose-6-phosphate} + \text{ADP}$$

The hexokinases in other cell types have a relatively low K_M for glucose (less than 0.1 mM), which means that the enzymes are saturated with substrate at physiological glucose concentrations. Glucokinase, in contrast, has a high K_M of 5–10 mM, *so it is never saturated and its activity is maximally sensitive to the concentration of available glucose* (**Fig. 19-7**).

Interestingly, the velocity versus substrate curve for glucokinase is not hyperbolic, as might be expected for a monomeric enzyme such as glucokinase. Instead, the curve is sigmoidal, which is typical of allosteric enzymes with multiple active sites operating cooperatively (see Section 7-2). The sigmoidal kinetics of glucokinase, which has only one active site, may be due to a substrate-induced conformational change such that at the end of the catalytic cycle, the enzyme briefly maintains a high affinity for the next glucose molecule. At high glucose concentrations, this would mean a high reaction velocity; at low glucose concentrations, the enzyme would operate more slowly because it reverts to a low-affinity conformation before binding another glucose substrate.

The role of glucokinase as a pancreatic glucose sensor is supported by the fact that mutations in the glucokinase gene cause a rare form of diabetes. However, other cellular factors may be involved, particularly in the mitochondria of the β cells. The glucose sensor responsible for triggering insulin release may also depend on the mitochondrial

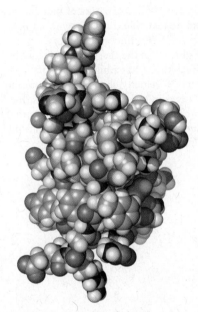

Figure 19-5 Structure of human insulin. This two-chain hormone is colored by atom type: C gray, O red, N blue, H white, and S yellow. [Structure (pdb 1AI0) determined by X. Chang, A. M. M. Jorgensen, P. Bardrum, and J. J. Led.]

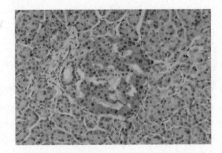

Figure 19-6 Pancreatic islet cells. The pancreatic islets of Langerhans (named for their discoverer) consist of two types of cells. The β cells produce insulin, and the α cells produce glucagon. Most other pancreatic cells produce digestive enzymes. [Carolina Biological Supply Co./Phototake.]

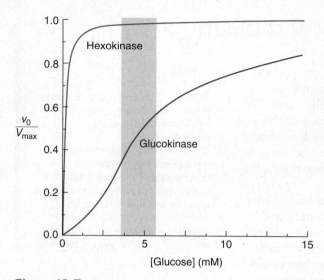

Figure 19-7 Activities of glucokinase and hexokinase. Both enzymes catalyze the ATP-dependent phosphorylation of glucose as the first step of glycolysis. Glucokinase has a high K_M, so its reaction velocity changes in response to changes in glucose concentrations. In contrast, hexokinase is saturated with glucose at physiological concentrations (shaded region).

$NAD^+/NADH$ or ADP/ATP ratios. For this reason, age-related declines in mitochondrial function may be a factor in the development of diabetes in the elderly.

Once released into the bloodstream, insulin can bind to receptors on cells in muscle and other tissues. Insulin binding to its receptor stimulates the tyrosine kinase activity of the receptor's intracellular domains (Section 10-3). These kinases phosphorylate each other as well as tyrosine residues in other proteins, including IRS-1 and IRS-2 (insulin receptor substrates 1 and 2). The IRS proteins then trigger additional events in the cell, not all of which have been fully characterized.

Insulin promotes fuel use and storage

Only cells that bear insulin receptors can respond to the hormone, and the cells' response is tissue specific. In general, *insulin signals fuel abundance: It decreases the metabolism of stored fuel while promoting fuel storage.* The effects of insulin on various tissues are summarized in Table 19-1.

In tissues such as muscle and adipose tissue, insulin stimulates glucose transport into cells by several fold. The V_{max} for glucose transport increases, not because insulin alters the intrinsic catalytic activity of the transporter but because insulin increases the number of transporters at the cell surface. These transporters, named GLUT4 to distinguish them from other glucose-transport proteins, are situated in the membranes of intracellular vesicles. When insulin binds to the cell, the vesicles fuse with the plasma membrane. *This translocation of transporters to the cell surface increases the rate at which glucose enters the cell* (Fig. 19-8). GLUT4 is a passive transporter, operating similarly to the erythrocyte glucose transporter (see Fig. 9-12). When the insulin stimulus is removed, endocytosis returns the transporters to intracellular vesicles.

Insulin stimulates fatty acid uptake as well as glucose uptake. When the hormone binds to its receptors in adipose tissue, it activates the extracellular protein lipoprotein lipase, which helps remove fatty acids from circulating lipoproteins so that they can be taken up for storage by adipocytes.

The insulin signaling pathway also alters the activity of glycogen-metabolizing enzymes. Glycogen metabolism is characterized by a balance between glycogen synthesis and glycogen degradation. Synthesis is carried out by the enzyme glycogen synthase, which adds glucose units donated by UDP–glucose to the ends of the branches of a glycogen polymer (see Section 13-3):

$$UDP–glucose + glycogen_{(n\ residues)} \rightarrow UDP + glycogen_{(n+1\ residues)}$$

Glycogen phosphorylase mobilizes glucose residues from glycogen by phosphorolysis (cleavage through addition of a phosphoryl group rather than water):

$$glycogen_{(n\ residues)} + P_i \rightarrow glycogen_{(n-1\ residues)} + glucose-1-phosphate$$

This reaction, followed by an isomerization reaction, yields glucose-6-phosphate, the first intermediate of glycolysis.

Glycogen synthase is a homodimer, and glycogen phosphorylase is a heterodimer. Both enzymes are regulated by allosteric effectors. For example, glycogen synthase is activated by glucose-6-phosphate. AMP activates glycogen phosphorylase and ATP inhibits it. These effects are consistent with the role of glycogen phosphorylase in making glucose available to boost cellular ATP production. However, *the primary mechanism for regulating glycogen synthase and glycogen phosphorylase is through covalent modification (phosphorylation and dephosphorylation) that is under hormonal control.* Both enzymes undergo reversible phosphorylation at specific Ser residues. Phosphorylation deactivates glycogen synthase and activates glycogen phosphorylase. Removal of the phosphoryl groups has the opposite effect:

[TABLE 19-1] Summary of Insulin Action

Target Tissue	Metabolic Effect
Muscle and other tissues	Promotes glucose transport into cells
	Stimulates glycogen synthesis
	Suppresses glycogen breakdown
Adipose tissue	Activates extracellular lipoprotein lipase
	Increases level of acetyl-CoA carboxylase
	Stimulates triacylglycerol synthesis
	Suppresses lipolysis
Liver	Promotes glycogen synthesis
	Promotes triacylglycerol synthesis
	Suppresses gluconeogenesis

Dephosphorylation activates glycogen synthase and deactivates glycogen phosphorylase (Fig. 19-9).

Covalent modification is a type of allosteric regulation (Section 7-3). The attachment or removal of the highly anionic phosphoryl group triggers a conformational shift between a more active (*a* or R) state and a less active (*b* or T) state. *The reciprocal regulation of glycogen synthase and glycogen phosphorylase promotes metabolic efficiency, since the two enzymes catalyze key reactions in opposing metabolic pathways.* The advantage of this regulatory system is that a single kinase can tip the balance between glycogen synthesis and degradation. Similarly, a single phosphatase can tip the balance in the opposite direction. Covalent modifications, such as phosphorylation and dephosphorylation, permit a much wider range of enzyme activities than could be accomplished solely through the allosteric effects of metabolites whose cellular concentrations do not vary much. Insulin signaling activates phosphatases that dephosphorylate (activate) glycogen synthase and dephosphorylate (deactivate) glycogen phosphorylase. As a result, glycogen synthesis accelerates and the rate of glycogenolysis decreases when glucose is abundant.

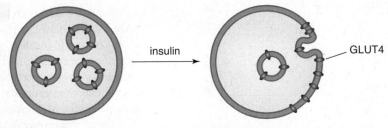

Figure 19-8 Effect of insulin on GLUT4. Insulin triggers vesicle fusion so that the glucose transport protein GLUT4 is translocated from intracellular vesicles to the plasma membrane. This increases the rate at which the cells take up glucose. ⊕ **See Animated Figure. GLUT4 activity.**

Glucagon and epinephrine trigger fuel mobilization

Within hours of a meal, dietary glucose has mostly been taken up by cells and consumed as fuel, stored as glycogen, or converted to fatty acids for long-term storage. At this point, the liver must begin mobilizing glucose in order to keep the blood glucose concentration constant. This phase of fuel metabolism is governed not by insulin but by other hormones, mainly glucagon and the catecholamines epinephrine and norepinephrine.

Glucagon, a 29-residue peptide hormone, is synthesized and released by the α cells of pancreatic islets when the blood glucose concentration begins to drop below about 5 mM (Fig. 19-10). Catecholamines are tyrosine derivatives (see Section 18-2) that are synthesized by the central nervous system as neurotransmitters and by the adrenal glands as hormones. Glucagon, epinephrine, and norepinephrine bind to receptors with seven membrane-spanning segments. Hormone binding triggers a conformational change that activates an associated G protein, which goes on to activate other cellular

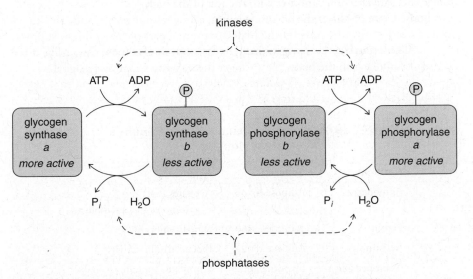

Figure 19-9 The reciprocal regulation of glycogen synthase and glycogen phosphorylase. Phosphorylation (transfer of a phosphoryl group from ATP to the enzyme) deactivates glycogen synthase and activates glycogen phosphorylase. Dephosphorylation has the opposite effect. The more active form of each enzyme is known as the *a* form (indicated in green), and the less active form is known as the *b* form (in red).

? **Would insulin signaling lead to activation of the kinases or of the phosphatases in this diagram?**

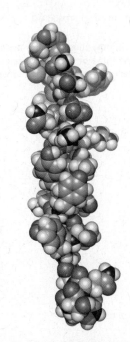

Figure 19-10 Structure of glucagon. The atoms of the 29-residue peptide are colored by type: C gray, O red, N blue, H white, and S yellow. [Structure (pdb 1GCN) determined by T. L. Blundell, K. Sasaki, S. Dockerill, and I. J. Tickle.]

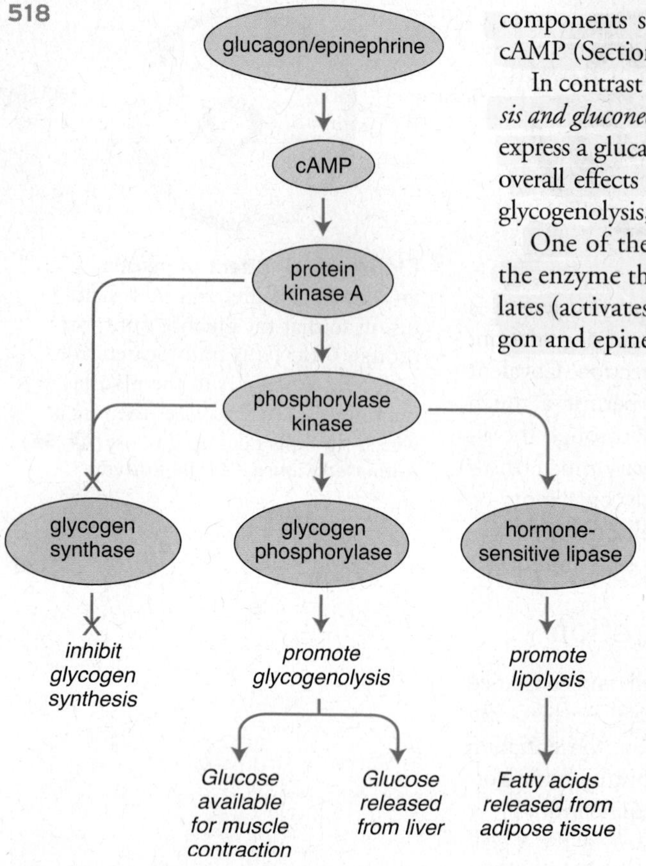

Figure 19-11 Effect of glucagon and epinephrine on fuel metabolism. Green arrows represent activation events, and red symbols represent inhibition. Both glucagon and epinephrine inhibit glycogen synthesis and promote the mobilization of glucose and fatty acids.

? **Explain how these events account for the fight-or-flight response triggered by epinephrine (adrenaline).**

components such as an adenylate cyclase, which produces the second messenger cAMP (Section 10-2). cAMP activates protein kinase A.

In contrast to insulin, *glucagon stimulates the liver to generate glucose by glycogenolysis and gluconeogenesis, and it stimulates* **lipolysis** *in adipose tissue.* Muscle cells do not express a glucagon receptor but do respond to catecholamines, which elicit the same overall effects as glucagon. Thus, epinephrine stimulation of muscle cells activates glycogenolysis, which makes more glucose available to power muscle contraction.

One of the intracellular targets of protein kinase A is phosphorylase kinase, the enzyme that phosphorylates (deactivates) glycogen synthase and phosphorylates (activates) glycogen phosphorylase. Consequently, hormones such as glucagon and epinephrine, which lead to cAMP production, promote glycogenolysis and inhibit glycogen synthesis. Although phosphorylase kinase is activated by protein kinase A, it is maximally active when Ca^{2+} ions are also present. Ca^{2+} concentrations increase during signaling via the phosphoinositide pathway, which responds to the catecholamine hormones (Section 10-2).

In adipocytes, protein kinase A phosphorylates an enzyme known as hormone-sensitive lipase, thereby activating it. This lipase catalyzes the rate-limiting step of lipolysis, the conversion of stored triacylglycerols to diacylglycerols and then to monoacylglycerols, which releases fatty acids. *Hormone stimulation not only increases the lipase catalytic activity, it also relocates the lipase from the cytosol to the fat droplet of the adipocyte.* Co-localization with its substrate, possibly through binding to a lipid-binding protein, boosts the rate at which fatty acids are mobilized. Thus, glucagon and epinephrine promote the breakdown of both glycogen and fat. These responses are summarized in **Figure 19-11**.

Additional hormones influence fuel metabolism

In addition to well-known endocrine organs such as the pancreas (the source of insulin and glucagon) and adrenal glands (the source of epinephrine and norepinephrine), many other tissues produce hormones that help regulate all aspects of food acquisition and use (Table 19-2). In fact, adipose tissue, once thought to be a relatively inert fat-storage site, actively communicates with the rest of the body.

Adipose tissue produces the hormone leptin, a 146-residue polypeptide that functions as a satiety signal. It acts on the hypothalamus, a part of the brain, to suppress appetite. The level of leptin is proportional to the amount of adipose tissue: The more fat that accumulates in the body, the stronger the appetite-suppressing signal.

[TABLE 19-2] Some Hormones that Regulate Fuel Metabolism

Hormone	Source	Action
Adiponectin	Adipose tissue	Activates AMPK (promotes fuel catabolism)
Leptin	Adipose tissue	Signals satiety
Resistin	Adipose tissue	Blocks insulin activity
Neuropeptide Y	Hypothalamus	Stimulates appetite
Cholecystokinin	Intestine	Suppresses appetite
Incretin	Intestine	Promotes insulin release; inhibits glucagon release
PYY_{3-36}	Intestine	Suppresses appetite
Amylin	Pancreas	Signals satiety
Ghrelin	Stomach	Stimulates appetite

Like leptin, adiponectin is released by adipose tissue, but this 247-residue polypeptide exists as an assortment of multimers with different receptor-binding properties. Adiponectin exerts its effects on a variety of tissues by activating an AMP-dependent protein kinase (see below). The effects of adiponectin include increased combustion of glucose and fatty acids. It also increases the sensitivity of tissues to insulin.

Adipocytes also release a 108-residue hormone called resistin, which blocks the activity of insulin. Levels of resistin increase during obesity, which would help explain the link between weight gain and decreased responsiveness to insulin (Section 19-3).

The digestive system produces at least 20 different peptide hormones with varying functions. Several of these signal that a meal has been consumed. For example, incretins are released by the intestine and enhance insulin secretion by the pancreas. The oligopeptide known as PYY$_{3-36}$, whose release is triggered especially by a high-protein meal, acts on the hypothalamus to suppress appetite. The level of ghrelin, a 28-residue peptide produced by the stomach, increases during fasting and decreases immediately following a meal; this is the only gastrointestinal hormone that stimulates appetite.

AMP-dependent protein kinase acts as a fuel sensor

So far we have looked at a variety of signals that regulate fuel intake, storage, and mobilization to help the body maintain homeostasis. Individual cells also have a fuel gauge to adjust their activity on a finer scale. The AMP-dependent protein kinase (AMPK) responds to the cell's balance of ATP, ADP, and AMP to activate and inhibit a number of enzymes involved in different metabolic pathways. *AMP and ADP, representing the cell's need for energy, activate AMPK, and ATP, representing a state of energy sufficiency, inhibits the kinase.*

AMPK is a highly conserved Ser/Thr kinase consisting of a catalytic subunit and a regulatory subunit linked by a "scaffolding" subunit (**Fig. 19-12**). Like many other kinases, AMPK is activated by phosphorylation of a specific Thr residue. ADP binding to the regulatory subunit of AMPK prevents dephosphorylation of this residue, thereby maintaining the kinase in an active state (about 200 times more active than its dephospho form). AMP also acts as an allosteric activator of the kinase so that overall its activity increases about 2000-fold. ATP, which competes with AMP and ADP for binding to the regulatory subunit, inhibits AMPK. This multipart regulatory scheme allows AMPK to respond to a wide range of cellular energy states.

In addition to responding to intracellular energy deficits, AMPK responds to hormones such as leptin and adiponectin. *As a result of AMPK activity, the cell switches off ATP-consuming anabolic pathways and switches on ATP-generating catabolic pathways. For example, in exercising muscle,* AMPK phosphorylates and activates the enzyme that produces fructose-2,6-bisphosphate, an allosteric activator of phosphofructokinase, so that glycolytic flux increases (Section 13-1). In adipose tissue, AMPK phosphorylates and inactivates acetyl-CoA carboxylase, the enzyme that generates malonyl-CoA, to suppress fatty acid synthesis. Since malonyl-CoA inhibits fatty acid transport into mitochondria, AMPK increases the rate of mitochondrial β oxidation in tissues such as muscle. AMPK activation also promotes the production of new mitochondria. Some metabolic effects of AMPK are listed in Table 19-3.

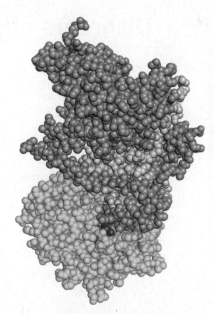

Figure 19-12 Structure of AMPK. A portion of the catalytic subunit (green) wraps around the scaffolding subunit (blue) and interacts with the regulatory subunit (yellow). An AMP bound to the regulatory subunit is shown in red. [Structure (pdb 2Y94) determined by B. Xiao, M. J. Sanders, E. Underwood, et al.]

[**TABLE 19-3**]

Effects of AMP-Dependent Protein Kinase

Tissue	Response
Hypothalamus	Increases food intake
Liver	Increases glycolysis
	Increases fatty acid oxidation
	Decreases glycogen synthesis
	Decreases gluconeogenesis
Muscle	Increases fatty acid oxidation
	Increases mitochondrial biogenesis
Adipose tissue	Decreases fatty acid synthesis
	Increases lipolysis

CONCEPT REVIEW
- Summarize the metabolic effects of insulin signaling on muscle cells and adipocytes.
- How does glucokinase differ from hexokinase?
- How does insulin increase the rate of glucose entry into cells?
- Explain how phosphorylation and dephosphorylation reciprocally regulate glycogen synthase and glycogen phosphorylase.
- Summarize the metabolic effects of glucagon and epinephrine on liver cells and adipocytes.
- Summarize the metabolic effects of leptin, adiponectin, resistin, incretins, PYY, and ghrelin.
- How does AMPK function as a cell's energy sensor?
- Which metabolic pathways are stimulated by AMPK? Which are suppressed?

KEY CONCEPTS
- The body breaks down glycogen, fats, and proteins to generate glucose, fatty acids, and ketone bodies during starvation.
- Obesity appears to result from metabolic, environmental, and genetic factors.
- In diabetes, either a lack of insulin or the inability to respond to it leads to hyperglycemia.
- The metabolic syndrome is characterized by obesity and insulin resistance.

The multifaceted regulation of mammalian fuel metabolism offers many opportunities for things to go wrong. Excessive intake and storage of fuel can cause obesity. Starvation results from insufficient food. The faulty regulation of carbohydrate and lipid metabolism can lead to diabetes. In this section we examine some of the biochemistry behind these conditions.

The body generates glucose and ketone bodies during starvation

Most tissues in the body use glucose as their preferred fuel and turn to fatty acids only when the glucose supply diminishes. Except in the intestine, amino acids are not a primary fuel. But when no food is available for an extended period, the body must make adjustments to mobilize different types of fuels. An average adult can survive a famine lasting up to a few months, an adaptation likely shaped by seasonal food shortages during human evolution. Starvation in children, of course, may severely impact development (Box 19-C).

The liver and muscles store less than a day's supply of glucose in the form of glycogen. As glycogen stores are depleted, muscle switches from burning glucose to burning fatty acids. Insulin secretion ceases with the drop in circulating glucose, so insulin-responsive tissues are not stimulated to take up glucose. This means that more glucose will be available for tissues such as the brain, which stores very little glycogen and cannot use fatty acids as fuel.

The liver and kidneys respond to the continued demand for glucose by increasing the rate of gluconeogenesis, using noncarbohydrate precursors such as amino acids (derived from protein degradation) and glycerol (from fatty acid breakdown). After several days, the liver begins to convert mobilized fatty acids to acetyl-CoA and then to ketone bodies. These small water-soluble fuels are used by a variety of tissues,

---

BOX 19-C ◆ BIOCHEMISTRY NOTE

Marasmus and Kwashiorkor

Chronic malnutrition takes a toll on human life in many ways. For example, it exacerbates infectious diseases that would not necessarily be fatal in well-nourished individuals. Severely malnourished children also fail to reach their full potential in terms of body size and cognitive development, even if their food intake later increases to normal levels. There are two major forms of severe malnutrition, **marasmus** and **kwashiorkor,** which may also occur in combination.

In marasmus, inadequate intake of metabolic fuels of all types causes wasting. Individuals with this condition are emaciated, with very little muscle mass and essentially no subcutaneous fat. Similar symptoms also develop in some chronic diseases such as cancer, tuberculosis, and AIDS.

Kwashiorkor results from inadequate protein intake, which may or may not be accompanied by inadequate energy intake. A child with kwashiorkor typically has thin limbs, reddish hair, and a swollen belly. Without an adequate supply of amino acids, the liver makes too little albumin, a protein that helps retain fluid inside blood vessels. When the concentration of albumin drops, fluid enters tissues by osmosis. This swelling (edema) also occurs in other diseases that impair liver function. The liver is enlarged in kwashiorkor due to the deposition of fat. Depigmentation of the hair and skin occurs because tyrosine, which is derived from the essential amino acid phenylalanine, is also the precursor of melanin, a brown pigment molecule.

◆ **Question:** Explain why kwashiorkor sometimes develops in infants who are fed rice milk rather than breast milk.

[TABLE 19-4] Source of Metabolic Fuels Under Different Conditions			
	Carbohydrates (%)	Fatty Acids (%)	Amino Acids (%)
Immediately after a meal	50	33	17
After an overnight fast	12	70	18
After a 40-day fast	0	95[a]	5

[a]This value reflects a high concentration of fatty acid–derived ketone bodies.

including the heart and brain. The gradual switch from glucose to ketone bodies prevents the body from using up its proteins to supply gluconeogenic precursors. During a 40-day fast, the concentration of circulating fatty acids varies about 15-fold, and the concentration of ketone bodies increases about 100-fold. In contrast, the concentration of glucose in the blood varies by no more than threefold. These patterns of fuel use are summed up in Table 19-4.

Obesity has multiple causes

Obesity has become an enormous public health problem. In addition to its impact on the quality of life, it is physiologically costly: Masses of fat prevent the lungs from fully expanding; the heart must work harder to circulate blood through a larger body; and the additional weight stresses hip, knee, and ankle joints. Obesity also increases the risk of cardiovascular disease, diabetes, and cancer. And it is not an exaggeration to describe obesity as an epidemic, since it affects an estimated one-third of the adult population of the United States.

Like many conditions, obesity has no single cause. It is a complex disorder involving appetite and metabolism and reflecting environmental as well as genetic factors. Despite a high degree of heritability for obesity, the relevant genes have not been clearly identified. There is an obvious link between overeating and the deposition of fat in adipose tissue, but metabolic regulation is so complex that—as dieters can attest—a simple adjustment such as eating less food may not be sufficient to correct a tendency to gain weight.

The human body appears to have a **set-point** for body weight that remains constant and relatively independent of energy intake and expenditure even over many decades. The hormone leptin may help establish the set-point, since the absence of leptin causes severe obesity in rodents and humans (**Fig. 19-13**). However, the majority of obese humans do not appear to lack leptin, so they may instead be suffering from leptin resistance due to a defect somewhere in the leptin signaling pathway. When leptin is less effective at suppressing the appetite, the individual gains weight. Eventually, the increase in leptin concentration resulting from the increase in adipose tissue mass succeeds in signaling satiety, but the result is a high set-point (a higher body weight that must be maintained). This may be one reason why overweight people who manage to shed a few pounds often regain the lost weight and return to the original set-point.

It turns out that humans have several types of fat, including subcutaneous fat (beneath the skin), visceral fat (surrounding the abdominal organs), and brown fat. **Brown adipose tissue,** named for its high mitochondrial content, is specialized for generating heat to maintain body temperature. Brown adipose tissue is prominent in newborns and hibernating mammals (see Box 15-B), but it also occurs at least in small amounts in adult humans, mainly in the neck and upper chest.

Developmentally and metabolically, brown adipose tissue more closely resembles muscle than ordinary white adipose tissue (**Fig. 19-14**). Instead of one large fat globule, brown adipose tissue contains many small fat droplets, which are a source of fatty acids that are oxidized to generate heat. The hormone norepinephrine binds to receptors on brown adipocytes, and signal transduction via protein kinase A activates a lipase that liberates fatty acids from triacylglycerols. The uncoupling protein (UCP) is expressed in the mitochondria of brown adipose tissue, so fuel

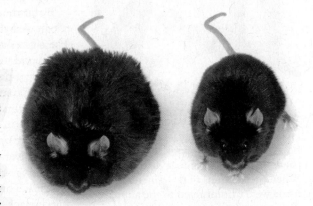

Figure 19-13 Normal and obese mice. The mouse on the left lacks a functional gene for leptin and is several times the size of a normal mouse (right). [The Rockefeller University/AP/© Wide World Photos.]

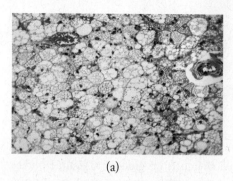
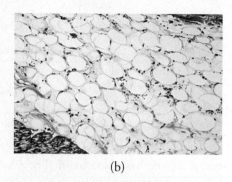

(a)   (b)

Figure 19-14 Brown and white adipose tissue. (a) In brown adipose tissue, cells contain relatively more mitochondria, and triacylglycerols are present as numerous small globules in the cytoplasm. (b) In white adipose tissue, each cell is occupied mostly by a single large fat globule, and there is little cytoplasm. [Biophoto Associates/ Photo Researchers, Inc.]

oxidation occurs without ATP synthesis. A compelling hypothesis is that lean individuals have a higher capacity to burn off excess fuel in this manner instead of storing it in white adipose tissue.

Diabetes is characterized by hyperglycemia

Another well-characterized disorder of fuel metabolism is diabetes mellitus, which affects about 10% of the population of the United States. Worldwide, the disease affects about 350 million people, killing about 3.5 million each year. The highest rates of diabetes are observed in middle-income countries, where around 20% of the population is affected.

The words *diabetes* (meaning "to run through") and *mellitus* ("honey") describe an obvious symptom of the disease. Diabetics excrete large amounts of urine containing high concentrations of glucose (the kidneys work to eliminate excess circulating glucose by excreting it in urine, a process that requires large amounts of water).

Type 1 diabetes (juvenile-onset or insulin-dependent diabetes) is an autoimmune disease in which the immune system destroys pancreatic β cells. Symptoms first appear in childhood as insulin production begins to drop off. At one time, the disease was invariably fatal. This changed dramatically in 1922, when Frederick Banting and Charles Best administered a pancreatic extract to save the life of a severely ill diabetic boy (Fig. 19-15). Since then, the treatment of type 1 diabetes with purified insulin has been refined, with delivery options including preloaded syringes and small pumps.

An ongoing challenge lies in tailoring the delivery of insulin to the body's needs over the course of a typical 24-hour cycle of eating and fasting. Diabetics typically measure the glucose concentration in a tiny sample of blood—often less than a microliter—several times a day. Freeing patients from frequent needle sticks, whether for monitoring blood glucose or injecting insulin, is the goal of pancreatic islet cell transplants, which have seen some success. Gene therapy to treat diabetes is an elusive goal because the insulin gene must be introduced into the body in such a way that the gene's expression is glucose-sensitive.

By far the most common form of diabetes, accounting for up to 95% of all cases, is type 2 diabetes (also known as adult-onset or non-insulin-dependent diabetes). These cases are characterized by **insulin resistance,** which is the failure of the body to respond to normal or even elevated concentrations of the hormone. Only a small fraction of patients with type 2 diabetes bear genetic defects in the insulin receptor, as might be expected; in the majority of cases, the underlying cause is not known.

The primary feature of untreated diabetes is chronic **hyperglycemia** (high levels of glucose in the blood). *The loss of responsiveness of tissues to insulin means that cells fail to take up glucose.* The body's metabolism responds as if no glucose were available, so liver gluconeogenesis increases, further promoting hyperglycemia. Glucose circulating at high concentrations can participate in nonenzymatic glycosylation of proteins. This process is slow, but the modified proteins may gradually accumulate and damage tissues with low turnover rates, such as neurons.

Tissue damage also results from the metabolic effects of hyperglycemia. Since muscle and adipose tissue are unable to increase their uptake of glucose in response to insulin, glucose tends to enter other tissues. Inside these cells, aldose reductase catalyzes the conversion of glucose to sorbitol:

▶▶ WHAT goes wrong in diabetes?

Figure 19-15 Banting and Best. Frederick Banting (right) and Charles Best (left) surgically removed the pancreas of dogs to induce diabetes. When preparations of the pancreatic tissue were administered to the animals, their symptoms improved. This work laid the foundation for treating human diabetes with pancreatic extracts, which contain insulin. [Hulton Archive/Getty Images, Inc.]

$$\text{Glucose} \quad \xrightarrow[\text{aldose reductase}]{\text{NADPH} + \text{H}^+ \quad \text{NADP}^+} \quad \text{Sorbitol}$$

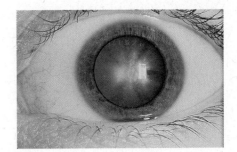

Figure 19-16 Photo of a diabetic cataract. The accumulation of sorbitol in the lens leads to swelling and precipitation of lens proteins. The resulting opacification can cause blurred vision or complete loss of sight. [Courtesy Dr. Manuel Datiles III, Cataract and Cornea Section, OGCSB, National Eye Institute, National Institutes of Health.]

Because aldose reductase has a relatively high K_M for glucose (about 100 mM), flux through this reaction is normally very low. But under hyperglycemic conditions, sorbitol accumulates and may alter the cell's osmotic balance. This may alter kidney function and may trigger protein precipitation in other tissues. Aggregation of lens proteins leads to cataracts (Fig. 19-16). Neurons and cells lining blood vessels may be similarly damaged, increasing the likelihood of neuropathies and circulatory problems that in severe cases result in kidney failure, heart attack, stroke, or the amputation of extremities.

Although commonly considered a disorder of glucose metabolism, *diabetes is also a disorder of fat metabolism,* since insulin normally stimulates triacylglycerol synthesis and suppresses lipolysis in adipocytes. Uncontrolled diabetics tend to metabolize fatty acids rather than carbohydrates, and the resulting production of ketone bodies may give the breath a sweet odor. Overproduction of ketone bodies leads to diabetic ketoacidosis.

A variety of drugs have been developed to help compensate for the physiological effects of insulin resistance; within each class of drugs there are multiple options with slightly different pharmacokinetics (Table 19-5). For example, metformin improves diabetic symptoms by activating AMPK in liver and other tissues. Liver glucose production is suppressed by the decreased expression of the gluconeogenic enzymes phosphoenolpyruvate carboxykinase and glucose-6-phosphatase (Section 13-2). Metformin also increases glucose uptake and fatty acid oxidation in muscle.

Drugs of the thiazolidinedione class, such as rosiglitazone (Avandia; see Box 7-A), act via intracellular receptors known as peroxisome proliferator–activated receptors. These receptors, which normally respond to lipid signals, are transcription factors that alter gene expression (Section 10-4). Thiazolidinediones increase adiponectin levels and decrease resistin levels (in fact, research on the pharmacology of these drugs led to the discovery of resistin). The net result is an increase in insulin sensitivity.

Many diabetic patients use a combination of drugs to help lower blood glucose levels. The most widely prescribed drugs can be taken orally, although they all have side effects. For example, the increased risk of heart attacks has severely restricted the use of rosiglitazone.

The metabolic syndrome links obesity and diabetes

In diabetes, the body behaves as if it were starving. Paradoxically, about 80% of patients with type 2 diabetes are obese, and obesity—particularly when abdominal

[**TABLE 19-5**] Some Antidiabetic Drugs

Class	Example	Mechanism of Action
Biguanides	Metformin (Glucophage)	Stimulates AMPK; reduces glucose release from liver; increases glucose uptake by muscle
Sulfonylureas	Glipizide (Glucotrol)	Blocks a K^+ channel in β cells, leading to increased production and secretion of insulin
Thiazolidinediones	Rosiglitazone (Avandia)	Binds to peroxisome proliferator–activated receptors to activate gene transcription; increases insulin sensitivity

fat deposits are large—is strongly correlated with the development of the disease. Some researchers use the term **metabolic syndrome** to refer to a set of symptoms, including obesity and insulin resistance, that appear to be related. As many as 40% of Americans over age 60 meet the criteria for a diagnosis of metabolic syndrome. Individuals with this disorder often develop type 2 diabetes, they may have atherosclerosis and hypertension (high blood pressure) that put them at risk for a heart attack, and they have a higher incidence of cancer. Several factors appear to underlie metabolic syndrome and to link obesity with diabetes.

Individuals with metabolic syndrome tend to have a relatively high proportion of visceral fat (assessed as a high waist-to-hip ratio). This type of fat exhibits a different hormone profile than subcutaneous fat. For example, visceral fat produces less leptin and adiponectin (hormones that increase insulin sensitivity) and more resistin (which promotes insulin resistance). Visceral fat also produces a hormone called tumor necrosis factor α (TNFα), which is a powerful mediator of inflammation, a normal part of the body's immune defenses. Chronic inflammation triggered by the visceral fat–derived TNFα may be responsible for some of the symptoms, such as atherosclerosis, that characterize metabolic syndrome. The TNFα signaling pathway in cells may lead to phosphorylation of IRS-1, a modification that prevents its activation by the insulin receptor kinase. This would explain the insulin resistance of metabolic syndrome and may also explain why disturbances in the intestinal microbiome that lead to inflammation (see Box 19-B) may trigger insulin resistance.

Another possible cause of metabolic syndrome, which may operate in concert with inflammation, is fat toxicity. High levels of dietary fatty acids promote fat accumulation in muscle tissue in addition to adipose tissue, impairing GLUT4 translocation and impeding glucose uptake. Circulating fatty acids also trigger gluconeogenesis in the liver, contributing to hyperglycemia. Pancreatic β cells respond to the hyperglycemia by increasing insulin secretion, which may stress the cells to the point of death, resulting in "β cell exhaustion."

Whatever its biochemical basis, the link between obesity and metabolic syndrome is underscored by the improvement of symptoms when the individual loses weight. If lifestyle changes related to diet and exercise are not effective, metabolic syndrome can be treated by the same drugs used to treat type 2 diabetes, since these increase insulin sensitivity.

CONCEPT REVIEW
- Summarize the metabolic changes that occur during starvation.
- Explain how a signaling molecule such as leptin could help determine the set-point for body weight.
- What are the differences between type 1 and type 2 diabetes?
- Explain why hyperglycemia is a symptom of diabetes.
- Describe the action of some antidiabetic drugs.
- How is obesity related to type 2 diabetes?

[SUMMARY]

19-1 Integration of Fuel Metabolism

- The liver is specialized to store glucose as glycogen, to synthesize triacylglycerols, to carry out gluconeogenesis, and to synthesize ketone bodies and urea. The muscles synthesize glycogen and can use glucose, fatty acids, and ketone bodies as fuel. Adipose tissue stores fatty acids as triacylglycerols.
- Pathways such as the Cori cycle and the glucose–alanine cycle link different organs.

19-2 Hormonal Control of Fuel Metabolism

- Insulin, which is synthesized by the pancreas in response to glucose, binds to a receptor tyrosine kinase. Cellular responses to insulin include increased uptake of glucose and fatty acids.

- The balance between glycogen synthesis and degradation depends on the relative activities of glycogen synthase and glycogen phosphorylase, which are controlled by hormone-triggered phosphorylation and dephosphorylation.
- Glucagon and catecholamines lead to the activation of cAMP-dependent protein kinase, which promotes glycogenolysis in liver and muscle, and lipolysis in adipose tissue.
- Adipose tissue is the source of the hormones leptin, adiponectin, and resistin, which help regulate appetite, fuel combustion, and insulin resistance. The stomach, intestines, and other organs also produce hormones that regulate appetite.
- AMP is an allosteric activator of AMPK, whose activity switches on pathways such as glycolysis and fatty acid oxidation.

19-3 Disorders of Fuel Metabolism

• In starvation, glycogen stores are depleted, but the liver makes glucose from amino acids and converts fatty acids into ketone bodies.
• The cause of obesity is not clear but may involve a failure in leptin signaling that raises the body weight set-point.

• The most common form of diabetes is characterized by insulin resistance, the inability to respond to insulin. The resulting hyperglycemia can lead to tissue damage.
• Metabolic disturbances resulting from obesity may lead to insulin resistance, a condition termed metabolic syndrome.

[GLOSSARY TERMS]

Warburg effect
Cori cycle
microbiome
glucose–alanine cycle
hormone

diabetes mellitus
lipolysis
marasmus
kwashiorkor
set-point

brown adipose tissue
insulin resistance
hyperglycemia
metabolic syndrome

[PROBLEMS]

19-1 Integration of Fuel Metabolism

1. Name the two small metabolites at the "crossroads" of metabolism. How are these metabolites connected to the metabolic pathways we have studied?

2. Glucose-6-phosphate (G6P) is a metabolite that is linked to several metabolic pathways in carbohydrate metabolism. Describe how glucose-6-phosphate is linked to these pathways.

3. Incubating brain slices in a medium containing ouabain (an Na,K-ATPase inhibitor) decreases respiration by 50%. What does this tell you about ATP use in the brain? What pathways are involved in producing ATP in the brain?

4. Red blood cells lack mitochondria. Describe the metabolic pathways involved in ATP production in the red blood cell. What is the ATP yield per glucose molecule?

5. Glycogen phosphorylase cleaves glucose residues from glycogen via a phosphorolytic rather than a hydrolytic cleavage.
(a) Write an equation for each process.
(b) What is the metabolic advantage of phosphorolytic cleavage?

6. Adenylate kinase catalyzes the reaction

$$AMP + ATP \rightleftharpoons 2\,ADP$$

(a) How can you tell whether this reaction is likely to be a near-equilibrium reaction?
(b) Explain why muscle adenylate kinase would be very active during vigorous exercise.

7. During exercise, the concentration of AMP in muscle cells increases (see Solution 6). AMP is a substrate for the adenosine deaminase reaction:

$$AMP + H_2O \rightarrow IMP + NH_4^+$$

AMP is subsequently regenerated by a process in which the amino group of aspartate becomes attached to the purine ring of IMP and fumarate is released (this set of reactions is known as the purine nucleotide cycle).
(a) What is the likely fate of the fumarate product?
(b) Why doesn't the muscle cell increase the concentration of citric acid cycle intermediates by converting aspartate to oxaloacetate by a simple transamination reaction?

8. Ammonium ion stimulates the activity of phosphofructokinase and pyruvate kinase. Use this information and your responses to Problems 6 and 7 to explain how adenosine deaminase activity could promote ATP production in active muscle.

9. What is the "energy cost" in ATP of running the Cori cycle? How is the ATP obtained?

10. In the Cori cycle, the muscle converts pyruvate into lactate, which then diffuses out of the muscle and travels via the bloodstream to the liver, where the reaction is reversed and lactate is converted back to pyruvate. Why is this extra step necessary? Why doesn't the muscle simply release pyruvate for uptake by the liver?

11. Explain how the reactions of the glucose–alanine cycle would operate during starvation.

12. What happens to plasma alanine levels in patients with inherited diseases of pyruvate or lactate metabolism that result in elevated plasma pyruvate levels? Explain.

13. An infant seemed normal at birth but was diagnosed at the age of three months with a pyruvate carboxylase deficiency. She suffered from lactic acidosis and ketosis. She had poor muscle tone and was experiencing seizures.
(a) Which metabolites would be elevated in this patient? Which metabolites would be deficient?
(b) Why does the patient suffer from lactic acidosis and ketosis?
(c) The poor muscle tone results from a lack of the neurotransmitter amino acids glutamate, aspartate, and γ-aminobutyric acid (GABA). Why would a pyruvate carboxylase deficiency decrease synthesis of these neurotransmitters?
(d) Acetyl-CoA was added to the patient's cultured fibroblasts to see whether pyruvate carboxylase activity could be detected. What was the rationale behind this experiment?

14. Physicians treating a second infant with a pyruvate carboxylase deficiency (see Problem 13) noted that the patient suffered from hyperammonemia and that plasma levels of citrulline were elevated. Provide an explanation for this observation.

15. Treating farm animals such as cows and chickens with low doses of antibiotics promotes weight gain. (a) Propose an explanation for this observation. (b) The widespread use of antibiotics in animals has been implicated in the rise of antibiotic resistance in human pathogenic bacteria. How might this happen?

16. The caterpillar (larva) of the blue *Morpho* butterfly contains about 20 mg of fatty acids, compared to about 7 mg in the adult butterfly. What does this tell you about the energy source for metamorphosis (when the insect does not eat)?

17. During early lactation, cows cannot eat enough to supply the nutrients required for milk production. A recent study showed that enzymes involved in glycogen synthesis and the citric acid cycle were decreased during the early lactation period, whereas glycolytic enzyme activity, lactate production, and fatty acid degradation activity increased. What metabolic strategies are used by the cows in order to obtain the nutrients required for milk production in the absence of sufficient food intake? Be sure to specify which tissues are involved.

18. The compound 2-deoxy-D-glucose is structurally similar to glucose and can be taken up by cells via glucose transporters.

(a) Once inside the cells, the compound is acted upon by hexokinase to produce 2-deoxy-D-glucose-6-phosphate. Write the balanced equation for this reaction.

(b) If 2-deoxy-D-glucose is added to cultured cancer cells, the intracellular concentration of ATP rapidly decreases. In a separate experiment, antimycin A (which prevents the transfer of electrons to cytochrome *c* in the electron transport chain) is added to the cells, but in this case there is no effect on ATP production. Explain these results.

19-2 Hormonal Control of Fuel Metabolism

19. Why does the sigmoidal behavior of glucokinase, the liver isozyme of hexokinase, help the liver adjust its metabolic activities to the amount of available glucose?

20. Why might small molecule allosteric activators of glucokinase be effective in treating diabetes?

21. Explain why a tyrosine phosphatase might be involved in limiting the signaling effect of insulin.

22. Why is insulin required for triacylglycerol synthesis in adipocytes?

23. Glycogen synthase kinase 3 (GSK3) can phosphorylate glycogen synthase in muscle cells. Activation of the insulin receptor leads to the activation of protein kinase B (Akt; see Section 10-2), which phosphorylates GSK3. How does insulin affect glycogen metabolism through GSK3?

24. Insulin resistance is characterized by the failure of insulin-sensitive tissues to respond to the hormone, which normally results in the uptake of glucose and its subsequent conversion to a storage form, such as glycogen or triacylglycerols. Why might glycogen synthase kinase 3 (GSK3, see Problem 23) be useful for treating diabetes?

25. In addition to its role in activating AMPK, adiponectin blocks the phosphorylation of glycogen synthase kinase 3 (GSK3, see Problem 23). How does the lack of adiponectin in an obese person predispose the individual to insulin resistance (see Problem 24)?

26. Inexperienced athletes might consume a meal high in glucose just before a race, but veteran marathon runners know that doing so would impair their performance. Explain.

27. The results of one study showed that glucagon leads to an increase in the rate of glucose-6-phosphate hydrolysis.

(a) How could this explain the following results: Phosphoenolpyruvate concentration increased twofold, glucose-6-phosphate concentration decreased by 60%, and hepatic glucose concentration was increased twofold in the presence of glucagon and exogenously administered dihydroxyacetone phosphate.

(b) Inhibition of glucose-6-phosphate hydrolysis resulted in both activation of gluconeogenesis and inhibition of glycolysis. Explain.

28. A 15-year-old male patient sees a physician because his parents are concerned about his inability to perform any kind of strenuous exercise without suffering painful muscle cramps. His liver is normal in size, but his muscles are flabby and poorly developed. Liver and muscle biopsies reveal that glycogen content in the liver is normal, but muscle glycogen content is elevated. The biochemical structure of glycogen in both tissues appears to be normal. A fasting glucose test shows that he is neither hypo- nor hyperglycemic. The patient's response to glucagon is tested by injecting a high dose of glucagon intravenously and then drawing samples of blood periodically and measuring the glucose content.

(a) After the glucagon injection, the patient's blood sugar rises dramatically. Is this the response you would expect in a normal person? Explain.

(b) Suggest an explanation for the results obtained from the liver and muscle biopsies. What type of glycogen storage disease does this patient have?

(c) The patient performs 30 minutes of ischemic (anaerobic) exercise and blood is withdrawn every few minutes and analyzed for alanine. In a normal person, the concentration of alanine in blood increases during ischemic exercise. But in the patient, alanine decreases during exercise, leading you to believe that his muscle cells are taking up alanine rather than releasing it. Why would blood alanine concentrations increase in a normal person? Why do blood alanine concentrations decrease in the patient?

(d) Explain why the patient does not suffer from either hypo- or hyperglycemia.

(e) The patient is advised to avoid strenuous exercise. If he does wish to perform light or moderate exercise, he is advised to consume sports drinks containing glucose or fructose frequently while exercising. Why would this help alleviate the muscle cramps suffered during exercise?

29. Phosphorylase kinase is the one of the most complex enzymes known. It consists of four copies each of four different subunits, denoted as $\alpha_4\beta_4\gamma_4\delta_4$. The γ subunit contains the catalytic site. The α and β subunits can be phosphorylated. The δ subunit is calmodulin (see Section 10-2). What does this information tell you about the regulation of this enzyme's activity?

30. Hyperthyroidism is a condition that occurs when the thyroid secretes an excess of hormones, leading to an increase in the metabolic rate of all the cells in the body. Hyperthyroidism increases the demand for glucose and for substrates for oxidative phosphorylation. Which metabolic pathways are active in the fasting state in the liver, muscle, and adipose tissue in order to meet this demand?

31. AMPK affects gene transcription. Predict the effect of AMPK activation on the expression of (a) muscle GLUT4 and (b) liver glucose-6-phosphatase.

32. How would phosphorylation by AMPK affect the activity of the following enzymes?

(a) glycogen synthase
(b) HMG-CoA reductase
(c) hormone-sensitive lipase
(d) phosphorylase kinase

33. The compound 5-aminoimidazole-4-carboxamide ribonucleoside (AICAR) activates AMPK. Adding AICAR to cancer cells in culture increases the concentration of reactive oxygen species (see Box 19-A). Explain why.

34. Insulin activates cAMP phosphodiesterase. Explain why this augments insulin's metabolic effect.

35. What would happen if IRS-1 were overexpressed in rat muscle cells in culture?

36. Why is the liver sometimes referred to as the body's "glucose buffer"?

19-3 Disorders of Fuel Metabolism

37. Explain why fasting results in an increase in the liver concentrations of phosphoenolpyruvate carboxykinase and glucose-6-phosphatase.

38. After several days of starvation, the ability of the liver to metabolize acetyl-CoA via the citric acid cycle is severely compromised. Explain why.

39. During a 24-hour fast, a person utilizes protein at a rate of 75 g/day. If a nonobese person has 6000 g of protein reserves and if death occurs when 50% of the protein reserves have been utilized, how prolonged can the fast be before death occurs?

40. In fact, during starvation, protein utilization does not progress at a rate of 75 g/day (Problem 39) but dramatically slows down to 20 g/day as the fast increases in duration. What body fuels are utilized during a prolonged fast in order to conserve body protein?

41. Why do dieters who follow the Atkins diet (a diet high in fat and protein and very low in carbohydrate) sometimes suffer from bad breath? (*Hint*: The odorous component of the breath is acetone.)

42. Individuals who are trying to lose weight are advised to consume fewer calories as well as to exercise. Why would exercising (keeping muscles active) help promote the loss of stored fat from adipose tissue?

43. Adipocytes secrete leptin, a hormone that suppresses appetite. Leptin exerts its effects through the central nervous system and also directly on target tissues by binding to specific receptors. Leptin can inhibit insulin secretion but can also act as an insulin mimic by activating some of the same intracellular signaling components as insulin. For example, leptin can induce tyrosine phosphorylation of insulin receptor substrate-1 (IRS-1). Using this information, predict leptin's effect on the following:
(a) glucose uptake by skeletal muscle
(b) hepatic glycogenolysis and liver glycogen phosphorylase activity
(c) cAMP phosphodiesterase

44. Adults have deposits of brown adipose tissue located mainly in the muscles of the lower neck and collarbone.
(a) Brown adipose tissue expresses greater amounts of cytochrome *c* than white adipose tissue. What is the purpose of the elevated cytochrome *c*?
(b) Investigators measured the uptake of labeled glucose into brown fat in volunteers under room-temperature conditions and while the volunteers placed one foot in 7–9°C water. There was a 15-fold increase in uptake of labeled glucose by brown fat when the subjects were exposed to the colder temperature. Explain.

45. Decreased carnitine acyltransferase activity along with reduced activity of the mitochondrial electron transport chain have been observed in obese individuals. Explain the significance of these observations.

46. The activity of acetyl-CoA carboxylase is stimulated by a fat-free diet and inhibited in starvation and diabetes. Explain.

47. The properties of acetyl-CoA carboxylase (ACC) were studied to see whether the enzyme might be a possible drug target to treat obesity. Mammals have two forms of acetyl-CoA carboxylase, termed ACC1 and ACC2, whose properties are summarized in the table.

	ACC1	**ACC2**
Molecular mass (D)	265,000	280,000
Tissue expression	Liver and adipose tissue	Heart and muscle
Cellular location	Cytosol	Mitochondrial matrix
Sensitive to regulation by malonyl-CoA	Yes	Yes

(a) How does malonyl-CoA regulate the activity of acetyl-CoA carboxylase?
(b) In ACC2 knockout mice, the ACC2 gene is missing, but the ACC1 gene is still expressed normally. The ACC2-knockout mice showed a 20% reduction in liver glycogen compared to control mice. Explain why.
(c) In the knockout mice, the concentration of fatty acids in the blood was lower but the concentration of triacylglycerols was higher than in the wild-type mice. Explain.
(d) Fatty acid oxidation was measured in muscle tissue samples collected from both knockout and control mice. Administration of insulin caused a 45% decrease in palmitate oxidation in muscle tissue from normal mice, but there was no change in the rate of palmitate oxidation in the knockout mice. Explain.
(e) Both knockout and normal mice were allowed access to as much food as they cared to eat. At the end of a 27-week period, the knockout mice had consumed 20–30% more food than the wild-type mice. Interestingly, despite the increased food intake, the knockout mice weighed about 10% less and accumulated less fat in the adipose tissue than normal mice. Explain.
(f) How might you design the next new "diet pill," based on these results?

48. Fatty acid synthase inhibitors have been investigated as possible candidates for weight-loss drugs. A fatty acid synthase (FAS) inhibitor called C75 was synthesized (this is "Inhibitor D" in Problem 17-34).
(a) Mice were injected intraperitoneally with C75 and radioactively labeled acetate. What is the fate of the label?
(b) Mice receiving intraperitoneal injections of C75 reduced their food intake by more than 90% and lost nearly one-third of their body weight, although they gained back the weight when the drug was withdrawn. The investigators measured brain concentrations of neuropeptide Y (NPY), a compound known to act on the hypothalamus to increase appetite during starvation. Based on the results presented here, predict the effect of C75 on brain levels of NPY.
(c) Because hepatic malonyl-CoA levels were high in the C75-treated mice but not in control mice, the investigators hypothesized that malonyl-CoA inhibits feeding. If their hypothesis is correct, predict what would happen if mice were pretreated with an acetyl-CoA carboxylase inhibitor prior to injection with C75.
(d) What other cellular metabolites accumulate when concentrations of malonyl-CoA rise? (These molecules are candidates for signaling molecules, which might stimulate a biochemical pathway that decreases appetite.)

49. Explain why some drugs used to treat type 1 diabetes are compounds that diffuse into the cell and activate tyrosine kinases.

50. PTP-1B is a phosphatase that dephosphorylates the insulin receptor and might also dephosphorylate IRS-1.

(a) After feeding, mice deficient in PTP-1B reduce circulating blood glucose levels, using half as much insulin as normal mice. Explain this observation.

(b) What intracellular changes are observed in muscle cells when insulin is injected into the PTP-1B–deficient mice?

(c) How would you use this information to design a drug to treat diabetes? Are there any concerns involved in using the drug you have designed?

51. Several studies have shown that the insulin insufficiency of type 1 diabetes is accompanied by a hypersecretion of glucagon. The excess glucagon leads to release of glucose from the liver, which exacerbates the hyperglycemia characteristic of untreated diabetes. This observation has led to the suggestion that a diabetic's treatment regimen should include the administration of a glucagon antagonist along with insulin. A glucagon antagonist binds to receptors on the surface of liver cells but cannot initiate signal transduction. The antagonist prevents endogenous glucagon from binding and stimulating glycogenolysis.

Construction of a glucagon antagonist involves modifying the peptide hormone in such a way that amino acid residues important for binding are retained, while residues important for signal transduction are modified. There is an aspartate residue in the glucagon receptor that is essential for glucagon binding.

Glucagon analogs were synthesized in which amino acid residues were modified at several positions. The resulting analogs were tested for their ability to bind to liver membrane receptors and to initiate signal transduction (determined by measuring an increase in cAMP). A true antagonist binds to receptors without eliciting a response. Analogs capable of binding with diminished activity are referred to as partial agonists.

Sequence of human glucagon

1	2	3	4	5	6	7	8	9	10
His	Ser	Gln	Gly	Thr	Phe	Thr	Ser	Asp	Tyr
11	12	13	14	15	16	17	18	19	20
Ser	Lys	Tyr	Leu	Asp	Ser	Arg	Arg	Ala	Gln
21	22	23	24	25	26	27	28	29	
Asp	Phe	Val	Gln	Trp	Leu	Met	Asn	Thr	

(a) Glucagon carries out its biological function by binding to hepatic receptors and putting into motion a series of events that leads to glycogenolysis. Draw a diagram showing the steps of this process.

(b) Why did the investigators choose to modify the amino acids at positions 1, 12, and 18?

(c) Some of the synthetic glucagon analogs are listed in the table below. The ability of the glucagon analogs to both bind to receptors and elicit a biological response was measured and compared to native glucagon. What is the effect of substituting or eliminating the amino acid at position 9? Of replacement or modification at position 12? Of replacement at position 18? What is the role of the histidine at position 1?

(d) Of the glucagon analogs presented here, which is the best glucagon antagonist?

Glucagon analog (The *des* prefix indicates a deleted amino acid residue)	Binding affinity (%)	% of maximum activity
Glucagon	100	100
Des-Asp^9	45	8.3
Lys^9	54	0
Ala^{12}	17.3	59.7
Glu^{12}	1.0	80.4
Ala^{18}	13	94.4
Leu^{18}	56	95
Glu^{18}	6.2	100
Des-His^1	63	44
Des-His^1-Des-Asp^9	7	0
Des-His^1-Lys^9	70	0

52. Some obese patients with type 2 diabetes have undergone gastric bypass surgery, in which the upper part of the stomach is reconnected to the lower part of the small intestine. In some patients, the surgery appears to cure the symptoms of diabetes even before the patient has lost any weight. Propose an explanation for this observation.

53. One target of AMPK is phosphofructokinase-2, the enzyme that catalyzes the synthesis of fructose-2,6-bisphosphate (see Section 13-1). How does the stimulation of AMPK assist in the treatment of diabetes?

54. There is convincing evidence that AMPK can phosphorylate acetyl-CoA carboxylase. It's also possible that AMPK phosphorylates protein kinase B (see Section 10-2), which increases the translocation of GLUT4 vesicles to the plasma membrane. Given this information, how does metformin treat the symptoms of metabolic syndrome?

[SELECTED READINGS]

Friedman, J. M., Causes and control of excess body fat, *Nature* **459**, 340–342 (2009). [A short question-and-answer summary of obesity and the role of leptin.]

Hardie, D. G., AMP-activated protein kinase—an energy sensor that regulates all aspects of cell function, *Genes Dev.* **25**, 1895–1908 (2011). [An extensive review of AMPK's effects on fuel metabolism and other aspects of cell growth.]

Rosen, E. D., and Spiegelman, B. M., Adipocytes as regulators of energy balance and glucose homeostasis, *Nature* **444**, 847–853 (2006). [Includes a summary of the actions of adipocyte hormones.]

Saltiel, A. R., and Kahn, C. R., Insulin signaling and the regulation of glucose and lipid metabolism, *Nature* **414**, 799–806 (2001). [Summarizes the mechanisms of insulin action and insulin resistance.]

Vander Heiden, M. G., Cantley, L. C., and Thompson, C. B., Understanding the Warburg effect: the metabolic requirements of cell proliferation, *Science* **324**, 1029–1033 (2009). [Explains why rapidly growing cells rely on glycolysis rather than oxidative metabolism.]

20

DNA REPLICATION AND REPAIR

[© Alfred Pasieka/Photo Researchers, Inc.]

▶▶ HOW does DNA fit inside the nucleus?

A human cell contains 46 separate DNA molecules—chromosomes—comprising over 6 billion base pairs. A single DNA helix of this size would be slightly longer than 2 m, but the average diameter of a mammalian nucleus, such as the one shown at left, is only 6 μm (0.000006 m). Fitting all the DNA into the nucleus would be like stuffing a 100-km strand of hair into your backpack. Even if the DNA were broken into 46 pieces, you would still have a hard time packing it up without breaking it or tangling it. Fortunately, cells have a more efficient strategy for packing DNA so that it stays intact, organized, and accessible.

THIS CHAPTER IN CONTEXT

Part 1 Foundations

Part 2 Molecular Structure and Function

Part 3 Metabolism

Part 4 Genetic Information

  20 DNA Replication and Repair

Do You Remember?

- A DNA molecule contains two antiparallel strands that wind around each other to form a double helix in which A and T bases in opposite strands, and C and G bases in opposite strands, pair through hydrogen bonding (Section 3-1).

- Double-stranded nucleic acids are denatured at high temperatures; at lower temperatures, complementary polynucleotides anneal (Section 3-1).

- A DNA molecule can be sequenced or amplified by using DNA polymerase to make a copy of a template strand (Section 3-4).

- A reaction that breaks a phosphoanhydride bond in ATP occurs with a large change in free energy (Section 12-3).

(a) (b)

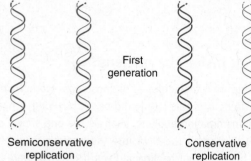

Parental DNA

First generation

Semiconservative replication

Conservative replication

Figure 20-1 Semiconservative versus conservative DNA replication. (a) Experiments performed by Meselson and Stahl demonstrated that new DNA molecules contain one parental (heavy) and one new (light) polynucleotide strand. Thus, DNA replication is semiconservative. (b) If DNA replication were conservative, the parental DNA (both strands heavy) would persist, while new DNA would consist of two light strands.

⊕ **See Animated Figure. Meselson and Stahl experiment.**

? **What would the second generation of DNA look like for each mode of replication?**

When Watson and Crick described the complementary, double-stranded nature of DNA in 1953, they recognized that DNA could be duplicated by a process involving separation of the strands followed by the assembly of two new complementary strands. This mechanism of copying, or **replication,** was elegantly demonstrated by Matthew Meselson and Franklin Stahl in 1958. They grew bacteria in a medium containing the heavy isotope ^{15}N in order to label the cells' DNA. The bacteria were then transferred to fresh medium containing only ^{14}N, and the newly synthesized DNA was isolated and sedimented according to its density in an ultracentrifuge. Meselson and Stahl found that the first generation of replicated DNA had a lower density than the parental DNA but a higher density than DNA containing only ^{14}N. From this, they concluded that newly synthesized DNA is a hybrid containing one parental (heavy) strand and one new (light) strand. In other words, DNA is replicated **semiconservatively.** Because Meselson and Stahl did not observe any all-heavy DNA in the first generation, they were able to discount the possibility that DNA was copied in a way that left intact—or conserved—the original double-stranded molecule (**Fig. 20-1**).

Although simple in principle, DNA replication is a process of many steps, involving many enzymes and accessory factors that may also be called into action to restore damaged DNA. The complexity of DNA replication and repair mechanisms reflects the relatively large size of DNA molecules, the need to copy them quickly and accurately, and the importance of maintaining the integrity of the genome. In addition, the very structure of DNA—its helical twisting and its extraordinary length—presents challenges to the cellular machinery for replication and other processes. In this chapter we examine the topology of DNA, the enzymes that replicate DNA and repair damaged DNA, and the packaging of DNA molecules in eukaryotic cells.

20-1 DNA Supercoiling

KEY CONCEPTS
- The DNA helix is slightly underwound so that it is negatively supercoiled.
- Topoisomerases alter DNA supercoiling by transiently cutting one or both strands of DNA.

⊕ **See Guided Exploration. DNA supercoiling.**

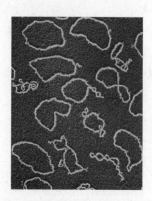

Before DNA can be replicated, its two strands, which are coiled around each other in the form of a double helix, must be separated. Inside cells, the DNA helix is already slightly unwound. However, in order to maintain a conformation close to the stable B form, the DNA molecule twists up on itself, much like the cord of an old-fashioned telephone. This phenomenon, termed **supercoiling,** is readily apparent in small circular DNA molecules (**Fig. 20-2**).

The geometry of a DNA molecule can be described by the branch of mathematics known as topology. For example, consider a strip of paper. When the strip is looped once, it is said to have one writhe. If the ends of the paper are gently pulled in opposite directions, the strip is deformed into a shape called a twist.

A writhe ⇌ A twist

Twisting the strip further introduces more writhes; twisting it in the opposite direction removes writhes, or introduces negative writhes. The same topological terms

Figure 20-2 Supercoiled DNA molecules. The circular DNA molecules are slightly underwound, so they coil up on themselves, forming supercoils. [Dr. Gopal Murti/Photo Researchers Inc.]

apply to DNA: *Each writhe, or supercoil, in DNA is the result of overtwisting or under-twisting the DNA helix.* The twisted molecule, like the piece of paper, prefers to writhe, since this is more energetically favorable.

To demonstrate supercoiling for yourself, cut a flat rubber band to obtain a linear piece a few inches long. Hold the ends apart, twist one of them, and then bring the ends closer together. You will see the twists collapse into writhes (supercoils). The same thing happens if you let the rubber band relax and then twist it in the opposite direction. Note that the twisted rubber band (representing a double-stranded DNA molecule) adopts a more compact shape. In fact, supercoiling is essential for packaging DNA efficiently into a small space inside a bacterial cell or in a eukaryotic nucleus.

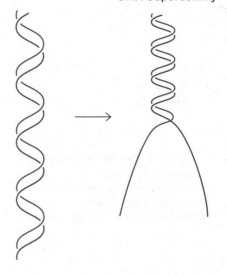

Naturally occurring DNA molecules are negatively supercoiled. This means that if the DNA were stretched out (that is, its writhes converted to twists), the two strands would unwind, in effect loosening the double helix. Consequently, *the negative supercoiling of DNA helps the replication machinery gain access to the two template strands during synthesis of two new complementary DNA strands.* In the absence of negative supercoils, strand separation would force additional twisting ahead of the separation point (shown at right).

Topoisomerases alter DNA supercoiling

Normal replication and transcription require that cells actively maintain the supercoiling state of DNA by adding or removing supercoils. It is difficult to imagine how such a long molecule could be pulled taut and its ends twisted or untwisted. Instead, topological changes in DNA molecules occur when an enzyme called a **topoisomerase** cuts one or both strands of the supercoiled DNA, alters the structure, and then reconnects the broken strands. Type I topoisomerases cut one DNA strand; the type II enzymes cut both strands of DNA and require ATP.

Type I topoisomerases occur in all cells and alter supercoiling by altering the DNA's helical twisting. Type IA enzymes **nick** the DNA (cut the backbone of one strand), pass the intact strand through the break, then seal the broken strand in order to unwind the DNA by one turn (**Fig. 20-3**). Type IB enzymes also cut one strand but hold the DNA on one side of the nick while allowing the DNA on the other side of the nick to rotate one or more turns before the broken strand is sealed. In both cases, unwinding is driven by the strain in the supercoiled DNA, so *a type I topoisomerase can "relax" both negatively and positively supercoiled DNA.*

Human topoisomerase I, a type IB enzyme, has four domains that encircle the DNA (**Fig. 20-4**). The surface of the protein that contacts the DNA is rich in positively charged groups that can interact with the backbone phosphate groups of about 10 base pairs. When a type I topoisomerase cleaves one strand of the DNA, an active-site Tyr residue forms a covalent bond with the backbone phosphate at one side of the nick, for example, see the structure at right. Formation of this diester linkage conserves the free energy of the broken phosphodiester bond of the

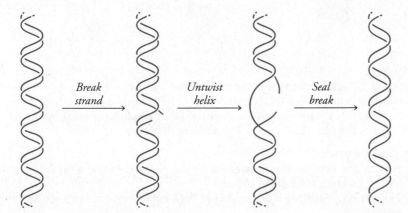

Figure 20-3 Action of a type I topoisomerase. In this scheme, the supercoiling of a DNA segment is decreased when a DNA strand is broken, the helix untwisted by passing the other strand through the break, and then sealed.

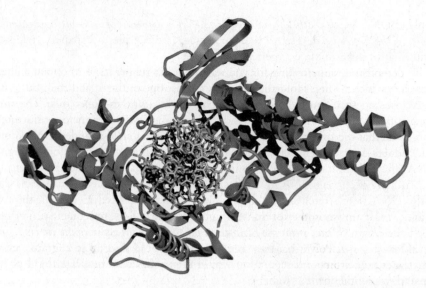

Figure 20-4 Human topoisomerase I.
The protein (blue) surrounds a DNA
molecule (shown end-on in this model).
[Structure (pdb 1A36) determined by L. Stewart,
M. R. Redinbo, X. Qiu, J. J. Champoux, and
W. G. J. Hol.]

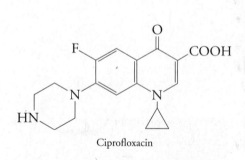

Ciprofloxacin

DNA strand, so strand cleavage and subsequent sealing do not require any other
source of free energy.

Type II topoisomerases directly alter the number of writhes in a supercoiled
DNA molecule by passing one DNA segment through another, a process that requires a double-strand break (Fig. 20-5). The type II enzymes are dimeric, with two
active-site Tyr residues that form covalent bonds to the 5′ phosphate groups of the
broken DNA strands. The enzyme must undergo conformational changes in order
to cleave a DNA molecule and hold the ends apart while another DNA segment
passes through the break. *ATP hydrolysis appears to produce the free energy to restore
the enzyme to its starting conformation.* Type II topoisomerases can relieve both negative and positive supercoiling.

Bacteria contain a type II topoisomerase called DNA gyrase, which can introduce additional negative supercoils into DNA, so its net effect is to further under-
wind the DNA helix. (Eukaryotic cells lack DNA gyrase but maintain negative
supercoiling by wrapping DNA in nucleosomes; Section 20-5.) A number of antibiotics inhibit DNA gyrase without affecting eukaryotic type II topoisomerases.
Drugs such as ciprofloxacin (*left*) act on DNA gyrase to enhance the rate of DNA
cleavage or reduce the rate of sealing broken DNA. The result is a large number of
DNA breaks that interfere with the transcription and replication required for
normal cell growth and division. Ciprofloxacin is prescribed as an antibiotic for a
wide variety of bacterial infections.

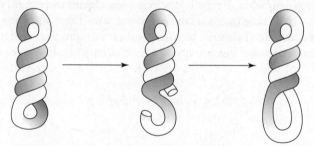

Figure 20-5 Action of a type II topoisomerase. The twisted worm shape represents
a supercoiled double-stranded DNA molecule. A type II topoisomerase breaks both
strands of the DNA, passes another segment of DNA through the break, then seals the
break. In this diagram, the degree of supercoiling decreases.

CONCEPT REVIEW
• What is the relationship between a writhe (supercoil) and a twist? How does
helical twisting alter supercoiling?
• Explain why negative supercoiling of DNA assists processes such as
replication.
• Describe how type I and type II topoisomerases operate. When is a source
of free energy required?

DNA polymerase, the enzyme that catalyzes the polymerization of deoxynucleotides, is just one of the proteins involved in replicating double-stranded DNA. The entire process—separating the two template strands, initiating new polynucleotide chains, and extending them—is carried out by a complex of enzymes and other proteins. In this section we examine the structures and functions of the major players in DNA replication.

Replication occurs in factories

In the circular chromosomes of bacteria, DNA replication begins at a particular site called the origin. Here, proteins bind to the DNA and melt it open in an ATP-dependent manner. Polymerization then proceeds in both directions from this point until the entire chromosome (4.6×10^6 bp in *E. coli*) has been replicated. The point where the parental strands separate and the new strands are synthesized is known as the **replication fork.**

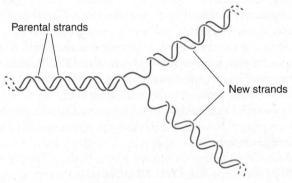

Parental strands

New strands

Replication fork

Replication begins at numerous sites in the 46 chromosomes in the nucleus of each human cell. These sites do not appear to have a defined sequence, as in prokaryotic origins, but instead have a structure that is located by an origin recognition complex.

At one time, complexes of DNA polymerase and other replication proteins were thought to move along DNA like a train on tracks. This "locomotive" model of DNA replication requires that the large replication proteins move along the relatively thin template strand, rotating around it while generating a double-stranded helical product. In fact, cytological studies indicate that DNA replication (as well as transcription) occurs in "factories" at discrete sites. For example, in bacteria, DNA polymerase and associated factors appear to be immobilized in one or two complexes near the plasma membrane. In the eukaryotic nucleus, newly synthesized DNA appears at 100 to 150 spots, each one representing several hundred replication forks (**Fig. 20-6**). *According to the **factory model of replication,** the protein machinery is stationary and the DNA is reeled through it.* In eukaryotes, this organization presumably facilitates the synchronous elongation of many DNA segments, which allows for efficient replication of enormous eukaryotic genomes.

Helicases convert double-stranded DNA to single-stranded DNA

One of the jobs of the origin recognition complex in eukaryotes is to recruit an enzyme called a helicase, which catalyzes unwinding of the DNA helix. The helicase is not yet active but awaits a signal (through kinase-catalyzed phosphorylation) that the cell is ready to copy its DNA. Then, in an ATP-dependent process, the helicase separates the parental DNA strands so that each one can serve as a template for DNA polymerase. This unwinding is aided by the negative supercoiling of DNA. Prokaryotic DNA replication also depends on the activity of helicases.

KEY CONCEPTS
- DNA replication is carried out by stationary protein complexes.
- The proteins required for DNA replication include helicase, single-strand binding protein, primase, DNA polymerase, a sliding clamp, RNase, and DNA ligase.
- DNA polymerase synthesizes the leading strand continuously and the lagging strand as a series of Okazaki fragments.
- The accuracy of DNA polymerase is enhanced by its ability to sense correctly paired nucleotides and to hydrolytically excise mispaired nucleotides.

⊕ **See Guided Exploration. DNA replication in *E. coli*.**

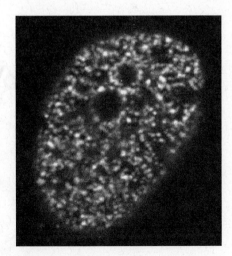

Figure 20-6 Replication foci. The fluorescent patches (foci) in the nucleus of a eukaryotic cell mark the presence of newly synthesized DNA. These sites of DNA replication remain stationary, probably due to attachment of the replication machinery to the nucleoskeleton. [Courtesy A. Pombo. From *Science* 284, 1790–1795 (1999).]

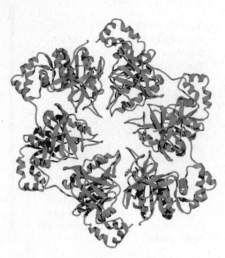

Figure 20-7 Structure of a hexameric helicase. This helicase, from the bacteriophage T7, forms a hexameric ring around a single strand of DNA and pushes double-stranded DNA apart. [Structure (pdb 1E0J) determined by M. R. Singleton, M. R. Sawaya, T. Ellenberger, and D. B. Wigley.]

Some helicases are hexameric proteins that encircle a single strand of DNA (Fig. 20-7). As the helicase pulls the strand through its ring, it unwinds the double-stranded DNA ahead of it. For each ATP hydrolyzed, up to five base pairs of the helix are separated. The helicase appears to operate in a rotary fashion, with conformational changes driven by ATP hydrolysis. This mechanism is reminiscent of the binding change mechanism of the F_1 component of ATP synthase, another hexameric ATP-binding protein (see Section 15-4).

As replication proceeds, the parental DNA strands are continuously separated by the action of the helicase, working in concert with DNA polymerases. A second helicase, which may be a dimer rather than a hexamer, may bind to the other strand of DNA to help open up the replication fork. A topoisomerase working ahead of the fork helps relieve the strain of helix unwinding. Although eukaryotic genomes include numerous origins, each portion of DNA is replicated only once because after DNA synthesis commences, no additional helicases can be loaded onto the origins.

As single-stranded DNA is exposed at the replication fork, it associates with a protein known as single-strand binding protein (SSB). SSB coats the DNA strands to protect them from nucleases and to prevent them from reannealing or forming secondary structures that might impede replication. *E. coli* SSB is a tetramer with a positively charged cleft that accommodates loops of single-stranded—but not double-stranded—DNA. The eukaryotic SSB, called replication protein A, is a larger protein that includes four DNA-binding domains separated by flexible regions (Fig. 20-8). Both prokaryotic and eukaryotic SSBs can adopt different conformations and bind DNA in several different ways. Studies of replication protein A suggest that DNA binding occurs in stages such that the first protein–DNA interaction is relatively weak (with a K_d in the μM range) but allows the other domains to "unroll" along the DNA to form a stable complex that protects about 30 nucleotides and has an overall dissociation constant of about 10^{-9} M. As the template DNA is reeled through the polymerase, SSB is displaced and is presumably redeployed as helicase exposes more single-stranded DNA.

DNA polymerase faces two problems

The mechanism of DNA polymerase and the double-stranded structure of DNA present two potential obstacles to the efficient replication of DNA. First, *DNA polymerase can only extend a preexisting chain; it cannot initiate polynucleotide synthesis.* However, RNA polymerase can do this, so DNA chains *in vivo* begin with a short stretch of RNA that is later removed and replaced with DNA:

(a)   (b)

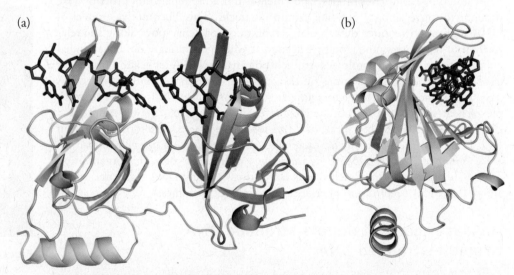

Figure 20-8 DNA-binding domains of replication protein A. Two of the four DNA-binding domains are shown in yellow and green in this model, with a bound octanucleotide (polydeoxycytidine, shown in purple). (a) Front view. (b) Side view. [Structure (pdb 1JMC) determined by A. Bochkarev, R. Pfuetzner, A. Edwards, and L. Frappier.]

? Which types of amino acids would you expect to find in the DNA-binding site of this protein?

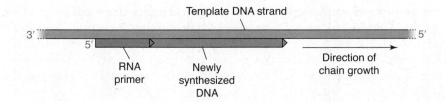

This stretch of RNA, about 10 to 60 nucleotides long, is known as a **primer,** and the enzyme that produces it during DNA replication is known as a primase. As we will see, primase is required throughout DNA replication, not just at the start. The active site of the primase is narrow (about 9 Å in diameter) at one end, where the single-stranded DNA template is threaded through. The other end of the active site is wide enough to accommodate a DNA–RNA hybrid helix (which has an A-DNA–like conformation; see Fig. 3-6).

The second problem facing DNA polymerase is that the antiparallel template DNA strands are replicated simultaneously by a pair of polymerase enzymes. But each DNA polymerase catalyzes a reaction in which the 3′ OH group at the end of the growing DNA chain attacks the phosphate group of a free nucleotide that base pairs with the template DNA strand **(Fig. 20-9)**. For this reason, *a polynucleotide chain is said to be synthesized in the 5′ → 3′ direction.* Because the two template DNA strands are antiparallel, the synthesis of two new DNA strands would require that the template strands be pulled in opposite directions through the replication machinery so that the DNA polymerases could continually add nucleotides to the 3′ end of each new strand.

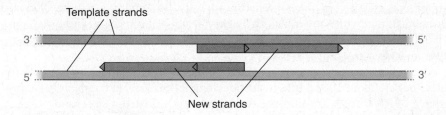

This awkward situation does not occur in cells. Instead, *the two polymerases work side-by-side, and one template DNA strand periodically loops out.* In this scenario, one strand of DNA, called the **leading strand,** can be synthesized in one continuous

Figure 20-9 Mechanism of DNA polymerase. The 3′ OH group (at the 3′ end of a growing polynucleotide chain) is a nucleophile that attacks the phosphate group of an incoming deoxynucleoside triphosphate (dNTP) that base pairs with the template DNA strand. Formation of a new phosphodiester bond eliminates PP$_i$. The reactants and products of this reaction have similar free energies, so the polymerization reaction is reversible. However, the subsequent hydrolysis of PP$_i$ makes the reaction irreversible *in vivo*. RNA polymerase follows the same mechanism.

Figure 20-10 A model for DNA replication. Two DNA polymerase enzymes (green) are positioned at the replication fork to make two complementary strands of DNA. The leading and lagging strands both start with RNA primers (red) and are extended by DNA polymerase in the $5' \rightarrow 3'$ direction. The replication machinery is stationary, and the template DNA is reeled through it. Because the two template strands are antiparallel, the lagging-strand template loops out (the single-stranded DNA becomes coated with SSB). The leading strand is therefore synthesized continuously, while the lagging strand is synthesized as a series of Okazaki fragments.

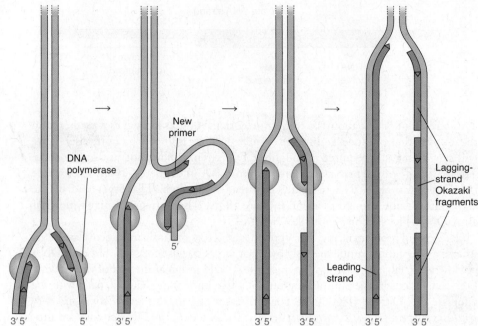

piece. It is initiated by the action of a primase, then extended in the $5' \rightarrow 3'$ direction by the action of one DNA polymerase enzyme. The other strand, called the **lagging strand,** is synthesized in pieces, or **discontinuously.** Its template is repeatedly looped out so that its polymerase can also operate in the $5' \rightarrow 3'$ direction. Thus, *the lagging strand consists of a series of polynucleotide segments, which are called **Okazaki fragments** after their discoverer* (Fig. 20-10).

Bacterial Okazaki fragments are about 500 to 2000 nucleotides long; in eukaryotes, they are about 100 to 200 nucleotides long. Each Okazaki fragment has a short stretch of RNA at its 5' end, since each segment is initiated by a separate priming event. This explains why primase is required throughout replication: Although the leading strand theoretically requires only one priming event, the discontinuously synthesized lagging strand requires multiple primers.

The mechanism for continuous synthesis of the leading strand and discontinuous synthesis of the lagging strand means that the lagging-strand template must periodically be repositioned. Each time an Okazaki fragment is completed, a polymerase begins extending the RNA primer of the next Okazaki fragment. Other protein components of the replication complex assist in this repositioning in order to coordinate the activities of the two DNA polymerases at the replication fork. In *E. coli* and possibly other organisms, three polymerases localize to the replication fork. One synthesizes the leading strand, and the other two probably take turns synthesizing Okazaki fragments; sharing the work appears to increase replication efficiency.

DNA polymerases share a common structure and mechanism

Most DNA polymerases are shaped somewhat like a hand, with domains corresponding to palm, fingers, and thumb. These structure are likely the result of convergent evolution, as only the palm domains exhibit strong homology. One of the best known polymerases is *E. coli* DNA polymerase I, the first such enzyme to be characterized (Fig. 20-11; this is the enzyme used for sequencing DNA; Section 3-4). The template strand and the newly synthesized DNA strand, which form a double helix, lie across the palm, in a cleft lined with basic residues.

The polymerase active site, at the bottom of the cleft, contains two Mg^{2+} ions about 3.6 Å apart. These metal ions are coordinated by Asp side chains of the enzyme and by the phosphate groups of the substrate nucleoside triphosphate. One of the Mg^{2+} ions interacts with the 3' O atom at the end of the primer or growing DNA strand to enhance its nucleophilicity as it attacks the incoming nucleotide:

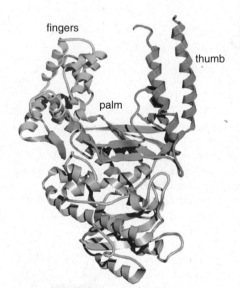

Figure 20-11 *E. coli* DNA polymerase I. This model shows the so-called Klenow fragment of DNA polymerase I (residues 324 to 928). The palm, fingers, and thumb domains are labeled. A loop at the end of the thumb is missing in this model.
[Structure (pdb 1KFD) determined by L. S. Beese, J. M. Friedman, and T. A. Steitz.]
⊕ **See Interactive Exercise.** *E. coli* DNA **polymerase I.**

(a)

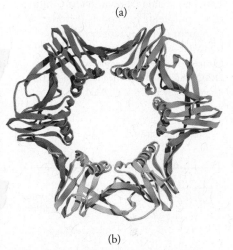

(b)

Figure 20-12 DNA polymerase-associated clamps. In *E. coli*, the β subunit of DNA polymerase III forms a dimeric clamp. (b) In humans, the clamp is a trimer, called the proliferating cell nuclear antigen (PCNA). The inner surface of each clamp is positively charged and encloses a space with a diameter of about 35 Å, more than large enough to accommodate double-stranded DNA or a DNA–RNA hybrid helix with a diameter of 26 Å. Both structures enhance the processivity of their respective DNA polymerases, thereby increasing the efficiency of DNA replication. [Structure of the β clamp (pdb 2POL) determined by X.-P. Kong and J. Kuriyan; structure of PCNA (pdb 1AXC) determined by J. M. Gulbis and J. Kuriyan.]
⊕ **See Interactive Exercise. PCNA.**

? **What structural changes must occur each time the clamp is loaded onto DNA?**

The deoxy carbon at the 2′ position of the nucleotide lies in a hydrophobic pocket. This binding site allows the polymerase to discriminate against ribonucleotides, which bear a 2′ OH group.

After each polymerization event, the enzyme must advance the template strand by one nucleotide. Most DNA polymerases are **processive** enzymes, which means that they undergo several catalytic cycles (about 10 to 15 for *E. coli* DNA polymerase *in vitro*) before dissociating from their substrates. *E. coli* DNA polymerase is even more processive *in vivo*, polymerizing as many as 5000 nucleotides before releasing the DNA. *This enhanced processivity is due to an accessory protein that forms a sliding clamp around the DNA and helps hold DNA polymerase in place.* In *E. coli* the clamp is a dimeric protein, and in eukaryotes the clamp is a trimer. Both types of protein have a hexagonal ring structure and similar dimensions, consistent with a common function **(Fig. 20-12).** Additional proteins help assemble the clamp around the DNA. For the lagging-strand polymerase, the clamp must be reloaded at the start of each Okazaki fragment. This occurs about once every second in *E. coli*.

E. coli has five different DNA polymerase enzymes (I–V), and there are at least 13 different eukaryotic DNA polymerases (designated by Greek letters), not including those found in mitochondria and chloroplasts. Why so many? In eukaryotes and at least some prokaryotes, two different types of polymerases work in concert to synthesize the leading and lagging strands, and other polymerases perform specialized roles. For example, in eukaryotes, DNA polymerase α initiates Okazaki fragment synthesis but is soon replaced with polymerase δ, which is responsible for synthesizing most of the lagging strand, while DNA polymerase ε synthesizes the leading strand. Many polymerases—such as *E. coli* polymerases II, IV, and V and the majority of the eukaryotic polymerases—appear to function primarily in DNA repair pathways, some of which involve the excision and replacement of damaged DNA (Section 20-4).

DNA polymerase proofreads newly synthesized DNA

During polymerization, the incoming nucleotide base pairs with the template DNA so that the new strand will be complementary to the original. The polymerase accommodates base pairs snugly—recall that all possible pairings (A:T, T:A, C:G, and G:C) have the same overall geometry (see Section 3-1). The tight fit minimizes the chance of mispairings. In fact, structural studies indicate that DNA polymerase shifts between an "open" conformation and a "closed" conformation, analogous to the fingers moving closer to the thumb, when a nucleotide substrate is bound **(Fig. 20-13).** This conformational change may facilitate the polymerization of a tightly bound (correctly paired) nucleotide, or it could reflect a mechanism for quickly releasing a mismatched nucleotide before catalysis occurs.

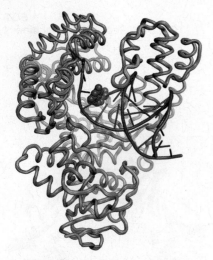

Figure 20-13 Open and closed conformations of a DNA polymerase. The structure of the polymerase from *Thermus aquaticus* was determined in the absence (magenta trace) and presence (green trace) of a nucleotide substrate analog (shown in space-filling form, purple). The models include a segment of DNA representing the template and primer strands (purple). [Structures of the open conformation (pdb 2KTQ) and closed conformation (pdb 3KTQ) determined by Y. Li and G. Waksman.]

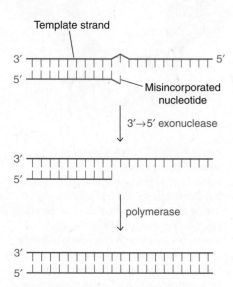

Figure 20-14 Proofreading during polymerization. DNA polymerase detects a distortion in double-stranded DNA that results from the incorporation of a mismatched nucleotide. The $3' \rightarrow 5'$ exonuclease activity hydrolyzes the nucleotide at the 3' end of the new strand (recall that an exonuclease removes residues from the end of a polymer; an endonuclease cleaves within the polymer). The polymerase then resumes its activity, generating an accurately base-paired DNA product.

If the wrong nucleotide does become covalently linked to the growing chain, the polymerase can detect the distortion it creates in the newly generated double helix. Many DNA polymerases contain a second active site that catalyzes hydrolysis of the nucleotide at the 3' end of the growing DNA strand. *This $3' \rightarrow 5'$ exonuclease excises misincorporated nucleotides, thereby acting as a* **proofreader** *for DNA polymerase* (Fig. 20-14). In *E. coli* DNA polymerase I, the $3' \rightarrow 5'$ exonuclease active site is located about 25 Å from the polymerase active site, indicating that the enzyme–DNA complex must undergo a large conformational change in order to shift from polymerization to nucleotide hydrolysis.

Proofreading during polymerization limits the error rate of DNA polymerase to about one in 10^6 bases. Misincorporated bases can also be removed following replication, through various DNA repair mechanisms, which further reduces the error rate of replication. This high degree of fidelity is absolutely essential for the accurate transmission of biological information from one generation to the next.

An RNase and a ligase are required to complete the lagging strand

Replication is not complete until the lagging strand—which is synthesized one Okazaki fragment at a time—is made whole. *As the polymerase finishes each Okazaki fragment, its RNA primer, along with some of the adjoining DNA, is hydrolytically removed and replaced with DNA; then the backbone is sealed to generate a continuous DNA strand.* This process also increases the accuracy of DNA replication: Primase has low fidelity, so the RNA primers tend to contain errors, as do the first few deoxynucleotides added to the primer by the action of the DNA polymerase. For example, in eukaryotes, DNA polymerase α lacks exonuclease activity and therefore cannot proofread its work.

In many cells, an exonuclease known as RNase H (H stands for hybrid) operates in the $5' \rightarrow 3'$ direction to excise nucleotides at the primer end of an Okazaki fragment. Nucleotide hydrolysis may continue until DNA polymerase, now in the process of completing another Okazaki fragment, "catches up" with RNase H (the polymerase is faster than the exonuclease). In extending the newer Okazaki fragment, the polymerase replaces the excised ribonucleotides of the older Okazaki fragment with deoxyribonucleotides, leaving a single-strand nick between the two lagging-strand segments (Fig. 20-15).

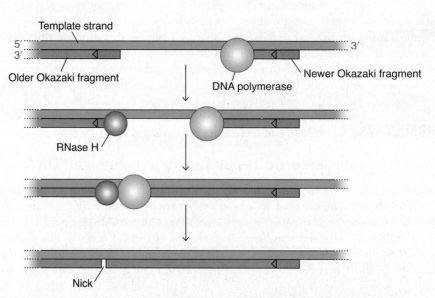

Figure 20-15 Primer excision. RNase H removes the RNA primer and some of the adjoining DNA in the older Okazaki fragment, allowing DNA polymerase to accurately replace these nucleotides. The nick can then be sealed.

In the event that DNA polymerase reaches the older Okazaki fragment before its primer has been completely removed, the polymerase displaces the primer, creating a single-stranded "flap." RNase H or another protein then acts as an endonuclease to cut away the primer (Fig. 20-16). This flap endonuclease appears to recognize the junction between the RNA and DNA portions of a single polynucleotide strand.

The *E. coli* DNA polymerase I polypeptide actually includes a $5' \rightarrow 3'$ exonuclease activity (this is in addition to the $3' \rightarrow 5'$ proofreading endonuclease), so that, at least *in vitro*, a single protein can remove ribonucleotides from the previous Okazaki fragment as it extends the next Okazaki fragment. The combined activities of removing and replacing nucleotides have the net effect of moving the nick in the $5' \rightarrow 3'$ direction. This phenomenon is known as **nick translation**. DNA polymerase cannot seal the nick; this is the function of yet another enzyme.

The discontinuous segments of the lagging strand are joined by the action of DNA ligase. The reaction, which results in the formation of a phosphodiester bond, consumes the free energy of a similar bond in a nucleotide cofactor. Prokaryotes use NAD^+ for the reaction, yielding as products AMP and nicotinamide mononucleotide; eukaryotes use ATP and produce AMP and PP_i.

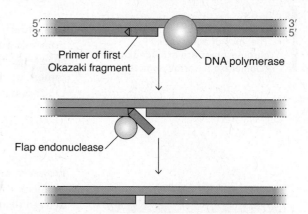

Figure 20-16 Function of the flap endonuclease. If DNA polymerase displaces the RNA primer of the preceding Okazaki fragment, an endonuclease can remove the single-stranded flap.

Eukaryotes   Adenine — Ribose — (P) — (P) — (P)

ATP

Adenine — Ribose — (P)   +   (P) — (P)

AMP PP_i

Prokaryotes   Adenine — Ribose — (P) — (P) — Ribose — Nicotinamide

Nicotinamide adenine dinucleotide (NAD^+)

Adenine — Ribose — (P)   +   (P) — Ribose — Nicotinamide

AMP   Nicotinamide mononucleotide

Ligation of Okazaki fragments yields a continuous lagging strand, completing the process of DNA replication.

CONCEPT REVIEW
- Explain why the factory model for DNA replication is superior to the locomotive model.
- Summarize the roles of helicase and SSB during replication.
- Explain why DNA polymerization must be primed by RNA.
- Describe how two DNA polymerases, operating in the $5' \rightarrow 3'$ direction, replicate the two antiparallel template strands.
- What protein enhances the processivity of DNA polymerase?
- How does the polymerase proofread its mistakes?
- Explain why an RNase, a polymerase, and a ligase are required to complete lagging-strand synthesis.

20-3 Telomeres

In a circular bacterial DNA molecule, replication terminates where the two replication forks meet, at a point more or less opposite the replication origin. In a linear eukaryotic chromosome, replication must proceed to the very end of the chromosome. Copying the 5′ ends of the two parental DNA strands is straightforward, since DNA polymerase extends a new complementary strand in the 5′ → 3′ direction.

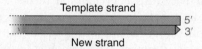

However, replication of the extreme 3′ ends of the parental DNA strands presents a problem, for the same reason. Even if an RNA primer were paired with the 3′ end of a template strand, DNA polymerase would not be able to replace the ribonucleotides with deoxynucleotides (*left*). The 3′ end of each parental (template) DNA strand would then extend past the end of each new strand and would be susceptible to nucleases. Consequently, *each round of DNA replication would lead to chromosome shortening.*

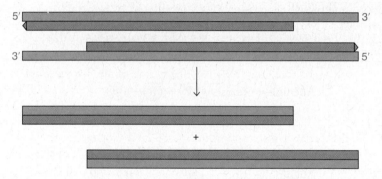

Eukaryotic cells counteract the potential loss of genetic information by actively extending the ends of chromosomes. The resulting structure, which consists of short tandem (side-by-side) repeats, is called a **telomere.** The proteins associated with the telomeric DNA protect the end of the chromosome from nucleolytic degradation and prevent the joining of two chromosomes through end-to-end ligation (a reaction that normally occurs to repair broken DNA molecules).

Telomerase extends chromosomes

The telomere proteins include an enzyme known as telomerase, which was first described by Elizabeth Blackburn. *Telomerase repeatedly adds a sequence of six nucleotides to the 3′ end of a DNA strand, using an enzyme-associated RNA molecule as a template.* The catalytic subunit of telomerase is a **reverse transcriptase,** a homolog of a viral enzyme that copies the viral RNA genome into DNA (Box 20-A).

Although the telomerase catalytic core is highly conserved among eukaryotes, its RNA subunit varies from ~150 to ~1300 nucleotides in different species. In humans, the RNA is a sequence of 451 bases, including the six that serve as a template for the addition of DNA repeats with the sequence TTAGGG. Similar G-rich sequences extend the 3′ ends of chromosomes in all eukaryotes. In order to synthesize telomeric DNA, the template RNA must be repeatedly realigned with the end of the preceding hexanucleotide extension. The precise alignment depends on base pairing between telomeric DNA and a nontemplate portion of the telomerase RNA. When the 3′ end of a DNA strand has been extended by telomerase, it can then serve as a template for the conventional synthesis of a C-rich extension of the complementary DNA strand (Fig. 20-17).

In humans, telomeric DNA is 2 to 10 kb long, depending on the tissue and the age of the individual, and includes a 3′ single-strand overhang of 100–300 nucleotides. This single strand of DNA appears to fold back on itself to form a structure

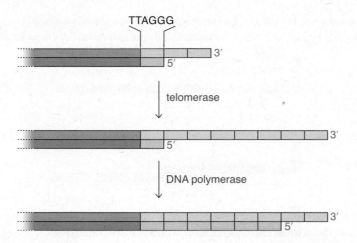

TTAGGG

telomerase

DNA polymerase

Figure 20-17 Synthesis of telomeric DNA. Telomerase extends the 3′ end of a DNA strand by adding TTAGGG repeats (yellow segments). DNA polymerase can then extend the complementary strand by the normal mechanism for lagging-strand synthesis (orange segments). Note that this process still leaves a 3′ overhang (up to 300 nucleotides in humans), but the chromosome has been lengthened.

BOX 20-A ⬡ BIOCHEMISTRY NOTE

HIV Reverse Transcriptase

The human immunodeficiency virus (HIV, introduced in Box 7-B) is a retrovirus, a virus whose RNA genome must be copied to DNA inside the host cell. After entering a cell, the HIV particle disassembles. The 9-kb viral RNA is then transcribed into DNA by the action of the viral enzyme reverse transcriptase. Another viral enzyme, an integrase, incorporates the resulting DNA into the host genome. Expression of the viral genes produces 15 different proteins, some of which must be processed by HIV protease to achieve their mature forms. Eventually, new viral particles are assembled and bud off from the host cell, which dies. Because HIV preferentially infects cells of the immune system, cell death leads to an almost invariably fatal immunodeficiency.

HIV reverse transcriptase, which synthesizes DNA from an RNA template (a contradiction of the central dogma outlined in Section 3-2), resembles other polymerases in having fingers, thumb, and palm domains. These parts of the protein (colored red below) comprise a polymerase active site that can use either DNA or RNA as a template (no other enzyme has this dual specificity). A separate domain (green) contains an RNase active site that degrades the RNA template.

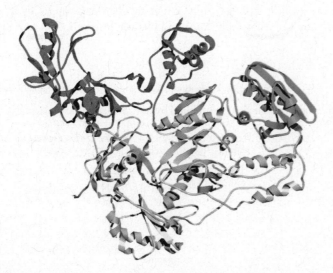

[Structure of HIV reverse transcriptase (pdb 1BQN) determined by Y. Hsiou, K. Das, and E. Arnold.]
⊕ **See Interactive Exercise. HIV reverse transcriptase.**

Reverse transcription occurs as follows: The enzyme binds to the RNA template and generates a complementary DNA strand. In a host cell, DNA synthesis is primed by a transfer RNA molecule. As polymerization proceeds, the RNase active site degrades the RNA strand of the RNA–DNA hybrid molecule, leaving a single DNA strand that then serves as a template for the reverse transcriptase polymerase active site to use in synthesizing a second strand of DNA. The result is a double-stranded DNA molecule.

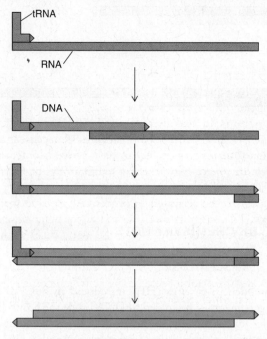

Viral reverse transcriptase has proved to be more than a biological curiosity. It has become a valuable laboratory tool, allowing researchers to purify messenger RNA transcripts from cells, transform them to DNA (called **cDNA** for complementary DNA), and then clone the DNA or express its protein product.

Reverse transcriptase activity can be blocked by two different types of drugs. Nucleoside analogs such as AZT and ddC readily enter cells and are phosphorylated. The resulting nucleotides bind in the reverse transcriptase active site and are linked, via their 5′ phosphate group, to the growing DNA chain. However, because they lack a 3′ OH group, further addition of nucleotides is impossible. Such **chain terminators** are also used for DNA sequencing (Section 3-4).

HOCH$_2$ O Thymine	HOCH$_2$ O Cytosine
H H	H H
H H	H H
$\overset{-}{N}=\overset{+}{N}=N$ H	H H
AZT	**ddC**
(3′-Azido-2′,3′-dideoxythymidine, Zidovudine)	(2′,3′-Dideoxycytidine, Zalcitabine)

Reverse transcriptase can also be inhibited by non-nucleoside analogs such as nevirapine:

Nevirapine

This noncompetitive inhibitor binds to a hydrophobic patch on the surface of reverse transcriptase near the base of the thumb domain. This does not interfere with RNA or nucleotide binding, but it does inhibit polymerase activity, probably by restricting thumb movement.

● **Question:** Explain why the drugs described here interfere only minimally with nucleic acid metabolism in the human host.

called a T-loop (Fig. 20-18). Multiple copies of six different proteins, collectively known as shelterin, bind to the telomere. These proteins play a critical role in regulating telomere length, which may be an indicator of the cells' longevity.

Is telomerase activity linked to cell immortality?

Cells that normally undergo a limited number of cell divisions appear to contain no active telomerase. Consequently, the size of the telomeres decreases with each replication cycle, until the cells reach a senescent stage and no longer divide. For most human cells, this point is reached after about 35–50 cell divisions. In contrast, cells that are "immortal"—such as unicellular organisms, stem cells, and the reproductive cells of multicellular organisms—appear to have active telomerase. These findings are consistent with the role of telomerase in maintaining the ends of chromosomes over many rounds of replication.

Telomerase appears to be activated in cancer cells derived from tissues that are normally senescent. This suggests that shutting down telomerase activity might be an effective treatment for cancer. However, for many cells, telomerase activity by itself is not an indicator of the cells' potential for immortality. Rather, the ability of the cells to undergo repeated divisions without chromosome shortening appears to be a more complicated function of telomerase activity, telomere length, and the integrity of the telomere, which, like other portions of DNA, is susceptible to damage.

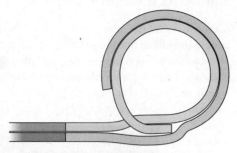

Figure 20-18 A T-loop. Telomeric DNA folds back on itself, and the G-rich single strand invades the double helix. Shelterin proteins (not shown here) associate with both single-stranded and double-stranded DNA to form a cap protecting the chromosome end.

> **CONCEPT REVIEW**
> - Why does DNA replication lead to chromosome shortening?
> - Describe the structure of telomeric DNA.
> - What is the function of the RNA component of telomerase?
> - Explain the relationship between telomeres and cell immortality.

20-4 DNA Damage and Repair

An alteration in a cell's DNA, whether from a polymerization error or another cause, becomes permanent—that is, a **mutation**—unless it is repaired. In a unicellular organism, altered DNA is replicated and passed on to the daughter cells when the cell divides. In a multicellular organism, a mutation is passed to offspring only if the altered DNA is present in the reproductive cells. Mutations that arise in other types of cells affect only the progeny of those cells within the parent organism.

A genetic change can affect the expression of genes positively or negatively (or it may have no measurable effect on the organism). The accumulation of genetic changes over an individual's lifetime may contribute to the gradual loss of functionality associated with aging. Cells with badly damaged DNA tend to be so impaired that they undergo **apoptosis,** or programmed cell death, making room for their replacement by healthy cells. But in some cases, cells with damaged DNA do not die but instead escape the normal growth-control mechanisms and proliferate excessively, resulting in cancer. For this reason, cancer can be considered a disease of the genes (Box 20-B). In this section, we examine different types of DNA damage and the mechanisms cells use to restore DNA integrity.

KEY CONCEPTS
- A mutation in DNA can arise from DNA replication errors, from nonenzymatic cellular processes, and from environmental factors.
- A variety of enzymes, including nucleases, polymerases, and ligases, participate in DNA repair pathways.

DNA damage is unavoidable

DNA damage, even if it does not lead to cancer, is a fact of life. Even with its proofreading activity, DNA polymerase makes mistakes by introducing mismatched bases or incorporating uracil rather than thymine. DNA polymerase may occasionally add or delete nucleotides, producing bulges or other irregularities in the DNA that lead to insertion or deletion mutations.

Cellular metabolism itself exposes DNA to the damaging effects of reactive oxygen species (for example, the superoxide anion $\cdot O_2^-$, the hydroxyl radical $\cdot OH$, or H_2O_2)

BOX 20-B ● CLINICAL CONNECTION

Cancer Is a Genetic Disease

Cancer is one of the most common diseases, affecting one in three people and killing one in four. The clinical picture of cancer varies tremendously, depending on the tissue affected, but all cancer cells share an ability to proliferate uncontrollably and ignore the usual signals to differentiate or undergo apoptosis. Eventually, cancer may kill by invading and eroding the surrounding normal tissue.

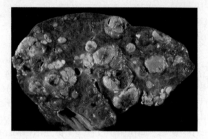

Tumors in human liver. The white bulges are masses of cancerous cells growing in the liver. [CNRI/Science Photo Library/ Photo Researchers.]

Many cancers appear to result from genetic predisposition, environmental factors, or infections. Only about 1% of cancers are classified as hereditary, but there are over 20 different kinds of these diseases, and they shed considerable light on specific molecular mechanisms of **carcinogenesis** (cancer development). The linkage between environmental factors and cancer comes mostly from epidemiological studies, which have shown, for example, that sunlight increases the risk of cutaneous melanoma (skin cancer) and smoking and asbestos exposure promote the development of lung cancer. Viral infections contribute to certain types of cancer, for example, liver cancer from hepatitis B virus and cervical cancer from human papillomaviruses. Chronic bacterial infections may also lead to cancer.

All the known causes of cancer are linked in some way to damage of the cell's DNA. Thus, *cancer is fundamentally a disease of the genes.* We have already seen that oncogenic mutations in growth signaling pathways can produce constitutively active receptors and kinases so that cells grow and divide even in the absence of a growth signal (see Box 10-B).

Inactivating genetic events also contribute to cancer. For example, the childhood cancer known as retinoblastoma (which is characterized by retinal tumors) occurs in an inherited and a sporadic (noninherited) form. The age of tumor onset varies widely, which is consistent with a scenario in which both copies (alleles) of a gene must be inactivated in order to develop the disease. Children with the inherited form of retinoblastoma succumb early because they already bear one defective allele (inherited from one of their parents). Inactivation of the second allele leads to transformation. The retinoblastoma gene is known as a **tumor suppressor gene,** *since its loss leads to the development of cancer.*

Genetic changes—whether they activate a growth-promoting oncogene or inactivate a tumor suppressor gene—can take the form of point mutations, large or small deletions, chromosomal rearrangements, or inappropriate methylation that leads to gene silencing (Section 20-5). Because cell growth and division are tightly regulated, a single genetic change is unlikely to send a cell into a pattern of uncontrolled proliferation that leads to cancer.

The ability of a cell to grow and divide depends on numerous factors, including the balance between growth-promoting signals and apoptotic signals, the state of the telomeres, contacts with neighboring cells, and the delivery of oxygen and nutrients to support expansion. In general, several regulatory pathways must be overridden by separate genetic events. This is known as the multiple-hit hypothesis for carcinogenesis.

There is no defined sequence of events for transforming a normal cell to a tumor, so cancer is not a single disease. Efforts to identify common mutations in human cancers, the goal of the Cancer Genome Atlas project (http://cancergenome.nih.gov/), promise to extend understanding of cancer and uncover new therapeutic targets. Techniques for profiling a tumor's unique features using DNA microarrays (see Box 12-B) are already used to predict its growth potential and to select the most appropriate regimen for halting it. Part of a gene expression profile of breast tumors is shown here. Each horizontal band corresponds to a different gene used in the microarray, and the columns represent different tumors. The green spots indicate levels of gene expression below the mean, red spots indicate levels above the mean, and intermediate shades indicate levels closer to the mean (gray spots indicate no data).

[Courtesy David Botstein and Therese Sorlie, Stanford University. From *Nature* 406, 747–752 (2000).]

In the absence of specific information about a tumor, cancer therapy can only target the features common to all forms of the disease. At present, this mainly means suppressing cell proliferation. Most drugs that interfere with mitosis and cell division are useful but are not specific for cancer cells; they also kill tissues that normally undergo regular renewal. Radiation is effective in killing cancer cells, although it may act indirectly by destroying the radiation-sensitive endothelial cells of the blood vessels that provide nutrients to the rapidly growing cancer cells. Drugs that inhibit angiogenesis (the growth of new blood vessels) can also slow tumor growth.

Because cancer is so difficult to cure, preventive efforts should also be considered. Epidemiological studies suggest that deaths from cancer could be significantly reduced (by as much as one-half) by eliminating known risk factors, including smoking and exposure to certain chemicals and microorganisms. The remaining cases of cancer might someday be treated with agents designed to rectify specific defects at the molecular level.

DNA repair pathways are closely linked to cancer

Identifying specific genetic lesions in cancer cells is tricky because a single tumor may contain several different lineages of cancer cells and because cancers are genetically unstable. In addition to small mutations, entire chromosomes may be lost or gained, and parts of chromosomes may be translocated to other chromosomes. The state of the DNA, which seems to go from bad to worse, apparently reflects the failure of the mechanisms that detect damaged DNA and promote its restoration. These steps require a host of proteins, including kinases and other intracellular signaling components that respond rapidly (usually within minutes) to DNA damage. We discuss some of these proteins here.

One key player in the damage response pathway is the protein kinase known as ATM, which is defective in ataxia telangiectasia. This disease is characterized by neurodegeneration, premature aging, and a propensity to develop cancer. ATM is a large protein (350,000 D) that includes a DNA-binding domain. It is activated in response to DNA damage, such as double-strand breaks. Among the substrates for ATM's kinase activity are proteins involved in initiating cell division, the protein BRCA1 (whose gene is commonly mutated in breast cancer), and the tumor suppressor known as p53.

Breast cancer strikes about one in nine women in developed countries. About 10% of breast cancers have a familial form, and about half of these exhibit mutations in the *BRCA1* or *BRCA2* genes. A woman bearing one of these mutated genes has a 70% chance of developing breast cancer. BRCA1 and BRCA2 are large multidomain proteins that function in part as scaffolding proteins to link proteins that detect DNA damage to proteins that can repair the damage or halt the cell cycle. For example, BRCA2 binds to Rad51, a protein necessary for the recombination repair pathway. The loss of BRCA1 or BRCA2 increases the likelihood that a cell would attempt to divide without first repairing damaged DNA.

The tumor suppressor gene p53 is found to be mutated in at least half of all human tumors. The level of p53 in the cell is controlled by its rate of degradation (it is ubiquitinated and targeted to a proteasome for destruction; Section 12-1). The concentration of p53 increases when its degradation is slowed. This can occur by a decrease in the rate at which ubiquitin is attached to it or when it is phosphorylated by the action of a kinase such as ATM. Thus, *DNA damage, which activates ATM, leads to an increase in the cellular concentration of p53.*

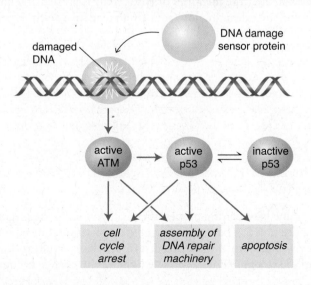

Other modifications, such as acetylation and glycosylation, also increase the activity of p53, which responds not just to DNA damage but also to other forms of cellular stress such as low oxygen and elevated temperatures. Covalent modification of

p53 triggers a conformational change that allows the protein to bind to specific DNA sequences in order to promote the transcription of several dozen different genes. p53 binds as a tetramer (a dimer of dimers) to DNA, encircling it.

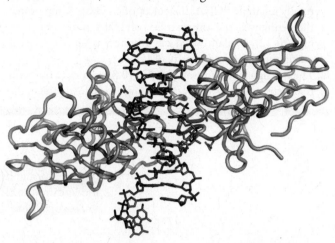

A p53 dimer (gray) bound to DNA (blue). The six residues that are most commonly mutated in p53 from cancer cells are shown in red. These residues either interact directly with the DNA or are involved in stabilizing p53 structure. [Structure (pdb 2GEQ) determined by W. C. Ho, M. X. Fitzgerald, and R. Marmorstein.]
⊕ **See Interactive Exercise. Human p53 in complex with its target DNA.**

p53 stimulates production of a protein that blocks the cell's progress toward cell division. This regulatory mechanism buys time for the cell to repair DNA using enzymes whose synthesis is also stimulated by p53. Moreover, activated p53 turns on the gene for a subunit of ribonucleotide reductase (see Section 18-3), which promotes synthesis of the deoxynucleotides required for DNA repair.

Some of p53's other target genes encode proteins that carry out apoptosis, a multistep process in which the cell's contents are apportioned into membrane-bounded vesicles that are subsequently engulfed by macrophages. For multicellular organisms, death by apoptosis is often a better option than allowing a malfunctioning cell to persist.

It is possible that p53's ultimate effects are dose-dependent, with lower doses (reflecting mild DNA damage) pausing the cell cycle and higher doses (reflecting severe, irreparable DNA damage) leading to cell death. *The position of p53 at the interface of pathways related to DNA repair, cell cycle control, and apoptosis indicates why the loss of the p53 gene is so strongly associated with the development of cancer.*

⬢ **Questions:**

1. From what you know about the causes of cancer, explain why the risk of developing cancer increases with age.
2. The product of the retinoblastoma gene is a protein that binds to and inhibits the activity of a transcription factor that induces the expression of genes required for the cell to synthesize DNA. Explain how a mutation in the retinoblastoma gene promotes cancer.
3. Explain why inheriting a defective gene for the retinoblastoma protein, BRCA1, or p53 does not guarantee that an individual will develop cancer.

(*continued on the next page*)

4. The human immune system can recognize markers on the surface of cancerous cells so that these cells can be selectively killed. What feature of a transformed cell would change its appearance?

5. One of the proteins that senses damaged DNA may structurally resemble PCNA, the eukaryotic sliding-clamp protein that enhances the processivity of DNA polymerase (see Fig. 20-12). Explain why such a protein is suited for monitoring the chemical integrity of DNA.

6. The Ras signaling pathway (Section 10-3) activates a transcription factor called Myc, which turns on a gene for a protein that blocks the activity of the protein that adds ubiquitin to p53. Is this mechanism consistent with the ability of excessive Ras signaling to promote cell growth?

7. In the 1920s, Otto Warburg noted that cancer cells primarily use glycolysis rather than aerobic respiration to produce ATP (see Box 19-A). In 2006, researchers discovered that p53 stimulates the expression of cytochrome c oxidase. Explain how the connection between p53 and cytochrome c oxidase explains Warburg's observation.

8. Why do unicellular organisms have no need for apoptosis?

that are normal by-products of oxidative metabolism. Over 100 different oxidative modifications of DNA have been catalogued. For example, guanine can be oxidized to 8-oxoguanine (oxoG):

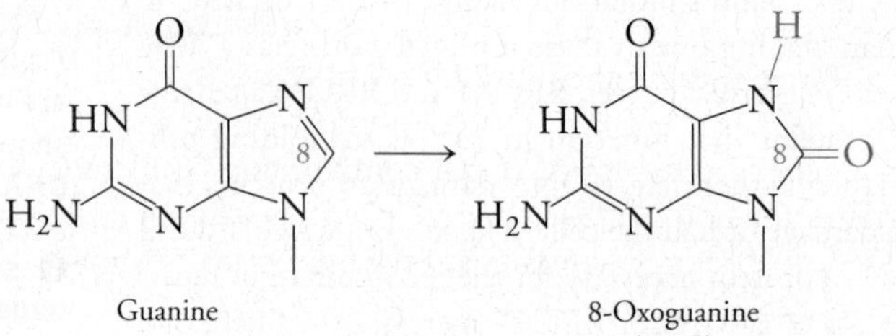

Guanine   8-Oxoguanine (oxoG)

When the modified DNA strand is replicated, the oxoG can base pair with either an incoming C or A. Ultimately, the original G:C base pair can become a T:A base pair. A nucleotide substitution such as this is called a **point mutation.** Changing a purine (or pyrimidine) to another purine (or pyrimidine) is known as a **transition mutation;** a **transversion** occurs when a purine replaces a pyrimidine or vice versa.

Other nonenzymatic reactions disintegrate DNA under physiological conditions. For example, hydrolysis of the *N*-glycosidic bond linking a base to a deoxyribose group yields an **abasic site** (also called an apurinic or apyrimidinic or AP site).

Deamination reactions can alter the identities of bases. This is particularly dangerous in the case of cytosine, since deamination (actually an oxidative deamination) yields uracil:

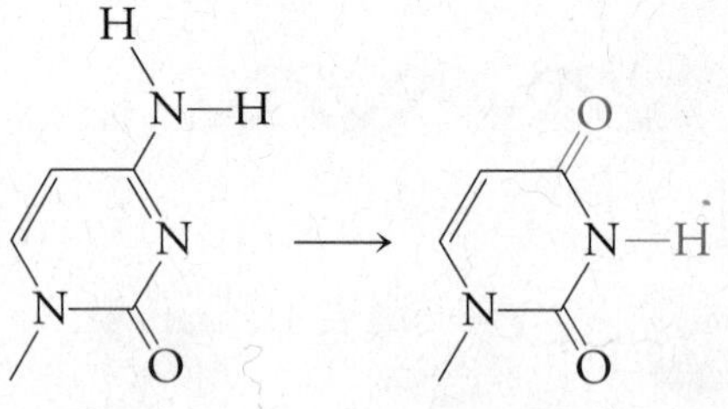

Cytosine   Uracil

Recall that uracil has the same base-pairing propensities as thymine, so the original C:G base pair could give rise to a T:A base pair after DNA replication. Since DNA has evolved to contain thymine rather than uracil, a uracil base resulting from cytosine deamination can be recognized and corrected before the change becomes permanent.

In addition to the cellular factors described above, *environmental agents such as ultraviolet light, ionizing radiation, and certain chemicals can physically damage DNA.* For example, ultraviolet (UV) light induces covalent linkages between adjacent thymine bases:

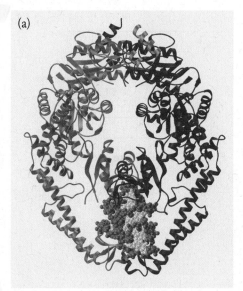

This brings the bases closer together, which distorts the helical structure of DNA. Thymine dimers can thereby interfere with normal replication and transcription.

Ionizing radiation also damages DNA either through its direct action on the DNA molecule or indirectly by inducing the formation of free radicals, particularly the hydroxyl radical, in the surrounding medium. This can lead to strand breakage. Many thousands of chemicals, both natural and man-made, can potentially react with the DNA molecule and cause mutations. Such compounds are known as **mutagens,** or as **carcinogens** if the mutations lead to cancer.

The unavoidable nature of many DNA lesions has driven the evolution of mechanisms to detect and remedy errors. A cell containing a point mutation or a small insertion or deletion may suffer no ill effects, particularly if the mutation is in a part of the genome that does not contain an essential gene. However, more serious lesions, such as single- or double-strand breaks, usually bring replication or translation to a halt, and the cell must deal with these show-stoppers immediately.

Repair enzymes restore some types of damaged DNA

In a few cases, repair of damaged DNA is a simple process involving one enzyme. For example, in bacteria and some other organisms (but not mammals), UV-induced thymine dimers can be restored to their monomeric form by the action of a light-activated enzyme called DNA photolyase.

Mammals can reverse other simple forms of DNA damage, such as the methylation of a guanine residue, which yields O^6-methylguanine (this modified base can pair with either cytosine or thymine):

O^6-Methylguanine

A methyltransferase removes the offending methyl group, transferring it to one of its Cys residues. This permanently inactivates the protein. Apparently, the expense of sacrificing the methyltransferase is justified by the highly mutagenic nature of O^6-methylguanine.

In bacteria as well as eukaryotes, *nucleotide mispairings are corrected shortly after DNA replication by a **mismatch repair** system.* A protein, called MutS in bacteria, monitors newly synthesized DNA and binds to the mispair. Although it binds only 20 times more tightly to the mispair than to a normal base pair, MutS undergoes a conformational change and causes the DNA to bend (**Fig. 20-19**). These changes apparently induce an endonuclease to cleave the strand with the incorrect base at a site as far as 1000 bases away. A third protein then unwinds the helix so that the defective segment of DNA can be destroyed and replaced with accurately paired nucleotides by DNA polymerase. How does the endonuclease know which strand

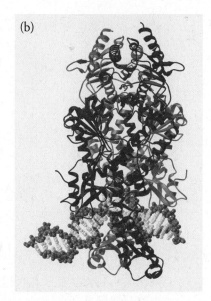

Figure 20-19 The mismatch repair protein MutS bound to DNA. Two subunits of the comma-shaped MutS protein encircle the DNA at the site of a mispaired nucleotide. Binding causes the DNA to bend sharply, compressing the major groove and widening the minor groove. (a) Front view. (b) Side view. [Courtesy Wei Yang/NIH. From *Nature* 407, 703–710 (2000).]

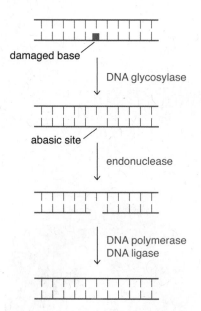

damaged base

DNA glycosylase

abasic site

endonuclease

DNA polymerase
DNA ligase

Figure 20-20 Base excision repair.

contains the incorrect base? Cellular DNA is normally methylated (Section 20-5), and the endonuclease can select the newly synthesized strand because it has not yet undergone methylation. A defect in the human homolog of MutS increases the mutation rate of DNA and predisposes individuals with the defect to a form of hereditary colon cancer.

Base excision repair corrects the most frequent DNA lesions

*Modified bases that cannot be directly repaired can be removed and replaced in a process known as **base excision repair**.* This pathway begins with a glycosylase, which removes the damaged base. An endonuclease then cleaves the backbone, and the gap is filled in by DNA polymerase (Fig. 20-20).

The structures and mechanisms of several DNA glycosylases have been described in detail. For example, there is a glycosylase that recognizes oxoG. Uracil-DNA glycosylase recognizes and removes the uracil bases that are mistakenly incorporated into DNA during replication or that result from cytosine deamination. When a glycosylase binds DNA, the damaged base flips out from the helix so that it can bind in a cavity on the protein surface, and various protein side chains take the place of the damaged base and form hydrogen bonds with its complement on the other DNA strand (Fig. 20-21).

After excising the offending base, DNA glycosylases appear to remain bound to the DNA at the abasic site, possibly to help recruit the next enzyme of the repair pathway. This enzyme, usually called the AP endonuclease (APE1 in humans), nicks the DNA backbone at the 5′ side of the abasic ribose. The nuclease inserts two protein loops into the major and minor grooves of the DNA and bends the DNA by about 35° to expose the abasic site. A backbone structure with a base attached cannot enter the active-site pocket. During the hydrolysis reaction, an Mg^{2+} ion in the active site stabilizes the anionic leaving group (Fig. 20-22). Like the glycosylases, the AP endonuclease remains with its product. For most enzymes, rapid product dissociation is the rule, but in DNA repair, the continued association of the endonuclease with the broken DNA strand may be favored because it prevents unwanted side reactions.

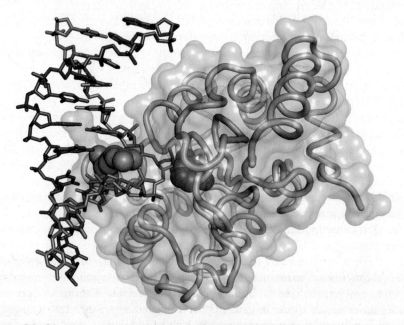

Figure 20-21 Uracil-DNA glycosylase bound to DNA. The enzyme is shown in gray, and the DNA substrate is shown with blue and purple strands. The flipped-out uracil (whose glycosidic bond has already been hydrolyzed) is shown in red. An Arg side chain that takes the place of the uracil is shown in orange. [Structure (pdb 4SKN) determined by G. Slupphaug, C. D. Mol, B. Kavil, A. S. Arvai, H. E. Krokan, and J. A. Tainer.]

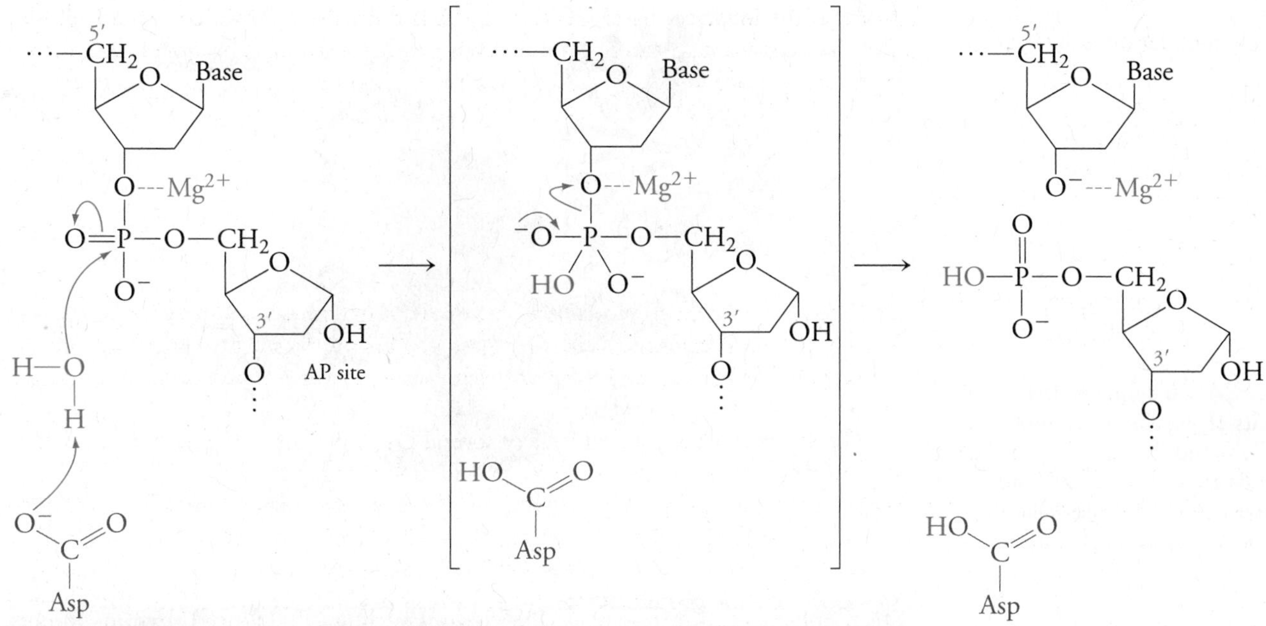

Figure 20-22 The AP endonuclease reaction.

In the final step of base excision repair, a DNA polymerase (such as DNA polymerase β in eukaryotes) fills in the one-nucleotide gap, and a DNA ligase seals the nick. In some cases, DNA polymerase may replace as many as 10 nucleotides. The displaced single strand can then be cleaved off by a flap endonuclease (Section 20-2).

Nucleotide excision repair targets the second most common form of DNA damage

Nucleotide excision repair, as its name suggests, is similar to base excision repair but mainly targets DNA damage resulting from insults such as ultraviolet light or oxidation. In nucleotide excision repair, a segment containing the damaged nucleotide and about 30 of its neighbors is removed, and the resulting gap is filled in by a DNA polymerase that uses the intact complementary strand as a template (**Fig. 20-23**). Many of the 30 or so proteins that are involved in this pathway in humans have been identified through mutations that are manifest as two genetic diseases.

The rare hereditary disease Cockayne syndrome is characterized by neural underdevelopment, failure to grow, and sensitivity to sunlight. It results from a mutation in any of several genes that encode proteins participating in a pathway for recognizing an RNA polymerase that has stalled in the act of transcribing a gene into messenger RNA. Stalling occurs when the DNA template is damaged and distorted so that it blocks the progress of the RNA polymerase. The polymerase must be removed so that the damage can be addressed by the nucleotide excision repair system. Defects in the Cockayne syndrome genes prevent the cell from recognizing and removing the stalled RNA polymerase. Consequently, *the DNA never has a chance to be repaired, and the cell undergoes apoptosis.* The death of transcriptionally active cells may account for the developmental symptoms of Cockayne syndrome.

Like Cockayne syndrome, the disease xeroderma pigmentosum is characterized by high sensitivity to sunlight, but individuals with xeroderma pigmentosum are about 1000 times more likely to develop skin cancer and do not suffer from developmental problems. Xeroderma pigmentosum is caused by a mutation in one of the genes that participate directly in nucleotide excision repair. Whereas Cockayne syndrome gene products appear to detect DNA damage that prevents transcription, the xeroderma pigmentosum proteins are responsible for repairing the damage. Thus, *in xeroderma pigmentosum, apoptosis is not triggered, but damaged DNA cannot be repaired.* The failure to repair UV-induced lesions explains the high incidence of skin cancer.

When the cell attempts to replicate damaged DNA that has not been repaired, it may rely on one of its nonstandard DNA polymerases. For example, eukaryotic

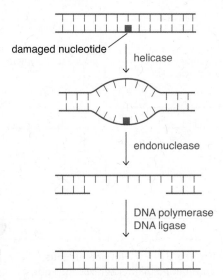

Figure 20-23 Nucleotide excision repair.

? **Compare this process to base excision repair (Fig. 20-20).**

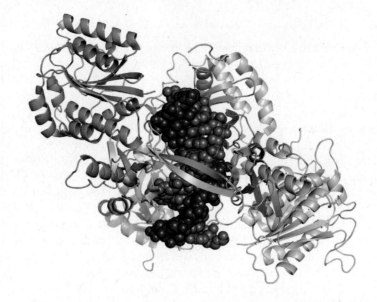

Figure 20-24 Ku bound to DNA. The two subunits of the Ku heterodimer are shown in light and dark green, and the DNA strands are shown in blue and purple. [Structure (pdb 1JEY) determined by J. R. Walker, R. A. Corpina, and J. Goldberg.]

DNA polymerase η can bypass DNA lesions such as UV-induced thymine dimers by incorporating two adenine bases in the new strand. Although it is useful as a translesion polymerase, DNA polymerase η is relatively inaccurate and has no proofreading exonuclease activity. It inserts an incorrect base on average every 30 nucleotides. This may not be problematic, as the errors can be detected and corrected by the mismatch repair system described above.

The existence of error-prone polymerases provides a fail-safe mechanism for replicating stretches of DNA that cannot be navigated by the standard replication machinery. In fact, synthesis of these alternative polymerases increases when bacterial cells experience DNA damage. Evidently, the possibility of introducing small errors during replication is acceptable when the only other option is cell death.

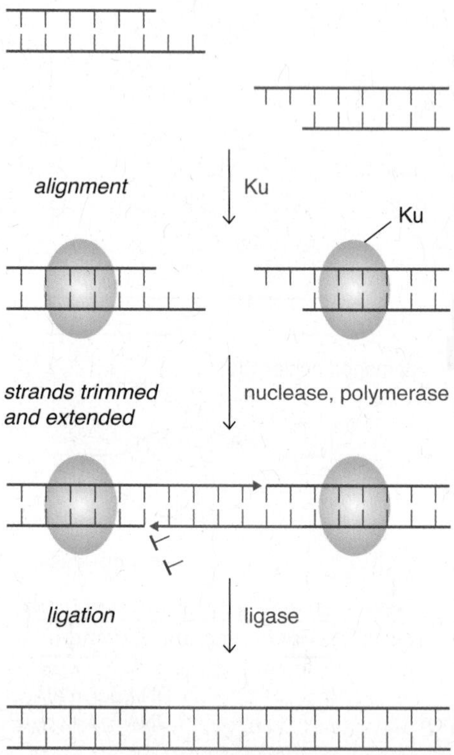

Double-strand breaks can be repaired by joining the ends

Segments of a DNA double helix that have been completely severed by the effects of radiation or free radicals can be rejoined through recombination (a process that requires another DNA molecule with a similar sequence; discussed below) or by **nonhomologous end-joining,** a process that does not require the presence of a homologous DNA molecule. In mammals, nearly all double-strand breaks are repaired by nonhomologous end-joining.

The first step in this repair pathway is the recognition of the broken DNA ends by a dimeric protein called Ku (**Fig. 20-24**). When Ku binds the cut DNA, it undergoes a conformational change so that it can recruit a nuclease, which trims up to 10 residues from the ends of the DNA molecule. The protein–DNA complex may then be joined by a DNA polymerase such as polymerase μ, which can extend the ends of the DNA either with or without a template. Template-independent polymerization, along with a tendency for polymerase μ to slip, means that the break site may end up with additional nucleotides that were not present in the DNA before it broke. A DNA ligase finishes the repair job by joining the two backbones of the DNA segments (**Fig. 20-25**).

Two Ku–DNA complexes associate with each other so that, ideally, the proper halves of a broken DNA molecule can be stitched back together. However, nonhomologous end-joining is also responsible for combining DNA segments that do not belong together, leading to chromosomal rearrangements. In addition, the Ku–DNA complex can interact with the nuclease, polymerase, and ligase in any order, so a DNA break could potentially be repaired in many different ways, with or without the addition and removal of nucleotides. Consequently, *nonhomologous end-joining is inherently mutagenic,* but, as in other forms of DNA repair, this may be a small price to pay for restoring a continuous double-stranded DNA.

alignment   Ku

Ku

strands trimmed and extended   nuclease, polymerase

ligation   ligase

Figure 20-25 Nonhomologous end-joining. Ku recognizes the ends of broken DNA and aligns them. The activities of a nuclease, polymerase, and ligase generate an unbroken DNA molecule that may differ in sequence from the original.

Recombination also restores broken DNA molecules

In some organisms, double-strand breaks are repaired through **recombination,** a process that also occurs in the absence of DNA damage as a mechanism for shuffling genes between homologous chromosomes in meiosis. Recombination repair of a broken chromosome can occur at any time in a diploid organism (which has two sets of homologous chromosomes) but can occur only after DNA replication in an organism with only one chromosome. Recombination requires another intact homologous double-stranded molecule as well as nucleases, polymerases, ligases, and other proteins (Fig. 20-26).

In recombination repair, a single strand from the damaged DNA molecule changes places with a homologous strand in another DNA molecule. In order for a single strand of DNA to "invade" a double-stranded DNA molecule (step 2 in Fig. 20-26), the single strand must first be coated with an ATP-binding protein, called RecA in *E. coli* and Rad51 in humans. RecA binds to a DNA strand in a cooperative fashion, beginning at the break point. This binding unwinds and stretches the DNA by about 50%, but not uniformly: Sets of three nucleotides retain a near-standard conformation (with about 3.4 Å between bases) but are separated from the next triplet by about 7.8 Å (Fig. 20-27). During recombination, the RecA–DNA filament aligns with double-stranded DNA containing a complementary strand. The elongated structure

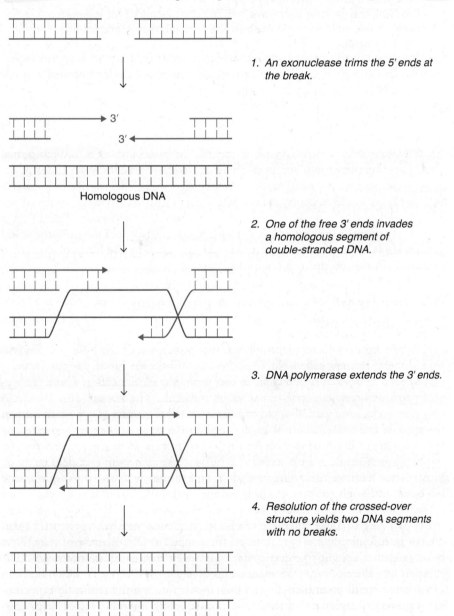

1. *An exonuclease trims the 5' ends at the break.*

Homologous DNA

2. *One of the free 3' ends invades a homologous segment of double-stranded DNA.*

3. *DNA polymerase extends the 3' ends.*

4. *Resolution of the crossed-over structure yields two DNA segments with no breaks.*

Figure 20-26 Homologous recombination to repair a double-strand break.

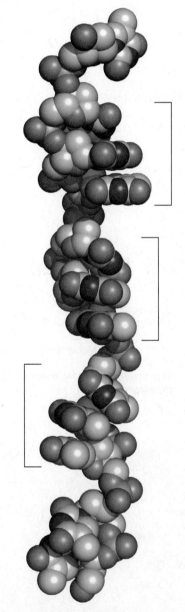

Figure 20-27 Conformation of DNA bound to RecA. The single strand of DNA is shown in space-filling form with atoms color coded: C gray, N blue, O red, and P orange. Brackets indicate sets of three nucleotides that retain a near–B-DNA conformation. The DNA backbone stretches in between these triplets so that the strand extends to about 1.5 times its original length. The RecA protein subunits are not shown. [Structure (pdb 3CMW) determined by N. P. Pavletich.]

of the RecA filament may induce a similar shape change in the double-stranded DNA so that base pairs in the double-stranded DNA are disrupted and bases become unstacked. These changes would facilitate the strand swapping that occurs during recombination. At this point, hydrolysis of the ATP bound to RecA allows the protein to release the displaced single strand and a new double-stranded DNA.

The mechanism of recombination is highly conserved among prokaryotes and eukaryotes, which argues for its essential role in maintaining the integrity of DNA as a vehicle of genetic information. The proteins that carry out recombination appear to function **constitutively** (that is, the genes are always expressed), which is consistent with proposals that DNA strand breaks are unavoidable. However, *many DNA repair mechanisms are induced only when the relevant form of DNA damage is detected.* This makes sense, since the repair enzymes might otherwise interfere with normal replication. In fact, activation of the repair pathways usually halts DNA synthesis, an advantage when error-prone DNA polymerases—which would be a liability in normal replication—are active.

> **CONCEPT REVIEW**
> - Provide examples of how intracellular and environmental agents can damage DNA.
> - Can all forms of DNA damage be attributed to one of these agents?
> - How does a point mutation arise?
> - Describe the general pathways and enzymes required for mismatch repair, base excision repair, nucleotide excision repair, end-joining, and recombination.
> - Describe the advantages and disadvantages of error-prone DNA polymerases.

20-5 DNA Packaging

> **KEY CONCEPTS**
> - Eukaryotic DNA is wrapped around histones to form nucleosomes, which can form higher-order structures.
> - Both histones and DNA may be chemically modified

At some point after it is replicated, eukaryotic DNA assumes a highly condensed form. This is advantageous for a cell about to divide, since elongated DNA molecules would become hopelessly tangled rather than segregating neatly to form two equivalent sets of chromosomes. Even when the cell is not dividing, much of the DNA is packaged in a form that compresses its length considerably. This DNA is known as **heterochromatin**, which is transcriptionally silent. **Euchromatin** is less condensed and appears to be transcribed at a higher rate. The two forms of chromatin are distinguishable by electron microscopy (**Fig. 20-28**).

The fundamental unit of DNA packaging is the nucleosome

Both heterochromatin and euchromatin contain structural units known as **nucleosomes**, which are complexes of DNA and protein. *The core of a nucleosome consists of eight **histone** proteins: two each of the histones known as H2A, H2B, H3, and H4. Approximately 146 base pairs of DNA wind around the histone octamer* (**Fig. 20-29**). A complete nucleosome contains the core structure plus histone H1, a small protein that appears to bind outside the core. Neighboring nucleosomes are separated by short stretches of DNA of variable length.

The histone proteins interact with DNA in a sequence-independent manner, primarily via hydrogen bonding and ionic interactions with the sugar–phosphate backbone. Although prokaryotes lack histones, other DNA-binding proteins may help package DNA in bacterial cells.

The winding of DNA in the nucleosome (it makes approximately 1.65 turns around the histone octamer) generates negative supercoils. In other words, the DNA in nucleosomes is slightly underwound; this is why eukaryotes do not need DNA gyrase to introduce negative supercoils into their DNA. During DNA replication, nucleosomes are disassembled as the DNA spools through the replication machinery. A protein complex called the replication-coupling assembly factor helps reassemble nucleosomes on newly replicated DNA, using the displaced histones as well as newly synthesized histones imported from the cytoplasm to the nucleus.

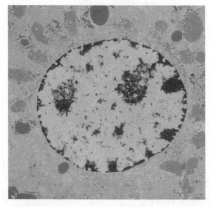

Figure 20-28 A eukaryotic nucleus. Heterochromatin is the darkly staining material (red in this colorized electron micrograph); euchromatin stains more lightly (yellow). [Gopal Murti/Science Photo Library/Photo Researchers.] ⊕ See Guided Exploration. Nucleosome structure.

? Identify the euchromatin and heterochromatin in the photo at the beginning of this chapter.

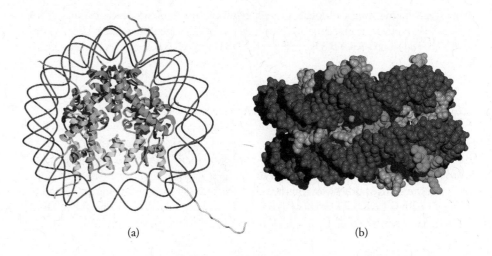

(a)   (b)

Figure 20-29 Structure of the nucleosome core. (a) Top view. (b) Side view (space-filling model). The DNA (dark blue) winds around the outside of the histone octamer. [Structure (pdb 1AOI) determined by K. Luger, A. W. Maeder, R. K. Richmond, D. F. Sargent, and T. J. Richmond.]

As the question at the beginning of the chapter suggests, cells have a mechanism for efficiently packaging their DNA. Nucleosomes compact the DNA only by a factor of about 30 to 40, but the chain of nucleosomes itself coils into a solenoid (coil) with a diameter of about 30 nm (**Fig. 20-30**). The DNA in the 30-nm fiber is presumably well protected from nuclease attack and is inaccessible to the proteins that carry out replication and transcription. During cell division, the chromatin condenses even further so that each chromosome has an average length of about 10 μm and a diameter of 1 μm.

Histones are covalently modified

The histones are among the most highly conserved proteins known, in keeping with their essential function in packaging the genetic material in all eukaryotic cells. Each histone pairs with another, and the set of eight forms a compact structure. However, the tails of the histones, which are flexible and charged, extend outward from the core of the nucleosome (see Fig. 20-29). These histone tails are subject to covalent modification, including acetylation, methylation, and phosphorylation (**Fig. 20-31**).

The addition or removal of the various histone-modifying groups can potentially introduce considerable variation in the fine structure of chromatin, offering a mechanism for promoting or preventing gene expression. For example, a Lys residue of a histone is positively charged and could interact strongly with the negatively charged DNA backbone. Acetylation of the Lys side chain would neutralize it and weaken its interaction with the DNA, possibly destabilizing the nucleosome or allowing other proteins access to the DNA. In fact, acetylation of histones is associated with transcriptionally active chromatin, and deacetylation appears to repress transcription. Histone phosphorylation appears to be a prelude to the chromosome condensation that occurs during cell division.

Some of the enzymes that modify histones act on many residues in different histone proteins; others are highly specific for one residue in one histone. The modifications themselves are interdependent. For example, in histone H3, methylation of Lys 9 inhibits phosphorylation of Ser 10, but phosphorylation of Ser 10 promotes acetylation of Lys 14. The correlation between the pattern of histone modifications and the DNA's readiness for transcription is believed to act as a **histone code** that can be read by various proteins. *Alterations in histone–histone and histone–DNA contacts could alter the structure of the nucleosomes as well as provide binding sites for proteins such as transcription factors* (these are discussed in greater detail in Section 21-1).

Figure 20-30 Levels of chromatin structure. The DNA helix (blue) is wrapped around a histone octamer (orange) to form a nucleosome; nucleosomes aggregate to form the 30-nm fiber; this packs into loops in the fully condensed chromosome. The approximate diameter of each structure is given in parentheses.

▶▶ **HOW** does DNA fit inside the nucleus?

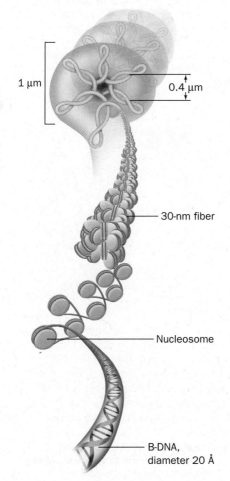

1 μm

0.4 μm

30-nm fiber

Nucleosome

B-DNA, diameter 20 Å

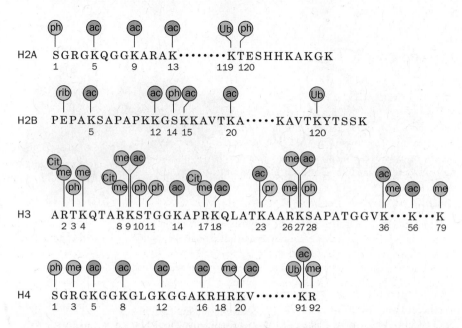

Figure 20-31 Histone modifications. Partial sequences of the four histones are shown using one-letter amino acid codes. The added groups are represented by colored symbols (ac = acetyl, me = methyl, ph = phosphoryl, pr = propionyl, rib = ADP-ribose, and ub = ubiquitin). Cit represents citrulline (deiminated arginine). Note that some residues can be modified in multiple ways. This diagram is a composite; not all modifications occur in all organisms. [After a diagram by Ali Shilatifard, St. Louis University School of Medicine.]

? Which histone residues undergo acetylation? Methylation? Phosphorylation?

DNA also undergoes covalent modification

5-Methylcytosine residue

In many organisms, including plants and animals, DNA methyltransferases add methyl groups to cytosine residues (*left*). The methyl group projects out into the major groove of DNA and could potentially alter interactions with DNA-binding proteins.

In mammals, the methyltransferases target C residues next to G residues, so that about 80% of CG sequences (formally represented as CpG) are methylated. CpG sequences occur much less frequently than statistically predicted, but clusters of CpG (called **CpG islands**) are often located near the starting points of genes. Interestingly, these CpG islands are usually unmethylated. This suggests that *methylation may be part of the mechanism for marking or "silencing" DNA that contains no genes.*

After DNA is replicated, methyltransferases can modify the new strand, using the methylation pattern of the parent DNA strand as a guide. Cells also contain enzymes that methylate DNA without a template as well as enzymes that demethylate DNA.

Variations in DNA methylation may be responsible for **imprinting,** in which the level of expression of a gene depends on its parental origin. Recall that an individual receives one copy of a gene from each parent. Normally, both copies (alleles) are expressed, but a gene that has been methylated may not be expressed. The gene behaves as if it has been "imprinted" and knows its parentage. Imprinting explains why some traits are transmitted in a maternal or paternal fashion, contrary to the Mendelian laws of inheritance. Methylation and other DNA modifications (such as the addition of telomeres) are said to be **epigenetic** (from the Greek *epi*, "above") because they represent heritable information that goes beyond the sequence of nucleotides in the DNA.

CONCEPT REVIEW
- How does heterochromatin differ from euchromatin in structure and function?
- What is the role of histones in packaging DNA?
- How could covalent modification of histones or DNA alter gene expression?

[SUMMARY]

20-1 DNA Supercoiling

- In order to be replicated, the two strands of DNA must be separated, or unwound. This unwinding is facilitated by the negative supercoiling (underwinding) of DNA molecules. Enzymes called topoisomerases can add or remove supercoils by temporarily introducing breaks in one or both DNA strands.

20-2 The DNA Replication Machinery

- DNA replication requires a host of enzymes and other proteins located in a stationary factory. Helicases separate the two DNA strands at a replication fork, and SSB binds to the exposed single strands.
- DNA polymerase can only extend a preexisting chain and therefore requires an RNA primer synthesized by a primase. Two polymerases operate side-by-side to replicate DNA, so the leading strand of DNA is synthesized continuously while the lagging strand is synthesized discontinuously as a series of Okazaki fragments. The RNA primers of the Okazaki fragments are removed, the gaps are filled in by DNA polymerase, and the nicks are sealed by DNA ligase.
- DNA polymerase and other polymerases catalyze a reaction in which the 3′ OH group of the growing chain nucleophilically attacks the phosphate group of an incoming nucleotide that base pairs with the template strand. A sliding clamp promotes the processive activity of DNA polymerase.

- Many DNA polymerases contain a second active site that hydrolytically excises a mismatched nucleotide.

20-3 Telomeres

- In eukaryotes, the extreme 3′ end of a DNA strand cannot be replicated, so the enzyme telomerase adds repeating sequences to the 3′ end to create a structure known as a telomere.

20-4 DNA Damage and Repair

- Normal replication errors, spontaneous deamination, radiation, and chemical damage can cause mutations in DNA.
- Mechanisms for repairing damaged DNA include direct repair, mismatch repair, base excision repair, nucleotide excision repair, end-joining, and recombination.

20-5 DNA Packaging

- Eukaryotic DNA is wound around a core of eight histone proteins to form a nucleosome, which represents the first level of DNA compaction in the nucleus.
- The histone components of nucleosomes may be modified by acetylation, phosphorylation, and methylation as a potential mechanism for regulating gene expression. The DNA may also be covalently modified, often by methylation of C residues in "silent" regions of the genome.

[GLOSSARY TERMS]

replication
semiconservative replication
supercoiling
topoisomerase
nick
replication fork
factory model of replication
primer
leading strand
lagging strand
discontinuous synthesis
Okazaki fragment
processivity
proofreading
nick translation

telomere
reverse transcriptase
cDNA
chain terminator
mutation
apoptosis
carcinogenesis
tumor suppressor gene
point mutation
transition mutation
transversion mutation
abasic site
mutagen
carcinogen
mismatch repair

base excision repair
nucleotide excision repair
nonhomologous end-joining
recombination
constitutive
heterochromatin
euchromatin
nucleosome
histone
histone code
CpG island
imprinting
epigenetics

[PROBLEMS]

20-1 DNA Supercoiling

1. Why do type II topoisomerase enzymes require ATP whereas type I topoisomerases do not?

2. A topoisomerase in *E. coli*, called topoisomerase IV, separates the two newly replicated circular DNA molecules from one another just before cell division. Is topoisomerase IV a member of the type I or type II class of topoisomerases?

3. A variety of compounds inhibit topoisomerases. Novobiocin is an antibiotic and, like ciprofloxacin, it inhibits DNA gyrase. Doxorubicin and etoposide are anticancer drugs that inhibit eukaryotic topoisomerases. What properties distinguish the antibiotics from the anticancer drugs?

4. Deoxyribonuclease (DNase) processively nicks the phosphodiester backbone of double-stranded DNA to yield small oligonucleotide fragments and can be used therapeutically to treat patients with cystic fibrosis (CF). The enzyme is inhaled into the lungs, where it hydrolyzes the DNA contained in the sputum to decrease its viscosity and improve lung function. Genetic engineering techniques have been used to synthesize hyperactive DNase variants that produce nicks in the DNA more rapidly than the wild-type enzyme. A more efficient DNase can be used to achieve the same results at a lower enzyme concentration.

(a) The engineered hyperactive DNase variants are listed in the table. (A note on nomenclature: Q9R means that the Gln at position 9 in the wild-type DNase I has been changed to an Arg.) What structural feature do all of the DNase variants have in common? Why might these changes improve the catalytic efficiency of DNase?

(b) The enzymatic activities of the DNase variants were tested using a DNA hyperchromicity assay in which the absorbance of a solution of intact DNA is first measured at 260 nm. Then the enzyme is added, and the solution is monitored for an increase in absorbance that occurs when the concentration of single-stranded DNA in the solution increases. Why is the hyperchromicity assay a useful tool to assess the activity of the DNase variants?

(c) The DNA hyperchromicity assay was used to measure the K_M and V_{max} values for each variant. The results are shown in the table. How do the amino acid changes affect the activity of the enzyme variants as compared to the wild-type? Which variant(s) is(are) most efficient?

DNase variant	Charge (relative to wild-type)	K_M (μg/mL DNA)	V_{max} (A_{260} units · min^{-1} · mg^{-1})
Wild-type		1.0	1.0
N74K	+1	0.77	3.6
E13R/N74K	+2	0.20	5.3
E13R/N74K/T205K	+3	0.18	7.7
E13R/T14K/N74K/ T205K	+4	0.37	3.5
Q9R/E13R/H44K/ N74K/T205K	+5	0.11	2.4
Q9R/E13R/T14K/ H44R/N74K/T205K	+6	0.20	2.5

(d) The DNase variants were evaluated for their ability to cut or nick DNA. A *cut* refers to the hydrolysis of phosphodiester bonds on both strands, whereas a *nick* is the hydrolysis of just one strand. This was assessed by using a circular plasmid substrate, which is most stable in its supercoiled form. If one strand is nicked, the plasmid forms a relaxed circle, but if the backbones of both strands are cut, the circle linearizes, as shown below.

Supercoiled, relaxed circular, and linear DNA can be detected by their different rates of migration during electrophoresis through agarose gels. In a series of experiments, the plasmid was incubated with DNase and then the products were analyzed by agarose gel electrophoresis. The results are shown above right. Describe the results for each lane. Compare the selected variants to the wild-type DNase with regard to their ability to cut or nick the DNA. (C, control; WT, wild-type)

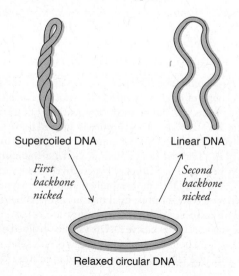

Supercoiled DNA   Linear DNA

First backbone nicked   *Second backbone nicked*

Relaxed circular DNA

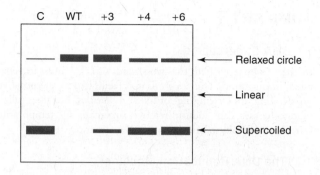

Relaxed circle   ←

Linear   ←

Supercoiled   ←

(e) The variants were tested for their ability to act on DNA at high and low substrate concentration. High-molecular-weight (MW) DNA is present in the lung secretions of CF patients in fairly high concentrations. DNase might also be effective as a therapeutic agent to treat the autoimmune disease systemic lupus erythematosus, in which DNA is present in the serum at low concentrations. What variant would be a good drug candidate to treat a CF patient? A lupus patient? Explain.

	Nicking activity (compared to wild type)			
	Low DNA concentration		High DNA concentration	
	Low MW	High MW	Low MW	High MW
Wild-type	1	1	1	1
N74K	26		10.4	
E13R/N74K	211	31	24.3	13
E13R/N74K/ T205K	7		1.3	
E13R/T14K/ N74K/T205K	7		0.7	

20-2 The DNA Replication Machinery

5. Semiconservative replication is shown in Figure 20-1. Draw a diagram that illustrates the composition of DNA daughter molecules for the second, third, and fourth generations.

6. What is the role of DNA gyrase during bacterial DNA replication?

7. DNA replication of the circular chromosome in *E. coli* begins when a protein called DnaA binds to the replication origin on the DNA. DnaA initiates replication only when the DNA that constitutes the DNA replication origin is negatively supercoiled. Why?

8. Would the replication origin described in Problem 7 likely be richer in G:C or A:T base pairs?

9. DNA helicases can be considered to be molecular motors that convert the chemical energy of NTP hydrolysis into mechanical energy for separating DNA strands. The bacteriophage T7 genome encodes a protein that assembles into a hexameric ring with helicase activity.

(a) In order for the helicase to unwind DNA, it requires two single-stranded DNA tails at one end of a double-stranded DNA segment. However, the helicase appears to bind to only one of the single strands, apparently by encircling it. Is this consistent with its ability to unwind a double-stranded DNA helix?

(b) Kinetic measurements indicate that the T7 helicase moves along the DNA at a rate of 132 bases per second. The protein hydrolyzes 49 dTTP per second. What is the relationship between dTTP hydrolysis and DNA unwinding?

(c) What does the structure of T7 helicase suggest about its processivity?

10. At eukaryotic origins of replication, helicase cannot be activated until the polymerase is also positioned on the DNA. Explain what would happen if the helicase became active in the absence of DNA polymerase.

11. Temperature-sensitive mutations are defined as mutations that allow a protein to function at a low temperature (the permissive temperature) but not a high (nonpermissive) temperature. Temperature-sensitive mutations in some replication proteins result in the immediate cessation of bacterial growth; for other mutant proteins, growth comes to a halt more gradually when the bacteria are exposed to a nonpermissive temperature. What happens to DNA replication and bacterial growth when the temperature is suddenly increased and the temperature-sensitive mutation is in **(a)** helicase and **(b)** DnaA (see Problem 7)?

12. Based on your knowledge of replication proteins, compare DNA polymerase and single-strand binding protein (SSB) with respect to **(a)** affinity for DNA and **(b)** cellular concentration.

13. In the chain-termination method of DNA sequencing (described in Section 3-4), the Klenow fragment of DNA polymerase I (see Fig. 20-11) is used to synthesize a complementary strand using single-stranded DNA as a template. Along with a primer and the four dNTP substrates, a small amount of $2',3'$-dideoxynucleoside triphosphate (ddNTP) is added to the reaction mixture. When the ddNTP is incorporated into the growing nucleotide chain, polymerization stops. Explain.

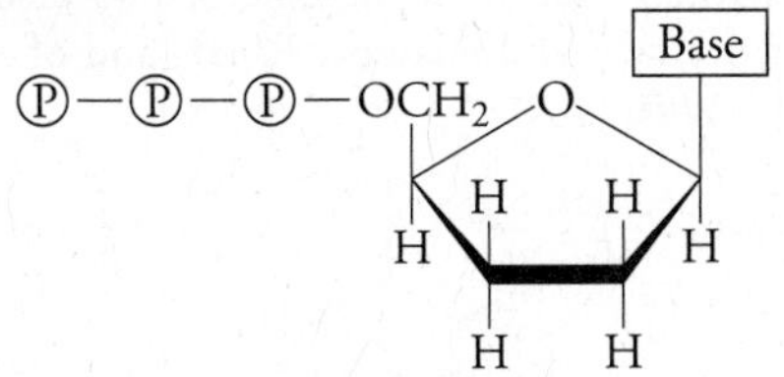

2′,3′-Dideoxynucleoside
triphosphate (ddNTP)

14. You have discovered a drug that inhibits the activity of the enzyme inorganic pyrophosphatase. What effect would this drug have on DNA synthesis?

15. Which eukaryotic DNA polymerase would you expect to have greater processivity: polymerase α or polymerase ε? Explain.

16. An *in vitro* system was developed in which simian virus 40 (SV40) can be replicated using purified mammalian proteins. Make a list of the proteins required to replicate SV40 DNA in a cell-free system.

17. Describe the ways in which a cell minimizes the incorporation of mispaired nucleotides during DNA replication.

18. DNA polymerases include two Mg^{2+} ions in the active site. **(a)** Describe how Mg^{2+} could enhance the nucleophilicity of the 3′ OH group that attacks the α-phosphate of an incoming nucleotide.
(b) The polymerization mechanism includes the formation of a pentacovalent phosphorus. Sketch this structure. How could an Mg^{2+} ion contribute to transition state stabilization during DNA polymerization?

19. A reaction mixture contains purified DNA polymerase, the four dNTPs, and one of the DNA molecules whose structures are represented here. Which reaction mixture produces PP_i?

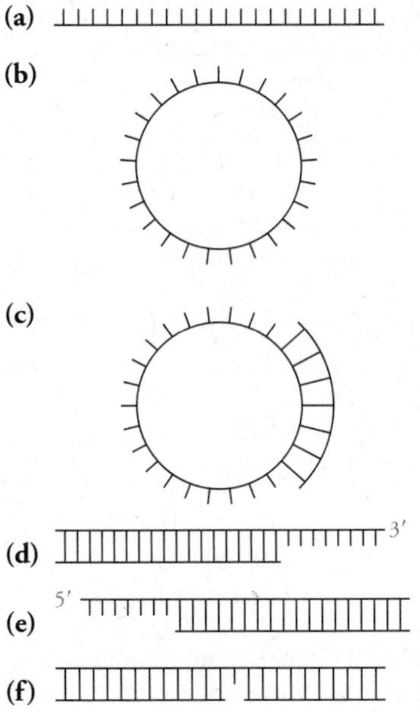

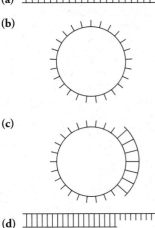

20. The "flap" endonuclease recognizes the junction between RNA and DNA near the 5′ end of an Okazaki fragment. What feature(s) of this structure might the enzyme recognize?

21. Why would it not make sense for the cell to wait to combine Okazaki fragments into one continuous lagging strand until the entire DNA molecule has been replicated?

22. Explain how DNase (the endonuclease described in Problem 4 that cleaves the backbone of one strand of a DNA molecule), *E. coli* DNA polymerase I (which includes 5′ → 3′ exonuclease activity), and DNA ligase could be used in the laboratory to incorporate radioactive nucleotides into a DNA molecule.

23. The mechanism of *E. coli* DNA ligase involves the transfer of the adenylyl (AMP) group of NAD^+ to the ε-amino group of the side chain of an essential lysine residue on the enzyme. The adenylyl group is subsequently transferred to the 5′ phosphate group of the nick. The first step is shown in the figure below. Draw the mechanism of *E. coli* DNA ligase.

$$Enzyme—(CH_2)_4—NH_2$$

Nicotinamide—Ribose—O—P—O—P—O—Ribose—Adenine

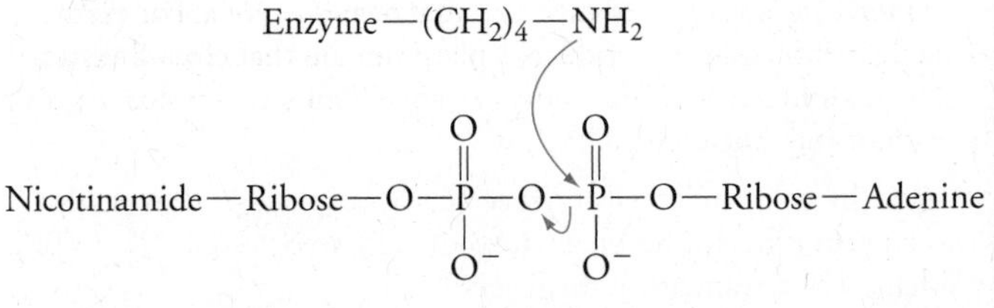

24. Why would ligase enzymes that used NAD^+ as a cofactor be attractive drug targets for treating diseases caused by bacteria?

20-3 Telomeres

25. Give the name of the enzyme that catalyzes each of the following reactions:
(a) makes a DNA strand from a DNA template
(b) makes a DNA strand from an RNA template
(c) makes an RNA strand from a DNA template

26. Human cells infected by HIV increase the expression of SAMDH1, a phosphohydrolase that removes triphosphate groups from dNTP substrates. Explain how SAMDH1 inhibits HIV replication.

27. In some species, G-rich telomeric DNA folds up on itself to form a four-stranded structure. In this structure, four guanine residues assume a hydrogen-bonded planar arrangement with an overall geometry that can be represented as

(a) This is called a G quartet, and it may play a role as a negative regulator of telomerase activity. Draw the complete structure of this G quartet, including the hydrogen bonds between the purine bases.

(b) Show schematically how a single strand of four repeating TTAGGG sequences can fold to generate a structure with three stacked G quartets linked by TTA loops.

28. Why might a drug that induced the formation of the G quartet (see Problem 27) be effective as an antitumor agent?

29. An experiment is carried out in which the AAUCCC RNA template on the telomerase (see Fig. 20-17) is mutated. Will there be a change in the telomeric sequence as a result of this mutation? Explain.

30. What cellular changes occur when the RNA template on the telomerase is mutated as described in Problem 29?

20-4 DNA Damage and Repair

31. Mutations in the genes that code for repair enzymes can often lead to the transformation of a normal cell into a cancerous cell. Explain why.

32. Although DNA damage is a cause of cancer, some cancer chemotherapies are based on drugs that damage DNA. For example, the drug carboplatin introduces a platinum ion that cross-links adjacent guanine residues in a DNA strand. Explain why this leads to death of the cancer cell.

33. In eukaryotic cells, a specific triphosphatase cleaves deoxy-8-oxoguanosine triphosphate (oxo-dGTP) to oxo-dGMP + PP_i. What is the advantage of the reaction?

34. Draw the structure of an oxoguanine:adenine base pair. (*Hint:* The oxoguanine base pivots around the glycosidic bond in order to form two hydrogen bonds with adenine.)

35. What change does the methylation of a guanine residue make in the succeeding generations of DNA?

36. Compounds such as ethidium bromide, proflavin, and acridine orange (above right) can be used to stain DNA bands in a gel. These compounds are known as intercalating agents, and they interact with the DNA by slipping in between stacked base pairs. This interaction with the DNA increases the fluorescence of the intercalating agent and allows for visualization of the DNA bands in a gel under ultraviolet light. Care must be taken when working with these compounds, however, because they are powerful mutagens. Explain why.

Ethidium bromide

Proflavin

Acridine orange

37. The compound 5-bromouracil is a thymine analog and can be incorporated into DNA in the place of thymine. 5-Bromouracil readily converts to an enol tautomer, which can base pair with guanine. (The keto and enol tautomers freely interconvert through the movement of a hydrogen between an adjacent nitrogen and oxygen.) Draw the structure of the base pair formed by the enol form of 5-bromouracil and guanine. What kind of mutation can 5-bromouracil induce?

5-Bromouracil

38. Experiments with cells are performed in a tissue culture hood, which filters the air to minimize the chance of contaminating the cells with airborne bacteria. When the experiments have been completed, the cells are removed from the hood, and an ultraviolet light is switched on until the hood is used again. What is the rationale for this procedure?

39. As discussed in the text, deamination of cytosine produces uracil. Deamination of other DNA bases can also occur. For example, deamination of adenine produces hypoxanthine.

(a) Draw the structure of hypoxanthine.

(b) Hypoxanthine can base pair with cytosine. Draw the structure of this base pair.

(c) What is the consequence to the DNA if this deamination is not repaired?

40. Deamination of guanine produces xanthine.

(a) Draw the structure of xanthine.

(b) Xanthine base pairs with cytosine. Would this cause a mutation? Explain.

41. The fact that DNA has evolved to contain the bases A, C, G, and T makes the DNA molecule easy to repair. For example, deamination of adenine produces hypoxanthine, deamination of guanine produces xanthine, and deamination of cytosine produces uracil. Why are these deaminations repaired quickly?

42. Studies of bacteria and other organisms indicate that mutations occur more frequently at certain positions. These "hotspots" are due to the presence of 5-methylcytosine, which may undergo oxidative deamination.

(a) Draw the structure of deaminated 5-methlycytosine.

(b) By what other name is the base known?

(c) What kind of mutation results from 5-methylcytosine deamination?

(d) Why is the cell unable to repair the altered base?

43. The adenine analog 2-aminopurine (shown below) is a potent mutagen in bacteria. The 2-aminopurine substitutes for adenine during DNA replication and gives rise to mutations because it pairs with cytosine instead of thymine. Structural studies show that there is an equilibrium between a "neutral wobble" base pair (so-called because it is the dominant structure at neutral pH) and a "protonated Watson–Crick" structure, which forms at lower pH.

(a) Two hydrogen bonds form between cytosine and 2-aminopurine in the "neutral wobble" pair. Draw the structure of this base pair.

(b) At lower pH, either the cytosine or the 2-aminopurine can become protonated. The two protonated forms are in equilibrium in the "protonated Watson–Crick" base pair structure; the proton essentially "shuttles" from one base to the other in the pair, and the hydrogen bond is maintained. Draw the two possible structures for the base pair, one with a protonated 2-aminopurine, and one with a protonated cytosine.

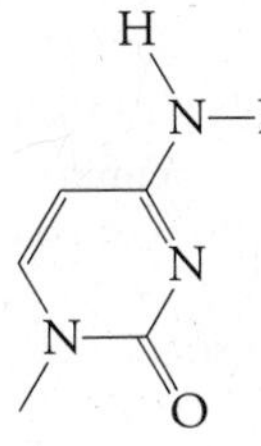

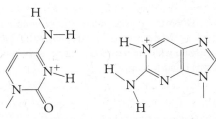

Cytosine 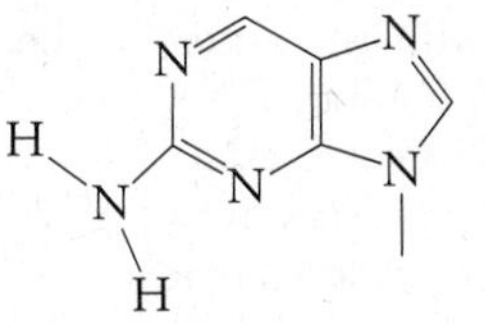   2-Aminopurine

Protonated cytosine   Protonated 2-aminopurine

44. Chemical methylating agents can react with guanine residues in DNA to produce O^6-methylguanine, which causes mutations because it can pair with thymine. The structure of O^6-methylguanine is shown in the text. This type of chemical damage is difficult to repair because the O^6-methylguanine:T base pair is structurally similar to a normal G:C base pair and the mismatch repair system has difficulty recognizing it. The resulting mutations can generate oncogenes. The O^6-methylguanine can form a "wobble" pair with thymine (similar to the 2-aminopurine:C base pair in Problem 43). The methylated guanine can also form a base pair with protonated cytosine at lower pH values.

(a) At physiological pH, the O^6-methylguanine:T wobble base pair predominates. Two hydrogen bonds form between the two bases. Draw the structure of this base pair. Does a mutation result?

(b) At lower pH values, cytosine becomes protonated (see Problem 43b). Draw the structure of the O^6-methyguanine: protonated cytosine base pair (three hydrogen bonds form between the two bases). Does a mutation result?

45. What is the most common DNA lesion in individuals with the disease xeroderma pigmentosum?

46. Thymine dimers can be restored to their original form by DNA photolyases that cleave the dimer. The photolyases are so named because they are activated by absorption of light. Why is this a good biochemical strategy?

47. A strain of mutant bacterial cells lacks the enzyme uracil-DNA glycosylase. What is the consequence for the organism?

48. In many cases, a point mutation in DNA has no effect on the encoded amino acid sequence. Explain.

49. During replication in *E. coli*, certain types of DNA damage, such as thymine dimers, can be bypassed by DNA polymerase V. This polymerase tends to incorporate guanine residues opposite the damaged thymine residues and has a higher overall error rate than other polymerases. Thymine dimers can be bypassed by other polymerases, such as DNA polymerase III (which carries out most DNA replication in *E. coli*). DNA polymerase III incorporates adenine residues opposite the damaged thymine residues, but much more slowly than DNA polymerase V. Polymerase III is a highly processive enzyme, whereas polymerase V adds only 6–8 nucleotides before dissociating from the DNA. Explain how DNA polymerases III and V together carry out the efficient replication of UV-damaged DNA with minimal errors.

50. Eukaryotic cells contain a number of DNA polymerases. Several of these enzymes were tested for their ability to cleave nucleotides from the 3′ end of a DNA chain (3′ → 5′ exonuclease activity). The enzymes were also tested for the accuracy of DNA polymerization, expressed as the rate of base substitution. The results are summarized in the table.

Polymerase	3′ → 5′ exonuclease activity	Base substitution rate ($\times 10^{-5}$)
α	No	16
β	No	67
δ	Yes	1
ε	Yes	1
η	No	3500

(a) Is there a correlation between the presence of 3′ → 5′ exonuclease activity and the error rate (base substitutions) during DNA polymerization?

(b) Express the error rate of each polymerase in terms of how often a wrong base is incorporated.

(c) Polymerization errors can result from the ability to insert an incorrect (mispaired) base or from the inability to efficiently insert the correct (template-matched) base. To tell which mechanism accounts for the high error rate of polymerase η, the catalytic efficiency (k_{cat}/K_M) of the polymerization reaction was measured for matched and unmatched bases. The results were compared to the catalytic efficiency of another polymerase, HIV reverse transcriptase (HIV RT). The data are summarized in the table.

Polymerase	Template base	Incoming base	k_{cat}/K_M ($\mu M \cdot min^{-1} \cdot 10^3$)
Polymerase η	T	A	420
	T	G	22
	T	C	1.6
	G	C	760
	G	G	8.7
HIV RT	T	A	800
	T	G	0.07

Compare the efficiency of DNA polymerase η and HIV RT in incorporating correct bases and incorrect bases. What do these results reveal about the cause of errors during polymerization by polymerase η?

(d) The overexpression of genes that code for DNA polymerases similar to polymerase η has been observed in bacteria. What effect would this have on the mutation rate in these bacteria?

20-5 DNA Packaging

51. The percentages of arginine and lysine residues in the histones of calf thymus DNA are shown in the table below. Why do histones have a large number of Lys and Arg residues?

	% Arg	% Lys
H1	1	29
H2A	9	11
H2B	6	16
H3	13	10
H4	14	11

52. Why is 0.5 M NaCl effective at dissociating histones from DNA in a sample of chromatin?

53. Draw the structures of the side chains that correspond to the following histone modifications: **(a)** acetylation of Lys, **(b)** phosphorylation of Ser, **(c)** phosphorylation of His, **(d)** methylation of Lys, and **(e)** methylation of Arg. How do these modifications change the character of their respective amino acid side chains?

54. The histone protein H4 from cows and the H4 from peas, representing groups that diverged from one another over a billion years ago, differ by only two amino acid residues out of 102 total residues. The changes are conservative—Val 60 in calf thymus H4 is replaced by an Ile in peas; Lys 77 in the calf H4 is replaced by an Arg in peas. Propose a hypothesis that explains why the H4 histone protein is so highly evolutionarily conserved.

55. DNA methylation requires the methyl group donor S-adenosylmethionine, which is produced by the condensation of methionine with ATP. The sulfonium ion's methyl group is used in methyl-group transfer reactions.

S-Adenosylmethionine

(a) The demethylated S-adenosylmethionine is then hydrolyzed to produce adenosine and a nonstandard amino acid. Draw the structure of this amino acid. How does the cell convert this compound back to methionine to regenerate S-adenosylmethionine?

(b) The proper regulation of gene expression requires methylation as well as demethylation of cytosine residues in DNA. If a demethylase carries out a hydrolytic reaction to restore cytosine residues, what is the other reaction product?

56. Mice that are homozygous for a mutation that renders the DNA methyltransferase enzyme nonfunctional usually die *in utero*. Why does this mutation have such serious consequences?

57. Proteins destined for the proteasome are tagged with ubiquitin as described in Section 12-1. How does this modification compare with the modification of a histone with ubiquitin? Are ubiquitinated histones marked for proteolytic destruction by the proteasome?

58. Enzymes that catalyze the acetylation of histones (histone acetyltransferases, or HAT) are closely associated with transcription factors, which are proteins that promote transcription. Why is this a good biochemical strategy?

59. During sperm development, about 95% of the cell's DNA is associated with small proteins known as protamines rather than histones. Protamine–DNA complexes pack more compactly than histone–DNA complexes.

(a) Explain the advantage of replacing histones with protamines during sperm development.

(b) What type of genes would you expect to find in the remaining nucleosomes? (*Hint*: these genes are not transcribed until after fertilization.)

60. *In vivo*, DNA undergoes transient local melting, producing small "bubbles" of single-stranded DNA. Does this DNA "breathing" activity explain why the rate of C → T mutations is lower when the DNA is part of a nucleosome?

[SELECTED READINGS]

Deweese, J. E., Osheroff, M. A., and Osheroff, N., DNA topology and topoisomerases, *Biochem. Mol. Biol. Educ.* **37,** 2–10 (2009). [Includes discussion of topoisomerase action and inhibitors.]

Dunn, B., Solving an age-old problem, *Nature* **483,** S2–S6 (2012). [Discusses major advances in research on the causes, genetic basis, treatment, and prevention of cancer.]

Hakem, R., DNA-damage repair; the good, the bad, and the ugly, *EMBO J.* **27,** 589–605 (2008). [Summarizes the different repair pathways and diseases associated with their defects.]

Kunkel, T. A., and Burgers, P. M., Dividing the workload at a eukaryotic replication fork, *Trends Cell Biol.* **18,** 521–527 (2008).

[Discusses experimental evidence supporting the involvement of several polymerases in eukaryotic DNA replication.]

Luger, K., and Hansen, J. C., Nucleosome and chromatin fiber dynamics, *Curr. Opin. Struct. Biol.* **15,** 188–196 (2005).

O'Donnell, M., Replisome architecture and dynamics in *E. coli*, *J. Biol. Chem.* **281,** 10653–10656 (2006).

O'Sullivan, R. J., and Karlseder, J., Telomeres: protecting chromosomes against genome instability, *Nat. Rev. Mol. Cell Biol.* **11,** 171–181 (2010). [Discusses telomere structure as well as the roles of telomere-binding proteins.]

chapter 21

TRANSCRIPTION AND RNA

[ittipol nampochai/iStockphoto.]

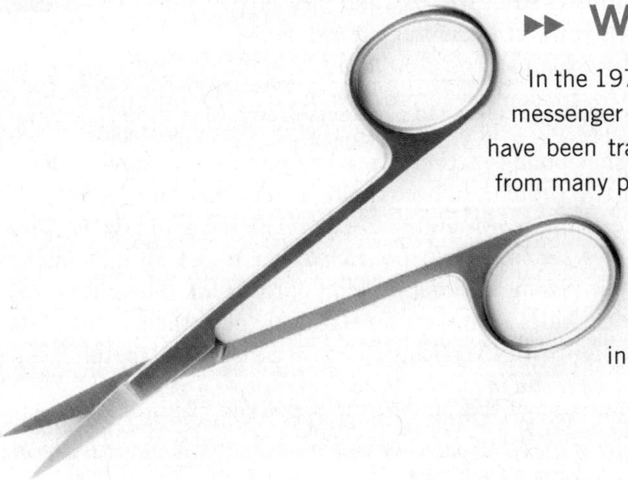

▸▸ WHY do genes include introns?

In the 1970s, Richard Roberts and Phillip Sharp discovered that some eukaryotic messenger RNA molecules are much shorter than the genes from which they have been transcribed. More surprising still, the messenger RNAs are assembled from many pieces that are precisely cut from a longer RNA and joined together. In humans, virtually all genes are discontinuous and require complicated machinery to carry out mRNA splicing. Natural selection would have long ago eliminated such an inefficient process for expressing genetic information—unless there were advantages in arranging genes in bits and pieces.

THIS CHAPTER IN CONTEXT

Part 1 Foundations

Part 2 Molecular Structure and Function

Part 3 Metabolism

Part 4 Genetic Information

21 Transcription and RNA

Do You Remember?

- **DNA and RNA are polymers of nucleotides, each of which consists of a purine or pyrimidine base, deoxyribose or ribose, and phosphate (Section 3-1).**
- **The biological information encoded by a sequence of DNA is transcribed to RNA and then translated into the amino acid sequence of a protein (Section 3-2).**
- **Genes can be identified by their nucleotide sequences (Section 3-3).**
- **DNA replication is carried out by stationary protein complexes (Section 20-2).**
- **Both histones and DNA may be chemically modified (Section 20-5).**

Transcription is the fundamental mechanism by which a gene is expressed. It is the conversion of stored genetic information (DNA) to a more active form (RNA). The information contained in the sequence of deoxynucleotides in DNA is preserved in the sequence of ribonucleotides in the RNA transcript. Like DNA replication, transcription is characterized by template-directed nucleotide polymerization that requires a certain degree of fidelity. However, unlike DNA synthesis, RNA synthesis takes place selectively, on a gene-by-gene basis.

In describing RNA transcription, we invoke the idea of **gene,** a segment of DNA that is transcribed for the purpose of expressing the encoded genetic information or transforming it to a form that is more useful to the cell. This definition of a gene requires some qualification:

1. *For protein-coding genes, the RNA transcript (called **messenger RNA** or **mRNA**) includes all the information specifying the sequence of amino acids in a polypeptide.* But keep in mind that not all RNA molecules are translated into protein. **Ribosomal RNA (rRNA), transfer RNA (tRNA),** and other types of RNA molecules carry out their functions without undergoing translation.

2. *Most RNA transcripts correspond to a single functional unit,* for example, one polypeptide. But some RNAs, particularly in prokaryotes, code for multiple proteins and result from the transcription of an **operon,** a set of contiguous genes whose products have related metabolic functions. In a few rare cases, a single mRNA carries information for two proteins in overlapping sequences of nucleotides.

3. *RNA transcripts typically undergo **processing**—the addition, removal, and modification of nucleotides—before becoming fully functional.* For nearly all mammalian protein-coding genes, processing includes splicing out **introns.** These long segments of noncoding nucleotides are not expressed but are nevertheless part of the gene. Because of variations in mRNA splicing and post-translational modification, *several different forms of a protein may be derived from a single gene.*

4. Finally, *the proper transcription of a gene may depend on DNA sequences that are not transcribed but help position the RNA polymerase at the transcription start site or are involved in the regulation of gene expression.*

Most of this chapter focuses on the transcription of eukaryotic protein-coding genes. As introduced in Section 3-3, the human genome includes an estimated 21,000 such genes, which have an average length of about 27,000 bp. The actual fraction of protein-coding sequences is quite small, however (about 1.5% of the genome), because a typical gene consists of eight **exons** (protein-coding segments) with an average length of 145 bp that are separated by introns with an average length of 3365 bp. Nevertheless, an estimated 80% of the human genome may undergo transcription to produce **noncoding RNAs (ncRNAs)** of various sizes. Some of these are listed in Table 21-1.

The function of some types of ncRNA molecules have been elucidated and are described later in this chapter, but others remain mysterious. It is possible that they represent transcriptional "noise," or random synthesis, which is consistent with the observation that they are degraded soon after their synthesis. However, evidence that some of these sequences are conserved among species indicates that they may play a role in large-scale regulation of transcriptional activity.

[TABLE 21-1] Some Noncoding RNAs

Type	Size (nucleotides)	Function
Ribosomal RNA (rRNA)	120–4718	Translation (ribosome structure and catalytic activity)
Transfer RNA (tRNA)	54–100	Delivery of amino acids to ribosome during translation
Small interfering RNA (siRNA)	20–25	Sequence-specific inactivation of mRNA
Micro RNA (miRNA)	20–25	Sequence-specific inactivation of mRNA
Large intervening noncoding RNA (lincRNA)	Up to 17,200	Transcriptional control
Small nuclear RNA (snRNA)	60–300	RNA splicing
Small nucleolar RNA (snoRNA)	70–100	Sequence-specific methylation of rRNA

Although the functions of the more familiar mRNA, rRNA, and tRNA are well understood, their synthesis is far from simple. Transcription of these RNAs requires a mechanism for selecting sites where RNA polymerase initiates transcription. Numerous protein cofactors interact with each other, with the polymerase, and with the DNA template to regulate where and when transcription takes place, to accurately transcribe the DNA, and to convert the initial RNA product to its mature, functional form.

21-1 Transcription Initiation

Like DNA replication, *RNA transcription is apparently carried out by immobile protein complexes that reel in the DNA.* These transcription factories in eukaryotic nuclei can be visualized by immunofluorescence microscopy and are distinct from the replication factories where DNA is synthesized (Fig. 21-1). If the RNA polymerase were free to track along the length of a DNA molecule, rotating around the helical template, the newly synthesized RNA strand would become tangled with the DNA. In fact, except for a short 8- to 9-bp hybrid DNA–RNA helix at the polymerase active site, the newly synthesized RNA is released as a single-stranded molecule.

KEY CONCEPTS
- Transcription in eukaryotes may require alteration of histones and chromatin structure.
- Promoters are DNA sequences where RNA polymerase initiates transcription.
- Eukaryotic transcription factors interact with DNA, with RNA polymerase, and with each other.
- Additional DNA-binding proteins may regulate transcription.
- In the *lac* operon, transcription of several genes is regulated by one repressor.

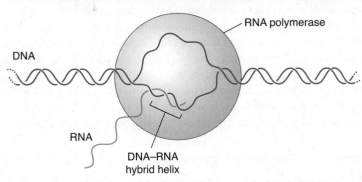

Synthesis of a new RNA molecule is a multistep process, which includes locating the transcription start site, melting apart the double-stranded DNA, and initiating RNA synthesis.

Chromatin remodeling may precede transcription

As discussed in Section 20-5, eukaryotic DNA is packaged in nucleosomes, which associate closely to form higher-order structures. Transcriptionally inactive DNA is maintained in a condensed state by mutually interdependent factors such as the degree of DNA methylation, the presence of variant histones, and the pattern of histone modification (the histone code).

Some histone modifications associated with "silent" and "active" chromatin are shown in Table 21-2. The histone code also involves phosphorylation of Ser and Thr residues, methylation of Arg residues, and other covalent changes (see Fig. 20-31). The combinatorial possibilities for all the different histone modifications may be as great as 10^{11}, and even if not all of them are realized, there is much to be learned about how the histone code operates.

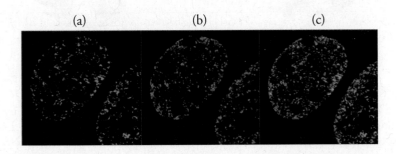

Figure 21-1 Spatial separation of transcription and replication. In these fluorescence microscopy images of mouse cells in early S phase, sites of DNA replication are green (a) and sites of RNA transcription are red (b). The merged images are shown in (c). A single nucleus may contain 2000 to 3000 transcription sites or "factories." [Courtesy Ronald Berezney. From Wei, X. et al., *Science* 281, 1502–1505 (1998).]

[TABLE 21-2]   Transcriptional Activity and Histone Modification

Transcriptionally Silent Chromatin

H3K9me2	Lys 9 of histone H3 is dimethylated
H3K27me3	Lys 27 of histone H3 is trimethylated

Transcriptionally Active Chromatin

H3K4me3	Lys 4 of histone H3 is trimethylated
H3K9ac	Lys 9 of histone H3 is acetylated
H4K16ac	Lys 16 of histone H4 is acetylated

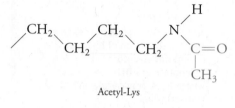

Acetyl-Lys

A key change in the transformation of silent to active chromatin is the acetylation of histones. Enzymes known as histone acetyltransferases (HATs) add acetyl groups from acetyl-CoA to the side chains of Lys residues (*left*). This modification can be reversed later by the action of a histone deacetylase. Other changes, including demethylation (of histones and DNA) and histone exchange, promote transcriptional activity.

Decondensing chromatin to make it more accessible to the transcription machinery also involves remodeling the chromatin so that critical DNA sequences are exposed rather than wrapped tightly around histones in the nucleosome core. The dynamics of chromatin therefore include a process in which nucleosomes change their positions, appearing to slide along the DNA. Chromatin-remodeling complexes contain multiple protein subunits and require the free energy of ATP. A model for one of these complexes, which neatly grasps the nucleosome, is shown in **Figure 21-2**.

How do these complexes work? Complete dissociation of the DNA from the histone octamer would be energetically costly. Instead, the complexes appear to operate by loosening a segment of DNA from its histone attachments. The resulting loop of DNA travels around the nucleosome as a wave so that a segment of DNA that was part of the nucleosome core is now available to interact with transcription factors (**Fig. 21-3**). In eukaryotes, transcription starts in nucleosome-free stretches of DNA.

Transcription begins at promoters

Prokaryotes typically have compact genomes without much nontranscribed DNA (see Fig. 3-13a), whereas in eukaryotes, protein-coding genes may be separated by large tracts of DNA (see Fig. 3-13b). But in both types of organisms, the efficient expression of genetic information involves the initiation of RNA synthesis at a particular site, known as a **promoter,** near the protein-coding sequence. The DNA sequence at the promoter is recognized by specific proteins, which either are part of the RNA polymerase protein or subsequently recruit the appropriate RNA polymerase to the DNA to begin RNA synthesis.

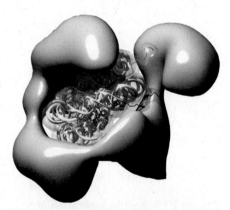

Figure 21-2 Model of a nucleosome bound to a chromatin-remodeling complex. The X-ray structure of the nucleosome (see Fig. 20-29) was modeled into the structure of the yeast RSC complex (gray) determined by cryoelectron microscopy. RSC stands for Remodels the Structure of Chromatin. [Courtesy of Andres Leschziner, Harvard University.]

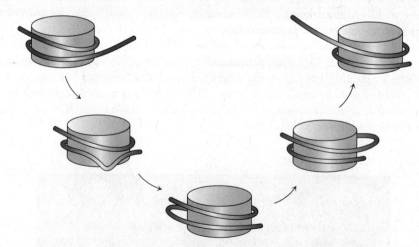

Figure 21-3 Nucleosome sliding. Chromatin-remodeling complexes may act by loosening a portion of DNA (blue) wound around a histone octamer (orange). The nucleosome appears to slide along the DNA, exposing different segments of DNA.

Transcription
start site
+1

−35 region   −10 region

5′···NNNTTGACANNNNNNNNNNNNNNNNNNTATAATNNNNNNNNNNNN···3′

Figure 21-4 The *E. coli* promoter. The first nucleotide to be transcribed is position +1. The two consensus promoter sequences, centered around positions −10 and −35, are shaded. N represents any nucleotide.

> **?** Approximately how many turns of the DNA helix separate the −10 region from the −35 region and the transcription start site?

In bacteria such as *E. coli,* the promoter comprises a sequence of about 40 bases on the 5′ side of the transcription start site. By convention, such sequences are written for the coding, or nontemplate, DNA strand (see Section 3-2) so that the DNA sequence has the same sense and 5′→3′ directionality as the transcribed RNA. The *E. coli* promoter includes two conserved segments, whose **consensus sequences** (sequences indicating the nucleotides found most frequently at each position) are centered at positions −35 and −10 relative to the transcription start site (position +1; **Fig. 21-4**).

E. coli RNA polymerase is a five-subunit enzyme with a mass of about 450 kD and a subunit composition of $\alpha_2\beta\beta'\omega\sigma$. The σ subunit, or σ factor, recognizes the promoter, thereby precisely positioning the RNA polymerase enzyme to begin transcribing. Although bacterial cells contain only one core RNA polymerase ($\alpha_2\beta\beta'\omega$), they contain multiple σ factors, each specific for a different promoter sequence. Since different genes may have similar promoters, *bacterial cells can regulate patterns of gene expression through the use of different σ factors.* Once transcription is under way, the σ factor is jettisoned, and the remaining subunits of RNA polymerase extend the transcript.

In eukaryotes, some promoters for protein-coding genes include a **TATA box,** an element that resembles the prokaryotic promoter. Most eukaryotic genes lack this sequence but may instead exhibit a variety of other conserved promoter elements located both upstream (on the 5′ side) and downstream (on the 3′ side) of the transcription initiation site (**Fig. 21-5**). A given gene may contain several—although not all—of these elements, which may act synergistically. For example, the DPE (downstream promoter element) is always accompanied by an Inr (initiator) element. In a large eukaryotic genome, a given sequence of 6–8 nucleotides is expected to occur many times by chance, so the multipart nature of eukaryotic promoters may help guard against random transcription initiation at a single one of these sequences.

Approximately two-thirds of vertebrate genes lack well-defined promoters with a precise transcription start site, so transcription initiates over a stretch of 50 to 100 nucleotides. This DNA typically contains unmethylated CpG islands (see Section 20-5), but little is known about how transcription factors and RNA polymerase interact with this DNA. Eukaryotic RNA polymerase does not include a subunit corresponding to the prokaryotic σ factor. Instead, RNA polymerase locates the promoter through a series of complex protein–protein and protein–DNA interactions.

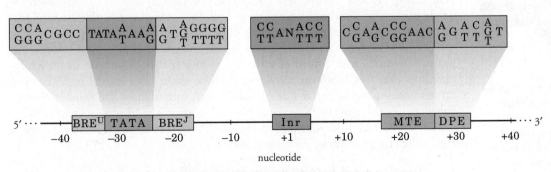

Figure 21-5 Sequences of some eukaryotic promoter elements. The consensus sequence for each element in humans is given. Some positions can accommodate two or three different nucleotides. N represents any nucleotide.

> **?** Compare these sequences to the *E. coli* promoter (see Fig. 21-4).

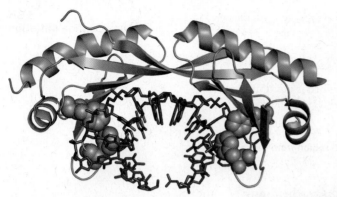

Figure 21-6 Structure of TBP bound to DNA. The TBP polypeptide (green) forms a pseudosymmetrical structure that straddles a segment of DNA containing a TATA box (the DNA is shown in blue and viewed end-on). The insertion of TBP Phe residues (orange) bends the DNA in two places. [Structure (pdb 1YTB) determined by Y. Kim, J. H. Geiger, S. Hahn, and P. B. Sigler.] ⊕ See Interactive Exercise. TATA-binding protein in complex with TATA box.

Transcription factors recognize eukaryotic promoters

*In eukaryotes, the initiation of transcription typically requires a set of five highly conserved proteins known as **general transcription factors.*** They are abbreviated as TFIIB, TFIID, TFIIE, TFIIF, and TFIIH (the II indicates that these transcription factors are specific for RNA polymerase II, the enzyme that transcribes protein-coding genes). Some of these general transcription factors interact specifically with some of the promoter elements shown in Figure 21-5. For example, TFIIB binds to the BRE sequences, and the TATA-binding protein (TBP), a subunit of TFIID, binds to the TATA box. Other TFIID subunits, called TAFs (TBP-associated factors), interact with the Inr and DPE promoter elements. There is no set of universal transcription factors required for transcribing every gene. In fact, TFIID has multiple forms with variable subunit compositions, which may suit it for recognizing different combinations of promoter elements.

The various transcription factors play numerous roles in preparing the DNA for transcription and recruiting RNA polymerase. TBP, a saddle-shaped protein about $32 \times 45 \times 60$ Å, consists of two structurally similar domains that sit astride the TATA box DNA at an angle (Fig. 21-6). This protein–DNA interaction introduces two sharp kinks into the DNA. The kinks are caused by the insertion of two Phe side chains like a wedge between a T and an A residue at each end of the TATA box. There are other sequence-specific interactions based on hydrogen bonding and van der Waals interactions. TBP may be an important player in transcription initiation even at promoters without a TATA box.

Once TBP is in place at the promoter, the conformationally altered DNA may serve as a stage for the assembly of additional proteins, including RNA polymerase and other transcription factors (Fig. 21-7). TFIIB, for example, helps position the DNA near the polymerase active site. TFIIE joins the complex and recruits TFIIH, a helicase that unwinds the DNA in an ATP-dependent manner. The result is an open structure called a transcription bubble that is stabilized in part by the binding of TFIIF, which interacts with the nontemplate DNA strand.

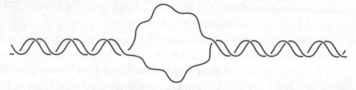

Transcription bubble

The structural changes in the DNA may extend beyond the immediate site of the transcription bubble. For instance, the TFIID component known as TAF1 has histone acetyltransferase activity, which may allow it to alter nucleosome packing by neutralizing Lys side chains. TAF1 may also diminish histone H1 cross-linking of neighboring nucleosomes by helping to link the small protein ubiquitin to H1, thereby marking it for proteolytic destruction by a proteasome (see Section 12-1).

TAF1 and some other transcription factors include structural motifs known as bromodomains. These consist of a bundle of four α helices surrounding a hydrophobic pocket that binds an acetylated Lys side chain (Fig. 21-8). In TAF1, two bromodomains are arranged side by side, which might allow the protein to bind cooperatively to a multiply acetylated histone protein. The histone acetyltransferase activity of the TAF itself may lead to localized hyperacetylation, a positive feedback mechanism for promoting gene transcription.

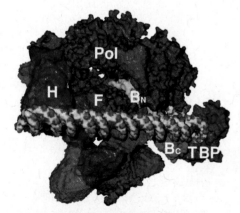

Figure 21-7 Model of RNA polymerase and some transcription factors. This composite model is based on models of the various proteins. RNA polymerase is gray, and the various transcription factors (TBP, TFIIF, TFIIH, and the N- and C-terminal portions of TFIIB) are shown in different colors. The model of DNA is colored by electrostatic potential (red negative, blue positive). [Courtesy of Roger Kornberg, Stanford University School of Medicine.]

Enhancers and silencers act at a distance from the promoter

Additional sets of protein–protein and protein–DNA interactions may participate in the highly regulated expression of many eukaryotic genes. Whereas the rate of prokaryotic gene transcription varies about 1000-fold between the most and least frequently expressed genes, the gene transcription rate in eukaryotes may vary by as much as 10^9-fold. Some of this fine control is due to **enhancers,** DNA sequences that range from 50 to 1500 bp and are located up to 120 kb upstream or downstream of a promoter. A single gene may have more than one enhancer functionally associated with it, and hundreds of thousands of these elements may be scattered around the human genome.

The proteins that bind to enhancers are commonly called **activators,** although they are also known simply as transcription factors (hence the term *general transcription factor* for the ones that bind to promoters). By facilitating (or inhibiting) transcription, these DNA-binding proteins can shape an organism's gene expression patterns in response to internal or external signals. We have already seen how some transcription factors are linked to signal transduction pathways (Chapter 10). For example, steroid hormones directly activate DNA-binding proteins, and Ras signaling indirectly activates several transcription factors. These proteins interact with DNA in a variety of ways (Box 21-A).

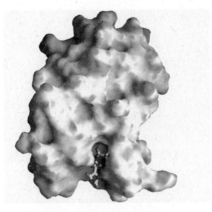

Figure 21-8 A bromodomain. The protein surface is shown in this model. A single bromodomain (~110 amino acids) includes a cavity for an acetyl-Lys group (ball-and-stick model). [Courtesy Ming-Ming Zhou, Mt. Sinai School of Medicine. From *Science* 285, 1201 (1999).]

⊕ **See Guided Exploration. Transcription factor–DNA interactions.**

BOX 21-A ⬡ BIOCHEMISTRY NOTE

DNA-Binding Proteins

The proteins that directly participate in or regulate processes such as DNA replication, repair, and transcription must interact intimately with the DNA double helix. In fact, many of the proteins that promote or suppress transcription recognize and bind to particular sequences in the DNA. However, there do not seem to be any strict rules by which certain amino acid side chains pair with certain nucleotide bases. In general, interactions are based on van der Waals interactions and hydrogen bonds, often with intervening water molecules.

An examination of the structures of a large variety of protein–DNA complexes in prokaryotic and eukaryotic cells reveals that the DNA-binding proteins fall into a limited number of classes, depending on the structural motif that contacts the DNA. Many of these motifs are likely the result of convergent evolution and may therefore represent the most stable and evolutionarily versatile ways for proteins to interact with DNA.

By far the most prevalent mode of protein–DNA interaction involves a protein α helix that binds in the major groove of DNA. This DNA-binding motif may take the form of a helix–turn–helix (HTH) structure in which two perpendicular α helices are connected by a small loop of at least four residues. The HTH motifs are colored red in the structure shown at right, and the DNA is blue. In most cases, the side chains of one helix insert into the major groove and directly contact the exposed edges of bases in the DNA. Residues in the other helix and the turn may interact with the DNA backbone.

In prokaryotic and eukaryotic proteins, the HTH helices are usually part of a larger bundle of several α helices, which form a stable domain with a hydrophobic core. Prokaryotic transcription factors tend to be homodimeric proteins (as in the previous example) that recognize palindromic DNA sequences. In contrast, eukaryotic transcription factors more commonly are heterodimeric or contain multiple domains that recognize a nonsymmetrical series of binding sites. For this reason, eukaryotic DNA-binding proteins are able to interact with a wider variety of target DNA sequences.

Many eukaryotic transcription factors include a DNA-binding motif in which one zinc ion (sometimes two) is tetrahedrally coordinated by Cys or His side chains. The metal ion stabilizes a small protein domain (which is sometimes involved in protein–protein rather than protein–DNA interactions). In most cases, the DNA-binding motif, known as a zinc finger (see Fig. 4-15), consists of two antiparallel β strands followed by an α helix. In the structure shown on the next page, one of the three zinc fingers is colored red. The Zn^{2+} ions are represented by purple spheres. As in the HTH proteins, the helix of each zinc finger motif inserts into the major groove of DNA, where it interacts with a three-base-pair sequence.

(continued on the next page)

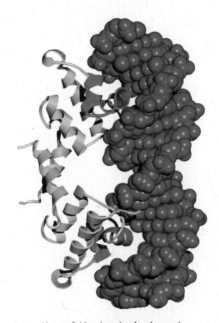

A portion of the bacteriophage λ repressor bound to DNA. [Structure (pdb 1LMB) determined by L. J. Beamer and C. O. Pabo.]

Zinc finger domains from the mouse transcription factor Zif268.
[Structure (pdb 1AAY) determined by M. Elrod-Erickson, M. A. Rould, and C. O. Pabo.]

Some homodimeric DNA-binding proteins in eukaryotes include a leucine zipper motif that mediates protein dimerization. Each subunit has an α helix about 60 residues long, which forms a coiled coil with its counterpart in the other subunit (see Section 5-2). Leucine residues, appearing at every eighth position, or about every two turns of the α helix, mediate hydrophobic contacts between the two helices (they do not actually interdigitate, as the term *zipper* might suggest). The DNA-binding portions of a leucine zipper protein are extensions of the dimerization helices that bind in the major groove.

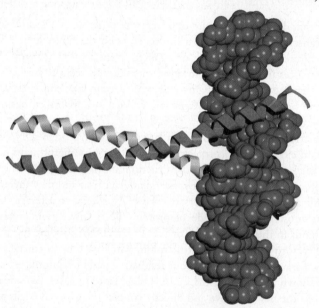

A portion of the yeast transcription factor GCN4. [Structure (pdb 1DGC) determined by W. Keller, P. Konig, and T. J. Richmond.]

In a few proteins, a β sheet interacts with the DNA (TBP is one such protein; see Fig. 21-6). In a few cases, two antiparallel β strands constitute the DNA-binding segment, fitting into the major groove so that protein side chains can form hydrogen bonds with the functional groups on the edges of the DNA bases.

The DNA-binding proteins described here interact with a limited portion of the DNA (typically just a few base pairs), marking the positions of regulatory DNA sequences and making additional protein–protein contacts that control processes such as gene transcription. Proteins that carry out catalytic functions (polymerases, for example) interact with the DNA much more extensively—but in a sequence-independent manner—and they tend to envelope the entire DNA helix.

⬡ **Question:** Explain why sequence-specific DNA-binding proteins usually interact with the major groove rather than with the minor groove.

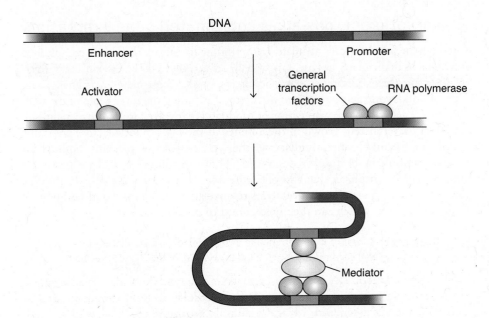

DNA

Enhancer   Promoter

Activator   General transcription factors   RNA polymerase

Mediator

Figure 21-9 Overview of enhancer function. An activator protein binds to a gene's enhancer sequence. Mediator, binding to the activator as well as to the general transcription factors and RNA polymerase at the gene's promoter, transmits a transcription-activating signal to RNA polymerase, thereby promoting gene expression. The negative regulation of gene expression may occur through the binding of a repressor protein to a gene's silencer sequence. In both cases, protein–protein interactions cause the DNA to loop out between the regulatory sequence and the promoter. This simple diagram does not convey the complexity of many activator and repressor pathways, which may involve competition for binding sites among the many protein factors.

When an activator binds to the enhancer, a protein complex known as Mediator links the enhancer-bound activator to the transcription machinery poised at the promoter. Note that this interaction requires that the DNA form a loop in order to connect the enhancer and promoter (**Fig. 21-9**). The packaging of DNA in nucleosomes may facilitate this long-range interaction by minimizing the length of the intervening DNA loop. In addition to enhancers, *a gene may have associated **silencer** sequences that bind proteins known as **repressors**.* Mediator may also relay silencer–repressor signals to the transcription machinery in order to repress gene transcription.

The Mediator complex, which contains about 20 polypeptides in yeast and as many as 30 in mammals, is a discrete particle visible by electron microscopy. Mediator interacts with RNA polymerase as well as with the general transcription factors (**Fig. 21-10**). As many as 60 proteins may congregate at a transcription initiation site. Mediator complexes with different polypeptide compositions could potentially recognize different activators and repressors. In more complicated organisms, the variability among Mediator-like complexes, coupled with multiple enhancers and silencers, could constitute a sophisticated system for fine-tuning gene expression, leading to the different patterns of gene expression that distinguish the 200 or so different cell types found in the human body.

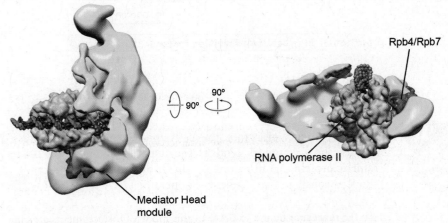

Rpb4/Rpb7

RNA polymerase II

Mediator Head module

Figure 21-10 Model of yeast Mediator bound to RNA polymerase. Mediator is shown in yellow and RNA polymerase in orange, with two subunits (Rpb4/Rpb7) that interact closely with Mediator in red. The promoter DNA is blue and green. This model is based on results from electron microscopy and X-ray crystallography. [Courtesy Francisco Asturias, the Scripps Research Institute, La Jolla, CA.]

Prokaryotic operons allow coordinated gene expression

Although prokaryotic cells regulate transcription initiation using less complicated mechanisms than eukaryotes use, functionally related genes may be organized in operons to ensure their coordinated expression in response to some metabolic signal. About 13% of prokaryotic genes are found in operons, including the three genes of the well-studied *E. coli lac* operon, whose protein products are involved in lactose metabolism.

When bacterial cells grown in the absence of the disaccharide lactose are transferred to a medium containing lactose, the synthesis of two proteins required for lactose catabolism is quickly enhanced. These proteins are lactose permease (Fig. 9-14), a transporter that allows lactose to enter the cells, and β-galactosidase (Section 11-2), an enzyme that catalyzes the hydrolysis of lactose to its component monosaccharides, which can then be oxidized to produce ATP:

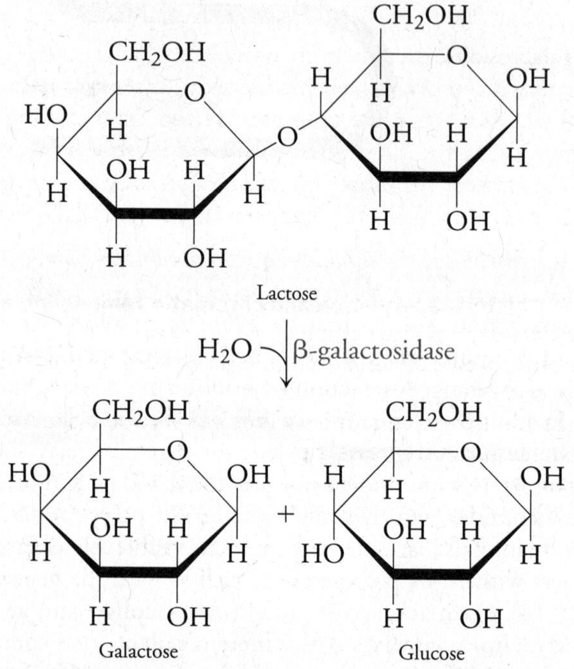

Lactose

H_2O → β-galactosidase

Galactose + Glucose

See Guided Exploration. Regulation of gene expression by the *lac* repressor system.

β-Galactosidase and lactose permease, along with a third enzyme (thiogalactoside transacetylase, whose physiological function is not clear), form the three-gene *lac* operon. The three genes (designated *Z*, *Y*, and *A*) are transcribed as a single unit from a promoter, *P*. Near the start of the β-galactosidase gene is a regulatory site called the operator, *O*, which binds a protein called the *lac* repressor. The repressor itself is a product of the *I* gene, which lies just upstream of the *lac* operon:

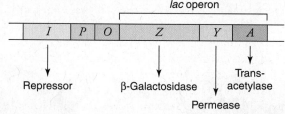

lac operon

| *I* | *P* | *O* | *Z* | *Y* | *A* |

Repressor   β-Galactosidase   Trans-acetylase

Permease

In the absence of lactose, the *Z*, *Y*, and *A* genes are not expressed because the *lac* repressor binds to the operator. The repressor is a tetrameric protein that functions as a dimer of dimers so that it can bind simultaneously to two segments of operator DNA (**Fig. 21-11**). The

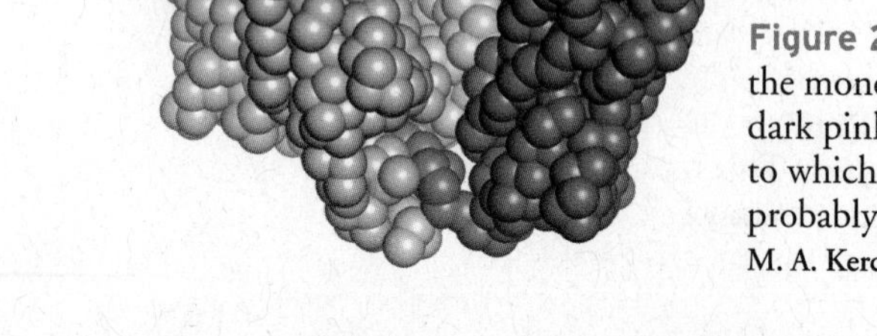

Figure 21-11 *lac* repressor bound to DNA. In this model, DNA segments are blue, the monomeric units of the repressor are light green, light blue, light pink, and dark pink (each amino acid is represented by a sphere). The two operator sequences to which the repressor binds are separated by a stretch of DNA of 93 bp, which probably forms a loop. [Structure (pdb 1LBG) determined by M. Lewis, G. Chang, N. C. Horton, M. A. Kercher, H. C. Pace, and P. Lu.]

repressor does not interfere with RNA polymerase binding, but it prevents the polymerase from initiating transcription at the promoter.

When the cell is exposed to lactose, a lactose isomer called allolactose

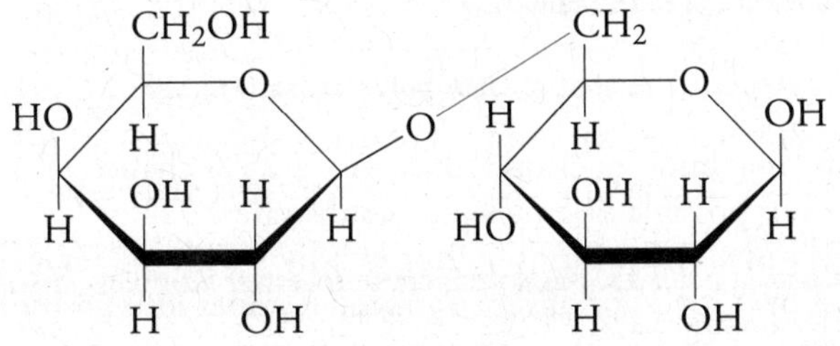

Allolactose

acts as an inducer of the *lac* operon (allolactose is generated from lactose by trace amounts of β-galactosidase present in the bacterial cell). The inducer binds to the *lac* repressor, triggering a conformational change that causes it to release its grip on the operator sequences. As a result, the promoter is freed for transcription, and the production of β-galactosidase and lactose permease can increase a thousandfold within a few minutes. This simple regulatory system ensures that the proteins required for metabolizing lactose are synthesized only when lactose is available as a metabolic fuel.

> **CONCEPT REVIEW**
> - Explain why histone acetylation and chromatin remodeling are prerequisites for eukaryotic gene transcription.
> - Compare prokaryotic and eukaryotic promoters in terms of sequence, variability, and recognition by proteins.
> - Summarize the activities of the various eukaryotic transcription factors and how these facilitate transcription.
> - Summarize the roles of enhancers, silencers, activators, repressors, and Mediator in regulating gene transcription.
> - Describe how the *lac* operon is switched on and off.

21-2 RNA Polymerase

Bacterial cells contain just one type of RNA polymerase, but eukaryotic cells contain three (plus additional polymerases for chloroplasts and mitochondria). Eukaryotic RNA polymerase I is responsible for transcribing rRNA genes, which are present in multiple copies. RNA polymerase III mainly synthesizes tRNA molecules and other small RNAs. Protein-coding genes are transcribed by RNA polymerase II, which is the main focus of this section.

The structure of RNA polymerase II was determined by Roger Kornberg (Fig. 21-12). The enzyme has a mass of over 500 kD and at first glance has the same fingers, thumb, and palm domains as DNA polymerase (see Fig. 20-11). The highly conserved sequences of eukaryotic RNA polymerases indicate that they have virtually identical structures. The core structure of RNA polymerase and its catalytic mechanism are also very similar between eukaryotes and prokaryotes. The differences are mainly at the enzyme surface, where the proteins interact with transcription factors and other regulatory proteins.

The active site of RNA polymerase is located at the bottom of a positively charged cleft between the two largest subunits. The DNA to be transcribed enters the active-site cleft of RNA polymerase, and the two DNA strands separate to form a transcription bubble that extends to 12 or 14 nucleotides. The location of the nontemplate strand has not been delineated, but the template strand threads through the polymerase, making an abrupt right-angle turn where it encounters a wall of protein. Here, the template base points away from the standard B-DNA conformation and toward the floor of the active-site cleft. This geometry allows the deoxynucleotide residue to base pair with an incoming ribonucleoside triphosphate, which enters the active site through a channel in the floor (Fig. 21-13).

KEY CONCEPTS
- During transcription, the DNA strands separate, and RNA polymerase constructs an RNA molecule that forms a short hybrid helix with the template strand.
- RNA polymerase is a processive enzyme that proofreads its work.
- The shift from initiation to elongation involves structural changes in RNA polymerase, including phosphorylation of its C-terminal domain.
- The mechanisms for transcription termination differ in prokaryotes and eukaryotes.

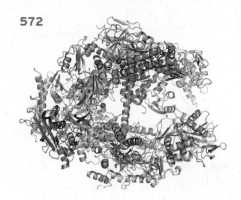

Figure 21-12 Structure of eukaryotic RNA polymerase II. The model includes 10 of the 12 subunits of the yeast enzyme (the two small subunits that are missing are not essential for transcription), with subunits shown in different colors. A magenta sphere represents an Mg^{2+} ion in the active site. [Structure (pdb 1I50) determined by P. Cramer, D. A. Bushnell, and R. D. Kornberg.]

? **Compare this structure to that of DNA polymerase (Fig. 20-11).**

Note that the incoming nucleotide has only a 25% chance of pairing correctly (since there are four possible ribonucleotides substrates: ATP, CTP, GTP, and UTP). When the correct nucleotide forms hydrogen bonds with the template base, a loop of protein closes over it and brings the catalytic residues into position. This mechanism is believed to enhance the accuracy of transcription since only a correctly paired nucleotide can be efficiently added to the growing chain.

Catalysis requires two metal ions (Mg^{2+}) coordinated by negatively charged side chains. Like DNA polymerase, *RNA polymerase catalyzes nucleophilic attack of the 3′ OH group of the growing polynucleotide chain on the 5′ phosphate of an incoming nucleotide* (see Fig. 20-9). The RNA molecule is therefore extended in the $5′{\rightarrow}3′$ direction. No primer is needed, so the RNA chain begins with the joining of two ribonucleotides. As the RNA strand is synthesized, it forms a hybrid double helix with the DNA template strand for eight or nine base pairs. The conformation of the hybrid is intermediate to the A form (as in double-stranded RNA) and the B form (as in double-stranded DNA).

⊕ **See Interactive Exercise. An RNA–DNA hybrid helix.**

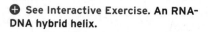

A-DNA   DNA–RNA Hybrid   B-DNA

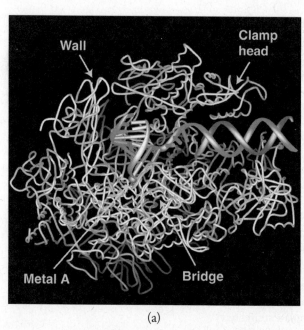

(a)

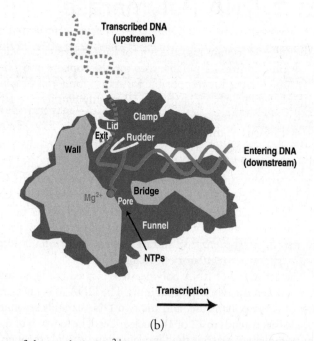

(b)

Figure 21-13 RNA polymerase with bound DNA and RNA.
(a) X-Ray structure of RNA polymerase (gray and orange) with DNA (coding strand green and template strand blue) and RNA (red). Part of the protein has been cut away to better reveal the active site region. (b) Cutaway diagram. DNA enters the enzyme at the right. A magenta sphere marks the position of one of the catalytic Mg^{2+} ions. Newly synthesized RNA forms a short hybrid helix with the template strand before exiting the enzyme. The two DNA strands separate in advance of the polymerization site and reanneal just beyond the RNA exit site. [Courtesy Roger Kornberg. From *Science* 292, 1876, 1844 (2001).] ⊕ See **Interactive Exercise. The RNA polymerase II elongation complex.**

RNA polymerase is a processive enzyme

During transcription, a portion of RNA polymerase known as the clamp (the orange structure in Fig. 21-13) rotates by about 30° to close down snugly over the DNA template. *Clamp closure appears to promote the high processivity of RNA polymerase.* In experiments where RNA polymerase was immobilized and a magnetic bead was attached to the DNA, up to 180 rotations (representing thousands of base pairs at 10.4 bp per turn) were observed before the RNA polymerase slipped. This processivity is essential, since genes are usually thousands—sometimes millions—of nucleotides long, and the largest ones may require many hours to transcribe.

With each reaction cycle, a helix located near the active site (the long green "bridge" helix visible in Figures 21-12 and 21-13a) appears to oscillate between a straight and bent conformation. This alternating movement appears to act as a ratchet to aid translocation of the template so that the next nucleotide can be added to the growing RNA chain. Throughout transcription, the sizes of the transcription bubble and the DNA–RNA hybrid helix remain constant. A protein loop known as the rudder (see Fig. 21-13b) may help separate the RNA and DNA strands so that a single RNA strand is extruded from the enzyme as the template and nontemplate DNA strands reanneal to restore the double-stranded DNA.

Like DNA polymerase, *RNA polymerase carries out proofreading.* If a deoxynucleotide or a mispaired ribonucleotide is mistakenly incorporated into RNA, the DNA–RNA hybrid helix becomes distorted. This causes polymerization to cease, and the newly synthesized RNA "backs out" of the active site through the channel by which ribonucleotides enter (**Fig. 21-14**). Transcription factor TFIIS binds to the RNA polymerase and stimulates it to act as an endonuclease to trim away the RNA containing the error. Transcription may resume if the 3′ end of the truncated transcript is then repositioned at the active site.

Transcription elongation requires a conformational change in RNA polymerase

One of the puzzles of RNA polymerase action is that the enzyme appears to initiate RNA synthesis repeatedly, producing and releasing many short transcripts (up to about 12 nucleotides) before committing to elongating a transcript. This suggests that *the transcription machinery must undergo a transition from initiation mode to elongation mode.* Several structural changes must occur at this point. While the first few ribonucleotides are polymerized, the transcription machinery remains firmly associated with the promoter. As a result, the template DNA enters the RNA polymerase active site but has nowhere to go after being transcribed. In prokaryotes, the buildup of strain—termed DNA scrunching—may provide the driving force for the polymerase to eventually escape the promoter and discard its σ factor. In eukaryotes, TFIIB occupies part of the RNA polymerase active site and must be displaced to accommodate an RNA longer than a few residues. Moreover, the exit channel for the RNA is initially partially blocked. The shift to an elongation conformation relieves these constraints and allows the polymerase to advance beyond the promoter.

In eukaryotes, the shift in RNA polymerase involves the C-terminal domain of the largest subunit of RNA polymerase (a structure that is disordered and therefore not visible in the models shown in Figs. 21-12 and 21-13). The C-terminal domain of mammalian RNA polymerase contains 52 seven–amino acid pseudorepeats with the consensus sequence

$$\text{Tyr–Ser–Pro–Thr–Ser–Pro–Ser}$$
$$1 \quad 2 \quad 3 \quad 4 \quad 5 \quad 6 \quad 7$$

Serine residues 2 and 5 (and possibly 7) of each heptad can potentially be phosphorylated. During the initiation phase of transcription, the C-terminal domain of RNA polymerase is not phosphorylated, but elongating RNA polymerase bears numerous phosphate groups. The Ser 5 phosphorylation that triggers the switch from an initiating to an elongating RNA polymerase is carried out by the kinase activity of TFIIH. Other kinases continue the phosphorylation process, mainly targeting Ser 2.

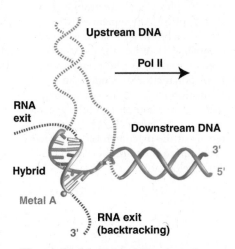

Figure 21-14 Schematic view of backtracking RNA in RNA polymerase. If polymerization stops due to a polymerization error, the 3′ end of the RNA transcript may back up into the channel for incoming nucleotides. The enzyme can then cleave off the 3′ end of the RNA and resume transcription. [Courtesy Roger Kornberg. From *Science* 292, 1879 (2001).]

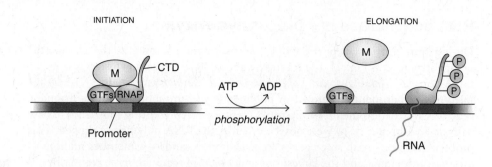

INITIATION

CTD

M

GTFs RNAP

Promoter

ATP → ADP

phosphorylation

ELONGATION

M

P
P
P

GTFs

RNA

Figure 21-15 The transition from transcription initiation to elongation. During initiation, the nonphosphorylated C-terminal domain (CTD) of RNA polymerase (RNAP) serves as a binding site for a Mediator complex (M). When the C-terminal domain undergoes phosphorylation, RNA polymerase switches to an elongation mode, dissociating from the Mediator complex and the general transcription factors (GTFs) that remain at the promoter. Other proteins may bind to the phosphorylated CTD during elongation.

When RNA polymerase becomes phosphorylated at its C-terminal domain, it can no longer bind a Mediator complex; this would allow the polymerase to abandon transcription-initiating factors and advance along the template DNA. In fact, when RNA polymerase "clears" the promoter, it leaves behind some general transcription factors, including TFIID (Fig. 21-15). These proteins, along with the Mediator complex, can reinitiate transcription by recruiting another RNA polymerase to the promoter. Consequently, *the first RNA polymerase to transcribe a gene acts as a "pioneer" polymerase that helps pave the way for additional rounds of transcription.* Histone acetyltransferases associated with the pioneer RNA polymerase may alter the nucleosomes of a gene undergoing transcription. However, the histone octamer never entirely dissociates from the transcribed DNA (Fig. 21-16).

During transcription elongation, other proteins bind to the phosphorylated C-terminal domain of RNA polymerase II, taking the place of the jettisoned initiation factors. Although RNA polymerase by itself can transcribe a DNA sequence *in vitro*, the presence of these additional factors accelerates transcription. Interestingly, the general transcription factors TFIIF and TFIIH, which participate in transcription initiation, remain associated with RNA polymerase during elongation. The phosphorylated domain of an elongating RNA polymerase also serves as a docking site for proteins that begin processing the nascent (newly made) RNA transcript. Transcription may not terminate until after these enzymes have completed their tasks.

Transcription is terminated in several ways

In both prokaryotes and eukaryotes, transcription termination involves cessation of RNA polymerization, release of the complete RNA transcript, and dissociation of the polymerase from the DNA template. In prokaryotes such as *E. coli,* termination occurs mainly by one of two mechanisms. In about half of *E. coli* genes, the 3′ end includes palindromic sequences followed by a stretch of T residues. The RNA corresponding to the gene can form a stem–loop or hairpin structure followed by a stretch of U residues. The other half of *E. coli* genes lack a hairpin sequence, and their termination depends on the action of a protein such as Rho, a hexameric helicase that may act by prying the nascent RNA away from the DNA and pushing the polymerase off the template. Both types of termination mechanisms can be explained in terms of

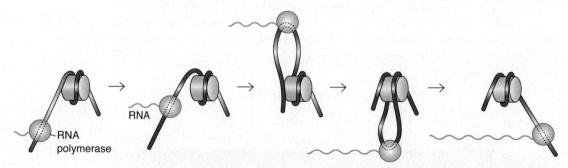

RNA polymerase

RNA

Figure 21-16 Transcription through a nucleosome. In this model, which shows RNA polymerase advancing along a DNA molecule (blue) and synthesizing RNA (red), the DNA loops out so that it never entirely leaves the histone octamer (orange). [After G. Orphanides and D. Reinberg, *Nature* 407, 472 (2000).]

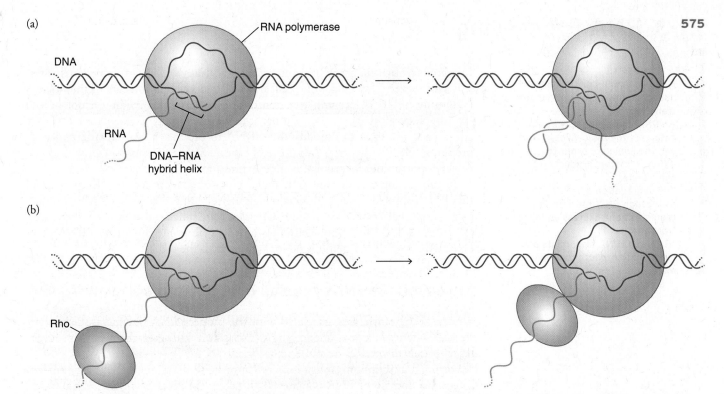

Figure 21-17 Mechanisms for transcription termination in prokaryotes. (a) Formation of an RNA hairpin shortens the length of the DNA–RNA hybrid helix, which is rich in easily separated A:U base pairs. (b) The movement of Rho along the RNA transcript pushes the polymerase forward, leaving a short hybrid helix from which the RNA more easily dissociates.

destabilizing the DNA–RNA hybrid helix that forms in the transcription bubble during elongation. Hairpin formation or the ATP-dependent action of Rho exerts a force that causes the RNA polymerase to advance, extending the leading end of the transcription bubble without extending the RNA (**Fig. 21-17**). The hybrid helix, now shortened and consisting of relatively weak U:A base pairs, is easily disrupted, freeing the RNA transcript.

In eukaryotic protein-coding genes, termination appears to be an imprecise process. In response to poorly defined signals in the DNA sequence, RNA polymerase may periodically pause during elongation; these pauses may provide opportunities for transcription termination. Pause sites occur following a polyadenylation signal, the stretch of nucleotides that marks the site where the growing mRNA is cleaved and then polyadenylated (see below). While paused, the RNA polymerase may undergo a conformational change that allows regulatory proteins to bind to its phosphorylated C-terminal domain and trigger termination of elongation. Alternatively, after the mRNA has been cut, an exonuclease may "eat away" the tail of the RNA still being synthesized, until it catches up with the polymerase and causes it to stop transcribing. In either scenario, the termination point is imprecise, but it doesn't really matter because the coding portion of the mRNA has already been completed.

CONCEPT REVIEW

- Describe the overall structure of RNA polymerase and the arrangement of DNA and RNA within the protein.
- Compare RNA polymerase and DNA polymerase in terms of mechanism, accuracy, and processivity.
- What would happen if the DNA–RNA hybrid helix were too long or too short?
- Describe the role of the C-terminal domain during transcription initiation, elongation, and termination.
- Describe the events that accompany the switch from transcription initiation to elongation.
- What does the "pioneer" polymerase accomplish?

21-3 RNA Processing

In a prokaryotic cell, an mRNA transcript is typically translated immediately after it is synthesized. In a eukaryotic cell, however, transcription occurs in the nucleus (where the DNA is located), but transcription takes place in the cytosol (where ribosomes are located). The separation of these processes gives eukaryotic cells two advantages: (1) They can modify the mRNA to produce a greater variety of gene products, and (2) the extra steps of RNA processing and transport provide additional opportunities for regulating gene expression.

In this section we examine some major types of RNA processing. Keep in mind that RNA is probably never found alone in the cell but rather interacts with a variety of proteins that covalently modify the transcript, splice out unneeded sequences, export the RNA from the nucleus, deliver it to a ribosome (if it is an mRNA), and eventually degrade it when it is no longer useful to the cell.

Eukaryotic mRNAs receive a 5' cap and a 3' poly(A) tail

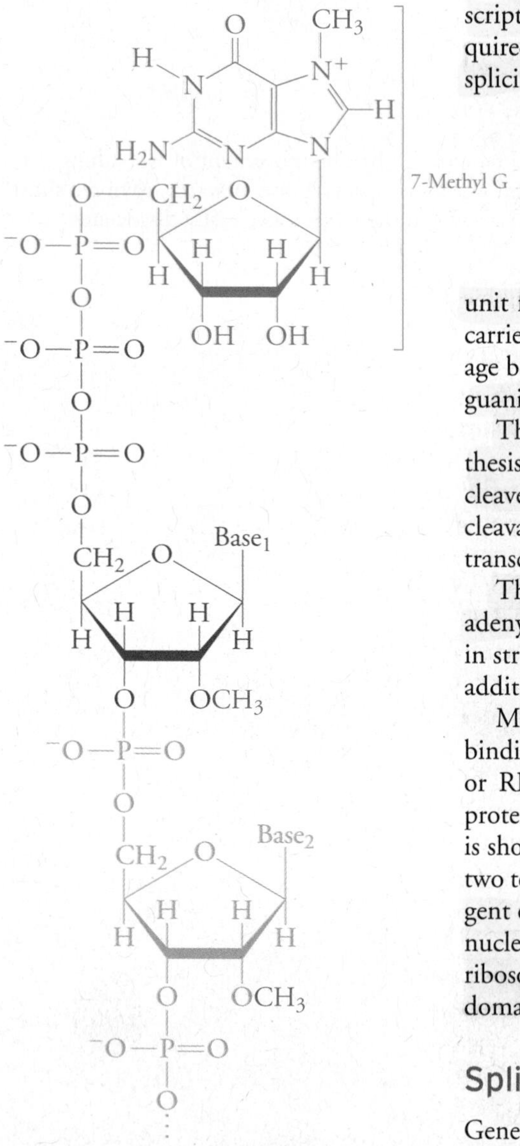

mRNA processing begins well before transcription is complete, as soon as the transcript begins to emerge from RNA polymerase. Many of the various enzymes required for capping the 5' end of the mRNA, for extending the 3' end, and for splicing are recruited to the phosphorylated domain of RNA polymerase, *so processing is closely linked to transcription.* In fact, the presence of processing enzymes may actually promote transcriptional elongation.

At least three enzyme activities modify the 5' end of the emerging mRNA to produce a structure called a **cap** *that protects the polynucleotide from 5' exonucleases.* First, a triphosphatase removes the terminal phosphate from the 5' triphosphate end of the mRNA. Next, a guanylyltransferase transfers a GMP unit from GTP to the remaining 5' diphosphate. These two reactions, which are carried out by a bifunctional enzyme in mammals, create a 5'–5' triphosphate linkage between two nucleotides. Finally, methyltransferases add a methyl group to the guanine and to the 2' OH group of ribose residues (Fig. 21-18).

The 3' end of an mRNA is also modified. Processing begins following the synthesis of the RNA sequence AAUAAA, which is a signal for a protein complex to cleave the transcript and extend it by adding adenosine residues. In fact, the RNA cleavage reaction, which occurs while the RNA polymerase is still operating, causes transcription termination.

The enzyme poly(A) polymerase generates a 3' **poly(A) tail** (also called a polyadenylate tail) of about 200 A residues. The enzyme resembles other polymerases in structure and catalytic mechanism, but it does not need a template to direct the addition of nucleotides.

Multiple copies of a binding protein associate with the mRNA tail. The poly(A)-binding protein consists of four copies of an RNA-binding domain (called an RBD, or RRM for RNA recognition motif) plus a C-terminal domain that mediates protein–protein contacts. A portion of the poly(A)-binding protein bound to RNA is shown in Figure 21-19. Each domain of about 80 amino acids can interact with two to six RNA nucleotides. Hence an mRNA's poly(A) tail can carry a large contingent of binding proteins, which help protect the 3' end of the transcript from exonucleases. They may also provide a "handle" for proteins that deliver mRNA to ribosomes. Other RNA-binding proteins with different types of nucleotide-binding domains participate in RNA processing and other events.

Splicing removes introns from eukaryotic genes

Genes were once believed to be continuous stretches of DNA, but experimental work initiated by Roberts and Sharp showed that hybridized DNA and mRNA molecules included large loops of unpaired DNA (Fig. 21-20). As the gene is being transcribed, portions of the sequence called introns (intervening sequences) are cut out, and the remaining portions (expressed sequences, or exons) are joined together.

7-Methyl G

Figure 21-18 An mRNA 5' cap.

? **How many phosphoanhydride bonds are cleaved during cap construction?**

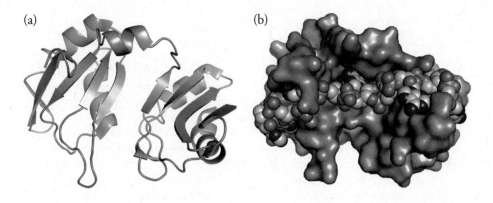

Figure 21-19 Poly(A)-binding protein bound to poly(A). (a) Two of the RNA-binding domains from human poly(A)-binding protein are shown in ribbon form. (b) Surface view of the protein with a 9-residue poly(A) nucleotide with atoms color-coded: C gray, O red, N blue, and P gold. [Structure (pdb 1CVJ) determined by R. C. Deo, J. B. Bonanno, N. Sonenberg, and S. K. Burley.]

These **splicing** reactions do not occur in prokaryotes, whose genes are continuous, but are the rule in complex eukaryotes. In simple organisms such as yeast, only a few genes contain introns, but in humans, almost all genes contain at least one intron.

Like capping, splicing commences before RNA polymerase has finished transcribing a gene, and some components of the splicing machinery assemble on the phosphorylated C-terminal domain of RNA polymerase. Most mRNA splicing is carried out by a **spliceosome,** a complex of five small RNA molecules (called **snRNAs,** for **small nuclear RNAs**) and hundreds of proteins. The spliceosome recognizes conserved sequences at the 5′ intron/exon junction and at a conserved A residue within the intron, called the branch point (Fig. 21-21). Recognition depends on base pairing between the conserved mRNA sequences and snRNA sequences. However, the sequence conservation is relatively weak, and introns can be enormous, ranging from less than a hundred nucleotides to a record of 2.4 million nucleotides (the average is ~3400). These factors make it difficult to identify introns and exons in genomic DNA sequences.

Splicing is a two-step transesterification process (Fig. 21-22). Each step requires an attacking nucleophile (a ribose OH group) and a leaving group (a phosphoryl group). A catalytically essential Mg^{2+} ion enhances the nucleophilicity of the attacking hydroxyl group and stabilizes the phosphate leaving group. Because there is no net change in the number of phosphodiester bonds, the splicing reaction needs no external source of free energy.

Some types of introns, particularly in protozoan rRNA genes, undergo self-splicing; that is, they catalyze their own transesterification reactions without the aid of proteins. These rRNA molecules were the first RNA enzymes (**ribozymes**) to be described, in 1982. The activity of these self-splicing RNA molecules, along with other evidence that protein-free snRNAs can carry out splicing, indicates that the catalytic activity of the spliceosome is a property of its snRNA components, not its

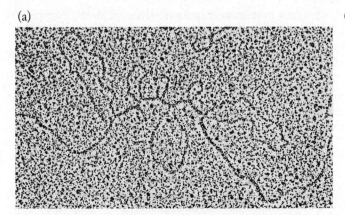

Figure 21-20 Hybridization of ovalbumin DNA and mRNA. The template DNA strand for the chicken ovalbumin gene was allowed to hybridize with the corresponding mRNA. Complementary sequences, representing exons, have annealed, while single-stranded DNA, coding for introns that have

been spliced out of the mRNA, forms loops. (a) Electron micrograph. (b) Interpretive drawing. The mRNA is shown as a dashed line, the introns of the single-stranded DNA (blue line) are labeled I–VII, and the exons are labeled 1–7. [Courtesy Pierre Chambon and Fabienne Perrin.]

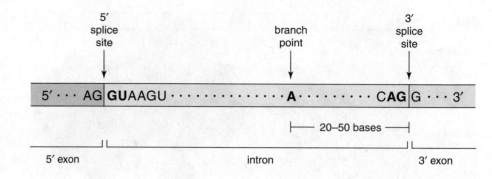

Figure 21-21 Consensus sequence at eukaryotic mRNA splice sites. Nucleotides shown in bold are invariant.

proteins. One hypothesis for the evolutionary origin of splicing suggests that introns, and the splicing machinery itself, are the result of RNA molecules that spliced themselves *into* mRNA molecules, which were converted to DNA by the action of a reverse transcriptase (see Box 20-A) and then incorporated into the genome through recombination.

▶▶ **WHY** do genes include introns?

Introns typically comprise over 90% of a gene's total length, which means that a lot of RNA must be transcribed and then discarded. Moreover, a cell must spend energy to synthesize the RNA and proteins that make up the spliceosome and that destroy intronic RNA and incorrectly spliced transcripts. Finally, the complexity of the splicing process creates many opportunities for things to go wrong: A majority of mutations linked to inherited diseases involve defective splicing. So just what is the advantage of arranging a gene as a set of exons separated by introns?

One answer to this question is that splicing allows cells to increase variation in gene expression through alternative splicing. At least 95% of human protein-coding genes exhibit splice variants. Variation may result from selecting alternative sequences to serve as 5' or 3' splice sites, from skipping an exon, or from retaining an intron. Thus, certain exons present in the gene may or may not be included in the mature RNA transcript (Fig. 21-23). The signals that govern exon selection and splice sites probably involve RNA-binding proteins that recognize sequences or secondary structures within introns as well as exons. As a result of alternative splicing, *a given gene can generate more than one protein product,* and gene expression can be

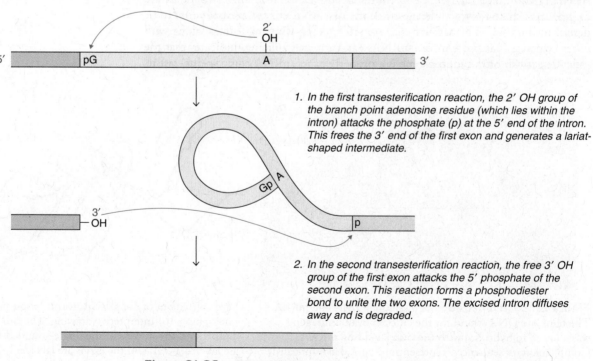

1. In the first transesterification reaction, the 2' OH group of the branch point adenosine residue (which lies within the intron) attacks the phosphate (p) at the 5' end of the intron. This frees the 3' end of the first exon and generates a lariat-shaped intermediate.

2. In the second transesterification reaction, the free 3' OH group of the first exon attacks the 5' phosphate of the second exon. This reaction forms a phosphodiester bond to unite the two exons. The excised intron diffuses away and is degraded.

Figure 21-22 mRNA splicing.

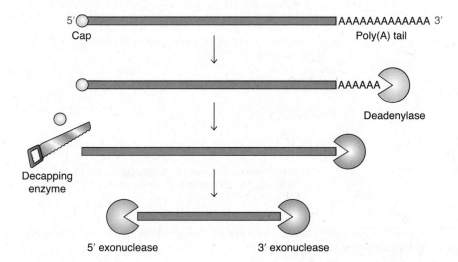

gene

mRNA transcripts

Striated muscle

Smooth muscle

Nonmuscle/
fibroblast

Brain

Figure 21-23 Alternative splicing. The rat gene for the muscle protein α-tropomyosin (top) encodes 12 exons. The mature mRNA transcripts in different tissues consist of different combinations of exons (some exons are found in all transcripts), reflecting alternative splicing pathways. [After Breitbart, R. E., Andreadis, A., and Nadal-Ginard, B., *Annu. Rev. Biochem.* 56, 481 (1987).]

finely tailored to suit the needs of different types of cells. The evolutionary advantage of this regulatory flexibility clearly outweighs the cost of making the machinery that cuts and pastes RNA sequences. Alternative splicing also explains why humans are vastly more complex than organisms such as roundworms, which contain a comparable number of genes (see Table 3-4).

mRNA turnover and RNA interference limit gene expression

Although mRNA accounts for only about 5% of cellular RNA (rRNA accounts for about 80% and tRNA for about 15%), it continuously undergoes synthesis and degradation. The life span of a given mRNA molecule is another regulated aspect of gene expression: mRNA molecules decay at different rates. In mammalian cells, mRNA life spans range from less than an hour to about 24 hours.

The rate of mRNA turnover depends in part on how rapidly its poly(A) tail is shortened by the activity of deadenylating exonucleases. Poly(A) tail shortening is followed by decapping, which allows exonucleases access to the 5′ end of the transcript and eventually leads to destruction of the entire message (Fig. 21-24). *In vivo,* the RNA cap and tail are close together because a protein involved in translation binds to both ends of the mRNA, effectively circularizing it. RNA-binding regulatory proteins are almost certainly involved in monitoring RNA integrity. For example, transcripts with a premature stop codon are preferentially degraded, thereby avoiding the waste of synthesizing a nonfunctional truncated polypeptide.

5′ Cap   AAAAAAAAAAAAA 3′ Poly(A) tail

AAAAAA   Deadenylase

Decapping enzyme

5′ exonuclease   3′ exonuclease

Figure 21-24 mRNA decay. A mature mRNA bears a 5′ cap and a 3′ poly(A) tail. After a deadenylase has shortened the tail, a decapping enzyme removes the methylguanosine cap at the 5′ end. The mRNA can then be degraded by exonucleases from both ends.

Sequence-specific degradation of certain RNAs, a phenomenon called **RNA interference (RNAi),** provides another mechanism for regulating gene expression after transcription has occurred. RNA interference was discovered by researchers who were attempting to boost gene expression in various types of cells by introducing extra copies of genetic information in the form of RNA. They observed that instead of increasing gene expression, the RNA—particularly if it was double-stranded—actually blocked production of the gene's product. This interference or gene-silencing effect results from the ability of the introduced RNA to target a complementary cellular mRNA for destruction. Endogenously produced RNAs, known as **small interfering RNAs (siRNAs)** and **micro RNAs (miRNAs),** appear to mediate RNA interference in virtually all types of eukaryotic cells, including human cells.

For siRNA, the RNA interference pathway begins with the production of double-stranded RNA, which may result when a single polynucleotide strand folds back on itself in a hairpin. A ribonuclease called Dicer cleaves the double-stranded RNA to generate segments of 20 to 25 nucleotides with a 2-nucleotide overhang at each 3' end **(Fig. 21-25)**. These siRNAs bind to a multiprotein

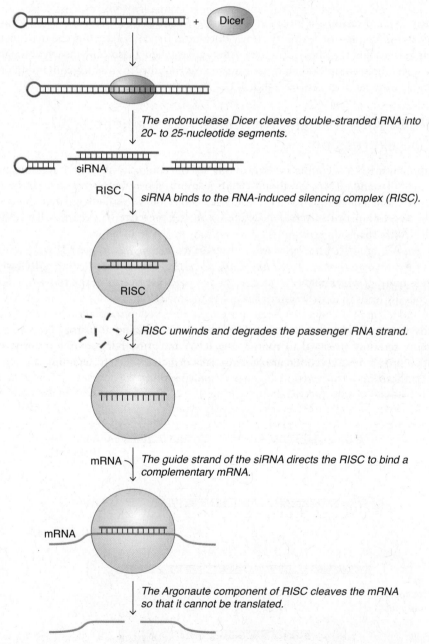

Figure 21-25 RNA interference. The steps involving siRNA are shown. The miRNA pathway for inactivating mRNA is similar.

complex called the RNA-induced silencing complex (RISC), where one strand of the RNA (the "passenger" strand) is separated from the other by a helicase and/or degraded by a nuclease. The remaining strand serves as a guide for the RISC to identify and bind to a complementary mRNA molecule. The "Slicer" activity of the RISC, a protein known as Argonaute, then cleaves the mRNA, rendering it unfit for translation.

Like RNA splicing, RNA interference at first appears wasteful, but it provides cells with a mechanism for eliminating mRNAs—which could otherwise be translated into protein many times over—in a highly specific manner. It is believed that RNA interference originally evolved as an antiviral defense, since many viral life cycles include the formation of double-stranded RNAs.

In the miRNA pathway, RNA hairpins containing imperfectly paired nucleotides are processed by Dicer and other enzymes to double-stranded miRNAs that bind to the RISC. The passenger RNA strand is ejected and the remaining strand helps the RISC locate complementary target mRNAs. Whereas an siRNA specifically seeks and destroys an mRNA that is perfectly complementary, an miRNA can bind to a large number of target mRNAs—possibly hundreds—because it forms base pairs with a stretch of only 6 or 7 nucleotides. The captured mRNAs are unavailable for translation and are susceptible to the standard mechanisms for RNA degradation diagrammed in Figure 21-24.

In addition to serving as a powerful laboratory technique for silencing genes in order to explore their functions, the RNA interference system is being exploited for therapeutic reasons. One application uses siRNAs that repress expression of a protein, called vascular endothelial growth factor, to treat a form of macular degeneration characterized by the overgrowth of small blood vessels behind the retina. Clinical trials are under way to test whether siRNAs can turn off the expression of viral genes in order to block viral replication. Other diseases in which gene silencing would be desirable, such as cancer, are amenable to RNAi therapy, provided that the siRNA can be delivered selectively to cancerous cells. In general, introducing exogenous RNAs into cells presents a challenge, since nucleic acids don't easily cross cell membranes, and the presence of extracellular RNA may trigger the body's innate RNA-degrading antiviral defenses.

rRNA and tRNA processing includes the addition, deletion, and modification of nucleotides

rRNA transcripts, which are generated mainly by RNA polymerase I in eukaryotes, must be processed to produce mature rRNA molecules. rRNA processing and ribosome assembly take place in the **nucleolus,** a discrete region in the nucleus. The initial eukaryotic rRNA transcript is cleaved and trimmed by endo- and exonucleases to yield three rRNA molecules **(Fig. 21-26)**. The rRNAs are known as 18S, 5.8S, and 28S rRNAs for their sedimentation coefficients (large molecules have larger sedimentation coefficients, a measure of how quickly they settle in an ultra-high-speed centrifuge).

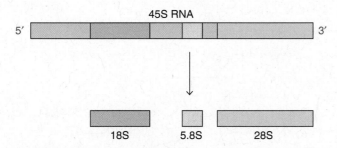

Figure 21-26 Eukaryotic rRNA processing. The initial transcript of about 13.7 kb has a sedimentation coefficient of 45S. Three smaller rRNA molecules (18S, 5.8S, and 28S) are derived from it by the action of nucleases.

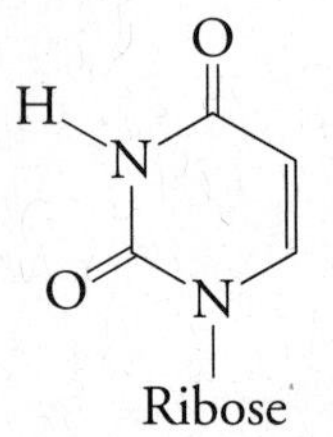

Uridine   Pseudouridine (ψ)

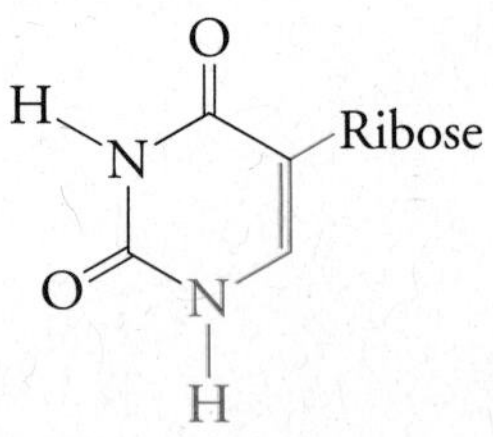

rRNA transcripts may be covalently modified (in both prokaryotes and eukaryotes) by the conversion of some uridine residues to pseudouridine (*left*) and by the methylation of certain bases and ribose 2′ OH groups. This last type of modification is guided by a multitude of **small nucleolar RNA molecules** (called **snoRNAs**) that recognize and pair with specific 15-base segments in the rRNA sequences, thereby directing an associated protein methylase to each site. *Without the snoRNAs to mediate sequence-specific ribose methylation, the cell would require many different methylases in order to recognize all the different nucleotide sequences to be modified.*

A rapidly growing mammalian cell may synthesize as many as 7500 rRNA transcripts each minute, each of which associates with about 150 different snoRNAs. The processed rRNAs then combine with some 80 different ribosomal proteins to generate fully functional ribosomes, a task that requires careful coordination between RNA synthesis and ribosomal protein synthesis.

tRNA molecules, produced by the action of RNA polymerase III in eukaryotes, undergo nucleolytic processing and covalent modification. The initial tRNA transcripts are trimmed by ribonuclease P (RNase P), a ribonucleoprotein enzyme whose catalytic activity resides in its RNA component (Box 21-B).

BOX 21-B   BIOCHEMISTRY NOTE

RNA: A Versatile Molecule

All cells contain two essential ribozymes: the tRNA-processing RNAse P and the ribosomal RNA that catalyzes peptide bond formation during protein synthesis. There are at least six other naturally occurring types of catalytic RNAs (such as those involved in splicing) and many more synthetic ribozymes. Although the discovery of catalytic RNA was initially greeted with skepticism, subsequent studies have amply demonstrated that *RNA has the same properties that allow proteins to function as catalysts, namely, complex tertiary structure and reactive functional groups.*

Unlike DNA, whose conformational flexibility is considerably constrained by its double-stranded nature, single-stranded RNA molecules can adopt highly convoluted shapes through base pairing between different segments. In addition to the standard (Watson–Crick) types of base pairs, RNA accommodates nonstandard base pairs as well as hydrogen-bonding interactions among three bases. A few of the possibilities are shown here. R represents the ribose–phosphate backbone. Base stacking stabilizes the RNA tertiary structure, achieving the same sort of balance between rigidity and flexibility exhibited by protein enzymes. A folded RNA molecule can then bind substrates, orient them, and stabilize the transition state of a chemical reaction.

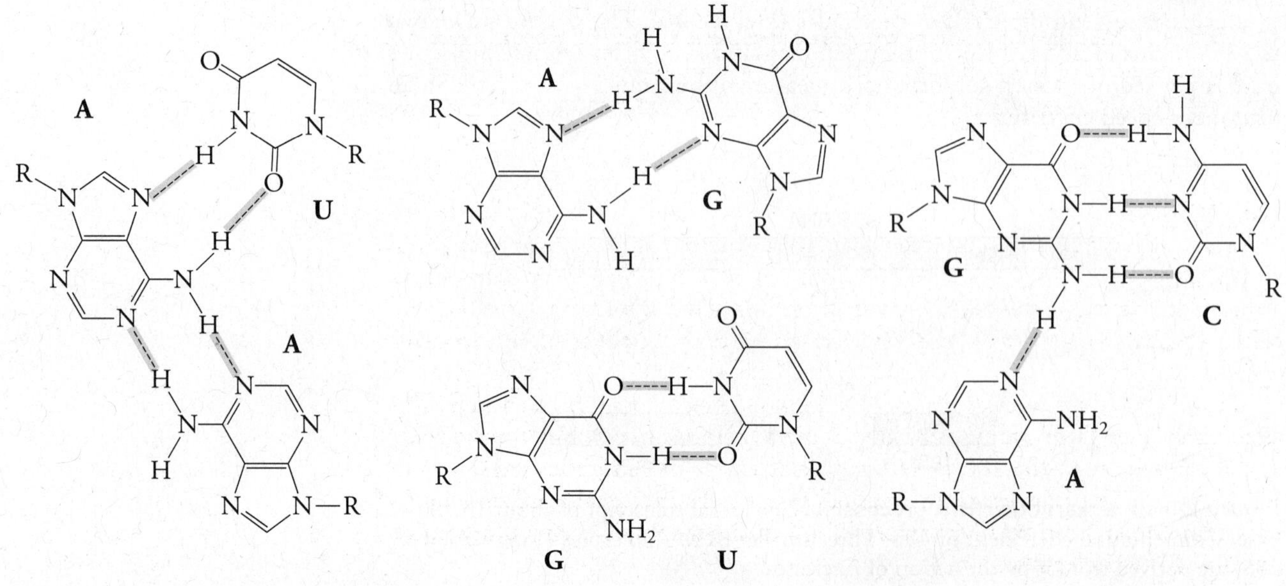

Among the first ribozymes to be thoroughly studied were RNAs that catalyze their own cleavage. However, these are not true catalysts, as they cannot participate in more than one reaction cycle. The RNA component of RNase P is a true catalyst. At one time it was believed that the enzyme's RNA molecule helped align the tRNA substrate for the protein to cleave, but the bacterial RNase P RNA is able to cleave its substrate in the absence of the RNase P protein. The structure of RNase P from *Thermotoga maritima* with its product tRNA (red) is shown here. The numerous base-paired stems of the 347-nucleotide RNA (gold) form a compact structure, much like a protein enzyme. The small protein component of the RNase P (117 amino acids) is shown in green. Eukaryotic RNase P enzymes contain a similar RNA component plus 9 or 10 protein subunits whose functions are not completely understood.

[Structure (pdb 3Q1Q) determined by N. J. Reiter, A. Osterman, A. Torres-Larios, K. K. Swinger, T. Pan, and A. Mondragon.]

The existence of RNA enzymes such as RNase P lends support to the theory of an early **RNA world** when RNA functioned as a repository of biological information (like modern DNA) as well as a catalyst (like modern proteins). Experiments with synthetic RNAs *in vitro* have demonstrated that RNA can catalyze a wide variety of chemical reactions, including the biologically relevant synthesis of glycosidic bonds (the type of bond that links the base and ribose in a nucleoside) and RNA-template-directed RNA synthesis. Apparently, most ribozymes that originated in an early RNA world were later supplanted by protein catalysts, leaving only a few examples of RNA's catalytic abilities.

● **Question:** In human mitochondria, RNase P consists entirely of protein. Does this disprove the RNA world theory?

Some tRNA transcripts undergo splicing to remove introns. In prokaryotes, tRNAs end with a 3′ CCA sequence, which serves as the attachment point for an amino acid that will be used for protein synthesis. In eukaryotes, the three nucleotides are added to the 3′ end of the immature molecule by the action of a nucleotidyltransferase.

Up to 25% of the nucleotides in tRNA molecules are covalently modified. The alterations range from simple additions of methyl groups to complex restructuring of the base. Some of the 100 or so known nucleotide modifications are shown in **Figure 21-27**. These are yet more examples of how cells alter genetic information as it is transcribed from relatively inert DNA to highly variable and much more dynamic RNA molecules.

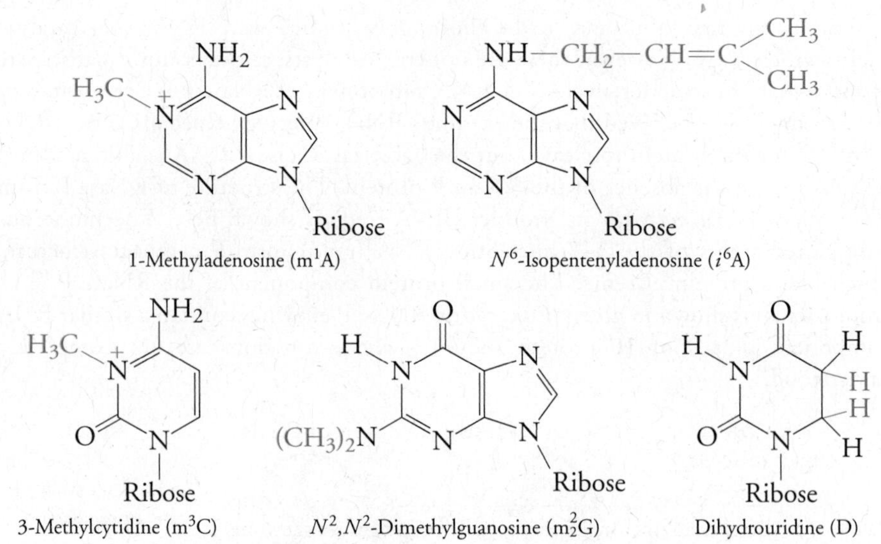

Figure 21-27 Some modified nucleotides in tRNA molecules. The parent nucleotide is in black; the modification is shown in red.

1-Methyladenosine (m^1A)

N^6-Isopentenyladenosine (i^6A)

3-Methylcytidine (m^3C)

N^2,N^2-Dimethylguanosine (m_2^2G)

Dihydrouridine (D)

CONCEPT REVIEW
- Summarize the types of covalent changes that occur in the maturation of a eukaryotic mRNA molecule, and describe their purpose.
- Explain why splicing does not require free energy input.
- What are the advantages and disadvantages of splicing protein-coding genes?
- Summarize the steps of RNA interference.
- Explain why the products of transcription exhibit much more variability than the genes that encode them.

[SUMMARY]

21-1 Transcription Initiation

- Transcription is the process of converting a segment of DNA into RNA. An RNA transcript may represent a protein-coding gene or it may participate in protein synthesis or other activities, including RNA processing.
- Alterations in histones and the rearrangement of nucleosomes may help expose DNA sequences for transcription.
- Transcription begins at a DNA sequence known as a promoter. A gene to be transcribed must be recognized by a regulatory factor such as the σ factor in prokaryotes.
- In eukaryotes, a set of general transcription factors interact with DNA at the promoter to form a complex that recruits RNA polymerase and may further alter chromatin structure.
- Regulatory DNA sequences may affect transcription through binding proteins that interact with RNA polymerase via the Mediator complex.
- The bacterial *lac* operon illustrates the regulation of transcription by a repressor protein.

21-2 RNA Polymerase

- Eukaryotic RNA polymerase II transcribes protein-coding genes. It requires no primer and polymerizes ribonucleotides to generate

an RNA chain that forms a short double helix with the template DNA.
- The polymerase acts processively along the DNA template but reverses to allow the excision of a mispaired nucleotide.
- The elongation phase of transcription in eukaryotes is triggered by phosphorylation of the C-terminal domain of RNA polymerase II.
- Transcription termination in prokaryotes involves destabilization of the DNA–RNA hybrid helix. In eukaryotes, transcription termination is linked to polymerase pausing and RNA cleavage.

21-3 RNA Processing

- mRNA transcripts undergo processing that includes the addition of a 5′ cap structure and a 3′ poly(A) tail. mRNA splicing, carried out by RNA–protein complexes called spliceosomes, joins exons and eliminates introns.
- RNA interference is a pathway for inactivating mRNAs according to their ability to pair with a complementary siRNA or miRNA.
- rRNA and tRNA transcripts are processed by nucleases and enzymes that modify particular bases.

[GLOSSARY TERMS]

transcription	tRNA	exon
gene	operon	ncRNA
mRNA	RNA processing	promoter
rRNA	intron	consensus sequence

TATA box	cap	RNAi
general transcription factor	poly(A) tail	siRNA
enhancer	splicing	miRNA
activator	spliceosome	nucleolus
silencer	small nuclear RNA (snRNA)	small nucleolar RNA (snoRNA)
repressor	ribozyme	RNA world

[PROBLEMS]

21-1 Transcription Initiation

1. Why does the genome contain so many more genes for rRNA than mRNA?

2. Why is it effective for a bacterial cell to organize genes for related functions as an operon? How do eukaryotes achieve the same benefits?

3. Proteins can interact with DNA through relatively weak forces, such as hydrogen bonds and van der Waals interactions, as well as through stronger electrostatic interactions such as ion pairs. Which types of interactions predominate for sequence-specific DNA-binding proteins and for sequence-independent binding proteins?

4. Certain proteins that stimulate expression of a gene bind to DNA in a sequence-specific manner and also induce conformational changes in the DNA. Describe the purpose of these two modes of interaction with the DNA.

5. The sense (coding strand) for the *E. coli* promoter for the *rrnA1* gene is shown below. The transcription initiation site is shown by +1. Identify the −35 and −10 region for this gene.

+1

AAAATAAATGCTTGACTCTGTAGCGGGAAGGCGTATTATCCAACACCC

6. Predict the effect of a mutation in one of the bases in either the −35 or the −10 region of the promoter.

7. Sp1 is a sequence-specific human DNA-binding protein that binds to a region on the DNA called the GC box, a promoter element with the sequence GGGCGG. Binding of Sp1 to the GC box enhances RNA polymerase II activity 50- to 100-fold. How would you use affinity chromatography (see Section 4-5) to purify Sp1?

8. The promoters for genes transcribed by eukaryotic RNA polymerase I exhibit little sequence variation, yet the promoters for genes transcribed by eukaryotic RNA polymerase II are highly variable. Explain.

9. Eukaryotic promoters include an AT-rich sequence just prior to the transcription start site. Why is this sequence composed of A:T base pairs, not G:C base pairs?

10. Identify possible eukaryotic promoter elements in the sequence of the mouse β globin gene shown below. The first nucleotide to be transcribed is indicated by +1.

+1

GAGCATATAAGGTGAGGTAGGATCAGTTGCTCCTCACATTT

11. A specific type of histone methyltransferase (HMT) catalyzes the methylation of a single lysine or a single arginine in a histone protein (usually H3 or H4). Draw the structures of methylated lysine and methylated arginine residues.

12. The reversal of histone Arg methylation converts the methylated Arg to citrulline in a reaction that consumes H_2O. Draw the resulting amino acid residue. What is the other product of the reaction?

13. The enzyme EZH2 is a histone lysine methyltransferase that has been shown to be upregulated in various types of cancers. How does this upregulation affect lysine residues 4 and 27 of histone 3? What is the effect on the transcription of genes associated with this histone?

14. Three human TBP-associated factors (TAFs) have been found to contain protein domains that are homologous to those found in histones H2B, H3, and H4. Why is this finding not surprising?

15. The RNA polymerase from bacteriophage T7 recognizes specific promoter sequences and melts open the DNA to form a transcription bubble. No other transcription factors are necessary for T7 RNA polymerase to initiate transcription.

(a) Dissociation constants (K_d) were measured for the interaction between the polymerase and DNA segments containing the promoter sequences. In some cases, the DNA contained a bulge, caused by a mismatch of one, four, or eight bases, to mimic the intermediates in the formation of a transcription bubble. The results are shown in the table. To which DNA segment does the polymerase bind most tightly? Explain in terms of the DNA structure.

DNA promoter segment	K_d (nM)
Fully base paired	315
One-base bulge	0.52
Four-base bulge	0.0025
Eight-base bulge	0.0013

(b) How can the data in part (a) be used to calculate the free energy change for the binding of T7 RNA polymerase to DNA? (Note that a dissociation constant is the inverse of an association constant.)

(c) Compare the ΔG values for T7 RNA polymerase binding to double-stranded DNA and to DNA with an eight-base bulge.

(d) What do these results reveal about the thermodynamics of melting open a DNA helix for transcription? What is the approximate free energy cost of forming a transcription bubble equivalent to eight base pairs?

(e) Which of the following DNA segments most likely represents the sequence where the transcription bubble begins to form during transcription initiation? Explain.

```
CTATA       GGGAG
GATAT       CCCTC
```

16. In bacteria, the core RNA polymerase binds to DNA with a dissociation constant of 5×10^{-12} M. The polymerase in complex with its σ factor has a dissociation constant of 10^{-7} M. Explain.

17. One of the genes expressed by the *lac* operon is the *lacY* gene, which encodes a lactose permease transporter that allows lactose to enter the cell. Why does the expression of this gene assist in the expression of the operon?

18. The genes of the *lac* operon are not expressed when the *lac* repressor binds to the operator. But removal of the *lac* repressor is not sufficient to allow gene expression—a protein called catabolite activator protein (CAP) is also required to assist RNA polymerase and facilitate transcription. CAP can bind to the operon only when its ligand, cAMP, is bound. In *E. coli*, the intracellular concentration of cAMP falls when glucose is present. Describe the activity of the *lac* operon in each of the following scenarios:

(a) Both lactose and glucose are present.
(b) Glucose is present but lactose is absent.
(c) Both glucose and lactose are absent.
(d) Lactose is present and glucose is absent.

19. Researchers have isolated bacterial cells with mutations in various segments of the *lac* operon. What is the effect on gene expression if a mutation in the operator occurs so that the repressor cannot bind? What happens when lactose is added to the growth medium of these mutants?

20. A bacterial strain expresses a mutant *lac* repressor protein that retains its ability to bind to the operator but cannot bind lactose. What is the effect on gene expression in these mutants? What happens when lactose is added to the growth medium?

21. The compound phenyl-β-D-galactose (phenyl-Gal) is not an inducer of the *lac* operon because it is unable to bind to the repressor. However, it can serve as a substrate for β-galactosidase, which cleaves phenyl-Gal to phenol and galactose. How can the addition of phenyl-Gal to growth medium distinguish between wild-type bacterial cells and cells that have a mutation in the *lacI* gene?

22. In bacterial cells, the genes that code for the enzymes of the tryptophan biosynthetic pathway are organized in an operon as shown below. Another gene encodes a repressor protein that binds tryptophan. How does the repressor protein control the expression of the genes in the *trp* operon?

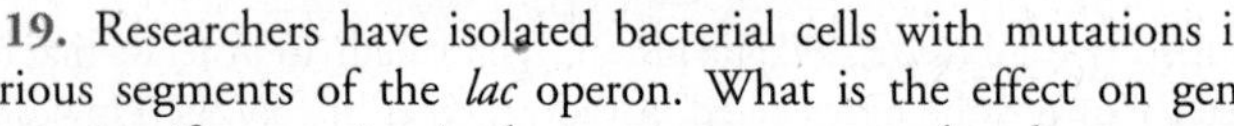

P	trpL	trpE	trpD	trpC	trpB	trpA

21-2 RNA Polymerase

23. RNA synthesis is much less accurate than DNA synthesis. Why does this not harm the cell?

24. Radioactively labeled γ-[^{32}P]GTP is added to a bacterial culture undergoing transcription. Is the resulting RNA labeled? If so, where?

25. Explain why the adenosine derivative cordycepin inhibits RNA synthesis.

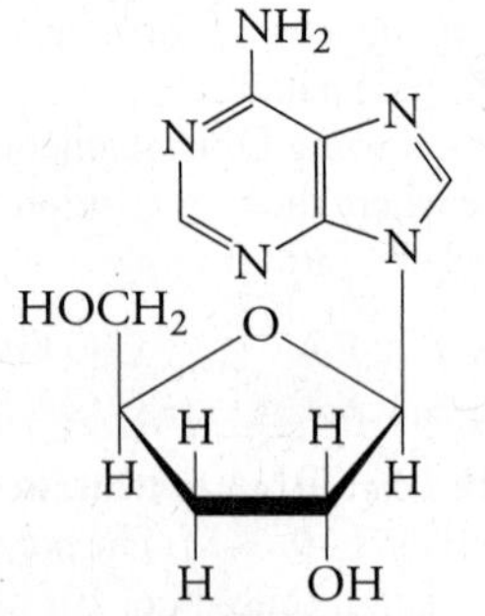

Cordycepin

26. How does your answer to Problem 25 provide evidence to support the hypothesis that transcription occurs in the $5' \rightarrow 3'$ direction, not the $3' \rightarrow 5'$ direction?

27. In the presence of the antibiotic rifampicin, bacterial cells in culture are capable of synthesizing only short RNA oligomers. At what point in the transcription process does rifampicin exert its inhibitory effect?

28. Rifampicin (see Problem 27) binds to the β subunit of bacterial RNA polymerase. Why is rifampicin used as an antibiotic to treat diseases caused by bacteria?

29. The coding strand of a gene has the sequence shown below. Write the sequence of the mRNA that corresponds to this DNA sequence.

$$G\ T\ C\ C\ G\ A\ T\ C\ G\ A\ A\ T\ G\ C\ A\ T\ G$$

30. The bacterial enzyme polynucleotide phosphorylase (PNPase) is a $3' \rightarrow 5'$ exoribonuclease that degrades mRNA.

(a) The enzyme catalyzes a phosphorolysis reaction, as does glycogen phosphorylase (see Section 13-3), rather than hydrolysis. Write an equation for the mRNA phosphorolysis reaction.
(b) *In vitro*, PNPase also catalyzes the reverse of the phosphorolysis reaction. What does this reaction accomplish and how does it differ from the reaction carried out by RNA polymerase?
(c) PNPase includes a binding site for long polyribonucleotides, which may promote the enzyme's processivity. Why would this be an advantage for the primary activity of PNPase *in vivo*?

31. The activity of RNA polymerase II is inhibited by the mushroom toxin α-amanitin ($K_d = 10^{-8}$ M). In contrast, RNA polymerase III is only moderately inhibited by α-amanitin ($K_d = 10^{-6}$ M), and RNA polymerase I is not affected at all. What would be the effect of adding 10 nM α-amanitin to cells in culture?

32. The three different eukaryotic RNA polymerases were discovered in the 1970s by researchers who solubilized whole cells and loaded the extracts onto a DEAE ion exchange column (see Section 4-5). An ammonium sulfate salt gradient was applied in order to elute (dislodge) the bound proteins. Each fraction eluting from the column was assayed for RNA polymerase activity (see the figure below). The samples were also analyzed for activity in the presence of Mg^{2+} ions and in the presence of the mushroom toxin α-amanitin. In addition to the difference in α-amanitin sensitivity described in Problem 31, the investigators noted that RNA polymerases II and III achieved only 50% activity in the presence of 5 mM Mg^{2+} ions, whereas RNA polymerase I was fully active under these conditions. How did the investigators use their results to conclude that the three peaks constituted three different forms of RNA polymerase?

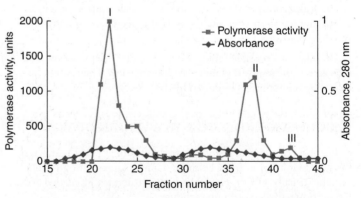

33. The C-terminal domain of RNA polymerase II projects away from the globular portion of the protein. Why?

34. Experiments were carried out to truncate RNA polymerase II so that its C-terminal domain (CTD) was missing. What are the consequences for the cells?

35. The DNA sequence of a hypothetical *E. coli* terminator is shown below. N stands for any of the four nucleotides.

```
5′···NNAAGCGCCGNNNNCCGGCGCTTTTTTNNN···3′
3′···NNTTCGCGGCNNNNGGCCGCGAAAAAANNN···5′
```

(a) Write the sequence of the mRNA transcript that is formed if the bottom strand is the noncoding (template) strand.
(b) Draw the RNA hairpin structure that would form in this RNA transcript.

36. In an experiment, inosine triphosphate (ITP) is added to bacterial cells in culture. ITP is used by the cells in place of GTP. Inosine (I) forms base pairs with cytidine (C), and the I:C base pairs form two hydrogen bonds.
(a) Write the sequence of the mRNA transcript that is formed from the gene shown in Problem 35 if the bottom strand is the noncoding (template) strand.
(b) Draw the RNA hairpin structure that would form in this RNA transcript. Compare the stability of this RNA hairpin with the RNA hairpin you drew in Problem 35b. How is termination of transcription affected by the ITP substitution?

37. Formation of an RNA hairpin cannot be the sole factor in the termination of transcription in prokaryotes. Why?

38. The addition of β,γ-imido nucleoside triphosphates to cells in culture has been shown to inhibit Rho-dependent termination. Explain why.

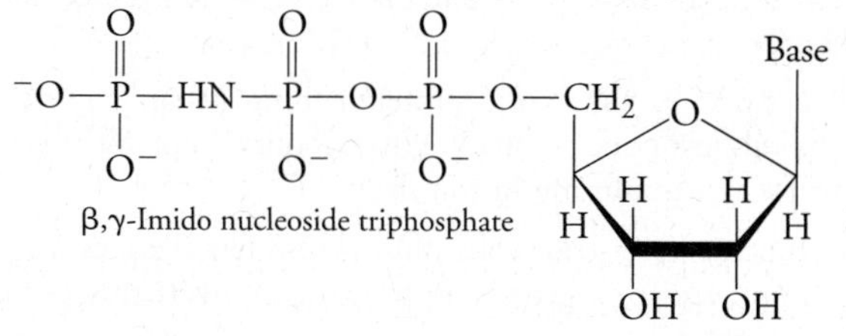

β,γ-Imido nucleoside triphosphate

39. In bacteria, the organization of functionally related genes in an operon allows the simultaneous regulation of expression of those genes. If the operon consists of genes encoding the enzymes for a biosynthetic pathway, then the pathway activity as a whole can be feedback inhibited when the concentration of the pathway's final product accumulates.
(a) In one mode of feedback regulation, a repressor protein binds to a site in the operon (called the operator) to decrease the rate of transcription only when the repressor has bound a molecule representing the operon's ultimate metabolic product. Draw a diagram showing how such a regulatory system would work.
(b) Feedback regulation of gene expression can also occur after RNA synthesis has begun. In this case, the presence of the operon's ultimate product causes transcription to terminate prematurely or leads to an mRNA that cannot be translated. Draw a diagram illustrating this control mechanism. Assume that the feedback mechanism includes a protein to which the product binds.
(c) How would the feedback inhibition system in part (b) differ if no protein were involved?

(d) In some bacteria, several genes required for the biosynthesis of the redox cofactor flavin adenine dinucleotide (FAD; see Fig. 3-3c) are arranged in an operon. Comparisons of the sequences of this operon in different species reveal a conserved sequence in the untranslated region at the 5′ end of the operon's mRNA. The tertiary structure of an RNA molecule typically includes regions of base pairing and unpaired loops (stem–loop structures). By examining an RNA sequence and noting which positions are most conserved, it is possible to predict the stem–loop structure of the RNA. A portion of a conserved mRNA sequence called RFN, which regulates the expression of the FAD-synthesizing operon, is shown below. Draw the stem–loop structure for this RNA segment.

··· G A U U C A G U U U A A G C U G A A G C ···

(e) In order to function as an FAD sensor, the RFN element (which consists of about 165 nucleotides) must alter its conformation when FAD binds. How could researchers assess RNA conformational changes?
(f) FAD can be considered as a derivative of flavin mononucleotide (FMN), which in turn is derived from riboflavin.

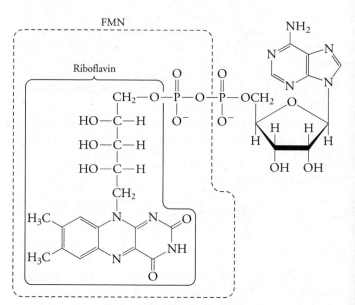

Flavin adenine dinucleotide (FAD)

The ability of FAD, FMN, and riboflavin to bind to the RFN element was measured as a dissociation constant, K_d.

Compound	K_d (nM)
FAD	300
FMN	5
Riboflavin	3000

Which compound would be the most effective regulator of FAD biosynthesis in the cell? What portion of the FAD molecule is likely to be important for interacting with the mRNA?

40. A number of human neurological diseases result from the presence of trinucleotide repeats in certain protein-coding genes. The severity of each disease is correlated with the number of repeats, which may increase due to the slippage of DNA polymerase during replication.

(a) The most common repeated triplet is CAG, which is almost always located within an open reading frame. What amino acid is encoded by this triplet (see Table 3-3), and what effect would the repeats have on the protein?

(b) To test the effect of CAG repeats on transcription, researchers used a yeast expression system with genes engineered to contain CAG repeats. In addition to the expected transcripts corresponding to the known lengths of the genes, RNA molecules up to three times longer were obtained. Based on your knowledge of RNA synthesis and processing, what factors could account for longer-than-expected transcripts of a given gene?

(c) Unexpectedly long transcripts could result from slippage of RNA polymerase II during transcription of the CAG repeats. In this scenario, the polymerase temporarily ceases polymerization, slides backward along the DNA template, then resumes transcription, in effect retranscribing the same sequence. Slippage may be triggered by the formation of secondary structure in the DNA template strand. Draw a diagram showing how a DNA strand containing CAG repeats could form a secondary structure that might prevent the advance of RNA polymerase.

41. In *E. coli*, replication is several times faster than transcription. Occasionally, the replication fork catches up to an RNA polymerase that is moving in the same direction as replication fork movement. When this occurs, translation stops and the RNA polymerase is displaced from the template DNA. DNA polymerase can use the existing RNA transcript as a primer to continue replication. Draw a diagram of this process, showing how such collisions would produce a discontinuous leading strand.

42. Refer to Problem 41 to explain why most *E. coli* genes are oriented such that replication and transcription proceed in the same direction.

21-3 RNA Processing

43. In *E. coli*, mRNA degradation is carried out by an endonuclease, but the mRNA must first be modified by a 5′ pyrophosphohydrolase. What reaction does this enzyme catalyze?

44. Complete the following table about nucleic acid polymerases.

Polymerase	Type of template	Substrates	Reaction product
DNA polymerase			
Human telomerase			
RNA polymerase			
Poly(A) polymerase			
Bacterial CCA-adding enzyme			

45. Why are only mRNAs capped and polyadenylated? Why do these post-transcriptional modifications not take place on rRNA or tRNA?

46. Explain why capping the 5′ end of an mRNA molecule makes it resistant to 5′ → 3′ exonucleases. Why is it necessary for capping to occur before the mRNA has been completely synthesized?

47. The poly(A) polymerase that modifies the 3′ end of mRNA molecules differs from other polymerases. The active sites of DNA and RNA polymerases are large enough to accommodate a double-stranded polynucleotide, but the active site of poly(A) polymerase is much narrower. Explain.

48. Explain how the substrate specificity of poly(A) polymerase differs from that of a conventional RNA polymerase.

49. A poly(A)-binding protein (PABP) has an affinity for RNA molecules with poly(A) tails. What is the effect of adding PABP to a cell-free system containing mRNA and RNases?

50. The only mRNA transcripts that lack poly(A) tails are those encoding histones. Why do these mRNA transcripts not require poly(A) tails?

51. ATP can be labeled with ^{32}P at any one of its three phosphate groups, designated α, β, and γ:

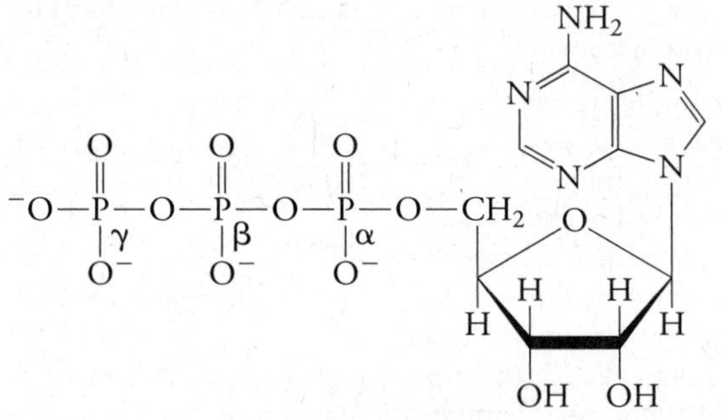

A eukaryotic cell carrying out transcription and RNA processing is incubated with labeled ATP. Where will the radioactive isotope appear in RNA if the ATP is labeled with ^{32}P at the **(a)** α position, **(b)** β position, and **(c)** γ position?

52. Genetic engineers must modify eukaryotic genes so that they can be expressed in bacterial host cells. Explain why the DNA from a eukaryotic gene cannot be placed directly into the bacteria but is first transcribed to mRNA and then reverse-transcribed back to cDNA.

53. Introns in eukaryotic protein-coding genes may be quite large, but almost none are smaller than about 65 bp. What are some reasons for this minimum intron size?

54. Biochemistry textbooks published a few decades ago often used the phrase "one gene, one protein." Why is this phrase no longer accurate?

55. Introns are removed co-transcriptionally rather than post-transcriptionally. Why is this a good cellular strategy?

56. A portion of the gene for the β chain of hemoglobin is shown below. The upper sequence includes the 5′ splice site and the lower sequence includes the 3′ splice site for the intron between exons 1 and 2. Identify the 5′ splice site and the 3′ splice site.

··· CCCTGGGCAGGTTGGTA ···

··· TTTCCCACCCTTAGGCTGCT ···

57. The ribozyme known as RNase P processes certain immature tRNA and rRNA molecules. Comparisons of RNase P RNAs from different species reveal conserved features that appear to be involved in endonuclease activity. For example, an unpaired uridine at position 69 is universally conserved. This residue does not pair with another nucleotide but forms a bulge in the RNA secondary structure. To test whether the identity or the geometry of U69 is critical for RNase P activity, several mutants were constructed and their endonuclease activity studied. The results for each mutant are given as a rate constant (k) for catalysis and a dissociation constant (K_d) for substrate binding.

RNAse P RNA	k (min^{-1})	K_d (nM)
Wild-type U69	0.26	1.7
U69 → A69	0.062	4
U69 → G69	0.0034	73
U69 → C69	0.0056	3
U69 deletion	0.0056	7
U69 + U70	0.0054	181

(a) According to these results, is the U69 bulge more important for substrate binding or catalysis?

(b) What is the effect of increasing the size of the bulge by adding a second U residue (the U69 + U70 mutant)?

58. Some tRNA molecules include sulfur-containing nucleotides. Draw the structures of 4-thiouridine and 2-thiocytidine.

59. RNA interference was investigated as a method to silence the gene for vascular endothelial growth factor (VEGF), a protein required for angiogenesis (development of blood vessels) in most cancers. The addition of siRNA targeted against the VEGF gene almost completely eliminated the secretion of VEGF from prostate cancer cells in culture. A portion of the gene sequence (bases 189–207) is shown below. Design an siRNA targeted to this region of the gene.

5′··· GGAGTACCCTGATGAGATC ··· 3′

60. Explain why the vast majority of nucleic acids with catalytic activity are RNA rather than DNA.

61. Like proteins, RNAs have primary, secondary, and tertiary structure. Use what you have learned about the primary, secondary, and tertiary structures of proteins to describe the structure of the ribozyme RNase P.

62. Based on the information in this chapter, give at least three reasons why a silent mutation in a gene (that is, a mutation that does not alter the amino acid sequence of the encoded protein) could decrease the amount of protein expressed.

[SELECTED READINGS]

Borukhov, S., and Nudler, E., RNA polymerase: the vehicle of transcription, *Trends Microbiol.* **16,** 126–134 (2008). [Discusses the structural changes that occur in bacterial RNA polymerase during initiation and elongation.]

Gilmour, D. S., and Fan, R., Derailing the locomotive: transcription termination, *J. Biol. Chem.* **283,** 661–664 (2008). [Outlines the mechanisms for transcription termination in prokaryotes and eukaryotes.]

Juven-Gershon, T., and Kadonaga, J. T., Regulation of gene expression via the core promoter and the basal transcription machinery, *Dev. Biol.* **339,** 225–229 (2010). [Describes some of the eukaryotic promoter elements.]

Kornberg, R. D., The molecular basis of eukaryotic transcription, *Proc. Nat. Acad. Sci.* **104,** 12955–12961 (2007). [A summary of many aspects of transcription by the scientist who won a Nobel Prize for his work on RNA polymerase.]

Sashital, D., and Doudna, J. A., Structural insights into RNA interference, *Curr. Opin. Struct. Biol.* **20,** 90–97 (2010). [Includes information on the structure of RISC.]

Sharp, P. A., The discovery of split genes and RNA splicing, *Trends Biochem. Sci.* **30,** 279–281 (2005). [A brief summary of splicing by one of its discoverers.]

Sikorski, T. W., and Buratowski, S., The basal initiation machinery: beyond the general transcription factors, *Curr. Opin. Cell Biol.* **21,** 344–351 (2009).

22

PROTEIN SYNTHESIS

▶▶ **WHY** do all tRNAs have the same size and shape?

RNA molecules exhibit a huge variety of structures, which allows them to carry out a range of essential cellular functions related to the storage and expression of genetic information. The small RNAs known as transfer RNA molecules deliver 20 different types of amino acids to the ribosome during protein synthesis. But the tRNA molecules are surprisingly uniform: They are limited to about 76 nucleotides, and they are invariably L-shaped with a characteristic arrangement of stems and loops. In this chapter we will see why this conformity is necessary for tRNAs to do their job.

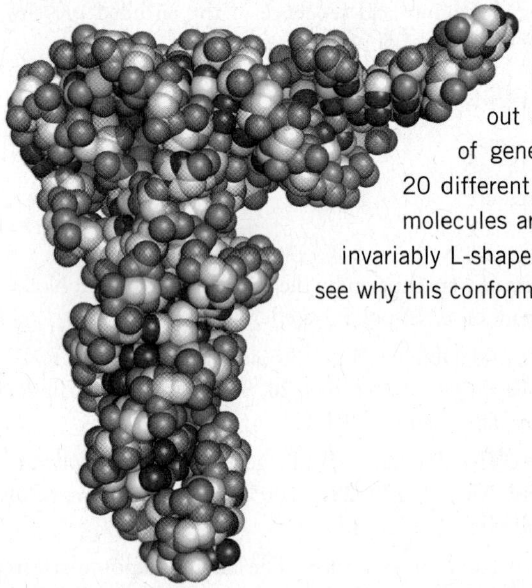

[Structure of tRNA^{Phe} (pdb 6TNA) determined by J. L. Sussman, S. R. Holbrook, R. W. Warrant, G. M. Church, and S. H. Kim.]

THIS CHAPTER IN CONTEXT

Part 1   Foundations

Part 2   Molecular Structure and Function

Part 3   Metabolism

Part 4 Genetic Information

Do You Remember?

- DNA and RNA are polymers of nucleotides, each of which consists of a purine or pyrimidine base, deoxyribose or ribose, and phosphate (Section 3-1).
- The biological information encoded by a sequence of DNA is transcribed to RNA and then translated into the amino acid sequence of a protein (Section 3-2).
- Amino acids are linked by peptide bonds to form a polypeptide (Section 4-1).
- Protein folding and protein stabilization depend on noncovalent forces (Section 4-3).
- rRNA and tRNA transcripts are modified to produce functional molecules (Section 21-3).

In the decade that followed Watson and Crick's 1953 elucidation of DNA structure, nearly all the components required for expressing genetic information were identified, including mRNA, tRNA, and ribosomes. Crick had already hypothesized that protein synthesis, or **translation,** required "adaptor" molecules (subsequently identified as tRNA) that carried an amino acid and recognized genetic information in the form of a sequence of nucleotides. The correspondence between DNA sequences and protein sequences was indisputable, but it required some biochemical detective work to discover the nature of the genetic code. Ultimately, *the genetic code was shown to be based on three-nucleotide* **codons** *that are read in a sequential and nonoverlapping manner.*

A triplet code is a mathematical necessity, since the number of possible combinations of three nucleotides of four different kinds (4^3, or 64) is more than enough to specify the 20 amino acids found in polypeptides (a doublet code, with 4^2, or 16, possibilities, would be inadequate). Genetic experiments with mutant bacteriophages demonstrated that triplet codons are read sequentially. For example, a mutation resulting from the deletion of a nucleotide within a gene can be corrected by a second mutation that inserts another nucleotide into the gene. The second mutation can restore gene function because it maintains the proper **reading frame** for translation. Since a given nucleotide sequence in an mRNA molecule can potentially have three different reading frames (**Fig. 22-1**), the selection of the proper one depends on the precise identification of a translation start site.

The genetic code, shown in Table 22-1, is said to be **degenerate** because *several mRNA codons may correspond to the same amino acid.* In fact, most amino acids are specified by two or more codons (Arg, Leu, and Ser each have six codons). Only Met and Trp have only one codon each (they are also among the amino acids that occur least frequently in polypeptides). The Met codon also functions as a translation initiation point. Three codons, known as stop or nonsense codons, signal translation termination. In Table 22-1, codons are shaded according to the overall hydrophobic, polar, or ionic character of the corresponding amino acid (using the scheme introduced in Fig. 4-2). Codons for chemically similar amino acids appear to cluster; for example, U at the second codon position invariably specifies a hydrophobic amino acid. This apparently nonrandom pattern of codon–amino acid correspondence suggests that the genetic code might have evolved from a simpler system involving only two nucleotides and a handful of amino acids.

The process of accurately translating a series of mRNA codons requires interactions among nucleic acids and proteins, beginning with the attachment of a specific amino

A C C A U C U C G A G A G U
Thr Ile Ser Arg
A C C A U C U C G A G A G U
Pro Ser Arg Glu
A C C A U C U C G A G A G U
His Leu Glu Ser

Figure 22-1 Reading frames. Even with a nonoverlapping genetic code based on nucleotide triplets, a given nucleotide sequence has three possible reading frames. An mRNA molecule can therefore potentially specify three different amino acid sequences.

[**TABLE 22-1**] The Standard Genetic Code

First Position (5′ end)	Second Position				Third Position (3′ end)
	U	C	A	G	
U	UUU Phe	UCU Ser	UAU Tyr	UGU Cys	U
	UUC Phe	UCC Ser	UAC Tyr	UGC Cys	C
	UUA Leu	UCA Ser	UAA Stop	UGA Stop	A
	UUG Leu	UCG Ser	UAG Stop	UGG Trp	G
C	CUU Leu	CCU Pro	CAU His	CGU Arg	U
	CUC Leu	CCC Pro	CAC His	CGC Arg	C
	CUA Leu	CCA Pro	CAA Gln	CGA Arg	A
	CUG Leu	CCG Pro	CAG Gln	CGG Arg	G
A	AUU Ile	ACU Thr	AAU Asn	AGU Ser	U
	AUC Ile	ACC Thr	AAC Asn	AGC Ser	C
	AUA Ile	ACA Thr	AAA Lys	AGA Arg	A
	AUG Met	ACG Thr	AAG Lys	AGG Arg	G
G	GUU Val	GCU Ala	GAU Asp	GGU Gly	U
	GUC Val	GCC Ala	GAC Asp	GGC Gly	C
	GUA Val	GCA Ala	GAA Glu	GGA Gly	A
	GUG Val	GCG Ala	GAG Glu	GGG Gly	G

acid to the appropriate tRNA molecule. The tRNAs must then align their **anticodons** with mRNA sequences on a ribosome so that peptide bonds can link amino acids in the order specified by the mRNA. In the following sections, we will examine these processes in detail. We will also look at some of the steps required to convert a newly made polypeptide to a fully functional protein.

22-1 tRNA Aminoacylation

KEY CONCEPTS
- tRNAs are compact L-shaped molecules with an anticodon at one end and an aminoacyl group at the other end.
- An aminoacyl-tRNA synthetase uses ATP to activate an amino acid and transfer it to the appropriate tRNA.
- Because of wobble pairing, a tRNA anticodon may pair with more than one mRNA codon.

⊕ **See Guided Exploration. The structure of tRNA.**

A bacterial cell typically contains 30 to 40 different tRNAs, and a mammalian cell as many as 150 (this is an obvious example of redundancy in biological systems, since only 20 different amino acids are routinely incorporated into polypeptides). tRNAs that bear the same amino acid but have different anticodons are called **iso-acceptor tRNAs.** The structures of all tRNA molecules are similar—even those that carry different amino acids.

Each tRNA molecule contains about 76 nucleotides (the range is 54 to 100), of which up to one-quarter are post-transcriptionally modified (the structures of some of these modified nucleotides are shown in Fig. 21-27). Many of the tRNA bases pair intramolecularly, generating the short stems and loops of what is commonly called a cloverleaf secondary structure **(Fig. 22-2a)**. A segment at the 5′ end of the tRNA pairs with bases near the 3′ end to form the acceptor stem (an amino acid attaches to the 3′ end). Several other base-paired stems end in small loops. The D loop often contains the modified base dihydrouridine (abbreviated D), and the TψC loop usually contains the indicated sequence (ψ is the symbol for the nucleotide pseudouridine; see Section 21-3). The variable loop, as its name implies, ranges from 3 to 21 nucleotides in different tRNAs. The anticodon loop includes the three nucleotides that pair with an mRNA codon.

The various elements of tRNA secondary structure fold into a compact L shape that is stabilized by extensive stacking interactions and nonstandard base pairs (Fig. 22-2b). Virtually all the bases are buried in the interior of the tRNA molecule, except for the anticodon triplet and the CCA sequence at the 3′ end. The narrow elongated structure

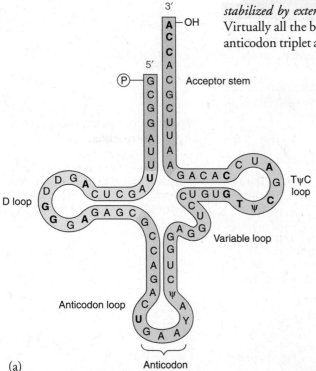

(a)

(b)

Figure 22-2 Structure of yeast tRNA^Phe. (a) Secondary structure. The 76 nucleotides of this tRNA molecule, which can carry a phenylalanine residue at its 3′ end, form four base-paired stems arranged in a cloverleaf pattern. Invariant bases are shown in boldface. ψ is pseudouridine and Y is a guanosine derivative. Some C and G residues in this structure are methylated. (b) Tertiary structure, with the various structures colored as in part (a). The long arm of the **L**

consists primarily of the anticodon loop and D loop, and the short arm is primarily made up of the TψC loop and acceptor stem. The anticodon and acceptor ends of the molecule are separated by about 75 Å. [Structure (pdb 4TRA) determined by E. Westhof, P. Dumas, and D. Moras.]

? **Why is it important that the bases in the anticodon loop point outward?**

of tRNA molecules allows them to align side-by-side so that they can interact with adjacent mRNA codons during translation. However, the tRNA anticodon is located a considerable distance (about 75 Å) from the 3' aminoacyl group, whose identity is specified by that anticodon.

tRNA aminoacylation consumes ATP

Aminoacylation, the attachment of an amino acid to a tRNA, is catalyzed by an aminoacyl–tRNA synthetase (AARS). *To ensure accurate translation, the synthetase must attach the appropriate amino acid to the tRNA bearing the corresponding anticodon.* As expected, most AARSs interact with the tRNA anticodon as well as the aminoacylation site at the other end of the tRNA molecule.

An AARS catalyzes the formation of an ester bond between an amino acid and an OH group of the ribose at the 3' end of a tRNA to yield an aminoacyl–tRNA:

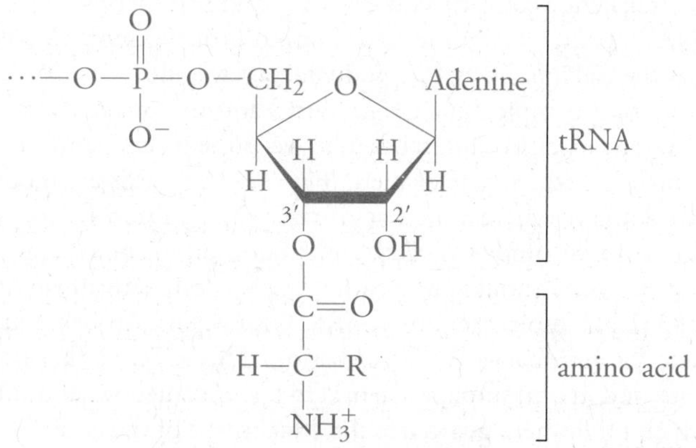

Aminoacyl–tRNA

The tRNA molecule is then said to be "charged" with an amino acid. The aminoacylation reaction has two steps and requires the free energy of ATP (**Fig. 22-3**). The overall reaction is

$$\text{amino acid} + \text{tRNA} + \text{ATP} \rightarrow \text{aminoacyl–tRNA} + \text{AMP} + \text{PP}_i$$

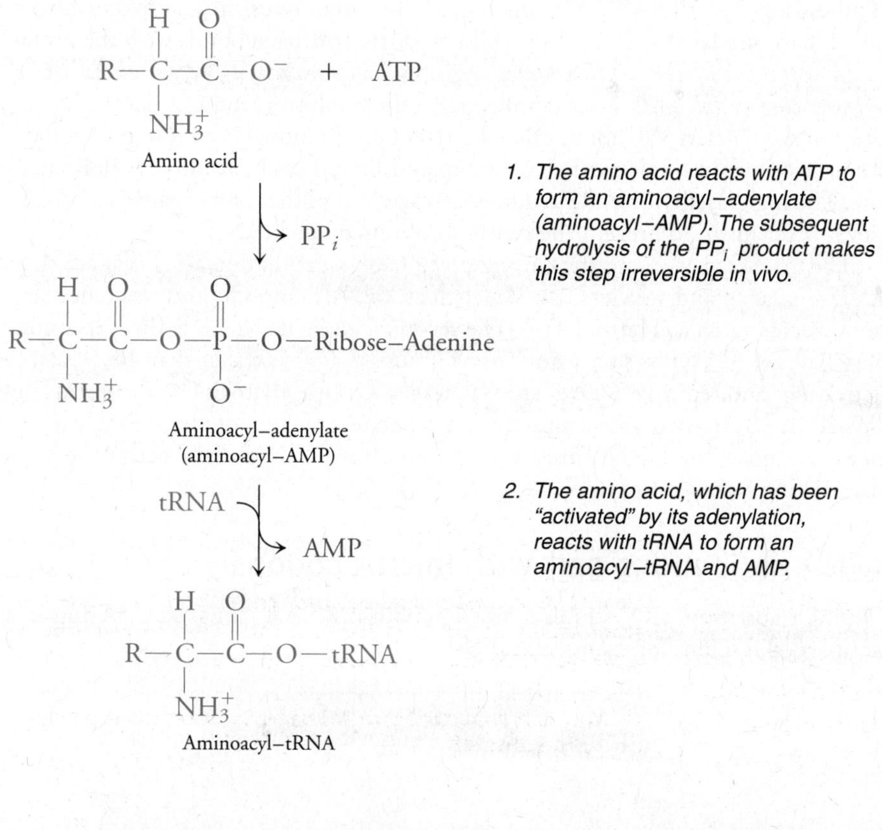

Amino acid

1. The amino acid reacts with ATP to form an aminoacyl–adenylate (aminoacyl–AMP). The subsequent hydrolysis of the PP_i product makes this step irreversible in vivo.

Aminoacyl–adenylate (aminoacyl–AMP)

2. The amino acid, which has been "activated" by its adenylation, reacts with tRNA to form an aminoacyl–tRNA and AMP.

Aminoacyl–tRNA

Figure 22-3 **The aminoacyl-tRNA synthetase reaction.**

? How many "high-energy" phosphoanhydride bonds break in this process? How many "high-energy" acyl-phosphate bonds form?

[**TABLE 22-2**]

Classes of Aminoacyl–tRNA Synthetases

	Amino Acids	
Class I	Arg	Leu
	Cys	Met
	Gln	Trp
	Glu	Tyr
	Ile	Val
Class II	Ala	Lys
	Asn	Pro
	Asp	Phe
	Gly	Ser
	His	Thr

Most cells contain 20 different AARS enzymes, corresponding to the 20 standard amino acids (isoacceptor tRNAs are recognized by the same AARS). Although all AARSs catalyze the same reaction, they do not exhibit a conserved size or quaternary structure. Nevertheless, the enzymes fall into two groups based on several shared structural and functional features (Table 22-2). For example, the class I enzymes attach an amino acid to the 2′ OH group of the tRNA ribose, whereas the class II enzymes attach an amino acid to the 3′ OH group (this distinction is ultimately of no consequence, as the 2′-aminoacyl group shifts to the 3′ position before it takes part in protein synthesis).

Some bacteria appear to lack the full complement of 20 AARSs. The enzymes most commonly missing are GlnRS and AsnRS (which aminoacylate tRNA^{Gln} and tRNA^{Asn}). In these organisms, Gln–tRNA^{Gln} and Asn–tRNA^{Asn} are synthesized indirectly. First, GluRS and AspRS with relatively low tRNA specificity charge tRNA^{Gln} and tRNA^{Asn} with their corresponding acids (glutamate and aspartate). Next, an amidotransferase converts Glu–tRNA^{Gln} and Asp–tRNA^{Asn} to Gln–tRNA^{Gln} and Asn–tRNA^{Asn} using glutamine as an amino-group donor. In some microorganisms, this is the only pathway for producing asparagine.

The structure of a complex of *E. coli* GlnRS and its cognate (matching) tRNA (tRNA^{Gln}) shows the extensive interaction between the protein and the concave face of the tRNA molecule (the inside of the L; Fig. 22-4). AARSs are modular proteins with a catalytic domain, where amino acid activation and transfer to a tRNA occur, as well as a domain that binds the tRNA anticodon or, in some cases, another part of the tRNA such as the variable loop. Most AARSs can activate an amino acid in the absence of a tRNA molecule, but GlnRS, GluRS, and ArgRS require a cognate tRNA molecule for aminoacyl–AMP formation. This suggests that the anticodon-recognition site and the aminoacylation active site somehow communicate with each other, which might help guarantee the attachment of the correct amino acid to the tRNA.

Some synthetases have proofreading activity

Studies of individual AARS enzymes indicate that the amino acid binding site may be tailored precisely to the geometry and electrostatic properties of a particular amino acid, making it less likely that one of the other 19 amino acids would be activated or transferred to a tRNA molecule. For example, TyrRS (the enzyme responsible for synthesizing Tyr–tRNA^{Tyr}) can distinguish between tyrosine and phenylalanine, which have similar shapes, by their ability to form hydrogen bonds with the protein.

In some cases, *the specificity of tRNA aminoacylation may be enhanced through proofreading by the AARS*. For example, IleRS almost always produces Ile–tRNA^{Ile} and only rarely transfers Val (about once every 50,000 reactions) even though Val differs from Ile only by a single methylene group and should easily fit into the IleRS active site. The high fidelity of IleRS requires two active sites that participate in a "double-sieve" mechanism to prevent the synthesis of mischarged tRNA^{Ile}.

The first active site activates Ile and presumably other amino acids that are chemically similar to and smaller than Ile (such as Val, Ala, and Gly) but excludes larger amino acids (such as Phe and Tyr). The second active site, which hydrolyzes aminoacylated tRNA^{Ile}, admits only aminoacyl groups that are smaller than Ile. Thus, the activating and editing active sites together ensure that IleRS produces only Ile–tRNA^{Ile}. The two active sites are on separate domains of the synthetase, so a newly aminoacylated tRNA must visit the proofreading hydrolytic active site before dissociating from the enzyme.

tRNA anticodons pair with mRNA codons

During translation, tRNA molecules align with mRNA codons, base pairing in an antiparallel fashion, for example

Figure 22-4 Structure of GlnRS with RNA^{Gln}. In this complex, the synthetase is green and the cognate tRNA is red. Both the 3′ (acceptor) end of the tRNA (top right) and the anticodon loop (lower left) are buried in the protein. ATP at the active site is shown in yellow. [Structure of the *E. coli* complex (pdb 1QRT) determined by J. G. Arnez and T. A. Steitz.]

$$\begin{array}{lc} \text{tRNA anticodon} & 3'-A-A-G-5' \\ & \quad\vdots\ \vdots\ \vdots \\ \text{mRNA codon} & 5'-U-U-C-3' \end{array}$$

At first glance, this sort of specific pairing would require the presence of 61 different tRNA molecules, one to recognize each of the "sense" codons listed in Table 22-1. In fact, *many isoacceptor tRNAs can bind to more than one of the codons that specify their amino acid.* For example, yeast tRNA^{Ala} has the anticodon sequence 3′–CGI–5′ (I represents the purine nucleotide inosine, a deaminated form of adenosine) and can pair with the Ala codons GCU, GCC, and GCA.

| tRNA anticodon | –C–G–I– | –C–G–I– | –C–G–I– |
| mRNA codon | –G–C–U– | –G–C–C– | –G–C–A– |

As Francis Crick originally proposed in the **wobble hypothesis,** the third codon position and the 5′ anticodon position experience some flexibility, or wobble, in the geometry of their hydrogen bonding. The base pairs permitted by wobbling are given in Table 22-3. The wobble hypothesis explains why many bacterial cells can bind all 61 codons with a set of less than 40 tRNAs (the reasons why mammalian cells contain over 150 tRNAs are not clear). Variations in tRNA anticodon sequences allow nonstandard amino acids to be occasionally incorporated into polypeptides at positions corresponding to stop codons (Box 22-A).

[TABLE 22-3]

Allowed Wobble Pairs at the Third Codon-Anticodon Position

5′ Anticodon Base	3′ Codon Base
C	G
A	U
U	A, G
G	U, C
I	U, C, A

BOX 22-A BIOCHEMISTRY NOTE

The Genetic Code Expanded

In addition to the 20 standard amino acids listed in Figure 4-2, some amino acid variants can be incorporated into proteins during translation (keep in mind that a mature protein may contain a number of modified amino acids, but these changes almost always take place *after* the protein has been synthesized). Addition of a nonstandard amino acid during protein synthesis requires a dedicated tRNA and a stop codon that can be reinterpreted. The expanded genetic code includes two naturally occurring amino acids, selenocysteine and pyrrolysine, plus a number of amino acids produced in the laboratory.

Selenocysteine occurs in a few proteins in both prokaryotes and eukaryotes, which explains why selenium is an essential trace element. Humans may produce as many as two dozen selenoproteins.

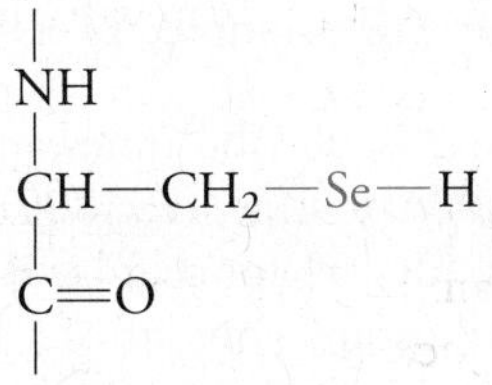

Selenocysteine (Sec) residue

Selenocysteine (Sec), which resembles cysteine, is generated from serine that has been attached to tRNA^{Sec} by the action of SerRS. A separate enzyme then converts Ser–tRNA^{Sec} to Sec–tRNA^{Sec}. This charged tRNA has an ACU anticodon (reading in the 3′ → 5′ direction), which recognizes a UGA codon. Normally, UGA functions as a stop codon, but a hairpin secondary structure in the selenoprotein's mRNA provides the contextual signal for selenocysteine to be delivered to the ribosome at that point.

A few prokaryotic species incorporate pyrrolysine (Pyl) into certain proteins. Synthesis of these proteins requires a 21st type of AARS that directly charges the tRNA^{Pyl} with pyrrolysine.

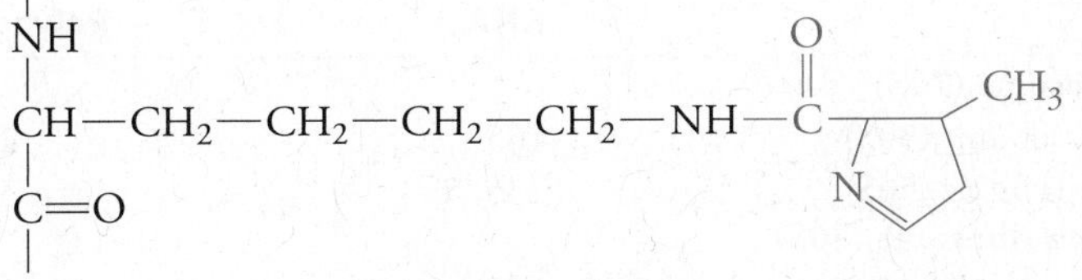

Pyrrolysine (Pyl) residue

(*continued on next page*)

The Pyl–tRNA^{Pyl} recognizes the stop codon UAG, which is reinterpreted as a Pyl codon with the help of a protein that recognizes secondary structure in the mRNA bound to the ribosome.

In the laboratory, proteins containing unnatural amino acids can be synthesized in bacterial, yeast, and mammalian cells. These experimental systems rely on a pair of genetically engineered components: a tRNA that can "read" a stop codon and an AARS that can attach the unnatural amino acid to the tRNA. When the cell translates an mRNA containing the stop codon, the novel amino acid is incorporated at that codon. Dozens of amino acid derivatives with fluoride, reactive acetyl and amino groups, fluorescent tags, and other modifications have been introduced into specific proteins using this technology. Because the novel amino acids are genetically encoded, they appear only at the expected positions in the translated protein—a more reliable outcome than chemically modifying a protein in a test tube.

⬡ **Question:** What is the disadvantage for a cell to have a variant tRNA that can insert an amino acid at a stop codon?

CONCEPT REVIEW
• Describe the free energy cost of aminoacyl-tRNA synthesis.
• Explain why accurate aminoacylation is essential for accurate translation.
• Why don't cells require 60 different codons?

22-2 Ribosome Structure

KEY CONCEPT
• The two subunits of the ribosome consist largely of rRNA that binds mRNA and three tRNAs during protein synthesis.

In order to synthesize a protein, genetic information (in the form of mRNA) and amino acids (attached to tRNA) must get together so that the amino acids can be covalently linked in the specified order. This is the job of the **ribosome,** a large complex containing both RNA and protein. At one time, ribosomal RNA (rRNA) was believed to serve as a structural scaffolding for ribosomal proteins, which presumably carried out protein synthesis, but it is now clear that rRNA itself is central to ribosomal function.

A bacterial cell may contain 20,000 ribosomes and a yeast cell, about 200,000. This accounts for the observation that at least 80% of a cell's RNA is located in ribosomes (tRNA comprises about 15% of cellular RNA; mRNA accounts for only a few percent of the total). A ribosome consists of a large and a small subunit containing rRNA molecules, all of which are described in terms of their sedimentation coefficients, S. Thus, the 70S bacterial ribosome has a large (50S) and a small (30S) subunit (the sedimentation coefficient indicates how quickly a particle settles during ultracentrifugation; it is related to the particle's mass). The 80S eukaryotic ribosome is made up of a 60S large subunit and a 40S small subunit. The compositions of prokaryotic and eukaryotic ribosomes are listed in Table 22-4. Regardless of its source, about two-thirds of the mass of a ribosome is due to the rRNA; the remainder is due to dozens of different proteins (over 80 in eukaryotes).

[TABLE **22-4**] Ribosome Components

	RNA	Polypeptides
E. coli ribosome (70S)		
Small subunit (30S)	16S	21
Large subunit (50S)	23S, 5S	31
Mammalian ribosome (80S)		
Small subunit (40S)	18S	33
Large subunit (60S)	28S, 5.8S, 5S	49

(a)

(b)

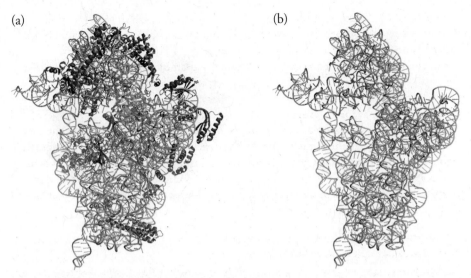

Figure 22-5 **Structure of the 30S ribosomal subunit from *Thermus thermophilus.***
(a) The 30S subunit with the rRNA in gray and the proteins in purple. (b) Structure
of the 16S rRNA alone. Note how the overall shape of the 30S subunit reflects the
structure of the rRNA. [Structure (pdb 1J5E) determined by B. T. Wimberly, D. E. Brodersen,
W. M. Clemons, Jr., R. J. Morgan-Warren, A. P. Carter, C. Vonrhein, T. Hartsch, and V. Ramakrishnan.]

The structures of intact ribosomes from both prokaryotes and eukaryotes have
been elucidated by X-ray crystallography—a monumental undertaking, given the
ribosome's large size (about 2600 kD in bacteria and about 4300 kD in eukaryotes). The small ribosomal subunit from the heat-tolerant bacterium *Thermus
thermophilus* is shown in **Figure 22-5**. The overall shape of the subunit is defined
by the 16S rRNA (1542 nucleotides in *E. coli*), which has numerous base-paired
stems and loops that fold into several domains. This multidomain structure appears to confer some conformational flexibility on the 30S subunit—a requirement for protein synthesis. Twenty small polypeptides dot the surface of the
structure.

Compared to the 30S subunit, the prokaryotic 50S subunit is solid and immobile. Its 23S rRNA (2904 nucleotides in *E. coli*) and 5S rRNA (120 nucleotides)
fold into a single mass **(Fig. 22-6)**. As in the small subunit, the ribosomal proteins
associate with the surface of the rRNA, but the surfaces of the large and small subunits that make contact in the intact 70S ribosome are largely devoid of protein.

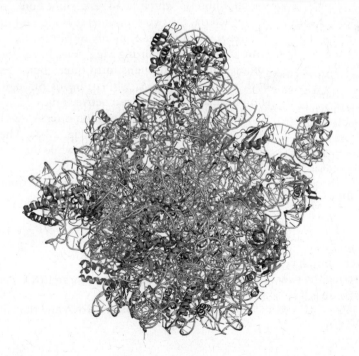

Figure 22-6 **Structure of the 50S
ribosomal subunit from *Haloarcula
marismortui.*** The 50S subunit with
rRNA in gray and proteins in green.
Most of the ribosomal proteins are not
visible in this view. The protein-free
central area forms the interface with
the 30S subunit. [Structure (pdb 1JJ2)
determined by D. J. Klein, T. M. Scheming,
P. B. Moore, and T. A. Steitz.]

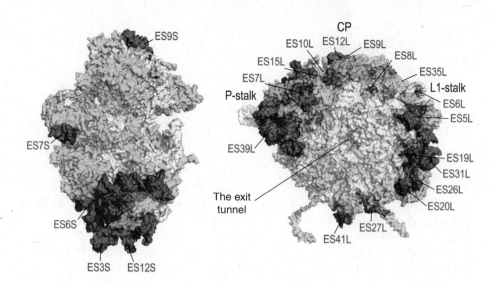

Figure 22-7 Eukaryotic ribosomal subunits. The solvent-exposed surfaces of the 40S subunit (left) and 60S subunit (right) are shown with the conserved ribosomal core structures in gray. Proteins that are unique to eukaryotes are shown in transparent yellow, and rRNA expansion segments are red. Newly synthesized proteins emerge from the ribosome through the exit tunnel. [From M. Yusupov, *Science* 334, 1524–1529 (2011). Reprinted with permission from AAAS. Courtesy Marat Yusupov.]

? **Which areas of the ribosome appear to be most conserved between bacteria and eukaryotes? Why?**

This highly conserved rRNA-rich subunit interface is the site where mRNA and tRNA bind during protein synthesis.

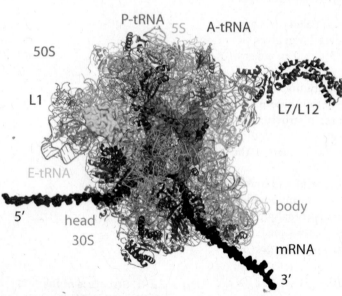

Figure 22-8 Model of the complete bacterial ribosome. The large subunit is shown in shades of gold (rRNA) and brown (proteins), and the small subunit in shades of blue (rRNA) and purple (proteins). The three tRNAs are colored magenta (A site), green (P site), and yellow (E site). An mRNA molecule is shown in dark gray. Note that the anticodon ends of the tRNAs contact the mRNA in the small subunit, while their aminoacyl ends are buried in the large subunit, where peptide bond formation occurs. [From M. Schmeing, *Nature* 461, 1234–1242 (2009). Reprinted by permission from Macmillan Publishers, Ltd. Photo Courtesy of M. Schmeing, McGill University.]
⊕ **See Interactive Exercise. Ribosome with tRNA and mRNA.**

Eukaryotic ribosomes are about 40–50% larger than those from bacteria and contain many additional proteins and more extensive rRNA. The RNA sequences that have no counterparts in bacterial ribosomes are known as expansion segments; these structures, along with the unique eukaryotic protein components, surround a core structure that is shared with the simpler bacterial ribosome (**Fig. 22-7**).

Up to three tRNA molecules may bind to the ribosome at a given time (**Fig. 22-8**). The binding sites are known as the **A site** (for *aminoacyl*), which accommodates an incoming aminoacyl–tRNA; the **P site** (for *peptidyl*), which binds the tRNA with the growing polypeptide chain; and the **E site** (for *exit*), which transiently binds a deacylated tRNA after peptide bond formation. The anticodon ends of the tRNAs extend into the 30S subunit to pair with mRNA codons, while their aminoacyl ends extend into the 50S subunit, which catalyzes peptide bond formation.

In bacteria, the two ribosomal subunits and the various tRNAs are held in place mainly by RNA–RNA contacts, with a number of stabilizing Mg^{2+} ions. In eukaryotic ribosomes, numerous proteins form intersubunit bridges. In both cases, the mRNA, which threads through the 30S subunit, makes a sharp bend between the codons in the A site and P site, where an Mg^{2+} ion interacts with mRNA backbone phosphate groups (see Fig. 22-8). The kink allows two tRNAs to fit side-by-side while interacting with consecutive mRNA codons. It may also help the ribosome maintain the reading frame by preventing it from slipping along the mRNA.

CONCEPT REVIEW
• Describe the overall structure of a ribosome and its three tRNA binding sites.
• Summarize the structural importance of ribosomal RNA and ribosomal proteins.

22-3 Translation

Like DNA replication and RNA transcription, protein synthesis can be divided into separate phases for initiation, elongation, and termination. These stages require an assortment of accessory proteins that bind to tRNA and to the ribosome in order to enhance the speed and accuracy of translation.

Initiation requires an initiator tRNA

In both prokaryotes and eukaryotes, protein synthesis begins at an mRNA codon that specifies methionine (AUG). In prokaryotic mRNAs, this initiation codon lies about 10 bases downstream of a conserved mRNA sequence called a Shine–Dalgarno sequence (**Fig. 22-9**). This sequence base pairs with a complementary sequence at the 3′ end of the 16S rRNA, thereby positioning the initiation codon in the ribosome. Eukaryotic mRNAs lack a Shine–Dalgarno sequence that can pair with the 18S rRNA. Instead, translation usually begins at the first AUG codon of an mRNA molecule.

The initiation codon is recognized by an initiator tRNA that has been charged with methionine. This tRNA does not recognize other Met codons that occur elsewhere in the coding sequence of the mRNA. In bacteria, the methionine attached to the initiator tRNA is modified by the transfer of a formyl group from tetrahydrofolate (see Section 18-2). The resulting aminoacyl group is designated fMet, and the initiator tRNA is known as $tRNA_f^{Met}$:

$$CH_3$$
$$|$$
$$S$$
$$|$$
$$CH_2$$
$$|$$
$$CH_2$$
$$HC-NH-CH-C-O-tRNA_f^{Met}$$

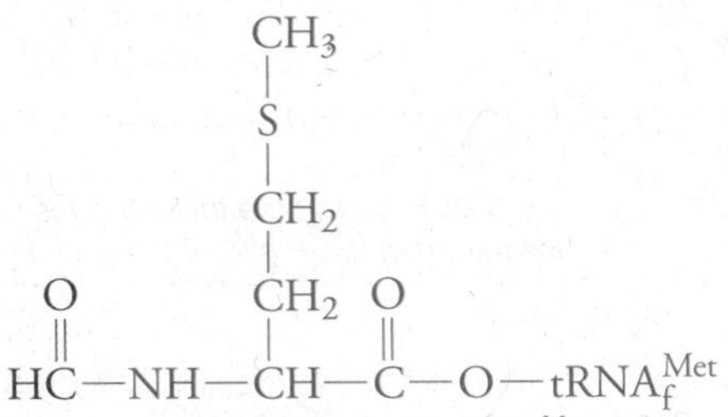

N-Formylmethionine–$tRNA_f^{Met}$
(fMet–$tRNA_f^{Met}$)

Because the amino group of fMet is derivatized, it cannot form a peptide bond. Consequently, fMet can be incorporated only at the N-terminus of a polypeptide. Later, the formyl group or the entire fMet residue may be removed. In eukaryotic cells, the initiator tRNA, designated $tRNA_i^{Met}$, is charged with Met but is not formylated.

Initiation in *E. coli* requires three **initiation factors** (IFs) called IF-1, IF-2, and IF-3. IF-3 binds to the small ribosomal subunit to promote the dissociation of the large and small subunits. fMet–$tRNA_f^{Met}$ binds to the 30S subunit with the assistance of IF-2, a GTP-binding protein. IF-1 sterically blocks the A site of the small subunit, thereby forcing the initiator tRNA into the P site. An mRNA molecule may bind to the 30S subunit either before or after the initiator tRNA has bound, indicating that a codon–anticodon interaction is not essential for initiating protein synthesis.

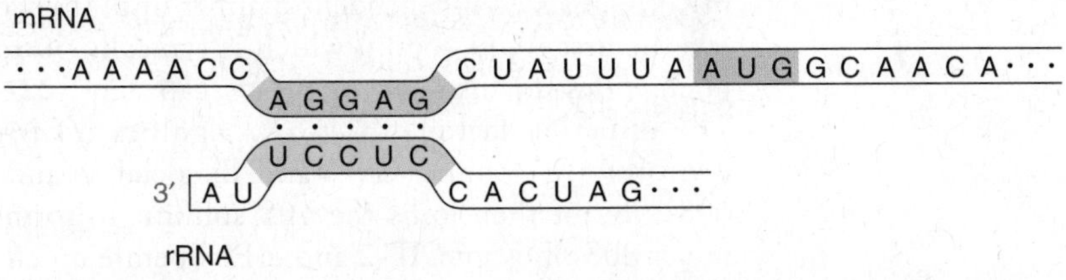

KEY CONCEPTS
- Translation begins with the binding of a methionine-bearing initiator tRNA to the P site of a ribosome.
- For each new aminoacyl-tRNA, the ribosome senses correct codon-anticodon pairing.
- Ribosomal RNA promotes peptide bond formation by a transpeptidation reaction.
- G proteins such as EF-Tu and EF-G use the free energy of GTP to increase the efficiency of translation.
- Translation terminates when a release factor rather than an aminoacyl-tRNA binds to a stop codon positioned in the A site.
- Multiple ribosomes can simultaneously translate an mRNA.

See Guided Exploration. Translational initiation.

Figure 22-9 Alignment of a Shine-Dalgarno sequence with 16S rRNA. The Shine–Dalgarno sequence (blue) in an mRNA pairs with a complementary region (purple) at the 3′ end of the 16S rRNA molecule. The initiation codon is shaded green. This mRNA–rRNA interaction helps position the mRNA at the start of translation. The mRNA shown here encodes the ribosomal protein S12.

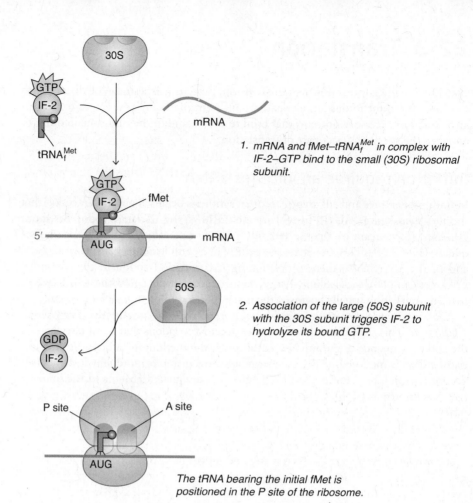

1. mRNA and fMet–tRNA$_f^{Met}$ in complex with IF-2–GTP bind to the small (30S) ribosomal subunit.

2. Association of the large (50S) subunit with the 30S subunit triggers IF-2 to hydrolyze its bound GTP.

The tRNA bearing the initial fMet is positioned in the P site of the ribosome.

Figure 22-10 Summary of translation initiation in *E. coli*. Similar events occur during translation initiation in eukaryotes, when a 40S and a 60S subunit associate following the binding of Met–tRNA$_i^{Met}$ to an initiation codon.

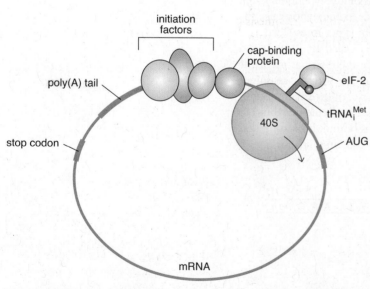

Figure 22-11 Circularization of eukaryotic mRNA at translation initiation. A number of initiation factors form a complex that links the 5′ cap and 3′ poly(A) tail of the mRNA. The small (40S) ribosomal subunit binds to the mRNA and locates the AUG start codon that is complementary to the anticodon of the initiator tRNA.

After the 30S–mRNA–fMet–tRNA$_f^{Met}$ complex has assembled, the 50S subunit associates with it to form the 70S ribosome. This change causes IF-2 to hydrolyze its bound GTP to GDP + P$_i$ and dissociate from the ribosome. The ribosome is now poised—with fMet–tRNA$_f^{Met}$ at the P site—to bind a second aminoacyl–tRNA in order to form the first peptide bond (**Fig. 22-10**).

In eukaryotes, translation initiation requires at least 12 distinct initiation factors. Among these are proteins that recognize the 5′ cap and poly(A) tail of the mRNA (see Section 21-3) and form a complex so that the mRNA actually forms a circle. Initiation may also require the helicase activity of the ribosome to remove secondary structure in the mRNA that would impede translation. The 40S subunit scans the mRNA in an ATP-dependent manner until it encounters the first AUG codon, which is typically 50 to 70 nucleotides downstream of the 5′ cap (**Fig. 22-11**). The initiation factor eIF2 (the *e* signifies *eukaryotic*) hydrolyzes its bound GTP and dissociates, and the 60S subunit then joins the 40S subunit to form the intact 80S ribosome. IF-2 and eIF2 operate much like the heterotrimeric **G proteins** that participate in intracellular signal transduction pathways (see Section 10-2). In each *case, GTP hydrolysis induces conformational changes that trigger additional steps of the reaction sequence.*

The appropriate tRNAs are delivered to the ribosome during elongation

All tRNAs have the same size and shape so that they can fit into small slots in the ribosome. In each reaction cycle of the elongation phase of protein synthesis, an aminoacyl–tRNA enters the A site of the ribosome (the initiator tRNA is the only one that enters the P site without first binding to the A site). After peptide bond formation, the tRNA moves to the P site, and then to the E site. As Figure 22-8 shows, there isn't much room to spare. In addition, all tRNAs must be able to bind interchangeably with protein cofactors.

Aminoacyl–tRNAs are delivered to the ribosome in a complex with a GTP-binding **elongation factor (EF)** known as EF-Tu in *E. coli*. EF-Tu is one of the most abundant *E. coli* proteins (about 100,000 copies per cell, enough to bind all the aminoacyl–tRNA molecules). An aminoacyl–tRNA can bind on its own to a ribosome *in vitro*, but EF-Tu increases the rate *in vivo*.

Because EF-Tu interacts with all 20 types of aminoacyl–tRNAs (representing more than 20 different tRNA molecules), it must recognize common elements of tRNA structure, primarily the acceptor stem and one side of the TψC loop (**Fig. 22-12**). A highly conserved protein pocket accommodates the aminoacyl group. Despite the differing chemical properties of their amino acids, *all aminoacyl–tRNAs bind to EF-Tu with approximately the same affinity* (uncharged tRNAs bind only weakly to EF-Tu). Apparently, the protein interacts with aminoacyl–tRNAs in a combinatorial fashion, offsetting less-than-optimal binding of an aminoacyl group with tighter binding of the acceptor stem and vice versa. This allows EF-Tu to deliver and surrender all 20 aminoacyl–tRNAs to a ribosome with the same efficiency.

The incoming aminoacyl–tRNA is selected on the basis of its ability to recognize a complementary mRNA codon in the A site. Due to competition among all the aminoacyl–tRNA molecules in the cell, this is the rate-limiting step of protein synthesis. Before the 50S subunit catalyzes peptide bond formation, the ribosome must verify that the correct aminoacyl–tRNA is in place. When tRNA binds to the A site of the 30S subunit, two highly conserved residues (A1492 and A1493) of the 16S rRNA "flip out" of an rRNA loop in order to form hydrogen bonds with various parts of the mRNA codon as it pairs with the tRNA anticodon. *These interactions physically link the two rRNA bases with the first two base pairs of the codon and anticodon so that they can sense a correct match between the mRNA and tRNA* (**Fig. 22-13**). Incorrect base pairing at the first or second codon position would prevent this three-way mRNA–tRNA–rRNA interaction. As expected from the wobble hypothesis (Section 22-1), the A1492/A1493 sensor does not monitor nonstandard base pairing at the third codon position.

As the rRNA nucleotides shift to confirm a correct codon–anticodon match, the conformation of the ribosome changes in such a way that the G protein EF-Tu is induced to hydrolyze its bound GTP. As a result of this reaction, EF-Tu dissociates

▶▶ **WHY** do all tRNAs have the same size and shape?

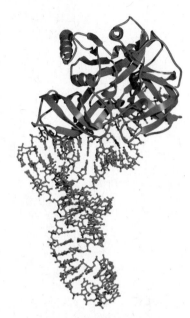

Figure 22-12 Structure of an EF-Tu-tRNA complex. The protein (blue) interacts with the acceptor end and TψC loop of an aminoacyl–tRNA (red). [Structure (pdb 1TTT) determined by P. Nissen, M. Kjeldgaard, S. Tharp, G. Polekhina, L. Reshetnikova, B. F. C. Clark, and J. Nyborg.]

⊕ **See Guided Exploration. Translational elongation.**

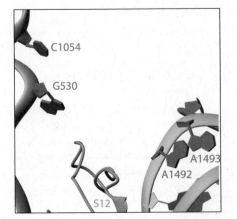
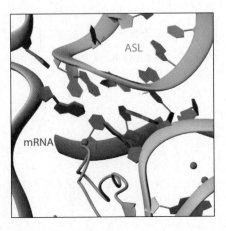

Figure 22-13 The ribosomal sensor for proper codon-anticodon pairing. These images show the A site of the 30S subunit in the (a) absence and (b) presence of mRNA and tRNA analogs. The rRNA is gray, with the "sensor" bases in red. The mRNA analog, representing the A-site codon, is purple, and the tRNA analog (labeled ASL) is gold. A ribosomal protein (S12) and two Mg^{2+} ions (magenta) are also visible. Note how rRNA bases A1492 and A1493 flip out to sense the codon–anticodon interaction. [Courtesy Venki Ramakrishnan. From *Science* 292, 897–902 (2001). Reproduced with permission of AAAS.]

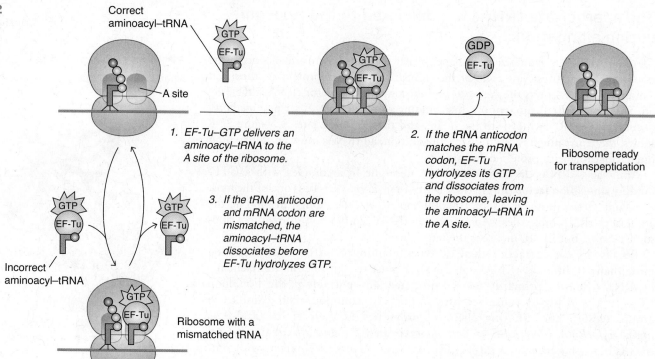

1. EF-Tu–GTP delivers an aminoacyl–tRNA to the A site of the ribosome.

2. If the tRNA anticodon matches the mRNA codon, EF-Tu hydrolyzes its GTP and dissociates from the ribosome, leaving the aminoacyl–tRNA in the A site.

3. If the tRNA anticodon and mRNA codon are mismatched, the aminoacyl–tRNA dissociates before EF-Tu hydrolyzes GTP.

Figure 22-14 Function of EF-Tu in translation elongation.

from the ribosome, leaving behind the tRNA with its aminoacyl group to be incorporated into the growing polypeptide chain.

However, if the tRNA anticodon is not properly paired with the A-site codon, the 30S conformational change and GTP hydrolysis by EF-Tu do not occur. Instead, the aminoacyl–tRNA, along with EF-Tu–GTP, dissociates from the ribosome. Because a peptide bond cannot form until after EF-Tu hydrolyzes GTP, *EF-Tu ensures that polymerization does not occur unless the correct aminoacyl–tRNA is positioned in the A site.* The energetic cost of proofreading at the decoding stage of translation is the free energy of GTP hydrolysis (catalyzed by EF-Tu). The function of EF-Tu is summarized in **Figure 22-14**. In eukaryotes, elongation factor eEF1α performs the same service as the prokaryotic EF-Tu. The functional correspondence between some prokaryotic and eukaryotic translation cofactors is given in Table 22-5.

The ribosome itself performs a bit of proofreading. The departure of EF-Tu–GDP leaves behind the aminoacyl–tRNA, whose acceptor end can now slip all the way into the A site of the 50S ribosomal subunit. The 30S subunit closes in around the tRNA, but at this point, the only interactions that hold the aminoacyl–tRNA in place are codon–anticodon contacts. If there is still a slight mismatch, such as a G:U base pair at the first or second position, not detectable by the A1492/A1493 sensor, the strain of not being able to form a perfect Watson–Crick pair will be felt by the

[**TABLE 22-5**] Prokaryotic and Eukaryotic Translation Factors

Prokaryotic Protein	Eukaryotic Protein	Function
IF-2	eIF2	Delivers initiator tRNA to P site of ribosome
EF-Tu	eEF1α	Delivers aminoacyl–tRNA to A site of ribosome during elongation
EF-G	eEF2	Binds to A site to promote translocation following peptide bond formation
RF-1, RF-2	eRF1	Binds to A site at a stop codon and induces peptide transfer to water

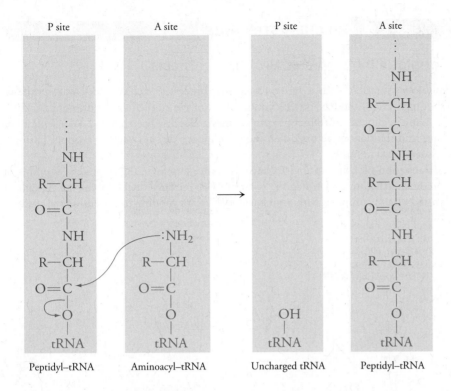

P site   A site   P site   A site

Peptidyl–tRNA   Aminoacyl–tRNA   Uncharged tRNA   Peptidyl–tRNA

Figure 22-15 The peptidyl transferase reaction. Note that the nucleophilic attack of the aminoacyl group on the peptidyl group produces a free tRNA in the P site and a peptidyl–tRNA in the A site.

? Compare this reaction to the condensation reaction of two amino acids shown in Section 4-1.

ribosome and possibly the tRNA itself, and the tRNA will slip out of the A site. In this way, the ribosome verifies correct codon–anticodon pairing twice for each aminoacyl–tRNA: when EF-Tu first delivers it to the ribosome and after EF-Tu departs. Ribosomal proofreading helps limit the error rate of translation to about 10^{-4} (one mistake for every 10^4 codons).

The peptidyl transferase active site catalyzes peptide bond formation

When the ribosomal A site contains an aminoacyl–tRNA and the P site contains a peptidyl–tRNA (or, prior to formation of the first peptide bond, fMet–tRNA$_f^{Met}$), the peptidyl transferase activity of the 50S subunit catalyzes a **transpeptidation** reaction in which the free amino group of the aminoacyl–tRNA in the A site attacks the ester bond that links the peptidyl group to the tRNA in the P site **(Fig. 22-15)**. This reaction lengthens the peptidyl group by one amino acid at its C-terminal end. Thus, *a polypeptide grows in the N → C direction.* No external source of free energy is required for transpeptidation because the free energy of the broken ester bond of the peptidyl–tRNA is comparable to the free energy of the newly formed peptide bond. (Recall, though, that ATP was consumed in charging the tRNA with an amino acid.)

The peptidyl transferase active site lies in a highly conserved region of the 50S subunit, and the newly formed peptide bond is about 18 Å away from the nearest protein. Thus, the ribosome is a ribozyme (an RNA catalyst). How does rRNA catalyze peptide bond formation? Two highly conserved rRNA nucleotides, G2447 and A2451 in *E. coli*, do not function as acid–base catalysts, as was initially proposed. Rather, these residues help position the substrates for reaction, an example of induced fit (Section 6-3). Binding of a tRNA at the A site triggers a conformational change that exposes the ester bond of the peptidyl–tRNA in the P site. At other times, the ester bond must be protected so that it does not react with water, a reaction that would prematurely terminate protein synthesis. Proximity and orientation effects in the ribosome increase the rate of peptide bond formation about 10^7-fold above the uncatalyzed rate. Some antibiotics exert their effects by binding to the peptidyl transferase active site to directly block protein synthesis (Box 22-B).

During transpeptidation, the peptidyl group is transferred to the tRNA in the A site, and the P-site tRNA becomes deacylated. The new peptidyl–tRNA then moves into the P site, and the deacylated tRNA moves into the E site. The mRNA, which is still base

BOX 22-B   BIOCHEMISTRY NOTE

Antibiotic Inhibitors of Protein Synthesis

Antibiotics interfere with a variety of cellular processes, including cell-wall synthesis, DNA replication, and RNA transcription. Some of the most effective antibiotics, including many in clinical use, target protein synthesis. Because bacterial and eukaryotic ribosomes and translation factors differ, these antibiotics can kill bacteria without harming their mammalian hosts.

Translation-blocking antibiotics have also proved useful in the laboratory as probes of ribosomal structure and function. For example, puromycin resembles the 3′ end of Tyr–tRNA and competes with aminoacyl–tRNAs for binding to the ribosomal A site.

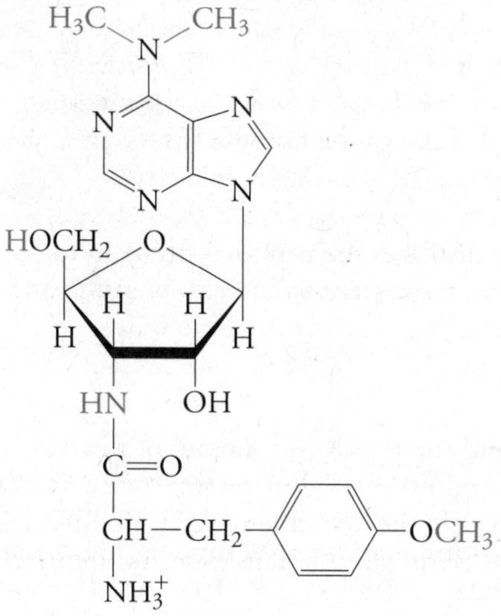

Puromycin

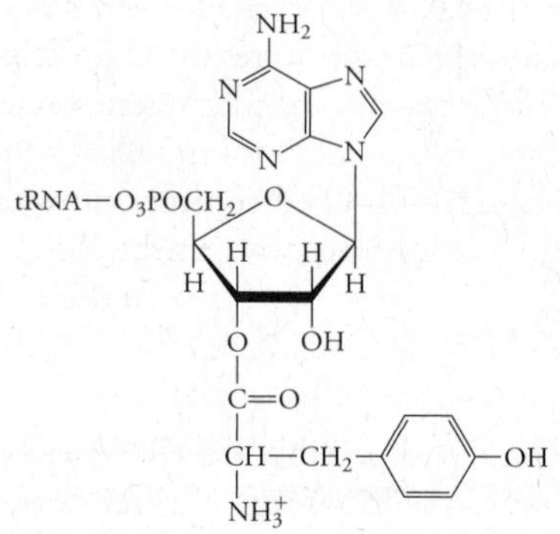

Tyrosyl–tRNA

Transpeptidation generates a puromycin–peptidyl group that cannot be further elongated because the puromycin "amino acid" group is linked by an amide bond rather than an ester bond to its "tRNA" group. As a result, peptide synthesis comes to a halt.

The antibiotic chloramphenicol blocks protein synthesis by binding at the peptidyl transferase active site.

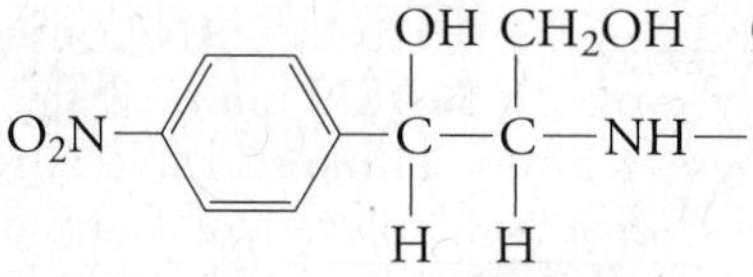

Chloramphenicol

X-Ray studies indicate that this relatively small compound interacts with the active-site nucleotides, including the catalytically essential A2451, to prevent transpeptidation.

Other antibiotics with more complicated structures interfere with protein synthesis through different mechanisms. For example, erythromycin physically blocks the tunnel that conveys the nascent polypeptide away from the active site. Six to eight peptide bonds form before the constriction of the exit tunnel blocks further chain elongation.

Streptomycin kills cells by binding tightly to the backbone of the 16S rRNA and stabilizing an error-prone conformation of the ribosome. In the presence of the antibiotic, the ribosome's affinity for aminoacyl–tRNAs increases, which increases the likelihood of codon–anticodon mispairing and therefore increases the error rate of translation. Presumably, the resulting burden of inaccurately synthesized proteins kills the cell.

The drugs described here, like all antibiotics, lose their effectiveness when their target organisms become resistant to them. Resistance can be acquired in a variety of ways. For example, mutations in ribosomal components can prevent antibiotic binding. In fact, mapping the locations of such mutations in ribosomal RNA has been instrumental in elucidating ribosomal function. Alternatively, an antibiotic-susceptible organism may acquire a gene, often present on an extrachromosomal plasmid, whose product inactivates the antibiotic. Acquisition of a gene encoding a particular acetyl-transferase, for example, leads to the addition of an acetyl group to chloramphenicol, which prevents its binding to the ribosome. Acquisition of a gene for an ABC transporter (Section 9-3) can hasten a drug's export from the cell, rendering it useless.

● **Question:** Explain why some of the side effects of antibiotic use in humans can be traced to impairment of mitochondrial function.

paired with the peptidyl–tRNA anticodon, advances through the ribosome by one codon. Experiments designed to assess the force exerted by the ribosome demonstrate that formation of a peptide bond causes the ribosome to loosen its grip on the mRNA, although it is not clear how events at the peptidyl transferase site in the large subunit are communicated to the mRNA decoding site in the small subunit. *The movement of tRNA and mRNA, which allows the next codon to be translated, is known as* **translocation**. This dynamic process requires the G protein called elongation factor G (EF-G) in *E. coli*.

EF-G bears a striking resemblance to the EF-Tu–tRNA complex (**Fig. 22-16**), and the ribosomal binding sites for the two proteins overlap. Structural studies show that the EF-G–GTP complex physically displaces the peptidyl–tRNA in the A site, causing it to translocate to the P site. This movement would also bump the deacylated tRNA from the P site to the E site. EF-G binding to the ribosome stimulates its GTPase activity. After EF-G hydrolyzes its bound GTP, it dissociates from the ribosome, leaving a vacant A site available for the arrival of another aminoacyl–tRNA and another round of transpeptidation.

The GTPase activity of G proteins such as EF-Tu and EF-G allows the ribosome to cycle efficiently through all the steps of translation elongation. Because GTP hydrolysis is irreversible, the elongation reactions—aminoacyl–tRNA binding, transpeptidation, and translocation—proceed unidirectionally. The *E. coli* ribosomal elongation cycle is shown in **Figure 22-17**. Eukaryotic cells contain elongation factors that function similarly to EF-Tu and EF-G. These G proteins are continually recycled during protein synthesis. In some cases, accessory proteins help replace bound GDP with GTP to prepare the G protein for another reaction cycle.

Figure 22-16 Structure of EF-G from *T. thermophilus*. [Structure (pdb 2BV3) determined by S. Hansson, R. Singh, A. T. Gudkov, A. Liljas, and D. T. Logan.]

[?] **Compare the size and shape of this complex to the size and shape of the complex containing EF-Tu and an aminoacyl-tRNA (see Fig. 22-12).**

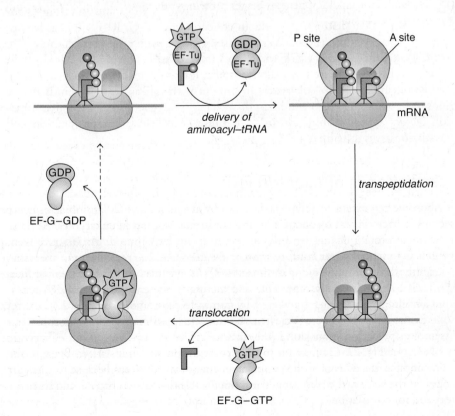

Figure 22-17 The ribosomal elongation cycle.

Release factors mediate translation termination

As the peptidyl group is lengthened by the transpeptidation reaction, it exits the ribosome through a tunnel in the center of the large subunit. The tunnel, about 100 Å long and 15 Å in diameter in the bacterial ribosome, shelters a polypeptide chain of up to 30 residues. The tunnel is defined by ribosomal proteins as well as the 23S rRNA. A variety of groups—including rRNA bases, backbone phosphate groups, and protein side chains—form a mostly hydrophilic surface for the tunnel. There are no large hydrophobic patches that could potentially impede the exit of a newly synthesized peptide chain.

Translation ceases when the ribosome encounters a stop codon (see Table 22-1). With a stop codon in the A position, the ribosome cannot bind an aminoacyl–tRNA but instead binds a protein known as a **release factor (RF).** In bacteria such as *E. coli*, RF-1 recognizes stop codons UAA and UAG, and RF-2 recognizes UAA and UGA. In eukaryotes, one protein—called eRF1—recognizes all three stop codons.

The release factor must specifically recognize the mRNA stop codon; it does this via an "anticodon" sequence of three amino acids, such as Pro–Val–Thr in RF-1 and Ser–Pro–Phe in RF-2, that interact with the first and second bases of the stop codon. At the same time, a loop of the release factor with the conserved sequence Gly–Gly–Gln projects into the peptidyl transferase site of the 50S subunit (**Fig. 22-18**). The amide group of the Gln residue promotes transfer of the peptidyl group from the P-site tRNA to water, apparently by stabilizing the transition state of this hydrolysis reaction. The product of the reaction is an untethered polypeptide that can exit the ribosome. At one time, release factors were believed to act by mimicking tRNA molecules, as EF-G does. However, it now appears that release factors undergo conformational changes upon binding to the ribosome, so they don't operate straightforwardly as tRNA surrogates.

In *E. coli*, an additional RF (RF-3), which binds GTP, promotes the binding of RF-1 or RF-2 to the ribosome. In eukaryotes, eRF3 performs this role. Hydrolysis of the GTP bound to RF-3 (or eRF3) allows the release factors to dissociate (**Fig. 22-19**). This leaves the ribosome with a bound mRNA, an empty A site, and a deacylated tRNA in the P site.

In bacteria, preparing the ribosome for another round of translation is the responsibility of a **ribosome recycling factor (RRF)** that works in concert with EF-G. RRF apparently slips into the ribosomal A site. EF-G binding then translocates RRF to the P site, thereby displacing the deacylated tRNA. Following GTP hydrolysis, EF-G and RRF dissociate from the ribosome, leaving it ready for a new round of translation initiation. In eukaryotes, a protein with ATPase activity binds to the ribosome containing eRF3 and helps dissociate the large and small subunits. Initiation factors appear to bind during this process, so translation termination and reinitiation are tightly linked.

Translation is efficient *in vivo*

A ribosome can extend a polypeptide chain by approximately 20 amino acids every second in bacteria and by about 4 amino acids every second in eukaryotes. At these rates, most protein chains can be synthesized in under a minute. As we have seen, various G proteins trigger conformational changes that keep the ribosome operating efficiently through many elongation cycles. Cells also maximize the rate of protein synthesis by forming **polysomes.** These structures represent a single mRNA molecule simultaneously being translated by multiple ribosomes (see **Fig. 22-20**). As soon as the first ribosome has cleared the initiation codon, a second ribosome can assemble and begin translating the mRNA. The circularization of eukaryotic mRNAs (see **Fig. 22-11**) may promote repeated rounds of translation. Because the stop codon at the 3′ end of the coding sequence may be relatively close to the start codon at the 5′ end, the ribosomal components released at termination can be easily recycled for reinitiation.

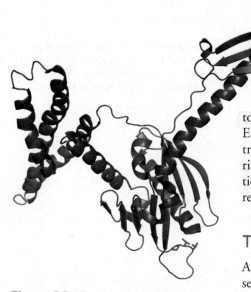

Figure 22-18 Structure of RF-1. The RF-1 protein is shown as a purple ribbon with its anticodon-binding Pro–Val–Thr (PVT) sequence in red and its peptidyl transferase–binding Gly–Gly–Gln (GGQ) sequence in orange. This image shows the structure of RF-1 as it exists when bound to a ribosome. [Structure of the *T. thermophilus* ribosome with RF-1 (pdb 3D5A) determined by M. Laurberg, H. Asahara, A. Korostelev, J. Zhu, S. Trakhanov, and H. F. Noller.]

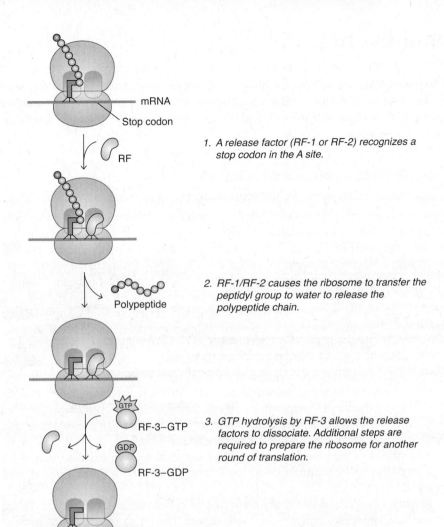

Figure 22-19 Translation termination in E. coli.

mRNA
Stop codon

RF

1. A release factor (RF-1 or RF-2) recognizes a stop codon in the A site.

Polypeptide

2. RF-1/RF-2 causes the ribosome to transfer the peptidyl group to water to release the polypeptide chain.

GTP
RF-3—GTP

GDP
RF-3—GDP

3. GTP hydrolysis by RF-3 allows the release factors to dissociate. Additional steps are required to prepare the ribosome for another round of translation.

(a)

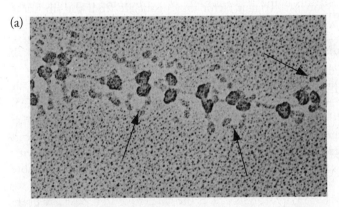

(b)

Figure 22-20 A polysome. (a) In this electron micrograph, a single mRNA strand encoding silkworm fibroin is studded with ribosomes. Arrows indicate growing fibroin polypeptide chains. [Courtesy Oscar L. Miller, Jr. and Steven L. McKnight, University of Virginia.]

(b) In this cryoelectron microscopy–based reconstruction of a polysome, ribosomes are blue and gold, and the emerging polypeptides are red and green. [Courtesy Wolfgang Baumeister and Julip Ortiz, from *Cell* 136, 261–271 (2009).]

CONCEPT REVIEW

- What determines the translation initiation and termination sites in prokaryotes and eukaryotes?
- Summarize the roles of initiation, elongation, and release factors in translation. Which of these proteins are G proteins?
- How does the ribosome prevent polymerization of the wrong amino acid?
- Explain how polysomes increase the rate of protein synthesis.

22-4 Post-Translational Events

The polypeptide released from a ribosome is not yet fully functional. For example, it must fold to its native conformation; it may need to be transported to another location inside or outside the cell; and it may undergo post-translational modification, or **processing.**

Chaperones promote protein folding

Studies of protein folding *in vitro* have revealed numerous insights into the pathways by which proteins (usually relatively small ones that have been chemically denatured) assume a compact globular shape with a hydrophobic core and a hydrophilic surface (see Section 4-3). Protein folding *in vivo* is only partly understood. For one thing, a protein can begin to fold as soon as its N-terminus emerges from the ribosome, even before it has been fully synthesized. In addition, a polypeptide must fold in an environment crowded with other proteins with which it might interact unfavorably. Finally, for proteins with quaternary structure, individual polypeptide chains must assemble with the proper stoichiometry and orientation. All of these processes may be facilitated in a cell by proteins known as **molecular chaperones.**

To prevent improper associations within or between polypeptide chains, chaperones bind to exposed hydrophobic patches on the protein surface (recall that hydrophobic groups tend to aggregate, which could lead to nonnative protein structure or protein aggregation and precipitation). *Many chaperones are ATPases that use the free energy of ATP hydrolysis to drive conformational changes that allow them to bind and release a polypeptide substrate while it assumes its native shape.* Chaperones were originally identified as heat-shock proteins (Hsp) because their synthesis is induced by high temperatures—conditions under which proteins tend to denature (unfold) and aggregate.

The first chaperone a bacterial protein meets, called trigger factor, is poised just outside the ribosome's polypeptide exit tunnel, bound to a ribosomal protein (**Fig. 22-21**). When trigger factor binds to the ribosome, it opens up to expose a hydrophobic patch facing the exit tunnel. Hydrophobic segments of the emerging polypeptide bind to this patch and are thereby prevented from sticking to each other or to other cellular components. Trigger factor may dissociate from the ribosome but remain associated with the nascent polypeptide until another chaperone takes over. Eukaryotes lack trigger factor, although they have other small heat-shock proteins that function in the same manner to protect newly made proteins. These chaperones are extremely abundant in cells, so there is at least one per ribosome.

Trigger factor may hand off a new polypeptide to another chaperone, such as DnaK in *E. coli* (it was named when it was believed to participate in DNA synthesis). DnaK and other heat-shock proteins in prokaryotes and eukaryotes interact with newly synthesized polypeptides and with existing cellular proteins and therefore can prevent as well as reverse improper folding. These chaperones, in a complex with ATP, bind to a short extended polypeptide segment with exposed hydrophobic groups (they do not recognize folded proteins, whose hydrophobic groups are sequestered in the interior). ATP hydrolysis causes the chaperone to release the polypeptide. As the polypeptide folds, the heat-shock protein may repeatedly bind and release it.

Ultimately, protein folding may be completed by multisubunit chaperones, called chaperonins, which form cagelike structures that physically sequester a folding polypeptide. The best-known chaperonin complex is the GroEL/GroES complex of *E. coli*. Fourteen GroEL subunits form two rings of seven subunits, with each ring enclosing a 45-Å-diameter chamber that is large enough to accommodate a folding polypeptide. Seven GroES subunits form a domelike cap for one GroEL chamber (**Fig. 22-22**). The GroEL ring nearest the cap is called the *cis* ring, and the other is the *trans* ring.

Each GroEL subunit has an ATPase active site. All seven subunits of the ring act in concert, hydrolyzing their bound ATP and undergoing conformational changes.

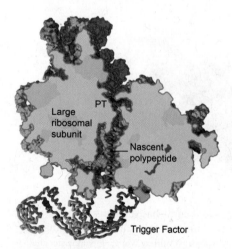

Figure 22-21 Trigger factor bound to a ribosome. The ribosome-binding portion of trigger factor is shown with its domains in different colors. It binds to the 50S ribosomal subunit where a polypeptide (magenta helix) emerges. PT represents the peptidyl transferase active site. [Courtesy Nenad Ban, Eidgenossische Technische Hochschule Honggerberg, Zurich.]

The two GroEL rings of the chaperonin complex act in a reciprocating fashion to promote the folding of two polypeptide chains in a safe environment (**Fig. 22-23**). Note that 7 ATP are consumed for each 10-second protein-folding opportunity. If the released substrate has not yet achieved its native conformation, it may rebind to the chaperonin complex. Only about 10% of bacterial proteins seem to require the GroEL/GroES chaperonin complex, and most of these range in size from 10 to 55 kD (a protein larger than about 70 kD probably could not fit inside the protein-folding chamber). Immunocytological studies indicate that some proteins never stray far from a chaperonin complex, perhaps because they tend to unfold and must periodically restore their native structures.

In eukaryotes, a chaperonin complex known as TRiC functions analogously to bacterial GroEL, but it has eight subunits in each of its two rings, and its fingerlike projections take the place of the GroES cap.

The signal recognition particle targets some proteins for membrane translocation

For a cytosolic protein, the journey from a ribosome to the protein's final cellular destination is straightforward (in fact, the journey may be short, since some mRNAs are directed to specific cytosolic locations before translation commences). In contrast, an integral membrane protein or a protein that is to be secreted from the cell follows a different route since it must pass partly or completely through a membrane.

In prokaryotic cells, membrane and secretory proteins are synthesized by ribosomes in the cytosol and are subsequently ushered to or through the plasma membrane. In eukaryotes, the insertion of most proteins into a membrane occurs cotranslationally, that is, as the polypeptide is being elongated by a ribosome. However, membrane translocation is fundamentally similar in all cells and usually requires a ribonucleoprotein known as the **signal recognition particle (SRP)**. *The SRP directs certain proteins to the plasma membrane (in bacteria) or the endoplasmic reticulum (in eukaryotes).*

Figure 22-22 The GroEL/GroES chaperonin complex. The two seven-subunit GroEL rings, viewed from the side, are colored red and yellow. A seven-subunit GroES complex (blue) caps the so-called *cis* GroEL ring. [Structure (pdb 1AON) determined by Z. Xu, A. L. Horwich, and P. B. Sigler.]

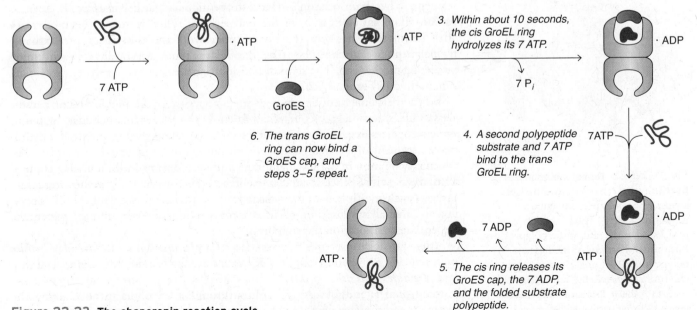

1. A GroEL ring binds 7 ATP and an unfolded polypeptide, which associates with hydrophobic patches on the GroEL subunits.

2. A GroES cap binds, triggering a conformational change that retracts the hydrophobic patches, thereby releasing the polypeptide into the GroEL chamber, where it can fold.

3. Within about 10 seconds, the cis GroEL ring hydrolyzes its 7 ATP.

4. A second polypeptide substrate and 7 ATP bind to the trans GroEL ring.

5. The cis ring releases its GroES cap, the 7 ADP, and the folded substrate polypeptide.

6. The trans GroEL ring can now bind a GroES cap, and steps 3–5 repeat.

7 ATP · ATP · ATP · GroES · 7 Pᵢ · ADP · 7ATP · ADP · 7 ADP · ATP

Figure 22-23 The chaperonin reaction cycle.

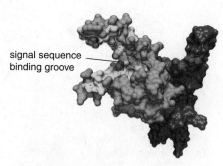

signal sequence
binding groove

Figure 22-24 **The SRP signal peptide binding domain.** This model shows the molecular surface of a portion of the *E. coli* signal recognition particle. The protein is magenta with hydrophobic residues in yellow. Adjacent RNA phosphate groups are red, and the rest of the RNA is dark blue. A signal peptide binds in the SRP groove, making hydrophobic as well as electrostatic contacts with the protein and RNA. [From R. Batey, *Science* 287, 1232–1239 (2000). Reprinted with permission from AAAS. Courtesy Robert Batey.]

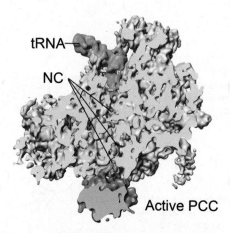

tRNA

NC

Active PCC

Figure 22-25 **Ribosome bound to Sec61.** In this cryoEM-based image, a yeast ribosome (partially cut away) is bound to Sec 61 (pink), which is positioned at the end of the polypeptide exit tunnel. Portions of the nascent polypeptide (NC, green) are visible in the exit tunnel. [From R. Beckmann, *Science* 326, 1369–1373 (2009). Reprinted with permission from AAAS. Courtesy Roland Beckmann.]

As in other ribonucleoproteins, including the ribosome and the spliceosome (see Section 21-3), the RNA component of the SRP is highly conserved and is essential for SRP function. The *E. coli* SRP consists of a single multidomain protein and a 4.5S RNA. The mammalian SRP contains a larger RNA and six different proteins, but its core is virtually identical to the bacterial SRP. The RNA component of the SRP includes several nonstandard base pairs, including G:A, G:G, and A:C, and interacts with several protein backbone carbonyl groups (most RNA–protein interactions involve protein side chains rather than the backbone).

How does the SRP recognize membrane and secretory proteins? Such proteins typically have an N-terminal **signal peptide** consisting of an α-helical stretch of 6 to 15 hydrophobic amino acids preceded by at least one positively charged residue. For example, human proinsulin (the polypeptide precursor of the hormone insulin) has the following signal sequence:

$$\text{M A L W M } \boxed{\text{R}} \boxed{\text{L L P L L A L L A L}} \text{ W G P D P A A A F V N} \cdots$$

The hydrophobic segment and a flanking Arg residue are shaded in green and pink.

The signal peptide binds to the SRP in a pocket formed mainly by a methionine-rich protein domain. The flexible side chains of the hydrophobic Met residues allow the pocket to accommodate helical signal peptides of variable sizes and shapes. In addition to the Met residues, the SRP binding pocket contains a segment of RNA, whose negatively charged backbone interacts electrostatically with the positively charged N-terminus and basic residue of the signal peptide (**Fig. 22-24**).

Electron microscopic studies indicate that in eukaryotes, the SRP binds to the ribosome at the polypeptide exit tunnel. When a signal peptide emerges from the ribosome, the SRP binds it and undergoes a conformational change that is transmitted to the rest of the ribosome so that translation elongation is halted. The ribosome–SRP complex then docks at a receptor on the endoplasmic reticulum membrane. When translation resumes, the growing polypeptide is translocated through the membrane. The prokaryotic SRP docks a full-length polypeptide (already free of the ribosome) to the membrane-translocating machinery, called a **translocon.** In all cases, GTP hydrolysis by the SRP is an essential step of this process.

The translocon proteins, called SecY in prokaryotes and Sec61 in eukaryotes, form a transmembrane channel with a constricted pore that limits the diffusion of other substances across the membrane (**Fig. 22-25**). During protein translocation, the pore must enlarge to a diameter of at least 20 Å to accommodate a polypeptide segment. When translocation occurs co-translationally, the driving force is provided by the ribosomal elongation of the polypeptide. When translocation takes place after the ribosome has completed polypeptide synthesis (as occurs in prokaryotes), the polypeptide chain must be driven through the translocon by an ATP-consuming ratcheting mechanism. The steps of translocating a eukaryotic secretory protein are summarized in **Figure 22-26**.

When the signal peptide emerges on the far side of the membrane, it may be cleaved off by an integral membrane protein known as a signal peptidase. This enzyme recognizes extended polypeptide segments such as those flanking the hydrophobic segment of a signal peptide, but it does not recognize α-helical structures, which are common in mature membrane proteins. In some integral membrane proteins, the signal sequence is not cleaved off and remains anchored in the membrane. Proteins with multiple membrane-spanning segments interact with the translocon in a way that allows insertion of discrete polypeptide segments into the membrane rather than translocation through it.

After translocation, *chaperones and other proteins in the endoplasmic reticulum may help the polypeptide fold into its native conformation, form disulfide bonds, and assemble with other protein subunits.* Extracellular proteins are then transported from the endoplasmic reticulum through the Golgi apparatus and to the plasma membrane via vesicles. The proteins may undergo processing (described next) en route to their final destination. Not all membrane and secretory proteins follow the SRP-mediated pathway described above. For example, some translocated proteins lack signal sequences. In addition, some eukaryotic proteins, particularly mitochondrial proteins, are translocated across membranes post-translationally, as occurs in bacteria.

Many proteins undergo covalent modification

Proteolysis is part of the maturation pathway of many proteins. For example, after it has entered the endoplasmic reticulum lumen and had its signal peptide removed and its Cys side chains cross-linked as disulfides, the insulin precursor undergoes proteolytic processing. The prohormone is cleaved at two sites to generate the mature hormone (Fig. 22-27).

Many extracellular eukaryotic proteins are **glycosylated** at Asn, Ser, or Thr side chains to generate glycoproteins (see Section 11-3). The short sugar chains (oligosaccharides) attached to glycoproteins may protect the proteins from degradation or mediate molecular recognition events. Methyl, acetyl, and propionyl groups may be added to various side chains. N-terminal groups are frequently modified by acetylation (up to 80% of human proteins), and C-terminal groups by amidation. Fatty acyl chains and other lipid groups are added to proteins to anchor them to membranes (Section 8-3). The addition and removal of phosphoryl groups is a powerful mechanism for allosterically regulating cellular signaling components (Section 10-2) and metabolic pathways (Section 19-2).

Modification of a protein by covalently attaching another protein occurs during protein degradation, when ubiquitin is covalently linked to a target protein (Section 12-1). A related protein, called SUMO (for *s*mall *u*biquitin-like *mo*difier), is also covalently attached to a target protein's Lys side chains, but rather than mark the protein for degradation, as ubiquitin does, SUMO is involved in various other processes, including protein transport into the nucleus.

All the post-translational modifications mentioned above are catalyzed by specific enzymes, which act more or less reliably depending on the nature of the modification and the cellular context. One consequence of post-translational processing therefore is that proteins may exhibit a great deal of variation beyond the sequence of amino acids that is specified by the genetic code.

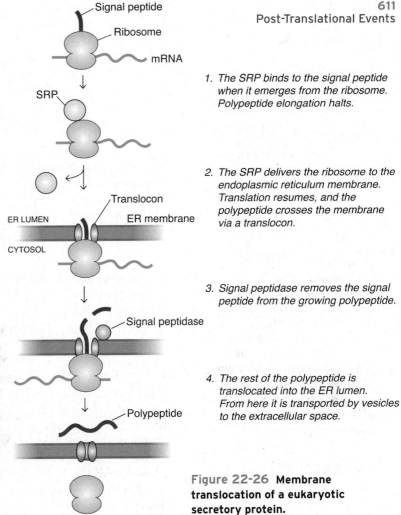

1. The SRP binds to the signal peptide when it emerges from the ribosome. Polypeptide elongation halts.

2. The SRP delivers the ribosome to the endoplasmic reticulum membrane. Translation resumes, and the polypeptide crosses the membrane via a translocon.

3. Signal peptidase removes the signal peptide from the growing polypeptide.

4. The rest of the polypeptide is translocated into the ER lumen. From here it is transported by vesicles to the extracellular space.

Figure 22-26 Membrane translocation of a eukaryotic secretory protein.

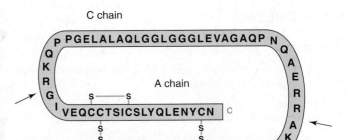

Figure 22-27 Conversion of proinsulin to insulin. The prohormone, with three disulfide bonds, is proteolyzed at two bonds (indicated by arrows) to eliminate the C chain. The mature insulin hormone consists of the disulfide-linked A and B chains.

> **CONCEPT REVIEW**
> - Compare protein folding as catalyzed by heat-shock proteins and by the chaperonin complex.
> - How does the SRP recognize a membrane or secretory protein?
> - What are the functions of the translocon and the signal peptidase?
> - Describe the types of reactions that a polypeptide may undergo following its synthesis.

[SUMMARY]

22-1 tRNA Aminoacylation

- The sequence of nucleotides in DNA is related to the sequence of amino acids in a protein by a triplet-based genetic code that must be translated by tRNA adaptors.
- tRNA molecules have similar L-shaped structures with a three-base anticodon at one end and an attachment site for a specific amino acid at the other end.
- Attachment of an amino acid to a tRNA is catalyzed by an aminoacyl–tRNA synthetase in a reaction that requires ATP. Various proofreading mechanisms ensure that the correct amino acid becomes linked to the tRNA.

22-2 Ribosome Structure

- The ribosome, the site of protein synthesis, consists of two subunits containing both rRNA and protein. The ribosome includes a binding site for mRNA and three binding sites (called the A, P, and E sites) for tRNA.

22-3 Translation

- Translation of mRNA requires an initiator tRNA bearing methionine (formyl-methionine in bacteria). Proteins known as initiation factors facilitate the separation of the ribosomal subunits and their reassembly with the initiator tRNA and an mRNA to be translated.

- During the elongation phase of protein synthesis, an elongation factor (EF-Tu in *E. coli*) interacts with aminoacyl–tRNAs and delivers them to the A site of the ribosome. Correct pairing between the mRNA codon and the tRNA anticodon allows the EF-Tu to hydrolyze its bound GTP and dissociate from the ribosome.
- Transpeptidation, or formation of a peptide bond, is catalyzed by rRNA in the large ribosomal subunit. The growing polypeptide chain becomes attached to the tRNA in the A site, which then moves to the P site. This movement is assisted by a GTP-binding protein elongation factor (EFG in *E. coli*).
- Translation terminates when a release factor recognizes an mRNA stop codon in the A site of the ribosome. Additional factors prepare the ribosome for another round of translation.

22-4 Post-Translational Events

- Chaperone proteins bind newly synthesized polypeptides to facilitate their folding. Large chaperonin complexes form a barrel-shaped structure that encloses a folding protein.
- Proteins to be secreted must pass through a membrane. An RNA–protein complex called a signal recognition particle directs polypeptides bearing an N-terminal signal sequence to a membrane for translocation.
- Additional modifications to newly synthesized proteins include proteolytic processing and the attachment of carbohydrate, lipid, or other groups.

[GLOSSARY TERMS]

translation	P site	polysome
codon	E site	processing
reading frame	initiation factor (IF)	molecular chaperone
degenerate code	G protein	signal recognition particle (SRP)
anticodon	elongation factor (EF)	signal peptide
isoacceptor tRNA	transpeptidation	translocon
wobble hypothesis	translocation	glycosylation
ribosome	release factor (RF)	
A site	ribosome recycling factor (RRF)	

[PROBLEMS]

22-1 tRNA Aminoacylation

1. How many combinations of four nucleotides are possible with a hypothetical quadruplet code?

2. Explain why a degenerate genetic code helps protect an organism from the effects of mutations.

3. Marshall Nirenberg and his colleagues deciphered the genetic code in the early 1960s. Their experimental strategy involved constructing RNA templates by using various ribonucleotides and polynucleotide phosphorylase, an enzyme that links available nucleotides together in random order. What protein sequence was obtained when the following templates were added to a cell-free translation system? **(a)** poly(U) **(b)** poly(C) **(c)** poly(A).

4. The experimental strategy described in Problem 3 was used to synthesize poly(UA). Note that the incorporation of nucleotides by polynucleotide phosphorylase is random; therefore, all possible codons containing U and A could occur in the RNA template.

(a) What amino acids would be incorporated into a polypeptide synthesized by a cell-free translation system using this template?
(b) What amino acids would be incorporated into a protein when poly(UC) is used as the template?
(c) Repeat the exercise from part (b) for poly(UG).
(d) How do these results show that the genetic code is redundant?

5. The elucidation of the genetic code was completed using H. Gobind Khorana's method of synthesizing polynucleotides with precise rather than random sequences.

(a) What polypeptides are synthesized from a nonrandom poly(UAUC) template?
(b) Explain why a single RNA template can yield more than one polypeptide.

6. What polypeptide is synthesized from a nonrandom poly(AUAG) template? Compare your results with your solution to Problem 5.

7. Organisms differ in their codon usage. For example, in yeast, only 25 of the 61 amino acid codons are used with high frequency. The cells contain high concentrations of the tRNAs that pair best with those codons, that is, form standard Watson–Crick base pairs. Explain how a point mutation in a gene, which does not change the identity of the encoded amino acid, could decrease the rate of synthesis of the protein corresponding to that gene.

8. (a) Would you expect that the highly expressed genes in a yeast cell would have sequences corresponding to the cell's set of 25 preferred codons (see Problem 7)? Would you expect this to be the case for genes that are expressed only occasionally?

(b) The genomes of many bacterial species appear to contain genes acquired from other species, including mammals. Even when a gene's function cannot be identified, the gene's nonbacterial origin can be recognized. Explain.

9. The 5′ nucleotide of a tRNA anticodon is often a nonstandard nucleotide such as a methylated guanosine. Why doesn't this interfere with the ribosome's ability to read the genetic code?

10. In this chapter, we have frequently compared eukaryotic translation to *bacterial* translation rather than to *prokaryotic* translation in general. This distinction is intentional, because the protein-synthesizing machinery in Archaea is more similar to the eukaryotic system than to the system in Bacteria. Is this consistent with the evolutionary scheme outlined in Figure 1-15?

11. Draw the "wobble" base pair that forms between inosine (I) and adenosine.

I

12. A new tRNA was discovered in *E. coli* in which a uridine on the tRNA was modified to form uridine-5′-oxyacetic acid (cmo^5U). The modified uridine can base pair with G, A, and U.

(a) What mRNA codons are recognized by $tRNA^{Leu}_{cmo^5UAG}$?

(b) How would the modified tRNA affect the sequence of a polypeptide being synthesized?

13. Some RNA transcripts are substrates for an adenosine deaminase. This "editing enzyme" converts adenosine residues to inosine residues, which can base pair with guanosine residues. Explain how the action of the deaminase could potentially increase the number of gene products obtained from a given gene.

14. In some types of cells, an aminoacyl–tRNA synthetase appears to play a secondary role in promoting certain events in RNA splicing (see Section 21-3).

(a) What structural feature of the synthetase is most likely involved in both the aminoacylation and splicing reactions?

(b) The aminoacyl–tRNA synthetases are among the oldest proteins. Why is this relevant to the presence of multiple activities in a single synthetase molecule?

15. Why doesn't GlyRS need a proofreading domain?

16. IleRS uses a double-sieve mechanism to accurately produce Ile–tRNA^{Ile} and prevent the synthesis of Val–tRNA^{Ile}. Which other pairs of amino acids differ in structure by a single carbon and might have AARSs that use a similar double-sieve proofreading mechanism?

17. An examination of AlaRS (the enzyme that attaches alanine to tRNA^{Ala}) suggests that the enzyme's aminoacylation active site cannot discriminate between Ala, Gly, and Ser.

(a) List all the products of the AlaRS reaction.

(b) A double-sieve mechanism like the one in IleRS cannot entirely solve the problem of misacylated tRNA^{Ala}. Explain.

(c) Many organisms express a protein called AlaXp, which is a soluble analog of the AlaRS editing domain that is able to hydrolyze Ser–tRNA^{Ala} but not Gly–tRNA^{Ala}. What is the purpose of AlaXp?

18. A tRNA molecule cannot be aminoacylated unless it bears a 3′ CCA sequence. Many tRNA precursors are synthesized without this sequence, so a CCA-adding enzyme must append the three nucleotides to the 3′ end of the immature tRNA molecule.

(a) The CCA-adding enzyme does not require a polynucleotide template. What does this imply about the mechanism for adding the CCA sequence?

(b) What can you conclude about the substrate specificity of the CCA-adding enzyme?

(c) Most CCA-adding enzymes consist of a single polymerase domain, but in one species of bacteria, the enzyme has two polymerase domains. Explain how this CCA-adding enzyme operates.

19. Predict the effect on a protein's structure and function for all possible nucleotide substitutions at the first position of a Lys codon in the gene encoding the protein.

20. Protein engineers who study the effects of nonstandard amino acids on protein structure and function can use a cell-based system for synthesizing proteins containing nonstandard amino acids. In theory, a polypeptide containing the amino acid norleucine could be produced by cells growing in media containing high concentrations of norleucine and lacking norleucine's standard counterpart, leucine.

Norleucine

Experimental results have shown that peptides containing norleucine in place of leucine were not produced unless the cells contained a mutant LeuRS. Explain these results.

22-2 Ribosome Structure

21. The sequences for all the ribosomal proteins in *E. coli* have been elucidated and have been found to contain large amounts of Lys and Arg residues. Why is this finding not surprising? What kinds of interactions are likely to form between the ribosomal proteins and the ribosomal RNA?

22. In eukaryotes, the primary rRNA transcript is a 45S rRNA that includes the sequences of the 18S, 5.8S, and 28S rRNAs separated by short spacers. What is the advantage of this operon-like arrangement of rRNA genes?

23. The sequence of ribosomal RNA is highly conserved, even though there are many rRNA genes in the genome. How does this observation argue for a functional (not just structural) role for rRNA?

24. Ribosomal inactivating proteins (RIPs) are RNA *N*-glycosidases found in plants. They catalyze the hydrolysis of specific adenine residues in RNA. RIPs are highly toxic but might be useful as antitumor drugs because ribosome synthesis is upregulated in transformed cells. Give a general explanation that describes how RIPs inactivate ribosomes.

25. What is the effect of adding EDTA, a chelating agent specific for divalent cations, to a bacterial cell extract carrying out protein synthesis?

26. In an experiment, >95% of the proteins are extracted from the 50S ribosomal subunit. Only the 23S rRNA and some protein fragments remain. Peptidyl transferase activity is unaffected. What can you conclude from these results? Propose a role for the protein fragments left behind. Why might it have been difficult to remove them?

27. Ribosomal proteins from the small subunit are designated "S" and those from the large subunit are designated "L." The S or L prefix is followed by a number that indicates their position (from upper left to lower right) on a two-dimensional electrophoretic gel (a technique that separates proteins based on both charge and size). One of the proteins in the large ribosomal subunit is sometimes acetylated at its N-terminus. Explain why two-dimensional electrophoresis of ribosomal proteins yields two spots corresponding to this protein.

28. Like their protein counterparts, RNA molecules fold into a variety of structural motifs. Ribosomal proteins contain a so-called RNA-recognition motif. Rho factor and the poly(A)-binding protein contain this same motif. Why is this observation not surprising?

22-3 Translation

29. In the diagram of a gene and its mRNA (see Fig. 21-20), indicate the approximate locations of the start and stop codons.

30. Complete the following table about eukaryotic replication, transcription, and translation.

Process	Replication	Transcription	Translation
Substrates			
Product			
Template or guide			
Primer			
Enzyme			
Cellular location			

31. The direction of protein synthesis was determined by carrying out an experiment in a cell-free system in which the mRNA consisted of a polymer of A residues with C at the 3′ end, as shown. What polypeptide was synthesized? What would be the result if the mRNA were read in the 3′ → 5′ direction? How does this directionality allow prokaryotes to begin translation before transcription is complete?

$$5'\ A\ A\ A\ A\ \cdots\ A\ A\ A\ C\ 3'$$

32. Mycobacteriophage genes occasionally begin with GUG or, even more rarely, UUG (rather than AUG). Which amino acids correspond to these codons?

33. The translation initiation sequence for the ribosomal protein L10 is shown below. Draw a diagram that shows how the Shine–Dalgarno sequence aligns with the appropriate sequence on the 16S rRNA. Identify the initiation codon.

$$5'–CUACCAGGAGCAAAGCUAAUGGCUUUA–3'$$

34. S1, a protein in the small ribosomal subunit, has a high affinity for single-stranded RNA and has been shown to be important in initiation. What role might S1 play during initiation?

35. What happens when colicin E3, which cleaves on the 5′ side of A1493 in the 16S rRNA, is added to a bacterial culture?

36. Explain why modification or mutagenesis of prokaryotic 16S rRNA at position 1492 or 1493 increases the error rate of translation.

37. The pairing of an aminoacyl–tRNA with EF-Tu offers an opportunity for proofreading during translation. EF-Tu binds all 20 aminoacyl–tRNAs with approximately equal affinity so that it can deliver them and surrender them to the ribosome with the same efficiency. Based on the experimentally determined binding constants for EF-Tu and correctly charged and mischarged aminoacyl–tRNAs (shown here), explain how the tRNA–EF-Tu recognition system could prevent the incorporation of the wrong amino acid in a protein.

Aminoacyl–tRNA	Dissociation constant (nM)
Ala–tRNA^{Ala}	6.2
Gln–tRNA^{Ala}	0.05
Gln–tRNA^{Gln}	4.4
Ala–tRNA^{Gln}	260

38. The affinity of a ribosome for a tRNA in the P site is about 50 times higher than for a tRNA in the A site. Explain why this promotes translational accuracy.

39. The bacterial elongation factors EF-Tu and EF-G are essential for translation *in vivo*, but bacterial ribosomes can translate mRNA into protein *in vitro* in the absence of EF-Tu and EF-G. Why are these factors not required *in vitro*? How does their absence affect the accuracy of translation?

40. Predict the effect on protein synthesis if EF-Tu were able to recognize and form a complex with fMet–tRNA$_f^{Met}$.

41. Identify the peptide encoded by the DNA sequence shown below (the lower strand is the template for RNA synthesis).

C G A T A A T G T C C G A C C A A G C G A T C T C G T A G C A
G C T A T T A C A G G C T G G T T C G C T A G A G C A T C G T

42. The sequence of a portion of the sense strand of a gene is shown below, along with the sequence of a mutant form of the gene.

wild-type	ACACCATGGTGCATCTGACT
mutant	ACACCATGGTTGCATCTGAC

(a) Give the polypeptide sequence that corresponds to the wild-type gene.
(b) How does the mutant gene differ from the wild-type gene, and how does the mutation affect the encoded polypeptide?

43. The gene for CFTR, the protein that is mutated in cystic fibrosis, contains 250,000 nucleotides. After splicing, the mature mRNA contains 6129 nucleotides. How many nucleotides are required to encode the 1480 residues of the protein? What is the function of the "extra" mRNA nucleotides?

44. Calculate the approximate energetic cost (in kJ) for the ribosomal synthesis of one mole of a 20-residue polypeptide. Why is the actual energetic cost *in vivo* probably higher than this value?

45. In prokaryotes, translation can begin even before an mRNA transcript has been completely synthesized. Why is co-transcriptional translation not possible in eukaryotes?

46. All cells contain an enzyme that hydrolyzes peptidyl–tRNA molecules that are not bound to a ribosome. Cells that are deficient in peptidyl–tRNA hydrolase grow very slowly. What is the function of this enzyme, which cleaves the peptidyl group from the tRNA? What does this tell you about the ability of ribosomes to carry out protein synthesis?

47. In an experimental system, the rate of the peptidyl transferase reaction varies with the identity of the amino acid residue attached to the tRNA in the P site, as shown in the table.

Peptidyl group	rate of peptidyl transfer (s^{-1})
Ala	57
Arg	90
Asp	8
Lys	100
Phe	16
Pro	0.14
Ser	44
Val	16

(a) Explain why transpeptidation involving Pro is much slower than for other amino acids.

(b) What can you conclude about the electrostatic environment of the peptidyl transferase active site?

(c) For the nonpolar amino acids, what factor seems to facilitate transpeptidation?

48. The peptidyl transferase center of the ribosome actually catalyzes two reactions involving the ester bond linking a polypeptide to the tRNA: aminolysis and hydrolysis. Explain.

49. The rate of the peptidyl transferase reaction increases as the pH increases from 6 to 8. Use your knowledge of the peptidyl transferase reaction mechanism to explain this result.

50. The events of transpeptidation resemble peptide bond hydrolysis in reverse (see Fig. 6-10).

(a) Draw the "tetrahedral intermediate" of the transpeptidation reaction.

(b) According to an early hypothesis about the catalytic action of rRNA, residue A2451 that has been protonated at position N1 could help stabilize the reaction intermediate. Draw this protonated adenine and explain how it might stabilize the tetrahedral intermediate. Would this catalytic mechanism be enhanced as the pH increases?

51. A "nonsense suppressor" mutation results from a mutation in a tRNA anticodon sequence so that the tRNA can pair with a stop codon (also called a nonsense codon).

(a) What would be the effect of a nonsense suppressor mutation on protein synthesis in the cell?

(b) Would all of the cell's proteins be affected by the mutation?

(c) Could the aminoacyl–tRNA synthetase that aminoacylates the nonmutated tRNA play a role in minimizing the effects of a nonsense suppressor mutation?

52. Many naturally occurring antibiotics target bacterial ribosomes in order to inhibit protein synthesis. In theory, any of the events of translation—including ribosome assembly, tRNA binding, codon–anticodon pairing, and peptide bond formation—are potential targets for antibiotics. The synthetic antibiotics known as oxazolidinones inhibit protein synthesis. Experiments show that these compounds do not directly inhibit peptide bond formation. They do inhibit translation initiation, but only at very high concentrations. This suggests that the oxazolidinones act by interfering with another aspect of translation. To test different possibilities, researchers used an *E. coli lacZ* expression system (*lacZ* codes for the enzyme β-galactosidase; see Section 21-1) and measured β-galactosidase enzymatic activity in the presence and absence of an oxazolidinone.

(a) First, the researchers engineered a stop codon into the N-terminal region of the polypeptide sequence and then measured β-galactosidase activity. What is the expected effect of the stop codon? Why was the stop codon introduced near the N-terminus of the protein rather than the C-terminus?

(b) The researchers observed that the level of β-galactosidase expressed from *lacZ* containing a stop codon was extremely low but increased about eightfold in the presence of the oxazolidinone. What does this suggest about the action of the oxazolidinone?

(c) Next, the researchers used a *lacZ* gene containing either a one-nucleotide insertion or a one-nucleotide deletion. How would these mutations affect the measured β-galactosidase activity?

(d) When the insertion and deletion mutations were tested in the presence of the antibiotic, β-galactosidase activity levels increased 15–25 times relative to levels in the absence of the antibiotic. What does this suggest about the action of the oxazolidinone?

(e) Next, the researchers measured β-galactosidase activity using *lacZ* genes in which the codon for an active-site Glu residue was mutated. Which codons specify Glu (see Table 22-1)? Describe how a single-base substitution in a Glu codon can give rise to codons specific for Ala, Gln, Gly, and Val.

(f) All of the genes with mutations at the Glu codon yielded very low levels of β-galactosidase activity, and the oxazolidinone had no effect. What does this tell about the action of the oxazolidinone?

(g) Based on your understanding of oxazolidinone action, explain why the compound inhibits bacterial growth.

(h) Oxazolidinones bind to a single site on the ribosome that is in the 50S subunit near the peptidyl transferase active site. Is this information consistent with the experimental results described above?

22-4 Post-Translational Events

53. In 1957, Christian Anfinsen carried out a denaturation experiment *in vitro* with ribonuclease, a pancreatic enzyme consisting of a single chain of 124 amino acids cross-linked by four disulfide bonds (see Problem 4-41). Urea (a denaturing agent) and 2-mercaptoethanol (a reducing agent) were added to a solution of purified ribonuclease, resulting in protein unfolding with a concomitant loss of biological activity. When urea and 2-mercaptoethanol were removed, the ribonuclease spontaneously folded back up to its native conformation and regained its full enzymatic activity. Why could proper protein folding occur in this experiment in the absence of molecular chaperones?

54. In another set of experiments, the lysine side chains on the surface of ribonuclease (see Problem 53) were covalently attached to an eight-residue chain of polyalanine. The presence of these polyalanine chains did not affect the ability of the ribonuclease to fold properly. What do these experiments reveal about the driving force for protein folding?

55. Chaperones are located not just in the cytosol but in the mitochondria as well. What is the role of mitochondrial chaperones?

56. Hsp90 is a chaperone that interacts with the tyrosine kinase domain of a growth factor receptor that has lost its ligand-binding domain but retains its tyrosine kinase domain (see Box 10-B). The antibiotic geldanamycin inhibits Hsp90 function. What is the effect of adding geldanamycin to cells expressing the abnormal growth factor receptor?

57. Multidomain proteins tend to fold better inside cagelike chaperonin structures (such as GroEL/GroES in *E. coli*) than with cytosolic chaperones. Explain why.

58. In immature red blood cells, globin synthesis is carefully regulated. The genes for α and β globin (see Section 5-1) are on separate chromosomes, and there are two α globin genes for every β globin gene. If too many β chains are produced, they form a functionally useless tetrameric hemoglobin. Excess α chains tend to precipitate and damage red blood cells.

(a) Explain why it is advantageous for the cell to synthesize a slight excess of α chains.

(b) Red blood cells express a protein that appears to stabilize the α chains and prevents their precipitation. Why is the stabilizing protein necessary?

(c) The disease β thalassemia results from a defect in a β globin gene. Heterozygotes (who have one normal and one abnormal gene) may develop a mild anemia, but homozygotes (who have two defective β globin genes) exhibit severe anemia. Explain why an extra copy of an α globin gene in an individual lacking one β globin gene would result in more severe anemia.

(d) Explain why a mutation in an α globin gene would reduce the severity of anemia in β thalassemia.

(e) Immature red blood cells produce a kinase that phosphorylates the ribosomal initiation factor eIF2. The phosphorylated eIF2 is unable to exchange bound GDP for GTP. How does this affect the rate of protein synthesis in the cell?

(f) In the presence of heme, the kinase is inactive. How does this mechanism regulate hemoglobin synthesis?

59. The N-terminal sequence of the secreted protein bovine pro-albumin is shown below. Identify the essential features of the signal peptide in this protein.

signal peptidase
cleavage site
↓

MKWVTFISLLLLFSSAYSRGV

60. The mammalian signal recognition particle (SRP) consists of one molecule of RNA and six proteins. In addition to interacting with the signal peptide, what might be the role of the RNA in the SRP?

61. One of the six proteins in the mammalian signal recognition particle (SRP) contains a cleft lined with hydrophobic amino acids. Propose a role for this protein in the SRP.

62. Why does the mammalian signal recognition particle (SRP) bind to the nascent polypeptide as it emerges from the ribosome? Why doesn't the SRP wait until translation is complete before escorting the new polypeptide to the ER membrane?

63. A eukaryotic "cell-free" translation system contains all the components required for protein synthesis—ribosomes; tRNAs; aminoacyl–tRNA synthetases; initiation, elongation, and termination factors; amino acids; GTP; and Mg^{2+} ions. An exogenous mRNA added to this mixture can direct protein synthesis *in vitro*. When an mRNA encoding a secretory protein is added to this cell-free system, along with the SRP, the entire protein is synthesized. When microsomes (sealed vesicles derived from ER membranes) are subsequently added, the protein is not translocated into the microsomal lumen and the signal sequence is not removed. What does this observation reveal about the role of SRP in the synthesis of secretory proteins?

64. The elongation factor EF-Tu serves a proofreading role by monitoring codon–anticodon base pairing during translation. When a match is confirmed, a conformational change occurs and EF-Tu hydrolyzes its bound GTP. Could the SRP use a similar mechanism to perform a proofreading function?

65. Polyglutamine diseases are neurodegenerative disorders caused by mutations in the DNA that produce triplet CAG repeats. Proteins translated from the mutated genes contain long stretches of glutamines that interfere with protein folding. Polyglutamine proteins tend to aggregate and form cytosolic and nuclear inclusion bodies. The result is a loss of neuron function, although the mechanism is unknown. Recent studies show that post-translational modification of the polyglutamine proteins may play a role in the progression of the disease. Interestingly, some post-translational modifications are neurotoxic, while others are protective.

(a) Protein kinase B (see Section 10-2) phosphorylates a polyglutamine protein on an essential Ser, resulting in decreased toxicity. Draw the structure of a phosphorylated serine residue.

(b) Proteins are marked for degradation by the attachment of the protein ubiquitin to a lysine on the condemned protein (see Section 12-1). An isopeptide bond forms between the lysine side chain and the carboxyl terminal group of the ubiquitin. Ubiquitination enhances the degradation of the inclusion body protein. Draw the structure of the linkage between ubiquitin and a polyglutamine protein.

(c) Polyglutamine proteins interact with a histone acetyltransferase (see Section 20-5), sequestering the enzyme in inclusion bodies and hastening its degradation. What are the cellular consequences of this interaction?

66. The protein c-Myc is a leucine zipper protein that regulates gene expression in cell proliferation and differentiation. Its activity was known to be regulated by phosphorylation on a specific threonine, but later studies showed that this same threonine could be modified by an *N*-acetylglucosamine residue and that phosphorylation and glycosylation were competitive processes. The specific threonine is mutated in some human lymphomas. Draw the structure of the *O*-glycosylated threonine residue.

67. In the *N*-myristoylation process, myristic acid (14:0) is attached to an N-terminal glycine residue of a protein during translation. Draw the structure of a myristoylated N-terminal glycine residue.

68. In the palmitoylation process, palmitate (16:0) is attached to the side chain of an internal Cys residue of a protein. Draw the structure of a palmitoylated Cys residue. What proteins involved in cell-signaling pathways include this modification, and what is the role of the palmitate?

[SELECTED READINGS]

Hartl, F. U., Bracher, A., and Hayer-Hartl, M., Molecular chaperones in protein folding and proteostasis, *Nature* **475**, 324–332 (2011). [Explains why constant surveillance is necessary to maintain protein homeostasis.]

Ibba, M., and Söll, D., Aminoacyl-tRNAs: setting the limits of the genetic code, *Genes Dev.* **18**, 731–738 (2004). [Discusses strategies organisms use to produce all types of aminoacyl–tRNAs.]

Moore, P. B., The ribosome returned, *J. Biol.* **8**, 8 (2009). [Reviews many aspects of translation, including the accuracy of decoding.]

Nirenberg, M., Historical review: deciphering the genetic code—a personal account, *Trends Biochem. Sci.* **29**, 46–54 (2004). [Describes the experimental approaches used to elucidate the genetic code.]

Rapoport, T. A. Protein translocation across the eukaryotic endoplasmic reticulum and bacterial plasma membranes, *Nature* **450**, 663–669 (2007). [Reviews the different ways for moving proteins into or through membranes.]

Schmeing, T. M., and Ramakrishnan, V., What recent ribosome structures have revealed about the mechanism of translation, *Nature* **461**, 1234–1242 (2009). [Discusses initiation, elongation, release, and recycling, including the proteins that interact with the ribosome at each step.]

A site. The ribosomal binding site that accommodates an aminoacyl–tRNA.

Abasic site. The deoxyribose residue remaining after the removal of a base from a DNA strand.

ABC transporter. A member of a family of structurally similar transmembrane proteins that use the free energy of ATP to drive conformational changes that move substances across the membrane.

Acid. A substance that can donate a proton.

Acid catalysis. A catalytic mechanism in which partial proton transfer from an acid lowers the free energy of a reaction's transition state.

Acid dissociation constant (K_a). The dissociation constant for an acid in water.

Acid–base catalysis. A catalytic mechanism in which partial proton transfer from an acid or partial proton abstraction by a base lowers the free energy of a reaction's transition state.

Acidic solution. A solution whose pH is less than 7.0 ($[H^+] > 10^{-7}$ M).

Acidosis. A pathological condition in which the pH of the blood drops below its normal value of 7.4.

Action potential. The momentary reversal of membrane potential that occurs during transmission of a nerve impulse.

Activation energy (free energy of activation, $\Delta G^{\ddagger}$). The free energy of the transition state minus the free energies of the reactants in a chemical reaction.

Activator. A protein that binds at or near a gene so as to promote its transcription.

Active site. The region of an enzyme in which catalysis takes place.

Active transport. The transmembrane movement of a substance from low to high concentrations by a protein that couples this endergonic transport to an exergonic process such as ATP hydrolysis. See also secondary active transport.

Acyl group. A portion of a molecule with the formula —COR, where R is an alkyl group.

Adipose tissue. Tissue consisting of cells that are specialized for the storage of triacylglycerols. See also brown adipose tissue.

A-DNA. A conformation of DNA in which the double helix is wider than the standard B-DNA helix and in which base pairs are inclined to the helix axis.

Aerobic. Occurring in or requiring oxygen.

Affinity chromatography. A procedure in which a molecule is isolated by its ability to bind specifically to a second immobilized molecule.

Agonist. A substance that binds to a receptor so as to evoke a cellular response.

Aldose. A sugar whose carbonyl group is an aldehyde.

Alkalosis. A pathological condition in which the pH of the blood rises above its normal value of 7.4.

Allele. An alternate form of a gene; a diploid organism may contain two alleles for each gene.

Allosteric protein. A protein in which the binding of ligand at one site affects the binding of other ligands at other sites. See also cooperative binding.

Allosteric regulation. Binding of an activator or inhibitor to one subunit of a multisubunit enzyme, which increases or decreases the catalytic activity of all the subunits. See also positive effector and negative effector.

α anomer. A sugar in which the OH substituent of the anomeric carbon is on the opposite side of the ring from the CH_2OH group of the chiral center that designates the D or L configuration.

α carbon. See Cα.

α helix. A regular secondary structure of polypeptides, with 3.6 residues per right-handed turn and hydrogen bonds between each backbone C=O group and the backbone N—H group that is four residues further.

α-amino acid. See amino acid.

Amido group. A portion of a molecule with the formula —CONH—.

Amino acid (α-amino acid). A compound consisting of a carbon atom to which are attached a primary amino group, a carboxylate group, a side chain (R group), and an H atom.

Amino group. A portion of a molecule with the formula —NH_2, —NHR, or —NR_2, where R is an alkyl group.

Amino terminus. See N-terminus.

Amphiphilic (amphipathic). Having both polar and nonpolar regions and therefore being both hydrophilic and hydrophobic.

Amyloid deposit. An accumulation of certain types of insoluble protein aggregates in tissues (e.g., in the brain in Alzheimer's disease).

Anabolism. The reactions by which biomolecules are synthesized from simpler components.

Anaerobic. Occurring independently of oxygen.

Anaplerotic reaction. A reaction that replenishes the intermediates of a metabolic pathway.

Anemia. A condition caused by the insufficient production of or loss of red blood cells.

Anneal. To allow base pairing between complementary single polynucleotide strands so that double-stranded segments form.

Anomers. Sugars that differ only in the configuration around the carbonyl carbon that becomes chiral when the sugar cyclizes.

Antagonist. A substance that binds to a receptor but does not elicit a cellular response.

Antenna pigment. A molecule that transfers its absorbed energy to other pigment molecules and eventually to a photosynthetic reaction center.

Anticodon. The sequence of three nucleotides in a tRNA that recognizes an mRNA codon through complementary base pairing.

Antiparallel. Running in opposite directions.

Antiparallel β sheet. See β sheet.

Antiport. Transport that involves the simultaneous transmembrane movement of two molecules in opposite directions.

Antisense strand. See noncoding strand.

Apoptosis. Programmed cell death that results from extracellular or intracellular signals and involves the activation of enzymes that selectively degrade cellular structures.

Archaea. One of the two major groups of prokaryotes.

Atherosclerosis. A disease characterized by the formation of cholesterol-containing fibrous plaques in the walls of blood vessels.

ATPase. An enzyme that catalyzes the hydrolysis of ATP to ADP + P_i.

Autoactivation. A process by which the product of an activation reaction also acts as a catalyst for the same reaction, so that it appears that the compound catalyzes its own activation.

Autophosphorylation. The phosphorylation of a kinase by another molecule of the same kinase.

Axon. The extended portion of a neuron that conducts an action potential from the cell body to a synapse with a target cell.

Backbone. The atoms that form the repeating linkages between successive residues of a polymeric molecule, exclusive of the side chains.

Bacteria. One of the two major groups of prokaryotes.

Bacteriophage. A virus specific for bacteria. Also known as a phage.

Base. (1) A substance that can accept a proton. (2) A purine or pyrimidine component of a nucleoside, nucleotide, or nucleic acid.

Base catalysis. A catalytic mechanism in which partial proton abstraction by a base lowers the free energy of a reaction's transition state.

Base excision repair. A DNA repair pathway in which a damaged base is removed by a glycosylase so that the resulting abasic site can be repaired.

Base pair. The specific hydrogen-bonded association between nucleic acid bases. The standard base pairs are A:T and G:C. See also bp.

Basic solution. A solution whose pH is greater than 7.0 ($[H^+] < 10^{-7}$ M).

B-DNA. The standard conformation of double-helical DNA.

β anomer. A sugar in which the OH substituent of the anomeric carbon is on the same side of the ring as the CH_2OH of the chiral center that designates the D or L configuration.

β barrel. A protein structure consisting of a β sheet rolled into a cylinder.

β oxidation. A series of enzyme-catalyzed reactions in which fatty acids are progressively degraded by the removal of two-carbon units as acetyl-CoA.

β sheet. A regular secondary structure in which extended polypeptide chains form interstrand hydrogen bonds. In parallel β sheets, the polypeptide chains all run in the same direction; in antiparallel β sheets, neighboring chains run in opposite directions.

Bilayer. An ordered, two-layered arrangement of amphiphilic molecules in which polar segments are oriented toward the two solvent-exposed surfaces and the nonpolar segments associate in the center.

Bile acid. A cholesterol derivative that acts as a detergent to solubilize lipids for digestion and absorption.

Bimolecular reaction. A reaction involving two molecules, which may be identical or different.

Binding change mechanism. The mechanism whereby the subunits of ATP synthase adopt three successive conformations to convert ADP + P_i to ATP as driven by the dissipation of the transmembrane proton gradient.

Biofilm. A complex of bacterial cells and a protective extracellular matrix containing polysaccharides.

Bioinformatics. The use of computers in collecting, storing, accessing, and analyzing biological data, such as molecular sequences and structures.

Bisubstrate reaction. An enzyme-catalyzed reaction involving two substrates.

Blue-white screening. A technique for distinguishing cells containing a β-galactosidase gene that has been interrupted by a foreign DNA segment such that it cannot give rise to an active enzyme that converts a substrate to a blue-colored product.

Blunt ends. The fully base-paired ends of a DNA fragment that are generated by a restriction endonuclease that cuts both strands at the same point.

Bohr effect. The decrease in O_2 binding affinity of hemoglobin in response to a decrease in pH.

bp. Base pair, the unit of length used for DNA molecules.

Brown adipose tissue. A type of adipose tissue in which fatty acid oxidation is uncoupled from ATP production so that the free energy of the fatty acids is released as heat.

Buffer. A solution of a weak acid and its conjugate base, which resists changes in pH upon the addition of acid or base.

Cα. The alpha carbon, the carbon of an amino acid whose substituents are an amino group, a carboxylate group, an H atom, and a variable R group.

Calvin cycle. The sequence of photosynthetic reactions in which ribulose-5-phosphate is carboxylated, converted to three-carbon carbohydrate precursors, and regenerated.

cAMP. Cyclic AMP, an intracellular second messenger.

Cap. A 7-methylguanosine residue that is post-transcriptionally added to the 5′ end of a eukaryotic mRNA.

Carbanion. A compound that bears a negative charge on a carbon atom.

Carbohydrate. A compound with the formula $(CH_2O)_n$, where $n \geq 3$. Also called a saccharide.

Carbon fixation. The incorporation of CO_2 into biologically useful organic molecules.

Carbonyl group. A portion of a molecule with the formula $\geq C = O$.

Carboxyl group. A portion of a molecule with the formula —COOH.

Carboxyl terminus. See C-terminus.

Carcinogen. An agent that causes a mutation in DNA that leads to cancer.

Carcinogenesis. The process of developing cancer.

Catabolism. The degradative metabolic reactions in which nutrients and cell constituents are broken down for energy and raw materials.

Catalyst. A substance that promotes a chemical reaction without undergoing permanent change. A catalyst increases the rate at which a reaction approaches equilibrium but does not affect the free energy change of the reaction.

Catalytic constant (k_{cat}). The ratio of the maximal velocity (V_{max}) of an enzyme-catalyzed reaction to the enzyme concentration. Also called a turnover number.

Catalytic perfection. A state achieved by an enzyme that operates at the diffusion-controlled limit.

Catalytic triad. The hydrogen-bonded Ser, His, and Asp residues that participate in catalysis in serine proteases.

cDNA. See complementary DNA.

CF. See cystic fibrosis.

C₄ pathway. A photosynthetic process used in some plants to concentrate CO_2 by incorporating it into oxaloacetate (a C_4 compound).

Chain terminator. A nucleotide lacking a 3′ OH group that is incorporated into a polynucleotide but cannot support further polymerization.

Channeling. The transfer of an intermediate product from one enzyme active site to another in such a way that the intermediate remains in contact with the protein.

Chaperone. See molecular chaperone.

Chemical labeling. A technique for identifying functional groups in a macromolecule by treating the molecule with a reagent that reacts with those groups.

Chemiosmotic theory. The postulate that the free energy of electron transport is conserved in the formation of a transmembrane proton gradient that can be subsequently used to drive ATP synthesis.

Chemoautotroph. An organism that obtains its building materials and free energy from inorganic compounds.

Chirality. The asymmetry or "handedness" of a molecule such that it cannot be superimposed on its mirror image.

Chloroplast. The plant organelle in which photosynthesis takes place.

Chromatin. The complex of DNA and protein that comprises the eukaryotic chromosomes.

Chromatography. A technique for separating the components of a mixture of molecules based on their partition between a mobile solvent phase and a porous matrix (stationary phase), often performed in a column.

Chromosome. The complex of protein and a single DNA molecule that comprises some or all of an organism's genome.

Citric acid cycle. A set of eight enzymatic reactions, arranged in a cycle, in which energy in the form of ATP, NADH, and QH_2 is recovered from the oxidation of the acetyl group of acetyl-CoA to CO_2.

Clinical trial. A three-phase series of tests of a drug's safety and effectiveness in human subjects.

Clone. An organism or collection of identical cells derived from a single parental cell.

Cloning vector. A DNA molecule, such as a plasmid, that can accommodate a segment of foreign DNA for cloning.

Coagulation. The process of forming a blood clot.

Coding strand. The DNA strand that has the same sequence (except for the replacement of U with T) as the transcribed RNA; it is the nontemplate strand. Also called the sense strand.

Codon. The sequence of three nucleotides in DNA or RNA that specifies a single amino acid.

Coenzyme. A small organic molecule that is required for the catalytic activity of an enzyme. A coenzyme may be tightly associated with the enzyme as a prosthetic group.

Cofactor. A small organic molecule (coenzyme) or metal ion that is required for the catalytic activity of an enzyme.

Coiled coil. An arrangement of polypeptide chains in which two α helices wind around each other.

Competitive inhibition. A form of enzyme inhibition in which a substance competes with the substrate for binding to the enzyme active site and thereby appears to increase K_M.

Complement. (1) A molecule that pairs in a reciprocal fashion with another. (2) A set of circulating proteins that sequentially activate each other and lead to the formation of a pore in a microbial cell membrane.

Complementary DNA (cDNA). A segment of DNA synthesized from an RNA template.

Condensation reaction. The formation of a covalent bond between two molecules, during which the elements of water are lost.

Conformation. The three-dimensional shape of a molecule that it attained through rotation of its bonds.

Conjugate acid. The compound that forms when a base accepts a proton.

Conjugate base. The compound that forms when an acid donates a proton.

Consensus sequence. A DNA or RNA sequence showing the nucleotides most commonly found at each position.

Conservative substitution. A change of an amino acid residue in a protein to one with similar properties (e.g., Leu to Ile or Asp to Glu).

Constitutive. Being expressed at a continuous, steady rate rather than induced.

Convergent evolution. The independent development of similar characteristics in unrelated species.

Cooperative binding. A situation in which the binding of a ligand at one site on a macromolecule affects the affinity of other sites for the same ligand. See also allosteric protein.

Cori cycle. A metabolic pathway in which lactate produced by glycolysis in the muscles is transported via the bloodstream to the liver, where it is used for gluconeogenesis. The resulting glucose returns to the muscles.

Covalent catalysis. A catalytic mechanism in which the transient formation of a covalent bond between the catalyst and a reactant lowers the free energy of a reaction's transition state.

CpG island. A cluster of CG sequences that often marks the beginning of a gene in a mammalian genome.

Cristae. The invaginations of the inner mitochondrial membrane.

Cross-talk. The interactions of different signal transduction pathways through activation of the same signaling components.

Cryoelectron microscopy. A variation of electron crystallography in which electron diffraction data from a molecular structure are collected at very low temperatures.

C-terminus. The end of a polypeptide that has a free carboxylate group.

Cyclic electron flow. The light-driven circulation of electrons between Photosystem I and cytochrome b_6f, which leads to the production of ATP but not NADPH.

Cystic fibrosis (CF). A genetic disease that is caused by a mutation in a gene for a membrane transport protein and is characterized by thick mucus and bacterial lung infections.

Cytochrome. A protein that carries electrons via a prosthetic Fe-containing heme group.

Cytokinesis. The splitting of the cell into two following mitosis.

Cytoplasm. The entire contents of a cell excluding the nucleus.

Cytoskeleton. The network of intracellular fibers that gives a cell its shape and structural rigidity.

Cytosol. The contents of a cell (cytoplasm) minus its nucleus and other membrane-bounded organelles.

D sugar. A monosaccharide isomer in which the asymmetric carbon farthest from the carbonyl group has the same spatial arrangement as the chiral carbon of D-glyceraldehyde.

Dark reactions. The photosynthetic reactions in which NADPH and ATP produced by the light reactions are used to incorporate CO_2 into carbohydrates.

ddNTP. A dideoxynucleoside triphosphate.

Deamination. The hydrolytic removal of an amino group.

Degenerate code. A code in which more than one "word" encodes the same entity.

$\Delta G^{\ddagger}$. See activation energy.

$\Delta G^{\circ\prime}$. See standard free energy change.

$\Delta G_{reaction}$. The difference in free energy between the reactants and products of a chemical reaction; $\Delta G_{reaction} = \Delta G_{products} - \Delta G_{reactants}$.

Denaturation. The loss of ordered structure in a polymer, such as the disruption of native conformation in an unfolded polypeptide or the unstacking of bases and separation of strands in a nucleic acid.

Denitrification. The conversion of nitrate (NO_3^-) to nitrogen (N_2).

Deoxyhemoglobin. Hemoglobin that does not contain bound oxygen or is not in the oxygen-binding conformation.

Deoxynucleotide. A nucleotide in which the pentose is 2′-deoxyribose.

Deoxyribonucleic acid. See DNA.

Desensitization. A cell's adaptation to long-term stimulation through a reduced response to the stimulus.

Diabetes mellitus. A disease caused by a deficiency of insulin or the inability to respond to insulin, and characterized by elevated levels of glucose in the blood.

Diazotroph. A bacterium that carries out nitrogen fixation, the conversion of N_2 to NH_3.

Dideoxy DNA sequencing. A technique for determining the nucleotide sequence of a DNA using dideoxy nucleotides so as to yield a collection of strands of all possible lengths.

Dielectric constant. A measure of the ability of a substance to interfere with electrostatic interactions; a solvent with a high dielectric constant is able to dissolve salts by shielding the attractive electrostatic forces that would otherwise bring the ions together.

Diffraction pattern. The record of the radiation scattered from an object, for example, in X-ray crystallography.

Diffusion-controlled limit. The theoretical maximum rate of an enzymatic reaction in solution, about 10^8 to 10^9 $M^{-1} \cdot s^{-1}$.

Dimer. An assembly consisting of two monomeric units.

Diploid. Having two equivalent sets of chromosomes.

Dipole–dipole interaction. A type of van der Waals interaction between two strongly polar groups.

Disaccharide. A carbohydrate consisting of two monosaccharides.

Discontinuous synthesis. A mechanism whereby the lagging strand of DNA is synthesized as a series of fragments that are later joined.

Dissociation constant (K). The ratio of the products of the concentrations of the dissociated species to those of their parent compounds at equilibrium.

Disulfide bond. A covalent —S—S— linkage, often between two Cys residues in a protein.

DNA. A polymer of deoxynucleotides whose sequence of bases encodes genetic information in all living cells.

DNA chip. See microarray.

DNA fingerprinting. A technique for distinguishing individuals on the basis of DNA polymorphisms, such as the number of short tandem repeats.

DNA ligase. An enzyme that catalyzes the formation of a phosphodiester bond to join two DNA strands.

DNA marker. An element of DNA structure, such as a gene or other known sequence, whose position on a chromosome is known.

dNTP. A deoxyribonucleoside triphosphate.

Domain. A stretch of polypeptide residues that fold into a globular unit with a hydrophobic core.

$\mathcal{E}$. See reduction potential.

$\mathcal{E}^{\circ\prime}$. See standard reduction potential.

E site. The ribosomal binding site that accommodates a deacylated tRNA before it dissociates from the ribosome.

Edman degradation. A procedure for the stepwise removal and identification of the N-terminal residues of a polypeptide.

EF. See elongation factor.

Ehlers–Danlos syndrome. A genetic disease characterized by elastic skin and joint hyperextensibility, caused by mutations in genes for collagen or collagen-processing proteins.

EI complex. The noncovalent complex that forms between an enzyme and a reversible inhibitor.

Eicosanoids. Compounds derived from the C_{20} fatty acid arachidonic acid, which act in or near the cells that produce them and mediate pain, fever, and other physiological responses.

Electron crystallography. A technique for determining molecular structure by analyzing the pattern of diffraction of a beam from an electron microscope. See also cryoelectron microscopy.

Electron tomography. A technique for reconstructing three-dimensional structures by analyzing electron micrographs of consecutive tissue slices.

Electron transport chain. A series of membrane-associated electron carriers that pass electrons from reduced coenzymes to molecular oxygen so as to recover free energy for the synthesis of ATP.

Electronegativity. A measure of an atom's affinity for electrons.

Electrophile. A compound containing an electron-poor center. An electrophile (electron-lover) reacts readily with a nucleophile (nucleus-lover).

Electrophoresis. A procedure in which macromolecules are separated on the basis of charge or size by their differential migration through a gel-like matrix under the influence of an applied electric field. In polyacrylamide gel electrophoresis (PAGE), the matrix is cross-linked polyacrylamide. In SDS-PAGE, the detergent sodium dodecyl sulfate is used to denature proteins.

Electrostatic catalysis. A catalytic mechanism in which sequestering the reacting groups away from the aqueous solvent lowers the free energy of a reaction's transition state.

Elongation factor (EF). A protein that interacts with tRNA and/or the ribosome during polypeptide synthesis.

Epigenetics. The inheritance of genetic information that does not depend on the sequence of the DNA.

Enantiomers. Stereoisomers that are nonsuperimposable mirror images of one another.

Endergonic reaction. A reaction that has an overall positive free energy change (a nonspontaneous process).

Endocytosis. The inward folding and budding of the plasma membrane to form a new intracellular vesicle.

Endonuclease. An enzyme that catalyzes the hydrolysis of the phosphodiester bonds between two nucleotide residues within a polynucleotide strand.

Endopeptidase. An enzyme that catalyzes the hydrolysis of a peptide bond within a polypeptide chain.

Enhancer. A eukaryotic DNA sequence located some distance from the transcription start site, where an activator of transcription may bind.

Enthalpy (H). A thermodynamic quantity that is taken to be equivalent to the heat content of a biochemical system.

Entropy (S). A measure of the degree of randomness or disorder of a system.

Enzyme. A biological catalyst. Most enzymes are proteins; a few are RNA.

Epimers. Sugars that differ only by the configuration at one C atom (excluding the anomeric carbon).

Equilibrium. The point in a process at which the forward and reverse rates are exactly balanced so that it undergoes no net change.

Equilibrium constant (K_{eq}). The ratio, at equilibrium, of the product of the concentrations of reaction products to that of the reactants.

ES complex. The noncovalent complex that forms between an enzyme and its substrate in the first step of an enzyme-catalyzed reaction.

Essential compound. An amino acid, fatty acid, or other compound that an animal cannot synthesize and must therefore obtain in its diet.

Ester group. A portion of a molecule with the formula —COOR, where R is an alkyl group.

Ether. A molecule with the formula ROR', where R and R' are alkyl groups.

Euchromatin. The transcriptionally active, relatively uncondensed chromatin in a eukaryotic cell.

Eukarya. See eukaryote.

Eukaryote. An organism consisting of a cell (or cells) whose genetic material is contained in a membrane-bounded nucleus.

Exciton transfer. A mode of decay of an energetically excited molecule, in which electronic energy is transferred to a nearby unexcited molecule.

Exergonic reaction. A reaction that has an overall negative free energy change (a spontaneous process).

Exocytosis. The fusion of an intracellular vesicle with the plasma membrane in order to release the contents of the vesicle outside the cell.

Exon. A portion of a gene that appears in both the primary and mature mRNA transcripts.

Exonuclease. An enzyme that catalyzes the hydrolytic excision of a nucleotide residue from the end of a polynucleotide strand.

Exopeptidase. An enzyme that catalyzes the hydrolytic excision of an amino acid residue from one end of a polypeptide chain.

Extrinsic protein. See peripheral membrane protein.

$\mathcal{F}$. See Faraday.

F-actin. The polymerized form of the protein actin. See also G-actin.

Factory model of replication. A model for DNA replication in which DNA polymerase and associated proteins remain stationary while the DNA template is spooled through them.

Faraday ($\mathcal{F}$). The charge of one mole of electrons, equal to 96,485 coulombs $\cdot$ mol^{-1} or 96,485 J $\cdot$ V^{-1} $\cdot$ mol^{-1}.

Fatty acid. A carboxylic acid with a long-chain hydrocarbon side group.

Feedback inhibitor. A substance that inhibits the activity of an enzyme that catalyzes an early step of the substance's synthesis.

Feed-forward activation. The activation of a later step in a reaction sequence by the product of an earlier step.

Fermentation. An anaerobic catabolic process.

Fibrous protein. A protein characterized by a stiff, elongated conformation, that tends to form fibers.

First-order reaction. A reaction whose rate is proportional to the concentration of a single reactant.

Fischer projection. A graphical convention for specifying molecular configuration in which horizontal lines represent bonds that extend above the plane of the paper and vertical bonds extend below the plane of the paper.

5' end. The terminus of a polynucleotide whose C5' is not esterified to another nucleotide residue.

Flip-flop. See transverse diffusion.

Flippase. See translocase.

Fluid mosaic model. A model of biological membranes in which integral membrane proteins float and diffuse laterally in a fluid lipid layer.

Fluorescence. A mode of decay of an excited molecule, in which electronic energy is emitted in the form of a photon.

Flux. The rate of flow of metabolites through a metabolic pathway.

Fractional saturation (Y). The fraction of a protein's ligand-binding sites that are occupied by ligand.

Free energy (G). A thermodynamic quantity whose change indicates the spontaneity of a process. For spontaneous processes, $\Delta G < 0$, whereas for a process at equilibrium, $\Delta G = 0$.

Free energy of activation. See activation energy.

Free radical. A molecule with an unpaired electron.

Futile cycle. Two opposing metabolic reactions that function together to provide a control point for regulating metabolic flux.

G. See free energy.

G-actin. The monomeric form of the protein actin. See also F-actin.

G protein. A guanine nucleotide–binding and –hydrolyzing protein, involved in a process such as signal transduction or protein synthesis, that is inactive when it binds GDP and active when it binds GTP.

Gas constant (R). A thermodynamic constant equivalent to 8.3145 J $\cdot$ K^{-1} $\cdot$ mol^{-1}.

Gated channel. A transmembrane channel that opens and closes in response to a signal such as changing voltage, ligand binding, or mechanical stress.

Gel filtration chromatography. See size-exclusion chromatography.

Gene. A unique sequence of nucleotides that encodes a polypeptide or RNA; it may include nontranscribed and nontranslated sequences that have regulatory functions.

Gene expression. The transformation by transcription and translation of the information contained in a gene to a functional RNA or protein product.

Gene therapy. The transfer of genetic material to the cells of an individual in order to produce a therapeutic effect.

General transcription factor. One of a set of eukaryotic proteins that are typically required for the synthesis of mRNAs.

Genetic code. The correspondence between the sequence of nucleotides in a nucleic acid and the sequence of amino acids in a polypeptide; a series of three nucleotides (a codon) specifies an amino acid.

Genetic engineering. See recombinant DNA technology.

Genome. The complete set of genetic instructions in an organism.

Genome map. A reconstruction of an organism's genome, based on DNA sequences and physical DNA markers.

Genome-wide association study (GWAS). An attempt to correlate genetic variations with a trait such as a particular disease.

Genomics. The study of the size, organization, and gene content of organisms' genomes.

Genotype. An organism's genetic characteristics.

Globin. The polypeptide component of myoglobin and hemoglobin.

Globular protein. A water-soluble protein characterized by a compact, highly folded structure.

Glucogenic amino acid. An amino acid whose degradation yields a gluconeogenic precursor. See also ketogenic amino acid.

Gluconeogenesis. The synthesis of glucose from noncarbohydrate precursors.

Glucose–alanine cycle. A metabolic pathway in which pyruvate produced by glycolysis in the muscles is converted to alanine and transported to the liver, where it is converted back to pyruvate for gluconeogenesis. The resulting glucose returns to the muscles.

Glycan. See polysaccharide.

Glycerophospholipid. An amphipathic lipid in which two fatty acyl groups and a polar phosphate derivative are attached to a glycerol backbone.

Glycogen storage disease. An inherited defect in an enzyme or transporter that affects the formation, structure, or degradation of glycogen.

Glycogenolysis. The enzymatic degradation of glycogen to glucose-1-phosphate.

Glycolipid. A lipid to which carbohydrate is covalently attached.

Glycolysis. The 10-reaction pathway by which glucose is broken down to 2 pyruvate with the concomitant production of 2 ATP and the reduction of 2 NAD^+ to 2 NADH.

Glycomics. The systematic study of the structures and functions of carbohydrates, including large glycans and the small oligosaccharides of glycoproteins.

Glycoprotein. A protein to which carbohydrate is covalently attached.

Glycosaminoglycan. An unbranched polysaccharide consisting of alternating residues of an amino sugar and a sugar acid.

Glycosidase. An enzyme that catalyzes the hydrolysis of glycosidic bonds.

Glycoside. A molecule containing a saccharide linked to another molecule by a glycosidic bond to the anomeric carbon.

Glycosidic bond. The covalent linkage between two monosaccharide units in a polysaccharide, or the linkage between the anomeric carbon of a saccharide and an alcohol or amine.

Glycosylation. The attachment of carbohydrate chains to a protein through *N*- or *O*-glycosidic linkages.

Glycosyltransferase. An enzyme that catalyzes the addition of a monosaccharide residue to a polysaccharide.

Glyoxylate pathway. A variation of the citric acid cycle in plants that allows acetyl-CoA to be converted quantitatively to gluconeogenic precursors.

Glyoxysome. A membrane-bounded plant organelle in which the reactions of the glyoxylate pathway take place.

Gout. An inflammatory disease, usually caused by impaired uric acid excretion and characterized by painful deposition of uric acid in the joints.

GPCR. G protein–coupled receptor, a transmembrane protein that binds an extracellular ligand and transmits the signal to the cell interior by interacting with an intracellular G protein.

GWAS. See genome-wide association study.

H. See enthalpy.

Half-reaction. The single oxidation or reduction process, involving the reduced and oxidized forms of a substance, that must be combined with another half-reaction to form a complete oxidation–reduction reaction.

Haploid. Having one set of chromosomes.

Haworth projection. A drawing of a sugar ring in which ring bonds that project in front of the plane of the paper are represented by heavy lines and ring bonds that project behind the plane of the paper are represented by light lines.

Helicase. An enzyme that unwinds DNA.

Heme. A protein prosthetic group that binds O_2 (in myoglobin and hemoglobin) or undergoes redox reactions (in cytochromes).

Henderson–Hasselbalch equation. The mathematical expression of the relationship between the pH of a solution of a weak acid and its pK: pH = pK + log ([A^-]/[HA]).

Hetero-. Different. In a heteropolymer, the subunits are not all identical.

Heterochromatin. Highly condensed, nonexpressed eukaryotic DNA.

Heterotroph. An organism that obtains its building materials and free energy from organic compounds produced by otsxher organisms.

Heterozygous. Having one each of two gene variants.

Hexose. A six-carbon sugar.

Highly repetitive DNA. Clusters of short DNA sequences that are repeated side-by-side and are present at millions of copies in the human genome.

High-performance liquid chromatography (HPLC). An automated chromatographic procedure for fractionating molecules using high pressure and computer-controlled solvent delivery.

Histone code. The correlation between patterns of covalent modification of histone proteins and the transcriptional activity of the associated DNA.

Histones. Highly conserved basic proteins that form a core to which DNA is bound in a nucleosome.

HIV. The human immunodeficiency virus, the causative agent of acquired immunodeficiency syndrome (AIDS).

Homo-. The same. In a homopolymer, all the subunits are identical.

Homologous genes. Genes that are related by evolution from a common ancestor.

Homologous proteins. Proteins that are related by evolution from a common ancestor.

Homozygous. Having two identical copies of a particular gene.

Horizontal gene transfer. The transfer of genetic material between species.

Hormone. A substance that is secreted by one tissue into the bloodstream and that induces a physiological response in other tissues.

Hormone response element. A DNA sequence to which an intracellular hormone–receptor complex binds so as to regulate gene expression.

HPLC. See high-performance liquid chromatography.

Hydration. The molecular state of being surrounded by and interacting with solvent water molecules; that is, solvated by water.

Hydrogen bond. A partly electrostatic, partly covalent interaction between a donor group such as O—H or N—H and an electronegative acceptor atom such as O or N.

Hydrolase. An enzyme that catalyzes a hydrolytic reaction.

Hydrolysis. The cleavage of a covalent bond accomplished by adding the elements of water; the reverse of a condensation.

Hydronium ion. A proton associated with a water molecule, H_3O^+.

Hydrophilic. Having high enough polarity to readily interact with water molecules. Hydrophilic substances tend to dissolve in water.

Hydrophobic. Having insufficient polarity to readily interact with water molecules. Hydrophobic substances tend to be insoluble in water.

Hydrophobic effect. The tendency of water to minimize its contacts with nonpolar substances, thereby inducing the substances to aggregate.

Hydroxyl group. A portion of a molecule with the formula —OH.

Hyperglycemia. Elevated levels of glucose in the blood.

IF. See initiation factor.

Imine. A molecule with the formula $>C{=}NH$.

Imino group. A portion of a molecule with the formula $\text{>C}\text{=}\text{NH}$.

Imprinting. A heritable variation in the level of expression of a gene according to its parental origin.

In vitro. In the laboratory (literally, in glass).

***In vitro* mutagenesis.** See site-directed mutagenesis.

In vivo. In a living organism.

Induced fit. An interaction between a protein and its ligand that induces a conformational change in the protein that enhances the protein's interaction with the ligand.

Inhibition constant (K_I). The dissociation constant for the complex between an enzyme and a reversible inhibitor.

Initiation factor (IF). A protein that interacts with mRNA and/or the ribosome and that is required to initiate translation.

Insulin resistance. The inability of cells to respond to insulin.

Integral membrane protein. A membrane protein that is embedded in the lipid bilayer. Also called an intrinsic protein.

Intermediate filament. A 100-Å-diameter cytoskeletal element consisting of coiled-coil polypeptide chains.

Intermembrane space. The compartment between the inner and outer mitochondrial membranes, which is equivalent to the cytosol in ionic composition.

Intrinsic protein. See integral membrane protein.

Intrinsically unstructured protein. A protein whose tertiary structure includes highly flexible extended segments that can adopt different conformations.

Intron. A portion of a gene that is transcribed but excised by splicing prior to translation.

Invariant residue. A residue in a protein that is the same in all evolutionarily related proteins.

Ion exchange chromatography. A fractionation procedure in which charged molecules are selectively retained by a matrix bearing oppositely charged groups.

Ion pair. An electrostatic interaction between two ionic groups of opposite charge.

Ionic interaction. An electrostatic interaction between two groups that is stronger than a hydrogen bond but weaker than a covalent bond.

Ionization constant of water (K_w). A quantity that relates the concentrations of H^+ and OH^- in pure water: $K_w = [H^+][OH^-] = 10^{-14}$.

Irregular secondary structure. A segment of a polymer in which each residue has a different backbone conformation; the opposite of regular secondary structure.

Irreversible inhibitor. A molecule that binds to and permanently inactivates an enzyme.

Isoacceptor tRNA. A tRNA that carries the same amino acid as another tRNA but has a different codon.

Isoelectric point (pI). The pH at which a molecule has no net charge.

Isomerase. An enzyme that catalyzes an isomerization reaction.

Isoprenoid. A lipid constructed from five-carbon units with an isoprene skeleton. Also called a terpenoid.

Isozymes. Different proteins that catalyze the same reaction.

k. See rate constant.

K. See dissociation constant.

K_a. See acid dissociation constant.

kb. Kilobase pairs; 1000 base pairs.

k_{cat}. See catalytic constant.

k_{cat}/K_M. The apparent second-order rate constant for an enzyme-catalyzed reaction; it indicates the enzyme's overall catalytic efficiency.

K_{eq}. See equilibrium constant.

Ketogenesis. The synthesis of ketone bodies from acetyl-CoA.

Ketogenic amino acid. An amino acid whose degradation yields compounds that can be converted to fatty acids or ketone bodies but not to glucose. See also glucogenic amino acid.

Ketone bodies. Compounds (acetoacetate and 3-hydroxybutyrate) that are produced from acetyl-CoA by the liver and used as metabolic fuels in other tissues when glucose is unavailable.

Ketose. A sugar whose carbonyl group is a ketone.

K_I. See inhibition constant.

Kinase. An enzyme that transfers a phosphoryl group between ATP and another molecule.

Kinetics. The study of chemical reaction rates.

K_M. See Michaelis constant.

K_w. See ionization constant of water.

Kwashiorkor. A form of severe malnutrition resulting from inadequate protein intake; marked by abdominal swelling and reddish hair. See also marasmus.

L sugar. A monosaccharide isomer in which the asymmetric carbon farthest from the carbonyl group has the same spatial arrangement as the chiral carbon of L-glyceraldehyde.

Lagging strand. The DNA strand that is synthesized as a series of discontinuous fragments that are later joined.

Lateral diffusion. The movement of a lipid within one leaflet of a bilayer.

Le Châtelier's principle. The observation that a change in concentration, temperature, volume, or pressure in a system at equilibrium causes the equilibrium to shift in order to counteract the change.

Leading strand. The DNA strand that is synthesized continuously during DNA replication.

Ligand. (1) A small molecule that binds to a larger molecule. (2) A molecule or ion bound to a metal ion.

Ligase. An enzyme that catalyzes bond formation coupled with the hydrolysis of ATP.

Light reactions. The photosynthetic reactions in which light energy is absorbed and used to generate NADPH and ATP.

Light-harvesting complex. A pigment-containing protein that collects light energy in order to transfer it to a photosynthetic reaction center.

Lineweaver–Burk plot. A rearrangement of the Michaelis–Menten equation that permits the determination of K_M and V_{max} from a linear plot.

Lipid. Any member of a broad class of macromolecules that are largely or wholly hydrophobic and therefore tend to be insoluble in water but soluble in organic solvents.

Lipid bilayer. See bilayer.

Lipid-linked protein. A protein that is anchored to a biological membrane via a covalently attached lipid.

Lipolysis. The degradation of a triacylglycerol so as to release fatty acids.

Lipoprotein. A globular particle, containing lipids and proteins, that transports lipids between tissues via the bloodstream.

Lock-and-key model. An early model of enzyme action, in which the substrate fit the enzyme like a key in a lock.

Locus. The chromosomal location of a gene or other DNA marker.

London dispersion forces. The weak van der Waals interactions between nonpolar groups as a result of fluctuations in their electron distributions that create a temporary separation of charge (polarity).

Loop. A segment of a polypeptide that joins two elements of secondary structure; usually found on the protein surface.

Low-barrier hydrogen bond. A short, strong hydrogen bond in which the proton is equally shared by the donor and acceptor atoms.

Lyase. An enzyme that catalyzes the elimination of a group to form a double bond.

Lysosome. A membrane-bounded organelle in a eukaryotic cell that contains a battery of hydrolytic enzymes and that functions to digest ingested material and to recycle cell components.

Major groove. The wider of the two grooves on a DNA double helix.

Marasmus. Body wasting due to inadequate intake of all types of foods. See also kwashiorkor.

Mass action ratio. The ratio of the product of the concentrations of reaction products to that of the reactants.

Mass spectrometry. A technique for identifying molecules by measuring the mass-to-charge ratios of gas-phase ions, such as peptide fragments.

Melting temperature (T_m). The midpoint temperature of the melting curve for the thermal denaturation of a macromolecule. For a lipid, the temperature of transition from an ordered crystalline state to a more fluid state.

Membrane potential ($\Delta\psi$). The difference in electrical charge across a membrane.

Messenger RNA (mRNA). A ribonucleic acid whose sequence is complementary to that of a protein-coding gene in DNA.

Metabolic acidosis. A low blood pH caused by the overproduction or retention of hydrogen ions.

Metabolic alkalosis. A high blood pH caused by the excessive loss of hydrogen ions.

Metabolic fuel. A molecule that can be oxidized to provide free energy for an organism.

Metabolic pathway. A series of enzyme-catalyzed reactions by which one substance is transformed into another.

Metabolic syndrome. A set of symptoms related to obesity, including inulin resistance, atherosclerosis, and hypertension.

Metabolically irreversible reaction. A reaction whose value of ΔG is large and negative so that the reaction cannot proceed in reverse.

Metabolism. The total of all degradative and biosynthetic cellular reactions.

Metabolite. A reactant, intermediate, or product of a metabolic reaction.

Metabolome. The complete set of metabolites produced by a cell or tissue.

Metabolomics. The study of all the metabolites produced by a cell or tissue.

Metal ion catalysis. A catalytic mechanism that requires the presence of a metal ion to lower the free energy of a reaction's transition state.

Micelle. A globular aggregate of amphiphilic molecules in aqueous solution that are oriented such that polar segments form the surface of the aggregate and the nonpolar segments form a core that is out of contact with the solvent.

Michaelis constant (K_M). For an enzyme that follows the Michaelis–Menten model, $K_M = (k_{-1} + k_2)/k_1$; K_M is equal to the substrate concentration at which the reaction velocity is half-maximal.

Michaelis–Menten equation. A mathematical expression that describes the activity of an enzyme in terms of the substrate concentration ([S]), the enzyme's maximal velocity (V_{max}), and its Michaelis constant (K_M): $v_0 = V_{max}[S]/(K_M + [S])$.

Micro RNA (miRNA). A 20- to 25-nucleotide double-stranded RNA that binds to and inactivates a number of complementary mRNA molecules in RNA interference.

Microarray. A collection of DNA sequences that hybridize with RNA molecules and that can therefore be used to identify active genes. Also called a DNA chip.

Microbiome. The collection of microorganisms that live in or on the human body.

Microenvironment. A group's immediate neighbors, whose chemical and physical properties may affect the group.

Microfilament. A 70-Å-diameter cytoskeletal element composed of polymerized actin subunits.

Microtubule. A 240-Å-diameter cytoskeletal element consisting of a hollow tube of polymerized tubulin subunits.

Minor groove. The narrower of the two grooves on a DNA double helix.

(−) end. The end of a polymeric filament where growth is slower. See also (+) end.

miRNA. See micro RNA.

Mismatch repair. A DNA repair pathway that removes and replaces mispaired nucleotides on a newly synthesized DNA strand.

Mitochondrial matrix. The gel-like solution of enzymes, substrates, cofactors, and ions in the interior of the mitochondrion.

Mitochondrion (pl. mitochondria). The double-membrane-enveloped eukaryotic organelle in which aerobic metabolic reactions occur, including those of the citric acid cycle, fatty acid oxidation, and oxidative phosphorylation.

Mixed inhibition. A form of enzyme inhibition in which an inhibitor binds to the enzyme such that it cause the apparent V_{max} to decrease and the apparent K_M to increase or decrease.

Mobilization. The process in which polysaccharides, triacylglycerols, and proteins are degraded to make metabolic fuels available.

Moderately repetitive DNA. Sequences of DNA that are present at hundreds of thousands of copies in the human genome.

Molecular chaperone. A protein that binds to unfolded or misfolded proteins in order to promote their normal folding.

Molecular cloning. See recombinant DNA technology.

Molecular weight. See M_r.

Monomer. A structural unit from which a polymer is built up.

Monosaccharide. A carbohydrate consisting of a single sugar molecule.

Motor protein. An intracellular protein that couples the free energy of ATP hydrolysis to molecular movement relative to another protein that often acts as a track for the linear movement of the motor protein.

M_r. Relative molecular mass. A dimensionless quantity that is defined as the ratio of the mass of a particle to 1/12th the mass of a ^{12}C atom. Also known as molecular weight.

mRNA. See messenger RNA.

Multienzyme complex. A group of noncovalently associated enzymes that catalyze two or more sequential steps in a metabolic pathway.

Multifunctional enzyme. A protein that carries out more than one chemical reaction.

Mutagen. An agent that induces a mutation in an organism.

Mutase. An enzyme that catalyzes the transfer of a functional group from one position to another on a molecule.

Mutation. A heritable alteration in an organism's genetic material.

Myelin sheath. The multilayer coating of sphingomyelin-rich membranes that insulates a mammalian neuron.

Native structure. The fully folded conformation of a macromolecule.

Natural selection. The evolutionary process by which the continued existence of a replicating entity depends on its ability to survive and reproduce under the existing conditions.

ncRNA. See noncoding RNA.

Near-equilibrium reaction. A reaction whose ΔG value is close to zero, so that it can operate in either direction depending on the substrate and product concentrations.

Negative effector. A substance that diminishes an enzyme's activity through allosteric inhibition.

Nernst equation. An expression of the relationship between the actual ($\mathcal{E}$) and standard reduction potential ($\mathcal{E}°'$) of a substance A: $\mathcal{E} = \mathcal{E}°' - (RT/n\mathcal{F}) \ln([A_{reduced}]/[A_{oxidized}])$.

Neurotransmitter. A substance released by a nerve cell to alter the activity of a target cell.

Neutral solution. A solution whose pH is equal to 7.0 ([H^+] = 10^{-7} M).

Nick. A single-strand break in a double-stranded nucleic acid.

Nick translation. The progressive movement of a single-strand break (nick) in DNA through the actions of an exonuclease that removes residues followed by a polymerase that replaces them.

Nitrification. The conversion of ammonia (NH_3) to nitrate (NO_3^-).

Nitrogen cycle. A set of reactions, including nitrogen fixation, nitrification, and denitrification, for the interconversion of different forms of nitrogen.

Nitrogen fixation. The process by which atmospheric N_2 is converted to a biologically useful form such as NH_3.

N-linked oligosaccharide. An oligosaccharide linked to the amide group of a protein Asn residue.

NMR. See nuclear magnetic resonance spectroscopy.

Noncoding RNA (ncRNA). An RNA molecule that is not translated into protein.

Noncoding strand. The DNA strand that has a sequence complementary (except for the replacement of U with T) to the transcribed RNA; it is the template strand. Also called the antisense strand.

Noncompetitive inhibition. A form of enzyme inhibition in which an inhibitor binds to an enzyme such that the apparent V_{max} decreases but K_M is not affected.

Noncyclic electron flow. The light-driven linear path of electrons from water through Photosystems II and I, which leads to the production of O_2, NADPH, and ATP.

Nonessential amino acid. An amino acid that an organism can synthesize from common intermediates.

Nonhomologous end joining. A ligation process that repairs a double-stranded break in DNA.

Nonreducing sugar. A saccharide with an anomeric carbon that has formed a glycosidic bond and cannot therefore act as a reducing agent.

Nonspontaneous process. A thermodynamic process that has a net increase in free energy ($\Delta G > 0$) and can occur only with the input of free energy from outside the system. See also endergonic reaction.

NMR spectroscopy. See nuclear magnetic resonance spectroscopy.

N-terminus. The end of a polypeptide that has a free amino group.

Nuclear magnetic resonance (NMR) spectroscopy. A spectroscopic method in which the signals emitted by atomic nuclei in a magnetic field can be used to determine the three-dimensional molecular structure of a protein or nucleic acid.

Nuclease. An enzyme that degrades nucleic acids.

Nucleic acid. A polymer of nucleotide residues. The major nucleic acids are deoxyribonucleic acid (DNA) and ribonucleic acid (RNA). Also known as a polynucleotide.

Nucleolus. The region of the eukaryotic nucleus where rRNA is processed and ribosomes are assembled.

Nucleophile. A compound containing an electron-rich group. A nucleophile (nucleus-lover) reacts with an electrophile (electron-lover).

Nucleoside. A compound consisting of a nitrogenous base linked to a five-carbon sugar (ribose or deoxyribose).

Nucleosome. The disk-shaped complex of a histone octamer and DNA that represents the fundamental unit of DNA organization in eukaryotes.

Nucleotide. A compound consisting of a nucleoside esterified to one or more phosphate groups. Nucleotides are the monomeric units of nucleic acids.

Nucleotide excision repair. A DNA repair pathway in which a damaged single-stranded segment of DNA is removed and replaced with normal DNA.

Okazaki fragments. The short segments of DNA formed in the discontinuous lagging-strand synthesis of DNA.

Oligonucleotide. A polynucleotide consisting of a few nucleotide residues.

Oligopeptide. A polypeptide consisting of a few amino acid residues.

Oligosaccharide. A polymeric carbohydrate containing a few monosaccharide residues.

O-linked oligosaccharide. An oligosaccharide linked to the hydroxyl group of a protein Ser or Thr side chain.

Omega-3 fatty acid. A fatty acid with a double bond starting at the third carbon from the methyl (omega) end of the molecule.

Oncogene. A mutant gene that interferes with the normal regulation of cell growth and contributes to cancer.

Open reading frame (ORF). A portion of the genome that potentially codes for a protein.

Operon. A prokaryotic genetic unit that consists of several genes with related functions that are transcribed as a single mRNA molecule.

Ordered reaction. A multisubstrate reaction with a compulsory order of substrate binding to the enzyme.

Orientation effects. See proximity and orientation effects.

ORF. See open reading frame.

Orphan gene. A gene that appears to have no counterpart in the genome of another species.

Osmosis. The movement of solvent from a region of low solute concentration to a region of high solute concentration.

Osteogenesis imperfecta. A disease caused by mutations in collagen genes and characterized by bone fragility and deformation.

Oxidant. See oxidizing agent.

Oxidation. A reaction in which a substance loses electrons.

Oxidative phosphorylation. The process by which the free energy obtained from the oxidation of metabolic fuels is used to generate ATP from $ADP + P_i$.

Oxidizing agent. A substance that can accept electrons, thereby becoming reduced. Also called an oxidant.

Oxidoreductase. An enzyme that catalyzes an oxidation–reduction reaction.

Oxyanion hole. A cavity in the active site of a serine protease that accommodates the reactants during the transition state and thereby lowers its energy.

Oxyhemoglobin. Hemoglobin that contains bound oxygen or is in the oxygen-binding conformation.

P site. The ribosomal binding site that accommodates a peptidyl–tRNA.

PAGE. Polyacrylamide gel electrophoresis. See electrophoresis and SDS-PAGE.

Palindrome. A segment of DNA that has the same sequence on each strand when read in the $5' \rightarrow 3'$ direction.

Parallel β sheet. See β sheet.

Partial oxygen pressure (pO_2). The concentration of gaseous O_2 in units of torr.

Passive transport. The thermodynamically spontaneous protein-mediated transmembrane movement of a substance from high to low concentration.

Pasteur effect. The greatly increased sugar consumption of yeast grown under anaerobic conditions compared to that of yeast grown under aerobic conditions.

PCR. See polymerase chain reaction.

Pentose. A five-carbon sugar.

Pentose phosphate pathway. A pathway for glucose degradation that yields ribose-5-phosphate and NADPH.

Peptide. A short polypeptide.

Peptide bond. An amide linkage between the α-amino group of one amino acid and the α-carboxylate group of another. Peptide bonds link the amino acid residues in a polypeptide.

Peptide group. The planar —CO—NH— group that encompasses the peptide bond between amino acid residues in a polypeptide.

Peptidoglycan. The cross-linked polysaccharides and polypeptides that form bacterial cell walls.

Peripheral membrane protein. A protein that is weakly associated with the surface of a biological membrane. Also called an extrinsic protein.

Periplasmic compartment. The space between the inner membrane and the outer membrane of gram-negative bacteria.

Peroxisome. A eukaryotic organelle with specialized oxidative functions, including fatty acid degradation.

p_{50}. The ligand concentration (or pressure for a gaseous ligand) at which a binding protein such as hemoglobin is half-saturated with ligand.

pH. A quantity used to express the acidity of a solution, equivalent to $-\log[H^+]$.

Phage. See bacteriophage.

Pharmacokinetics. The behavior of a drug in the body, including its metabolism and excretion.

Phenotype. An organism's physical characteristics.

Phosphatase. An enzyme that hydrolyzes phosphoryl ester groups.

Phosphodiester bond. The linkage in which a phosphate group is esterified to two alcohol groups (e.g., two ribose units that join the adjacent nucleotide residues in a polynucleotide).

Phosphoinositide signaling system. A signal transduction pathway in which hormone binding to a cell-surface receptor induces phospholipase C to catalyze the hydrolysis of phosphatidylinositol bisphosphate to yield the second messengers inositol trisphosphate and diacylglycerol.

Phospholipase. An enzyme that hydrolyzes one or more bonds in a glycerophospholipid.

Phosphorolysis. The cleavage of a chemical bond by the substitution of a phosphate group rather than water.

Phosphoryl group. A portion of a molecule with the formula $-PO_3H_2$.

Photoautotroph. An organism that obtains its building materials from inorganic compounds and its free energy from sunlight.

Photon. A packet of light energy.

Photooxidation. A mode of decay of an excited molecule, in which oxidation occurs through the transfer of an electron to an acceptor molecule.

Photophosphorylation. The synthesis of ATP from ADP + P$_i$ coupled to the dissipation of a proton gradient that has been generated through light-driven electron transport.

Photoreceptor. A light-absorbing molecule, or pigment.

Photorespiration. The consumption of O$_2$ and evolution of CO$_2$ by plants (a dissipation of the products of photosynthesis), a consequence of the competition between O$_2$ and CO$_2$ for ribulose-5-phosphate carboxylase.

Photosynthesis. The light-driven incorporation of CO$_2$ into organic compounds.

P$_i$. Inorganic phosphate or a phosphoryl group: HPO$_3^-$ or PO$_3^{2-}$.

pI. See isoelectric point.

Ping pong reaction. An enzymatic reaction in which one or more products are released before all the substrates have bound to the enzyme.

pK. A quantity used to express the tendency for an acid to donate a proton (dissociate); equal to $-\log K$, where K is the dissociation constant.

Planck's law. An expression for the energy (E) of a photon: $E = hc/\lambda$, where c is the speed of light, λ is its wavelength, and h is Planck's constant (6.626×10^{-34} J · s).

Plasmid. A small circular DNA molecule that autonomously replicates and may be used as a cloning vector.

(+) end. The end of a polymeric filament where growth is faster. See also (−) end.

P:O ratio. The ratio of the number of molecules of ATP synthesized from ADP + P$_i$ to the number of atoms of oxygen reduced.

Point mutation. The substitution of one base for another in DNA, arising from mispairing during DNA replication or from chemical alterations of existing bases.

Polarity. Having an uneven distribution of charge.

Poly(A) tail. The sequence of adenylate residues that is post-transcriptionally added to the 3′ end of eukaryotic mRNAs.

Polyacrylamide gel electrophoresis (PAGE). See electrophoresis.

Polymer. A molecule consisting of numerous smaller units that are linked together in an organized manner.

Polymerase. An enzyme that catalyzes the addition of nucleotide residues to a polynucleotide.

Polymerase chain reaction (PCR). A procedure for amplifying a segment of DNA by repeated rounds of replication centered between primers that hybridize with the two ends of the DNA segment of interest.

Polynucleotide. See nucleic acid.

Polypeptide. A polymer consisting of amino acid residues linked in linear fashion by peptide bonds.

Polyprotein. A polypeptide that undergoes proteolysis after its synthesis to yield several separate protein molecules.

Polyprotic acid. A substance that has more than one acidic proton and therefore has multiple ionization states.

Polysaccharide. A polymeric carbohydrate containing multiple monosaccharide residues. Also called a glycan.

Polysome. An mRNA transcript bearing multiple ribosomes in the process of translating the mRNA.

Porin. A β barrel protein in the outer membrane of bacteria, mitochondria, or chloroplasts that forms a weakly solute-selective pore.

Positive effector. A substance that boosts an enzyme's activity through allosteric activation.

Post-translational modification (processing). The removal or derivatization of amino acid residues following their incorporation into a polypeptide.

pO$_2$. See partial oxygen pressure.

PP$_i$. A pyrophosphoryl group: H$_3$P$_2$O$_6$, H$_2$P$_2$O$_6^-$, HP$_2$O$_6^{2-}$, or P$_2$O$_6^{3-}$.

Primary structure. The sequence of residues in a polymer.

Primer. An oligonucleotide that base pairs with a template polynucleotide strand and is extended through template-directed polymerization.

Prion. An infectious protein that causes its cellular counterparts to misfold and aggregate, thereby leading to the development of a disease such as transmissible spongiform encephalopathy.

Probe. A labeled single-stranded DNA or RNA segment that can hybridize with a DNA or RNA of interest in a screening procedure.

Processing. See RNA processing and post-translational modification.

Processivity. A property of a motor protein or other enzyme that undergoes many reaction cycles before dissociating from its track or substrate.

Product inhibition. A form of enzyme inhibition in which the reaction product acts as a competitive inhibitor.

Prokaryote. A unicellular organism that lacks a membrane-bounded nucleus. All bacteria are prokaryotes.

Promoter. The DNA sequence at which RNA polymerase binds to initiate transcription.

Proofreading. An additional catalytic activity of an enzyme, which acts to correct errors made by the primary enzymatic activity.

Prosthetic group. An organic group (such as a coenzyme) that is permanently associated with a protein.

Protease. An enzyme that catalyzes the hydrolysis of peptide bonds.

Protease inhibitor. An agent, often a protein, that reacts incompletely with a protease so as to inhibit further proteolytic activity.

Proteasome. A multiprotein complex with a hollow cylindrical core in which cellular proteins are degraded to peptides in an ATP-dependent process.

Protein. A macromolecule that consists of one or more polypeptide chains.

Protein kinase. An enzyme that catalyzes the transfer of a phosphoryl group from ATP to the OH group of a protein Ser, Thr, or Tyr residue.

Proteoglycan. An extracellular aggregate of protein and glycosaminoglycans.

Proteome. The complete set of proteins synthesized by a cell.

Proteomics. The study of all the proteins synthesized by a cell.

Protofilament. One of the 13 linear polymers of tubulin subunits that forms a microtubule.

Proton jumping. The rapid movement of a proton among hydrogen-bonded water molecules.

Proton wire. A series of hydrogen-bonded water molecules and protein groups that can relay protons from one site to another.

Protonmotive force. The free energy of the electrochemical proton gradient that forms during electron transport.

Proximity and orientation effects. A catalytic mechanism in which reacting groups are brought close together in an enzyme active site to accelerate the reaction.

Purine. A derivative of the compound purine, such as the nucleotide base adenine or guanine.

Pyrimidine. A derivative of the compound pyrimidine, such as the nucleotide base cytosine, uracil, or thymine.

Pyrosequencing. A procedure for determining the sequence of nucleotides in DNA by detecting a flash of light generated by the addition of a new nucleotide to a growing DNA strand.

Q cycle. The cyclic flow of electrons involving a semiquinone intermediate in Complex III of mitochondrial electron transport and in photosynthetic electron transport.

qPCR. See quantitative PCR.

Quantitative PCR (qPCR). A variation of the polymerase chain reaction in which the level of gene expression can be measured. Also known as real-time PCR.

Quantum yield. The ratio of carbon atoms fixed or oxygen molecules produced to the number of photons absorbed by the photosynthetic machinery.

Quaternary structure. The spatial arrangement of a macromolecule's individual subunits.

Quorum sensing. The ability of cells to monitor population density by detecting the concentrations of extracellular substances.

R. See gas constant.

R group. A symbol for a variable portion of a molecule, such as the side chain of an amino acid.

R state. One of two conformations of an allosteric protein; the other is the T state.

Raft. An area of a lipid bilayer with a distinct lipid composition and near-crystalline consistency.

Random reaction. A multisubstrate reaction without a compulsory order of substrate binding to the enzyme.

Rate constant (*k*). The proportionality constant between the velocity of a chemical reaction and the concentration(s) of the reactant(s).

Rate equation. A mathematical expression for the time-dependent progress of a reaction as a function of reactant concentration.

Rate-determining reaction. The slowest step in a multistep sequence, such as a metabolic pathway, whose rate determines the rate of the entire sequence.

Rational drug design. The synthesis of more effective drugs based on detailed knowledge of the target molecule's structure and function.

Reactant. One of the starting materials for a chemical reaction.

Reaction center. A chlorophyll-containing protein where photooxidation takes place.

Reaction coordinate. A line representing the progress of a reaction, part of a graphical presentation of free energy changes during a reaction.

Reaction specificity. The ability of an enzyme to discriminate between possible substrates and to catalyze a single type of chemical reaction.

Reading frame. The grouping of nucleotides in sets of three whose sequence corresponds to a polypeptide sequence.

Real-time PCR. See quantitative PCR.

Receptor. A binding protein that is specific for its ligand and elicits a discrete biochemical effect when its ligand is bound.

Receptor tyrosine kinase. A cell-surface receptor whose intracellular domain becomes active as a Tyr-specific kinase as a result of extracellular ligand binding.

Recombinant DNA technology. The set of techniques used to construct a DNA molecule containing DNA segments from different sources. Also called genetic engineering and molecular cloning.

Recombination. The exchange of polynucleotide strands between separate DNA segments; recombination is one mechanism for repairing damaged DNA by allowing a homologous segment to serve as a template for replacement of the damaged bases.

Redox center. A group that can undergo an oxidation–reduction reaction.

Redox reaction. A chemical reaction in which one substance is reduced and another substance is oxidized.

Reducing agent. A substance that can donate electrons, thereby becoming oxidized. Also called a reductant.

Reducing sugar. A saccharide with an anomeric carbon that has not formed a glycosidic bond and can therefore act as a reducing agent.

Reductant. See reducing agent.

Reduction. A reaction in which a substance gains electrons.

Reduction potential ($\mathcal{E}$). A measure of the tendency of a substance to gain electrons.

Regular secondary structure. A segment of a polymer in which the backbone adopts a regularly repeating conformation; the opposite of irregular secondary structure.

Release factor (RF). A protein that recognizes a stop codon and causes a ribosome to terminate polypeptide synthesis.

Renaturation. The refolding of a denatured macromolecule so as to regain its native conformation.

Replication. The process of making an identical copy of a DNA molecule. During DNA replication, the parental polynucleotide strands separate so that each can direct the synthesis of a complementary daughter strand, resulting in two complete DNA double helices.

Replication fork. The point in a replicating DNA molecule where the two parental strands separate in order to serve as templates for the synthesis of new strands.

Repressor. A protein that binds at or near a gene so as to prevent its transcription.

Residue. A term for a monomeric unit after it has been incorporated into a polymer.

Resonance stabilization. The effect of delocalization of electrons in a molecule that cannot be depicted by a single structural diagram.

Respiration. The metabolic phenomenon whereby organic molecules are oxidized, with the electrons eventually transferred to molecular oxygen.

Respiratory acidosis. A low blood pH caused by insufficient elimination of CO_2 (carbonic acid) by the lungs.

Respiratory alkalosis. A high blood pH caused by the excessive loss of CO_2 (carbonic acid) from the lungs.

Restriction digest. The generation of a set of DNA fragments by the action of a restriction endonuclease.

Restriction endonuclease. A bacterial enzyme that cleaves a specific DNA sequence.

Restriction fragment. A segment of DNA produced by the action of a restriction endonuclease.

Reverse transcriptase. A DNA polymerase that uses RNA as its template.

RF. See release factor.

Ribonucleic acid. See RNA.

Ribosomal RNA (rRNA). The RNA molecules that provide structural support for the ribosome and catalyze peptide bond formation.

Ribosome. The RNA-and-protein particle that synthesizes polypeptides under the direction of mRNA.

Ribosome recycling factor (RRF). A protein that binds to a ribosome after protein synthesis to prepare it for another round of translation.

Ribozyme. An RNA molecule that has catalytic activity.

RNA. A polymer of ribonucleotides, such as messenger RNA (mRNA), transfer RNA (tRNA), and ribosomal RNA (rRNA).

RNA interference (RNAi). A phenomenon in which short RNA segments direct the degradation of complementary mRNA, thereby inhibiting gene expression.

RNA processing. The addition, removal, or modification of nucleotides in an RNA molecule that is necessary to produce a fully functional RNA.

RNA world. A hypothetical time before the evolution of DNA or protein, when RNA stored genetic information and functioned as a catalyst.

RRF. See ribosome recycling factor.

rRNA. See ribosomal RNA.

S. See entropy.

Saccharide. See carbohydrate.

Salvage pathway. A pathway that reincorporates an intermediate of nucleotide degradation into a new nucleotide, thereby minimizing the need for the nucleotide biosynthetic pathways.

Saturated fatty acid. A fatty acid that does not contain any double bonds in its hydrocarbon chain.

Saturation. The state in which all of a macromolecule's ligand-binding sites are occupied by ligand.

Schiff base. An imine that forms between an amine and an aldehyde or ketone.

Scissile bond. The bond that is to be cleaved during a proteolytic reaction.

SDS-PAGE. A form of polyacrylamide gel electrophoresis in which denatured polypeptides are separated by size in the presence of the detergent sodium dodecyl sulfate. See electrophoresis.

Second messenger. An intracellular ion or molecule that acts as a signal for an extracellular event such as ligand binding to a cell-surface receptor.

Secondary active transport. Transmembrane transport of one substance that is driven by the free energy of an existing gradient of a second substance.

Secondary structure. The local spatial arrangement of a polymer's backbone atoms without regard to the conformations of its substituent side chains.

Second-order reaction. A reaction whose rate is proportional to the square of the concentration of one reactant or to the product of the concentrations of two reactants.

Selection. A technique for distinguishing cells that contain a particular feature, such as resistance to an antibiotic.

Semiconservative replication. The mechanism of DNA duplication in which each new molecule contains one strand from the parent molecule and one newly synthesized strand.

Sense strand. See coding strand.

Serine protease. A peptide-hydrolyzing enzyme that has a reactive Ser residue in its active site.

Set-point. A body weight that is maintained through regulation of fuel metabolism and that resists change when an individual attempts to alter fuel consumption or expenditure.

Signal peptide. A short sequence in a membrane or secretory protein that binds to the signal recognition particle in order to direct the translocation of the protein across a membrane.

Signal recognition particle (SRP). A complex of protein and RNA that recognizes membrane and secretory proteins and mediates their binding to a membrane for translocation.

Signal transduction. The process by which an extracellular signal is transmitted to the cell interior by binding to a cell-surface receptor such that binding triggers a series of intracellular events.

Silencer. A DNA sequence some distance from the transcription start site, where a repressor of transcription may bind.

Single-nucleotide polymorphism (SNP). A nucleotide sequence variation in the genomes of two individuals from the same species.

siRNA. See small interfering RNA.

Site-directed mutagenesis. A technique in which a cloned gene is mutated in a specific manner. Also called *in vitro* mutagenesis.

Size-exclusion chromatography. A procedure in which macromolecules are separated on the basis of their size and shape. Also called gel filtration chromatography.

Small interfering RNA (siRNA). A 20- to 25-nucleotide double-stranded RNA that targets for destruction a fully complementary mRNA molecule in RNA interference.

Small nuclear RNA (snRNA). Highly conserved RNAs that participate in eukaryotic mRNA splicing.

Small nucleolar RNA (snoRNA). RNA molecules that direct the sequence-specific methylation of eukaryotic rRNA transcripts.

snoRNA. See small nucleolar RNA.

SNP. See single-nucleotide polymorphism.

snRNA. See small nuclear RNA.

Solute. The substance that is dissolved in water or another solvent to make a solution.

Solvation. The state of being surrounded by solvent molecules.

Specificity pocket. A cavity on the surface of a serine protease, whose chemical characteristics determine the identity of the substrate residue on the N-terminal side of the bond to be cleaved.

Sphingolipid. An amphipathic lipid containing an acyl group, a palmitate derivative, and a polar head group attached to a serine backbone. In sphingomyelins, the head group is a phosphate derivative.

Sphingomyelin. See sphingolipid.

Spliceosome. A complex of protein and snRNA that carries out the splicing of immature mRNA molecules.

Splicing. The process by which introns are removed and exons are joined to produce a mature RNA transcript.

Spontaneous process. A thermodynamic process that has a net decrease in free energy ($\Delta G < 0$) and occurs without the input of free energy from outside the system. See also exergonic reaction.

SRP. See signal recognition particle.

Stacking interactions. The stabilizing van der Waals interactions between successive (stacked) bases in a polynucleotide.

Standard conditions. A set of conditions including a temperature of 25°C, a pressure of 1 atm, and reactant concentrations of 1 M. Biochemical standard conditions include a pH of 7.0 and a water concentration of 55.5 M.

Standard free energy change ($\Delta G°'$). The force that drives reactants to reach their equilibrium values when the system is in its biochemical standard state.

Standard reduction potential ($\mathcal{E}°'$). A measure of the tendency of a substance to gain electrons (to be reduced) under standard conditions.

Steady state. A set of conditions under which the formation and degradation of individual components are balanced such that the system does not change over time.

Stereocilia. Microfilament-stiffened cell processes on the surface of cells in the inner ear, which are deflected in response to sound waves.

Sticky ends. Single-stranded extensions of DNA that are complementary, often because they have been generated by the action of the same restriction endonuclease.

Stroma. The gel-like solution of enzymes and small molecules in the interior of the chloroplast; the site of carbohydrate synthesis.

Substrate. A reactant in an enzymatic reaction.

Substrate-level phosphorylation. The transfer of a phosphoryl group to ADP that is directly coupled to another chemical reaction.

Subunit. One of several polypeptide chains that make up a protein.

Sugar–phosphate backbone. The chain of (deoxy)ribose groups linked by phosphodiester bonds in a polynucleotide chain.

Suicide substrate. A molecule that chemically inactivates an enzyme only after undergoing part of the normal catalytic reaction.

Sulfhydryl group. A portion of a molecule with the formula —SH.

Supercoiling. A topological state of DNA in which the helix is underwound or overwound so that the molecule tends to writhe or coil up on itself.

Symport. Transport that involves the simultaneous transmembrane movement of two molecules in the same direction.

Synaptic vesicle. A vesicle loaded with neurotransmitters to be released from the end of an axon.

T state. One of two conformations of an allosteric protein; the other is the R state.

TATA box. A eukaryotic promoter element with an AT-rich sequence located upstream from the transcription start site.

Tautomer. One of a set of isomers that differ only in the positions of their hydrogen atoms.

Telomerase. An enzyme that uses an RNA template to polymerize deoxynucleotides and thereby extend the 3′-ending strand of a eukaryotic chromosome.

Telomere. The end of a linear eukaryotic chromosome, which consists of tandem repeats of a short G-rich sequence on the 3′-ending strand and its complementary sequence on the 5′-ending strand.

Terpenoid. See isoprenoid.

Tertiary structure. The entire three-dimensional structure of a single-chain polymer, including the conformations of its side chains.

Tetramer. An assembly consisting of four monomeric units.

Tetrose. A four-carbon sugar.

Thalassemia. A hereditary disease caused by insufficient synthesis of hemoglobin, which results in anemia.

Thermogenesis. The process of generating heat by muscular contraction or by metabolic reactions.

Thick filament. A muscle cell structural element that is composed of several hundred myosin molecules.

Thin filament. A muscle cell structural element that consists primarily of an actin filament.

Thioester. A compound containing an ester linkage to a sulfur rather than an oxygen atom.

3′ end. The terminus of a polynucleotide whose C3′ is not esterified to another nucleotide residue.

Thylakoid. The membranous structure in the interior of a chloroplast that is the site of the light reactions of photosynthesis.

T_m. See melting temperature.

Topoisomerase. An enzyme that alters DNA supercoiling by breaking and resealing one or both strands.

Trace element. An element that is present in small quantities in a living organism.

Transamination. The transfer of an amino group from an amino acid to an α-keto acid to yield a new α-keto acid and a new amino acid.

Transcription. The process by which RNA is synthesized using a DNA template, thereby transferring genetic information from the DNA to the RNA.

Transcription factor. A protein that promotes the transcription of a gene by binding to DNA sequences at or near the gene or by interacting with other proteins that do so.

Transcriptome. The set of all the RNA molecules produced by a cell.

Transcriptomics. The study of the genes that are transcribed in a certain cell type or at a certain time.

Transfer RNA (tRNA). The small L-shaped RNAs that deliver specific amino acids to ribosomes according to the sequence of a bound mRNA.

Transferase. An enzyme that catalyzes the transfer of a functional group from one molecule to another.

Transgenic organism. An organism that stably expresses a foreign gene.

Transition mutation. A point mutation in which one purine (or pyrimidine) is replaced by another purine (or pyrimidine).

Transition state. The point of highest free energy, or the structure that corresponds to that point, in the reaction coordinate diagram of a chemical reaction.

Transition state analog. A stable substance that geometrically and electronically resembles the transition state of a reaction and that therefore may inhibit an enzyme that catalyzes the reaction.

Translation. The process of transforming the information contained in the nucleotide sequence of an RNA to the corresponding amino acid sequence of a polypeptide as specified by the genetic code.

Translocase. An enzyme that catalyzes the movement of a lipid from one bilayer leaflet to another. Also called a flippase.

Translocation. The movement of tRNA and mRNA, relative to the ribosome, that occurs following formation of a peptide bond and that allows the next mRNA codon to be translated.

Translocon. The complex of membrane proteins that mediates the transmembrane movement of a polypeptide.

Transmissible spongiform encephalopathy (TSE). A fatal neurodegenerative disease caused by infection with a prion.

Transpeptidation. The ribosomal process in which the peptidyl group attached to a tRNA is transferred to the aminoacyl group of another tRNA, forming a new peptide bond and lengthening the polypeptide by one residue at its C-terminus.

Transposable element. A segment of DNA, sometimes including genes, that can move (be copied) from one position to another in a genome.

Transverse diffusion. The movement of a lipid from one leaflet of a bilayer to the other. Also called flip-flop.

Transversion mutation. A point mutation in which a purine is replaced by a pyrimidine or vice versa.

Treadmilling. The addition of monomeric units to one end of a polymer and their removal from the opposite end such that the length of the polymer remains unchanged.

Triacylglycerol. A lipid in which three fatty acids are esterified to a glycerol backbone. Also called a triglyceride.

Triglyceride. See triacylglycerol.

Trimer. An assembly consisting of three monomeric units.

Triose. A three-carbon sugar.

Triple helix. The right-handed helical structure formed by three left-handed helical polypeptide chains in collagen.

Trisaccharide. A carbohydrate consisting of three monosaccharides.

tRNA. See transfer RNA.

TSE. See transmissible spongiform encephalopathy.

Tumor suppressor gene. A gene whose loss or mutation may lead to cancer.

Turnover number. See catalytic constant.

Uncompetitive inhibition. A form of enzyme inhibition in which an inhibitor binds to an enzyme–substrate complex such that the apparent V_{max} and K_M are both decreased to the same extent.

Uncoupler. A substance that allows the proton gradient across a membrane to dissipate without ATP synthesis so that electron transport proceeds without oxidative phosphorylation.

Unimolecular reaction. A reaction involving one molecule.

Uniport. Transport that involves transmembrane movement of a single molecule.

Unsaturated fatty acid. A fatty acid that contains at least one double bond in its hydrocarbon chain.

Urea cycle. A cyclic metabolic pathway in which amino groups are converted to urea for disposal.

Usher syndrome. A genetic disease characterized by profound deafness and retinitis pigmentosa that leads to blindness, caused in some cases by a defective myosin protein.

v. Velocity (rate) of a reaction.

van der Waals interaction. A weak noncovalent association between molecules that arises from the attractive forces between polar groups (dipole–dipole interactions) or between nonpolar groups whose fluctuating electron distribution gives rise to temporary dipoles (London dispersion forces).

van der Waals radius. The distance from an atom's nucleus to its effective electronic surface.

Variable residue. A position in a polypeptide that is occupied by different residues in evolutionarily related proteins; its substitution has little or no effect on protein function.

Vesicle. A fluid-filled sac enclosed by a lipid-bilayer membrane.

Vitamin. A metabolically required substance that cannot be synthesized by an animal and must therefore be obtained from the diet.

V_{max}. Maximal velocity of an enzymatic reaction.

Voltage-gated channel. A membrane transport channel that opens and closes in response to a change in membrane potential.

v_0. Initial velocity of an enzymatic reaction.

Warburg effect. The increased rate of glycolysis observed in cancerous tissues.

Wobble hypothesis. An explanation for the nonstandard base pairing between tRNA and mRNA at the third codon position, which allows a tRNA to recognize more than one codon.

X-Ray crystallography. A method for determining three-dimensional molecular structures from the diffraction pattern produced by exposing a crystal of a molecule to a beam of X-rays.

$\Delta\psi$. See membrane potential.

Y. See fractional saturation.

Z. The net charge of an ion.

Zinc finger. A protein structural motif consisting of 20–60 residues, including Cys and His residues to which one or two Zn^{2+} ions are tetrahedrally coordinated.

Z-scheme. A Z-shaped diagram indicating the electron carriers and their reduction potentials in the photosynthetic electron transport system of plants and cyanobacteria.

Zymogen. The inactive precursor (proenzyme) of a proteolytic enzyme.

SOLUTIONS

Chapter 1

1. (a)

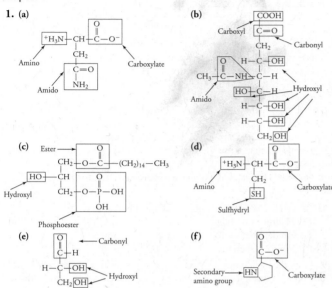

3. Amino acids, monosaccharides, nucleotides, and lipids are the four types of biological small molecules. Amino acids, monosaccharides, and nucleotides can form polymers of proteins, polysaccharides, and nucleic acids, respectively.

5. (a) Mainly C and H plus some O
(b) C, H, and O
(c) C, H, O, and N plus small amounts of S

7. (a) You should measure the nitrogen content, since this would indicate the presence of protein (neither lipids nor carbohydrates contain nitrogen).
(b) You could add the compound that contains the most nitrogen, compound B, which is melamine. (Melamine is a substance that has been added to some pet foods and milk products from China so that they would appear to contain more protein. Melamine is toxic to pets and children.)
(c) Compound C is an amino acid, so it would already be present in protein-containing food.

9. All amino acids have carboxylate groups. All have primary amino groups except for proline, which has a secondary amino group.

11. Asn has an amido group and Cys has a sulfhydryl group.

13.

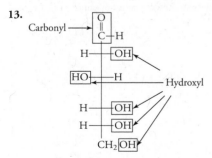

15. Uracil has a carbonyl functional group, whereas cytosine has an amino functional group.

17. As described in the text, palmitate and cholesterol are highly nonpolar and are therefore insoluble in water. Both are highly aliphatic. Alanine is water soluble because its amino group and carboxylate group are ionized, which render the molecule "saltlike." Glucose is also water soluble because its aldehyde group and many hydroxyl groups are able to form hydrogen bonds with water.

19. DNA forms a more regular structure because DNA consists of only four different nucleotides, whereas proteins are made up of as many as 20 different amino acids. In addition, the 20 amino acids have much more individual variation in their structures than do the four nucleotides. Both of these factors result in a more regular structure for DNA. The cellular role of DNA relies on the *sequence* of the nucleotides that make up the DNA, not on the overall shape of the DNA molecule itself. Proteins, on the other hand, fold into unique shapes, as illustrated by endothelin in Figure 1-4. The ability of proteins to fold into a wide variety of shapes means that proteins can also serve a wide variety of biochemical roles in the cell. According to Table 1-2, the major roles of proteins in the cell are to carry out metabolic reactions and to support cellular structures.

21. The pancreatic amylase is unable to digest the glycosidic bonds that link the glucose residues together in cellulose. Figure 1-6 shows the structural differences between starch and cellulose. Pancreatic amylase binds to starch prior to catalyzing the hydrolysis of the glycosidic bond; thus, the enzyme and the starch must have shapes that are complementary. The enzyme would be unable to bind to the cellulose, whose structure is much different from that of starch.

23. A positive entropy change indicates that the system has become more disordered; a negative entropy change indicates that the system has become more ordered.

(a) negative   **(b)** positive
(c) positive   **(d)** positive
(e) negative

25. The polymeric molecule is more ordered and thus has less entropy. A mixture of constituent monomers has a large number of different arrangements (like the balls scattered on a pool table) and thus has greater entropy.

27. The dissolution of ammonium nitrate in water is a highly endothermic process, as indicated by the positive value of ΔH. This means that when ammonium nitrate dissolves in water, the system absorbs heat from the surroundings and the surroundings become cold. The plastic bag containing the ammonium nitrate becomes cold and can be used as a cold pack to treat an injury.

29. First, calculate ΔH and ΔS, as described in Sample Calculation 1-1:

$$\Delta H = H_B - H_A$$
$$\Delta H = 60 \text{ kJ} \cdot \text{mol}^{-1} - 54 \text{ kJ} \cdot \text{mol}^{-1}$$
$$\Delta H = 6 \text{ kJ} \cdot \text{mol}^{-1}$$
$$\Delta S = S_B - S_A$$
$$\Delta H = 43 \text{ J} \cdot \text{K}^{-1} \cdot \text{mol}^{-1} - 22 \text{ J} \cdot \text{K}^{-1} \cdot \text{mol}^{-1}$$
$$\Delta H = 21 \text{ J} \cdot \text{K}^{-1} \cdot \text{mol}^{-1}$$

(a) $\Delta G = (6000 \text{ J} \cdot \text{mol}^{-1}) - (4 + 273 \text{ K})(21 \text{ J} \cdot \text{K}^{-1} \cdot \text{mol}^{-1})$
$\Delta G = 180 \text{ J} \cdot \text{mol}^{-1}$
The reaction is not favorable at 4°C.

(b) $\Delta G = (6000 \text{ J} \cdot \text{mol}^{-1}) - (37 + 273 \text{ K})(21 \text{ J} \cdot \text{K}^{-1} \cdot \text{mol}^{-1})$
$\Delta G = -510 \text{ J} \cdot \text{mol}^{-1}$
The reaction is favorable at 37°C.

31. $0 > -14.3 \text{ kJ} \cdot \text{mol}^{-1} - (273 + 25 \text{ K})(\Delta S)$
$14.3 \text{ kJ} \cdot \text{mol}^{-1} > -(273 + 25 \text{ K})(\Delta S)$
$-48 \text{ J} \cdot \text{mol}^{-1} > \Delta S$
ΔS could be any positive value, or it could have a negative value smaller than $-48 \text{ J} \cdot \text{mol}^{-1}$.

33. Process (d) is never spontaneous.

35. The dissolution of urea in water is an endothermic process and has a positive ΔH value. In order to be spontaneous, the process must also have a positive ΔS value in order for the free energy change of the process to be negative. Solutions have a higher order of entropy than the solvent and solute alone.

37. **(a)** The conversion of glucose to glucose-6-phosphate is not favorable because the ΔG value for the reaction is positive, indicating an endergonic process.
(b) If the two reactions are coupled, the overall reaction would be the sum of the two individual reactions. The ΔG value would be the sum of the ΔG values for the two individual reactions.

$$\text{ATP + glucose} \rightleftharpoons \text{ADP + glucose-6-phosphate}$$
$$\Delta G = -16.7 \text{ kJ} \cdot \text{mol}^{-1}$$

Coupling the conversion of glucose to glucose-6-phosphate with the hydrolysis of ATP has converted an unfavorable reaction to a favorable reaction. The ΔG value of the coupled reaction is negative, which indicates that the reaction as written is favorable.

39. C (most oxidized), A, B (most reduced)

41. **(a)** oxidized   **(b)** oxidized
(c) oxidized   **(d)** reduced

43. **(a)** Palmitate's carbon atoms, which have the formula $-CH_2-$, are more reduced than CO_2, so their reoxidation to CO_2 can release free energy.
(b) Because the $-CH_2-$ groups of palmitate are more reduced than those of glucose ($-HCOH-$), their conversion to the fully oxidized CO_2 would be even more thermodynamically favorable (have a larger negative value of ΔG) than the conversion of glucose carbons to CO_2. Therefore, palmitate carbons could provide more free energy than glucose carbons.

45. The experiment was significant because it demonstrated that it was possible to synthesize the building blocks of biological macromolecules (amino acids, carbohydrates, and nucleic acids) using only inorganic gases as starting materials and lightning as an energy source, that is, the conditions that most likely existed in the prebiotic world.

47. Morphological differences, which are useful for classifying large organisms, are not useful for bacteria, which often look alike. Furthermore, microscopic organisms do not leave an easily interpreted imprint in the fossil record, as vertebrates do. Thus, molecular information is often the only means for tracing the evolutionary history of bacteria.

49. **(a)** H15 and H7 are closely related, as are H4 and H14.
(b) H4 and H14 are most closely related to H3.

Chapter 2

1. The water molecule is not perfectly tetrahedral because the electrons in the nonbonding orbitals repel the electrons in the bonding orbitals more than the bonding electrons repel each other. The angle between the bonding orbitals is therefore slightly less than 109°.

3. Ammonia is polar because it has one unshared electron pair. Its shape is trigonal pyramidal, and the molecule is not symmetrical. Nitrogen is more electronegative than hydrogen, so partial negative charges reside on the nitrogen and partial positive charges on the hydrogens.

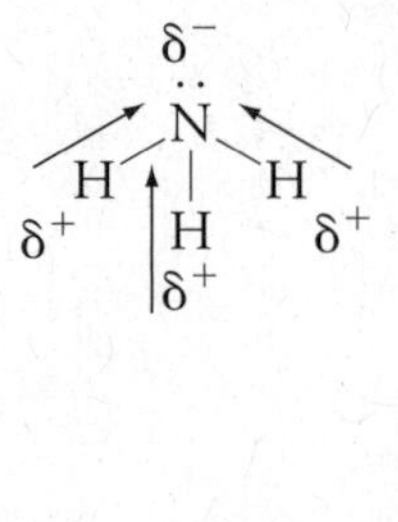

5. The arrows point toward hydrogen acceptors and away from hydrogen donors.

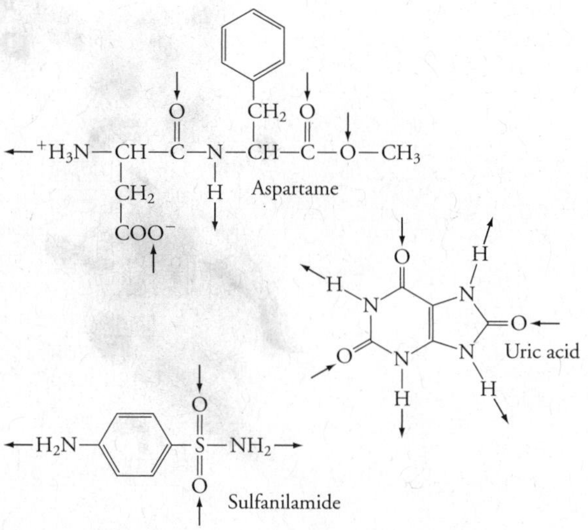

7. **(a)** van der Waals forces (dipole–dipole interactions)
(b) Hydrogen bonding
(c) van der Waals forces (London dispersion forces)
(d) Ionic interactions

9. From the highest melting point to the lowest melting point: C, B, E, A, D. Compound C (urea, melting point 133°C) has three functional groups that can serve as hydrogen donors and/or acceptors. Compound B (acetamide, melting point 80.16°C) has one less $-NH$ group than Compound C and therefore forms fewer hydrogen bonds. Compound E (propionaldehyde, melting point -80°C) has one functional group that can serve as a hydrogen bond acceptor, but it has no donors, so dipole–dipole forces are the strongest intermolecular forces in a sample of this compound. Compound A (methyl ethyl ether, melting point -113°C) also has one functional group that can serve as a hydrogen bond acceptor (it has no donors), but it has hydrocarbon portions that interact with one another via London dispersion forces. Compound D (pentane, melting point -139.67°C) is nonpolar and experiences only London dispersion forces, so it has the lowest melting point in the group.

11. Aquatic organisms that live in the pond are able to survive the winter. Since the water at the bottom of the pond remains in the liquid form instead of freezing, the organisms are able to move around. The ice on top of the pond also serves as an insulating layer from the cold winter air.

13. The positively charged ammonium ion is surrounded by a shell of water molecules that are oriented so that their partially negatively charged oxygen atoms interact with the positive charge on the ammonium ion. Similarly, the negatively charged sulfate ion is hydrated with water molecules oriented so that the partially positively charged hydrogen atoms interact with the negative charge on the sulfate anion. (Not shown in the diagram is the fact that the ammonium ions outnumber the sulfate ions by a 2:1 ratio. Also note that the exact number of water molecules shown is unimportant.)

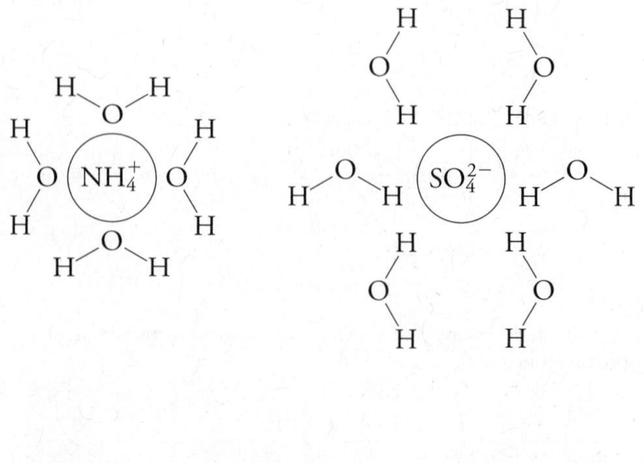

15. (a) Surface tension is defined as the force that must be applied to surface molecules in a liquid so that they may experience the same forces as the molecules in the interior of the liquid. Water's surface tension is greater than ethanol's because the strength and number of water's intermolecular forces (hydrogen bonds) are both greater. Ethanol's —OH group also forms hydrogen bonds, but the hydrocarbon portion of the molecule cannot interact favorably with water, and weaker London dispersion forces form instead.
(b) The kinetic energy of the water molecules increases when temperature increases. Intermolecular forces are weaker in strength as a consequence of the increased molecular motion. Because surface tension increases when the strength of intermolecular forces increases, as described in part (a), surface tension decreases when temperature increases.

17. Methanol, which has the highest dielectric constant, would be the best solvent for the cationic NH_4^+. The polarity of the alcohols, which all contain a primary —OH group, varies with the size of the hydrocarbon portion. 1-Butanol, with the largest hydrophobic group, is the least polar and therefore has the highest dielectric constant.

19. (a) First, calculate the number of moles of protein using Avogadro's number:

$$1000 \text{ molecules} \times \frac{1 \text{ mole}}{6.02 \times 10^{23} \text{ molecules}} = 1.66 \times 10^{-21} \text{ moles}$$

Next, calculate the volume of the cell, expressing r in centimeters:

$$\text{volume} = \frac{4\pi r^3}{3} = \frac{4\pi (5 \times 10^{-5} \text{ cm})^3}{3} = 5.2 \times 10^{-13} \text{ cm}^3$$

Since $1 \text{ cm}^3 = 1 \text{ mL}$, the volume is 5.2×10^{-13} mL, or 5.2×10^{-16} L. Therefore, the concentration of the protein is

$$\frac{1.66 \times 10^{-21} \text{ moles}}{5.2 \times 10^{-16} \text{ L}} = 3.3 \times 10^{-6} \text{ M, or } 3.2 \text{ } \mu\text{M}$$

(b) $\dfrac{5 \times 10^{-3} \text{ moles}}{\text{L}} \times \dfrac{6.02 \times 10^{23} \text{ molecules}}{\text{mole}}$
$\times 5.2 \times 10^{-16} \text{ L} = 1.6 \times 10^{6} \text{ molecules}$

21. Compound A is amphiphilic and has a polar head and a nonpolar tail as indicated and can form a micelle (see Fig. 2-10). Compound B is nonpolar and cannot form a micelle or a bilayer. Compound C is polar (ionic) and cannot form a micelle or a bilayer. Compound D is amphiphilic and has a polar head and two nonpolar tails as indicated and can form a bilayer (see Fig. 2-11). Compound E is polar and forms neither a micelle nor a bilayer.

23.

(c) The hydrophobic grease can move into the hydrophobic core of the water-soluble soap micelle. The "dissolved" grease can then be washed away with the micelle.

25. (a) The nonpolar core of the lipid bilayer helps prevent the passage of water since the polar water molecules cannot easily penetrate the hydrophobic core of the bilayer.
(b) Most human cells are surrounded by a fluid containing about 150 mM Na^+ and slightly less Cl^- (see Fig. 2-13). A solution containing 150 mM NaCl mimics the extracellular fluid and therefore helps maintain the isolated cells in near-normal conditions. If the cells were placed in pure water, water would tend to enter the cells by osmosis; this might cause the cells to burst.

27. (a) CO_2 is nonpolar and would be able to cross a bilayer.
(b) Glucose is polar and would not be able to pass through a bilayer because the presence of the hydroxyl groups means glucose is highly hydrated and would not be able to pass through the nonpolar tails of the molecules forming the bilayer.
(c) DNP is nonpolar and would be able to cross a bilayer.
(d) Calcium ions are charged and are, like glucose, highly hydrated and would not be able to cross a lipid bilayer.

29. Substances present at high concentration move to an area of low concentration spontaneously, or "down" a concentration gradient in a process that increases their entropy. The export of Na^+ ions out of the cell requires that the sodium ions be transported from an area of low concentration to an area of high concentration. The same is true for potassium transport. Thus, these processes are not spontaneous, and an input of cellular energy is required to accomplish the transport.

31. In a high-solute medium, the cytoplasm loses water; therefore its volume decreases. In a low-solute medium, the cytoplasm gains water and therefore its volume increases.

33. Since the molecular mass of H_2O is $18.0 \text{ g} \cdot \text{mol}^{-1}$, a given volume (for example, 1 L or 1000 g) has a molar concentration of $1000 \text{ g} \cdot \text{L}^{-1} \div 18.0 \text{ g} \cdot \text{mol}^{-1} = 55.5$ M. By definition, a liter of water at pH 7.0 has a hydrogen ion concentration of 1.0×10^{-7} M. Therefore, the ratio of $[H_2O]$ to $[H^+]$ is $55.5 \text{ M}/(1.0 \times 10^{-7} \text{ M}) = 5.55 \times 10^{8}$.

35. The HCl is a strong acid and dissociates completely. This means that the concentration of hydrogen ions contributed by the HCl is 1.0×10^{-9} M. But the concentration of the hydrogen ions contributed by the dissociation of water is 100-fold greater than this: 1.0×10^{-7} M. The concentration of the hydrogen ions contributed by the HCl is negligible in comparison. Therefore, the pH of the solution is equal to 7.0.

37. In aqueous solution, where virtually all biochemical reactions take place, an extremely strong acid such as HCl dissociates completely, so that all its protons are donated to water: $HCl + H_2O \rightarrow H_3O^+ + Cl^-$. This leaves H_3O^+ as the only acidic species remaining.

39. Since $pH = -\log[H^+]$, $[H^+] = 10^{-pH}$

For saliva, $[H^+] = 10^{-6.6} = 2.5 \times 10^{-7}$ M

For urine, $[H^+] = 10^{-5.5} = 3.2 \times 10^{-6}$ M

41. (a) The final concentration of HNO_3 is $\dfrac{(0.020 \text{ L})(1.0 \text{ M})}{0.520 \text{ L}} = 0.038$ M

Since HNO_3 is a strong acid and dissociates completely, the added $[H^+]$ is equal to $[HNO_3]$. (The existing hydrogen ion concentration in the water itself, 1.0×10^{-7} M, can be ignored because it is much smaller than the hydrogen ion concentration contributed by the nitric acid.)

$$pH = -\log[H^+]$$
$$pH = -\log(0.038)$$
$$pH = 1.4$$

(b) The final concentration of KOH is

$$\frac{(0.015 \text{ L})(1.0 \text{ M})}{0.515 \text{ L}} = 0.029 \text{ M}$$

Since KOH dissociates completely, the added $[OH^-]$ is equal to the $[KOH]$. (The existing hydroxide ion concentration in the water itself, 1.0×10^{-7} M, can be ignored because it is much smaller than the hydroxide ion concentration contributed by the KOH.)

$$K_w = 1.0 \times 10^{-14} = [H^+][OH^-]$$
$$[H^+] = \frac{1.0 \times 10^{-14}}{[OH^-]}$$
$$[H^+] = \frac{1.0 \times 10^{-14}}{(0.029 \text{ M})}$$
$$[H^+] = 3.4 \times 10^{-13} \text{ M}$$
$$pH = -\log[H^+]$$
$$pH = -\log(3.4 \times 10^{-13})$$
$$pH = 12.5$$

43. The stomach contents have a low pH due to the contribution of gastric juice (pH 1.5–3.0). When the partially digested material enters the small intestine, the addition of pancreatic juice (pH 7.8–8.0) neutralizes the acid and increases the pH.

45. (a) $C_2O_4^{2-}$ **(b)** SO_3^{2-} **(c)** HPO_4^{2-} **(d)** CO_3^{2-}
(e) AsO_4^{3-} **(f)** PO_4^{3-} **(g)** O_2^{2-}

47.

HO—C—COOH with CH2—COOH groups
Citric acid

Piperidine (H+ on N, H)

^+H_3N—CH—C—OH, H, CH2, CH2, CH2, CH2, H—NH2$^+$
Lysine

Oxalic acid: HO—C—C—OH

Barbituric acid (NH, N—H ring)

4-Morphine ethanesulfonic acid (MES)

49.

^+H_3N—CH—C—O$^-$ with CH2, CH2, CH2, CH2, NH3$^+$ chain

51. The pK of the fluorinated compound would be lower (it is 9.0); that is, the compound becomes less basic and more acidic. This occurs because the F atom, which is highly electronegative, pulls on the nitrogen's electrons, loosening its hold on the proton.

53. (a) 10 mM glycinamide buffer because its pK is closer to the desired pH.
(b) 20 mM Tris buffer because the higher the concentration of the buffering species, the more acid or base it can neutralize.
(c) Neither; each solution will contain an equilibrium mixture of the boric acid and its conjugate base (borate).

55. Because it is small and nonpolar (see Solution 2), CO_2 can quickly diffuse across cell membranes to exit the tissues and enter red blood cells.

57. (a) The three ionizable protons of phosphoric acid have pK values of 2.15, 6.82, and 12.38 (Table 2-4). The pK values are the midpoints of the titration curve.

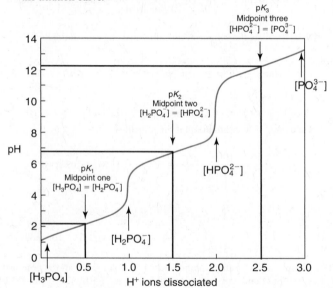

(b) The dissociation of the second proton has a pK of 6.82, which is closest to the pH of blood. Therefore, the weak acid present in blood is $H_2PO_4^-$ and the weak acid is HPO_4^{2-}.
(c) The dissociation of the third proton has a pK of 12.38. Therefore, a buffer solution at pH 11 would consist of the weak acid HPO_4^{2-} and its conjugate base, PO_4^{3-} (supplied as the sodium salts Na_2HPO_4 and Na_3PO_4).

59. Calculate the final concentrations of the weak acid ($H_2PO_4^-$) and conjugate base (HPO_4^{2-}). Note that K^+ is a spectator ion.

$$[H_2PO_4^-] = \frac{(0.025 \text{ L})(2.0 \text{ M})}{0.200 \text{ L}} = 0.25 \text{ M}$$

$$[HPO_4^{2-}] = \frac{(0.050 \text{ L})(2.0 \text{ M})}{0.200 \text{ L}} = 0.50 \text{ M}$$

Next, substitute these values into the Henderson–Hasselbalch equation using the pK values in Table 2-4:

$$pH = pK + \log \frac{[A^-]}{[HA]}$$
$$pH = 6.82 + \log(0.50 \text{ M})/(0.25 \text{ M})$$
$$pH = 6.82 + 0.30$$
$$pH = 7.12$$

61. First, determine the ratio of $[A^-]$ to $[HA]$:

$$pH = pK + \log\frac{[A^-]}{[HA]}$$

$$\log\frac{[A^-]}{[HA]} = pH - pK$$

$$\frac{[A^-]}{[HA]} = 10^{(pH-pK)}$$

Substitute the values for the desired pH (5.0) and the pK (4.76):

$$\frac{[A^-]}{[HA]} = 10^{(5.0-4.76)} = 10^{0.24} = 1.74$$

Calculate the number of moles of acetate (A^-) already present:

$$(0.50 \text{ L})(0.20 \text{ mol} \cdot \text{L}^{-1}) = 0.10 \text{ moles acetate}$$

Calculate the moles of acetic acid needed, based on the calculated ratio:

$$\frac{[A^-]}{[HA]} = 1.74$$

$$[HA] = \frac{0.10 \text{ moles}}{1.74}$$

$$[HA] = 0.057 \text{ moles}$$

Finally, calculate the volume of glacial acetic acid needed:

$$\frac{0.057 \text{ moles}}{17.4 \text{ mol} \cdot \text{L}^{-1}} = 0.0033 \text{ L, or } 3.3 \text{ mL}$$

The addition of 3.3 mL to a 500-mL solution dilutes the solution by less than 1%, which doesn't introduce significant error.

63. (a)

$$HO-(H_2C)_2-NH^+ \quad N-(CH_2)_2-SO_3^- + H_2O \rightleftharpoons$$

Weak acid (HA)

$$HO-(H_2C)_2-N \quad N-(CH_2)_2-SO_3^- + H_3O^+$$

Conjugate base (A)

(b) The pK for HEPES is 7.55; therefore, its effective buffering range is 6.55–8.55.

(c) $1.0 \text{ L} \times \dfrac{0.10 \text{ mole}}{\text{L}} \times \dfrac{260.3 \text{ g}}{\text{mol}} = 26 \text{ g}$

Weigh 26 g of the HEPES salt and add to a beaker. Dissolve in slightly less than 1.0 liter of water (leave "room" for the HCl solution that will be added in the next step).

(d) At the final pH,

$$\frac{[A^-]}{[HA]} = 10^{(pH-pK)} = 10^{(8.0-7.55)} = 10^{0.45} = 2.82$$

For each mole of HCl added, x, one mole of HEPES salt (A^-) will be converted to a mole of HEPES acid (HA). The starting amount of A^- is $(1.0 \text{ L})(0.10 \text{ mol} \cdot \text{L}^{-1}) = 0.10$ mole. After the HCl is added, the amount of A^- will be 0.10 mole $- x$, and the amount of HA will be x. Consequently,

$$\frac{[A^-]}{[HA]} = 2.82 = \frac{0.10 \text{ mole} - x}{x}$$

$$2.82x = 0.10 \text{ mol} - x$$

$$3.82x = 0.10 \text{ mol}$$

$$x = 0.10 \text{ mol}/3.82 = 0.0262 \text{ mol}$$

Calculate how much 6.0 M HCl to add:

$$\frac{0.0262 \text{ mol}}{6.0 \text{ mol} \cdot \text{L}^{-1}} = 0.0044 \text{ L, or } 4.4 \text{ mL}$$

To make the buffer, dissolve 26 g of HEPES salt [see part (c)] in less than 1.0 L. Add 4.4 mL of 6.0 M HCl, then add water to bring the final volume to 1.0 L.

65. (a) First, calculate the ratio of $[A^-]$ to $[HA]$. Rearranging the Henderson–Hasselbalch equation gives

$$\frac{[A^-]}{[HA]} = 10^{(pH-pK)} = 10^{(2.0-8.3)} = 10^{-6.3} = 5 \times 10^{-7}$$

Virtually all of the Tris is in the weak acid form. Therefore, the concentration of the weak acid, HA, is 0.10 M and the concentration of the conjugate base, A^-, is 5.0×10^{-8} M.

(b) The added HCl dissociates completely, so the amount of H^+ added is $(0.0015 \text{ L})(3.0 \text{ mol} \cdot \text{L}^{-1}) = 0.0045$ mol. In an effective buffer, the acid would convert some of the conjugate Tris base to weak acid. But the concentration of conjugate base is already negligible. Therefore, the moles of additional H^+ should be added to the concentration of hydrogen ions already present $(1.0 \times 10^{-2}$ M), for a total concentration of 0.0145 M.

$$pH = -\log[H^+] = \log(0.0145 \text{ M}) = 1.84$$

The buffer has not functioned effectively. There was not enough conjugate base to react with the additional hydrogen ions added. The result is a decrease in pH from 2.0 to 1.84.

(c) When NaOH is added, an equivalent amount of Tris acid (HA) is converted to Tris base (A^-). Let $x =$ moles of OH^- added $= (0.0015 \text{ L})(3.0 \text{ mol} \cdot \text{L}^{-1}) = 0.0045$ moles $= 4.5$ mmol.
The final amount of A^- is 5.0×10^{-8} mol $+ 4.5$ mmol $= 4.5$ mmol.
The final amount of HA is 100 mmol $- 4.5$ mmol $= 95.5$ mmol.
The new pH is determined by substituting the new concentrations of H^- and HA into the Henderson–Hasselbalch equation:

$$pH = pK + \log\frac{[A^-]}{[HA]}$$

$$pH = 8.3 + \log\frac{(4.5 \text{ mmol})}{(95.5 \text{ mmol})}$$

$$pH = 8.3 + (-1.3) = 7.0$$

Tris is not an effective buffer at pH = 2.0, a pH more than 6 units lower than its pK value. Virtually all of the Tris is in the weak acid form at this pH. If acid is added, there is not enough base to absorb the excess added hydrogen ions, and the pH decreases. If base is added, some of the weak acid is converted to the conjugate base and the pH approaches the value of the pK.

67. Ammonia and ammonium ions are in equilibrium, as represented by the following equation:

$$NH_4^+ \rightleftharpoons H^+ + NH_3$$

Carbonic acid and bicarbonate ions are in equilibrium, as represented by the following equation:

$$H_2CO_3 \rightleftharpoons H^+ + HCO_3^-$$

Phosphate ions are in equilibrium, according to the following equation:

$$H_2PO_4^- \rightleftharpoons H^+ + HPO_4^{2-}$$

In metabolic acidosis, the concentration of protons increases, so the equilibrium shifts to form $H_2PO_4^-$, carbonic acid and ammonium ions. In order to bring the pH back to normal, the kidney will excrete $H_2PO_4^-$ and ammonium ions and bicarbonate ions will be reabsorbed. The result is a decrease in the concentration of protons and an increase in blood pH.

Chapter 3

1. The heat treatment destroys the polysaccharide capsule of the wild-type *Pneumococcus*, but the DNA survives the heat treatment. The DNA then "invades" the mutant *Pneumococcus* and supplies the genes encoding the enzymes needed for the capsule synthetic pathway that the mutant lacks. The mutant is now able to synthesize a capsule and has the capacity to cause disease, which results in the death of the mice and the appearance of encapsulated *Pneumococcus* in the mouse tissue.

3. Some of the labeled "parent" DNA appears in the progeny, but none of the labeled protein appears in the progeny. This indicates that the bacteriophage DNA is involved in the production of progeny bacteriophages, but bacteriophage protein is not required.

5.

5-Methylcytosine

7. The base, 5-chlorouracil, is a substitute for thymine (5-methyluracil).

9. Thymine (5-methyluracil) contains a methyl group attached to C5 of the pyrimidine ring of uracil.

11.

Phosphodiester bond

If the dinucleotide were DNA, it would lack OH groups at each ribose C2′ position.

13. The total amount of purines (A + G) in DNA must equal the total amount of pyrimidines (C + T) because each base pair in the double-stranded DNA molecule consists of a purine and a pyrimidine. This is not true for RNA, which is single-stranded.

15. (a) Using Chargaff's rules (see Solution 14), the number of C residues must also be 24,182. Subtracting 2 × 24,182) from 97,004 yields 48,640 (A + T) residues. Dividing this number by 2 yields 24,320 residues each of A and T.

(b) GenBank reports only the sequence of a single strand of DNA, since the sequence of the complementary strand can easily be deduced using Chargaff's rules.

17. It is a G:C base pair.

19. The statement is false because the greater stability of GC-rich DNA is due to the stronger stacking interactions involving G:C base pairs and does not depend on the number of hydrogen bonds in the base pairs.

21. The sugar–phosphate backbone is found on the outside of the molecule. The polar sugar molecules can form hydrogen bonds with the surrounding water molecules. The negatively charged phosphate groups interact favorably with positively charged ions. The nonpolar nitrogen bases are found on the inside of the molecule and interact favorably via stacking interactions. In this way, contact with the aqueous solution is minimized, as described by the hydrophobic effect.

23.

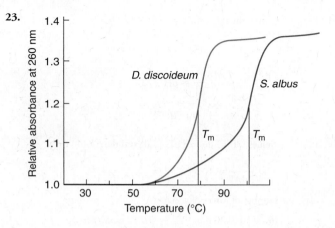

25. The DNA from the organisms that thrive in hot environments would contain more G and C than DNA from species living in a more temperate environment. The higher GC content increases the stability of DNA at high temperatures.

27. You should increase the temperature to melt out imperfect matches between the probe and the DNA.

29. (a) An inherited characteristic could be determined by more than one gene.

(b) Some sequences of DNA encode RNA molecules that are not translated into protein (for example, rRNA and tRNA).

(c) Some genes are not transcribed during a cell's lifetime. This can occur if the gene is expressed only under certain environmental conditions or in certain specialized cells in a multicellular organism.

31. (a) TGTGGTACCACGTAGACTGA
(b) ACACCAUGGUGCAUCUGACU

33. (a) A poly(Phe) polypeptide was produced.
(b) Poly(A) produces poly(Lys); poly(C) yields poly(Pro); and poly(G) yields poly(Gly).

35.
First reading frame:

AGG TCT TCA GGG AAT GCC TGG CGA GAG GGG AGC AGC
Ser-Ser-Ser-Gly-Asn-Ala-Trp-Arg-Glu-Gly-Ser-Ser-

TGG TAT CGC TGG GCC CAA AGG C
Trp-Tyr-Arg-Trp-Ala-Gln-Arg

Second reading frame:

A GGT CTT CAG GGA ATG CCT GGC GAG AGG GGA GCA
Gly-Leu-Gln-Gly-Met-Pro-Gly-Glu-Arg-Gly-Ala-

GCT GGT ATC GCT GGG CCC AAA GGC
Ala-Gly-Ile-Ala-Gly-Pro-Lys-Gly

Third reading frame:

AG GTC TTC AGG GAA TGC CTG GCG AGA GGG GAG CAG
Val-Phe-Arg-Glu-Cys-Leu-Ala-Arg-Gly-Glu-Gln-

CTG GTA TCG CTG GGC CCA AAG GC
Leu-Val-Ser-Leu-Gly-Pro-Lys

37. Asparagine has two codons, AAU and AAC (see Table 3-3). An A → G mutation at the second position could generate a codon for serine (AGU or AGC).

39. The genetic code (shown in Table 3-3) is redundant. Since there are 64 different possibilities for 3-base codons and only 20 amino acids, some amino acids have more than one codon. If a mutation just happens to occur in the third position (3′ end), the mutation might not alter the protein sequence. For example, GUU, GUC, GUA, and GUG all code for valine. A mutation in the third position of a valine codon would still result in the selection of valine and would have no effect on the amino acid sequence of the protein.

41. First, identify the translation start site, the Met residue whose codon is AUG in the mRNA (see Table 3-3) or ATG in the DNA. Translation stops at the DNA sequence TAA, which corresponds to the stop codon UAA in the mRNA. Use Table 3-3 to decode the intervening codons, substituting U for T.

CTCAGAGTTCACC ATG GGC TCC ATC GGT GCA GCA AGC ATG GAA
              Met Gly Ser Ile Gly Ala Ala Ser Met Glu

... 1104 bp ... UUC UUU GGC AGA UGU GUU UCC CCU UAA AAAGAA
.... Phe Phe Gly Arg Cys Val Ser Pro *

43. *C. ruddii*, with such a small genome and only 182 genes, must be some sort of parasite rather than a free-living bacterium. (In fact, *C. ruddii* is an insect symbiont.)

45. The 35 million differences out of 3.2 billion total nucleotides represent approximately 1%, or a bit less than the original claim. (This number reflects single-base differences and does not account for insertions and deletions of multiple bases.)

47. **(a)** The first reading frame is the longest ORF.

First reading frame:

TAT GGG ATG GCT GAG TAC AGC ACG TTG AAT GAG GCG
Tyr-Gly-Met-Ala-Glu-Tyr-Ser-Ser-Leu-Tyr-Glu-Ala-

ATG GCC GCT GGT GAT G
Met-Ala-Ala-Gly-Asp-

Second reading frame:

T ATG GGA TGG CTG AGT ACA GCA CGT TGA ATG AGG
  Met-Gly-Trp-Leu-Ser-Thr-Ala-Arg-<u>stop</u>-Met-Arg-

CGA TGG CCG CTG GTG ATG
Arg-Trp-Pro-Leu-Val-Met

Third reading frame:

TA TGG GAT GGC TGA GTA CAG CAC GTT GAA TGA GGC
   -Tyr-Asp-Gly-<u>stop</u>-Leu-Gln-His-Val-Glu-<u>stop</u>-Gly-

GAT GGC CGC TGG TGA TG
Gly-Arg-Trp-<u>stop</u>-

(b) Assuming the reading frame has been correctly identified, the most likely start site is the first Met residue in the first ORF.

49. If a SNP occurs every 300 nucleotides or so, and if there are about 3 million kb in the human genome (see Table 3-4), then a SNP occurs every $(3 \times 10^6/300) = 10,000$ kb or so. [Source: http://ghr.nlm.nih.gov/handbook/genomicresearch/snp]

51. **(a)** The strongest associations are located between positions 67,400,000 and 67,450,000.
(b) Gene B contains SNPs associated with the disease whereas genes A and B do not. [From Duerr, R. H., et al., *Science* **314**, 1461–1463 (2006).]

53. Polymerization occurs in the $5' \rightarrow 3'$ direction and a 3' OH group must be available, so the primer must be complementary to the sequence as shown.

5'-AGTCGATCCCTGATCGTACGCTACGGTAACGT-3'
                    3'-TGCCATTGCA-5'

55. A restriction endonuclease is often used to prepare fragments of DNA for insertion into a cloning vector. Since the cloned DNA contains the recognition site (whose sequence is known), this sequence can be used as a starting point to sequence the unknown DNA segment.

57. You can use a DNA polymerase that is not heat-stable. You would have to cool the reaction mixture to a temperature at which the polymerase works best, and you would have to add the enzyme at each reaction cycle because it would be destroyed every time the temperature was raised to melt the double-stranded DNA.

59. MspI, AsuI, EcoRI, PstI, SauI, and NotI generate sticky ends. AluI and EcoRV generate blunt ends.

61. The restriction enzyme with the longer recognition sequence would be a rare cutter because it is likely to encounter this sequence less often and therefore will cleave the DNA less frequently than a restriction enzyme with a shorter recognition sequence.

63. If you were to use plasmids, you would need at least 150,000 clones to accommodate the 3×10^6–kb genome in fragments of 20 kb each (see

Table 3-6). This is an almost unmanageable number. However, if you were to use yeast artificial chromosomes, you would need only a minimum of 3000 clones to accommodate the 3×10^6–kb genome in fragments of 1000 kb.

65. A stop codon would need to follow the Leu residue (instead of the Ser that is present in the wild-type protein). A mismatched primer could substitute a UCA (Ser) codon for a UGA stop codon. There are several correct answers due to the redundancy of the genetic code. One possible sequence is GTTTTCGCTGTTCTTTCAUGA. [From Yue, L., *J. Virol.* **83**, 11588–11598 (2009).]

Chapter 4

1. **(a)** The chiral carbon is marked with an asterisk. D-Alanine, the mirror image isomer, is shown.

L-Alanine   D-Alanine

(b) Since the majority of proteins contain L-amino acids, the presence of D-amino acids in the bacterial cell wall renders the cell wall less susceptible to digestion by proteases (enzymes produced by certain organisms to destroy bacteria).

3. **(a)** His, Phe, Pro, Tyr, Trp
(b) His, Phe, Tyr, Trp
(c) His, Cys, Ser, Thr, Tyr
(d) Gly
(e) Arg, Lys
(f) Asp, Glu
(g) Cys, Met

5. From least soluble to most soluble: Trp, Val, Thr, Ser, Arg. You can use Table 4-3 as a guide, but you should also be able to do this type of problem without using the table.

7. This combination cannot occur in significant amounts at any pH. An unprotonated amino group cannot exist with a protonated carboxylate group because the amino group's pK is much greater than the carboxyl group's pK (therefore the carboxyl group ionizes at a lower pH than the amino group).

9. In a free amino acid, the charged amino and carboxylate groups, which are separated only by the alpha carbon, electronically influence each other. When the amino acid forms a peptide bond, one of these groups is neutralized, thereby altering the electronic properties of the remaining group.

11. At pH 6.0, the N-terminus is protonated (+1 charge), and the C-terminus is unprotonated (−1 charge). The four His side chains (pK = 6.0) are half-protonated (see Fig. 2-17) so that each has a charge of +0.5. The tetrapeptide therefore has a net charge of +2.

13. **(a)** The three amino acids are Ser, Tyr, and Gly.
(b) Cyclization of the polypeptide backbone occurs between the carbonyl carbon of Ser and the amide nitrogen of Gly.
(c) Oxidation results in a double bond in the Tyr side chain between Cα and Cβ (the second carbon of the side chain).

15.

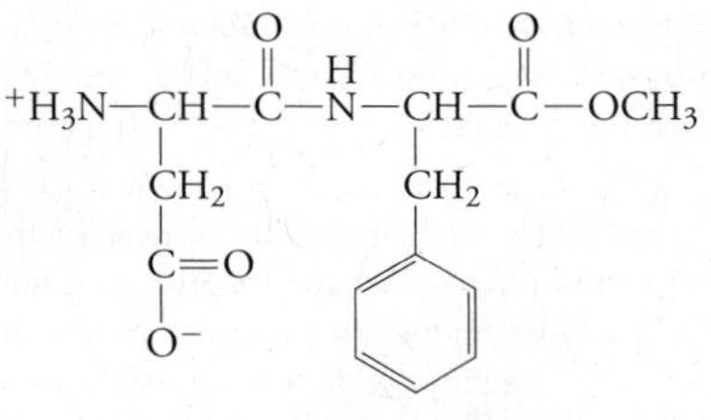

Aspartame (Asp–Phe–OMe)

17. (a)

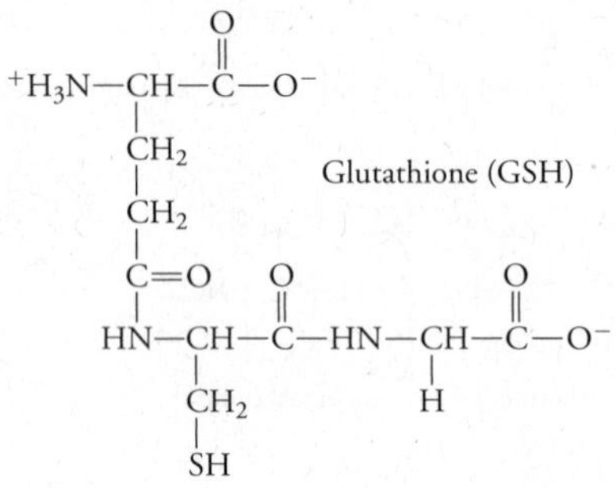

Glutathione (GSH)

(b)

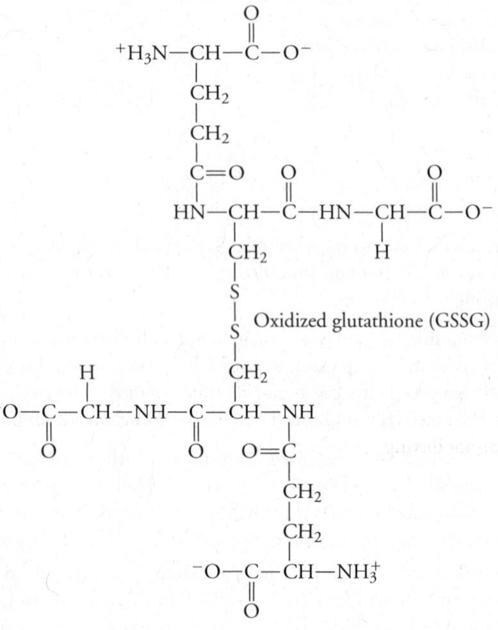

Oxidized glutathione (GSSG)

19.

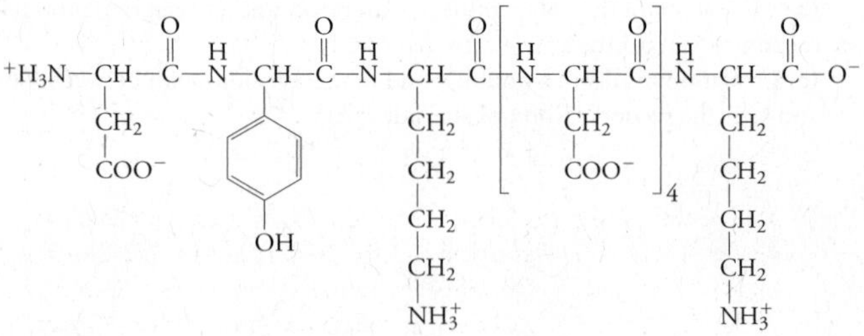

21. There are six possible sequences: HPR, HRP, PHR, PRH, RHP, RPH.

23. (a) tertiary
   (b) quaternary
   (c) primary
   (d) secondary

25.

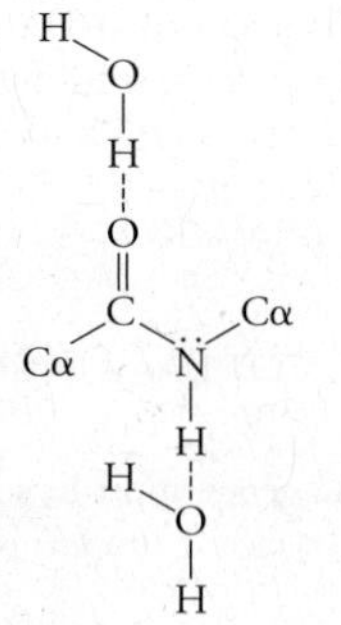

27. Both the DNA helix and the α helix turn in the right-handed direction. Both helices have tightly packed interiors: In DNA, the interior of the helix is occupied by nitrogenous bases; in the α helix, the atoms of the polypeptide backbone contact one another. In the α helix, the side chains extend outward from the helix; no such structure exists in the DNA helix.

29. The amino group of Pro is linked to its side chain (see Fig. 4-2), which limits the conformational flexibility of a peptide bond involving the amino group. The geometry of this peptide bond is incompatible with the bond angles required for a polypeptide to form an α helix.

31.

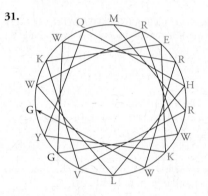

The polar amino acid residues are shown in red; the nonpolar residues are shown in blue. The polar residues are mainly on one side of the helix, while the nonpolar residues are on the other side. Quite a few of the polar side chains are positively charged. [This is an example of an amphipathic helix. From Martoglio, B., Graf, R., and Dobberstein, B., *EMBO J.* **16,** 6636–6645 (1997).]

33. Triose phosphate isomerase is an example of an α/β protein.

35. It's possible that the ligand has a positive charge and forms an ion pair with the negatively charged Glu on the receptor. When the Glu is mutated to an Ala, the negative charge on the receptor is lost and the ion pair between the receptor and the ligand can no longer form.

37. There are many possible answers for this question. An example is shown for each (below).

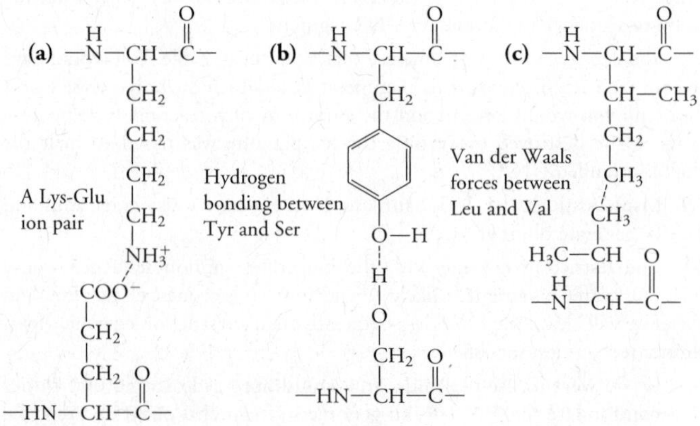

(a) A Lys-Glu ion pair

(b) Hydrogen bonding between Tyr and Ser

(c) Van der Waals forces between Leu and Val

39. A polypeptide synthesized in a living cell has a sequence that has been optimized by natural selection so that it folds properly (with hydrophobic residues on the inside and polar residues on the outside). The random sequence of the synthetic peptide cannot direct a coherent folding process, so hydrophobic side chains on different molecules aggregate, causing the polypeptide to precipitate from solution.

41. Anfinsen's ribonuclease experiment demonstrated that a protein's primary structure dictates its three-dimensional structure. Although some proteins, like ribonuclease, can renature spontaneously *in vitro,* most proteins require the assistance of molecular chaperones to fold properly *in vivo.*

43. When the temperature increases, the vibrational and rotational energy of the atoms making up the protein molecules also increases, which increases the chance that the proteins will denature. Increasing the synthesis of chaperones under these conditions allows the cell to renature, or refold, proteins that have been denatured by heat.

45. Proline does not fit well into the structure of the α helix, because of both its geometry (see Problem 29) and the absence of a peptide —NH to contribute to hydrogen bonding (see Problem 37). This amino acid substitution would produce a protein with decreased stability, which would affect the ability of red blood cells to deform their shape in order to squeeze through capillaries. The cells would become damaged and would be removed from circulation, causing anemia. [From Johnson, C. P., Gaetani, M., Ortiz, V., Bhasin, N., Harper, S., Gallagher, P. G., Speicher, D. W., and Discher, D. E., *Blood* **109,** 3538–3543 (2007).]

47. If the proteins were homodimers, they would be more likely to have two identical sites to interact with their palindromic recognition sites in the DNA. Heterodimeric proteins would likely lack the necessary symmetry. (In fact, these enzymes are homodimeric.)

49. The arginine residues, with their positively charged side chains, and the aspartate residues, with their negatively charged side chains, are likely to be found on the surface of the monomer. These residues likely form ion pairs that stabilize the dimeric form. When these residues were mutated to neutral amino acid side chains, the ion pairs could not form, the dimer could not form, and the equilibrium shifted in favor of the monomers. [From Huang, Y., Misquitta, S., Blond, S. Y., Adams, E., and Colman, R. F., *J. Biol. Chem.*, **283,** 32800–32888 (2008).]

51. The only ionizable groups in the dipeptide are the C-terminus ($pK = 3.5$) and the N-terminus ($pK = 9.0$). $pI = ½ (3.5 + 9.0) = 6.25$.

53. The protein must contain groups that undergo protonation/deprotonation at pH values near 4.3. The only amino acids with side chain pK values in this range are Asp and Glu (Table 4-1), so the protein likely contains an abundance of these residues.

55. At pH 7.0, the peptide likely has a net positive charge since Arg (R) and Lys (K) outnumber Asp (D) and Glu (E). Therefore, the peptide is likely to bind to CM groups but not to DEAE groups.

57. The amino terminal residue is Ala. The carboxyl terminal residue must be Met, since the dodecapeptide was not cleaved when CNBr was added. Chymotrypsin cleaves after Phe. Fragment II contains the Asp, so it appears in the sequence first, and Phe must be the cleavage site. Trypsin cleaves after Lys. Since fragment III contains Asp, Lys must be the cleavage site. Elastase cleaves after Gly, Val and Ser. Val must occupy the second position followed by a Pro, and this Val was not cleaved.

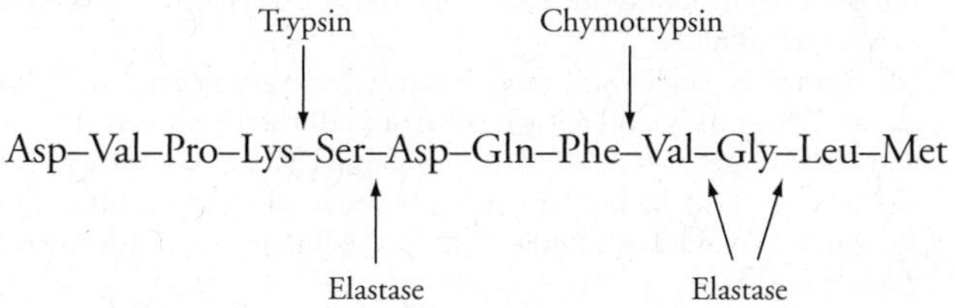

Trypsin
Chymotrypsin
Asp–Val–Pro–Lys–Ser–Asp–Gln–Phe–Val–Gly–Leu–Met
Elastase
Elastase

[Based on Anastasi, A., Montecucchi, P., Erspamer, V., and Visser, J., *Experientia* **33,** 857–858 (1977).]

59. Edman degradation of a polypeptide with a disulfide cross-link would not work properly when the first Cys became exposed at the N-terminus of the polypeptide (the Cys would not be released in the next reaction since it would still be covalently linked to a Cys residue farther along the polypeptide chain). Reduction before sequencing breaks the disulfide bond, and alkylation of the two free Cys groups prevents re-formation of the bond.

61. (a) If only one proteolytic cleavage is carried out, the sequences of the fragments could be determined but it would not be possible to place the fragments in the proper order. If different enzymes are used to generate two sets of fragments, overlapping peptides would allow ordering of all the sequences.

(b) Trypsin cleaves after Lys and Arg residues. The fragments resulting from digestion of the heavy and light chains are shown below, identified by residue number.

Light chain	Heavy chain
6	1–7
7–9	8–18
10–11	19–31
12	32–35
13–20	36–38
21–28	39–42
29–38	43–57
	58–61
	62–71
	78–85
	86–91

(c) Chymotrypsin would be a good choice for the second enzyme because it cleaves after Phe, Tyr, and Trp. The following fragments would be obtained:

Light chain	Heavy chain
6–13	1–63
14–25	64–83
26–38	84–91

63. Since each codon corresponds to an amino acid, the error rate is

$$\frac{5 \times 10^{-4} \text{ error}}{\text{residue}} \times 500 \text{ residues} = 0.25$$

About one-quarter of the polypeptides would contain a substitution.

65. Leu and Ile are isomers and have the same mass; therefore, mass spectrometry cannot distinguish them.

67.

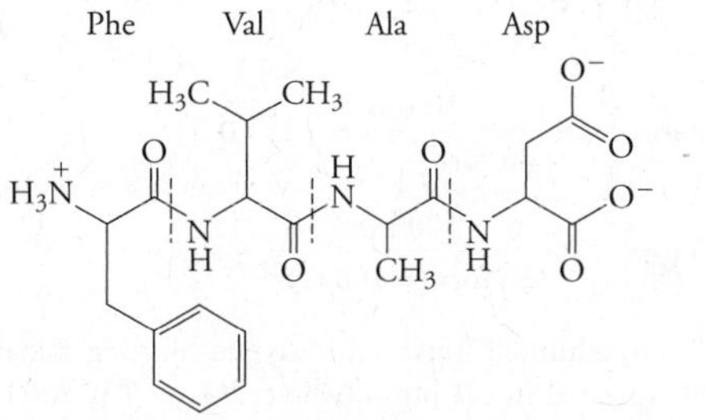

Phe
Val
Ala
Asp

Dashed lines indicate broken bonds. The smallest charged fragment is the N-terminal residue (Phe), which has a mass of approximately 149 D (9 C + 1 N + 1 O + 11 H).

69. In a protein crystal, the residues at the end of a polypeptide chain may experience fewer intramolecular contacts and therefore tend to be less ordered (more mobile in the crystal). If their disorder prevents them from generating a coherent diffraction pattern, it may be impossible to map their electron density.

Chapter 5

1. Globin lacks an oxygen-binding group and therefore cannot bind O_2. Heme alone is easily oxidized and therefore cannot bind O_2. The bound heme gives a protein such as myoglobin the ability to bind O_2. In turn, the protein helps prevent oxidation of the heme Fe atom.

3. Myoglobin facilitates oxygen diffusion in the cell by acting as a "molecular bucket brigade." It accepts oxygen delivered to the cell by hemoglobin and then transfers the oxygen to proteins in the mitochondrion. Myoglobin is 50% saturated with oxygen when the intracellular oxygen concentration is equal to its p_{50} value, so it functions most effectively under these conditions. At intracellular oxygen concentrations greater than p_{50}, oxygen remains bound to hemoglobin and is not transferred; at oxygen concentrations less than p_{50}, myoglobin doesn't bind sufficient oxygen. In either case, the transfer of oxygen from hemoglobin to mitochondrial proteins is compromised.

5. (a) The interior of the roast is purple because it does not have access to oxygen and thus the myoglobin in the meat is primarily in the deoxygenated form, which is purple. When sliced, the myoglobin is exposed to oxygen and becomes oxygenated, turning red.

(b) When meat is cooked, the globin chains are denatured and the iron is oxidized to Fe^{3+}, converting the myoglobin to met-myoglobin, which is brown.

(c) In a vacuum package, the cells rapidly use up any available oxygen in metabolism, which results in mostly deoxygenated myoglobin, which is purplish. Meat packaged in oxygen-permeable packaging continues to be metabolically active so that most of the myoglobin is oxygenated and red, which is more appealing to consumers. [Source: "Color Changes in Cooked Beef," James Claus, National Cattlemen's Association.]

7. Isoleucine has a larger side chain than valine and will decrease the size of the pocket. This steric hindrance results in decreased binding of oxygen. [From Olson, J. S., and Phillips, Jr., G. N., *J. Biol. Chem.* **271**, 17593–17596 (1996).]

9. In the arteries, nearly all the hemoglobin is oxygenated and therefore takes on the color of the Fe(II), in which the sixth coordination site is occupied by O_2. Blood that has passed through the capillaries and given up some of its oxygen contains a mixture of oxy- and deoxyhemoglobin. Deoxyhemoglobin, in which the Fe(II) has only five ligands, imparts a bluish tinge to venous blood.

11. (a) Position 6 (Gly) and position 9 (Val) appear to be invariant.

(b) Conservative substitutions occur at position 1 (Asp and Lys, both charged), position 10 (Ile and Leu, similar in structure and hydrophobicity), and position 2 (all uncharged bulky side chains). Positions 5 and 8 appear to tolerate some substitution.

(c) The most variable positions are 3, 4, and 7, where a variety of residues appear.

13. Use Equation 5-4 to calculate the fractional saturation (Y) for hyperbolic binding, letting $K = 26$ torr:

$$Y = \frac{pO_2}{K + pO_2}$$

At 30 torr, $Y = \dfrac{30 \text{ torr}}{26 \text{ torr} + 30 \text{ torr}} = 0.54$

At 100 torr, $Y = \dfrac{100 \text{ torr}}{26 \text{ torr} + 100 \text{ torr}} = 0.79$

Therefore, if hemoglobin exhibited hyperbolic oxygen-binding behavior, it would be only 79% saturated in the lungs (where $pO_2 \approx 100$ torr) and would exhibit a loss of saturation of only 25% (79% − 54%) in the tissues (where $pO_2 \approx 30$ torr). Hemoglobin's sigmoidal binding behavior allows it to bind more oxygen in the lungs so that it can deliver relatively more oxygen to the tissues (for an overall change in saturation of about 40%; see Fig. 5-7).

15. $Y = \dfrac{(pO_2)^n}{(p_{50})^n + (pO_2)^n}$

$Y = \dfrac{(25 \text{ torr})^3}{(40 \text{ torr})^3 + (25 \text{ torr})^3}$

$Y = 0.20$

$Y = \dfrac{(pO_2)^n}{(p_{50})^n + (pO_2)^n}$

$Y = \dfrac{(120 \text{ torr})^3}{(40 \text{ torr})^3 + (120 \text{ torr})^3}$

$Y = 0.96$

17. At the high altitude where the bar-headed goose resides, less oxygen is available to bind to hemoglobin in the lungs. The bar-headed goose hemoglobin has a lower p_{50} value and a higher oxygen affinity than the plains-dwelling grelag goose hemoglobin, so the bar-headed goose hemoglobin can more easily bind oxygen in order to deliver it to the tissues. [From Jessen, T.-H., Weber, R. E., Fermi, G., Tame, J., and Braunitzer, G., *Proc. Natl. Acad. Sci.* **88**, 6519–6522 (1991).]

19. The increased O_2 release is the result of the Bohr effect. The increase in $[H^+]$ promotes the shift from the oxy to the deoxy conformation of hemoglobin. The decrease in oxygen affinity improves oxygen delivery to the muscle, where it is needed.

21. (a) Asp 94 and His 146, which is protonated, form an ion pair in the deoxy form of hemoglobin. The protons are produced by cellular respiration. As described by the Bohr effect, an increase in hydrogen ion concentration favors the deoxy form of hemoglobin so that oxygen can be delivered to the tissues.

(b) The presence of the negatively charged Asp 94 increases the pK value of the imidazole ring of His and promotes the formation of the ion pair between the two side chains. The increased pK value means that the imidazole ring's affinity for protons is increased. [From Berenbrink, M., *Resp. Physiol. Neurobiol.* **154**, 165–184 (2006).]

23. Negatively charged glutamate side chains on the surface of the oxygenated lamprey monomer would resist association due to the charge–charge repulsion. But when the pH decreases, excess protons would bind to the glutamate side chains, neutralizing them. The monomers would associate to form the deoxygenated tetramer. In this manner, oxygen is delivered to lamprey tissue when the pH decreases in metabolically active tissue. [From Qiu, Y., Maillett, D. H., Knapp, J., Olson, J. S., and Riggs, A. F., *J. Biol. Chem.* **275**, 13517–13528 (2000).]

25. (a) Curve A represents fetal hemoglobin, which has a higher oxygen affinity than adult hemoglobin (curve B). Curve A shows greater fractional saturation than curve B at any given oxygen concentration.

(b) In the γ chain, the positively charged His 21 in the β chain is replaced by the neutral Ser, resulting in a central cavity with fewer positive charges. The allosteric effector BPG cannot bind to Hb F as effectively as it binds to Hb A because the negatively charged BPG forms fewer ion pairs in the central cavity of Hb F. BPG plays a role in decreasing oxygen-binding affinity, so the inability of BPG to bind to Hb F means that fetal hemoglobin has a higher oxygen-binding affinity. Oxygen transfer from the maternal circulation to the fetal circulation can be accomplished more efficiently if the fetal hemoglobin has a higher oxygen affinity than maternal hemoglobin.

27. (a) Normal hemoglobin has a sigmoidal curve, which means that the binding and release of oxygen from normal hemoglobin is cooperative. The hyperbolic binding curve of Hb Great Lakes indicates that there is little cooperativity in oxygen binding and release.

(b) Hb Great Lakes has a higher affinity for oxygen. More than 60% of the mutant hemoglobin has bound oxygen. Hb A is about 30% oxygenated.

(c) Both hemoglobins are essentially 100% oxygenated and therefore have equal affinities.

(d) Normal hemoglobin is more efficient at oxygen delivery. It delivers about 70% of its bound oxygen (since it is 100% oxygenated at 75 torr and 30% oxygenated at 20 torr). Hb Great Lakes is less efficient since less than 40% of its bound oxygen is delivered to the tissues. [From Rahbar, S., Winkler, K., Louis, J., Rea, C., Blume, K., and Beutler, E., *Blood* **58**, 813–817 (1981).]

29. (a) A person hyperventilates in order to obtain more oxygen, and the blood pH increases as a result, as shown by the equations below. Excessive removal of $CO_2(g)$ from the lungs during hyperventilation causes the third equation to shift to the right. This depletes $CO_2(aq)$, which causes the second equation to shift right, thus depleting carbonic acid. This causes the first equation to shift right as hydrogen ions and bicarbonate

ions combine to form more carbonic acid. Depletion of hydrogen ions results in a basic pH.

$$H^+(aq) + HCO_3^-(aq) \rightleftharpoons H_2CO_3(aq)$$
$$H_2CO_3(aq) \rightleftharpoons CO_2(aq) + H_2O(l)$$
$$CO_2(aq) \rightleftharpoons CO_2(g)$$

(b) The decrease in alveolar pCO_2 concentration can be explained by the hyperventilation, as described in part (a). The concentration of 2,3-BPG increases in order to convert more of the hemoglobin molecules to the low-affinity T form so that oxygen can be effectively delivered to the cells.

31. The substitution at the C-terminus could affect the position of the F8 histidine in such a way that oxygen either binds more readily or dissociates less readily. The substitution is quite near the His 146 involved in binding protons (see Problem 21) and may decrease this side chain's proton-binding affinity. Since His 146 is located in the central cavity, BPG would bind less readily, which favors the oxygenated or R form of hemoglobin.

33. Both types of molecules are proteins and consist of polymers of amino acids. Both contain elements of secondary structure. But globular proteins are water soluble and nearly spherical in shape. Examples include proteins such as hemoglobin and myoglobin as well as enzymes. Their cellular role involves participating in the chemical reactions of the cell in some way. In contrast, fibrous proteins tend to be water insoluble and have an elongated shape. Their cellular role is structural—as elements of the cytoskeleton of the cell and the connective tissue matrix between cells.

35. Microfilaments and microtubules consist entirely of subunits that are assembled in a head-to-tail fashion, so the polarity of the subunits (actin monomers in microfilaments and tubulin dimers in microtubules) is preserved in the fully assembled fiber. In intermediate filament assembly, only the initial step (dimerization of parallel helices) maintains polarity. In subsequent steps, subunits align in an antiparallel fashion, so in a fully assembled intermediate filament, each end contains heads and tails.

37. Because phalloidin binds to F-actin but not to G-actin, the addition of phalloidin fixes actin in the filamentous form. This impairs cell motility because cell movement requires both actin polymerization at the leading edge of the cell and depolymerization at the trailing edge. In the presence of phalloidin, depolymerization does not occur and cell movement is not possible.

39. During rapid microtubule growth, β-tubulin subunits containing GTP accumulate at the (+) end because GTP hydrolysis occurs following subunit incorporation into the microtubule. In a slowly growing microtubule, the (+) end will contain relatively more GTP that has already been hydrolyzed to GDP. A protein that preferentially binds to (+) ends that contain GTP rather than GDP could thereby distinguish fast- and slow-growing microtubules.

41. Polymers composed of β-tubulin molecules allowed to polymerize in the presence of a nonhydrolyzable analog of GTP are more stable. When the β-tubulin subunits are exposed to GTP in solution, the GTP binds to the β-tubulin and then is hydrolyzed to GDP, which remains bound to the β-tubulin. Additional αβ heterodimers are then added. The microtubule ends with GDP bound to the β-tubulin are less stable than those bound to GTP because protofilaments with GDP bound are curved rather than straight and tend to fray. If a nonhydrolyzable analog is bound, it will resemble GTP and the protofilament will be straight rather than curved. It is less likely to fray and the resulting protofilament is more stable as a result.

43. Microtubules form the mitotic spindle during cell division. Because cancer cells are rapidly dividing cells, and hence undergo mitosis at a rate more rapid than in most other body cells, drugs that target tubulin and thus interfere with the formation of the mitotic spindle in some way will slow the growth of cancerous tumors.

45. Colchicine, which promotes microtubule depolymerization, inhibits the mobility of the neutrophils because cell mobility results from polymerization and depolymerization of microtubules.

47. As shown in Figure 5-22, microtubules link replicated chromosomes to two points at opposite sides of the cell. Vinblastine's ability to stabilize the microtubules at the (+) end while destabilizing the (−) end disrupts this linkage. Mitosis slows down or completely halts as a result. [From Panda, D., Jordan, M. A., Chu, K. C., and Wilson, L., *J. Biol. Chem.*, **271**, 29807–29812 (1996).]

49. (a) The first and fourth side chains are buried in the coiled coil, but the remaining side chains are exposed to the solvent and therefore tend to be polar or charged.

(b) Although the residues at positions 1 and 4 in both sequences are hydrophobic, Trp and Tyr are much larger than Ile and Val and would therefore not fit as well in the area of contact between the two polypeptides in a coiled coil (see Fig. 5-25).

51. The reducing agent breaks the disulfide bonds (—S—S—) between keratin molecules. Setting the hair brings the reduced Cys residues (with their —SH groups) closer to new partners on other keratin chains. When the hair is then exposed to an oxidizing agent, new disulfide bonds form between the Cys residues and the hair retains the shape of the rollers.

53. (a) Actin's primary structure is its amino acid sequence. Its secondary structure includes its α helices, β sheets, and other conformations of the polypeptide backbone. Its tertiary structure is the arrangement of its backbone and all its side chains in a globular structure. Monomeric actin by definition has no quaternary structure. However, when actin monomers associate to form a microfilament, the arrangement of subunits becomes the filament's quaternary structure. Thus, actin is an example of a protein that has quaternary structure under certain conditions.

(b) Collagen's primary structure is its amino acid sequence. Its secondary structure is the left-handed helical conformation characteristic of the Gly–Pro–Hyp repeating sequence. Its tertiary structure is essentially the same as its secondary structure, since most of the protein consists of one type of secondary structure. Collagen's quaternary structure is the arrangement of its three chains in a triple helix. It is also possible to view the triple helix as a form of tertiary structure, with quaternary structure referring to the association of collagen molecules.

55. The bacterial enzymes degrade collagen, the major protein in connective tissue. Treatment of the tissue with these enzymes degrades the collagen in the extracellular matrix without harming the cells themselves and thus facilitates the preparation of cells for culturing. [Source: Worthington Biochemical Corporation.]

57. (a) Collagen B is from rat, and collagen A is from the sea urchin.

(b) The stability of each of these collagens is correlated with their hydroxyproline content. The higher the percentage of hydroxyproline, the more regular the structure and the more difficult it is to melt, resulting in more stable collagen. The rat has a more stable collagen, and the sea urchin, which lives in cold water, has a less stable collagen. It is important to note that the melting temperatures of each collagen molecule are higher than the temperature at which each organism lives. Thus, each organism has stable collagen at the temperature of its environment. [From Mayne, J., and Robinson, J. J., *J. Cell. Biochem.* **84**, 567–574 (2001).]

59. (a) (Pro–Pro–Gly)$_{10}$ has a melting temperature of 41°C, while (Pro–Hyp–Gly)$_{10}$ has a melting temperature of 60°C. (Pro–Hyp–Gly)$_{10}$ and (Pro–Pro–Gly)$_{10}$ both have an imino acid content of 67%, but (Pro–Hyp–Gly)$_{10}$ contains hydroxyproline, whereas (Pro–Pro–Gly)$_{10}$ does not. Hydroxyproline therefore has a stabilizing effect relative to proline.

(b) (Pro–Pro–Gly)$_{10}$ and (Gly–Pro–Thr(Gal))$_{10}$ have the same melting point, indicating that they have equal stabilities. This is interesting because (Pro–Pro–Gly)$_{10}$ has an imino acid content of 67%, whereas (Gly–Pro–Thr(Gal))$_{10}$ has an imino acid content of only 33%. The glycosylated threonine must have an effect similar to that of proline. It is possible that the galactose, which contains many hydroxyl groups, provides additional sites for hydrogen bonding and would thus contribute to the stability of the triple helix.

(c) The inclusion of (Gly–Pro–Thr)$_{10}$ is important because the results show that this molecule doesn't form a triple helix. This molecule is included as a control to show that the increased stability of the (Gly–Pro–Thr(Gal))$_{10}$ is due to the galactose, not to the threonine residue itself. [From Bann, J. G., Peyton, D. H., and Bächinger, H. P., *FEBS Lett.* **473**, 237–240 (2000).]

61.

63. Because collagen has such an unusual amino acid composition (almost two-thirds consists of Gly and Pro or Pro derivatives), it contains relatively fewer of the other amino acids and is therefore not as good a source of amino acids as proteins containing a greater variety of amino acids. In particular, gelatin lacks tryptophan and contains only small amounts of methionine.

65. (a) The patients all suffer from scurvy, a disease resulting from the lack of vitamin C, or ascorbate, in the diet.
(b) Ascorbic acid is necessary for the formation of hydroxyproline residues in newly synthesized collagen chains. Underhydroxylated collagen is less stable, so tissues containing the defective collagen are less sound, leading to bruising, joint swelling, fatigue, and gum disease.
(c) Patients with a gastrointestinal disease may actually be consuming foods with vitamin C, but the disease impairs absorption. Patients suffering from poor dentition and alcoholism may have overall difficulties with food intake. Patients following various food fads might consume diets that are so unusual or restrictive that their intake of vitamin C is insufficient to support healthy collagen synthesis. [From Olmedo, J. M., Yiannias, J. A., Windgassen, E. B., and Gornet, M. K., *Int. J. Dermatol.*, **45**, 909–913 (2006).]

67.

69. (a) Minoxidil inhibits the lysyl hydroxylase enzyme. In the presence of minoxidil, fewer Lys residues are hydroxylated, as demonstrated by the decrease of [³H]-lysine incorporated into collagen.
(b) Since minoxidil inhibits lysyl hydroxylase, procollagen chains would be underhydroxylated. Lysines lacking hydroxyl groups decrease the stability of the collagen and increase the likelihood that the collagen will be degraded once it is secreted from the fibroblast cell. This would be effective in reducing collagen concentrations in patients with fibrosis.
(c) A similar explanation indicates that long-term minoxidil use in patients without fibrosis has the potential to compromise collagen synthesis in fibroblasts of the skin. The underhydroxylated collagen synthesized in the presence of minoxidil will be less stable, and skin structure might be affected as a result. The medical literature reports only scalp irritation, dryness, scaling, itching, and redness as a side effect in some men who received topical minoxidil treatments for nearly two years, however. [From Murad, S., Walker, L. C., Tajima, S., and Pinnell, Sr. R., *Arch. Biochem. Biophys.* **308**, 42–47 (1994) and Price, V. H., *N. Engl. J. Med.* **341**, 964–973 (1999).]

71. Myosin is both fibrous and globular. Its two heads are globular, with several layers of secondary structure. Its tail, however, consists of a single fibrous coiled coil.

73. (a) Diffusion is a random process. It tends to be slow (especially for large substances and over long distances). Because it is random, it operates in three dimensions (not linearly) and has no directionality.
(b) An intracellular transport system must have some sort of track (for linear movement of cargo) and an engine that moves cargo along the track by converting chemical energy to mechanical energy. The engine must operate irreversibly to promote rapid movement in one direction. Finally, some sort of addressing system is needed to direct cargo from its source to a certain destination.

75. When muscles contract, myosin heads bind and release actin in a process that requires ATP for the physical movement of myosin along the actin filament. At the time of death, cellular processes that generate ATP cease. Myosin heads remain bound to actin, but in the absence of ATP, the conformational change that causes myosin to release the actin does not occur, and stiffened muscles are the consequence.

77. Normal bone development involves the formation of bone tissue in response to stresses placed on the bone. When muscle activity is impaired, as in muscular dystrophy, the forces that shape bone development are also abnormal, leading to abnormal bone growth.

Chapter 6

1. A globular protein can bind substrates in a sheltered active site and can support an arrangement of functional groups that facilitates the reaction and stabilizes the transition state. Most fibrous proteins are rigid and extended and therefore cannot surround the substrate to sequester it or promote its chemical transformation.

3. The rate enhancement is calculated as the ratio of the catalyzed rate to the uncatalyzed rate:

$$\frac{61 \text{ s}^{-1}}{1.3 \times 10^{-10} \text{ s}^{-1}} = 4.7 \times 10^{11}$$

[From Bryant, R. A. R., and Hansen, D. E., *J. Am. Chem. Soc.*, **118**, 5498–5499 (1996).]

5. For adenosine deaminase:

$$\frac{370 \text{ s}^{-1}}{1.8 \times 10^{-10} \text{ s}^{-1}} = 2.1 \times 10^{12}$$

For triose phosphate isomerase:

$$\frac{4300 \text{ s}^{-1}}{4.3 \times 10^{-6} \text{ s}^{-1}} = 1.0 \times 10^9$$

The rate of the uncatalyzed reaction is slower for the adenosine deaminase reaction than for the triose phosphate isomerase reaction. But adenosine deaminase is able to catalyze its reaction so that it occurs more quickly than the reaction catalyzed by triose phosphate isomerase. Therefore, the rate enhancement for the adenosine deaminase reaction is greater.

7.

9. (a) Pyruvate decarboxylase is a lyase. During the elimination of the carboxylate group (—COO⁻) of pyruvate, a double bond is formed in CO_2 (O=C=O).
(b) Alanine aminotransferase is a transferase. The amino group is transferred from alanine to α-ketoglutarate.
(c) Alcohol dehydrogenase is an oxidoreductase. Acetaldehyde is reduced to ethanol or ethanol is oxidized to acetaldehyde.
(d) Hexokinase is a transferase. The phosphate group is transferred from ATP to glucose to form glucose-6-phosphate.
(e) Chymotrypsin is a hydrolase. Chymotrypsin catalyzes the hydrolysis of peptide bonds.

11. Succinate dehydrogenase is an oxidoreductase.

13. A kinase transfers a phosphate group from ATP to a substrate.

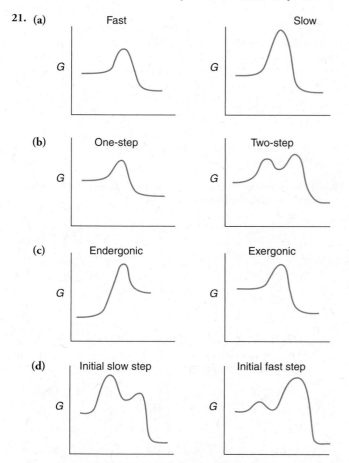

15. (a) Reaction 4; (b) Reaction 1; (c) Reaction 3; (d) Reaction 2

17. $2\,H_2O_2 \rightleftharpoons O_2 + 2\,H_2O$

19. Every tenfold increase in rate corresponds to a decrease of about 5.7 kJ · mol^{-1} in $\Delta G^{\ddagger}$. For the nuclease, with a rate enhancement on the order of 10^{14}, $\Delta G^{\ddagger}$ is lowered about 14×5.7 kJ · mol^{-1}, or about 80 kJ · mol^{-1}.

21. (a) Fast / Slow

(b) One-step / Two-step

(c) Endergonic / Exergonic

(d) Initial slow step / Initial fast step

23. Yes. An enzyme decreases the activation energy barrier for both the forward and the reverse directions of a reaction.

25. (a) Gly, Ala, and Val have side chains that lack the functional groups required for acid–base or covalent catalysis.
(b) Mutating one of these residues may alter the conformation at the active site enough to disrupt the arrangement of other groups that are involved in catalysis.

27. (a) In order for any molecule to act as an enzyme, it must be able to recognize and bind a substrate specifically, it must have the appropriate functional groups to effect a chemical reaction, and it must be able to position those groups for reaction.
(b) Functional groups on the nitrogen bases can participate in chemical reactions in much the same way as amino acid side chains on proteins. For example, the amino groups on adenine, guanine, and cytosine bases could act as nucleophiles and could also act as proton donors.
(c) DNA, as a double-stranded molecule, has limited conformational freedom. RNA, which is single-stranded, is able to assume a greater

range of conformations. This flexibility allows it to bind to substrates and carry out chemical transformations.

29. His 57 abstracts a proton from Ser 195, thus rendering the serine oxygen a better nucleophile. When Ser 195 is modified by formation of a covalent bond with DIP, the proton is no longer available and Ser 195 is unable to function as a nucleophile.

31.

33. His residues are often involved in proton transfer. A carboxymethylated His would be unable to donate or accept protons.

[From Shapiro, R., Weremowicz, S., Riordan, J. F., and Vallee, B., *Proc. Natl. Acad. Sci.* **84**, 8783–8787 (1987).]

35. (a)

(b) Since benzoate resembles the substrate, it is likely that benzoate binds to the active site of the enzyme. Under these conditions, FDNP does not have access to the active site and will be unable to react with the tyrosine (which is also assumed to be part of the enzyme's active site because of its unusual reactivity). [From Nishino, T., Massey, V., and Williams, C. H., *J. Biol. Chem.,* **255**, 3610–3616 (1979).]

37. At very low pH values, His would be protonated and unable to form a hydrogen bond with Ser. Asp would also be protonated and unable to form a hydrogen bond with His. At very high pH values, His would be unprotonated and unable to form a hydrogen bond with Asp.

39. (a) Glu 35 has a pK of 5.9 and Asp 52 has a pK of 4.5.
(b) Lysozyme is inactive at pH 2.0 because both the Glu and the Asp are protonated. The Asp is no longer negatively charged and cannot nucleophilically attack the carbocation intermediate. Lysozyme is inactive

at pH 8.0 because both the Glu and the Asp are unprotonated. The Glu would be unable to donate a hydrogen to cleave the bond between the sugar residues.

(c)

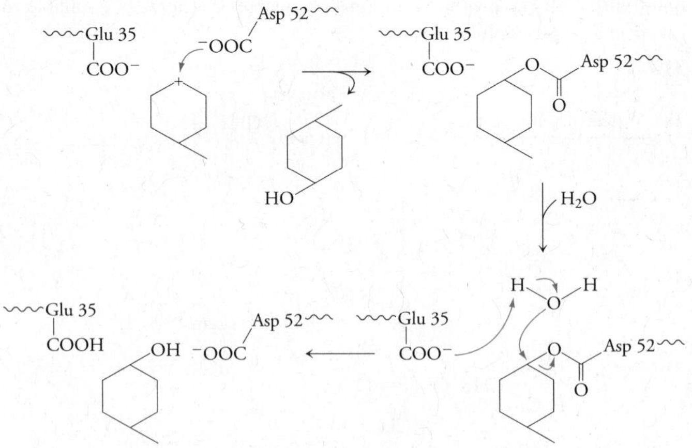

41. (a) In the first part of the reaction, the ester bond is cleaved and the chymotrypsin is acetylated. The *p*-nitrophenolate ion is quickly released, which accounts for the rapid increase in absorbance seen at 410 nm. The enzyme must be regenerated before a second round of catalysis can begin, which requires a deacetylation step. This step is much slower than the first step. Once the acetate is released, the enzyme is regenerated and another molecule of substrate can bind and react. Thus a steady state is reached and the absorbance increases at a uniform rate until the substrate is depleted.

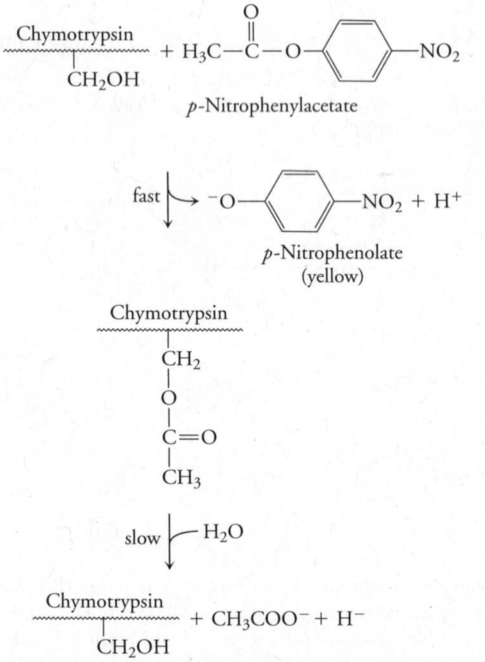

(b) The reaction coordinate diagram will look like the one in Figure 6-7, since this is a two-step reaction. Each step has a characteristic activation energy. The acetylated chymotrypsin is the intermediate.

(c) Yes, chymotrypsin and trypsin use the same catalytic mechanism, so trypsin can act as an esterase as well as a protease.

43. This is an example of convergent evolution, in which unrelated proteins evolve similar characteristics.

45. Cys 278 is highly exposed and unusually reactive compared to other cysteines in creatine kinase. Cys 278, because of its high reactivity, is probably one of the catalytic residues in the enzyme. The other cysteine residues

are not as reactive because they are not directly involved in catalysis and/or because they are shielded in some way that prevents them from reacting with NEM.

47.

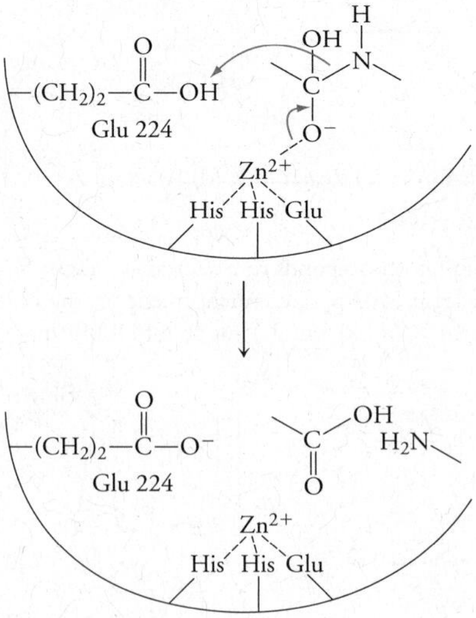

[From Li, L., Binz, T., Niemann, H., and Singh, B. R., *Biochemistry* **39**, 2399–2405 (2000).]

49. (a) The deamidation reaction for asparagine is shown. The deamidation reaction for glutamine is similar.

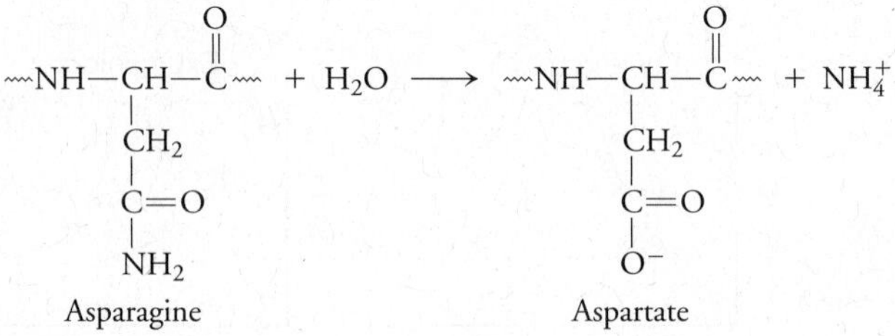

(b)

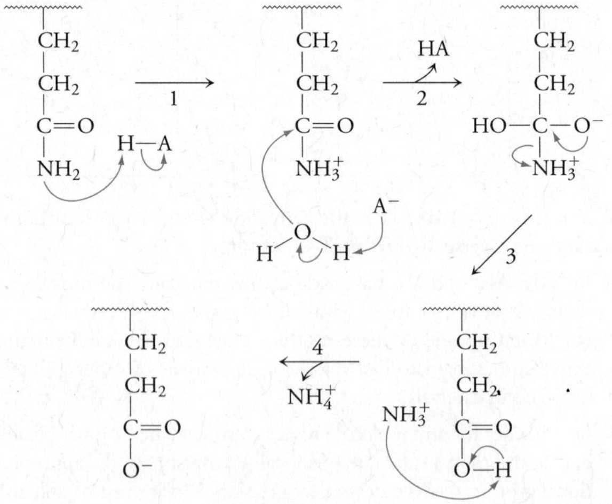

(c) Ser and Thr residues could stabilize the transition state. They could also serve as bases (if unprotonated) and accept a proton from water to form a hydroxide ion that would act as the attacking nucleophile. Ser and Thr (in the unprotonated form) could also act as attacking nucleophiles themselves.

(d) The mechanism for the deamidation of an amino terminal Gln residue is shown. Amino terminal Asn residues are not deamidated because a four-membered ring, which is unstable, would result.

$$+ NH_3 + HA$$

(e) Water is a substrate in the reaction. The Asn and Gln residues on the surface of the protein have much greater access to water molecules than interior Asn and Gln residues. [From Wright, H. T., *Crit. Rev. Biochem. Mol. Biol.* **26**, 1–52 (1991).]

51. The ability of an enzyme to accelerate a reaction depends on the free energy difference between the enzyme-bound substrate and the enzyme-bound transition state. As long as this free energy difference is less than the free energy difference between the unbound substrate and the uncatalyzed transition state, the enzyme-mediated reaction proceeds more quickly.

53. In a serine protease, there is no need to exclude water from the active site, since it is a reactant for the hydrolysis reaction catalyzed by the enzyme.

55. The zinc ion participates in catalysis by polarizing the water molecule so that its proton is more easily abstracted by Glu 224. The positively charged zinc ion stabilizes the negatively charged oxygen in the transition state.

57. The transition state structure is likely tetrahedral at position 6 on the purine ring, since adenosine is planar whereas 1,6-dihydropurine is tetrahedral at this position. Enzymes bind the transition state much more tightly than the substrate.

59. A mutation can increase or decrease an enzyme's catalytic activity, depending on how it affects the structure and activity of groups in the active site.

61. **(a)** Trypsin cleaves peptide bonds on the carboxyl side of Lys and Arg residues, which are positively charged at physiological pH. These residues fit into the specificity pocket and interact electrostatically with Asp 189.
(b) A mutant trypsin with a positively charged Lys residue in its specificity pocket would no longer prefer basic side chains because the like charges would repel one another. The mutant trypsin might instead prefer to cleave peptide bonds on the carboxyl side of negatively charged residues such as Glu and Asp, whose side chains could interact electrostatically with the positively charged Lys residue.
(c) If the substrate specificity pocket does not include a positively charged Lys residue, then there would be no reason to expect the mutant enzyme to prefer substrates with acidic side chains. Instead, the mutant enzyme is more likely to prefer substrates with nonpolar side chains such as Leu or Ile. [From Graf, L., Craik, C. S., Patthy, A., Roczniak, S., Fletterick, R. J., and Rutter, W. J., *Biochemistry* **26**, 2616–2623 (1987).]

63. During chymotrypsin activation, chymotrypsin cleaves other chymotrypsin molecules at a Leu, a Tyr, and an Asn residue. Only one of these (Tyr) fits the standard description of chymotrypsin's specificity. Clearly, chymotrypsin has wider substrate specificity, probably determined in part by the identities of residues near the scissile bond.

65. No, the compound shown in Problem 8 would not be hydrolyzed by chymotrypsin. The side chain on the carboxyl side of the amide bond is an arginine side chain which would not fit into the chymotrypsin's specificity pocket.

67. **(a)** Persistent activation of trypsinogen to trypsin also results in the activation of chymotrypsinogen to chymotrypsin (see Problem 64) and causes proteolytic destruction of the pancreatic tissue.

(b) Since trypsin is at the "top of the cascade," it makes sense to inactivate it by using a trypsin inhibitor. [From Hirota, M., Ohmuraya, M., and Baba, H., *Postgrad. Med. J.* **82**, 775–778 (2006).]

69. A protease with extremely narrow substrate specificity (that is, a protease with a single target) would pose no threat to nearby proteins because these proteins would not be recognized as substrates for hydrolysis.

Chapter 7

1. The hyperbolic shape of the velocity versus substrate curve suggests that the enzyme and substrate physically combine so that the enzyme becomes saturated at high concentrations of substrate. The lock-and-key model describes the interaction between an enzyme and its substrate in terms of a highly specific physical association between the enzyme (lock) and the substrate (key).

3. $\dfrac{k_{cat}}{K_M} = \dfrac{4.0 \text{ s}^{-1}}{1.4 \times 10^{-4} \text{ M}} = 2.8 \times 10^4 \text{ s}^{-1} \cdot \text{M}^{-1}$

5. $v = \dfrac{d[P]}{dt} = \dfrac{25 \times 10^{-6} \text{ M}}{50 \text{ d} \times 24 \text{ h} \cdot \text{day}^{-1} \times 3600 \text{ s} \cdot \text{h}^{-1}}$

$v = 5.8 \times 10^{-12} \text{ M} \cdot \text{s}^{-1}$

[From Bryant, R. A. R., and Hansen, D. E., *J. Am. Chem. Soc.*, **118**, 5498–5499 (1996).]

7. $(5.8 \times 10^{-12} \text{ M} \cdot \text{s}^{-1})(4.7 \times 10^{11}) = 2.7 \text{ M} \cdot \text{s}^{-1}$

9. $v = -\dfrac{d[S]}{dt}$

$v = -\dfrac{0.065 \text{ M}}{60 \text{ s}}$

$v = -1.1 \times 10^{-3} \text{ M} \cdot \text{s}^{-1}$

11.

Reaction	Molecularity	Rate equation
A → B + C	Unimolecular	Rate = k [A]
A + B → C	Bimolecular	Rate = k [A][B]
2 A → B	Bimolecular	Rate = k [A]^2
2 A → B + C	Bimolecular	Rate = k [A]^2

Units of k	Reaction velocity proportional to …	Order
s^{-1}	[A]	First
M$^{-1} \cdot$ s^{-1}	[A] and [B]	Second
M$^{-1} \cdot$ s^{-1}	[A] squared	Second
M$^{-1} \cdot$ s^{-1}	[A] squared	Second

13. rate = k [sucrose]

rate = $(5.0 \times 10^{-11} \text{ s}^{-1})(0.050 \text{ M})$

rate = $2.5 \times 10^{-12} \text{ M} \cdot \text{s}^{-1}$

15. **(a)** The reaction is a second-order reaction because the units of the rate constant k are M$^{-1} \cdot$ s^{-1}.
(b) Convert the partial pressure of CO_2 into units of molar concentration by using the ideal gas law:

$$PV = nRT$$

$$\frac{n}{V} = \frac{P}{RT}$$

$$\frac{n}{V} = \frac{40 \text{ torr} \times \dfrac{1 \text{ atm}}{760 \text{ torr}}}{\dfrac{0.0821 \text{ L} \cdot \text{atm}}{\text{K} \cdot \text{mol}} \times 310 \text{ K}}$$

$$\frac{n}{V} = 0.0021 \text{ M}$$

Next, substitute values for concentration and k into the rate law:

$$\text{rate} = k[\text{RNH}_2][\text{CO}_2]$$

$$= 4950 \text{ M}^{-1} \cdot \text{s}^{-1} \times 0.6 \times 10^{-3} \text{ M} \times 0.0021 \text{ M}$$

$$= 6.2 \times 10^{-3} \text{ M} \cdot \text{s}^{-1}$$

(c) The rate constant k would increase with increasing pH because an amino group is more likely to be unprotonated (and thus able to react) at a higher pH.

[From Gros, G., Forster, E., and Lin, L., *J. Biol. Chem.* **251**, 4398–4407 (1976).]

17.

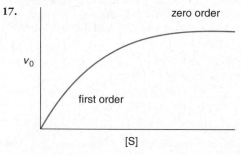

19. The apparent K_M would be greater than the true K_M because the experimental substrate concentration would be less than expected if some of the substrate has precipitated out of solution during the reaction.

21. (a) Velocity measurements can be made using any convenient units. K_M is by definition a substrate concentration, so its value does not reflect how the velocity is measured.

(b) It's not necessary to know the enzyme concentration in order to determine K_M or V_{max}; only [S] and v_0 must be known so that a Lineweaver–Burk plot can be constructed. The value of $[E]_T$ is required to calculate k_{cat} since $k_{cat} = V_{max}/[E]_T$, according to Equation 7-23.

23. $v_0 = \dfrac{V_{max}[S]}{K_M + [S]}.$

$$v_0 = \frac{7.5 \ \mu\text{mol} \cdot \text{min}^{-1} \cdot \text{mg}^{-1} \times 0.15 \text{ mM}}{0.5 \text{ mM} + 0.15 \text{ mM}}$$

$$v_0 = \frac{7.5 \ \mu\text{mol} \cdot \text{min}^{-1} \cdot \text{mg}^{-1} \times 0.15 \text{ mM}}{0.65 \text{ mM}}$$

$$v_0 = 7.5 \ \mu\text{mol} \cdot \text{min}^{-1} \cdot \text{mg}^{-1} \times 0.23$$

$$v_0 = 1.7 \ \mu\text{mol} \cdot \text{min}^{-1} \cdot \text{mg}^{-1}$$

[From Phillips, R. S., Parniak, M. A., and Kaufman, S., *J. Biol. Chem.* **259**, 271–277 (1984).]

25. The V_{max} is approximately 30 μM $\cdot$ s^{-1} and the K_M is approximately 5 μM.

27. (a) $v_0 = 0.75 \ V_{max}$, so substitute in:

$$0.75 \ V_{max} = \frac{V_{max}[S]}{[S] + K_M}$$

V_{max} cancels out on both sides.

$$0.75 = \frac{[S]}{[S] + K_M}$$

$$0.75 \ ([S] + K_M) = [S]$$

$$0.75 \ K_M = 0.25[S]$$

$$3 \ K_M = [S]$$

Thus, the substrate concentration is three times as high as the K_M.

(b) $v_0 = 0.9 \ V_{max}$, so substitute in:

$$0.9 \ V_{max} = \frac{V_{max}[S]}{[S] + K_M}$$

V_{max} cancels out on both sides.

$$0.9 = \frac{[S]}{[S] + K_M}$$

$$0.9 \ ([S] + K_M) = [S]$$

$$0.9 \ K_M = 0.1[S]$$

$$9 \ K_M = [S]$$

Thus, the substrate concentration is nine times as high as the K_M.

29. $$k_{cat} = \frac{V_{max}}{[E]_T}$$

$$k_{cat} = \frac{4.0 \times 10^{-7} \text{ M} \cdot \text{s}^{-1}}{1.0 \times 10^{-7} \text{ M}}$$

$$k_{cat} = 4.0 \text{ s}^{-1}$$

The k_{cat} is the turnover number, which is the number of catalytic cycles per unit time. Each molecule of the enzyme therefore undergoes 4 catalytic cycles per second.

31.

Reaction	$1/V_{max}$ (s $\cdot$ M^{-1})	V_{max} (M $\cdot$ s^{-1})	$-1/K_M$ (M^{-1})	K_M (M)
1	4	0.25	−4	0.25
2	2	0.50	−1	1.0
3	2	0.50	−2	0.5

Reaction 1 has the lowest K_M; Reactions 2 and 3 are tied for the highest V_{max}.

33. (a) *N*-Acetyltyrosine ethyl ester, with its lower K_M value, has a higher affinity for the chymotrypsin enzyme. The aromatic tyrosine residue more easily fits into the nonpolar "pocket" on the enzyme than does the smaller aliphatic valine residue.

(b) The value of V_{max} is not related to the value of K_M, so no conclusion can be drawn.

35. The maximum rate at which two molecules can collide with one another in solution is 10^8 to 10^9 M$^{-1} \cdot$ s^{-1}. Enzymes with k_{cat}/K_M values in this range can be considered to be diffusion-controlled, which means the reaction is catalyzed as rapidly as the two reactants can encounter each other in solution. Thus, enzymes B and C are diffusion-controlled but enzyme A is not.

Enzyme	K_M	k_{cat}	k_{cat}/K_M
A	0.3 mM	5000 s^{-1}	1.7×10^7 M$^{-1} \cdot$ s^{-1}
B	1 nM	2 s^{-1}	2×10^9 M$^{-1} \cdot$ s^{-1}
C	2 μM	850 s^{-1}	4.2×10^8 M$^{-1} \cdot$ s^{-1}

37. (a) The reaction is a trisubstrate reaction and therefore does not obey Michaelis–Menten kinetics.

(b) The K_M value for one substrate is obtained by varying its concentration while holding the concentrations of the other two substrates constant at saturating levels.

(c) V_{max} is achieved by saturating the enzyme with each substrate. Therefore, the concentration of each substrate must be much greater than its K_M value. [From Brekken, D. L., and Phillips, M. A., *J. Biol. Chem.* **273**, 26317–26322 (1998).]

39. (a) The enzyme catalyzes the hydrolysis of the peptide bond on the carboxyl side of the Phe residue. One of the products, *p*-nitrophenolate, is bright yellow, and the rate of its appearance was monitored spectrophotometrically.

(b) The K_M values are nearly identical, which means that each enzyme has the same affinity for its substrate. The k_{cat} value for the Leu

31 enzyme is nearly six times greater than the k_{cat} value for the wild-type enzyme, which means that the mutant enzyme has a greater catalytic efficiency and a higher turnover rate of substrate converted to product per minute. The k_{cat}/K_M ratio reflects the specific reactivity of the substrate AAPF with the enzyme, and this ratio is larger in the mutant enzyme than in the wild type.

(c) The mutant enzyme is nearly threefold more active toward the casein substrate than the wild-type enzyme. Thus the improvement in catalytic activity seen in the mutant is not just toward an artificial substrate but toward a natural substrate as well.

(d) Ile 31 is near the Asp–His–Ser catalytic triad of the subtilisin E enzyme. Because Leu 31 improves the catalytic activity, this residue must somehow improve the function of the catalytic triad. Residue 31 is especially close to Asp, so it's possible that this residue plays a role in assisting Asp in its function in the catalytic triad. Histidine acts as a base catalyst and abstracts a proton from serine. The imidazole ring of histidine is positively charged as a result, and the role of Asp is to stabilize the positively charged imidazole ring. Thus leucine must somehow enable the Asp to fulfill this function better than isoleucine. Another possibility is that the substitution of leucine for isoleucine has altered the three-dimensional protein structure such that the catalytic triad residues are closer to one another and thus proton transfer is facilitated.

(e) Subtilisin would remove protein stains by hydrolyzing the peptide bonds of the protein and producing amino acids or short peptides as products. These products could be easily washed away from the clothing. [From Takagi, H., Morinaga, Y., Ikemura, H., and Inouye, M., *J. Biol. Chem.* **263**, 19592–19596 (1988).]

41. (a) If an irreversible inhibitor is present, the enzyme's activity would be exactly 100 times lower when the sample is diluted 100-fold. Dilution would not change the degree of inhibition.

(b) If a reversible inhibitor is present, dilution would lower the concentrations of both the enzyme and the inhibitor enough that some inhibitor would dissociate from the enzyme. The enzyme's activity would therefore not be exactly 100 times less than the diluted sample; it would be slightly greater because the proportion of uninhibited enzyme would be greater at the lower concentration.

43. (a) Since the structures are similar (both have choline groups), the inhibitor is competitive. Competitive inhibitors compete with the substrate for binding to the active site, so the structures of the inhibitor and the substrate must be similar.

(b) Yes, the inhibition can be overcome. If large amounts of substrate are added, the substrate will be able to effectively compete with the inhibitor such that very little inhibitor will be bound to the active site. The substrate "wins" the competition when it is in excess.

(c) Like all competitive inhibitors, the inhibitor binds reversibly.

45. It is difficult to envision how an inhibitor that interferes with the catalytic function (represented by k_{cat} or V_{max}) of amino acid side chains at the active site would not also interfere with the binding (represented by K_M) of a substrate to a site at or near those same amino acid side chains.

47. (a) NADPH is structurally similar to $NADP^+$ and is likely to be a competitive inhibitor.

(b) The V_{max} is the same in the presence and absence of the inhibitor since inhibition can be overcome at high substrate concentrations. The K_M increases because a higher concentration of substrate is needed to achieve half-maximal activity in the presence of an inhibitor.

(c) The K_M is 400 times greater for NAD^+, indicating that the enzyme prefers $NADP^+$ as a cofactor. The differences in V_{max} are not as great. [From Hansen, T., Schicting, B., and Schonheit, P., *FEMS Microbiol. Lett.* **216**, 249–253 (2002).]

49. The compound is a transition state analog (it mimics the planar transition state of the reaction) and therefore acts as a competitive inhibitor.

51. The structure of coformycin structurally resembles the proposed transition state for adenosine deaminase (see Section 7-3), and this supports the proposed structure. However, 1,6-dihydroinosine has a K_I of 1.5×10^{-13} M whereas coformycin's K_I is about 0.25 μM; thus, 1,6-dihydroinosine more closely resembles the transition state than does coformycin.

53.

$$\alpha = \frac{K_M \text{ (with I)}}{K_M \text{(no I)}}$$

$$\alpha = \frac{40 \text{ μM}}{10 \text{ μM}} = 4$$

$$\alpha = 1 + \frac{[I]}{K_I}$$

$$4 = 1 + \frac{30 \text{ μM}}{K_I}$$

$$K_I = 10 \text{ μM}$$

[From Gross, R. W., and Sobel, B. E., *J. Biol. Chem.* **258**, 5221–5226 (1983).]

55. The inhibitor is a mixed inhibitor. The V_{max} is decreased and the K_M value is increased in the presence of the inhibitor.

The V_{max} in the absence of inhibitor can be calculated by taking the reciprocal of the y intercept:

$$V_{max} = \frac{1}{y \text{ int}}$$

$$V_{max} = \frac{1}{1.51 \text{ (OD}^{-1} \cdot \text{min)}}$$

$$V_{max} = 0.66 \text{ OD} \cdot \text{min}^{-1}$$

The V_{max} in the presence of inhibitor can be calculated similarly:

$$V_{max} = \frac{1}{y \text{ int}}$$

$$V_{max} = \frac{1}{4.27 \text{ (OD}^{-1} \cdot \text{min)}}$$

$$V_{max} = 0.23 \text{ OD} \cdot \text{min}^{-1}$$

The K_M value in the absence of inhibitor can be determined by first calculating the x intercept and then by taking its reciprocal:

$$x \text{ int} = -\frac{b}{m}$$

$$x \text{ int} = -\frac{1.51 \text{ OD}^{-1} \cdot \text{min}}{1.52 \text{ min} \cdot \text{OD}^{-1} \cdot \text{mM}}$$

$$x \text{ int} = -0.99 \text{ mM}^{-1}$$

$$K_M = -\frac{1}{x \text{ int}}$$

$$K_M = -\frac{1}{-0.99 \text{ mM}^{-1}}$$

$$K_M = 1.0 \text{ mM}$$

The K_M value in the presence of inhibitor can be similarly determined:

$$x \text{ int} = -\frac{b}{m}$$

$$x \text{ int} = -\frac{4.27 \text{ OD}^{-1} \cdot \text{min}}{1.58 \text{ min} \cdot \text{OD}^{-1} \cdot \text{mM}}$$

$$x \text{ int} = -2.70 \text{ mM}^{-1}$$

$$K_M = -\frac{1}{x \text{ int}}$$

$$K_M = -\frac{1}{-2.70 \text{ mM}^{-1}}$$

$$K_M = 0.37 \text{ mM}$$

Dodecyl gallate is an uncompetitive inhibitor. In the presence of the inhibitor, the V_{max} and the K_M values decreased to a similar extent. The slopes of the lines in the Lineweaver–Burk plot are nearly the same. [From Kubo, I., Chen, Q.-X., and Nihei, K.-I., *Food Chem.* **81,** 241–247 (2003).]

57. (a) The Lineweaver–Burk plot is shown. The K_M is calculated from the x intercept, the V_{max} from the y intercept.

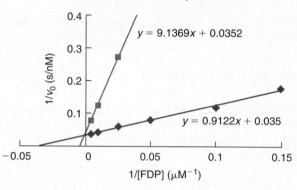

Inhibition of PTP1B by vanadate

$y = 9.1369x + 0.0352$

$y = 0.9122x + 0.035$

◆ Without inhibitor
■ With inhibitor

	Without vanadate	With vanadate
x intercept (μM^{-1})	−0.038	−0.0039
K_M (μM)	26	260
y intercept (s · nM^{-1})	0.035	0.035
V_{max} ($nM \cdot s^{-1}$)	28.5	28.5

(b) The inhibitor is a competitive inhibitor. The V_{max} is the same in the presence and absence of the inhibitor, but the K_M has increased tenfold, indicating that the vanadate is competing with the substrate for binding to the active site of the enzyme.

59. (a) The Lineweaver–Burk plot is shown. The K_M is calculated from the x intercept, the V_{max} from the y intercept.

	Without inhibitor	With inhibitor
y intercept, $(mM/min)^{-1}$	0.704	1.90
V_{max} (mM/min)	1.42	0.52
x intercept $(mM)^{-1}$	−0.949	−2.54
K_M (mM)	1.05	0.39

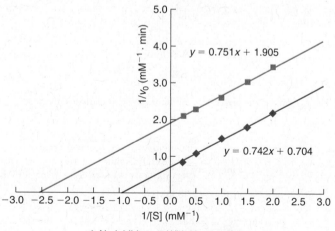

$y = 0.751x + 1.905$

$y = 0.742x + 0.704$

◆ No inhibitor ■ With homoarginine

(b) Homoarginine is an uncompetitive inhibitor. The slopes of the lines in the Lineweaver–Burk plot are nearly identical. A proportional decrease in V_{max} and K_M occurs in the presence of the inhibitor.

(c) Because homoarginine is an uncompetitive inhibitor, it does not bind to the active site of the alkaline phosphatase enzyme but to another site that interferes with the activity of the enzyme in some way. The intestinal alkaline phosphatase catalyzes the same reaction as the bone alkaline phosphatase, so the active sites of the two enzymes are likely to be similar, but the structures of the enzymes may be sufficiently different that the intestinal enzyme lacks the binding site for homoarginine. [From Lin, C.-W., and Fishman, W. H., *J. Biol. Chem.* **247,** 3082–3097 (1972).]

61. (a) ATCase is an allosteric enzyme because its activity versus [S] curve is sigmoidally shaped.

(b) CTP is a negative effector, or inhibitor, because when CTP is added, the K_M increases and thus the affinity of the enzyme for the substrate decreases. CTP is the eventual product of the pyrimidine biosynthesis pathway; thus, when the concentration of CTP is sufficient for the needs of the cell, CTP inhibits an early enzyme in the synthetic pathway, ATCase, by feedback inhibition.

(c) ATP is a positive effector, or activator, because when ATP is added, the K_M decreases and thus the affinity of the enzyme for its substrate increases. ATP is a reactant in the reaction sequence, so it serves as an activator. ATP is also a purine nucleotide, whereas CTP is a pyrimidine nucleotide. Stimulation of ATCase by ATP encourages CTP synthesis when ATP synthesis is high, thus balancing the cellular pool of purine and pyrimidine nucleotides.

63. The formation of a disulfide bond under oxidizing conditions, or its cleavage under reducing conditions, could act as an allosteric signal by altering the conformation of the enzyme in a way that affects the groups at the active site.

Chapter 8

1. (a)

$$H_3C—(CH_2)_{12}—COO^-$$
Myristate (14:0)

(b) $H_3C—(CH_2)_5—CH{=}CH—(CH_2)_7—COO^-$
Palmitoleate (16:1 *n*-7)

(c) $H_3C—CH_2—(CH{=}CH—CH_2)_3—(CH_2)_6—COO^-$
α-Linolenate (18:3 *n*-3)

(d) $H_3C—(CH_2)_7—CH{=}CH—(CH_2)_{13}—COO^-$
Nervonate (24:1 *n*-9)

3.

$$H_3C—(CH_2)_4—CH{=}CH—CH_2—CH{=}CH—(CH_2)_4—CH{=}CH—(CH_2)_3—COO^-$$
Sciadonate (20:3$\Delta^{5, 11, 14}$)

[From Sayanova, O., Haslam, R., Venegas Caleron, M., and Napier, J. A., *Plant Physiol.* **144,** 455–467 (2007).]

5.

(a) $H_3C—(CH_2)_{13}—CH{=}CH—(CH_2)_2—CH{=}CH—(CH_2)_3—COO^-$
24:2$\Delta^{5, 9}$

(b)

$$H_3C-(CH_2)_4-(CH=CH-CH_2)_2-(CH_2)_3-CH=CH-(CH_2)_2-CH=CH-(CH_2)_3-COO^-$$

$$24:2\Delta^{5, 9, 15, 18}$$

7.

$$H_3C-(CH_2)_7-\overset{\overset{\textstyle H}{|}}{C}=\overset{\overset{\textstyle }{\underset{\underset{\textstyle H}{|}}{C}}}{}-(CH_2)_7-COOH$$

Elaidic acid

The melting point of elaidic acid is higher than the melting point of oleic acid, since the *trans* double bond of elaidic acid gives it an elongated shape, whereas the *cis* double bond oleic acid gives it a bent shape.

9. (a) Because it has a head group with just one anionic charge, SQDG is likely to substitute for phosphatidylglycerol (net charge −1) rather than phosphatidylethanolamine (with a positive and a negative charge).
(b) When phosphorus is limited, the organism produces relatively more of the sulfur-containing SQDG to substitute for phospholipids in the cell membrane.

11.

13.

DPPC

15.

17. All except phosphatidylcholine have hydrogen-bonding head groups.

19. Both DNA and phospholipids have exposed phosphate groups that are recognized by the antibodies.

21. The spicy ingredient in the food is a powder made from peppers that contains the hydrophobic compound capsaicin (see Section 8-1). Yogurt containing whole milk also contains hydrophobic ingredients that can cleanse the palate of the irritating capsaicin. Water is polar, so it does not dissolve the capsaicin and cannot cleanse the palate.

23. Vitamin A, and the compound from which it is derived, β-carotene, are lipid-soluble molecules. The vegetables in a typical salad do not contain large amounts of lipid. The addition of the lipid-rich avocado provided a means to solubilize the β-carotene and thus increase its absorption. [From Unlu, N. Z., Bohn, T., Clinton, S. K., and Schwartz, S. J., *J. Nutr.* **135,** 431–436 (2005).]

25. (b) is polar, (d) is nonpolar, and (a), (c), and (e) are amphipathic.

27. (a) A hydrocarbon chain is attached to the glycerol backbone at position 1 by a vinyl ether linkage. In a glycerophospholipid, an acyl group is attached by an ester linkage.
(b) The presence of this plasmalogen would not have a great effect since it has the same head group and same overall shape as phosphatidylcholine.

29. Lipids that form bilayers are amphiphilic, whereas triacylglycerols are nonpolar. Amphiphilic molecules orient themselves so that their polar head groups face the aqueous medium on the inside and outside of the cell. Also, triacylglycerols are cone-shaped rather than cylindrical and thus would not fit well in a bilayer structure, as shown in Figure 8-4.

31. Two factors that influence the melting point of a fatty acid are the number of carbons and the number of double bonds. Double bonds are a more important factor than the number of carbons, since a significant change in structure (a "kink") occurs when a double bond is introduced. An increase in the number of carbons increases the melting point, but the change is not nearly as dramatic. For example, the melting point of palmitate (16:0) is 63.1°C, whereas the melting point of stearate (18:0) is only slightly higher at 69.1°C. However, the melting point of oleate (18:1) is 13.2°C, a dramatic decrease with the introduction of a double bond.

33. In general, animal triacylglycerols must contain longer and/or more saturated acyl chains than plant triacylglycerols, since these chains have higher melting points and are more likely to be in the crystalline phase at room temperature. The plant triacylglycerols must contain shorter and/or less saturated acyl chains in order to remain fluid at room temperature.

35. The lipids from the meat of the reindeer slaughtered in February contained fewer unsaturated acyl chains than their healthy counterparts. Unsaturated fatty acids have a lower melting point due to the presence of double bonds that prevent them from packing together tightly. These lipids therefore help membranes remain fluid even at low temperatures as the reindeer walks through the snow. Saturated fatty acids, which pack together more efficiently and have higher melting points, decrease membrane fluidity at low temperatures. The decreased percentage of lipids with unsaturated fatty acyl chains result in decreased membrane fluidity and may compromise the ability of the animal to survive a cold winter. [From Suppela, P., and Nieminen, M., *Comp. Biochem. Physiol.* **128,** 53–72 (2001).]

37. The cyclopropane ring in lactobacillic acid produces a bend in the aliphatic chain and thus its melting point should be closer to the melting point of oleate, which also has a bend due to the double bond. The presence of bends decreases the opportunity for van der Waals forces to act among neighboring molecules. Less heat is required to disrupt the intermolecular forces, resulting in a melting point that is lower than that of a saturated fatty acid with a similar number of carbons. Therefore stearate has the highest melting point (69.6°C), followed by lactobacillic acid (28°C) and oleate (13.4°C).

39. Cholesterol's planar ring system interferes with the movement of acyl chains and thus tends to decrease membrane fluidity. At the same time, cholesterol prevents close packing of the acyl chains, which tends to prevent their crystallization. The net result is that cholesterol helps the membrane resist melting at high temperatures and resist crystallization at low temperatures. Therefore, in a membrane containing cholesterol, the shift from the crystalline form to the fluid form is more gradual than it would be if cholesterol were absent.

41. No. Higher temperatures increase fatty acid fluidity. To counter the effect of temperature, the plants make relatively more fatty acids with higher melting points. Dienoic acids have higher melting points than trienoic acids because they are more saturated. Therefore, the plants convert fewer dienoic acids into trienoic acids.

43. **(a)** PS and PE both contain amino groups.
(b) PC and SM both contain choline groups.
(c) PE, PC, and SM are all neutral, but PS carries an overall negative charge. Since PS is exclusively found on the cytosolic-facing leaflet, this side of the membrane is more negatively charged than the other side.

45. **(a)** Detergents are required to solubilize a transmembrane protein because the protein domains that interact with the nonpolar acyl chains are highly hydrophobic and would not form favorable interactions with water.
(b) A schematic diagram of the detergent SDS interacting with a transmembrane protein is shown here. The polar head group of SDS is represented by a circle and the nonpolar tail as a wavy line. The nonpolar tail of the SDS interacts with the nonpolar regions of the protein, effectively masking these regions from the polar solvent. The polar head group of the SDS interacts favorably with water. The presence of the detergent effectively solubilizes the transmembrane protein so that it can be purified.

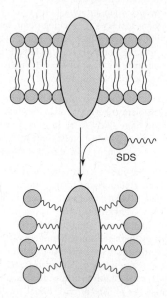

47. A. Fatty acyl–anchored protein (the acyl group is myristate)
B. Prenyl-anchored protein
C. Glycosylphosphatidylinositol (GPI)-anchored protein

49. The membrane-spanning segment is a stretch of 19 residues that are all uncharged and mostly hydrophobic.

LSTTEVAMHTTTSSSVSKSYISSQTNDTHKRDTYAATPRAHEV-
SEISVRTVYPPEEETGERVQLAHHFSEPEITLIIFGVMAGVIGTILLI-
SYGIRRLIKKSPSDVKPLPSPDTDVPLSSVEIENPETSDQ

51. A steroid is a hydrophobic lipid that can easily cross a membrane to enter the cell. It does not require a cell-surface receptor, as does a polar molecule such as a peptide.

53. **(a)** Glycosphingolipids pack together loosely because their very large head groups do not allow tight association.
(b) The lipid raft is less fluid because of both the presence of cholesterol and the saturated fatty acyl chains, which pack together more tightly than unsaturated acyl chains. [From Pike, L., *J. Lipid Res.* **44**, 655–667 (2003).]

55. **(a)** Alcohols, ether, and chloroform are nonpolar molecules and can easily pass through the nonpolar portion of the lipid bilayer, the aliphatic acyl chains of the phospholipids. Salts, sugars, and amino acids are highly polar and would not be able to traverse the nonpolar portion of the membrane.
(b) Cells contain proteins that serve as transporters. Proteins that transport water, known as aquaporins, have been identified. [From Kleinzeller, A., *News Physiol. Sci.* **12**, 49–54 (1997).]

57. After fusion, the green and red markers were segregated because they represent cell-surface proteins derived from two different sets of cells. Over time, the cell-surface proteins that could diffuse in the lipid bilayer became distributed randomly over the surface of the hybrid cell, so the green and red markers were intermingled. At 15°C, the lipid bilayer was in a gel-like rather than a fluid state, which prevented membrane protein diffusion. Edidin's experiment supported the fluid mosaic model by demonstrating the ability of proteins to diffuse through a fluid membrane.

Chapter 9

1.
$$\Delta\psi = 0.058 \log \frac{[Na^+]_{in}}{[Na^+]_{out}}$$

$$-0.070 = 0.058 \log \frac{[Na^+]_{in}}{[Na^+]_{out}}$$

$$-1.20 = \log \frac{[Na^+]_{in}}{[Na^+]_{out}}$$

$$10^{-1.20} = \frac{[Na^+]_{in}}{[Na^+]_{out}}$$

$$\frac{0.063}{1} = \frac{[Na^+]_{in}}{[Na^+]_{out}}$$

3. $\Delta G = RT \ln \frac{[Na^+]_{in}}{[Na^+]_{out}} + Z\mathcal{F}\Delta\psi$

$\Delta G = (8.3145 \times 10^{-3}\,\text{kJ} \cdot \text{K}^{-1} \cdot \text{mol}^{-1})(310\,\text{K})\ln\frac{0.063}{1}$
$\qquad + (+1)(96,485 \times 10^{-3}\,\text{kJ} \cdot \text{V}^{-1} \cdot \text{mol}^{-1})(-0.070\,\text{V})$
$\Delta G = -7.12\,\text{kJ} \cdot \text{mol}^{-1} - 6.75\,\text{kJ} \cdot \text{mol}^{-1}$
$\Delta G = -13.9\,\text{kJ} \cdot \text{mol}^{-1}$

At the resting potential, the movement of Na^+ ions into the cell is a favorable process.

5. $\Delta G = RT \ln \frac{[Na^+]_{in}}{[Na^+]_{out}} + Z\mathcal{F}\Delta\psi$

$\Delta G = (8.3145 \times 10^{-3}\,\text{kJ} \cdot \text{K}^{-1} \cdot \text{mol}^{-1})(20 + 273\,\text{K})$

$\qquad \ln\frac{40\,\text{mM}}{450\,\text{mM}} + (1)(96,485 \times 10^{-3}\,\text{kJ} \cdot \text{V}^{-1} \cdot \text{mol}^{-1})(-0.070\,\text{V})$

$\Delta G = -5.90\,\text{kJ} \cdot \text{mol}^{-1} - 6.75\,\text{kJ} \cdot \text{mol}^{-1}$
$\Delta G = -12.64\,\text{kJ} \cdot \text{mol}^{-1}$

$\Delta G = RT \ln \frac{[Ca^{2+}]_{in}}{[Ca^{2+}]_{out}} + Z\mathcal{F}\Delta\psi$

$\Delta G = (8.3145 \times 10^{-3}\,\text{kJ} \cdot \text{K}^{-1} \cdot \text{mol}^{-1})(20 + 273\,\text{K})$

$$\ln \frac{0.0001 \text{ mM}}{4 \text{ mM}} + (2)(96{,}485 \times 10^{-3} \text{ kJ} \cdot \text{V}^{-1} \cdot \text{mol}^{-1})(-0.070 \text{ V})$$

$$\Delta G = -25.8 \text{ kJ} \cdot \text{mol}^{-1} - 13.5 \text{ kJ} \cdot \text{mol}^{-1}$$

$$\Delta G = -39.3 \text{ kJ} \cdot \text{mol}^{-1}$$

Since the concentrations of both ions are greater outside the cell than inside and the cell potential is negative, the passive movement of the ions will be from outside the cell to inside. In order to maintain the ion concentrations given in the problem, energy-consuming active transport processes are required.

7. Use Equation 9-4 and let $Z = 2$ and $T = 310$ K:

(a) $\Delta G = RT \ln \dfrac{[\text{Ca}^{2+}]_{in}}{[\text{Ca}^{2+}]_{out}} + Z\mathcal{F}\Delta\psi$

$$\Delta G = (8.3145 \text{ J} \cdot \text{K}^{-1} \cdot \text{mol}^{-1})(310 \text{ K})\ln\frac{10^{-7}}{10^{-3}}$$

$$+ (2)(96{,}485 \text{ J} \cdot \text{V}^{-1} \cdot \text{mol}^{-1})(-0.05 \text{ V})$$

$$\Delta G = -23{,}700 \text{ J} \cdot \text{mol}^{-1} - 9600 \text{ J} \cdot \text{mol}^{-1}$$

$$\Delta G = -33{,}300 \text{ J} \cdot \text{mol}^{-1} = -33.3 \text{ kJ} \cdot \text{mol}^{-1}$$

The negative value of ΔG indicates a thermodynamically favorable process.

(b) $\Delta G = RT \ln \dfrac{[\text{Ca}^{2+}]_{in}}{[\text{Ca}^{2+}]_{out}} + Z\mathcal{F}\Delta\psi$

$$\Delta G = (8.3145 \text{ J} \cdot \text{K}^{-1} \cdot \text{mol}^{-1})(310 \text{ K})\ln\frac{10^{-7}}{10^{-3}}$$

$$+ (2)(96{,}485 \text{ J} \cdot \text{V}^{-1} \cdot \text{mol}^{-1})(+0.05 \text{ V})$$

$$\Delta G = -23{,}700 \text{ J} \cdot \text{mol}^{-1} + 9600 \text{ J} \cdot \text{mol}^{-1}$$

$$\Delta G = -14{,}100 \text{ J} \cdot \text{mol}^{-1} = -14.1 \text{ kJ} \cdot \text{mol}^{-1}$$

The negative value of ΔG indicates a thermodynamically favorable process, but not as favorable as in part (a).

9. (a) Since all the terms on the right side of Equation 9-1 are constant, except for T, the following proportion for the two temperatures (310 K and 313 K) applies:

$$\frac{-70 \text{ mV}}{310 \text{ K}} = \frac{\Delta\psi}{313 \text{ K}}$$

$$\Delta\psi = -70.7 \text{ mV}$$

The difference in membrane potential at the higher temperature would not significantly affect the neuron's activity.

(b) It is more likely that an increased temperature would increase the fluidity of cell membranes. This in turn might alter the activity of membrane proteins, including ion channels and pumps, which would have a more dramatic effect on membrane potential than temperature alone.

11. (a) $\Delta G = RT \ln \dfrac{[\text{glucose}]_{in}}{[\text{glucose}]_{out}}$

$$\Delta G = (8.3145 \times 10^{-3} \text{ kJ} \cdot \text{K}^{-1} \cdot \text{mol}^{-1})(310 \text{ K})\ln\frac{0.5 \text{ mM}}{15 \text{ mM}}$$

$$\Delta G = -8.8 \text{ kJ} \cdot \text{mol}^{-1}$$

(b) $\Delta G = RT \ln \dfrac{[\text{glucose}]_{in}}{[\text{glucose}]_{out}}$

$$\Delta G = (8.3145 \times 10^{-3} \text{ kJ} \cdot \text{K}^{-1} \cdot \text{mol}^{-1})(310 \text{ K})\ln\frac{0.5 \text{ mM}}{4 \text{ mM}}$$

$$\Delta G = -5.4 \text{ kJ} \cdot \text{mol}^{-1}$$

13. The less polar a substance, the faster it can diffuse through the lipid bilayer. From slowest to fastest: C, A, B.

15. (a) Glucose has a slightly larger permeability coefficient than mannitol and therefore moves across the synthetic bilayer more easily.

(b) Both solutes have higher permeability coefficients for the red blood cell membrane, indicating that transport is occurring via a protein transporter rather than diffusion through the membrane. The transporter binds glucose specifically and transports it rapidly across the membrane, whereas it is less specific for mannitol and transports it less effectively.

17. (a) Phosphate ions are negatively charged, and lysine side chains most likely carry a full positive charge at physiological pH. It is possible that an ion pair forms between the phosphate and the lysine side chains and that the lysine side chains serve to funnel the phosphate ions through the porin.

(b) If the hypothesis described in part (a) is correct, the replacement of lysines with the negatively charged glutamates would abolish phosphate transport by the porin, due to charge–charge repulsion. Possibly the mutated porin might even transport positively charged ions instead of phosphate. [From Sukhan, A., and Hancock, R. E. W., *J. Biol. Chem.* **271**, 21239–21242 (1996).]

19. (a) Acetylcholine binding triggers the opening of the channel, an example of a ligand-gated transport protein.

(b) Na^+ ions flow into the muscle cell, where their concentration is low.

(c) The influx of positive charges causes the membrane potential to increase.

21. The transfer to pure water increases the influx of water by osmosis, and the cell begins to swell. Swelling, which puts pressure on the cell membrane, causes mechanosensitive channels to open. As soon as the cell's contents flow out, the pressure is relieved and the cell can return to its normal size. Without these channels, the cell would swell up and burst.

23. The hydroxyl and amido groups act as proton donors to coordinate the negatively charged chloride ion. Cations could not interact with the protons and so would be excluded.

25. (a) A transport protein, like an enzyme, carries out a chemical reaction (in this case, the transmembrane movement of glucose) but is not permanently altered in the process. Because the transport protein binds glucose, its rate does not increase in direct proportion to increases in glucose concentration, and it becomes saturated at high glucose concentrations.

(b) The transport protein has a maximum rate at which it can operate (corresponding to $V_{\max}$, the upper limit of the curve). It also binds glucose with a characteristic affinity (corresponding to K_M, the glucose concentration at half-maximal velocity). The estimated $V_{\max}$ for this transporter is about 0.8×10^6 mM $\cdot$ cm $\cdot$ s^{-1}, and the K_M is about 0.5 mM.

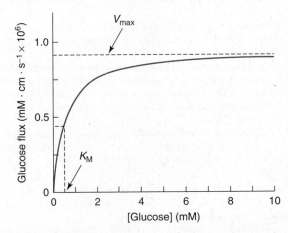

27. Intracellular exposure of the glucose transporter to trypsin indicates that there is at least one cytosolic domain of the transport protein that is essential for glucose transport. Hydrolysis of one or more peptide bonds in this domain(s) abolishes glucose transport. But extracellular exposure of the ghost transporter to trypsin has no effect, so there is no trypsin-sensitive extracellular domain that is essential for transport. This experiment also shows that the glucose transporter is asymmetrically arranged in the erythrocyte membrane.

29. As the glutamate (charge -2) enters the cell, four positive charges also enter (3 Na$^+$, 1 H$^+$) for a total of two positive charges. Since 1 K$^+$ exits the cell at the same time, a total of one positive charge is added to the cell for each glutamate transported inside.

31. CO$_2$ produced by respiring tissues enters the red blood cell and combines with water to form carbonic acid, which then dissociates to form H$^+$ and HCO$_3^-$ ions. The HCO$_3^-$ ions are transported out of the cell by Band 3 in exchange for Cl$^-$ ions, which enter the cell. The HCO$_3^-$ ions travel through the circulation to the lungs, where they recombine with H$^+$ ions to form carbonic acid, which subsequently dissociates to form water and CO$_2$. The CO$_2$ is then exhaled in the lungs.

33.

Aspartyl phosphate

35. **(a)** A transporter similar to a porin would be inadequate since even a large β barrel would be far too small to accommodate the massive ribosome. Likewise, a transport protein with alternating conformations would not be up to the task due to its small size relative to the ribosome. In addition, neither type of protein would be suited for transporting a particle across two membranes. In fact, ribosomes and other large particles move between the nucleus and cytoplasm via nuclear pores, which are constructed from many different proteins and form a structure, even larger than the ribosome, that spans both nuclear membranes.
(b) Initially, one might expect ribosomal transport to be a thermodynamically favorable process, since the concentration of ribosomes is greater in the nucleus, where they are synthesized. However, free energy would ultimately be required to establish a pore (which would span two membrane thicknesses) for the ribosome to pass through. In fact, the nucleocytoplasmic transport of all but very small substances requires the activity of GTPases that escort particles through the nuclear pore assembly and help ensure that transport proceeds in one direction.

37. **(a)** Glucose uptake increases as sodium concentration increases in pericytes. In endothelial cells, glucose uptake is constant regardless of sodium ion concentration.
(b) The shape of the curve for the pericytes indicates that a protein transporter is involved. Glucose uptake initially increases as sodium ion concentration increases, then reaches a plateau at high sodium ion concentration, indicating that the transporter is saturated and is operating at its maximal capacity.
(c) It is likely that the pericytes use secondary active transport to import glucose. Sodium ions and glucose molecules enter the cell in symport. The sodium ions are then ejected from the cell by the Na,K-ATPase transporter.

39. **(a)** The maximal velocity is estimated to be between 6 and 7 pmol choline · mg^{-1} · mL^{-1}. The K_M is the substrate concentration at half-maximal velocity and is about 40 mM.
(b) The choline transporter responds to increased choline concentrations by increasing its rate of transport so that efficient uptake occurs over a range of choline concentrations near the K_M. At low concentrations of choline (10 μM), the transporter operates at 20% of its maximal velocity, whereas at high concentrations of choline (80 μM), the transporter operates at nearly 100% of its maximal velocity. The K_M value of 40 μM is between the low and high physiological concentrations of choline.

(c) It is possible that the choline transporter cotransports hydrogen ions and choline. A hydrogen ion might be exported when choline is imported. This is an example of antiport transport.
(d) TEA is structurally similar to choline and acts as a competitive inhibitor. TEA might bind to the choline transporter, preventing choline from binding. In this manner, TEA is brought into the cell and choline transport is inhibited. [From Sinclair, C. J., Chi, K. D., Subramanian, V., Ward, K. L., and Green, R. M., *J. Lipid Res.* **41**, 1841–1847 (2000).]

41. The ABC transporters bind ATP, then undergo a conformational change as the ATP is hydrolyzed and P$_i$ is released, leaving ADP. Vanadate, a phosphate analog, might serve as a competitive inhibitor by binding to the phosphate portion of the ATP binding site. With ATP unable to bind, the necessary conformational change cannot occur and the transporter is inhibited.

43. Both transporters are examples of secondary active transport. The H$^+$/Na$^+$ exchanger uses the free energy of the Na$^+$ gradient (established by the Na,K-ATPase) to remove H$^+$ from the cell as Na$^+$ enters. Similarly, a preexisting Cl$^-$ gradient (see Fig. 2-13) allows the cell to export HCO$_3^-$ as Cl$^-$ enters.

45. Di- and tripeptides enter the cell in symport with H$^+$ ions. The H$^+$ ions leave in exchange for Na$^+$ via the antiport protein. The Na$^+$ ions are ejected via the Na,K-ATPase. This is an example of secondary active transport in which the expenditure of ATP by the Na,K-ATPase pump is the driving force for peptide entry into the cell.

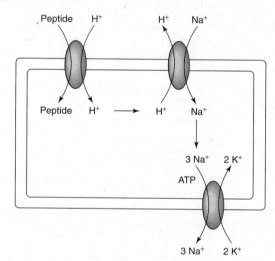

47. Acetylcholinesterase inhibitors will prevent the enzyme from breaking down acetylcholine (see Section 9-4). This increases the concentration of acetylcholine in the synaptic cleft and increases the chances that acetylcholine will bind to a dwindling number of receptors in the postsynaptic cell. [From Thanvi, B. R., and Lo, T. C. N., *Postgrad. Med. J.*, **80**, 690–700 (2004).]

49.

51. The tetanus toxin cleaves the SNAREs, which are required for the fusion of synaptic vesicles with the neuronal plasma membrane. This prevents the release of acetylcholine, interrupting communication between nerves and muscles and causing paralysis.

53. By adding a phosphate group, the kinase increases the size and negative charge of the lipid head group, which then occupies a larger volume and more strongly repels neighboring negatively charged lipid head groups. The phosphatidylinositol would become more cone-shaped, thereby increasing bilayer curvature, which is a necessary step in the formation of a new vesicle by budding.

55. Two lipid bilayers separate the damaged organelle from the rest of the cell.

Chapter 10

1. Signal molecules that are lipids (cortisol and thromboxane) or very small (nitric oxide) can diffuse through lipid bilayers and do not need a receptor on the cell surface.

3. Let $[R \cdot L] = x$ and $[R] = 0.010 - x$

$$K_d = \frac{[R][L]}{[R \cdot L]}$$

$$[R \cdot L] = x = \frac{[R][L]}{K_d}$$

$$x = \frac{(0.010 - x)(0.0025)}{0.0015}$$

$$x = \frac{(0.000025 - 0.0025x)}{0.0015}$$

$$0.0015x = 0.000025 - 0.0025x$$

$$0.0040x = 0.000025$$

$$x = 0.00625 = 6.25 \text{ mM} = [R \cdot L]$$

The percentage of receptors occupied by ligand is 6.25 mM/10 mM, or 62.5%.

5.

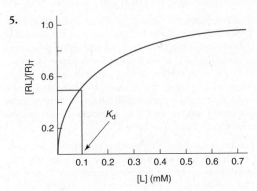

The K_d estimated from the curve is about 0.1 mM.

7.

$$K_d = \frac{[R][L]}{[RL]}$$

$$[R]_T = [R] + [RL]$$

$$[R] = [R]_T - [RL]$$

$$K_d = \frac{([R]_T - [RL])[L]}{[RL]}$$

$$K_d[RL] = [R]_T[L] - [RL][L]$$

$$(K_d[RL]) + ([RL][L]) = [R]_T[L]$$

$$[RL](K_d + [L]) = [R]_T[L]$$

$$\frac{[RL]}{[R]_T} = \frac{[L]}{K_d + [L]}$$

9.

$$\frac{[RL]}{[R]_T} = \frac{[L]}{[L] + K_d}$$

$$\frac{100}{1000} = \frac{[L]}{[L] + 1.0 \times 10^{-10} \text{ M}}$$

$$0.10([L] + 1.0 \times 10^{-10} \text{ M}) = [L]$$

$$0.10[L] + 1.0 \times 10^{-11} \text{ M} = [L]$$

$$1.0 \times 10^{-11} \text{ M} = 0.9[L]$$

$$[L] = 1.11 \times 10^{-11} \text{ M}$$

11. **(a)** The binding site with a K_d of 0.35 μM is the high-affinity binding site, and the one with a K_d of 7.9 μM is the low-affinity binding site. K_d is the ligand concentration at which the receptor is half-saturated with ligand; therefore, the lower the K_d, the lower the concentration of ligand required to achieve half-saturation.
(b) The high-affinity binding site with a K_d of 0.35 μM is most effective in the 0.1–0.5 μM range because at the upper limit of this range the high-affinity binding sites will be more than 50% occupied, whereas the low-affinity sites will be less than 50% occupied.
(c) Both of these agonists can compete at high concentrations, but the methylthio-ADP has a lower K_d and will be a more effective inhibitor at low concentrations. [From Jefferson, J. R., Harmon, J. T., and Jamieson, G. A., *Blood* **71**, 110–116 (1988).]

13. The K_d is obtained from the x intercept in the double-reciprocal plot.

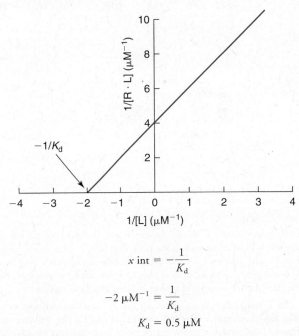

$$x \text{ int} = -\frac{1}{K_d}$$

$$-2 \text{ μM}^{-1} = \frac{1}{K_d}$$

$$K_d = 0.5 \text{ μM}$$

15. Cell-surface receptors are difficult to purify because they are usually integral membrane proteins and require the addition of detergents to dissociate them from the membrane. The receptor proteins constitute a very small proportion of all of the proteins in the cell; this makes it difficult for the experimenter to isolate the receptor protein from other cellular proteins.

17. The different types of G protein–linked receptors are found in different types of cells. The cellular response elicited when a ligand binds to a receptor depends on how that particular cell integrates and processes the signal. Different cells have different intracellular components, which results in different responses to what appears to be the same signal.

19. If receptors are removed from the cell surface, the ligand cannot bind and an intracellular response cannot occur. If a receptor is phosphorylated, it binds to arrestin, which blocks the ligand from binding.

21.

23. Epinephrine and norepinephrine are ligands that bind to β_2-adrenergic receptors. When these ligands bind to their receptors, the signal transduction process produces a number of effects, including increases in heart rate, muscle contraction, and blood pressure. In addition, the smooth muscle in the bronchial tubes relaxes, making it easier to expand the lungs. These physiological effects are all necessary components of the fight-or-flight response but are harmful to a person suffering from high blood pressure. β-Blockers are antagonists that bind to the same β_2-adrenergic receptors but do not elicit a response. By occupying the receptors, the antagonists prevent epinephrine and norepinephrine from doing so. The result is that the heart rate and blood pressure decrease and the heart contractions are less intense.

25. Stimulation of GTPase activity by RGS accelerates the hydrolysis of GTP to GDP, converting the receptor-associated G protein into its inactive form more rapidly. This will shorten the duration of signaling.

27. (a)

(b)

29. Because of their structural similarity to diacylglycerol, phorbol esters will stimulate protein kinase C, as diacylglycerol does. Increased protein kinase C activity will lead to an increase in the phosphorylation of the kinase's cellular targets. Because protein kinase C phosphorylates proteins involved in cell division and growth, the addition of phorbol esters will have profound effects on the rates of cell division and growth if added to cells in culture.

31. The T cell is stimulated when an extracellular ligand binds to a G protein–linked receptor and activates phospholipase C. The activated phospholipase C catalyzes the hydrolysis of phosphatidylinositol bisphosphate, yielding diacylglycerol and inositol trisphosphate. The inositol trisphosphate binds to channel proteins in the endoplasmic reticulum and allows calcium ions to flow into the cytosol. Calcium ions then bind to calmodulin, causing a conformational change that allows it to bind and activate calcineurin. The activated calcineurin then activates NFAT, as described in the problem.

33. Overexpression of PTEN in mammalian cells would promote apoptosis. PTEN removes a phosphate group from inositol trisphosphate; when this occurs, inositol trisphosphate is no longer able to activate protein kinase B. In the absence of protein kinase B, cells are not stimulated to grow and proliferate and instead undergo apoptosis.

35. (a) Upon stimulation by an action potential, acetylcholine-containing synaptic vesicles in neurons fuse with the plasma membrane and release their contents into the synaptic cleft (see Section 9-4). Acetylcholine then diffuses across the synaptic cleft to the endothelial cell.
(b) Acetylcholine binds to G protein–linked cell-surface receptors in the endothelial cell and activates phospholipase C, which hydrolyzes phosphatidylinositol bisphosphate to diacylglycerol and inositol trisphosphate. The inositol trisphosphate binds to calcium channels in the endoplasmic reticulum, which opens the channels and floods the cell with Ca^{2+}. Calcium ions bind to calmodulin, changing its conformation and allowing it to bind to NO synthase to activate the enzyme.
(c) Cyclic GMP is formed, along with pyrophosphate, from GTP. The enzyme that catalyzes the reaction is guanylate cyclase.

(d) It's possible that cGMP activates protein kinase G in a manner analogous to that of cAMP activating protein kinase A; that is, cGMP binding could displace regulatory subunits from protein kinase G to release active catalytic subunits. The active protein kinase G would next phosphorylate proteins involved in the muscle contraction process, perhaps myosin or actin, resulting in smooth muscle relaxation.

37. If the NO enzyme is missing, the signaling pathway described in Problem 35 cannot be completed. NO cannot be synthesized in the absence of the NO synthase enzyme, and subsequent steps, including the production of the second messenger cGMP and activation of protein kinase G, do not occur. Protein kinase G acts on muscle in such a way that the muscle relaxes. If this does not occur, muscles lining the blood vessels are constricted, resulting in high blood pressure. This makes it more difficult for the heart to pump blood through the circulatory system, leading to an increased heart rate and increased size of the ventricular heart chambers.

39. Nitroglycerin decomposes to form NO, which passes through cell membranes in tissues of the tongue to enter the bloodstream. NO activates guanylate cyclase in smooth muscle cells, as described in Problem 35, producing cyclic GMP, which subsequently activates protein kinase G. The kinase phosphorylates proteins involved in muscle contraction, which leads to the relaxation of the smooth muscle cell. This increases blood flow to the heart and relieves the pain associated with angina.

41. (a) Adenylate cyclase generates the second messenger cAMP in response to activation of G proteins by G protein–coupled receptors. The EF toxin would generate large amounts of cAMP in the absence of any specific hormone signal.
(b) When Ca^{2+} calmodulin is bound to EF, it is not available to activate any other Ca^{2+}-sensitive proteins that might be involved in normal cell signaling.

43. Since the growth factor stimulates kinase activity, the H_2O_2 second messenger is likely to produce similar responses, so it must inactivate the phosphatases.

45. In the presence of GEF, the activity of the signaling pathway increases, since GEF promotes dissociation of bound GDP, and Ras · GDP is inactive. Once GDP has dissociated, GTP can bind and activate Ras. The opposite is true in the presence of GAP. Ras · GTP is active, but when the GTP is hydrolyzed to GDP, Ras is converted from the active to the inactive form.

47. Phosphatases that remove phosphate groups from the insulin receptor would turn the insulin signaling pathway off and protein kinases B and C would not be activated. Without active protein kinase B, glycogen synthase is inactive and glycogen cannot be synthesized from glucose. Without protein kinase C, glucose transporters are not translocated to the membrane and glucose is not brought into the cell but remains in the blood. Drugs that act as inhibitors of these phosphatases would potentiate the action of the insulin receptor, allowing the receptor to remain active with a lower concentration of ligand, and thus are potentially effective treatments for diabetes.

49. As noted in Problem 29, phorbol esters are diacylglycerol analogs that can activate protein kinase C. According to the information given in this problem, protein kinase C activates the MAP kinase cascade, which leads to the phosphorylation of proteins that influence gene expression. When these genes are expressed, progression through the cell cycle is altered and cells are stimulated to grow and proliferate, a characteristic of tumor cells.

51. In order to become activated, two inactive PKR proteins must come close enough to phosphorylate each other (autophosphorylation). A long RNA molecule can bind two PKR proteins simultaneously, holding them in close proximity so that they can activate each other. Short RNA molecules prevent PKR activation because when a short RNA molecule occupies the PKR RNA-binding site, the PKR cannot bind to another RNA where it might encounter a second PKR and get phosphorylated. [From Nallagatla, S. R., Toroney, R., and Bevilacqua, P. C., *Curr. Opin. Struct. Biol.* **21,** 119–127 (2011).]

53. Substances cannot enter the nucleus unless they possess a nuclear localization signal, a sequence that interacts with the nuclear pore and allows entry into the nucleus. The nuclear localization signal on the progesterone receptor must be exposed, even when ligand is not bound. But the nuclear localization signal on the glucocorticoid receptor must be masked. When ligand binds, a conformational change occurs that unmasks the nuclear localization signal, and the complex can pass through the nuclear pore and enter the nucleus.

55. Arachidonate is the substrate for the production of prostaglandins, many of which have inflammatory properties. Stimulating the release of arachidonate from the membrane by C1P increases the concentration of substrate available for prostaglandin synthesis. One of the enzymes that catalyzes the first step in the production of prostaglandins is COX-2, which is stimulated by S1P. Both C1P and S1P can potentially increase production of prostaglandins, which accounts for their inflammatory properties, as shown in the diagram.

57. S1P might use a variety of mechanisms to activate Ras, either through receptor tyrosine kinases or activation of protein kinase C. Ras then activates the MAP kinase pathway (see Problem 49), which leads to the phosphorylation of transcription factors that promote the expression of proteins involved in the cell cycle, ultimately leading to cell survival.

59.

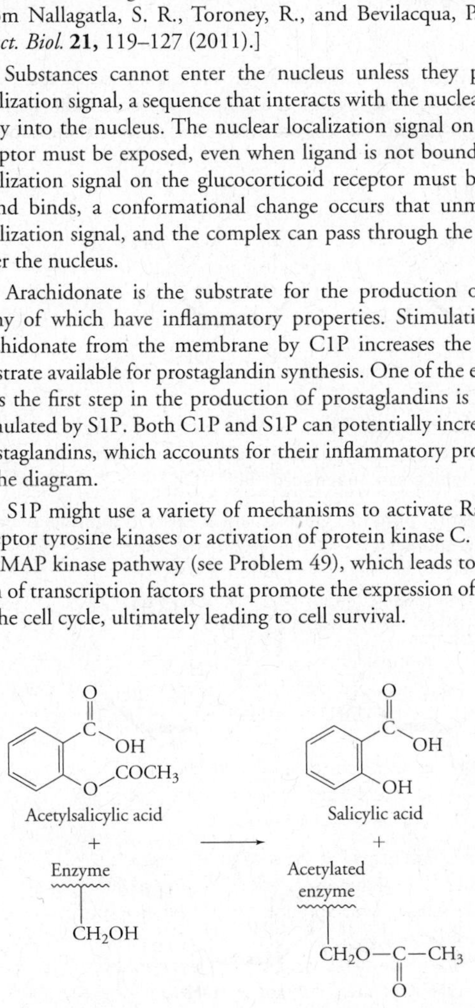

Without knowing the mechanism of the enzyme, it is not possible to say for certain why acetylating the serine inhibits cyclooxygenase activity. But it's possible that the acetylation alters the structure of the active site such that the arachidonic acid substrate is unable to bind. It's also possible that serine participates in catalysis, possibly as a nucleophile, as in chymotrypsin. An acetylated serine would be unable to function as a nucleophile, which would explain why the modified enzyme is catalytically inactive.

61. Phospholipase A_2 catalyzes the release of arachidonate from membrane phospholipids. Blocking this reaction would prevent the COX-catalyzed conversion of arachidonate to proinflammatory prostaglandins.

63.

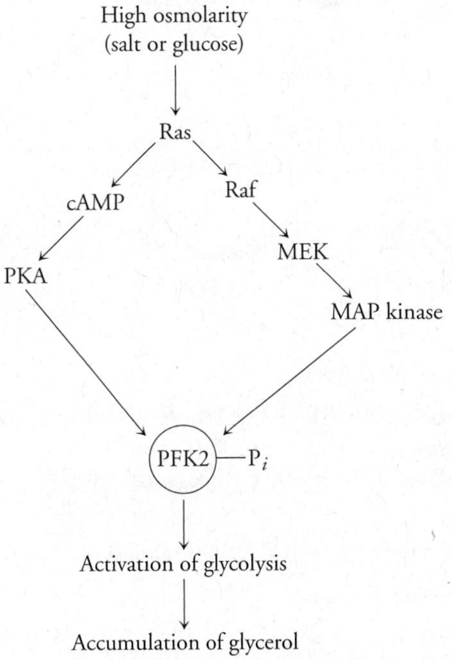

Chapter 11

1. (a) aldose (b) ketose (c) aldose

3. Chiral carbons are indicated with an asterisk:

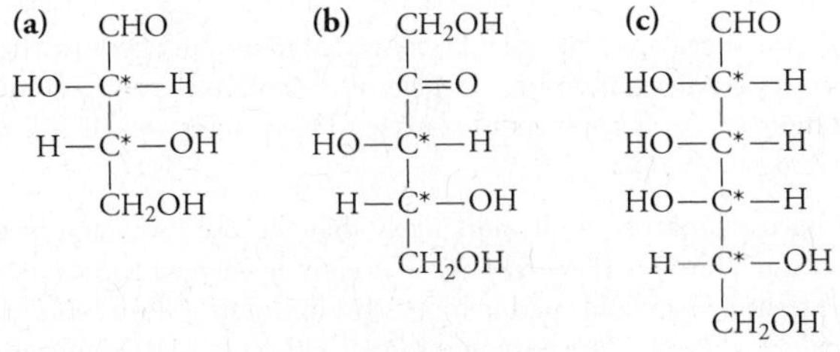

5. (a) D (b) L (c) D (d) L

7. (a) D-Psicose and D-sorbose are epimers.
   (b) D-Sorbose and D-fructose are structural isomers.
   (c) D-Fructose and L-fructose are enantiomers.
   (d) D-Ribose and D-ribulose are structural isomers.

9. Fructose and galactose are isomers of glucose.

11. (a)

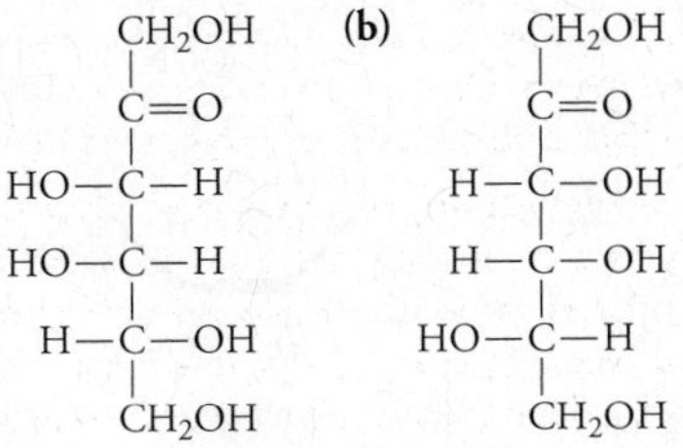

(c) Tagatose is less effectively absorbed in the small intestine because the transport proteins in the epithelial cells lining the small intestine do not bind and transport tagatose as efficiently as they do sugars that are more naturally and commonly present in the diet.

13. The β anomer is more stable because most of the bulky hydroxyl substituents are in the equatorial position.

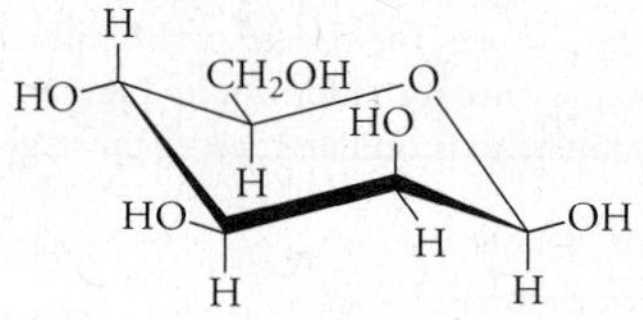

β-Mannose

15.

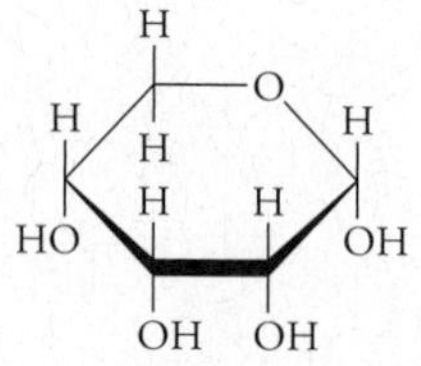

α-D-Ribose 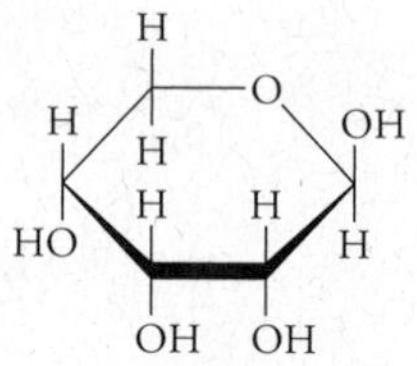   β-D-Ribose

17. (a) A five-membered ring results.

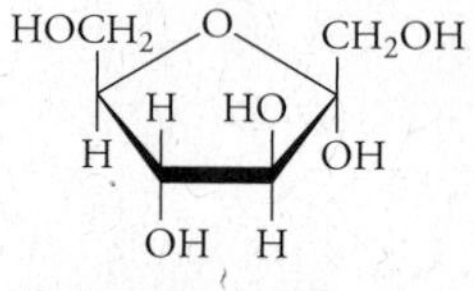

α-D-Fructose 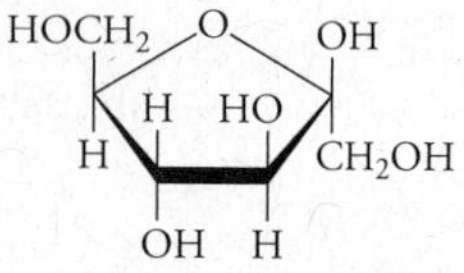   β-D-Fructose

(b) A six-membered ring results.

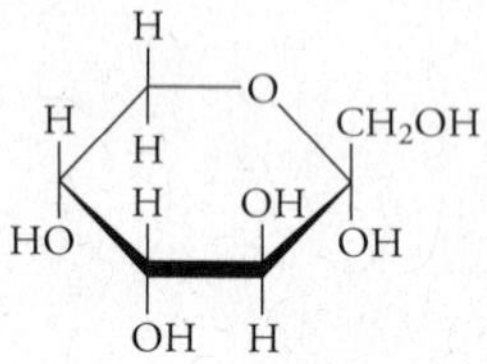

α-D-Fructose 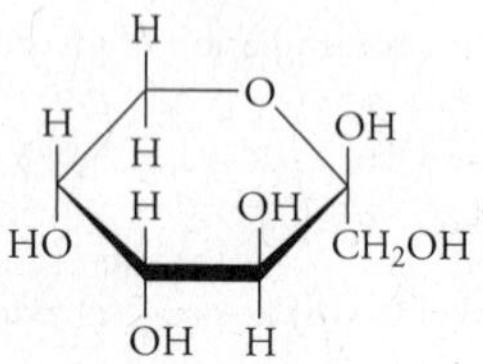   β-D-Fructose

19. All the sugar molecules will be converted to product because the α and β anomers are in equilibrium. Depletion of molecules in the α form will cause more of the β anomers to convert to α anomers, which will then be converted to product.

21. Glucose-6-phosphate is more likely than glucose to remain in the cell because the phosphorylated glucose is negatively charged and cannot easily cross the nonpolar lipid bilayer by passive diffusion. Glucose-6-phosphate cannot exit the cell via transport proteins either, as these transporters are specific for glucose, not glucose-6-phosphate.

23. Glucose is a reducing sugar because the cyclization reaction can occur in reverse to form the straight-chain structure that contains an aldehyde group (see Fig. 11-1). Fructose contains a ketone functional group that cannot be oxidized by Cu^{2+}, so it is not a reducing sugar. (However, in the presence of the strong base in Benedict's solution, fructose undergoes isomerization to an aldose, which does react with Cu^{2+}.)

25.

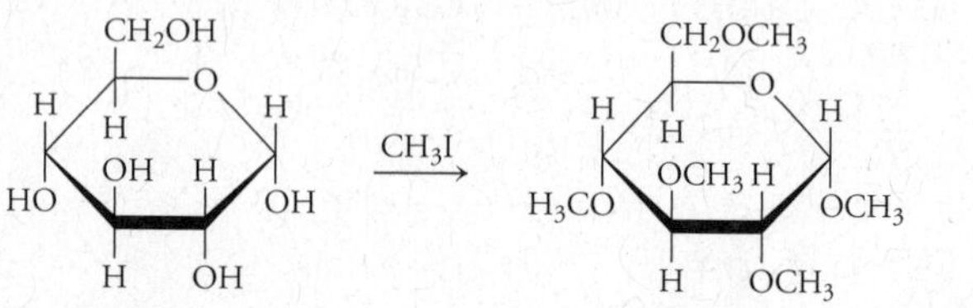

$CH_3I \longrightarrow$

α-D-Glucose   1,2,3,4,6-Penta-O-methyl-D-glucose

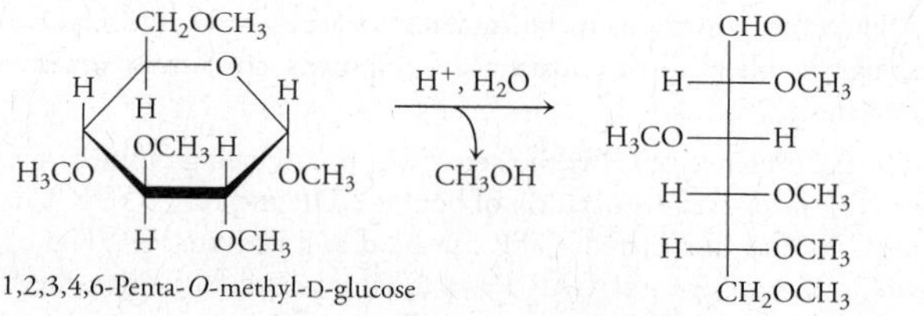

CH_2OCH_3 $\xrightarrow{H^+, H_2O}$ CHO

CH_3OH

1,2,3,4,6-Penta-O-methyl-D-glucose   2,3,4,6-Tetra-O-methyl-D-glucose

27.

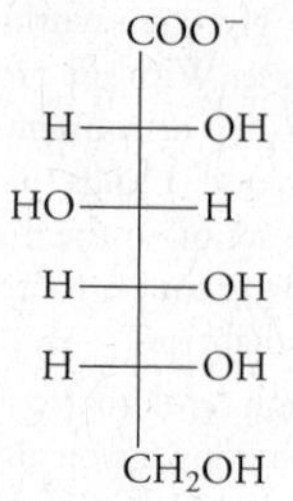

Gluconate

29.

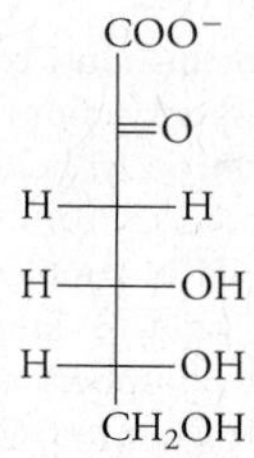

KDG

31. Lactose is a reducing sugar because it has a free anomeric carbon (C1 of the glucose residue). Sucrose is not a reducing sugar because the anomeric carbons of both glucose and fructose are involved in the glycosidic bond.

33.

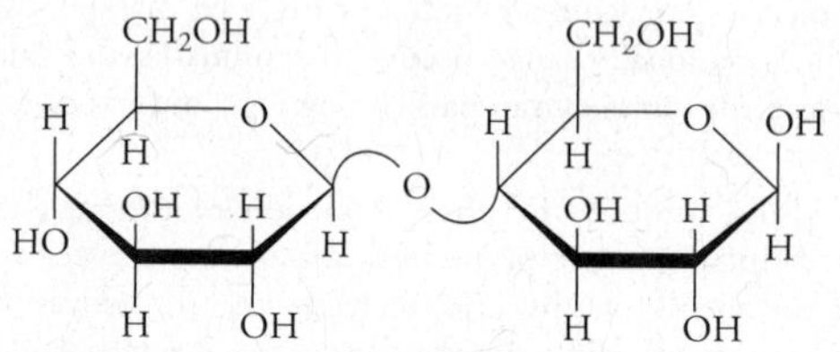

Cellobiose is a reducing sugar. The anomeric carbon of the glucose on the right side is free to reverse the cyclization reaction to re-form the aldehyde functional group, which can be reduced.

35. Trehalase digestion produces glucose, which exists in solution as a mixture of the α and β anomers.

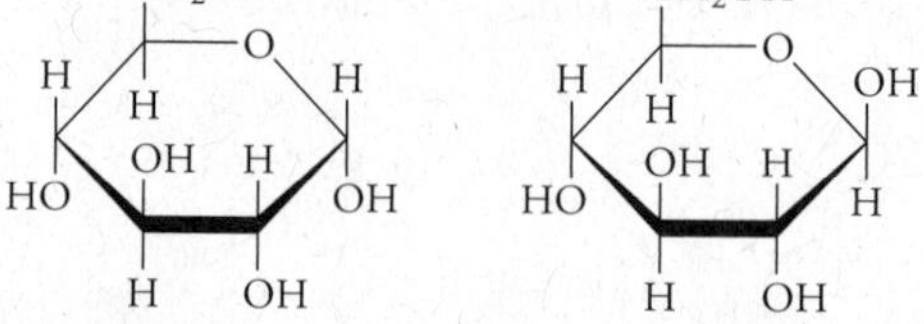

37.

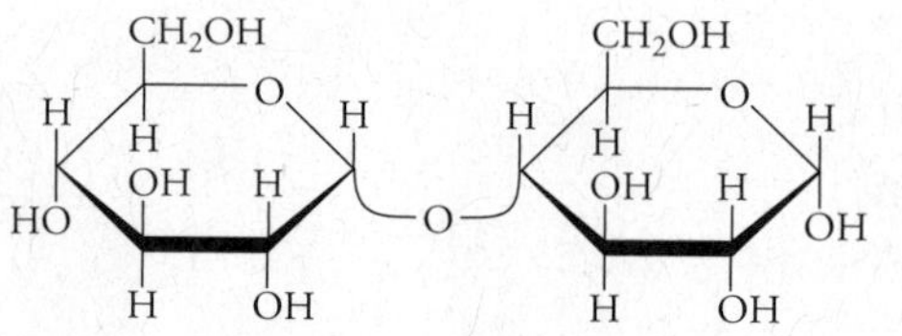

39. In order for sorbitol to be catabolized to yield energy, it would need to enter the same catabolic pathway as glucose. The enzymes that catalyze glucose catabolism are specific for glucose and do not bind to sorbitol, so the sugar alcohol is not metabolized and passes through the body undigested. In this manner, sorbitol contributes no calories to the food containing it, but it is nearly identical to its parent monosaccharide in taste and sweetness.

41. Humans do not have one of the enzymes needed to hydrolyze the glycosidic bonds in order to convert the trisaccharide raffinose to its constituent monosaccharides. The glycosidic linkages in raffinose are an $\alpha(1 \rightarrow 6)$ linkage between galactose and glucose and an $\alpha(1 \rightarrow 2)$ glycosidic linkage between glucose and fructose. Humans have the enzymes to digest the second glycosidic bond but not the first.

43. Starch, glycogen, cellulose, and chitin are homopolymers. Peptidoglycan and chondroitin sulfate are heteropolymers.

45. (a)

(b) Humans do not have the enzymes to digest $\beta(2 \rightarrow 1)$ glycosidic bonds (although the bacteria that inhabit the small intestine possess the necessary enzymes and do have this capability). Nondigestible carbohydrates are often classified by food manufacturers as "fiber"; thus, inulin extracted from chicory root is often added to processed foods to boost their fiber content.

47. Pectin is a highly hydrated polysaccharide, so it thickens the fruit preparation and helps turn it into a gel.

49. The cellulose-based plant cell wall is strong and rigid, but it must be remodeled as the plant cell grows. The cell uses cellulase to weaken the cell wall so that the cell can expand.

51.

53. The *N*-linked saccharide is *N*-acetylglucosamine, and the bond has the β configuration. The *O*-linked oligosaccharide is *N*-acetylgalactosamine, and the bond has the α configuration.

55. The residues of the disaccharide are glucuronate linked by a $\beta(1 \rightarrow 3)$ glycosidic bond to *N*-acetylgalactosamine-4-sulfate. Disaccharides are linked to each other by $\beta(1 \rightarrow 4)$ bonds.

57.

59. The monosaccharide is *N*-acetylglucosamine.

61. The amide bond forms between Ala and the carboxylate group on the C4 substituent in the disaccharide.

Chapter 12

1. (a) chemoautotroph **(b)** photoautotroph
   (c) chemoautotroph **(d)** heterotroph
   (e) heterotroph **(f)** chemoautotroph
   (g) photoautotroph

3. The pH of the stomach is ~2. At this pH, the salivary amylase is denatured and can no longer catalyze the hydrolysis of glycosidic bonds in dietary carbohydrates.

5. Maltase is required to hydrolyze the $\alpha(1 \rightarrow 4)$ glycosidic bonds in maltotriose and maltose (see Solution 4). Isomaltase is needed to hydrolyze the $\alpha(1 \rightarrow 6)$ glycosidic bonds in the limit dextrins because α-amylase only catalyzes the hydrolysis of $\alpha(1 \rightarrow 4)$ glycosidic bonds and cannot accommodate branch points (see Problem 4). These enzymes are required to completely hydrolyze starch to its component monosaccharides, since only monosaccharides can be absorbed.

7. Sugar alcohols are not present in abundance naturally, which explains the absence of transporters for these molecules. Passive diffusion is less effective than passive transport (as shown in Problem 9-15).

9. Because the products of nucleic acid digestion are relatively large, charged nucleotides, a transport protein is required to facilitate their movement across the cell membrane. The transport protein most likely uses the free energy of a Na^+ gradient (active transport), as is the case for intestinal monosaccharide and amino acid transporters.

11. The low pH denatures the protein, unfolding it so peptide bonds are more accessible to proteolytic digestion by stomach enzymes.

13. The pH optimum for pepsin is 2, which is the pH of the stomach. The pH optimum for trypsin and chymotrypsin is 7–8, as the small intestine is slightly basic (see Table 2-3). Each enzyme functions optimally in the conditions of its environment.

15. Amino acids enter the cells lining the small intestine via secondary active transport. This system is similar to the process for glucose absorption shown in Figure 9-18.

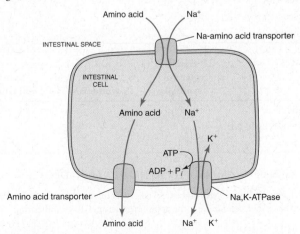

17. (a)

Triacylglycerol → Diacylglycerol + Fatty acid (via Gastric lipase, H_2O)

(b) Both diacylglycerol and fatty acids are amphipathic molecules—they have both hydrophilic and hydrophobic domains. These molecules can form micelles that emulsify the dietary triacylglycerols, which are nonpolar and are unable to form micelles.

19.

Cholesteryl ester

Cholesteryl esterase

Fatty acid

Cholesterol

21. (a) The polar glycogen molecule is fully hydrated, so its weight reflects a large number of closely associated water molecules. Fat is stored in anhydrous form. Therefore, a given weight of fat stores more free energy than the same weight of glycogen.

(b) Because it must be hydrated, a glycogen molecule occupies a large effective volume of the cytoplasm, which it shares with other glycogen molecules, enzymes, organelles, and so on. Because hydrophobic fat molecules are sequestered from the bulk of the cytoplasm, they do not have the same potential for interfering with other cellular constituents, so their collective volume is virtually unlimited.

23. The phosphorylated glucose molecule is not recognized by the glucose transporter. Removal of the phosphate group allows the glucose to more easily leave the cell.

25.

	Glycolysis	Citric acid cycle	Fatty acid metabolism
Acetyl-CoA		✓	✓
Glyceraldehyde-3-P	✓		
Pyruvate	✓		

	Triacylglycerol synthesis	Photosynthesis	Transamination
Acetyl-CoA			
Glyceraldehyde-3-P	✓	✓	
Pyruvate			✓

27. (a) NAD^+ **(b)** NADPH **(c)** NADH **(d)** $NADP^+$

29. Individuals with gastrointestinal disorders might have a gastrointestinal tract that is not colonized by the appropriate vitamin B_{12}–synthesizing bacte-

ria. A deficiency in haptocorrin or intrinsic factor would be manifested as a vitamin B_{12} deficiency, since these proteins are essential for absorption of the vitamin. Vegetarians and vegans who consume no animal products would also be at risk for a deficiency of vitamin B_{12}.

31. (a)

γ-Carboxyglutamate

(b) The additional carboxylate group on the glutamate residue confers a -2 charge on the side chain, generating a high-affinity binding site for the Ca^{2+} ions essential for blood clotting.

33. (a) Vitamin C **(b)** biotin **(c)** pyridoxine **(d)** pantothenic acid

35. Because K_{eq} is the ratio of the product concentration to the reactant concentration at equilibrium, the reaction with the larger K_{eq} will have a higher concentration of product. Therefore, the concentration of B in Tube 1 will be greater than the concentration of D in Tube 2.

37. (a) Since $K_{eq} = 1$, $\ln K_{eq} = 0$ and $\Delta G^{\circ\prime}$ is also equal to zero (Equation 12-2).

(b) Since $K_{eq} = 1$, the concentrations of reactants and products must be equal at equilibrium. If the reaction started with 1 mM F, the equilibrium concentrations will be 0.5 mM E and 0.5 mM F.

39.
$$\Delta G = \Delta G^{\circ\prime} + RT \ln \frac{[B]}{[A]}$$

$$\Delta G = -5.9 \text{ kJ} \cdot \text{mol}^{-1} + (8.3145 \times 10^{-3} \text{ kJ} \cdot \text{K}^{-1} \cdot \text{mol}^{-1})$$
$$(310 \text{ K})\ln \left(\frac{0.1 \times 10^{-3} \text{ M}}{0.9 \times 10^{-3} \text{ M}} \right)$$

$$\Delta G = -11.6 \text{ kJ} \cdot \text{mol}^{-1}$$

The reaction will proceed as written, with A converted to B until the ratio of [B]/[A] = 10/1.

41. (a) The equilibrium constant can be determined by rearranging Equation 12-2 (see Sample Calculation 12-2):

$$K_{eq} = e^{-\Delta G^{\circ\prime}/RT}$$
$$K_{eq} = e^{-10 \text{ kJ} \cdot \text{mol}^{-1}/(8.3145 \times 10^{-3} \text{ kJ} \cdot \text{K}^{-1} \cdot \text{mol}^{-1})(298 \text{ K})}$$
$$K_{eq} = e^{-4.04}$$
$$K_{eq} = 0.018$$
$$K_{eq} = e^{-\Delta G^{\circ\prime}/RT}$$
$$K_{eq} = e^{-20 \text{ kJ} \cdot \text{mol}^{-1}/(8.3145 \times 10^{-3} \text{ kJ} \cdot \text{K}^{-1} \cdot \text{mol}^{-1})(298 \text{ K})}$$
$$K_{eq} = e^{-8.07}$$
$$K_{eq} = 0.000313$$

Small changes in $\Delta G^{\circ\prime}$ result in large changes in K_{eq}. Doubling the $\Delta G^{\circ\prime}$ value (a positive, unfavorable value) leads to a 60-fold decrease in K_{eq}.

(b)
$$K_{eq} = e^{-\Delta G^{\circ\prime}/RT}$$
$$K_{eq} = e^{-(-10 \text{ kJ} \cdot \text{mol}^{-1})/(8.3145 \times 10^{-3} \text{ kJ} \cdot \text{K}^{-1} \cdot \text{mol}^{-1})(298 \text{ K})}$$
$$K_{eq} = e^{4.04}$$
$$K_{eq} = 56.6$$
$$K_{eq} = e^{-\Delta G^{\circ\prime}/RT}$$
$$K_{eq} = e^{-(-20 \text{ kJ} \cdot \text{mol}^{-1})/(8.3145 \times 10^{-3} \text{ kJ} \cdot \text{K}^{-1} \cdot \text{mol}^{-1})(298 \text{ K})}$$
$$K_{eq} = e^{8.07}$$
$$K_{eq} = 3200$$

The same conclusion can be made: Small changes in $\Delta G^{\circ\prime}$ lead to large changes in K_{eq}. Doubling a (favorable) $\Delta G^{\circ\prime}$ results in a K_{eq} value that is nearly 60 times as large.

43. The complete reaction is $ATP + H_2O \rightarrow ADP + P_i$. Use Equation 12-3 and the value of $\Delta G^{\circ\prime}$ from Table 12-4. The concentration of water is assumed to be equal to 1.

$$\Delta G = \Delta G^{\circ\prime} + RT \ln \frac{[ADP][P_i]}{[ATP]}$$

$$\Delta G = -30.5 \text{ kJ} \cdot \text{mol}^{-1} + (8.3145 \times 10^{-3} \text{ kJ} \cdot \text{K}^{-1} \cdot \text{mol}^{-1})$$
$$(310 \text{ K}) \ln \frac{(0.001)(0.005)}{(0.003)}$$

$$\Delta G = -30.5 \text{ kJ} \cdot \text{mol}^{-1} - 16.5 \text{ kJ} \cdot \text{mol}^{-1}$$

$$\Delta G = -47 \text{ kJ} \cdot \text{mol}^{-1}$$

45. First, convert Calories to joules: $72,000 \text{ cal} \times 4.184 \text{ J/cal} = 300,000 \text{ J}$ or 300 kJ. Since the $ATP \rightarrow ADP + P_i$ reaction releases $30.5 \text{ kJ} \cdot \text{mol}^{-1}$, the apple contains $300 \text{ kJ}/30.5 \text{ kJ} \cdot \text{mol}^{-1}$ or the equivalent of about 9.8 moles of ATP.

47.

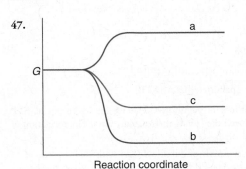

Reaction coordinate

49. (a) The phosphate groups on the ATP molecule would be less negative at a lower pH. Therefore, there would be less charge–charge repulsion and therefore less energy released upon hydrolysis. The ΔG° would be less negative at a lower pH.
(b) Magnesium ions are positively charged and form ion pairs with the negatively charged phosphate groups. Thus, magnesium ions serve to decrease the charge–charge repulsion associated with the phosphate groups. In the absence of magnesium ions, the charge–charge repulsion is greater; thus, more free energy is released upon the removal of one of the phosphate groups. This results in a $\Delta G^{\circ\prime}$ value that is more negative.

51. (a) The synthesis of ATP from ADP requires $30.5 \text{ kJ} \cdot \text{mol}^{-1}$ of energy:
$$ADP + P_i \rightarrow ATP + H_2O \quad \Delta G^{\circ\prime} = +30.5 \text{ kJ} \cdot \text{mol}^{-1}$$
$$\frac{2850 \text{ kJ} \cdot \text{mol}^{-1}}{30.5 \text{ kJ} \cdot \text{mol}^{-1}} \times 0.33 = 30.8 \text{ ATP}$$
(b) $\dfrac{9781 \text{ kJ} \cdot \text{mol}^{-1}}{30.5 \text{ kJ} \cdot \text{mol}^{-1}} \times 0.33 = 105.8 \text{ ATP}$

(c) For glucose, 30.8 ATP/6 carbons = 5.1 ATP/carbon. For palmitate, 105.8 ATP/16 carbons = 6.6 ATP/carbon. Most of the carbon atoms of fatty acids are fully reduced —CH_2— groups. Most of the carbon atoms of glucose have hydroxyl groups attached to them (—CHOH—) and are therefore already partly oxidized. Consequently, more free energy is available from a carbon in a triacylglycerol than from a carbon in a glycogen molecule.

53. The apple provides 9.8 moles of ATP (see Solution 45). The moderately active female described in Problem 52 requires 100 moles ATP daily. Therefore, 100 moles ATP/9.8 apples mol^{-1} = 10.2 apples. Keeping the 33% efficiency in mind, 10.2/0.33 = 31 apples would be required.

55. Reactions involving phosphoenolpyruvate, 1,3-bisphosphoglycerate, and phosphocreatine could drive the synthesis of ATP because transfer of a phosphoryl group from one of these compounds occurs with a greater change in free energy than the transfer of a phosphoryl group to ADP.

57. (a) The equilibrium constant can be determined by rearranging Equation 12-2 (see Sample Calculation 12-2):

$$K_{eq} = e^{-\Delta G^{\circ\prime}/RT}$$
$$K_{eq} = e^{-5 \text{ kJ} \cdot \text{mol}^{-1}/(8.3145 \times 10^{-3} \text{ kJ} \cdot \text{K}^{-1} \cdot \text{mol}^{-1})(298 \text{ K})}$$
$$K_{eq} = e^{-2.02}$$
$$K_{eq} = 0.133$$

(b) Since

$$K_{eq} = \frac{[\text{isocitrate}]}{[\text{citrate}]} = 0.133$$

$$[\text{isocitrate}] = 0.133 \, [\text{citrate}]$$

The total concentration of isocitrate and citrate is 2 M, so
$$[\text{isocitrate}] = 2 \text{ M} - [\text{citrate}]$$

Combining the two equations gives

$$0.133 \, [\text{citrate}] = 2 \text{ M} - [\text{citrate}]$$
$$1.133 \, [\text{citrate}] = 2 \text{ M}$$
$$[\text{citrate}] = 1.77 \text{ M}$$
$$[\text{isocitrate}] = 2 \text{ M} - 1.77 \text{ M} = 0.23 \text{ M}$$

(c) The preferred direction under standard conditions is toward the formation of citrate.
(d) The reaction occurs in the direction of isocitrate synthesis because standard conditions do not exist in the cell. Also, the reaction is the second step of an eight-step pathway, so isocitrate is removed as soon as it is produced in order to serve as the reactant for the next step of the pathway.

59.

Mechanism 1: glutamate + $NH_3 \rightarrow$ glutamine
$$ATP + H_2O \rightarrow ADP + P_i$$

glutamate + NH_3 + ATP + $H_2O \rightarrow$ glutamine + ADP + P_i

Mechanism 2: glutamate + ATP $\rightarrow$ γ-glutamylphosphate + ADP
γ-glutamylphosphate + H_2O + $NH_3 \rightarrow$ glutamine + P_i

glutamate + NH_3 + ATP + $H_2O \rightarrow$ glutamine + ADP + P_i

Mechanism 2 is the more likely mechanism because it proceeds through a phosphorylated intermediate that "captures" the energy of the phosphoanhydride bond of ATP. In the first mechanism, the two reactions are not linked to a common intermediate. ATP is hydrolyzed, but this energy is not "harnessed" in any way and is simply dissipated as heat.

61. The equilibrium constant can be determined by rearranging Equation 12-2 (see Sample Calculation 12-2):

(a) $K_{eq} = e^{-\Delta G^{\circ\prime}/RT}$
$$K_{eq} = e^{-47.7 \text{ kJ} \cdot \text{mol}^{-1}/(8.3145 \times 10^{-3} \text{ kJ} \cdot \text{K}^{-1} \cdot \text{mol}^{-1})(298 \text{ K})}$$
$$K_{eq} = e^{-19.2}$$
$$K_{eq} = 4.4 \times 10^{-9}$$

$$K_{eq} = \frac{[\text{fructose-1,6-bisphosphate}]}{[\text{fructose-6-phosphate}][P_i]}$$

$$4.4 \times 10^{-9} = \frac{[\text{fructose-1,6-bisphosphate}]}{[\text{fructose-6-phosphate}](5.0 \times 10^{-3})}$$

$$\frac{[\text{fructose-1,6-bisphosphate}]}{[\text{fructose-6-phosphate}]} = 2.2 \times 10^{-11}$$

(b)

fructose-6-phosphate + P$_i$ $\rightleftharpoons$ fructose-1,6-bisphosphate

$$\Delta G°' = 47.7 \text{ kJ} \cdot \text{mol}^{-1}$$

ATP + H$_2$O $\rightleftharpoons$ ADP + P$_i$ $\Delta G°' = -30.5 \text{ kJ} \cdot \text{mol}^{-1}$

fructose-6-phosphate + ATP + H$_2$O $\rightleftharpoons$

fructose-1,6-bisphosphate + ADP $\Delta G°' = 17.2 \text{ kJ} \cdot \text{mol}^{-1}$

(c)

$$K_{eq} = e^{-\Delta G°'/RT}$$

$$K_{eq} = e^{-17.2 \text{ kJ} \cdot \text{mol}^{-1}/(8.3145\times10^{-3} \text{ kJ} \cdot \text{K}^{-1} \cdot \text{mol}^{-1})(298 \text{ K})}$$

$$K_{eq} = e^{-6.94}$$

$$K_{eq} = 1.0 \times 10^{-3}$$

$$K_{eq} = \frac{[\text{fructose-1,6-bisphosphate}][\text{ADP}]}{[\text{fructose-6-phosphate}][\text{ATP}]}$$

$$1.0 \times 10^{-3} = \frac{[\text{fructose-1,6-bisphosphate}][\text{ADP}]}{[\text{fructose-6-phosphate}][\text{ATP}]}$$

$$\frac{(1.0 \times 10^{-3})[\text{ATP}]}{[\text{ADP}]} = \frac{[\text{fructose-1,6-bisphosphate}]}{[\text{fructose-6-phosphate}]}$$

$$\frac{(1.0 \times 10^{-3})(3.0 \times 10^{-3})}{(1.0 \times 10^{-3})} = \frac{[\text{fructose-1,6-bisphosphate}]}{[\text{fructose-6-phosphate}]}$$

$$3.0 \times 10^{-3} = \frac{[\text{fructose-1,6-bisphosphate}]}{[\text{fructose-6-phosphate}]}$$

(d) The conversion of fructose-6-phosphate to fructose-1,6-bisphosphate is unfavorable. The ratio of products to reactants at equilibrium is 2.2×10^{-11} under standard conditions. But if the conversion of fructose-6-phosphate to fructose-1,6-bisphosphate is coupled with the hydrolysis of ATP, the reaction becomes more favorable and the ratio of fructose-1,6-bisphosphate to fructose-6-phosphate increases to 3×10^{-3}, a change of eight orders of magnitude.

(e) The second mechanism is biochemically feasible because the two steps are coupled via a common phosphorylated intermediate that "captures" the energy of ATP. In the first mechanism, the two steps are not coupled. The ATP is hydrolyzed and the energy is lost as heat instead of being used to assist the conversion of fructose-6-phosphate to fructose-1,6-bisphosphate.

63. (a) The equilibrium constant can be determined by rearranging Equation 12-2 (see Sample Calculation 12-2):

$$K_{eq} = e^{-\Delta G°'/RT}$$

$$K_{eq} = e^{-31.5 \text{ kJ} \cdot \text{mol}^{-1}/(8.3145\times10^{-3} \text{ kJ} \cdot \text{K}^{-1} \cdot \text{mol}^{-1})(298 \text{ K})}$$

$$K_{eq} = e^{-12.7}$$

$$K_{eq} = 3.0 \times 10^{-6}$$

$$K_{eq} = \frac{[\text{palmitoyl-CoA}]}{[\text{palmitate}][\text{CoA}]}$$

$$3.0 \times 10^{-6} = \frac{[\text{palmitoyl-CoA}]}{[\text{palmitate}][\text{CoA}]}$$

Therefore the ratio of products to reactants is 3.0×10^{-6}:1. The reaction is not favorable.

(b) Coupling the synthesis of palmitoyl-CoA with ATP hydrolysis to ADP produces a standard free energy change of $1.0 \text{ kJ} \cdot \text{mol}^{-1}$ for the coupled process [31.5 kJ $\cdot$ mol^{-1} + (-30.5 kJ $\cdot$ mol^{-1}) = 1.0 kJ $\cdot$ mol^{-1}].

palmitate + CoA + ATP $\rightarrow$ palmitoyl-CoA + ADP + P$_i$

$$\Delta G°' = 1.0 \text{ kJ} \cdot \text{mol}^{-1}$$

$$K_{eq} = e^{-\Delta G°'/RT}$$

$$K_{eq} = e^{-1.0 \text{ kJ} \cdot \text{mol}^{-1}/(8.3145\times10^{-3} \text{ kJ} \cdot \text{K}^{-1} \cdot \text{mol}^{-1})(298 \text{ K})}$$

$$K_{eq} = e^{-0.40}$$

$$K_{eq} = 0.67$$

$$K_{eq} = \frac{[\text{palmitoyl-CoA}][\text{ADP}][\text{P}_i]}{[\text{palmitate}][\text{CoA}][\text{ATP}]}$$

$$0.67 = \frac{[\text{palmitoyl-CoA}][\text{ADP}][\text{P}_i]}{[\text{palmitate}][\text{CoA}][\text{ATP}]}$$

Coupling the synthesis of palmitoyl-CoA with the hydrolysis of ATP to ADP has improved the [product]/[reactant] ratio considerably, but the formation of products is still not favored.

(c) Coupling the synthesis of palmitoyl-CoA with ATP hydrolysis to AMP produces a standard free energy change of $-14.1 \text{ kJ} \cdot \text{mol}^{-1}$ for the coupled process [31.5 kJ $\cdot$ mol^{-1} + (-45.6 kJ $\cdot$ mol^{-1}) = -14.1 kJ $\cdot$ mol^{-1}].

palmitate + CoA + ATP $\rightarrow$ palmitoyl-CoA + AMP + PP$_i$

$$\Delta G°' = -14.1 \text{ kJ} \cdot \text{mol}^{-1}$$

$$K_{eq} = e^{-\Delta G°'/RT}$$

$$K_{eq} = e^{-(-14.1 \text{ kJ} \cdot \text{mol}^{-1})/(8.3145\times10^{-3} \text{ kJ} \cdot \text{K}^{-1} \cdot \text{mol}^{-1})(298 \text{ K})}$$

$$K_{eq} = e^{5.7}$$

$$K_{eq} = 296$$

$$296 = \frac{[\text{palmitoyl-CoA}][\text{AMP}][\text{PP}_i]}{[\text{palmitate}][\text{CoA}][\text{ATP}]}$$

Coupling the synthesis of palmitoyl-CoA with the hydrolysis of ATP to AMP has improved the [product]/[reactant] ratio. The formation of products is now favored.

(d) Coupling the synthesis of palmitoyl-CoA with ATP hydrolysis to AMP and PP$_i$ followed by PP$_i$ hydrolysis produces a standard free energy of -34.2 kJ $\cdot$ mol^{-1} for the coupled process [-14.1 kJ $\cdot$ mol^{-1} + (-19.2 kJ $\cdot$ mol^{-1}) = -33.3 kJ $\cdot$ mol^{-1}].

palmitate + CoA + ATP + H$_2$O $\rightarrow$ palmitoyl-CoA + AMP + 2 P$_i$
$$\Delta G°' = -33.3 \text{ kJ} \cdot \text{mol}^{-1}$$

$$K_{eq} = e^{-\Delta G°'/RT}$$

$$K_{eq} = e^{-(-33.3 \text{ kJ} \cdot \text{mol}^{-1})/(8.3145\times10^{-3} \text{ kJ} \cdot \text{K}^{-1} \cdot \text{mol}^{-1})(298 \text{ K})}$$

$$K_{eq} = e^{13.4}$$

$$K_{eq} = 6.9 \times 10^5$$

$$6.9 \times 10^5 = \frac{[\text{palmitoyl-CoA}][\text{AMP}][\text{P}_i]^2}{[\text{palmitate}][\text{CoA}][\text{ATP}]}$$

Coupling the activation of palmitate to palmitoyl-CoA with the hydrolysis of ATP to AMP, with subsequent hydrolysis of pyrophosphate, is a thermodynamically effective means of accomplishing the reaction. Coupling the reaction with hydrolysis of ATP to ADP is not effective.

Chapter 13

1. (a) Reactions 1, 3, 7, and 10; **(b)** Reactions 2, 5, and 8; **(c)** Reaction 6; **(d)** Reaction 9; **(e)** Reaction 4.

3. (a)

$$\Delta G°' = -RT \ln \frac{[\text{glucose-6-phosphate}][\text{ADP}]}{[\text{glucose}][\text{ATP}]}$$

$$-16.7 \text{ kJ} \cdot \text{mol}^{-1} = -(8.3145 \times 10^{-3} \text{ kJ} \cdot \text{K}^{-1} \cdot \text{mol}^{-1})$$

$$(298 \text{ K})\ln \frac{[\text{glucose-6-phosphate}][\text{ADP}]}{[\text{glucose}][\text{ATP}]}$$

$$16.7 \text{ kJ} \cdot \text{mol}^{-1} = 2.48 \text{ kJ} \cdot \text{mol}^{-1}$$

$$\ln \frac{[\text{glucose-6-phosphate}][\text{ADP}]}{[\text{glucose}][\text{ATP}]}$$

$$6.73 = \ln \frac{\text{[glucose-6-phosphate][ADP]}}{\text{[glucose][ATP]}}$$

$$e^{6.73} = \frac{\text{[glucose-6-phosphate][ADP]}}{\text{[glucose][ATP]}}$$

$$840 = \frac{\text{[glucose-6-phosphate][ADP]}}{\text{[glucose][ATP]}}$$

$$840 = \frac{\text{[glucose-6-phosphate]}(1)}{\text{[glucose]}(10)}$$

$$8.4 \times 10^3 = \frac{\text{[glucose-6-phosphate]}}{\text{[glucose]}}$$

(b)

$$\Delta G^{\circ\prime} = -RT \ln \frac{\text{[glucose][ADP]}}{\text{[glucose-6-phosphate][ADP]}}$$

$$16.7 \text{ kJ} \cdot \text{mol}^{-1} = -(8.3145 \times 10^{-3} \text{ kJ} \cdot \text{K}^{-1} \cdot \text{mol}^{-1})$$

$$(298 \text{ K}) \ln \frac{\text{[glucose][ATP]}}{\text{[glucose-6-phosphate][ADP]}}$$

$$-16.7 \text{ kJ} \cdot \text{mol}^{-1} = 2.45 \text{ kJ} \cdot \text{mol}^{-1} \ln \frac{\text{[glucose][ATP]}}{\text{[glucose-6-phosphate][ADP]}}$$

$$-6.73 = \ln \frac{\text{[glucose][ATP]}}{\text{[glucose-6-phosphate][ADP]}}$$

$$e^{-6.73} = \frac{\text{[glucose][ATP]}}{\text{[glucose-6-phosphate][ADP]}}$$

$$1.2 \times 10^{-3} = \frac{\text{[glucose][ATP]}}{\text{[glucose-6-phosphate][ADP]}}$$

$$1.2 \times 10^{-3} = \frac{\text{[glucose]}(10)}{\text{[glucose-6-phosphate]}(1)}$$

$$1.2 \times 10^{-4} = \frac{\text{[glucose]}}{\text{[glucose-6-phosphate]}}$$

In order to reverse the reaction, the ratio of glucose-6-phosphate to glucose would have to be 8.3×10^3:1.

5. (a) Because the brain relies on glucose from the blood, it stores very little glucose in the form of glycogen. Therefore, glucose rather than phosphorylated glucose is the substrate that enters the glycolytic pathway. The first step of glucose catabolism in the brain is catalyzed by hexokinase, so this step is the rate-determining step of the pathway. In other tissues that break down glycogen for glycolysis, the hexokinase step is bypassed.
(b) The low K_M means that the enzyme will be saturated with glucose and will therefore operate at maximum velocity. Even if the concentration of glucose were to fluctuate slightly, the brain's ability to catabolize glucose would not be affected.

7. One might expect the product of a reaction to inhibit the enzyme that catalyzes the reaction, while the reactant would act as an activator. Although it is true that ADP is a direct product of the PFK reaction, PFK is sensitive to the ATP needs of the cell as a whole. Rising ADP concentrations are an indication that ATP is needed; the subsequent stimulation of PFK increases glycolytic flux and generates ATP as a final pathway product.

9. In the presence of the inhibitor, the curve is sigmoidal and the K_M increases dramatically (nearly 10-fold, to 200 μM), indicating that a greater quantity of substrate is required to achieve ½ V_{max}. PEP stabilizes the T form of PFK (see Solution 8).

11. Glycerol can serve as an energy source because glycerol can be converted to glyceraldehyde-3-phosphate, which can then enter the glycolytic pathway "below" the phosphofructokinase step. The mutants cannot grow on glucose because glucose enters the glycolytic pathway by first being converted to glucose-6-phosphate, then fructose-6-phosphate. The next step, conversion to fructose-1,6-bisphosphate, requires the phosphofructokinase enzyme. Thus, glycerol is a suitable substrate for this mutant, but glucose is not.

13. In the presence of iodoacetate, fructose-1,6-bisphosphate accumulates, which suggests that iodoacetate inactivates the enzyme that uses fructose-1,6-bisphosphate as a substrate. Because iodoacetate reacts with Cys residues, the inactivation of the enzyme by the reagent suggests that a Cys residue is in the active site. Later studies showed that Cys was not part of the active site; instead, the reaction of the Cys with iodoacetate probably caused a conformational change that rendered the enzyme inactive.

15.

$$\Delta G = \Delta G^{\circ\prime} + RT \ln \frac{\text{[GAP]}}{\text{[DHAP]}}$$

$$4.4 \text{ kJ} \cdot \text{mol}^{-1} = 7.9 \cdot \text{mol}^{-1}$$

$$+ (8.3145 \times 10^{-3} \text{ kJ} \cdot \text{K}^{-1} \cdot \text{mol}^{-1})(310 \text{ K}) \ln \frac{\text{[GAP]}}{\text{[DHAP]}}$$

$$-3.5 \text{ kJ} \cdot \text{mol}^{-1} = 2.58 \text{ kJ} \cdot \text{mol}^{-1} \ln \frac{\text{[GAP]}}{\text{[DHAP]}}$$

$$-1.36 = \ln \frac{\text{[GAP]}}{\text{[DHAP]}}$$

$$e^{-1.36} = \frac{\text{[GAP]}}{\text{[DHAP]}}$$

$$0.26 = \frac{\text{[GAP]}}{\text{[DHAP]}}$$

The ratio of [GAP] to [DHAP] is 0.26:1, which seems to indicate that the formation of DHAP, not the formation of GAP, is favored. However, GAP, the product of the triose phosphate isomerase reaction, is the substrate for the glyceraldehyde-3-phosphate dehydrogenase reaction. The continuous removal of the product GAP by the action of the dehydrogenase shifts the equilibrium toward formation of GAP from DHAP.

17. Arsenate is a metabolic poison, and the cells eventually die. In the presence of arsenate, 1,3-bisphosphoglycerate is not formed; instead, this step is essentially skipped. Two moles of ATP per mole of glucose are normally generated at this step. If ATP is not generated at this step, the net ATP yield for the glycolytic pathway is zero, and the cells die because they are unable to meet their energy requirements.

19. Phosphoglycerate kinase catalyzes the conversion of 1,3-bisphosphoglycerate to 3-phosphoglycerate with concomitant production of ATP from ADP. The kinase can generate the ATP required by the ion pump, and the ADP produced when the pump is phosphorylated can serve as a substrate in the kinase reaction.

21. Phosphate levels increase because phosphate is a reactant for the GAPDH enzyme. ATP levels decrease, since ATP is a product of the GAPDH reaction. Levels of 2,3-BPG decrease as well, since levels of 1,3-BPG decrease as a result of GAPDH inhibition.

23. Fluoride inhibits the enzyme enolase. If the enzyme is inactive, its substrate, 2-phosphoglycerate, will accumulate. The previous reaction is at equilibrium, so 3-phosphoglycerate will accumulate as 2-phosphoglycerate accumulates.

25. In cells with a pyruvate kinase deficiency the [ADP]/[ATP] ratio increases and the [NAD$^+$]/[NADH] ratio decreases. Pyruvate kinase catalyzes the second ATP-generating step in glycolysis; in the absence of this reaction, ATP levels decrease and the [ADP]/[ATP] ratio increases. The substrate PEP accumulates in a pyruvate kinase deficiency, stimulating PFK and increasing the concentration of F16BP. This ultimately leads to an increase in glyceraldehyde-3-phosphate, which reacts with NAD$^+$ to form NADH. In the absence of pyruvate kinase, pyruvate is not produced, so lactate cannot subsequently re-oxidize NADH to NAD$^+$; therefore, the [NAD$^+$]/[NADH] decreases.

27. The alcohol dehydrogenase enzyme catalyzes the reduction of acetaldehyde to ethanol. Concomitantly, NADH is oxidized to NAD$^+$. The NADH reactant is produced by glycolysis in the GAPDH reaction. The NAD$^+$ produced in the alcohol dehydrogenase reaction can serve as a reactant for the glycolytic GAPDH reaction, allowing glycolysis to continue.

29. (a) Methanol reacts with alcohol dehydrogenase (as ethanol does) to produce formaldehyde (Section 13-1).

$$H_3C-OH \xrightleftharpoons[\text{alcohol dehydrogenase}]{NAD^+ \quad NADH, H^+} H-\overset{\displaystyle O}{\overset{\displaystyle \|}{C}}-H$$

Methanol ⟶ Formaldehyde

(b) Administering ethanol is a good antidote because ethanol will compete with methanol for binding to alcohol dehydrogenase and will produce the less harmful acetaldehyde. This allows time for methanol to be eliminated from the system. [From Cooper, J. A., and Kini, M., *Biochem. Pharmacol.* **11**, 405–416 (1962).]

31. (a) One mole of ATP is invested when KDG is converted to KDPG. One mole of ATP is produced when 1,3-BPG is converted to 3PG. One mole of ATP is produced when phosphoenolpyruvate is converted to pyruvate. Therefore, the net yield of this pathway (per mole of glucose) is one mole of ATP.

(b) In order to keep the pathway going, subsequent reactions would need to reoxidize the NADPH that is produced when glucose is converted to gluconate and the NADH that is produced by GAPDH. [From Johnsen, U., Selig, M., Xavier, K. B., Santos, H., and Schönheit, P., *Arch. Microbiol.* **175**, 52–61 (2001).]

33. (a) Acetyl-CoA produced from pyruvate is a substrate for the citric acid cycle, an energy-producing pathway. When the cell's need for energy is low, acetyl-CoA accumulates and activates pyruvate carboxylase, which catalyzes the first step of gluconeogenesis. As a result, the cell can synthesize glucose when the need to catabolize fuel is low.

(b) Following its deamination, alanine is a substrate for gluconeogenesis. By inhibiting pyruvate kinase, alanine suppresses glycolysis so that flux through the shared steps of glycolysis and gluconeogenesis will favor gluconeogenesis.

35. Insulin, the hormone of the fed state, might be expected to suppress the transcription of the gluconeogenic enzymes pyruvate carboxylase, PEPCK, fructose-1,6-bisphosphatase, and glucose-6-phosphatase. [In fact, insulin has been shown to suppress the transcription of PEPCK and glucose-6-phosphatase.]

37. (a) The phosphatase activity is active under fasting conditions. The phosphatase removes the phosphate group from F26BP, forming fructose-6-phosphate. Thus F26BP is not present to stimulate glycolysis (or to inhibit gluconeogenesis); therefore, gluconeogenesis is active.

(b) The hormone of the fasted state is glucagon.

(c) When glucagon binds to its receptors, cellular cAMP levels rise as described in Section 10-2. This activates protein kinase A, which phosphorylates the bifunctional enzyme, resulting in the activation of the phosphatase activity and the inhibition of the kinase activity.

39. (a) Increasing the activity of the enzyme that produces F26BP would increase the concentration of this metabolite, which stimulates PFK and inhibits fructose-1,6-bisphosphatase. This would have the effect of stimulating glycolysis and inhibiting gluconeogenesis. Stimulation of PFK would increase the concentration of fructose-1,6-bisphosphate, which activates pyruvate kinase via feed-forward activation. Additional stimulation of pyruvate kinase by brazilin would lead to an increase in flux through the glycolytic pathway.

(b) If brazilin is able to act on the liver to increase glycolysis and decrease gluconeogenesis, this could help alleviate the high blood glucose concentrations that occur in diabetes. An active gluconeogenic pathway in the liver would result in efflux of glucose from the liver, which is not desirable in the diabetic patient. [From You, E.-J., Khill, L.-Y., Kwak, W.-J., Won, H.-S., Chae, S.-H., Lee, B.-H., and Moon, C.-K., *J. Ethnopharmacol.* **102**, 53–57 (2005).]

41. A diagram of this pathway, referred to as the Cori cycle, is shown in Figure 19-3. Lactate is released from the muscle as a result of anaerobic glycolytic activity. It travels via the bloodstream to the liver, where it is taken up, converted to pyruvate, and then transformed back to glucose via gluconeogenesis. The cost of running this cycle is 4 ATP, since 2 ATP are generated in glycolysis and the cost of running gluconeogenesis is 6 ATP.

43. The starch in the grains must be converted to glucose because the yeast that carry out fermentation use glucose as their starting material.

45. Production of glucose-1-phosphate requires only an isomerization reaction catalyzed by phosphoglucomutase to convert it to glucose-6-phosphate, which can enter glycolysis. This skips the hexokinase step and saves a molecule of ATP. Hydrolysis, which produces glucose, would require expenditure of an ATP to phosphorylate glucose to glucose-6-phosphate.

47. This observation revealed that the pathways for glycogen degradation and synthesis must be different, since a defect in the degradative pathway has no effect on the synthetic pathway.

49. Normally, muscle glycogen is degraded to glucose-6-phosphate, which enters glycolysis to be oxidized to yield ATP for the active muscle. In anaerobic conditions, pyruvate, the end product of glycolysis, is converted to lactate, which is released from the muscle into the blood and enters the liver to be converted back to glucose via gluconeogenesis. The patient's muscle cells are unable to degrade glycogen to glucose-6-phosphate; thus, there is no glucose-6-phosphate to enter glycolysis and lactate formation does not occur. [From Stanbury, J. B., Wyngaarden, J. B., and Fredrickson, D. S., *The Metabolic Basis of Inherited Disease*, pp. 151–153, McGraw-Hill, New York (1978).]

51. Glucose-6-phosphatase catalyzes the last reaction in gluconeogenesis (and glycogenolysis) in the liver. Glucose-6-phosphate is converted to glucose, and the glucose transporters export glucose to the circulation, where it is available to other body tissues that do not carry out gluconeogenesis and that do not store glycogen. In the absence of this enzyme, glucose-6-phosphate cannot be converted to glucose and instead accumulates in the liver and is converted to glucose-1-phosphate, which is used for glycogen synthesis. Glycogen synthesis is therefore elevated in the livers of patients with this disease. The accumulation of glycogen enlarges the liver and causes the abdomen to protrude.

53. In order for the enzyme to be active, the serine in the active site must be phosphorylated. This phosphate group is donated to C1 of the glucose-6-phosphate in the first step of the conversion of glucose-6-phosphate to glucose-1-phosphate. If glucose-1,6-bisphosphate dissociates prematurely, the serine is not phosphorylated, the enzyme is not regenerated, and further rounds of catalysis cannot occur.

55. The ATP yield for the pathway is 2 ATP. *T. tenax* stores energy in the form of glycogen, which is phosphorolyzed to glucose-1-phosphate. The glucose-1-phosphate is converted to glucose-6-phosphate, an isomerization reaction that does not require an investment of ATP. Glucose-6-phosphate is then converted to fructose-6-phosphate. The next step, in which fructose-6-phosphate is converted to fructose-1,6-bisphosphate, also does not require an investment of ATP, since the phosphofructokinase reaction in *T. tenax* is reversible and uses pyrophosphate rather than ATP. Since the GAPDH reaction does not produce 1,3-bisphosphoglycerate, there is no ATP produced until the last step, when phosphoenolpyruvate is converted to pyruvate. Since this reaction occurs twice for each glucose molecule, the pathway yields two moles ATP per mole of glucose. [From Brunner, N. A., Brinkmann, H., Siebers, B., and Hensel, R., *J. Biol. Chem.* **273**, 6149–6156 (1998).]

57. (a) The first committed step of the pentose phosphate pathway is the first reaction, which is catalyzed by glucose-6-phosphate dehydrogenase and is irreversible. Once glucose-6-phosphate has passed this point, it has no other fate than conversion to a pentose phosphate.

(b) The hexokinase reaction does not commit glucose to the glycolytic pathway, since the product of the reaction, glucose-6-phosphate, can also enter the pentose phosphate pathway.

59. The pentose phosphate pathway in the red blood cell generates NADPH, which is used to regenerate oxidized glutathione. Glucose-6-phosphate dehydrogenase is the enzyme that catalyzes the first step of the oxidative branch of the pathway. Its deficiency results in a decreased output of NADPH from the pathway. As a result, glutathione remains in the oxidized form and cannot fulfill its roles of decreasing the concentrations of organic peroxides, maintaining red blood cell shape, and keeping the iron ion of hemoglobin in the +2 form. Hemolytic anemia is the likely result.

61.

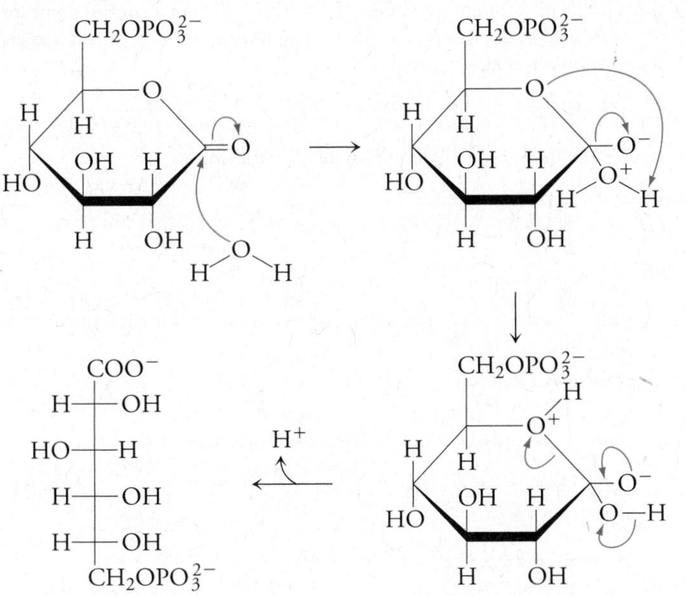

$CH_2OPO_3^{2-}$ (hexose structure with reaction arrows) → intermediate →

$CH_2OPO_3^{2-}$ intermediate

$$\begin{array}{c} COO^- \\ H\!-\!OH \\ HO\!-\!H \\ H\!-\!OH \\ H\!-\!OH \\ CH_2OPO_3^{2-} \end{array} \quad \xleftarrow{\;H^+\;} $$

63. G16BP inhibits hexokinase but stimulates PFK and pyruvate kinase. This means that glycolysis will be active, but only if the substrate is glucose-6-phosphate, since glucose cannot be phosphorylated in the absence of hexokinase activity. The pentose phosphate pathway is inactive, since 6-phosphogluconate dehydrogenase is inhibited. Phosphoglucomutase is activated, which converts glucose-1-phosphate (the product of glycogenolysis) to glucose-6-phosphate. Thus, in the presence of G16BP, glycogenolysis is active and produces substrate for glycolysis but not the pentose phosphate pathway. This is a more efficient process than using glucose taken up from the blood, which would need to be phosphorylated at the expense of ATP. [From Beitner, R., *Trends Biol. Sci.* **4**, 228–230 (1979).]

Chapter 14

1. In mammalian cells, pyruvate can be converted to lactate by lactate dehydrogenase. Pyruvate can also be transformed into oxaloacetate; this reaction is catalyzed by pyruvate carboxylase. Pyruvate can be converted to acetyl-CoA by the pyruvate dehydrogenase complex. Pyruvate can be converted to alanine by transamination.

3. The purpose of steps 4 and 5 is to regenerate the enzyme. In step 3, the product acetyl-CoA is released, but the lipoamide prosthetic group of E2 is reduced. In step 4, the E3 reoxidizes the lipoamide group by accepting the protons and electrons from the reduced lipoamide. In step 5, the enzyme is reoxidized by NAD^+. The product NADH then diffuses away.

5. Arsenite reacts with the reduced lipoamide group on E2 of the pyruvate dehydrogenase complex to form a compound with the structure shown in the figure. The enzyme cannot be regenerated and can no longer catalyze the conversion of acetyl-CoA to pyruvate. The α-ketoglutarate dehydrogenase complex has a lipoamide group on its E2 subunits and will be inhibited as well. The entire citric acid cycle cannot function, glucose cannot be oxidized aerobically, and respiration comes to a halt, which explains why these compounds are so toxic.

7. In both cases, the activity of the pyruvate dehydrogenase complex decreases, as both NADH and acetyl-CoA are products of the reaction. Rising concentrations of NADH and acetyl-CoA decrease pyruvate dehydrogenase activity by competing with NAD^+ and CoASH for binding sites on the enzyme.

9. The E1 subunit of the pyruvate dehydrogenase complex requires TPP, the phosphorylated form of thiamine, as a cofactor. Administering large doses of thiamine might be a successful treatment option if the E1 mutation happens to occur in the thiamine-binding site.

11. The phosphofructokinase reaction is the major rate-control point for the pathway of glycolysis. Inhibiting phosphofructokinase slows the entire pathway, so the production of acetyl-CoA by glycolysis followed by the pyruvate dehydrogenase complex can be decreased when the citric acid cycle is operating at maximum capacity and the citrate concentration is high.

13. In the citrate synthase mechanism, a proton is removed from the acetyl group of acetyl-CoA. The histidine then forms a hydrogen bond to stabilize the resulting enediolate intermediate. The alanine side chain cannot form this hydrogen bond; thus, the reaction cannot continue. [From Pereira, D. S., Donald, L. J., Hosfield, D. J., and Duckworth, H. W., *J. Biol. Chem.* **269**, 412–417 (1994).]

15.

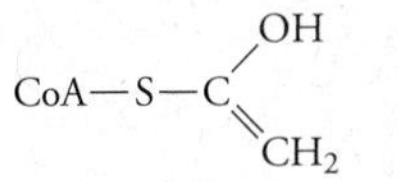

$$CoA\!-\!S\!-\!C\!\overset{OH}{\underset{CH_2}{\Big\langle}}$$

17. (a) Aconitase is the enzyme that catalyzes the reversible isomerization of citrate to isocitrate. Because this reaction is followed by and preceded by irreversible reactions, the inhibition of aconitase leads to an accumulation of citrate. The concentrations of other citric acid cycle intermediates will be decreased.

(b) If the citric acid cycle and mitochondrial respiration are not functioning, the cell turns to glycolysis to produce the ATP required for its energy needs. Consequently, flux through glycolysis increases. The increase in the rate of the pentose phosphate pathway is required to meet the increased demand for reducing equivalents during hyperoxia. [From Allen, C. B., Guo, X. L., and White, C. W., *Am. J. Physiol.* **274 (3 Pt. 1)**, L320–L329 (1998).]

19. *Cis*-aconitate is an intermediate in the reaction when citrate is converted to isocitrate by aconitase. *Trans*-aconitate structurally resembles *cis*-aconitate and would be expected to compete with *cis*-aconitate for binding to the enzyme. But because *trans*-aconitate is a noncompetitive inhibitor when citrate is used as the substrate, the citrate binding site must be distinct from the aconitate binding site. Citrate and *trans*-aconitate do not compete for binding and can bind to the enzyme simultaneously, but when both substrate and inhibitor are bound, the substrate cannot be converted to product. [From Villafranca, J. J., *J. Biol. Chem.* **249**, 6149–6155 (1974).]

21. See Sample Calculation 12-2. K_{eq} can be calculated by rearranging Equation 12-2:

$$K_{eq} = e^{-\Delta G^{\circ\prime}/RT}$$
$$K_{eq} = e^{-(-21\ kJ\cdot mol^{-1})/(8.3145\times10^{-3}\ kJ\cdot K^{-1}\cdot mol^{-1})(298\ K)}$$
$$K_{eq} = e^{8.5}$$
$$K_{eq} = 4.8\times10^3$$

23. (a) Usually, phosphorylation of an enzyme causes a conformational change in the protein that subsequently alters its activity. For the bacterial isocitrate dehydrogenase, however, phosphorylation of an active site Ser residue introduces negative charges that repel the negatively charged isocitrate and prevent it from binding.

(b) Construction of the mutant supports this hypothesis. The introduction of the negatively charged Asp residue in place of the Ser residue similarly introduces a negative charge to the active site and prevents isocitrate binding in the same manner. [From Dean, A. M., Lee, M. H. I., and Koshland, D. E., *J. Biol. Chem.* **264**, 20482–20486 (1989).]

25. *Step 1.* In the first step, α-ketoglutarate is decarboxylated, a process that requires TPP. The carbon of the carbonyl group becomes a carbanion, which forms a bond with TPP.

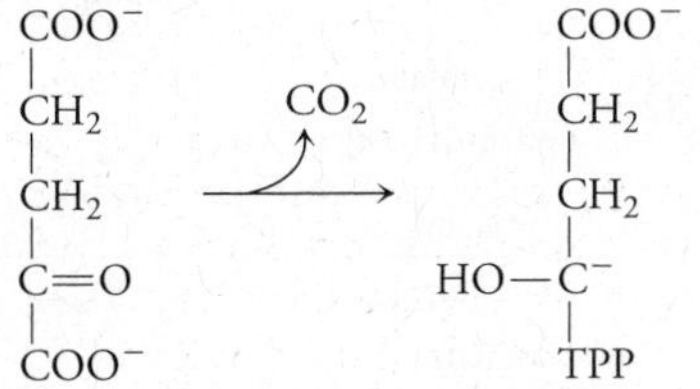

$$\begin{array}{c} COO^- \\ CH_2 \\ CH_2 \\ C\!=\!O \\ COO^- \end{array} \xrightarrow{\;CO_2\;} \begin{array}{c} COO^- \\ CH_2 \\ CH_2 \\ HO\!-\!C \\ TPP \end{array}$$

α-Ketoglutarate

Step 2. The succinyl group is then transferred to the lipoamide prosthetic group of E2 of the α-ketoglutarate dehydrogenase complex.

Lipoamide

Step 3. The succinyl group is transferred to coenzyme A, and the lipoamide group is reduced.

reduced lipoamide

Coenzyme A

Succinyl-CoA

Steps 4 and 5. The last two steps are the same as for the pyruvate dehydrogenase complex. E3 reoxidizes the lipoamide when its disulfide group accepts two protons and two electrons. The NAD$^+$ reoxidizes the enzyme, and the NADH and H$^+$ products diffuse away.

27. Succinyl-CoA resembles acetyl-CoA sufficiently that it is able to compete with acetyl-CoA for binding to the active site of citrate synthase. Likewise, succinyl-CoA competes with CoASH binding to the active site of α-ketoglutarate dehydrogenase. Both cases are examples of feedback inhibition: Succinyl-CoA inhibits the enzyme that produces it and inhibits an earlier enzyme in the pathway.

29. When operating in reverse, succinyl-CoA synthetase catalyzes a kinase-type reaction, the transfer of a phosphoryl group from a nucleotide triphosphate (GTP or ATP).

31. Succinate accumulates because it cannot be converted to fumarate. Succinyl-CoA also accumulates because the succinyl-CoA synthetase reaction is reversible. However, the succinyl-CoA ties up some of the cell's CoA supply, so the α-ketoglutarate dehydrogenase reaction, which requires CoA, slows. As a result, α-ketoglutarate accumulates.

33. $\Delta G = \Delta G^{\circ\prime} + RT \ln \dfrac{[\text{malate}]}{[\text{fumarate}]}$

$0 = -3.4 \text{ kJ} \cdot \text{mol}^{-1} + (8.3145 \times 10^{-3} \text{ kJ} \cdot \text{K}^{-1} \cdot \text{mol}^{-1})$

$(310 \text{ K}) \ln \dfrac{[\text{malate}]}{[\text{fumarate}]}$

$3.4 \text{ kJ} \cdot \text{mol}^{-1} = 2.58 \text{ kJ} \cdot \text{mol}^{-1} \ln \dfrac{[\text{malate}]}{[\text{fumarate}]}$

$1.32 = \ln \dfrac{[\text{malate}]}{[\text{fumarate}]}$

$e^{1.32} = \dfrac{[\text{malate}]}{[\text{fumarate}]}$

$3.7 = \dfrac{[\text{malate}]}{[\text{fumarate}]}$

The ratio of malate to fumarate is 3.7 to 1, indicating that the reaction proceeds in the direction of formation of malate. This is not a control point for the citric acid cycle because the ΔG is close to zero, indicating it is a near-equilibrium reaction.

35. (a) malate + NAD$^+$ → oxaloacetate + NADH + H$^+$
$$\Delta G^{\circ\prime} = 29.7 \text{ kJ} \cdot \text{mol}^{-1}$$
oxaloacetate + acetyl-CoA → citrate + coenzyme A
$$\Delta G^{\circ\prime} = -31.5 \text{ kJ} \cdot \text{mol}^{-1}$$

---

malate + NAD$^+$ + acetyl-CoA → citrate + NADH + H$^+$ + CoA
$$\Delta G^{\circ\prime} = -1.8 \text{ kJ} \cdot \text{mol}^{-1}$$

(b) The equilibrium constant for the coupled reaction is 3.4×10^5 times greater than the equilibrium constant for the uncoupled reaction.

Reactions 1 and 8

$K_{eq} = e^{-\Delta G^{\circ\prime}/RT}$

$K_{eq} = e^{-(-1.8 \text{ kJ} \cdot \text{mol}^{-1})/(8.3145 \times 10^{-3} \text{ kJ} \cdot \text{K}^{-1} \cdot \text{mol}^{-1})(298 \text{ K})}$

$K_{eq} = e^{0.73}$

$K_{eq} = 2.1$

Uncoupled Reaction 8

$K_{eq} = e^{-\Delta G^{\circ\prime}/RT}$

$K_{eq} = e^{-(29.7 \text{ kJ} \cdot \text{mol}^{-1})/(8.3145 \times 10^{-3} \text{ kJ} \cdot \text{K}^{-1} \cdot \text{mol}^{-1})(298 \text{ K})}$

$K_{eq} = e^{-12.0}$

$K_{eq} = 6.2 \times 10^{-6}$

37. (a) The isotopic label on C4 of oxaloacetate is released as $^{14}CO_2$ in the α-ketoglutarate dehydrogenase reaction.
(b) The isotopic label on C1 of acetyl-CoA is scrambled at the succinyl-CoA synthetase step. Because succinate is symmetrical, C1 and C4 are chemically equivalent, so in a population of molecules, both C1 and C4 would appear to be labeled (half the label would appear to be at C1 and half at C4). Consequently, one round of the citric acid cycle would yield oxaloacetate with half the labeled carbon at C1 and half at C4. Both of these labeled carbons would be lost as $^{14}CO_2$ in a second round of the citric acid cycle.

39. (a) Substrate availability: Acetyl-CoA and oxaloacetate levels regulate citrate synthase activity.
(b) Product inhibition: Citrate inhibits citrate synthase, NADH inhibits isocitrate dehydrogenase and α-ketoglutarate dehydrogenase, and succinyl-CoA inhibits α-ketoglutarate dehydrogenase.
(c) Feedback inhibition: NADH and succinyl-CoA inhibit citrate synthase.

41. Pyruvate carboxylase converts pyruvate to oxaloacetate, one of the reactants for the first reaction of the citric acid cycle. If the first reaction of the cycle cannot take place, the remaining reactions cannot proceed.

43. (a) The citric acid cycle is a multistep catalyst. Degrading an amino acid to a citric acid cycle intermediate boosts the catalytic activity of the cycle but does not alter the stoichiometry of the overall reaction (acetyl-CoA → 2 CO$_2$).
(b) Pyruvate derived from the degradation of an amino acid can be converted to acetyl-CoA by the pyruvate dehydrogenase complex; these amino acid carbons can then be completely oxidized by the citric acid cycle.

45. Any metabolite that can be converted to oxaloacetate can enter gluconeogenesis and serve as a precursor for glucose. Biological molecules that are degraded to acetyl-CoA cannot be used as glucose precursors because acetyl-CoA enters the citric acid cycle and its two carbons are oxidized to carbon dioxide. Thus, glyceraldehyde-3-phosphate, tryptophan, phenylalanine, and pentadecanoate can serve as gluconeogenic substrates because at least one of their breakdown products can be converted to oxaloacetate. Palmitate and leucine are not glucogenic because their breakdown products are acetyl-CoA or one of its derivatives. Acetyl-CoA cannot be converted to pyruvate in mammals.

47. Alanine can be converted to pyruvate via the transamination reaction shown in Problem 44. In gluconeogenesis, pyruvate is converted to oxaloacetate via the pyruvate carboxylase reaction, then oxaloacetate is converted to phosphoenolpyruvate, and so on to produce glucose. But if pyruvate carboxylase is deficient, alanine is converted to pyruvate, but the gluconeogenic pathway can go no further.

49. Pyruvate carboxylase requires biotin as a cofactor (see Table 12-2). If the nature of the defect involves a mutated pyruvate carboxylase with a decreased affinity for biotin, administering large doses of the vitamin might help treat the disease. Biotin treatment is ineffective for more severe forms of the disease in which the mutation does not occur in the biotin-binding region or in which the enzyme is expressed at extremely low levels or not at all.

51. The exercising muscle required greater concentrations of ATP to power it, so the rates of glycolysis and the citric acid cycle increased. Phosphoenolpyruvate was converted to pyruvate more rapidly, so the concentration of phosphoenolpyruvate decreased. Some of the pyruvate was converted to acetyl-CoA via the pyruvate dehydrogenase reaction. Since equimolar amounts of acetyl-CoA and oxaloacetate are required for the first step of the citric acid cycle, some of the pyruvate was converted to oxaloacetate via the pyruvate carboxylase reaction. This explains why oxaloacetate concentrations increased. The concentration of pyruvate did not increase because a steady state was reached: The rate of production of pyruvate from phosphoenolpyruvate was equal to the rate of consumption of pyruvate.

53. (a) Isocitrate is a branch point between the citric acid cycle and the glyoxylate pathway in organisms such as *E. coli*, which have the glyoxylate pathway. When acetate is the only food source, isocitrate dehydrogenase is inactive and acetate enters the glyoxylate pathway (as acetyl-CoA). This pathway produces glucose (by gluconeogenesis) and other intermediates that can be used as precursors in other biosynthetic reactions.

Citrate $\longrightarrow$ Isocitrate
$\quad\quad\quad$ isocitrate $\downarrow$ $\longrightarrow$ Succinate
$\quad\quad\quad$ lyase
$\quad\quad\quad\quad\quad$ Glyoxylate
$\quad\quad\quad\quad\quad\quad$ +
Acetate $\longrightarrow$ Acetyl-CoA $\xrightarrow{\text{malate synthase}}$ Malate $\xrightarrow{\text{malate dehydrogenase}}$ Oxaloacetate

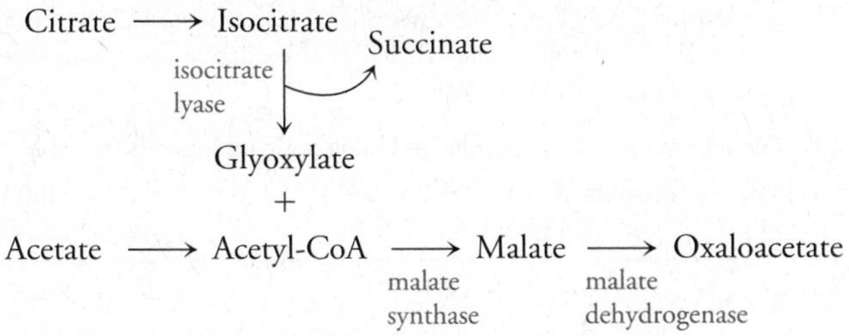

(b) When glucose is the substrate for the cultured *E. coli*, the glyoxylate pathway, which leads to glucose, is no longer necessary. Isocitrate dehydrogenase then becomes active so that glucose enters glycolysis and then the citric acid cycle.

55. If the fatty acid is broken down, the resulting acetyl-CoA units can enter the citric acid cycle for further oxidation. Because the labeled carbons are not lost as CO_2 in the first round of the cycle (see Fig.14-10), some of the label will appear in oxaloacetate. If some of this oxaloacetate is siphoned from the citric acid cycle for gluconeogenesis, some of the label will appear in glucose. This sequence of reactions does not violate the statement that fatty acids cannot be converted to carbohydrates because two carbons have already been lost as CO_2. Thus, there is no net conversion of fatty acid carbons to glucose carbons.

57. The concentration of AMP in muscle cells rises during periods of high muscle activity, indicating the need for ATP production via glycolysis and the citric acid cycle. AMP is converted to IMP by adenosine deaminase, as shown in the figure. Breakdown of muscle protein produces aspartate, which joins with IMP to produce adenylosuccinate. This substrate is lysed to form AMP and fumarate. Fumarate is a citric acid cycle intermediate, and increasing its concentration leads to greater citric acid cycle activity. Thus, the purine nucleotide cycle acts as an anaplerotic mechanism for the citric acid cycle (at the expense of muscle protein).

59. (a) NADH, citrate, and succinyl-CoA inhibit citrate synthase in mammals but do not inhibit citrate synthase in *H. pylori*. Isocitrate dehydrogenase is NADP⁺-dependent rather than NAD⁺-dependent and is regulated differently (by higher concentrations of its substrates NADP⁺

and isocitrate instead of NADH). *H. pylori* lacks α-ketoglutarate dehydrogenase and instead has α-ketoglutarate oxidase. The enzyme succinyl-CoA synthetase is missing in *H. pylori*. This enzyme catalyzes the only substrate-level phosphorylation reaction in the citric acid cycle; therefore, no GTP is produced in the citric acid cycle of this organism. Succinate dehydrogenase is missing. Fumarate reductase is present. Mammals do not have the glyoxylate pathway, but most bacteria do. *H. pylori* has only one step, catalyzed by malate synthase (isocitrate lyase is not present).

(b) High K_M values indicate a low affinity of the enzyme for the substrate. The high values indicate that substrate concentrations must be relatively high for the enzyme to attain half-maximal velocity. This indicates that the pathway does not operate unless the concentrations of the citric acid cycle intermediates are relatively high, which occurs when the *H. pylori* is in a nutrient-rich environment with plentiful resources (which would occur when the human "host" ingested a meal). Since the primary purpose of the citric acid cycle in *H. pylori* is to provide biosynthetic intermediates, it makes sense that the pathway operates only when metabolic resources are plentiful.

(c) *H. pylori* citrate synthase is inhibited by ATP but is not affected by NADH or any of the other citric acid cycle intermediates. Since the citric acid cycle in this organism is not used to produce metabolic energy in the form of reducing equivalents for oxidative phosphorylation, it makes sense that NADH would not serve as an inhibitor.

(d) Citrate synthase, isocitrate dehydrogenase, and α-ketoglutarate oxidase may serve as regulatory control points since these enzymes catalyze irreversible reactions and are subject to activation and inhibition by allosteric modulators.

(e) Enzymes unique to *H. pylori* would be good therapeutic targets: α-ketoglutarate oxidase, fumarate reductase, and malate synthase. [From Pitson, S. M., Mendz, G. L., Srinivasan, S., and Hazell, S. L., *Eur. J. Biochem.* **260**, 258–267 (1999).]

61. Enzymes of the glyoxylate pathway, particularly malate dehydrogenase and isocitrate lyase (which are unique to this pathway), would be inactivated. The glyoxylate pathway produces glucose from noncarbohydrate sources but is not required when glucose is available. Enzymes required for gluconeogenesis that are not involved in glycolysis would also be inactivated, mainly phosphoenolpyruvate carboxykinase and fructose-1,6-bisphosphatase.

63. (a) This reaction is an anaplerotic reaction in bacteria and plants, analogous to the pyruvate carboxylase reaction in animals. PPC produces oxaloacetate for the citric acid cycle to ensure its continued operation as a pathway for oxidizing fuel molecules and for producing intermediates for biosynthetic reactions.

(b) Acetyl-CoA and oxaloacetate are required in equimolar amounts as substrates for the citrate synthase reaction that begins the citric acid cycle. If the concentration of acetyl-CoA rises, the concentration of oxaloacetate will need to increase as well, so acetyl-CoA stimulates the enzyme that produces its cosubstrate. The activation by fructose-1,6-bisphosphate appears to be a feed-forward mechanism to ensure that sufficient oxaloacetate is present to condense with the acetyl-CoA produced by glycolysis and the pyruvate dehydrogenase reaction.

65. (a) Glutamine is converted to glutamate by deamidation, and glutamate is converted to α-ketoglutarate by transamination.

(b) α-Ketoglutarate dehydrogenase, succinyl-CoA synthetase, succinate dehydrogenase, fumarase, and malic enzyme.

Chapter 15

1. Reverse the NADH half-reaction and the sign of its $\mathcal{E}^{\circ\prime}$ value to indicate oxidation, then combine the half-reactions and their $\mathcal{E}^{\circ\prime}$ values.

pyruvate + 2 H⁺ + 2 e⁻ → lactate	$\mathcal{E}^{\circ\prime} = -0.185$ V
NADH → NAD⁺ + H⁺ + 2 e⁻	$\mathcal{E}^{\circ\prime} = 0.315$ V
NADH + pyruvate + H⁺ → NAD⁺ + lactate	$\Delta\mathcal{E}^{\circ\prime} = 0.130$ V

Use Equation 15-4 to calculate $\Delta G^{\circ\prime}$ for this reaction:

$$\Delta G^{\circ\prime} = -n\mathcal{F}\Delta\mathcal{E}^{\circ\prime}$$
$$\Delta G^{\circ\prime} = -(2)(96{,}485\ \text{J}\cdot\text{V}^{-1}\cdot\text{mol}^{-1})(0.130\ \text{V})$$
$$\Delta G^{\circ\prime} = -25.1\ \text{kJ}\cdot\text{mol}^{-1}$$

The reduction of pyruvate by NADH (Section 13-1) is spontaneous under standard conditions.

3. The relevant reactions and their $\mathcal{E}^{\circ\prime}$ values are obtained from Table 15-1:

Dihydrolipoic acid $\rightarrow$ lipoic acid + 2 H$^+$ + 2 e^- $\mathcal{E}^{\circ\prime} = 0.29$ V
NAD$^+$ + H$^+$ + 2 $e^- \rightarrow$ NADH $\mathcal{E}^{\circ\prime} = -0.315$ V

Dihydrolipoic acid + NAD$^+$ $\rightarrow$ lipoic acid + NADH $\Delta\mathcal{E}^{\circ\prime} = -0.025$ V

Use Equation 15-4 to calculate $\Delta G^{\circ\prime}$ for this reaction:

$$\Delta G^{\circ\prime} = -n\mathcal{F}\Delta\mathcal{E}^{\circ\prime}$$
$$\Delta G^{\circ\prime} = -(2)(96{,}485\ \text{J}\cdot\text{V}^{-1}\cdot\text{mol}^{-1})(-0.025\ \text{V})$$
$$\Delta G^{\circ\prime} = 4.8\ \text{kJ}\cdot\text{mol}^{-1}$$

5. Consult Table 15-1 for the relevant half-reactions involving acetaldehyde and NAD$^+$. Reverse the acetaldehyde half-reaction and the sign of its $\mathcal{E}^{\circ\prime}$ value to indicate oxidation, then combine the half-reactions and their $\mathcal{E}^{\circ\prime}$ values.

acetaldehyde + H$_2$O $\rightarrow$ acetate + 3 H$^+$ + 2 e^- $\mathcal{E}^{\circ\prime} = 0.581$ V
NAD$^+$ + H$^+$ + 2 $e^- \rightarrow$ NADH $\mathcal{E}^{\circ\prime} = -0.315$ V

acetaldehyde + H$_2$O + NAD$^+$ $\rightarrow$ NADH + acetate + 2 H$^+$ $\Delta\mathcal{E}^{\circ\prime} = 0.266$ V

Use Equation 15-4 to calculate $\Delta G^{\circ\prime}$ for this reaction:

$$\Delta G^{\circ\prime} = -n\mathcal{F}\Delta\mathcal{E}^{\circ\prime}$$
$$\Delta G^{\circ\prime} = -(2)(96{,}485\ \text{J}\cdot\text{V}^{-1}\cdot\text{mol}^{-1})(0.266\ \text{V})$$
$$\Delta G^{\circ\prime} = -51.3\ \text{kJ}\cdot\text{mol}^{-1}$$

The oxidation of acetaldehyde by NAD$^+$ is spontaneous, as shown by the negative $\Delta G^{\circ\prime}$ value.

7. Consult Table 15-1 for the relevant half-reactions involving ubiquinol and cytochrome c_1. Reverse the ubiquinol half-reaction and the sign of its $\mathcal{E}^{\circ\prime}$ value to indicate oxidation, multiply the coefficients in the cytochrome c_1 equation by 2 so that the number of electrons transferred will be equal, then combine the half-reactions and their $\mathcal{E}^{\circ\prime}$ values.

QH$_2 \rightarrow$ Q + 2 H$^+$ + 2 e^- $\mathcal{E}^{\circ\prime} = -0.045$ V
2 cyt c_1 (Fe^{3+}) + 2 $e^- \rightarrow$ 2 cyt c_1 (Fe^{2+}) $\mathcal{E}^{\circ\prime} = 0.220$ V

QH$_2$ + 2 cyt c_1 (Fe^{3+}) $\rightarrow$ Q + 2 cyt c_1 (Fe^{2+}) $\Delta\mathcal{E}^{\circ\prime} = 0.175$ V

Use Equation 15-4 to calculate $\Delta G^{\circ\prime}$ for this reaction:

$$\Delta G^{\circ\prime} = -n\mathcal{F}\Delta\mathcal{E}^{\circ\prime}$$
$$\Delta G^{\circ\prime} = -(2)(96{,}485\ \text{J}\cdot\text{V}^{-1}\cdot\text{mol}^{-1})(0.175\ \text{V})$$
$$\Delta G^{\circ\prime} = -33.8\ \text{kJ}\cdot\text{mol}^{-1}$$

The reaction is spontaneous under standard conditions.

9. (a) Use Equation 15-2 to determine the $\mathcal{E}$ values for these two half-reactions.

$$\text{QH}_2 \rightarrow \text{Q} + 2\ \text{H}^+ + 2\ e^- \quad \mathcal{E}^{\circ\prime} = -0.045\ \text{V}$$
$$\mathcal{E} = \mathcal{E}^{\circ\prime} - \frac{0.026\ \text{V}}{n}\ln\frac{[\text{QH}_2]}{[\text{Q}]}$$
$$\mathcal{E} = -0.045\ \text{V} - \frac{0.026\ \text{V}}{2}\ln 10$$
$$\mathcal{E} = -0.075\ \text{V}$$

$$2\ \text{cyt}\ c\ (\text{Fe}^{3+}) + 2\ e^- \rightarrow 2\ \text{cyt}\ c\ (\text{Fe}^{2+}) \quad \mathcal{E}^{\circ\prime} = 0.235\ \text{V}$$

$$\mathcal{E} = \mathcal{E}^{\circ\prime} - \frac{0.026\ \text{V}}{n}\ln\frac{[\text{cyt}\ c\ (\text{Fe}^{2+})]}{[\text{cyt}\ c\ (\text{Fe}^{3+})]}$$
$$\mathcal{E} = 0.235\ \text{V} - \frac{0.026\ \text{V}}{2}\ln\frac{1}{5}$$
$$\mathcal{E} = 0.235\ \text{V} + 0.021\ \text{V} = 0.256\ \text{V}$$

QH$_2 \rightarrow$ Q + 2 H$^+$ + 2 e^- $\mathcal{E} = -0.075$ V
2 cyt c (Fe^{3+}) + 2 $e^- \rightarrow$ 2 cyt c (Fe^{2+}) $\mathcal{E} = 0.256$ V

QH$_2$ + 2 cyt c (Fe^{3+}) $\rightarrow$ Q + 2 H$^+$ + 2 cyt c (Fe^{2+}) $\Delta\mathcal{E} = 0.181$ V

(b) Use Equation 15-4 to calculate ΔG for this reaction:

$$\Delta G = -n\mathcal{F}\Delta\mathcal{E}$$
$$\Delta G = -(2)(96{,}485\ \text{J}\cdot\text{V}^{-1}\cdot\text{mol}^{-1})(0.181\ \text{V})$$
$$\Delta G = -34.9\ \text{kJ}\cdot\text{mol}^{-1}$$

11. Consult Table 15-1 for the relevant half-reactions involving O$_2$ and NADH. Reverse the NADH half-reaction and the sign of its $\mathcal{E}^{\circ\prime}$ value to indicate oxidation, then combine the half-reactions and their $\mathcal{E}^{\circ\prime}$ values.

½ O$_2$ + 2 H$^+$ + 2 $e^- \rightarrow$ H$_2$O $\mathcal{E}^{\circ\prime} = 0.815$ V
NADH $\rightarrow$ NAD$^+$ + H$^+$ + 2 e^- $\mathcal{E}^{\circ\prime} = 0.315$ V

NADH + ½ O$_2$ + H$^+$ $\rightarrow$ NAD$^+$ + H$_2$O $\Delta\mathcal{E}^{\circ\prime} = 1.13$ V

Use Equation 15-4 to calculate $\Delta G^{\circ\prime}$ for this reaction:

$$\Delta G^{\circ\prime} = -n\mathcal{F}\Delta\mathcal{E}^{\circ\prime}$$
$$\Delta G^{\circ\prime} = -(2)(96{,}485\ \text{J}\cdot\text{V}^{-1}\cdot\text{mol}^{-1})(1.13\ \text{V})$$
$$\Delta G^{\circ\prime} = -218\ \text{kJ}\cdot\text{mol}^{-1}$$

The synthesis of 2.5 ATP requires a free energy investment of 2.5 $\times$ 30.5 kJ $\cdot$ mol^{-1}, or 76.3 kJ $\cdot$ mol^{-1}. The efficiency of oxidative phosphorylation is therefore 76.3/218 = 0.35, or 35%.

13. The relevant reactions and their $\mathcal{E}^{\circ\prime}$ values are obtained from Table 15-1:

succinate $\rightarrow$ fumarate + 2 H$^+$ + 2 e^- $\mathcal{E}^{\circ\prime} = -0.031$ V
Q + 2 H$^+$ + 2 $e^- \rightarrow$ QH$_2$ $\mathcal{E}^{\circ\prime} = 0.045$ V

succinate + Q $\rightarrow$ fumarate + QH$_2$ $\Delta\mathcal{E}^{\circ\prime} = 0.014$ V

Use Equation 15-4 to calculate $\Delta G^{\circ\prime}$ for this reaction:

$$\Delta G^{\circ\prime} = -n\mathcal{F}\Delta\mathcal{E}^{\circ\prime}$$
$$\Delta G^{\circ\prime} = -(2)(96{,}485\ \text{J}\cdot\text{V}^{-1}\cdot\text{mol}^{-1})(0.014\ \text{V})$$
$$\Delta G^{\circ\prime} = -2.7\ \text{kJ}\cdot\text{mol}^{-1}$$

15. **(a)** Since all of these inhibitors interfere with electron transfer somewhere in the electron transport chain, oxygen consumption will decrease when any of the inhibitors are added to a suspension of respiring mitochondria. Adding any of these inhibitors prevents electrons from being transferred to the oxygen, the final electron acceptor.
(b) In rotenone- or amytal-blocked mitochondria, NADH and Complex I redox centers are reduced while components from ubiquinone on are oxidized. In antimycin A–blocked mitochondria, NADH, Complex I redox centers, ubiquinol, and Complex III redox centers are reduced while cytochrome c and Complex IV redox centers are oxidized. In cyanide-blocked mitochondria, all of the electron transport components are reduced and only oxygen remains oxidized.

17. The donation of a pair of electrons to Complex IV will result in the synthesis of about 1.3 ATP per atom of oxygen (½ O$_2$). Therefore, the P:O ratio of this compound is 1.3 (see Solution 12c).

19. Adding tetramethyl-p-phenylenediamine to rotenone-blocked and antimycin A–blocked mitochondria effectively bypasses the block as the compound donates its electrons to Complex IV and electron transport resumes. Adding tetramethyl-p-phenylenediamine is not an effective bypass for cyanide-blocked mitochondria because cyanide inhibits electron transport in Complex IV. Similarly, ascorbate, which donates its electrons to cytochrome c and then to Complex IV, can act as an effective bypass for antimycin A–blocked mitochondria but not cyanide-blocked mitochondria.

21. Cyanide binds to the Fe^{2+} in the Fe–Cu center of cytochrome a_3 (see Problem 15). When the iron in hemoglobin is oxidized from Fe^{2+} to Fe^{3+}, cytochrome a_3 can donate an electron to reoxidize the hemoglobin to Fe^{2+}. This oxidizes the iron in cytochrome a_3 to Fe^{3+}. Cyanide does not bind to Fe^{3+}, so it is released and Complex IV can again function normally. The cyanide binds to the Fe^{2+} in hemoglobin, where it does not interfere with respiration (although it does interfere with oxygen delivery).

$$HbO_2\ (Fe^{2+})$$
$$\downarrow$$
$$HbO_2\ (Fe^{3+})$$
$$\downarrow \quad \text{cytochrome } a_3\ (Fe^{2+}\text{—Cu)—CN}^-$$
$$\downarrow \quad \text{cytochrome } a_3\ (Fe^{3+}\text{—Cu)}$$
$$HbO_2\ (Fe^{2+})$$

23. All these enzymes catalyze reactions in which electrons are transferred from reduced substances, such as NADH, to ubiquinone, a compound with a higher reduction potential. The flavin group, whose reduction potential is lower than that of ubiquinone (Table 15-1), is ideally suited to shuttle electrons between the reduced NADH and the ubiquinone.

25. Like the lipids that compose the membrane, coenzyme Q is amphiphilic, with a hydrophilic head and a hydrophobic tail. "Like dissolves like," and coenzyme Q literally dissolves in the membrane, which facilitates rapid diffusion.

27. Cytochrome c is a water-soluble, peripheral membrane protein and is easily dissociated from the membrane by adding salt solutions that interfere with the ionic interactions that tether it to the inner mitochondrial membrane. Cytochrome c_1 is an integral membrane protein and is largely water insoluble due to the nonpolar amino acids that interact with the acyl chains of the membrane lipids. Detergents are required to dissociate cytochrome c_1 from the membrane because amphiphilic detergents can disrupt the membrane and coat membrane proteins, acting as substitute lipids in the solubilization process.

29. The dead algae are a source of food for aerobic microorganisms lower in the water column. As the growth of these organisms increases, the rates of respiration and O_2 consumption increase to the point where the concentration of O_2 in the water becomes too low to sustain larger aerobic organisms.

31. In the absence of myoglobin function, the mice developed several compensatory mechanisms to ensure adequate oxygen delivery to tissues. The symptoms described all involve increasing the amount of available hemoglobin. In this manner, hemoglobin takes over some of the functions usually performed by myoglobin.

33. Myoglobin functions in muscle cells to facilitate oxygen diffusion throughout the cell and possibly assumes this same role in tumor cells. The resulting increase in oxygen concentration may allow the tumor cells to oxidize a higher percentage of the available glucose aerobically and thus obtain more ATP per molecule of glucose than if glucose were oxidized anaerobically. [From Kristiansen, G., et al., *J. Biol. Chem.,* **286,** 43417–43428 (2011).]

35. As a result of the exercise program, the number of mitochondria in the muscle cells of the participants increased, as indicated by an increase in DNA content. The increase in Complex II activity is of similar magnitude, which might be the result of the increased number of mitochondria. However, the total electron transport chain activity is twofold greater after the exercise intervention, indicating that mitochondrial function increased as well. These results suggest that even though oxidative damage to mitochondria increases with age, exercise can help maintain or increase mitochondrial function. [From Menshikova, E. V., Ritov, V. B., Fairfull, L., Ferrell, R. E., Kelley, D. E., and Goodpaster, B. H., *J. Gerontol. A Biol. Sci. Med. Sci.* **61,** 534–540 (2006)].

37. The free energy change for generating the electrical imbalance is calculated using Equation 15-6:

$$\Delta G = Z\mathcal{F}\Delta\psi$$
$$\Delta G = (1)(96{,}485\ J \cdot V^{-1} \cdot mol^{-1})(0.081\ V)$$
$$\Delta G = 7.8\ kJ \cdot mol^{-1}$$

39. **(a)** The pH of the intermembrane space is lower than the pH of the mitochondrial matrix because protons are pumped out of the matrix, across the inner membrane, and into the intermembrane space. The increase in concentration of protons in the intermembrane space decreases the pH; the deficit of protons in the matrix results in an increase in pH.

(b) Detergents disrupt membranes. An intact inner mitochondrial membrane is required for oxidative phosphorylation to take place. Without an intact membrane, an electrochemical gradient, which is the energy reservoir that drives ATP synthesis, cannot be established, and ATP synthesis does not occur.

(c) Uncouplers such as DNP ferry protons across the inner mitochondrial membrane and dissipate the proton gradient established by electron transport. In the presence of DNP, electron transport still occurs, but the free energy released by the process is dissipated as heat instead of being harnessed to synthesize ATP.

41. (a)

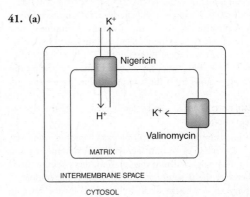

(b) Potassium ions enter the matrix with the assistance of valinomycin. These ions are then exported by nigericin in exchange for protons. Importing protons into the mitochondrial matrix dissipates the proton gradient. Since the proton gradient serves as the energy reservoir that drives ATP synthesis, no ATP can be synthesized.

43. A total of 32 ATP are obtained from the exergonic oxidation of glucose under aerobic conditions:

glycolysis	2 ATP	2 ATP
	2 NADH	$2 \times 2.5 = 5$ ATP
2 pyruvate $\rightarrow$ 2 acetyl-CoA	2 NADH	$2 \times 2.5 = 5$ ATP
citric acid cycle (2 rounds)	2×3 NADH	$6 \times 2.5 = 15$ ATP
	2×1 QH_2	$2 \times 1.5 = 3$ ATP
	2×1 GTP	$2 \times 1 =$ ATP
Total		**32 ATP**

A total of 2 ATP per glucose are obtained when glucose is oxidized in the absence of oxygen by conversion to lactate or ethanol (Section 13-1).

45. Aerobic respiration yields 32 ATP (see Problem 43), each of which requires 30.5 kJ · mol⁻¹ to synthesize:

$$\frac{32\ \text{ATP} \times 30.5\ kJ \cdot mol^{-1}\ \text{ATP}}{2850\ kJ \cdot mol^{-1}} \times 100 = 34\%$$

Lactate fermentation yields 2 ATP:

$$\frac{2\ \text{ATP} \times 30.5\ kJ \cdot mol^{-1}\ \text{ATP}}{196\ kJ \cdot mol^{-1}} \times 100 = 31\%$$

Alcoholic fermentation yields 2 ATP:

$$\frac{2\ \text{ATP} \times 30.5\ kJ \cdot mol^{-1}\ \text{ATP}}{235\ kJ \cdot mol^{-1}} \times 100 = 26\%$$

47. (a) Aerobic oxidation of glucose yields 32 ATP per glucose, whereas alcoholic fermentation of glucose by the yeast yields only 2 ATP per glucose. Assuming that the energy needs of the yeast cell remain constant under both aerobic and anaerobic conditions, the catabolism of glucose by the yeast will be 16-fold greater in the absence of oxygen than in the presence of oxygen in order to produce the same amount of ATP. Thus, the rate of consumption of glucose decreases when the cells are exposed to oxygen because fewer glucose molecules must be oxidized to yield the same amount of ATP.

(b) Both ratios will initially increase, as the citric acid cycle (which does not operate under anaerobic conditions) produces more NADH equivalents for electron transport. The [ATP]/[ADP] ratio will also increase, since aerobic oxidation of glucose produces more ATP per mole of glucose than anaerobic oxidation [as described in part (a)]. ATP and NADH will "reset" the equilibrium by inhibiting the regulatory enzymes of glycolysis and the citric acid cycle, slowing down these processes. Eventually, the [NADH]/[NAD$^+$] and [ATP]/[ADP] ratios return to their "original" values.

49. (a) The import of ADP (net charge –3) and the export of ATP (net charge –4) represent a loss of negative charge inside the mitochondria. This decreases the difference in electrical charge across the membrane, since the outside is positive due to the translocation of protons during electron transport. Consequently, the gradient is diminished by the activity of the adenine nucleotide translocase.

(b) The activity of the P_i-H^+ symport protein diminishes the proton gradient by allowing protons from the intermembrane space to reenter the matrix.

(c) Both transport systems are driven by the free energy of the electrochemical proton gradient.

51. Inactivation of one c subunit by DCCD blocks all proton translocation by F_0 since the movement of protons across the membrane requires continuous rotation of the c ring. Without this rotation, the γ subunit of F_1 cannot move, and therefore the β subunits cannot undergo the conformational changes necessary to synthesize or hydrolyze ATP by the binding change mechanism.

53. DNP uncouples the electron transport chain from oxidative phosphorylation by dissipating the proton gradient. Electron transport still occurs, but the energy released by electron transport is dissipated as heat instead of being harnessed to synthesize ATP. One might think that DNP would be an effective diet aid because the sources of the electrons that go down the electron transport chain are dietary carbohydrates and fatty acids. If the energy of these compounds is dissipated as heat instead of used to synthesize ATP (which would then be used for, among other processes, the synthesis of fatty acids in adipocytes), weight gain from the ingestion of food could theoretically be prevented.

55. Organic compounds are oxidized by oxygen through the activity of the electron transport complexes of mitochondria. This activity generates the proton gradient that is dissipated during phosphorylation of ADP. When there is no more ADP, ATP synthase is inactive and unable to dissipate the proton gradient. As a result of the tight coupling between oxidative phosphorylation and electron transport, electron transport and oxygen consumption also come to a halt.

57. F_0 acts as a proton channel as the c ring rotates, feeding protons through the a subunit (see Fig. 15-23). The addition of F_1 blocks proton movement because the γ shaft rotates along with the c ring. In this system, the γ subunit and the c ring can rotate only when the binding change mechanism is in operation, that is, when the β subunits are binding and releasing nucleotides. ATP or ADP + P_i must be added to the system in order for the γ subunit to move.

59. Since it has 10 c subunits, the bacterial enzyme can theoretically produce 3 ATP for every 10 protons translocated. In the chloroplast, 3 ATP are synthesized for every 14 protons. Thus, the bacterium is more efficient in its use of the proton gradient established during electron transport and has a higher ratio of ATP produced per oxygen consumed.

61. α-Ketoglutarate is an intermediate of the citric acid cycle, so increasing its concentration can increase flux through the citric acid cycle, contributing to ATP synthesis by substrate-level phosphorylation. However, the α-ketoglutarate dehydrogenase reaction generates NADH, which cannot be efficiently reoxidized when electron transport and oxidative phosphorylation are not operating normally. The added aspartate helps to eliminate the excess NADH: Aspartate is transaminated to oxaloacetate (Section 14-3), which is then converted to malate by malate dehydrogenase in a reaction that consumes the NADH. [From Sgarbi, G., Casalena, G. A., Baracca, A., Lenaz, G., DiMauro, S., and Solaini, G., *Arch. Neurol.* **66**, 951–957 (2009).]

63. During anaerobic fermentation, only 2 ATP molecules can be obtained from each glucose molecule. Thus, in order to get enough ATP to satisfy the energy needs of the cell, glucose consumption by the glycolytic pathway predominates, and only a small amount is left over for the pentose phosphate pathway. But during aerobic respiration, 32 ATP are available per glucose molecule. Thus, less glucose is needed to synthesize the same amount of ATP, so a larger fraction of glucose can enter the pentose phosphate pathway.

65. (a) In the root system of the skunk cabbage, starch is broken down to yield glucose, which is catabolized via glycolysis, the citric acid cycle, and the electron transport chain. The oxidation of glucose provides the reduced cofactors (NADH and QH$_2$) required to keep electron transport going so that thermogenesis can occur.

(b) In the skunk cabbage, thermogenesis increases as the temperature decreases. Thus, the rate of aerobic oxidation of glucose increases to increase the flux of NADH and QH$_2$ through the electron transport chain. Since oxygen is the final electron acceptor, an increase in electron transport will also increase oxygen consumption. When the ambient air temperature is warmer, the need for heat production is less, so there is less flux through the electron transport chain. Thus, oxygen consumption decreases during the day.

67. (a) Glutamate enters the mitochondrion through a specific transporter and is then oxidized to α-ketoglutarate by glutamate dehydrogenase in the mitochondrial matrix. Concomitantly, NAD$^+$ is reduced to NADH, which can then enter electron transport.

(b) Ceramide could inhibit the glutamate transporter, the glutamate dehydrogenase enzyme, any of the three complexes involved in the electron transport chain, ATP synthase, or the ATP translocase.

(c) Ceramide doesn't interfere with ATP synthase. Ceramide must block respiration at an earlier step because the rate of respiration did not increase in the presence of the uncoupler.

(d) Ceramide does not inhibit the glutamate transporter or glutamate dehydrogenase. Ceramide must inhibit one of the three complexes involved in electron transport. (In fact, ceramide inhibits the activity of Complex III.) [From Gudz, T. I., Tserng, K.-Y., and Hoppel, C. L., *J. Biol. Chem.* **272**, 24154–24158 (1997).]

Chapter 16

1.

proton translocation	C, M
photophosphorylation	C
photooxidation	C
quinones	C, M
oxygen reduction	M
water oxidation	C
electron transport	C, M
oxidative phosphorylation	M
carbon fixation	C
NADH oxidation	M
Mn cofactor	C
heme groups	C, M
binding change mechanism	C, M
iron–sulfur clusters	C, M
NADP$^+$ reduction	C

3.

5. **(a)** Use Planck's law multiplied by Avogadro's number (N) to calculate the energy of the photons:

$$E = \frac{hc}{\lambda} \times N$$

$$E = \frac{(6.626 \times 10^{-34}\ \text{J} \cdot \text{s})(2.998 \times 10^{8}\ \text{m} \cdot \text{s}^{-1})}{4 \times 10^{-7}\ \text{m}}$$
$$\times\ (6.022 \times 10^{23}\ \text{photons} \cdot \text{mol}^{-1})$$

$$E = 300\ \text{kJ} \cdot \text{mol}^{-1}$$

(b) $E = \dfrac{hc}{\lambda} \times N$

$$E = \frac{(6.626 \times 10^{-34}\ \text{J} \cdot \text{s})(2.998 \times 10^{8}\ \text{m} \cdot \text{s}^{-1})}{7 \times 10^{-7}\ \text{m}}$$
$$\times\ (6.022 \times 10^{23}\ \text{photons} \cdot \text{mol}^{-1})$$

$$E = 170\ \text{kJ} \cdot \text{mol}^{-1}$$

7. The difference in reduction potential between P680* and P680 is $-0.8\ \text{V} - 1.15\ \text{V} = -1.95\ \text{V}$.

$$\Delta G^{\circ\prime} = -n\mathcal{F}\Delta\mathcal{E}^{\circ\prime}$$
$$= -(1)(96,485\ \text{J} \cdot \text{V}^{-1} \cdot \text{mol}^{-1})(-1.95\ \text{V})$$
$$= 188,000\ \text{J} \cdot \text{mol}^{-1} = 188\ \text{kJ} \cdot \text{mol}^{-1}$$

9. Because the algae appear red, red light is transmitted rather than absorbed. Therefore, the photosynthetic pigments in the red algae do not absorb red light but absorb light of other wavelengths.

11.

- The central metal ion in chlorophyll a is Mg^{2+}, whereas in heme a the central metal ion is Fe^{2+}.
- In chlorophyll a, there is a cyclopentanone ring fused to ring C.
- Ring B in chlorophyll a has an ethyl side chain; the chain in heme b is unsaturated.
- The propionyl side chain in ring D of chlorophyll a is esterified to a long, branched-chain alcohol.

13. The buildup of the proton gradient indicates a high level of activity of the photosystems. A steep gradient could therefore trigger photoprotective activity to prevent further photooxidation when the proton-translocating machinery is operating at maximal capacity.

15. The order of action is water–plastoquinone oxidoreductase (Photosystem II), then plastoquinone–plastocyanin oxidoreductase (cytochrome $b_6 f$), then plastocyanin–ferredoxin oxidoreductase (Photosystem I).

17. If electrons cannot be transferred to Photosystem I, then Photosystem II remains reduced and cannot be reoxidized. The photosynthetic production of oxygen ceases. No proton gradient is generated, so ATP synthesis does not occur in the presence of DCMU.

19. Like the lipids that compose the membrane, plastoquinone is amphiphilic, with a hydrophilic head and a hydrophobic tail. "Like dissolves like," so plastoquinone literally dissolves in the membrane, which facilitates rapid diffusion.

21. Use Equation 15-7, as applied in Sample Calculation 15-3.
The matrix and the stroma are both *in*.

$$\Delta G = 2.303\ RT\,(\text{pH}_{in} - \text{pH}_{out}) + Z\mathcal{F}\Delta\psi$$
$$\Delta G = 2.303(8.3145\ \text{J} \cdot \text{K}^{-1} \cdot \text{mol}^{-1})(298\ \text{K})(3.5)$$
$$+\ (1)(96,485\ \text{J} \cdot \text{V}^{-1} \cdot \text{mol}^{-1})(-0.05\ \text{V})$$

$$\Delta G = 20,000\ \text{J} \cdot \text{mol}^{-1} - 4800\ \text{J} \cdot \text{mol}^{-1}$$
$$\Delta G = 15.2\ \text{kJ} \cdot \text{mol}^{-1}$$

23. Consult Table 15-1 for the reduction potentials of the relevant half-reactions, reversing the sign for the water oxidation half-reaction.

$H_2O \rightarrow \frac{1}{2}\,O_2 + 2\,H^+ + 2\,e^-$	$\mathcal{E}^{\circ\prime} = -0.815\ \text{V}$
$NADP^+ + H^+ + 2\,e^- \rightarrow NADPH$	$\mathcal{E}^{\circ\prime} = -0.320\ \text{V}$
$H_2O + NADP^+ \rightarrow \frac{1}{2}\,O_2 + NADPH + H^+$	$\Delta\mathcal{E}^{\circ\prime} = -1.135\ \text{V}$

Use Equation 15-4 to calculate $\Delta G^{\circ\prime}$:

$$\Delta G^{\circ\prime} = -n\mathcal{F}\Delta\mathcal{E}^{\circ\prime}$$
$$\Delta G^{\circ\prime} = -(2)(96,485\ \text{J} \cdot \text{V}^{-1} \cdot \text{mol}^{-1})(-1.135\ \text{V})$$
$$\Delta G^{\circ\prime} = 219,000\ \text{J} \cdot \text{mol}^{-1}$$

Divide by Avogadro's number to obtain the free energy change per molecule:

$$\frac{219,000\ \text{J} \cdot \text{mol}^{-1}}{6.022 \times 10^{23}\ \text{molecules} \cdot \text{mol}^{-1}} = 3.6 \times 10^{-19}\ \text{J} \cdot \text{molecule}^{-1}$$

25. The final electron acceptor in photosynthesis is $NADP^+$. The final electron acceptor in mitochondrial electron transport is oxygen.

27. **(a)** An uncoupler dissipates the transmembrane proton gradient by providing a route for translocation other than ATP synthase. Therefore, chloroplast ATP production would decrease.
(b) The uncoupler would not affect $NADP^+$ reduction since light-driven electron-transfer reactions would continue regardless of the state of the proton gradient.

29. **(a)** More c subunits means that more protons are required to rotate the ATP synthase through one ATP-synthesizing step. Therefore, more photons must be absorbed to drive the translocation of more protons, so the quantum yield decreases.
(b) Cyclic electron flow contributes to the proton gradient and therefore leads to ATP synthesis. However, carbon fixation by the Calvin cycle requires NADPH also, so the additional photons that drive cyclic flow do not lead to more carbon fixed. Consequently, the quantum yield decreases.

31. This statement is false. The "dark" reactions do not require darkness in order to proceed. Sometimes the "dark" reactions are called "light-independent" reactions in order to specify that these reactions do not directly require light energy. This term is also misleading because the "dark" reactions of the Calvin cycle do require the products of the light reactions—ATP and NADPH—in order to proceed. Thus, for a majority of plants, the "dark" reactions actually occur during the day when the light reactions are operational and can produce the needed ATP and NADPH.

33. 3-Phosphoglycerate is the first stable radioactive intermediate that forms when algal cells are exposed to $^{14}CO_2$. The radioactive label is found on the carboxyl group of the compound.

35. The increase in mass comes from carbon dioxide. CO_2 is the carbon source for cellulose, a major structural component of the tree. Water also contributes to the increase in mass. Soil nutrients contribute a very small percentage of the mass of the full-grown oak tree.

37. Normally, plants must synthesize large quantities of rubisco, a protein whose constituent amino acids all contain nitrogen. If rubisco had greater catalytic activity, the plant might produce less of the enzyme, thereby decreasing its need for nitrogen.

39. The unprotonated Lys side chain serves as a nucleophile when reacting with CO_2. At high pH, a higher percentage of ε-amino groups are in the unprotonated form.

41. Grasses turn brown because they undergo photorespiration in hot, dry conditions. Rubisco reacts with oxygen to form 2-phosphoglycolate, which subsequently consumes large amounts of ATP and NADPH. CO_2 concentrations are low because the plants close their stomata in order to avoid loss of water when the weather is hot and dry (see Box 16-A). Without CO_2, photosynthesis does not occur and the grass turns brown. But

C_4 plants such as crab grass can generate CO_2 from oxaloacetate, which can enter the Calvin cycle. Carbon fixation occurs, and the crab grass thrives even in hot, dry weather.

43.

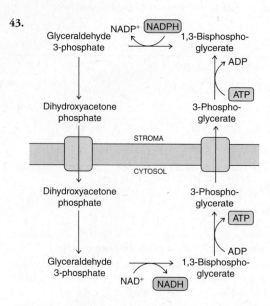

45. (a) PEPC catalyzes the formation of oxaloacetate, one of the two reactants for the first reaction of the citric acid cycle. Anaplerotic reactions are important because they replenish citric acid cycle intermediates (see Fig. 14-18). If oxaloacetate is unavailable, the citric acid cycle cannot continue.
(b) Acetyl-CoA is an allosteric activator of PEPC. When the concentration of acetyl-CoA rises, additional oxaloacetate will be required to react with it in the first reaction of the citric acid cycle. Activation of PEPC by acetyl-CoA will lead to increased production of the required oxaloacetate.

47. The primers should have the following sequences: GTAGTGGGATT-GTGCGTC and GCTCCTACAAATGCCATC (see Fig. 3-17).

49. Glyphosate herbicides are effective at killing weeds because glyphosate inhibits the plant EPSPS enzyme required for the synthesis of aromatic amino acids. The transgenic crops are protected from this inhibitor because these crops contain the bacterial enzyme that is not subject to inhibition by glyphosate. Using this strategy allows for weed eradication while preserving the desired crop.

Chapter 17

1. The lipoproteins increase in density as the percentage of protein content increases and the percentages of lipid content decreases. Thus, chylomicrons have the lowest density and HDL have the highest density.

3. The products are one palmitoyl methyl ester, two oleoyl methyl esters, and one glycerol:

$$H_3C-(CH_2)_{14}-\overset{\displaystyle O}{\overset{\displaystyle \|}{C}}-O-CH_3$$

$$H_3C-(CH_2)_7-CH=CH-(CH_2)_7-\overset{\displaystyle O}{\overset{\displaystyle \|}{C}}-O-CH_3$$

$$\begin{array}{ccc} H_2C-CH-CH_2 \\ | \quad\; | \quad\; | \\ HO \quad OH \quad OH \end{array}$$

5. The hydrolysis of pyrophosphate releases -19.2 kJ $\cdot$ mol^{-1}.
(a)

fatty acid + CoA + ATP $\rightleftharpoons$ acyl-CoA + AMP + PP$_i$
$$\Delta G^{\circ\prime} = 0 \text{ kJ} \cdot \text{mol}^{-1}$$

PP$_i$ + H$_2$O $\rightarrow$ 2 P$_i$
$$\Delta G^{\circ\prime} = -19.2 \text{ kJ} \cdot \text{mol}^{-1}$$

fatty acid + CoA + ATP + H$_2$O $\rightarrow$ acyl-CoA + AMP + 2 P$_i$
$$\Delta G^{\circ\prime} = -19.2 \text{ kJ} \cdot \text{mol}^{-1}$$

(b)
$$K_{eq} = e^{-\Delta G^{\circ\prime}/RT}$$
$$K_{eq} = e^{-(-19.2 \text{ kJ}\cdot\text{mol}^{-1})/(8.3145\times10^{-3}\text{ kJ}\cdot\text{K}^{-1}\cdot\text{mol}^{-1})(310 \text{ K})}$$
$$K_{eq} = e^{7.4}$$
$$K_{eq} = 1.7 \times 10^3$$

7. If carnitine is deficient, fatty acid transport from the cytosol to the mitochondrial matrix (the site of β oxidation) is impaired. Fatty acid oxidation generates a great deal of ATP to power the muscle, so in the absence of fatty acid oxidation the muscle must rely on stored glycogen or uptake of circulating glucose to obtain the necessary ATP. Muscle cramping is exacerbated by fasting because the concentration of circulating glucose is decreased and glycogen stores are depleted. Exercise also increases muscle cramping because the demand for ATP by the muscle is greater.

9. Medium-chain acyl-CoA (4–12 carbons) would accumulate in individuals with MCAD deficiency, since the conversion of fatty acyl-CoA to enoyl-CoA is blocked. Acyl-carnitine esters would also accumulate.

11. The conversion of fatty acyl-CoA to enoyl-CoA is similar to the conversion of succinate to fumarate because both reactions involve oxidation of the substrate and concomitant reduction of FAD to FADH$_2$ (see Section 14-2). The conversion of enoyl-CoA to hydroxyacyl-CoA is similar to the conversion of fumarate to malate because both reactions involve the addition of water across a *trans* double bond. The conversion of hydroxyacyl-CoA to ketoacyl-CoA is similar to the conversion of malate to oxaloacetate because both reactions involve the oxidation of an alcohol to a ketone with concomitant reduction of NAD$^+$ to NADH.

β Oxidation	Citric acid cycle
Fatty acyl-CoA	Succinate
↓	↓
Enoyl-CoA	Fumarate
↓	↓
Hydroxyacyl-CoA	Malate
↓	↓
Ketoacyl-CoA	Oxaloacetate

13. (a) Benzoate was produced when the dogs were fed phenylpropionate.
(b) Phenylacetate was produced when the dogs were fed phenylbutyrate.

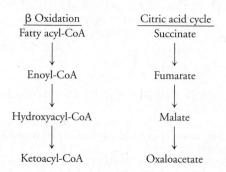

15. (a) Palmitate goes through seven cycles of β oxidation. The first six cycles produce 1 QH_2, 1 NADH, and 1 acetyl-CoA. The seventh cycle produces 1 QH_2, 1 NADH, and 2 acetyl-CoA. Each QH_2 generates 1.5 ATP in the electron transport chain, each NADH generates 2.5 ATP in the electron transport chain, and each acetyl-CoA generates a total of 10 ATP (1 $QH_2 \times 1.5 = 1.5$ ATP; 3 NADH $\times$ 2.5 = 7.5 ATP; 1 GTP = 1 ATP for a total of 10 ATP per acetyl-CoA). The total is 108 ATP. Two ATP must be subtracted from this total to account for the ATP spent in activating palmitate to palmitoyl-CoA. This gives a total of 106 ATP. **(b)** The same logic is used for stearate, except that stearate goes through eight cycles of β oxidation. The total is 120 ATP.

17. A C_{17} fatty acid goes through seven cycles of β oxidation. The first six cycles produce 1 QH_2, 1 NADH, and 1 acetyl-CoA. The seventh cycle produces 1 QH_2, 1 NADH, 1 acetyl-CoA, and 1 propionyl-CoA. Each QH_2 generates 1.5 ATP in the electron transport chain, each NADH generates 2.5 ATP, and each acetyl-CoA generates a total of 10 ATP (1 $QH_2 \times 1.5 = 1.5$ ATP; 3 NADH $\times$ 2.5 = 7.5 ATP; 1 GTP = 1 ATP for a total of 10 ATP per acetyl-CoA). The total is 98 ATP. Propionyl-CoA is metabolized to succinyl-CoA (at a cost of 1 ATP; see Fig. 17-7) and enters the citric acid cycle. Conversion of succinyl-CoA to succinate yields 1 GTP (which offsets the cost of the propionyl-CoA → succinyl-CoA conversion), and conversion of succinate to fumarate yields 1 QH_2 (equivalent to 1.5 ATP). Fumarate is converted to malate, then malate is converted to pyruvate, yielding 1 NADPH (equivalent to 2.5 ATP). The pyruvate dehydrogenase reaction converts pyruvate to acetyl-CoA (which is subsequently oxidized by the citric acid cycle to yield 10 ATP) and 1 NADH, which yields 2.5 ATP. Therefore, oxidation of propionyl-CoA yields an additional 16.5 ATP. The total is 98 ATP + 16.5 ATP = 114.5 ATP. Two ATP must be subtracted from this total to account for the ATP spent in activating the C_{17} fatty acid to fatty acyl-CoA. This gives a final total of 112.5 ATP. Note that a C_{17} fatty acid yields more ATP than palmitate (106 ATP) and less than oleate (118.5 ATP).

19. The patient could be treated with injections of vitamin B_{12} directly into the bloodstream. Alternatively, the patient could be treated with high doses of oral vitamin B_{12}. In the presence of high concentrations of the vitamin, sufficient amounts may be absorbed even in the absence of intrinsic factor.

21. A fatty acid cannot be oxidized until it has been activated by its attachment to coenzyme A in an ATP-requiring step. The first phase of glycolysis also requires the investment of free energy in the form of ATP. Consequently, neither β oxidation nor glycolysis can produce any ATP unless some ATP is already available to initiate these catabolic pathways.

23.

	Fatty acid degradation	Fatty acid synthesis
Cellular location	Mitochondrial matrix	Cytosol
Acyl-group carrier	Coenzyme A	Acyl-carrier protein
Electron carrier(s)	Ubiquinone and NAD$^+$ *accept* electrons to become ubiquinol and NADH	NADPH *donates* electrons and becomes oxidized to NADP$^+$
ATP requirement	One ATP (two high-energy phospho-anhydride bonds) required to activate the fatty acid	Consumes one ATP per two carbons incorporated into the growing fatty acyl chain.
Unit product/unit donor	Two-carbon acetyl units (acetyl-CoA)	C_3 intermediate (malonyl-CoA)
Configuration of hydroxyacyl intermediate	L	D
Shortening/growth occurs at which end of the fatty acyl chain?	Thioester end	Thioester end

25.

Phase I

27. All three enzymes involve the ATP-dependent addition of a carboxylate group (donated by bicarbonate) to a substrate. The bicarbonate is activated via its attachment to biotin, a cofactor required by all three enzymes. The carboxy-biotin, a prosthetic group of all three enzymes, transfers the carboxylate group to the substrate, increasing the number of carbons in the substrate by one.

29. The K_I is lower when the enzyme is phosphorylated. Phosphorylation decreases the activity of the enzyme (see Solution 28). Palmitoyl-CoA (a pathway product) allosterically inhibits the enzyme; a lower concentration is required for inhibition if the enzyme is already phosphorylated. A low K_I value indicates more efficient inhibition (see Section 7-3); therefore, palmitoyl-CoA binds more tightly to the phosphorylated enzyme. The allosteric inhibition of acetyl-CoA carboxylase by palmitoyl-CoA is a method of "fine-tuning" the regulation of the enzyme by phosphorylation.

31. The synthesis of palmitate from acetyl-CoA costs 42 ATP. Seven rounds of the synthase reaction are required. ATP is required to convert each of 7 acetyl-CoA to malonyl-CoA for a total of 7 ATP. Two NADPH are required for seven rounds of synthesis, which is equivalent to $2 \times 7 \times 2.5 = 35$ ATP.

33. Mammalian fatty acid synthase is structurally different from bacterial fatty acid synthase; thus, triclosan can act as an inhibitor of the bacterial enzyme but not the mammalian enzyme. The mammalian fatty acid synthase is a multifunctional enzyme made up of two identical polypeptides. In bacteria, the enzymes of the fatty acid synthetic pathway are separate proteins. Triclosan actually inhibits the bacterial enoyl-ACP reductase. The enzymes of the mammalian multifunctional enzyme must be arranged in such a way as to preclude the binding of triclosan to the active site of the enoyl-ACP reductase.

35. The shorthand form for fatty acids is described in Problem 8-1. Fatty acids that cannot be synthesized from palmitate using cellular elongases and desaturases are essential fatty acids and must be obtained from the diet. Mammals do not have a desaturase enzyme that can introduce double bonds beyond C9. Oleate and palmitoleate, with a double bond at the 9,10 position, are not essential fatty acids. Linoleate has a second double bond at the 12,13 position. Because a double bond could not be introduced at the 12,13 position, linoleate is an essential fatty acid. α-Linolenate has double bonds at positions 9,10, 12,13, and 15,16 and is also essential.

Oleate   Linoleate   α-Linolenate   Palmitoleate

37. In gluconeogenesis, the input of free energy is required to undo the exergonic pyruvate kinase reaction of glycolysis. Pyruvate is carboxylated to produce oxaloacetate, and then oxaloacetate is decarboxylated to produce phosphoenolpyruvate. Each of these reactions requires the cleavage of one phosphoanhydride bond (in ATP and GTP, respectively). In fatty acid synthesis, ATP is consumed in the acetyl-CoA carboxylase reaction, which produces malonyl-CoA. The decarboxylation reaction is accompanied by cleavage of a thioester bond, which has a similar change in free energy to cleaving a phosphoanhydride bond.

39. (a) Acetoacetate is a ketone body. It is converted to acetyl-CoA, which can be oxidized by the citric acid cycle to supply free energy to the cell.
(b) Intermediates of the citric acid cycle are also substrates for other metabolic pathways. Unless they are replenished, the catalytic activity of the cycle is diminished. Ketone bodies are metabolic fuels, but they cannot be converted to citric acid cycle intermediates. A three-carbon glucose-derived compound such as pyruvate can be converted to oxaloacetate to increase the pool of citric acid cycle intermediates and keep the cycle operating at a high rate.

41. The synthesis of the ketone body acetoacetate does not require the input of free energy (the thioester bonds of 2 acetyl-CoA are cleaved; see Fig. 17-16). Conversion of acetoacetate to 3-hydroxybutyrate consumes NADH (which could otherwise generate 2.5 ATP by oxidative phosphorylation). However, the conversion of 3-hydroxybutyrate back to 2 acetyl-CoA regenerates the NADH (see Fig. 17-17). This pathway also requires a CoA group donated by succinyl-CoA. The conversion of succinyl-CoA to succinate by the citric acid cycle enzyme succinyl-CoA synthetase generates GTP from GDP + P$_i$, so the conversion of ketone bodies to acetyl-CoA has a free energy cost equivalent to one phosphoanhydride bond.

43. In the absence of pyruvate carboxylase, pyruvate cannot be converted to oxaloacetate. Without sufficient oxaloacetate to react with acetyl-CoA in the first reaction of the citric acid cycle, acetyl-CoA accumulates. The excess acetyl-CoA is converted to ketone bodies.

45. All cells can obtain glycerol-3-phosphate from glucose because glucose enters glycolysis to form dihydroxyacetone phosphate, which is subsequently transformed to glycerol-3-phosphate via the glycerol-3-phosphate dehydrogenase reaction. Cells capable of carrying out gluconeogenesis, such as liver cells, can transform pyruvate to dihydroxyacetone phosphate. (Interestingly, adipocytes do not carry out gluconeogenesis but do express PEPCK and are thus able to convert pyruvate to glycerol-3-phosphate.)

47. (a) Fumonisin inhibits ceramide synthase. The concentration of the final product, ceramide, decreased, while other lipid synthetic pathways were not affected. The first enzyme in the pathway, serine palmitoyl transferase, was not the target of fumonisin B$_1$ because there was no significant decrease in the production of its product, 3-ketosphinganine. The second enzyme in the pathway, 3-ketosphinganine reductase, was also not a target because if it were, the substrate of this reaction would have accumulated in the presence of fumonisin. Accumulation of sphinganine indicates that ceramide synthase is inhibited. When ceramide synthase is inhibited, sphinganine cannot be converted to dihydroceramide and instead accumulates.
(b) Fumonisin likely acts as a competitive inhibitor. It is structurally similar to sphingosine and its derivatives and thus can bind to the active site and prevent substrate from binding. The fumonisin may form a covalent bond with the enzyme, or it may bind to the enzyme's active site noncovalently with high affinity. Alternatively, fumonisin may act as a substrate and be converted to a product that cannot be subsequently converted to ceramide. [From Wang, E., Norred, W. P., Bacon, C. W., Riley, R. T., and Merrill, A. H., *J. Biol. Chem.* **266**, 14486–14490 (1991).]

49. A hydrolysis reaction removes a fatty acyl group from a triacylglycerol, leaving a diacylglycerol. The intent is to reduce the total fatty acid content of the oil, thereby reducing its caloric value, without drastically altering the fluidity of the product.

51. (a) The enzyme transfers a methyl group from a donor molecule three times in order to convert phosphoethanolamine to phosphocholine.

$$^{2-}O_3P-O-CH_2-CH_2-NH_3^+ \longrightarrow ^{2-}O_3P-O-CH_2-CH_2-N(CH_3)_3^+$$

Phosphoethanolamine   Phosphocholine

(b) The methyltransferase would be a suitable drug target only if the enzyme were unique to the parasite and was not expressed in humans. [In fact, this is the case.]

53. (a) Because cholesterol is water insoluble, it is commonly found associated with other lipids in cell membranes. Only an integral membrane protein would be able to recognize cholesterol, which has a small OH head group and is mostly buried within the lipid bilayer.
(b) Proteolysis releases a soluble fragment of the SREBP that can travel from the cholesterol-sensing site to other areas of the cell, such as the nucleus.
(c) The DNA-binding portion of the protein might bind to a DNA sequence near the start of certain genes to mark them for transcription. In this way, the absence of cholesterol could stimulate the expression of proteins required to synthesize or take up cholesterol.

55. (a) Insufficient apolipoprotein B-100 affects the production of LDL. Without enough of the protein portion of the lipoprotein, these individuals are less able to export triacylglycerols and other lipids from the liver in the form of LDL, and the lipids remain in the liver.
(b) LDL are a major vehicle for cholesterol transport from the liver to other tissues, so low levels of LDL would cause hypocholesterolemia.

Chapter 18

1. The ATP-induced conformational change must decrease the $\mathcal{E}^{\circ\prime}$, from -0.29 V to about -0.40 V. The decrease in reduction potential allows the protein to donate electrons to N$_2$, since electrons flow spontaneously from a substance with a lower reduction potential to a substance with a higher reduction potential. Without the conformational change, nitrogenase could not reduce N_2.

3. Leghemoglobin, like myoglobin, is an O_2-binding protein. Its presence decreases the concentration of free O_2 that would otherwise inactivate the bacterial nitrogenase.

5. (a)

$$Glu + ATP + NH_4^+ \rightarrow Gln + P_i + ADP$$
$$Gln + \alpha\text{-ketoglutarate} + NADPH \rightarrow 2\ Glu + NADP^+$$
$$Glu + \alpha\text{-keto acid} \rightleftharpoons \alpha\text{-ketoglutarate} + amino\ acid$$
_____
$$NH_4^+ + ATP + NADPH + \alpha\text{-keto acid} \rightarrow$$
$$ADP + P_i + NADP^+ + amino\ acid$$

(b)

$$\alpha\text{-ketoglutarate} + NH_4^+ + NAD(P)H \rightleftharpoons$$
$$Glu + H_2O + NAD(P)^+$$
$$Glu + \alpha\text{-keto acid} \rightleftharpoons \alpha\text{-ketoglutarate} + amino\ acid$$
_____
$$\alpha\text{-keto acid} + NAD(P)H + NH_4^+ \rightleftharpoons$$
$$NAD(P)^+ + amino\ acid + H_2O$$

7. Cancer cells upregulate glutamine membrane transport proteins, which results in the effective uptake of glutamine from the circulation. Another strategy used by cancer cells is upregulation of the glutamine synthetase enzyme, which converts glutamate to glutamine. Glutamine analogs (compounds that are structurally similar to glutamine) could be used to either block the transport proteins or to bind to the enzyme and interfere with substrate binding.

9. (a)

(b)

(c)

(d)

11. (a) Leucine   **(b)** Methionine   **(c)** Tyrosine   **(d)** Valine

13. In the glutamine synthetase reaction, ATP donates a phosphoryl group to glutamate, which is then displaced by an ammonium ion, producing glutamine and phosphate. The ammonium ion is the nitrogen source. The asparagine synthetase reaction also requires ATP as an energy source, but the nitrogen donor is glutamine, not an ammonium ion. Aspartate is converted to asparagine, and the glutamine becomes glutamate after donating an amino group. ATP is hydrolyzed to AMP and pyrophosphate instead of ADP and phosphate as in the glutamine synthetase reaction.

15. Phosphoglycerate dehydrogenase converts 3-phosphoglycerate to 3-phosphohydroxypyruvate, which then undergoes transamination with glutamate to yield 3-phosphoserine and α-ketoglutarate, a citric acid cycle intermediate. The additional α-ketoglutarate allows greater flux through the cycle to meet the cell's energy needs. [From Possemato, R. et al., *Nature* **476**, 346–350 (2011).]

17.

19. Cysteine is the source of taurine. The cysteine sulfhydryl group is oxidized to a sulfonic acid, and the amino acid is decarboxylated.

21. If an essential amino acid is absent from the diet, then the rate of protein synthesis drops significantly, since most proteins contain an assortment of amino acids, including the deficient one. The other amino acids that would normally be used for protein synthesis are therefore broken down and their nitrogen excreted. The decrease in protein synthesis, coupled with the normal turnover of body proteins, leads to the excretion of nitrogen in excess of the intake.

23. (a) The sigmoidal shape of the velocity versus substrate concentration plot indicates that threonine deaminase binds its substrate in a positively cooperative manner. As the threonine concentration increases, threonine binds to the enzyme with increasing affinity.

(b) Isoleucine is an allosteric inhibitor of threonine deaminase and binds to the T form of the enzyme. Velocity decreases by about 15%, but the nearly 10-fold increase in K_M is more dramatic. The decrease in velocity and increase in K_M indicate that isoleucine, the end product of the pathway, acts as a negative allosteric inhibitor of the enzyme that catalyzes an early, committed step of its own synthesis. The velocity versus substrate concentration curve obtained for threonine deaminase in the presence of isoleucine has greater sigmoidal character, which means that binding of threonine is even more cooperative in the presence of the inhibitor.

(c) Valine stimulates threonine deaminase by binding to the R form. The maximal velocity is somewhat increased, but the K_M is decreased, indicating that the threonine substrate has a higher affinity for the enzyme in the presence of valine. The cooperative binding of threonine to threonine deaminase is abolished in the presence of valine, however, as indicated by the hyperbolic shape of the curve.

25. In an autoimmune disease, the body's own white blood cells become activated to mount an immune response that leads to pain, inflammation, and tissue damage. The activity of the white blood cells, which proliferate rapidly, can be diminished by methotrexate, as occurs in rapidly dividing cancer cells.

27. (a) ADP and GDP both serve as allosteric inhibitors of ribose phosphate pyrophosphokinase.

(b) PRPP, the substrate of the amidophosphoribosyltransferase, stimulates the enzyme by feed-forward activation. AMP, ADP, ATP, GMP, GDP, and GTP are all products and inhibit the enzyme by feedback inhibition.

29. Inhibiting HGPRT would block production of IMP, which is a precursor of AMP and GMP. In order to be an effective drug target, HGPRT must be essential for parasite growth; that is, the parasite cannot synthesize its own purine nucleotides *de novo* but instead relies on salvage reactions using the host cell's hypoxanthine.

31. Lymphocytes that have not fused with a myeloma cell are unable to use the *de novo* synthetic pathway because it is blocked. These cells are still able to use the HGPRT salvage pathway, but the cells will not survive beyond 7–10 days. Myeloma cells cannot survive in HAT medium because the aminopterin blocks the *de novo* pathway and these cells lack HGPRT and cannot use the salvage pathway. Only hybridomas that result from the fusion of a lymphocyte (which can carry out the salvage pathway) and a myeloma cell (which can divide in culture indefinitely) will survive in HAT medium.

33. (a) Arginine residues are converted to citrulline residues by a process of deamination (water is a reactant and ammonia is a product). Note that free citrulline produced by the urea cycle or in the generation of nitric oxide is not incorporated into polypeptides by ribosomes since there is no codon for this nonstandard amino acid.

(b) The nonstandard amino acid citrulline is not normally incorporated into polypeptides, so its presence appears foreign to the immune system, increasing the risk of triggering an autoimmune response.

35. Pyruvate can be transaminated to alanine, carboxylated to oxaloacetate, or oxidized to acetyl-CoA to enter the citric acid cycle. α-Ketoglutarate, succinyl-CoA, fumarate, and oxaloacetate are all citric acid cycle intermediates; they can also all enter gluconeogenesis. Acetyl-CoA can enter the citric acid cycle, be converted to acetoacetate, or be used for fatty acid synthesis. Acetoacetate is a ketone body and can be converted to acetyl-CoA for the citric acid cycle or fatty acid synthesis.

37. Threonine catabolism yields glycine and acetyl-CoA. The acetyl-CoA is a substrate for the citric acid cycle, which ultimately provides ATP for the rapidly dividing cell. Glycine is a source of one-carbon groups, which become incorporated into methylene-tetrahydrofolate via the glycine cleavage system. THF delivers one-carbon groups for the synthesis of purine nucleotides and for the methylation of dUMP to produce dTMP; nucleotides are needed in large amounts in rapidly dividing cells. [From Wang, J., Alexander, P., Wu, L., Hammer, R., Cleaver, O., and McKnight, S. L., *Science* **325**, 435–439 (2009).]

39. The fate of propionyl-CoA produced upon degradation of isoleucine is identical to that of propionyl-CoA produced in the oxidation of odd-chain fatty acids (see Fig. 17-7). Propionyl-CoA is converted to (*S*)-methylmalonyl-CoA by propionyl-CoA carboxylase. A racemase converts the (*S*)-methylmalonyl-CoA to the (*R*) form. A mutase enzyme converts the (*R*)-methylmalonyl-CoA to succinyl-CoA, which enters the citric acid cycle.

41. (a) Acetyl-CoA can enter the citric acid cycle if sufficient oxaloacetate is available; if not, excess acetyl-CoA is converted to ketone bodies. Leucine differs from isoleucine in that leucine is exclusively ketogenic, generating a ketone body and a ketone body precursor upon degradation. Isoleucine produces propionyl-CoA along with acetyl-CoA; the former can be converted to succinyl-CoA (see Solution 39) and then to glucose. Thus, isoleucine is glucogenic as well as ketogenic.

(b) Persons deficient in HMG-CoA lyase are unable to degrade leucine and must restrict this amino acid in their diets. A low-fat diet is also recommended because this same enzyme is involved in the production of ketone bodies (see Reaction 3 in Fig. 17-16). A diet high in fat would generate a high concentration of acetyl-CoA, which would not be able to be converted to ketone bodies in the absence of this enzyme.

43. (a) Insulin inhibits the enzyme, whereas glucagon stimulates the enzyme.

(b) In the presence of Phe, the activity of the enzyme increases dramatically, more so in the presence of glucagon. Phe acts as an allosteric activator of phenylalanine hydroxylase and plays a role in converting the enzyme from the inactive dimeric form to the active tetrameric form.

(c) The incorporation of phosphate into the active form of phenylalanine hydroxylase indicates that the enzyme is regulated by phosphorylation as well as allosteric control. Glucagon signaling must lead to phosphorylation of the enzyme.

(d) Phenylalanine hydroxylase is most active when the glucagon concentration is high, corresponding to the fasting state. Under these circumstances, phenylalanine can be degraded to produce acetoacetate (a ketone body) and fumarate (which can be converted to glucose); both compounds provide necessary resources in the fasting state.

45. Persons with NKH lack a functioning glycine cleavage system. This is the major route for the disposal of glycine, and in its absence, glycine accumulates in body fluids. The presence of excessive glycine, a neurotransmitter, in the cerebrospinal fluid explains the effects on the nervous system.

47. The glutamate dehydrogenase reaction converts α-ketoglutarate to glutamate. In the presence of excess ammonia, α-ketoglutarate in the brain could be depleted, diminishing flux through the citric acid cycle.

49. Glutamine is degraded to glutamate with the release of ammonia, which can bind protons to form NH_4^+. This helps counteract the acidosis that occurs when the concentration of acidic ketone bodies in the blood increases during starvation.

51. The reaction of serine and homocysteine to produce cysteine and α-ketobutyrate, the catabolism of the pyrimidine breakdown products β-ureidopropionate and β-ureidoisobutyrate, the reaction catalyzed by asparaginase, the conversion of serine to pyruvate, the conversion of cysteine to pyruvate, the glycine cleavage system, and the glutamate dehydrogenase reaction all generate free ammonia.

53. (a) A urea cycle enzyme deficiency decreases the rate at which nitrogen can be eliminated as urea. Since the sources of nitrogen for urea synthesis include free ammonia, low urea cycle activity may lead to high levels of ammonia in the body.

(b) A low-protein diet might reduce the amount of nitrogen to be excreted.

55. Adding arginine, the product of the argininosuccinase reaction, would increase flux through the urea cycle.

57. (a) The condensation of ammonia and α-ketoglutarate produces the amino acid glutamate.

(b) The reverse reaction, in which glutamate is deaminated to produce α-ketoglutarate, replenishes the citric acid cycle intermediate.

59. The amino acids released by protein degradation are used as metabolic fuels. The amino groups are removed and the nitrogens are eventually excreted in the form of urea. The carbon skeletons can be completely catabolized to CO_2 by muscle cells to produce ATP via the citric acid cycle and oxidative phosphorylation, or they can be partially broken down and transported to the liver to be used in gluconeogenesis, which indirectly supplies the muscle cells with glucose.

61. (a) UTase is stimulated by α-ketoglutarate and ATP, both substrates (either directly or indirectly) of glutamine synthetase. UTase is inhibited by glutamine and inorganic phosphate, both products of the reaction. UR is stimulated by glutamine, a product of the glutamine synthetase reaction. High concentrations of glutamine decrease the activity of glutamine synthetase.

(b) Histidine, tryptophan, carbamoyl phosphate, glucosamine-6-phosphate, AMP, CTP, and NAD^+ are all end products of glutamine metabolic pathways. Alanine, serine, and glycine reflect the overall cellular nitrogen level. When nitrogen levels are adequate, glutamine synthetase activity is inhibited.

(c) The first batch of glutamine synthetase was adenylylated and was in its less active form and more susceptible to inhibition by allosteric modulators. The enzyme was in the adenylylated form because the growth medium contained glutamate, and under these conditions, glutamine synthetase activity is inhibited. The second batch of enzyme was not adenylylated and was in its fully active form, which is what would be expected when NH_4^+ is the sole nitrogen source. Because the second batch of enzyme was not adenylylated, it was not susceptible to inhibition by the allosteric modulators. [From Stadtman, E. R., *J. Biol. Chem.* **276**, 44357–44363 (2001).]

Chapter 19

1. The two main metabolites at the "crossroads" are pyruvate and acetyl-CoA. Pyruvate is the main product of glycolysis. It can be converted to acetyl-CoA by pyruvate dehydrogenase. Pyruvate is produced from a transamination reaction involving alanine. Pyruvate can be carboxylated to oxaloacetate for gluconeogenesis. Acetyl-CoA is a product of fatty acid degradation and one of the reactants in the citric acid cycle. Acetyl-CoA is a product of the degradation of ketogenic amino acids. Acetyl-CoA can be used to synthesize fatty acids and ketone bodies.

3. The Na,K-ATPase pump requires ATP to expel Na^+ ions while importing K^+ ions, both against their concentration gradients. Inhibition by oubain reveals that the brain devotes half of its ATP production solely to power this pump. Because the brain does not store much glycogen, it must obtain glucose from the circulation. Glucose is oxidized aerobically in order to maximize ATP production.

5. **(a)** *Phosphorolytic cleavage*

glycogen$_{(n \text{ residues})}$ + P$_i$ → glycogen$_{(n-1 \text{ residues})}$ + glucose-1-phosphate

Hydrolytic cleavage

glycogen$_{(n \text{ residues})}$ + H$_2$O → glycogen$_{(n-1 \text{ residues})}$ + glucose

(b) Phosphorolytic cleavage yields glucose-1-phosphate, which is negatively charged due to its phosphate group and cannot exit the cell via the glucose transporter. In addition, glucose-1-phosphate can be isomerized to glucose-6-phosphate (and can enter glycolysis) without the expenditure of ATP. Hydrolytic cleavage yields neutral glucose, which can leave the cell via the glucose transporter. Converting free glucose to glucose-6-phosphate so that it can enter glycolysis requires the expenditure of ATP in the hexokinase reaction.

7. **(a)** The fumarate is used to boost the level of citric acid cycle intermediates to increase flux through this pathway and increase ATP synthesis by oxidative phosphorylation.

(b) Transamination of aspartate would produce oxaloacetate, but another α-keto acid, such as pyruvate or α-ketoglutarate, would become an amino acid in the process. Consequently, there would be no net change in the level of citric acid cycle intermediates.

9. Glycolysis produces two moles of ATP per mole of glucose. Synthesis of one mole of glucose via gluconeogenesis costs six moles of ATP. Therefore, the cost of running one round of the Cori cycle is four ATP. The extra ATP is generated from the oxidation of fatty acids in the liver.

11. During starvation, muscle proteins are broken down to produce gluconeogenic precursors. The amino groups of the amino acids are transferred to pyruvate via transamination reactions. The resulting alanine travels to the liver, which can dispose of the nitrogen via the urea cycle and produce glucose from the alanine skeleton (pyruvate) and other amino acid skeletons. This glucose circulates not just to the muscles but to all tissues that need it, so the metabolic pathway is not truly a cycle involving just the liver and muscles.

13. **(a)** Since pyruvate carboxylase catalyzes the carboxylation of pyruvate to oxaloacetate, a deficiency of the enzyme would result in increased pyruvate levels and decreased oxaloacetate levels. Some of the excess pyruvate would also be converted to alanine, so alanine levels would be elevated.

(b) Some of the excess pyruvate is converted to lactate, which explains why the patient suffers from lactic acidosis. Decreased oxaloacetate levels decrease the activity of the citrate synthase reaction, the first step of the citric acid cycle. This causes the accumulation of acetyl-CoA, which forms ketone bodies that accumulate in the blood to cause ketosis.

(c) A pyruvate carboxylase deficiency results in decreased oxaloacetate levels. Aspartate is formed by transamination of oxaloacetate, so aspartate levels are decreased as well. Low levels of oxaloacetate decrease the levels of all the citric acid cycle intermediates due to the decreased activity of the citrate synthase reaction. This results in lower levels of α-ketoglutarate, which can be transaminated to glutamate. Therefore, glutamate levels are low, as are GABA levels, since GABA is produced from glutamate (see Section 18-2).

(d) Acetyl-CoA stimulates pyruvate carboxylase activity. Adding acetyl-CoA would allow the investigators to determine whether there was a slight amount of pyruvate carboxylase activity that could be detected by adding this activator. [From Stanbury, J. B., Wyngaarden, J. B., Fredrickson, D. S., Goldstein, J. L., and Brown, M. S., *The Metabolic Basis of Inherited Disease*, pp. 196–198, McGraw-Hill Book Company, New York (1983).]

15. **(a)** The antibiotics affect the composition of the intestinal microbiome, favoring species that allow the animal to gain more weight from a given amount of food.

(b) The presence of antibiotics selects for the growth of microbial species that can resist the antibiotic. The result is a large number of bacteria with antibiotic-resistance genes that can be transferred to other species, including those that cause disease in humans.

17. The increase in glycolytic activity and the decrease in citric acid cycle activity indicate that aerobic metabolism is decreased. Instead, metabolic intermediates are funneled into pathways that will ultimately produce the glucose and

amino acids required for milk production. The main source of these intermediates is the muscle. Amino acids released from the muscle travel via the bloodstream to the liver, where they are converted to glucose via gluconeogenesis. Rather than returning to the muscle (which would occur during the Cori cycle), the glucose and amino acids are diverted to the mammary gland for milk production. [From Khula, B., Nuernberg, G., Albrecht, D., Goers, S., Hammon, H. M., and Metges, C. C., *J. Proteome Res.*, **10**, 4252–4262 (2011).]

19. Because glucokinase is not saturated at physiological glucose concentrations, it can respond to changes in glucose availability with an increase or decrease in reaction velocity. Consequently, the entry of glucose into glycolysis and subsequent metabolic pathways depends on the glucose concentration. Because hexokinase is saturated at physiological glucose concentrations, its rate does not change with changes in glucose concentration.

21. Insulin binding to its receptor stimulates the tyrosine kinase activity of the receptor. Proteins whose tyrosine residues are phosphorylated by the receptor tyrosine kinase can then interact with additional components of the signaling pathway. These interactions could not occur if a tyrosine phosphatase removed the phosphoryl groups attached to the Tyr residues.

23. Phosphorylation of glycogen synthase by GSK3 inactivates the enzyme so that glycogen synthesis does not occur. But when insulin activates protein kinase B, the kinase phosphorylates GSK3. Phosphorylated GSK3 is inactive and unable to phosphorylate glycogen synthase. In the dephosphorylated state glycogen synthase is active and glycogen synthesis can occur.

25. If the phosphorylation of GSK3 is blocked, GSK3 remains active and can phosphorylate glycogen synthase, rendering it inactive (see Solution 23). This decreases the ability of the cell to take up glucose and convert it to a storage form and exacerbates insulin resistance (see Problem 24).

27. **(a)** Stimulation of the rate of phosphate removal from glucose-6-phosphate explains why the glucose concentration increased and the glucose-6-phosphate concentration decreased. The phosphoenolpyruvate concentration increased because flux from dihydroxyacetone phosphate to phosphoenolpyruvate increased (the glycolytic reactions are all near equilibrium).

(b) Removing the phosphate from glucose-6-phosphate is the final step of gluconeogenesis (and glycogenolysis) and allows glucose to leave the liver. Increasing the rate of this step enhances the overall rate of gluconeogenesis. At the same time, glycolysis is inhibited because the very first step of glycolysis is the phosphorylation of glucose to glucose-6-phosphate. If phosphate groups are continually detached, glucose-6-phosphate cannot enter glycolysis and thus glycolysis is effectively inhibited. [From Ichai, C., Guignot, L., El-Mir, M. Y., Nogueira, V., Guigas, B., Chauvin, C., Fontaine, E., Mithieux, G., and Leverve, X. M., *J. Biol. Chem.* **276**, 28126–28133 (2001).]

29. Phosphorylase kinase is regulated by phosphorylation, which causes a conformational change that activates the enzyme. The enzyme is phosphorylated by protein kinase A (see Figure 19-11). The activity of the enzyme therefore depends somewhat on the G protein–coupled receptor pathway activated by either glucagon or epinephrine. But phosphorylase kinase is not fully active until calmodulin (a calcium-binding protein that is part of its structure) binds calcium and undergoes its own conformational change. The intracellular concentration of calcium increases when the phosphoinositide signaling system is activated (see Section 10-2), so the activation of phosphorylase kinase depends on this signaling pathway as well.

31. **(a)** AMPK increases the expression of GLUT4, since this will increase ATP production via glucose catabolism.

(b) AMPK decreases glucose-6-phosphatase expression, since this is a gluconeogenic enzyme that would contribute to cellular ATP consumption.

33. AMPK stimulates pathways involved in ATP production, so the rates of glycolysis and fatty acid oxidation increase. Cancer cells normally rely on anaerobic metabolism to meet their energy needs, so the stimulation of AMPK has the effect of increasing aerobic metabolism, which increases the production of reactive oxygen species.

35. Overexpression of IRS-1 would increase the rate of translocation of GLUT4 transporters to the cell surface. IRS-1 triggers the downstream activation of proteins such as the phosphatase enzyme that removes phosphate

groups from glycogen phosphorylase (which inactivates the enzyme) and glycogen synthase (which activates the enzyme). The result is that glycogen synthesis in the cultured muscle cells would increase.

37. These two enzymes are part of the gluconeogenic pathway. Their concentrations increase when dietary fuels are not available so that the liver can supply other tissues with newly synthesized glucose.

39. If 3,000 g of fat are utilized at a rate of 75/g day, the fast can last for 40 days before death occurs.

41. After a few days of a diet low in carbohydrate, glycogen stores are depleted and the liver converts fatty acids to ketone bodies to be used as fuel for muscle and other tissues. Acetone is produced from the nonenzymatic decarboxylation of the ketone body acetoacetate. The relatively nonpolar acetone passes from blood capillaries into the lung alveoli, and its smell can be detected in exhaled breath.

43. (a) Leptin stimulates glucose uptake by skeletal muscle.

(b) Glycogenolysis is inhibited, probably by direct inhibition of glycogen phosphorylase, which catalyzes the committed step in glycogenolysis.

(c) Leptin increases the activity of cAMP phosphodiesterase; the result is that cAMP cellular concentration decreases. In this way, leptin acts as a glucagon antagonist in the same manner that insulin does; glucagon's signal transduction pathway leads to an increase in cAMP concentration.

45. Both of these observations indicate a decrease in the ability of the mitochondrion to oxidize fatty acids. Carnitine acyltransferase transports fatty acids from the cytosol to the mitochondrial matrix, the site of β oxidation. Products of β oxidation, the reduced coenzymes NADH and QH₂, are reoxidized by the electron transport chain, with concomitant production of ATP. If fatty acid oxidation cannot occur, fatty acids will be used to synthesize triacylglycerols for storage in adipose tissue instead.

47. (a) Malonyl-CoA, the product of the acetyl-CoA carboxylase reaction (the committed reaction of fatty acid synthesis), inhibits simultaneous fatty acid oxidation by inhibiting carnitine acyltransferase, an enzyme required for shuttling fatty acids into the mitochondrial matrix for oxidation. This mechanism prevents the oxidation of newly synthesized fatty acids because they cannot enter the mitochondrion.

(b) In the absence of ACC2, heart and muscle are unable to synthesize fatty acids. This increases the demand on the liver to provide fatty acids for heart and muscle. In the knockout mice, liver glycogen is degraded to glucose, then oxidized to pyruvate and acetyl-CoA in order to provide acetyl-CoA for fatty acid synthesis.

(c) Fatty acid levels decrease because of the lack of ACC2. Triacylglycerols are released by adipose tissue and travel to the muscle and heart for oxidation, since the muscle and heart cannot synthesize their own fatty acids. This accounts for the increased blood levels of triacylglycerols.

(d) Insulin stimulates the activity of ACC2 in the muscle cells of normal mice, which promotes fatty acid synthesis and inhibits fatty acid oxidation (due to increased malonyl-CoA levels). The muscle cells in the knockout mice lack ACC2 and are not subject to insulin-mediated control. Fatty acid synthesis does not occur, malonyl-CoA levels do not rise, and fatty acid oxidation proceeds normally, even in the presence of insulin.

(e) Knockout mice are leaner because their heart and muscle tissue cannot synthesize fatty acids, so triacylglycerols are mobilized to provide fatty acids for these tissues, as described in part (d). Knockout mice have a higher rate of fatty acid oxidation and a lower rate of synthesis, as described in part (c), which also accounts for their lower weight gain despite the increased caloric intake.

(f) Molecular modeling techniques could be used to design a drug that inhibits the enzyme activity of ACC2 but not ACC1. The drug would have to be targeted in such a way that it would be delivered to the mitochondrial matrix, where ACC2 is located. [From Abu-Elheiga, L., Matzuk, M. M., Abo-Hashema, K. A. H., and Wakhil, S. J., *Science* **291,** 2613–2616 (2001).]

49. The drugs can activate the intracellular tyrosine kinase domains of the insulin receptor, bypassing the need for insulin to bind to the receptor.

51. (a)

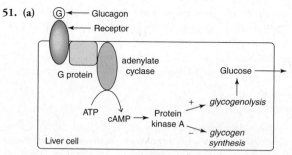

(b) These amino acids are positively charged. Since a negatively charged aspartate residue in the glucagon receptor has been shown to be essential for binding, it's possible that an ion pair forms between a positively charged amino acid side chain (His, Lys, or Arg) and the essential aspartate. This hypothesis can be tested by modifying His¹, Lys¹², and Arg¹⁸ to neutral or negatively charged side chains and assessing the resulting analogs' binding and signal-transducing capabilities.

(c) Eliminating the Asp at position 9 results in an analog with decreased affinity for the receptor and with little biological activity, indicating that the Asp plays a role in both binding and signal transduction. Substituting the Asp with a positively charged Lys decreases the binding affinity by about half but completely eliminates the biological activity. The Asp evidently plays an important role in binding, but conservation of the negative charge does not seem to be critical, since a positive charge does not abolish binding. Hence some other aspect of the Asp side chain structure is important for binding. The Asp at position 9 does seem to be important for biological activity, since deletion or substitution of the Asp greatly decreases biological activity.

Abolishing the positive charge at position 12 greatly decreases binding affinity. But once the analogs are bound, they are still capable of eliciting a biological response. The addition of a negative charge at position 12 virtually abolishes binding, so it's possible that the positively charged group at position 12 forms an ion pair with a negatively charged amino acid on the glucagon receptor.

Leu¹⁸ binds more effectively to receptors than does Ala¹⁸, supporting the hypothesis that hydrophobic interactions between the hormone and the receptor are important, since leucine has a more hydrophobic side chain than alanine. Substitution with a Glu residue also decreases binding, but not as much as at position 12. The positive charge is important, since replacement with the negatively charged Glu abolishes more than 90% of the binding ability of the analog.

The des-His¹-glucagon has decreased binding affinity and has a greater decrease in biological activity, indicating that the histidine at position 1 is important for binding but plays a greater role in signal transduction. This is also supported by the other des-His¹ analogs. The des-His¹-des-Asp⁹ analog does not bind well (only 7% of the control) and has no biological activity. Interestingly, the des-His¹-Lys⁹ derivative binds well (70%) but has no biological activity. This indicates that the substitution of aspartate for lysine at position 9 preserves characteristics that are important for binding. However, once bound, the analog does not trigger signal transduction.

(d) The des-His¹-Lys⁹ is the best antagonist because it binds to the receptor with 70% of the affinity of the native hormone but has no biological activity. In this derivative, the two amino acids important in signal transduction have been modified, while the positively charged residues at positions 12 and 18, which are critical for binding, have been retained. [From Unson, C. G., et al., *J. Biol. Chem.* **266,** 2763–2766 (1991); and Unson, C. G., et al., *J. Biol. Chem.* **273,** 10308–10312 (1998).]

53. AMPK phosphorylates and activates phosphofructokinase-2, the enzyme that catalyzes the synthesis of fructose-2,6-bisphosphate. This metabolite is a potent activator of the glycolytic enzyme phosphofructokinase and an inhibitor of fructose-1,6-bisphosphatase, which catalyzes the opposing reaction for gluconeogenesis. Stimulation of AMPK would increase the concentration of fructose-2,6-bisphosphate and would therefore stimulate glycolysis and inhibit gluconeogenesis. The increase in glucose utilization and decrease in glucose production would lower the level of glucose in the blood in the diabetic patient. [From Hardie, D. G., Hawley, S. A., and Scott, J. W., *J. Physiol.* **574,** 7–15 (2006).]

Chapter 20

1. Topoisomerase I reactions are driven by the free energy change of DNA shifting from a supercoiled conformation to a relaxed conformation, so no external source of free energy is needed. The enzyme merely accelerates a reaction that is already favorable. Topoisomerase II reactions involve more extensive mechanical intervention because both strands of the DNA are cleaved and held apart while another segment of DNA passes through the break. This process requires the free energy of ATP, since it is not thermodynamically favorable on its own.

3. Novobiocin and ciprofloxacin are useful as antibiotics because they inhibit prokaryotic DNA gyrase but not eukaryotic topoisomerases. They can kill disease-causing prokaryotes without harming host eukaryotic cells. Doxorubicin and etoposide inhibit eukaryotic topoisomerases and can be used as anticancer drugs. Although these drugs inhibit topoisomerases from both cancer cells and normal cells, cancer cells have a higher rate of DNA replication and are more susceptible to the effects of the inhibitors than are normal cells.

5. Parental ^{15}N-labeled DNA strands are shown in black, and newly synthesized ^{14}N DNA strands are shown in gray. The original ^{15}N-labeled parental DNA strands persist throughout succeeding generations, but their proportion of the total DNA decreases as new DNA is synthesized.

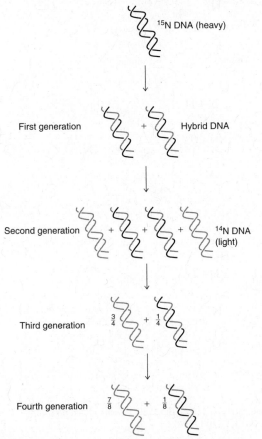

7. Negatively supercoiled DNA is more easily unwound, and thus the separation of the DNA strands occurs more easily, which facilitates the replication process.

9. (a) Yes. By moving along a single DNA strand, the helicase can act as a wedge to push apart the double-stranded DNA ahead of it.
(b) The free energy of dTTP hydrolysis is similar to the free energy of ATP hydrolysis. Each hydrolysis reaction drives the helicase along two to three bases of DNA.
(c) The T7 helicase is probably a processive enzyme. Its hexameric ring structure is reminiscent of the clamp structure that promotes the processivity of DNA polymerase (see Fig. 20–12). [From Kim, D.-E., Narayan, M., and Patel, S. S., *J. Mol. Biol.* **321**, 807–819 (2002).]

11. (a) DNA replication (and hence bacterial growth) halts immediately at the nonpermissive temperature because the DNA cannot be unwound ahead of the replication fork in the absence of the helicase.

(b) Bacterial growth slows and then stops because the role of DnaA is to locate the replication origin (see Problem 7). When the temperature shifts to the nonpermissive temperature, DNA replication already under way is not affected (those cells can complete cell division), but another round of replication cannot begin in the absence of functioning DnaA.

13. The ddNTP lacks the 3′ hydroxyl group that serves as the attacking nucleophile for the incoming dNTP.

15. DNA polymerase ε would need to have greater processivity because it synthesizes the leading strand continuously. Polymerase α can be less processive because it synthesizes only a short DNA segment at the start of each Okazaki fragment.

17. First, the cell contains roughly equal concentrations of the four deoxynucleotide substrates for DNA synthesis; this minimizes the chance for an overabundant dNTP to take the place of another or for the wrong dNTP to take the place of a scarce dNTP. Second, DNA polymerase requires accurate pairing between the template base and the incoming base. Third, the 3′ → 5′ exonuclease proofreads the newly formed base pair. Fourth, the removal of the RNA primer and some of the adjacent DNA helps minimize errors introduced by primase and by the DNA polymerase at the 5′ end of a new DNA segment. Finally, DNA repair mechanisms can excise mispaired or damaged nucleotides.

19. PP_i is the product of the polymerization reaction catalyzed by DNA polymerase. This reaction also requires a template DNA strand and a primer with a free 3′ end.
(a) There is no primer strand, so no PP_i is produced.
(b) There is no primer strand, so no PP_i is produced.
(c) PP_i is produced.
(d) No PP_i is produced because there is no 3′ end that can be extended.
(e) PP_i is produced.
(f) PP_i is produced.

21. The DNA molecule (chromosome) is much longer than an Okazaki fragment and must be condensed and packaged in some way to fit inside the nucleus (in a eukaryote) or cell (in a prokaryote). If the cell waited until the entire DNA molecule had been replicated, the newly synthesized lagging strand, in the form of many Okazaki fragments, might already be packaged and inaccessible to the endonuclease, polymerase, and ligase necessary to produce a continuous lagging strand.

23.

25. (a) DNA polymerase, (b) reverse transcriptase or telomerase, (c) primase or RNA polymerase.

27. (a)

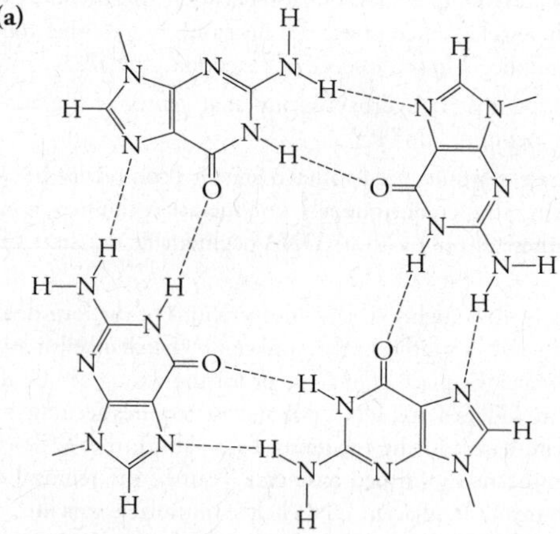

(b)

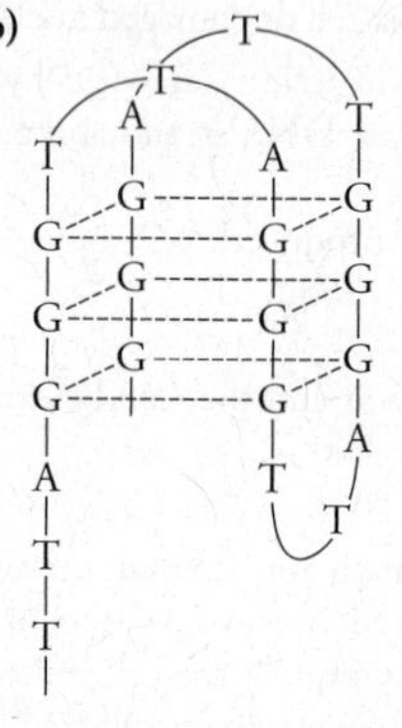

29. The resulting telomeres will have a sequence complementary to the mutated sequence of the telomerase-associated RNA template. This experiment was important because it established the mechanism of the enzyme and verified the role of the RNA template in extending chromosome length.

31. Without functional DNA repair enzymes, additional mutations may arise in genes that are involved in regulating cell growth. In the absence of proper growth controls, cells may begin to proliferate at an accelerated rate.

33. The triphosphatase removes nucleotides containing the modified base before they can be incorporated into DNA during replication.

35. The O^6-methylguanine produced by the methylation of guanine produces a residue that can base pair with either cytosine or thymine. If the O^6-methylguanine residue base pairs with thymine, the G:C base pair will eventually be changed to an A:T base pair.

37. Bromouracil causes an A:T to G:C transition.

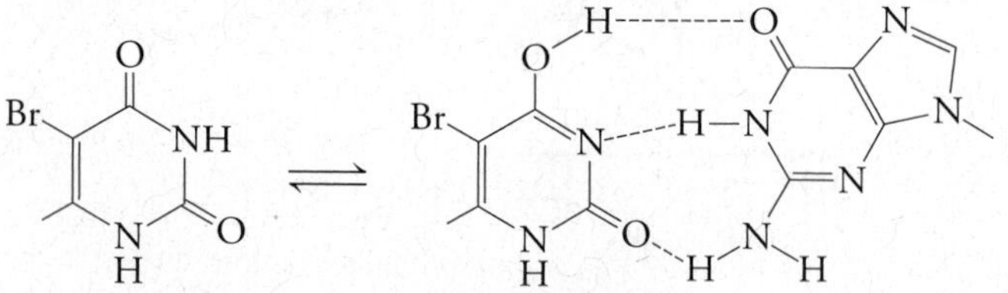

5-Bromouracil (keto tautomer) ⇌ 5-Bromouracil (enol tautomer)   Guanine

39. (a)

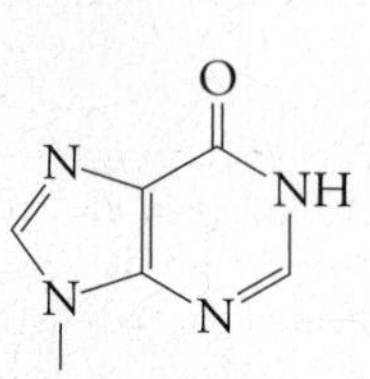

Hypoxanthine

(b)

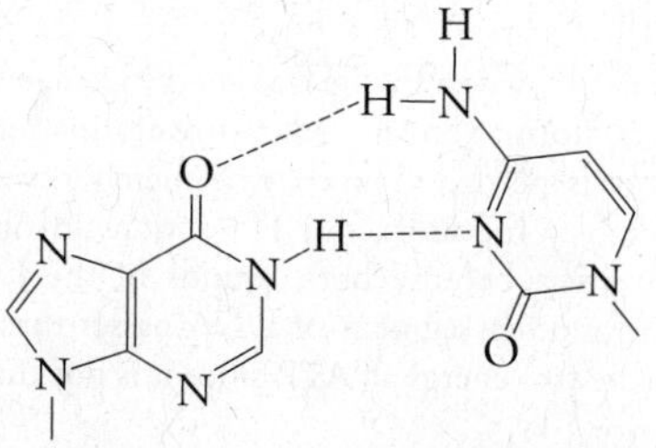

Hypoxanthine   Cytosine

(c) An A:T base pair is converted to a C:G base pair.

41. All of these deaminations produce bases that are foreign to DNA; therefore, they can be quickly spotted and repaired before DNA has replicated and the damage is passed on to the next generation.

43. (a)

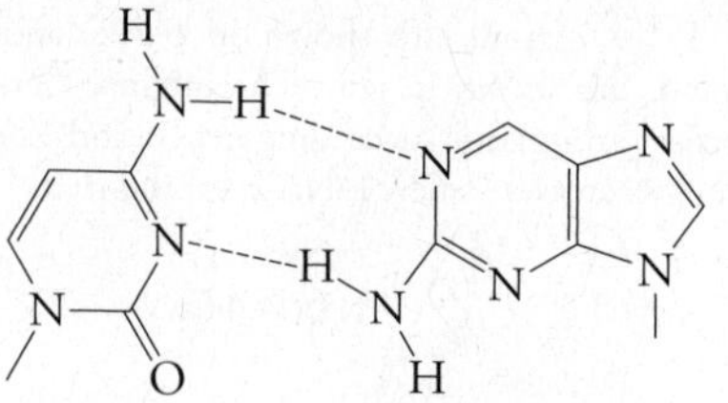

Cytosine   2-Aminopurine

(b)

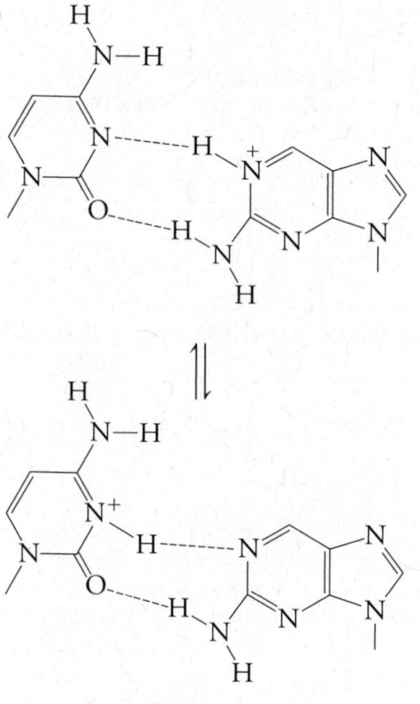

[From Sowers, L. C., Boulard, Y., and Fazakerley, G. V., *Biochemistry* **29,** 7613–7620 (2000).]

45. Most likely, the thymine–thymine dimer, since this lesion forms upon exposure of the DNA to ultraviolet light.

47. The mutant bacteria are unable to repair deaminated cytosine (uracil). In these cells, the rate of change of G:C base pairs to A:T base pairs is much greater than normal.

49. DNA polymerase III replicates DNA until a thymine dimer is encountered. Polymerase III is accurate but cannot quickly bypass the damage. Polymerase V, which can more quickly proceed through the damaged site, does so, but at the cost of misincorporating G rather than A opposite T. Thus, replication can continue at a high rate. The tendency for DNA polymerase V to continue to introduce errors is minimized by its low processivity: Soon after passing the thymine dimer, it dissociates, and the more accurate polymerase III can continue replicating the DNA with high fidelity.

51. The side chains of lysine and arginine residues have high pK values and are positively charged at physiological pH. The positively charged groups can form ion pairs with the negatively charged phosphate groups on the backbone of the DNA molecule.

53. (a)

```
|
CH2
|
CH2
|
CH2
|
CH2
|
NH
|
C=O
|
CH3
```

(b)

```
|
CH2
|
O
|
-O-P=O
|
O-
```

(c)

```
        |
        CH2
      /    \
   N        
   ||        
   N
    \
     -O-P-O-
        ||
        O
```

(d)

```
|
CH2
|
CH2
|
CH2
|
CH2
|
NH
|
CH3
```

(e)

```
|
CH2
|
CH2
|
CH2
|
NH
|
C-NH-CH3
||
NH2+
```

Both acetylation and methylation of the lysine produce a neutral side chain and remove the lysine residue's positive charge; phosphorylation of the serine and histidine side chains produces a side chain with two negative charges; methylation of the arginine residue increases the size of the side chain somewhat but does not alter the charge of the side chain.

55. (a) The nonstandard amino acid is homocysteine, which can accept a methyl group donated by methyl-tetrahydrofolate to regenerate methionine (see Section 18-2).

```
        COO-
        |
+H3N-C-H
        |
        CH2
        |
        CH2
        |
        SH
```

(b) The other product is methanol, CH_3OH.

57. No, histones modified with ubiquitin are not marked for proteolytic destruction by the proteasome because the amino acid side chains of the histone proteins have only one ubiquitin attached. In order to enter the proteasome for degradation, a chain of at least four ubiquitin molecules is required (see Section 12-1).

59. (a) Because the protamine–DNA complexes occupy less volume than nucleosomes, DNA can be more easily packed into the small volume of the sperm nucleus.

(b) DNA in nucleosomes is less compact and therefore more accessible for transcription, so one would expect nucleosomal DNA to contain genes that would need to be expressed immediately following fertilization, that is, genes essential for early embryogenesis.

Chapter 21

1. Messenger RNA is translated into protein, and a single mRNA can be used to translate many proteins. In this way, the mRNA is "amplified." Ribosomal RNA performs a structural role and is not amplified, so many more rRNA genes are needed to express sufficient rRNA to meet the needs of the cell.

3. Sequence-specific interactions require contact with the bases of DNA, which can participate in hydrogen bonding and van der Waals interactions with protein groups. Electrostatic interactions involve the ionic phosphate groups of the DNA backbone and are therefore sequence independent.

5. The promoter region is shaded.

$$^{+1}$$

AAAATAAATGCTTGACTCTGTAGCGGGAAGGCGTATTATCCAACACCC

7. Affinity chromatography takes advantage of the ability of the desired protein to bind to a specific ligand (see Section 4-5). To purify Sp1, the GGGCGG oligonucleotide is covalently attached to tiny beads, which constitute the stationary phase of a chromatography column. A cellular extract containing the Sp1 protein is loaded onto the column and a buffer (the mobile phase) is washed through the column to elute proteins that do not bind to the oligonucleotide ligand. Next, a high-salt buffer is applied to the column to disrupt the strong interactions between the Sp1 and the GC box, and the Sp1 protein is eluted. [From Kadonaga, J. T., et al., *Trends Biochem. Sci.* **11**, 20–23 (1986) and Kadonaga, J. T., and Tjian, R., *Proc. Natl. Acad. Sci.* **83**, 5889–5893 (1986).]

9. A:T base pairs have weaker stacking interactions (see Section 3-1) and are easier to melt apart than G:C base pairs, which have stronger stacking interactions. This facilitates DNA unwinding to expose the template for transcription.

11.

```
        O                          O
        ||                         ||
-NH-CH-C-                 -NH-CH-C-
        |                          |
        CH2                        CH2
        |                          |
        CH2                        CH2
        |                          |
        CH2                        CH2
        |                          |
        CH2                        NH
        |                          |
        NH2+                       C=N-CH3
        |                            |
        CH3                          H+
                                   NH2
```

13. Upregulation of the methylase will increase the extent of methylation of Lys 4 (K4) and Lys 27 (K27) in histone 3. Since H3K4me3-associated genes normally tend to be in transcriptionally active chromatin (see Table 21-2), these genes are hyperactivated in cancer cells. The opposite occurs with H3K27me3-associated genes, which are normally transcriptionally inactive and will be hypersilenced in cancer cells. [From Stark, G. R., Wang, Y., and Lu, T., *Cell Res.*, **21**, 375–380 (2011).]

15. (a) The polymerase binds most tightly to the DNA segment with the largest bulge. This DNA mimics the transcription bubble, in which the DNA strands have already been separated.

(b) Since K_d is a dissociation constant, the apparent equilibrium constant for the binding reaction is $1/K_d$. Equation 12-2 gives the relationship between ΔG and K:

$$\Delta G = -RT \ln K \text{ or } \Delta G = -RT \ln(1/K_d)$$

(c) For double-stranded DNA:

$$\Delta G = -(8.3145 \text{ J} \cdot \text{K}^{-1} \cdot \text{mol}^{-1})(298 \text{ K})\ln(1/315 \times 10^{-9})$$
$$= -37 \text{ kJ} \cdot \text{mol}^{-1}$$

For the eight-base bulge,

$$\Delta G = -(8.3145 \text{ J} \cdot \text{K}^{-1} \cdot \text{mol}^{-1})(298 \text{ K})\ln(1/1.3 \times 10^{-12})$$
$$= -68 \text{ kJ} \cdot \text{mol}^{-1}$$

Polymerase binding to the melted DNA is more favorable than binding to double-stranded DNA.

(d) Melting open a DNA helix is thermodynamically unfavorable. Some of the favorable free energy of binding the polymerase to the DNA is spent in forming the transcription bubble. When the transcription bubble is preformed (for example, in the DNA with an eight-base bulge), this energy is not spent and is reflected in the apparent energy of polymerase binding. The difference in ΔG values for polymerase binding to double-stranded DNA and to the eight base bulge is $-68 - (-37) = -31 \text{ kJ} \cdot \text{mol}^{-1}$. This value estimates the free energy cost ($+31 \text{ kJ} \cdot \text{mol}^{-1}$) of melting open eight base pairs of DNA.

(e) The AT-rich sequence is easier to melt open than the GC-rich sequence because GC-rich DNA experiences stronger stacking interactions. [From Bandwar, R. P., and Patel, S. S., *J. Mol. Biol.* **324**, 63–72 (2002).]

17. The lactose permease allows lactose to enter the cell, which increases the intracellular lactose concentration. Allolactose can then bind to the repressor protein to remove it from the operator. The presence of additional lactose assists in the full expression of the operon.

19. If the repressor cannot bind to the operator, the genes of the *lac* operon are constitutively expressed; that is, the genes are expressed irrespective of whether lactose is present or absent in the growth medium. Adding lactose has no effect on gene expression.

21. Wild-type cells cannot grow in the presence of phenyl-Gal. The wild-type cells produce a small amount of β-galactosidase in the absence of *lac* operon expression, but not in sufficient amounts to be able to cleave phenyl-Gal to phenol and galactose. The *lacI* mutants, however, will thrive in this growth medium. The mutation in the *lacI* gene results in the expression of a nonfunctional repressor (or perhaps no repressor); in any case, the *lac* operon is constitutively expressed and β-galactosidase is produced in sufficient amounts to act on phenyl-Gal to release galactose. The use of this growth medium permits selection of repressor mutants, since the mutants survive while the wild type cells do not.

23. The accurate transmission of genetic information from one generation to the next requires a high degree of fidelity in DNA replication. A higher rate of error in RNA transcription is permitted because the cell's survival usually does not depend on accurately synthesized RNA. If translated, an RNA transcript containing an error may lead to a defective protein, which is likely to be destroyed by the cell before it can do much damage. The gene can be transcribed again and again to generate accurate transcripts.

25. Cordycepin, which resembles adenosine, can be phosphorylated and used as a substrate by RNA polymerase. However, it blocks further RNA polymerization because it lacks a 3′ OH group.

27. Rifampicin inhibits the transition from RNA chain initiation to elongation. Normally, RNA polymerase initiates RNA synthesis repeatedly, releasing many short transcripts before committing to elongation. In the presence of rifampicin, the RNA polymerase cannot convert from initiation mode to elongation mode and remains bound to its promoter. Synthesis of longer RNA transcripts is not possible in the presence of the drug.

29. G U C C G A U C G A A U G C A U G

31. If α-amanitin were added to cells in culture, the synthesis of mRNA would be inhibited, but the synthesis of all other types of RNA would be relatively unaffected. RNA polymerase II is responsible for mRNA synthesis and is the most sensitive to inhibition by α-amanitin. Experiments with this toxin permitted investigators to determine the types of RNA synthesized by each polymerase.

33. The C-terminal domain (CTD) is phosphorylated on multiple serine residues when RNA polymerase transitions to elongation mode. The presence of many negatively charged phosphate groups causes charge–charge repulsions that cause this domain to be positioned away from the globular domain of the RNA polymerase as well as away from the negatively charged DNA.

35. (a) 5′ · · · NNAAGCGCCGNNNNCCGGCGCUUUUUUNNN · · · 3′

(b)

```
        N─N
      N      N
      N      C
       G---C
       C---G
       C---G
       G---C
       C---G
       G---C
       A---U
       A---U
  5′···NN   UUUU···3′
```

37. As transcription proceeds, the nascent RNA forms a variety of secondary structures as portions of the transcript form complementary base pairs. The formation of these secondary structures may cause transcription to pause but not necessarily terminate.

39. (a)   *Product absent*

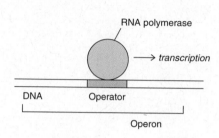

Product present

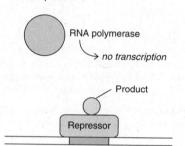

(b)   *Product absent*

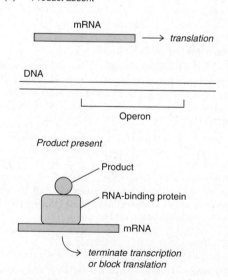

(c) If no protein were involved, the operon's product would have to interact directly with the mRNA.

(d)
```
        U A
      U     A
      U     G
      G---C
      A---U
      C---G
      U---A
· · · GAU---AGC · · ·
```

(e) Researchers could probe the conformation of an RNA molecule in the presence and absence of FAD by monitoring a conformation-dependent property of the RNA molecule, such as its susceptibility to an endonuclease.

(f) The FMN component of FAD is the most effective, since it has the lowest dissociation constant. The phosphate group is important for RNA binding, since riboflavin, which lacks a phosphate group, binds much less tightly. [From Winkler, W. C., Cohen-Chalamish, S., and Breaker, R. B., *Proc. Nat. Acad. Sci.* **99**, 15908–15913 (2002).]

41.

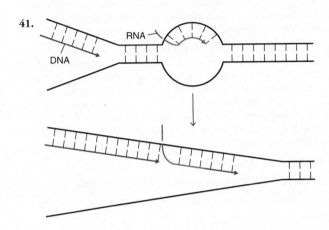

43. Bacterial mRNAs have a 5′ triphosphate group. The pyrophospho-hydrolase removes two of the phosphoryl groups as pyrophosphate (PP_i), leaving a 5′ monophosphate (this apparently makes the mRNA a better substrate for the endonuclease).

45. Messenger RNAs are transcribed only by RNA polymerase II. The phosphorylated tail of RNA polymerase II recruits the enzymes needed for capping and polyadenylation. Other types of RNAs are synthesized by different RNA polymerases that do not have phosphorylated tails and cannot recruit enzymes involved in post-transcriptional modification. Thus, only mRNAs are capped and polyadenylated.

47. The active site of poly(A) polymerase is narrower because it does not need to accommodate a template strand.

49. The PABP binds to the poly(A) tails and protects the mRNA from degradation by the nucleases. Increasing the concentration of PABP extends the half-lives of the mRNAs bound to this protein.

51. **(a)** The phosphate groups of the phosphodiester backbone of RNA will be labeled wherever α-[^{32}P]-ATP is used as a substrate by RNA polymerase.
(b) ^{32}P will appear only at the 5′ end of RNA molecules that have A as the first residue (this residue retains its α and β phosphates). In all other cases where β-[^{32}P]-ATP is used as a substrate for RNA synthesis, the β- and γ-phosphates are released as PP_i (see Fig. 20-9).
(c) No ^{32}P will appear in the RNA chain. During polymerization, the β- and γ-phosphates are released as PP_i. The terminal (γ) phosphate of an A residue at the 5′ end of an RNA molecule is removed during the capping process.

53. The splicing reactions are mediated by the spliceosome, a large RNA–protein complex. The intron must be large enough to include spliceosome binding site(s). In addition, the formation of a lariat-shaped intermediate (see Fig. 21-22) requires a segment of RNA long enough to curl back on itself without strain.

55. Introns are removed co-transcriptionally rather than post-transcriptionally because the former strategy makes it less likely that exons will be skipped during the intron-removal process.

57. **(a)** U69 is more important for catalysis, since deleting this residue or substituting it with another residue generally resulted in a dramatic decrease in the catalytic constant (k) but had a modest effect on K_d.
(b) Increasing the size of the bulge dramatically decreased substrate binding affinity, suggesting that although the presence of the bulge is necessary for catalysis, the geometry of the bulge is important for substrate binding. [From Kaye, N. M., Zahler, N. H., Christian, E. L., and Harris, M. E., *J. Mol. Biol.* **324**, 429–442 (2002).]

59.
```
5′ · · · GGAGUACCCUGAUGAGAUC · · · 3′
3′ · · · CCUCAUGGGACUACUCUAG · · · 5′
```

[From Takei, Y., Kadomatsu, K., Yuzawa, Y., et al., *Cancer Res.* **64**, 3365–3370 (2004).]

61. The primary structure of a protein refers to its sequence of amino acids; in the RNase P RNA this corresponds to the sequence of 417 nucleotides. Secondary structure in proteins refers to regular, repeating structural motifs such as α helices and β sheets. In the ribozyme, secondary structure refers to the base-paired stem and loop structures. Tertiary structure in proteins refers to the overall three-dimensional shape of the macromolecule; similarly for the ribozyme, the tertiary structure refers to the three-dimensional shape of the molecule.

Chapter 22

1. A hypothetical quadruplet code would have 4^4, or 256, possible combinations.

3. **(a)** Poly(Phe), **(b)** poly(Pro), and **(c)** poly(Lys).

5. **(a)** A polypeptide consisting of a repeating Tyr–Leu–Ser–Ile tetrapeptide will be produced.
(b) Depending on the reading frame, the polypeptide may begin with Tyr, Ile, or Ser.

7. Because the tRNAs that match the common codons are most abundant in the yeast cell, protein synthesis is normally efficient. If a mutation alters a codon so that it is not one of the 25 commonly used codons, it is likely that the isoacceptor tRNA for that codon is relatively scarce. Consequently, waiting for the appropriate tRNA to deliver the amino acid to the ribosome would result in a lower rate of protein synthesis, even though the sequence of the protein is unchanged.

9. The 5′ nucleotide is at the wobble position, which can participate in non–Watson–Crick base pairings with the 3′ nucleotide of an mRNA codon. Because the first two codon positions are more important for specifying an amino acid (see Table 22-1), wobble at the third position may not affect translation.

11.

I:A

13. If the adenosine of an mRNA codon were changed to inosine, then the codon could pair with a tRNA anticodon containing guanosine rather than uridine. This could result in the incorporation of a different amino acid at that codon position. A given gene could give rise to several different polypeptide products, depending on how many A residues were edited to I residues and how many of these changes resulted in amino acid substitutions.

15. Gly is the smallest amino acid, so the aminoacylation site in GlyRS can be small enough to prevent the entry of any other amino acid.

17. **(a)** AlaRS can generate Ala–tRNA^{Ala}, Gly–tRNA^{Ala}, and Ser–RNA^{Ala}.
(b) An editing active site that admits amino acids smaller than alanine would destroy Gly–tRNA^{Ala}, but it would not get rid of Ser–RNA^{Ala}, since it cannot fit into the editing active site (serine is larger than alanine). **(c)** AlaXp functions as a free-standing editor for AlaRS by eliminating Ser–tRNA^{Ala} after it has been synthesized by AlaRS.

19. The two Lys codons are AAA and AAG. Substitution with C would yield CAA and CAG, which code for Gln; substitution with G would yield GAA and GAG, which code for Glu; and substitution with U would yield AUU and UAG, which are stop codons. Replacing a Lys codon with a stop codon would terminate protein synthesis prematurely, most likely producing a non-functional protein. Replacing Lys with Glu or Gln could disrupt the protein's structure and therefore its function if the Lys residue was involved in a structurally essential interaction such as an ion pair in the protein interior. If the Lys residue was on the surface of the protein, replacing it with Glu or Gln, both of which are hydrophilic, might not have much impact on the protein's structure or function.

21. Like other nucleic acid–binding proteins we have studied (histones are an example), proteins containing the positively charged Lys and Arg residues interact favorably with the polyanionic RNA. The most important interactions between the protein and the nucleic acid are likely to be ion pairs.

23. In protein-coding genes, a high degree of conservation indicates a segment of protein that is intolerant of substitution. Mutations that produce amino acid substitutions in these regions are likely to produce nonfunctional proteins. It is the same for ribosomal RNA. Mutations in the rRNA that disrupt its function are likely to be lethal and will not be passed on to progeny. Consequently, the sequences of the rRNAs are intolerant of substitution and are highly conserved.

25. The small and large ribosomal subunits are held together by RNA–RNA contacts that are stabilized by Mg^{2+} ions. The tRNAs that bind to the A and P sites in the ribosome contact the rRNA; these interactions have the same structural features. Adding EDTA chelates the magnesium ions and causes loss of ribosomal stability; this may be accompanied by the unfolding of the ribosomal RNA and the dissociation of the ribosomal proteins, so translation cannot occur.

27. Acetylation removes the positive charge on the N-terminus, as shown below. If some copies of the protein are acetylated whereas others are not, two spots will appear on the gel, which separates proteins based on charge as well as size (the difference in size is not significant). [From Kaltschmidt, E., and Wittmann, H. G., *Proc. Natl. Acad. Sci.* **67**, 1276–1282 (1970).]

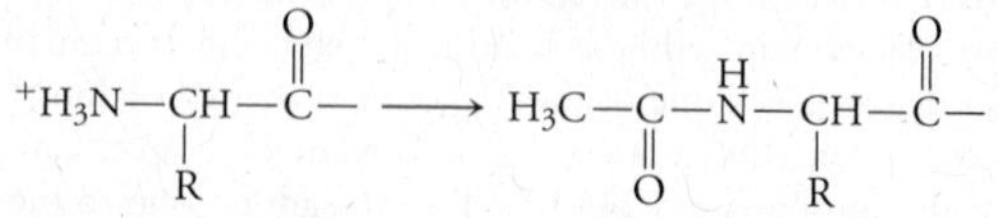

$$^{+}H_3N-CH-C- \longrightarrow H_3C-C-N-CH-C-$$

29. The start codon is likely near the 5′ end of the mRNA, and the stop codon near the 3′ end (neither is at the very end of the mRNA molecule).

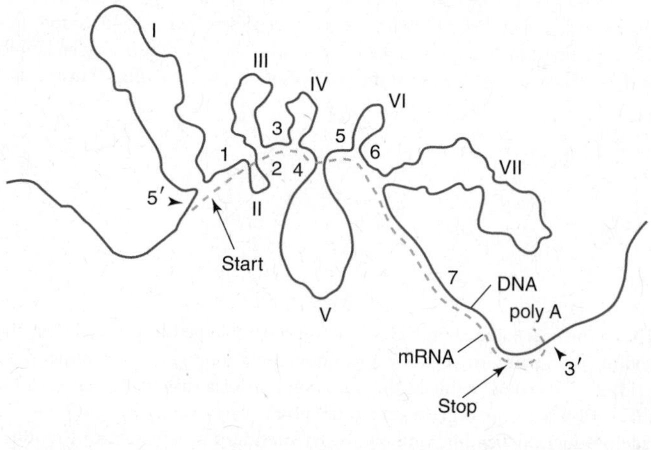

31. The polypeptide would contain all Lys residue with an Asn at the C-terminus. If the mRNA were read from 3′ → 5′, the polypeptide would consist of an N-terminal Gln followed by a series of Lys residues. Transcription and translation both take place in the 5′ → 3′ direction; this allows bacterial cells to begin translating nascent mRNA before transcription is complete. If translation took place in the 3′ → 5′ direction, the ribosome would have to wait for mRNA synthesis to be completed before translation could begin.

33. The sequence on the 16S rRNA that aligns with the Shine–Dalgarno sequence is shown. The initiation codon is highlighted in gray.

5′ · · · CUACCAGGAGCAAAGCUAAUGGCUUUA · · · 3′
3′ · · · UCCUC · · · 5′

35. Colicin E3 is lethal to the cells because it prevents accurate and efficient translation. Cleavage of the 16S rRNA at A1493 destroys the part of the 30S ribosomal subunit that verifies codon–anticodon pairing. As a result, the ribosome is less able to incorporate the correct aminoacyl group into a growing polypeptide. In addition, EF-Tu hydrolysis of GTP is slow because EF-Tu does not receive a signal from the ribosome that an mRNA–tRNA match has occurred, so the speed of translation decreases.

37. The correctly charged tRNAs (Ala–tRNA^{Ala} and Gln–tRNA^{Gln}) bind to EF-Tu with approximately the same affinity, so they are delivered to the ribosomal A site with the same efficiency. The mischarged Ala–tRNA^{Gln} binds to EF-Tu more loosely, indicating that it may dissociate from EF-Tu before it reaches the ribosome. The mischarged Gln–tRNA^{Ala} binds to EF-Tu much more tightly, indicating that EF-Tu may not be able to dissociate from it at the ribosome. These results suggest that either a higher or a lower binding affinity could affect the ability of EF-Tu to carry out its function, which would decrease the rate at which mischarged aminoacyl–tRNAs bind to the ribosomal A site during translation.

39. In a living cell, EF-Tu and EF-G enhance the rate of protein synthesis by rendering various steps of translation irreversible. They also promote the accuracy of protein synthesis through proofreading. In the absence of the elongation factors, translation would be too slow and too inaccurate to support life. These constraints do not apply to an *in vitro* translation system, which can proceed in the absence of EF-Tu and EF-G. However, the resulting protein is likely to contain more misincorporated amino acids than a protein synthesized in a cell.

41. The mRNA has the sequence

CGAUAAUGUCCGACCAAGCGAUCUCGUAGCA

The start codon and stop codon are highlighted. The encoded protein has the sequence

Met–Ser–Asp–Gln–Ala–Ile–Ser.

43. To encode 1480 amino acids, 1480 codons or 4440 nucleotides (1480 × 3) are needed, plus a stop codon. The 1686 additional mRNA nucleotides (6129 − 4443) include segments at the 3′ and 5′ ends where translation factors and the ribosome bind.

45. In prokaryotes, both mRNA and protein synthesis take place in the cytosol, so a ribosome can assemble on the 5′ end of an mRNA even while RNA polymerase is synthesizing the 3′ end of the transcript. In eukaryotes, RNA is produced in the nucleus, but ribosomes are located in the cytosol. Because transcription and translation occur in separate compartments, they cannot occur simultaneously. A eukaryotic mRNA must be transported from the nucleus to the cytosol before it can be translated.

47. **(a)** The ribosome positions the peptidyl group for reaction with the incoming aminoacyl group, so a peptidyl group with a constrained geometry, like Pro, is unable to react optimally.
(b) Because Arg and Lys (both with positively charged side chains) react much faster than Asp (negatively charged side chain), the active site must be more accommodating of cationic groups than anionic groups.
(c) Transpeptidation of Ala is faster than for Phe or Val, so for nonpolar amino acids, small size is more favorable. [From Wohlgemuth, I., Brenner, S., Beringer, M., and Rodnina, M. V., *J. Biol. Chem.* **283**, 32229–32235 (2008).]

49. **(a)** Transpeptidation involves the nucleophilic attack of the amino group of the aminoacyl–tRNA on the carbonyl carbon of the peptidyl–tRNA (see Fig. 22-15). The higher the pH, the more nucleophilic the amino group (the less likely it is to be protonated).

51. **(a)** The mutation would allow the aminoacylated tRNA rather than a release factor to enter the ribosome and pair with a stop codon. The result would be incorporation of an amino acid into a polypeptide rather than translation termination, so the ribosome would continue to read mRNA codons and produce elongated polypeptides. The inability of the mutated tRNA to recognize its amino acid–specifying codons would have a minor impact on protein synthesis, since the cell likely contains other isoacceptor tRNAs that can recognize the same codons.

(b) Not all proteins would be affected. Only proteins whose genes include the stop codon that is read by the mutated tRNA would be affected. Proteins whose genes include one of the other two stop codons would be synthesized normally.

(c) Aminoacyl–tRNA synthetases usually recognize both the anticodon and acceptor ends of their tRNA substrates. A mutation in the tRNA anticodon, such as a nonsense suppressor mutation, might interfere with tRNA recognition, so the mutated tRNA molecule might not undergo aminoacylation. This would minimize the ability of the mutated tRNA to insert the amino acid at a position corresponding to a stop codon.

53. Anfinsen's ribonuclease experiment demonstrated that a protein's primary structure dictates its three-dimensional structure. The purified ribonuclease was able to refold without the assistance of chaperones because other cellular components were absent. Chaperones are required *in vivo* because they prevent the interaction and aggregation of the many proteins and other components that exist in the cell. Molecular chaperones assist in the protein-folding process and do not contribute any additional information regarding the tertiary structure of the protein; that information is contained within the primary structure of the protein, as demonstrated by Anfinsen.

55. Mitochondria contain ribosomes that synthesize proteins encoded by mitochondrial DNA. Like cytosolic proteins, mitochondrial proteins require the assistance of chaperones for proper folding. Other proteins are synthesized in the cytosol and are transported partially unfolded through pores in the mitochondrial membranes; these proteins also require the assistance of chaperones to fold properly once they reach their destination.

57. The different domains in a multidomain protein associate with one another via van der Waals forces, since the domain interfaces eventually end up in the interior of the protein. A cagelike chaperonin structure allows these proteins to fold in a protected environment where the hydrophobic regions of the protein are not exposed to other intracellular proteins with which they could potentially aggregate.

59. The basic residue is highlighted in gray; the hydrophobic core is underlined.

MKWVT<u>FISLLLLF</u>SSAYSRGV

61. The hydrophobic cleft of this particular protein might allow it to recognize the hydrophobic core of the signal sequence (see Problem 59). Other proteins in the SRP might be involved in pausing translation and promoting translocation.

63. In the cell-free system, SRP can bind to the exit tunnel of the ribosome, but translation is not arrested when no membrane is present. This indicates that the SRP must interact with both the nascent polypeptide and the ER membrane in order to pause translation. When microsomal membranes are subsequently added, the protein is not translocated, indicating that translocation must occur co-translationally, not post-translationally. Proteins that are not translocated retain their signal sequences because they do not have access to the signal peptidase, which is located in the microsomal lumen.

65. (a) **(b)**

Polyglutamine protein

(c) The acetyltransferase acetylates Lys residues in histones, neutralizing the positive charge of the Lys side chain and weakening its interaction with the DNA, so transcriptional activity is increased. If the acetyltransferase is inactive, the DNA will be less transcriptionally active and certain genes will not be expressed. The loss of transcriptional activity could contribute to the progression of the polyglutamine disease. [From Pennuto, M., Palazzolo, I., and Poletti, A., *Hum. Mol. Gen.* **18**, R40–R47 (2009).]

67.

Chapter 1

1. Since ΔH is positive, heat is absorbed during the reaction.

2. Because entropy increases, ΔS is positive.

3. $\Delta G = \Delta H - T\Delta S$

$$= -15{,}000 \text{ J} \cdot \text{mol}^{-1} - 298 \text{ K}(-75 \text{ J} \cdot \text{K}^{-1} \cdot \text{mol}^{-1})$$
$$= -15{,}000 + 22{,}400 \text{ J} \cdot \text{mol}^{-1}$$
$$= 7400 \text{ J} \cdot \text{mol}^{-1} \text{ or } 7.4 \text{ kJ} \cdot \text{mol}^{-1}$$

Because ΔG is greater than 0, the reaction is not spontaneous.

4. ΔG would be zero when the ΔH and $T\Delta S$ terms are equal:

$$\Delta H = T\Delta S$$
$$T = \Delta H/\Delta S$$
$$= (15{,}000 \text{ J} \cdot \text{mol}^{-1})/(75 \text{ J} \cdot \text{K}^{-1} \cdot \text{mol}^{-1})$$
$$= 200 \text{ K or } -73°\text{C}$$

Chapter 2

1. HCl dissociates completely, so the added $[\text{H}^+]$ is equal to [HCl]:

$$[\text{H}^+] = \frac{(0.05 \text{ L})(0.025 \text{ M})}{0.55 \text{ L}} = 0.0023 \text{ M}$$

$$\text{pH} = -\log[\text{H}^+] = -\log 0.0023 = 2.6$$

2. NaOH dissociates completely, so the added $[\text{OH}^-]$ is equal to [NaOH]:

$$[\text{OH}^-] = \frac{(0.25 \text{ L})(0.005 \text{ M})}{0.5 \text{ L}} = 0.0025 \text{ M}$$

$$K_w = 10^{-14} = [\text{H}^+][\text{OH}^-]$$

$$[\text{H}^+] = \frac{10^{-14}}{0.0025} = 4 \times 10^{-12} \text{ M}$$

$$\text{pH} = -\log[\text{H}^+] = 11.4$$

3. The final volume is 1 L + 25 mL + 25 mL = 1.05 L

$$[\text{acetic acid}] = [\text{HA}] = \frac{(0.025 \text{ L})(0.01 \text{ M})}{1.05 \text{ L}} = 2.4 \times 10^{-4} \text{ M}$$

$$[\text{acetate}] = [\text{A}^-] = \frac{(0.025 \text{ L})(0.03 \text{ M})}{1.05 \text{ L}} = 7.1 \times 10^{-4} \text{ M}$$

$$\text{pH} = \text{p}K + \log\frac{[\text{A}^-]}{[\text{HA}]}$$

$$= 4.76 + \log\frac{7.1 \times 10^{-4}}{2.4 \times 10^{-4}}$$

$$= 4.76 + 0.47 = 5.23$$

4. The final volume is 500 mL + 10 mL + 20 mL = 0.53 L

$$[\text{boric acid}] = [\text{HA}] = \frac{(0.01 \text{ L})(0.05 \text{ M})}{0.53 \text{ L}} = 9.4 \times 10^{-4} \text{ M}$$

$$[\text{borate}] = [\text{A}^-] = \frac{(0.02 \text{ L})(0.02 \text{ M})}{0.53 \text{ L}} = 7.5 \times 10^{-4} \text{ M}$$

$$\text{pH} = \text{p}K + \log\frac{[\text{A}^-]}{[\text{HA}]}$$

$$= 9.24 + \log\frac{7.5 \times 10^{-4}}{9.4 \times 10^{-4}}$$

$$= 9.24 - 0.10 = 9.14$$

5. $$\text{pH} = \text{p}K + \log\frac{[\text{A}^-]}{[\text{HA}]}$$

$$\log\frac{[\text{A}^-]}{[\text{HA}]} = \text{pH} - \text{p}K = 5.0 - 4.76 = 0.24$$

$$\frac{[\text{A}^-]}{[\text{HA}]} = 1.74 \text{ or } [\text{A}^-] = 1.74[\text{HA}]$$

$$[\text{A}^-] = 1.74[\text{HA}] = 1.74(0.05 \text{ M} - [\text{A}^-])$$

$$[\text{A}^-] = 0.087 - 1.74[\text{A}^-]$$

$$2.74[\text{A}^-] = 0.087 \text{ M}$$

$$[\text{A}^-] = 0.032 \text{ M or } 32 \text{ mM}$$

6. $$\text{pH} = \text{p}K + \log\frac{[\text{A}^-]}{[\text{HA}]}$$

$$\log\frac{[\text{A}^-]}{[\text{HA}]} = \text{pH} - \text{p}K = 3.0 - 2.15 = 0.85$$

$$\frac{[\text{A}^-]}{[\text{HA}]} = 7.08 \text{ or } [\text{HA}] = \frac{[\text{A}^-]}{7.08}$$

$$[\text{HA}] = \frac{[\text{A}^-]}{7.08} = \frac{(0.05 \text{ M} - [\text{HA}])}{7.08}$$

$$7.08[\text{HA}] = 0.05 \text{ M} - [\text{HA}]$$

$$8.08[\text{HA}] = 0.05 \text{ M}$$

$$[\text{HA}] = 0.0062 \text{ M or } 6.2 \text{ mM}$$

7. At pH 6, pH < pK_2, so the H_2PO_4^- form predominates.

8. At pH 8, pH > pK_2, so the HPO_4^{2-} form predominates.

9. The pK of the ammonium ion is 9.25, so at pH 9.25, $[\text{NH}_3] = [\text{NH}_4^+]$.

10. Since the dicarboxylic acid has a pK of 4.2 and the monocarboxylic acid has a pK of 5.64, the monocarboxylate form predominates between pH 4.2 and pH 5.64 (above pH 5.64, the dicarboxylate form predominates).

11. $$\text{pH} = \text{p}K + \log\frac{[\text{A}^-]}{[\text{HA}]}$$

$$\log\frac{[\text{A}^-]}{[\text{HA}]} = \text{pH} - \text{p}K$$

$$\frac{[\text{A}^-]}{[\text{HA}]} = 10^{(\text{pH}-\text{p}K)}$$

$$\frac{[\text{A}^-]}{[\text{HA}]} = 10^{(9.6-9.24)} = 10^{0.36} = 2.29$$

Since the starting solution contains $(0.2 \text{ L})(0.05 \text{ mol} \cdot \text{L}^{-1}) = 0.01$ mole of boric acid (HA), the amount of borate (A^-) needed is $2.29(0.01 \text{ mol}) = 0.023$ moles. The stock sodium borate is 5.0 M, so the volume of sodium borate to be added is

$$\frac{0.023 \text{ mol}}{5.0 \text{ mol} \cdot \text{L}^{-1}} = 0.0046 \text{ L or } 4.6 \text{ mL}$$

12.
$$\text{pH} = \text{p}K + \log\frac{[A^-]}{[HA]}$$

$$\log\frac{[A^-]}{[HA]} = \text{pH} - \text{p}K$$

$$\frac{[A^-]}{[HA]} = 10^{(\text{pH}-\text{p}K)}$$

$$\frac{[A^-]}{[HA]} = 10^{(6.5-7.0)} = 10^{-0.5} = 0.316$$

Since the starting solution contains $(0.5 \text{ L})(0.01 \text{ mol} \cdot \text{L}^{-1}) = 0.005$ mole of imidazole (A^-), the amount of imidazolium chloride (HA) needed is $0.005 \text{ mol}/0.316 = 0.016$ moles. The stock imidazolium chloride is 1 M, so the volume of imidazolium chloride to be added is

$$\frac{0.016 \text{ mol}}{1.0 \text{ mol} \cdot \text{L}^{-1}} = 0.016 \text{ L or } 16 \text{ mL}$$

Chapter 4

1. At pH 6.0, groups with pK values less than 6.0 are mostly deprotonated, and groups with pK values greater than 6.0 are mostly protonated. The dipeptide has a net charge of -1.

Group	Charge
N-terminus	+1
Glu	−1
Tyr	0
C-terminus	−1
Net charge	**−1**

2. At pH 7.0, groups with pK values less than 7.0 are mostly deprotonated, and groups with pK values greater than 7.0 are mostly protonated. The tripeptide has a net charge of -3.

Group	Charge
N-terminus	+1
3 Asp	−3
C-terminus	−1
Net charge	**−3**

3. At pH 8.0, groups with pK values less than 8.0 are mostly deprotonated, and groups with pK values greater than 8.0 are mostly protonated. The tripeptide has a net charge of 0.

Group	Charge
N-terminus	+1
His	0
Lys	+1
Glu	−1
C-terminus	−1
Net charge	**0**

4. In order for alanine to have no net charge, its α-carboxyl group (p$K \approx 3.5$) must be unprotonated (negatively charged) and its α-amino group (p$K \approx 9.0$) must be protonated (positively charged):

$$\text{pI} = \tfrac{1}{2}(3.5 + 9.0) = 6.25$$

5. In order for glutamate to have no net charge, its α-carboxyl group must be unprotonated (negatively charged), its side chain must be protonated (neutral), and its α-amino group must be protonated (positively charged). Because deprotonation of the side chain or the α-amino group would change the amino acid's net charge, the pK values of these groups (4.0 and 9.0) should be used to calculate the pI:

$$\text{pI} = \tfrac{1}{2}(4.0 + 9.0) = 6.5$$

Chapter 5

1.
$$Y = \frac{pO_2}{K + pO_2}$$

$$= \frac{5.6}{2.8 + 5.6} = 0.67$$

2. Y increases.

$$Y = \frac{pO_2}{K + pO_2}$$

$$= \frac{5.6}{1.4 + 5.6} = 0.80$$

3.
$$Y = 0.75 = \frac{pO_2}{2.8 + pO_2}$$

$$Y = 0.75(2.8 + pO_2) = pO_2$$

$$2.1 + 0.75\,pO_2 = pO_2$$

$$2.1 = 0.25\,pO_2$$

$$pO_2 = 8.4 \text{ torr}$$

Chapter 7

1. Since $v = k[X][Y]$, $k = v/([X][Y])$

$$k = (5 \text{ μM} \cdot \text{s}^{-1})/[(5 \text{ μM})(5 \text{ μM})]$$

$$= 0.2 \text{ μM}^{-1} \cdot \text{s}^{-1}$$

2. $v = k[X][Y]$

$$= (0.2 \text{ μM}^{-1} \cdot \text{s}^{-1})(20 \text{ μM})(10 \text{ μM})$$

$$= 40 \text{ μM} \cdot \text{s}^{-1}$$

3. Since $v = k[X][Y]$, $[X][Y] = [X]^2 = v/k$

$$[X]^2 = (8 \text{ mM} \cdot \text{s}^{-1})/(0.5 \text{ mM}^{-1} \cdot \text{s}^{-1}) = 16 \text{ mM}^2$$
$$[X] = [Y] = 4 \text{ mM}$$

4. $v_0 = \dfrac{(5 \text{ nM} \cdot \text{s}^{-1})(1.5 \text{ mM})}{(1 \text{ mM}) + (1.5 \text{ mM})}$

$$= \frac{7.5}{2.5} \text{ nM} \cdot \text{s}^{-1}$$

$$= 3 \text{ nM} \cdot \text{s}^{-1}$$

5. $v_0 = \dfrac{(5 \text{ nM} \cdot \text{s}^{-1})(10 \text{ mM})}{(1 \text{ mM}) + (10 \text{ mM})}$

$$= \frac{50}{11} \text{ nM} \cdot \text{s}^{-1} = 4.5 \text{ nM} \cdot \text{s}^{-1}$$

6. $v_0 = \dfrac{V_{\max}[S]}{K_M + [S]}$

$$K_M = \frac{V_{\max}[S]}{v_0} - [S]$$

$$K_M = \frac{(7.5 \text{ μM} \cdot \text{s}^{-1})(1 \text{ μM})}{5 \text{ μM} \cdot \text{s}^{-1}} - 1 \text{ μM}$$

$$= 1.5 \text{ μM} - 1 \text{ μM} = 0.5 \text{ μM}$$

7.
$$v_0 = \frac{V_{max}[S]}{K_M + [S]}$$

$$[S] = \frac{V_{max}[S]}{v_0} - K_M$$

$$V_{max}[S] = v_0 K_M + v_0[S]$$

$$V_{max}[S] - v_0[S] = v_0 K_M$$

$$[S](V_{max} - v_0) = v_0 K_M$$

$$[S] = \frac{v_0 K_M}{V_{max} - v_0}$$

$$= \frac{(2.5\ \mu M \cdot s^{-1})(0.5\ \mu M)}{7.5\ \mu M \cdot s^{-1} - 2.5\ \mu M \cdot s^{-1}}$$

$$= 1.25\ \mu M / 5 = 0.25\ \mu M$$

8. Calculate the reciprocals of [S] and v_0 and construct a plot of $1/v_o$ versus $1/[S]$. The intercept on the $1/[S]$ axis is -0.1 mM^{-1}, which is equal to $-1/K_M$. Therefore, $K_M = 10$ mM. The intercept on the $1/v_0$ axis is 0.05 mM$^{-1} \cdot$ s, which is equal to $1/V_{max}$. Therefore, $V_{max} = 20$ mM $\cdot$ s^{-1}.

9. Since $K_M^{app} = 3K_M$, $\alpha = 3$.

$$K_I = \frac{[I]}{\alpha - 1} = \frac{10\ \mu M}{3 - 1} = \frac{10\ \mu M}{2} = 5\ \mu M$$

10. First, calculate α using Equation 7-29:

$$\alpha = 1 + \frac{[I]}{K_I} = 1 + \frac{4\ \mu M}{2\ \mu M} = 1 + 2 = 3$$

The apparent K_M is $\alpha K_M = (3)(10\ \mu M) = 30\ \mu M$.

11. For inhibitor A, $\alpha = 2$ and $K_I = \frac{[I]}{\alpha - 1} = \frac{2\ \mu M}{2 - 1} = 2\ \mu M$

For inhibitor B, $\alpha = 4$ and $K_I = \frac{[I]}{\alpha - 1} = \frac{9\ \mu M}{4 - 1} = 3\ \mu M$

The ratio is 3 μM/2 μM or 1.5.

Chapter 9

1.
$$\Delta\psi = 0.058 \log \frac{[Na^+]_{in}}{[Na^+]_{out}}$$

$$\log[Na^+]_{in} = \frac{\Delta\psi}{0.058} + \log[Na^+]_{out}$$

$$\log[Na^+]_{in} = \frac{-0.100}{0.058} + \log(0.160)$$

$$\log[Na^+]_{in} = -1.72 - 0.796$$
$$\log[Na^+]_{in} = -2.52$$
$$[Na^+]_{in} = 0.003\ M = 3\ mM$$

2. $\Delta\psi = 0.058 \log \frac{[Na^+]_{in}}{[Na^+]_{out}}$

$$= 0.058 \log \frac{(0.010)}{(0.100)}$$

$$= 0.058 \log (0.10)$$
$$= -0.058\ V = -58\ mV$$

3. $\Delta\psi = 0.058 \log \frac{[Na^+]_{in}}{[Na^+]_{out}}$

$$= 0.058 \log \frac{(0.040)}{(0.025)}$$

$$= 0.058 \log (1.6)$$
$$= 0.012\ V = 12\ mV$$

4. $\Delta G = RT(-4.6)$
$$= (8.3145\ J \cdot K^{-1} \cdot mol^{-1})(293\ K)(-4.6)$$
$$= -11,200\ J \cdot mol^{-1} = -11.2\ kJ \cdot mol^{-1}$$

5. $\Delta G = RT \ln \frac{[glucose]_{in}}{[glucose]_{out}}$

$$= RT \ln \frac{(0.0005)}{(0.005)}$$

$$= (8.3145\ J \cdot K^{-1} \cdot mol^{-1})(293\ K)(-2.3)$$
$$= -5610\ J \cdot mol^{-1} = -5.61\ kJ \cdot mol^{-1}$$

6. $\Delta G = RT \ln \frac{[glucose]_{in}}{[glucose]_{out}}$

$$= RT \ln \frac{(0.005)}{(0.0005)}$$

$$= (8.3145\ J \cdot K^{-1} \cdot mol^{-1})(293\ K)(2.3)$$
$$= +5610\ J \cdot mol^{-1} = +5.61\ kJ \cdot mol^{-1}$$

7. $\Delta G = RT \ln \frac{[X]_{in}}{[X]_{out}} + Z\mathcal{F}\Delta\psi$

$$= (8.3145\ J \cdot K^{-1} \cdot mol^{-1})(293\ K) \ln \frac{(0.050)}{(0.010)}$$

$$+ (1)(96,485\ J \cdot V^{-1} \cdot mol^{-1})(-0.05\ V)$$
$$= 3920\ J \cdot mol^{-1} - 4820\ J \cdot mol^{-1}$$
$$= -900\ J \cdot mol^{-1} = -0.9\ kJ \cdot mol^{-1}$$

This process is spontaneous.

8. $\Delta G = RT \ln \frac{[X]_{in}}{[X]_{out}} + Z\mathcal{F}\Delta\psi$

$$= (8.3145\ J \cdot K^{-1} \cdot mol^{-1})(293\ K) \ln \frac{(0.025)}{(0.100)}$$

$$+ (1)(96,485\ J \cdot V^{-1} \cdot mol^{-1})(0.05\ V)$$
$$= -3380\ J \cdot mol^{-1} + 4820\ J \cdot mol^{-1}$$
$$= 1440\ J \cdot mol^{-1} = 1.44\ kJ \cdot mol^{-1}$$

This process is not spontaneous.

Chapter 10

1. Because 40% of the receptors are occupied, $[R \cdot L] = 9.6\ \mu M$ and $[R] = 14.4\ \mu M$.

$$K_d = \frac{[R][L]}{[R \cdot L]}$$

$$= \frac{(14.4 \times 10^{-6})(10 \times 10^{-6})}{(9.6 \times 10^{-6})}$$

$$= 15 \times 10^{-6}\ M = 15\ \mu M$$

2. $K_d = \frac{[R][L]}{[R \cdot L]}$

$$[R \cdot L] = \frac{[R][L]}{K_d}$$

$$= \frac{(0.005)(0.018)}{(0.003)}$$

$$= 0.03\ M = 30\ mM$$

3. Let $[R \cdot L] = x$ and $[R] = 20\ mM - x$.

$$K_d = \frac{[R][L]}{[R \cdot L]}$$

$$[R \cdot L] = x = \frac{[R][L]}{K_d}$$

$$x = \frac{(0.02 - x)(0.005)}{(0.01)}$$

$$x = \frac{0.0001 - 0.005x}{0.01}$$

$$0.01x = 0.0001 - 0.005x$$
$$0.015x = 0.0001$$
$$x = 0.0067 \text{ M} = 6.7 \text{ mM}$$

The percentage of receptors occupied by ligand is 6.7 mM/20 mM or 33%.

Chapter 12

1. $\Delta G^{\circ\prime} = -RT \ln K_{eq}$

$$= -(8.3145 \text{ J} \cdot \text{K}^{-1} \cdot \text{mol}^{-1})(298 \text{ K}) \ln 0.25$$
$$= 3400 \text{ J} \cdot \text{mol}^{-1} = 3.4 \text{ kJ} \cdot \text{mol}^{-1}$$

2. $\Delta G^{\circ\prime} = -RT \ln K_{eq}$

$$= -(8.3145 \text{ J} \cdot \text{K}^{-1} \cdot \text{mol}^{-1})(310 \text{ K}) \ln 0.25$$
$$= 3600 \text{ J} \cdot \text{mol}^{-1} = 3.6 \text{ kJ} \cdot \text{mol}^{-1}$$

Increasing the temperature makes the change in free energy slightly more positive.

3. Since $\Delta G^{\circ\prime} = -RT \ln K_{eq}$,

$$\ln K_{eq} = -\Delta G^{\circ\prime}/RT$$
$$K_{eq} = e^{-\Delta G^{\circ\prime}/RT}$$
$$= e^{-(-10,000 \text{ J} \cdot \text{mol}^{-1})/(8.3145 \text{ J} \cdot \text{K}^{-1} \cdot \text{mol}^{-1})(310 \text{ K})}$$
$$= e^{3.88} = 48$$

4. $\Delta G = \Delta G^{\circ\prime} + RT \ln \dfrac{[\text{glucose-6-phosphate}]}{[\text{glucose-1-phosphate}]}$

$$= -7100 \text{ J} \cdot \text{mol}^{-1}$$
$$\quad + (8.3145 \text{ J} \cdot \text{K}^{-1} \cdot \text{mol}^{-1})(310 \text{ K}) \ln(0.020/0.005)$$
$$= -7100 \text{ J} \cdot \text{mol}^{-1} + 3600 \text{ J} \cdot \text{mol}^{-1}$$
$$= -3500 \text{ J} \cdot \text{mol}^{-1} = -3.5 \text{ kJ} \cdot \text{mol}^{-1}$$

Because $\Delta G < 0$, the reaction is spontaneous

5. $$K_{eq} = 17.6 = \frac{[\text{glucose-6-phosphate}]}{[\text{glucose-1-phosphate}]}$$

$$[\text{glucose-1-phosphate}] = [\text{glucose-6-phosphate}]/17.6$$
$$= 0.035/17.6$$
$$0.002 \text{ M} = 2 \text{ mM}$$

6. $$\Delta G = \Delta G^{\circ\prime} + RT \ln \frac{[\text{glucose-6-phosphate}]}{[\text{glucose-1-phosphate}]}$$

$$\ln \frac{[\text{glucose-6-phosphate}]}{[\text{glucose-1-phosphate}]} = \frac{\Delta G - \Delta G^{\circ\prime}}{RT}$$

$$\ln \frac{[\text{glucose-6-phosphate}]}{[\text{glucose-1-phosphate}]} = \frac{-2000 - (-7100 \text{ J} \cdot \text{mol}^{-1})}{(8.3145 \text{ J} \cdot \text{K}^{-1} \cdot \text{mol}^{-1})(310 \text{ K})}$$

$$\ln \frac{[\text{glucose-6-phosphate}]}{[\text{glucose-1-phosphate}]} = \frac{5100}{2600}$$

$$\ln \frac{[\text{glucose-6-phosphate}]}{[\text{glucose-1-phosphate}]} = 2.0$$

$$\frac{[\text{glucose-6-phosphate}]}{[\text{glucose-1-phosphate}]} = 7.4$$

Chapter 15

1. Use Equation 15-1.

$$\mathcal{E} = \mathcal{E}^{\circ\prime} - \frac{RT}{n\mathcal{F}} \ln \frac{[\text{A}_{reduced}]}{[\text{A}_{oxidized}]}$$

$$= 0.031 \text{ V} - \frac{(8.3145 \text{ J} \cdot \text{K}^{-1} \cdot \text{mol}^{-1})(310 \text{ K})}{(2)(96,485 \text{ J} \cdot \text{V}^{-1} \cdot \text{mol}^{-1})} \ln \frac{(1 \times 10^{-4})}{(8 \times 10^{-5})}$$

$$= 0.031 \text{ V} - 0.003 \text{ V} = 0.028 \text{ V}$$

2. Rearrange Equation 15-2:

$$\mathcal{E} = \mathcal{E}^{\circ\prime} - \frac{0.026 \text{ V}}{n} \ln \frac{[\text{A}_{reduced}]}{[\text{A}_{oxidized}]}$$

$$\mathcal{E}^{\circ\prime} = \mathcal{E} + \frac{0.026 \text{ V}}{n} \ln \frac{[\text{A}_{reduced}]}{[\text{A}_{oxidized}]}$$

$$= 0.50 \text{ V} + \frac{0.026 \text{ V}}{2} \ln \frac{(5 \times 10^{-6})}{(2 \times 10^{-4})}$$

$$= 0.50 \text{ V} + (-0.048 \text{ V}) = 0.452 \text{ V}$$

3. The relevant half-reactions are

malate $\rightarrow$ oxaloacetate $+ 2 \text{ H}^+ + 2 \; e^-$	$\mathcal{E}^{\circ\prime} = +0.166 \text{ V}$
ubiquinone $+ 2 \text{ H}^+ + 2 \; e^- \rightarrow$ ubiquinol	$\mathcal{E}^{\circ\prime} = +0.045 \text{ V}$

net: malate $+$ ubiquinone $\rightarrow$ $\Delta\mathcal{E}^{\circ\prime} = 0.211 \text{ V}$
oxaloacetate $+$ ubiquinol

$$\Delta G^{\circ\prime} = -n\mathcal{F}\Delta\mathcal{E}^{\circ\prime}$$
$$= -(2)(96,485 \text{ J} \cdot \text{V}^{-1} \cdot \text{mol}^{-1})(0.211 \text{ V})$$
$$= -40,700 \text{ J} \cdot \text{mol}^{-1} = -40.7 \text{ kJ} \cdot \text{mol}^{-1}$$

4. The relevant half-reactions are

acetaldehyde $+ 2 \text{ H}^+ + 2 \; e^- \rightarrow$ ethanol	$\mathcal{E}^{\circ\prime} = -0.197 \text{ V}$
NADH $\rightarrow$ NAD$^+$ $+ \text{ H}^+ + 2 \; e^-$	$\mathcal{E}^{\circ\prime} = +0.315 \text{ V}$

net: acetaldehyde $+$ NADH $\rightarrow$ $\Delta\mathcal{E}^{\circ\prime} = 0.118 \text{ V}$
ethanol $+$ NAD$^+$

$$\Delta G^{\circ\prime} = -n\mathcal{F}\Delta\mathcal{E}^{\circ\prime}$$
$$= -(2)(96,485 \text{ J} \cdot \text{V}^{-1} \cdot \text{mol}^{-1})(0.118 \text{ V})$$
$$= -22,800 \text{ J} \cdot \text{mol}^{-1} = -22.8 \text{ kJ} \cdot \text{mol}^{-1}$$

5. For the reverse reaction,
cytochrome $c_1(\text{Fe}^{3+}) + e^- \rightarrow$ cytochrome $c_1(\text{Fe}^{2+})$ $\mathcal{E}^{\circ\prime} = 0.22 \text{ V}$
cytochrome $c(\text{Fe}^{2+}) \rightarrow$ cytochrome $c(\text{Fe}^{3+}) + e^-$ $\mathcal{E}^{\circ\prime} = -0.235 \text{ V}$

$$\Delta\mathcal{E}^{\circ\prime} = 0.22 \text{ V} + (-0.235 \text{ V}) = -0.015 \text{ V}$$
$$\Delta G^{\circ\prime} = -n\mathcal{F}\Delta\mathcal{E}^{\circ\prime}$$

$$= -(1)(96,485 \text{ J} \cdot \text{V}^{-1} \cdot \text{mol}^{-1})(-0.015 \text{ V})$$

$$= 1400 \text{ J} \cdot \text{mol}^{-1} = 1.4 \text{ kJ} \cdot \text{mol}^{-1}$$

6. Use the equation derived in Sample Calculation 15-3:

$$\Delta G = 2.303 \; RT(\text{pH}_{in} - \text{pH}_{out}) + Z\mathcal{F}\Delta\psi$$

$$\Delta G = 2.303 \; (8.3145 \text{ J} \cdot \text{K}^{-1} \cdot \text{mol}^{-1})(310 \text{ K})(7.6 - 7.3)$$
$$\quad + (1)(96,485 \text{ J} \cdot \text{V}^{-1} \cdot \text{mol}^{-1})(0.170 \text{ V})$$
$$= 1780 \text{ J} \cdot \text{mol}^{-1} + 16,400 \text{ J} \cdot \text{mol}^{-1}$$
$$= 18,200 \text{ J} \cdot \text{mol}^{-1} = 18.2 \text{ kJ} \cdot \text{mol}^{-1}$$

7. Use the equation derived in Sample Calculation 15-3:

$$\Delta G = 2.303\ RT\ (pH_{in} - pH_{out}) + Z\mathcal{F}\Delta\psi$$

$$(pH_{in} - pH_{out}) = \frac{\Delta G - Z\mathcal{F}\Delta\psi}{2.303\ RT}$$

$$= \frac{30{,}500\ \text{J}\cdot\text{mol}^{-1} - (1)(96{,}485\ \text{J}\cdot\text{V}^{-1}\cdot\text{mol}^{-1})(0.170\ \text{V})}{2.303\ (8.3145\ \text{J}\cdot\text{K}^{-1}\cdot\text{mol}^{-1})(298\ \text{K})}$$

$$= \frac{30{,}500\ \text{J}\cdot\text{mol}^{-1} - 16{,}400\ \text{J}\cdot\text{mol}^{-1}}{5700\ \text{J}\cdot\text{mol}^{-1}}$$

$$= 2.5$$

The pH of the matrix must be 2.5 units greater than the pH of the cytosol.

Chapter 16

1. $E = (3.6 \times 10^{-19}\ \text{J})(6.022 \times 10^{23}\ \text{mol}^{-1})$
$= 217{,}000\ \text{J}\cdot\text{mol}^{-1} = 217\ \text{kJ}\cdot\text{mol}^{-1}$

2. $E = \dfrac{hc}{\lambda} \times N$

$\lambda = \dfrac{hcN}{E}$

$$= \frac{(6.626 \times 10^{-34}\ \text{J}\cdot\text{s})(2.998 \times 10^{8}\ \text{m}\cdot\text{s}^{-1})(6.022 \times 10^{23}\ \text{mol}^{-1})}{250 \times 10^{3}\ \text{J}\cdot\text{mol}^{-1}}$$

$= 479 \times 10^{-9}\ \text{m} = 479\ \text{nm}$

Box 2-D Acid–Base Balance in Humans

1. $H^+(aq) + HCO_3^-(aq) \rightleftharpoons H_2CO_3(aq)$
$$\rightleftharpoons H_2O(l) + CO_2(aq)$$

Failure to eliminate CO_2 in the lungs would cause a buildup of $CO_2(aq)$. This would shift the equilibrium of the above equations to the left. The increase in $CO_2(aq)$ would lead to the increased production of carbonic acid, which would in turn dissociate to form additional hydrogen ions, causing acidosis.

2. (a) Mechanical hyperventilation removes CO_2 from the patient's lungs. Carbonic acid in the blood would produce more water and CO_2 to make up for the loss of CO_2. This in turn would cause additional hydrogen ions and bicarbonate ions to form more carbonic acid. The loss of hydrogen ions would result in an increased pH, bringing the patient's pH back to normal.

(b) The additional bicarbonate would combine with hydrogen ions to form carbonic acid. The additional carbonic acid would dissociate to form water and carbon dioxide. This helps alleviate the acidosis because the bicarbonate combines with excess hydrogen ions, thus decreasing the hydrogen ion concentration and increasing the pH. However, it is not acceptable for use in patients with ALI because of the increased production of aqueous CO_2 in the blood. The CO_2 produced would need to be exhaled in the lungs, which would be difficult in patients with ALI.

(c) Tris becomes protonated to form its conjugate base. This removes H^+ from circulation and increases the pH back to normal. The protonated form of Tris is excreted in the urine. This method of acidosis treatment does not involve exhalation of CO_2 and is therefore an acceptable treatment for patients with ALI.

3. During hyperventilation, too much CO_2 (which is equivalent to H^+ in the form of carbonic acid) is given off, resulting in respiratory alkalosis. By repeatedly inhaling the expired air, the individual can recover some of this CO_2 and restore acid–base balance.

4. (a) The NH_3 molecule is small and uncharged and can diffuse through the cell membrane. The cationic ammonium ion cannot diffuse through the membrane and so must travel through an anion exchange protein.

(b) Na^+, which is present at high concentrations outside the cell, moves into the cell as NH_4^+ moves out.

5. The concentrations of both Na^+ and Cl^- are greater outside the cell than inside (see Fig. 2-13). Therefore, the movement of these ions into the cell is thermodynamically favorable. Na^+ movement into the cell drives the exit of H^+ via an exchange protein in the plasma membrane (the favorable movement of Na^+ into the cell "pays for" the unfavorable movement of H^+ out of the cell). Similarly, the movement of Cl^- into the cell drives the movement of HCO_3^- out of the cell through another exchange protein.

6. Acetoacetate and 3-hydroxybutyrate are acids (they are ionized at physiological pH). The accumulation of the ketone bodies therefore causes metabolic acidosis. The body attempts to compensate by increasing the breathing rate in order to eliminate more CO_2.

7. The cell-surface carbonic anhydrase can catalyze the conversion of $H^+ + HCO_3^-$ to CO_2, which can then diffuse into the cell (the ionic H^+ and HCO_3^- cannot cross the hydrophobic lipid bilayer on their own). Inside the cell, carbonic anhydrase converts the CO_2 back to $H^+ + HCO_3^-$.

8. The lungs can compensate for metabolic acidosis through an increase in the breathing rate in order to eliminate more CO_2. The kidneys can compensate for respiratory acidosis by increasing the breakdown of glutamine to produce NH_3 (excreted in urine as NH_4^+); however, this mechanism requires the synthesis of enzymes, which takes several hours at least.

Box 3-A Discovery of the Cystic Fibrosis Gene

1. Whereas the deletion of an amino acid dramatically affects a protein's structure and function, other defects, such as an amino acid substitution, may not have a large effect on the protein. Likewise, a modest diminution in the amount of functional CFTR protein may not impair mucus secretion as much as the near-complete absence of the protein.

2. The mutated DNA sequence is missing two nucleotides (T and A). The deletion alters the sequence of codons so that past this point, all the codons specify different amino acids. The resulting protein is likely to be nonfunctional.

3. Individuals with two copies of the defective CF gene develop CF and die before passing on their genes. Individuals with two copies of the normal CF gene do not develop CF but could die from infectious disease. Individuals with one normal and one defective gene would not die from CF and would be more resistant to infection, since the mutant CFTR would help prevent pathogen entry. These individuals would be most likely to survive and pass on their genes (including the defective CF gene) to the next generation.

4. Because CF is caused by a single-gene defect, it would be a good candidate for correction by replacing the defective gene with a copy of the normal gene. However, CFTR is located in the membranes of certain types of cells; in other words, it does not circulate. Consequently, CFTR function would need to be restored in a tissue-specific manner; the protein would not be able to reach all tissues in the body after being introduced at one site.

Box 4-C Protein Misfolding and Disease

1. The extra copy of chromosome 21 increases the amount of the precursor protein and therefore contributes to a higher concentration of the amyloid-β fragment. As a result, amyloid fibers begin forming in the brains of these individuals at a younger age.

2. Since the mice without PrP^C do not develop TSE, the disease must require the presence of PrP^C, not just PrP^{Sc}. These results also suggest that PrP^C is not an essential protein since mice that lack the PrP gene are normal.

3. (a) The extensive α-helical secondary structure in myoglobin makes it unlikely to easily adopt the all-β conformation necessary for amyloid formation.

(b) This result suggests that any polypeptide—even one whose native conformation is all α-helical—can assume the β conformation if conditions permit.

4. The experiment with myoglobin, which suggests that any polypeptide can form a β sheet, and the model for PrP misfolding indicate that the ability to form an amyloid fiber is a property of the polypeptide backbone, not its side chains. However, it is possible that amino acid side chains, particularly if they were hydrophobic, would promote the aggregation of individual amyloid fibers into larger structures.

Box 5-C Hemoglobin Mutations

1. The decreased stability of the β146His–β94Asp ion pair decreases the stability of the deoxy form of hemoglobin. Thus, the oxygenated form of Hb Ohio is more stable, resulting in a decreased p_{50} value. Hb Ohio's increased oxygen affinity means that it does not deliver oxygen to cells as effectively as Hb A. In fact, Hb Ohio has a p_{50} of 16.8 torr, whereas Hb A has a p_{50} of 27.2 torr.

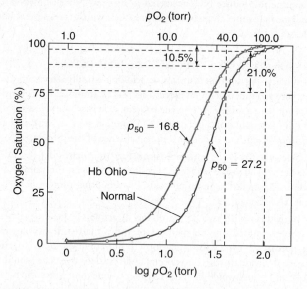

2. Hb S migrates more slowly because the Glu → Val mutation makes the protein less negatively charged. The Ala → Asp mutation in Hb Ohio makes the protein more negatively charged, so it migrates faster.

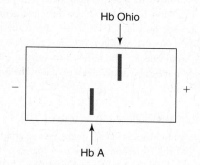

3. Hb Milledgeville has an increased oxygen affinity because a mutation has occurred at the $\alpha_1\beta_2$ interface. Leu is larger and more elongated than Pro and may interfere with important contacts at this interface. Since these contacts occur in the deoxy form, the substitution of Leu for Pro destabilizes the deoxy form and thus the oxygenated form is more stable. Therefore, the oxygen affinity of Hb Milledgeville is greater than that of Hb A.

4. (a) The two codons for Lys are AAA and AAG (see Table 3-3). The two codons for Asn are AAU and AAC. Therefore, there has been a mutation in the coding strand of the DNA in the third position of the codon. Either an A or a G has been mutated to a T or C.

(b) The oxygen affinity of Hb Providence is increased. The substitution of the neutral Asn for the positively charged Lys results in decreased binding of BPG in the central cavity of hemoglobin, since Lys forms

an ion pair with the negatively charged BPG. BPG binds to the T form but not the R form of hemoglobin. Therefore, decreased BPG binding means the R form is favored and the oxygen affinity of the mutant is increased.

(c)

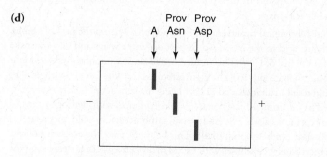

(d)

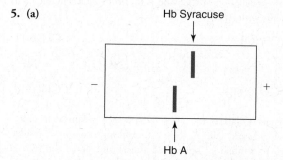

(e) The oxygen affinity of Hb Providence Asp is even greater than that of Hb Providence Asn. The presence of the negatively charged Asp repels the negatively charged phosphate groups of BPG, resulting in an even greater decrease in affinity for BPG. Since BPG binds only to deoxyhemoglobin, the inability of BPG to bind to Hb Providence Asp results in a stabilization of the oxygenated form of hemoglobin and an increase in its oxygen affinity.

5. (a)

Hb Syracuse

Hb A

(b) Hb Syracuse has increased oxygen affinity. Proline cannot accept protons, as His 146 does. His 146 on the β chain of hemoglobin forms ion pairs in the deoxy form. The inability of Hb Syracuse to form these ion pairs results in the greater stability of the oxygenated form of hemoglobin, so increased oxygen affinity is the result.

Box 5-E Genetic Collagen Diseases

1. Delivery of nutrients is slow in avascular tissues such as ligaments, so tissue regeneration is slow. Even though bone has a high mineral content, the cells are supplied with blood, so bone remodeling (removal and replacement of the tissue) is more rapid.

2. Bone becomes stronger when stressed, so muscles pulling on bones during exercise will lead to remodeling that counters the osteoporosis.

3. Since individuals with a severe case of osteogenesis imperfecta do not survive to reproductive age, their particular genetic defect is not passed on. Hence, most cases arise from new mutations.

4. First, the collagen would have to travel from the skin surface through the hard keratinized layers of dead cells, even as those cells are being shed. Second, even if the exogenous collagen molecules were able to reach the living dermal layers, it is not clear how they would be incorporated into newly synthesized collagen fibers, which are laid down in an organized fashion by fibroblasts.

Box 6-B Blood Coagulation Requires a Cascade of Proteases

1. The function of factor IXa is to activate factor X in order to promote thrombin activation and fibrin formation. Factor VIIa can also activate factor X, so the physiological effect is similar.

2. Factor IXa leads to the activation of thrombin, so the absence of factor IX delays clot formation, causing bleeding. Although factor XIa also leads to thrombin production, factor XI plays no role until it is activated by thrombin itself. By this point, coagulation is already well under way, so a deficiency of factor XI may not significantly delay coagulation.

3. Because factor IX is not the only trigger for blood coagulation, partial restoration of factor IX function is enough to supplement factor VIIa–tissue factor. Even if initiation of coagulation is slower than normal, once thrombin has been generated, fibrin production is guaranteed.

4. Thrombin must have been the first protease to evolve, since it acts directly on fibrinogen to produce a fibrin clot. Over time, additional proteases evolved, appearing in the inverse order of their action in the modern coagulation pathway.

5. In DIC, the high rate of coagulation actually leads to depletion of platelets and the various coagulation factors. This prevents normal coagulation from occurring, so the patient bleeds.

6. The thrombin variant can promote coagulation but cannot be inhibited normally, so it would lead to an increased risk of clotting.

7. Heparin must enter the bloodstream (via intravenous administration) in order to act with antithrombin as an anticoagulant. If consumed orally, it will not enter the circulation but will be degraded to its monosaccharide components.

8. The leech is a parasite that feeds on blood. In order to maintain the fluidity of the blood during a meal, the leech produces anticoagulants to prevent the host's blood from clotting.

Box 7-A Drug Development

1. A drug candidate's small size and limited hydrogen-bonding capacity indicate that the compound would be able to diffuse across biological membranes in order to enter cells to exert its effects.

2. With a mass of about $357 \, \text{g} \cdot \text{mol}^{-1}$, only one hydrogen bond donor group, and only six hydrogen bond acceptor groups, the drug obeys the rules.

3. Digestive enzymes in the stomach and small intestine may destroy the drug before it has a chance to be absorbed by the body. Therefore, some drugs must bypass the digestive system and be delivered directly to the bloodstream.

4. The liver is damaged by acetamidoquinone because the cytochrome P450 enzymes that convert acetaminophen to its more toxic derivative are located in the liver.

5. Because cytochrome P450 enzymes can modify warfarin to hasten its excretion, it is helpful to know which P450 variants are present. The clinician can then use this information to predict how quickly the drug will be modified and can select the appropriate dose to achieve the desired anticoagulant effect.

6. (a) Hypertension (high blood pressure) can be treated with enalapril (or a similar drug), which blocks angiotensin II production and thereby keeps blood vessels in a more relaxed (lower-pressure) state.

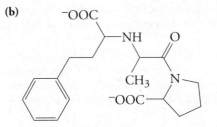

Box 8-B The Lipid Vitamins A, D, E, and K

1. Oxidation of retinal (an aldehyde) yields retinoic acid (a carboxylic acid).

2. Inflammatory diseases may result from overactivation of the immune system such that normal tissues (not just pathogens) are attacked or damaged. By further stimulating immune system cells, retinoic aid can worsen the inflammation.

3. Because vitamin D_3 can be produced from cholesterol, it is not strictly a vitamin. However, its production does require ultraviolet light, which an individual must obtain through exposure to the sun and which may not always be available.

4. At higher latitudes, the intensity of ultraviolet light from the sun is less than at lower latitudes, leading to lower rates of production of vitamin D.

5. The antibiotics suppress the growth of the intestinal bacteria that produce vitamin K, leading to vitamin K deficiency.

6. In obese individuals, the excess body fat acts as a sink for the fat-soluble vitamins. Higher dietary intake is therefore required to provide the necessary amounts of vitamins needed for normal metabolism.

7. Vitamins are natural products and therefore cannot be patented. Without the protection of a patent, market competition keeps the prices of vitamins low. Consequently, the sale of vitamins cannot generate enough money to offset the cost of conducting large clinical trials.

Box 9-C Antidepressants Block Serotonin Transport

1. Because serotonin is derived from tryptophan, a protein-rich diet that supplied ample tryptophan could promote serotonin production.

2. Serotonin recycling depends on a transporter that uses the free energy of the Na^+ gradient to move serotonin back into the cell. The Na^+ gradient is established through the action of the Na,K-ATPase.

3. The reduction in serotonin transport capacity increases the concentration of serotonin in synapses, prolonging the mood-elevating effects of serotonin.

4. No; preventing serotonin reuptake would prolong its signaling potential, but blocking its receptor would be expected to have the opposite effect. [In fact, due to the complexity of serotonin signaling pathways, blocking a receptor may have an antidepressive effect.]

5. Together, the drugs could contribute to excessive serotonin signaling, the MAO inhibitor by slowing serotonin's degradation and the SSRI by prolonging its time in the synaptic space.

Box 13-B Alcohol Metabolism

1. Normally, the skin is cooler than the core of the body, which helps conserve body heat. Ethanol triggers vasodilation, which increases the flow of blood in the skin and allows a greater loss of heat from the body. In situations where conserving body heat is important (staying warm when outdoors during the winter), hypothermia may result.

2. Ethanol acts as a diuretic, an agent that increases urination and leads to dehydration. Consuming extra water can help offset the water loss from the kidneys.

3. Increased expression of cytochrome P450 would allow more rapid modification and excretion of the drug molecules, thereby making them less effective.

4. Both enzymes consume NAD^+. Since the lactate dehydrogenase reaction that converts lactate to pyruvate requires NAD^+, this process is slowed. Less pyruvate means less glucose produced by gluconeogenesis, and less glucose is released from the liver into the bloodstream, producing hypoglycemia.

5. Production of acetate consumes NAD^+, which decreases the ability of the cell to produce ATP by glycolysis. As a result, there is less ATP available to convert acetate to acetyl-CoA for further catabolism.

Box 13-C Glycogen Storage Diseases

1. Chronic hypoglycemia apparently limits growth and development. Because the glycogen storage diseases are genetic, they are present at birth and therefore may have lifelong consequences.

2. Glucose-6-phosphatase is normally present in both the liver and the kidneys, the two gluconeogenic tissues. In the absence of the enzyme, both the liver and the kidneys swell as glycogen accumulates, since glycogen-derived glucose-6-phosphate is unable to exit the cell as glucose.

3. Yes; the corn starch is a source of glucose. Frequent feedings would increase the amount of glucose leaving the intestine and entering the circulation, thereby relieving the hypoglycemia caused by the inability of the liver to stockpile glucose as glycogen.

4. Because the defect (lack of glycogen debranching enzyme) affects all types of tissues, a liver transplant would cure only the disease in the liver. Muscles and other tissues that synthesize and degrade glycogen would not be directly affected by the transplant, although they might benefit from the restoration of normal liver function.

5. GLUT2 normally functions to allow glycogen-derived glucose to leave the liver and enter the circulation. The same transporter must also facilitate galactose transport, because galactose cannot enter cells and remains in the blood (hypergalactosemia) when the transporter is defective.

Box 14-A Mutations in Citric Acid Cycle Enzymes

1. Pyruvate increases because the patient relies on glycolysis in the absence of proper citric acid cycle function. Fumarate increases because it cannot be converted to malate in the absence of fumarase. This also explains why the concentration of malate decreases.

2. The reactions catalyzed by these two enzymes are part of the succinate → oxaloacetate regeneration phase of the citric acid cycle (Reactions 6–8), so both deficiencies have the effect of slowing the cycle by preventing the production of oxaloacetate.

3. In the absence of succinate dehydrogenase, the substrate succinate accumulates. This in turn slows the production of succinate from succinyl-CoA, leading to a buildup of succinyl-CoA and making less coenzyme A available for other cellular processes.

4. The low activity of the citric acid cycle, caused by the fumarase deficiency, means that the cycle cannot accommodate the usual flux of carbon produced by glycolysis. As a result, the pyruvate generated by glycolysis is converted to lactate (lactic acid), which causes the blood pH to decrease (see Box 2-D).

5.

$$^-OOC - \underset{\underset{OH}{|}}{CH} - CH_2 - CH_2 - COO^-$$

2-Hydroxyglutarate

6. Arginine is encoded by four codons: CGU, CGC, CGA, and CGG. A G → A point mutation in the first or second codon would generate a His codon: CAU or CAC (Table 3-3).

7. The conversion of α-ketoglutarate to 2-hydroxyglutarate is a reduction reaction, so it is accompanied by the conversion of NADPH to $NADP^+$. This shifts the cell's balance of cofactors to more oxidized.

Box 17-B Inhibitors of Fatty Acid Synthesis

1. Mutations that alter active-site residues in enoyl-ACP reductase could make bacteria resistant to triclosan. Such mutations might introduce bulkier side chains that would prevent triclosan from slipping into the active site. Amino acid substitutions might reduce the ability of the enzyme to form hydrogen bonds with triclosan, without affecting substrate binding.

2. In mammals, fatty acid desaturation requires O_2 for dehydrogenation. Because bacteria use a different enzyme to introduce a double bond, inhibition of that enzyme could block bacterial growth without harming the mammalian host.

3.

$$H_3C - (CH_2)_{46} - \underset{\beta}{\overset{\overset{\displaystyle HO}{|}}{C}} - \underset{\alpha}{\overset{\overset{\displaystyle COO^-}{|}}{C}} - (CH_2)_{25} - CH_3$$

4. **(a)** Approximately 10,000 (10^4) cells will be resistant.
(b) The probabilities are multiplied:

$$\frac{1}{10^8} \times \frac{1}{10^8} \times \frac{1}{10^8} = \frac{1}{10^{24}}$$

In theory, it is highly unlikely that even one cell could survive the combination of three drugs.

5. Yes; by blocking fatty acid synthesis and by promoting fatty acid oxidation, cerulenin decreases the levels of fatty acids available in the cell, thereby preventing fungal growth.

Box 19-A Cancer Metabolism

1.

2-Deoxy-2-fluoroglucose

2. **(a)** The negatively charged phosphate group makes glucose unable to fit in the glucose transporter.
(b) The lack of a hydroxyl group at C2 prevents the sugar from being isomerized to fructose-6-phosphate and continuing through the steps of glycolysis.

3. The brain maintains a high metabolic rate and relies on glucose to supply its energy needs, so a significant portion of the FDG tracer accumulates in the brain. Since FDG is injected into the bloodstream, some of it is filtered out by the kidneys and is stored in the bladder until it can be eliminated. (For these reasons, PET scans using FDG are not suitable for monitoring brain or bladder cancers.)

4. Pyruvate can be transaminated to alanine, and 3-phosphoglycerate is used to produce serine, which is also a precursor of glycine (Section 18-2).

5. Glutamine is a precursor for the synthesis of both purine and pyrimidine nucleotides. It is also a source of two amino groups: One is removed as free ammonia and the other is part of glutamate, which can donate its amino group in transamination reactions. The carbons, as α-ketoglutarate, can be used in other biosynthetic reactions.

6. Nucleotide biosynthesis requires aspartate (derived from oxaloacetate) and glutamine (derived from α-ketoglutarate via glutamate; Section 18-3).

7. The result of the pathway is that pyruvate, a precursor for anabolic reactions, is produced without the ATP that is normally produced by the pyruvate kinase reaction. In a cancer cell, the need for molecular building blocks is greater than the need to make additional ATP.

Box 20-B Cancer Is a Genetic Disease

1. Cancer results from damage to DNA, for example, from environmental agents (UV light or hydroxyl radicals) or viral infections. DNA damage most likely accumulates over time (more quickly if there is already an inherited genetic defect). Consequently, the cells of older individuals have more DNA mutations and are more likely than the cells of younger individuals to undergo transformation.

2. Normally, the retinoblastoma protein acts as a tumor suppressor by preventing the cell from synthesizing DNA, a prerequisite for cell division. A mutation in the retinoblastoma gene would allow the cell to synthesize DNA and proceed with cell division, possibly leading to cancer.

3. According to the multiple-hit hypothesis, more than one genetic change is required to initiate tumor growth. In addition, individuals with a defective copy of an oncogene or tumor suppressor gene also have another normal copy of the gene that helps protect them from carcinogenesis.

4. The same genetic changes that alter the growth behavior of a cancerous cell may also affect genes for many proteins, resulting in the display of abnormal proteins on the cell surface.

5. The ring shape of PCNA allows it to slide along the DNA helix without making sequence-specific contacts. A protein with a similar structure could likewise slide along the DNA. Distortions in the DNA helix caused by nicks, gaps, missing bases, or bulky chemical adducts could halt the progress of the sensor and allow it to recruit DNA repair proteins.

6. According to the pathway described in the question, overactivation of Ras would lead to less ubiquitinated p53. p53 would be less rapidly degraded and so would be more available to halt the cell cycle. This would actually counteract the growth-promoting activity of the Ras pathway.

7. p53 increases the production of cytochrome c oxidase, the terminal enzyme of the electron transport chain, which consumes oxygen and contributes to the proton gradient that powers ATP synthesis (Section 15-3). In the absence of p53, less cytochrome c oxidase is made, so the cell relies less on aerobic respiration as a source of ATP and switches to anaerobic glycolysis.

8. A damaged unicellular organism does not harm others of its kind; it can go on to proliferate or die according to natural selection. In a multicellular organism, a runaway cell could threaten the well-being of the whole organism. Thus, apoptosis is necessary only in multicellular organisms.

INDEX

Page references followed by T indicate tables. Page references followed by F indicate figures.

AMINO ACID STRUCTURES AND ABBREVIATIONS

Hydrophobic amino acids

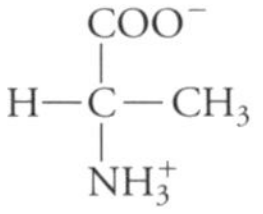

Alanine (Ala, A)

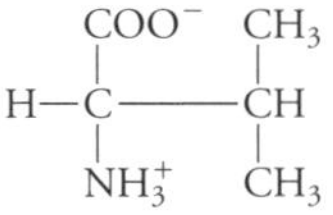

Valine (Val, V)

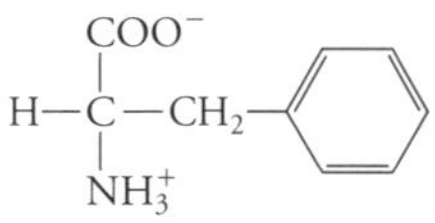

Phenylalanine (Phe, F)

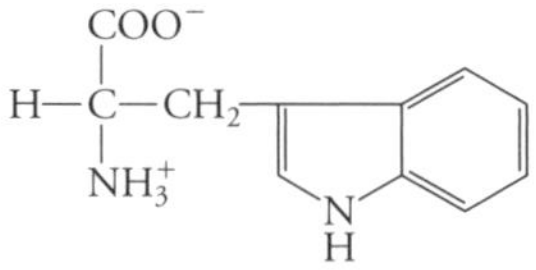

Tryptophan (Trp, W)

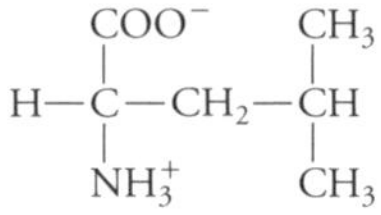

Leucine (Leu, L)

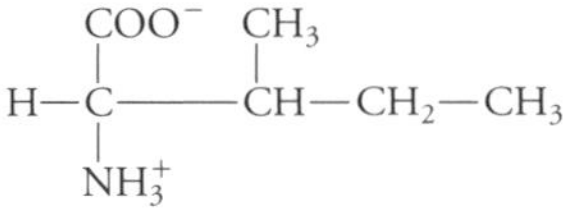

Isoleucine (Ile, I)

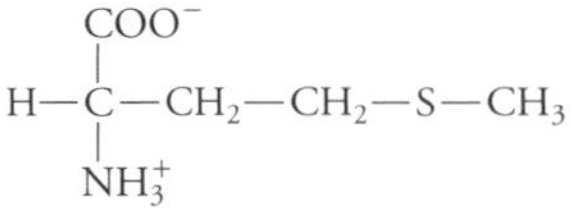

Methionine (Met, M)

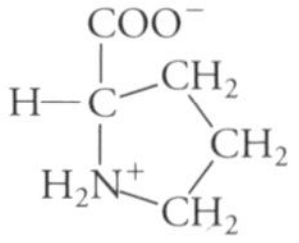

Proline (Pro, P)

Polar amino acids

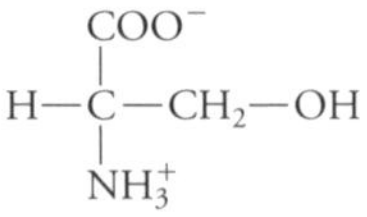

Serine (Ser, S)

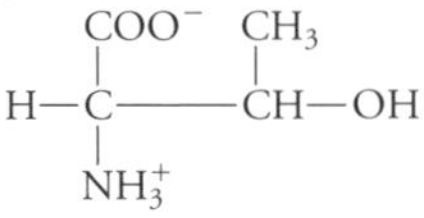

Threonine (Thr, T)

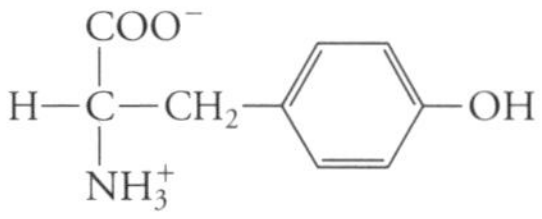

Tyrosine (Tyr, Y)

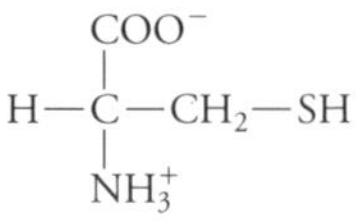

Cysteine (Cys, C)

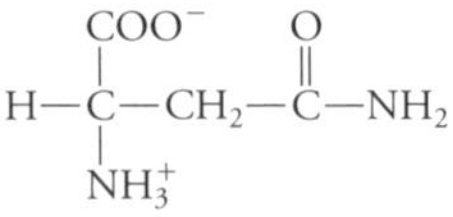

Asparagine (Asn, N)

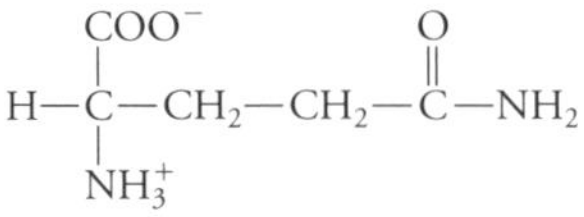

Glutamine (Gln, Q)

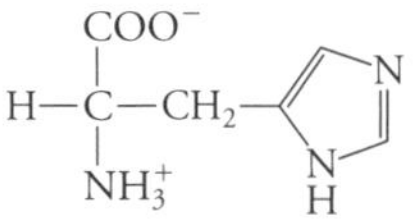

Histidine (His, H)

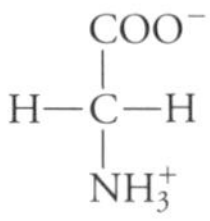

Glycine (Gly, G)

Charged amino acids

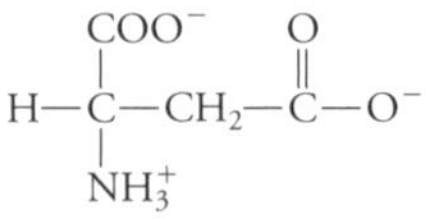

Aspartate (Asp, D)

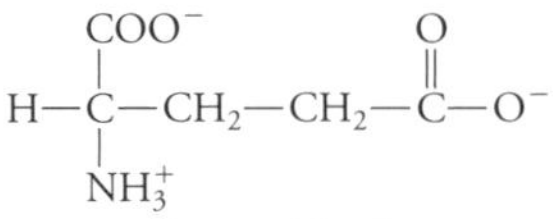

Glutamate (Glu, E)

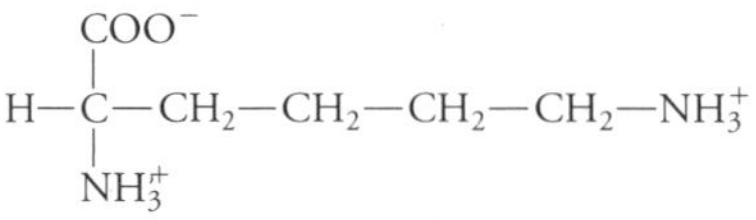

Lysine (Lys, K)

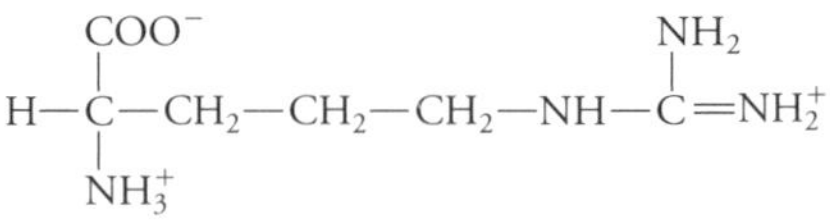

Arginine (Arg, R)